P9-DMB-703

Modern Statistics

A Canadian Perspective

William M. Goodman
University of Ontario Institute of Technology

NELSON EDUCATION

NELSON EDUCATION

Modern Statistics: A Canadian Perspective
by William M. Goodman

Associate Vice President, Editorial Director:
Evelyn Veitch

Editor-in-Chief, Higher Education:
Anne Williams

Acquisitions Editor:
Shannon White

Marketing Manager:
Sean Chamberland

Developmental Editor:
Joanne Sutherland

Photo Researcher/Permissions Coordinator:
Melody Tolson

Content Production Managers:
Sabrina Mansour, Imoinda Romain

Production Service:
Lachina Publishing Services

Copy Editor:
June Trusty

Proofreader:
Lachina Publishing Services

Indexer:
Lachina Publishing Services

Production Coordinator:
Ferial Suleman

Design Director:
Ken Phipps

Managing Designer:
Katherine Strain

Interior Design:
Kyle Gell

Cover Design:
Peter Papayanakis

Cover Image:
Eastcott Momatiuk/Digital Vision/Getty Images

Compositor:
Lachina Publishing Services

Printer:
RR Donnelley

Printed and bound in the United States of America
1 2 3 4 11 10 09 08

For more information contact Nelson Education Ltd., 1120 Birchmount Road, Toronto, Ontario, M1K 5G4. Or you can visit our Internet site at http://www.nelson.com

Library and Archives Canada Cataloguing in Publication

Goodman, William Martin, 1950-

Modern statistics : a Canadian perspective / William M. Goodman.

Includes index.

ISBN 978-0-17-625179-6

1. Statistics—Textbooks. 2. Statistics—Data processing—Textbooks.

I. Title.

QA276.12.M37 2008 519.5 C2007-906300-4

ISBN-10: 0-17-625179-0
ISBN-13: 978-0-17-625179-6

Contents

Brief Contents

Preface

Introduction

Computers have revolutionized what can be done in the field of statistics, but many introductory textbooks on the subject have not kept pace with this progress. While most textbooks today do include some feature items or appendixes that mention relevant statistical software, this material is never fully integrated into the main body of the text. In general, other books follow conventions from the time when students had no tools available besides calculators and hard-copy formulas and tables.

Many courses today, however, also include a computer lab component in the course presentation, and perhaps even a laptop for each student. Also, there are online or online-enhanced courses for which all students have some access to computers. In all such cases, however, the existing conventional textbooks fall short.

Modern Statistics: A Canadian Perspective tailors the content and presentation of the introductory material provided to take full advantage of the computer resources now available to students of statistics and to fully integrate the coverage of the statistics curriculum with the related coverage of relevant statistical software. This book attempts to offer a fresh, experiential introduction to statistical concepts and calculations while maximizing the potential provided by computer-enriched statistics courses.

Special Features of *Modern Statistics: A Canadian Perspective*

The in-text and computer examples and procedures are integrated.
The clear focus of this book is statistics but its computer content is not merely an afterthought. If, for instance, a chapter example uses a certain set of numbers, all of the relevant computer instructions and illustrations follow through by using the same example. As well, *Modern Statistics* does not ignore discrepancies where traditional procedures and software default procedures do not give the same results; either the difference is acknowledged and explained or the instructions given in the text or on the accompanying CD are modified for consistency.

The book's approach to software usage encourages reflective problem solving.
Rather than reading like a computer manual for a software package, the computer-related sections of this book minimize sequences of "click on this, then click on that" instructions in favour of explanations of how the software input or output features can be related back to the in-text concepts. Students are encouraged and prepared to take a more experimental and exploratory approach in applying computer software to statistical problems.

Computer software usage is designed to provide new opportunities for statistical exploration and experience.
Including well over a hundred data sets and more than a thousand exercises in the book and on the CD, as well as at least two "Hands On" activities per chapter in the book,

Modern Statistics provides a great number of opportunities for students to solve statistical problems using real data and to experiment with statistical concepts such as computer simulations.

The focus is on up-to-date, not obsolescent, statistical procedures.
To make room for modern statistical procedures, conventions that have become obsolescent with the increased use of statistical computer software do not appear in this textbook. One example is the omission of tables such as those for t or χ^2. Despite their long, respected history, these tables were really just look-up devices for numbers calculated in advance. With today's computer software, there is an easier way to look up a value (e.g., a p-value, given a test statistic and other relevant information). None of the underlying theory is lost by the substitution of computerized look-up tools. Computational formulas, such as those for the standard deviation or for regression, have also been omitted. Instead, definitional formulas, which express the essence of a concept, are used in this book. The computational or "shortcut" versions (which are not very intuitive to look at) were not designed for pedagogy, but to simplify calculations—*assuming that the calculations would be done manually.* With computers, that advantage for introducing the transforms no longer exists.

***Modern Statistics* is a primary textbook for a one- or two-semester introductory statistics course at the university or college level.**
While innovative in much of its content and presentation, this book also meets the need for a solid course textbook, covering a standard set of statistical topics with all of the apparatus of explained formulas, solved problems, exercises, chapter summaries, and so on. By doing so, it is a valuable teaching tool for any postsecondary course in introductory statistics in programs ranging from business, to science, to the social sciences.

Pedagogy

Modern Statistics: A Canadian Perspective embodies the belief that students learn best by doing, whenever possible. The text maximizes the potential that today's computer software provides for students to perform experiments, simulations, and demonstrations that otherwise would not realistically have been possible. Solved problems in the text, along with the majority of chapter exercises, are based on real-world data with which the students themselves can experiment. Computer software coverage uses screen captures to encourage learning by exploring, rather than by offering long lists of "click on this, then click on that" instructions.

Learning Aids

"Learning Objectives"

Each chapter opens with a list of "**Learning Objectives**" that students can use to check their own progress. The numbering of the objectives follows the section numbering in the body of the chapter.

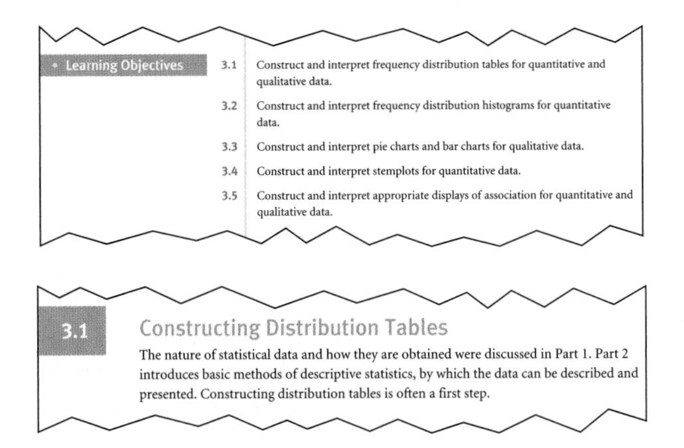

• Learning Objectives

3.1	Construct and interpret frequency distribution tables for quantitative and qualitative data.
3.2	Construct and interpret frequency distribution histograms for quantitative data.
3.3	Construct and interpret pie charts and bar charts for qualitative data.
3.4	Construct and interpret stemplots for quantitative data.
3.5	Construct and interpret appropriate displays of association for quantitative and qualitative data.

3.1 Constructing Distribution Tables

The nature of statistical data and how they are obtained were discussed in Part 1. Part 2 introduces basic methods of descriptive statistics, by which the data can be described and presented. Constructing distribution tables is often a first step.

Solved Examples

Every section in each chapter includes at least one solved example to illustrate the concepts and methods that are being discussed.

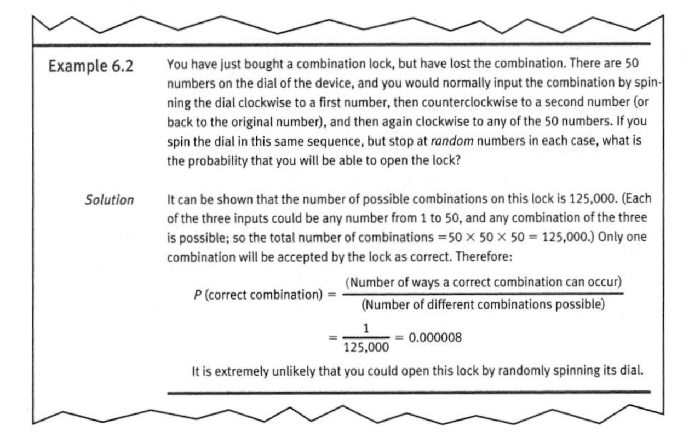

Example 6.2 You have just bought a combination lock, but have lost the combination. There are 50 numbers on the dial of the device, and you would normally input the combination by spinning the dial clockwise to a first number, then counterclockwise to a second number (or back to the original number), and then again clockwise to any of the 50 numbers. If you spin the dial in this same sequence, but stop at *random* numbers in each case, what is the probability that you will be able to open the lock?

Solution It can be shown that the number of possible combinations on this lock is 125,000. (Each of the three inputs could be any number from 1 to 50, and any combination of the three is possible; so the total number of combinations $= 50 \times 50 \times 50 = 125{,}000$.) Only one combination will be accepted by the lock as correct. Therefore:

$$P\,(\text{correct combination}) = \frac{(\text{Number of ways a correct combination can occur})}{(\text{Number of different combinations possible})}$$

$$= \frac{1}{125{,}000} = 0.000008$$

It is extremely unlikely that you could open this lock by randomly spinning its dial.

"In Brief" Features

Just-in-time introductions to statistical software techniques, integrated with the in-text discussions of statistical methods, are provided in "**In Brief**" features, where Excel, Minitab, and SPSS software procedures are detailed. The text minimizes keystroke instructions in favour of showing screen captures that invite the students to explore for themselves.

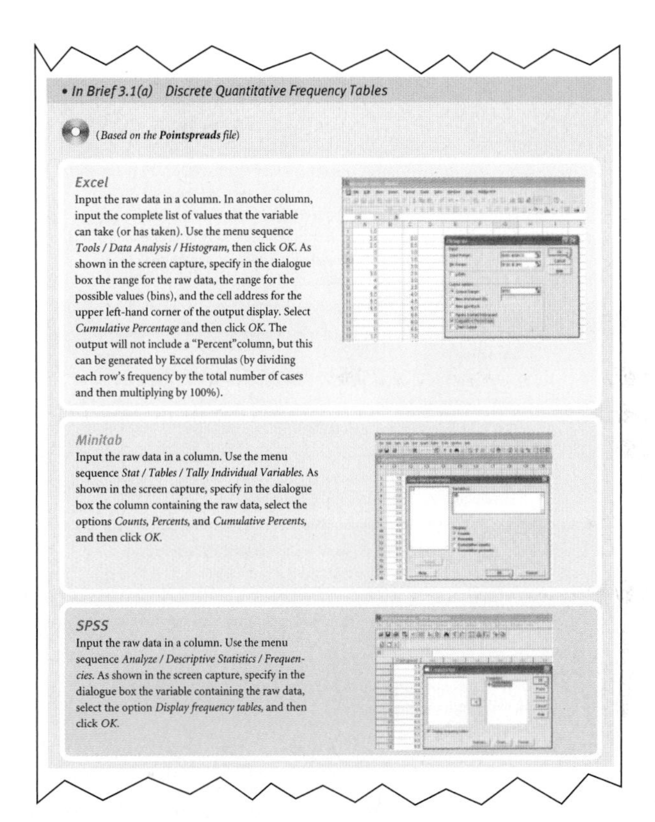

• *In Brief 3.1(a) Discrete Quantitative Frequency Tables*

(*Based on the* ***Pointspreads*** *file*)

Excel

Input the raw data in a column. In another column, input the complete list of values that the variable can take (or has taken). Use the menu sequence *Tools / Data Analysis / Histogram*, then click *OK*. As shown in the screen capture, specify in the dialogue box the range for the raw data, the range for the possible values (bins), and the cell address for the upper left-hand corner of the output display. Select *Cumulative Percentage* and then click *OK*. The output will not include a "Percent"column, but this can be generated by Excel formulas (by dividing each row's frequency by the total number of cases and then multiplying by 100%).

Minitab

Input the raw data in a column. Use the menu sequence *Stat / Tables / Tally Individual Variables*. As shown in the screen capture, specify in the dialogue box the column containing the raw data, select the options *Counts*, *Percents*, and *Cumulative Percents*, and then click *OK*.

SPSS

Input the raw data in a column. Use the menu sequence *Analyze / Descriptive Statistics / Frequencies*. As shown in the screen capture, specify in the dialogue box the variable containing the raw data, select the option *Display frequency tables*, and then click *OK*.

"Hands On" Activities

At least two sections per chapter end with a "**Hands On**" activity. Students are coached to use software, sometimes including Internet resources, to visualize and creatively experiment with statistical concepts introduced in the section.

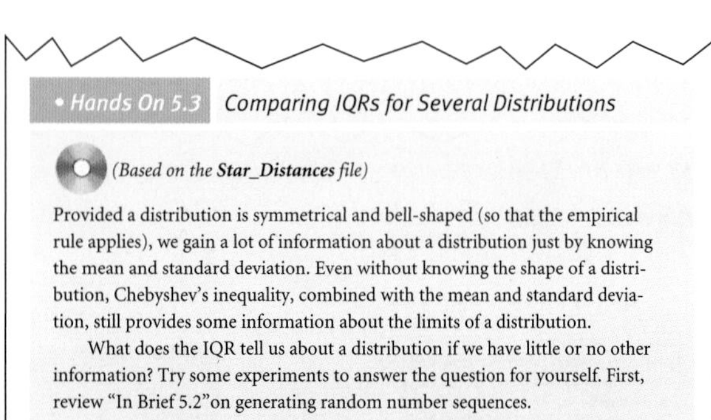

• Hands On 5.3 *Comparing IQRs for Several Distributions*

*(Based on the **Star_Distances** file)*

Provided a distribution is symmetrical and bell-shaped (so that the empirical rule applies), we gain a lot of information about a distribution just by knowing the mean and standard deviation. Even without knowing the shape of a distribution, Chebyshev's inequality, combined with the mean and standard deviation, still provides some information about the limits of a distribution.

What does the IQR tell us about a distribution if we have little or no other information? Try some experiments to answer the question for yourself. First, review "In Brief 5.2"on generating random number sequences.

1. Use the computer to generate the following five random distributions:
   a) 1,000 numbers with a "normal" distribution: Mean = 10, Standard deviation = 2.
   b) 1,000 numbers with a "normal" distribution: Mean = 10, Standard deviation = 4.
   c) 1,000 numbers with a "normal" distribution: Mean = 10, Standard deviation = 6.
   d) 1,000 numbers with a "normal" distribution: Mean = 10, Standard deviation = 8.
   e) 1,000 numbers with a "uniform"distribution:from 0 to 10.
2. For each case, find the three quartiles and the IQR. How does the IQR compare with the standard deviation? For each distribution, answer the question: Fifty percent of the values are between what two numbers?
3. Next load the data set stored in the file **Star_Distances.** Find the quartiles and IQR. Then find the mean and standard deviation for the same data. If you knew only the quartiles, mean, and standard deviation, and had no other information, and you were a space traveller running low on fuel, which measure or measures might be most useful? Explain.

Seeing Statistics Resource

Other opportunities to creatively explore chapter concepts are provided by an Internet-based resource created by Duxbury Press: *Seeing Statistics.* Throughout the text, icons are used to identify the *Seeing Statistics* sections that complement what is covered in this book.

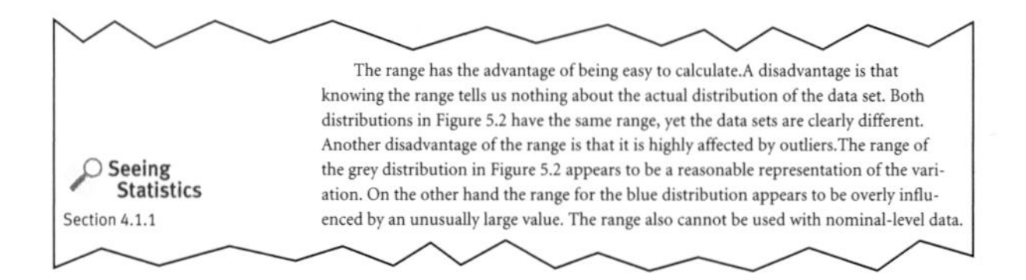

Seeing Statistics
Section 4.1.1

The range has the advantage of being easy to calculate.A disadvantage is that knowing the range tells us nothing about the actual distribution of the data set. Both distributions in Figure 5.2 have the same range, yet the data sets are clearly different. Another disadvantage of the range is that it is highly affected by outliers.The range of the grey distribution in Figure 5.2 appears to be a reasonable representation of the variation. On the other hand the range for the blue distribution appears to be overly influenced by an unusually large value. The range also cannot be used with nominal-level data.

Exercises

Numerous exercises are provided at the end of each chapter. For each section, there are **basic exercises** that review core concepts, definitions, and so on, and **applied exercises** based on real data, a large percentage of which are Canadian. The same is true for the "**Review Exercises**" section, which covers material in the whole chapter.

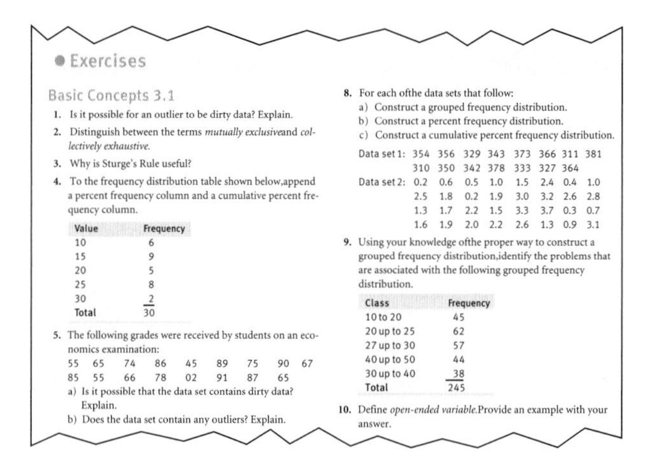

● Exercises

Basic Concepts 3.1

1. Is it possible for an outlier to be dirty data? Explain.
2. Distinguish between the terms *mutually exclusive*and *collectively exhaustive.*
3. Why is Sturge's Rule useful?
4. To the frequency distribution table shown below,append a percent frequency column and a cumulative percent frequency column.

Value	Frequency
10	6
15	9
20	5
25	8
30	2
Total	30

5. The following grades were received by students on an economics examination:
   55 65 74 86 45 89 75 90 67
   85 55 66 78 02 91 87 65
   a) Is it possible that the data set contains dirty data? Explain.
   b) Does the data set contain any outliers? Explain.

8. For each ofthe data sets that follow:
   a) Construct a grouped frequency distribution.
   b) Construct a percent frequency distribution.
   c) Construct a cumulative percent frequency distribution.

   Data set 1: 354 356 329 343 373 366 311 381
   310 350 342 378 333 327 364

   Data set 2: 0.2 0.6 0.5 1.0 1.5 2.4 0.4 1.0
   2.5 1.8 0.2 1.9 3.0 3.2 2.6 2.8
   1.3 1.7 2.2 1.5 3.3 3.7 0.3 0.7
   1.6 1.9 2.0 2.2 2.6 1.3 0.9 3.1
9. Using your knowledge ofthe proper way to construct a grouped frequency distribution,identify the problems that are associated with the following grouped frequency distribution.

Class	Frequency
10 to 20	45
20 up to 25	62
27 up to 30	57
40 up to 50	44
30 up to 40	38
Total	245

10. Define *open-ended variable.*Provide an example with your answer.

Chapter Summary

A "**Chapter Summary**" rounds out each chapter by referring back to the "Learning Objectives" at the beginning of the chapter and touching on all of the main concepts that were covered.

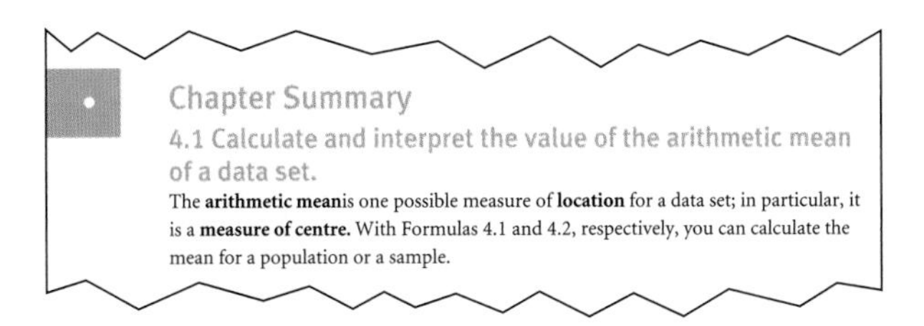

Chapter Summary

4.1 Calculate and interpret the value of the arithmetic mean of a data set.

The **arithmetic meanis** one possible measure of **location** for a data set; in particular, it is a **measure of centre.** With Formulas 4.1 and 4.2, respectively, you can calculate the mean for a population or a sample.

Key Terms

Located near the end of each chapter, the "**Key Terms**" list summarizes all of the new statistical terms that were introduced in that chapter.

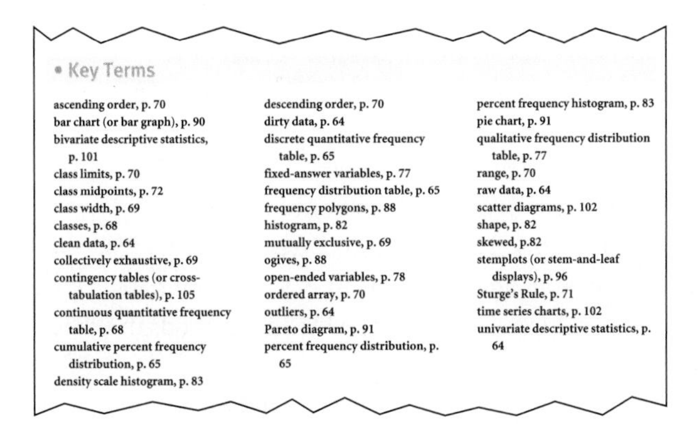

Key Terms

ascending order, p. 70
bar chart (or bar graph), p. 90
bivariate descriptive statistics, p. 101
class limits, p. 70
class midpoints, p. 72
class width, p. 69
classes, p. 68
clean data, p. 64
collectively exhaustive, p. 69
contingency tables (or cross-tabulation tables), p. 105
continuous quantitative frequency table, p. 68
cumulative percent frequency distribution, p. 65
density scale histogram, p. 83
descending order, p. 70
dirty data, p. 64
discrete quantitative frequency table, p. 65
fixed-answer variables, p. 77
frequency distribution table, p. 65
frequency polygons, p. 88
histogram, p. 82
mutually exclusive, p. 69
ogives, p. 88
open-ended variables, p. 78
ordered array, p. 70
outliers, p. 64
Pareto diagram, p. 91
percent frequency distribution, p. 65
percent frequency histogram, p. 83
pie chart, p. 91
qualitative frequency distribution table, p. 77
range, p. 70
raw data, p. 64
scatter diagrams, p. 102
shape, p. 82
skewed, p.82
stemplots (or stem-and-leaf displays), p. 96
Sturge's Rule, p. 71
time series charts, p. 102
univariate descriptive statistics, p. 64

Glossary

Following the text chapters, the "**Glossary**" provides a simple definition for each of the statistical terms introduced in the text, together with a chapter reference indicating where the term was introduced.

Key Formulas

Identified by their section numbers, all of the **key formulas** are listed, along with any helpful information.

Exercise Solutions

The **Solutions for Odd-Numbered Exercises** are provided at the back of the book. The solutions to all text exercises are in the *Instructor's Manual.*

List of "Hands On" Activities

The titles of all of the **"Hands On"** activities in the book are listed on the inside of the front cover of the book, citing the text section number and page number where each is located.

List of "In Brief" Features

As a handy reference for applying learned software techniques to new problems, the titles of all of the "**In Brief**" features in the book are listed on the inside of the back cover of the book, citing the text section number and page number where each is located.

Student Resources

- A **data CD-ROM** accompanies each new copy of the textbook. The CD contains all of the data sets referenced in the applied exercises in the text, presented in Excel, Minitab, and SPSS formats. It also includes a "data dictionary" that lists the names and descriptions of all of the data set variables provided on the CD.
- The ***Student Solutions Manual*** (ISBN 978-0-17-647380-8) includes solutions to the odd-numbered text exercises that are even more detailed than those provided in the textbook.
- The ***Modern Statistics* website** at www.modernstatistics.nelson.com incorporates chapter links, interactive quiz questions, case examples, and other valuable resources.
- The book incorporates references to ***Seeing Statistics*** by Duxbury Press—an online product located at www.seeingstatistics.com that uses over 150 Java applets to create an intuitive learning environment, including links to examples, exercises, definitions, and search capabilities.

Seeing Statistics

Instructor Resources

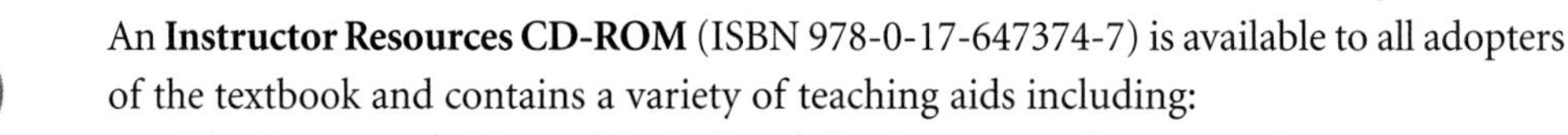

An **Instructor Resources CD-ROM** (ISBN 978-0-17-647374-7) is available to all adopters of the textbook and contains a variety of teaching aids including:

- The ***Instructor's Manual***, including full solutions to all text exercises.
- **PowerPoint slide presentations** that can be used as is or customized. These include:
  - A presentation provided for each chapter that offers a text outline to match the chapter's content, along with figures and tables from the text.
  - A separate presentation for each software version (Excel, Minitab, SPSS) that includes, in ordered sequence, all figures for all "In Brief" sections about that software, showing how the software can be applied to the methods described in the text.
- A **test bank** in Word format that contains a comprehensive selection of questions from which to choose. A computerized version of the test bank is available in ExamView testing software. ExamView is a Windows-based program that is easy to use. With this program, you can create, edit, store, print, and otherwise customize your quizzes, tests, and exams.

Downloadable versions of the *Instructor's Manual* and PowerPoint slides are provided for instructors on the *Modern Statistics* website at www.modernstatistics.nelson.com.

Acknowledgments

Many people have contributed greatly to this project, and to all I offer my sincere thanks. Special thanks to Richard Hird and Anthony Rahilly, who invited me to borrow some helpful turns of phrase and basic exercises from a now out-of-print book of theirs. This final product has also benefited a great deal from the suggestions of the technical editors and reviewers:

Jim Stallard, University of Calgary
Catherine Stanley, Acadia University
Gary Sneddon, Memorial University of Newfoundland
Robert Connolly, Algonquin College
Chris Kellman, British Columbia Institute of Technology
Jianan Peng, Acadia University
Peter Macdonald, McMaster University
Bob Sealy, Mount Allison University
Hong Wang, Memorial University of Newfoundland
Alison Weir, University of Toronto at Mississauga
Lang Wu, University of British Columbia

The team supporting me at Nelson Education Ltd. and Lachina Publishing Services have been outstanding. I interfaced most directly with Joanne Sutherland, June Trusty, Melody Tolson, Imoinda Romain, and Bonnie Briggle—but I know that many others also worked very hard to bring this book into print and I thank them all. A special thanks, too, to my assistant, Silke Martin, who provided tremendous help in a range of areas—from finding new data sets, to proofreading formulas and exercises, to creating draft solutions for exercises. And thanks to Laura Gannon for her work on SPSS data sets.

I also thank the countless webmasters who have posted and maintained sites in Canada and worldwide containing a wealth of information, and all of those data sources who have kindly given permission for use of some of their data in the book or on the data CD. Thank you, too, to the more than 400 UOIT students who worked with the prepublication version of this textbook in a statistics course and to the large number of these students who submitted signed or anonymous suggestions or corrections in relation to that version. All of your feedback has been very valuable.

On a personal note, I thank my wife Kathryn and my children Rachel and Aleksey for enduring my ever-present laptop—even when it showed up on vacation—and, most importantly, for putting all of this into perspective.

About the Author

Dr. William Goodman is an associate professor in the faculty of Business and Information Technology at the University of Ontario Institute of Technology (UOIT). He received his Ph.D. from the University of Waterloo, and was awarded the university's Masters Gold Medal for work related to computer modelling of a logical analysis.

Dr. Goodman has conducted extensive research into the distribution of incident-outcome severities in relation to industrial and radiological health and safety. His core model for analyzing severe-outcome accidents has been refined and applied commercially to data for the Ontario Power Generation company, the Atomic Energy Control Board, and the Bruce Nuclear Generating Station, among others.

In collaboration with researchers in the UOIT faculties of Education and Health Sciences, Dr. Goodman also applies statistical analyses to the evaluation of technology use in education. Recent research funded by the Public Health Agency of Canada has extended these analyses into program evaluation and health economics—an area he hopes to pursue further. His interest in updating student educational materials to take advantage of modern technologies has led to numerous publications, including educational simulations included with three editions of an economics textbook.

PART 1

Statistical Data

Chapter 1

Introduction to Statistical Data

Chapter Outline

Learning Objectives

1.1 Define the term *statistics,* and distinguish between descriptive and inferential statistics.

1.2 Describe the difference between a population and a sample, and explain some advantages of sampling.

1.3.1 Name and describe the basic types of data.

1.3.2 Explain the differences among the four levels of measurement.

1.1 Why Study Statistics?

The core subject of this book can be applied in almost any field or formal endeavour. You will learn techniques that can be used to:

- Clearly present numerical or other information.
- Draw or verify conclusions about populations based on sample information.
- Make reliable forecasts.
- Monitor and improve processes.

Contrary to what some people may tell you, statistics does not have to be viewed as a set of arcane procedures—to be struggled with in this course and forgotten as soon as possible. First and foremost, statistics can be viewed as a branch of critical thinking and problem solving. Suppose you are in the market to buy a house. How helpful would it be to download thousands of the most recent house prices? If you find that using a chart or table of the prices is more helpful, you are working with statistics. If you decide *not* to consider the prices of houses sold in Nunavut (unless you're moving there), you are making a statistical judgment. If you estimate how large a down payment that you can raise within a year, you are making a forecast. We all face these types of problems and, in many cases, statistics can help in their solution.

Statistics can be defined as the art of collecting, classifying, interpreting, and reporting numerical information related to a particular subject. In the house price example, decisions must be made about what price data to include and how to obtain it, and how (or whether) to distinguish the sizes of properties or types of houses; then, decisions must be made about the conclusions that can be drawn from the data. Some books define statistics not as an art but as a science, but this term possibly inflates its mystique. Yes, many powerful techniques have been developed in statistics—many of these are presented in this book. But eventually *you* must use your own judgment in each case about which techniques to apply and what limitations they may have for your application. (And there *will* be limitations.)

For conducting research in medicine, business, psychology, agriculture, and many other disciplines, the role of statistics is essential. In medicine, new drugs are tested using formal experiments. Business uses statistics in many areas, including forecasting, marketing research, accounting, quality control, and setting prices. Psychologists apply statistics in their studies of human behaviour, and in agriculture, new crop varieties and pesticides are tested experimentally. In every case, statistics provides a tool to help ensure that better decisions are made.

In one sense, *statistics* is just the plural of *statistic*—a numerical fact of interest in some context. Statistics Canada and the media regularly publish such numbers: Canada's population in 2000; the number of serious crimes committed that year; the number of days since the last rainfall in Goose Bay, Labrador; the pollution count in Sackville, New Brunswick; the goals-against average of the Calgary Flames.

If you are involved with a related topic (demographics, the weather, sports, etc.), these numbers can be interesting. But as soon as you are tempted to compare them (the record of the Calgary Flames versus the Edmonton Oilers, for instance) or to make predictions (such as next year's pollution count in Sackville), you are engaged in the *art* of statistics as defined above. If Calgary's goals-against average is lower by four compared to Edmonton's, can you say that Calgary is "better" in this regard? Or could this small difference be random and of no consequence, or else due to another factor (maybe

Edmonton was playing short-handed for longer periods of time)? Such questions go beyond the simple numbers as reported on TV, and require judgments about what data to consider and how to evaluate them.

This text introduces you to this art of statistics. Examples using real-world data will be drawn from a variety of disciplines. You will learn how to describe the characteristics of data and how to make inferences from the data that are on hand about a larger **population** (the total set of objects or measurements that are of interest to a decision maker) or to a projected future time. Guidelines for gathering the suitable data will also be discussed.

Description and Inference with Statistics

Traditionally, a distinction is made between two sets of techniques within statistics: the use of descriptive statistics and the use of inferential statistics. **Descriptive statistics** are focused on summarizing and presenting information. Typical examples might be a pie chart or graph comparing populations of different provinces, a table comparing the amounts of ozone over North America over a series of years, or a world map showing imports of Brazilian cherry wood by various countries. In the examples below, Table 1.1 shows how many U.K. mining accidents occurred due to various causes during a certain time period, and Figure 1.1 uses a pie chart to illustrate how many of the accidents happened on the surface and how many happened underground. Figure 1.2 graphs chemical-emission trends over a number of years.

Table 1.1 Types and Locations of Accidents in U.K. Coal Mines, 1999–2000

	Location of Incident		
Description of Incident	**Surface**	**Underground**	**Total**
Falls of ground—headings		3	**3**
Falls of ground—longwall face		7	**7**
Falls of ground—elsewhere underground		1	**1**
Machine accident		6	**6**
Stumbling, falling, or slipping	9		**9**
Transport—conveyors		4	**4**
Transport—others		8	**8**
Miscellaneous	9	48	**57**
Total	**8**	**77**	**95**

Source: U.K. Health and Safety Executive. Based on data in *Accidents & Do's at Coal Mines 1996–2001*. Retrieved from http://www.hse.gov.uk/mining/accident/inc96_0_2.htm

Each of these examples displays characteristics of an available **data set** (the set of all observations for a given project or purpose). No claims are made explicitly about the possible results if a larger population was examined or what might happen in some future year. There is some choice in how to present these data—whether using tables, different types of graphs, or other illustrations. The best choice is one that conveys the necessary information as clearly as possible, while reflecting the level of knowledge of the audience. (Obviously, keep it simple, especially if your audience will not have much statistical background.)

Inferential statistics goes beyond the data set that is at hand. Its techniques are designed for making estimates, or inferences, about the characteristics of a population, based on information found in one or more **samples** (representative subsets of the population). In practice, even descriptive statistics are hard to examine without making some inferences.

Figure 1.1
Locations of Accidents in U.K. Coal Mines, 1999–2000

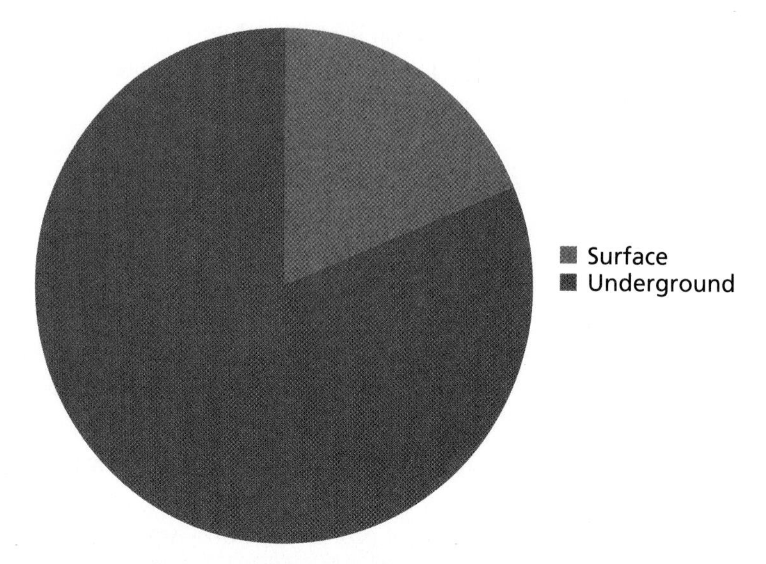

Source: U.K. Health and Safety Executive. Based on data in *Accidents & Do's at Coal Mines, 1996–2001.* Retrieved from http://www.hse.gov.uk/mining/accident/inc96_0_2.htm

Figure 1.2
Emissions of Manganese and Chromium to Water from Member Operations of the Canadian Chemical Producers' Association, 1992–2003

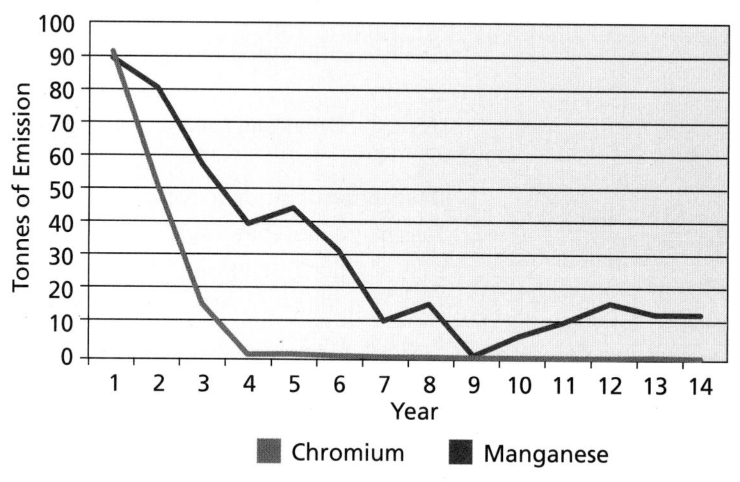

Source: Canadian Chemical Producers' Association. Found at: http://www.ccpa.ca/files/Library/Reports/NERM12full.pdf

Inspecting Table 1.1, for example, we see that for the observed time period, *all* stumbling and falling incidents happened on the mine surface. You may find yourself concluding that "this type of incident *generally,* or maybe *always,* happens on the surface of the mine" (i.e., even in *other mines* or in *other years*), but this type of inference takes you beyond the data. The techniques of inferential statistics help to ensure that the conclusions you reach of this type are supported by the evidence.

Two main types of problems call for inferential statistics:

1. *Estimate* a characteristic of a population, based on data from a sample.

   Example: A survey finds that 52% of questioned voters support a certain piece of legislation. Estimate the proportion of *all* voters that support the legislation.

2. *Test a claim or assumption* about a characteristic of a population, based on data from the sample.

Section 1.0

Example: A reporter claims that "a majority" of voters support a certain piece of legislation. A survey finds that 52% of questioned voters support that legislation. Does this evidence support the reporter's claim?

This textbook will help you to solve both types of inferential problems.

• *In Brief 1.1 Loading Up Your Data*

*(Based on the **Emissions_of_Metals** file)*

Each numbered section of a chapter includes at least one "In Brief" feature. Each of these features has subsections on three versions of statistical software: Excel, Minitab, and SPSS. If you are unsure about the basics of your software product, you are encouraged to consider other resources, such as Help screens, specialized manuals, and Internet-based support. These "In Brief" features are focused on the statistical content of each section being covered and help to ensure that you make the most use of your software to automate key procedures related to the section's content.

Excel

Consulting your manual if necessary, open your Excel software application now. Make sure the main worksheet screen is displayed. You will often be required in this textbook to load and open a data set that is stored on the CD provided with this book. The data file names listed in this textbook will not display extensions (the three letters following a period at the end of the file name). If you are using Excel, assume that each file's extension is ".xls" as in **"Emissions_of_Metals.xls"**—the file used for Figure 1.2.

Begin with the menu sequence *File / Open / Data;* make sure the box at the upper left of the resulting dialogue box displays the directory where the data files that you want can be found, as shown in the screen capture. If the correct directory is not displayed, you can click the down-arrow (circled in the screen capture) to display a drop-down menu from which you can select the correct disk drive and directory to find the data. Then input the name of the file you want and click *Open* (or, alternatively, double-click on the file name, which should be displayed).

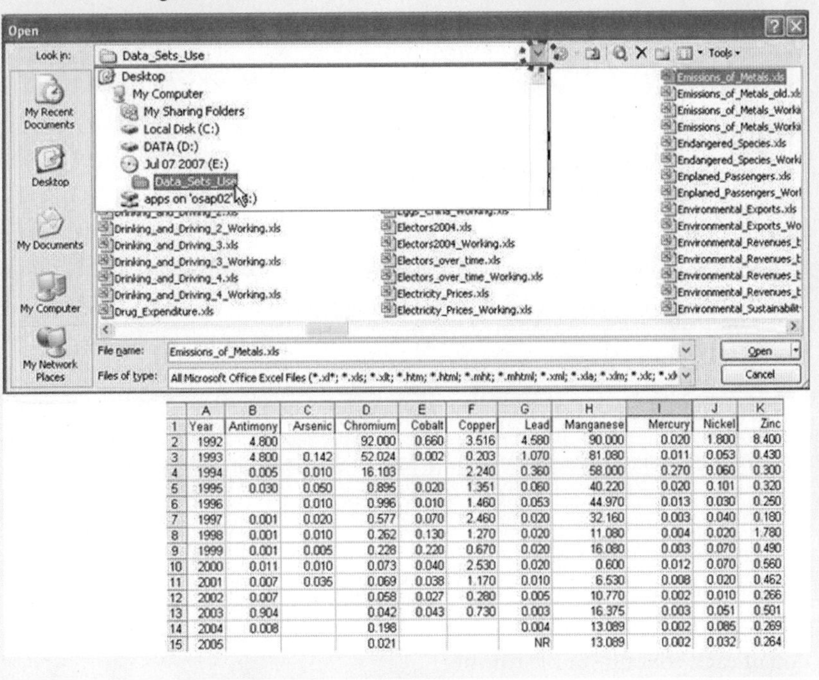

	A	B	C	D	E	F	G	H	I	J	K
1	Year	Antimony	Arsenic	Chromium	Cobalt	Copper	Lead	Manganese	Mercury	Nickel	Zinc
2	1992	4.800		92.000	0.660	3.516	4.580	90.000	0.020	1.800	8.400
3	1993	4.800	0.142	52.024	0.002	0.203	1.070	81.080	0.011	0.053	0.430
4	1994	0.005	0.010	16.103		2.240	0.360	58.000	0.270	0.060	0.300
5	1995	0.030	0.050	0.895	0.020	1.351	0.060	40.220	0.020	0.101	0.320
6	1996		0.010	0.996	0.010	1.460	0.053	44.970	0.013	0.030	0.250
7	1997	0.001	0.020	0.577	0.070	2.460	0.020	32.160	0.003	0.040	0.180
8	1998	0.001	0.010	0.262	0.130	1.270	0.020	11.080	0.004	0.020	1.780
9	1999	0.001	0.005	0.228	0.220	0.670	0.020	16.080	0.003	0.070	0.490
10	2000	0.011	0.010	0.073	0.040	2.530	0.020	0.600	0.012	0.070	0.560
11	2001	0.007	0.035	0.069	0.038	1.170	0.010	6.530	0.008	0.020	0.462
12	2002	0.007		0.058	0.027	0.280	0.005	10.770	0.002	0.010	0.266
13	2003	0.904		0.042	0.043	0.730	0.003	16.375	0.003	0.051	0.501
14	2004	0.008		0.198			0.004	13.089	0.002	0.085	0.269
15	2005			0.021			NR	13.089	0.002	0.032	0.264

An example of an opened worksheet file is shown at the bottom of the illustration. Note the standardized format: Each variable name, such as *Year,* is listed at the very top of each column, with the corresponding data lined up below. If you wanted to input your own data, you would use the same format.

Minitab

Consulting your manual if necessary, open your Minitab software application now. Make sure a worksheet screen is displayed. You will often be required in this textbook to load and open a data set that has been stored on the CD provided with this book. The data file names listed in this textbook will not display extensions (the three letters following a period at the end of the file name). If you are using Minitab, assume that each file's extension is ".MTW" as in "**Emissions_of_Metals.MTW**"—the file used for Figure 1.2.

Begin with the menu sequence *File / Open Worksheet;* make sure the box at the upper left of the resulting dialogue box displays the directory where the data files that you want can be found, as shown in the screen capture. If the correct directory is not displayed, you can click the down-arrow (circled in the screen capture) to display a drop-down menu from which you can select the correct disk drive and directory to find the data. Then input the name of the file you want and click *Open* (or, alternatively, double-click on the yellow icon to the left of the file name, which should be displayed). The program will advise you that the worksheet will be added to the current project; click *OK.*

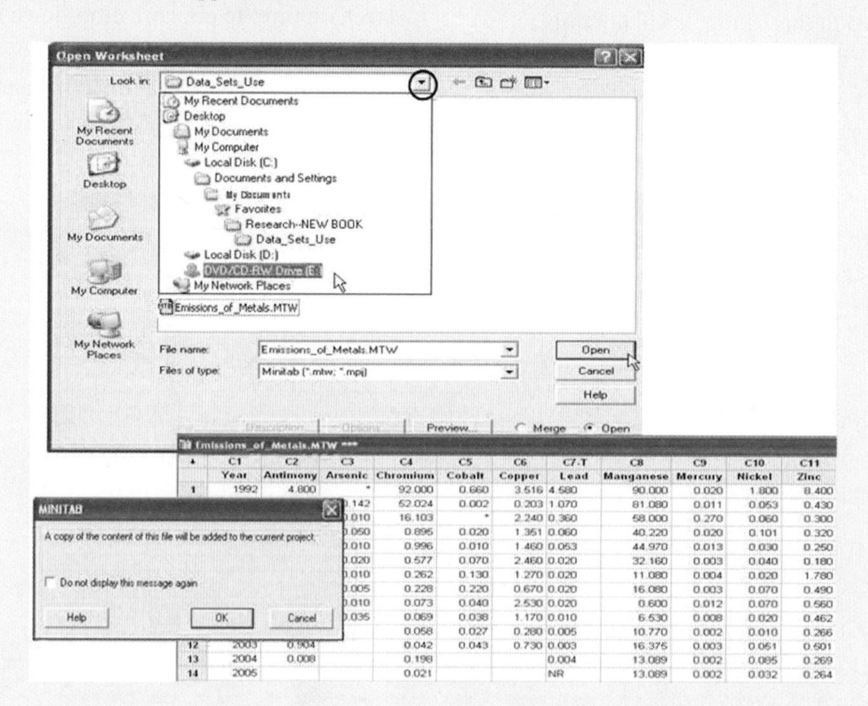

An example of an opened worksheet file is shown at the bottom of the illustration. Note the standardized format: Each variable name, such as *Year,* is listed at the very top of each column (in the row of grey cells), with the corresponding data lined up underneath. If you wanted to input your own data, you would use the same format.

SPSS

Consulting your manual if necessary, open your *SPSS* software application now. Make sure that *Data View* is selected by using the tabs at the bottom of the screen. You will often be required in this textbook to load and open a data set that has been stored on the CD provided with this book. The data file names listed in this textbook will not display extensions (the three letters following a period at the end of the file name). If you are using SPSS, assume that each file's extension is ".sav" as in **"Emissions_of_Metals.sav"**—the file used for Figure 1.2.

Begin with the menu sequence *File / Open / Data;* make sure the box at the upper left of the resulting dialogue box displays the directory where the data files can be found, as shown in the screen capture. If the correct directory is not displayed, you can click the down-arrow (circled in the screen capture) to display a drop-down menu from which you can select the correct disk drive and directory to find the data. Then you input the name of the file you want and click *Open* (or, alternatively, double-click on the icon to the left of the file name, which should be displayed).

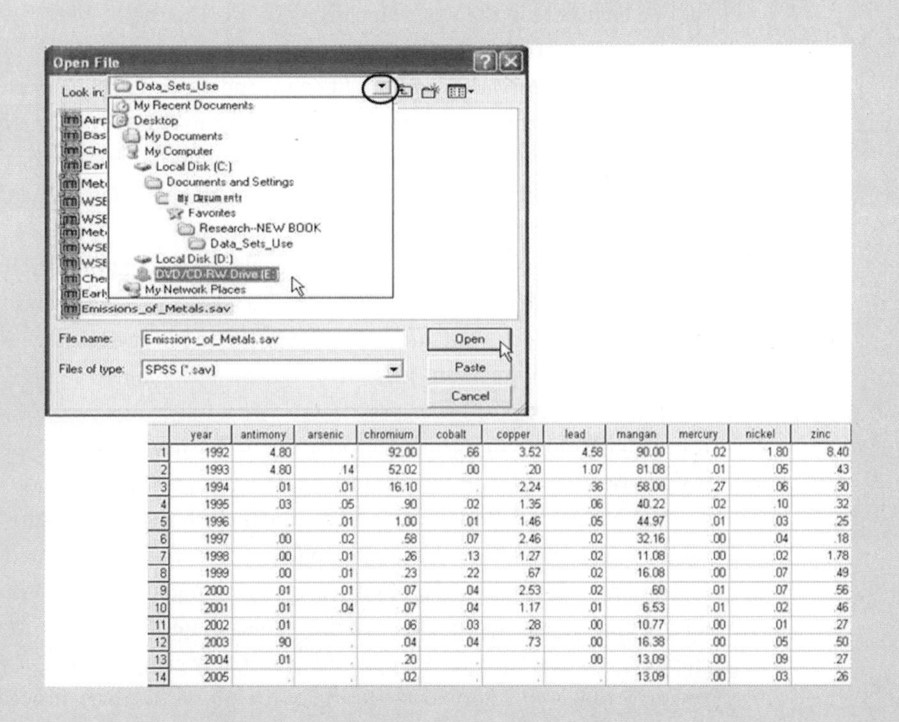

	year	antimony	arsenic	chromium	cobalt	copper	lead	mangan	mercury	nickel	zinc
1	1992	4.80	.	92.00	.66	3.52	4.58	90.00	.02	1.80	8.40
2	1993	4.80	.14	52.02	.00	.20	1.07	81.08	.01	.05	.43
3	1994	.01	.01	16.10	.	2.24	.36	58.00	.27	.06	.30
4	1995	.03	.05	.90	.02	1.35	.06	40.22	.02	.10	.32
5	1996	.	.01	1.00	.01	1.46	.05	44.97	.01	.03	.25
6	1997	.00	.02	.58	.07	2.46	.02	32.16	.00	.04	.18
7	1998	.00	.01	.26	.13	1.27	.02	11.08	.00	.02	1.78
8	1999	.00	.01	.23	.22	.67	.02	16.08	.00	.07	.49
9	2000	.01	.01	.07	.04	2.53	.02	.60	.01	.07	.56
10	2001	.01	.04	.07	.04	1.17	.01	6.53	.01	.02	.46
11	2002	.01	.	.06	.03	.28	.00	10.77	.00	.01	.27
12	2003	.90	.	.04	.04	.73	.00	16.38	.00	.05	.50
13	2004	.01	.	.20	.	.	.00	13.09	.00	.09	.27
14	2005	.	.	.02	.	.	.	13.09	.00	.03	.26

An example of an opened worksheet file is shown at the bottom of the illustration. Note the standardized format: Each variable name, such as *Year,* is listed at the very top of each column (in the row of grey cells), with the corresponding data lined up underneath. If you wanted to input your own data, you would use the same format. However, to input the column headings manually requires an extra step: Click on the *Variable View* tab at the bottom of the worksheet. Manually input the variable names (i.e., column headings) at the left of each row (one row per column heading). When you click on the *Data View* tab at the bottom of the worksheet, you will see that the headings have been changed.

1.2 Populations, Samples, and Inference

The fundamental concepts of populations, samples, and inference were all introduced in Section 1.1. In this section, these and other terms commonly used in statistical analysis will be defined more precisely.

As already defined, a population represents the total set of objects or measurements that are of interest to a decision maker. Each object may have one or more features or characteristics that are of interest. Suppose you are in the market to buy a new car. The population may be every car that is potentially available to you for purchase. This could include cars displayed currently at accessible dealerships and cars there or elsewhere that could be ordered. Each car in this population has features you will want to consider: Price, special financing, colour, number of seats, storage space, delivery costs, operating costs, and so on.

In a statistical sense, the population just described is one type of **infinite population.** It is not "infinite" in the sense having no limit (since the number of cars for sale in the world is less than, say, a limit of one billion). But in practice, it is not realistically possible to identify and *count* every member of the population, especially if you include cars for sale anywhere in the world, including on the Internet.

On the other hand, a scientist who is tracking an endangered species of large mammal in a preservation area might count and tag every single member of that species. In a **finite population,** it is possible and realistic to count every member. Similar to the car example, more than one feature of each population member may be of interest. The scientist may be tracking values for weight, length, gender, and approximate age, for example.

A set of observations taken from every member of a population is called a **census.** Every five years, for example, Statistics Canada takes a census of all households in Canada. Ideally, information is obtained for everyone living in the country. A census is possible only for a finite population.

It is often not possible or practical to take a full census even for a finite population. If General Motors ran a crash test on every brand-new truck, this would be a census. It would also ruin a lot of trucks. Instead, the company takes a representative subset (a sample) of the truck population and conducts crash tests only on this subset. The idea is for the information from the sample to tell us something about the overall population.

You may not realize how often you use sampling in daily life. When cooking a meal, do you ever taste a spoonful to see if the whole pot needs more salt? If you get your blood checked, you don't give *all* of your blood; you only give a portion. If you road-test a car, you drive only *some* kilometres—not all of the kilometres that will accumulate in the car's lifetime. In each case, you are interested in the whole population, but you are relying on a sample for your data.

The relationship between a population and a sample is illustrated in Figure 1.3. The population is the nationalities of the 55 winners of the Nobel Prize in Literature from 1951 to 2003. Twelve of those nationalities were sampled, as shown in the lower portion of the figure.

A **parameter** is a value for summarizing the measurements of a quantifiable feature of the population. For example, the proportion of winners of the 1951–2003 Nobel Prize in Literature who came from the former U.S.S.R. is a particular feature of that population; its exact value (i.e., the parameter) is 3/55. If the corresponding feature is meas-

Figure 1.3
Nationalities of Winners of the Nobel Prize in Literature, 1951–2003

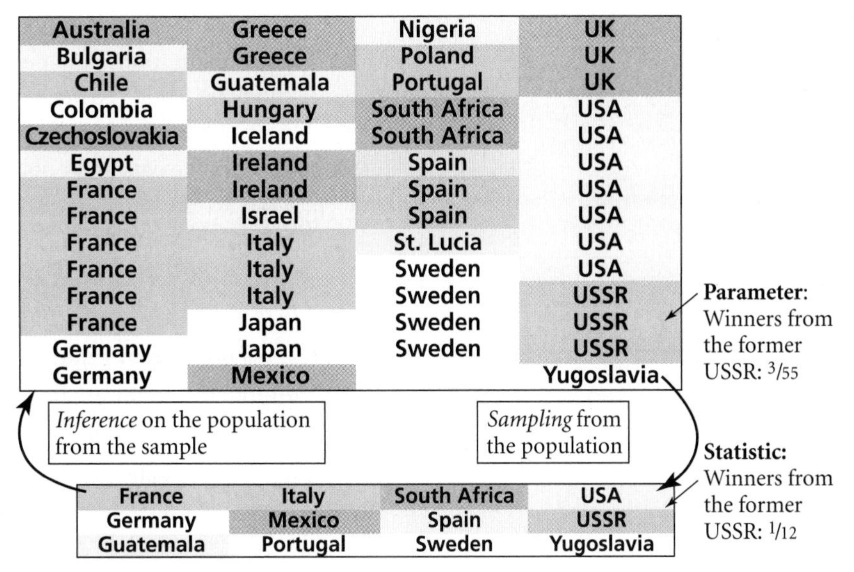

Source: Nobel Foundation. Nobel Prize for Literature winners (1951–2003). Found at: http://nobelprize.org/nobel_prizes/literature/laureates

ured in the sample, it is called a **sample statistic.** Based on the sample, we might estimate (or **infer**) that the sample statistic is "close to" or "approximately equal to" the population parameter. For example, one of the 12 nationalities for Nobel Prize winners included in the sample (1/12) is the U.S.S.R. We might therefore infer that *in the population* about 1/12 of all winners of the 1951–2003 Nobel Prize in Literature came from countries in the former U.S.S.R. In this case, the estimate appears to be reasonable but not perfect. However, if you look next at St. Lucia, shown in bright yellow in the population list, you will find that 1 of the 55 winners (1/55) came from there, yet *none* of the winners in the sample (0/12) happened to come from St. Lucia.

Why Use Samples?

The St. Lucia example illustrates a risk of using sample statistics to estimate population parameters: The sample statistic may not equal precisely the parameter to be measured. Indeed, some elements in the population (such as Nobel Prize winners from St. Lucia) may be missing from the sample. So why not always analyze the full population in a census? Why do we ever rely on a sample? Several valid reasons for using samples, in spite of these risks, are listed below.

1. **Time constraints:** It may be too time-consuming to survey or observe all of the members of the population. Several possible reasons for this include:
   a) **Decision deadlines:** Every summer, city beaches are inspected for levels of fecal coliform bacteria. If levels at any beach are too high, the public is warned to avoid swimming in those waters until further notice. Suppose it were possible to test every litre of water within swimming distance of the beach. Surely this would take a lot of time and, meanwhile, people would keep swimming. Sampling allows inspectors to make their close/don't-close decisions for each beach in a timely fashion, and to post the necessary signs.

Another example is public opinion polling by newspapers and television networks. These organizations have deadlines for publishing their results. They may be tracking weekly changes in support for election candidates. Again, the benefits of approaching every possible voter are negated if it is impossible to collect the data and publish or broadcast on time.

b) **Trends:** Suppose you are expanding your business internationally and would like to know the proportions of households in Argentina, Brazil, Mexico, and Colombia that use cellphones. These rates have been increasing each year (i.e., there is an "upward trend"), which means that if you take a long time to conduct a census, the population parameters will be changing even as you collect the data. (For more details on trends, see Chapter 17.)

c) **Seasonality:** Some variables increase and decrease in value over time in a cyclical pattern. Sales of bathing suits go up in early summer and are lower in the late fall. Depending on where you live, the local population of robins increases and decreases periodically, as a result of migration. Knowing that such "seasons" soon pass is similar to having a decision deadline: If you take too long counting robins in the spring, the count is out of date in the fall. (For more details on seasonality, see Chapter 17.)

2. **Cost constraints:** Not surprisingly, it costs less to survey fewer people or make fewer observations than to survey or observe a whole population. Even the Canadian census process acknowledges this fact. There is a short form that is expected to be completed by every household, so it is a true census. But there is a set of additional questions that is distributed to a minority of households; this part is really a sample, designed to increase the overall information provided. The costs of printing, distributing, and analyzing the larger version, plus ensuring that people comply, are reduced, compared to requiring everyone to answer the long version.
3. **Unknown population size:** Infinite populations are in this category. Suppose you are deciding whether to buy advertising time during the next Grey Cup broadcast. Your product is sportswear. To study the potential buyers of sportswear among Grey Cup viewers, who is in the population and where do they live? How many people are in that population? It is not possible to answer these questions—especially before the game. Instead, you might decide to survey a sample of individuals who are known to buy sportswear and to watch football on TV.
4. **Destructive tests:** How fast can a vehicle be driven into a barrier without denting the vehicle's bumper? How much heat and flame can a child's blanket endure without igniting? Companies need to know such information to ensure the safety of their products. Obviously, destroying every bumper or blanket to test these limits makes no sense. Even bumpers or blankets that pass the tests will probably be scratched or scorched, and could not then be sold. Clearly, then, only a sample of the product can reasonably be subjected to destructive tests.
5. **Greater accuracy of some samples:** You may be getting the impression that sampling is a necessary evil. It can introduce error, but for reasons of cost or time, you go ahead and use it. Less obvious is that sampling can possibly yield more accurate results than a census.

A small team of trained individuals may be able to conduct a very reliable survey based on a sample, whereas a census of the population, for example, requires more people to administer it, possibly including part-timers who may have inadequate training, experience, or motivation. This could lead to errors in how questions are worded or to mechanical errors in collecting the data. With so many people working on the survey, there may be inconsistencies in how different people interpret the questions and answers or code people's responses, or communications mix-ups. A well-planned survey by a qualified team may avoid some of these problems.

Representative Samples

A good sample should be **representative** of the population that it is drawn from. The term is hard to define exactly, but Figure 1.4 should convey the basic idea. Ideally, *every* sample statistic would approximate its corresponding population parameter. In the figure, this ideal is close to being met; we would say that this sample is representative of the population.

Figure 1.4
A Representative Sample

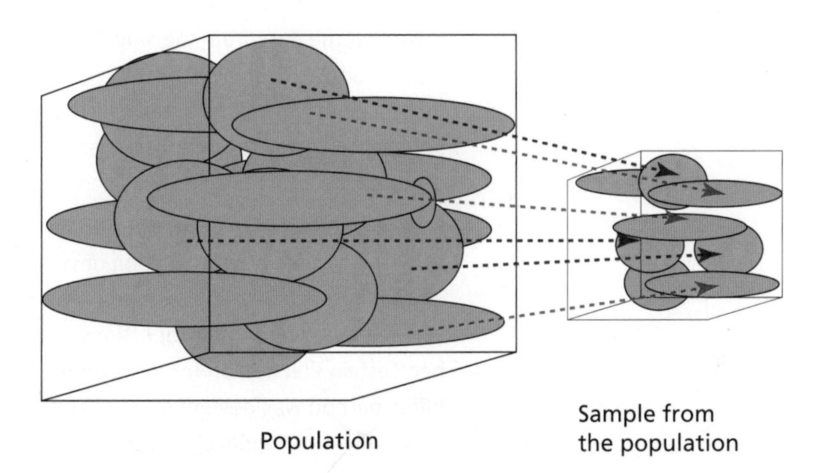

For example, in *both* the population and the sample shown in Figure 1.4, about half of the elements are balls, about half are blue, the vertical and horizontal distributions of elements in the boxes are similar, and so on. In practice, it is not important that every aspect of the population be matched in the sample, as long as those aspects that are the focus of the analysis are represented. For instance, the sample appears to have a smaller proportion of blue circles near the centre of the cube, but if our analysis is unconcerned with this property, the sample remains a good one.

Example 1.1

1. You would like to know the class average for the test you just wrote. You ask your friends for their marks so you can calculate the average.
2. A public company is being audited for possible misuse of funds. An auditor randomly picks 30 expense reports that were filed by staff on business trips; she examines the proportion of those reports that declared over $85 as the cost of an individual business lunch.

3. Throughout the 2005 federal election campaign, SES Research conducted daily polling of eligible voters' electoral preferences. Samples were selected so as to reflect the geographical distributions of the Canadian population.

For each of the above cases, indicate the following: (a) What is the population and what is the sample? (b) What is the inference to be made? (c) Is it reasonable to assume that the sample will be representative of the population? Why or why not?

Solution

1. (a) The population is the full set of the test marks for every student in the class who wrote the test. (Typically, the class average ignores the marks for those who did not write the test.) The sample is the less complete set of marks that you collected from your friends. (b) Based on the marks scored by your friends, you plan to infer the class average for the whole class. (c) No, it is not reasonable to assume that the marks scored by your friends are representative of the marks scored by the whole class. Your friends may happen to be people who tend to do especially well—or poorly—in such tests.
2. (a) The population is the complete set of expense reports filed by staff at the company. The sample is the 30 expense reports that were randomly selected. (b) Based on the data from the sampled reports, the auditor plans to infer the proportion of reports overall that declared over $85 for a business lunch. (c) As will be discussed in the next chapter, the auditor's method of randomly selecting the sample helps to ensure that the sample is reasonably representative of the population.
3. (a) The population is the set of all eligible Canadian voters at the time of the election. The sample is the selection of voters who were approached by SES Research and took the survey. (b) Based on the data from the sampled voters, inferences can be made about the proportions of voters, overall, as well as by region, that had certain election preferences on the day of sampling. (c) The described sampling method was designed to help ensure a representative sample, at least in terms of voters' regional distributions.

• *In Brief 1.2 Samples and Populations in Your Statistical Software*

Specific techniques related to sampling with the help of your computer will be discussed in the next chapter. A more general observation is made below.

Excel

Excel does not build in any assumptions about whether a stored data set is to be interpreted as a sample or a population. The software includes functions for working with either case. Therefore, if you work with Excel, it is your responsibility to determine whether you are working with sample or population data, and to key in (when it makes a difference) the appropriate formulas for the applicable case.

Minitab and SPSS

Both Minitab and SPSS are designed to presume that the data you load are for a sample. Where needed, some ways around this assumption will be described in subsequent "In Brief" features. The population-based calculations may tend to be less convenient to execute than the built-in sample versions in the software.

• *Hands On 1.1* *Representative Samples*

*(Based on the **Accident_Coal_Mines** file)*

In Section 1.2, the importance of representative sampling was stressed. Some formal techniques for sampling will be discussed in the next chapter. The goal of this exercise is to try out, informally, some possible schemes for sampling and to see how each scheme fares in terms of generating representative samples.

First, open the data file **Accident_Coal_Mines.** A description of each incident is stored in the second data column (labelled "Classif"). To see the total number of incidents of each type in the full population, compare the left and right columns of Table 1.1. We see that there were three cases of "Falls of ground—headings" and more than twice as many cases (seven) of "Falls of ground—longwall face," etc. A sample that is representative should roughly preserve these relative proportions among the incident types.

Try these five schemes for selecting a sample from all of the recorded incident descriptions:

1. Select the first 15 descriptions in the data set.
2. Select the 3rd description in the second column, then the 9th description, then the 15th description, and then every 6th description thereafter until you have 15 descriptions in your sample.
3. Arbitrarily write 15 numbers on a piece of paper. Each number should be between 1 and 95. Try your best *not* to think of any pattern—whether the numbers are high or low, or close or far from each other does not matter. Then pick out your sample from the ordered column of descriptions in this fashion: Use each of your 15 numbers in turn to select *positions* in the second column. The description in that position becomes part of the sample. (For example, if one of your 15 numbers was 5, then pick the 5th value down in the second column. That is "Falls of ground—longwall face" becomes part of the sample. Continue with the same procedure based on the next of 15 numbers you wrote down.)

4. Repeat procedure 3, above, but *reselect* 15 arbitrary numbers in the range of 1 to 95. Again try not to aim for any particular order, and it is OK to use again a number that you used in (3).
5. Think of another scheme of *your own* choosing to select a sample of 15 incident descriptions from the population.

For each of the above cases, write or input the sample that results. How does the sample compare with the original population? Would you say that your sample is reasonably representative of the population? Why or why not? Can you explain how the sampling method you used contributed to the apparent representativeness or nonrepresentativeness of the result?

1.3 Data Types and Levels of Measurement

The relationship between data and information is not always clear-cut, but as defined on a NATO-sponsored website, *information* is "data that have been transformed through analysis and interpretation into a form useful for drawing conclusions and making decisions."[1] **Data** therefore are the raw materials for these analyses. They are sets of numeric or nonnumeric facts that represent records of observations. A single observation is called a **datum.** The set of all observations for a given project or purpose is called our *data set.*

Consider the locations of accidents in U.K. coal mines in 1999–2000. Safety officials observed and recorded the location of each accident. Each recorded location is a datum. Your data file **Accidents_Coal_Mines** is a data set, recording the data for all the accident locations (plus other observations). Figure 1.1, on the other hand, represents information. Statistics have played the role of summarizing and displaying the data in a new way that is easier to interpret and apply to decision making. Because the statistics start from data, it is important to understand data's nature and distinctions.

Two basic forms of data are constants and variables. Both of these refer to characteristics being observed for each member of a sample or population. A **constant** is an observed characteristic whose value does not change over time or in different experiments. The ratio of a circle's circumference to its diameter is a constant—the number pi ($\approx$3.1416). Suppose an archeologist has discovered an old structure that features repeated circular patterns. When recording data about the circles, the archeologist does not need to record repeatedly for each circle that 3.1416 is the ratio of its circumference to its diameter. This value is constant for every circle in the pattern. Similarly, a baseball statistician does not need to track the number of defensive players on the field during play; that number will always be 9, a constant.

A **variable** also refers to an observed characteristic. However, the values of a variable can change from observation to observation. In the archeologist example, the diameters of all the observed circles may differ from one another. In baseball, statisticians may track the number of pitchers used in a game or the number of strikeouts thrown in the game. The value of this variable can easily vary from game to game.

A particular type of variable—the **random variable**—is the focus for many statistical procedures. "Random" refers to something that happens by chance or is unplanned. So, informally, a random variable is a variable whose value cannot be known in advance, prior to its discovery by observation or experiment. Examples could include the number of hits in the next Blue Jays game, or the number of chromosomes counted in a newly discovered cell.

Figure 1.5 illustrates the types of data to be discussed in this section. For completeness, random variables are contrasted with nonrandom variables. For example, if we have a time series that starts with the years 2006, 2007, . . ., we *do* have knowledge of the next (nonrandom) value in the series; namely, 2008. The random variables, in turn, can be subdivided—into qualitative or quantitative random variables.

Figure 1.5
Types of Data

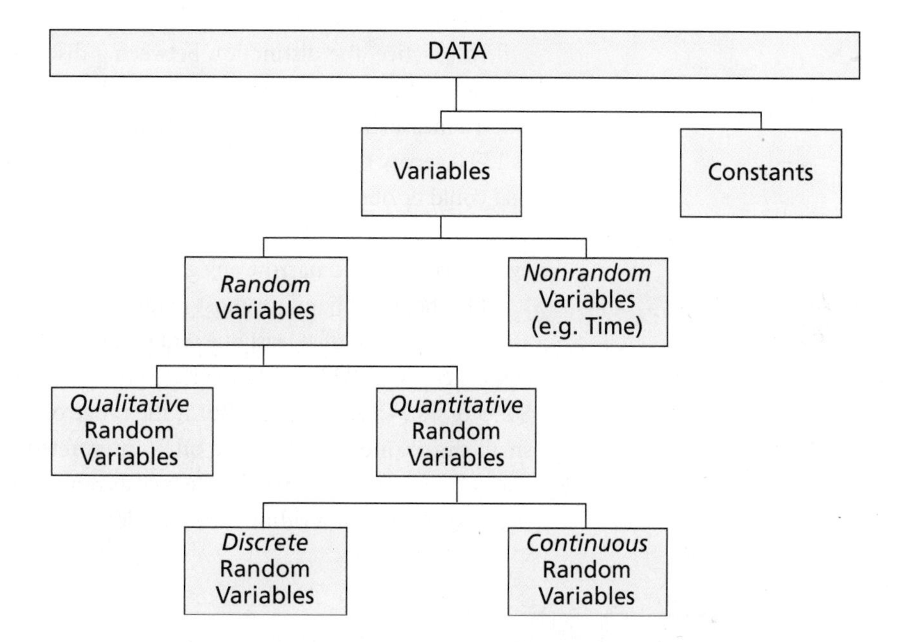

An observation of a **qualitative** (or **categorical) random variable** does not result in a numerical value. Instead, one observes a trait or characteristic that can be classified into one of a number of **categories.** Suppose you are sampling your classmates to identify their political preferences. Responses might be classified into categories such as "Conservative," "Liberal," "NDP," "Other party," or "No preference." No numbers are associated with these values. It is possible to arbitrarily code "Conservative" as "1," "Liberal" as "2," and so on, but these codes should be interpreted as being only labels.

If a random variable takes values that vary in magnitude, it is classified as a **quantitative** (or **numerical) random variable.** The number of circular patterns on a wall, the number of hits in a ball game, and the proportion of bumpers that crumple in a crash test are all quantitative random variables.

A quantitative random variable can be either discrete or continuous. There are restraints on the values that could be observed for a **discrete random variable.** If you are counting all functioning cars on Toronto's Don Valley Parkway, 1, 2, 30, or 4,000 cars might be involved; there cannot be 1.5 or 20.75 cars. (We are ignoring any wrecked vehicles on the shoulder!) Or all items' prices in a dollar store might be one of $1, $0.50 (i.e., two for a dollar), or $0.25 (four for a dollar). Observe that decimal values are possible for a discrete variable—but there are gaps between the possible values that the variable can assume.

On the other hand, a **continuous random variable** can possibly assume *any* value over a particular range of values: There are no gaps between the possible values. Variables with values obtained by measurement, such as heights or weights, are this type of variable. Possible values of a randomly observed person's height could be 179.00 cm, or 179.01 cm, or 179.005 cm; there is no gap in the possible values.

There may be limits to the *range* of possible values for a continuous variable: According to the website http://www.famousheights.com, the shortest height on record for a human adult is 45.7 cm and the tallest is 2.72 m.[2] It would be hard to exceed these extreme values, but biologically it appears possible for a randomly observed height to fall anywhere between those bounds.

In practice, the distinction between a discrete and continuous variable is relative. Few metre sticks could precisely distinguish a height of 179.00 cm from a height of 179.01 cm. If a measuring instrument is capable of **precision** only to four significant digits (such as 179.0 versus 179.1), then in practice there is a small gap between the measured values that could be observed. The preciseness of the measurement depends on the instrument used for the measuring. A measured variable is generally interpreted as continuous, however, since we could narrow any gaps by using a more precise instrument.

On the other hand, in most contexts, the possible values of a money-related variable will have gaps of at least one cent ($0.01). An item purchased at Canadian Tire cannot cost exactly $0.33333—not even if a sign says "3 for $1." The possible cost values are only $0.33 or $0.34 for one item. But if the range of possible costs is large compared to this small gap, money amounts are often interpreted as if they were continuous. At Canadian Tire, for example, possible item costs range from pennies for nuts and bolts to thousands of dollars for a riding mower. Price "gaps" of one penny are scarcely noticed.

Section 1.1

Example 1.2

You are collecting various data in the process of building your first-ever shed. Identify the types of data listed below, using the appropriate classifications from Figure 1.5.

1. The *type of screwdriver* required for each screw in the construction kit. (Note: The kit includes some slotted, Phillips, Robertson, and hex screws.)
2. The *speed of falling, after 0.3 seconds,* for any screws that you drop.
3. The *clock time* when you begin your next half-hour break. (Note: You are using a precision 24-hour digital clock, set to Greenwich Mean Time.)
4. The *clock time* when you are ready to begin installation of the roof.
5. *The number of hammer blows* required to secure the first roof shingle.

Solution

1. Several types of screwdrivers are required, so this is a *variable.* Until you observe each screw, you do not know which type of screwdriver it requires, so this is a *random* variable. Values such as "slotted" or "Phillips" are not numerical; they merely classify the available types. Therefore, the type of screwdriver required is a *qualitative random variable.*
2. As Galileo demonstrated 400 years ago when he dropped various objects off the Leaning Tower of Pisa, all objects (discounting air resistance, etc.) fall with the same rate of acceleration, regardless of their weight or mass. Therefore, the speed that a dropped object reaches after 0.3 seconds is a *constant,* and does not need to be measured by experiment.

3. The clock time is always changing, so this is a *variable*. But if you know your start time and begin your breaks on each subsequent half-hour, then no experiment is needed to know the time for your first break, your second break, and so on. This is a *nonrandom variable*.
4. As in (3), clock time is a *variable*. But since you have never built a shed before, you do not know in advance how long it will take to get to the point where you can install the roof. So the clock time when you reach that point is a *random variable*. The value to be displayed is numerical (in units of hours plus minutes), so it is a *quantitative random variable*. Limited only by your clock's precision, the instrument might display *any* time between the range of 00:00 to 24:00 when you begin the roof. Therefore, this is a *continuous random variable*.
5. Since not all objects require the same number of hammer blows to be secured, the number of blows is a *variable*. Until you try it, you will not know how many blows it takes to secure the first shingle. So this is a *quantitative random variable* because the value will be some number of blows. Only a limited number of values are possible: one blow, or two, or three, etc. This is therefore a *discrete random variable*.

Levels of Measurement

When performing statistical analyses, we choose procedures that are appropriate to the types of data being studied. We have just examined the main types of statistical data that may be available. Similarly, not all data convey or contain the same amount of information. The higher the **level of measurement** of data, the more information these data contain. The levels of measurement and their relationships are illustrated in Figure 1.6.

Figure 1.6
Levels of Measurement

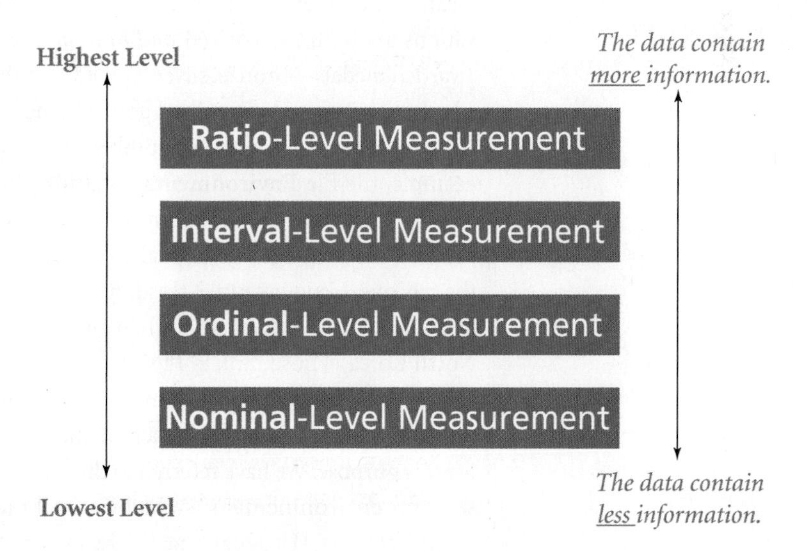

Nominal-Level Measurement

Nominal-level measurement is the lowest level of measurement. Qualitative data are generally measured at this level, unless the data values (categories) imply some kind of ranking (such as "Good," "Better," and "Best"). For example, in the data underlying Figure 1.1, the locations of the accidents in U.K. coal mines are categorized into either "Surface" or

"Underground." There is no order to the categories. Although for data entry on a computer we could *code* the category names to look like numbers (e.g., "Surface" is "1" and "Underground" is "2"), there is no quantitative interpretation for the numbers; they are just symbols for categories.

If you plan to use nominal-level data, two important rules must be observed when you are designing the categories:

1. *The full set of categories should cover all possible observations.* Mine-worker accidents, for example, might happen in a wide variety of places. If "Surface" and "Underground" are the only available categories, then it must be possible to classify *any* accident under one of these headings. What if a worker has a car accident while driving on company business? If the locations "Surface" or "Underground" do not apply, then there must be an additional location category (perhaps a category "Other") to handle such cases. In political polls, there is often an "Undecided" category for respondents who cannot be classified under a particular party name or viewpoint.
2. *All of the categories should be mutually exclusive.* That is, an observation cannot be classified into more than one category. If a mine accident occurs in an elevator before it begins moving to the lower levels, there must be no confusion about whether to classify it as a surface or underground incident. Mistakes are often made because such borderline cases are not anticipated when designing the categories or because rules that everyone will follow are not documented to guide decisions on how such cases will be handled.

If these rules are followed, then every observation will be classified into one and only one category.

Ordinal-Level Measurement

Ordinal-level measurement conveys more information about each observation. Observations are both categorized *and ranked.* The three top swimmers in the Olympics are awarded medals—bronze, silver, or gold. But these are not equal classifications: Silver has a higher rank than bronze, and gold a higher rank than silver.

It is very common to use numbers as the values for ordinal-level data. Consider, for example, the file **Environmental_Stability,** based on an environmental stability index (ESI) published by Yale and Columbia university researchers. Based on ESI calculations, countries' overall environmental performances are ranked. Canada ranks sixth among the nations; Uruguay ranks third. By this measure (on a lower-is-better scale), Canada is in a less favourable position than Uruguay—but both countries are doing better than North Korea, whose rank is 146.

This use of numbers for ranked data is convenient for data input and manipulation on a computer, but there is also a potential for misunderstanding when numbers are used for this purpose. We have referred to three countries ranked at 3, 6, and 146 on the ordinal scale for environmental sustainability. All that we are entitled to conclude is that the rank-3 country (Uruguay) has more sustainability than the rank-6 country (Canada), and Canada has more sustainability than the rank-146 country (North Korea). Only the ordering property of the rank numbers is relevant. In terms of real improvement in the environment, there may well be a tremendous gap between a rank-3 and a rank-6 country, while the gap between a rank-6 country down to a rank-146 could possibly be relatively small. In short: *The mathematical differences between rank numbers have no interpretation; only the relative order of the numbers conveys information.*

Interval-Level Measurement

Data measured at the interval-level are numbers. As for ordinal-level data, these numbers can be compared meaningfully to distinguish higher from lower ranks. But interval-level measurements convey an extra amount of information: Equal differences between numbers on the measurement scale correspond to equal differences between the real-world characteristics being measured. Put it another way: Any one-unit change anywhere on the scale corresponds to a known constant unit of change in the actual characteristic.

Common examples of measurements at this level are based on temperature data. For example, the temperatures at two pairs of cities were recorded on a certain July date:

1. Iqaluit: 13° C; Whitehorse: 14° C
2. Edmonton: 21° C; Moncton: 22° C

For both pairs of cities, the differences between the temperatures are the same: For Whitehorse compared to Iqaluit, 14° C − 13° C = 1° C. For Moncton compared to Edmonton, 22°C − 21°C = 1° C. A difference of 1° C represents a known, real amount of extra heat—the amount defined as 1/100th of the increase in temperature required to bring pure water at sea level from the freezing point to the boiling point. Every 1° difference in measured temperatures between two objects or places represents this same, known difference in the amount of heat.

Index values, such as measures of inflation or stock market indexes, are also examples of data measured at the interval level. Compare the daily "close" prices listed in the file **Dow_Jones_Industrial_Average_Index.** The index provides information about the prices of stocks being traded on the U.S. stock market. Compare the following two pairs of index values:

1. January 2004: 10488.07 May 2004: 10188.45
2. February 2005: 10766.23 May 2005: 10467.48

For both pairs of dates, the differences between the earlier and later index values are almost identical: ≈300. These two equal decreases on the scale represent equal decreases in actual stock values for the two periods.

In other words, for data measured at the interval level, both *the relative orders of the numbers and the mathematical differences between the numbers convey information.* However, *the ratios between numbers on these scales have no interpretation.* A temperature of 28° C *does not* represent twice as much heat as a temperature of 14° C, even though, as numbers, 28 = 2 × 14. The reason is that for interval scales, there is *no true zero point.* 0° C does not mean "no heat"; it is just an arbitrary starting point. Without a true zero, ratios on a scale have no interpretation.

Ratio-Level Measurement

Data measured at the ratio level convey the most information. At this level, *the relative orders of the numbers, the mathematical differences between the numbers, and the ratios between numbers on the scale* all apply meaningfully to the characteristic being measured. Compare three objects weighing 20 grams, 40 grams, and 60 grams, respectively. The increasing order of measured values reflects a genuine increase in weight from one object to the other. And just as the intervals from 20 to 40 and from 40 to 60 both have the same value (i.e., 20), so too the actual differences in weight between objects 1 and 2, and between objects 2 and 3, are the same. Lastly, the ratio of 2/1 between 2nd and 1st numbers, respectively, reflects a genuine property of the measured objects' weights: The second object has twice as much weight as the first object.

This last property of meaningful ratios can apply only when a scale originates at a "true" of zero. The scale for weight does have this property ("0 grams" literally means no weight), whereas the scale for temperature does not ("0° C" does *not* mean "no heat"). In performing statistical analysis on a data set, note carefully that not all statistical techniques apply validly for all levels of measurement.

Levels of measurement can be related to the types of data discussed earlier in this section.

- Qualitative random variables can be measured at one of two levels:
  1. Nominal level (e.g., "Ford," "Chrysler," "GM") ***or***
  2. Ordinal level (e.g., "very bad," "somewhat bad," "neutral," "good"). Note that qualitative categories may have the *appearance* of being quantitative (e.g., "That course ranks 4 on a scale from 1 to 5.")
- Discrete quantitative random variables can be measured at one of two levels:
  1. Interval level (e.g., "This dryer can be set to only 20°, 30°, or 40° C.")
  2. Ratio level (e.g., "The coin dryer can be set for 5, 10, 15, 20, or 25 minutes.")

     (Note: A sequence of ordinal variables with apparently numerical values, such as "0, 1, 2, 3, or 4" may also *appear* to belong in this list, but actually these are qualitative, rather than quantitative, random variables.)
- Continuous quantitative random variables can be measured at one of two levels:
  1. Interval level (e.g., "His temperature is 37° C.")
  2. Ratio level (e.g., "That vehicle weighs 1,400 kg.")

Example 1.3

Continuing Example 1.2, you are collecting various data in the process of building your first-ever shed. Identify the appropriate levels of measurement for each variable:

1. The *type of screwdriver required* for each screw in the construction kit.
2. The *shed model* on a scale: Basic, Deluxe, or Super Deluxe.
3. The *speed of falling, after 0.3 seconds,* for any screws that you drop.
4. The *clock time* (in Greenwich Mean Time) displayed on a 24-hour digital clock when you are ready to begin installation of the roof.
5. The *number of hammer blows* required to secure the first roof shingle.

Solution

1. This qualitative variable is measured at the nominal level.
2. This qualitative variable is measured at the ordinal level.
3. This constant is measured at the ratio level (since there is a true zero—i.e., there could be *zero* speed of falling toward the ground).
4. This continuous quantitative variable is measured at the interval level, since "00:00.00" (i.e., midnight Greenwich Mean Time) on the clock does not mean an absence of time.
5. This discrete quantitative variable is measured at the ratio level (since it is possible to have zero hammer blows).

• *In Brief 1.3 Types of Data and Levels of Measurement in Statistical Software*

For data input in the standard format described in "In Brief 1.1," each column conventionally represents values of a variable. Alternatively, you could repeat the value of a single constant down an entire column. The software does not distinguish automatically between a constant and a random or nonrandom variable. Your software *can* distinguish between qualitative and quantitative data, as discussed in the Excel, Minitab, and SPSS subsections below.

With regard to levels of measurement, the computer may actually contribute to errors. If you input qualitative data as text, the computer will not distinguish between ordinal data (e.g., "good," "better," "best") and nominal data (e.g., "red," "white," "blue"). If you input data (such as "100") as numbers, then the computer will *assume* that all numbers are ratio numbers, whether you actually intend them as ranks (e.g., 100 on a scale of 1 to 100), or as interval level (e.g., 100 to signify a consumer price index), or even as just a category. Be careful to avoid making calculations or misinterpreting computer outputs that are not appropriate to the level of measurement.

Excel

If you start inputting the numbers 4, 6, 32, and so on into an Excel column, Excel will presume that these are quantitative values. Observe in the illustration how these numerical values line up on the right side of the cells. If you input words such as "slotted," "hex," and "Phillips," Excel presumes that these are qualitative values. Observe in the illustration how these text values line up on the left side of the cells. (These formats can be overwritten.) If you input a number such as 999 as a code for a category, Excel will not recognize this automatically. To avoid problems, you should instruct Excel that the number is only a text code: Highlight any code numbers in your data and use the menu sequence *Format / Cells* to call up the dialogue box shown in the screen capture. Click *Text* then *OK* to make the format change. (All of the other illustrated formats are for different ways of showing quantitative data—with dollar signs, percentage notations, date formats, and so on.)

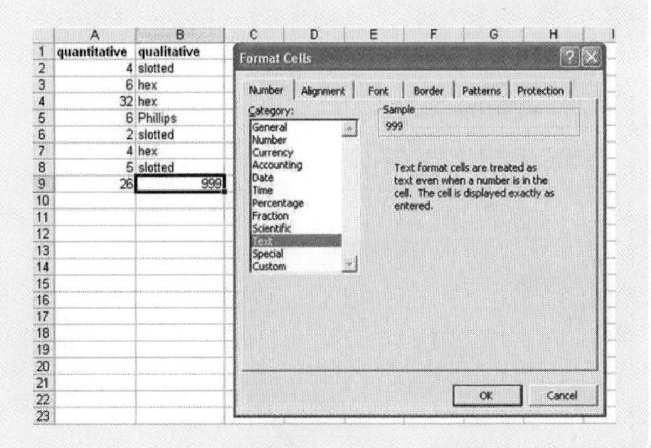

Minitab

If you start inputting the numbers 4, 6, 32, and so on into a Minitab column, Minitab will presume that these are quantitative values. Observe in the illustration how these numerical values line up on the right side of the cells. If you input words such as "slotted," "hex," and "Phillips," Minitab presumes that these are qualitative values. Observe in the illustration how these text values line up on the left side of the cells. Also note the changed column

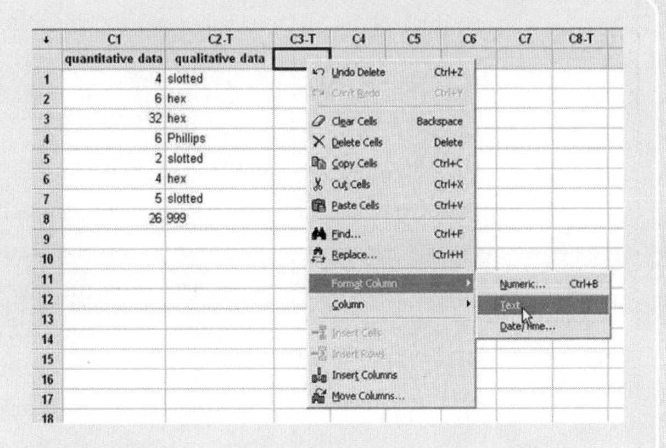

heading ("C2-T"); the "-T" indicates text (qualitative) data. If you input a number such as 999 in a column identified as text, Minitab will interpret the apparent number as just a string of characters.

Suppose you wanted to input the code 999 at the top of a *new* column. You would have to instruct Minitab that this will be a text column: Right-click anywhere on the intended new column, and then right-click *Format Column* on the menu. Then click to choose *Text*. Everything you now input into that column will be interpreted as text.

SPSS

If you start inputting the numbers 4, 6, 32, and so on into an SPSS column, SPSS will presume that these are quantitative values. Observe in the illustration how these numerical values line up on the right side of the cells. If you input words such as "slotted," "hex," and "Phillips," SPSS presumes that these are qualitative values. Observe in the illustration how these text values line up on the left side of the cells. If you input a number such as 999 into a column identified as text, SPSS will interpret the apparent number as just a string of characters.

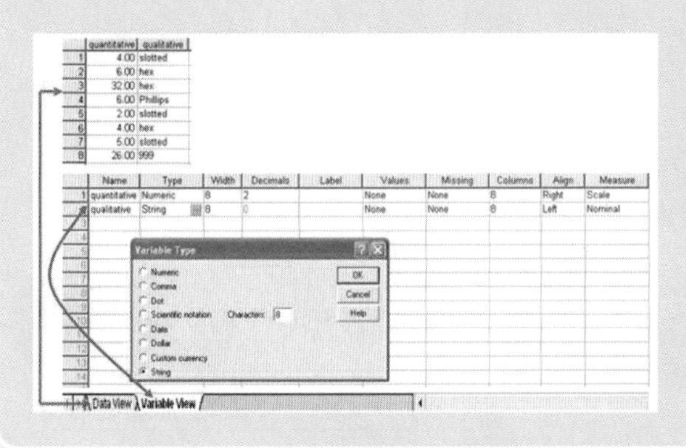

Suppose you wanted to input the code 999 at the top of a *new* column. You would have to instruct SPSS that this will be a text column: Click on the lower right-hand tab labelled *Variable View* (see the illustration). Each row sets the properties for a column in the data view. (Note that the illustration shows names input for the column/variables.) Right-click on the right side of the "Type" column for the variable to be interpreted as qualitative. In the dialogue box shown in the screen capture, click on *String*, and optionally specify how many characters can be used for each category name. Click *OK*. Everything you now input into that column will be interpreted as text. (All of the other illustrated formats are for different ways of showing quantitative data—with dollar signs, percentage notations, date formats, and so on.)

• Hands On 1.2 *Questioning Internet Data Sources*

Without question, the Internet is a great source of information, but you should question and approach this resource critically. A great deal of what we find posted includes already-prepared summaries or presentations of information, and accessing the raw data is not always possible. Even if the raw data are unpublished, you can at least ask what types of data would have to be consulted to back up the published claim. Then make your judgment: Is this Internet source likely (1) to have accurate access to the needed data and (2) to present and manipulate it competently and fairly? Compare these two examples:

1. On July 25, 2005, the Canadian Football League (CFL) published some midseason statistics for the Saskatchewan Roughriders, one of which was that the "PCT complete" rate for the Roughrider quarterbacks was 63.9% (i.e., 63.9% of attempted passes were completed).[3] What data would be needed to calculate or prove this? Records of every offensive play of that team would be required, with a variable to indicate which plays were attempted passes. This would be a categorical random variable. (Other values might be "attempted lateral," "quarterback sneak," and so on.) Also required would be a categorical variable (with values like "yes" and "no") to indicate whether an attempted pass was completed. If the CFL had these data, it could easily tally the numbers of attempts and successes and compare the two. Critical questions: (a) Is the CFL likely to have access to this data? (b) Is it reasonable to presume the accuracy of its conclusion? It is probable that, given this source, the answer to both questions is yes, but make your own judgment.
2. An Internet newsletter that claims to help people make money by buying and selling the right stocks at the right times published an "average gain" figure of over 200% for its stock pick suggestions in 2003. Ideally, the raw data would be a list of stocks actually traded by the Internet source, with the variables "net amount invested" and "net amount returned" for each stock. But the Internet vendors of such advice might not even buy the stock themselves—they might just recommend the transaction. Who then are the people in the population who actually bought and sold stocks on the published advice? Were these folks able to buy and sell at exactly the same prices as the newsletter recommends (and how much commission did they have to pay)? Did all clients invest an equal amount, and did all invest equally in of all the recommended stocks—or could some have invested more in the losers than in the winners? In short: We really need a database of every buy/sell transaction of every actual client of the advice service. The data could include random qualitative variables for stock names and investor IDs, plus random quantitative variables for amounts paid and amounts received per transaction. Did the Internet source have this data? Are you prepared to trust its judgment?

Your first "Hands On" activity is to find websites that present information on topics that interest you. (Repeat the exercise for four separate websites, related to four different topics.) Using the above examples as a guide, pick out at least one summary conclusion that is made on each site. (Possible examples are claims that concert sales are going down or that the world's temperature is going up, and by how much in each case.) Identify the types of variables that would be needed to support that particular claim. Do you think the Internet source has actually looked at all required data and has reported accurately and objectively? Or, if you are doubtful, why?

A second "Hands On" activity continues the exercise begun in the previous paragraph. For each of the possible variables that you identified, write down the level of measurement that would be suitable for that variable. Once again: In your judgment, do you think the Internet source would have the necessary access to reliable information or the capability to obtain that needed level of measurement for all of the presumed variables? Explain.

Chapter Summary

1.1 Define the term *statistics*, and distinguish between descriptive and inferential statistics.

Statistics was defined as "the art and process of collecting, classifying, interpreting, and reporting numerical information with respect to a particular subject." In **descriptive statistics,** the focus is on *summarizing* and *presenting* information. In **inferential statistics,** we estimate (or we **infer**) the characteristics of a population, based on information in one or more samples. Two main types of inferential statistics are: (1) *estimating* a characteristic of a population, based on data from the sample, and (2) *testing a claim or assumption* about a characteristic of a population, based on data from the sample.

1.2 Describe the difference between a population and a sample, and explain some advantages of sampling.

A **population** is the total set of objects that are of interest to a decision maker. In a **finite population,** it is possible to list or count every member of the population. In an **infinite population,** it is either not possible to count every member (for example, the population of all points on a line), or not realistic to count all members (for example, the full population of flies in Ontario on June 15 of a particular year). The collection of data for *all* members of a population is called a **census.** A value for summarizing the measurements of a quantifiable aspect of the population (such as a proportion) is a **parameter.**

A **sample** is a subset of the population that is observed and studied to provide information about the overall population. A **sample statistic** is a summary measure for the sample that corresponds to (and can help to estimate) a parameter of the population. Making a conclusion about a parameter based on a statistic is called *making an inference.* A good sample should be **representative** of the population from which it is drawn; ideally, every statistic taken from the sample would approximate its corresponding population parameter.

Sample-based inferences are not always accurate, yet there are many reasons for using samples as opposed to attempting a full census of the population. These may include time and cost constraints, an infinite population, the necessity to use destructive tests for sampling, and the possibility that a sample may be *more* accurate than a census, especially if trained researchers and a more manageable data set are used for the sample.

1.3.1 Name and describe the basic types of data.

The raw materials for statistics are **data.** These are sets of facts (numerical or nonnumerical) that record the results of observations. A single observation is called a **datum,** and the set of all observations for a given purpose is called a **data set.** Two basic forms of data are **constants,** whose values do not change over time, and **variables,** whose values may change from observation to observation. For a **random variable,** its next observed value is not known until it is observed.

Random variables can be categorized in several important ways. For a **qualitative** (or **categorical**) **random variable,** the observed values are not numerical; they are traits or characteristics that can be classified into one of a number of **categories.** For a **quantitative** (or **numerical**) **random variable,** the observed values are numbers that can vary in magnitude. There are two types of quantitative random variables: A **continuous random variable** can possibly assume *any* value over a particular range of values. (Based on the **precision** of the observed measurements, however, there can be apparent "gaps" between possible observed values—such as between 10.1 cm and 10.2 cm—for measurements precise to only three significant digits.) For a **discrete random variable,** the numerical values that could be observed are constrained (beyond just the issue of precision); for example, possible values may be limited to just whole numbers.

1.3.2 Explain the differences among the four levels of measurement.

The higher the **level of measurement** of the data, the more information these data contain. The lowest of the four levels of measurement is *nominal-level measurement.* Data measured at this level are all of the categorical type (e.g., "M" or "F"). Every observation should be classifiable into only one category, and there should be no order or rank (implicit or explicit) in the classifications.

The next level of measurement above the nominal level is *ordinal-level measurement.* Data measured at this level may appear to be categories (e.g., "Never," "Sometimes," "Always") or numbers (e.g., "1," "2," "3"). In either case, the data values can be arranged into a meaningful order. However, if numbers are used, only the relative *orders* of the numbers convey meaningful information; any apparent mathematical *differences* between the numbers *have no interpretation* at this level of measurement.

The third level of measurement is *interval-level measurement.* All values are numbers, and the numbers' relative order *and* any mathematical *differences* between the numbers *both* convey meaningful information. However, because there is no true zero point, the apparent *ratios between numbers* on this scale *have no interpretation.* Examples are temperatures in degrees Celsius or values of the Dow Jones Index.

The highest level of measurement is *ratio-level measurement.* Again, all values are numbers, and *all* of the following properties of the measured numbers convey meaningful information: *the relative orders of the numbers, the mathematical differences between the numbers, and the ratios between the numbers on the scale.* Examples include weights measured in kilograms and prices in dollars.

• Key Terms

categories, p. 17
census, p. 10
constant, p. 16
continuous random variable, p. 18
data, p. 16
data set, p. 5
datum, p. 16
descriptive statistics, p. 5
discrete random variable, p. 17
finite population, p. 10
infer, p. 11
inferential statistics, p. 5
infinite population, p. 10
level of measurement, p. 19
parameter, p. 10

population, p. 5
precision, p. 18
qualitative (or categorical) random variable, p. 17
quantitative (or numerical) random variable, p. 17
random variable, p. 16
representative, p. 13
sample, p. 5
sample statistic, p. 11
statistics, p. 4
variable, p. 16

Exercises

Basic Concepts 1.1

1. Explain the difference between the term *statistics*—when it means the plural of *statistic*—and the concept of statistics, which is the subject of this course.
2. Give a definition for the art of statistics.
3. What is the difference between descriptive statistics and inferential statistics? Explain.
4. Describe how knowledge of statistical techniques may be applied in *three* different areas of study or work (such as market research or weather forecasting).
5. List three examples of the use of descriptive statistics.
6. What are the two main types of problems that call for inferential statistics? Explain.
7. For each of the following statements, indicate whether the statistics being used are descriptive statistics or inferential statistics:
   a) The class average for the first midterm statistics exam was 66%.
   b) Based on a review of 200 randomly selected real estate transactions in Calgary, the value of house prices in the city has increased since last year.
   c) Of all the species of lichen in Japan, 4.5% are included on the endangered list.
   d) Over 75% of NBA players are at least 183 cm tall.
   e) Based on the results of a survey, a majority of employers attempt to monitor the websites visited by employees during work hours.

Basic Concepts 1.2

8. Discuss the advantages and disadvantages of sampling, compared to surveying an entire population.
9. If you are given a set of numbers with no explanation, can you tell whether it represents a sample or a population? Explain.
10. A naturalist is monitoring the health of white sturgeon living in the Kootenay River in British Columbia.
    a) Describe the population of interest to this researcher.
    b) Is the population finite or infinite? Explain.
    c) Discuss a possible sample of sturgeon that could be used by the naturalist to make an inference about the population.
11. Does a census always provide better information than a sample? Explain.
12. A large retailer in Whitby, Ontario, had to recall some defective products that had been sold by mistake. In order to assess how the customers affected by the recall felt about the company's response to the problem, the owner mails everyone on the recall list a questionnaire about the issue.
    a) Describe the population of interest to the store owner.
    b) Is the population finite or infinite?
    c) Recommend at least one way that the store owner could improve the data collection process to ensure that good data are gathered.
13. Look in a recent newspaper for an article that reports on the results of a survey or provides other numerical information.
    a) Do the data represent a sample or a population? Explain.
    b) Do the data represent descriptive or inferential statistics?
14. Desperate to score a run in the final inning of a baseball game, the manager replaces the next scheduled batter with a pinch hitter. Up to this point in the season, the pinch hitter has gotten on base 55% of the time that he has gone to bat. With respect to the baseball season as a whole, is the above ratio a sample statistic or a population parameter? Explain.
15. The following headline is published just before an election: "Poll results are in: 60% of voters support Franklin." What is the implied inference of this statement? Under what circumstances might the inference be mistaken?

Applications 1.2

16. Each month, approximately 54,000 households are surveyed by Statistics Canada as part of the *Labour Force Survey.* One objective of this survey is to determine the proportion of Canadians (15 years old or older) who are in the labour force. In the survey report for October 2006, this proportion was reported to be 67.1%.[4]
    a) Describe the population of interest for that survey.
    b) Describe the sample.
    c) What was the value of the sample statistic? What was the estimated value of the population parameter?
    d) Explain why for this monthly report, Statistics Canada did not utilize a complete census to collect the data.

17. The file **Religions** shows the stated religions of Canadians according to the 2001 Canadian census. Data for British Columbia are shown below.[5]

Religious Affiliation	Christian: Catholic	Christian: Protestant	Christian: Orthodox	Christian: Other	Muslim	Jewish	Buddhist	Hindu	Sikh	Other Eastern Religion	Other	No Religious Affiliation
Frequencies	675,320	1,213,295	35,655	200,345	56,220	21,230	85,540	31,500	135,310	9,970	16,205	1,388,300

a) Do these data represent descriptive or inferential statistics?
b) From these data you can calculate the proportion of British Columbians who have no religious affiliation. Is this value a parameter or a statistic? Explain.

18. According to a letter written by the Secretary-General of the United Nations in 2006, the countries shown in the table below were the top 10 contributors to UN funding in 2004.[6]

Country	U.S.	Japan	U.K.	Germany	Netherlands	Canada	Norway	France	Sweden	Italy
Amount contributed in 2004 (millions of US$)	3,651.6	1,590.7	1,163.9	755.7	664.2	518.1	516.0	512.3	505.8	453.9

a) Do these data represent descriptive or inferential statistics?
b) From these data you can calculate the proportion of funding that was given by Canada. Is this value a parameter or a statistic? Explain.

19. On his website, a home inspector ranks the top 10 problems he encountered in 400 homes that he inspected.[7]

Problem Rank	Problem Category
1	Surface grading and drainage: Not correctly installed
2	Exteriors: Penetration by water and air
3	Poor ventilation
4	Incorrect electrical wiring
5	Roof damage
6	Heating system problems
7	Plumbing problems
8	Structure-related problems
9	Poor maintenance
10	Minor interior items, not affecting the integrity of the house

a) This ranking is based on a sample of 400 home inspections. Describe the population.
b) Can we be confident that the sample is representative of the population? Why or why not?
c) Suppose that these 400 house inspections were all of houses in a rundown part of the city. Would this affect your answer to (b)? Explain.

20. In 2002, Natural Resources Canada completed interviews with a sample of 1,500 randomly selected adult Canadians. Participants were asked to assess which of the following industries were causing significant damage to the environment. The percentage of those who believed an industry was causing such damage is shown in parentheses in each case: forestry (43%); oil and gas (43%); coal (35%); mining (25%); hydroelectric (21%).[8]

Would you believe that in 2002, the percentage of all adult Canadians who believed that the coal industry was causing significant damage to the environment was *exactly* 35%? Explain.

21. According to a report published by a major Canadian accounting firm, 19% of Canadians polled would hide their income from the Canada Revenue Agency, if they thought they could get away with it.[9]
a) Describe the population of interest in the reported study.
b) Make an inference from the sample result to the population parameter.

22. A data set on coal production in the United States was compiled by an energy planner based in Alabama. (See the file **Coal Production.**) She is comparing the 2004 ratio of active surface mines to active underground mines in the state. Judging from the data set, is the above-mentioned ratio in Alabama representative of the ratio of those same types of mines in other states? Why or why not?

23. In the summer of 2003, SES Canada Research Inc. conducted a Canada Day survey of 500 randomly selected beer drinkers. Each was asked what one Canadian item they'd most like to have if stranded on a desert island. (For the set of results, see the file **Desert_Island.**)
a) Describe the population of interest for this study.
b) Is the population finite or infinite?
c) 61% of those polled choose Canadian beer as the item of choice. Can you infer that 61% of *all* Canadian adults would choose Canadian beer? Why or why not?

Basic Concepts 1.3

24. What is meant by the term *variable?* What is the difference between a constant and a variable?

25. What is the difference between a *quantitative* random variable and a *qualitative* random variable? Provide an example of each of these types of variables.

26. What is the difference between a *discrete* random variable and a *continuous* random variable? Provide an example of each of these types of variables.

27. Determine which of the following random variables are discrete and which are continuous:
a) The number of students in your class.
b) The weight of your statistics book.
c) The height of the CN Tower.
d) The number of people who watched the Grey Cup game on television.

28. Explain the following statement: "The variable price for items listed in the Sears catalogue is, technically speaking, a discrete variable. But, in practice, it can be treated as if it were a continuous variable."

29. Determine which of the following random variables are quantitative variables and which are qualitative variables:
a) The salaries of members of Parliament.
b) The modes of transportation that students take to class.
c) The colour of one's eyes.
d) The number of herring gulls that eat fish from Lake Ontario.

30. Indicate the level of the measurement for the following sets of observations:
a) Daily values of the Dow Jones Industrial stock index.
b) Daily prices of McDonald's stock on the New York Stock Exchange.
c) The jersey numbers assigned to hockey players.
d) Degrees in Celsius.
e) Blood types of ambulance patients.
f) Ranking of a university in Maclean's annual survey results.
g) Weights of patients in an outpatient clinic.
h) The number of defects per car coming off an assembly line.
i) The amount of money invested by foreign companies in Canada in 2006.
j) The proportion of restaurant customers who are vegetarians.
k) Viewers' feedback on a new TV show, on a 1–5 scale.

31. Define the ordinal level of measurement. Give an example of a typical survey question whose answers are at the ordinal level.

32. Are data measured at the nominal level continuous or discrete? Explain.

33. Are data measured at the nominal level qualitative or quantitative? Explain.

34. You are developing a questionnaire on students' attitudes toward recycling. Provide examples of questions whose responses would be measured at these levels:
a) At the nominal level.
b) At the ordinal level.
c) At the ratio level.

35. A school's administrators are concerned about absenteeism at their school and about its effect on students' grades. The grade of each student in the study (measured as "low," "medium," or "high") was compared with his or her level of absenteeism (measured as "absent infrequently," "moderate absenteeism," or "absent frequently"). What levels of measurement are being used?

Applications 1.3

36. Some of the variables in a national hospital discharge survey are listed below.[10]
a) Specify which of the variables are quantitative and which are qualitative.
b) Specify the level of measurement for each of the variables.
i) Name of the medical diagnosis of the person leaving the hospital.
ii) Numerical code for the medical diagnosis of the person leaving the hospital.
iii) Age of the person leaving the hospital.
iv) Gender of the person leaving the hospital.
v) Type of admission (e.g., emergency) of the person into the hospital.
vi) Number of days of care before the person left the hospital.

37. A data set on coal production in the United States was compiled by an energy planner. (See the file **Coal Production.**)
a) Specify which of the three variables listed below are discrete and which are continuous. Explain your answers.
b) Specify the levels of measurement of each of these variables.
i) The number of surface mines in a certain state in 2004.
ii) The increase in the number of surface mines in a certain state from 2003 to 2004.
iii) The ratio of the number of surface mines in a state in 2004 to the number of surface mines in the same state in 2003.

38. A Canadian online source for diamond shopping provides data about diamonds for sale. (See the file **Diamonds.**) Specify which of the following variables from the data set are quantitative and which are qualitative. What is the level of measurement for each variable?
a) Name for how the diamond is cut, such as "round brilliant," "princess cut," or "emerald cut."
b) Numerical code for how the diamond is cut.
c) Weight of the diamond (measured in carats).

d) Alphabetical code for the colour of the diamond.
e) Price of the diamond in Canadian dollars.

39. In July 2006, *The Globe and Mail* reported on free agent trades in the NHL. Which, if any, of the variables listed below are continuous quantitative variables?
a) Team the player was traded *from*.
b) Total value of the contract in millions of U.S. dollars.
c) The player's new annual salary in millions of U.S. dollars.
d) Duration of the contract in complete years.

40. Are any of the variables listed in the previous exercise a *discrete quantitative* variable? If so, which one(s)?

41. Engineering Dynamics Corporation provides automobile-simulation software for use by those who engineer or test the safety of vehicles.[11] For analyzing tire performance, the program may include data on the level of tire inflation, cornering stiffness, radial tire stiffness, rolling resistance, and the ratio of a tire's circumference to its diameter. Which, if any, of these listed items are constants?

42. IQ (intelligence quotient) scores are interpreted by some as comparative measures of people's intelligence.[12] For a given age or reference group, the average score is assumed to be 100, and scores of, say, 25 or more above or below that value are taken to indicate notably high or low levels of intelligence. Given this information, can we say that a person with a score of 120 is "twice as intelligent" as a person with a score of 60? Answer this question with reference to the level of measurement of the data.

43. The questions below are taken from a medical survey on hearing loss used by Veterans Affairs Canada.[13] For each question, indicate the expected level of measurement of the response:

a) _______

Is this diagnosis: ○ confirmed or ○ provisional?

b) _______

Please choose the most appropriate statement:
○ Occasional tinnitus, present less than once a week affecting one or both ears.
○ Occasional tinnitus, present at least once a week affecting one or both ears.
○ Intermittent tinnitus, present daily, but not all day long, affecting one or both ears.
○ Continuous tinnitus, present all day and all night, everyday, affecting one or both ears, but does not require use of prescribed masking device ...
○ Continuous tinnitus, present all day and all night, everyday, affecting one or both ears, and requiring the ongoing use of prescribed masking device ...

c) _______

Please choose the most appropriate statement:
○ History of Vertigo/Disequilibrium but no current symptoms.
○ Intermittent symptoms of Vertigo/Disequilibrium with or without objective findings, such as nystagmus or ataxia.
○ Continuous symptoms of Vertigo/Disequilibrium are present with supportive objective findings, such as nystagmus or ataxia.

d) _______ e) _______

Height _______ Blood Pressure _______

Review Applications

For Exercises 44–51, use data from the file **National_Parks.**

44. Natural Resources Canada publishes a list of all Canada's national parks. Do these data represent a sample or a population? Were these data collected by taking a census or collecting a sample?

45. A tourist wants to quickly assess the typical size of national parks in Canada. She proposes to use just the parks in Ontario as a sample on which to base her estimate. Is this a good idea? Why or why not?

46. Which, if any, of the variables in the **National_Parks** file data are qualitative?

47. Which, if any, of the variables are discrete quantitative?

48. Which, if any, of the variables are continuous quantitative?

49. Which, if any, of the variables are measured at the nominal level?

50. Which, if any, of the variables are measured at the ordinal level?

51. Which, if any, of the variables are measured at the ratio level?

For Exercises 52–60, use data from the file **Nobel_Prizes_Literature.**

52. An aspiring author in Sweden hopes to someday win the Nobel Prize in Literature. To estimate the likelihood of a Swedish author actually winning this award in some future year, she collects the data in the file **Nobel_Prizes_Literature.** Given the author's intention to estimate a future possibility, do the data represent a sample or a population? Explain your answer.

53. A history major plans to write biographies of all Nobel Prize winners in Literature from 1951 to 2003. He collects the data in the file **Nobel_Prizes_Literature** with the intention of learning the names of all of the winners during that period. Given the history major's goals for the data, do the collected data represent a sample or a population? Explain your answer.

54. Can the proportions among the winners' nationalities in the data set be safely assumed to be representative of the proportions for winners that will occur over the next 50 years? Explain why or why not.

55. Which, if any, of the variables in the **Nobel_Prizes_Literature** file are qualitative?

56. Which, if any, of the variables are discrete quantitative?

57. Which, if any, of the variables are continuous quantitative?

58. Which, if any, of the variables are measured at the nominal level?

59. Which, if any, of the variables are measured at the ordinal level?

60. Which, if any, of the variables are measured at the ratio level?

For Exercises 61–67, use data from the file **Major_Bridges.**

61. A student of architecture is studying designs of the world's major of bridges. Find the major bridges—that exist *now*—that are located in Canada. Would it be reasonable to use these current data as a sample to make an inference about the length of the next major bridge to be built in Canada? Why or why not?
62. Which, if any, of the variables in the **Major_Bridges** file are qualitative?
63. Which, if any, of the variables are discrete quantitative?
64. Which, if any, of the variables are continuous quantitative?
65. Which, if any, of the variables are measured at the nominal level?
66. Which, if any, of the variables are measured at the ordinal level?
67. Which, if any, of the variables are measured at the ratio level?

Chapter 2

Obtaining the Data

Chapter Outline

Learning Objectives

2.1 Name and describe the basic types of research and the types of data sources.

2.2.1 Distinguish between statistical and nonstatistical sampling, and describe the basic types of statistical sampling.

2.2.2 Describe the sources of error associated with sampling.

2.3 Explain the importance of following accepted moral standards in both the collecting and the presenting of statistical data.

2.1 Sources of Data

In most of this book, we explore how to use a variety of techniques to utilize, display, and get the most out of data. But where do we find the data for these tasks? It is better to ask first: Why do we need the data?

As discussed in Chapter 1, we use statistics—and statistical data—to solve problems. Will there be enough rain next August for my crops? How will rising oil prices affect truck sales? You might take a guess (use intuition) or ask an expert (appeal to an authority) for answers. These may have their place in research, but if you are committed to a more **scientific approach** to problem solving, statistics can play an important role.

For a scientific approach, the search for data begins with the recognition of a problem to investigate. We develop a research plan. This plan will call for evidence and will identify the steps for obtaining relevant data. Figure 2.1 displays the generic steps in this research process, which is discussed in more detail in the next section. Many books offer variations of this chart, but most would agree on its basic principles.

Figure 2.1
The Research Process

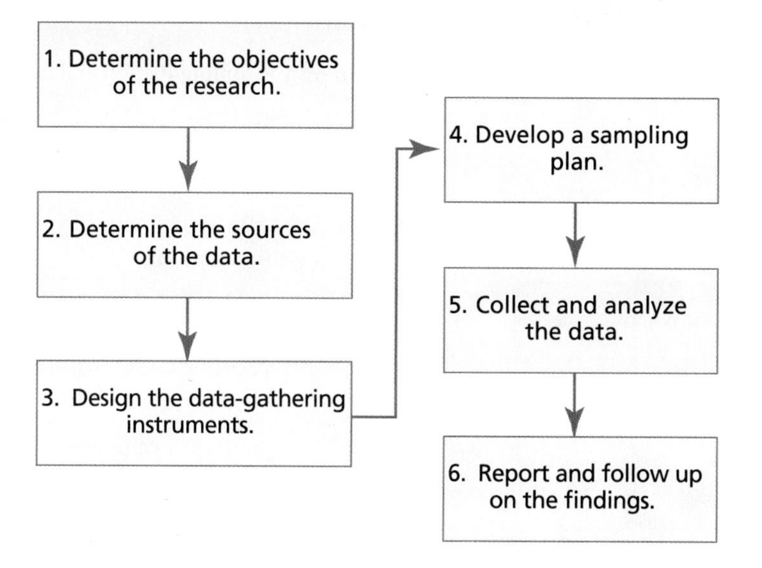

The first step is to formulate a clear statement of your **research objectives.** Time and money are required to conduct research and collect data, so if you cannot state the purpose for your endeavour in clear, unambiguous terms, you might want to reconsider the project. Stakeholders who could be affected by the research might also want to participate in the discussion.

In this first phase, you might brainstorm specific questions to be answered. A farmer may have many questions about how changing climate will affect his crops. He might formulate some educated guesses (research hypotheses) about what the answers might be. The researcher may think of more questions and hypotheses than a single study can address. Clear, well-worded objectives (there can be more than one) are therefore required to determine what this particular project will accomplish. For example, the farmer's objective may be to learn whether a change in crop is advisable in response to new climate conditions.

Notice that objectives reflect the nature of the decision to be made with the data. The farmer needs data to project future temperatures, plus data on how different crops respond to climatic conditions, plus data on the markets for alternative crops. A biologist may need data to correctly classify a new species in relation to other, more familiar species. The search for data will be guided by the objectives.

Types of Research

There are two basic types of research, depending on what is already known about the problem to be studied. The selected research type can impact how the data are collected.

1. **Exploratory research** is preliminary research, conducted when very little is known as yet about the problem being studied. The problem may be new, or perhaps familiar but ambiguous and in need of clarification. Examples of this type of research are **pilot studies** or **focus groups.** These projects or interactive sessions can help in better formulating the research problems, fine-tuning the questions in survey instruments, or defining other procedures for data gathering.
2. **Conclusive research** can be conducted when the research objectives are clear and the problems are unambiguous.
   - **Descriptive research** is used to describe the characteristics of a population. Data could be collected from either the full population (a census) or from a sample, with the goal to make predictions or inferences about a population. Many business applications, such as marketing studies and forecasting, are of this type.
   - **Causal research** designs (also called *experimental research* designs) go beyond describing a population as it is, to exploring possible cause-and-effect relationships among the observed factors. This approach is common in the sciences. For example, medical researchers may test whether increased exposure of mice to a chemical agent causes asthmatic reactions among normally healthy individuals.

Types of Data Sources

There are two main types of data sources: *primary data* and *secondary data.* Each type can be subdivided as shown in Figure 2.2.

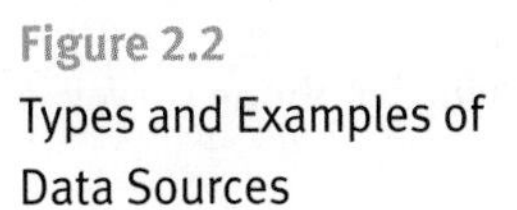

Figure 2.2
Types and Examples of Data Sources

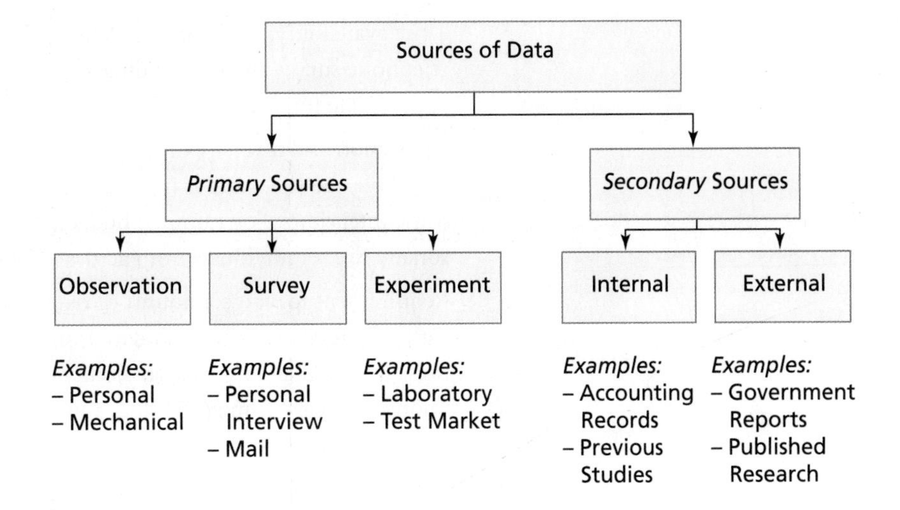

Primary Data

Primary data are not available prior to the research. Your organization or research group will have to find it or delegate the task. There are three main methods of collecting these data: observation, surveys, and experiments. The choice depends on your research objectives and the nature of the data required to answer the research questions.

1. Observation involves actually observing what is happening in order to gather the data. This can be done by personal or mechanical means. You may have seen people taking notes on clipboards at an intersection. They may be counting all of the cars that pass or stop or turn at an intersection that is being considered for a stoplight. Planners need to know the actual activity at that intersection, and obtaining the data from personal observation is often the easiest way to proceed.

On the other hand, a physicist studying the decay of atomic particles called *kaons* (which persist for only about 12 billionths of a second) is obviously unable to observe this phenomenon directly. Instead, she would use a large specialized particle detector. This is an example of using mechanical means for observation.

2. Surveys obtain data by asking people questions about their experiences, opinions, preferences, backgrounds, and many other variables. Surveys are often used in research for business and the social sciences.

A good survey depends on finding a way to contact a *representative sample* of individuals, as discussed in Chapter 1. Common methods are personal (face to face) and telephone interviews and questionnaires distributed by various means, including traditional and electronic mail.

Many books have been written specifically about the do's and don'ts for good survey design; the following are some highlights:

a) As for any research, be sure that you have clarified your objectives.

b) Identify the population that can answer your questions.

c) Choose a survey approach to reliably elicit the data you need from a representative sample of the population (or from the full census).

d) You will face these constraints on your choices:

- Budget and costs: For example, personal interviews are more expensive to conduct than mail, e-mail, or telephone-based surveys. Costs can also limit the size of your sample.
- Time availability: For example, data can generally be collected faster with a telephone survey than by sending a mail survey, provided a phone survey is appropriate for your research.
- Dispersion and diversity of the population: For example, for a population based on one site, such as in a factory, it may be feasible to conduct personal interviews. For a limited budget, it would be more difficult to personally reach individuals from across an entire province or country.
- Requirement for large amounts of data or interactivity: For example, a trained interviewer can elicit data from nonverbal clues and from interactions during dialogue. The possibility of getting the same data from a telephone interview is more limited, and from a written questionnaire, severely limited.

e) Regardless of how you plan to deliver your questions, ensure that they are unambiguous; clearly state an operational definition of how you intend to use the information that they elicit. Are your questions on-topic for the objectives? (It is generally useful to pilot a survey instrument with a small group of participants prior to administering it to the full sample. If the participants do not respond to the terminology and questions the way you intended, there is still time for adjustments.)

f) Beware of introducing bias and other errors. (This topic is discussed in more detail later in this chapter.) For example, nonrepresentative sampling; leading questions—where the question's wording influences the response; and interviewer bias—where the words, emphasis, body language, etc., of the interviewer influence the response.

Section 1.1

3. Experiments comprise the third method of collecting primary data. This method plays a critical role in causal research. From data collected by other methods, you might observe apparent connections between variables, and hypothesize a cause-and-effect relationship between them. Formal testing of a causal hypothesis requires an experiment.

In an experiment, the researcher attempts to control (keep constant) all factors except those factors being investigated. For example, medical researchers testing whether increased exposure of mice to a chemical agent causes asthmatic reactions would have to keep constant all factors other than the exposure to the agent while monitoring each animal's reaction, the initial health of each mouse, all components of the air in the test area, the duration and method of each exposure, and so on. If induced changes in one variable (e.g., exposure to the agent) lead to significant changes in another (e.g., the asthmatic reactions) when all other factors are kept constant, some evidence has been provided of a causal relation between those two factors.

Note that this approach is not always feasible in studies of human behaviour or in studies where human beings are affected by the outcomes (such as studies of industrial accidents). Deliberately exposing human subjects to risks obviously poses ethical issues. Also, it could be difficult to identify and control all of the factors that could influence the outcome of the experiment.

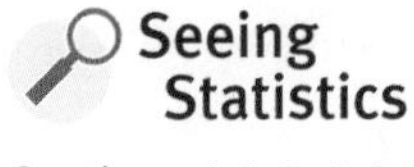

Sections 1.1.2–1.1.3

Secondary Data

All data are originally collected by someone as primary data—an expensive and time-consuming process. When planning your own research, begin by exploring what is already available. When you use previously collected data as input to your own research, it is called **secondary data.**

Secondary data does have some disadvantages. The original research objectives for which the data were collected may not exactly match your own objectives. Not all of the variables you need may be included, the data may be incomplete for your purposes, or the values may have slightly different interpretations. In addition there is often a financial cost for secondary data. Statistics Canada, for example, provides some free daily summaries of its data sets, but if you want more information or access to the raw data, there is often a charge. When secondary data are apparently free, such as data about herbal remedies published on the Internet, be cautious: The site may actually be promoting sales. Always check the reliability of your sources.

Nevertheless, secondary data can be very useful. Besides the time and cost advantages, it can offer a basis for comparison with any new primary data that are collected. Also, during exploratory analysis, the information provided may help when formulating the research questions and hypotheses for new primary studies.

As was shown in Figure 2.2, there are two main subtypes of secondary data: internal and external.

1. Internal secondary data were originally collected in one's own organization for other purposes. In business, much data is collected routinely to meet legal requirements or to assist in management. Examples are employee records, financial data, health and safety reports, quality control data, and so on. Similarly, the Canadian Avalanche Association routinely publishes an annual report with data pertaining to avalanches and related fatalities.[1] A marketing manager may use the company's past financial data when preparing a sales forecast, while a Canadian Avalanche Association member may use data from the association's annual report when studying needs for first-aid facilities.

2. External secondary data were originally collected outside of one's own organization. As part of her research, a Canadian Avalanche Association member might consult data published by the American Avalanche Association, or she might review published papers on the topic. A marketing forecaster might study trend reports published by Statistics Canada or consumer information published on the Industry Canada website.

Example 2.1

Following an accident at a local airport in which a plane overshot the runway, investigators are studying the general safety of the airport. Some of the variables to be studied are:

1. Annual numbers of accidents of any kind at this airport.
2. Annual numbers of all aviation accidents in Canada.
3. Exact distances between (a) where planes landing at this airport generally come to a halt on the runway and (b) the physical end of the runway. (Note: These data were never previously collected.)
4. Number of employees at the airport who have concerns about airport safety.
5. Minimum stopping distance required for a Boeing 747 on a runway if there is a tail wind of 20 knots.

Referring to the classification of data types in Figure 2.2, determine what types of data are required to determine values for each of these variables.

Solution

1. By law, the airport would have to keep records of all accidents that occur there. These internal records could be consulted as *internal secondary data.*
2. Aviation safety data are regularly collected and published by Transport Canada. Those reports could be consulted as *external secondary data.*
3. Since these distances were not previously recorded, primary data must now be collected. It would be possible to observe, either manually or mechanically (e.g., by radar or photographically) where each plane came to a stop, and then measure the distance to the end of the runway. This is *primary data* by *observation.*
4. For this variable, some instrument for getting feedback from employees is needed. This is *primary data* by *survey.*

Section 1.1.4

5. This question goes beyond descriptions of what has actually happened previously. It is asking for causal predictions that relate to the aircraft design. Probably experiments would have to be conducted in wind tunnels or with computer simulations, etc., to answer this question. This involves *primary data* by *experiment*.

• *In Brief 2.1 Sources of Data for Statistical Software*

Your statistical software cannot distinguish data by source. All of the data files stored on this textbook's CD are, from your viewpoint, secondary data—unless by chance you created or commissioned some of that data personally. Often, primary data have to be transcribed or manipulated in some way before they end up in a statistical spreadsheet. Survey data, for example, may first arrive in the form of hand-completed questionnaires. If there are open-ended questions (e.g., where the respondent writes a one-line answer), you may have to use some judgment to read difficult handwriting or to restate long-winded answers in shorter terms. This point should be remembered: The spreadsheet that is eventually input into the computer and thought of later as the raw data may already include some preprocessing, which could lead to error. Sometimes data that looks out of place can be explained by going back to the source documents or other mechanical records.

• *Hands On 2.1 Sources of Data Displayed on the Internet*

All or most of the data you find on the Internet are secondary data. The only exception would be pages that you or your organization (or your delegates) created to meet a particular need. Everything else is created by teams or individuals following their own plans and objectives.

Sometimes the data found on the Internet are secondary data even for the website creators. For example, the Canada Health Portal website[2] may report data that were originally collected by Statistics Canada.[3] Statistics Canada, in turn, is a very large unit. If you are viewing data from the Canadian Community Health Survey, only a portion of Statistics Canada resources and personnel were actually involved with that project.

For this section, find five unrelated websites that present data on topics of interest to you. These data may be formatted as graphs or tables or simply as numerical claims about a topic. Figure 2.3 shows an example of such a page, with the displayed data highlighted in the red oval.

Separately, for each of the pages you choose, ask yourself the questions listed below. If you can track down sure answers to some of the questions, using the Internet or other resources, all the better. Otherwise, take your best guess, but be ready to explain and justify your answers. (Note: A single Internet page may display data from more than one source. In that case, try to answer the questions in relation to each source that is represented.)

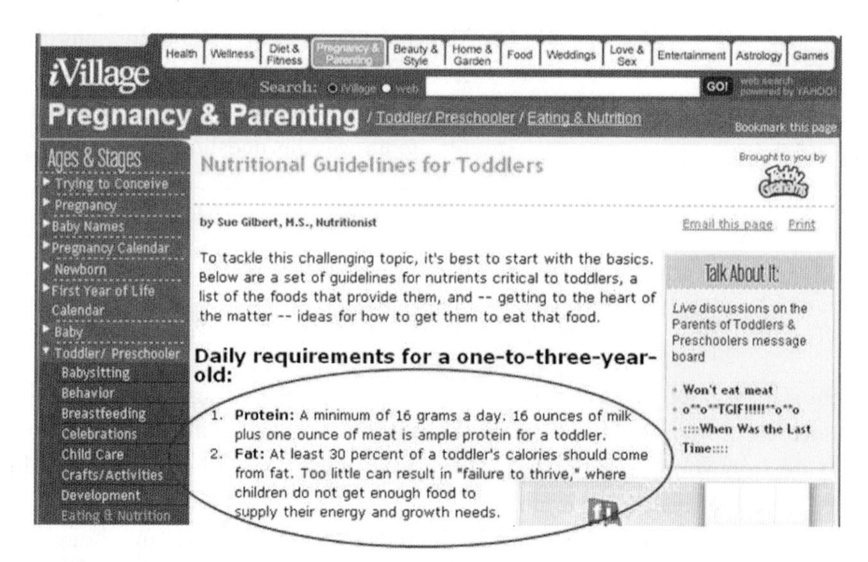

Figure 2.3
Example of Internet-Presented Data

Source: iVillage, *Pregnancy & Parenting,* "Nutritional Guidelines for Toddlers." Found at: http://parenting.ivillage.com/tp/tpnutrition/0,,6w,00.html

1. From the perspective of the organization that is displaying the Internet page, is this primary or secondary data?
2. If secondary, can you find out who collected the original primary data? If you do not know, try to guess at least *what type* of individual or organization might have collected the primary data.
3. Regardless of who collected it, is the original primary data based on observation, a survey, or an experiment? Explain.

2.2 Designing a Sampling Plan

As defined in Chapter 1, a sample is a selection of objects or individuals from a population. There are two basic reasons for sampling: (1) To estimate the values of characteristics in a population based on their values in the sample, and (2) to evaluate previous assumptions or hypotheses that have been made about a population. Both of these applications are part of inferential statistics. A "good" sample is selected in such a way that it is legitimate to draw these sorts of inferences.

Statistical Sampling

To apply the inferential techniques presented in this textbook, the data must be collected by means of **statistical sampling** (also called *probability sampling*). This approach uses **random selection** to best ensure that the collected samples are *representative* of the population, as illustrated in Figure 1.4 in Chapter 1. When samples are representative in this way, we can have some confidence that patterns and relationships found in the sample are also present in the population (or *will* be present, in the case of forecasts). Even the best statistical sample can, by pure chance, fail to be representative, but if sampling is conducted using random selection, then the likelihood or probability of obtaining a non-representative sample can be estimated and taken into account.

The core idea behind random selection is that no one could predict exactly which objects will be collected in the sample. It is hoped that objects of varying characteristics will be selected, by chance, in roughly the same proportions as in the full population—with no systematic tendency to favour larger or smaller values, etc., for any characteristic of interest. In practice, there are several strategies for the random selection of a statistical sample, four of which (simple random, stratified random, simple cluster, and systematic) are discussed later in this section.

Whichever technique you use, you must specify the **sampling frame**—a list of all individuals or objects from which the sample will be drawn. Ideally, a sample would be selected from the entire population—but we cannot realistically select objects that we do not know about or cannot find a way to reach. Suppose you want to sample from the population composed of "all employees at my university or workplace." Can you obtain a list of all of these people, complete with contact information? If so, you could sample from the list. The sampling frame is the list of all individuals or objects from which the sample will be drawn. Unfortunately, your list may not exactly correspond to the real population. Some part-time or contract workers may not appear on the list; some who have retired may still have their names listed. It is important to recognize the limitations of your sampling frame, and if possible to fill in the gaps and remove any false entries.

Simple Random Sampling

In **simple random sampling,** a sample is selected in such a way that (1) each *object* in the sampling frame has an equal chance of being selected, and (2) each possible *combination of objects* (i.e., every possible sample) of a given size has an equal chance of being selected from the sampling frame. These two requirements are not the same.

Suppose you want to sample four words that are printed on this page, and your technique is to pick a starting point randomly and then select the next four words in sequence as your sample. (Note: If you run out of words on the page, continue your selection of words starting with the first word on this page. Treat the next word in sequence after the end of the page as returning to the first word at the top.) This method meets condition (1): Because the starting point is random, any word on the page has just as good a chance as another to fall into the selected group. However, condition (2) is not met: Because you are selecting words in four-word sequences, you could never get a sample that, for instance, contains the 3rd word on the page, as well as the 10th, 12th, and 39th words on the page, respectively. So, not all possible combinations of objects could be selected by this method.

The strategy of simple random sampling meets both required conditions. If you participate in a raffle, you might buy one ticket and place it in a container along with everyone else's ticket. If the container's contents are thoroughly mixed and someone picks four tickets, one at a time, without looking, that person has selected a simple random sample of four tickets. Any one ticket has the same chance as another to be selected *and* any combination of four tickets is as likely as any other to be drawn.

For a very large population, such as all the adults in Quebec, it would be cumbersome to accomplish simple random sampling by literally drawing names from a hat. But if you could access a suitable sampling frame (a phone book might be a start, although it is not a complete list), a computer could help to select any arbitrary number of random names from the list (refer to "In Brief 2.2" on page 50).

Stratified Random Sampling

Sometimes a researcher wants to know characteristics not only of the full population, but also of certain subgroups of the population. A scientist studying exposures of Canadians to *E. coli* bacteria in their drinking water might want to examine breakdowns of people's exposure levels by province and territory.

Simple random sampling can be impractical in such cases. According to Statistics Canada 2005 estimates, only 0.13% of Canada's population lived in the Northwest Territories.[4] A simple random sample of, say, 1,000 Canadians would be unlikely to include by chance more than two or three people from that region—too small a sample to make valid inferences about the territory. Although a very large random sample of 40,000 Canadians would likely improve the numbers included from the Northwest Territories, this would be an expensive solution—and would provide far more data than needed for the large provinces.

Therefore, if the data contain subpopulations that are relatively small and you want to ensure that all subpopulations have reasonable representation in the sample, **stratified random sampling** can be used. Here, the population is divided into subpopulations (called *strata*) of interest, and simple random samples are then taken *from each subpopulation.* Perhaps the scientist would sample about 100 people from the Northwest Territories, about 500 from British Columbia, and so on.

You need to be careful if you also want to make estimates about the entire population. In the above example, the stratified sampling collected a larger-than-proportional sample of some subpopulations (such as the Northwest Territories) compared to others. To get the *overall* national picture, you cannot simply combine all of the samples, because the smaller subpopulations would be overrepresented. Instead, for the combining of results, you would need to take a weighted average (discussed in detail in Chapter 4): If you sampled 1 of every 425 households in the Northwest Territories, you would need to weight the Northwest Territories results by 425; if you sampled 1 of every 8,000 British Columbians, you would need to weight the British Columbia results by 8,000; and so on. This method will compensate for the unequal sampling rates among the strata.

Example 2.2

A biologist is studying mercury levels in the fish from two lakes (see Figure 2.4). She plans to analyze results for both lakes individually, as well as combined. Recommend an appropriate sampling strategy. Are there other factors that need to be considered?

Solution

Lake Small appears to contain a relatively small number of fish compared to Lake Big, so simple random sampling may fail to include enough fish from Lake Small for analysis. The biologist would be advised to use stratified random sampling. Perhaps she could sample four fish from Lake Small (2/3 of all of the fish) and ten fish from Lake Big (5/13 of all fish). When recombining results from the subpopulations, she would weight Lake Small's results by 3/2 and Lake Big's results by 13/5.

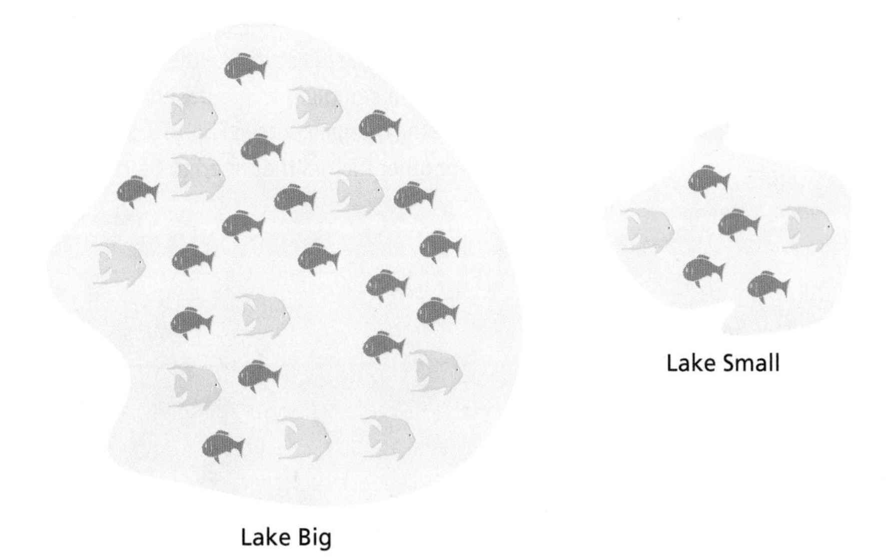

Figure 2.4
Stratified Random Sampling of Fish from Two Lakes

We have not yet discussed how large a sample should be to improve the chance of representativeness (this will be discussed in Chapter 11). Also, do we have a sampling frame for the fish in each lake? How would we get one? Or, if we cannot, what should we do? A pilot study might be useful to begin with, to get a picture of what is in each lake. Also, if the biologist plans to break down the results by type of fish (e.g., green versus yellow), then more strata may be needed, based on that additional subdivision of the data.

Simple Cluster Sampling

Suppose that a large retail company with outlets across Canada wants to personally interview its employees regarding their views about closed-circuit surveillance systems. The managers are not interested in breaking down the results by province, territory, or other characteristic, so stratified random sampling does not apply. But travel budgets for the company are limited. To take a simple random sample of its employees, the company might have to send interviewers to sites scattered widely across the country, which could be very expensive and time-consuming. **Simple cluster sampling** provides an alternative.

In simple cluster sampling, the population is divided into clusters, or groups. A selection of these clusters is chosen by simple random sampling. Then, if budgets permit, every individual in every selected cluster is sampled. Alternatively (less time-consuming and less expensive), a simple random sample of individuals is selected from each of the selected clusters for the final sample.

Unlike "strata," the different "clusters" are not distinguished by having different characteristics (such as strata based on different locations, genders, or species, etc.). In fact, ideally, every cluster would have the same mix of characteristics in its subpopulation as every other. That assumption would make it reasonable to sample from some of the clusters and ignore the others. Clusters are rarely ideal in that sense. If some clusters are in rural areas and others are in urban areas, the mix of attitudes in the two sets of clusters may be somewhat different. However, if you sample enough clusters and select them randomly, a final mix of individuals that reflects the whole population should result.

Example 2.3 An environmentalist is studying the aftereffects of vegetation fires on the environment. He develops a sampling frame based on areas recorded to have had vegetation fires in 2005 (see the orange spots in Figure 2.5). Recommend an appropriate sampling strategy. Are there other factors that need to be considered?

Figure 2.5
Simple Cluster Sampling of Fire Hotspots (Simplified Example)

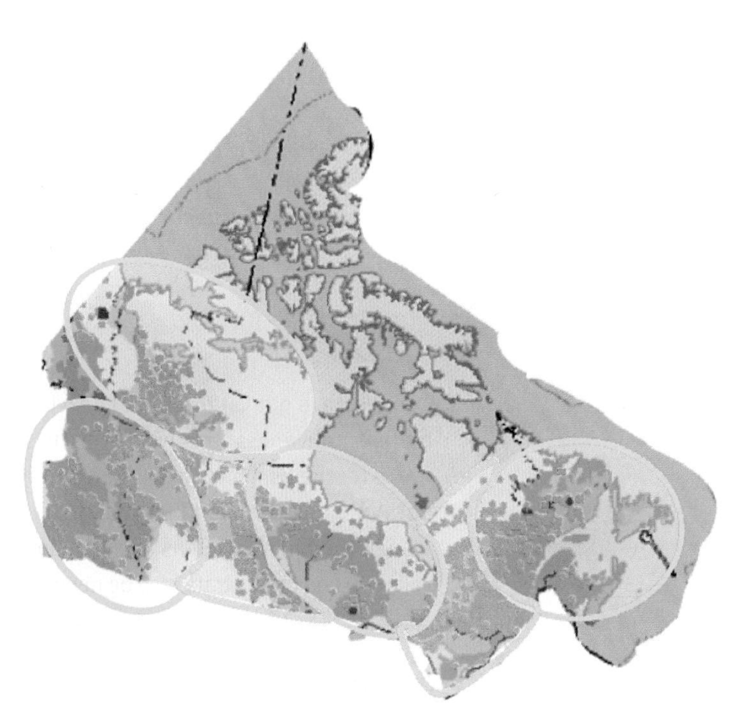

Source: Original map data provided by *The Atlas of Canada* http://atlas.gc.ca © 2007. Produced under licence from Her Majesty the Queen in Right of Canada, with permission of Natural Resources Canada.

Solution If there are time or budget constraints, a simple random sample would be impractical. Selected locations might be widely spread and hard to reach. Nor does stratified random sampling apply. Simple cluster sampling could be used instead.

In Figure 2.5, all of the areas where fires were detected are divided into six groups or clusters. If a given cluster happens to be randomly selected by the technique, then the areas within it are reasonably accessible. The clusters are large enough that each contains a mix of different types of vegetation and ecological environments. The clusters shown in the figure are just for illustration. In practice, the environmentalist should design a set of clusters with care, so that all are relatively similar in their mixes of environmental characteristics. At the very least, a random selection of the clusters should reasonably be expected to have a representative mix of characteristics.

This type of sampling could produce misleading results if the clusters are not properly defined or if it is not possible to prevent some clusters from being too homogeneous. In that case, a selected cluster might overrepresent certain features of the population and underrepresent others.

Systematic Sampling

Under certain conditions **systematic random sampling** is an easy-to-use alternative to simple random sampling. It can be applied when data for the sampling frame are available in a randomly ordered and conveniently accessible list. For example, the telephone directory might be such a list when sampling individuals to contact for a telephone interview. Of course, the telephone directory is not random in every respect. It is ordered very rigidly by alphabetical order of last names. But this is acceptable—provided there is no connection at all between the ordering of items in the list (e.g., the order of last names) and the characteristics you are hoping to measure.

When such a list is available, the procedure is to first determine a sampling interval, k. Assign $k =$ (the number of items in the sampling frame) divided by (the intended sample size). Suppose there are 5,000 names in the phone book, and we want to sample 50 of them. The sampling interval is $k = 5{,}000/50 = 100$.

Next, choose randomly a starting position in the list between 1 (the beginning of the list) and the k th position on the list. For example, randomly select the 29th name in the phone book as your first selected item. From this point, select every k th subsequent item in the list. Continuing the phone survey example, select the name listed 129th (since it is the 100th name following the name listed 29th), then select the 229th name (which is the 100th name following the name listed 129th), and so on, until you reach the end of the list. If you follow this technique, the procedure ends when you have selected exactly the desired number of items from the list, spaced evenly throughout the list. Since the starting place was random, and we presume that there is no meaningful order in the list, then every possible item will have had an equal chance at the outset to have been chosen. However, not all possible combinations will have had a chance to occur. For example, if two people's names are next to each other in the phone book, they could not be both selected by this method.

Example 2.4

The owner of a pet store is concerned about the health of the *Girella punctata* species (Green Fish) in her inventory. They are all placed sequentially along a spiral walkway (see Figure 2.6). She decides to collect a sample of six of the fish to take to a lab for analysis. Recommend an appropriate sampling strategy. Are there other factors that need to be considered?

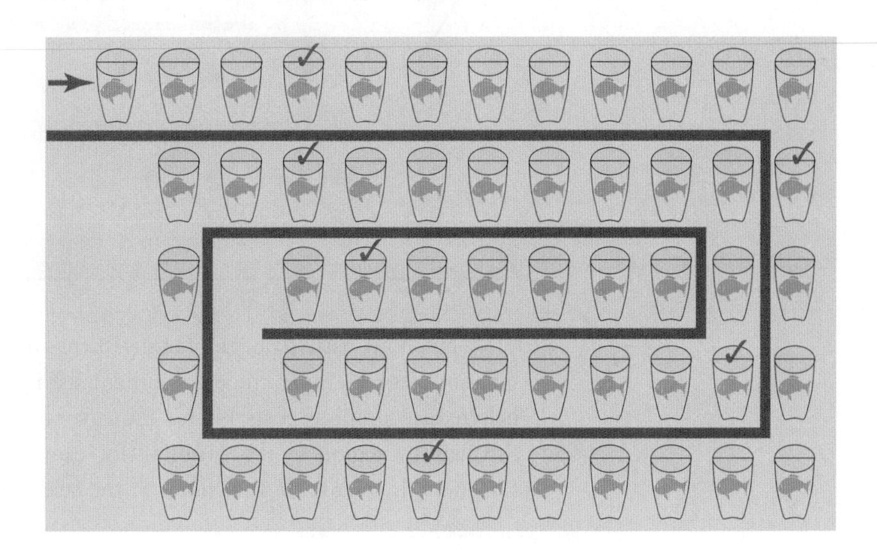

Figure 2.6
Systematic Random Sampling of Fish in a Pet Store

Solution

If the owner has a checklist of all of the Green Fish identified by a name or number (a sampling frame), she could randomly select six of their IDs, and then walk up and down the aisles trying to find the selected specimens. This would be a simple random sample. But, given the narrow walkway in this case, it might be inconvenient to use this method. We can eliminate grouping the items by strata or clusters, since there is no obvious basis for this.

Systematic random sampling appears to be the best strategy. The ordered position numbers of each fish provide an implicit sampling frame, whether or not the numbers have been printed separately. There are 54 fish, and the owner wants to sample 6 of them. Therefore, the sampling interval is $k = 54/6 = 9$. (If this quotient had not been an integer, you could round k to the nearest whole number.) Randomly select a number from 1 to k (i.e., 1 to 9); suppose it is 4. Note in Figure 2.6 how the 4th fish is selected. After that, walk down the aisle and select fish in intervals of 9. This completes your sample.

If the walkway had branches, you would need an explicit list for the sampling frame, so there was no ambiguity about the position number of each object. There could also be a problem if there is an unknown *nonrandom* relationship between the order of the objects and the characteristic to be measured. For example, suppose the aerial view of the pet store walkway is as shown in Figure 2.7. The light area represents a glass skylight that has some long cracks, allowing a liquid film of pollutant to leak down on the fish below. Note that, by chance, *none* of the contaminated fish will be included in the sample. This was possible because the sequence used for the sampling frame (the physical positions of the fish) is *not* unrelated to the measured characteristic (the fish's health). If simple random sampling had been used, it would have been less likely to have so many continuous blocks of fish left unsampled.

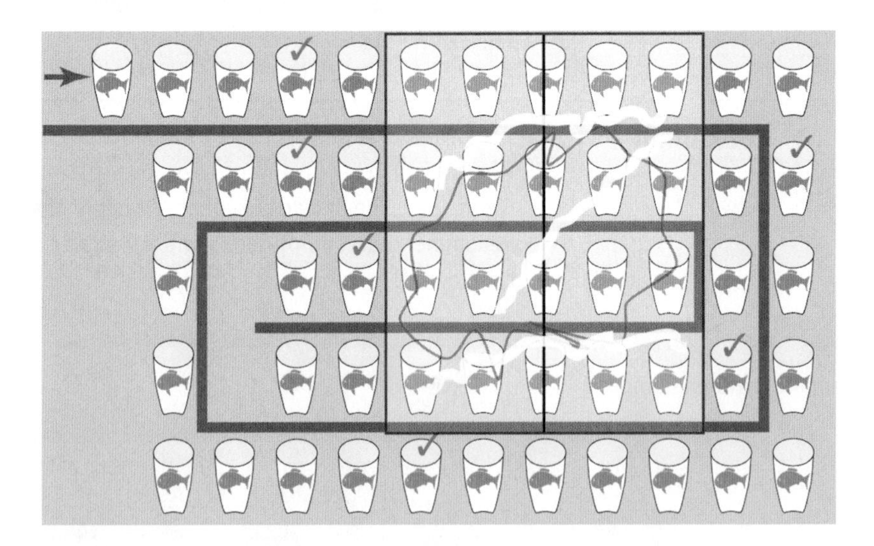

Figure 2.7
Aerial View of the Pet Store Walkway

These four basic strategies for statistical sampling can be used, as needed, in combinations. For example, the environmentalist in Example 2.3 used cluster sampling to help reduce sampling costs, but his strategy was not designed to give a breakdown of the fire data by region in the country (for example, "Western Canada" and "Northern Canada"). It is possible that none of the selected clusters would include a particular region. To avoid this risk, he could start with stratified sampling, with each national region being a stratum. These subpopulations are still quite extensive geographically, so the environmentalist could choose to use cluster sampling *within* each regional stratum. Then, to further save time and costs, he may investigate only a simple random sample of sites from each of the selected clusters within each region, rather than investigate every site within selected clusters. This illustrates a sampling strategy that combines stratified, cluster, and simple random sampling.

Nonstatistical Sampling Techniques

Using nonstatistical sampling techniques, data can be collected in many ways without using the randomizing strategies that are fundamental for statistical sampling. Some of those alternatives, such as convenience sampling, judgment sampling, and quota sampling, are discussed below. From the viewpoint of **quantitative statistics,** these alternatives are, at best, compromises to be used with caution and, at worst, error-prone techniques. This introductory textbook is, like many others in statistics, quantitatively oriented, focusing on what can be counted or measured in the data and on what can then validly be inferred from those counts or measurements. Students are warned against sampling methods that are viewed as less likely to produce representative samples and are therefore less reliable as the basis for inferences.

Students in the social sciences or business, however, may encounter another viewpoint in some future courses. You may have heard of focus groups used in marketing or other approaches involving strategic selection for intensive interviews. These methods can play a role in **qualitative statistics,** which aims for a deeper view of how people are thinking in certain contexts and how they relate their ideas. While these methods cannot claim inferences in the statistical sense, sometimes they can give a human face to more quantitative findings. For example, instead of concluding statistically that a majority of people favour a certain product, qualitative research might provide insights about why they feel that way.

Convenience Sampling

"Man on the street" interviews conducted on TV broadcasts are an example of **convenience sampling.** As crowds mingle for the big sports playoff or political rally, the reporter asks some individuals to voice their opinions about a sport or a political candidate. This is like a meat inspector taking all her samples from the local store where her family buys groceries—there is no way to know if these convenient-to-get samples are representative of the entire population.

Judgment Sampling

Judgment sampling has many variations with names like *critical case sampling, typical case sampling,* and *extreme case sampling.* The common theme is that *in the judgment of the researcher,* these particular individuals can provide some key information that might be missed or diluted in a broader random sample. The importance placed on thorough accident investigation for data to use in industrial safety research can be viewed partly in this light: If relatively few accidents occur, they may or may not be representative of all of the unsafe acts and near-misses that occur in a workplace. Yet, by learning all of the factors that contributed to each specific case, remedial possibilities for enhanced industrial safety might be identified.

Another example is the selection of a test market for introducing a new product. Rather than selecting the location randomly, marketers may learn more by choosing a city that has more people who can afford the product and who have used similar products.

This approach might also be useful for infinite samples, where it is not possible to enumerate the whole population. For example, if you plan to invest in a display ad on a popular website, the exact population of viewers is not likely known, but perhaps you could make a judgment of the types of people who would view the ad and find a way to talk to them.

Unfortunately, a judgment sample is only as good as the judgment of the researcher. Does the person allow biases to affect the selection method? Might this person tend to favour individuals who support his or her preconceptions? One way that researchers try to avoid these traps is through *triangulation:* Use different methods, with different samples, and see if they yield corresponding results.

Quota Sampling

Quota sampling is a special case of judgment sampling. Perhaps an advertiser is examining people's reaction to a new display ad he has purchased on a website. Based on secondary data, he might expect a certain mix of characteristics among the visitors to the website—so many men compared to so many women, a mix of income groups and political views, and so on. With quota sampling, he deliberately chooses a number of individuals to question from each gender, income group, and so on, based on his expectations of their exposure to the ad. Quota sampling can often be observed at shopping malls, where a questioner waits to approach a certain number of people with specific characteristics.

Errors Associated with Sampling

As we learned in Chapter 1, there can be advantages to taking a sample from a population of interest instead of attempting a full census. Time, cost, and quality are three important considerations. Nonetheless, using samples will always introduce some risk of error (see Figure 2.8). It is important to know the source of an error in order to minimize the chance of it recurring and to possibly identify other areas where it might arise.

Figure 2.8
Errors Associated with Sampling

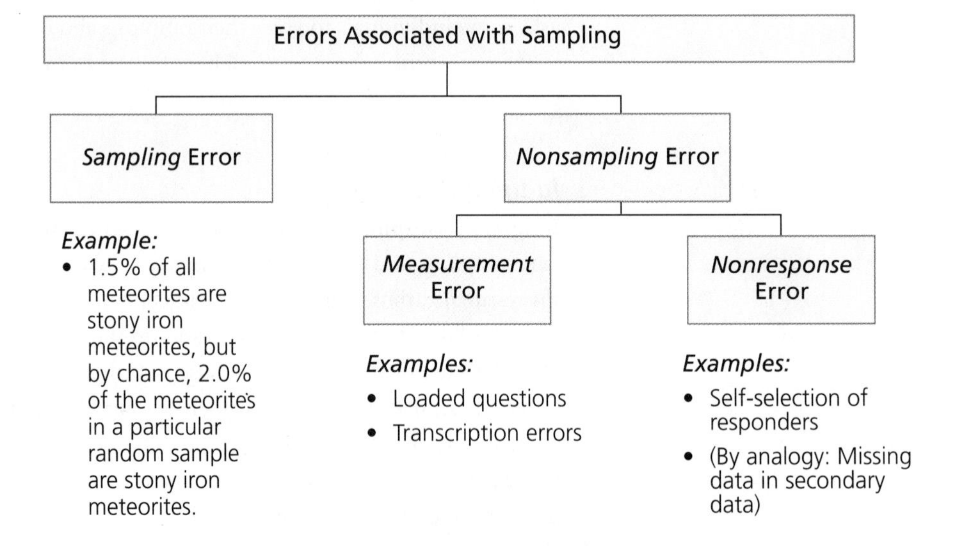

Sampling Error

Even the best-collected sample is bound to have results that differ slightly from the actual properties of the population. **Sampling error** is defined as the difference between the information in the sample and the information in the population that occurs simply because the sample is a subset of the population. For example, the statement has been made that 1.5% of meteorites can be classified as stony iron meteorites.[5] Assuming that this is true of the population, you diligently collect a random sample from all of the meteorites that have fallen. Even if you take all steps to ensure that your sample is representative, however, you are highly likely, by chance alone, to encounter in your sample a few more or a few less stony iron types than exactly 1.5% of your sample. Sampling error is *not* a type of "mistake"—it is an inevitable consequence of the sampling process itself.

Fortunately, if you select an appropriate random sample and meet some other conditions, you can estimate the size of the sampling error that will result. This estimate depends in part on the size of the sample that you have chosen. While you cannot completely eliminate error, knowing how exact (or inexact) your sample results are can be a guide to decision making.

Nonsampling Error

Nonsampling error, on the other hand, is the difference between information in the sample and the information in the population that is due to missing data or incorrect measurement. Unlike the case for sampling error, the occurrence of nonsampling error may suggest that a mistake of some kind was made. Naturally, no one is perfect—not even researchers—but the occurrence of any specific nonsampling error is not inevitable and precautions can be taken.

Nonsampling errors come in a variety of forms, many of which can be called **measurement errors.** Think of how you might obtain the wrong result when trying to measure someone's height. Perhaps your ruler is improperly calibrated. In a survey, this may correspond to poorly worded questions or answer choices that do not measure what they are intended to. Or you may misread information or incorrectly record some measurements from a ruler that is accurate. Analogous errors can occur when transcribing results from a well-designed survey or copying the correct values displayed on a screen.

In a survey approach to data collection, respondents may deliberately withhold or falsify information if, for example, they think that the interviewer would have a negative opinion of their preferred answer. In loaded questions, the question wording itself might influence people's responses. If either the questioner or respondent does not fully understand the terms or concepts contained in the question, this could also lead to collecting inaccurate information.

Nonresponse Error

Nonresponse error is a different class of error. Many people who receive questionnaires choose not to reply or refuse to take part in a telephone survey. Assuming that the survey answers would be the same for those who respond and those who do not respond may not be correct. Perhaps those who have a complaint to make on the subject will return the questionnaire or stay on the line, and those who are quietly satisfied will not want to be bothered. Such response patterns could **bias** the results; that is, cause the collected data to tend misleadingly more toward one possible conclusion than another. To counter this problem, surveyors often take measures to increase survey response rates, perhaps by sending follow-up letters or offering gifts for participating.

The bias of allowing self-selection for a survey has a similar effect to nonresponse bias. For example, if you are visiting a particular website voluntarily (e.g., not for a course) and are engaged enough to respond to a survey posted there, you have already demonstrated certain interests that might not be shared by nonrespondents.

Issues similar to those faced in survey-based research may arise for research that depends on secondary data. In industrial safety research, for example, the data are often collected from companies, safety agencies, or governments regarding accidents that have already happened. By definition, we have no information about incidents that did not happen or were not recorded, so there is a risk of bias in the analytical results. Similar to using triangulation in qualitative studies, it can be helpful to look at the research problem from different angles—with possibly different data sets—to look for confirmation of results.

• *In Brief 2.2 Using the Computer for Random Selection*

(*Based on the* **BC_Votes** *file*)

Provided you can input a column of data that corresponds to the sampling frame, the computer can assist with the process of random selection. Sometimes, you can input the entire frame—for example, if you are working from a directory of names and addresses. On other occasions, you might just input reference numbers for each member of the population (such as student numbers, if your population is students); for those selected, you can look up the required, additional information about the individuals later.

The steps described here and illustrated using the file **BC_Votes** can also be applied for cluster sampling. In that case, input reference names or numbers for the clusters, instead for individuals, and make a random selection.

Excel

Once the sampling frame data have been input, decide on the sample size you require. There are now four remaining steps, as shown in the illustration.

1. Create a new column and label it "Random." In the top cell below that heading, key in the Excel formula

(1) Enter this formula in this top cell

C2 =RAND()

	A	B	C
1	Riding	Tot_Vote	Random
2	Abbotsford	48139	0.168132
3	Burnaby-Douglas	45263	0.1477804
4	Burnaby-New Westminster	41732	0.5627567
5	Cariboo-Prince George	42120	0.0452736
6	Chilliwack-Fraser Canyon	44862	0.8434345
7	Delta-Richmond-East	46727	0.5812705
8	Dewdney-Alouette	48000	0.8691236
9	Esquimalt-Juan de Fuca	54756	0.8743111
10	Fleetwood-Port Kells	39180	0.8203362
11	Kamloops-Thompson	51176	0.7334465
12	Kelowna	53238	0.3889251
13	Kootenay-Columbia	41119	0.8443396
14	Langley	50956	0.470048
15	Nanaimo-Alberni	58770	0.5525339
16	Nanaimo-Cowichan	56032	0.6466193
17	Newton-North Delta	41215	0.6775848
18	New Westminster-Coquitlam	47615	0.7716919
19	North Okanagan-Shuswap	51853	0.9973824
20	North Vancouver	56645	0.5040415
21	Okanagan-Coquihalla	48812	0.4027774
22	Port Moody-Westwood-Port Coquitlam	45527	0.8880795
23	Prince George-Peace River	36237	0.0403845
24	Richmond	40782	0.8278625
25	Saanich-Gulf Islands	63773	0.6488982
26	Skeena-Bulkley Valley	37025	0.0404087
27	Southern Interior	46268	0.0648215
28	South Surrey-White Rock-Cloverdale	53283	0.8232759
29	Surrey North	34454	0.6043231
30	Vancouver Centre	52510	0.2684696
31	Vancouver East	41594	0.8515538

(2) Then copy it down the column

(3) Convert random numbers to values, then sort

Riding	Tot_Vote	Random
Prince George-Peace River	36237	0.040384537
Skeena-Bulkley Valley	37025	0.040408665
Cariboo-Prince George	42120	0.045273608
Southern Interior	46268	0.0648215
Burnaby-Douglas	45263	0.147780412
Abbotsford	48139	0.168132
Vancouver Quadra	56699	0.212999186
Victoria	58127	0.241380309
Vancouver Centre	52510	0.268460629
Vancouver Island North	52449	0.32116083
Vancouver Kingsway	42648	0.331840104
Kelowna	53238	0.38892506
Okanagan-Coquihalla	48812	0.402777391
Langley	50956	0.47004798
North Vancouver	56645	0.504041488
Nanaimo-Alberni	58770	0.552533876
Burnaby-New Westminster	41732	0.56275674
Delta-Richmond-East	46727	0.581270522
Surrey North	34454	0.604323090
Nanaimo-Cowichan	56032	0.646519322
Saanich-Gulf Islands	63773	0.648898186
Newton-North Delta	41215	0.677584804
Kamloops-Thompson	51176	0.733446451
New Westminster-Coquitlam	47615	0.77169189
West Vancouver-Sunshine Coast	60543	0.807344012
Fleetwood-Port Kells	39180	0.82033623
South Surrey-White Rock-Cloverdale	53283	0.823275839
Richmond	40782	0.827862496
Chilliwack-Fraser Canyon	44862	0.84343471
Kootenay-Columbia	41119	0.84433955

(4) The top *n* items are your sample

=**Rand**(). This causes a pseudo-random number to be generated that will appear in that location. That multidigit number will have some value between zero and almost one.

2. Using standard Excel procedures, copy and paste the =**Rand**() formula down the remainder of the new column for as many rows as there are data.
3. The first part of this step is required to stabilize the random numbers you have created in step 2. Otherwise, you will find that the numbers change as you perform other operations on the worksheet. Click and drag the cell pointer to highlight the column of random numbers. Then use the menu sequence *Edit / Copy* and *leave the column of numbers highlighted.* Immediately, use the menu sequence *Edit / Paste Special* and then, in the dialogue box, select *Values* and then click on *OK.* Now the visible random numbers have replaced the underlying formulas.

   The second part of step 3 is to sort the sampling frame data based on the random numbers. Click and drag the cell pointer to highlight the full block of data, including the new "Random" column. (Be sure to include all data rows.) Use the menu sequence *Data / Sort* and then use the drop-down menu to select *Random* as the *Sort by* variable; then click *OK.*
4. Your sampling frame data are now sorted in a totally random order. Count down to include as many rows as the sample size you require. Your sample has now been selected.

One other formula in Excel can be very useful for random selection. Suppose you wanted to collect a systematic random sample. Recall that you need a sampling interval k, and that you should start at some random position between 1 and k. (In our previous example, $k = 9$.) In any empty cell, you could input a formula of the form =**Randbetween(1,9).** This returns a random integer between the first number (**1** in this case) and the second number (**9**), inclusive.

Minitab

All of Minitab's functions that are accessible through menus, which we usually introduce in this text, are also available through a command language—using session commands. If you use Minitab for your course, your instructor may opt to give some instructions on this alternative approach.

Once the sampling frame data have been input, decide on the sample size you require. Begin with the menu sequence *Calc / Random Data / Sample from Columns.* In the dialogue box, perform the following three steps, as illustrated in the screen capture.

1. In the *Sample* box, input the desired size of the sample.
2. From the list of columns provided originally at the left, click to select all of the column headings (i.e., variables) that you want moved to the *Sample rows . . . from column(s)* box. These variables will be included in the sample.
3. Lastly, in the *Store samples in* box, input column headings for where the sample data should be output; then click *OK.* (The illustration shows a sample output from this procedure in columns C4 and C5.)

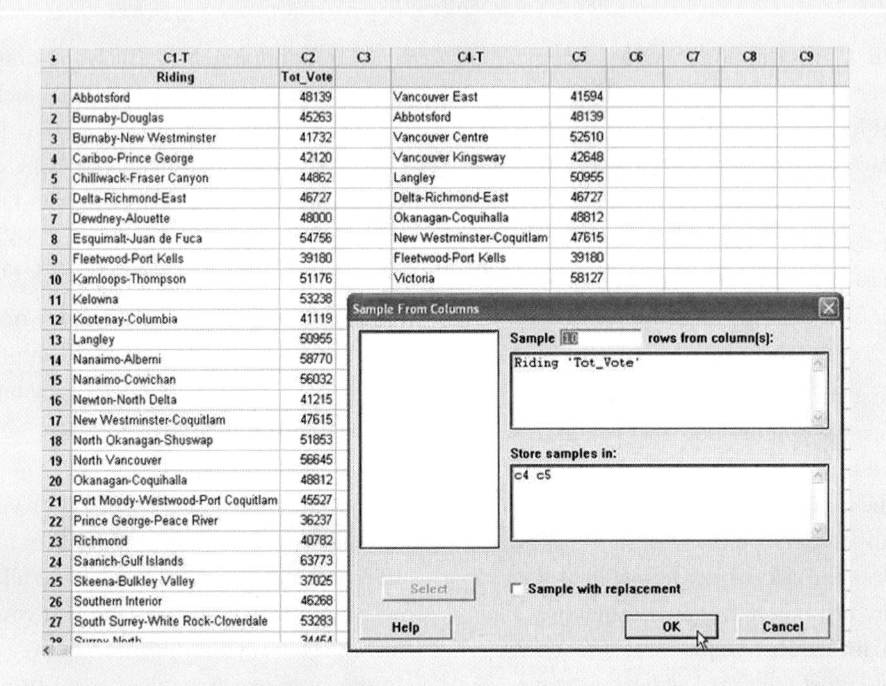

On some occasions, you need to generate only one random integer. Suppose, for example, that you wanted to collect a systematic random sample. Recall that you need a sampling interval k, and that you should start at some random position between 1 and k. (In our previous example, $k = 9$.) In Minitab, begin with the menu sequence *Calc / Random Data / Integer.* In the dialogue box shown in the second screen capture, perform the following steps.

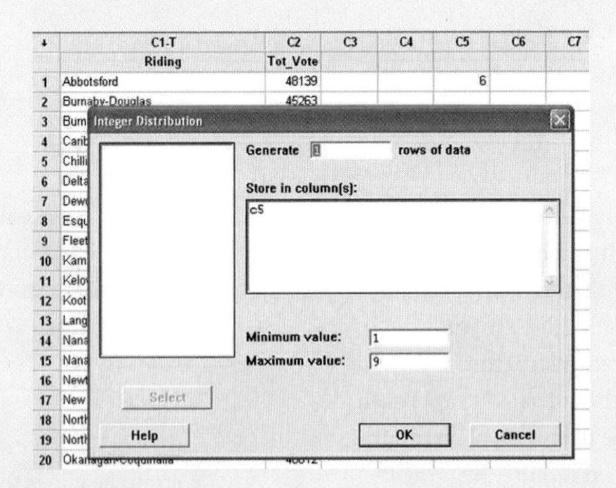

1. In the *Generate* box, input "1" as the size of the sample.
2. Specify in the *Store in column(s)* box a column heading under which the result is to be placed.
3. Input the desired minimum and maximum for the selected random integer, then click *OK*. (The illustration shows a sample output from this procedure in column C5.)

SPSS

Once the sampling frame data have been input, decide on the sample size you require. Begin with the menu sequence *Data / Select Cases.* In the first dialogue box, select *Random sample of cases,* then click the *Sample* button under it (see the arrow in the screen capture). In the second, smaller dialogue box, select *Exactly* and then specify the sample size (left box) and the number of records in the sampling frame (right box). Click on *Continue* and then *OK* to exit the two dialogue boxes.

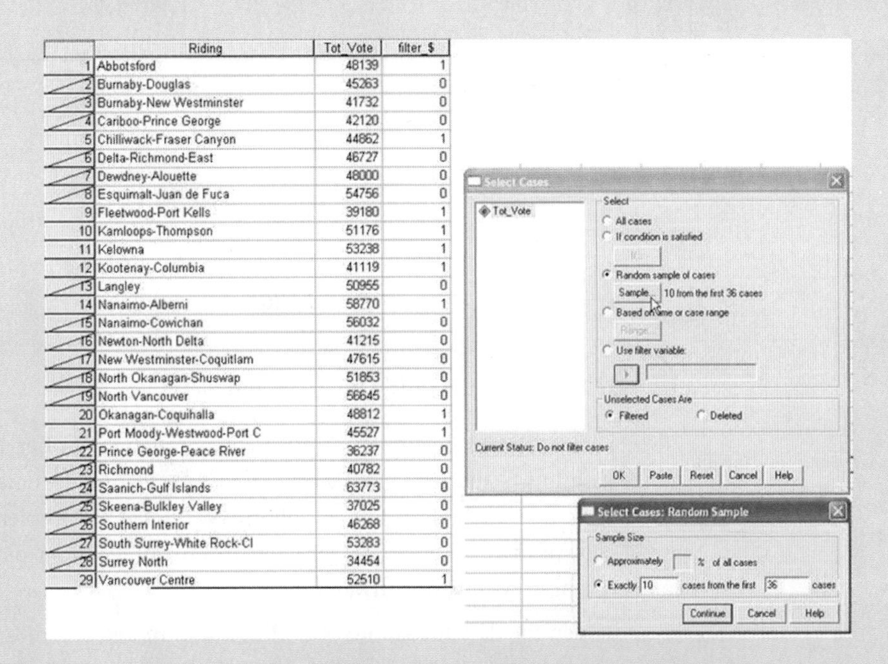

The illustration also shows the result of using these procedures: The number "1" in the added column labelled "filter_$" indicates that a record has been randomly selected. Note that the record numbers of nonselected records have been crossed out in the first column at the left of the page. This indicates that when you later perform statistics, only the selected records' data will be included. An alternative procedure is to select *Deleted* (instead of *Filtered*) in the first dialogue box, which would totally erase all unselected records from the data set.

On some occasions, you need to generate only one random integer. Suppose you wanted to collect a systematic random sample. Recall that you need a sampling interval k and that you should start at some random position between 1 and k. (In our previous example, $k = 9$.) We can use SPSS as a "calculator" to generate the random integer. Use the menu sequence *File / New / Data* to open a new data editor screen, and input any value in the upper left cell. Then use the menu sequence *Transform / Compute.* In the *Target Variable* box at the upper left of the resulting dialogue box (see the second screen capture below), input a name for the new column into which SPSS will place the answer. In the box *Numeric Expression,* input a formula in this format: **TRUNC(RV.UNIFORM[2,10)],** where **2** is a number equal to the desired beginning of the range *plus one* and **10** is a number equal to the desired end of the range *plus one.* The formula randomly produces a decimal number anywhere between just over the beginning of the range to some fraction above the highest number in the range, and then removes the decimal portion. (The illustration shows a sample output from the procedure.)

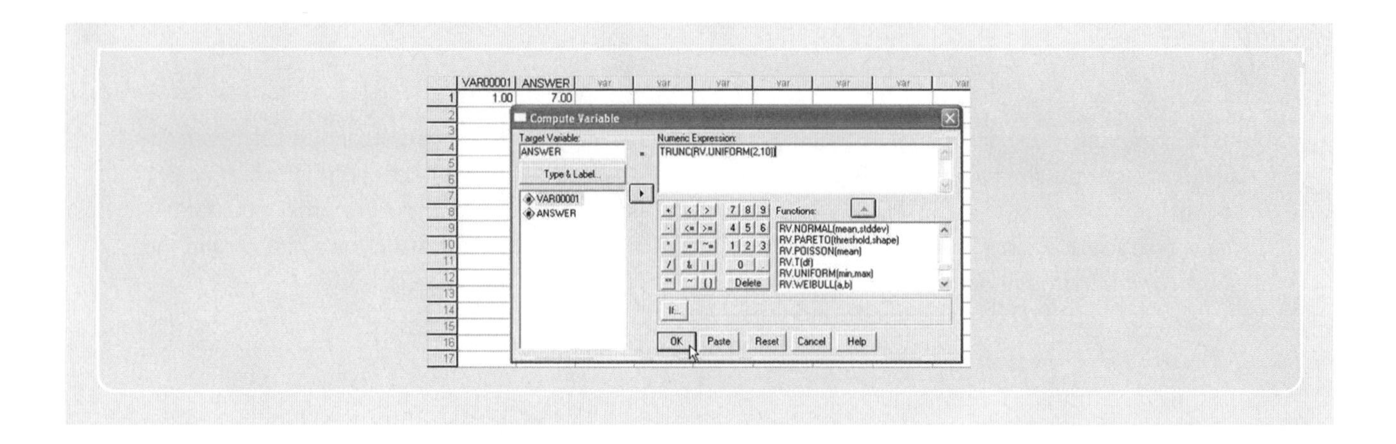

• Hands On 2.2 *Conceiving and Implementing a Sampling Plan*

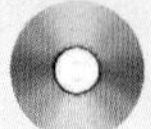
(Based on the ***Hands_On_2_2_Data*** *file)*

To perform each task below, your software may or may not offer a fully automated solution. Sometimes simple operations like sorting the data and copying and pasting relevant subsets to new columns can solve the problem. (Be sure to save a copy of the unaltered data set.) You are encouraged to explore what you can accomplish with your software.

1. Open the data file **Hands_On_2_2_Data.** In two columns, the IDs and region numbers of 120 fictitious employees of a company are shown. Using the concepts provided in this chapter and with the help of the computer, create the following samples of employees:
   a) A simple random sample of 30 employees (regardless of region).
   b) A sample of 20 employees—10 chosen from each of two randomly selected clusters.
   c) A systematic random sample of 10 employees from the SE region.
   d) A stratified sample that includes about 20% of the employees from each region.
   e) A stratified sample that includes exactly 15 employees from each region.

2. Find a complete copy of a modern fashion magazine, such as *GQ* or *Cosmopolitan.* Randomly select 30 pages from that magazine using any two of the four random sampling methods we have discussed. From each sample, count how many of the 30 sampled pages are full-page advertisements.
   a) What is the population of interest?
   b) What is the variable of interest?
   c) What is the parameter being estimated, and what is the statistic that you have created? (Save your data.)

2.3 Planning an Ethical Study

To reach conclusions from data can involve many challenges. Among the first is the necessity to obtain data that are relevant and meaningful. If you use secondary data, you first need to know where to find the data. Data that already exist may or may not be fully representative of the population you want to research. Can you trust the source of the data, and can you get permission to use it? And if you need to merge data from two or more data sets, how compatible are the sets with respect to, for example, their data structures or to their interpretations of variables?

If you collect primary data for the research, you may be able to avoid some of these concerns. At least, you would know the data source and structure, and have some input about the sampling frame that is used. But, as we learned in Section 2.1, not all primary data collection methods meet experimental standards; that is, they observe or take measurements of the population as it is, without imposing controls on the experimental conditions. Relationships found in such cases may turn out to be accidents due to unmeasured common influences, instead of true indications of causal relationships among the observed variables. (If experimental methods are not possible, approaching the research problem with multiple methods may help to limit this risk.)

Whatever the method of data collection, the researcher has a responsibility to be **ethical** in all aspects of the study. This means following accepted moral standards in both the collecting and the presenting of the data. Honest mistakes will happen, but to present findings that are misleading due to a careless study design or sampling plan, or due to poor training of the data collectors, can lead to disastrous results—no less than if someone deliberately falsifies conclusions. **Due diligence** requires that you plan and execute each step very carefully and critically assess the possible limitations of every method that is chosen. Take extra precautions if they are available, and admit openly the types of errors that might be present in your findings.

For those whose studies are conducted for hospitals, universities, and other agencies receiving government research funding, ethical research standards are particularly high: Any research plans involving humans, human tissues and fluids, and human remains must be thoroughly evaluated by a Research Ethics Board, made up of dedicated individuals both on- and off-site. Similar bodies exist to review research proposals that would involve animals, radiation, or environmental risks.

In the following chapters, the limitations and assumptions built into various methods will be discussed. In Parts 5 and 6, some positive guidelines and suggestions for good research methodology will be presented. It would be impossible to catalogue every error that could be made, or every experimental technique designed to avoid them. Whenever you practise statistics, whether collecting data, analyzing, or presenting the data, your ability to think critically, reflectively, and ethically at every step is as important as all of your knowledge of techniques.

Chapter Summary

2.1 Name and describe the basic types of research and the types of data sources.

A **scientific approach** to research was outlined, with an emphasis on clear **research objectives.** Preliminary types of studies called **exploratory research,** such as **pilot studies** or

focus groups, are distinguished from **conclusive research,** which can be conducted when the research objectives are clear and the problems are unambiguous. The latter includes **descriptive research,** used to describe the characteristics of a population, and **causal** or **experimental research,** which explores possible cause and effect relationships among the observed factors.

The two main types of data sources are **primary data,** which are not available prior to the research, and **secondary data,** which originally were collected for some other purpose. Primary data can be obtained by personal or mechanical **observation** of what is actually happening; by **surveys,** such as telephone interviews and mail-in questionnaires; or by **experiments.** Formal testing of a causal hypothesis requires an experimental approach. Secondary data can be subdivided into **internal secondary data,** originally collected in one's own organization, and **external secondary data,** originally collected outside of one's own organization.

2.2.1 Distinguish between statistical and nonstatistical sampling, and describe the basic types of statistical sampling.

To validly apply statistical techniques for making inferences, sample data must be collected by **statistical sampling** (or **probability sampling**) methods. These are based on **random selection** of the collected data, to ensure that the samples are *representative* of the population. Care is required in determining the **sampling frame** (i.e., the list the individuals or objects from which the sample will be drawn).

In **simple random sampling,** each *object* in the sampling frame has an equal chance of being selected, as does each possible *combination of objects* of a given size. If the data contain subpopulations that are relatively small, you can use **stratified random sampling** to ensure that all of the subpopulations are reasonably represented. If a population is widely dispersed, **simple cluster sampling** provides a way to group the population into clusters and samples are taken from randomly selected clusters. Finally, **systematic random sampling** can be applied when data for the sampling frame are available in a conveniently accessible list that is randomly ordered in relation to the variable of interest.

Qualitative statistics follow *nonstatistical sampling techniques,* which may provide an insight into how people are thinking in certain contexts. Example types are **convenience sampling** (such as "man on the street" interviews), and **judgment sampling** and **quota sampling** (which both focus on cases or individuals who the researcher believes can provide some key information).

2.2.2 Describe the sources of error associated with sampling.

If the information in the sample differs from the information in the population, there is an error. In the case of **sampling error,** this difference between sample and population information is unavoidable, and occurs because the sample is only a subset of the population. But in the case of **nonsampling error,** the difference between information in the sample and the population is due to missing data or incorrect measurement. Three common types of nonsampling error are **measurement errors, nonresponse errors** (where cases that were not included may have different properties from included cases), and **bias** (where the collected data tend to point—misleadingly—more toward one possible conclusion than another).

2.3 Explain the importance of following accepted moral standards in both the collecting and the presenting of statistical data.

It is important—and legally required—to be **ethical** in the collecting and the presenting of data. Although risks of error always exist, **due diligence** requires that you plan and execute each step of your research carefully, assess the limitations of methods used, and report on possible gaps and biases.

Key Terms

bias, p. 49
causal research (or experimental research), p. 35
conclusive research, p. 35
convenience sampling, p. 47
descriptive research, p. 35
due diligence, p. 55
ethical, p. 55
experiments, p. 37
exploratory research, p. 35
external secondary data, p. 38
focus groups, p. 35
internal secondary data, p. 38
judgment sampling, p. 47
measurement errors, p. 49
nonresponse error, p. 49
nonsampling error, p. 49
observation, p. 36
pilot studies, p. 35
primary data, p. 36
qualitative statistics, p. 47
quantitative statistics, p. 47
quota sampling, p. 48
random selection, p. 40
research objectives, p. 34
sampling error, p. 49
sampling frame, p. 41
scientific approach, p. 34
secondary data, p. 37
simple cluster sampling, p. 43
simple random sampling, p. 41
statistical sampling (or probability sampling), p. 40
stratified random sampling, p. 42
surveys, p. 36
systematic random sampling, p. 45

EXERCISES

Basic Concepts 2.1

1. Discuss some records that a business might keep that could be used as internal sources for secondary data.
2. Discuss some records that might be kept by the municipal Water Department that could be used as internal sources for secondary data.
3. List several examples of external sources of secondary data. Include in your list, at the minimum, a newspaper, a periodical, and a website.
4. Why are clearly stated research objectives important for any study?
5. Describe, in general terms, the ordered steps of the research process.
6. Discuss three main approaches to obtaining primary data.
7. Mail-in questionnaires and telephone interviews are examples of what approach to obtaining primary data?
8. Distinguish between descriptive and causal research studies.
9. Distinguish between exploratory research and conclusive research.

Applications 2.1

10. The following statements describe the original collection of primary data. In each case, decide whether the data are being collected by observation, survey, or experiment:
    a) A home inspector records and summarizes the most common problems encountered when he inspects a home.
    b) Natural Resources Canada asks Canadians about their attitudes toward natural resource issues.
    c) For a study on behalf of Canada's International Development Research Centre, the revised Model 3 of an "expeller" device (for squeezing oil from various types of seeds) is field-tested on cotton seeds, and the results are recorded.
    d) When a person makes a financial claim regarding a hearing problem to Veterans Affairs Canada, a form is

completed that summarizes the person's past conditions and other related diagnoses.

e) To compare the yields of different varieties of soybeans, the Minnesota Crop Improvement Association planted seeds of different varieties in pots containing comparable types of soil.

11. The following statements describe an application using secondary data. In each case, decide whether the data are internal or external secondary data.

a) An analyst for the Minnesota Crop Improvement Association is testing a new hypothesis on soybean yield. In the process, she uses data collected previously by the association as part of a different study.

b) A science journalist prepares for an article by analyzing the data in *Canadians' Attitudes Toward Natural Resources Issues, 2002,* a document published by Natural Resources Canada.

c) In order to prepare a forecast of future interest rates, a bank manager consults recent reports published by the Bank of Canada.

d) An official at Veterans Affairs Canada wants to prepare a profile of the typical veteran who has filed a claim for a hearing problem. All past forms relating to veterans' claims in this area have been gathered for the analysis.

e) In order to choose a suitable wardrobe when travelling to an environmental conference in the north, a scientist consults weather data published by Environment Canada for the Iqaluit A weather station in Nunavut.

12. The following statements describe research that has been conducted. In each case, categorize the research as primarily exploratory, descriptive, or causal:

a) A home inspector has recorded and summarized the most common problems encountered when he inspects a home.

b) Natural Resources Canada conducts research leading to the report *Canadians' Attitudes Toward Natural Resources Issues, 2002.*

c) In a study for Canada's International Development Research Centre, the relative effectiveness of the Model 3 "expeller" device (for squeezing oil from cotton seeds) is tested under field conditions.

d) Veterans Affairs Canada conducts a focus group with hearing-impaired veterans to obtain feedback on how they feel about the organization's services.

e) To compare the yields of different varieties of soybeans, the Minnesota Crop Improvement Association planted seeds of different varieties in pots containing comparable types of soil.

13. A criminologist is studying the relative lengths of jail sentences given in Canada over several years, compared by gender of the prisoner. What type of research is she conducting?

14. A movie fan wants to buy a subwoofer for his home theatre. (A subwoofer is a speaker that specializes in very low frequencies—such as the deep rumble sound of approaching enemy spacecraft.) He has narrowed his search to the Velodyne SPL-1000 subwoofer and, using secondary sources, he found that a typical price for such a unit online is about US$1,175. What kind of research is he conducting?

Basic Concepts 2.2

15. What does it mean to say that a sample is a *random* sample? Why are random samples used?

16. What is meant by a sampling frame? Why is it important that the sampling frame be chosen carefully?

17. Explain the difference between simple cluster sampling and stratified random sampling. Include examples in your answer.

18. Explain the differences between simple random sampling and systematic sampling. What are some advantages and disadvantages of using each?

19. Explain why—from the viewpoint of a statistician—random samples are preferable to judgment samples.

20. Describe a research objective for which it may be appropriate to use judgment samples or other nonstatistical methods.

21. The registrar at your university has access to the student IDs for all current students. Describe procedures she could use to select a simple random sample of those students to take part in a research project.

22. In inferential statistics, which problem is generally considered more serious: sampling errors or nonsampling errors? In your answer, describe both kinds of errors, and also give examples.

23. Which can occur in a census: Sampling errors? Nonsampling errors? Both? Explain.

24. What can be done to minimize sampling errors?

25. What can be done to minimize nonsampling errors?

Applications 2.2

26. You plan to survey the residents of your community about the location of a new, high-tech garbage incinerator planned for your region. What are some of the difficulties likely to be encountered if you use the telephone directory to derive the sampling frame? If you do use the telephone directory, explain how you could use it to obtain a systematic sample.

27. Natural Resources Canada recognizes the environmental drawbacks of prolonged idling by motor vehicles. In 2001, the department published the results of a residential telephone survey on the topic, conducted in Mississauga, Ontario.[6] Describe some alternative ways the department might have obtained its information on residents' attitudes toward this issue.

28. In 2005, the Ontario Universities' Application Centre published results of a survey of graduates from Ontario universities.[7] Respondents were asked about their employment positions and salaries following graduation. Do you perceive any risk of bias in asking graduates themselves about these apparent measures of their success? The report acknowledges that over 75% of those contacted did not respond. Is there reason to question whether salary levels and employment successes are basically the same for both the responders and the nonresponders? Explain. *If* there were some differences between the groups, would this lead to sampling error or nonsampling error?

29. In a questionnaire published by British Columbia's Thompson Rivers University for its agents who represent the institution internationally, an open-ended question is asked: "What types of services do you provide to the students you serve?"[8] As an alternative, the questionnaire could have listed an explicit set of typical services offered and asked respondents to check off those that are provided. Can you think of some advantages and disadvantages for each of these approaches, in terms of the risk of nonsampling errors?

30. After a big sporting event, reporters sometimes stand near one of the exits and question passersby about how they liked the game. What type of sampling is this? Explain how this method could lead to nonsampling error.

31. An investor using the website http://www.GlobeFund.com regularly monitors the performances of various stocks. In particular, she wants to compare the performances of Canadian equity stocks versus European equity stocks. What kind of sampling should be used to ensure a good-sized sample for these two categories?

32. The zipper is often the first part of a jacket to disintegrate. It may start opening from the wrong side or fail to zip. A jacket manufacturer wants to test different brands of zippers to select the most reliable brand for use in the company's product. Devise a plan whereby the jacket manufacturer can determine the proportion of ZipALot brand zippers that are defective.

33. Describe a realistic sampling situation where it may be advisable to use a nonstatistical sampling technique.

Review Applications

34. Give an example of an actual *secondary* data source that might be used by each of the following. (Hint: You might browse the Internet to find some suitable data sources for the different examples.)

a) Provincial police official preparing a forecast of future crime rates.

b) Person responsible for setting and publicizing avalanche warnings for a popular ski resort.

c) Sales manager for an automobile retailer who is preparing a forecast of sales of SUVs for the next calendar year.

d) Public health official investigating the prevalence of work-related stress.

e) Bridge designer in Saskatchewan who wants to analyze the patterns of previous earthquakes in that province.

35. Give an example of how each of the following might use a sample to gather *primary* data.

a) Provincial police official who wants to assess the steps that businesses are taking to reduce their exposure to crime.

b) Person responsible for setting and publicizing avalanche warnings for a popular ski resort.

c) Sales manager for an automobile retailer who is preparing a forecast of sales of SUVs for the next calendar year.

d) Public health official investigating the prevalence of work-related stress.

e) Engineer in Saskatchewan who wants to analyze the resistance of a new type of bridge design to damage from earthquakes.

36. A concerned member of Gamblers Anonymous counted and recorded information on hundreds of slot machines in Alberta gambling facilities. (Compare the file **Casinos_Alberta.**) Was the member engaging in exploratory research, descriptive research, or causal research?

37. A member of Gamblers Anonymous is conducting focus groups with those who quit gambling in Alberta, to learn how they feel about the number of the slot machines in the province. Is the member engaging in exploratory research, descriptive research, or causal research?

38. Every morning Jacob trains for the next International Marathon to be held in Seoul, South Korea. Afterward, before leaving for work, he drinks a high-caffeine energy drink. Jacob wants to know which of the many competing brands of energy drinks would be most successful in keeping him awake at work until lunch time. Is Jacob planning to engage in exploratory research, descriptive research, or causal research? Will it require primary or secondary research?

For Exercises 39–41, use data from the file **Chem_Element_Discoverers.**

39. A sampling frame for the population of discoverers of known chemical elements is given in the file **Chem_Element_Discoverers.** What type of sampling would you use if the goal of a study was to compare the elements discovered by French scientists with the elements discovered by British scientists?

40. If you started on a random row of the above sampling frame, what type of sampling would you be using if you selected data from every 9th row in the data file?

41. Suppose you base your sample of element discoverers on the names of elements you happen to remember from high school. This would be what type of nonstatistical sampling?

42. Suppose your research question was this: How fondly do new high-school graduates remember their time studying chemistry in high school? You gather together a small group of recent grads to question them on their experiences. What type of research is this?

43. The large Internet backbone service provider Global Crossing handles about 1.26 million hits per minute. Presume that this rate is constant throughout the day, and that you are studying the countries of origin of the hits. Explain how you might take a random sample of hits during a day without favouring hits from any particular time period.

44. After learning in the previous exercise the countries of origin of hits on the site Global Crossing, you decide to visit and collect data from some of the countries involved. Describe how with cluster sampling you could limit your travel costs and still visit a reasonably representative selection of the countries.

For Exercises 45 and 46, use data from the file **Meteors_N_Amer.**

45. Given the file **Meteors_N_Amer** as the sampling frame, a geologist wants to compare the diameters of meteor craters in Canada versus those in the United States. What type of sampling should she use?

46. Even if the average diameters of meteor craters are about the same in Canada and the United States, it is possible that, by chance, the samples from the two countries will have different average diameters (see Exercise 45). This is an example of what kind of error? Explain.

PART 2

Descriptions of Data

Chapter 3

Displaying Data Distributions

Chapter Outline

Learning Objectives

3.1 Construct and interpret frequency distribution tables for quantitative and qualitative data.

3.2 Construct and interpret frequency distribution histograms for quantitative data.

3.3 Construct and interpret pie charts and bar charts for qualitative data.

3.4 Construct and interpret stemplots for quantitative data.

3.5 Construct and interpret appropriate displays of association for quantitative and qualitative data.

3.1 Constructing Distribution Tables

The nature of statistical data and how they are obtained were discussed in Part 1. Part 2 introduces basic methods of descriptive statistics, by which the data can be described and presented. Constructing distribution tables is often a first step.

Part 2 focuses primarily on **univariate descriptive statistics,** which are descriptive statistics for just one variable. Examples of statements that express univariate descriptive statistics are:

- 41% of out-of-province tourists to Alberta learned about the province from the Internet.[1]
- Over 7,800 firearms licences have been revoked under the Canadian Firearms Program.[2]
- The average annual amount spent by Canadian retirees on gifts is $822.[3]

Descriptive statistics characterize available data in terms of patterns, clusters, and other features. Even if the analyzed data are a sample and one's goal is to make inferences about a larger population, it is useful to become familiar with the characteristics of each variable. For example, this knowledge could influence one's choice when selecting more advanced techniques. Inspecting data for individual variables can also uncover data that may not be suitable, as they stand, for further analysis.

It is important to ensure that **clean data** are used for analysis; that is, the recorded numbers or computer codes must be correct. This is not the case if someone incorrectly enters "688 kg" for a person's weight, when "68.8 kg" was intended. Or perhaps an untrained staff member inputs the number of grams—68,800—instead of the number of kilograms. In another example, one could encounter an impossible code, such as a "5" for data whose scales go up to only "4." All of these cases illustrate **dirty data.** If at all possible, you should track these cases back to the original questionnaire responses or instrument readings. If the errors cannot be corrected, you should discard the corrupted data.

Inspecting univariate descriptive statistics can also help to uncover **outliers** in the data. Outliers are values that appear remote from all or most of the other values for that variable. An example is a display of an MLS real estate listing for house prices in Richmond Hill, Ontario, as of March 2005. The top-priced house (at $3.68 million) appears to cost over a million dollars more than the second most expensive house.[4] It is prudent to double-check whether this is a proper part of the data set. In this example, based on a photograph and description of the most expensive house—a mansion with 10 bathrooms—there is no reason to suspect that the data value is dirty. But depending on the goal of the analysis (for example, to market houses to middle-class buyers), it may be wise to footnote this extreme value and remove it from subsequent descriptions. Outliers will be discussed in more detail in Chapter 5.

Where to Begin

Generally, the data that you originally collect (**raw data**) are not in a form that is easy to understand. For example, if you were planning a vacation flight out of Toronto in March 2005, and you were flexible about your destination, you might have encountered the following list of "flight deal" prices posted on the Red Label Vacations Inc. website (http://www.redtag.ca).[5]

$199 $189 $198 $198 $259 $197 $197 $197 $268 $197 $448 $224 $368

One approach to interpreting this raw data would be to characterize the data numerically by finding a single representative value (discussed in detail in Chapter 4) or a measure of how dispersed the data are (discussed in detail in Chapter 5). However, a common place to start is to construct a table called a **frequency distribution table.** A frequency distribution counts how many times (the frequency) that a specific data value reappears in the dataset; or, if the data values are grouped into ranges, the distribution counts how many values fall into each range. The same information contained in a table can also be represented in a graph, as discussed later in this chapter in Sections 3.2 and 3.3.

The details of interpreting and constructing frequency distributions vary, based on whether the data are qualitative or quantitative, discrete or continuous. Frequency distributions for discrete quantitative data are introduced in the following section.

Discrete Quantitative Frequency Tables

Many fans of National Football League (NFL) games follow the point spread data that are published for each game. For an upcoming game, these data represent forecasts by experts (or bookmakers) of the projected difference in points between the stronger team's score and the weaker team's score. Point spread data are quantitative and discrete, because the projections are limited to specific numeric values—those ending in ".0" or ".5."

The actual distribution of assigned point spreads for three seasons of NFL games (based on data in the file **Pointspreads**) is displayed in Table 3.1. The first column provides an exhaustive list of the point spreads that were assigned. The column headed "Frequency" displays the exact number of times each specific assigned value appears in the data set. For example, the experts predicted a tie game (Point spread = 0) exactly 20 times over the three seasons. Seven times, they forecast the stronger team to win by 14 points (i.e., by two converted touchdowns). This table contains the basic elements of a **discrete quantitative frequency table.** Any specific data value that occurred can be examined to determine how often it appeared in the overall data set. The total for the "Frequency" column represents the total number of recorded cases.

Also displayed is the **percent frequency distribution** of the data, which is in the column labelled "Percent." This column puts the frequency data into perspective. Just looking at the frequencies, it may not be obvious whether a point spread of, say, 10, which appeared 13 times, could be described as a common occurrence. But after converting raw frequencies into percentages (proportions can also be used), we see that a point spread of exactly 10 is relatively rare—it occurs in just 1.9% of the cases. For each distinct value x of the variable that is observed, calculate its percentage relative to the total number of observations n, as follows:

$$\text{Percentage}_x = \frac{\text{Frequency}_x}{n} \times 100\%$$

In the table, align the calculated percentages with the corresponding observed values of the variable. For example, we see in the table that the percentage of games for which a 14-point spread was predicted was 1.04%, based on (7/672) × 100%. The total for the "Percent" column is 100%, although it may vary slightly due to round-off.

The final column in the table, "Cumulative Percent," shows the **cumulative percent frequency distribution** for each distinct value of the variable. For any row on the table, the

Table 3.1 **Frequency Distribution for NFL Game Point Spreads**

Point Spread	Frequency	Percent	Cumulative Percent
0.0	20	2.98	2.98
1.0	22	3.27	6.25
1.5	27	4.02	10.27
2.0	41	6.10	16.37
2.5	48	7.14	23.51
3.0	77	11.46	34.97
3.5	39	5.80	40.77
4.0	43	6.40	47.17
4.5	17	2.53	49.70
5.0	27	4.02	53.72
5.5	23	3.42	57.14
6.0	46	6.85	63.99
6.5	42	6.25	70.24
7.0	46	6.85	77.09
7.5	25	3.72	80.81
8.0	13	1.93	82.74
8.5	7	1.04	83.78
9.0	26	3.87	87.65
9.5	14	2.08	89.73
10.0	13	1.93	91.66
10.5	7	1.04	92.70
11.0	5	0.74	93.44
11.5	4	0.60	94.04
12.0	11	1.64	95.68
12.5	4	0.60	96.28
13.0	8	1.19	97.47
13.5	6	0.89	98.36
14.0	7	1.04	99.40
14.5	1	0.15	99.55
16.5	1	0.15	99.70
17.5	1	0.15	99.85
19.5	1	0.15	100.00
Total	**672**	**100.00**	

Source: Hal Stern, Harvard University. Published in the Statlib Datasets Archive. Found at: http://Lib.stat.cmu.edu/datasets/profb

cumulative percent is the percentage of all cases having values up to or including the value displayed at the left of that row. It can be calculated manually by this procedure:

Cumulative percent for the first data row = Percent frequency for that data row

Cumulative percent for any subsequent row = Cumulative percent for the previous row + percent frequency for the current row

In other words, the cumulative percent for a given data value is the sum of all valid percentages up to and including that value. With this useful information, you can now answer questions such as: 90% of the predicted point spreads go up to what value? (*Answer:* 9.5) What percentage of the predicted point spreads are less than a converted touchdown (i.e., 6.5 points or less). (*Answer:* 70.2%)

• *In Brief 3.1(a) Discrete Quantitative Frequency Tables*

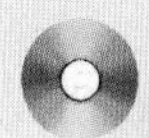

(*Based on the* ***Pointspreads*** *file*)

Excel

Input the raw data in a column. In another column, input the complete list of values that the variable can take (or has taken). Use the menu sequence *Tools / Data Analysis / Histogram,* then click *OK.* As shown in the screen capture, specify in the dialogue box the range for the raw data, the range for the possible values (bins), and the cell address for the upper left-hand corner of the output display. Select *Cumulative Percentage* and then click *OK.* The output will not include a "Percent" column, but this can be generated by Excel formulas (by dividing each row's frequency by the total number of cases and then multiplying by 100%).

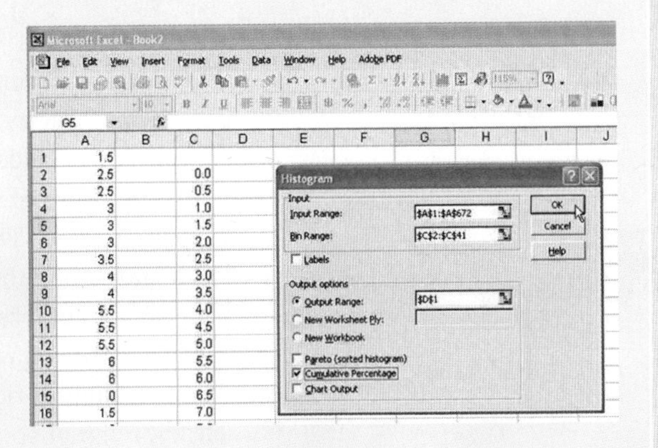

Minitab

Input the raw data in a column. Use the menu sequence *Stat / Tables / Tally Individual Variables.* As shown in the screen capture, specify in the dialogue box the column containing the raw data, select the options *Counts, Percents,* and *Cumulative Percents,* and then click *OK.*

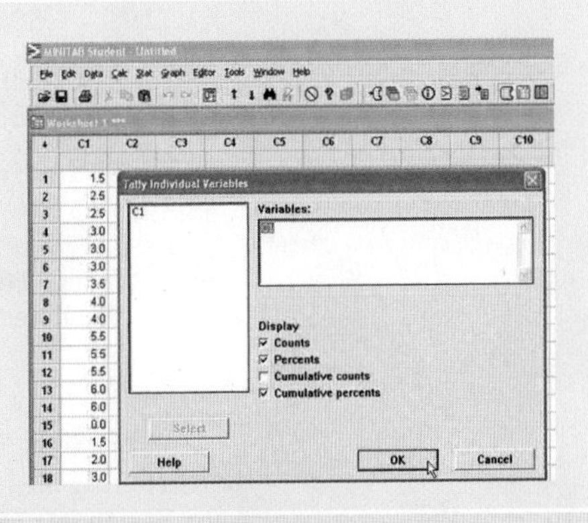

SPSS

Input the raw data in a column. Use the menu sequence *Analyze / Descriptive Statistics / Frequencies.* As shown in the screen capture, specify in the dialogue box the variable containing the raw data, select the option *Display frequency tables,* and then click *OK.*

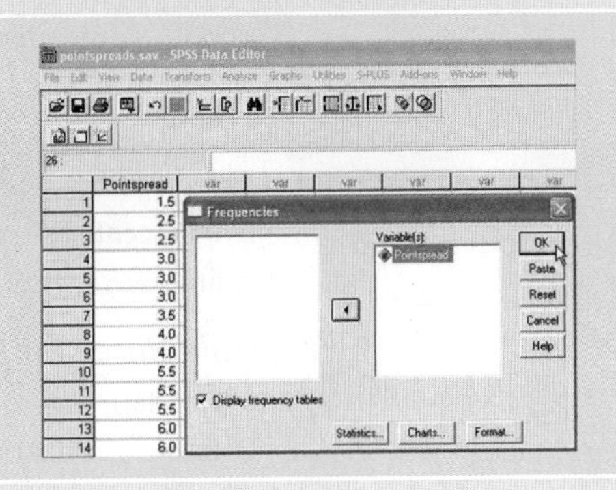

After constructing Table 3.1, we can get a sense of how the data are distributed. Almost all of the possible point spreads from 0 to 14.5 are represented in the data in increments of 0.5. The "Cumulative Percent" column tells us that virtually all the data (99.6%) fall below the three largest values—16.5, 17.5, and 19.5, which could be considered outliers. At the least, we know to recheck those high values, if possible, to ensure that they have been correctly input.

Table 3.1, however, includes so many distinct values that it is quite large, making it difficult to get a clear sense of how the values are distributed. The techniques described in the next section can be applied to both continuous quantitative data and to discrete quantitative data, such as point spreads, *if* there are many distinct values in the data set.

Continuous Quantitative Frequency Tables

When the data are continuous, the number of possible values is potentially infinite. Even if you could line up every value that occurred and display its frequency, the sheer number of values included would make it difficult to comprehend the information in that format. Instead, values displayed in a **continuous quantitative frequency table** are grouped into ranges (called **classes**). The frequencies in the table represent how many individual data values fall into each of the classes.

Suppose, for instance, a political scientist is examining the voting patterns of British Columbians in the Canadian federal election of 2004. The raw numbers of votes cast in each British Columbia riding are shown in Table 3.2. Although fractional votes are not possible, the range of possible values of votes cast per riding is very large (in the tens of thousands) so that the data can be interpreted as nearly continuous.

Table 3.2 Total Votes Cast in BC Ridings, 2004 Federal Election

48,139	45,263	41,732	42,120	44,862	46,727	48,000
41,119	50,955	58,770	56,032	41,215	47,615	51,853
40,782	63,773	37,025	46,268	53,283	34,454	52,510
40,763	58,127	60,543	54,756	39,180	51,176	53,238
41,594	52,449	42,648	56,645	48,812	45,527	36,237
55,599						

Source: "Federal Election Results, 2004." *The Globe and Mail.* Wednesday, June 30, 2004. Pages A12–A13. Reprinted with permission from *The Globe and Mail.*

Table 3.3 is a frequency distribution table for the variable. The basic principles of constructing and interpreting this table are the same as for a discrete frequency distribution table. However, counts and percentages are now related to the indicated *ranges* of data values, instead of to individual, discrete values. We observe, for example, that it was relatively rare for a riding to have fewer than 35,000 votes—only one (2.8%) of the 36 ridings was in the first class. The numbers of votes per riding that were most common lie in three classes that range from 40,000 up to almost 55,000. Based on the "Cumulative Percent" column, we see that the numbers of votes cast in each of over 80% of the ridings were fewer than 55,000.

Guidelines for Manually Constructing the Tables

Many statistical software packages can now automate the construction of various types of frequency distribution tables and graphs. It is still useful to understand the following guidelines so that you understand what the computer is doing—and, sometimes, to eval-

Table 3.3 **Total Votes Cast in BC Ridings, 2004 Federal Election**

Classes	Frequency	Percent	Cumulative Percent
30,000–34,999	1	2.8%	2.8%
35,000–39,999	3	8.3%	11.1%
40,000–44,999	9	25.0%	36.1%
45,000–49,999	8	22.2%	58.3%
50,000–54,999	8	22.2%	80.6%
55,000–59,999	5	13.9%	94.4%
60,000–64,999	2	5.6%	100.0%
TOTALS	**36**	**1**	

Source: "Federal Election Results 2004." *The Globe and Mail.* Wednesday, June 30, 2004. Pages A12–A13. Reprinted with permission from *The Globe and Mail.*

uate or revise the computer's automatic "decisions." Before elaborating on particular steps to create continuous frequency tables, some guidelines are presented here to ensure that the tables are both useful and valid. There can be more than one "right" answer to constructing a table, so there is room to use your own discretion.

1. *The displayed classes should be* ***mutually exclusive.*** Any specific data value should fall into one, and only one, of the classes. For instance, the number of votes in a riding can be either "30,000–34,999" or "35,000–39,999," but it cannot fall into both classes. An alternative way of labelling the classes in Table 3.3 is "30,000 up to 35,000" followed by "35,000 up to 40,000," etc., but this "up to" notation must be interpreted as meaning "up to *but not including,*" so that the highest possible value falling into the class "30,000 up to 35,000" would be 34,999 (or, if fractional values were possible, 34,999.99, etc.).
2. *The classes should be* ***collectively exhaustive,*** meaning that there is a class to which every data value can be assigned. There were 48,000 votes cast in the riding of Dewdney–Alouette, British Columbia. If the number had been 88,000, Table 3.3 would no longer meet this second guideline, since the table does not contain a class to which that high data value could be assigned.
3. *Try to use the same* ***class width*** *for all classes.* Because Table 3.3 follows this guideline, it is easy to interpret the larger frequency for the second class (compared to the first class) as reflecting more occurrences of values in a same-size range. It would be harder to interpret the difference in frequencies if the two classes stretched over unequal ranges of value, such as "30,000 up to 35,000" in the first class but "35,000 up to 60,000" in the second class. The next point explains an occasional exception to this third guideline.
4. *Try to include all classes, even if the frequency is zero.* A well-crafted frequency distribution should display the pattern of where the data values lie, between the upper and lower limits of the table. Omitting a class because its frequency is zero would hide the fact that the data contain a gap in one or more ranges. This hidden information might prove to be important.

A possible exception to the last two guidelines can occur if there are extreme values or outliers in the data. Suppose 88,000 votes had been cast in Dewdney–Alouette. Keeping all class widths equal, and displaying all classes, would require the addition of four zero-frequency classes between 65,000 and 84,999, until a non-empty class is reached. It is not uncommon, for such cases to instead include one open-ended class such as "65,000 or above."

5. *Select convenient numbers for **class limits.*** Each class in Table 3.3 has a nice, even lower limit: 30,000, 35,000, 40,000, and so on. There is no rule that classes must go up by 5s or 10s or 5,000s, but imagine yourself in the shoes of whoever will read the table you create. Select limits that appear to be reasonable and easy to interpret yet are consistent with the other guidelines.
6. *Choose an appropriate number of classes.* Choosing too few classes makes it difficult to distinguish patterns in the data. Having too many classes may overly stretch the display, so it will be difficult to comprehend. Typically, the number of classes is between 5 and 15, depending on the amount of data available, and there are techniques (for example, Sturge's Rule, described a little later in this section) that can help to guide your choice.
7. *The sum of the class frequencies must equal the number of original data values.* This serves as a check on your work. If the classes are mutually exclusive and collectively exhaustive, and if every data value has been correctly assigned to exactly one class, then this condition should be met automatically.
8. *Combine these guidelines—with a touch of common sense.* When following the steps to construct a frequency distribution (see the next section), you will have to balance the concerns for reasonable class limits with issues of class width, number of classes, or dealing with outliers. The important thing is to design a table that clearly communicates information about the distribution, rather than hiding or distorting the information.

Steps for Manually Constructing the Tables from Raw Data

The following steps should be applied with some flexibility, keeping the preceding guidelines in mind.

1. **Sort the data:** The first step is to put the data into an **ordered array,** which means that the data are reordered into either **ascending order** (lowest number to highest) or **descending order** (highest number to lowest). Ascending order is most common. The ordered array for the British Columbia votes data is shown here (reading left to right, by row):

```
34,454 36,237 37,025 39,180 40,763 40,782 41,119 41,215 41,594 41,732
42,120 42,648 44,862 45,263 45,527 46,268 46,727 47,615 48,000 48,139
48,812 50,955 51,176 51,853 52,449 52,510 53,238 53,283 54,756 55,599
56,032 56,645 58,127 58,770 60,543 63,773
```

This step alone makes the data look more meaningful. We can easily see some values of votes cast per riding that are typical (such as in the 40,000s), and other values that occur less frequently (such as in the 60,000s). We can also identify immediately the **range** of the full data set, defined as the spread or difference between the highest and the lowest value. For these data, the range is 63,773 – 34,454 = 29,319. For a data set that is not too large, one can identify by eye additional details such as (a) the approximate distances between successive values and (b) areas in the data where there are gaps or clusters of values.

• *In Brief 3.1(b) Sorting the Data Using the Computer*

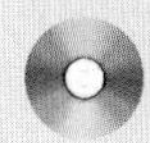

(Based on the ***BC_votes*** *file)*

Excel

Highlight the column containing the data. If the data column is part of a larger block of data, highlight the full block (otherwise the sorted data will go "out of sync" with other data fields it is related to). Use the menu sequence *Data / Sort.* Specify (1) whether you want to sort in ascending or descending order and (2) whether the highlighted column or block of data includes column titles in the first row. Then click *OK.*

Minitab

Use the menu sequence *Data / Sort.* In the dialogue box, highlight and select all of the columns to be sorted. Generally, if the column to be sorted is part of a larger block of data, you would select all of the columns in the data block. Then indicate the specific column to be sorted. An ascending sort is assumed unless you indicate otherwise. Next, click *OK.*

SPSS

Right-click on the heading for the column that contains the data to be sorted. Select *Sort Ascending* or *Sort Descending* in the drop-down menu. SPSS will automatically sort in tandem all of the columns in the full data set.

2. **Choose a tentative number of classes:** As a general rule, more classes are needed when more raw data are available. Numbers of classes typically range from 5 to 15, although some flexibility is permitted. A useful starting point is to estimate a number of classes by using **Sturge's Rule:**

$$\text{Estimated number of classes} = 1 + 3.3\log(n)$$

   where n is the number of values in the data set, $\log(n)$ is the base-10 logarithm, and the answer is rounded to the nearest whole number. For example, given a set of 100 data points, the formula recommends $1 + (3.3)(2) \approx 8$ classes.

   The output of this rule is only a guide and is subject to modification. In the British Columbia votes data set there are 36 values, for which Sturge's Rule would recommend $1 + (3.3)(1.5563) \approx 6$ classes.

3. **Choose a tentative class width:** Once you have chosen a tentative number of classes, you can estimate how wide each class should be. The class width is defined as the difference between the lower limit of one class and the lower limit of the next class. We can approximate class width by dividing the *range* of the full data set (found in step 1) by the tentative number of classes (step 2):

$$\text{Tentative class width} = (\text{Range of full data set})/(\text{Tentative number of classes})$$

For the British Columbia votes data, we can calculate class width = (29,319)/6 ≈ 4,887. This outcome is only a guideline and can be rounded up or down to create a more readable table. Since the range for the votes is very large—nearly 30,000—it seems reasonable to round the class widths to at least the nearest 100, if not to an even 5,000. This choice may impact on the final number of classes, remembering that every data point has to be assignable to exactly one class.

4. **Define all classes:** Considering the results from steps 1–3 and, keeping in mind the guidelines, establish specific limits for all of the classes. To satisfy yourself that you have the best results, you may want to revise some original estimates.

   It is also advisable for the lower limit of the first class to be a multiple of the class width. Table 3.3, for instance, uses a class width of 5,000, so its first lower limit is 30,000. This is not ideal, since the smallest data point (34,454) is actually near the top of its class range. Had a class width of 4,800 been selected, a lower class limit of 33,600 (an even multiple of 4,800) could have been appropriate.

   Once you have determined the lower limit for the first class, each successive class begins exactly one class width larger in value. Continue until the upper limit of the largest class is large enough to include the largest data value. Be mindful of guideline (1) when labelling the upper value in each class, so that this value does not overlap with the lowest value of the next class.

   Although not strictly needed for constructing a frequency distribution, it is often useful to calculate **class midpoints.** These can be used in subsequent calculations that are based on frequency distributions (discussed in the next chapter). A class's midpoint is

   $$\text{Midpoint} = (\text{Lower limit of class} + \text{Upper limit of class})/2$$

   For example, the midpoint of the first British Columbia votes class is (30,000 + *almost* 35,000)/2 ≅ 32,500. Note that the midpoint for the last votes class is (60,000 + *almost* 65,000)/2 ≅ 62,500.

5. **Construct the table:** Manually, this step includes grouping and counting the data, so that each value is assigned to one class; then the total counts (class frequencies) of values to be assigned to particular classes are determined. For large data sets in particular, software programs are generally used for this step.

   "In Brief 3.1(c)" shows how to construct a table similar to Table 3.3 using Excel, Minitab, or SPSS. It is presupposed that steps 1 to 4 and the guidelines have been followed before you reached this stage.

• *In Brief 3.1(c) Constructing a Continuous Quantitative Frequency Table*

*(Based on the **BC_votes** file)*

Excel

Open a data file or input the raw data into a column. (Only the "Total Votes" column is shown in the illustration.) In another column, input a complete, ordered list of upper values for the proposed class ranges. These must each be actual numerical values (such as 34,999 or 34,999.99) rather than word descriptions (such as "up to but not including 35,000"), and they should include as many decimal places as are found in the data.

Next, use the menu sequence *Tools / Data Analysis / Histogram,* then click *OK.* As shown in the screen capture, specify in the dialogue box the range for the raw data, the range for the sequence of upper class values (bins), and the cell address for the upper left-hand corner of the output display. Select *Cumulative Percentage,* then click *OK.* The output will not include a "Percent" column, but this can be generated by Excel formulas (by dividing each row's frequency by the total number of cases and then multiplying by 100%). The output will also not include the sequence of class labels (such as "30,000–34,999"), but these can be input manually into a new column.

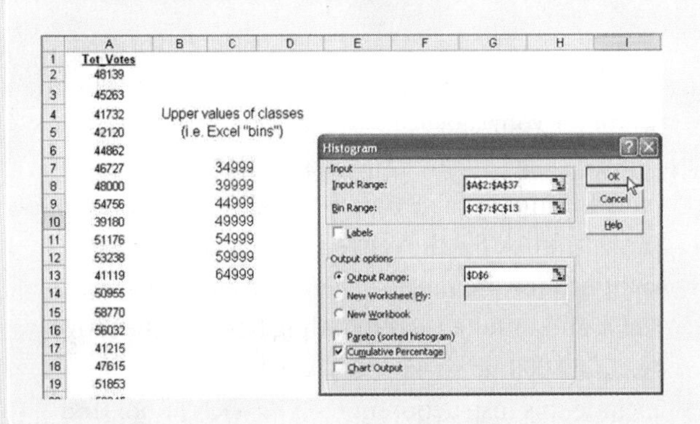

Minitab

Open a data file or input the raw data into a column. (Only the "Total Votes" data are shown in the illustration.) Next, the quantitative data must be converted into text labels corresponding to the selected classes. For this, use the menu sequence *Data / Code / Numeric to Text.* As shown in the screen capture, specify in the dialogue box the column that contains the data, as well as a column to which to send the created labels. Complete the encoding table in this manner: In each entry in the left column, show a range of raw data values that should be encoded into a single class label (e.g., "30,000:34,999"). In the right column, input the corresponding label for each range (e.g., "30,000 up to 35,000"). Click *OK* to complete the coding. (Partial output is illustrated in column C2 in the illustration.)

If you need to code for more than eight classes, you must run a small program. Use the menu sequence *Edit / Command Line Editor.* In the command window, input a sequence in this format:

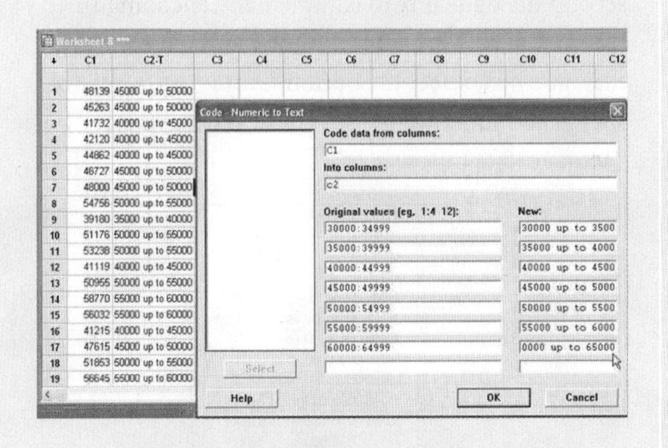

Code (30000:34999) "30000 up to 35000" &

(35000:39999) "35000 up to 40000" &

(40000:44999) "40000 up to 45000" &

(45000:49999) "45000 up to 50000" &

(50000:54999) "50000 up to 55000" &

(55000:59999) "55000 up to 60000" &

(60000:64999) "60000 up to 65000" &

C1 C2

Following the word "Code," each line shows a range of original values in parentheses followed by the desired corresponding label in quotation marks. The "&" symbol shows the continuation of the command. The final line shows the columns holding the original data and the encoded data, respectively. Click *Submit commands* to execute the program.

After coding the labels, proceed as for a discrete quantitative frequency data table: Use the menu sequence *Stat / Tables / Tally Individual Variables.* Specify in the dialogue box the column that contains the encoded (not the raw) data, and select the options *Counts, Percents,* and *Cumulative Percents;* then click *OK.*

SPSS

Open a data file or input the raw data into a column. (Only the "Total Votes" data are shown in the illustration.) Next, the quantitative data must be converted into text labels corresponding to the selected classes. For this, use the menu sequence *Transform / Recode / Into Different Variables.* As shown in the screen capture, specify in the dialogue box the column that contains the data and, in the box for *Output Variable, Name:* input a name such as "Encoded." Click on *Old and New Values.* Use the second dialogue box to complete the encoding in this manner: On the lower right-hand panel of the dialogue box, select the option *Output variables are strings* and then input a sufficiently long string *width* for your labels. In *Old Value,* select *Range* and input a range of raw data values that should be encoded into a single class label (e.g., input "60,000" in the left box for the range and "64,999" in the right box for the range). Then, in *New Value,* select *Value* and input a corresponding label for the range (e.g., "60,000 up to 65,000"). Click on *Add* to include this instruction in the *Old* → *New* box and then repeat the above sequence.

When all code instructions have been added, click *Continue* and then *OK* to execute the code. (Partial output is shown in the column labelled "encoded" in the illustration.)

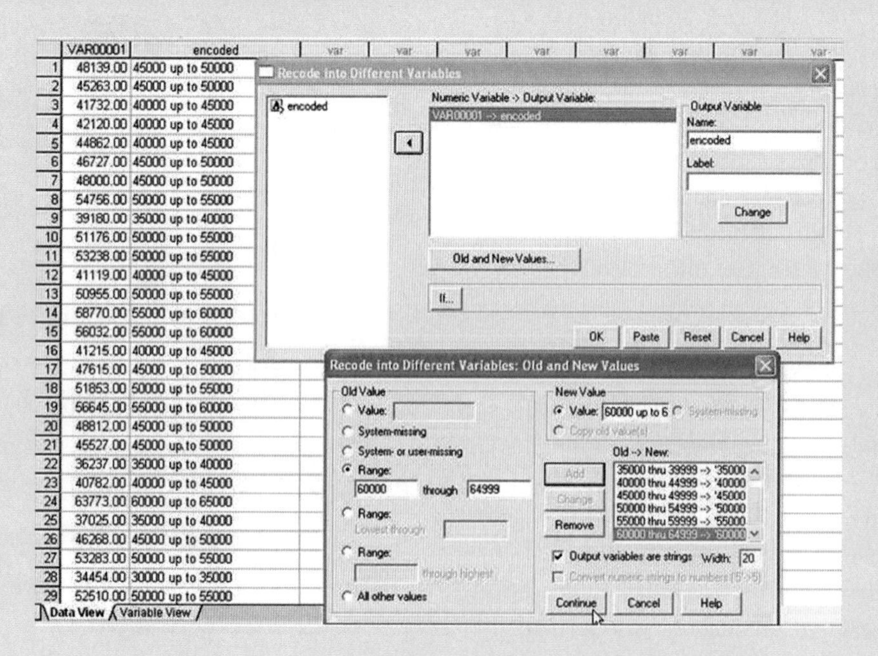

Next proceed as for a discrete quantitative frequency data table: Use the menu sequence *Analyze / Descriptive Statistics / Frequencies.* Specify in the dialogue box the variable containing the encoded (not the raw) data, and select the option *Display frequency tables;* then click *OK.*

Recall that the purpose of a frequency distribution is to provide useful information to the eventual user of the table. It is important to display a meaningful title for the distribution and, if appropriate, additional notes to ensure that the source and nature of the data are understood. If any open-ended classes (such as "65,000 and above") are included, it is also important to give the user a sense of how large (or small) the contained values actually are. Informed users may choose to handle exceptional cases as outliers when performing their own analyses.

Example 3.1

(Based on the ***Purchasing Power_World*** *file)*

An export consultant seeks to expand worldwide markets for locally produced consumer goods. She knows that, realistically, a country's suitability as a market depends on the general purchasing power of its citizens. To get a sense of what those levels are in different countries, she has compiled the data set **Purchasing Power_World** (based on data in the U.S. Central Intelligence Agency's 2003 edition of *The World Factbook*). The complete file contains data for 230 countries, plus summary data for "World" in the second row. Selected values (in Canadian dollars) from the data set are given below:

952 5,984 7,344 10,880 25,840 2,312 11,696 14,960 14,280 4,896

38,080 36,584 37,944 5,032 20,808 20,536 2,448 20,400 11,832 39,712

Construct a frequency distribution table for the full database, which is available on the textbook CD.

Solution

1. **Sort the data:** Use the computer software to sort the data into an array (do not include the summary row for "World"). A condensed version of the full 230-member array is displayed here:

   680 952 1,224 26,520 30,056 35,632 39,848 66,504

   Note that the largest value is 66,504 and the smallest is 680. Therefore:

   $$\text{Range} = 66{,}504 - 680 = 65{,}824$$

2. **Choose a tentative number of classes:** A first estimate can be made using Sturge's Rule.

   $$\text{Tentative number of classes} = 1 + 3.3\log(230) \approx 9$$

3. **Choose a tentative class width:** We have already found the range and a provisional number of classes.

   $$\text{Tentative class width} = 65824/9 = 7314$$

4. **Define all classes:** Given the large range of the data (in the tens of thousands), it would be reasonable to set the class width as a number of thousands (e.g., 7,000). Choosing a more precise class width such as 7,300 or 7,320 would result in visually confusing class limits. Following the guidelines to include the lowest raw data value in a class (680), and to ensure that the lowest class limit is a multiple of the class width, we set the lowest class limit at 0. Using the "up to" notation for the upper class limits, we get the following classes:

   0 up to 7,000
   7,000 up to 14,000
   14,000 up to 21,000
   21,000 up to 28,000
   28,000 up to 35,000
   35,000 up to 42,000
   42,000 up to 49,000
   49,000 up to 56,000
   56,000 up to 63,000
   63,000 up to 70,000

5. **Construct the table:** At this point, follow the steps in "In Brief 3.1(c)" to input both the raw data and your chosen class structure into the software you will be using.
   An example of the output is shown in Table 3.4.

Table 3.4 Frequency Distribution for Countries' Purchasing Power per Capita (Canadian Dollars)

C2	Count	Percent	CumPct
0 up to 7,000	114	49.57	49.57
7,000 up to 14,000	46	20.00	69.57
14,000 up to 21,000	18	7.83	77.39
21,000 up to 28,000	18	7.83	85.22
28,000 up to 35,000	11	4.78	90.00
35,000 up to 42,000	16	6.96	96.96
42,000 up to 49,000	5	2.17	99.13
49,000 up to 56,000*	1	0.43	99.57
63,000 up to 70,000*	1	0.43	100.00
Totals	230	1.00	

*See the comments in the text about the missing class "56,000 up to 63,000."

Source: U.S. Central Intelligence Agency. Based on data in the *The World Factbook, 2003*. Found at: https://www.cia.gov/cia/publications/factbook/fields/2004.html

Remember that you input the class labels using strings of text (e.g., "28,000 up to 35,000"), yet the classes in the finished table must appear to be sorted in correct *numerical* order. When coding the classes in Minitab or SPSS, include extra blank spaces or zeros before any small numbers (e.g., 7,000 in "7,000 up to 14,000," as shown in Table 3.4) so that they are properly sorted in front of numerically larger numbers such as 14,000 in "14,000 up to 21,000" in the table. Also observe that some computer outputs (e.g. Table 3.4, which simulates a Minitab output) do not obey the guideline to display *all* classes, including a class (such as "56,000 up to 63,000") of zero frequency. It is not uncommon that statistical computer program outputs need some modification—either by using advanced tools in the software itself or by exporting to another file—to be suitable for inclusion in a formal report to your manager or the public.

Qualitative Frequency Tables

As described in Chapter 1, qualitative (or categorical) data are nonnumerical in character. The possible values of a qualitative variable are words or categories, like "bus," "car," "subway," and so on. You can count the numbers of these answers and compare them in various ways. You can even encode them, arbitrarily, as "1," "2," "3," etc., but the actual values of the variable have no mathematical significance.

Like a frequency distribution for a quantitative variable, a **qualitative frequency distribution table** can help you visualize how the data values are distributed and shows how often each particular value or category of the qualitative variable has occurred. Table 3.5 illustrates the different categories of occupational disease for which claims have been paid by the Saskatchewan Workers' Compensation Board (WCB). The frequencies show how many claims there have been from each category.

Table 3.5 **Qualitative Frequency Table for Allowed Claims: Saskatchewan WCB**

Disease Category	Frequency	Percent	Cumulative Percent
Infectious & parasitic diseases	446	5.14	5.14
Neoplasms, tumours, & cancer	11	0.13	5.26
Prosthetic devices	259	2.98	8.25
Skin diseases	567	6.53	14.78
Systemic diseases & disorders	2,261	26.04	40.82
Traumatic injuries & disorders	456	5.25	46.07
Other	4,683	53.93	100.00
Total	**8,683**	**100.00**	

Source: Saskatchewan's Workers' Compensation Board. Found at: http://www.wcbsask.com/Facts_&_Figures/occupational_statistics.html

The column of frequencies in a qualitative frequency distribution table can be supplemented by additional columns for percent and cumulative percent frequencies. The percent frequency column shows what percentage of all claims occurred in relation to a particular disease category. For each distinct category x of the variable that is observed, calculate its percentage relative to the total number of observations n, as follows:

$$\text{Percentage}_x = \frac{\text{Frequency}_x}{n} \times 100\%$$

A cumulative percent frequency can also be calculated in the same manner as for a quantitative variable, but because the value categories have no inherent order, it may sometimes be hard to interpret. Nonetheless, in Table 3.5, the cumulative percent frequency could help answer a question such as "What percentage of the claims fell into the six main categories that are specifically named in the table?" (*Answer:* About 46%)

Constructing Qualitative Frequency Tables

Two basic steps are required to construct a frequency distribution table for qualitative data. A software program can facilitate the second step, but the first step is a prerequisite.

1. **Define and code the categories:** For **fixed-answer variables,** this step occurs before the data are collected. In a study on the mode of transportation used by commuters, a survey may list specific options such as "bus," "car," "subway,"

"walk," and so on. A preliminary study may be needed to assess what set of categories is most appropriate for the variable. Should the survey make separate entries for "car," "pickup truck," and "hitchhiking"? It is generally advisable to not have too many categories, which would thin out the frequencies for particular choices. On the other hand, having too few categories could violate the principle that, for every data value, there should be a category to which it can be assigned.

An alternative approach allows for **open-ended variables.** Simply asking commuters what mode of transportation they use, with a space to write in their answers, would solve the problem of leaving something out. But for constructing a table, the results can be confusing, containing many unique phrases and words that may be synonymous. For example, if four responses are "bus," "GO bus," "Oshawa Transit," and "Greyhound," respectively, should these be added to the frequency for just category "bus" or should they be distinguished in some way? The researcher will have to decide. An advantage of the open-ended approach is that it may elicit some unexpected responses that add to the researcher's understanding of the variable.

The category "Other" is used in Table 3.5 to join the frequencies for a number of other unlisted categories that individually had small frequencies. In a different sense, "Other" can appear in a survey as a way to combine the benefits of fixed and open answers. Six or seven specific choices may be listed, plus an opportunity to check "Other" and write in an open-ended response.

Coding the categories is a technical step, to prepare the data for input into the software program. Depending on the software used, the text-explanation version of a category (e.g., "intercity public transportation") may need to be coded into a number or a short abbreviation to facilitate processing. These codes are arbitrary and merely distinguish the categories that are included in the analysis.

2. **Construct the table:** Constructing the qualitative frequency table is in practice very similar to constructing a discrete quantitative frequency table. In both cases, there is generally a finite number of values that the variable can take, and these values are aligned in a single column. The number of occurrences for each value are tallied (by the software program) and displayed in the frequency column. As desired, a percent frequency and cumulative percent frequency column can be added, using the same calculations for either data type.

• *In Brief 3.1(d) Qualitative Frequency Tables*

*(Based on the **Baseball_hitting** file)*

For more information about the data and outputs in the illustrations below, see Example 3.2.

Excel

Open a data file or input the raw data into a column. (Only the "Event" data are needed; see the illustration.) In a second, currently unused column (it does *not* have to be a neighbour of the data column), input the complete list of category names that the variable can take. To the right of the top category name in that second column, input an Excel formula of this format:

= COUNTIF($A$2:$A$101,B2), as shown in the screen capture. Revise the absolute address for the first input (**$A$2:$A$101**) to point to the column that contains the raw data. Revise the second input (shown as the address **B2**) to refer to the cell for top entry in the column displaying the category names. The formula counts how often the value in **B2** appears within the full list of data. Copy the formula and insert it next to each of the values listed in the second column; this new, third column will display the frequencies corresponding to each category. Create appropriate labels and a percent frequency column as desired.

	A	B	C	D
1	Event (All Data)	Event_Category	Frequency	Formula used in the Frequency Column
2	BB	2B	5	=COUNTIF($A$2:$A$101,B2)
3	HR	AB	8	↓
4	RBI	BB	14	↓
5	RBI	G	9	↓
6	BB	H	6	{Copy the formula down the column}
7	RBI	HR	24	↓
8	R	R	13	↓
9	G	RBI	15	↓
10	AB	SB	6	=COUNTIF($A$2:$A$101,B10)
11	R			
12	HR	TOTAL	100	
13	HR			
14	↓			
15				
16	↓			
17	{List continues to Row 101}			
18	↓			
19	↓			
20	↓			

Minitab

Open a data file or input the raw data into a column. (Only the "Event" data are needed; see the illustration.) Use the menu sequence: *Stat / Tables / Tally Individual Variables.* As shown in the screen capture, specify in the dialogue box the column that contains the raw data, and select the options for *Counts, Percents,* and *Cumulative Percents;* then click *OK.*

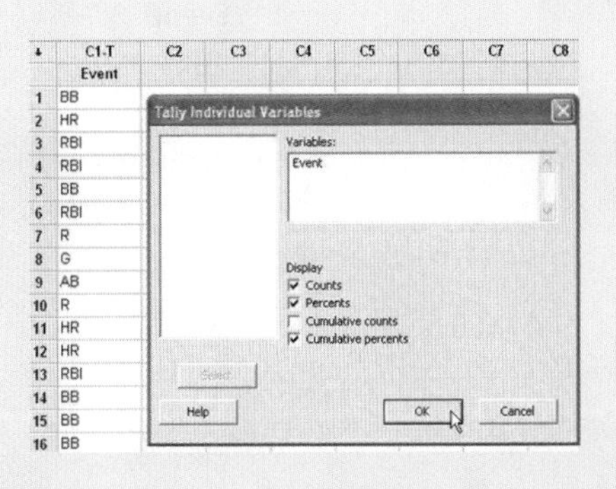

SPSS

Open a data file or input the raw data into a column. (Only the "Event" data are needed; see the illustration.) Use the menu sequence *Analyze / Descriptive Statistics / Frequencies.* As shown in the screen capture, specify in the dialogue box the variable containing the raw data, and select the option *Display frequency tables;* then click *OK.*

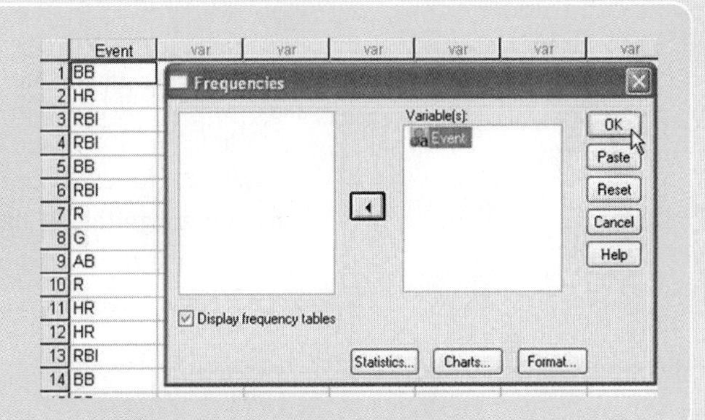

Example 3.2

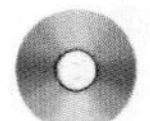

*(Based on the **Baseball_hitting** file)*

As any fan is aware, baseball is a sport of records and statistics, eagerly cited by its announcers during the game. The Major League Baseball website records no fewer than 400 milestones reached by hitters during the 2004 season—from Rafael Palmeiro passing Mike Schmidt for 9th place in career home runs, to Craig Biggio passing Doc Cramer to become 50th overall in career at-bats. For milestones related to hitting, are there any particular events (such as home runs [HR], runs batted in [RBI]), or bases on balls [BB]) that appear in the list of records more often than others? Construct a frequency distribution table for the qualitative variable "Event," using the 100-record file **Baseball_hitting** to answer the question.

Solution

1. **Define and code the categories:** The categories are the labels of batting-related events (such as HR, RBI, and BB) that were considered noteworthy by those who compiled the data set.
2. **Construct the table:** Whatever your software, input the full data set of 100 events that were considered record-setting. Procedures for generating the frequency distribution from this data are described in "In Brief 3.1(d)." The resulting output is shown as Table 3.6.

Table 3.6 Frequency Distribution for Record-Setting Batting Events

Event	Count	Percent	CumPct
2B	5	5.00	5.00
AB	8	8.00	13.00
BB	14	14.00	27.00
G	9	9.00	36.00
H	6	6.00	42.00
HR	24	24.00	66.00
R	13	13.00	79.00
RBI	15	15.00	94.00
SB	6	6.00	100.00
	$N = 100$		

Source: MLB Advanced Media. Found at: http://mlb.mlb.com/NASApp/mlb/mlb/stats/mlb_milestones.jsp?msTimeFrame=Achieved&statType=1&msStat=all&noHighlight=true&team=all&baseballScope=mlb&§ion1=1&sortByStat=GAME_ID. Major League Baseball trademarks and copyrights are used with permission of MLB Advanced Media, L.P. All rights reserved.

We see, for example, that nearly one-quarter of the noteworthy events were home runs. Merely getting at bat gets someone onto the list only if, for example, somebody else's lifetime record is surpassed; 8% of the milestones are of this type.

• Hands On 3.1 *What a Difference a Limit Makes*

*(Based on the **Purchasing Power_World** file)*

The ideal frequency distribution table should reveal actual patterns in the data, and the data values should be reasonably spread throughout the class ranges, rather than clustered near the class limits. Otherwise, a small shift in selected class limits, or in class width, could change which classes whole groups of values fall into, dramatically altering the appearance of the table.

This risk cannot be totally avoided, but software programs make it relatively easy to try out a variety of possible class structures as rough drafts. This can give a more complete picture of the data and, following guidelines and your judgment, you can then decide on the "best" class limits to use for constructing your final copy.

Figure 3.1 shows parts of three quite different tables that could be constructed based on the same data used for Table 3.4 (Frequency Distribution for Countries' Purchasing Power per Capita). The data are stored in the file **Purchasing Power_World.**

In the left-hand version of the distribution, frequencies appear to rise and then fall almost randomly as you look up or down the table. (The arrows

Figure 3.1

What Is the "Real" Frequency Distribution Pattern for This Data?

Based on Set of Class Limits A	Frequency	*Frequency Patterns?*
	82	
	39	
	27	
	20	
	5	
	9	
	14	
	5	
	10	
	11	
	2	
	4	
	1	
	0	
	0	
	0	
	1	
	0	

Based on Set of Class Limits B	Frequency	*Frequency Patterns?*
	128	
	42	
	22	
	19	
	14	
	4	
	0	
	1	

Based on Set of Class Limits C	Frequency	*Frequency Patterns?*
	148	
	34	
	29	
	17	
	1	
	1	

Source: U.S. Central Intelligence Agency. Based on data in the *The World Factbook, 2003.* Found at: https://www.cia.gov/cia/publications/factbook/fields/2004.htmlwww.cia.gov/cia/publications/factbook/fields/2004.html

highlight where frequencies are increasing [up arrows] or decreasing [down arrows] as you read up the table.) But in the right-hand version, the frequencies fall continuously as you read down the classes. The middle table is similar, but shows a small surge in the bottom class. None of these distributions is more "real," necessarily, than the other, but which is more informative depends on the purposes of your report. Do your readers need or want to know the fine detail in smaller classes, or is a broader-view pattern sufficient?

Load the data from the file **Purchasing Power_World** into your software program and try to fill in the blanks in Figure 3.1, determining what class limits were used to generate each of the three versions. (Hint: For each version, the class width is some multiple of 1,000.) It might take some trial and error to set up an alternative set of classes and then run the program to generate the corresponding frequencies. See if you can develop strategies with your software for making these trials rather quickly. Then try some alternative class structures that are not shown in the figure. Of all these trials, which version do you think is best, and why?

3.2 Graphing Quantitative Data

We have seen how a frequency table can improve our understanding of a data set and of how its values are distributed. A graph can represent the same information visually, making it easier to grasp or to compare with other data of interest. A graph is especially useful for revealing the **shape** of the distribution of quantitative data, which (if the future is like the past) may give a sense of approximately into what values the data tend "naturally" to fall. In this section, we will look at charts in the form of histograms, polygons (i.e., line graphs), and cumulative frequency charts.

The Histogram

Table 3.3 displayed the frequency distribution of numbers of votes cast in British Columbia ridings for the 2004 federal election. This same information (for the frequency column) is presented in Figure 3.2 in the form of a histogram. A **histogram** is a graph that uses the lengths of bars to represent the frequency of values in each class of a frequency distribution. All of the bars should have equal widths, to suggest to the eye the equal width of the classes. The height of each bar is proportional to the frequency of the corresponding class. The class values are shown along the horizontal axis.

Figure 3.2
Total Votes Cast in BC Ridings, 2004

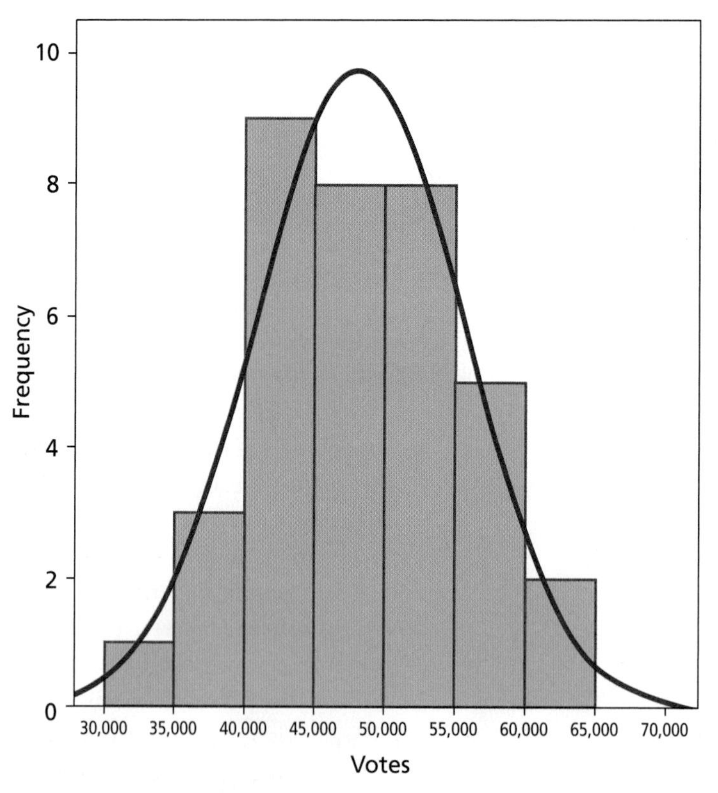

Source: "Federal Election Results, 2004." *The Globe and Mail.* Wednesday, June 30, 2004. Pages A12–A13. Reprinted with permission from *The Globe and Mail.*

The bell-shaped line in the Figure 3.2 is not part of the histogram, but it makes it easy to determine whether the data are clustered more or less symmetrically around some central value. Another notable shape for a distribution is **skewed.** The distribution of countries' purchasing power, for example (see left graph in Figure 3.3) is skewed to the right—meaning that while most countries have values in the lower class ranges, a notable (although diminishing) number of countries have values in the higher class ranges. (Distribution shape is discussed further near the end of Chapter 5.)

Figure 3.3
Histograms of Countries' Purchasing Power per Capita

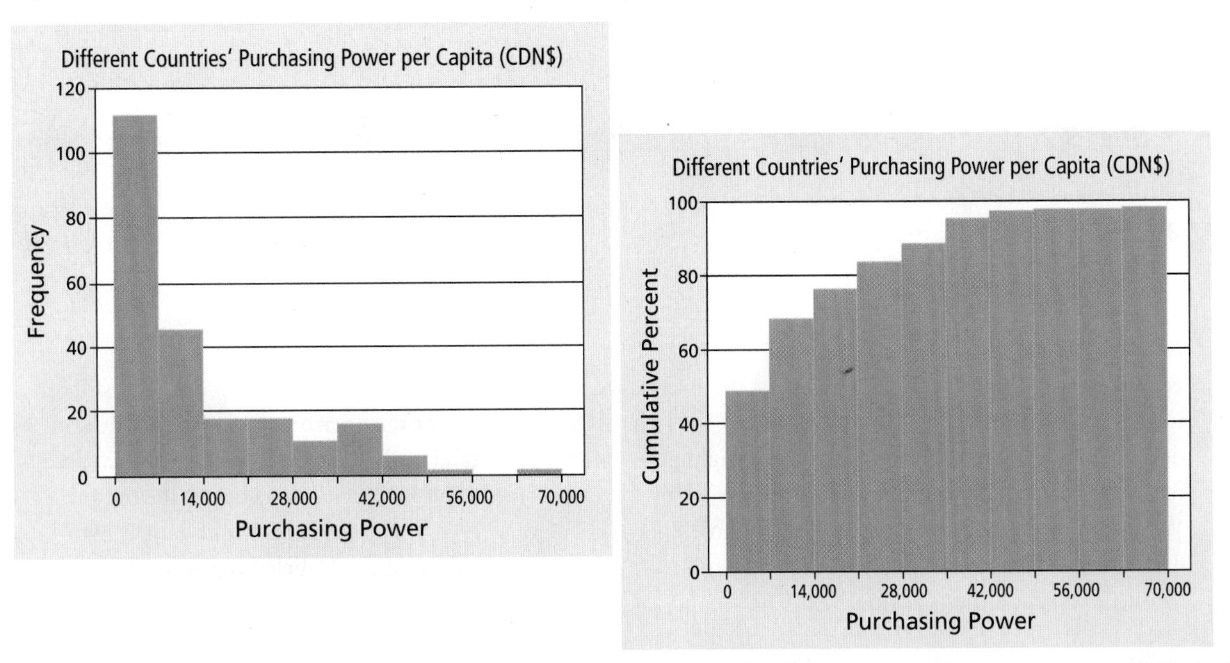

Source: U.S. Central Intelligence Agency. Based on data in the *The World Factbook, 2003*. Found at: https://www.cia.gov/cia/publications/factbook/fields/2004.html

The frequencies on the vertical axis of the histogram can be converted into relative or percent frequencies, to create a relative or **percent frequency histogram;** this is analogous to converting the frequencies in a frequency table to percent frequencies. The relative lengths of the bars in the figure are unaffected. Another variation is a **density scale histogram,** in which the *area* encompassed by each bar represents the percentage of the total number of observations that fall within the class boundaries for that bar; the total area under the histogram is defined as 1. In practice, the word *histogram* is often used to refer any of these forms of graph—a frequency histogram, a relative or percent frequency histogram, or a density scale histogram. The labels for the vertical axis make it clear which type of histogram is being used.

Traditional Steps for Constructing a Histogram

1. **Define all classes:** This step is the same as step 4 for constructing a frequency distribution table. It requires that the number of classes and the class width be determined, if this has not occurred already. The lower and upper class limits, starting from lowest class, must be decided before proceeding.

   Note that with some statistical software, you can skip this step. Both Minitab and SPSS offer a menu sequence *Graph / Histogram* that provides a quick snapshot of the data's distribution, which can be useful. The software itself chooses the number of classes and their boundaries; although the choices may not always conform to the guidelines recommended earlier. If you aim to produce a polished graph, to be published or circulated among professionals, consider carefully whether the software's decisions meet your needs.

2. **Construct and label the figure:** The tools provided by your statistical software can greatly simplify this step, as detailed in "In Brief 3.2(a)."

• *In Brief 3.2(a) Constructing a Histogram*

(Based on the ***BC_votes*** *file)*

Excel

For the cleanest, most flexible output, first use Excel to generate a frequency distribution table, as described in Section 3.1. From this output, highlight the column of frequencies, then click on the *Chart Wizard* icon (see the pointer in the illustration). The first dialogue box offers a choice of chart types. Select *Column* and then click on *Next.* The second dialogue box will display the frequencies, but still requires data for the horizontal axis labels. Click on the upper tab labelled *Series.* In the next display (see the screen capture), click on the box at the bottom *Category (X) axis labels* and click and drag to highlight the column of class labels from your table.

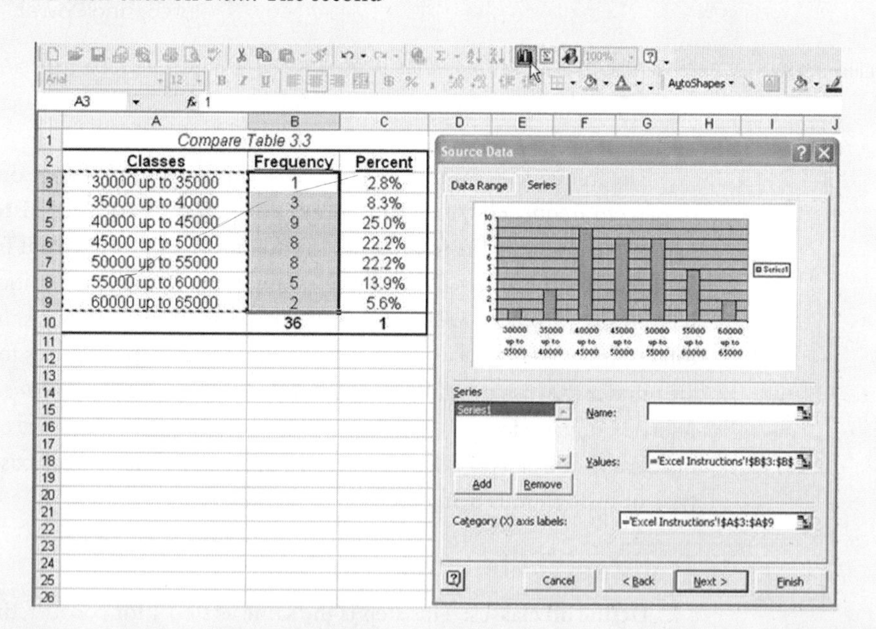

To improve the appearance of the chart, click *Next* and then input suitable labels for each of the axes and input a title for the chart itself. Click on *Finish.* Finally, right-click on one of the histogram bars and select *Format Data Series.* Select the upper tab labelled *Options* in the dialogue box, and adjust the *Gap Width* down to zero, so that the bars all touch.

Minitab

With the data set in a column, use the menu sequence *Graph / Histogram.* Click *OK* in the first dialogue box that appears to accept *Simple* for the histogram's style. From the data list that appears in the second dialogue box, highlight the column with the data and click *Select* and then *OK.* A graph is output, using Minitab's choices for the class limits.

To revise Minitab's choices, right-click somewhere on exactly the horizontal (*X*) axis of the graph; then select *Edit X Scale* in the displayed dialogue box. (Alternatively, double-click on any of the grey bars in the histogram.) In the dialogue box that then appears, click the tab at the top for *Binning;* then, as shown in the screen capture, select *Cutpoint* and then *Midpoint/Cutpoint positions.* Then input the desired sequence of class limits. The screen capture shows the histogram *after* this adjustment of class limits has been made.

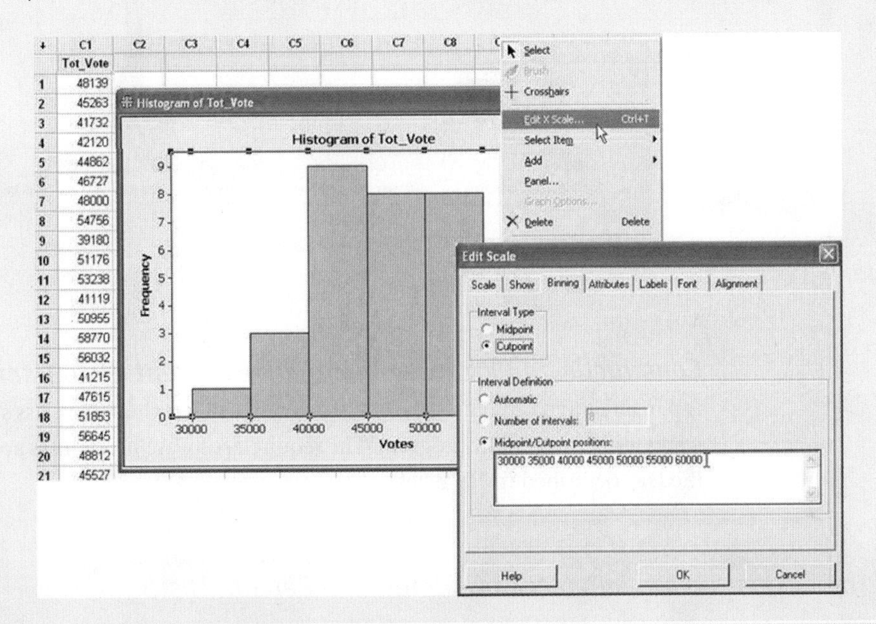

SPSS

With the data set in a column, use the menu sequence *Graph / Interactive / Histogram.* Following the steps indicated by the circled numbers in the illustration, in the dialogue box, (1) click and drag the choice of variable for the *X*-axis, placing it where shown in the screen capture. To control the choice of class boundaries, (2) click the *Histogram* tab. In the next dialogue box, (3) deselect the option *Set interval size automatically,* and (4) input the desired class width.

Now, (5) click the *Options* tab and, in *Scale Range*, at the lower left of the *Options* dialogue box, (6) select the name of *X*-variable for the analysis, (7) deselect the option *Auto* for *Scale Range,* and (8) input values that (relative to the data's level of precision) are *just below* the lower boundary of the lowest class and the upper boundary of the largest class, respectively, as shown in the screen capture.

As is clear from the partial output included at the bottom of the illustration (9), the displayed class boundaries will not automatically be rounded off. For this refinement, you can use graph editing features, which are beyond the scope of this book.

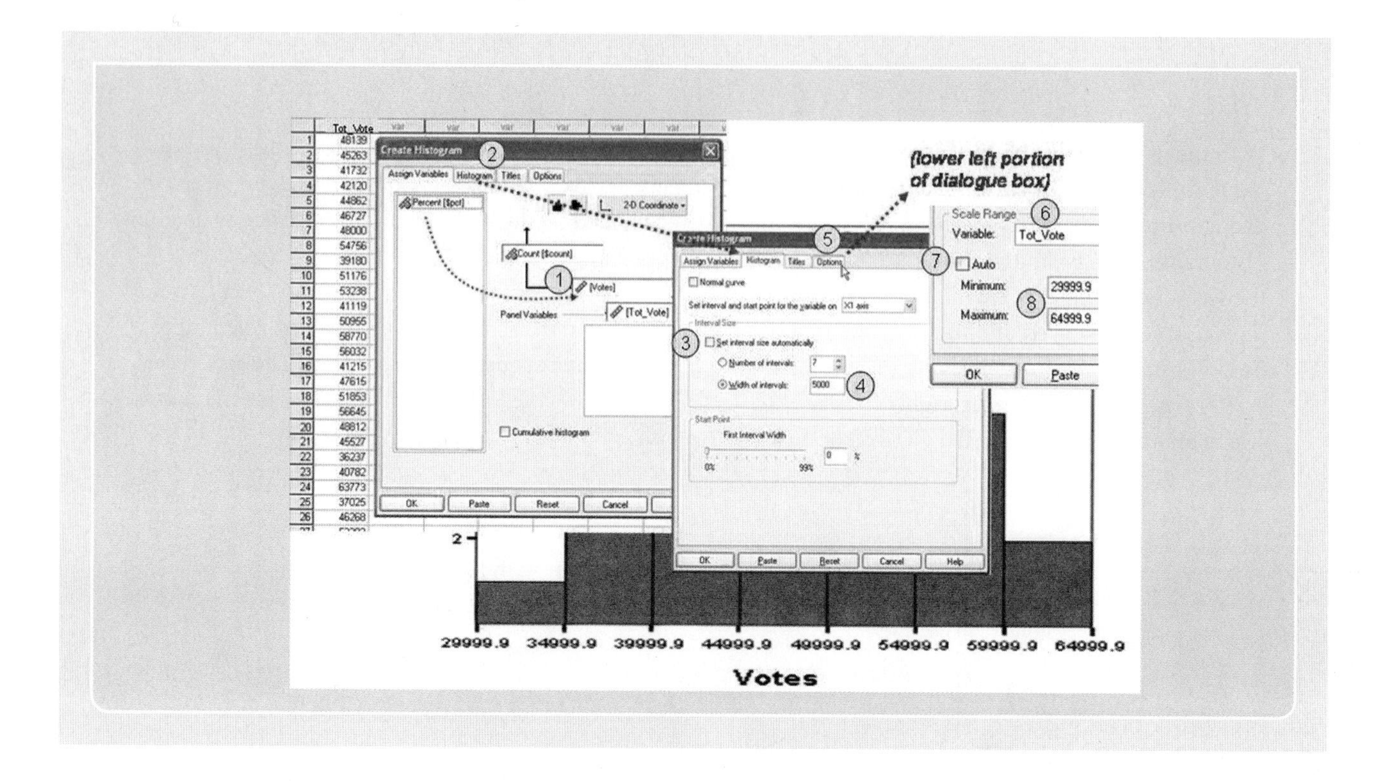

Seeing Statistics

Section 2.3

Constructing a Percent or Cumulative Percent Histogram

A percent and cumulative percent histogram can use the same class structure as the corresponding frequency histogram. The required software procedures are only slightly modified, as described in "In Brief 3.2(b)."

• *In Brief 3.2(b)* *Constructing a Percent or Cumulative Percent Histogram*

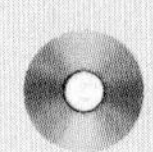
(Based on the ***BC_votes*** *file)*

Excel

Refer to "In Brief 3.2(a)" and follow the same procedures for Excel except, instead of highlighting the frequency column from a preexisting frequency table, choose instead the preexisting column for percent frequency or cumulative percent frequency. Specify the same column as previously for the class labels and label the *Y*-axis appropriately for percent frequency or cumulative percent frequency.

Minitab

Refer to "In Brief 3.2(a)" and follow all of the procedures for Minitab. Then, on the resulting graph, right-click somewhere on the vertical (*Y*) axis and then select *Edit Y Scale* in the displayed dialogue box. In the next dialogue box that appears, select the *Type* tab at the top and then, as shown in the screen capture, select *Percent,* if desired, and/or *Accumulate values across bins* for the cumulative percent frequency. The screen capture shows part of the histogram *after* this adjustment in the *Type* dialogue box.

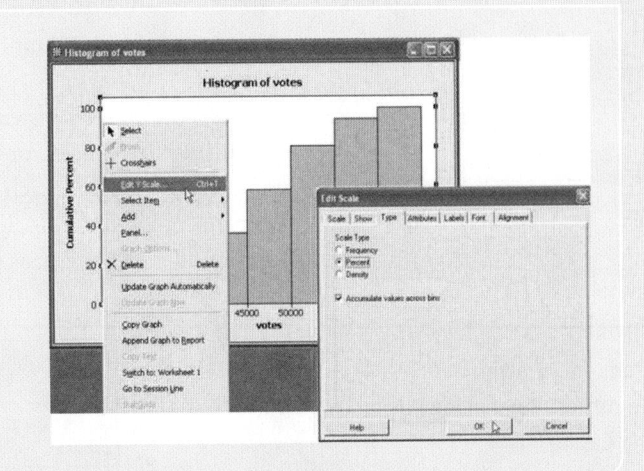

SPSS

Refer to "In Brief 3.2(a)" and follow the procedures for SPSS. The only change is in the first dialogue box: Drag the label "Percent[$Pct]" into the *Y*-position of the L-shaped diagram, as shown in the screen capture. As before, drag the variable name into the *X*-position. Also, if desired, select *Cumulative histogram* and then click *OK.*

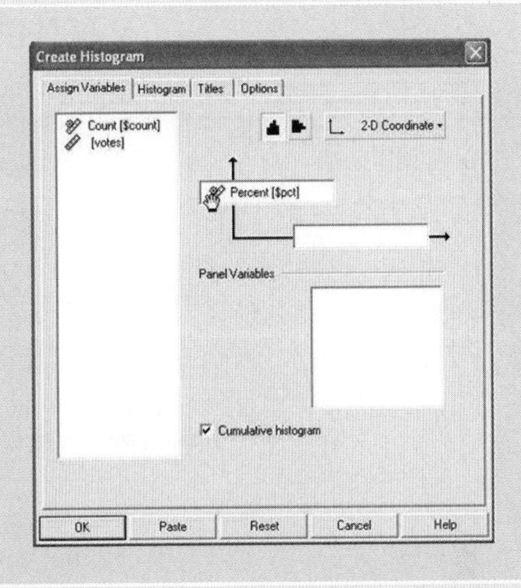

Example 3.3

*(Based on the **Purchasing Power_World** file)*

Construct a frequency histogram and cumulative percent frequency histogram for the per-capita countries' purchasing power data (in Canadian dollars) used in Example 3.1.

Solution

1. **Define all classes:** Having already constructed a table in Example 3.1, these classes have been defined: 0 up to 7,000; 7,000 up to 14,000; 14,000 up to 21,000; 21,000 up to 28,000; 28,000 up to 35,000; 35,000 up to 42,000; 42,000 up to 49,000; 49,000 up to 56,000; 56,000 up to 63,000; 63,000 up to 70,000.
2. **Construct and label the histogram:** Use the procedures described in "In Brief 3.2(a)" and "In Brief 3.2(b)" to construct the frequency histogram and the cumulative percent frequency histogram, respectively, for these data. Be sure to specify the class limits determined in step 1. With your software, you can customize the labels for the axes and title of the overall graph. (Refer back to Figure 3.3.)

Alternatives to the Histogram

Alternatives to histograms for displaying key features of a frequency distribution include stem-and-leaf diagrams (discussed briefly in Section 3.4), **frequency polygons** (or line graphs), and **ogives.** Instead of showing bars whose heights reflect the frequencies or cumulative frequencies for the classes, frequency polygons and ogives show *points* at the same relative heights. Each of the points (all of which are connected) is placed directly over the midpoint for its corresponding class. These figures create a smoother image for the overall data pattern, as shown in Figure 3.4.

Figure 3.4
Frequency Polygon of Countries' Purchasing Power per Capita

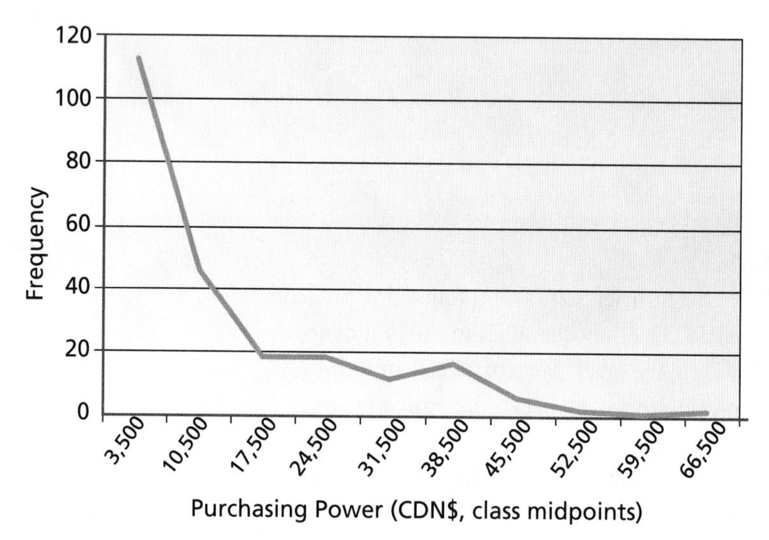

Source: U.S. Central Intelligence Agency. Based on data in the *The World Factbook, 2003*. Found at: https://www.cia.gov/cia/publications/factbook/fields/2004.html

We do not cover the procedures for building frequency polygons and ogives in this textbook, but if you see an illustration of either of these alternatives to a histogram, you should be able to interpret it. In your own reports, it is generally acceptable to use the histogram approach for displaying a distribution.

• Hands On 3.2 *Internet Download Times*

In today's culture, delays in downloading from the Internet have joined the weather as something everyone shares complaints about. How long does it really take to access a website? The answer depends on many variables, including the speed of your Internet connection, the reliability of your Internet provider, the amount and type of graphics on the Internet page (including pop-ups), and so on. But what if you limited a study to the computer and Internet connection you usually have access to and randomized the websites that you accessed? How long would it take to access the sites that you select? Here's an opportunity to make a study.

1. Make a sampling plan for selecting, with minimal bias, a sample of 100 websites to access. If you cannot devise a plan for true random sampling among the billions of available sites, try to make as diverse and random a sample as possible.
2. Document the steps in your plan and take note of ways in which your plan might be subject to bias. For example: If you have filters against pop-ups or adult content on your computer, you may not wish to remove those filters for the sake of the experiment. That is probably reasonable, but document this in your experiment notes because such filtering will potentially bias the study and because the filtered sites may tend to download slower or faster than nonfiltered sites. Why might that happen? Are there other sites that your method would entionally? How could that impact results?
3. Follow your sampling plan to call up sites that you have selected. Use a clock with a second hand or a stopwatch to keep track of time. For each selected site, copy or input the address (URL) into the location area of your browser. Record the time in seconds between (a) your *Enter* or *Click* to begin seeking the site and (b) how long it takes to fully access the site, as denoted by "*Done*" appearing on the browser.
4. If a site does not download properly, skip that site and try the next one that you have in your plan. This study is *not* addressing the issue of unsuccessful attempts to access a site; it is studying the downloading time for sites that *do* open (eventually).

   Question: Can you always be certain that a site that did not download before you gave up the attempt would not have downloaded if you had given it more time? If your answer is no, could this lead to potential bias of the study? Explain.
5. Once your 100 sample times have been collected, use the principles discussed in this chapter to construct a frequency distribution table, including a cumulative percent frequency column, and then a corresponding percent frequency histogram.
6. Use your results to estimate answers to the following questions, supposing that (despite the issues raised above) your sample of site download times is reasonably representative of site download times in general:

   Questions:

   a) Sixty percent of sites take less than how many seconds to download?
   b) Is there a range of site download times that occurs more frequently than other ranges of similar width? If so, what is that range? What is the shape of the distribution of site download times?

7. Most likely you found a variety of site download times in your sample. Were you able to detect any patterns regarding features of sites that downloaded more slowly and those that downloaded more quickly? How do you think you could test more formally whether you detected a real pattern in the data?

3.3 Graphing Qualitative Data

In this section, we focus on two types of charts designed to display the distribution of qualitative data: bar charts and pie charts.

Bar and Pie Charts

A **bar chart** (or **bar graph**) is similar to a histogram. All bars have equal widths and their lengths are proportional to the frequencies for the different categories. For qualitative data, it is possible to list all of the values of the variable (the categories) on an axis of the chart. The bars show the corresponding frequencies for the categories. These bars can be oriented vertically, similar to a histogram, or horizontally, with the category names shown on the left of each bar. This technique improves readability for categories with long names. Unlike a histogram, the bars have gaps between them (see Figure 3.5).

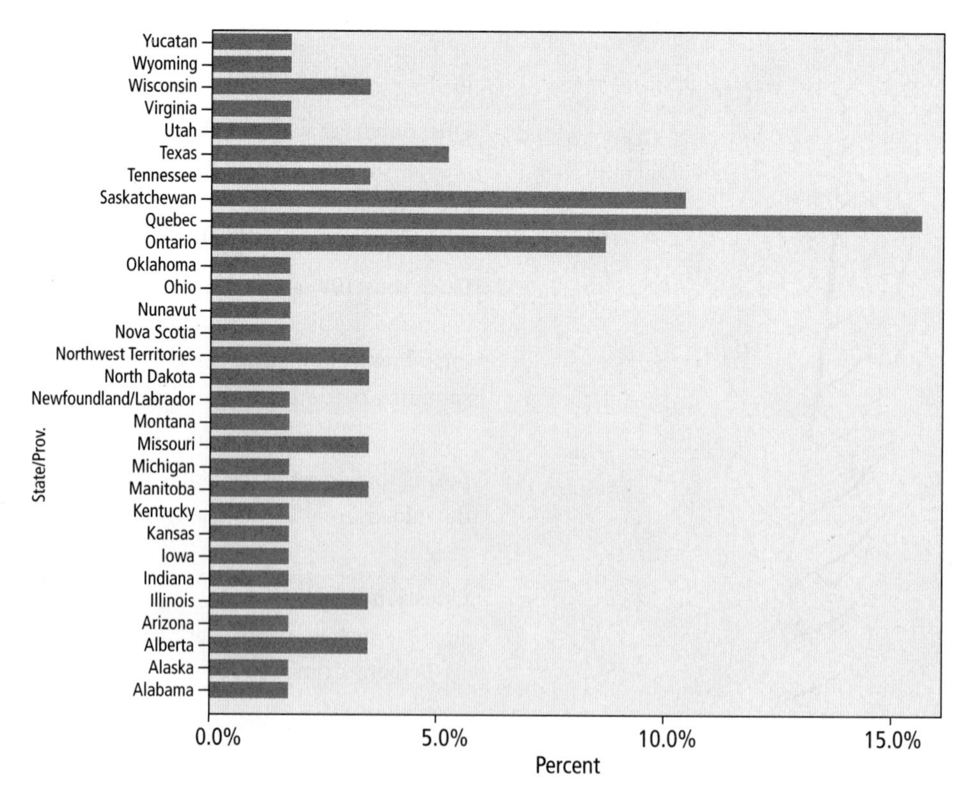

Figure 3.5
Bar Chart for North American Meteor Crater Locations by Province, Territory, State

Source: Earth Impact Database, 2006. Found at: http://www.unb.ca/passc/ImpactDatabase. Accessed 8 April 2005.

Figure 3.5 shows North American provinces, territories, and states in which meteor craters have been found. For each location, the figure displays the percent frequencies of identified meteor craters. Absolute frequencies could also have been shown.

Comparisons of percent frequencies can also be displayed on a **pie chart,** as illustrated in Figure 3.6. This figure shows the relative frequencies of meteor crater sites in each of the three countries of North America. Theoretically, a pie chart could have been also used to display the data for Figure 3.5; however, most "slices" would have been too thin to interpret. A pie chart works best when the number of categories is limited. It is also useful when categories can be interpreted as components of a whole. For example, Human Resources and Social Development Canada published a pie chart for fall 2004 that showed the breakdown of Employment Insurance claimants by age group: 15–29 years (16.1%), 30–44 years (42.9%), and so on.

Figure 3.6
Pie Chart for North American Meteor Crater Locations by Country

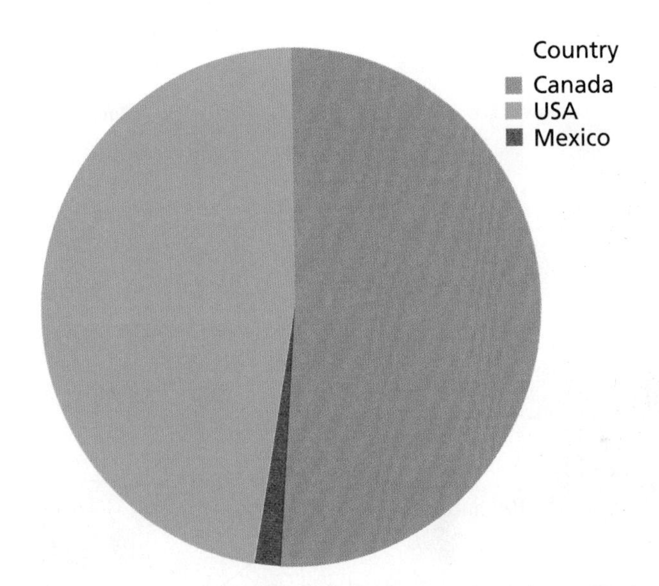

Source: Earth Impact Database, 2006. Found at: http://www.unb.ca/passc/ImpactDatabase. Accessed 8 April 2005.

Many common applications of pie charts are actually bivariate—comparing values of two variables. A typical example would be a published chart of "How your tax dollars are spent," comparing the total dollars spent in each of several tax categories. These applications are discussed in Section 3.5.

Pareto Diagrams

A **Pareto diagram** is just a bar chart in which categories are arranged in order of decreasing frequency to highlight the few categories that may account for most of the cases of interest. A quality control manager, for example, may be interested in what equipment parts are failing most often. A Pareto diagram approach to graphing the meteor crater locations (compare Figure 3.5) is shown in Figure 3.7.

Figure 3.7

Pareto Diagram for North American Meteor Crater Locations: Count by Province, Territory, State

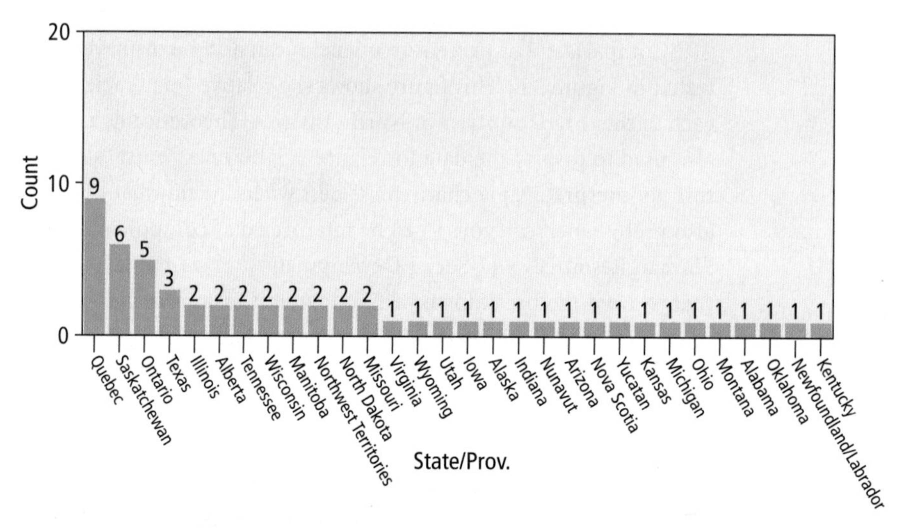

Source: Earth Impact Database, 2006. Found at: http://www.unb.ca/passc/ImpactDatabase. Accessed 8 April 2005.

• *In Brief 3.3 Constructing a Bar or Pie Chart*

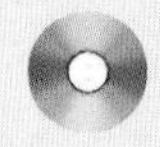

*(Based on the **Baseball_hitting** file)*

Excel

First use Excel to generate a frequency distribution table, as described in Section 3.1. From this output, highlight the "Frequencies" column (or percent frequencies—"PercentFreq"—as desired), and then click on the Chart Wizard icon (see the pointer in the illustration). The resulting dialogue box will display choices for chart type. Select *Column* (for vertical bars), *Bar* (for horizontal bars), or *Pie,* and then click on *Next.* The dialogue box will now display the frequencies, but still awaits data for the labels of the categories. Click on the upper tab labelled *Series.* In the next dialogue box, click on the option at the bottom for *Category (X) axis labels,* and click and drag to highlight the column of class labels from your table. The resulting bar chart should be as shown in the lower left-hand corner of the illustration.

To improve the appearance, click *Next* and then input suitable labels for each of the axes and an overall title for the chart. Click *Finish.* If there are many categories, some of their names may not be displayed. Right-click on the category axis and select *Format Axis.* Click the upper tab in the dialogue box for *Scale* and input 1 for the number of categories between tick-marked labels.

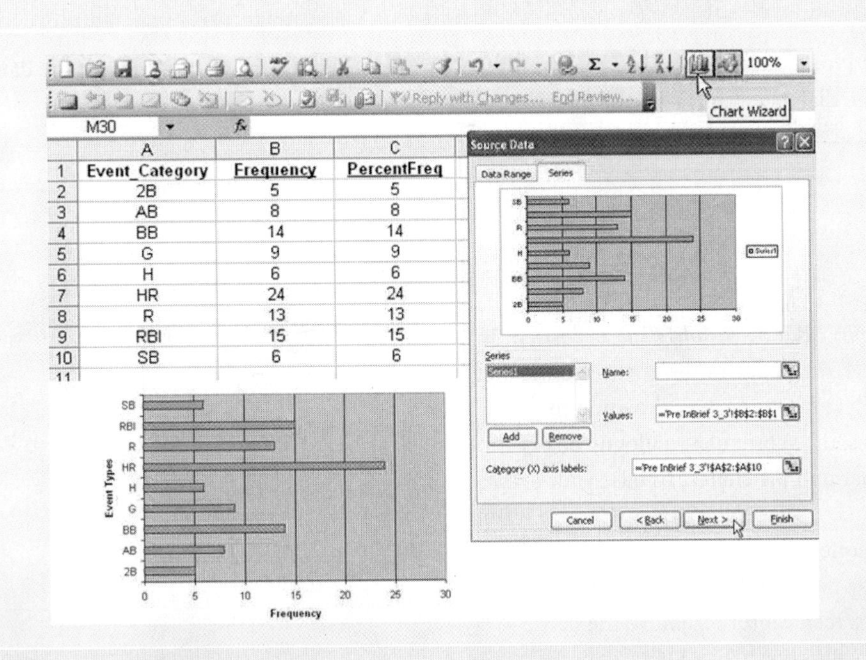

Minitab

With the data set in a column, use the menu sequence *Graph / Bar Chart* or *Graph / Pie Chart*, as desired. For the bar chart, click on *OK* to accept *Counts of unique values* as the interpretation of the bars, and then click *OK*. From the data list displayed in the next dialogue box, highlight the column with the data and choose *Select*. To output a frequency bar chart, click *OK* at this point. For a percent frequency display instead, select *Chart Options* (see the screen capture). In the next dialogue box, select both *Default* and *Show Y as a Percent* and then click *OK*. Now click *OK* again.

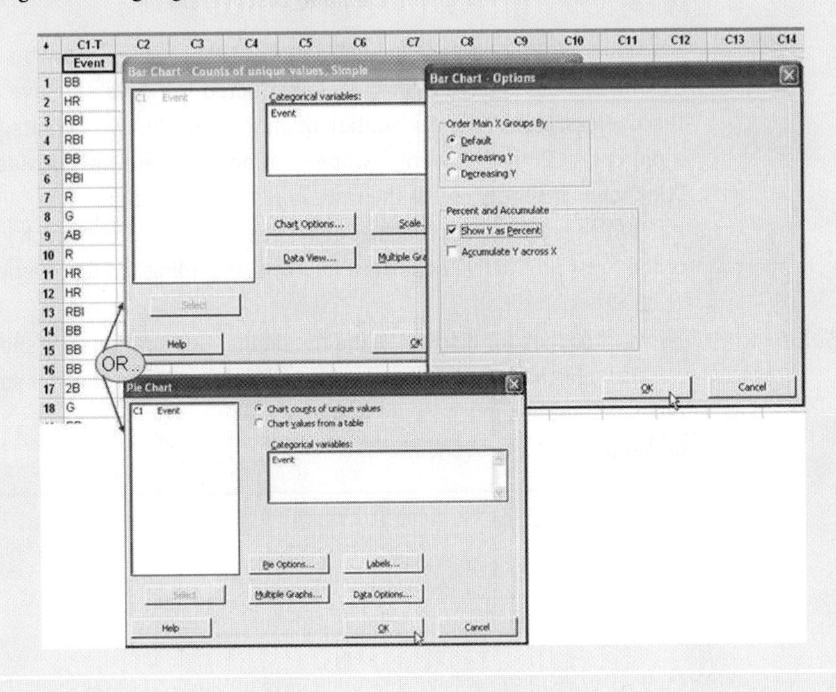

If you opted for a pie chart, select *Categorical Variables* in the dialogue box. From the data list now displayed at the left-hand side of the dialogue box, highlight the column with the data and choose *Select;* then click *OK.*

SPSS

With the data set in a column, use the menu sequence *Graphs / Bar* or *Graphs / Pie* as desired. In either case, click on *Define* in the next dialogue box to accept the default options. As shown in the screen captures, the subsequent dialogue box is similar for both bar and pie charts. To base your graph on the frequencies, select *N of cases* or to base it on the percent frequencies, select *% of cases.* Select the data column by name and click the arrow highlighted in the screen capture to move the name over to the *Category Axis* box. Click *OK.*

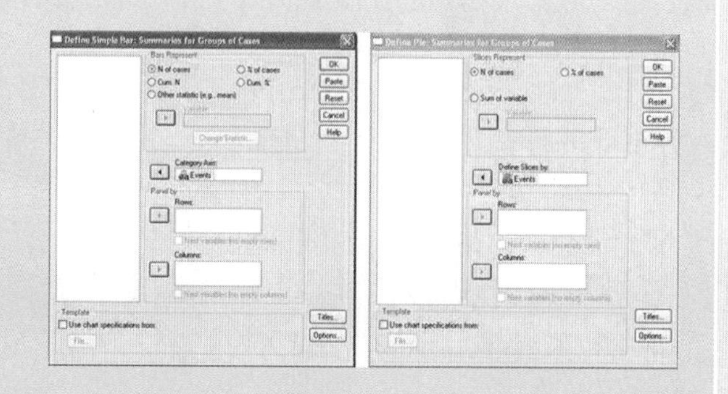

Example 3.4

*(Based on the **Chem_Element_Discoverers** file)*

A science reporter is researching nationalities of scientists who have led discoveries of new atomic elements. She has constructed the data set saved as **Chem_Element_Discoverers.** Depict the distribution of nationalities in two formats: as a bar chart and as a pie chart. Omit discoverers whose nationalities are not provided (i.e., left blank or "Unknown"); simply delete the rows in which this occurs.

Which graphing approach appears easier to interpret? Which gives the better sense of the most productive countries, in terms of leading the discoveries of new elements?

Solution

Two approaches for displaying the nationality distributions are shown in Figure 3.8.

The pie chart is clearly the more difficult to interpret given so many categories. In the bar chart, it is much easier to identify which countries are associated with the larger percentages of discoveries.

Figure 3.8
Graphic Displays of the Nationalities of Discoverers of Atomic Elements

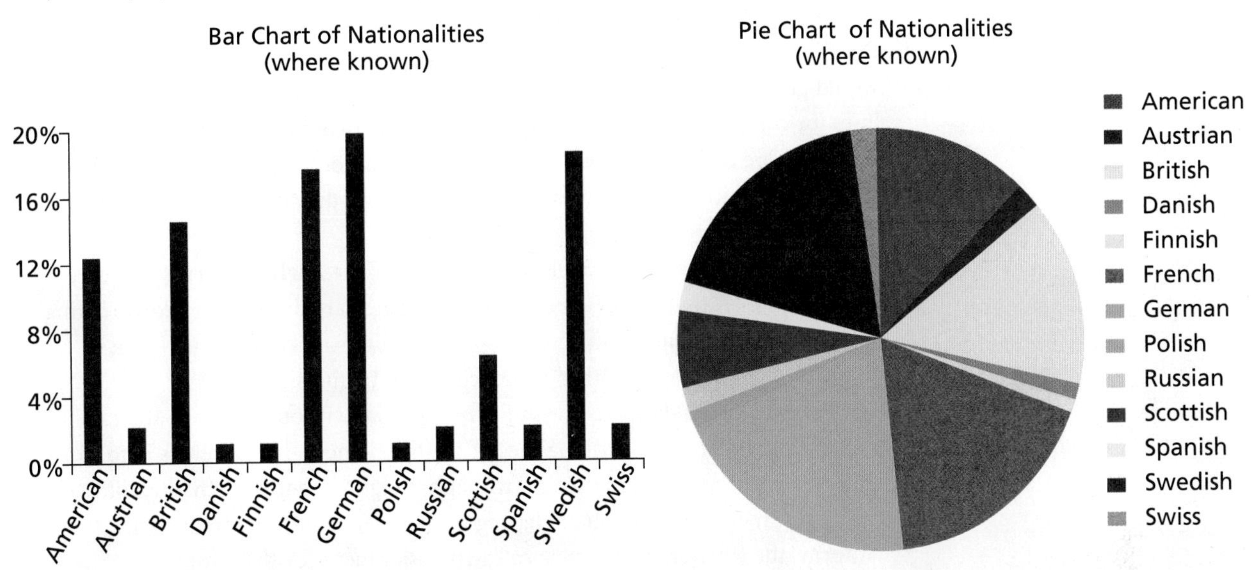

Source: Dr. John Andraos. Department of Chemistry, York University. Found at: http://www.careerchem.com/NAMED/Elements-Discoverers.pdf. 1.

• Hands On 3.3 When Should You Group?

*(Based on the **Pointspreads** and **Early_Returns** files)*

In Section 3.2, the focus was on graphs for *continuous quantitative* data. Guidelines were given for grouping the data into classes, then graphing the frequencies for those classes. This section has discussed graphing *qualitative* data, where the frequency of each possible category can be charted. *Discrete quantitative* variables have not yet been discussed specifically, but if the discrete values are treated as categories, a bar chart or Pareto diagram could easily be constructed.

Can categories be grouped? And can the values of discrete quantitative variables be grouped? The answer to both questions is yes, wherever this aids the reader's understanding, provided the following conditions are met:

1. Each new category should be inclusive of one or more original categories. (For example, "Atlantic Provinces" could be used to group frequencies for categories "PEI," "Nova Scotia," "New Brunswick," and "Newfoundland and Labrador.")
2. The categories actually used should be mutually exclusive and exhaustive.
3. The sum of the grouped categories' frequencies should exactly equal the sum of the original categories' frequencies.

Consider the CFL point spread data shown in Table 3.1. Although these values are technically discrete, the large number of possible values between 0 and 19.5 suggests that these values could be grouped *as if they were continuous.* Starting from the raw data in the file **Pointspreads,** graph these data first as if the discrete values were individual categories. Then group the values as you would group continuous quantitative variables. (Remember to use the guidelines for constructing the classes.) Compare these two graphs. Which graph appears to be easier to interpret? Could there be an occasion when graphing the frequencies of individual values is more useful than the alternative? Explain your answers.

Now consider the federal election data in the file **Early_Returns.** The data for the variable "number of candidates" (No_cand) are discrete. Try constructing two charts for this data—one grouped and one ungrouped. Can you think of contexts when one or the other approach would be most suitable?

For categorical data, we saw in Figure 3.8 that, without grouping the categories, a pie chart of the nationalities of atomic element discoverers is hard to read. Try creating an alternative chart that highlights the top five nationalities of the discoverers. (Hint: If you are letting the software count the frequencies, first "inform" the software of any new category assignment for the sample.)

3.4 Constructing Stemplots

An additional way to visualize distributions of quantitative data is by using **stemplots** (also called **stem-and-leaf displays**), which were developed by U.S. statistician John Tukey. The value of stemplots as a quick and easy technique for exploring data has been displaced somewhat, since computers can now generate traditional distribution tables and charts so quickly. But another feature of these charts remains useful: Although the data are grouped into classes, as in a frequency distribution, the displays retain original information about the raw, ungrouped data.

To create a stemplot, part of each value, on the left part of the number, must be identified as the "stem" and the next digit to the stem's right is identified as the "leaf." (Other, more complex schemes are less common and are beyond the scope of this textbook.) The clearest examples occur when the data are all two-digit numbers with a common decimal position. Consider the following list of approximate latitudes (in degrees north) for meteor craters found in North America.[6]

36	71	35	45	46	42	58	48	37	21	56	56
43	60	38	38	56	42	50	51	36	41	44	56
75	38	44	51	41	57	51	43	50	31	37	56
43	61	49	63	32	60	50	48	45	52	39	31
49	60	47	38	50	47	36	50	33			

In this case, the left-hand digits all represent the number of 10s in each value (three 10s in "36," seven 10s in "71," and so on). These left-hand digits become the stem. The leaves are the corresponding numbers of 1s in each original number (six 1s in the value "36," one 1 in "71," etc.). As shown below, each raw value is represented by its stem and leaf in the final diagram. Reading from the table, we see that the first data value represented is "21" (stem = 2, leaf = 1); the next data value is "31" (stem = 3, leaf = 1); and the final data value at the bottom corresponds to "75" (stem = 7, leaf = 5).

Stem	Leaf
2	1
3	112356667788889
4	112233344556778899
5	000001112666667 8
6	00013
7	15

By eye, the display can be interpreted as being similar to a histogram, oriented sideways: The implied classes are 20–29, 30–39, and so on; the lengths of the leaf sequences are similar to the heights of a histogram's bars. But unlike a histogram, it also is possible to view the distributions of values *within* each class. For example, we see that all of the values in the 60–69 class are actually in the bottom third of the range. Prior to easy access to computers, these charts could be constructed as quick variations to histograms.

Once the stems are decided, the framework of the diagram can be drawn. Scanning across the data, we can add the leaves in the correct positions, one at a time. For example, seeing "36," we add a "6" among the leaves for the stem "3"; seeing a "71," we add the leaf "1" for the stem "7." As a final step, the chart can be redrawn, sorting the leaves for each stem into ascending order.

Observe that for different data, the display shown above could *also* represent different data values such as "2.1," "3.1," "3.1," "3.2," . . ., *or* such as "21,000," "31,000," "31,000," "32,000." In our base-10 number system, the stem represents the "place" (as a power of 10) of the left digits of the data values (i.e., the 10's place, or the 100's, or the 0.0001's, etc.). Generally, the leaf represents the "place" for the next power to the right in the data values (the 1's place if the stems are 10s, or the 10's place if the stems are 100s, and so on).

Example 3.5

As part of a comparison of 41 Canadian cities to find the "best city for conducting business," *Canadian Business* magazine looked at the cost of living index for each city.[7] These values were:

69.63	65.79	70.27	68.09	72.2	74.47	73.89	75.41	76.4	71.44	
77.65	78.85	74.34	74.96	75.82	75.31	77.55	76.58	74.6	73.93	
79.12	79.47	78.23	79.05	78.41	77.5	77.44	77.19	82.39	82.42	
79.83	77.6	77.19	78.66	85.14	80.6	93.18	77.95	100	93.8	88.42

Construct a stemplot for these data.

Solution

The left-most digit for most of these numbers is in the 0's place ("6," "7," "8," etc.), so this place is the basis for the stem. The value "100" is an apparent exception—but it can be handled as a number whose stem is ten 10s. The leaves will be based on the next digit to the right of the 10's place (i.e., the 1's place), but because the raw data have additional decimal places, we can utilize for the leaves a rounded-off version of the 1's place, not necessarily the original values. (However, your software may use the original values without round-off.)

The display below is helpful, but not necessary, in preparing for the next steps. The data are pre-sorted, and the numbers have been rounded off.

66	68	70	70	71	72	74	74	74	74	
75	75	75	75	76	76	77	77	77	77	
78	78	78	78	78	78	78	79	79	79	
79	79	80	81	82	82	85	88	93	94	100

The corresponding stemplot is shown below. For each number, determine which stem it corresponds to and then identify the value of the leaf. Append the value of the leaf at the right-hand side of the growing list of leaves that correspond to the identified stem.

Stem	**Leaf**
6	68
7	0012444455556677778888888899999
8	912258
9	34
10	0

Figure 3.9 shows a variation of the stemplot that might be produced by a computer. To increase the number of implicit classes, a stem can be listed more than once. In this illustration, each stem appears twice—first followed by leaves of values 0 to 4 and then followed by leaves of values 5 to 9. The display also identifies possible "extreme values" or outliers.

Figure 3.9
Stemplot of Latitudes of North American Meteor Craters (Degrees North)

```
Stem & Leaf
   2 . 1
   2 .
   3 . 1123
   3 . 56667788889
   4 . 112233344
   4 . 556778899
   5 . 000001112
   5 . 6666678
   6 . 00013
   6 .
   7 . 1
1 Extremes . (>=75)
```

Source: University of New Brunswick. Based on data in *Earth Impact Database, 2003.* Retrieved April 8, 2005, http://www.unb.ca/passc/ImpactDatabase/NAmerica.html

• *In Brief 3.4 Constructing a Stemplot*

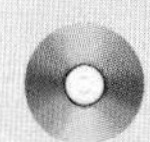

*(Based on the file **Meteors_N_Amer, "Latitude" column**)*

Excel

Excel does not provide a Chart Wizard or any other function to actually construct stemplots, but in preparing for manual construction of the chart, general commands such as *Data / Sort* can be useful.

Minitab

With the data set in a single column, use the menu sequence *Graph / Stem-and-Leaf.* Select the column with the data to be analyzed. In the box labelled *Increment,* input the class width to be used by the display (see the screen capture). For example, if the numbers 1, 2, 3 in the stems column represent data values in the ranges of 10s, 20s, 30s, then the increment is "10." If the stems are counting up in 100s, the increment is "100." In Figure 3.9, although the stems represent the 10's place in the original numbers, the leaves for each class increment only by 5s. Click *OK.*

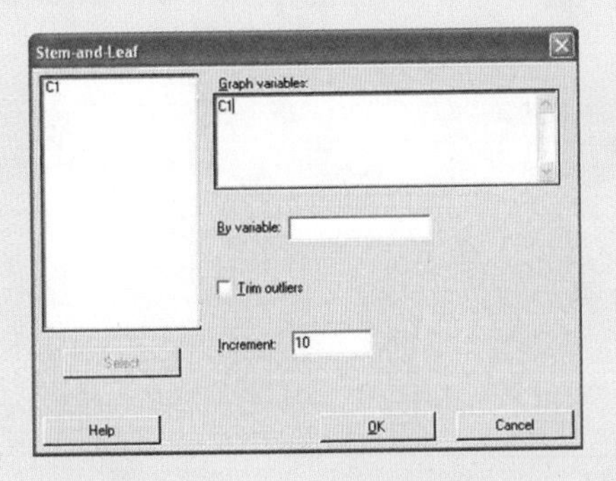

SPSS

With the data set in a single column, use the menu sequence *Analyze / Descriptive Statistics / Explore.* As shown in the screen capture, select the variable for analysis and then click *OK.* The output will display several tables and figures, including a stemplot.

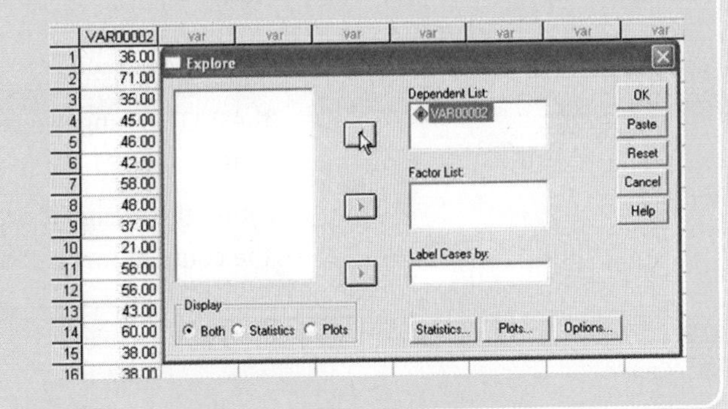

Example 3.6

*(Based on the **BC_votes** file)*

To illustrate the frequency distribution for votes cast in British Columbia ridings in the 2004 federal election, a histogram was used earlier in Figure 3.2. Create a stemplot for these same data, which are in the **BC_votes** file. Compare the two approaches. Use one of the graphs to answer each of the following questions. In each case, identify which graph approach was used.

1. Are the raw data that are in the range from 30,000 to 39,000 spread evenly across that range, or are the values clustered in some way?
2. How many raw data values are less than 40,000?

Solution

The new figure, to compare with the histogram, can be created automatically in Minitab or SPSS. Figure 3.10 is based on the SPSS version.

Figure 3.10
Stemplot of Total Votes in BC Ridings in the 2004 Federal Election

Stem &	Leaf
3 .	4
3 .	679
4 .	001111224
4 .	55667888
5 .	01122334
5 .	56688
6 .	03

Stem width: 10000
Each leaf: 1 case(s)

Source: "Federal Election Results, 2004." *The Globe and Mail.* Wednesday, June 30, 2004. Pages A12–A13. Reprinted with permission from *The Globe and Mail.*

1. The stemplot is more useful in answering this question. From the "leaves" corresponding to the stem "3" (= 30,000), we can reconstruct the approximate raw data values in the 30,000 range: The one leaf equal "4" represents a value around 30,000 + 4,000 = 34,000. The leaves (in the next row) equal "6" represents a value around 30,000 + 6,000 = 36,000, and so on. Scanning all of the leaves in the 30,000 range suggests that the values are clustered in or near the upper half of the range; they are not spread evenly across the range.
2. Either the stemplot or the frequency histogram could be used to determine that the count of raw data values less than 40,000 is four.

• Hands On 3.4 *Describe the Distribution*

*(Based on the files **Early_Returns, Meteors_N_Amer, Purchasing Power_World,** and **Star_Distances**)*

Starting with Part 4, much of this textbook introduces key concepts and techniques in inferential statistics—that is, techniques for estimating population parameters or testing hypotheses about them based on sample data. Not every technique presented there, however, can be applied to every data set. Many techniques can be applied meaningfully only if the data have a specific shape of frequency distribution. The further the real distribution is from this requirement, the less accurate the results generated by formulas or the computer.

Some of the most commonly used techniques for inferential statistics presuppose a "normal" distribution of the data. Figure 3.11 conveys the general pattern for this distribution. (This is discussed in more detail in Chapter 8.) Notice the largest frequency occurs in the middle of the diagram, with frequencies decreasing at about the same rate as you move farther from (below *or* above) the middle value.

Figure 3.11
Example of a "Normal" Distribution

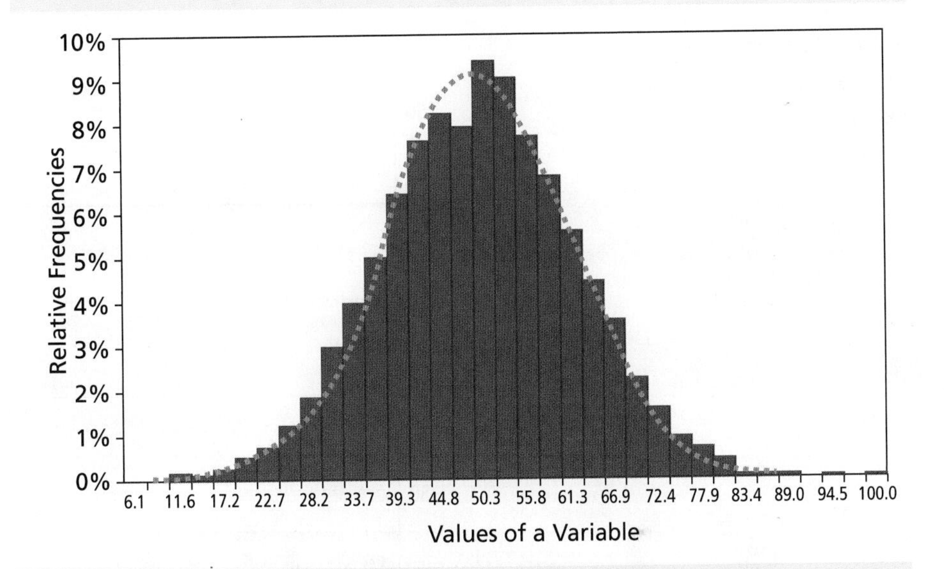

Suppose a friend who has reached Part 4 of this textbook is beginning to perform inferential statistics on the following data sets used in this chapter: the winning margin column (W_Margin) in the file **Early_Returns,** the approximate age in millions of years column (Av_Age) in the file **Meteors_N_Amer,** the purchasing power per capita column (GDP_CAD) in the file **Purchasing Power_World,** and the distances column (Distance) in the file **Star_Distances.** The techniques to be performed all presuppose that the data are normally distributed. For each data set, generate the corresponding histogram and stemplot, and visually assess the distribution. Which, if any of these sets, appear to have the distribution that your friend's methods will require?

3.5 Displays of Association

The previous sections focused on univariate descriptive statistics, that is, descriptive statistics for a single variable at a time. This section introduces **bivariate descriptive statistics,** which apply when the values of *two* variables characterize the individuals in a population. Growing children, for example, have both heights and weights. With bivariate descriptive statistics, the children's heights and weights can be compared. Similarly, galaxies have both ages and distances from our own Milky Way. The relationship between galaxies' ages and distances is an object of study in astronomy. Bivariate comparisons can be divided into three types of cases: (1) both variables are quantitative, (2) both variables are qualitative, and (3) the variables are a combination of qualitative and quantitative.

Scatter Diagrams

Scatter diagrams can be used for comparing two quantitative variables. Suppose a geologist is comparing the sizes of meteor craters in North America with their approximate ages. She creates the scatter diagram shown in Figure 3.12. Each dot displays simultaneously for *one* individual crater the values of *both* variables.

Figure 3.12
Scatter Diagram Comparing the Diameters and Ages of North American Meteor Craters

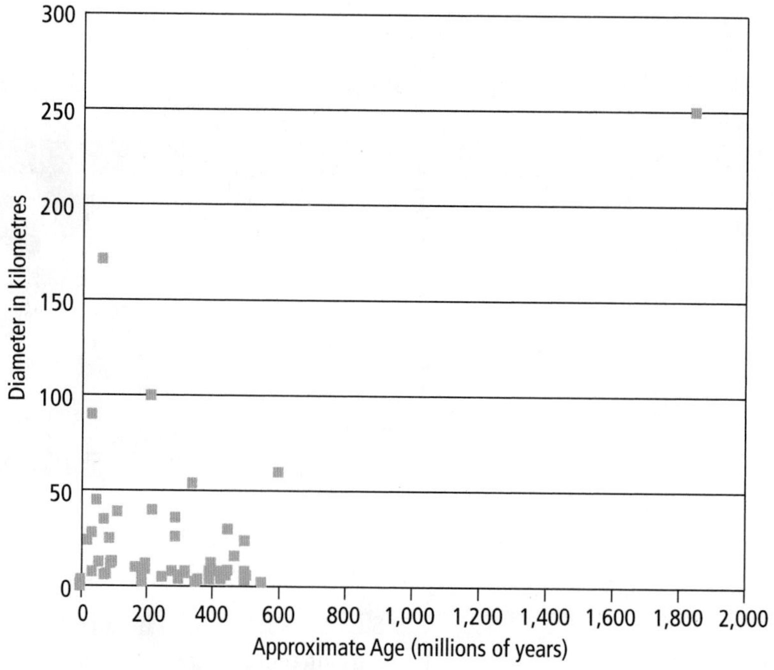

Source: Earth Impact Database, 2006. Found at: http://www.unb.ca/passc/ImpactDatabase. Accessed 8 April 2005.

In this example, the scatter diagram shows no particular relationship between the two variables: Regardless of age, most craters have diameters in the same range: from less than 15 km to (less commonly) up to 50 km. The two widest craters (more than 150 km) have drastically different ages—one quite recent geologically and the other over 1.8 million years old. Such figures convey useful information in an intuitive format. To assess the relationship between the variables with more mathematical precision, see Chapter 15.

Time Series Charts

A **time series chart** is a type of scatter diagram in which the variable measured horizontally is time (for example, periods of months or years) and the second variable is either a discrete count of some quantity or a continuous value. For example, Figure 3.13 shows the number of chemical elements discovered each year, starting from 1500. The years themselves are the variables mapped horizontally; the corresponding counts of discoveries are the second variable. Successive points are connected by lines.

Figure 3.13
Time Series (From 1500) for Discoveries of Chemical Elements

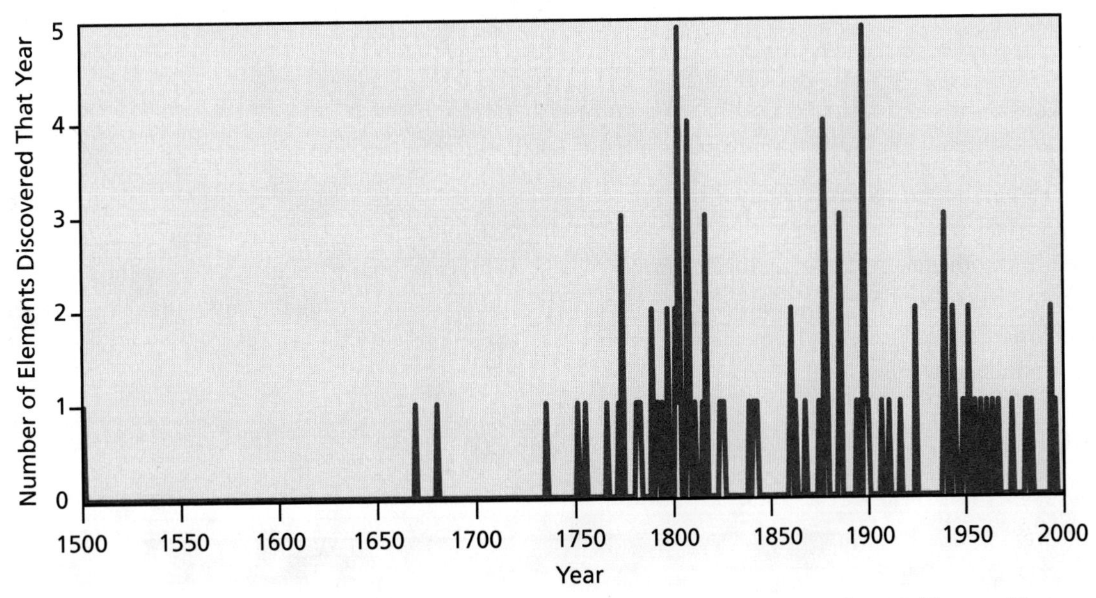

Source: Dr. John Andraos. Department of Chemistry, York University. Found at: http://www.careerchem.com/NAMED/Elements-Discoverers.pdf. 1.

Another typical example of a time series chart is shown in Figure 3.14. Each point represents one specific year's consumption of energy in Far East countries and in Oceania (in quadrillions of British thermal units). Clear trends over successive years for the continuous variable consumption can be observed. Again, successive points are connected by lines. Formal procedures to analyze time series are presented in Chapter 17.

Figure 3.14
Time Series for Primary Energy Consumption in the Far East and Oceania

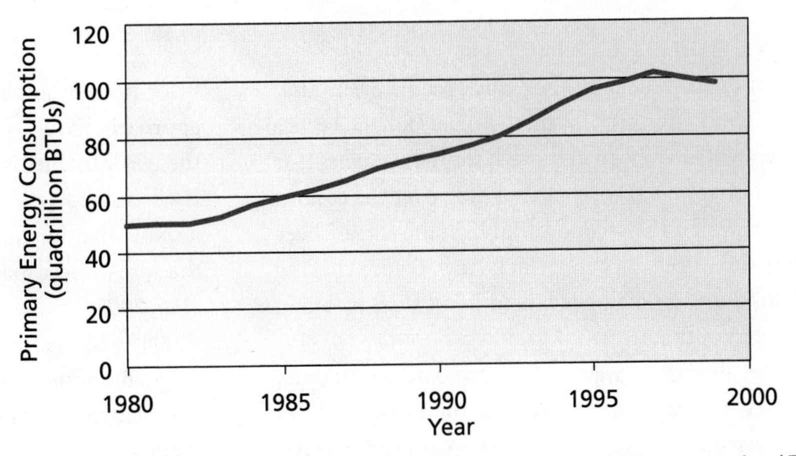

Source: Energy Information Administration, U.S. Dept of Energy. Based on data in "International Total Primary Energy and Related Information." Found at: http://www.eia.doe.gov/emeu/international/total.html #Consumption

• *In Brief 3.5(a)* *Constructing a Scatter Diagram or Time Series Chart*

*(Based on the **Meteors_N_Amer** file)*

Excel

There should be two columns of data—one for each of the variables. The columns do not need to be next to each other. Select and drag the column of values to be graphed on the vertical axis, then click on the Chart Wizard icon on the menu bar (see the arrow in the illustration). In the first dialogue box, select the *XY* (*Scatter*) option. Also select either the dots-only display (preferred when the horizontal values are in no particular order) or the second option of connecting the dots with a line (preferred for a time series chart). Click *Next.*

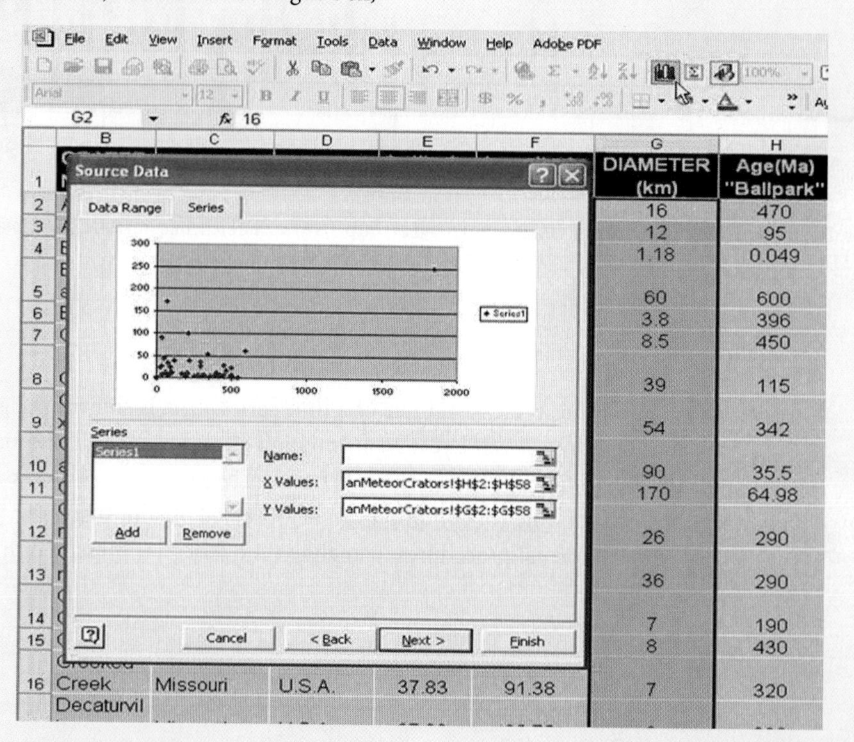

In the next dialogue box, click on the *Series* tab. The Excel addresses for the *Y-values* (for the vertical axis) will already be displayed; for the *X-values* (for the horizontal axis), click and drag over the column of values for the second variable in the spreadsheet (see the screen capture). Click *Next.* Before clicking *Finish,* use the dialogue box to input titles or other desired options.

An alternative way to input the data for graphing is to copy and paste the data for the *Y*- and *X*-variables, in that order, into two available neighbouring columns, and then highlight both columns prior to opening the Chart Wizard. Using this approach, Excel will automatically know which of the two variables is the *X*-variable and which is the *Y*-variable.

Note that for time series charts, the time data in the spreadsheet may be characters (e.g., "Dec. 2004," "Jan. 2005") rather than numbers (e.g., "2004," "2005"). In this case, choose a line graph from the original options instead of an *XY* (*Scatter*) graph; the dialogue boxes will guide you through the subsequent steps, which are similar to those presented above.

Minitab

There should be two columns of data—one for each of the variables. The columns do not need to be next each other. Use the menu sequence *Graph / Scatterplot.* In the first dialogue box, select either the *Simple* (dots-only) display (preferred when the horizontal values are in no particular order) or the option below it: *With Connect Line* (preferred for a time series chart). Click *OK.*

In the next dialogue box, Minitab is awaiting the name of the variable to display on the *Y* (vertical) axis. Click its name at the left and then click *Select.* Now click the name of the second variable, then *Select.* Click *OK.*

Note that for time series charts, Minitab also has a more specialized dialogue box sequence, which is discussed in Chapter 17.

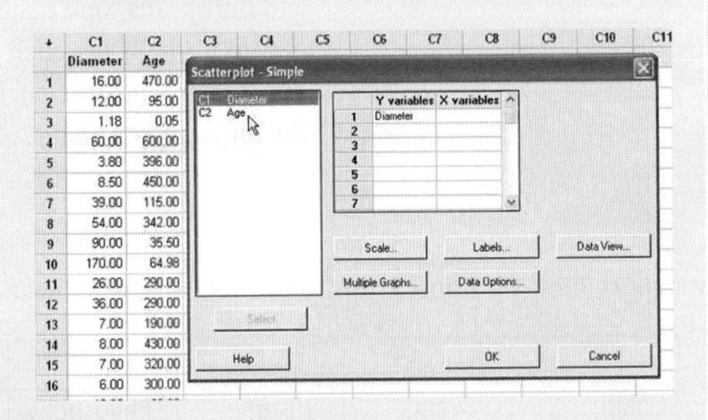

SPSS

There should be two columns of data—one for each of the variables. The columns do not need to be next each other. Use the menu sequence *Graphs / Scatter / Dot.* In the first dialogue box, select *Define* to accept the default option of *Simple* (dots-only) display.

On the next screen (see the screen capture), select the variable to be displayed on the *Y* (vertical) axis. Click the arrow to move the name over to the right. Select the name of the second variable and then click the arrow to move its name over. Click *OK.*

For time series charts, use instead the menu sequence *Graphs / Sequence* and indicate the time variable (which has to be sorted in ascending order) and one other variable. More specialized SPSS time series options are discussed in Chapter 17.

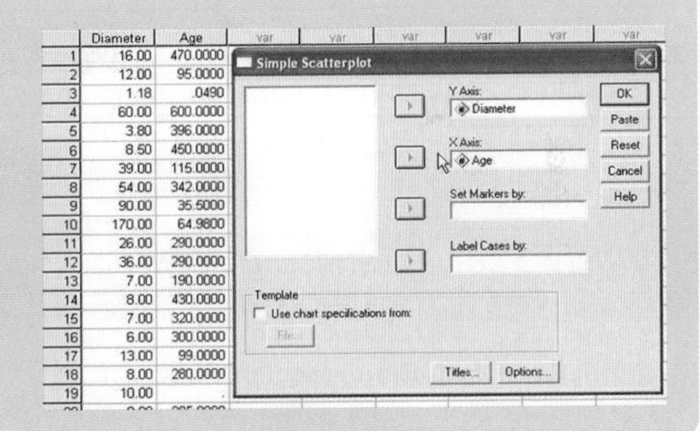

Contingency Tables

To display associations between two *categorical* variables, **contingency tables** (also called **cross-tabulation tables**) can be used. In effect, these tables are intersecting frequency distribution tables for two variables, simultaneously. Frequencies or relative frequencies are displayed both for the individual variables and also for combinations of the variables' values.

Suppose a candidate for the Nobel Prize in Literature is unsuccessful and charges that the awards committee is biased in its distribution of awards. In an effort to evaluate such a charge, it would be useful to construct a table such as the one in Figure 3.15. (The table would be equivalent if the rows and columns were interchanged.) The "Grand Total" column on the right is, in effect, the frequency distribution for the *Continents* variable. We see that, for whatever reason, candidates from Europe and the Americas receive the most prizes. The "Grand Total" row at the bottom is, in effect, the frequency distribution for the Rubric (i.e., prize category) variable. Clearly, most prizes are awarded for either poetry or prose fiction. If the complainant is a poet from Asia, one might understand some doubt about having a fair chance. Over a third of the successful Americans and Europeans were poets. Why not a third of the Asian authors? (Chapter 15 discusses how to formally test such questions about consistency of distributions.) Note that in this case, we have insufficient information about, for example, numbers of submissions from the different continents and about the quality of submissions.

Figure 3.15
Contingency Table for Nobel Prizes for Literature by Continent

	Rubric						
Contin	Drama	History	Philosophy	Poetry	Prose	Prose Fiction	Grand Total
Africa	1					3	4
America				5		7	12
Asia						3	3
Australia						1	1
Europe	1	1	1	12	1	19	35
Grand total	2	1	1	17	1	33	55

Source: Nobel Foundation. Nobel Prize for Literature winners (1951–2003). Found at: http://nobelprize.org/nobel_prizes/literature/laureates

• *In Brief 3.5(b) Constructing a Contingency Table*

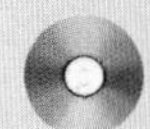

*(Based on the **Nobel_Prizes_Literature** file)*

Excel

Use the command sequence *Data / Pivot Table and Pivotchart Report.* In the first dialogue box, accept the default options by clicking *Next.* The following three dialogue boxes to be used are shown in the illustration. First, enter the range of the full data set where indicated and then click *Next.* In the next dialogue box, select *New worksheet* and then click the *Layout* button. This calls up the dialogue box shown at the bottom of the screen capture. One at a time, click and drag the names of the two variables for comparison into the template provided, in the indicated positions. Also click and drag one of the variable names into the *Data* area. Note the arrow in this area that points to *Count of Rubric.* If instead the corresponding box in your template is labelled *Sum of* [*variable name*], double-click on that box and choose *Count of Rubric* from the list that will appear at the left. Click *OK* and then *Finish.*

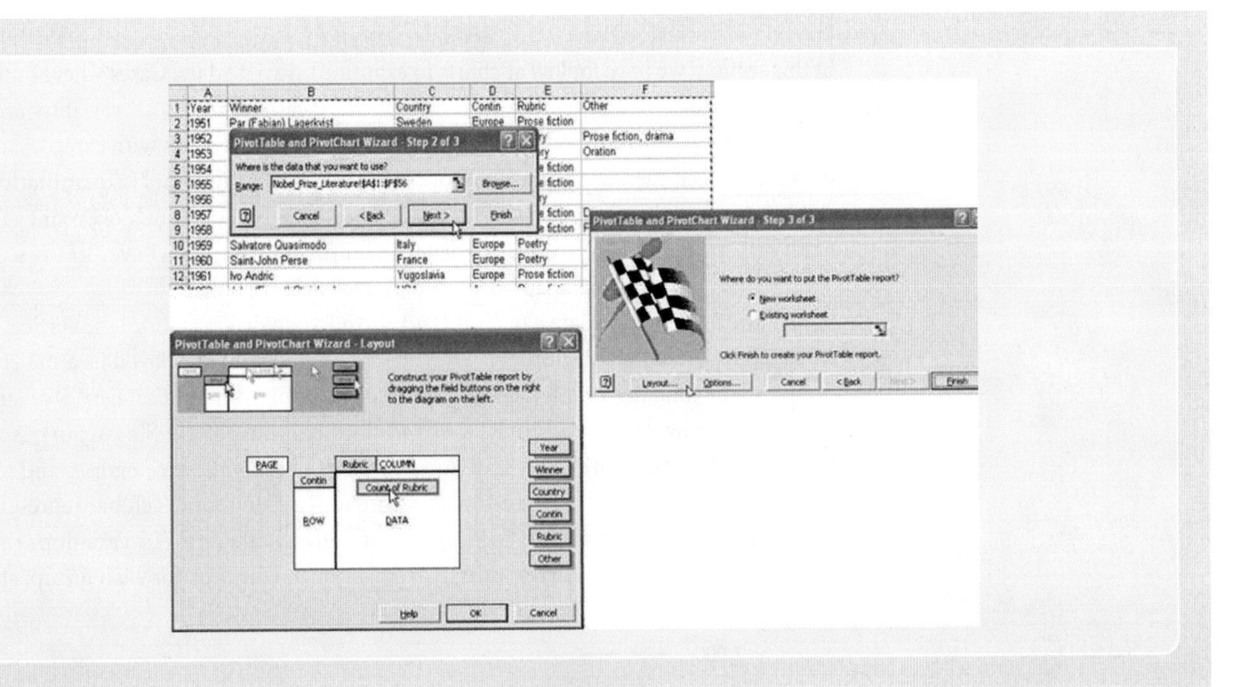

Minitab

Begin with the command sequence *Stat / Tables / Cross Tabulation and Chi-square.* As shown in the screen capture, select the variables in the dialogue box whose values should appear on the rows and columns of the table. Then click *OK.*

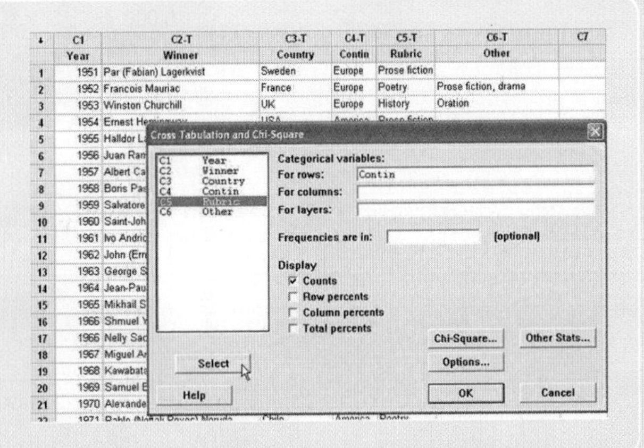

SPSS

Begin with the command sequence *Analyze / Descriptive Statistics / Crosstabs.* As shown in the screen capture, select the variables in the dialogue box whose values should appear on the rows and columns of the table. Then click *OK.*

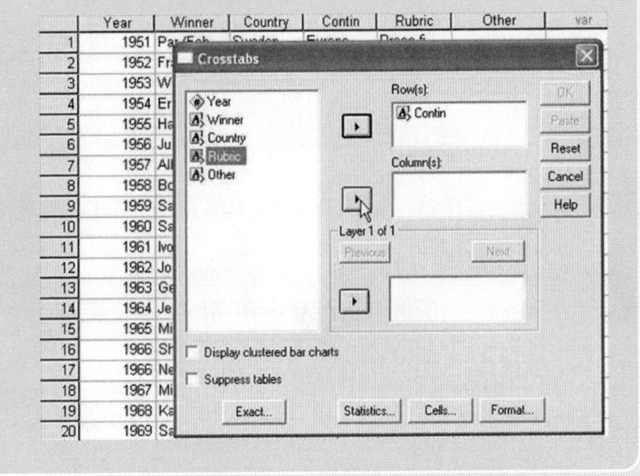

Bivariate Bar and Pie Charts

In this section, we have looked at charts to examine bivariate data. Cases where both variables are quantitative have been discussed, as have cases where both variables are categorical. Here, we introduce displays that link quantitative variables with categorical ones. In practice, such cases are quite common: What are people's incomes (a quantitative variable) if grouped by profession (a qualitative variable)? Or, how much tax money (quantitative) is spent by each government department (qualitative)? You have already seen that bar and pie charts can be adapted to display such relationships.

Figure 3.16 is a bivariate bar chart (and also a Pareto diagram, since the vertical values are arranged in decreasing order). These data could also be displayed as a pie chart. It compares the total revenue-tons of cargo enplaned at four different categories of airports. Examine the raw data found in the first and third columns of the file **Airport_Activity.** You will see that there are many individual airports in each of the categories and that for most airports, a value of revenue-tons is recorded. The length of each bar represents the sum of all of the revenue-tons for airports of a given category. (In variations of these charts, the lengths could correspond to averages or maximums for each group, etc.)

Figure 3.16
Bar Chart for Total Revenue-Tons by Airport Category

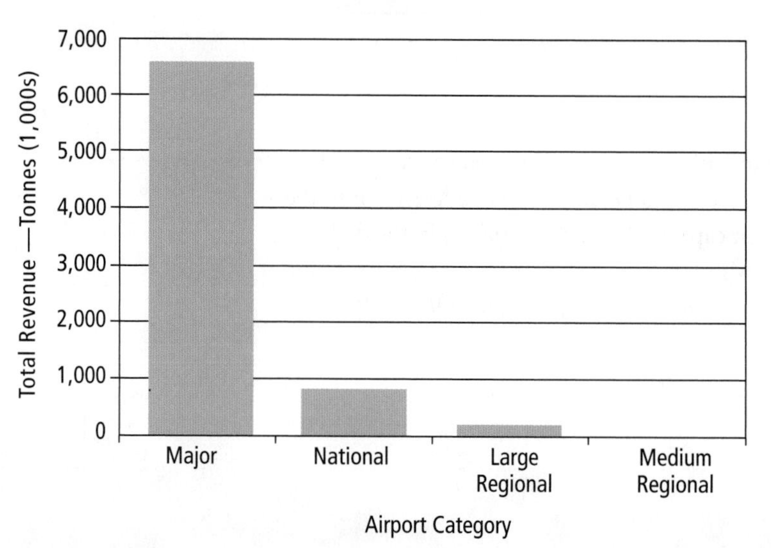

Source: U.S. Department of Transportation, Bureau of Transportation Statistics. Based on data in *Airport Activity Statistics of Certificated Air Carriers,* "Summary Tables: Twelve Months Ending December 3, 2000." Report BTS01-05. Found at: http://www.bts.gov/publications/airport_activity_statistics_of_certificated_air_carriers/2000/index.html

• *In Brief 3.5(c) Constructing a Bivariate Bar or Pie Chart*

*(Based on the **Airport_Activity** file)*

Excel

The relevant graphing commands in Excel are the same as for frequency-based bar and pie charts. (Refer back to the Excel section in "In Brief 3.3.") The only difference is that you start from a table that relates each category of the qualitative variable *not* to a frequency, but instead to a sum (or other desired statistic) from an associated quantitative variable. The first two columns in the illustration were used as the basis for Figure 3.16. (The right-hand column shows an Excel formula that *could* be used to automate summing of revenue-tons associated with each category. It might be easier to just sort the table by the category names and sum the revenue-tons for airports of each category.)

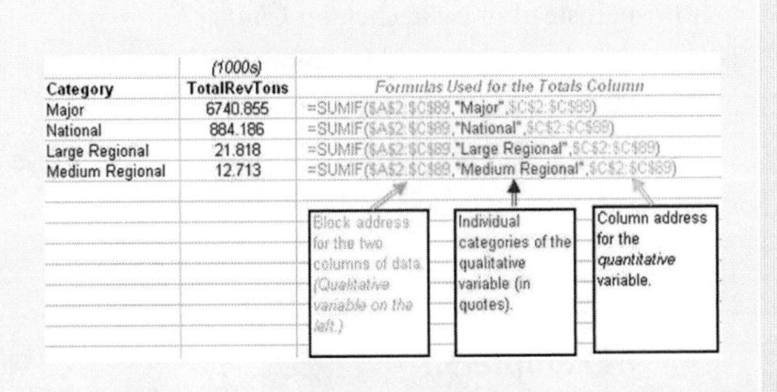

Category	(1000s) TotalRevTons	*Formulas Used for the Totals Column*
Major	6740.855	=SUMIF($A$2:$C$89,"Major",$C$2:$C$89)
National	884.186	=SUMIF($A$2:$C$89,"National",$C$2:$C$89)
Large Regional	21.818	=SUMIF($A$2:$C$89,"Large Regional",$C$2:$C$89)
Medium Regional	12.713	=SUMIF($A$2:$C$89,"Medium Regional",$C$2:$C$89)

Minitab

As in the Minitab section in "In Brief 3.3", begin with the menu sequence *Graph / Bar Chart.* In the first dialogue box that appears (see the screen capture), select *A function of a variable* from the drop-down menu. Click *OK.* The next dialogue box is also shown in the illustration. In the drop-down menu for *Function,* select *Sum* (or another function, if desired). Click in the box for *Graph Variables* and then select the quantitative variable to be summed. Click in the box for *Categorical Variable* then select the categorical variable. For a bar chart, click *OK* at this point. (Minitab does not have a bivariate pie chart function.)

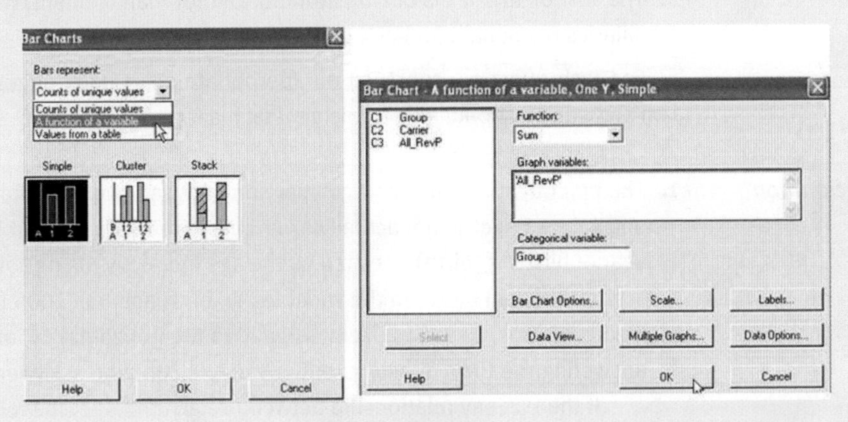

SPSS

The graphing commands in SPSS are essentially the same as for the frequency-based bar and pie charts described in the SPSS section in "In Brief 3.3." Input the categorical variable at the bottom of the bar or pie dialogue box, as before. The only change is that you will select *Other summary function* as the basis for the bars or slices, instead of *N of cases* or *% of cases* (see the screen capture). Then select the variable to be summed. If *Mean* or another function is

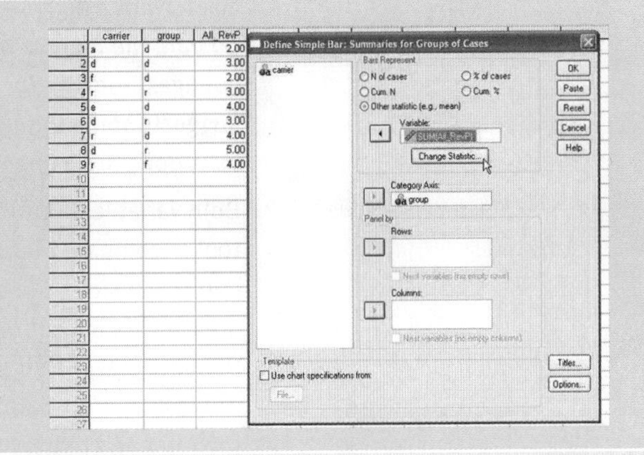

showing instead of *Sum,* click on *Change Summary* and then select the desired function. Click *Continue,* then *OK.* For a pie chart, the relevant portions of the dialogue boxes are similar to the bar chart version shown in the screen capture.

Example 3.7

*(Based on the **Players_Toronto_Blue_Jays** file)*

In 2005, the general manager of the Toronto Blue Jays was reviewing some player statistics in preparation for the trading season. He handed the file **Players_Toronto_Blue_Jays** to his team statistician for comments. The following are some questions that concerned him.

1. Does the age of his players appear to have any relationship to their salaries? For example, are the older players generally being paid more or less than the younger ones?
2. Each player has a preference in terms of batting as a "lefty" or a "righty" and in terms of throwing with the left or right hand. How do these two variables compare?
3. The cost of salaries is obviously important for management. What is the total in salaries being paid for each position on the team?

For each question, construct an appropriate chart and suggest what answers the team statistician should give to the general manager.

Solution

1. The first question calls for a comparison between two quantitative variables: age and salary. "Age" is not actually a variable in the data file, but knowing a player's year of birth (Yr_birth) does tell us the age indirectly. Using software, construct a scatter diagram between the variables Yr_birth and Sal_2005 (i.e., Salary). In Figure 3.17, Yr_birth has been assigned to the horizontal (X) axis, but for simply answering the GM's question, either variable could be assigned to X.

   If there is any relationship between salary and age, the relationship is obviously not linear, that is, an earlier birth date (greater age) is not consistently associated with a higher—or a lower—salary. It is interesting, however, that those who *do* earn the highest salaries are all in the *middle*-aged group—neither the youngest rookies nor those nearing retirement age. Notice how the scatterplot suggests some thought-provoking patterns. Without a more formal analysis (such as is done in Chapter 15), any conclusions must be tentative.
2. Both variables—batting approach and hitting approach—are qualitative. A good tool to compare them is a contingency table, as shown in Figure 3.18. Some tentative conclusions can be reached based on this figure. Two apparent relationships are that if a player *throws left,* that player also *bats left,* and if a player *bats right,* then the player also *throws right.* But the reverse statements do not hold: Two-thirds of the left-handed batters actually throw right-handed, and many of the right-handed batters throw left-handed.

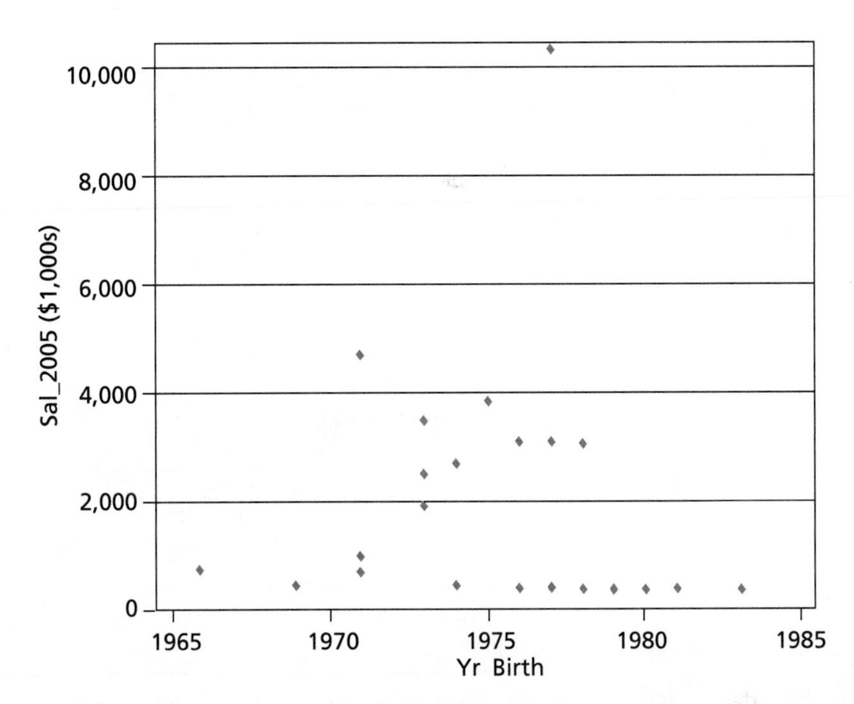

Figure 3.17
Scatterplot of Sal_2005 (in $1,000s) vs. Yr_Birth

Source: USA Today. 2005. Reprinted with permission. Found at: http://asp.usatoday.com/sports/baseball/salaries/teamdetail.aspx?team=14&year=2005

Figure 3.18
Contingency Table for Left/Right Preferences in Batting and Throwing

Count		Throws		
		Left	Right	Total
Bats	Both	0	2	2
	Left	3	6	9
	Right	0	15	15
Total		3	23	26

Source: USA Today. 2005. Reprinted with permission. Found at: http://asp.usatoday.com/sports/baseball/salaries/teamdetail.aspx?team=14&year=2005

Similar to the scatter diagram, we see that the contingency table by itself provides some promising ideas. For a more thorough analysis, techniques discussed in Chapter 6 (probability) and Chapter 15 (tests for association) should also be considered.

3. The sum of players' salaries by position requires bivariate data—qualitative (position) and quantitative (salary) combined. The Pareto diagram (Figure 3.19) is really just a bar chart that is sorted by decreasing sum of salary. Clearly, the pitching position is the most expensive to staff. From this figure alone, however, we cannot tell if the pitching expense is due to the larger number of pitchers, or to their relative salaries, or to both. (We will revisit this example in a later chapter.)

Figure 3.19
Pareto Diagram for the Sum of Salaries (2005, in $1,000s) by Position Played

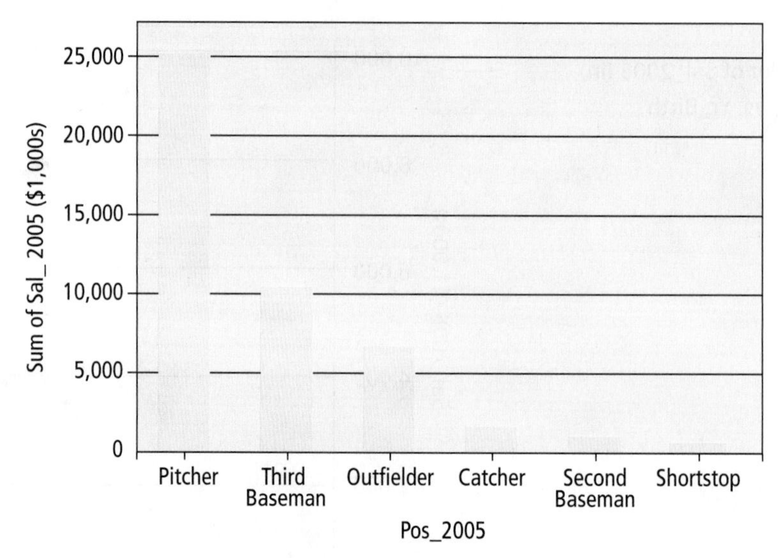

Source: USA Today. 2005. Reprinted with permission. Found at: http://asp.usatoday.com/sports/baseball/salaries/teamdetail.aspx?team=14&year=2005

• Hands On 3.5 *A Matter of Scale*

*(Based on the files **Bike_Car_Production** and **BC_votes**)*

At this point, many textbooks present a section of do's and don'ts for appropriate use and interpretation of statistical figures that can be briefly summarized: When creating a figure, the goal is to present information clearly, honestly, and with minimal chance for misinterpretation. For example, in a bar chart about oil production, avoid replacing bars with clever pictures of oil cans; the problem is that if you draw one object bigger than another, the reader does not know whether to compare them by height, area, or even volume. When reading other people's charts, be a critical thinker. Check the scale they've used and how they are indicating relative frequencies, and also ensure that they have not suppressed or distorted information.

A common way to "tell lies with statistics" is to intentionally or unintentionally choose a misleading scale for an axis. As shown in Figure 3.20 (based on the file **Bike_Car_Production**), a too-large scale for an axis can artificially compress the apparent range and variation of the data. On the other hand, we see in Figure 3.21 that a too-narrow scale can mislead about the degree of variation in the data. (For the purpose of this illustration, the file **BC_votes** was graphed as if the data were reported in the order in which they appear in the file.)

Some believe that a scale for numerical data (and frequencies) should always start at zero to avoid the above problems. This can be impractical if the data are all large numbers and there is insufficient room for a very tall or wide feature in the graph. Many traditional textbooks have required that if the scale does not start at zero, as in the right-hand graph in Figure 3.21, a small gap

Figure 3.20
Reasonable and Misleading Scales for Car Production

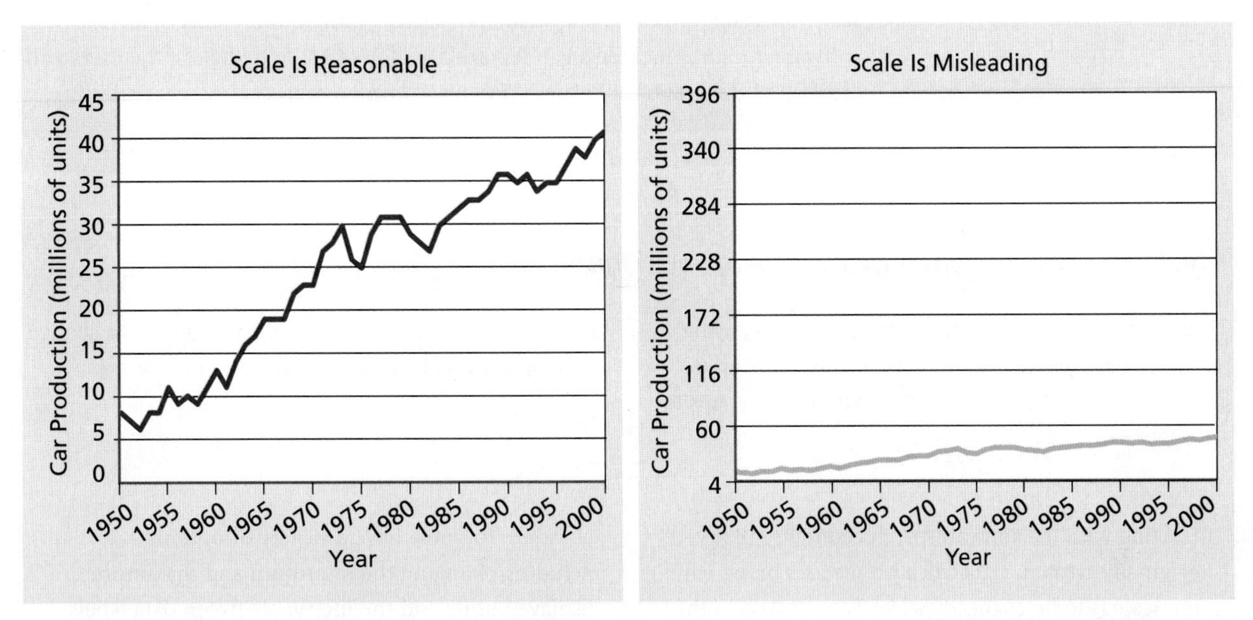

Source: "Figure: World Automobile Production" and "Figure: World Bicycle Production," from *Vital Signs 2002: The Trends that Are Shaping Our Future* by Lester R. Brown, et al. Copyright © 2002 by Worldwatch Institute. Used by permission of W.W. Norton & Company, Inc. Found at: http://www.ibike.org/library/statistics-data.htm

Figure 3.21
Reasonable and Misleading Scales for Total Votes

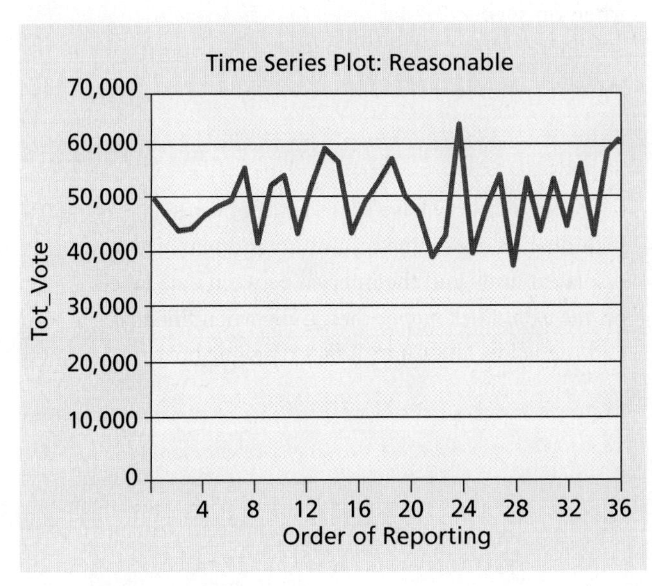

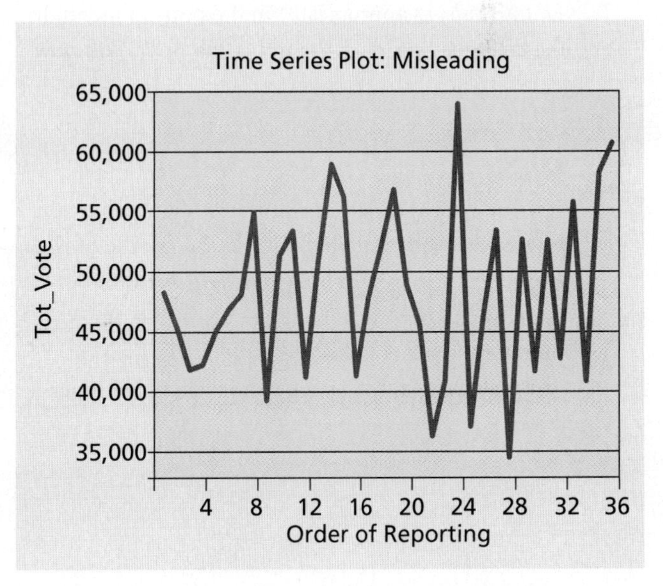

Source: "Federal Election Results, 2004." *The Globe and Mail.* Wednesday, June 30, 2004. Pages A12–A13. Reprinted with permission from *The Globe and Mail.*

should be drawn on the axis, near the origin, to highlight this compression of scale. This rule, although reasonable, is largely ignored today since statistical software does not include a gap feature for such cases; nonetheless, the users and creators of the graphs are expected to use common sense.

• *In Brief 3.5(d) Changing the Scale of Your Figure*

This "In Brief" feature addresses the case where, using methods discussed above, you have created a bivariate graph with at least one quantitative axis and, perhaps to avoid creating a misleading figure, you decide to change the displayed scale for one of the quantitative axes.

Excel

Starting with the software-produced figure that you originally created, right-click on the axis on which the scale is to be changed. Select *Format Axis* in the pop-up menu. In the Chart Wizard, click the upper tab *Scale*. You now have a variety of options, including changing the minimum and maximum displayed units and the interval between data labels on the axis. Click *OK* when finished.

Minitab

Starting with the software-produced figure that you originally created, right-click on the axis on which the scale is to be changed. Click on *Edit Y Scale* (or *Edit X Scale*, as appropriate) in the pop-up menu. In the dialogue box, click the upper tab *Scale*. You now have a variety of options, including changing the minimum and maximum displayed units and the interval between data labels on the axis. Click *OK* when finished.

SPSS

Starting with the software-produced figure that you originally created, double-click in the middle of the graph; this will call up the *Chart Editor*. In the upper menu bar, click the large *X* to adjust the *X* (horizontal) axis or click the large *Y* to adjust the *Y* (vertical) axis. In the dialogue box, click the upper tab *Scale*. You now have a variety of options, including changing the minimum and maximum displayed units and the interval between data labels on the axis. Click *Apply*, then *Close* when finished, and then close the *Chart Editor* dialogue box.

Using the above procedures for changing the numerical scales on a graph, revisit all of the bar, scatter, and other charts constructed in this chapter. Explore how you can make the scale clearer (if possible) and less clear (certainly possible!) than the graph you have just generated.

Chapter Summary

3.1 Construct and interpret frequency distribution tables for quantitative and qualitative data.

Most of this chapter is concerned with **univariate descriptive statistics,** that is, descriptive statistics for just one variable. Beginning with the **raw data** obtained from research, we must first ensure that the data are **clean** and not **dirty;** i.e., the recorded numbers or codes must be correct and free of mistyped or poorly measured values, etc. If there are **outliers** in the data, they will appear to be remote from all or most of the other values for the variable. When constructing a **frequency distribution table** for quantitative data, we count how many times the variable either (1) takes specific values from a finite list of possible values (i.e., in a **discrete quantitative frequency table**) or (2) takes values that fall within specified ranges or **classes** of values (i.e., in a **continuous quantitative frequency table**). We explored methods for constructing both types of table.

The frequency column of a frequency distribution table can be converted to show the **percent frequency distribution** of the variable. For each distinct value of the variable that is observed, we display its percentage relative to the total number of observations. Another column can be added to show the **cumulative percent frequency distribution,** displaying for each distinct value of the variable (if discrete) the cumulative percentage of all cases having values up to or including the value displayed at the left of that row, or (for continuous data) the cumulative percentage of cases with values up to or including the upper limit of the displayed class. For qualitative data, we learned to construct a **qualitative frequency distribution table,** which shows how often each particular value or category of the qualitative variable has occurred.

3.2 Construct and interpret frequency distribution histograms for quantitative data.

A **histogram** is a graph that uses the lengths of bars to represent the frequency (or relative frequency) of values in each class of a continuous frequency distribution. The class values are shown on the horizontal axis. Another variation of this type of graph is a **density scale histogram,** in which the *area* encompassed by each bar represents the percentage of the total number of observations that fall within the class boundaries for that bar. Histograms can often help to reveal the shape of a distribution of quantitative data—in particular, whether it is symmetrical or **skewed.**

Practical techniques for constructing histograms were discussed. **Frequency polygons** (or line graphs) or **ogives** were mentioned as additional alternatives to histograms that can create a smoother image for the data pattern.

3.3 Construct and interpret pie charts and bar charts for qualitative data.

A **bar chart** is similar to a histogram in that all bars have equal widths and their lengths are proportional to the frequencies (or relative frequencies) of occurrence of the different data values (i.e., categories). The bars (which have gaps between them) can be oriented vertically, similar to a histogram, or horizontally, with the category names shown on the left of each bar.

A **pie chart** displays relative frequencies in terms of the relative sizes of "slices" of the pie diagram. **Pareto diagrams** are similar to histograms, but designed to display all of the relative frequencies in decreasing order.

3.4 Construct and interpret stemplots for quantitative data.

Stemplots (or **stem-and-leaf displays**) group quantitative data into classes, similar to a histogram, but the displays retain original information about the raw, ungrouped data. To create a stemplot, part of each value in the left part of the number is identified as the "stem" and the next digit to the stem's right is identified as the "leaf."

3.5 Construct and interpret appropriate displays of association for quantitative and qualitative data.

In **bivariate descriptive statistics,** we compare the values for *two* variables that are observed for individuals in the population. **Scatter diagrams** can be used to compare two quantitative variables. Each dot displays simultaneously for *one* individual the values of *both* variables. **Time series charts** are a special case of scatter diagram (for date data that are numerical). The variable measured horizontally is time (for example, in periods of months or years) and the second variable can be either a discrete count of some quantity or else a continuous value.

Contingency tables (or **cross-tabulation tables**) can be used to display associations between two *categorical* variables. Frequencies or relative frequencies are displayed both for the individual variables and also for combinations of the variables' values. Bivariate bar charts and pie charts are useful when categorical and numerical data are combined; for example, in a pie chart showing the relative expenditures of government on health care, education, security, and so on.

• Key Terms

ascending order, p. 70
bar chart (or bar graph), p. 90
bivariate descriptive statistics, p. 101
class limits, p. 70
class midpoints, p. 72
class width, p. 69
classes, p. 68
clean data, p. 64
collectively exhaustive, p. 69
contingency tables (or cross-tabulation tables), p. 105
continuous quantitative frequency table, p. 68
cumulative percent frequency distribution, p. 65
density scale histogram, p. 83
descending order, p. 70
dirty data, p. 64
discrete quantitative frequency table, p. 65
fixed-answer variables, p. 77
frequency distribution table, p. 65
frequency polygons, p. 88
histogram, p. 82
mutually exclusive, p. 69
ogives, p. 88
open-ended variables, p. 78
ordered array, p. 70
outliers, p. 64
Pareto diagram, p. 91
percent frequency distribution, p. 65
percent frequency histogram, p. 83
pie chart, p. 91
qualitative frequency distribution table, p. 77
range, p. 70
raw data, p. 64
scatter diagrams, p. 102
shape, p. 82
skewed, p. 82
stemplots (or stem-and-leaf displays), p. 96
Sturge's Rule, p. 71
time series charts, p. 102
univariate descriptive statistics, p. 64

• Exercises

Basic Concepts 3.1

1. Is it possible for an outlier to be dirty data? Explain.
2. Distinguish between the terms *mutually exclusive* and *collectively exhaustive.*
3. Why is Sturge's Rule useful?
4. To the frequency distribution table shown below, append a percent frequency column and a cumulative percent frequency column.

Value	Frequency
10	6
15	9
20	5
25	8
30	2
Total	30

5. The following grades were received by students on an economics examination:

55 65 74 86 45 89 75 90 67
85 55 66 78 02 91 87 65

a) Is it possible that the data set contains dirty data? Explain.
b) Does the data set contain any outliers? Explain.

6. To the grouped frequency distribution table shown below, append a percent frequency column and a cumulative percent frequency column.

Class	Frequency
0 up to 10	3
10 up to 20	6
20 up to 30	9
30 up to 40	7
40 up to 50	8
50 up to 60	9
60 up to 70	8
Total	50

7. For each of the data sets that follow:
a) Determine the appropriate number of classes that could be used to convert the data set into a grouped frequency distribution.
b) Determine the approximate class width.
c) Construct a grouped frequency distribution.
d) Construct a percent frequency distribution.
e) Construct a cumulative percent frequency distribution.
f) Identify all of the class midpoints.

Data set 1: 15 16 22 35 13 29 36 24
10 12 24 27 33 17 16

Data set 2: 2 6 5 10 15 24 4 10
25 18 2 19 30 32 26 28
13 17 22 15 33 37 3 7
16 19 20 22 26 13 9 31

8. For each of the data sets that follow:
a) Construct a grouped frequency distribution.
b) Construct a percent frequency distribution.
c) Construct a cumulative percent frequency distribution.

Data set 1: 354 356 329 343 373 366 311 381
310 350 342 378 333 327 364

Data set 2: 0.2 0.6 0.5 1.0 1.5 2.4 0.4 1.0
2.5 1.8 0.2 1.9 3.0 3.2 2.6 2.8
1.3 1.7 2.2 1.5 3.3 3.7 0.3 0.7
1.6 1.9 2.0 2.2 2.6 1.3 0.9 3.1

9. Using your knowledge of the proper way to construct a grouped frequency distribution, identify the problems that are associated with the following grouped frequency distribution.

Class	Frequency
10 to 20	45
20 up to 25	62
27 up to 30	57
40 up to 50	44
30 up to 40	38
Total	245

10. Define *open-ended variable.* Provide an example with your answer.
11. The eye colours for each person in a group of 30 individuals are presented below.

brown blue grey green grey blue
brown brown brown blue grey brown
green brown brown blue grey green
green brown brown brown blue grey
brown brown blue brown grey brown

a) Construct a qualitative frequency distribution for these data.
b) Add a column for the percent frequency.

Applications 3.1

12. In a certain month, the Dovercourt Recreation Association in Ottawa, Ontario, sells 109 plans for its adult/senior fitness program. The price schedule for the plans (as published in 2006) is shown below (see the file **Price_Schedule**). The breakdown of the plans that were purchased is as follows:

Plan A: 15 adults; 20 seniors

Plan B: 12 adults; 10 seniors

Plan C: 5 adults; 7 seniors

Plan D: 3 adults; 2 seniors

Plan E: 35 "drop ins"

Plan	Description	Months	Adult Fee	Senior Fee
A	For 1 month	1	$28	$25
B	For 3 months	3	$77	$65
C	For 6 months	6	$145	$130
D	For 1 year	12	$277	$249
E	One time ("drop in")	0.033	$7	$7

a) Based on these data, construct a discrete frequency distribution table that shows the frequency of purchases at each available price level.
b) Expand the table created in (a) to include a column for percent frequency.
c) Which fee amounts were collected most often? Least often?

13. Two volunteers collected data, over several summers, on eruptions of the Grotto Geyser in Yellowstone National Park. The table below is derived from their data (see the file **Grotto_Geyser**). The table shows, for each of 83 days that were studied, the number of eruptions recorded on that day.

1 2 3 4 2 4 2 2 1 3 4 2 3 3 3 1 2 3
2 2 1 3 3 4 3 3 4 3 2 1 3 3 2 2 1 4
4 2 1 2 3 3 4 1 4 1 4 3 2 3 2 3 1 4
4 3 2 2 3 2 2 3 3 3 3 3 2 2 1 3 2 2
3 4 3 2 3 1 4 3 1 3 2

a) Construct a frequency distribution table that shows the frequency of days having 1, 2, 3 or 4 eruptions, respectively.
b) Expand the table created in (a) to include a column for percent frequency.
c) Based on the table, does there appear to be a "typical" number, or range of numbers, of eruptions each day? If so, what are they? Explain.

14. In 2001, the Canadian federal government and eight banks signed agreements to facilitate lower-cost banking services for consumers. The Financial Consumer Agency of Canada has reported some of the results. The following are among the low-cost options for monthly fees that have become available (see file **Low_Fee_Accounts**).

3.50	5.00	4.00	2.00	4.00	2.95	2.50
3.50	5.00	4.00	3.95	2.50	3.95	0.00

a) You can see there is repetition in the fee values that are set for these accounts. Construct a discrete frequency distribution table based on the specific values that have been chosen.
b) Expand the table created in (a) to include a column for percent frequency.
c) Does there appear to be a "typical" number, or range of numbers, that is selected for these low-cost accounts? If so, what is it? Explain.

15. During the course of each hockey season, *The Sports Network* (TSN) updates and publishes lists of the NHL's 25 top individual goal scorers. The data below were final as of April 4, 2004, for the 2003–2004 season. The list includes each athlete's number of *game-tying goals* for that season (see the file **NHL_Scoring_Leaders**).

0 0 1 0 1 0 1 0 1 0 0 0 0
2 1 1 0 0 2 0 1 0 0 3 1

a) Construct a frequency distribution table that shows the frequency of top scorers who scored 0, 1, 2, or 3 game-tying goals, respectively.
b) Expand the table created in (a) to include a column for percent frequency.
c) Does scoring a game-tying goal appear to be a major criterion for being considered a "top goal scorer" by TSN? Explain.

16. A chemist is experimenting with solvents to use in extreme-temperature environments, such as an open cargo bay on a NASA space shuttle. She will select from a list of 47 solvents for which the melting and boiling points are known. (See the tables below, adapted from the file **Freezing_Boiling_Data.**)

Melting Point (°C)						
80	17	−95	−6	73	5	−31
8	−87	177	−112	−23	−46	−64
31	35	6	25	4	9	−117
−105	−20	53	19	12	27	−114
−84	3	8	−91	31	−98	26
80	6	41	−10	−126	−21	−42
27	0	−95	0	13		

Boiling Point (°C)						
221	118	56	184	220	80	156
150	80	204	46	77	132	61
191	202	81	161	298	131	35
42	166	113	189	101	258	78
77	211	101	98	206	56	82
218	211	182	106	97	140	115
285	244	111	100	138		

For *each* of the two data sets:

a) Construct a grouped frequency distribution. The distribution should have a reasonable number of classes, an appropriate class width, and begin at a suitable value.
b) Add columns for (i) percent frequency and (ii) cumulative (less than) percent frequency.

c) How are the melting points and boiling points distributed among all of the solvents? For example, do values tend to fall evenly across the whole range of values, or do they cluster within particular classes? Are the distributions similar for the melting points and the boiling points?

17. The population density of a country is a measure of how many people live in the country relative to the surface area within its borders. A demographer claims that population density varies widely among countries, and has produced the list below (see the file **Population_Density**). Each value shows the population density of a particular country, in persons (on average) per square kilometre.

3.1	2.5	96.7	336.1	123.9	15.3	107.3
230.3	2.7	53.9	189.9	168.7	390.3	14.3
13.9	79.1	19.7	173.9	29.4		

a) Construct a grouped frequency distribution for this data set. Ensure that the distribution has a reasonable number of classes, an appropriate class width, and begins at a suitable value.

b) Add columns for (i) percent frequency, and (ii) cumulative (less than) percent frequency.

c) Does the table you created appear to back up the demographer's claim? Explain. Do the values tend to cluster within particular classes, or do they fall relatively evenly across the whole range of values?

18. Royal LePage, the realtor, periodically publishes a survey of Canadian house prices (in $CDN). Displayed below are representative prices for a standard two-storey house in a selected major urban area (see the file **Standard_2Storey_House_Prices**).

295,000	214,000	210,000	230,000
325,000	325,000	260,000	185,000
255,000	651,450	133,561	120,000
133,000	196,250	480,000	280,000
530,000	285,000	355,000	1,150,000
650,000	1,250,000	285,000	635,000
300,000	840,000	286,000	700,000
440,000	229,000	254,000	236,000
233,000	282,000	192,750	189,000
167,250	145,000	233,100	220,400
229,800	300,600	225,800	389,000
435,000	745,000	540,000	

a) Construct a grouped frequency distribution for this data set. Ensure that the distribution has a reasonable number of classes, an appropriate class width, and begins at a suitable value.

b) Add columns for (i) percent frequency, and (ii) cumulative (less than) percent frequency.

c) Comment on the distribution of these prices for the different areas. Suppose that, in a lottery, you win the cash value of a typical two-storey house in one of the areas that were surveyed; it will be determined by chance *which* area will be selected as the basis for the prize. Approximately how much money are you likely to win? What amount of money are you least likely to win? How were you able to answer these questions?

19. A mayor in a Prince Edward Island community is preparing to receive a guest from Australia. She has downloaded a price list for Australian wines from the website of the PEI Liquor Control Commission. Displayed below are prices for 750-mL bottles of wine (see the file **Australian_Wines**).

13.05	13.05	13.05	11.99	15.99	14.04	15.49
18.78	14.98	14.98	15.36	14.98	14.98	15.11
13.49	15.49	16.50	12.95	18.21	22.74	23.87
15.84	14.98	13.70	14.98	13.65	14.98	15.99
14.12	24.18					

a) Construct a grouped frequency distribution for this data set. Ensure that the distribution has a reasonable number of classes, an appropriate class width, and begins at a suitable value.

b) Add columns for (i) percent frequency, and (ii) cumulative (less than) percent frequency.

c) Based on the table, is there a range of wine prices that you would describe as "typical"? If so, what is the approximate price range? Are there any prices that are, comparatively, "higher than usual"? If yes, what are these values? Explain the reasons for your answers.

20. A mineral collector has identified 100 metallic minerals and recorded for each the values of density and hardness. He is examining the distributions of these two variables (see the file **Metallic_Minerals**).

a) Construct a suitable grouped frequency distribution for each of these variables and include columns for percent frequency and cumulative (less than) percent frequency.

b) Compare the two distributions. Which distribution is more symmetrical? Which variable has a greater spread of possible values? If you were to randomly pick one of the minerals, would you feel more confident guessing its hardness or its density, based on knowing their distributions? Explain your answer.

21. The University of New Brunswick's *Earth Impact Database* provides data on meteor craters in North America. The province, territory, or state for each crater is listed in the table below (see the file **Meteors_N_Amer**).

Oklahoma	Alaska	Arizona
Montana	Ontario	Michigan
Saskatchewan	Quebec	Virginia
NWT	Quebec	Quebec
Missouri	Missouri	Nova Scotia
Illinois	Alberta	Alabama
Saskatchewan	Illinois	Wisconsin
Saskatchewan	Nunavut	Kansas

Ontario	Quebec	Indiana
Quebec	Quebec	Iowa
Saskatchewan	Texas	Kentucky
NW T	Quebec	Quebec
N. Dakota	Wisconsin	Texas
Texas	Quebec	N. Dakota
Saskatchewan	Manitoba	Ohio
Saskatchewan	Ontario	Alberta
Ontario	Utah	Tennessee
Ontario	Tennessee	Manitoba
Nfld/Labrador	Wyoming	Yucatan

a) Construct a qualitative frequency distribution table for these data.

b) Append to the table created in (a) a column for percent frequency.

c) Based on the table, does there appear to be a province, territory, or state that has been particularly prone to large meteor falls? What parts of Canada, if any, do not have any known meteor craters?

22. A young taxonomist has identified 106 species of pine tree. As shown below, the great majority of pine species belong to the genus *Pinus.* Construct a frequency distribution table for the genera to which pine species belong. Include a column for percent frequency. What are the second and third most common genera? (See the file **Pine.**)

Abies	Calocedrus	Pinus	Larix	Larix
Larix	Picea	Picea	Picea	Picea
Pinus	Pinus	Pinus	Pinus	Pinus
Pinus	Pinus	Pinus	Pinus	Pinus
Pinus	Pinus	Pinus	Pinus	Pinus
Pinus	Pinus	Pinus	Pinus	Pinus
Pinus	Pinus	Pinus	Pinus	Pinus
Pinus	Pinus	Cryptomeria	Pinus	Pinus
Pinus	Pinus	Pinus	Pinus	Pinus
Pinus	Pinus	Pinus	Pinus	Pinus
Pinus	Pinus	Pinus	Pinus	Pinus
Pinus	Pinus	Pinus	Pinus	Pinus
Pinus	Pinus	Pinus	Pinus	Pinus
Pinus	Pinus	Pinus	Pinus	Pinus
Pinus	Pinus	Pinus	Pinus	Pinus
Pinus	Pinus	Pinus	Pinus	Pinus
Pinus	Pinus	Pinus	Pinus	Pinus
Pinus	Pinus	Pinus	Pinus	Pinus
Pinus	Pseudotsuga	Pinus	Pinus	Pinus
Pinus	Pinus	Pinus	Pinus	Pinus
Pinus	Pinus	Pinus	Sequoia	Taxus
Torreya				

23. According to the *Frequently Asked Questions* page on the Tim Hortons website, the company is often asked what people win during its "Roll Up the Rim to Win" promotion. The following statement was published in February 2006: "In 2005, we gave away 30 . . . GMC Canyons, 84 Panasonic plasma televisions, 460 cash prizes of $1,000 and 6,644 Coleman Camping Packages, as well as millions of Tim Hortons food prizes. . . ."[8] Suppose the number of food prizes was actually 3 million. Construct a qualitative frequency distribution that is based on this data. What percentage of awarded prizes were GMC Canyons? What percent were cash prizes and what percent were Coleman Camping Packages, respectively? If your "Roll Up the Rim" cup showed you won a prize in 2005, what type of prize did you most likely win? Least likely win?

24. As part of a yearly audit of licensed casinos in the province, the Alberta Gaming and Liquor Commission is reviewing the locations of all such facilities. These locations are listed below. (See the file **Casinos_Alberta.**)

Edmonton	Edmonton	Edmonton	Edmonton
Calgary	Calgary	Calgary	Calgary
Calgary	Red Deer	Red Deer	Fort McMurray
Grande Prairie	Medicine Hat	St. Albert	Lethbridge

a) Construct a qualitative frequency distribution table for these data and add a column for the percent frequency.

b) Based on the table, what city currently has the most licensed casinos in Alberta?

c) Compare the distribution of cities with a map of Alberta. Describe the distribution in terms of relative locations in the province (for example, are most licensed facilities in the east or north of the province, or are they evenly distributed?).

Basic Concepts 3.2

25. Construct a frequency histogram and a cumulative percentage frequency histogram for each of the grouped frequency distributions that you constructed in Exercise 7. The following were the original data for Exercise 7:

Data set 1:	15	16	22	35	13	29	36	24
	10	12	24	27	33	17	16	
Data set 2:	2	6	5	10	15	24	4	10
	25	18	2	19	30	32	26	28
	13	17	22	15	33	37	3	7
	16	19	20	22	26	13	9	31

26. Construct a frequency histogram or a frequency polygon corresponding to the following grouped frequency distribution.

Class	Frequency
10 up to 20	15
20 up to 30	25
30 up to 40	20
40 up to 50	20
50 up to 60	25
Total	105

27. Construct a cumulative frequency histogram or polygon that corresponds to the frequency table given in Exercise 24.

Applications 3.2

28. A chemist is experimenting with solvents to use in extreme-temperature environments. (See the data in Exercise 16, adapted from the file **Freezing_Boiling_Data.**)

a) Construct a frequency histogram for *each* of the two data sets.

b) Compare the two distributions. Include reference to the shapes of the distributions. List three respects in which the distribution patterns are different from each other.

c) Construct a cumulative percent histogram for just the boiling point data. Based on this graph, approximate the percentage of solvent types that would boil if the environment reaches 200°C.

29. The population density of a country is a measure of how many people live in the country relative to the surface area within its borders. (See the data in Exercise 17, based on the file **Population_Density.**) Population densities are given in persons per square kilometre.

a) Construct a frequency histogram, and a percent frequency histogram, for this data set.

b) Describe the shape of the distribution.

c) Construct a cumulative percent histogram for these data. Based on this figure, approximately what percentage of countries have a population density less than 300 people per square kilometre?

d) About what percentage of countries have a population density of *at least* 300 people per square kilometre?

30. Representative prices for buying a standard two-storey house in a selected urban location were displayed in Exercise 18. (See the file **Standard_2Storey_House_Prices.**)

a) Construct a frequency histogram and a relative frequency histogram for this data set.

b) Do any of the prices appear to be "outliers," that is, set apart from most of the values? Explain. Is the distribution symmetrical?

c) Based on the histograms you created in (a), what would you say is the typical house price for a standard two-storey house in an urban Canadian centre? Would you say that most other prices are arranged symmetrically around the typical value, or not? Explain.

31. To entertain a guest, a mayor has downloaded a price list for Australian wines from the website of the PEI Liquor Control Commission. (See the data in Exercise 19, based on the file **Australian_Wines.**)

a) Construct a frequency histogram, and a percent frequency histogram, for this data set.

b) Describe the shape of the distribution.

c) Construct a cumulative percent histogram for these data. Based on this figure, approximately what percentage of the wine prices are less than $20?

d) About what percentage of wines cost *at least* $20?

32. A mineral collector is examining the values for density and hardness among 100 metallic minerals. (See Exercise 20, based on the file **Metallic_Minerals.**)

a) Construct a frequency histogram and a relative frequency histogram for each of these two variables.

b) Describe the shape of the distributions. How do they differ?

c) Construct a cumulative percent histogram for the variable density. Approximately what percentage of the densities are less than 20?

d) About what percentage of the densities are *at least* 20?

33. An energy planner is comparing the oil reserves among the leading producer countries over a three-year period. (See the file **Oil_Reserves.**) The file records each country's reserves in billions of barrels at the end of three time periods—2000, 2003, and 2004.

a) For *each* of the three time periods, construct a relative frequency histogram and describe the shape of the distribution.

b) Compare the three distributions. Does the distribution of the countries' oil production appear to have changed significantly from 2000 to 2004?

c) If you randomly picked a barrel produced in 2004, which country or countries would it most likely have come from? Where would it least likely have been produced? Explain your reasons.

d) Construct a cumulative percent histogram for the 2004 data. Based on this data, approximately what percentage of the countries produced less than 30 billion barrels?

e) About what percentage of the countries produced *at least* 30 billion barrels?

Basic Concepts 3.3

34. Construct a bar chart (for frequencies) and a pie chart for each of the following tables:

a)

City	Frequency
Toronto	623
Montreal	592
Vancouver	433
Winnipeg	225
Regina	127
Total	2000

b)

Eye Colour	Frequency
Brown	52
Green	25
Blue	15
Grey	8
Total	100

c)

Mode of Transportation	Frequency
Air	12
Rail	6
Automobile	6
Bicycle	3
Total	27

35. Construct bar charts *for percentages* for each of the data sets shown in Exercise 34.

36. Construct a pie chart for the following data. The data represent the subject concentration for students in a business program.

accounting	marketing	human resources
accounting	accounting	marketing
human resources	accounting	marketing
human resources	accounting	accounting
accounting	marketing	accounting
marketing	human resources	marketing
accounting	marketing	marketing
accounting	accounting	marketing
human resources	accounting	human resources
accounting	marketing	marketing
marketing	accounting	accounting
marketing	marketing	

37. Construct a bar chart for the data in Exercise 36.

38. Based on the data in Exercise 36, construct a pie chart to compare the numbers of accounting and nonaccounting students in the program.

39. What type of chart would best represent the following data? Explain.

Type of Automobile	Frequency
Chevrolet	33
Honda	27
Toyota	15
Ford	14
Chrysler	12
Nissan	11
Total	112

Applications 3.3

40. The University of New Brunswick's *Earth Impact Database* provides data on the locations of meteor craters in North America. (See the data for Exercise 21, based on the file **Meteors_N_Amer.**)
    a) Construct a bar chart (for frequencies) for these data.
    b) Construct a bar chart (for percent frequencies) for the data.
    c) Construct a pie chart for the data.
    d) Which of the above three representations appears to communicate the data distribution most clearly? Explain.

41. A taxonomist has identified 106 species of pine tree. (See the data in Exercise 22, based on the file **Pine.**)
    a) Construct a bar chart for these data.
    b) Construct a pie chart for the data.
    c) Which of the above two representations appears to communicate the data distribution more clearly? Explain.

42. The distribution of winnings for the Tim Hortons "Roll Up the Rim to Win" promotion is given in Exercise 23.
    a) Construct a bar chart for these data.
    b) Construct a pie chart for the data.

43. An awards committee is reviewing hitting milestones reached by individual Major League Baseball players in 2004. For example, Jeff Bagwell attained the 25th-highest record for bases on balls. (See the file **Baseball_hitting.**)
    a) For *each* of the data columns "Team" and "Event," construct a bar chart.
    b) Are there any teams whose members account for a comparatively large percentage of the milestones? Are there events that have tended to occur most often? Would you expect these patterns to be repeated in future years? Why or why not?

44. An auditor is reviewing the locations of licensed casinos in Alberta. (See the data in Exercise 24, based on the file **Casinos_Alberta.**)
    a) Construct a bar chart for these data.
    b) Construct a pie chart for the data.

Basic Concepts 3.4

45. Create stem-and-leaf plots for each of the following data sets:

a) 23 24 45 55 23 47 34 24
21 33 44 56 25 46 43 32

b) 56 78 67 98 54 104 78 109
110 86 98 99 100 95

46. Create a stem-and-leaf plot for the following data set, listing the stems more than once:

22 33 44 35 26 22 48 26 37
25 45 41 21 39 31 32 46 48

47. Create a stem-and-leaf plot for the following data sets:

a) 112 125 132 119 134 125 116 117
122 123 138 131 114

b) 44 45 67 75 33 49 41
67 68 71 39 32 78 65
47 42 37 46 66

48. The following batting averages were obtained by members of the Toronto Blue Jays during the 2005 season:

0.269	0.291	0.256	0.262	0.262
0.271	0.251	0.301	0.269	0.274
0.249	0.216	0.250	0.207	0.194
0.308	0.083			

Create a stem-and-leaf plot for the batting averages.

Applications 3.4

49. The population density of a country is a measure of how many people live in the country relative to the surface area within its borders. (See the data in Exercise 17, based on the file **Population_Density.**) Population densities are given in persons per square kilometre.
    a) Construct a stemplot for this data set.
    b) Compare the stemplot with the frequency histogram that you created for the same data in Exercise 29. Which of these two displays seems best suited to this data set? Explain your answer.

50. Representative prices for buying a standard two-storey house in a selected urban location were displayed in Exercise 18. (See the file **Standard_2Storey_House_Prices.**)
    a) Construct a stemplot for this data set.
    b) Compare the stemplot with the frequency histogram that you created for the same data in Exercise 30. Name one advantage of the stemplot display for this data, compared to the histogram approach.

51. To entertain a guest, a mayor has downloaded a price list for Australian wines from the website of the PEI Liquor Control Commission. (See the data in Exercise 19, based on the file **Australian_Wines.**)
    a) Construct a stemplot for this data set.
    b) Compare the stemplot with the frequency histogram that you created for the same data in Exercise 31. Name one advantage of the histogram approach to displaying this data, compared to making a stemplot.

52. An economist is comparing the purchasing power of individuals in different countries. Her unit of measure is the per-capita GDP, in Canadian dollars, of each country. (See the file **Purchasing Power_World.**)
    a) Construct a stemplot for this data set.
    b) Compare the stemplot with a frequency histogram for these data, displayed in Figure 3.3 in this chapter. Suppose you wanted to know how many countries have a per-capita GDP between $32,000 and $33,000. Which of the two figures can be used to provide that answer? What is the answer?

Basic Concepts 3.5

53. Create a scatter diagram for the following data set:

Price	$9.65	$8.75	$9.25	$8.90	$8.00	$9.00
Quantity	50	60	55	58	40	65

54. Create a scatter diagram for the following data set:

X	15	25	34	19	16	22	30	35	28
Y	22	35	14	17	26	10	37	28	40

55. Create a contingency table from the following information: In a small village there are 75 men and 75 women. The village has 100 residents who state that they go to church regularly. Forty of the men go to church regularly.

56. Create a contingency table from the following information: In a statistics class 30 students have blue eyes, 50 students have brown eyes, and 10 students have green eyes. There are 40 males in the class. Twenty of the females have brown eyes and two males have green eyes.

57. An investor is tracking the monthly value of a stock. Display the following data as a time series:

Month:	1	2	3	4	5	6	7	8	9	10
$Value:	32	35	34	37	36	41	37	38	40	36

58. A chemist records the temperature of a certain process every hour. Display these data as a time series:

Time:	Temp. (°C)	Time:	Temp. (°C)
7 a.m.	5.1	11 a.m.	20.6
8 a.m.	6.3	Noon	39.9
9 a.m.	9.8	1 p.m.	43.7
10 a.m.	13.7	2 p.m.	43.6

59. A chipmunk made 10 trips to the patio of a cottage and each time returned with a handout from an agreeable child. These are the results of all the trips:

Trip Number:	Item Received:	How Many Received:
1	peanuts	2
2	apple	1
3	peanuts	2
4	peanuts	3
5	apple	0.5
6	apple	0.5
7	peanuts	1
8	cashews	9
9	peanuts	2
10	peanuts	2

Construct a bivariate bar chart to show how many of each type of item were received.

60. Following a citizen complaint, the police have been monitoring a certain location where a truck has been seen dumping refuse. So far, the police have documented seven cases of illegal dumping, as shown below. Construct a bivariate pie chart that compares how many of each type of object were dumped illegally.

Dumping Incident	Item Unloaded:	Quantity Unloaded:
1	mattress	1
2	garbage bag	3
3	mattress	2
4	garbage bag	2
5	garbage bag	3
6	furniture	1
7	furniture	2

Applications 3.5

61. As part of an international study on health, a researcher is comparing the life expectancies of newborns in different countries with the rates of infant mortality in those countries (measured in deaths per 1,000 births). (See the file **Infant_Mortality.**) Create a scatter diagram to compare those two variables. Do they appear to be related?

62. Complaints about airline service are being audited by a regulator. Numbers of complaints have been grouped based on airline company data in the file **Airline_Complaints.** The variable Problem refers to criticisms received due to cancellations, delays, or changes to the schedule. The variable Ov_Sale refers to complaints due to overbooking of the available seats. Create a scatter diagram to compare those two variables. Is there a clear relationship between an airline's number of overbooking complaints and its number of scheduling-related problems?

63. In the **Airline_Complaints** file, the variable Res_Tkt refers to complaints about mistakes in issuing or honouring tickets or boarding passes. The variable Refunds refers to perceived difficulties in receiving refunds. Compare those two variables by creating a scatter diagram. Is there any apparent relationship between the numbers of complaints due to these two causes? Why do you think that might be?

64. Governments in Canada continually reassess the levels of immigration. The numbers of people immigrating each year, over a 125-year period, are recorded in the file **Immigration_Time_Series.** Create a time series diagram to illustrate the pattern of immigration over this time. Did the level decrease, increase, or generally remain constant? In what year or years did exceptional numbers of people immigrate into Canada?

65. The *Chicken Data Handbook,* distributed by the Chicken Farmers of Canada, compares the inflation of chicken-related prices to the inflation of prices for other food items. As shown below, the price of a standard quantity of chicken or other item is set to 100 for 1992 (the base year). We can then see how prices vary before and after the base year. (See the file **Consumer_Price_Index.**) Create two time series diagrams—one each for the chicken prices and for the beef prices. Is there evidence of inflation (that is, do the prices tend to increase over time)? Does the inflation effect appear to be similar or different for these two categories of food?

Year	1983	1984	1985	1986	1987
Chicken	73.1	77.9	74.7	81.3	86.3
Beef	76.7	81.7	83.8	85.6	93.4

Year	1988	1989	1990	1991	1992
Chicken	87.4	97.5	102.6	100.4	100.0
Beef	94.8	96.9	100.6	101.3	100.0

Year	1993	1994	1995	1996	1997
Chicken	102.1	97.0	96.7	105.6	109.0
Beef	105.4	106.2	105.9	101.3	102.3

Year	1998	1999	2000	2001	2002	2003
Chicken	108.2	110.4	110.6	116.2	116.7	123.6
Beef	102.2	104.4	111.3	128.2	132.8	133.8

66. A senior student of engineering is especially interested in tunnel construction. He is hoping to find a job placement in the country that has the most impressive tunnels. During this search, he has found the list stored as the file **Notable_Tunnels.** Use this data to construct a contingency table. Base the rows on the list of countries (or country-pairs) and the columns on the list of tunnel usages. Then answer the following questions:

a) Which is the more common usage of notable tunnels—for railways or road vehicles?

b) What country has the largest share of notable tunnels?

67. A sociological study examined people's opinions on the question: *If* marijuana was decriminalized, what should be the minimum age permissible to possess this drug. The data in the file **Age_of_Marijuana_Possession** compare the ages of respondents and their respective opinions. Construct a contingency table based on this data. Then answer the following questions:

a) Overall, what is the minimum age most commonly suggested?

b) Judging by eye, are there any obvious differences of opinion based on the ages of the respondents? Explain.

68. A naturalist is studying the surface areas of Canadian lakes. A random selection of her data is stored in the file **Canadian_Lakes.**

a) Construct a bivariate bar chart that shows the *sum* of surface areas of the selected lakes, by province.

b) Construct a bivariate Pareto diagram to show the *sum* of surface areas of the selected lakes, by province.

c) Answer the following question and explain which of the above figures is most helpful for finding the answer: About 75% of the total surface area of the selected lakes is found in what two provinces and/or territories?

69. The Town of Wasaga Beach publishes information on all properties in the town. (See the file **Wasaga_Beach_Properties.**) A realtor is studying this list to get a sense of the market in the Wasaga Beach area.

a) Construct a bivariate bar chart that shows the *sum* of counts for properties for each of the different codes (e.g., Residential).

b) Construct a bivariate pie chart to show the *sum* of counts for properties for each of the different codes (e.g., Residential).

c) Answer the following question, and explain which of the above figures is most helpful for finding the answer: The total number of residential-code properties is approximately _____?

Review Applications

For Exercises 70–75, use data from the file **Brilliants.**

70. A dealer in fine diamonds is studying the data published by *Canada Diamonds* magazine to learn the factors that contribute to high or low prices for these jewels. A first step is to analyze some of the individual variables.

a) Construct a grouped frequency table for the variable Price. Add columns for (i) the percent frequency and (ii) the cumulative percent frequency.
b) Describe the distribution of the Price variable.
c) Construct a grouped frequency table for the variable Carat. Add columns for (i) the percent frequency and (ii) the cumulative percent frequency.
d) Describe the shape of the distribution of the Carat variable.

71. To supplement the findings from Exercise 70, the diamond dealer also creates the graphs listed below. For graphs (a) through (d), she uses the boundaries and class widths from the Exercise 70 answers.

a) For the variable Price, construct a histogram.
b) Construct a cumulative (less than) histogram for the variable Price.
c) For the variable Carat, construct a histogram.
d) Construct a cumulative (less than) histogram for the variable Carat.
e) Construct a stemplot for the variable Carat.

72. Next, the diamond dealer analyzes some qualitative variables.

a) Construct qualitative frequency tables for each of these three variables: (i) Colour, (ii) Clarity, and (iii) Certific (i.e., pertaining to certification).
b) For each of the above variables, identify the most common value and the least common value.
c) Construct qualitative *percent* frequency tables for each of these two variables: (i) Symmetry and (ii) Polish.
d) About what percentage of the time has a diamond's symmetry been judged Ideal? Excellent? VG (very good)?

73. Construct charts to supplement the tables that were created for Exercise 72.

a) For each of the variables Colour and Clarity, construct a bar chart.
b) For each of the variables Symmetry and Polish, construct a pie chart.

74. The dealer in fine diamonds now creates some charts to study comparisons between individual variables.

a) Construct a scatter diagram using the variables Price and Carat.
b) If a diamond has a high carat value, what if anything can you predict about the price?
c) If two diamonds have low carat values, can you predict confidently their difference in price based on the scatter diagram? Explain your answer.
d) Construct a contingency table, based on the variables Colour and Clarity.
e) What are the two most common Colour values for the variables? The two most common Clarity values? What two *combinations* of Colour and Clarity occur more often than any other combination?

75. Construct a bivariate bar chart to show, for each Colour value, the sum of all of the prices for sales of that colour. What two colour values account for most of the sales dollars?

For Exercises 76–81, use data from the file **Canadian_ Equity_ Funds.**

76. Following your decision to invest in Canadian equity mutual funds, your financial advisor sends you the **Canadian_Equity_Funds** data set to illustrate some alternatives. Your first step is to analyze some individual variables.

a) Construct a grouped frequency table for the variable Assets (fund assets in millions of Canadian dollars). Add columns for (i) the percent frequency and (ii) the cumulative percent frequency.
b) Describe the distribution of the Assets variable.
c) Construct a grouped frequency table for the variable NAVPS (net asset value per share as of the end of the previous month, in Canadian dollars). Add columns for (i) the percent frequency and (ii) the cumulative percent frequency.
d) Construct a grouped frequency table for the variable MER (management expense ratio—the percentage ratio between a fund's management fee plus expenses versus the fund's assets).

77. Create graphs to supplement your findings from Exercise 76. For (a) through d), use the boundaries and class widths from your answers to Exercise 76.

a) For the variable Assets, construct a histogram.
b) Construct a cumulative (less than) histogram for the variable Assets.
c) For the variable NAVPS, construct a histogram.
d) Construct a cumulative (less than) histogram for the variable MER.
e) Construct a stemplot for the variable MER.

78. Perform these analyses on the ordinal or qualitative variables in the **Canadian_Equity_Funds** data set.

a) Construct a qualitative frequency table for the variable Rating. Note that funds less than two years old are not rated.
b) For the variable LoadFees, construct a qualitative *percent* frequency table. (Note: "N" = no sales fee, "F" = front-end loaded, "D" = deferred load, "O" = optional, "R" = redemption fee, and "B" = both.)
c) Based on your answer for (b), what are the two most common fee approaches? The two least common?

79. Construct the following charts to supplement the tables generated for Exercise 78.
a) For the variable Rating, construct a bar chart.
b) For the variable LoadFees, construct a pie chart.

80. Create some charts to study comparisons between individual variables.
a) Construct a scatter diagram using the variables MER and Ret_3Yr.
b) Based on your answer to (a), would you be confident that a fund with a higher management expense ratio will provide a higher return over three years? Explain your answer.
c) Construct a contingency table based on the variables Rating and LoadFees.
d) What is the most common load fee arrangement for stocks rated at "3"?

81. Construct a bivariate bar chart to show, for each Rating value, the sum of all the Assets for funds having that rating. Does any one rating represent more than a third of the total assets?

For Exercises 82–86, use data from the file **Grace_Alumni_World.**

82. A globetrotting graduate of Grace University hopes to meet fellow alumni during her travels. She has found on the university website a list that shows the number of grads known to have moved to various countries.
a) Construct a grouped frequency table to show the distribution of alumni numbers in the various non-United States countries.
b) Add columns to table: (i) for the percent frequency and (ii) the cumulative percent frequency.
c) Describe the distribution of the Alumni variable.

83. Create the following graphs to supplement your findings from Exercise 82. Use boundaries and class widths from the previous table.
a) Construct a histogram.
b) Construct a cumulative (less than) histogram.

84. The data set also shows the continents to which the alumni have moved. Construct the charts described in (a) below and then use whichever chart is appropriate in order to answer (b) and (c).
a) Construct (i) a qualitative frequency table for the continent variable, (ii) a bivariate bar chart that shows the total number of alumni who have settled in each of the continents, and (iii) a qualitative *percent* frequency table for the continent variable.
b) Of the alumni who do not live in the United States, on what continent have most of them settled?
c) About what percentage of the alumni who do not live in the United States have settled in Africa? In Australia/Oceania? In Europe?

85. Construct a pie chart to illustrate your results for Exercise 84(c).

86. Construct a Pareto diagram that illustrates your results for Exercise 84(a).

For Exercises 87–92, use data from the file **National_Parks.**

87. Natural Resources Canada publishes a list of all of Canada's national parks. An ecologist is reviewing the list, with the goal of recommending expansions. A first step is to analyze some of the individual variables.
a) Construct a grouped frequency table for the variable Area. Add columns for (i) the percent frequency and (ii) the cumulative percent frequency.
b) Describe the distribution of data for that variable.

88. For the variable Area, construct a histogram, percent histogram, and cumulative percent histogram. Use the boundaries and class widths from your results for Exercise 87.

89. Construct a stemplot for the areas of Canada's national parks.

90. The ecologist would like to know how the national parks are distributed among the different provinces. Construct the charts shown in (a) below, and then use whichever chart is appropriate in order to answer (b) and (c).
a) Construct (i) a qualitative frequency table for the provinces, (ii) a bivariate Pareto diagram that shows the total area of the national parks within each province, and (iii) a qualitative *percent* frequency table for the province variable.
b) Out of the total number of parks, what percentage are located in Ontario? In Nunavut?
c) Of the total area of the national park system, what percentage is found in Ontario? In Nunavut?

91. Construct a scatter diagram using the variables Year and Area. Over the years, have new national parks been larger or smaller than parks created in earlier years? Or is there no clear pattern?

92. Construct a bivariate Pareto diagram to show, for each year, the sum of all of the areas for the parks created that year. In what five years did most of the growth in the Canadian national parks system occur?

Chapter 4

Measures of Location

• Chapter Outline

• Learning Objectives

4.1 Calculate and interpret the value of the arithmetic mean of a data set.

4.2 Calculate and interpret the values of the median and the mode (if any) of a data set, and for continuous quantitative data, distinguish between the mode and modal class.

4.3 Calculate a weighted mean, given appropriate weights or class frequencies.

4.4 Calculate and interpret the value of the geometric mean of a data set.

4.5 Explain what is meant by *quantiles*, and calculate values for particular quartiles and percentiles.

4.1 Arithmetic Mean

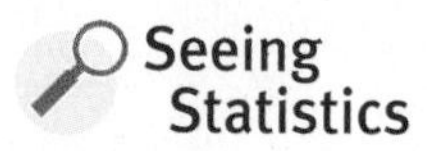

Section 3.1

Chapter 3 described methods for displaying the overall distribution pattern of a data set, but it sometimes is useful instead to summarize the same data set with one or two numerical values. If a friend asks you to compare the hitting skills of two batters, you probably cannot produce charts or tables on the spot. Even if you could, it might be difficult to compare all of the information found in two complex charts or tables. On the other hand, you could easily reply by comparing one number for each batter—in this case, their batting averages.

What is commonly called an *average* can more accurately be described as the ***arithmetic mean.*** This is the measure used in calculating, for example, the average student mark on a test, or the average wage rate in a contract settlement, or the average mercury level in Lake Erie's fish. The arithmetic mean is one indicator of the **location** of a data set; in particular, it is a **measure of centre.** Compared to a histogram for a data set, the arithmetic mean summarizes all of the data into one approximate, central location, which can be found on the horizontal axis. *If* the distribution is roughly symmetrical, then the arithmetic mean will be near the middle of the distribution.

Calculating the Arithmetic Mean

The arithmetic mean takes all of the data values into account. Put simply: Add all of the values in the data set and then divide by the number of values in the set. If you have ever calculated your 'average' quiz mark for a course, you have probably used this procedure. Using symbols for the procedure, we differentiate between the mean of a population (where you have values for every population member) and the mean of a *sample* of values in the population.

Formula 4.1 Arithmetic Mean for a Population

$$\mu = \frac{\sum_{i=1}^{N} x_i}{N}$$

where μ = Arithmetic mean for a population (pronounced 'mu')

x_i = Individual values of the population

[**Interpretation of the notation:** Suppose there are three values in the population data set. x_i would actually represent the following three values: x_1 (the first value in the data set); x_2 (the second value in the data set); x_3 (the third value in the data set).]

N = Number of values in the population (population size)

$\sum_{i=1}^{N}$ is a mathematical operator for summation (adding). Put simply, the notation

$\sum_{i=1}^{N} x_i$ means: Add all of the x_i-values in a data set or column, from the first value ($i = 1$) to the last ($i = N$).

[**Interpretation of the notation:** Suppose there are three values in the population data set. The notation calls for a procedure that is similar to a computer program: Identify all of the individual data values in the data set (for example, the values corresponding to x_1, x_2, and x_3). Add all of these values. If you are working with a spreadsheet, it can be helpful to think of the summation operator as calling for the *totalling* of a column of data.]

Formula 4.2 Arithmetic Mean for a Sample

$$\bar{x} = \frac{\sum_{i=1}^{n} x_i}{n}$$

where $\bar{x}$ = Arithmetic mean for a sample (The symbol is pronounced '*x* bar.' The *x* can be replaced with the name of any variable to be averaged.)

x_i = Individual values of the *sample*

[**Interpretation of the notation:** Same as in Formula 4.1.]

n = Number of values in the sample (sample size)

$\sum_{i=1}^{n} x_i$ means: Add all of the values (in this case, the *sample* values) in a data set or column from the first value ($i = 1$) to the last ($i = n$).

In Chapter 3, we constructed a frequency distribution histogram for the total votes cast in each British Columbia riding in the 2004 federal election (refer back to Figure 3.2). To visualize the relationship between the calculated mean and the entire frequency distribution, imagine the *x*-axis of Figure 3.2 as a beam or seesaw, and imagine the bars as having weights proportional to their lengths. As shown in Figure 4.1, the arithmetic mean would be located at the fulcrum (balance point) position—at which point, the two halves of the histogram are equally balanced. (This image is most accurate when the data points in every class are evenly distributed within their classes.)

Figure 4.1
The Arithmetic Mean as a Balance Point

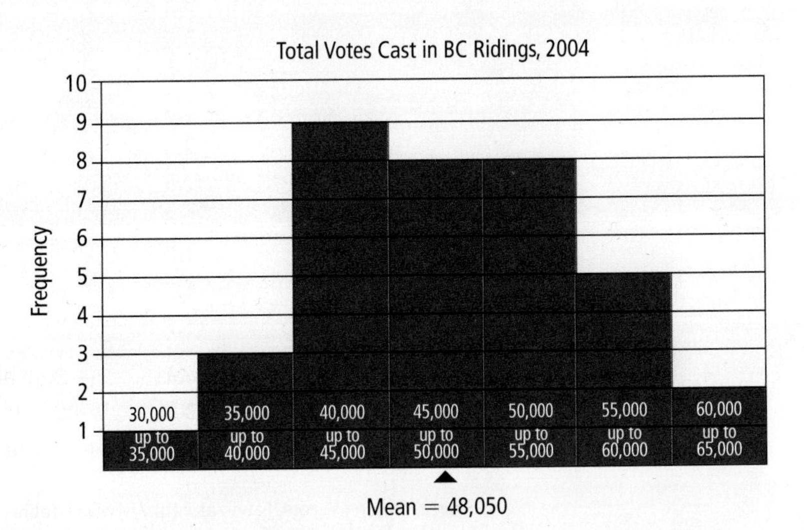

Source: "Federal Election Results, 2004." *The Globe and Mail.* Wednesday, June 30, 2004. Pages A12–A13. Reprinted with permission from *The Globe and Mail.*

The following example explains how to manually calculate the mean. The solution also illustrates how to interpret the summation (Σ) notation.

Example 4.1

 *(Based on the **Birds_Largest_Wingspan** file)*

A biologist has collected data on the wingspans of 13 birds. The file **Birds_Largest_Wingspan** displays the longest recorded wingspan for each of a selection of bird species.

1. Presume the data set is a sample of all species' longest wingspans on record; calculate the arithmetic mean.
2. Suppose that the 13 birds are actually caged in the biologist's personal zoo (these are the only birds in the collection). Calculate the arithmetic mean of wingspans for the birds in the zoo.

Solution

The formulas for solving the two problems look different, but the actual calculations are identical (see Figure 4.2). In both cases, add the 13 recorded lengths and divide by 13. The arithmetic mean is 2.65 metres. For problem (1), $\bar{x} = 2.65$, meaning that this is only a sample statistic. We can estimate—but not be sure—that the mean of longest wingspans taken from *all* bird species would be about the same value. For problem (2), $\mu = 2.65$, which signifies a specific measure of centre for this one fully represented population.

Figure 4.2
Calculation of the Mean Wingspans

	Wingspan (in m.)	
	3.6	
	3.6	
	3.3	
	3.2	
	3.0	
	3.0	
	2.8	
	2.5	
	2.5	
	1.8	
	1.7	
	1.7	
	1.7	
$\Sigma x_i =$	34.4	← Sum all the *x* values.
n *or* N =	13	← Divide by the number of values.
	2.65	← Calculated value of the mean.

Source: factophile.com. Found at: http://www.factophile.com/show.content?action=view&pageid=12 by Blackdog Media, 1081 Manawagonsih Rd, Saint John, NB E2M 3X5.

Round-Off Precision

In Example 4.1, a more precise answer (which computer software might yield) is 2.646153846. In the example's solution, this value was rounded to 2.65. Are there guidelines to use when rounding off such numbers? Different books and instructors may vary on the details, but the key guideline is: A computed answer cannot have more precision than the input numbers on which it is based.

Precision defines the smallest level of difference that can be measured. To distinguish between a wingspan of 2.646153846 m and 2.646153847 m (the next size larger) would require a precision level of 0.000000001! Yet the wingspan data were far less precise: They were rounded to the nearest 0.1 m, and the final answer should be on a similar scale. (The display of one extra decimal place for a calculated statistic, such as a mean, is generally allowed.)

Some books describe levels of precision in terms of displayed decimal places. This works well with calculators and computers, since you can set a preferred number of decimal places for display. Science texts often look at precision in terms of the number of **significant digits** displayed in the inputs and outputs. The significant digits in a number are all digits counting from the left-most nonzero digit to the right-most nonzero digit. All of the inputs for Example 4.1 have two digits, so the output should be comparable.

In any case, it is important to not round off numbers until you reach the very end of a calculation. Otherwise, in a chain of calculations, round-off errors will accumulate from one partial answer to the next. If possible, work with unrounded values in all of the intermediate calculations, although you may round these off for display purposes in a report or written solution.

Computer Calculations of the Mean

A computer can generate most of the measures for this chapter in a single operation. "In Brief 4.1" applies not only for the mean, but also for the median and mode, which are discussed later in this chapter.

• *In Brief 4.1 Computing the Mean, Median, and Mode*

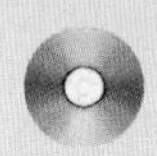

*(Based on the **Birds_Largest_Wingspan** file)*

Excel

The following instructions presuppose that your Excel program has the add-in program Analysis Toolpak installed. Otherwise, you may need to perform a one-time install operation from the Microsoft website.

Use the menu sequence *Tools / Data Analysis,* then select *Descriptive Statistics* and then click *OK.*

As shown in the screen capture, input the range for the data being analyzed and specify a cell for the upper left-hand cell address of the output. Select the *Summary Statistics* box and then click *OK.* The resulting output is shown in the illustration.

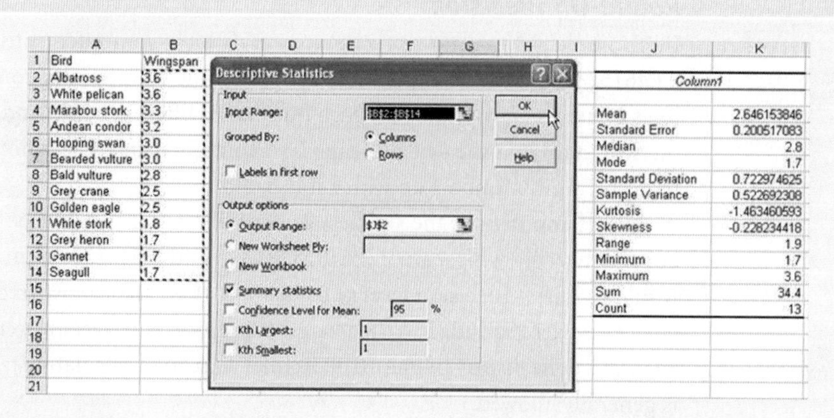

An alternative approach is to find a blank cell and input a formula of the form =**AVERAGE(range)**, replacing the word **range** with the Excel range containing the data. The mean of those data will be displayed.

Minitab

For flexible options, use the menu sequence *Stat / Basic Statistics / Display Descriptive Statistics.* In the first dialogue box (see the screen capture), select the variable to be analyzed. An option is to click on *Statistics* and then, as shown in the second dialogue box, select *Mean* and *Median,* or other desired statistics. (Several options are pre-selected by default.) Click *OK* in both dialogue boxes.

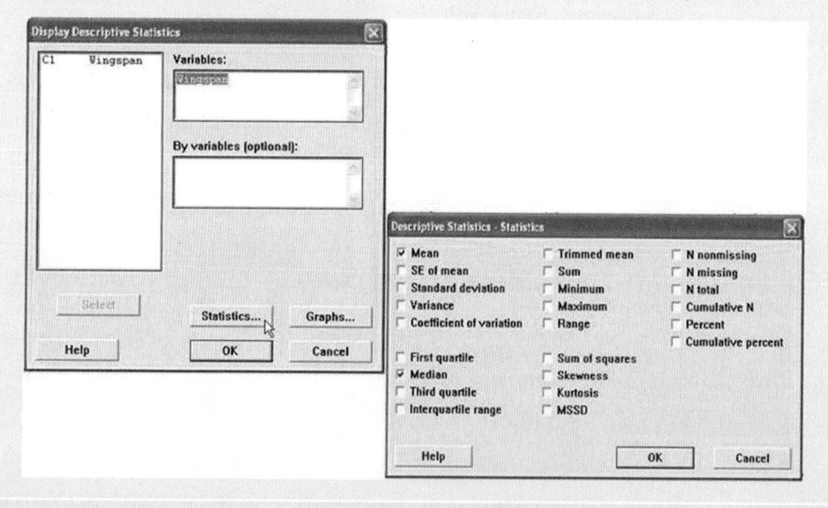

SPSS

For flexible options, use the menu sequence *Analyze / Descriptive Statistics / Frequencies.* In the first dialogue box (shown in the screen capture), select the variable to be analyzed (by highlighting its name in the left panel and clicking the arrow symbol near the centre); the variable name will shift to the centre panel of the dialogue box. It is now optional to click on *Statistics* and then, as shown in the second dialogue box, select *Mean, Median, Mode,* or other desired statistics. Click *Continue* to exit from the second dialogue box and then *OK* in the first dialogue box.

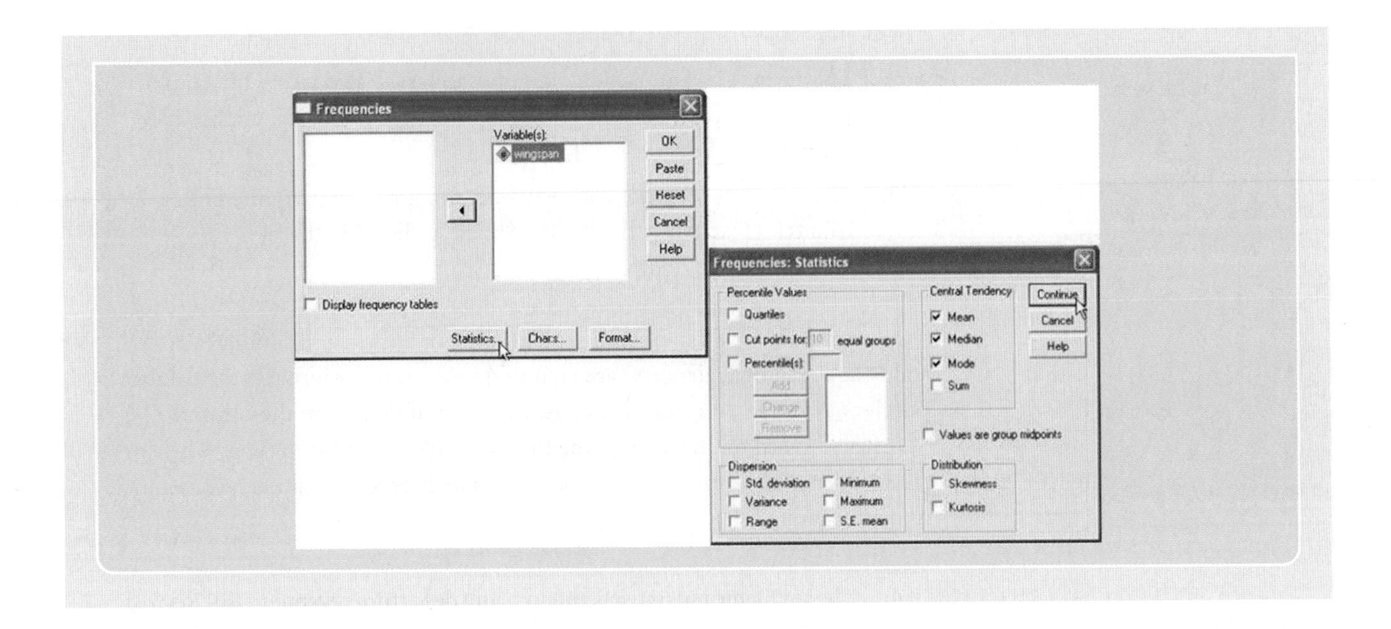

Example 4.2

*(Based on the **Birds_Largest_Wingspan** file)*

Redo Example 4.1 but use the computer to calculate the means.

Solution The computer-based solution for this example is identical to the manual solution for Example 4.1. Note, however, that the computer does not know (or care) whether the data set represents a sample or a full population. If you are using the output to write a report, be sure to write the correct symbols for the type of data source.

• Hands On 4.1 How Central Is the Mean?

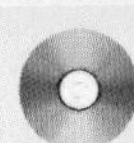

*(Based on the files **Animal_Weights, Diamonds, Grace_Alumni_World,** and **Carriers_revenue_tons**)*

As a one-number indicator of data location, the mean works best if there is a natural (such as genetic) or manufactured "target value" for a variable and if the variation of actual values around that target is more or less random. Mink bred for their fur are genetically programmed to be about a kilogram in weight. One kilogram, therefore, is close to the centre of the distribution for the size of adult specimens; larger and smaller individuals are roughly balanced around that centre.

Suppose you needed to know the centres of the six (*italicized*) variables in the following:

1. An animal lover has a collection of one each of the animals named in the file **Animal_Weights,** and each animal has the *weight* indicated (in grams).

2. The current inventory of a diamond dealer is listed in the file **Diamonds,** and each diamond has the *price* listed in the right-most column.
3. The numbers of Grace University *alumni* now living in different countries (excluding the United States), as listed in the file **Grace_Alumni_World.**

(Note: The data for the next three cases (4: major airports; 5: national airports; and 6: large regional airports) are all found in the same column in the data set **Carriers_revenue_tons.** Each case uses a subset of data from the column "All_RevS," based on the label in the first data column of the data set. The simplest way to analyze each subset separately is to copy and paste each subset into its own separate file.)

4. The total amounts of scheduled plane departure revenues (All_RevS) from major airports.
5. The total amounts of scheduled plane departure revenues (All_RevS) from national airports.
6. The total amounts of scheduled plane departure revenues (All_RevS) from large regional airports.

For each of these variables:

a) Calculate the arithmetic mean.
b) Using the techniques discussed in Chapter 3, assess the distribution shapes of these variables. (For example, construct a histogram. Is each distribution symmetrical or skewed?)
c) Decide whether just telling someone the mean in each case would successfully communicate the "centre" of the data. Who might be interested in knowing the centre, in practice, and what might be the consequences if they were unaware of the actual distribution? Short of giving someone a full table or graph, is there other information you could provide to help convey a clearer picture?

4.2 The Median and Mode

Sometimes, the arithmetic mean is not the most suitable measure for the centre of a data set. Suppose a young chemistry professor is offered university positions in a number of Canadian cities. One consideration is the price of housing, so she consults a Royal LePage survey of prices for standard two-storey houses in those cities. The results, based on the file **Standard-2Storey_House_Prices,** are as shown in Figure 4.3.

Because the arithmetic mean is influenced by every data value—including outliers—Figure 4.3 shows that the mean has been pulled up (by two houses costing more than a million dollars) to \$366,935. If the professor cannot afford this amount, learning the arithmetic mean may be very discouraging. But look again at the figure. Is the arithmetic

Figure 4.3
Distribution of Prices for Standard Two-Storey Houses in Various Cities

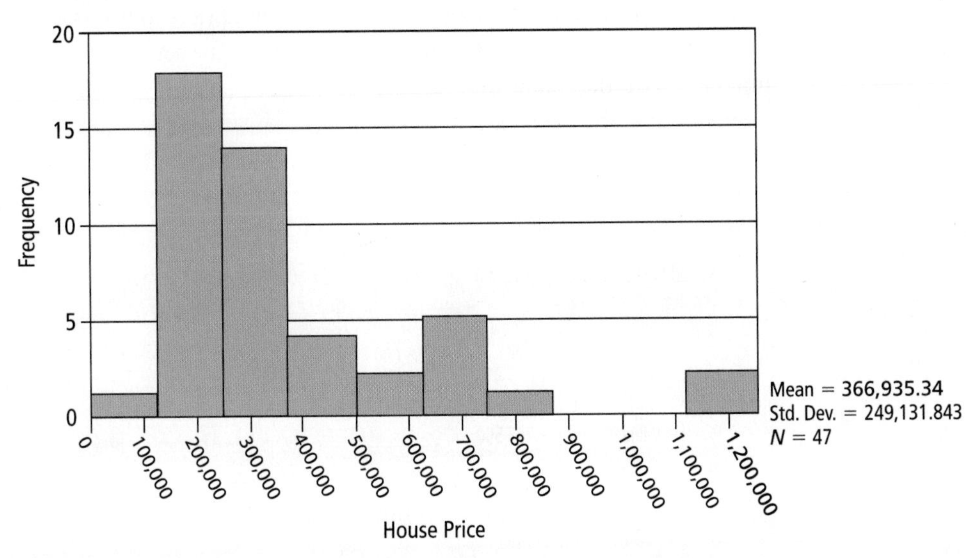

Source: Royal LePage Realty. Found at: http://www.royallepage.ca/schp/pdf/Q1_Survey_2005_Eng.pdf

mean really representative of what she probably will have to pay? *Most* of the house prices cited are, in fact, *below* that figure.

In cases like this, an alternative measure of location called the **median** can be used. If the data are arrayed from the lowest value to the highest, the median identifies the middle value in the array; or more precisely, a value such that half of the raw data have values less than or equal to the median, and half of the raw data have values greater than or equal to it. As shown in Figure 4.4, the middle value in the ordered array of house prices, which is the median, is somewhat more affordable: $282,000. This value is less influenced by the extreme values and gives a better picture of what a typical buyer might pay.

Figure 4.4
Sorted House Prices and the Median

Sorted Prices	120,000	133,000	133,561	145,000	167,250	185,000	189,000	192,750	196,250	210,000	214,000	220,400
Position in Sorted Array	1	2	3	4	5	6	7	8	9	10	11	12
		225,800	229,000	229,800	230,000	233,000	233,100	236,000	254,000	255,000	260,000	280,000
		13	14	15	16	17	18	19	20	21	22	23
Median Value:	282,000											
Middle Position:	24											
	285,000	285,000	286,000	295,000	300,000	300,600	325,000	325,000	355,000	389,000	435,000	440,000
	25	26	27	28	29	30	31	32	33	34	35	36
		480,000	530,000	540,000	635,000	650,000	651,450	700,000	745,000	840,000	1,150,000	1,250,000
		37	38	39	40	41	42	43	44	45	46	47

Source: Royal LePage Realty. Found at: http://www.royallepage.ca/schp/pdf/Q1_Survey_2005_Eng.pdf

If there is an even number of values in a data set and if you order the data from the lowest to highest value, then there is no single number in the middle. If the professor accepts the position in Montreal, there are 10 house prices to use as a guide (see Figure 4.5). If you rank-order the 10 values, the middle of the set is *between* the 5th and 6th numbers. Use the value midway between them (i.e., use (lower number + upper number)/2) as the median.

Figure 4.5
Median for House Prices in Montreal

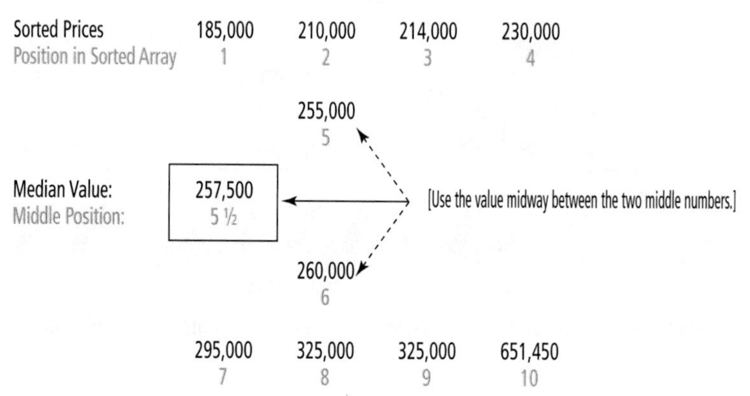

Source: Royal LePage Realty. Found at: http://www.royallepage.ca/schp/pdf/Q1_Survey_2005_Eng.pdf

For data at the ordinal level, the assumptions for calculating a mean are not met; so again the median is a preferred measure of centre. Suppose 10 students each rank the quality of a textbook on the ordinal scale: "unacceptable" (1); "poor" (2); "moderately good" (3); and "excellent" (4). There is no clearly defined, standard distance between these rankings. We know only that a "4" is better than a "3," and a "3" is better than a "2," and so on. If the students' rankings are as shown below, what is the measure of centre?

3 4 2 3 3 2 4 4 3 4

Sorting the values, we can easily see that the median (midway between the two innermost values) is 3. Students' rankings were centred around a rating of "moderately good." The arithmetic mean, which would be 3.2, is harder to interpret.

2 2 3 3 3 ^ 3 4 4 4 4

• *In Brief 4.2 Computing the Median (continued)*

For computing the median, see the procedures for your software shown in "In Brief 4.1." Excel and SPSS can also calculate a mode for the same data.

Example 4.3

 *(Based on the **Voter_Turnouts** file)*

A returning officer for the next election needs to plan the size and location of polling stations in his riding. Although he knows the number of eligible voters in the riding, he also knows that usually not everyone votes, so maybe a smaller space will be sufficient. Consulting data from Elections Canada (see the file **Voter_Turnouts**), he finds the percentages of eligible voters who voted in every federal election from 1900 to 2004. Calculate the median *and* the arithmetic mean for these data. How do they compare? Which would you rely on more for planning the sizes of polling stations?

Solution

Either manually or using computer software, you can find the median (75.0% turnout) and the arithmetic mean (74.2%, if displayed to one decimal place) (see Figure 4.6). These values are very close, which suggests that, over time, voter turnouts have been distributed rather evenly around these approximate values. That is, the mean has not been pulled away from the median value one way or the other due to extremely small or large cases. Given the variability of exact voter turnouts from election to election, there is little practical difference between using either of the two measures in this case.

Figure 4.6
Measures of Central Location for Percentage Voter Turnout

Worksheet 1

↓	C1	C2	C3	C4
	Year	Turnout		
1	1900	79		
2	1904	84		
3	1908	79		
4	1911	72		
5	1917	90		
6	1921	71		
7	1925	69		
8	1926	70		
9	1930	76		

Session

```
Worksheet size 10,000 cells

Welcome to Minitab, press F1 for help.
MTB> Describe 'Turnout';
SUBC> Mean;
SUBC> Median.
```

Descriptive Statistics: Turnout

```
Variable  Mean   Median
Turnout   74.20  75.00

MTB>
```

Source: Nodice Canada. Based on data in *Voter Turnout,* which was based on Elections Canada data. Found at: http://www.nodice.ca/election2004/voterturnout.html

Seeing Statistics

Sections 3.2–3.6

The difference between the mean and median becomes proportionally larger if (1) the data distribution is not very symmetrical and (2) the sample size is relatively small. To get a feel for how these factors interact, try the exercises in "Hands On 4.2" on page 139.

The Mode

The only measure of centre that can be used with all levels of data measurement is the **mode.** The mode is the particular value that occurs most frequently in the data set. A data set may have more than one mode. When there are two modes, the distribution is said to be *bimodal.* A distribution with more than two modes is called *multimodal.*

For nominal data, the mode is easiest to identify if you first create a frequency distribution table. The category with the largest frequency has, by definition, occurred most often in the data. Based on two surveys, Table 4.1 shows the modes for the most preferred hockey teams among beer drinkers (based on data in the file **Favorite_Hockey_Team**).

Table 4.1 **Preferred Hockey Teams among Beer Drinkers (n = 500, each year)**

Preferred Team	Frequency (2004 Survey)	Frequency (2003 Survey)
Calgary Flames	80	20
Edmonton Oilers	40	35
Montreal Canadiens	**135**	145
No preference	40	15
Ottawa Senators	20	50
Toronto Maple Leafs	130	**150**
Unsure	20	5
Vancouver Canucks	40	65

Source: SES Research Inc. SES Research Poll. Found at: http://www.sesresearch.com/news/press_releases/Moosehead%20PR.pdf

The formally defined *mode* is often not very meaningful if continuous quantitative data are being used. Depending on the round-off precision, specific data values may rarely be identical—and if there is a match in values, it may be just a coincidence. A more useful concept for continuous data is the **modal class.** If the data are grouped into a frequency distribution, the class having the largest frequency is the modal class. If there is a second large-frequency class elsewhere in the distribution, then the frequency distribution can be called *bimodal.*

Example 4.4

 *(Based on the **Meteors_N_Amer** file)*

A geologist is studying the diameters of meteor craters in North America. She compiles the data saved in the file **Meteors_N_Amer.** Based on that data, identify the modal class. Explain why there can be more than one right answer for this question.

Solution

Start by constructing a frequency distribution table or histogram (see Figure 4.7). We see that the class with the highest frequency is "0 km up to 10 km," so this is the modal class. Unless due to round-off, it is unlikely that any two or more craters are *exactly* the same diameter, as required for having a unique mode value.

Figure 4.7
Mode for Diameters of North American Meteor Craters

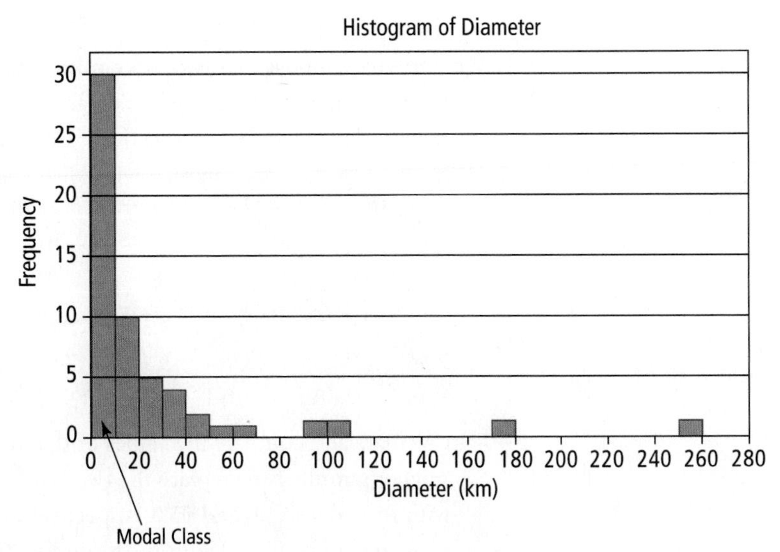

Source: Earth Impact Database, 2006. Found at: http://www.unb.ca/passc/ImpactDatabase/NAmerica.html. Accessed 8 April 2005.

Recall that using the guidelines given in Chapter 3, there can be more than one right way to define classes for a frequency distribution. Modal classes are determined with reference to a class's frequencies. Therefore, changing the class boundaries will affect the answer for modal class, as well.

• Hands On 4.2 *Sensitivities of Means and Medians to Outliers*

*(Based on the **Canadian_Equity_Funds** file)*

We have learned that outliers are values that lie beyond the typical range of values for a variable. Such outliers can be detected in the data by studying figures such as histograms and looking for values far off to one side or the other. (Boxplots, to be introduced in the next chapter, can be especially helpful as well.)

If outliers are present, we must decide whether to calculate measures of centre with the outliers included or with the outliers excluded (on the premise that the values are special cases). What difference does it make? The exercises in this section are intended to give you a feel for the impact of including or excluding outliers, when calculating the mean or median.

Suppose you are analyzing the management expense ratios (MERs) of a variety of mutual funds. MERs indicate the percentage ratio between a fund's management fee plus expenses versus the fund's assets. MERs are listed for a sample of 60 funds in the file **Canadian_Equity_Funds.** Visually inspect the distribution of MERs to assess if there may be outliers. Then calculate and compare the values of the mean and median for the MERs.

Now suppose that two more funds are added to the funds list, both headed by "big-name" managers, who demand large fees. The two extra funds have MERs of 11.0 and 15.0, respectively. Calculate and compare the mean and median for the MERs in the revised list. Did the two outliers have a large impact?

Suppose now that you take a smaller sample of funds, and obtain the following MERs.

2.80	0.97	2.62	3.69	2.71	2.43	2.47	2.54	3.06
2.56	2.74	2.96	3.11	2.51	3.15	1.25	3.27	3.15
3.65	2.68	2.14	1.75	3.65	3.61	1.62		

Calculate and compare the values of the mean and median for the MERs for this smaller sample. Now replace the two last values in the sample (3.61 and 1.62) with the values 11.0 and 15.0, respectively. (That is, presume that these outliers happened to be included in your sample.) Calculate and compare the mean and median for the MERs in the revised sample. Did the two outliers have a large impact? Data values that have a large impact on a summary statistic are called *influencing* data points. How did the impact compare with adding the outliers to the original, larger sample?

4.3 Weighted Mean

If you are reading this textbook as part of a statistics course, then you were probably given a weighting scheme to calculate your final mark. Perhaps your final exam is worth 40% of the mark; a midterm, 30%; a test, 20%; and a written assignment, 10%. This means that once you know all your test and assignment marks, you cannot just take their mean as was done earlier in this chapter. To find an arithmetic mean of a set of data where the values are assigned unequal importance, calculate the **weighted arithmetic mean.**

Formula 4.3 Weighted Arithmetic Mean

3a) For a Population

$$\mu = \frac{\sum_{i=1}^{N} x_i w_i}{\sum_{i=1}^{N} w_i}$$

3b) For a Sample

$$\overline{x} = \frac{\sum_{i=1}^{n} x_i w_i}{\sum_{i=1}^{n} w_i}$$

where μ = Arithmetic mean for a population ***or*** $\overline{x}$ = Arithmetic mean for a sample

x_i = Individual values of the population or sample

w_i = Corresponding weights assigned to the data values

N = Number of values in the population (population size) ***or*** n = Number of values in the sample (sample size)

$\sum_{i=1}^{N}$ *or* $\sum_{i=1}^{n}$ is the mathematical operator for summation (adding).

The interpretation of the formula is easiest to explain with reference to a spreadsheet, especially if you are using statistical software. A student's full set of marks (so, a population of marks) for a statistics course is listed in the second column of Figure 4.8. Each mark is to be weighted as shown to its right (weights are in percentages). The summation operator in the numerator of the formula is calling for a sum—shown in the figure as the sum of an appended column. Note that the heading of the new column corresponds to the expression to the right of the Σ operator ($x_i\ w_i$), (multiplication is the implied operator for the $x_i\ w_i$ pairs). Perform the (x_i times w_i) operation for each row (for row $i = 1$ to row $i = N$, the last row of data). Then sum.

Figure 4.8

Application of Formula for Weighted Arithmetic Mean of Population

	x_i	w_i	$x_i \times w_i$
Course Component	Mark (%)	Weight of Component	
Final Exam	84	40	3360
Midterm Exam	76	30	2280
Test	68	20	1360
Assignment	93	10	930
		100	7930

$$\mu = \frac{\sum_{i=1}^{N} x_i w_i}{\sum_{i=1}^{N} w_i} = \boxed{79.3}$$

For the summation operator in the denominator of the formula, there is already a column corresponding to the expression to be summed (w_i); simply find the sum of the column of weights. Divide the sum of products (x_i times w_i) by the sum of weights.

The same procedures would be followed if the data was only a sample. Suppose a course had several exams, tests, and assignments, and you sampled only some of each. In this case, the included individual values would comprise a *sample* of size *n*, rather than a population of size *N*. The calculations based on Formula 4.3(b) would, in practice, have been the same as shown in Figure 4.8.

Another application of weighted means is to combine the means from different strata when nonproportional stratified sampling is used, as discussed in Chapter 2. Suppose, for example, you obtain the mean ages of Canadians from each province, based on equal-sized samples from each province (strata). To find a national mean age, it would be an error to give equal weight to samples taken from all provinces while ignoring their population sizes. Instead, the procedure is to take a weighted average—giving the largest weight to the strata means that represent the largest number of individuals in the population.

Special Application of the Weighted Mean: Mean for Grouped Data

When data have been grouped into a frequency table, it is possible to *estimate* the arithmetic mean of the raw data, which may not be readily available. For example, you may be viewing the frequency table in a published report. Each class midpoint can be interpreted as an x-value, which approximates all of the raw data values falling in that range. Each midpoint is then *weighted* by how often that (approximate) x-value occurred—i.e., the frequencies can be used as the weights.

Sometimes data are already in a grouped format when they are collected. For example, a survey may ask people to indicate their ages, not as specific numbers, but by checking from a set of choices, such as "10 to 19," "20 to 29," "30 to 39," etc. In this case, we could compile a frequency table, but we would not have access to more precise data about individuals' ages.

Recall that the midpoint of each class is calculated as the sum of the class's lower and upper limits divided by 2. (For continuous data, the upper class limit $\approx$ the next class's lower limit). Treating the midpoints as the x-values and the frequencies as the w-values, use the manual procedures illustrated in Figure 4.8 or the computer procedures described in "In Brief 4.3."

• *In Brief 4.3 Computing the Weighted Mean*

(Based on the data shown in the box in Figure 4.8)

Excel

Although Excel does not have a function explicitly for the weighted mean, its other functions can be used to easily automate the calculation. As shown in the table at right, the data can be arranged with the x- and w-values aligned in columns (they do not need to be next to each other). The formula shows the use of the =**SUMPRODUCT( )** function, which performs the Σxw operation for the specified columns and divides this by the sum (=**SUM( ) function**) of the weights column.

Course Component	Mark (%)	Weight of Component
Final exam	84	40
Midterm exam	76	30
Test	68	20
Assignment	93	10

Weighted Arithmetic Mean
= SUMPRODUCT(address for marks column, address for weights column)/SUM(address for weights column)
= 79.3

Minitab

Although Minitab does not have a function explicitly for the weighted mean, its *Calculator* function can be used to easily automate the calculation. As shown in the illustration, the data can be arranged with the x- and w-values aligned in columns (they do not need to be next to each other). Open the calculator with the menu sequence *Calc / Calculator.* In the resulting dialogue box, first input in the upper right-hand box a name for a new column to contain the solution; in this case, the column's name is "WeightedAve." In the *Expression* area of the dialogue box, input an expression of the format **Sum(Mark*Weight)/Sum(Weight),** where **Mark** and **Weight** are your names for the x and w columns, respectively. Then click *OK* for the solution. Note that the Minitab calculator "thinks" in terms of column operations—so **Mark*Weight** automatically means to multiply each row's values for the "Mark" and "Weight" columns. **Sum(Mark*Weight)** adds all of those products, and **Sum(Weight)** adds all values in the weights column.

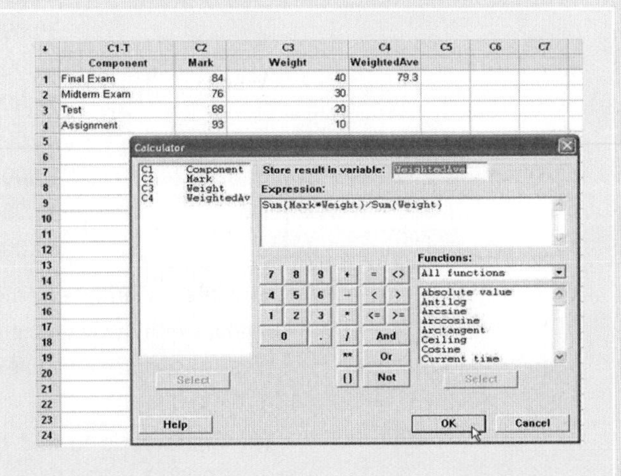

SPSS

SPSS allows you to declare a variable as representing weights and then perform a variety of possible analyses—including calculation of the weighted mean. First use the menu sequence *Data / Weight Cases.* In the dialogue box shown in the screen capture, click on *Select* and choose the name of the variable representing weights; then click *OK.* Then follow a procedure normally used for calculating the mean of the x-variable (such as *Analyze / Descriptive Statistics / Frequencies,* as detailed in "In Brief 4.1"). As shown on the right-hand side of the illustration, the weighted mean is calculated. Also shown is a table, which is applicable if the weights are interpreted as frequencies, as in Example 4.5.

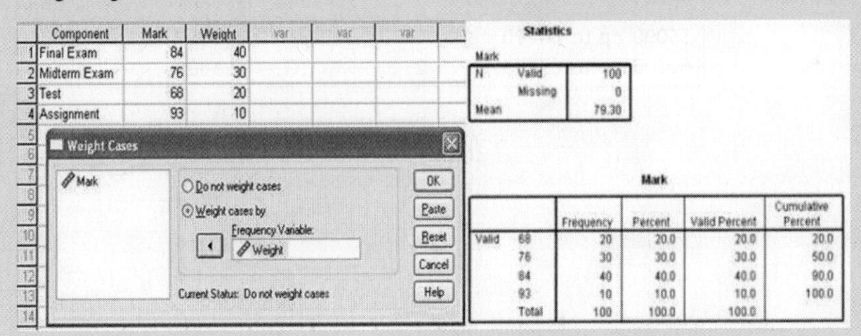

Two particular features in SPSS's approach should be noted: (1) Unless you turn off SPSS or rerun *Data / Weight Cases* and select *Do not weight cases,* the program will continue to weight *all* subsequent calculations by the variable you indicated. (2) The values in the "Weight" column must be whole numbers. If three weights are 0.333, 0.222, and 0.445, for example, you might multiply all weights by 1,000 to convert the weights to whole numbers 333, 222, and 445.

Example 4.5

*(Based on the **BC_votes** file)*

The total votes cast in British Columbia ridings in the 2004 federal election can be summarized as follows.

Classes	Frequency
30,000 up to 35,000	1
35,000 up to 40,000	3
40,000 up to 45,000	9
45,000 up to 50,000	8
50,000 up to 55,000	8
55,000 up to 60,000	5
60,000 up to 65,000	2

Estimate the arithmetic mean for the number of votes cast in BC ridings. Compare this number with the actual mean, based on the raw data in the file **BC_votes.** How close are these answers? Can you explain any discrepancy?

Solution

The first step is to find the midpoints for each class. The first midpoint is (30,000 + 34,999)/2 ≈ 32,500, and so on. Interpreting the column of midpoints as the *x*-values and the frequencies as the weights, the mean can be estimated by Formula 4.3: Mean ≅ 48,333 (see Figure 4.9). Since we happen to have access to the raw data, we can also calculate the actual mean directly, for comparison: Actual mean = 48,050. The actual mean is a bit less than the estimate.

Figure 4.9
Arithmetic Mean for Numbers of Voters in BC Ridings: Grouped Data

Classes	$f = w_i$ Frequency	x_i Class Midpoints	$x_i \times w_i$
30000 up to 35000	1	32500	32500
35000 up to 40000	3	37500	112500
40000 up to 45000	9	42500	382500
45000 up to 50000	8	47500	380000
50000 up to 55000	8	52500	420000
55000 up to 60000	5	57500	287500
60000 up to 65000	2	62500	125000

36 1740000

$$\mu = \frac{\sum_{i=1}^{N} x_i w_i}{\sum_{i=1}^{N} w_i} = \boxed{48333}$$

Raw Data: Mean = 48050
Compare the Stem and Leaf Diagram:

Frequency	Stem & Leaf
1.00	3 . 4
3.00	3 . 679
9.00	4 . 001111224
8.00	4 . 55667888
8.00	5 . 01122334
5.00	5 . 56688
2.00	6 . 03
Stem width:	10000.00
Each leaf:	1 case(s)

What might explain the small difference between the weighted-mean estimate and the true mean of the full data set? Taking advantage of the computer, we can explore a possible explanation by generating a stem-and leaf-diagram, as discussed in Chapter 3. When we used the class midpoints to estimate the x_i-values, our assumption was that the data in each class are spread evenly across their respective ranges—but that is not exactly the case for this data. For example, in the third class, the leaves show us that 8 of the 9 raw data values are closer to the low end of the range (near 41–42 thousand) than to the high end of the range (near 43–44 thousand). This effect could make the actual mean slightly different from the estimate.

• Hands On 4.3 *Estimating the Mean from Published Tables*

Because of the Internet and online libraries, the amount of information publicly available today is remarkable. Yet, finding raw uninterpreted data as originally output by primary research and experimentation can still be difficult and time-consuming.

Suppose you would like to calculate the mean income for all Canadians earning *under $75,000* who were employed in the year 2000. Eventually, you might track down the Statistics Canada website page simulated in Figure 4.10. Statistics Canada has already grouped the data into a frequency table, but note that you require only some of the rows and columns in the table. (Note also the department's choice of inconsistent class widths.) The public site does not offer access to the raw data, but you can estimate the mean by weighting the midpoints for the salary ranges under $75,000 (which are highlighted in the figure) by the corresponding frequencies for 2000 (also highlighted in the figure).

Figure 4.10
Incomes of Employed Canadians Earning Less Than $75,000

Title	Years (2)	
	1995	2000
Total—Employment income groups	22,628,925	23,901,360
Without employment income	7,632,810	7,485,580
With employment income	14,996,115	16,415,780
Under $2,000	1,244,705	1,142,860
$2,000-$4,999	1,202,615	1,171,875
$5,000-$6,999	755,920	727,515
$7,000-$9,999	904,410	895,665
$10,000-$11,999	643,790	622,895
$12,000-$14,999	708,945	808,745
$15,000-$19,999	1,261,505	1,289,845
$20,000-$24,999	1,123,460	1,338,805
$25,000-$29,999	1,201,510	1,248,255
$30,000-$34,999	1,242,275	1,313,260
$35,000-$39,999	844,790	1,024,760
$40,000-$44,999	848,755	974,055
$45,000-$49,999	589,620	688,575
$50,000-$59,999	1,020,430	1,140,765
$60,000-$74,999	722,415	1,028,485
$75,000 and over	680,965	999,430
Average employment income $	28,838	31,757
Median employment income $	22,901	25,052
Standard error of average employment income $	17	23

Source: Statistics Canada. Employment Income Groups (22) in Constant (2000) Dollars and Sex (3) for Population 15 Years and Over, for Canada, Provinces, Territories, Census Metropolitan Areas and Census Agglomerations, 1995 and 2000–20% Sample Data. Found at: http://www12.statcan.ca/english/census01/products/standard/themes/RetrieveProductTable.cfm?Temporal=2001&PID=56050&APATH=3&GID=431515&METH=1&PTYPE=55440&THEME=53&FOCUS=0&AID=0&PLACENAME=0&PROVINCE=0&SEARCH=0&GC=0&GK=0&VID=0&FL=0&RL=0&FREE=0. Accessed 10 August 2007.

But what if you wanted to estimate the mean for *all* of the data, including the highest salary range? Figure 4.10 is not unusual in showing an open-ended highest class ("$75,000 and over"). What's the midpoint of such a class? Depending on your need for accuracy in your estimate, you might try to make a "best guess" for the midpoint for the open-ended class, perhaps finding out the upper boundary from another source. For an open-ended upper class with a small frequency (*not* the case here), you might view it as containing only outliers, and exclude it in an estimate of the "usual" mean.

Use the Internet or the library to find an additional nine distinct published frequency distribution tables for which the raw data are not made available. How many of these include an open-ended upper and/or lower class? Estimate the mean for all of the nine tables that you find, plus the one in Figure 4.10. How have you handled any open-ended classes? How would you justify your method to a supervisor or peer reviewer? Can you also use *relative* frequency distribution tables to estimate the mean? Why or why not? Can you calculate the mean for discrete quantitative data using frequency distribution tables?

4.4 Geometric Mean

For a remarkable 24-year period, from 1950 to 1974, world annual production of bicycles increased steadily, without apparent reversals, from 11 million to 52 million units per year (see Table 4.2). For production during that period, what was the average annual rate of increase (more formally, the annual **growth factor**)? What was the average annual **rate of change**? (The rate of change is like an interest rate. It tells by what proportion of an initial value the *next* value increases [or decreases].)

For questions of this sort, we cannot simply calculate an arithmetic mean for the annual growth factors or rates of change, because values build on each other from year to year. The growth factor over a period can be defined by:

Growth factor = (New value/Old value)

Therefore, solving for the new value:

New value = Growth factor × Old value

From 1951 to 1952, for example, production increased by 1 million units, from 11 to 12 million. The growth factor = 12/11 = 1.09. From 1952 to 1953, the absolute amount of change was the same as for the previous year: A million more bikes were produced. But *due to the previous year's growth, the growth factor changed* to 13/12 = 1.08. Because growth factors accumulate their effects in this way, the **geometric mean** must be used for obtaining an appropriate measure of central location. Other data that can require this type of calculation are compound interest examples, as well as data on the rates of growth or decline in species' populations.

Table 4.2 **Annual Growth and Rates of Change in World Bicycle Production, 1950–1974 (Bike production shown in millions of units)**

Year	Bike Production	Growth Factor (from the previous year)	Rate of Change (from the previous year)
1950	11		
1951	11	1.00	0.00
1952	12	1.09	0.09
1953	13	1.08	0.08
1954	14	1.08	0.08
1955	15	1.07	0.07
1956	16	1.07	0.07
1957	17	1.06	0.06
1958	18	1.06	0.06
1959	19	1.06	0.06
1960	20	1.05	0.05
1961	20	1.00	0.00
1962	20	1.00	0.00
1963	20	1.00	0.00
1964	21	1.05	0.05
1965	21	1.00	0.00
1966	22	1.05	0.05
1967	23	1.05	0.05
1968	24	1.04	0.04
1969	25	1.04	0.04
1970	36	1.44	0.44
1971	39	1.08	0.08
1972	46	1.18	0.18
1973	52	1.13	0.13
1974	52	1.00	0.00

Source: "Figure: World Automobile Production" and "Figure: World Bicycle Production," from *Vital Signs 2002: The Trends that Are Shaping Our Future* by Lester R. Brown, et al. Copyright © 2002 by Worldwatch Institute. Used by permission of W.W. Norton & Company, Inc. Found at: http://www.ibike.org/library/statistics-data.htm

Formula 4.4 The Geometric Mean

$$GM = \sqrt[n]{x_1 \cdot x_2 \cdot x_3 \cdot \ldots \cdot x_n}$$

where GM = Geometric mean

n = Number of observations in the series, such as the number of years over which the growth rate is being assessed

x_i = Growth factor for a particular period in the series (e.g., x_2 = Growth factor for the second period)

For the data in Table 4.2, $n = 24$. (Although 25 years are listed, growth could not be measured until a year was completed.) To manually calculate the geometric mean of the growth factors, you would multiply all 24 of the growth factors listed in the third column and then take the 24th root of the product. See Example 4.8 on page 150 for the solution.

Example 4.6 Calculate and interpret the geometric mean for the first five growth factors listed in Table 4.2. What is the average rate of change for that same time period?

Solution The formula below says that bicycle production increased over those five years with an average annual growth factor of 1.06. Starting from the initial production level: If it was increased times 1.06 for the first period and if the new level was increased times 1.06 to reach the next year's level and so on for five periods, this would result in the net increase of production that actually occurred.

$$GM = \sqrt[5]{1.00 \cdot 1.09 \cdot 1.08 \cdot 1.08 \cdot 1.07} = \sqrt[5]{1.36} = 1.06$$

The rate of change looks at an increase or decrease in terms of its percentage of the starting value. For any given period:

$$\text{Rate of change} = \text{Growth factor} - 1$$

Therefore, the average rate of change for the five years in the example $= 1.06 - 1 = 0.06$.

For financial problems involving rates of increase, the calculations are essentially the same, but the terminology and data presentation may vary slightly. If you are forecasting the value of your savings account, for example, you know the rates of change (i.e., the interest rates) right away (or you estimate them), but the future values in the account and the growth factors are generally not provided directly.

Example 4.7 The interest rates payable on a Canada Savings Bond can vary each year. Suppose you hold a Canada Savings Bond for five years, and that the interest rates paid each year, in sequence, are 5%, 4%, 5%, 6%, and 4%. What is the average for the **annual rate of return**?

Solution The annual interest rates are the annual rates of change. Five percent interest for a period means that the value of the bond increases by 0.05 of its prior value. The geometric mean cannot be directly calculated using these rates of change. We know, however, that Rate of change $=$ Growth factor -1 so, solving for growth factor:

$$\text{Growth factor} = \text{Rate of change} + 1$$

The sequence of interest rates can be converted to this sequence of growth factors: 1.05, 1.04, 1.05, 1.06, 1.04. The geometric mean of this sequence is:

$$GM = \sqrt[5]{1.05 \cdot 1.04 \cdot 1.05 \cdot 1.06 \cdot 1.04} = \sqrt[5]{1.264} = 1.048$$

The question asks for the average rate of return, which is equivalent to the rate of change, as defined above. The mean rate of change is the geometric mean for the growth factor minus 1. Therefore, the average rate of return is about 4.8% annually.

• *In Brief 4.4 Computing the Geometric Mean*

(Based on the data for Canada Savings Bonds returns given in Example 4.7)

Excel

To calculate the geometric mean in Excel does not require dialogue boxes or a menu sequence. On the spreadsheet containing the data, simply place the cell pointer in an empty cell and then input a formula in the format =**geomean(*range*)**. For **range,** input the cell address range for the data to be averaged.

Minitab

Minitab does not offer a command or menu sequence explicitly for the geometric mean. The *Calculator* function shown in the screen capture, however, can do the job. Arrange the data into a column. Open the calculator with the menu sequence *Calc / Calculator.* In the upper right of the dialogue box, first input a name for a new column to contain the solution. (The resulting column, "GeoMean," is shown in the illustration.)

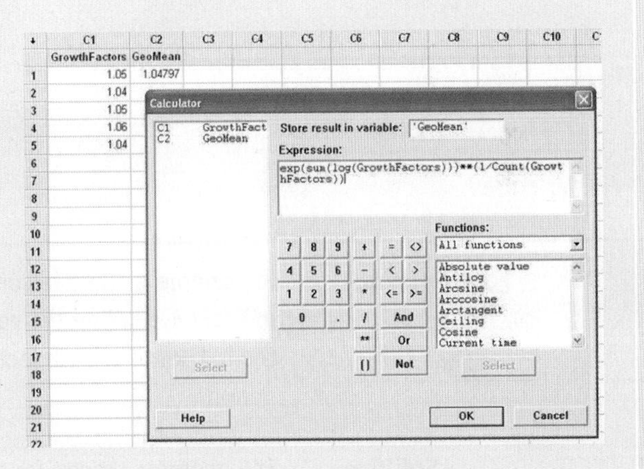

In the *Expression* area of the dialogue box, input an expression in this format: **exp(sum(log (GrowthFactors)))**(1/Count(GrowthFactors)),** replacing **GrowthFactors** with the name for the column of data. Then click *OK* for the solution.

Note that Minitab does not have a function to take the product of a column, but taking the sum of the logarithms for the data and then taking the antilog (**exp**) of the result is equivalent to finding the product of the data. "**" raises a value to a power, and raising the product of the data to the $1/n$ power is equivalent to taking the required n th root.

SPSS

Arrange the data in an SPSS column. Use the menu sequence *Analyze / Reports / Case Summaries.* Highlight and select the variable to be analyzed, then click on the *Statistics* button at the bottom of the dialogue box shown in the screen capture. A second dialogue box allows you to choose a mathematical function. There may already be a value other than the *Geometric Mean* showing in the *Cell Statistics* area on the right. If so, click its name and then click the left-pointing arrow in the middle to deselect the unwanted function. Select *Geometric Mean* in the list of possible *Statistics* on the left and then click the arrow (which at that time will be right-pointing) to select it. Click *Continue* and then click *OK* in the first dialogue box.

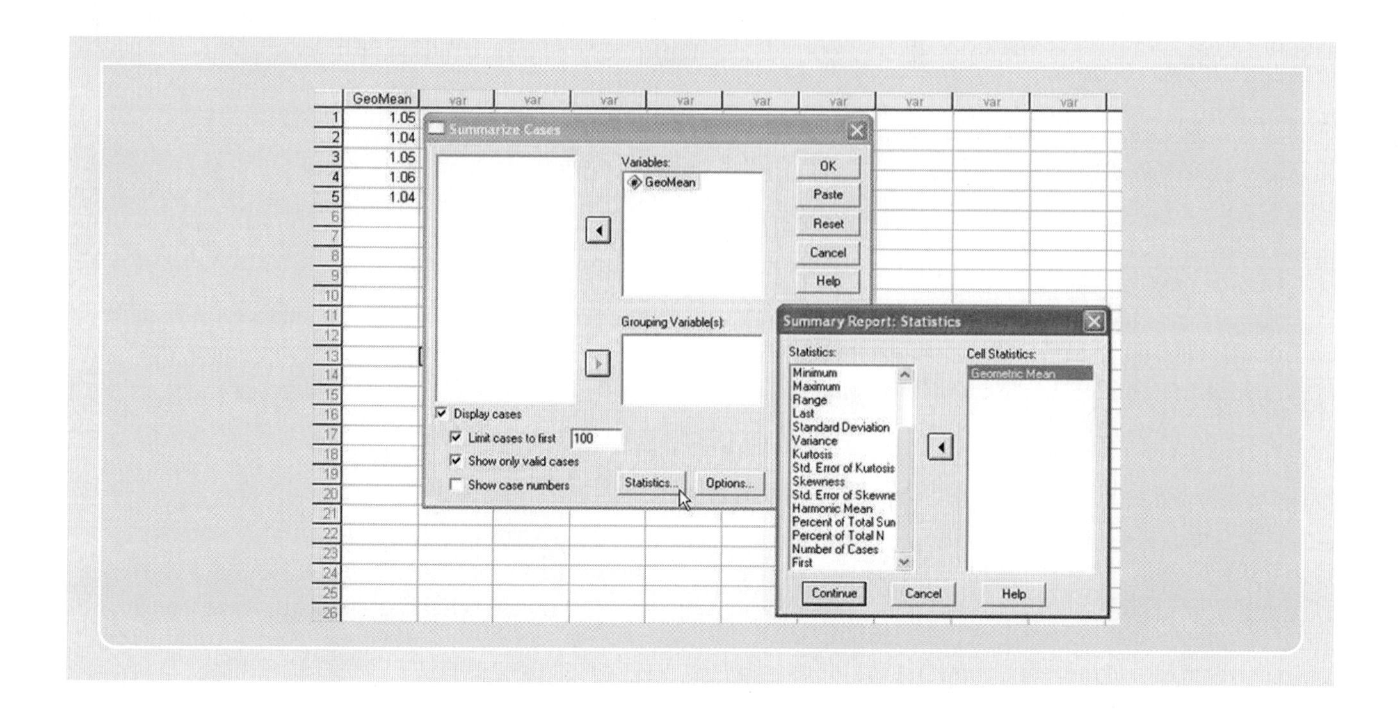

Example 4.8 In Example 4.6, we calculated the geometric mean for the first five growth factors listed in Table 4.2. Using the computer to handle more data, what is the geometric mean for the full set of growth factors in Table 4.2? What was the average rate of change for the period?

Solution The geometric mean for the growth factors can easily be calculated by the software as being 1.067. The geometric mean rate of change for the same period is therefore 1.067 − 1 = 0.067.

4.5 Quantiles

So far, we have looked at measures of location that focus on the *centre* of a data set: the mean, median, and mode, but other locations of the distribution may also be of interest. **Quantiles** present information about other key locations in the data. This topic also is a bridge to the next chapter, which discusses measures describing the spread of data relative to their general location.

The term *quantiles* actually refers to a family of measures. Each specific quantile (for example, quartiles or percentiles) divides a distribution into a number of equal parts. Technically, the median is a quantile dividing the ordered data set into two equal parts. **Quartiles** divide the data set into four equal parts; **quintiles** and **deciles** divide it into five or ten equal parts, respectively; and, finally, **percentiles** divide a data set into 100 equal parts. These measures can be applied to data of at least the ordinal level of measurement. The most frequently used quantiles are quartiles and percentiles.

Figure 4.11 illustrates the dividing of a data set into equal parts. There are 30 values in the file **Voter_Turnouts,** representing percentages of eligible voters who voted in particular federal elections. The data have been sorted in the left column, and the second column shows the position number of each value within the sorted data set. The third, fourth, and fifth columns illustrate the slicing of the full data set into two, four, or ten parts, respectively. Notice how the boundaries between the parts can be expressed in terms of the *positions* of the sorted values. For example, the median (dividing the two halves) occurs between the 15th and 16th positions. The first quartile (dividing the first quarter of the data from the second) occurs on or near the 8th sorted position. The first decile (dividing the first tenth of the data from the second tenth) occurs somewhere between the 3rd and 4th positions, and so on. This relationship between the boundaries of the parts and their positions in the sorted data forms the basis for manually calculating the various quantiles.

Figure 4.11
Example Divisions of Data Set for Voter Turnout (in percentages)

Turnout (sorted)	Ordered Positions of the Data Values	Turnout (in 2 parts)	Quartile	Turnout (in 4 parts)	Decile	Turnout (in 10 parts)
61	1	61		61		61
61	2	61		61		61
67	3	67		67	D1	67
68	4	68		68		68
69	5	69		69		69
69	6	69		69	D2	69
70	7	70		70		70
70	8	70	Q1	70		70
71	9	71		71	D3	71
71	10	71		71		71
71	11	71		71		71
72	12	72		72	D4	72
75	13	75		75		75
75	14	75		75		75
75	15	75	Q2	75	D5	75
75	16	75		75		75
76	17	76		76		76
76	18	76		76	D6	76
76	19	76		76		76
76	20	76		76		76
76	21	76		76	D7	76
76	22	76		76		76
77	23	77	Q3	77		77
79	24	79		79	D8	79
79	25	79		79		79
80	26	80		80		80
80	27	80		80	D9	80
81	28	81		81		81
84	29	84		84		84
90	30	90		90		90

Source: Nodice Canada. Based on data in *Voter Turnout,* which is based on Elections Canada data. Found at: http://www.nodice.ca/election2004/voterturnout.html

Specific calculations are shown below for quartiles and percentiles. (Medians were discussed in Section 4.2.) The procedures used in some textbooks and software may vary slightly. For example, the procedures used in Excel sometimes yield slightly different results from those shown here. Discuss with your instructor which approach to use.

Calculating Quartiles

If all of the data are sorted from lowest to highest, quartiles are values that divide the array into four equal parts. Just glancing at the fourth column in Figure 4.11, why not conclude that the first quartile occurs at (exactly) 70 and that the third quartile (which separates the 3rd from the 4th quarter of the data array) occurs (exactly) at 77? This intuitive approach might be sufficient if we cared only about the data in front of us. However, if these data are viewed as a sample, then a more formal method is preferable—one that allows us to infer where the quartile might fall for the full population.

The general formulas below do *not* directly calculate the quartiles; they calculate the quartile *positions*. Once these are known, refer back to the original ordered data to determine the value at the position indicated. The position number may be a fraction (e.g., $9^3/_4$). In that case, find the value that is the indicated distance (e.g., in this case, three-quarters of the way between the 9th and 10th number in an ordered array).

$$Q1 \text{ position} = {}^1/_4\,(n + 1)$$

$$Q2 \text{ position} = {}^2/_4\,(n + 1)$$

$$Q3 \text{ position} = {}^3/_4\,(n + 1)$$

where Q1, Q2, and Q3 refer to the first, second, and third quartiles, respectively.
n is the number of values in the data set.

Example 4.9

(Based on the ***Voter_Turnouts*** *file)*

Calculate the first, second, and third quartiles for the data set of voter turnouts in 30 federal elections (saved in the file **Voter_Turnouts**).

Solution Refer to the position numbers of the sorted data in Figure 4.11 and use the following formulas:

***Q*1:** *Q*1 position $= {}^1/_4\,(30 + 1) = 7.75$

The value in the 7th position = 70, and the value in the 8th position = 70.

The value that is 0.75 of the way from 70 to 70 is 70.

***Q*1 = 70**

***Q*2:** The median:

*Q*2 position $= {}^2/_4\,(30 + 1) = 15.5$

The value in the 15th position = 75, and the value in the 16th position = 75.

The value that is 0.50 of the way from 75 to 75 is 75.

***Q*2 (= median) = 75**

***Q*3:** *Q*3 position $= {}^3/_4\,(30 + 1) = 23.25$

The value in the 23rd position = 77, and the value in the 24th position = 79.

The value that is 0.25 of the way from 77 to 79 is $(77 + [0.25 \cdot (79 - 77)] = 77.5$.

***Q*3 = 77.5**

The values of the quartiles comprise an important part of a data-summary diagram called the *boxplot*, which is discussed in the next chapter. As we will see, the boxplot can be very useful for assessing the shape of a distribution and also for detecting outliers.

Calculating Percentiles and Other Quantiles

The basic principles are the same as above for calculating any other quantile: First you find the position of the quantile in the ordered array, and then you find or estimate the value at that position. You would rarely want to find every percentile for a data set, but often certain values are of special importance. For example, a graduate program at a university may only accept candidates who score at or above the 80th percentile in a certain standard test.

If you can calculate percentiles, you can easily calculate any of the other common quantiles—by converting them into percentiles. For example, the first decile (dividing the data array into tenths) is equivalent to the 10th percentile (since $1/10 = 10/100$). Similarly, the first quintile (dividing the array into fifths) is equivalent to the 20th percentile (since $1/5 = 20/100$), and so on.

The general format of a formula to calculate the *position* of a percentile is shown below. Again, remember that the actual value of the percentile has to be calculated separately.

$$\textbf{Percentile}_i \textbf{ position} = {}^{i}/_{100}\,(n + 1)$$

where Percentile$_i$ refers to the *i*th percentile (*i* being a number from 1 to 99).
n is the number of values in the data set.

Example 4.10

(Based on the ***Voter_Turnouts*** *file)*

For the data set of voter turnouts in 30 federal elections (saved in the file **Voter_Turnouts**), calculate (1) the 1st percentile, (2) the 47th percentile, (3) the 2nd quintile, and (4) the 9th decile.

Solution

Refer to the position numbers of the sorted data in Figure 4.11 and use the following formulas:

1. $\textbf{Percentile}_1 \textbf{ position} = {}^{1}/_{100}\,(30 + 1) = 0.31$

   The 0th position in the ordered data array is not defined. For any position that calculates to less than 1, interpret it as the first position (minimum value) in the data set.

   The value in the first position of the data is 61.

   $\textbf{Percentile}_1 = 61$

2. $\textbf{Percentile}_{47} \textbf{ position} = {}^{47}/_{100}\,(30 + 1) = 14.57$

   The value in the 14th position = 75, and the value in the 15th position = 75.

   The value that is 0.57 of the way from 75 to 75 is 75.

   $\textbf{Percentile}_{47} = 75$

3. $\textbf{Quintile}_2 \textbf{ position} = \textbf{Percentile}_{40} \textbf{ position (because } 2/5 = 40/100)$

   $\textbf{Percentile}_{40} \textbf{ position} = {}^{40}/_{100}\,(30 + 1) = 12.4$

   The value in the 12th position = 72, and the value in the 13th position = 75.

The value that is 0.4 of the way from 72 to 75 is $(72 + [0.4 \cdot (75 - 72)] = 73.2$.

$\text{Quintile}_2 = \text{Percentile}_{40}$ $= 73.2$

4. Decile_9 position = Percentile_{90} position (because 9/10 = 90/100)

   Percentile_{90} position = $^{90}/_{100}\,(30 + 1) = 27.9$

   The value in the 27th position = 80, and the value in the 28th position = 81.

   The value that is 0.9 of the way from 80 to 81 is 80.9.

   $\text{Decile}_9 = \text{Percentile}_{90} = 80.9$

• *In Brief 4.5 Computing the Quantiles*

*(Based on the **Voter_Turnouts** file)*

Not all software packages or options use the same procedures for calculating quantiles as described in this textbook, which can lead to minor differences from the solutions provided for the above examples. Check with your instructor to find out which tools and procedures are preferred.

Excel

Input the data to be analyzed into a spreadsheet column. Place the cell pointer in an empty cell. For a quartile, input a formula in the format shown below, replacing **N2:N31** with the actual range for the data and, if needed, replacing the **1** that follows the comma with a 2 for the 2nd quartile (median) or a 3 for the 3rd quartile. Then press *Enter.*

=QUARTILE(N2:N31,1)

For a percentile (or using percentiles, another quantile), input a formula in the format shown below, replacing **N2:N31** with the actual range for the data and, if needed, replacing **0.01** with 0.02 for the 2nd percentile, or with 0.03 for the 3rd percentile, or up to 0.99 for the 99th percentile. Then press *Enter.*

=PERCENTILE(N2:N31,0.01)

Minitab

The quartiles can be readily displayed in Minitab along with other common statistics, such as the mean and median. Begin with the menu sequence *Stat / Basic Statistics / Display Descriptive Statistics.* In the first dialogue box (see the screen capture), select the name of the variable to be analyzed. Click on *Statistics* to bring up the next dialogue box and then select the desired quartile and other statistics. Click *OK* and then *OK* again to exit both dialogue boxes.

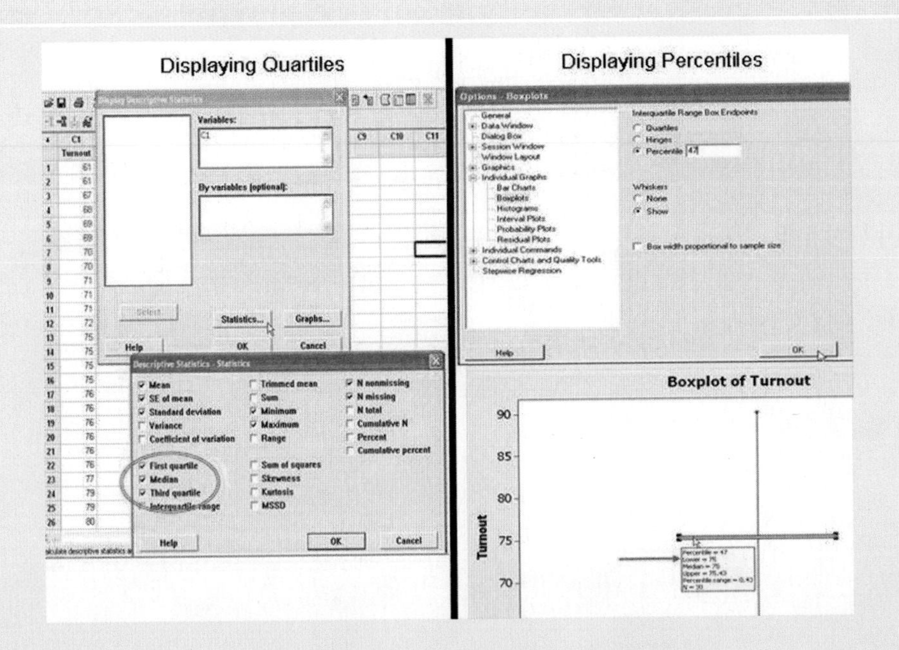

For the percentile—and using the percentiles, for other quantiles—a less direct technique is required. In effect, you instruct Minitab to modify a boxplot to draw a line at the desired percentile. Use the menu sequence *Tools / Options.* On the left side of the first dialogue box, select *Individual Graphs* and then *Boxplots.* As shown in the screen capture at the upper right, select the *Percentiles* option and input the desired percentile. If you want a percentile greater than 50, input a value that equals 100 – (the desired percentile). Leave other defaults unchanged.

Skip ahead in this textbook and use the procedures in "In Brief 5.4" to generate a boxplot for the data; then rest the arrow on the middle part of the rectangular (box) portion for the answers to be displayed (see the boxplot in the lower right-hand corner of the illustration). If the desired percentile in the boxplot is less than 50, read off the *Lower* value shown; if the desired percentile is more than 50, read off the *Upper* value shown.

SPSS

The quartiles can be readily displayed in SPSS, along with other common statistics such as the mean and median. Begin with the menu sequence *Analyze / Descriptive Statistics / Frequencies.* In the first dialogue box (see the screen capture), select the name of the variable to be analyzed and then click on *Statistics.*

Under the heading "Percentile Values" in the second dialogue box, for quartiles, simply select *Quartiles* and then click *Continue.* For a percentile (or using percentiles, another quantile), select *Percentile(s)* and input the desired percentile number to the right of the word *Percentile(s).* Click on *Add.* Repeat the two steps of inputting percentile numbers and clicking *Add* until all desired percentiles have been input; then click on *Continue* and then on *OK.*

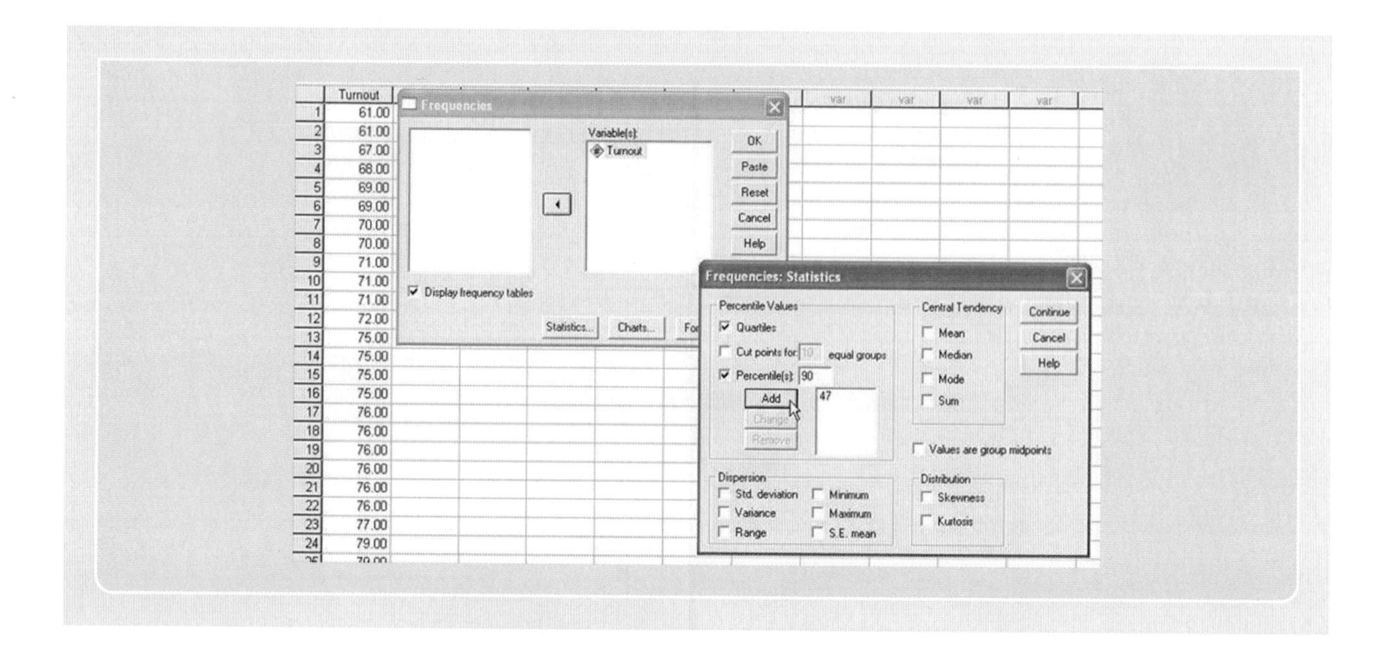

• Hands On 4.4 *Just Average*

Whatever our interests, we all talk loosely from time to time about "averages." Perhaps you follow the batting averages of first basemen or want to know what the average rents are in a new area you'd like to move to. Or, what's the average weight for someone of your size, the average house price in Winnipeg, or the average return for those who buy a certain mutual fund? These are all typical uses of the word *average.*

We saw in this chapter that *average* is a fuzzy way to refer to several possible measures of location for a data set. Often, more than one such measure can be applied, and each tells us something a little different about the data set. Other times, it is mathematically inappropriate to use certain methods, even if your software will let you try.

Examine your own uses of the word *average.* Do you follow the sports stats or house prices? Dig deeper into your references to find what actual measures of location the sources are using. Ask: Could they use alternative measures, and what would be the advantages and disadvantages if they did so?

Pick at least one area where you are interested in the average. Try to obtain as much raw data for the area as you can. (Random sampling would be ideal, but you might be restricted to published secondary data that you can find in such sources as magazines or on the Internet.) For example, if you follow apartment prices, you could draw from newspaper or online advertisements to obtain a diverse sample of prices for individual units.

Once you have the data, calculate a full set of the applicable measures of location discussed in this chapter; for example, mean, median, modal class, the quartiles. What do they tell you about the data? If you could know just one of these figures, which would be the most useful to you, and why? Can you envision other circumstances when another one or a set of the measures might be more helpful? Explain. Use a software program to construct a frequency distribution table or frequency histogram for the data, and compare where all of these measures of location fall in relation to the whole pattern.

Chapter Summary

4.1 Calculate and interpret the value of the arithmetic mean of a data set.

The **arithmetic mean** is one possible measure of **location** for a data set; in particular, it is a **measure of centre.** With Formulas 4.1 and 4.2, respectively, you can calculate the mean for a population or a sample.

4.2 Calculate and interpret the values of the median and the mode (if any) of a data set, and for continuous quantitative data, distinguish between the mode and modal class.

An alternative measure of centre is the **median,** which identifies the middle value if the data are arranged in an ordered array. The **mode,** which can be used with all levels of data measurement, is the particular value that occurs most frequently in the data set. A data set may also be *bimodal* or *multimodal.* If continuous quantitative data are grouped to create a frequency distribution, the class having the largest frequency is the **modal class.**

4.3 Calculate a weighted mean, given appropriate weights or class frequencies.

With the **weighted arithmetic mean,** we can calculate a mean for a set of data where the values are assigned unequal importance (see Formula 4.3). By using the frequencies for grouped data as weights, we can also use the technique to calculate the mean from a frequency distribution table.

4.4 Calculate and interpret the value of the geometric mean of a data set.

A **rate of change,** like an interest rate, tells by what proportion of an initial value the *next* value increases (or decreases). The **growth factor** (equals rate of change plus 1) tells by how much to multiply an initial value to calculate the next value. If these changes are compounded over time, the appropriate measure for centre for a series of these values is the **geometric mean** (see Formula 4.4).

4.5 Explain what is meant by quantiles, and calculate values for particular quartiles and percentiles.

Quantiles present information about other key locations in the data. Each specific quantile divides a distribution into a number of equal parts. For example, quartiles divide the data set into four equal parts; **quintiles** and **deciles** divide it into five or ten equal parts, respectively; and **percentiles** divide a data set into 100 equal parts. We learned to calculate specific values for quartiles and percentiles.

Key Terms

annual rate of return, p. 148
arithmetic mean, p. 128
deciles, p. 150
geometric mean, p. 146
growth factor, p. 146
location, p. 128
measure of centre, p.128
median, p. 135
modal class, p. 138
mode, p. 137
percentiles, p. 150
precision, p. 131
quantiles, p. 150
quartiles, p. 150
quintiles, p. 150
rate of change, p. 146
significant digits, p. 131
weighted arithmetic mean, p. 140

Exercises

Basic Concepts 4.1

1. The arithmetic mean is a measure of location. Of which location in the data set is the arithmetic mean a measure?

2. Determine the arithmetic mean for the following sample data sets:

Data set 1:	4	7	5	9	2
Data set 2:	4	7	5	9	25

3. Compare the two means calculated for the data sets in question 2. What impact does changing one number in the data set have? Does changing one number have a greater impact on the arithmetic mean for small data sets as opposed to larger data sets? Explain.

4. Determine the arithmetic mean for the following sample data sets:

Data set 1:	5	6	3	4	10	2	9	1	14
Data set 2:	6	3	2	8	7	3	2	1	6

5. According to the labels, each bottle of a certain soft drink holds 750 mL of liquid; however, the volume of liquid in each bottle may not be exactly 750 mL. From the following sample of 10 bottles, calculate the mean volume of liquid in the bottles. The values are expressed in millilitres.

749.5	748.0	749.7	748.6	750.1
751.3	750.5	749.7	750.1	750.8

6. Over the last 10 years, Shepitka Construction has built an average of 256 homes per year. The numbers of homes built in each of the first nine years of that period were as follows:

212 291 314 220 196 281 244 233 285

Determine the number of homes built in the tenth year.

Applications 4.1

7. Over several summers, two volunteers collected data on eruptions of the Grotto Geyser in Yellowstone National Park. The table below is derived from their data (see the file **Grotto_Geyser**). The table shows the number of eruptions recorded on each of the 83 days that were studied.

1	2	3	4	2	4	2	2	1	3	4	2	3	3	3	1	2	3
2	2	1	3	3	4	3	3	4	3	2	1	3	3	2	2	1	4
4	2	1	2	3	3	4	1	4	1	4	3	2	3	2	3	1	4
4	3	2	2	3	2	2	3	3	3	3	3	2	2	1	3	2	2
3	4	3	2	3	1	4	3	1	3	2							

a) Calculate the arithmetic mean for the number of eruptions each day.

b) In Chapter 3, Exercise 13, you constructed a frequency distribution for these same data. Does the mean value correspond to what appears to be a "typical" number of daily eruptions, judging from the frequency distribution? Explain your answer.

8. In 2001, the Canadian federal government and eight banks signed agreements to facilitate lower-cost banking services for consumers. The Financial Consumer Agency of Canada has reported some of the results. These are among the low-cost options for monthly fees. (See the file **Low_Fee_Accounts.**)

3.50	5.00	4.00	2.00	4.00	2.95	2.50
3.50	5.00	4.00	3.95	2.50	3.95	0.00

a) Calculate the arithmetic mean for these values of low-cost fees.

b) In Chapter 3, Exercise 14, you constructed a frequency distribution for these same data. Would you say that the mean value clearly expresses the "typical" value of low-cost fees? Why or why not?

9. A chemist is experimenting with solvents to use in extreme-temperature environments, such as the open cargo bay on a NASA space shuttle. She will select from a list of 47 solvents for which the melting and boiling points are known. (See the columns "Mpt" (Melting Point) and "Bpt" (Boiling Point) in the file **Freezing_Boiling_Data.**)

a) Calculate the arithmetic mean for (i) the melting points of the solvents and (ii) the boiling points of the solvents.

b) In Chapter 3, Exercise 16, you constructed frequency distributions for both of these variables. Compare those distributions with the means you have just calculated. If you randomly select one solvent, would the mean appear to be a better predictor of its melting point or of its boiling point? Explain your answer.

10. The population density of a country is a measure of how many people live in the country relative to the surface area within its borders. A demographer claims that population density varies widely among countries. (See the column "Resident" in the file **Population_Density.**) Each value shows the population density of a particular country in persons (on average) per square kilometre.
    a) Calculate the arithmetic mean for the population density of the countries.
    b) Consider the statement: "Sometimes the mean can be misleading as a measure of location." Compare the mean for the population density with the frequency distribution for the variable that you constructed in Chapter 3, Exercise 17. Would you agree that if you told someone just the mean in this case, it might give a misleading impression about the distribution of the data? Explain your answer.

11. A mineral collector has identified 100 metallic minerals and recorded the values of density and hardness for each. He is examining the distributions of these two variables. (See the file **Metallic_Minerals.**)
    a) Calculate the arithmetic means for density and hardness.
    b) Compare these means with the frequency distributions for these variables, which you constructed in Chapter 3, Exercise 20. For both variables, is a randomly selected value likely to be far greater than the mean (say, two times as large) or far smaller than the mean? If so, is this a desirable feature for a measure of central location? Explain your answer.

Basic Concepts 4.2

12. Find the median for each of the following data sets.

Data set 1:	4	7	5	9	2
Data set 2:	4	7	5	9	25

13. Compare the two medians calculated for the data sets in Exercise 12. What impact does changing one number in the data set have? Does changing one number have a greater impact on the arithmetic mean as a measure of location or on the median? Explain.

14. For each of following four data sets, find (a) the median and (b) the mode.

Data set 1:	16	12	19	10	14	15	18	
Data set 2:	8	9	6	7	8	5		
Data set 3:	10	11	17	19	20	15	26	17
	15	17	18	13	10			
Data set 4:	3	5	7	2	4	8	2	5
	9	5	7	1	6	3	2	5
	4	3	4	6	9			

15. The following batting averages were obtained from a sample of Toronto Blue Jays players during the 2005 season.

0.269	0.291	0.256	0.262	0.271	0.251	0.301
0.269	0.274	0.249	0.216	0.250	0.207	0.194
0.308	0.083	0.262				

    a) Determine the median batting average for the players in this sample.
    b) Determine the mode or modes of these batting averages.
    c) Explain this statement, with reference to the above data set: "For continuous data, such as batting averages, the mode may be misleading as a measure of central location."
    d) What is a modal class, and how might it be applied to a problem such as the preceding?

16. Over the last 10 years, the median number of homes built by Shepitka Construction has been 245. In each of the first nine years of that period the numbers of homes built in each year were as follows:

314	240	196	212	285	220	250
291	233					

Determine the minimum number of homes that could have been built in the tenth year.

17. Over the last 20 years, the mode for the number of homes built by Shepitka Construction was 250. In each of the first nine years of that period the numbers of homes built in each year were as follows:

314	240	196	212	285	220	250
291	233					

Determine the number of homes that were built in the tenth year.

18. In a small parking lot, the cars are the following colours:

red	burgundy	yellow	silver	white
silver	blue	burgundy	white	red
white	yellow	white	silver	green
brown	silver			

Determine the mode for this set of colours.

Applications 4.2

19. An amateur astronomer has been collecting facts about the first nine planets of our solar system. (There may be others.) She is particularly interested in their diameters, given below in kilometres. (See file **Planetary_Facts.**)

Planet	Mercury	Venus	Earth	Moon
Diameter (km)	4,879	12,104	12,756	3,475
Planet	Mars	Jupiter	Saturn	Uranus
Diameter (km)	6,794	142,984	120,536	51,118

Planet	Neptune	Pluto
Diameter (km)	49,528	2,390

a) Calculate the median for these diameters.

b) Construct a frequency distribution for these data and find their arithmetic mean. Which measure, the mean or median, gives a better picture of the "centre" of this data set? Explain your answer.

20. In 2001, the Canadian federal government and eight banks signed agreements to facilitate lower-cost banking services for consumers. The Financial Consumer Agency of Canada has reported some of the results. These are among the low-cost options for monthly fees. (See file **Low_Fee_Accounts.**)

3.50	5.00	4.00	2.00	4.00	2.95	2.50
3.50	5.00	4.00	3.95	2.50	3.95	0.00

a) Calculate the median for these values of low-cost fees.

b) In Chapter 3, Exercise 14, you constructed a frequency distribution for these same data. Would you say that the median value clearly expresses the "typical" value of low-cost fees? Why or why not?

21. The population density of a country is a measure of how many people live in the country relative to the surface area within its borders. A demographer claims that population density varies widely among countries. (See the column "Resident" in the file **Population_Density.**) Each value shows the population density of a particular country in persons (on average) per square kilometre.

a) Calculate the median for the population density of the countries.

b) In Exercise 10 in this chapter, you calculated the mean for this variable. For this data set, would you think that the mean or the median conveys a better sense of the central value? Explain your answer.

22. A mineral collector has identified 100 metallic minerals, and recorded for each the values of density and hardness. He is examining the distributions of these two variables. (See the file **Metallic_Minerals.**)

a) Calculate the medians for density and hardness.

b) Compare these medians with the variables' frequency distributions (see Chapter 3, Exercise 20) and their means (see Exercise 11 in this chapter). For both variables, is a randomly selected value more likely to be greater than the mean or greater than the median? Is a randomly selected value more likely to be less than the mean or less than the median? Explain your answers.

23. When asked about their plans for the next Canada Day holiday, in a SES Canada Research survey, people's answers included "Nothing," "At the cottage," "Working," and various other choices. (See the file **Celebration_Canada_Day.**) Suppose you are employed for the research company and are preparing a press release.

a) What is the mode for the data set?

b) Based on just the mode for the data set, put in clear words a statement for the press about what was found.

c) Suppose that everyone who answered "Unsure" changed their answers:

(i) If their new answers followed the same proportions as for those who originally expressed opinions, what would be the mode of the revised set of answers?

(ii) If everyone who had answered "Unsure" switched their answer to "Participate in Canada Day festivities," what would be the mode of the revised set of answers?

24. Given the diversity of Canada's population, it is a challenge for Statistics Canada to classify individuals' backgrounds in a sensitive and meaningful fashion. For example, in the 2001 Canadian census, the origin of a person identified as Aboriginal may in fact be one of several possibilities.[1] What is the mode of the following set of historical origins?

Historical Origin for Individuals Identified as Aboriginal	Number of Individuals
North American Indian	957,650
Métis	266,020
Inuit	51,390
More than one Aboriginal origin	44,835

25. To entertain a guest, a mayor has downloaded a price list for Australian wines from the website of the PEI Liquor Control Commission. (See the file **Australian_Wines.**)

a) What is the mode for the set of prices?

b) Compare the mode with the frequency histogram that you created for the same data in Chapter 3, Exercise 31. What is the modal class? Briefly state the similarities and differences in the information conveyed if you told a person the mode as opposed to showing them the distribution.

26. It would appear from data published in Vermont that many crimes are committed by relatively young people. In the file **Vermont_Crime,** the typical ages of apprehended individuals are shown.

a) What is mode of the ages shown?

b) The ages shown in the data set are themselves averages. Can we be sure that the mode for the ages of all apprehended individuals is the same as the mode of the average ages of individuals from various subgroups? Explain your answer.

Basic Concepts 4.3

27. Moran's Television Sales received a shipment of television sets, five of which were priced at $899 and six at $999. Find the weighted mean price for the television sets.

28. Mr. Lachance has been purchasing shares in a number of companies over the last year. His share holdings are as follows:

Company	Number of Shares	Share Price
ABC	200	$75.00
HRH	100	$22.00
PMG	500	$10.00
STU	300	$15.00

Determine the weighted mean share price for Mr. Lachance's holdings.

29. An environmental inspector has measured 12 trace compounds in the gas emissions from a waste landfill site. For two of the trace compounds (toluene and dichloromethane), the mean concentration was approximately 30,000 pbV (parts per billion, by volume). For the remaining 10 compounds in the sample, the mean concentration was approximately 3,500 pbV. Estimate the weighted mean concentration in the sample for all of the trace compounds combined.

30. In a certain week, the groundskeeper receives $15.00 per hour for her normal 40-hour workweek. She also receives 1.5 times her normal hourly rate for eight hours of overtime. What is the groundskeeper's mean hourly wage rate for that week?

31. Estimate the arithmetic mean for the following data.

Class	Frequency
10 to less than 20	5
20 to less than 30	15
30 to less than 40	3
40 to less than 50	8

32. Lajoie's French Cuisine has conducted a survey of the ages of its clientele. The results in grouped format are as follows:

Age	Number of Clients
10 to less than 20	22
20 to less than 30	55
30 to less than 40	76
40 to less than 50	63
50 to less than 60	55
60 to less than 70	23
70 to less than 80	14

Estimate the average age of Lajoie's clients.

Applications 4.3

33. According to a study published by the Public Health Agency of Canada, the number of patients who visit a physician for the treatment of diabetes in any given year is different for different age groups. Presuming a sample size of 999 patients across Canada, the following were the numbers of people visiting a physician for treatment of diabetes during the study.[2]

Age of Patient (Years)	Number of Visits
0 to 19	14
20 to 29	26
30 to 39	44
40 to 49	121
50 to 59	200
60 to 69	303
70 to 79	218
80 to 89	73

a) Calculate the arithmetic mean for the age of patients who visited a physician due to diabetes.

b) How might you use the solution to (a) if you were (i) selecting music for the waiting rooms used by these patients? (ii) Selecting TV shows and magazines in which to place health advisories and announcements for these patients?

34. The distribution of numbers of votes cast in each British Columbia riding in the 2004 federal election was as follows:[3]

Classes	Frequency
30,000–34,999	1
35,000–39,999	3
40,000–44,999	9
45,000–49,999	8
50,000–54,999	8
55,000–59,999	5
60,000–64,999	2

a) Calculate the weighted arithmetic mean for the number of votes cast per riding.

b) Given your answer to (a) and knowing that there were 36 ridings, quickly estimate the number of ballot slips that were needed in British Columbia during that election.

c) Suppose that each BC riding was initially supplied with only as many ballot slips as that mean number of votes cast per riding. Assuming that no ballot slips were spoiled, about how many ridings would have run short of slips during that election?

35. The University of New Brunswick's *Earth Impact Database* provides data on meteor craters in North America. Fifty-five of these are found below 65°N latitude, as shown below.[4]

Latitude (°N)	Frequency
30–34	5
35–39	10
40–44	9
45–49	10
50–54	9
55–59	7
60–64	5

a) Calculate the weighted arithmetic mean for the locations of these craters, in the °N latitude.

b) Suppose you drove to the position of the mean latitude. At your current position, would there be more meteors to your north or to your south? To answer this question, do you have to estimate the distribution within one of the classes? Explain.

36. Transport Canada safety regulations require that airlines conform to certain weight standards for each flight. The increasing weight, on average, of Canada's population is—literally—a growing concern. But airline passengers are not routinely weighed (as some critics recommend); instead, weight standards are applied to estimate the weights for different types of passengers. (Airlines still use imperial units for their weight measurements.) For a summer flight, it is assumed that each adult male adds 200 pounds to the cargo, each female adds 165 pounds, each child (over 2 years old) contributes 75 pounds, and every infant adds 30 pounds.[5]

a) Suppose that the human cargo (passengers and crew) on a Canadian flight include the following: 150 adult males; 135 adult females; 29 children over 2 years old; and 9 infants. Calculate the weighted mean for the weights of all of these passengers—assuming that Transport Canada's weight assumptions are correct.

b) Suppose that, in fact, adult men weigh 10 pounds more than Transport Canada assumes, adult women weigh 6 pounds more, and children over two weigh 3 pounds more. What is the actual mean weight of the human cargo on the flight? If the airplane can accommodate a maximum weight for human cargo of 56,000 pounds, is this flight safe to take off? Why or why not?

37. The management of the Calgary Stampeders football franchise is forecasting gate revenues for the 2006 football season. One component will be single ticket sales of adult seats. (Adjustments will be made later for special admission prices; e.g., tickets for students, seniors, and special game days). For a certain game, it is forecast that the following numbers of single adult tickets will be sold in the indicated categories.[6]

Seat Category	Price	Expected Number of Sales
Platinum	$80.00	1,576
Super Reds	$65.00	2,247
Reds	$60.00	2,918
Blue	$45.00	3,752
Super White	$35.00	3,367
White	$27.00	4,183

a) Based on the seat prices and expected numbers of sales, calculate the weighted mean of the adult ticket prices for the seats forecast to be sold that day.

b) Suppose the forecast is adjusted in this way: 200 more Reds are expected to be sold than shown in the table, and 400 fewer Whites are expected to be sold. Will the net change in the weighted mean be an increase or a decrease? How do you know?

Basic Concepts 4.4

38. A chemist is testing the reaction rate of a certain compound under experimental conditions. The concentration of the reactant in a beaker was reduced from 13 parts per million (ppm) to 2.9 ppm within 10 milliseconds.

Millisecond	0	1	2	3	4	5
Concentration	13	10.6	8.4	6	5	4.2
Millisecond	6	7	8	9	10	
Concentration	4	3.5	3.2	3	2.9	

What was the mean rate of decrease in concentration per millisecond?

39. The acceleration of a new sports car was tested by an auto magazine representative. The following table shows the speeds that were reached in two-second intervals, as the car accelerated from (nearly) a full stop to 100 kilometres per hour.

Seconds	0.1	2	4	6	8	10
Km/h	0.1	35	58	76	89	100

During this period of acceleration, what was the mean rate of growth in speed (per two-second interval)? In the data table, why is the first entry approximated at 0.1, instead of left at "0" for a standing stop?

40. The annual profits for a small business over the last five years were as follows: $425,000, $450,000, $430,000, $460,000, and $495,000. Determine the average percentage increase in profit over the five-year period.

41. The population in a small town was 950 in 2002, 975 in 2003, 970 in 2004, 1,000 in 2005, and 1,110 in 2006. Determine the average rate of growth in the town's population.

42. The annual interest rates for the last four years on a government bond have been 9%, 6%, 3%, and 4%. Use the geometric mean to determine the mean interest rate over the four-year period.

43. For a small country, the annual rates of inflation over a six-year period were as follows: 2%, 4%, 2.5%, 6%, 3%, and 1.5%. Determine the mean rate of inflation over this time period.

Applications 4.4

44. Expressing concern for environmental sustainability, Syncrude Canada reported the following levels of emission of greenhouse gases, in millions of tonnes:[7]

2000	2001	2002	2003	2004	2005 (forecast)
9.03	8.78	9.88	9.47	10.3	10.67

a) What has been the average rate of growth in the level of emissions?

b) What has been the average rate of change in level?

c) Suppose that following 2004, the rate of change in the level was reduced to one-half of the average rate as calculated in (b). On that basis, revise the forecast for levels of emission in 2005.

45. Although forest fires occur naturally, some people believe that for a variety of reasons, the extent and destruction of forest fires have been increasing. Based on the data in the file **Yukon_Wildfires,** find the rates of change for both the numbers of wildfires, and the numbers of hectares destroyed by wildfires in the Yukon from 1950 to 2003. (The 2004 value may be an outlier.) During that period, has there been an increasing rate of change for these variables?

46. Annual expenditures in Canada (by all sources) on research and development—in natural sciences, engineering, social sciences, and the humanities—have been increasing. Based on the following data published by Statistics Canada,[8] what was the average rate of growth of these expenditures from 1998 to 2004? Does the growth rate appear to have been constant, or did it increase or decrease in the last few years?

Year	$Millions Invested
1998	15,084
1999	16,469
2000	19,223
2001	21,344
2002	20,813
2003	21,557
2004	22,556

47. Although the past performance of a stock or mutual fund may not be a sure predictor of its future performance, investors commonly view past data when choosing a stock or fund to invest in. The following data show the annual returns on investment for the Investors Group-sponsored fund "Investors Mortgage."[9]

Year	2005	2004	2003	2002	2001
% Return	3.3	1.9	3.5	3.7	7.2
Year	**2000**	**1999**	**1998**	**1997**	**1996**
% Return	4.4	3.1	3.2	5.5	8.3
Year	**1995**	**1994**	**1993**	**1992**	**1991**
% Return	9.5	0.9	6.8	10.8	13.8

a) What was the average growth rate of this fund during the years shown?

b) What was the average, annual return on investment?

c) Recalculate the annual return on investment if the value for 1991 is omitted.

d) Replace the 1991 value in the data with "17" and recalculate the annual return from 1991 to 2005. Based on your answers to (c) and (d), does the geometric mean calculation appear relatively sensitive or insensitive to unusually high values?

48. The American Public Power Association has documented the typical rate of growth of "standard" nursery-raised trees.[10] As shown below, full growth is expected after 24 years. Complete the table to show the percentage rates of growth and of change, year by year. Then calculate the average annual rate of growth over the full period of development. During what portion or portions of the 24 years is the growth rate above the average? Below the average?

By End of Year	Cumulative Growth %	Percentage Rate of Growth	Percentage Rate of Change
1	1%		
2	2%	200%	100%
3	4%	200%	100%
4	6%	150%	50%
5	10%	167%	67%
6	14%		
7	18%		
8	23%		
9	29%		
10	35%		
11	41%		
12	47%		
13	53%		
14	59%		
15	65%		
16	71%		
17	77%		
18	82%		
19	86%		
20	90%		
21	94%		
22	96%		
23	98%		
24	100%		

Basic Concepts 4.5

49. Determine the first and third quartiles for the following data sets.

Data set 1: 4 6 2 7 8 3 5 9 1 6 7 4

Data set 2: 4 5 7 2 8 9 10 12 3 4 14 18 2 6 7 18

Data set 3: 15 13 19 18 15 16 20 11 14 16 17 12 21 22 23 10 17 13 24

50. For each of the data sets in the previous exercise, calculate the 20th and the 80th percentiles.

51. The batting averages below were obtained from a sample of Toronto Blue Jays players during the 2005 season.

0.269	0.291	0.256	0.262	0.262	0.271	0.251
0.301	0.269	0.274	0.249	0.216	0.250	0.207
0.194	0.308	0.083				

a) Determine the first, second, and third quartiles.

b) Determine the 5th, 50th, and 95th percentiles.

52. According to their labels, each bottle of a certain soft drink holds 750 mL of liquid. Ten bottles were sampled to measure their actual contents. From these sample values (in millilitres), calculate the first and third quartiles for the volume of liquid in the bottles.

749.5	748.0	749.7	748.6	750.1
751.3	750.5	749.7	750.1	750.8

53. For each of the data sets below, determine (a) the 5th percentile, (b) the 1st quartile, (c) the 3rd quartile, and (d) the 95th percentile.

Data set 1:	16	12	19	10	14	15	18	
Data set 2:	8	9	6	7	8	5		
Data set 3:	10	11	17	19	20	15	26	17
	15	17	18	13	10			
Data set 4:	3	5	7	2	4	8	2	5
	9	5	7	1	6	3	2	5
	4	3	4	6	9			

54. Which changes, if any, to data set 4 in the previous exercise would change the value of the 5th percentile: (a) If you make the smallest value smaller; (b) if you make the largest value larger; (c) if you change the largest value to a number equal to the lowest value? Explain.

Applications 4.5

55. An amateur astronomer has been collecting facts about the first nine planets of our solar system. She is particularly interested in their diameters (shown in the table below in kilometres). (See file **Planetary_Facts.**)

Planet	Mercury	Venus	Earth	Moon
Diameter (km)	4,879	12,104	12,756	3,475

Planet	Mars	Jupiter	Saturn
Diameter (km)	6,794	142,984	120,536

Planet	Uranus	Neptune	Pluto
Diameter (km)	51,118	49,528	2,390

a) Calculate the first and third quartiles for these diameters.

b) Calculate the 10th and 90th percentiles for the diameters.

56. In 2001, the Canadian federal government and eight banks signed agreements to facilitate lower-cost banking services for consumers. The following are among the low-cost options for monthly fees. (See file **Low_Fee_Accounts.**)

3.50	5.00	4.00	2.00	4.00	2.95	2.50
3.50	5.00	4.00	3.95	2.50	3.95	0.00

a) Calculate the first, second, and third quartiles for the fees.

b) Calculate the 5th and 95th percentiles for the fees.

c) In Chapter 3, Exercise 14, you constructed a frequency distribution for these same data. Would you say that the three quartiles, together, communicate the important features of the distribution? Is anything left out? Explain.

57. A chemist is experimenting with solvents to use in extreme-temperature environments, such as the open cargo bay of a NASA space shuttle. She will select from a list of 47 solvents for which the melting and boiling points are known. (See the columns "Mpt" (Melting Point) and "Bpt" (Boiling Point) in the file **Freezing_Boiling_Data.**)

a) Calculate the first and third quartiles for (i) the melting points of the solvents and (ii) the boiling points of the solvents.

b) Calculate the 15th and 85th percentiles for (i) the melting points of the solvents and (ii) the boiling points of the solvents.

c) If you randomly select one solvent, is its melting point more likely to be larger than the 3rd quartile or than the 85th percentile? Why?

58. The population density of a country is a measure of how many people live in the country relative to the surface area within its borders. In the column "Resident" in the file **Population_Density,** each value shows the population density of a particular country in persons (on average) per square kilometre.

a) Calculate the 1st and 3rd quartiles for the population density of the countries.

b) Calculate the 5th and 95th percentiles for the population density of the countries.

c) In Exercise 10 in this chapter, you calculated the mean for this variable. For this data set, why would communicating the quartiles along with the mean convey a better sense of the distribution? Explain.

59. A mineral collector has identified 100 metallic minerals and recorded for each the values of density and hardness. He is examining the distributions of these two variables. (See the file **Metallic_Minerals.**)

a) Calculate the first and third quartiles for (i) density and (ii) hardness.

b) Calculate the 10th and 90th percentiles for (i) density and (ii) hardness.

Chapter 5

Measures of Spread and Shape

• Chapter Outline

5.1 The Range, Variance, and Standard Deviation

5.2 The Empirical Rule and Interpreting the Standard Deviation

5.3 The Interquartile Range

5.4 Five-Number Summaries and Boxplots

5.5 Relative Variation

5.6 Assessing the Shape of a Distribution

• Learning Objectives

5.1 Explain measures of spread, and calculate the range, variance, and standard deviation for a population or sample.

5.2 Correctly select and apply the empirical rule or Chebyshev's inequality to interpret the standard deviation.

5.3 Calculate and interpret the interquartile range.

5.4 Construct five-number summaries and boxplots.

5.5 Calculate and interpret the coefficient of variation.

5.6 Describe the shape of a distribution.

5.1 The Range, Variance, and Standard Deviation

Chapter 4 introduced the notion of summarizing a data distribution with a small set of numerical values, called *measures of location.* We started by proposing a single number to do the job—the mean. Then we discovered that, by itself, the mean may present an incomplete, and possibly a misleading, picture of a distribution. Another problem with measures of location is that, by themselves, they do not measure the *variety* in values that make up a data set. Even among precision-manufactured parts, there is variety in the dimensions and properties of individual units.

Figure 5.1 illustrates the problem. Both distributions have identical measures of location. Their means, medians, and modal classes are all the same. Yet, obviously, the distributions are not identical. They are distinguished by how far individual values in the two populations tend to vary from that shared centre.

Figure 5.1
Two Frequency Distributions with the Same Mean

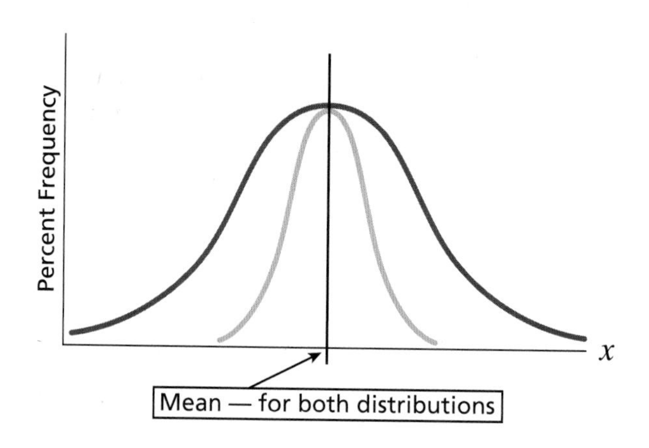

Seeing Statistics
Section 4.0.1

This chapter introduces some **measures of spread,** also called **measures of variation** or **measures of dispersion.** A measure of spread is a single number that describes the degree of variety among the values of a distribution.

The Range

The simplest measure of variation is the range. The **range** is the difference between the largest value in the data set and the smallest value. The formula for the range is

$$\mathbf{R = (Largest\ value) - (Smallest\ value)}$$

The range has the advantage of being easy to calculate. A disadvantage is that knowing the range tells us nothing about the actual distribution of the data set. Both distributions in Figure 5.2 have the same range, yet the data sets are clearly different. Another disadvantage of the range is that it is highly affected by outliers. The range of the grey distribution in Figure 5.2 appears to be a reasonable representation of the variation. On the other hand the range for the blue distribution appears to be overly influenced by an unusually large value. The range also cannot be used with nominal-level data.

Seeing Statistics
Section 4.1.1

The Mean Absolute Deviation

This section is included as a preview for the **standard deviation,** which is far more commonly used as a measure of variation. The standard deviation formula (given in the next

Figure 5.2
Two Frequency Distributions with the Same Range

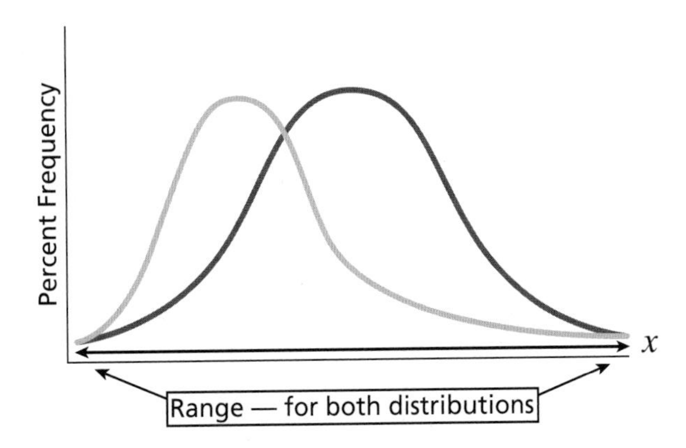

section) may look a bit complicated, but the basic concepts are illustrated here in our discussion of the **mean absolute deviation.**

Compare the distributions in Figure 5.1. We see that there is more variation "on average" in the grey distribution than in the blue distribution. Can we quantify this intuitive notion of a "greater average distribution"? A first attempt is to calculate a **mean deviation:**

$$\text{Mean Deviation} = \frac{\sum_{i=1}^{n}(x_i - \bar{x})}{n}$$

where x_i = Individual observation in the data set
$\bar{x}$ = Mean of the data set (use μ if the data set is a population rather than a sample)
n = Number of values in the data set (use N if the data set is a population rather than a sample)

The formula presumes that a mean has already been calculated for the data set. For each value in the data set (x_1, x_2, etc.), find its distance from that overall mean. The distance of a value from the mean is called its **deviation** from the mean. The formula finds the arithmetic mean of all of these deviations: All deviations are added, and the sum is divided by how many numbers (and so how many deviations) there are. Intuitively, this should tell us the typical amount by which actual x-values are varying from the mean.

Figure 5.3 illustrates the calculation of a mean deviation. It also demonstrates why the technique cannot be used without modification. The mean number of votes in each British Columbia riding in the 2004 federal election was calculated in Chapter 4 (equal to 48,049.75). The figure shows a column with the heading "x_i – mean." For each value in the data set (each x_i), a value equal to the x_i minus 48,049.75 (the mean) is calculated and displayed. The total of the "x_i – mean" column is the summation of the calculated differences. Divide this sum by n to get the mean deviation (which equals zero).

Unfortunately, the mean deviation is *always* equal to zero, because the formula for the mean was designed precisely to balance all of the plus and minus deviations of a data set! The mean deviation is not, in practice, very helpful for comparing distributions.

Figure 5.3
Mean Deviation for Numbers of Votes in BC Ridings, 2004 Federal Election

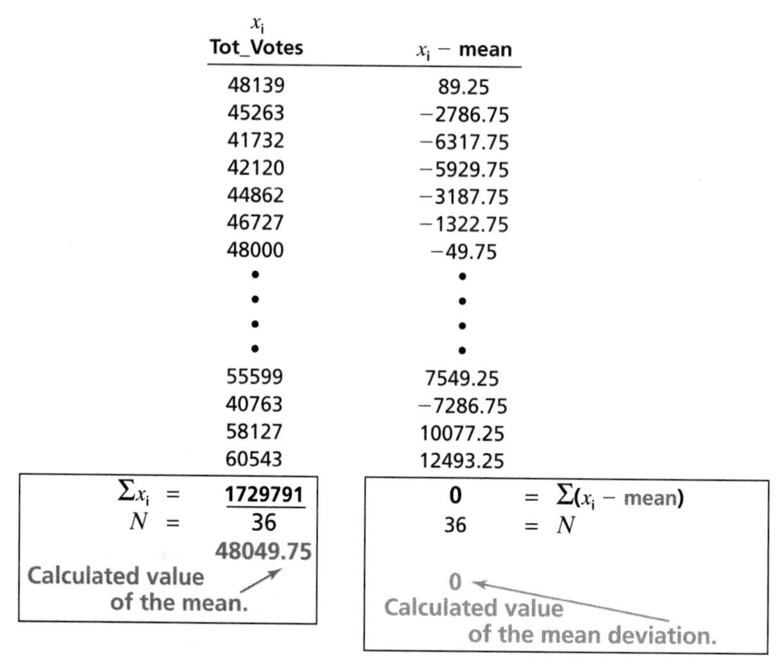

x_i Tot_Votes	x_i − mean
48139	89.25
45263	−2786.75
41732	−6317.75
42120	−5929.75
44862	−3187.75
46727	−1322.75
48000	−49.75
•	•
•	•
•	•
•	•
55599	7549.25
40763	−7286.75
58127	10077.25
60543	12493.25

Σx_i = 1729791
N = 36
48049.75
Calculated value of the mean.

0 = $\Sigma(x_i$ − mean)
36 = N
0
Calculated value of the mean deviation.

Source: "Federal Election Results, 2004." *The Globe and Mail.* Wednesday, June 30, 2004. Pages A12–A13. Reprinted with permission from *The Globe and Mail.*

As an alternative, we could calculate the mean absolute deviation. Figure 5.4 is very similar to Figure 5.3, but this time we ignore whether an x-value is above or below the mean for the data set; we are interested only in the *magnitudes* (also called *absolute values*) of deviations from the mean. In this revised formula, the numerator is the sum of all values' absolute distances from the mean—not caring whether a distance is positive or negative. Dividing this sum by n gives the mean distance by which values have varied from the mean.

$$\text{Mean Absolute Deviation} = \frac{\sum_{i=1}^{n} |x_i - \bar{x}|}{n}$$

where x_i = Individual observation in the data set
$\bar{x}$ = Mean of the data set (use μ if the data set is a population rather than a sample)
n = Number of values in the data set (use N if the data set is a population rather than a sample)
$|\ |$ is the absolute value operator. Consider only the magnitude of the expression within the lines, and treat the sign as positive.

Seeing Statistics
Section 4.2.1–4.2.2

The mean absolute deviation (MAD) is a reasonable measure of variation, but is not widely used. Mathematically, the absolute value operator is not suitable for certain advanced statistical calculations that have been developed.

Figure 5.4
Mean Absolute Deviation for Numbers of Votes in BC Ridings, 2004 Federal Election

x_i Tot_Votes	$\|x_i - \text{mean}\|$
48139	89.25
45263	2786.75
41732	6317.75
42120	5929.75
44862	3187.75
46727	1322.75
48000	49.75
•	•
•	•
•	•
•	•
55599	7549.25
40763	7286.75
58127	10077.25
60543	12493.25

Σx_i = 1729791
N = 36
48049.75
Calculated value of the mean.

219628.5 = $\Sigma |x_i - \text{mean}|$
36 = N
6100.79
Calculated value of the mean absolute deviation.

Source: "Federal Election Results, 2004." *The Globe and Mail.* Wednesday, June 30, 2004. Pages A12–A13. Reprinted with permission from *The Globe and Mail.*

The Variance and Standard Deviation of a Population

Another way to avoid the negative signs that make the mean deviation formula unhelpful is to *square* the differences between each data value and the mean—instead of taking the absolute values—before adding the adjusted differences. Then take a final square root of the result, to compensate for the squaring operation that was introduced. Formulas 5.1 and 5.2 illustrate the two-part sequence that results, as applied to a *population:*

Formula 5.1 Variance for a Population

$$\sigma^2 = \frac{\sum_{i=1}^{N} (x_i - \mu)^2}{N}$$

where σ (pronounced "sigma") is the population standard deviation, to be calculated.
σ^2 is called the *variance* of the population (discussed below).
x_i = Individual observation in the data set
μ = Mean of the population
N = Number of values in the population

Formula 5.2 Standard Deviation for a Population

$$\sigma = \sqrt{\sigma^2}$$

Observe how the right side of Formula 5.1 applies an approach similar to the mean absolute deviation: The deviation between each data value and the mean is calculated and adjusted to remove negative signs; this is done by squaring the magnitude of the deviation, instead of by taking the absolute value of the magnitude. Adding these squared differences and dividing by how many differences there are gives a mean—the mean of the *squared* deviations. Formula 5.2 "undoes" the squaring, to produce a result more in line with the "typical" distance between individual x-values and the mean.

Because of squaring of the differences, a new intermediate variable is created, called the **variance.** Note that the variance is not given its own symbol (such as "V," etc.) but is represented by σ^2. This symbol explicitly reminds us that we have not finished calculating the desired standard deviation (σ) until we take the square root of the variance. Formula 5.3 shows how the formula for a standard deviation can also be written as a single equation.

Formula 5.3 Standard Deviation for a Population

$$\sigma = \sqrt{\frac{\sum_{i=1}^{N} (x_i - \mu)^2}{N}}$$

In Figure 5.5, we see how the standard deviation can be calculated for the number of votes cast in BC ridings in the 2004 federal election. We have data for all of the ridings, so this is a population. A column is created to show, for each x_i-value, the value of $(x_i - 48049.75)^2$ (i.e., the square of the value's deviation from the mean of the x-values). Taking the sum of these squared deviations, and dividing this sum by N gives the mean of the squared deviations; so a final square root is needed to obtain the standard deviation.

Seeing Statistics

Sections 4.3.1– 4.3.2

Example 5.1

*(Based on the **Birds_Largest_Wingspan** file)*

An ornithologist keeps her own collection of unique birds. Each bird has an unusually large wingspan for its species, and the ornithologist is studying the variation among the wingspans of her birds. The inventory of birds and their wingspans are stored in the file **Birds_Largest_Wingspan.** Find the range and standard deviation for the population. How do these measures compare in terms of the information they give you?

Solution

Both questions can be solved manually, although the computer can simplify the repetitive operations of squaring the mean differences and so on. The necessary steps are shown in Figure 5.6.

The range is calculated as the largest value minus the smallest value = 3.6 – 1.7 = 1.9 metres. As shown in the figure, the standard deviation can be calculated in two parts: The mean of the squared deviations (which is the variance) works out to 6.2723/13 =

Figure 5.5
Standard Deviation for Numbers of Votes in BC Ridings, 2004 Federal Election

x_i Tot_Votes	$(x_i - \text{mean})^2$
48139	7965.56
45263	7765975.56
41732	39913965.1
42120	35161935.1
44862	10161750.1
46727	1749667.56
48000	2475.0625
•	•
•	•
•	•
•	•
55599	56991175.6
40763	53096725.6
58127	101550968
60543	156081296

Σx_i = 1729791
N = 36
48049.75
Calculated value of the mean.

1886764273 = $\Sigma(x_i - \text{mean})^2$
36 = N
52410118.7 = Variance
7239.48
Calculated value of the standard deviation.

Source: "Federal Election Results, 2004." *The Globe and Mail.* Wednesday, June 30, 2004. Pages A12–A13. Reprinted with permission from *The Globe and Mail.*

Figure 5.6
Measures of Variation for Wingspans of Birds

Bird	x_i Wingspan	$(x_i - \text{mean})^2$
Albatross	3.6	0.90982249
White pelican	3.6	0.90982249
Marabou stork	3.3	0.42751479
Andean condor	3.2	0.30674556
Hooping swan	3.0	0.1252071
Bearded vulture	3.0	0.1252071
Bald vulture	2.8	0.02366864
Grey crane	2.5	0.02136095
Golden eagle	2.5	0.02136095
White stork	1.8	0.71597633
Grey heron	1.7	0.8952071
Gannet	1.7	0.8952071
Seagull	1.7	0.8952071

Σx_i = 34.4
N = 13
2.6462
Calculated value of the mean.

6.27230769 = $\Sigma(x_i - \text{mean})^2$
13 = N
0.48248521 = Variance
0.69
Calculated value of the standard deviation.

Range = Largest − Smallest = 3.6−1.7 = 1.9

Source: factophile.com. Found at: http://www.factophile.com/show.content?action=view&pageid=12

0.4825 m. The standard deviation is the square root of the variance, and equals 0.69 metres.

The range gives a sense of the maximum of the variation: You will not find any birds farther apart in wingspan than the range—1.9 metres. This information might be useful if you were shopping for a common cage to hold two birds. For example: A cage that can handle only 1 metre of variation in wingspan will not be sufficient, but you need not pay for a cage that accommodates a 3-metre difference. On the other hand, the standard deviation gives a sense of the "typical" variation of the birds' wingspans from the mean value. A bird whose specific wingspan varied from the mean by 0.5 m or 0.9 m is probably not surprising, but a bird whose wingspan varied by 5.0 m would be quite unexpected.

The Variance and Standard Deviation of a Sample

In inferential statistics, we start from samples and the population parameters must be estimated. In Chapter 4, we introduced different notations for calculations with samples versus populations (for example, use $\overline{x}$ for the mean of a sample, but μ for the mean of a population). Until now, however, the actual calculations were the same. For the standard deviation, there is a slight difference in how the sample versus the population versions is calculated. For samples, compare Formulas 5.4, 5.5, and 5.6 with their counterparts for populations (Formulas 5.1, 5.2, and 5.3).

Formula 5.4 Variance for a Sample

$$s^2 = \frac{\sum_{i=1}^{n} (x_i - \overline{x})^2}{(n - 1)}$$

Formula 5.5 Standard Deviation for a Sample

$$s = \sqrt{s^2}$$

Formula 5.6 Standard Deviation for a Sample

$$s = \sqrt{\frac{\sum_{i=1}^{n} (x_i - \overline{x})^2}{(n - 1)}}$$

where
s is the *sample* standard deviation, to be calculated.
s^2 is called the variance of the sample.
x_i = Individual observation in the data set
$\overline{x}$ = Mean of the sample
n = Number of values in the sample

Apart from the symbols, the only difference between these formulas and the population versions is the denominator $n - 1$ (number of sample values *minus one*) as opposed to the denominator N (the *full* number of values in the population). Without this adjustment, the sample variance and standard deviation would be **biased estimators** for the population variance and standard deviation. This means that, starting from sample data, use of formulas without this adjustment would tend to systematically underestimate the corresponding parameters in the population.

This factor $n - 1$ is known as the **degrees of freedom** for the calculation. To calculate the variance, first the mean had to be estimated. But once the mean is given, only $n - 1$ of the remaining data points convey new information about the variance. (Once you know all of the data values but one, *and* you know the mean, the one remaining value is algebraically predetermined, with no "freedom" to convey new information.) In general, the degrees of freedom equal the sample size minus the number of parameters (such as the mean) that are estimated as part of calculating the statistic.

Seeing Statistics

Sections 4.3.3–4.3.5

Example 5.2

*(Based on the **Canadian_Lakes** file)*

A naturalist is studying a certain type of fish that lives only in large Canadian lakes. She randomly selects 29 of the lakes that the species is known to inhabit. (The list of sampled lakes is saved in the file **Canadian_Lakes.**) The surface area (in square kilometres) of each sampled lake has been recorded. Estimate the mean and standard deviation of the areas for *all* of the lakes that this type of fish inhabits. Can you see any problems with using the standard deviation as the sole measure of variation in this case?

Solution

The calculations for solving this problem are shown in Figure 5.7. Other than the denominator for calculating the variance (using $n - 1$), the process is the same as for the BC Votes example, where we had population data. Here, we have data for only a random selection of the lakes, yet we want to estimate the standard deviation for the full population. As explained above, we therefore need to adjust the calculation for the degrees of freedom. From the figure: The mean lake area is 3,705.2 square kilometres, with a standard deviation of 7,509.5 square kilometres.

The problem with the standard deviation for this example, as a measure of variation, is the same as the problem for the mean: They are both highly influenced by a handful of large values. Only 6 of the 29 lakes are even larger than 3,700 square kilometres in size. It seems odd to say that a value over twice that amount (7,509.5) estimates the "typical" variation in the lake sizes. The naturalist may be wise to consider another measure (such as the interquartile range, which will be discussed in Section 5.3), or else she might handle the two large outliers separately and just find the standard deviation for the other lakes.

• *In Brief 5.1 Calculating the Range and Standard Deviation*

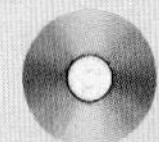
*(Based on the **Canadian_Lakes** file)*

Excel

If your data are a sample, review the Excel procedures in "In Brief 4.1." As shown there in the illustration, the output includes the range and standard deviation, as well as the variance.

Figure 5.7
Variance and Standard Deviation for the Areas of Selected Lakes

Lake	x_i Area	$(x_i - \text{mean})^2$
Atlin Lake	775	8586112.457
Aylmer Lake	847	8169346.663
Bras d´Or Lake	1099	6792314.388
Cedar Lake	1353	5532877.284
Clinton-Colden Lake	737	8810252.181
Contwoyto Lake	957	7552641.146
Doré Lake	640	9395493.319
Dubawnt Lake	3833	16331.07729
Ennadai Lake	681	9145827.353
Gods Lake	1151	6523972.87
Great Bear Lake	31328	763018698.8
Great Slave Lake	28568	618158480.9
Lac Bienville	1249	6032952.319
Lac de Gras	633	9438455.215
Lac La Rouge	1413	5254212.457
Lac Mistassini	2335	1877466.939
Lac Saint-Jean	1003	7301922.112
Lac Seul	1657	4195151.491
Lake Athabasca	7935	17891149.7
Lake Simcoe	744	8768746.284
Lake St. Clair	490	10337555.39
Lake Winnipegosis	5374	2784870.422
Playgreen Lake	657	9291565.284
Reindeer Lake	6650	8671806.422
Réservoir Cabonga	677	9170037.008
Réservoir Gouin	1570	4559108.491
Réservoir Pipmuacar	978	7437657.457
Tulemalu Lake	668	9224625.732
Yathkyed Lake	1449	5090469.56

Σx_i = 107451.0
n = 29
3705.2069
Calculated value of the mean.

1579030101 = $\Sigma(x_i - \text{mean})^2$
28 = $n - 1$
56393932.17 = Variance
7509.59
Calculated value of the standard deviation.

Source: Natural Resources Canada, Statistics Canada: GeoAccess Division. Found at: http://www40.statcan.ca/l01/cst01/phys05.htm

If you have data for a full population, you will have to use an Excel formula. Place the cell pointer in an empty cell and input a formula of this format: **=STDEVP(C2:C30)**. (Replace **C2:C30** with the range that contains the data.) Also, shown below are formulas that can be used as an alternative way for calculating a sample standard deviation and the range. For the variance, you can square the appropriate standard deviation.

Population Standard Deviation	7378.97783	= STDEVP(C2:C30)
Sample Standard Deviation	7509.589348	= STDEV(C2:C30)
Range	30838	= MAX(C2:C30)–MIN(C2:C30)

Minitab

If your data are a sample, use the menu sequence *Stat / Basic Statistics / Display Descriptive Statistics.* In the first dialogue box (illustrated in the Minitab procedures in "In Brief 4.1"), click on and select the variable to be analyzed. Click on the *Statistics* button at the bottom of the dialogue box, and then, as shown in the screen capture in this feature, select *Standard Deviation, Variance, Range,* and any other desired statistics. (Several are preselected by default.) Click *OK* to close this dialogue box, and then click *OK* in the first dialogue box.

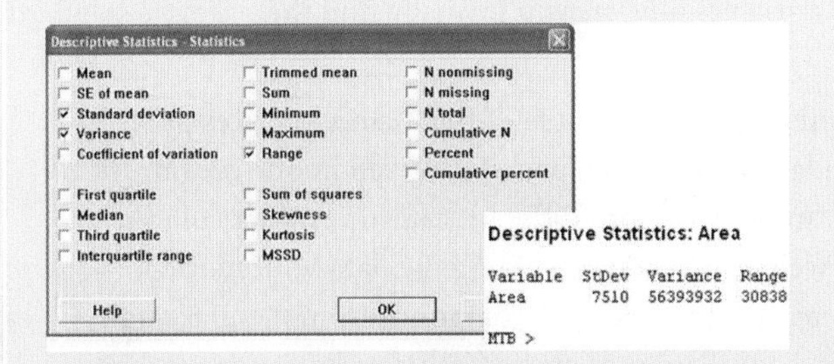

Minitab generally *assumes* that your raw data represent a sample. If your data represent a full population, you must take steps to convert the default answer: Use the menu sequence *Calc / Calculator* to open the calculator. As shown in the second screen capture in this feature, input a name for the solution column in the *Store result in variable* box. In the *Expression* box, input a formula of this format:

STDEV(Area)/SQRT(N(Area)/(N(Area)-1))

replacing **Area** with the name of your data column. The output (shown in the illustration) results from taking the sample standard deviation for the variable and converting it into the population version. For the population variance, square the population standard deviation.

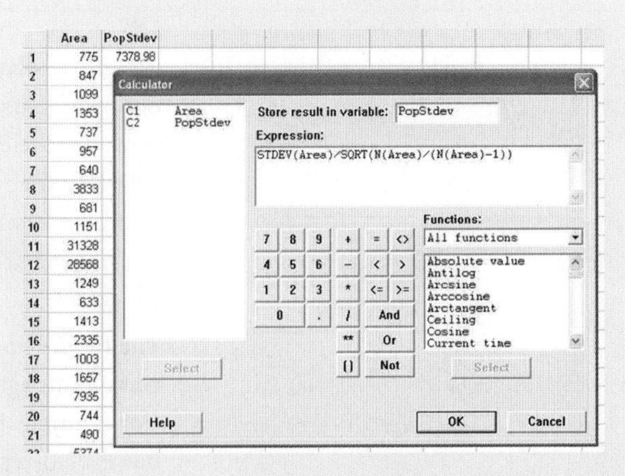

SPSS

If your data are a sample, use the menu sequence *Analyze / Descriptive Statistics / Frequencies.* In the first dialogue box (illustrated in the SPSS procedures in "In Brief 4.1"), select the variable to be analyzed. Click on the *Statistics* button at the bottom of this dialogue box and then, in the next dialogue box (see the screen capture in this feature), select *Standard Deviation, Variance, Range,* and any other desired statistics. Click *Continue* to exit from this dialogue box, and then click *OK* in the first dialogue box.

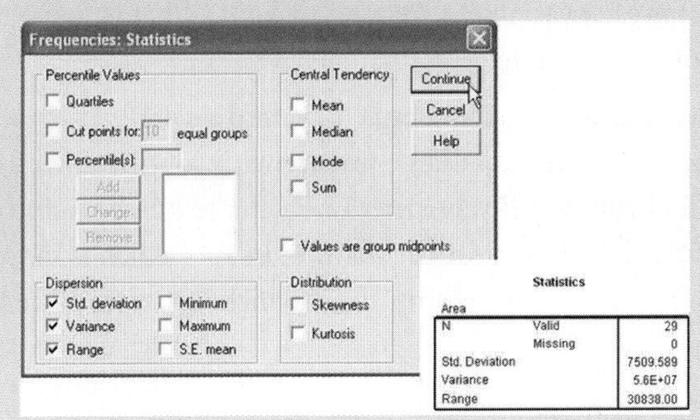

SPSS generally *assumes* that your raw data represent a sample. If your data represent a full population, you must take an extra step to convert the default answer: Use the formula $\sigma = \dfrac{s}{\sqrt{\dfrac{n}{(n-1)}}}$.

In the example, let SPSS calculate $s = 7{,}509.589$.

Then, calculate $\sigma = \dfrac{7{,}509.589}{\sqrt{\dfrac{29}{28}}} = 7{,}378.98$.

• Hands On 5.1 Getting a Feel for the Standard Deviation

 *(Based on the files **Brilliants** and **Airline_Complaints**)*

We saw in Example 5.2 that, like the mean, the standard deviation works best if the data set is approximately symmetrical and if more values are relatively near the mean than extremely far away. (The shape of a distribution is **symmetrical** when, if we could fold up a diagram of the distribution [e.g., its histogram] along a centre line, the diagram would line up with itself on each side.) That said, the standard deviation is the most commonly used measure of variation, and it would be useful to get a feel for what information it conveys under a variety of circumstances.

Open the following two files on the CD provided with your textbook and examine the described variables—*before* you formally find their sample standard deviations.

1. The file **Brilliants** contains a sample of the diamond prices published online by Canada Diamonds. All prices are given in one column, but note that the diamonds fall into different categories (first column) based on their weights in carats. Dividing the data set if needed, examine *separately* (a) the prices for diamonds from 0 to 0.5 carats, and (b) the prices for diamonds above 0.5 carats but no more than 1.0 carat.
2. The file **Airline_Complaints** lists the number of formally recorded complaints against different airline companies by complaint category (for example, flight problems, "over sales," fares, and so on). Ignore the Yes/No-type columns. Interpret the data as a sample that is representative of complaint rates and categories that might hold into the future.
3. For each of the described variables in the **Brilliants** and **Airline_Complaints** files:
   a) Use techniques from Chapter 3 to get a sense of the distributions of these variables. (For example, construct a histogram; don't worry about exact class widths, etc.)
   b) Before making any formal calculations, try to think of what answer you would give for each variable if someone asked: "What is the 'typical' variation for that variable?"
   c) Use the computer to calculate the sample standard deviation for each of the variables. How do these values compare with your intuitive guesses in (b)? Did you find it was easier or harder to guess a value close to the standard deviation for the diamond data (which is more symmetrical) or the airline complaints data (which has more extreme values)? Explain.

d) In practice, who might be interested in knowing the standard deviation of such data sets, and what might be the consequences if these people were unaware of the actual distribution? Short of providing a full table or figure, is there other information you could tell these people to help convey a clearer picture?

5.2 The Empirical Rule and Interpreting the Standard Deviation

Generally, you will encounter the standard deviation along with the mean. The latter, as we have seen, gives us a sense of where the data are located. The standard deviation adds to our understanding of the data, by providing a measure of the degree of variation in the data set. The larger the standard deviation, the more widely dispersed are the data values around the mean position. That is, a larger standard deviation suggests a greater dispersion among the data values.

The Empirical Rule (or the 68–95–99% Rule)

Regardless of whether the standard deviation is large or small, it can allow us to estimate the proportions of data that lie within certain distances of the mean. These estimates can be relatively precise if the population is bell-shaped and symmetrical, as in Figure 5.8. Under those conditions, there is a mathematical relationship between the distance of a value from the mean and the proportion of data values that lie within that distance from the mean. Three of the easiest-to-remember proportions are shown in this figure. If these guidelines are expressed in words, they are often referred to as the **empirical rule** or the **68–95–99% rule.**

Figure 5.8
The Empirical Rule (or the 68–95–99% Rule)

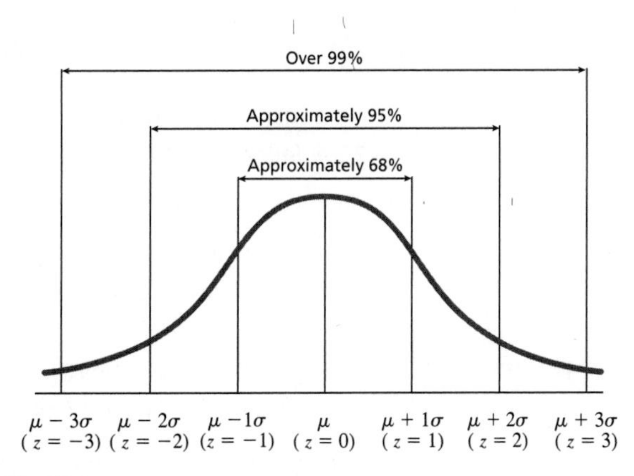

For normal distributions:
Proportion of values falling within one, two, or three standard deviations of the mean

Figure 5.8 shows that, for a suitably distributed variable, approximately 68% of all of the values for the variable lie within an area bounded by one standard deviation below the mean ($\mu - 1\sigma$) up to one standard deviation above the mean ($\mu + 1\sigma$). Suppose a population of bags to be lifted has a mean of 10 kg and a standard deviation of 1 kg. It follows from the empirical rule that about 68% of the bags' weights fall between (10 – 1) = 9 kg and (10 + 1) = 11 kg. If you are able to handle weights in this range from 9 to 11 kg, the rule gives some confidence that you can handle at least two-thirds of the bags.

The figure shows further that approximately 95% of all of the values for the variable have values from two standard deviations below the mean ($\mu - 2\sigma$) up to two standard deviations above the mean ($\mu + 2\sigma$). And, finally, over 99% of all values (in fact, about 99.7%) have values from three standard deviations below the mean ($\mu - 3\sigma$) up to three standard deviations above the mean ($\mu + 3\sigma$).

Continuing the weight example, you can be very confident that almost every bag (over 99%) will have a weight between (10 – 3) = 7 kg and (10 + 3) = 13 kg. If you are able to handle all of the weights in that range, the empirical rule assures you that you are prepared for almost every case. Notice that the curve never quite touches down to a frequency of zero, so there is always a remote chance of an occasional outlier. This possibility could be addressed by creating a contingency plan. (For example, maybe you could borrow a dolly if you encounter any unusually large bags.)

Figure 5.8 expresses the value ranges using two notations. In the upper row, values are expressed in terms of the *population* mean (μ) and standard deviation (σ). The rule is derived from the properties of a "normal" (symmetrical, bell-shaped) population distribution. If a sufficiently large random *sample* is taken from such a population, then the sample is likely to be representative of that same distribution. Under those conditions, the empirical rule can also be applied to the sample: You would expect 68% of the sample values to fall within one sample standard deviation (s) from the sample mean ($\bar{x}$), and so on.

The second notation used for representing value ranges in the figure is based on z-values. These will be looked at later in the section.

Example 5.3

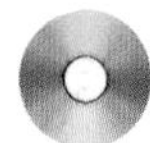 *(Based on the **BC_votes** file)*

We previously analyzed data for the numbers of votes cast in British Columbia ridings in the 2004 federal election. Find the raw data in the file **BC_votes.**

1. Apply the empirical rule to this data set, and interpret.
2. Using procedures from Chapter 3, determine the *actual* proportions of data values that occur within the regions identified in (1), and compare.

Solution

1. Using the computer, you can easily find that the mean for the numbers of votes cast per riding is 48,050. The standard deviation is approximately 7,342. From these parameters, the empirical rule tells us that approximately 68% of all of the data values lie between ($\mu - 1\sigma$) = 40,708 and ($\mu + 1\sigma$) = 55,392. This implies that about two-thirds of all of the ridings in British Columbia had numbers of votes cast in the range between 40,708 and 55,392. The empirical rule tells us further that approximately 95% of all the values lie between ($\mu - 2\sigma$) = 33,366 and ($\mu + 2\sigma$) = 62,734. We therefore expect it would be unusual to find a riding in BC whose number of votes cast was outside that range—although it does happen (about 5% of the time). Finding a riding with less than 26,024 votes cast ($\mu - 3\sigma$) or more than 70,076 ($\mu + 3\sigma$) would be quite exceptional. These expectations are shown in Figure 5.9.

Figure 5.9
An Approximate Application of the Empirical Rule

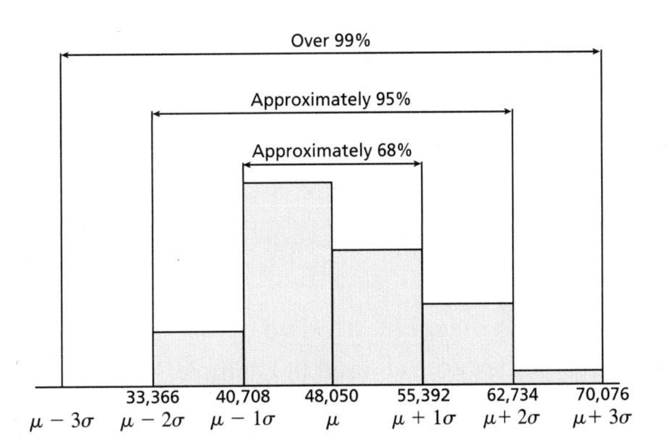

Source: "Federal Election Results, 2004." *The Globe and Mail.* Wednesday, June 30, 2004. Pages A12–A13. Reprinted with permission from *The Globe and Mail.*

2. Figure 5.10 shows that the predictions from the empirical rule can work quite well—even if a distribution is *not perfectly* symmetrical (see Figure 5.9). The table at the top of the figure reiterates the boundaries expected by the rule. The second table in the figure shows the relative frequency distribution for the dataset, based on the boundaries just specified. The two classes marked in green cover the range from ($\mu - 1\sigma$) up to ($\mu + 1\sigma$). If we add their relative frequencies, we account for $0.417 + 0.278 = 0.695$ of all of the values. Not too far from the expected 0.68! Adding in the relative frequencies for the next classes below (down to $\mu - 2\sigma$) and above (up to $\mu + 2\sigma$), respectively, the net proportion of values included is now 0.973; again, close to the estimate 0.95. Lastly, in this particular population *all* of the values fall within plus or minus three standard deviations of the mean.

Figure 5.10
A Comparison of Empirical Rule Estimates with Actual Values

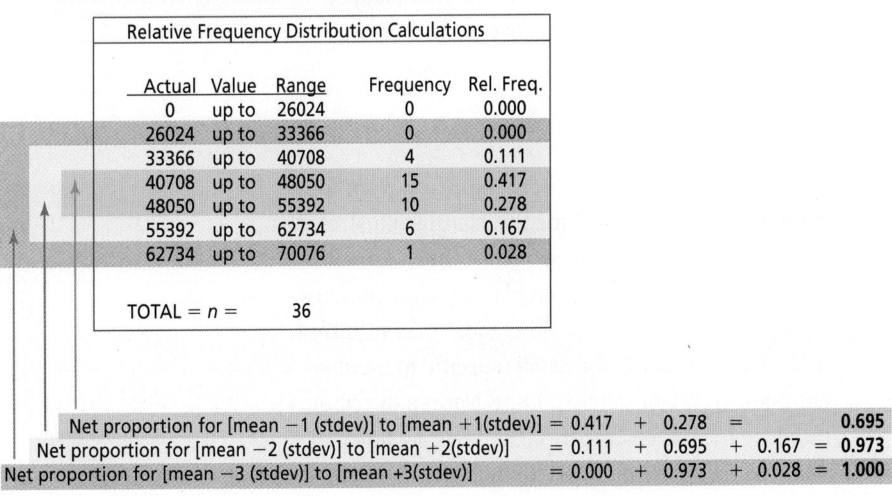

Mean −3(stdev) =	26024
Mean −2(stdev) =	33366
Mean −1(stdev) =	40708
Mean =	48050
Mean +1(stdev) =	55392
Mean +2(stdev) =	62734
Mean +3(stdev) =	70076

Relative Frequency Distribution Calculations

Actual	Value	Range	Frequency	Rel. Freq.
0	up to	26024	0	0.000
26024	up to	33366	0	0.000
33366	up to	40708	4	0.111
40708	up to	48050	15	0.417
48050	up to	55392	10	0.278
55392	up to	62734	6	0.167
62734	up to	70076	1	0.028
TOTAL = n =		36		

Net proportion for [mean −1 (stdev)] to [mean +1(stdev)]	= 0.417	+	0.278	=			**0.695**
Net proportion for [mean −2 (stdev)] to [mean +2(stdev)]	= 0.111	+	0.695	+	0.167	=	**0.973**
Net proportion for [mean −3 (stdev)] to [mean +3(stdev)]	= 0.000	+	0.973	+	0.028	=	**1.000**

Source: "Federal Election Results, 2004." *The Globe and Mail.* Wednesday, June 30, 2004. Pages A12–A13. Reprinted with permission from *The Globe and Mail.*

• *In Brief 5.2 Generating a Random Number Sequence*

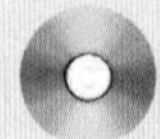

(Based on the ***BC_votes*** *file)*

There is no "In Brief" section for the empirical rule, since there are no associated computer functions to apply it. Procedures to find the mean, standard deviation, and relative frequencies were all used in Example 5.3, but procedures for these have already been addressed. This "In Brief" feature will be used, instead, to introduce a very useful procedure that is applied in the next "Hands On" section, as well as elsewhere in the text. The general task is to generate a sequence of pseudo-random numbers that conforms to a distribution you specify (such as every number being equally probable, called a *uniform distribution,* or a distribution having a bell or "normal" shape). In the procedures below, the task is to simulate a random sampling, with replacement, of 1,000 values based on **BC_Votes** data, which are approximately normally distributed with a mean = 48,050 and a standard deviation = 7,324.

Excel

Starting with an empty spreadsheet, use the menu sequence *Tools / Data Analysis / Random Number Generation.* In the resulting dialogue box (see the screen capture), specify 1 for the number of variables to generate. Next, input how many values you want to generate per variable (in this case, 1,000). Click to display the drop-down menu, and select the type of distribution you want to simulate. (Choose *Normal* for this example.) Then input, where labelled, the intended values for the distribution's mean (e.g., 48,050) and standard deviation (e.g., 7,324). The numbers you randomly generate can be interpreted as a random sample from the population you have described. Arbitrarily choose and input a "random seed." If you ever reuse the same seed number, you will find that you can regenerate exactly the same pseudo-random sequence on another occasion; however, if a new seed is chosen, no one can predict the numbers that will follow. Lastly, identify an output range (input the upper cell address) or else send the output to another worksheet page. Click *OK.*

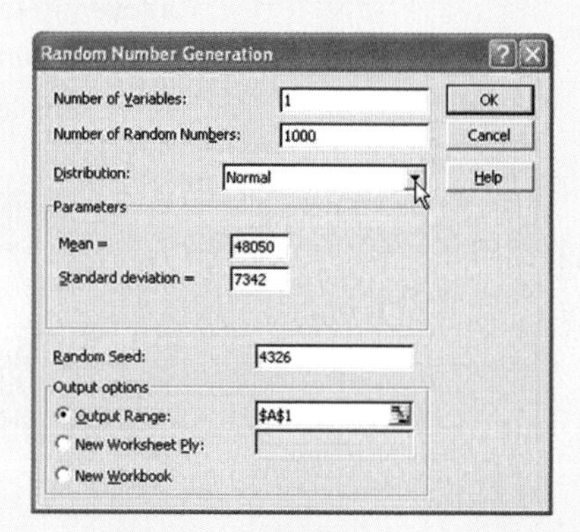

Minitab

Decide first what type of distribution you want to simulate (choose *Normal* for the example illustrated here). Then use the menu sequence *Calc / Random Data / {distribution name}* (replacing *{distribution name}* with the desired menu choice). The resulting dialogue box will ask for details specific to the distribution type you selected. For a *Normal* distribution (see the screen capture), first indicate how many rows (data values) are to be generated per variable (in this case, 1,000) and then input the

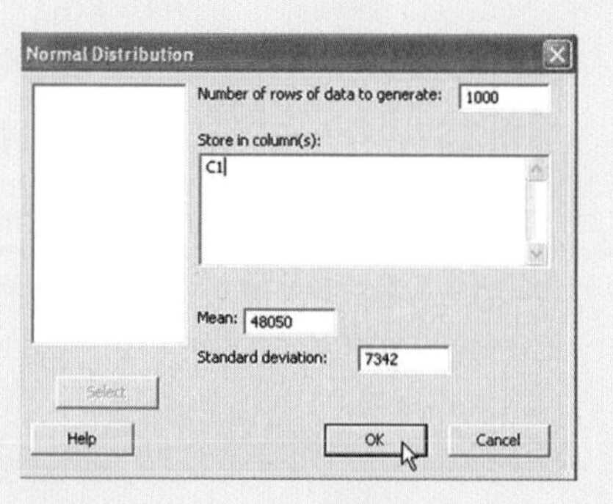

column number where the values are to be placed. Where specified at the bottom, input the intended values for the distribution's mean (e.g., 48,050) and standard deviation (e.g., 7,324). The set of numbers you randomly generate can be interpreted as a random sample from the population you have described. Click *OK.*

SPSS

In SPSS, a random number sequence can be generated by writing a small program, or "syntax" code. Unfortunately, the student version of this program, which many students use, does not support running syntax code. The following procedure offers an alternative.

Start by opening the SPSS file called **pre_random.sav.** Column 1 of the file contains 1,000 repetitions of 1.0 (see the spreadsheet behind the screen capture). To prepare for randomizing, use the menu sequence *Transform / Random Number Generators.* Select *Set Starting Point* and *Random* under *Active Generator Initialization.* This procedure will ensure to generate a fully random seed every time you start the following sequence.

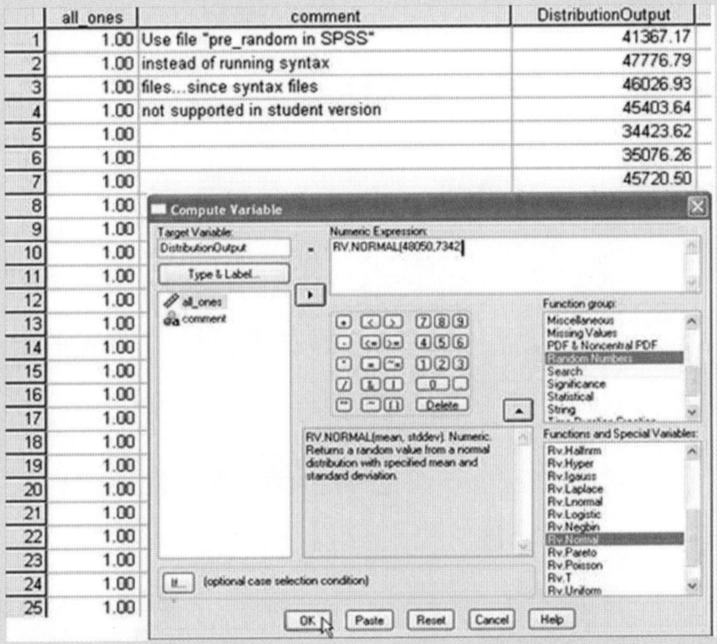

	all_ones	comment	DistributionOutput
1	1.00	Use file "pre_random in SPSS"	41367.17
2	1.00	instead of running syntax	47776.79
3	1.00	files...since syntax files	46026.93
4	1.00	not supported in student version	45403.64
5	1.00		34423.62
6	1.00		35076.26
7	1.00		45720.50
8	1.00		
9	1.00		
10	1.00		
11	1.00		
12	1.00		
13	1.00		
14	1.00		
15	1.00		
16	1.00		
17	1.00		
18	1.00		
19	1.00		
20	1.00		
21	1.00		
22	1.00		
23	1.00		
24	1.00		
25	1.00		

Compute Variable

Target Variable: DistributionOutput

Numeric Expression: RV.NORMAL(48050,7342)

RV.NORMAL(mean, stddev). Numeric. Returns a random value from a normal distribution with specified mean and standard deviation.

Now you are ready to use the menu sequence *Transform / Compute.* In the upper left of the dialogue box shown in the screen capture, input a name for the column to hold the data. (The output column "Distribution Output" is shown in the illustration.) In the *Numeric Expression* box, input the desired randomizing (RV) function from those available in SPSS. **RV.NORMAL** has been selected in the screen capture. Each randomizing function requires an appropriate input; **RV.NORMAL** requires the mean and standard deviation of the population to be simulated (in this example, 48,050 and 7,324, respectively). After clicking *OK,* a value will appear to the right of each of the 1.00 repetitions in the loaded file; each value will have been pseudo-sampled from the population of the specified mean and standard deviation.

• Hands On 5.2(a) Confirming the Empirical Rule

We saw in Example 5.3 a possible application of the empirical rule. The example distribution was not exactly symmetrical, and the proportions of values within one, two and three standard deviations of the mean were not exactly as specified in the rule. But such variation is typical in real-world examples.

Is it possible, without doing a formal mathematic proof, to informally check whether the rule really works? Yes, this can be done by simulation. Refer to "In Brief 5.2" for instructions on how to use the computer to generate a normally distributed sequence of 1,000 random numbers, with a given mean and standard deviation. Using the methods described in Chapter 3, code each of the 1,000 data values based on the range within it falls (see the bottom of Figure 5.11). Construct the frequency distribution and fill in the blank sections shown in the figure. How closely do the relative frequencies in the distribution match the expected proportions according to the empirical rule? Repeat the entire exercise four more times. Are you reasonably convinced that the rule is valid? Explain.

Figure 5.11
Confirming the Empirical Rule

Randomly Generate a Normal Distribution with These Parameters:

Mean	48050
Stdev	7342

Mean −3(stdev) =	26024
Mean −2(stdev) =	33366
Mean −1(stdev) =	40708
Mean =	48050
Mean +1(stdev) =	55392
Mean +2(stdev) =	62734
Mean +3(stdev) =	70076

For applying the empirical rule, these same boundaries are expected to apply. (Compare Figure 5.10)

Relative Frequency Distribution Calculations

Actual Value Range			Frequency	Rel. Freq.
0	up to	26024		
26024	up to	33366		
33366	up to	40708		
40708	up to	48050		
48050	up to	55392		
55392	up to	62734		
62734	up to	70076		
TOTAL = n =				

Complete the table based on your simulation.

Complete the calculations based on your table's relative frequencies.

Net proportion for [mean −1(stdev)] to [mean +1(stdev)] = ____ + ____ = ____
Net proportion for [mean −2(stdev)] to [mean +2(stdev)] = ____ + ____ + ____ = ____
Net proportion for [mean −3(stdev)] to [mean +3(stdev)] = ____ + ____ + ____ = ____

Counting Standard Deviations from the Mean: Finding the z-Value

In Figure 5.8, which introduced the empirical rule, there are two rows of labels for the horizontal (x) axis. Values along the axis represent possible values of a variable. Labels in the upper row express the variable values as equalling the mean value for the variable plus or minus a number of standard deviations. For example, if the mean of a population is 100 kg and the standard deviation is 10 kg, then the axis position labelled "$\mu + 2\sigma$" represents the value 100 + 2(10) = 120 kg.

However, the real-world values (in kilograms, parts per million, etc.) of points along the x-axis convey nothing about the proportions of data expected to lie above or below them in the distribution. To link the original data to the empirical rule or to similar rules (discussed in the next section), we need to convert data values into measurements of "how many standard deviations" they are from the mean. This unitless measure of "how many standard deviations from the mean" is called the **z-value** or **z-score** for a particular data value.

Whether the mean is 100 kg or 45 ppm, the mean is always *no* standard deviations from itself. So the z-value for a population's or sample's mean equals zero. Thus, the lower label for μ in Figure 5.8 is "$z = 0$." A value that is two standard deviations above the mean will always have a z-value = 2—regardless of whether, in the above examples, the raw value is 120 kg, or 71 ppm, or any other figure.

The empirical rule can now be restated in terms of z-values: For any population with a symmetrical, bell-shaped distribution, roughly 68% of the values (if first converted to z-values) lie between $z = -1$ and $z = +1$. Approximately 95% lie between $z = -2$ and $z = +2$ on that scale, and 99.7% lie between $z = -3$ and $z = +3$ on the scale. See Formulas 5.7 and 5.8 for converting a data value to a z-score.

Formula 5.7 Converting a Data Value to a z-Score (for a population)

$$z = \frac{x - \mu}{\sigma}$$

Formula 5.8 Converting a Data Value to a z-Score (for a sample)

$$z = \frac{x - \bar{x}}{s}$$

Example 5.4

A farmer in Avery County, North Carolina, wants to harvest her crops prior to the first frost of the year. She has contacted the North Carolina Cooperative Extension Service, from whom she learns that in Avery County the mean date for the first frost is day 225 of the year[1] (numbering January 1 as day 1). The standard deviation for the date distribution is 11 days. Presume that the distribution of dates for the first frost is symmetrical and bell-shaped.

1. Calculate the z-values for first-frost dates on the following three days: day 236 of the year; day 214; and day 203.
2. Estimate how often the date of the first frost has occurred between days 214 and 236 of the year.
3. Estimate how often the date of the first frost has occurred *prior to* day 203.
4. The farmer is very worried about a frost occurring prior to day 192. Given only the information provided above, is her concern justified? Explain.

Solution

1. Using the formula:

$$z_{\text{day236}} = (236 - 225)/11 = +1$$
$$z_{\text{day214}} = (214 - 225)/11 = -1$$
$$z_{\text{day203}} = (203 - 225)/11 = -2$$

2. We see that the interval between days 214 and 236 is equivalent to the interval between $z = -1$ and $z = +1$. Based on the empirical rule, first-frost dates would have occurred in this interval about 68% of the time.
3. The solution to this question is less direct. We found in (1) that the z-value for day 203 is $z = -2$. The empirical rule tells us that 95% of the first-frost dates occurred in the interval from $z = -2$ up to $z = +2$. That leaves 5% of the first-frost dates to be accounted for. Presuming that the distribution is symmetrical, one-half of the remaining 5% of first frosts would occur *after* the date corresponding to $z = +2$. The proportion of first frosts *earlier* than day 203 (i.e., below $z = -2$) is also $^1/_2$ of 5% = 2.5%.
4. We can calculate: $z_{\text{day192}} = (192 - 225)/11 = -3$. The empirical rule tells us that 99.7% of the first-frost dates occurred in the interval from $z = -3$ up to $z = +3$. That leaves 0.3% of the first-frost dates to be accounted for. Presuming a symmetrical distribution, the proportion of first frosts earlier than day 192 (i.e., below $z = -3$) is one-half of 0.3% = 0.15%. Unless she has reason to expect highly unusual weather this year, the farmer has no reason to worry. Past experience suggests that there is virtually never a frost earlier than day 192.

Chebyshev's Inequality

We have learned in this section that knowing the mean and standard deviation for a population (or estimating these from a sample), we can estimate the proportion of individual values that lie within a given distance of the mean. The empirical rule provides the best case of this possibility: *If* the raw data are bunched symmetrically around the mean, we can confidently estimate the proportions of data at various distances from the mean. We now look at a more general case when the conditions for using the empirical rule are not reasonably met.

The rule presented here was derived mathematically by P. L. Chebyshev (and independently by I. J. Bianaymé) from the formal definitions of the mean and standard deviation. It does not account for the frequency distribution. The rule, called **Chebyshev's inequality**, determines the *minimum* percentage of data points that could be expected to fall within k standard deviations away from the mean (on the condition that $k > 1$). The exact percentage of values in this range may be greater—depending on the actual shape of the distribution.

Using z notation, the rule can be expressed as follows:

$$\text{\% of values between } (z = -k) \text{ and } (z = +k) \geq [1 - (^1/_k)^2] \times 100\%$$

for any value of k *greater than* 1.

Consider the case $k = 2$. Where the empirical rule applies, we know that close to 95% of the distribution lies between $z = -2$ and $z = +2$. What if the distribution was more like Figure 5.12? We see that it is hard to know for sure, but *at least*

$$[1 - (^1/_2)^2] \times 100\% = (1 - ^1/_4) \times 100\% = 75\%$$

of the values are in that range, maybe more. For any possible distribution, the proportion of values between $z = -2$ and $z = +2$ is greater than or equal to 75%.

Figure 5.12
A Very Non-Normal Population

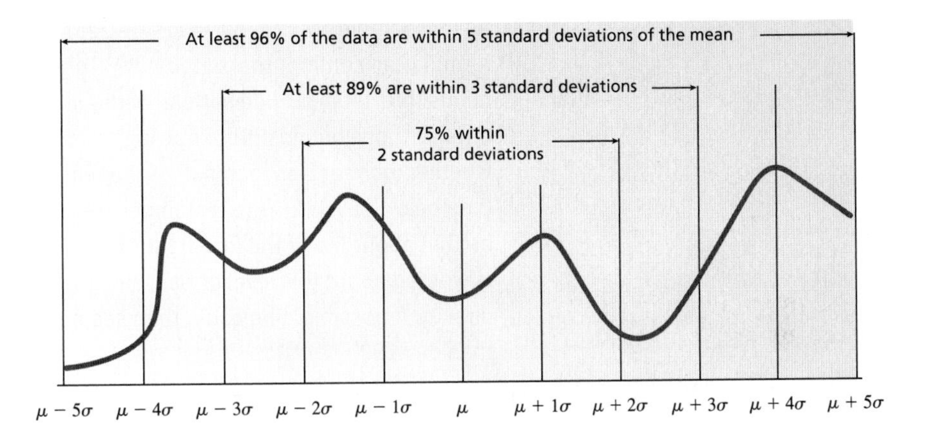

The inequality can be calculated for any value of *k*. Figure 5.12 shows two other commonly used estimates: At least 89% of all data values are within three standard deviations of the mean ($k = 3$), and at least 96% of all data values are within five standard deviations of the mean ($k = 5$). So even for the rarest distributions, we can estimate the upper and lower bounds for where most of the data lie.

Example 5.5

A farmer in Avery County, North Carolina, learned from the North Carolina Cooperative Extension Service that the mean date for the first seasonal frost in her county is day 225 of the year,[2] and the standard deviation is 11 days. We presumed in Example 5.4 that the distribution of dates for the first frost is symmetrical and bell-shaped, but this assumption was never confirmed. Solve the following questions *without* making any assumptions about the distribution of the raw data.

1. Estimate how often the date of the first frost has occurred between days 205 and 245 of the year.
2. Estimate how often the date of the first frost has occurred between days 214 and 236 of the year.
3. If the farmer is very worried about there being a frost prior to day 192, then given only the information provided above, is her concern justified? Explain.

Solution

1. We can calculate: $z_{day205} = (205 - 225)/11 = -1.82$ and $z_{day245} = (245 - 225)/11 = +1.82$. Setting $k = 1.82$:

$$(1 - ^1/_{1.82^2}) \times 100\% = (1 - ^1/_{3.31}) \times 100\% \approx 70\%$$

At least 70% of all the first-frost dates occurred between day numbers 205 and 245. (Note: "$1/1.82^2$" is equivalent to "$(\frac{1}{1.82})^2$".)

2. The interval between days 214 and 236 is equivalent to the interval between $z = -1$ and $z = +1$ (see Example 5.4). To use Chebyshev's inequality would require setting $k = 1$, but the Chebyshev formula *is not defined for $k \leq 1$.* Therefore there is no solution by Chebyshev. We do know, however, that the proportion of values between $z = -1$ and $z = +1$ cannot be more than the 68% proportion that the empirical rule predicts—and the proportion is probably less in this case.
3. As for Example 5.4: $z_{day192} = (192 - 225)/11 = -3$. Figure 5.12 shows (and the formula can confirm) that at least 89% of the data (maybe more) are within plus or minus three standard deviations of the mean. That leaves *at most* 11% of the data outside those bounds. Not knowing the distribution shape, we cannot be sure whether more or less than half of the remaining ($\leq$11%) cases are below $z = -3$ (i.e., below day 192). The remainder are above $z = +3$. A ballpark estimate might be that about 5% of the first-frost dates *could be* less than day 192. If the farmer is worried about this level of risk, she might consider inspecting the raw data directly, if possible. She could then see how often, if ever, a frost occurred earlier than day 192.

• Hands On 5.2(b) *Comparing Estimates from the Empirical Rule and Chebyshev's Inequality*

(Based on the files ***Animal_Weights, Diamonds, Grace_Alumni_World,*** *and* ***Carriers_revenue_tons****)*

In several of the "Hands On" features in Chapter 4, you were asked to examine these six variables:

1. *Weight,* from the file **Animal_Weights**
2. *Price,* from the file **Diamonds**
3. *Alumni,* from the file **Grace_Alumni_World**
4. *All_RevS* (scheduled plane departure revenues) from *major* airports, from the file **Carriers_revenue_tons**
5. *All_RevS* (scheduled plane departure revenues) from *national* airports, from the file **Carriers_revenue_tons**
6. *All_RevS* (scheduled plane departure revenues) from *large regional* airports, from the file **Carriers_revenue_tons**

You should now have quite a bit of information about these distributions. Do the following for each of the six variables in turn:

1. Use the empirical rule to estimate proportions of data between the values at:

   a) $z = -2$ and $z = +2$

   b) $z = -3$ and $z = +3$

2. Use Chebyshev's inequality to estimate proportions of data between the values at:

   a) $z = -2$ and $z = +2$

   b) $z = -3$ and $z = +3$

3. Find the actual proportions of values between those limits for the particular data set. Which of the above approaches gives the best estimates for this data set? Explain why that might be the case.

5.3 The Interquartile Range

The variance and standard deviation are not always the best measures of variation. Like the mean, they can be overly influenced by outliers or other types of nonsymmetrical distribution of the data. They also are not applicable for data whose level of measurement is not at least at the interval level. The **interquartile range** is a measure of variation that can avoid these limitations.

Consider the price distribution for standard two-storey houses illustrated in Figure 4.3, in Chapter 4. The computer-generated figure automatically included the mean ($366,935) and standard deviation ($249,132). For the same reasons that we introduced the median as perhaps a preferable measure of location for this data, the interquartile range (IQR) may be preferable as the measure of variation. By the way it is calculated, the IQR ignores the highest and lowest values in the data set, which might otherwise distort the measure.

We learned how to calculate the quartiles Q1, Q2, and Q3 in Section 4.5. Recall that these quartiles are based on sorting the data from the lowest to highest value. Q1 (the first quartile) separates the lowest quarter of the data values from the second quarter; Q2 (the second quartile = the median) separates the second quarter of the data (finishing off the first half of the data) from the third quarter; Q3 (the third quartile) separates the third quarter of the sorted data values from the final quarter of the values. These relationships are illustrated in Figure 5.13.

The interquartile range is simply calculated as Q3 – Q1. It comprises the middle 50% of the values in the data set, is not sensitive to extreme data points, and disregards any "noise" that may be present in data that fall below Q1 or above Q3. For the example in Figure 5.13, the IQR is $219,600, compared with the standard deviation of $249,132. For a house buyer of moderate means, the $30,000 difference in these numbers is quite significant. Unless a buyer inherits a fortune or signs with the NHL, the IQR probably conveys a better sense of the "typical" variation among the prices that the buyer will consider realistic.

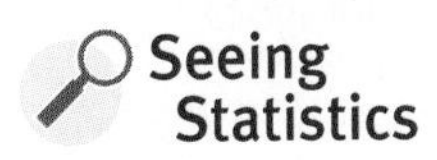

Section 4.1.2

• *In Brief 5.3 Computing the Interquartile Range*

If you compute quantiles Q1 and Q3 based on "In Brief 4.5," you need only to subtract Q1 from Q3.

Figure 5.13

Sorted House Prices and the Interquartile Range

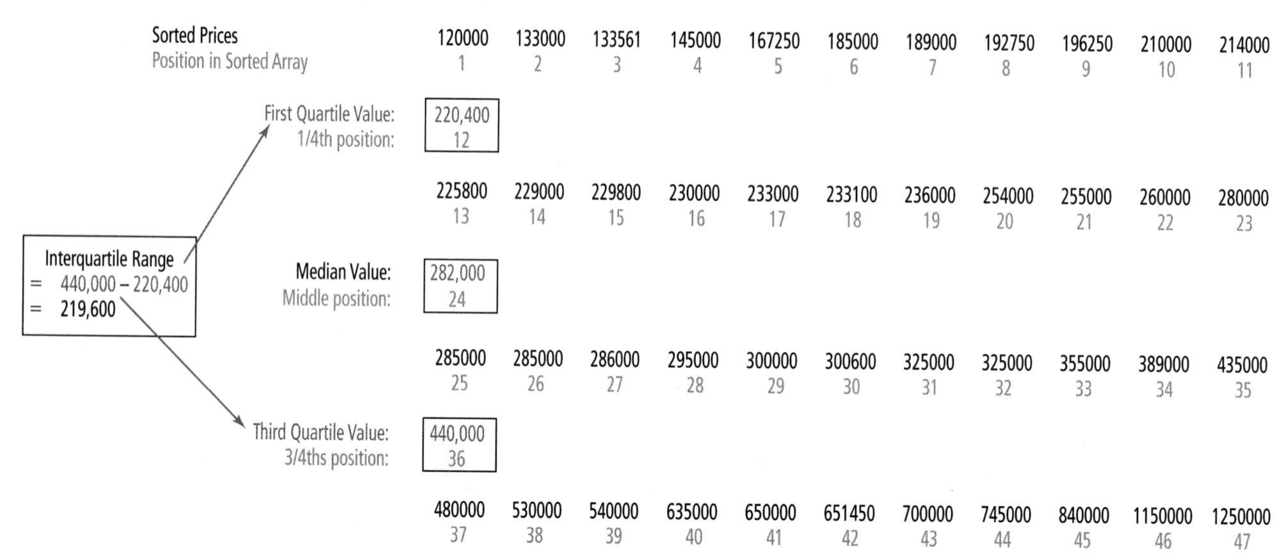

Source: Royal LePage Realty. Found at: http://www.royallepage.ca/schp/pdf/Q1_Survey_2005_Eng.pdf

Excel

Recall that Excel uses slightly different procedures for calculating the quartiles from those shown in this book. The result of subtracting Excel's Q1 from Excel's Q3 may vary slightly from an IQR result by SPSS or Minitab, but statistically the differences are negligible.

Minitab

The Minitab instructions for calculating the quartiles began with the menu sequence *Stat / Basic Statistics / Display Descriptive Statistics.* On clicking the *Statistics* button, you previously chose to display the quartiles. Note that in this dialogue box, you also have the option of selecting *Interquartile Range* to output the IQR directly.

SPSS

To output the IQR directly, use the menu sequence beginning *Analyze / Descriptive Statistics / Explore.* More details will be given on this sequence in "In Brief 5.4" on page 194.

Example 5.6

 *(Based on the* ***Environmental_Sustainability*** *file)*

The *Environmental Sustainability Index* (ESI) is a measure of a country's expected ability to sustain its environment over the next few decades. A ranking of 146 nations' performances on the basis of this index has been published in the report *Environmental Sustainability Index: Benchmarking National Environmental Stewardship* (go to http://www.yale.edu/esi and click on "Main Report" to access the most recent version).

A number of these countries have been randomly selected, and the file saved as **Environmental_Sustainability** lists their ESI results.

1. Find the interquartile ranges for the countries' ESI scores.
2. Find the interquartile ranges for the ESI ranks for the countries included in the sample.
3. Do we have any evidence of whether this sample's ESI distribution is representative of the full population's distribution? Explain.

Solution

1. Any statistical package could be used to find the quartiles $Q1$ and $Q3$. The interquartile range is $Q3 - Q1$. As shown in Figure 5.14, the IQR for the countries' ESI scores is 10.15.

Figure 5.14
Interquartile Range for Countries' ESI Scores and Ranks

Descriptive Statistics: ESIscore, ESI_Rank

Variable	Q1	Median	Q3	IQR
ESIscore	44.35	49.30	54.50	10.15
ESI_Rank	37.50	74.00	111.50	74.00

Source: 2005 Environmental Sustainability Index: Benchmarking National Environmental Stewardship. Yale Center for Environmental Law and Policy; Yale University. Center for International Earth Science Information Network; Columbia University. Main Report. Found at: http://www.yale.edu/esi/ESI2005_Main_Report.pdf

2. The IQR for the countries' ranks is 74. Note that the ranks are ordinal numbers, so using a standard deviation, for example, would not be strictly meaningful as an alternative measure of variation.
3. Yes, we have evidence that the sample's ESI distribution is representative of the population's ESI distribution. We are told that $N = 146$ for the population. If we lined up all of the rank numbers for the full population (1, 2, 3, . . ., 146), we would expect the first quartile to occur about one-quarter of the way from ranks 1 to 146 (≈ 37) and the third quartile to occur about three-quarters of the way from 1 to 146 (≈ 110). $Q1$ and $Q3$ for the *sampled* rank numbers occur at virtually these same values.

• Hands On 5.3 *Comparing IQRs for Several Distributions*

 *(Based on the* ***Star_Distances*** *file)*

Provided a distribution is symmetrical and bell-shaped (so that the empirical rule applies), we gain a lot of information about a distribution just by knowing the mean and standard deviation. Even without knowing the shape of a distribution, Chebyshev's inequality, combined with the mean and standard deviation, still provides some information about the limits of a distribution.

What does the IQR tell us about a distribution if we have little or no other information? Try some experiments to answer the question for yourself. First, review "In Brief 5.2" on generating random number sequences.

1. Use the computer to generate the following five random distributions:
   a) 1,000 numbers with a "normal" distribution: Mean = 10, Standard deviation = 2.
   b) 1,000 numbers with a "normal" distribution: Mean = 10, Standard deviation = 4.
   c) 1,000 numbers with a "normal" distribution: Mean = 10, Standard deviation = 6.
   d) 1,000 numbers with a "normal" distribution: Mean = 10, Standard deviation = 8.
   e) 1,000 numbers with a "uniform" distribution: from 0 to 10.
2. For each case, find the three quartiles and the IQR. How does the IQR compare with the standard deviation? For each distribution, answer the question: Fifty percent of the values are between what two numbers?
3. Next load the data set stored in the file **Star_Distances.** Find the quartiles and IQR. Then find the mean and standard deviation for the same data. If you knew only the quartiles, mean, and standard deviation, and had no other information, and you were a space traveller running low on fuel, which measure or measures might be most useful? Explain.

5.4 Five-Number Summaries and Boxplots

This section integrates a number of techniques and measures already introduced: the median (Section 4.2), quartiles (Section 4.5), and interquartile range (Section 5.3). Combining measures of location (median and quartiles, as well as minimum and maximum) plus spread (interquartile range), these methods can also provide information about the shape of a distribution.

The version of the **five-number summary** presented here summarizes a data set with these five numbers:

1. Min = Smallest (i.e., minimum) value
2. Q1 = First quartile
3. Q2 = Second quartile (= the median)

4. $Q3$ = Third quartile
5. Max = Largest (i.e., maximum) value

(An earlier variation, devised by John Tukey, replaces $Q1$ and $Q3$ with lower and upper "hinges"—which are values midway between Min and the median, and Max and the median, respectively. Following the hinge approach, the resulting five-number summaries may differ slightly from the versions presented here.)

These five points reveal some key proportions in the distribution: We know that half of the data are below the $Q2$-value, one-quarter of the data are between $Q1$ and $Q2$, and another quarter are between $Q2$ and $Q3$. If the smallest value Min or largest value Max is an outlier, we will see this—by the comparatively large distance between that extreme value and the nearest quartile, as compared to the distances between the quartiles.

The graphic method called a **boxplot** (or **box-and-whisker diagram**) is *sometimes* just a straightforward representation of the five-number summary. Suppose you wanted to analyze the distribution of residential electricity prices in the file **Electricity_Prices**—*excluding* the prices for Vermont, Hawaii, and New Hampshire. The results are shown in the graph portion of Figure 5.15. The vertical axis represents potential values of the price variable. Against that scale, we draw the "box" to range from the first quartile (bottom of the box) to the third quartile (top of the box). The line within the box shows the position of the median (i.e., second quartile). The "whisker" (vertical line) on the bottom traverses

Figure 5.15
Five-Number Summary and Boxplot: Simplest Case

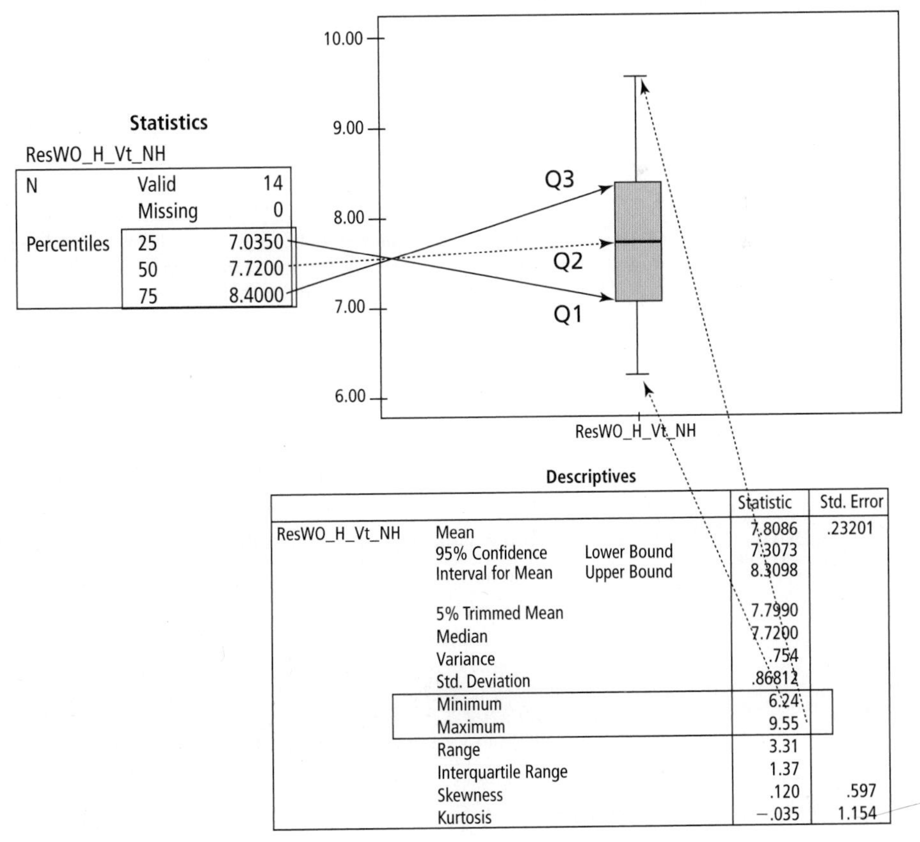

Statistics

ResWO_H_Vt_NH

N	Valid	14
	Missing	0
Percentiles	25	7.0350
	50	7.7200
	75	8.4000

Descriptives

			Statistic	Std. Error
ResWO_H_Vt_NH	Mean		7.8086	.23201
	95% Confidence Interval for Mean	Lower Bound	7.3073	
		Upper Bound	8.3098	
	5% Trimmed Mean		7.7990	
	Median		7.7200	
	Variance		.754	
	Std. Deviation		.86812	
	Minimum		6.24	
	Maximum		9.55	
	Range		3.31	
	Interquartile Range		1.37	
	Skewness		.120	.597
	Kurtosis		−.035	1.154

Source: U.S. Department of Energy, Energy Information Administration. Based on data in *Average Retail Price for Consumers by Sector and State, 2003* (a sample of 17 U.S. states). Found at: http://www.eia.doe.gov/neic/quickfacts/quickelectric.htm

from the minimum x-value to the first quartile; the upper whisker ranges from the third quartile up to the maximum value of the variable. (The same figure could also be oriented horizontally, with the variable scale increasing toward the right.)

Figure 5.16 provides a five-number summary and an alternative style of boxplot for the residential electricity prices—*including* the data for Vermont, Hawaii, and New Hampshire. As before, the box in the boxplot is bounded by the values for $Q1$ and $Q3$, and the inner line corresponds to $Q2$ (the median). Also, the lower "whisker" of the plot corresponds to the distance between $Q1$ and the minimum value in the data set. In this case, however, the upper whisker no longer reaches all the way to the maximum x-value.

Figure 5.16
Five-Number Summary and Boxplot: With Outliers

Descriptive Statistics: Price

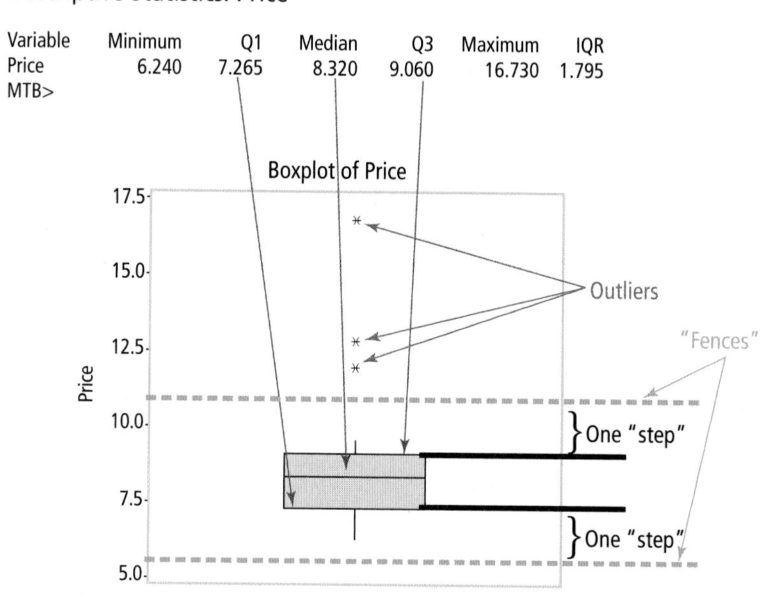

Variable	Minimum	Q1	Median	Q3	Maximum	IQR
Price	6.240	7.265	8.320	9.060	16.730	1.795

MTB>

Source: U.S. Department of Energy, Energy Information Administration. Based on data in *Average Retail Price for Consumers by Sector and State, 2003* (a sample of 17 U.S. states). Retrieved from http://www.eia.doe.gov/neic/quickfacts/quickelectric.htm

Figure 5.16 shows a common alternative convention for boxplots. The whiskers are drawn only as far as what some books call the **fence,** or a **one-step distance** from the nearest quartile. Using quartiles (rather than hinges), this maximum length for a whisker is 1.5 times the interquartile range. As in Figure 5.15, a whisker ends at the minimum or maximum data value *if* this value is not greater than one step from the nearest quartile. But a whisker continues only to the largest (or smallest) data value before the fence is reached; any more extreme values are shown as dots beyond the limit of the whisker. (The "fences" shown in the Figure 5.16 are for illustration only, and are not part of the output.)

Sections 5.1.1–5.1.2

• *In Brief 5.4 Constructing Five-Number Summaries and Boxplots*

*(Based on the **Electricity_Prices** file)*

We have already learned about the tools used to construct the five-number summary. See Section 4.5 for finding the quartiles and Section 5.3 for finding the interquartile range. Minimum and maximum values could be found, for example, by simply sorting the data. Instructions for drawing the boxplot are as follows.

Excel

Excel does not provide a function to construct these charts, per se, but a sketch can be made manually, using its outputs for quartiles, etc. Recall from Chapter 4 that Excel uses slightly different procedures for calculating the quartiles from those in the other software programs, so the boundaries of the drawn figure may vary slightly from solutions given in this book.

Minitab

With the data set in a column, use the menu sequence *Graph / Boxplot.* Accept the default option in the first dialogue box (for a simple one *Y*-variable—see the screen capture) by clicking *OK.* Then, from the list of variables in the second dialogue box, select the column with the data and then click *OK.* As shown in the illustration, the default boxplot style will show the outliers, if any. (For alternative options, click on *Data View* in the second dialogue box.)

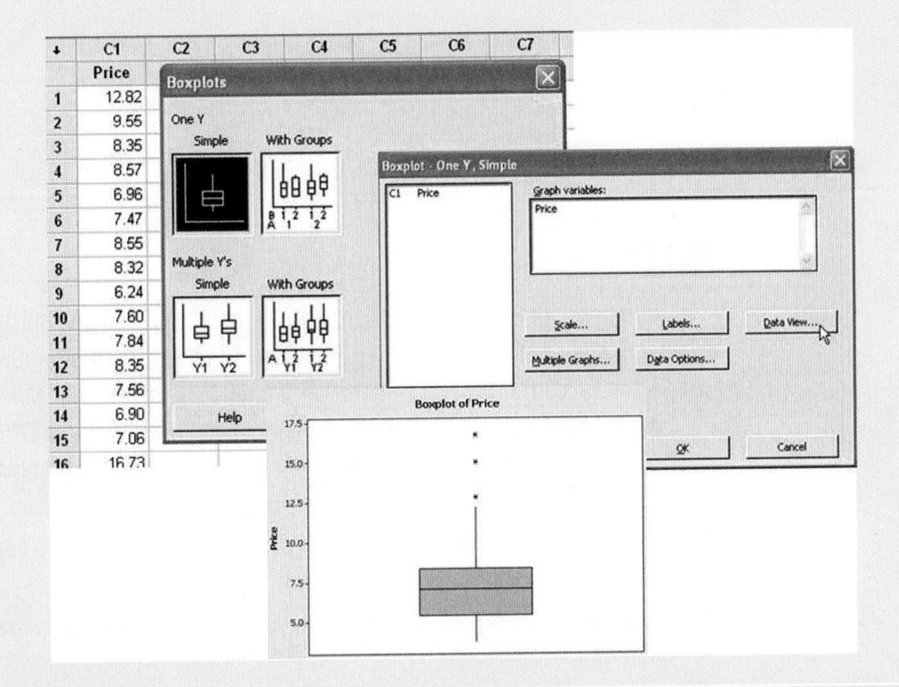

SPSS

With the data set in a column, use the menu sequence *Analyze / Descriptive Statistics / Explore.* As shown in the screen capture, select the variable for analysis and select *Both* as the option under *Display;* then click *OK.* The output will display several tables and figures, including a boxplot. The default display of the boxplot highlights possible outliers—shown lying beyond the "whiskers." Note that the labels on the outliers refer to their *positions* in the data column being analyzed; the labels *do not* display the actual data values.

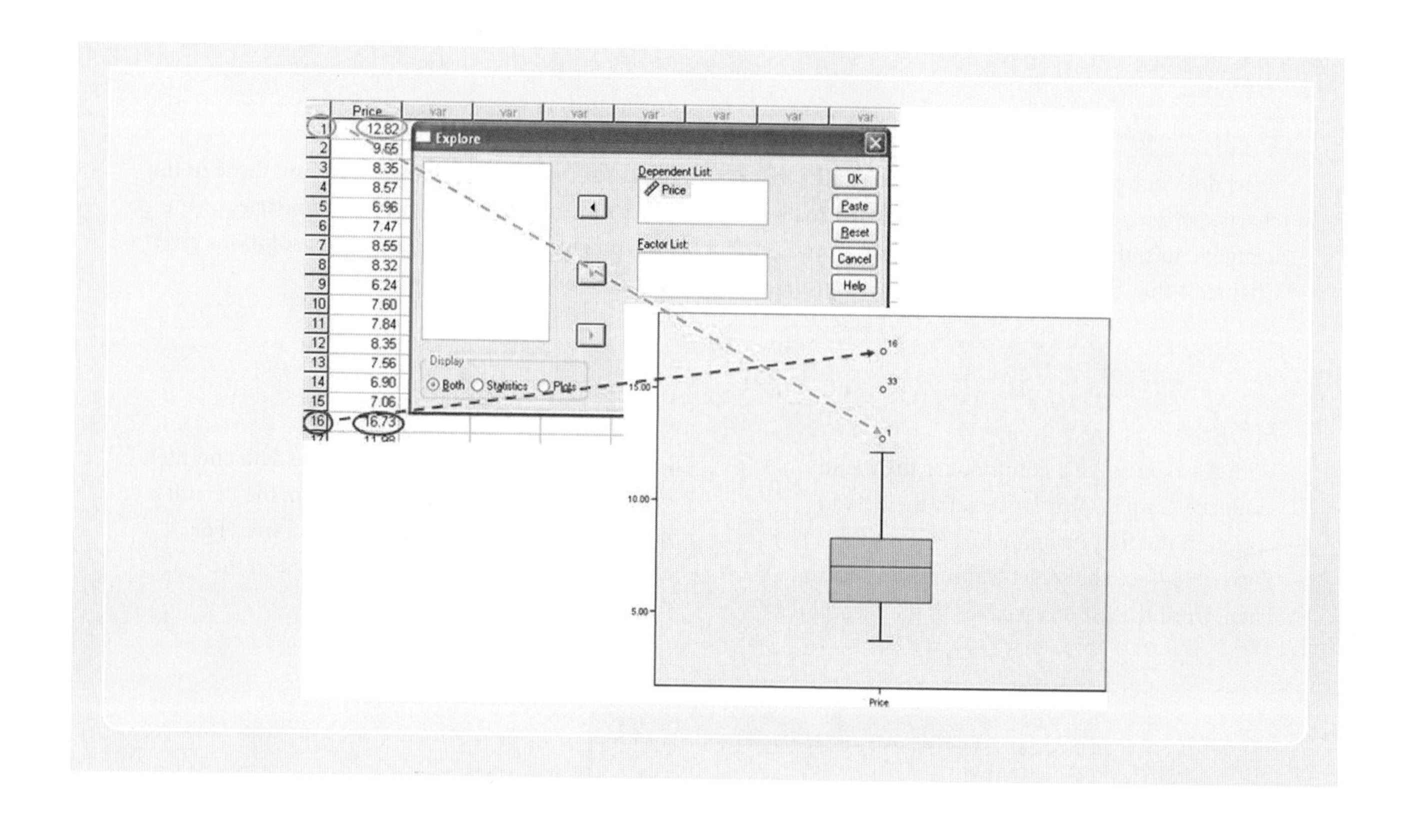

Example 5.7

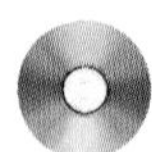 *(Based on the **Water_Mineral_Content** file)*

Responding to public concerns about the quality of commercially bottled mineral waters, an analyst randomly selected 30 examples of the product from around the world. Data on the mineral contents of these items are stored in the file **Water_Mineral_Content.** Find the columns relating to these specific minerals: lead (Pb); titanium (Ti); chloride (Cl); and arsenic (As). Chloride is measured in milligrams per litre; lead, titanium, and arsenic are measured in parts per billion.

1. Construct five-number summaries for the distributions of these elements among the samples.
2. To construct boxplots, determine the values of the fences for each of the elements.
3. Construct a boxplot of each element's distribution.
4. Using the boxplots as your only source of information, discuss which element or elements have:

   a) A symmetrical distribution.

   b) A very nonsymmetrical distribution.

   c) One or more outliers.

Solution

1. The components for five-number summaries can be found either manually or by using a computer. Each row of the following table gives the summary for one of the elements.

Element (units)	Minimum	*Q*1	*Q*2 (Median)	*Q*3	Maximum
Pb (ppb)	0.260	0.468	0.710	1.135	1.830
Ti (ppb)	0.270	1.025	1.585	2.223	4.580
Cl (mg/L)	0.000	3.390	9.300	45.500	591.100
As (ppb)	0.000	0.308	0.485	1.330	161.200

2. Based on the data in the above table, interquartile ranges (IQR) and the fences can be calculated for each element, as shown in the following table.

Element (units)	IQR (*Q*3 − *Q*1)	Lower Fence* [*Q*1 − (1.5 × IQR)]	Upper Fence [*Q*3 + (1.5 × IQR)]
Pb (ppb)	0.667	0*	2.136
Ti (ppb)	1.198	0*	4.020
Cl (mg/L)	42.110	0*	108.665
As (ppb)	1.022	0*	2.863

* Note that for these data, a negative value has no meaning. The lowest possible content of any mineral in the water is zero.

3. Boxplots can be drawn by hand based on the above information or computer-drawn (as shown in Figure 5.17), depending on your software.

Figure 5.17
Boxplots for the Distributions of Lead (Pb), Titanium (Ti), Chloride (Cl), and Arsenic (As) in Bottled Water

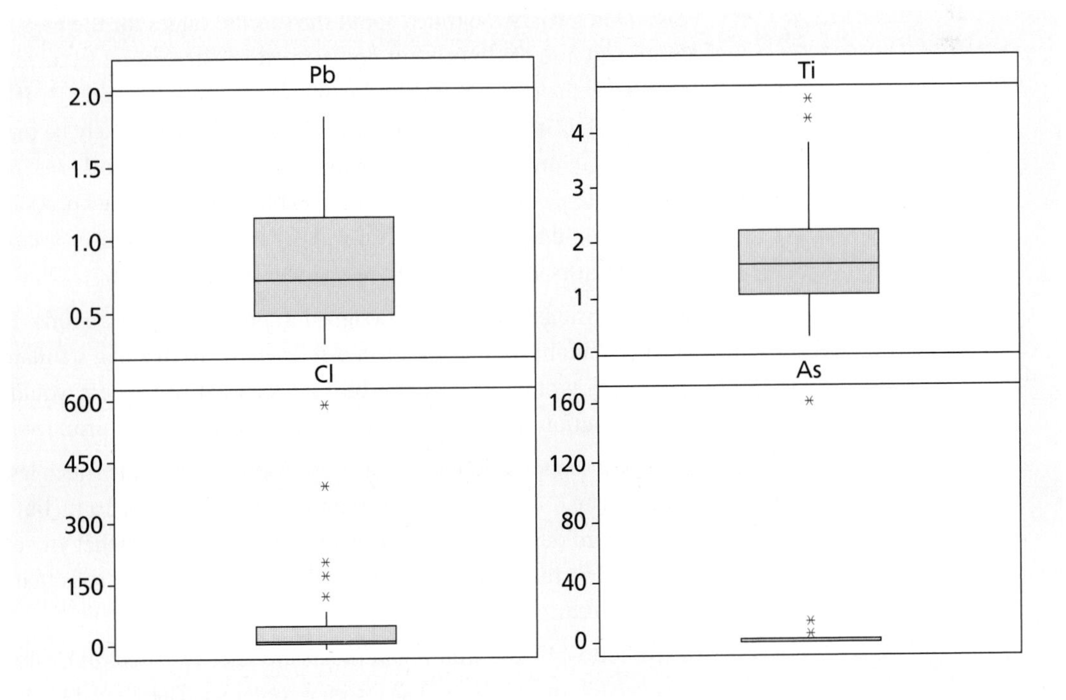

Source: Federal Agricultural Research Center (FAL). Found at: http://www.mineralwaters.org

4. Based on these boxplots:

   a) None of the distributions is altogether symmetrical. Otherwise, a boxplot would show (i) the median line equally far from both $Q1$ and $Q3$, and (ii) both whiskers having the same length. Lead and titanium are much closer to this pattern than are chloride and arsenic.

      It is notable that the titanium distribution is close to symmetrical *if the two outliers are discounted.* This *might* suggest that the distribution is usually symmetrical—except for special conditions that lead to the extreme values.

   b) Unlike titanium, the distributions of chlorine and arsenic are extremely non-symmetrical. We see from the medians that half of the values are close to zero. The next quarter of values grows noticeably; then the final quarter of values stretches to very high levels.

   c) The distribution for lead has no outliers; the titanium distribution has two; and the very nonsymmetrical distributions for chloride and arsenic have several outliers each.

• Hands On 5.4 *Anticipating Boxplots*

You will likely encounter boxplots in reports and studies that you use for research or for keeping up to date. The graphs will be most useful to you if you have developed an informed intuition of how they represent particular data distributions. Working through these exercises will help build this knowledge base.

1. Consider each of the following distributions. Without looking at any extra data, *sketch* what you *expect* the boxplot would look like in each case. Don't worry too much about the specific values for the mean or for the Q1 and Q3 lines—but do try to anticipate general features such as: Is the median likely to be midway between Q1 and Q3? If not, which is it likely to be closer to? Will the whiskers likely be the same length on both sides of the figure? If not, which whisker is likely to be the longer one? Will the whiskers be longer than the boxed area? Keep in mind that these answers will depend on the relative variations in data values within each of the quarters of a data set.

   a) A precision tool has been designed to produce 10-metre rods. The rods that it produces have a mean length of 10 m and a standard deviation of 0.3 m. Sketch a boxplot for what you expect would be the distribution of lengths of the next 1,000 rods that are produced.

   b) After the tool described in (a) was dropped, it became much less precise. The rods that it produces have a mean length of 10 m, but the standard deviation is now 1 m. Sketch a boxplot for what you expect would be the distribution of lengths of the next 1,000 rods that are produced.

   c) In Lotto 649, the winning numbers are selected based on the random selection of numbered balls from a container. There are 49 balls, each numbered 1 to 49, respectively. For the *first* number selected in each new draw, each of the 49 balls has an equal chance of being selected. Sketch a boxplot for what you expect would be the distribution of the variable "first number selected" over the course of 1,000 new draws.

2. Using simulations, generate approximations to the "real" distributions for the three cases in (1). First, review "In Brief 5.2" on generating random number sequences.

   a) For the first case, generate 1,000 numbers with a "normal" distribution: Mean = 10; Standard deviation = 0.3.

   b) For the second case, generate 1,000 numbers with a "normal" distribution: Mean = 10; Standard deviation = 1.0.

   c) The Lotto 649 case requires some adjustments to methods discussed previously. The goal is to use the computer to generate 1,000 numbers, each of which has an equal chance of being one of the integers 1 to 49.

   - **In Excel:** Follow the procedures in "In Brief 5.2" *with these changes:* When you reach the *Random Number Generation* screen, select *Uniform* as the type of distribution, and specify as the parameters between 0.5 and 49.5.

     Once you complete and execute the random number generation, the 1,000 numbers will all need to be rounded off. In the cell immediately to the right of the top number, input a formula of the form = **round(A1,0)** (where **A1** should be the address of the top number that was randomly generated). Copy and paste the formula so it fills the entire column to the right of the column of random numbers. This new column will now be random whole numbers in the range from 1 to 49.

   - **In Minitab:** Follow the procedures in "In Brief 5.2" *with these changes:* Following the *Calc / Random Data/* menu sequence, select *Integer* as the type of distribution. Specify 1 as the minimum value and 49 as the maximum value.

   - **In SPSS:** Follow the procedures in "In Brief 5.2." However, in the upper right-hand box of the *Compute* dialogue box, input **rnd(rv.uniform(0,1)*49+.5).**

     (**Explanation:** The function **rv.uniform(0,1)** randomly generates a decimal number between 0 and almost 1, and repeats this for all of the non-empty rows in the data file. If this decimal is multiplied by 49, the result is a number between 0 and almost 49; adding 0.5 to this product, each random number now falls between 0.5 and almost 49.5. The **rnd** function then rounds off the result to the nearest whole number, becoming an integer between 1 and 49.)

3. For each of the three 1,000-value distributions generated in step 2, construct the corresponding boxplot. Compare these boxplots with those you expected to obtain for the distributions, in step 1. How do they compare? If the actual boxplots are different from your expectations, try to explain the reasons.

5.5 Relative Variation

So far in this chapter, we have looked at ways to quantify the amount of variation in samples or populations. A commonly used measure is the standard deviation. To compare the amounts of variation in *two* distinct data sets, however, the standard deviation may not be suitable as it stands. For example, Figure 5.16 showed that among sampled U.S. states, there is considerable variation in the retail prices of electricity for residential customers. For the residential data in the file **Electricity_Prices,** the standard deviation is 2.65 cents per kilowatt hour. If you calculate the standard deviation for the sampled *industrial* data in the same file, the standard deviation is 2.25 cents per kilowatt hour.

Since 2.65 is larger than 2.25, should we conclude that there is "more variation" among residential prices than industrial prices? That would be misleading. The residential variation of 2.65 is in relation to a mean of 8.87 cents per kilowatt hour. For industrial customers, however, the mean is only 5.63 cents—so the standard deviation of 2.25 for that group represents a *comparatively* larger variation among the values of that group. In other words, when two groups' means are quite different, it is generally useful to compare each group's **relative variation,** rather than its individual standard deviations. The **coefficient of variation** (*CV*) is a unitless measure that expresses the relative size of the standard deviation as a percentage of the mean, for a given population or sample. The coefficient of variation can be calculated using Formula 5.9.

Formula 5.9 Coefficient of Variation

$$CV = \frac{\sigma}{\mu} \times 100\%$$

where CV is the coefficient of variation.
σ is the population standard deviation (or use s if working from a sample).
μ is the population mean (or use $\bar{x}$ if working from a sample).

In the example using prices of electricity,

$$CV_{\text{residential}} = 2.65/8.87 = 0.30 \times 100\% = 30\%$$

$$CV_{\text{industrial}} = 2.25/5.63 = 0.40 \times 100\% = 40\%$$

Since $CV_{\text{industrial}} > CV_{\text{residential}}$, we conclude that, relatively speaking, there is greater dispersion among industrial prices than among residential ones.

The coefficient of variation can also allow the variations of populations or samples to be compared—even if the two sets of data are not measured in the same units. This is illustrated in Example 5.8.

Example 5.8

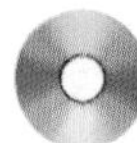
*(Based on the **Animal_Weights** file)*

During an environmental review of a piece of property, various animals were discovered to be living there. Each animal's species name, weight, and other characteristics were recorded and saved in the file **Animal_Weights.** Zoologists were called to the scene and asked for advice on feeding the animals. They noted a great variation among the beasts' food consumption per day, yet a child playing on the property pointed out that the various animals had quite different weights. Is the relative variation of the variable "daily grams of food consumed *per kilogram of body weight*" less than the relative variation of the variable "daily grams of food consumed"? Answer, and interpret the results.

Solution We can calculate the standard deviations for the two variables (their values are listed under the column headings "Food_day" and "Food_wgt" in the file):

$$s_{\text{(daily consumption)}} = 4{,}015 \text{ grams}$$

$$s_{\text{(daily consumption by weight)}} = 40.75 \text{ grams per kilogram}$$

Although 4,015 is larger than 40.75, these numbers have different units and cannot be compared directly. However, we can also calculate the means:

$$\bar{x}_{\text{(daily consumption)}} = 2{,}143 \text{ grams}$$

$$\bar{x}_{\text{(daily consumption by weight)}} = 63.50 \text{ grams per kilogram}$$

Next, calculate and compare the coefficients of variation for the two groups:

$$CV_{\text{(daily consumption)}} = 4{,}015\text{g}/2{,}143\text{g} \times 100\% = 187\%$$

$$CV_{\text{(daily consumption by weight)}} = 40.75\cancel{\text{g/kg}}/63.50\cancel{\text{g/kg}} = 64.2\%$$

Observe how the original units cancel out in the coefficient of variation calculations, leaving only percentage values, which can be compared. It does appear that there is less variation in the animals' food consumption per unit of weight compared to variation in simply how much they eat. This seems quite reasonable: At least some of the variation in how much the cows eat compared to the gerbils' consumption has to do with their diverse relative weights. The variable that accounts for weight shows relatively less variation than the variable that does not.

• *In Brief 5.5 Finding the Coefficient of Variation*

Neither Excel nor SPSS offer any special functions for the coefficient of variation.

Excel

Using techniques presented in "In Brief 4.1" and "In Brief 5.1," find the mean and standard deviation of the data set. Based on those values, apply Formula 5.9.

Minitab

If starting from a sample, follow the first paragraph of Minitab procedures described in "In Brief 5.1." In the dialogue box where you can select the standard deviation (and/or the mean), select instead the coefficient of variation. This result will be printed to the output.

If starting from a population, follow the second paragraph in "In Brief 5.1" for Minitab to find the standard deviation. Find the mean as described in "In Brief 4.1." Then, based on those values, apply Formula 5.9.

SPSS

Using techniques presented in "In Brief 4.1" and "In Brief 5.1," find the mean and standard deviation of the data set. Based on those values, apply Formula 5.9.

• Hands On 5.5 How Coefficients of Variation Compare for Varied Distributions

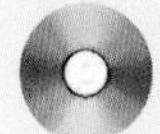 *(Based on the **Diamonds** file)*

This "Hands On" feature poses some questions—and encourages you to design and carry out some informal experiments that might suggest some answers. For data, you might devise some columns of numbers that have distribution properties of interest (e.g., symmetrical and bell-shaped, symmetrical but not bell-shaped, nonsymmetrical with large outliers, and so on). In some cases, it will be possible (but not necessary) to let the computer randomly generate suitable numbers (for example, for certain types of symmetrical distributions), or to reuse data sets whose distributions you have previously encountered in the CD provided with this textbook (such as the price data in the file **Diamonds**).

Use the computer to find the mean, standard deviation, and coefficient of variation for each data set you consider, and then "sketch" its frequency distribution (i.e., not worrying too much about exact class sizes). Then examine the relationship between the distributions and the parameters that you calculated. From these casual experiments, you cannot make firm rules about how coefficients of variation compare for varied distributions, but try to sharpen your intuition about what a data set looks like if you are given only its *CV* or its mean and standard deviation.

For each of the following questions, try to develop an answer by testing some relevant cases in the manner described above.

1. Do you think it is possible for a data distribution to be symmetrical if its *CV* is greater than 100%?
2. Does there seem to be a limit to how large the *CV* can be if a data distribution is symmetrical and bell-shaped?
3. When considering questions (1) and (2), did you allow the data to take on negative values? For many variables (such as those involving counts, weights, proportions, and so on), negative values are not really possible. Now answer questions (1) and (2) again, disallowing negative answers. How do the answers compare with the alternative of allowing negative values?
4. Deliberately try to find or create a variety of data sets with large *CV*s (say, 200% or larger) in each case. Then find or create a variety of data sets with relatively small *CV*s (say, less than 25%). Construct basic frequencies for each. Compare and try to identify any patterns.

5.6

Assessing the Shape of a Distribution

Sections 5.3.1–5.3.2

We briefly introduced the concept of distribution shape in Section 3.2. When we speak figuratively of the **shape of a distribution,** we are essentially visualizing the histogram that corresponds to the frequency distribution for the data. A histogram figure is symmetrical; if we could fold it along a centre line, like the fold of a birthday card, the figure would line up with itself on each side of the fold. We say a distribution is symmetrical if its histogram is symmetrical.

Skewness and *kurtosis* refer to two other aspects of a distribution's shape. **Skewness** measures the degree of nonsymmetry of a distribution. In several examples and "Hands On" features in this chapter, you have examined data sets that were nonsymmetrical. A clear example is the Prices variable in the **Diamonds** file. Compare its distribution with the distribution of the Power Play Goals variable ("PP") in the file **NHL_Scoring_Leaders.**

If the data, such as Prices in the **Diamonds** file, appear to be stretched toward values on the high side, then the distribution is said to be skewed positively (or to the right). If the data appear to be pulled toward the lower values, then the distribution is said to be skewed negatively (or to the left). For example, there *may* be some negative skewing in the Turnout variable in the **Voter_Turnouts** file (see Figure 5.18), but there the effect is small and could just be due to sampling error. Observe for the Power Play Goals variable in the **NHL_Scoring_Leaders** file that a distribution can be approximately symmetrical even if it is not altogether bell-shaped—the shape for which the empirical rule applies most accurately.

Figure 5.19 illustrates the relationship between skewing and the relative placements of the mean, median, and mode (actually, the modal class, if there is one predominant modal class). If the distribution is symmetrical, then the mean and median will fall at about the same location. If the distribution is skewed, then the more extreme values in the direction of the skewing will "pull" the mean toward that end.

Several mathematical measures for skewness have been developed, and statistical software generally incorporates one of these versions. But often it is sufficient to get a sense of the symmetry or skewness of a distribution by visually inspecting histograms or boxplots, and also by comparing the relative values of the mean and median.

Another measure of distribution shape is **kurtosis.** Kurtosis is the relative degree to which a distribution is tall or flat, compared to a smooth bell-shaped curve drawn on the same scale. Relatively tall distributions have positive kurtosis; these are characterized by high peaks and extended "fat" tails. In contrast, a distribution with negative kurtosis has a flatter shape, with only "thin" tails.

In practice, to distinguish these cases by eye is not always easy—especially if you are looking at a typical computer-drawn histogram. Compare the charts in Figure 5.20, based on player statistics for the Calgary Flames in 2003–2004 stored in the file **Calgary_Flames.** Notice how, as kurtosis varies from +0.84 down to –0.55, the bars tend to bunch up away from the extremes toward the mean. (We will not attempt to calculate the kurtosis values in this textbook.)

Figure 5.18

Skewed and Symmetrical Distributions

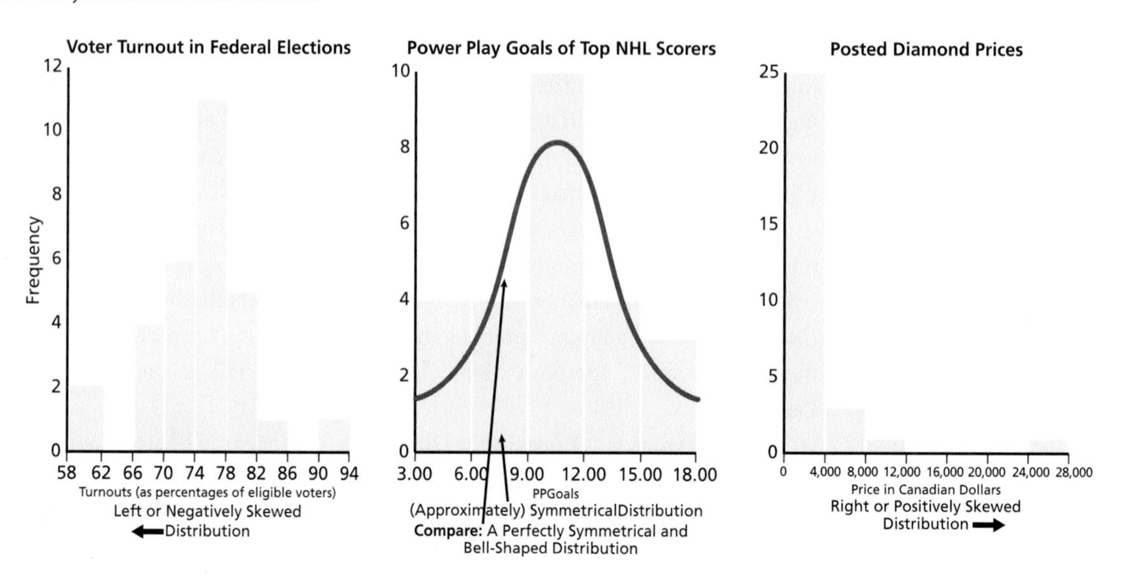

Sources: Nodice Canada. Based on data in *Voter Turnout,* which is based on Elections Canada data. Retrieved from www.nodice.ca/election2004/voterturnout.html; The Sports Network. Table: *Final 2003-2004 NHL Individual Scoring Leaders.* Retrieved from http://www.sportsnetwork.com/default.asp?c=sportsnetwork&page=nhl/stat/scoring-leaders.htm; and Canada Diamonds. Based on data retrieved from http://www.canadadiamonds.com/ round_brilliant_canadian_diamonds.php?page=650&limit=10&customer=CAD

Figure 5.19

Relationship of Mean/Median Positions and Skewing

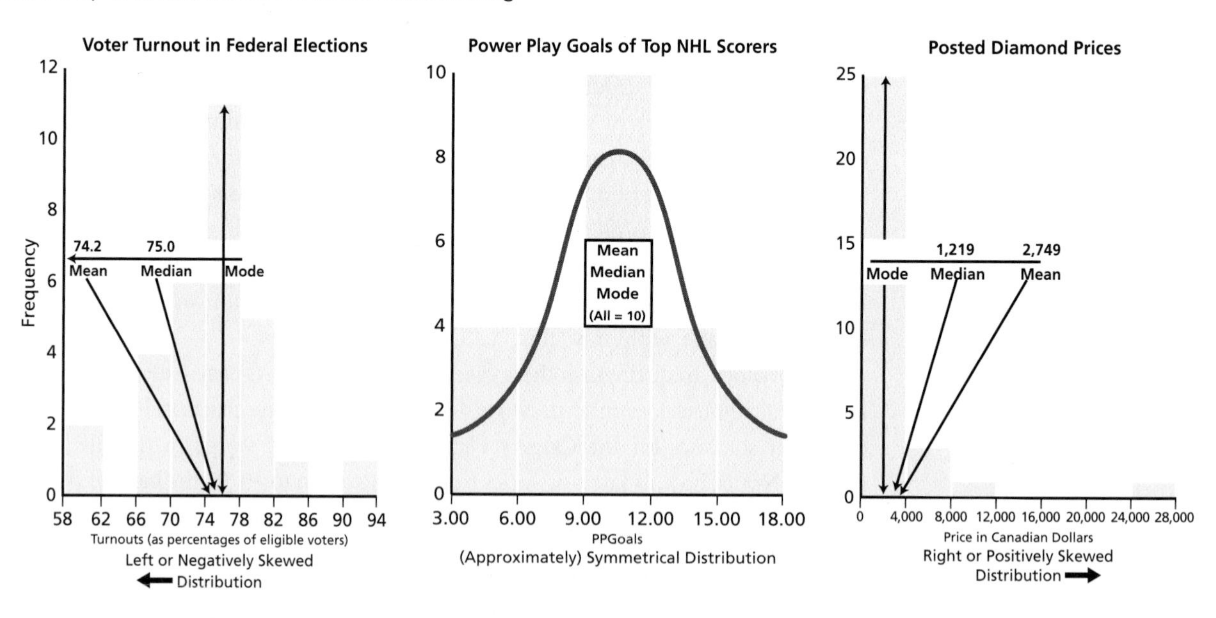

Sources: Nodice Canada. Based on data in *Voter Turnout,* which is based on Elections Canada data. Retrieved from www.nodice.ca/election2004/voterturnout.html; The Sports Network. Table: *Final 2003-2004 NHL Individual Scoring Leaders.* Retrieved from http://www.sportsnetwork.com/default.asp?c=sportsnetwork&page=nhl/stat/scoring-leaders.htm; and Canada Diamonds. Based on data retrieved from http://www.canadadiamonds.com/round_brilliant_canadian_diamonds.php?page=650&limit=10&customer=CAD

Figure 5.20

Examples of Positive and Negative Kurtosis: Hockey Player Statistics

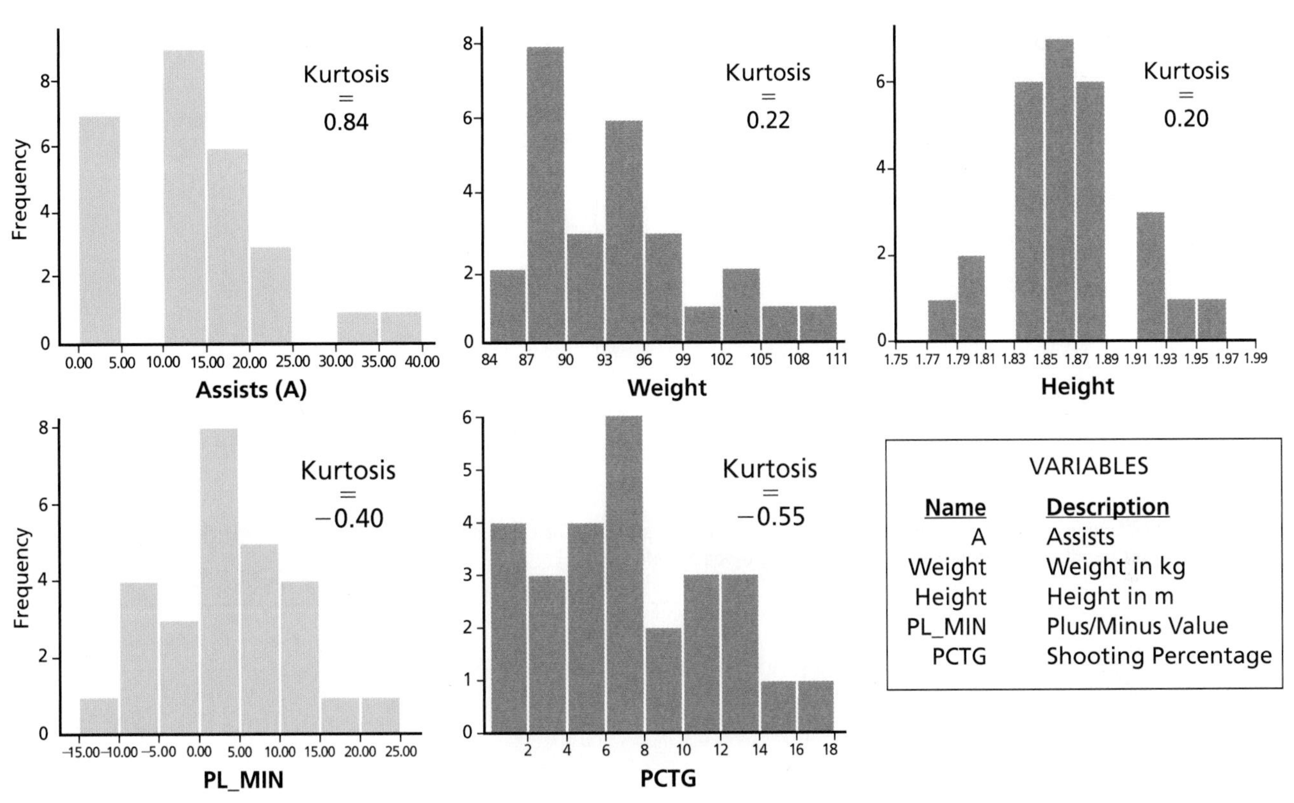

Source: The Sports Network. Found at: http://www.sportsnetwork.com/default.asp?c=sportsnetwork&page=nhl/stat/scoring-leaders.htm

Example 5.9

*(Based on the **Canadian_Equity_Funds** file)*

An investor is comparing the returns on various equity funds (see file **Canadian_Equity_Funds**). In particular, he is interested in the equity fund returns after one month (Ret_1Mth) and their returns after one year (Ret_1Yr). Compare the distributions for these two variables in relation to their skewness or symmetry.

Solution

With the computer, the distribution shapes can easily be visualized and compared. Figure 5.21 shows histograms and boxplots for the distributions. Also, for each distribution, its mean and median are compared.

By all of these indicators, both distributions have some negative skewing. For example, the mean has been pulled to a lower value than the median in each case. However, this skewing is proportionally less in the case of the one-year returns. Similarly, both the histogram and boxplot show less leftward skewing for the one-year returns, compared to the one-month returns.

Figure 5.21
Comparison of Two Distribution Shapes: Equity Fund Returns

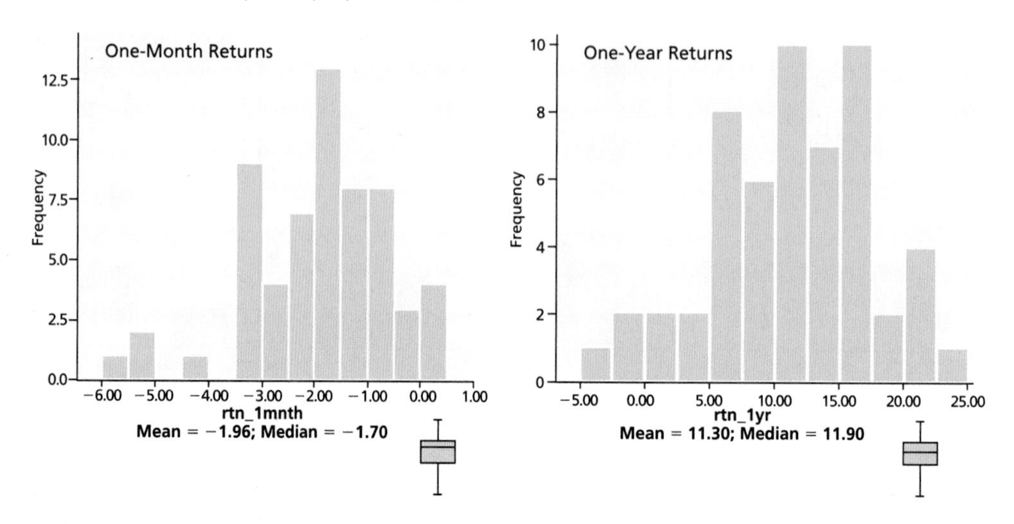

Source: Canadian Equities published on the Globefund.com website, sponsored by *The Globe and Mail.* Found at: http://www.globefund.com/static/romf/generic/tabceq.pdf. Reprinted with permission from *The Globe and Mail.*

Chapter Summary

5.1 Explain measures of spread, and calculate the range, variance, and standard deviation for a population or sample.

A **measure of spread** (or **measure of variation** or **dispersion**) is a single number that describes the degree of variety among the values of a distribution. The simplest measure of spread is the **range,** which is the difference between the largest and smallest data values. The **mean absolute deviation** is a measure based on the absolute distance (or **deviation**) of each data value from the mean. A more common measure of spread is the **standard deviation,** based on the squares of the deviations of each value from the mean. The standard deviation squared is the **variance.** Both population and sample versions for the formulas were given (see Formulas 5.1–5.6). The important concept of **degrees of freedom** was also introduced.

5.2 Correctly select and apply the empirical rule or Chebyshev's inequality to interpret the standard deviation.

For an approximately bell-shaped distribution, the **empirical rule** (or the **"68–95–99% rule"**) can be used to estimate the proportions of data that lie within certain distances of the mean. These "distances" are measured in standard deviations from the mean—a quantity that can be represented in terms of ***z*-values** (see Formulas 5.7–5.8). For distributions that are not close to bell-shaped, **Chebyshev's inequality** can be used to estimate the *minimum* proportions of values that can be expected to fall within a certain distance of the mean.

5.3 Calculate and interpret the interquartile range.

The **interquartile range** is a measure of spread that is not overly influenced by extreme values. It is calculated as the distance between the first and third quartiles for the data set.

5.4 Construct five-number summaries and boxplots.

The **five-number summary** uses the minimum and maximum data values, combined with the values for quartiles 1 through 3, to summarize a distribution of numbers. A **boxplot** (or **box-and-whisker diagram**) is a graphic method that can illustrate the five-number summary. Another variation of the boxplot uses **fences (a one-step distance from the nearest quartile)** as a way to identify and display possible outliers in the data.

5.5 Calculate and interpret the coefficient of variation.

The **coefficient of variation** is a unitless measure for comparing the **relative variations** of two or more data sets. This measure (see Formula 5.9) expresses the relative size of the standard deviation as a percentage of the mean for a given population or sample.

5.6 Describe the shape of a distribution.

The **shape of a distribution** can be characterized, in part, by whether it is **symmetrical,** that is whether we could fold up a picture of the distribution (e.g., its histogram) along a centre line and the figure would line up with itself on each side. **Skewness** is a measure of the nonsymmetry of a distribution. If the distribution is symmetrical, then the mean and median will fall at about the same location. **Kurtosis** is the relative degree to which a distribution is tall or flat, compared to a smooth bell-shaped curve drawn on the same scale.

• Key Terms

biased estimators, p. 174
boxplot (*or* box-and-whisker diagram), p. 193
Chebyshev's inequality, p. 186
coefficient of variation (CV), p. 200
degrees of freedom, p. 175
deviation, p. 169
empirical rule (*or* 68–95–99% rule), p. 179
fence (a one-step distance from the nearest quartile), p. 194
five-number summary, p. 192
interquartile range, p. 189
kurtosis, p. 203
mean absolute deviation, p. 169
mean deviation, p. 169
measure of spread (*or* measure of variation *or* measure of dispersion), p. 168
range, p. 168
relative variation, p. 200
shape of a distribution, p. 203
skewness, p. 203
standard deviation, p. 168
symmetrical, p. 178
variance, p. 172
z-value (*or* z-score), p. 185

• Exercises

Basic Concepts 5.1

1. When calculating the standard deviation, why are the differences between individual values and the mean squared before being summed?
2. For the following *populations,* determine the range, the variance, and the standard deviation.

Data set 1:	10	14	9	17	16
	5	13	15		
Data set 2:	9	4	6	8	
	9	5	7	3	8
Data set 3:	20	24	29	25	26
	27	22	20	24	

3. For the following *samples,* determine the variance and the standard deviation.

Data set 1:	10	14	9	17	16
	5	13	15		

Data set 2:	9	4	6	8	9
	5	7	3	8	
Data set 3:	20	24	29	25	26
	27	22	20	24	

4. The genders and years of service of 16 senior managers are:

Years:	6	8	9	12	15	3	14
Gender:	M	F	F	M	F	M	F
Years:	18	10	19	25	22	18	23
Gender:	M	F	M	M	M	F	M
Years:	5	16					
Gender:	F	M					

   a) Determine the variance and the standard deviation of the years of service for all of the senior managers.
   b) Determine the variance and the standard deviation for just the female managers.
   c) Determine the variance and the standard deviation for just the male managers.

5. Ten bottles of a certain soft drink were sampled; each claims to hold 750 mL of liquid. From the following sample of 10 bottles, calculate the standard deviation of the actual volumes of liquid in the bottles. Values are expressed in millilitres.

749.5	748.0	749.7	748.6	750.1
751.3	750.5	749.7	750.1	750.8

6. The number of houses built each year by Shepitka Construction is variable. Shown below is the number of houses built in each of the last 10 years. What is the variance and standard deviation for the number of houses built each year?

212	291	314	220	196	281
244	233	285	298		

Applications 5.1

7. Two volunteers collected data, over several summers, on eruptions of the Grotto Geyser in Yellowstone National Park. The table below shows, for each of 83 days that were studied, the number of eruptions recorded on that day. (Based on data in the file **Grotto_Geyser.**)

1	2	3	4	2	4
2	2	1	3	4	2
3	3	3	1	2	3
2	2	1	3	3	4
3	3	4	3	2	1
3	3	2	2	1	4
4	2	1	2	3	3
4	1	4	1	4	3
2	3	2	3	1	4
4	3	2	2	3	2
2	3	3	3	3	3
2	2	1	3	2	2
3	4	3	2	3	1
4	3	1	3	2	

   a) Calculate the standard deviation for number of eruptions each day.
   b) In Chapter 4, Exercise 7, you found the mean for the above data, and in Chapter 3, Exercise 13, you constructed its frequency distribution. Combine this information with your knowledge of the standard deviation for the data to answer this question: What percentage of the actual data values are in this range: larger than the value (mean minus one standard deviation) but less than the value (mean plus one deviation)? Does the standard deviation give a good sense of how the data are distributed? Explain.

8. In the file **World_Economic_Statistics,** the column labelled "Capita" shows a measure of purchasing power per person (in U.S. dollars) in many countries.
   a) Calculate the standard deviation for the population of values in the dataset.
   b) Using the computer, construct a quick histogram of the data set (do not worry about the titles or exact class widths). Based on just the standard deviation, and if you had not seen the histogram, would you have an accurate sense of how the data are distributed? Explain.

9. A chemist is experimenting with solvents to use in extreme-temperature environments. She will select from a sample of 47 solvents, for which the melting and boiling points are known. (See the columns "Mpt" (Melting Point) and "Bpt" (Boiling Point) in the file **Freezing_Boiling_Data.**)
   a) Calculate the standard deviations for (i) the melting points of the solvents and (ii) the boiling points of the solvents.
   b) In Chapter 4, Exercise 9, you found the mean for both of the variables in (a). For each variable, determine how many actual data points are below the value (mean minus two standard deviations) and how many data points are greater than the value (mean plus two standard deviations). For each of these variables, does the range from (mean minus two standard deviations) up to (mean plus two standard deviations) convey the "typical" range of values for the variable? Explain your answer.

10. An environmentalist is checking on the concentration of aluminum in the water supplies in Newfoundland. A sample of her data (in mg/L) is found in the file **Newfoundland_Water,** in the column labelled "Aluminum."
   a) Calculate the range, variance, and standard deviation for concentrations of aluminum in the sample.
   b) Use the computer to find the mean for this variable, then determine how many actual data points are below the value (mean minus two standard deviations) and

how many data points are greater than the value (mean plus two standard deviations). Does the range from (mean minus two standard deviations) up to (mean plus two standard deviations) convey the "typical" range of values for the concentrations of aluminum? Explain your answer.

11. A mineral collector has recorded values for density and hardness for a population of 100 metallic minerals. (See the file **Metallic_Minerals.**)

a) Calculate the range, variance, and standard deviation for each of the two variables.

b) In Chapter 4, Exercise 11, you found the means for both of the variables in (a). Answer this question for both variables: Is a randomly selected data value for this data set more likely to be more than two standard deviations above the mean or more than two standard deviations below the mean? What do your answers tell you about the shape of the distributions for these variables? Explain.

Basic Concepts 5.2

12. Unpaid bar bills at a local country club have a mean of \$505.27 and a standard deviation of \$100.45. Between what two amounts do:

a) Approximately 68% of all unpaid bar bills lie?
b) Approximately 95% of all unpaid bar bills lie?
c) Approximately 99% of all unpaid bar bills lie?

13. With reference to the data in Exercise 12, determine the z-value for unpaid bar bills of

a) \$600
b) \$400
c) \$590.76
d) \$495.21

14. Based on Chebyshev's inequality, between what two amounts do at least 75% of the unpaid bar bills in Exercise 12 lie?

15. The mean salary for support staff at a large publishing firm is \$16.00 per hour, and the standard deviation is \$2.50. If the salaries are normally distributed, then 95% of support staff salaries are between approximately what two dollar amounts? If the distribution of salaries is very non-normal, at least 96% of the salaries are between what two amounts?

16. With reference to the salary data in Exercise 15, determine the z-value for salaries of

a) \$17.00 per hour
b) \$15.00 per hour
c) \$21.50 per hour
d) \$14.10 per hour

17. Bartholomew owns two stocks. On randomly selected days, the two stocks' closing prices are:

Stock 1 Prices	Stock 2 Prices
\$26.07	\$23.30
\$24.28	\$20.35
\$19.15	\$22.50
\$18.60	\$20.80
\$25.78	\$25.87
\$23.24	\$21.29

a) Calculate the standard deviations for each set of prices.
b) Based on your answer to (a), which stock represents a more stable investment? Explain.

Applications 5.2

18. A team of health and environmental scientists has studied levels of fine particulate matter (PM_{10}) in the air around Mexico City. Over a five-year period, the following levels of PM_{10} were sampled in these three areas around the city.[3] Presume that the distributions are approximately normal.

Region	Mean Level (in $\mu g/m^3$)	Standard Deviation (in $\mu g/m^3$)
Xalostoc	94	53
Cero Estrella	65	35
Netzahualcoyotl	89	45

a) For what percentage of samples taken from Xalostoc is the PM_{10} level likely to be somewhere between 41 $\mu g/m^3$ and 147 $\mu g/m^3$?

b) For what percentage of samples taken from Cero Estrella is the PM_{10} level likely to be somewhere between 0 $\mu g/m^3$ and 135 $\mu g/m^3$?

c) In which of these three regions would it be *least unlikely* to be able to obtain an air sample with a PM_{10} concentration of 185 $\mu g/m^3$? Explain your answer.

d) Suppose one day's sample from Netzahualcoyotl has a PM_{10} concentration of 149 $\mu g/m^3$. What is the z-value for that sample?

19. An Australian psychologist developed an "Attitude to Authority" scale based on responses to various questions. In a 1971 paper based on the model, subjects scored an average of 88.34 on the scale, with a standard deviation of 14.14. It is not clear in the paper if the scores were normally distributed.[4]

a) For *at least* what percentage of subjects in the study was the score between 60.1 and 116.6?

b) For at least 89% of the subjects, the score was between what two values?

c) Suppose a participant in the study obtained a score of 92. What is the z-value for that person's score?

20. The two private online institutions that comprise the American Public University System (APUS) have published some learning outcomes of their programs on the Internet. In May 2006, they reported that their students in Criminal Justice obtained a mean score of 163.5, with a standard deviation of 12.9, on the *Major Field Test in Criminal Justice.*[5]

Presume that the scores are normally distributed.

a) About what percentage of the students obtained test scores between 124.8 and 202.2?
b) About 68% of the students scored between what two values?
c) Would it be unusual for a student to score 204 on the test? Explain why or why not.

21. A locker room is used primarily by males aged 65–74. For this age group, the mean weight is 74.8 kg, with a standard deviation of 12.8 kg.[6] Presume a normal distribution of weights for people in that demographic.

a) What percentage of those in the locker room are likely to weigh between 62.0 kg and 87.6 kg?
b) 99.6% of the people in the locker room are likely to weigh between what two values?
c) If a man in the locker room weighs 71.2 kg, what is z-value for his weight?
d) Suppose that, one day, together with the older male users, a number of pro football players and their young male children are using the locker room. Do you expect that the distribution of weights in the room is still normally distributed? Why or why not?

22. High noise levels, due to traffic, aircraft, construction, loud iPods, and many other causes are of serious concern to public health practitioners. Besides continuous noises, spikes of "impulse" noises, such as when something crashes, can pose health risks. One source suggests that the mean energy level at which impulse sounds become harmful to people is at 145 dB, with a standard deviation of 8 dB.[7]

Suppose 500 people are at a concert. What percentage of these people is at risk of harm if there is a loud impulse noise with an energy of

a) 145 dB?
b) 137 dB?
c) 153 dB?

Basic Concepts 5.3

For each of the data sets in Exercises 23–26, find the interquartile range.

23. Data set: 10 14 9 17 16 5 13 15 19 18 15 13 17 14 15 16 17

24. Data set: 9 4 6 8 9 5 7 3 8 3 5 8 8 6 4 9 4 5 6

25. Data set: 20 24 29 25 26 27 22 20 24 23 21 20 29 28 24

26. Data set: 101 122 113 143 194 136 129 148 179 155 132 145

27. Ten bottles of a certain soft drink were sampled; each claims to hold 750 mL of liquid. From the following sample of 10 bottles, calculate the interquartile range of the actual volumes of liquid in the bottles. Values are expressed in millilitres.

749.5	748.0	749.7	748.6	750.1
751.3	750.5	749.7	750.1	750.8

28. In each of the last 10 years, Shepitka Construction has built the number of houses shown below for each year. What is the interquartile range for the numbers of houses built each year?

212	291	314	220	196	281
244	233	285	298		

Applications 5.3

29. Two volunteers collected data, over several summers, on eruptions of the Grotto Geyser in Yellowstone National Park. The table below shows, for each of 83 days that were studied, the number of eruptions recorded on that day. (Based on data in the file **Grotto_Geyser.**)

1	2	3	4	2	4	2	2
1	3	4	2	3	3	3	1
2	3	2	2	1	3	3	4
3	3	4	3	2	1	3	3
2	2	1	4	4	2	1	2
3	3	4	1	4	1	4	3
2	3	2	3	1	4	4	3
2	2	3	2	2	3	3	3
3	3	2	2	1	3	2	2
3	4	3	2	3	1	4	3
1	3	2					

a) Calculate the interquartile range for number of eruptions each day.
b) In Exercise 7 of this chapter, you found the standard deviation for the data shown above. Explain how these two measures compare.

30. A list of the world's tallest buildings is given in the file **Tallest_Buildings.** The number of storeys in each building is indicated.

a) Calculate the interquartile range for the number of storeys in these buildings.
b) Using the computer, construct a quick histogram of the data set (do not worry about the titles or exact class widths). Based on just the interquartile range, and if you had not seen the histogram, would you have an accurate sense of how the data are distributed? Explain.

31. The file **Tallest_Buildings** also includes a column with the heights, in metres, of each building (including spires, but not antennae or flagpoles).

a) Calculate the interquartile range for the heights of these tall buildings.
b) Using the computer, construct a quick histogram of the data set. Does the interquartile range seem to be an

equally good indicator of distributions for the building heights and for the building storeys (see Exercise 30)? Explain.

32. An environmentalist is checking on the concentration of aluminum in the water supplies in Newfoundland. A sample of her data (in mg/L) is found in the file **Newfoundland_Water,** in the column labelled "Aluminum." Calculate the interquartile range for concentrations of aluminum in the samples.

33. A mineral collector has recorded values for density and hardness for a population of 100 metallic minerals. (See the file **Metallic_Minerals.**)

a) Calculate the interquartile range for each of the two variables.

b) Is a randomly selected data value more likely to be within the interquartile range or outside of the interquartile range, or neither? Would your answer change if the data set had a different distribution shape? Explain.

Basic Concepts 5.4

For each of the data sets in Exercises 34–37, construct the five-number summary and a boxplot.

34. Data set: 10 14 9 17 16 5 13 15 19 18 15 13 17 14 15 16 17

35. Data set: 9 4 6 8 9 5 7 3 8 3 5 8 8 6 4 9 4 5 6

36. Data set: 20 24 29 25 26 27 22 20 24 23 21 20 29 28 24

37. Data set: 101 122 113 143 194 136 129 148 179 155 132 145

38. Ten bottles of a certain soft drink were sampled; each claims to hold 750 mL of liquid. From the following sample of 10 bottles, construct a five-number summary and a boxplot of the actual volumes of liquid in the bottles. Values are expressed in millilitres.

749.5	748.0	749.7	748.6	750.1
751.3	750.5	749.7	750.1	750.8

39. In each of the last 10 years, Shepitka Construction has built the number of houses shown below in each year. Construct a five-number summary and a boxplot for the number of houses built each year.

212	291	314	220	196	281
244	233	285	298		

Applications 5.4

40. In Exercise 29, data were presented for the number of eruptions of the Grotto Geyser recorded on each of 83 days of observation. Use the same data to construct a five-number summary and a boxplot for the distribution.

For Exercises 41–43, use the data in the file **Tallest_ Buildings.**

41. Construct a five-number summary and a boxplot for the numbers of storeys in these tall buildings.

42. Construct a five-number summary and a boxplot for the heights of these tall buildings.

43. Compare your solutions to Exercises 41 and 42 with your solutions to Exercises 30 and 31, respectively, on the interquartile range. If you wanted to convey information about the distributions of the variables *storeys* and *height,* for which variable would it be *less* suitable to give only the interquartile range, rather than the full five-number summary? Why?

44. An environmentalist is checking on the concentration of aluminum in the water supplies in Newfoundland. A sample of her data (in mg/L) is found in the file **Newfoundland_Water,** in the column labelled "Aluminum." Construct a box plot and the five-number summary for concentrations of aluminum in the samples.

Basic Concepts 5.5

45. Suppose you want to compare the variability of two data sets. Under what condition would it be misleading to compare their standard deviations directly, and be clearer to compare their coefficients of variation?

46. For each of these data sets, calculate its coefficient of variation.

Data set 1:	10	14	9	17	16
	5	13	15		
Data set 2:	9	4	6	8	9
	5	7	3	8	
Data set 3:	20	24	29	25	26
	27	22	20	24	

47. The genders and years of service of 16 senior managers are:

Years:	6	8	9	12	15	3	14	18
Gender:	M	F	F	M	F	M	F	M
Years:	10	19	25	22	18	23	5	16
Gender:	F	M	M	M	F	M	F	M

a) Determine the coefficient of variation for all senior managers.

b) Compare coefficients of variation for the male managers and the female managers.

48. From the following sample of 10 bottles, each labelled "750 mL," calculate the coefficient of variation for the actual volume of liquid in the bottles. Values are expressed in millilitres.

749.5	748.0	749.7	748.6	750.1
751.3	750.5	749.7	750.1	750.8

49. In each of the last 10 years, the number of houses shown below has been built each year by Shepitka Construction. What is the coefficient of variation for the number of houses built each year?

212	291	314	220	196	281
244	233	285	298		

50. At East Breeding Farm, bulls live to a mean of 28 years, with a standard deviation of 6.5 years. At West Breeding Farm, the mean life span of its bulls has been 24 years, with a standard deviation of 6.2 years. At which breeding farm have the bulls' life spans been, relatively speaking, more variable?

Applications 5.5

51. A team of health and environmental scientists have studied levels of fine particulate matter (PM_{10}) in the air around Mexico City. Over a five-year period, the following levels of PM_{10} were sampled in these three areas around the city.[8] Presume approximately normal distributions.

Region	Mean Level (in $\mu g/m^3$)	Standard Deviation (in $\mu g/m^3$)
Xalostoc	94	53
Cero Estrella	65	35
Netzahualcoyotl	89	45

a) For each city, calculate the coefficient of variation for the level of PM_{10}.

b) Which city has the greatest absolute level of variability for PM_{10}?

c) Which city has the greatest relative variation in level of PM_{10}?

d) A person suffering from asthma wants to move to Xalostoc. He wants to estimate the upper levels that PM_{10} would likely reach, allowing for normal variation from the mean. Should he use the standard deviation or the coefficient of variation to make his estimate? Explain.

52. The two institutions that comprise the American Public University System (APUS) reported that their students in Criminal Justice obtained a mean score of 163.5, with a standard deviation of 12.9, on the *Major Field Test in Criminal Justice*.[9] Presuming that the scores are normally distributed, what is the coefficient of variation for the scores?

53. For males aged 65–74, the mean weight is 74.8 kg, with a standard deviation of 12.8 kg.[10] Presume a normal distribution of weights for men in that demographic.

a) For the weights of men in this group calculate the coefficient of variation.

b) Suppose that, one day, some males aged 65–74 are joined in a locker room by some pro football players and their young male children. Do you expect that the coefficient of variation of weights in the room will be larger, smaller, or no different from the *CV* calculated in (a)? Justify your answer.

54. To entertain a guest, a mayor has downloaded a price list for Australian wines from the website of the PEI Liquor Control Commission. (See the file **Australian_Wines.**) What is the coefficient of variation for this set of prices?

55. A political scientist is comparing candidates' winning margins in their ridings (i.e., the number of votes by which the winning candidate exceeded the votes cast for the candidate with the second-highest number of votes). For the 2004 federal election, find the winning margins of all winning candidates in the file **Early_Returns.**

a) For winning candidates in Quebec, find the coefficient of variation for their winning margins.

b) For winning candidates in Ontario, find the coefficient of variation for their winning margins.

c) In which of those provinces in 2004 was there a greater relative variability in the winning margins?

Basic Concepts 5.6

56. For each data set, determine whether its distribution is symmetrical, skewed to the left, or skewed to the right.

Data set 1:	4	6	2	7	8	3	5
	9	1	6	7	4		
Data set 2:	4	5	7	2	8	9	10
	12	3	4	14	18	2	6
	7	18					
Data set 3:	15	13	19	18	15	16	20
	11	14	16	17	12	21	22
	23	10	17	13	24		

57. What does it mean for a distribution to have positive kurtosis? Negative kurtosis?

58. The batting averages below were obtained by sampled members of the Toronto Blue Jays during the 2005 season.

0.269	0.291	0.256	0.262	0.262	0.271	0.251
0.301	0.269	0.274	0.249	0.216	0.250	0.207
0.194	0.308	0.083				

Determine whether this distribution is symmetrical, skewed to the left, or skewed to the right.

59. If a distribution has a value of kurtosis = 0, what is the shape of the distribution?

60. For each of the data sets below, comment on the shape of the distribution.

Data set 1: 16 12 19 10 14 15 18

Data set 2: 10 11 17 19 20 15 26 17 15 17 18 13 10

61. Comment on the shape of this distribution:

3 5 7 2 4 8 2 5 9 5 7 1 6 3 2 5 4 3 4 6 9

Applications 5.6

62. An amateur astronomer has collected facts about the diameters of the first nine planets of our solar system. (Data are in the file **Planetary_Facts.**) Assess the skewness of this data set, and justify your answer.

63. In 2001, the Canadian federal government and eight banks signed agreements to facilitate lower-cost banking services for consumers. For a sample of low-cost options for monthly fees, see the file **Low_Fee_Accounts.** Comment on the shape of this distribution.

64. A chemist is experimenting with solvents to use in extreme-temperature environments. The melting and boiling points of 47 solvents can be found in the file **Freezing_Boiling_ Data;** see the columns "Mpt" (Melting Point) and "Bpt" (Boiling Point). Comment on the shape of this distribution.

65. A mineral collector has identified 100 metallic minerals, and recorded for each the values of density and hardness. He is examining the distributions of these two variables. (See the file **Metallic_Minerals.**) Assess the level of skewness for each distribution. Justify your answer.

Review Applications

For Exercises 66–70, use data from the file **Petroleum_ Production_US.**

66. An energy planner has compiled U.S. data on petroleum production from 1949 to 2004. She is especially interested in the variability of the data set. The CD contains data for the crude oil and natural gas production, in 1,000s of barrels, during that period. Calculate the range, variance, standard deviation, and relative variation for (a) crude oil production and (b) natural gas production. Which distribution appears to have more variability? Explain.

67. Answer the following based on your answers to Exercise 66(a):
    a) About 95% of the annual production quantities, from 1949 to 2004, were between what two values?
    b) About 68% of the annual production quantities, from 1949 to 2004, were between what two values?
    c) What is the z-value for the quantity of crude oil produced in 1975?

68. For each of (a) crude oil and (b) natural gas, calculate the interquartile ranges.

69. Construct a five-number summary and a boxplot for the production of (a) crude oil and (b) natural gas.

70. Compare the skewness of the two variables' distributions.

For Exercises 71–76, use data from the file **Canadian_ Equity_ Funds.**

71. Following your decision to invest in Canadian Equity mutual funds, you began studying the fund data stored in the file **Canadian_Equity_Funds.** For the variable one-month returns ("Ret_1Mth"), calculate the range, variance, standard deviation, and relative variation.

72. The variable distribution for the management expense ratio (MER) variable is roughly normal.
    a) Between what range of actual values would you expect to find the management expense ratios for about 95% of the funds?
    b) What is the z-value for a MER of 0.43?

73. The variable distribution for the *one*-month returns is not altogether normal.
    a) Determine what range of values for one-month returns is bounded by (the mean minus 2 standard deviations) on the lower side, and (the mean plus 2 standard deviations) on the higher side.
    b) Complete this sentence, based on the information provided in Section 5.2: At least ___% and at most ___% of the values for one-month returns are in the range described in (a).

74. For each of (a) the three-month returns and (b) the MER data, calculate the interquartile range.

75. For each of (a) the three-month returns and (b) the MER data, construct a five-number summary and a boxplot.

76. Analyze the level of skewness for (a) the three-month returns and (b) the MER data.

For Exercises 77–81, use data from the file **National_ Parks.**

77. An ecologist is reviewing the land areas of all of Canada's national parks, as published by Natural Resources Canada.
    a) For the areas in the data file (in square kilometres), calculate the range, variance, and standard deviation.
    b) Do the areas of national parks appear to be highly variable or not particularly variable? Justify your answer.

78. The land areas for the national parks are not normally distributed.
    a) Nonetheless, you could expect that *at least* 89% of the parks' land areas are between what two values?
    b) What is the z-value for the land area of Kouchibouguac National Park in New Brunswick?

79. For the land areas of all the parks, calculate the interquartile range.

80. For the land areas of all the parks, construct a five-number summary and a boxplot.

81. Assess the skewness for the distribution of parks' areas.

For Exercises 82–86, use data from the file **Vehicle_ Ownership.**

82. Important considerations when buying a new car are (a) What are the likely costs to maintain it? and (b) How much will it cost for fuel? Preparing to buy a new car, you compile the data set **Vehicle_Ownership** and focus on two variables: typical annual repair costs in Canadian dollars for different models (variable: R_RepCst) and typical weekly costs for gas, for comparable driving distances (variable: R_FuelCs). For *both* these variables, calculate the range, variance, standard deviation, and relative variation.

83. The distribution of the repair costs variable is not altogether normal.

a) Determine what actual range of values for repair costs is bounded by (the mean minus 3 standard deviations) on the lower side, and (the mean plus 3 standard deviations) on the higher side.

b) Complete this sentence, based on the information provided in Section 5.2: At least ___% and at most ___% of the values for repair costs are in the range described in (a).

c) What is the *z*-value for the annual repair costs of a GMC Safari?

84. For each of (a) the annual repair costs and (b) the weekly fuel costs, calculate the interquartile range.

85. For each of (a) the annual repair costs and (b) the weekly fuel costs, construct a five-number summary and a boxplot.

86. Compare the skewness for the variables annual repair costs and weekly fuel costs.

PART 3

Probability and Distributions

Chapter 6

Concepts of Probability

Chapter Outline

Learning Objectives

6.1.1 Describe the core concepts of probability.

6.1.2 Explain and apply the three basic approaches to probability and explain some errors to avoid.

6.2 Explain and apply the concepts of conditional probability.

6.3 Define key terminology related to compound probabilities and apply the basic rules for compound probabilities.

6.4 Discuss the applications of Bayes' Rule and apply the rule to calculate posterior probabilities.

6.1 Basic Probability Calculations

In Part 2, our focus was primarily on descriptive statistics. Starting from data that may represent a census *or* a sample, we learned techniques for displaying and summarizing features of the data set, such as the central location and the shape of the distribution. In Part 3, we prepare the ground for inferential statistics. As described in Section 1.2, inferential statistics will allow us to make inferences about a population or about the next steps in a time series, based on the data in a sample.

The foundation for statistical inference is **probability.** Informally, we can view probability as a measure of the likelihood or chance that an event will occur. In any field, we rarely have complete information on which to base a decision. If a company produces a new car, the company cannot know for sure how many people will buy the car. When a geologist selects a location for digging an oil well or searching for diamonds, she cannot be sure that there is anything of value in that spot. There is always a degree of uncertainty or risk in making a final decision. Probability theory provides a method to analyze this uncertainty and to compare the risks of different choices.

As applied to inferential statistics, probability provides a tool to measure or quantify sampling error (see Section 2.2)—the unavoidable, but hopefully small chance that results obtained from a sample are simply off-base when compared to the population. Like the geologist choosing a digging site, the statistician is taking a risk if she concludes that the population has a similar value for the mean, variance, or other characteristic as does the sample. Again, probability theory helps us to identify and quantify this risk, and take steps to reduce it.

Core Concepts of Probability

The value that we assign for probability is really a hypothetical proportion. Suppose a geologist seeks a commercially viable source of raw diamond tailings, and estimates a probability of one-quarter that a particular site has that potential. This value implies that if she was to dig at many sites that had the same relevant characteristics, about one-quarter of them would prove to be commercially viable; the remainder would not. This model suggests a basic definition for probability (terms with an asterisk [*] are explained in the text that follows the definition):

> The probability of an event* is the expected proportion of times that the event would occur if the random experiment* that produces the event was repeated a large number of times.

The term **random experiment** refers here to a planned process of observation to obtain a value for a random variable. Digging for diamonds and observing what is found is one (very expensive) example of an experiment. Counting the number of Hummers in a parking lot is another example. Each distinct observation that results from an experiment is called an **outcome.** Having dug for diamonds, "finding no diamonds at all" could be an outcome. Having counted vehicles in the parking lot, "counting exactly four Hummers" is one possible outcome. For a given experiment, the set of all possible outcomes is called the **sample space.**

To describe an **event** is a bit fuzzier: It depends on which of the possible outcomes *counts as* the event, from the viewpoint of the researcher. For example, when a site is drilled for diamonds, it may yield anything from no diamond potential, to small traces of diamonds, up to very large quantities when a major diamond vein is discovered. Any one of these cases might describe a possible outcome of a drilling experiment. But which of the possible outcomes would count as the *event* "finding a commercially viable site"? This is a matter of the geologist's judgment.

Based on all known features of a drilling site, a geologist may expect that if many similar sites were drilled, about one-quarter of the *outcomes* (i.e., the actual diamond features and quantities discovered in each experiment) would meet the criteria for the *event* "being a commercially viable site." In that case, she could say (before drilling) that the probability is one-quarter that this type of site is commercially viable.

Because probability is related to an (expected) proportion, the following **conditions of probability** will always apply:

1. The assigned value of probability is always a number between 0 and +1. A probability 0 represents an impossible event: i.e., for any number n of experiments, we expect a proportion of $0/n = 0$ of the outcomes to be considered the event. A probability +1 signifies something that is certain to occur: For any number n of experiments, we expect a proportion of (all n)/n = 1 of the outcomes to be considered the event. Actual probabilities can fall anywhere in that range. (See Figure 6.1.)
2. The sum of probabilities for all of the possible outcomes always equals 1. The "event" in this case is any outcome whatsoever in the sample space; so, by definition, every experimental outcome (a proportion of 1/1) will count as the event.

Figure 6.1
The Range of Probability

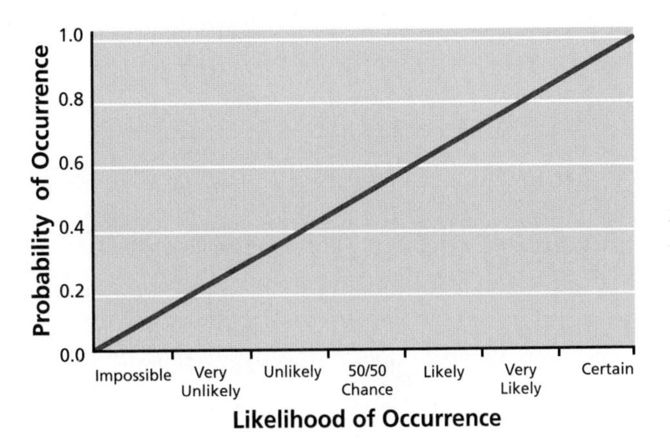

To calculate specific values for probabilities, there are three basic approaches or **types of probabilities:** (a) classical, (b) empirical, and (c) subjective (see Figure 6.2). All three approaches share the same core concept for probability as just discussed, and aim to quantify the chance or likelihood that a specific event will occur. In practice, the approaches can sometimes be combined.

Figure 6.2
Types of Probability

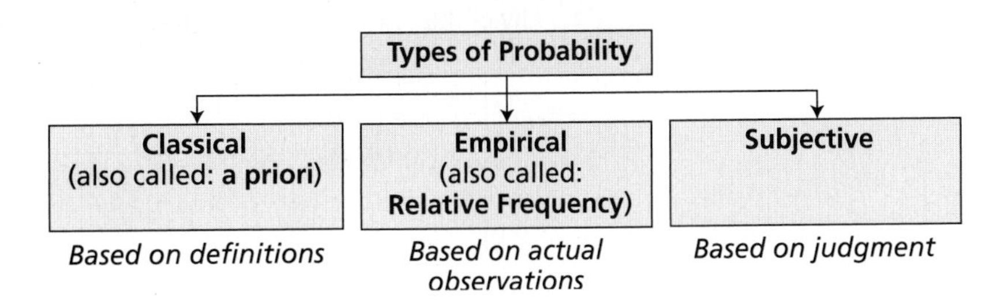

Classical Probability

In **classical probability,** the proportion of outcomes expected to count as an event of interest—i.e., the probability of that event—is determined from previously known definitions and properties of things. No actual observations (experiments) are required. This is also called an **a priori probability.**

For example, what is the probability of obtaining an odd number if you throw a conventional die with sides numbered (pictorially) from 1 to 6? From the definition of the die object itself, we determine that there are six possible outcomes: You throw a 1; or a 2, or a 3, and so on, up to a 6. From the properties of numbers, we know that only three of these outcomes (throwing a 1 or a 3 or a 5) count as the event "obtaining an odd number." Therefore, the probability equals the proportion of possible outcomes that count as the event, which is equal to 3/6 = 1/2. This approach can be given as a formula (see Formula 6.1).

Formula 6.1 Calculation of a Classical Probability

$$P(A) = \frac{\text{number of ways A can occur}}{\text{number of different outcomes possible}}$$

where A = An event of interest, ***and***

$P(A)$ = "The probability of the event A occurring," ***and***

$0 < P(A) < 1$, ***and***

"The number of ways A can occur" is the number of possible outcomes that would be considered cases of A occurring," ***and***

Both the number of ways A could occur (the numerator) and the number of different outcomes possible (the denominator) can be determined by the definitions of objects involved in the calculation.

Example 6.1 As the first step in a card game, you randomly draw one card from a standard full deck of 52 playing cards. Find the probabilities that the one card you draw is (1) the ace of spades, or (2) any ace, or (3) any spade.

Solution The first step in the solution is to review the object itself (the deck of cards) on which the probabilities will be based. The probabilities can be calculated from the known properties of that object. A standard deck of playing cards has a total of 52 cards, subdivided into four suits (clubs, diamonds, hearts, and spades) of 13 cards each. Each suit contains one card called an ace, so there are $4 \times 1 = 4$ aces in all. The ace of spades is the one ace card in the one suit spades; so there is only one ace of spades.

Therefore, if you randomly select one card from this deck, a number of things follow:

1. Any of the 52 cards could be drawn, so the number of outcomes possible is 52. We have seen that only one of the cards is an ace of spades. So:

$$P(\text{ace of spades}) = \frac{(\text{Number of ways ace of spades can occur})}{(\text{Number of different outcomes possible})}$$

$$= \frac{1}{52} = 0.0192$$

2. Of the 52 cards that could be drawn, we have seen that only four of the cards are aces. So:

$$P(\text{any ace}) = \frac{(\text{Number of ways any ace can occur})}{(\text{Number of different outcomes possible})}$$

$$= \frac{4}{52} = \frac{1}{13} = 0.0769$$

3. Of the 52 cards that could be drawn, we have seen that there are 13 ways to draw a spade (i.e., 13 of the cards would "count as" spades). So:

$$P(\text{any spade}) = \frac{(\text{Number of ways any spade can occur})}{(\text{Number of different outcomes possible})}$$

$$= \frac{13}{52} = \frac{1}{4} = 0.25$$

You may notice that examples of classical probability often refer to games of chance—dice, cards, roulette wheels, and so on. We are not trying to promote gambling, but these examples come naturally, since modern theories of mathematical probability got their start in early attempts (in the 17th and 18th centuries) to understand problems that arise in games of chance. Also, for teaching purposes, the events and outcomes in games of chance are very neatly defined: We know exactly how many sides are on a coin, and how many cards are in a deck.

Real-world examples can be more difficult for assigning classical probabilities. For example, if a child is randomly selected, what is the probability that he or she is blond? When trying to calculate the denominator of Formula 6.1 for this case, how many possible colours have to be included? Should we count only colours for which Clairol makes a hair dye? And for the numerator: How many specific colour shades can be considered a case of "blond"?

Nonetheless, some nongaming examples are also possible, as Example 6.2 illustrates.

Example 6.2

You have just bought a combination lock, but have lost the combination. There are 50 numbers on the dial of the device, and you would normally input the combination by spinning the dial clockwise to a first number, then counterclockwise to a second number (or back to the original number), and then again clockwise to any of the 50 numbers. If you spin the dial in this same sequence, but stop at *random* numbers in each case, what is the probability that you will be able to open the lock?

Solution

It can be shown that the number of possible combinations on this lock is 125,000. (Each of the three inputs could be any number from 1 to 50, and any combination of the three is possible; so the total number of combinations $= 50 \times 50 \times 50 = 125{,}000$.) Only one combination will be accepted by the lock as correct. Therefore:

$$P(\text{correct combination}) = \frac{(\text{Number of ways a correct combination can occur})}{(\text{Number of different combinations possible})}$$

$$= \frac{1}{125{,}000} = 0.000008$$

It is extremely unlikely that you could open this lock by randomly spinning its dial.

Some Helpful Counting Techniques

It can be easy to calculate classical probability—*if* you know the numbers for the denominator and the numerator of Formula 6.1 (i.e., the number of outcomes possible *and* the number of those outcomes interpreted as the event). The challenge is usually to find those two numbers. The following techniques can sometimes be helpful.

Combinations and Permutations

For some problems, you need to count the number of size k sequences of objects that can be created from a usually larger size n set of objects. For example, how many possible hands of five cards could you draw from a full deck of 52 cards? If the order of drawing the cards does not matter, the mathematics of **combinations** applies. If you also care about the *sequence* in which the five cards are drawn, the related formula for **permutations** applies. Although it is beyond the scope of this chapter to cover the underlying formulas (but see Formula 7.4 for a related formula), you can use the computer to help solve such problems.

Section 6.3

• *In Brief 6.1 Using the Computer to Count Combinations and Permutations*

Of the software discussed in this text, only Excel provides convenient formulas for calculating combinations and permutations. Many calculators can also perform these calculations, however.

Excel

The illustration shows how Excel can be used for these calculations. For the number of distinct size k combinations that could be generated as subsets of an original size n set, input an Excel formula of this form: **=COMBIN(*n*,*k*).** For the example shown, you can confirm the result: If you start with a set of three objects *a, b,* and *c,* there are three possible combinations of size 2 that can be generated (*a-b, a-c,* and *b-c*).

Similarly, for the number of size k arrangements or permutations (where order matters) that could be generated from an original size n set, input an Excel formula of this form: **=permut(*n*,*k*).** You can confirm the result shown in the illustration: If you start with a set of three objects *a, b,* and *c,* there are six possible permutations of size 2 that can be generated (*a-b, b-a, a-c, c-a, b-c,* and *c-b*).

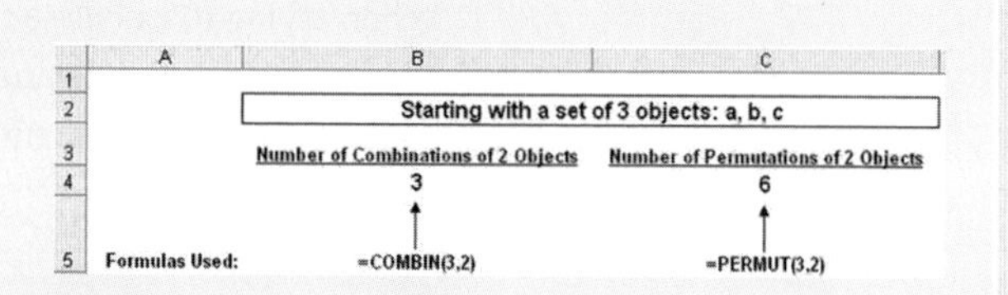

Example 6.3 A brand of candies comes in eight colours: red, yellow, orange, green, mauve, pink, brown, and blue. Suppose you have a bowl with one each of all of the colours and randomly pick two candies.

1. What is the probability that the two candies you draw (in whatever order) will be green and mauve?
2. What is the probability that you will draw, *in this order,* the green and the mauve candies?

Solution As in the previous examples, begin each solution with the basic formula for probability.

1. From the basic formula:

$$P(\text{the two candies are (in whatever order) green and mauve})$$

$$= \frac{(\text{Number of combinations that are green and mauve})}{(\text{Number of different combinations possible})}$$

Counting the denominator: There are eight colours available ($n = 8$). Any combination of two colours is a possible output ($k = 2$). Based on the Excel formula **=COMBIN(8,2),** the number for the denominator is 28. This can be confirmed manually by writing out every combination of two colours that could be selected. There are 28, as shown in the following table.

Possible Colour Combinations

red	&	yellow
red	&	orange
red	&	green
red	&	mauve
red	&	pink
red	&	brown
red	&	blue
yellow	&	orange
yellow	&	green
yellow	&	mauve
yellow	&	pink
yellow	&	brown
yellow	&	blue
orange	&	green

Possible Colour Combinations (continued)

	orange	&	mauve	
	orange	&	pink	
	orange	&	brown	
	orange	&	blue	
>>	green	&	mauve	<<
	green	&	pink	
	green	&	brown	
	green	&	blue	
	mauve	&	pink	
	mauve	&	brown	
	mauve	&	blue	
	pink	&	brown	
	pink	&	blue	
	brown	&	blue	

Counting the numerator: If order of selection is being ignored, the table shows that only one combination of colours counts as the event "green and mauve." So the number for the numerator is 1.

Therefore, $P(\text{the two candies are green and mauve}) = 1/28 = 0.0357$.

2. From the basic formula:

$$P(\text{draw green and mauve in exactly this order})$$

$$= \frac{(\text{Number of ways the event "draw green and mauve in this order" could occur})}{(\text{Number of different permutations possible})}$$

Counting the denominator: Since order matters, the possible outputs are all of the possible permutations of two colours, not just the combinations. There are eight colours of candies ($n = 8$). Any possible two-colour *sequence* is a possible output ($k = 2$). Based on the Excel formula **=PERMUT(8,2),** the number for the denominator is 56. This can be confirmed manually by reexamining the table drawn for solution 1—except that in this case, order matters, so double the number of displayed outcomes (e.g., the *combination* "red & green" corresponds to *two sequences*— "red → green" and "green → red," and so on for every pair displayed.) So the total number of sequences is $2 \times 28 = 56$.

Counting the numerator: There is only one way to "draw green and mauve in that exact order," so the count of the numerator equals 1.

Therefore, P(draw green and mauve in that exact order) $= 1/56 = 0.0179$

Tables of Outcomes

Creating a **table of outcomes** is another useful technique for finding the numbers needed for a basic probability calculation. Quite literally, you depict every possible outcome in a table format. (Compare, for example, the table included in the solution to the first part of Example 6.3.) For the denominator of the probability formula, simply count off the number of items listed; for the numerator, identify and count how many of the displayed outcomes would be considered as the event.

There are no hard-and-fast rules for how to construct such tables. Use your imagination to come up with a convenient way to display possible outcomes in a table format. (This works best if the number of possible outcomes is not very large.)

Example 6.4

Create tables of outcomes in the process of answering these two questions.

1. Suppose you randomly throw two standard six-sided dice and then add the numbers depicted on the two dice. What is the probability that the sum of the two numbers is greater than seven?
2. Suppose there are six items on a store shelf, numbered one to six. A customer enters the store and randomly removes one of the items. Then another customer enters the store and randomly removes one of the remaining items. What is the probability that the item numbered "two" has been removed from the shelf by one of the customers?

Solution

We create tables of outcomes to give us an idea of how to count the numerators and denominators for the basic probability formula.

1. An easy way to represent the possible outcomes is shown in Figure 6.3(a). Across the top are all of the possible values that the first die could land on; down the left are shown all of the possible values that the second die could land on. Each intersection of the corresponding rows and columns is a possible outcome of a throw of the dice. Of these, determine and check off the cases that correspond to the event (sum > 7).

$$P(\text{sum} > 7) = \frac{(\text{Number of outcome pairs whose sum is greater than 7})}{(\text{Number of possible outcomes of the dice throw})}$$

Counting the denominator: In Figure 6.3(a), we count 36 possible outcomes.
Counting the numerator: Counting only the checked cells in the table, we find 15 outcomes that correspond to "sum greater than 7."

Therefore, $P(\text{sum} > 7) = 15/36 = 0.417$.

2. For this example, the possible outcomes can be shown in a table such as Figure 6.3(b), which is similar to the table in Figure 6.3(a)—with one key difference: It is not possible for the second customer to pick an already-selected item, so the cells in the table for "1–1," "2–2," etc., are marked to indicate that these items should

Figure 6.3
Example Tables of Outcomes

a)

Number depicted on the second die thrown	Number depicted on the first die thrown: 1	2	3	4	5	6
1						
2						✓
3					✓	✓
4				✓	✓	✓
5			✓	✓	✓	✓
6		✓	✓	✓	✓	✓

✓ = The sum of these two numbers is greater than 7

b)

Numbered item selected by the second customer	Numbered item selected by the first customer: 1	2	3	4	5	6
1	✗	✓				
2	✓	✗	✓	✓	✓	✓
3		✓	✗			
4		✓		✗		
5		✓			✗	
6		✓				✗

✗ = These outcomes are not possible (don't count them)
✓ = One of the two customers took down item 2

not be taken into consideration. They are *not* possible outcomes that could actually occur. Every remaining cell is a possible outcome of item pairs that could be selected by the two customers. Of these, determine and check off the cases that correspond to the event (item 2 has been selected/removed by a customer).

$$P(\text{item 2 was selected/removed})$$

$$= \frac{(\text{Number of outcome pairs that include a 2})}{(\text{Number of item pairs that could be selected by the customers})}$$

Counting the denominator: In Figure 6.3(b), we count 30 possible item pairs that could be selected. (Remember not to count the cells marked with an "X.")

Counting the numerator: Counting only the checked cells in the figure, we find 10 outcomes that correspond to "item 2 was selected/removed."

Therefore, $P(\text{item 2 was selected/removed}) = 10/30 = 0.333$.

Contingency Tables

In Section 3.5, we introduced the **contingency table.** In some circumstances, the numbers needed for a basic probability calculation can be determined with the help of these tables. They can also be used for calculations of empirical probabilities.

Consider the data shown in Table 6.1, which illustrate the distributions of cards in a standard deck in relation to colour and ranks. From Chapter 3, we know that the set of column headings ("Face Card" or "Other") should list categories that are mutually

Table 6.1 **Contingency Table for Ranks and Colours of Cards**

		Ranks of Cards		
		Face Card	Other	Total
Colours of Cards	Red	6	20	26
	Black	6	20	26
	Total	12	40	52

exclusive and collectively exhaustive, as should the set of row headings ("Red" or "Black"). The values in the cells are frequencies. When applied to classical probabilities, these frequencies do not have to be found by observation: We know by the definition of a card deck, for example, that there are 26 red cards and 26 black cards, for a total of 52 cards. Once the data are presented this way, calculating probabilities is straightforward, as shown in Example 6.5.

Example 6.5

Suppose that one card is dealt randomly from a standard deck. Use Table 6.1 to calculate the following probabilities.

1. What is the probability that the selected card is a red card?
2. What is the probability that the selected card is a face card?
3. What is the probability that the selected card is a black face card?

Solution

Once again, base your solutions on the basic probability formula. The data needed for the counts can be easily read off Table 6.1.

1. $P(\text{draw a red card}) = \dfrac{(\text{Number of cards interpreted as red})}{(\text{Number of cards you could draw})}$

Counting the denominator: In the table, we see there are 52 cards in total.
Counting the numerator: We see that the total number of cards identified as "red" is 26.

Therefore, $P(\text{draw a red card}) = 26/52 = 0.5$.

2. $P(\text{draw a face card}) = \dfrac{(\text{Number of cards interpreted as face cards})}{(\text{Number of cards you could draw})}$

Counting the denominator: Again, we see there are 52 cards in total.
Counting the numerator: We see that the total number of cards identified as "face cards" is 12.

Therefore, $P(\text{draw a face card}) = 12/52 = 0.231$

3. $P(\text{draw a black face card}) = \dfrac{(\text{Number of cards interpreted as black face cards})}{(\text{Number of cards you could draw})}$

Counting the denominator: Again, we see there are 52 cards in total.
Counting the numerator: This time, look at the internal cell (left, bottom) for the frequency of cards that are black and face cards. The number is given as 6.

Therefore, $P(\text{draw a black face card}) = 6/52 = 0.115$.

Of course, you could also have used a table of outcomes for solving these classical probabilities. But if you are interested in only a few categories of outcomes, then a contingency table approach can be less cumbersome than individually displaying all possible outcomes in a table.

Empirical Probability

We have seen that, when using Formula 6.1 to calculate probability, the numbers needed for the denominator (how many outcomes are possible) and the numerator (how many outcomes count as the event) follow intrinsically from the definitions of objects involved. Our examples were based on some well-known artificial objects, such as standard card decks, dice, and well-defined combination locks. In many real-world applications, however, objects do not have the precise specifications that this approach requires, and the outcomes are not that predictable.

A common alternative is to calculate **empirical probability** (also called **relative frequency probability** or **experimental probability**). We learn from experience how often a particular event has occurred in the past, relative to all of the other alternatives, and then (with caution) we make this leap: We estimate that *if all relevant conditions remain the same,* the event will occur in roughly the same proportion in the future as it has in the past. Table 6.2 (a reinterpretation of some features of Table 3.3 in Chapter 3) shows how relative frequencies, based on historical data, can be interpreted as probabilities.

Table 6.2 Probabilities for Votes Cast in BC Ridings, 2004 Federal Election

Classes	Historical Percentages	Probabilities
30,000–34,999	2.8%	0.028
35,000–39,999	8.3%	0.083
40,000–44,999	25.0%	0.250
45,000–49,999	22.2%	0.222
50,000–54,999	22.2%	0.222
55,000–59,999	13.9%	0.139
60,000–64,999	5.6%	0.056
	100.0%	**1.000**

Source: "Federal Election Results, 2004." *The Globe and Mail.* Wednesday, June 30, 2004. Pages A12–A13. Reprinted with permission from *The Globe and Mail.*

In Table 3.3, we displayed the distribution of numbers of votes cast in British Columbia ridings in the 2004 federal election, showing the absolute frequencies for each number of votes cast as well as the relative frequencies. To find the probabilities shown in Table 6.2, the fact that the number of votes cast in a particular riding falls within a particular class is interpreted as a possible "event." The class's relative frequency (which is a historical fact) is interpreted as the probability of that event occurring in some possible future circumstance (such as in the next election). The probabilities are shown as decimal numbers between zero and one in Table 6.2, but they could be expressed as percentages as well. Thus, the probability for the event "30,000 up to 35,000 votes are cast in a randomly selected riding" is estimated as 0.028, with similar calculations for the other class "events."

Example 6.6 Table 3.1 in Chapter 3 displays the distribution of point spreads (i.e., the differences in teams' final scores) over the course of 672 NFL football games. Based on this historical data, estimate the probability that if one NFL game is randomly selected in the future, the point spread between the two teams' scores will be exactly 1. For this estimate to be reliable, why is it important that the game be randomly selected? Are there other factors that could affect whether the relative frequency provides a reasonable estimate of probability? Explain.

Solution According to Table 3.1, 3.3% of all of the NFL games had a point spread equal to 1. We can interpret this relative frequency as a probability. If one NFL game is randomly selected, the probability that the point spread equals 1 is 0.033.

This method presupposes that the game for which we want a probability will be randomly selected from all possible games. What if instead you are watching a game between the known best team and the known worst team in NFL, played late in the season? You would have good reason to suspect that the probability estimate—which was based on data for *all* teams over a whole season—may not apply in this situation.

In general, the historical relative frequency can be compared to a sample statistic. For a useful prediction or estimate, the original sample must be representative of the situation for which you want a probability. If the original conditions cease to apply in some way, it is risky to rely on the empirical probabilities (e.g., if the NFL changes its scoring rules or its officiating practices). The previous relative frequencies for point spreads may then not apply in future seasons.

To calculate empirical probability for one specific event, it is not necessary to create a complete relative frequency table. Formula 6.2 is sufficient.

Formula 6.2 Calculation of Empirical Probability

$$P(A) = \frac{\text{(number of ways } A \text{ did occur)}}{\text{(number of different outcomes possible)}}$$

where A = An event of interest, ***and***

$P(A)$ = "The probability of event A occurring," ***and***

One "outcome" is the result of one observation or experiment, among the set of n observations or experiments that were conducted, ***and***

The number of ways A did occur (the numerator) is the number of outcomes that historically met the conditions for being considered cases of "A occurred," ***and***

The size of the sample space (i.e., the number of different outcomes possible) (the denominator) equals the sample size n (i.e., the number of observations or experiments conducted).

Example 6.7 Based on data in the 2001 census, there are 7.14 million Canadians whose mother tongue is French, out of the total Canadian population of 29.64 million people.[1] If you randomly select one Canadian *in the present year,* what is the probability that his or her mother tongue is French?

Solution

Provided the proportions found in 2001 are reasonably representative for future years, we can use the formula:

$$P(\text{mother tongue is French}) = \frac{(\text{Number of people identified as "mother tongue is French"})}{(\text{Total number of people whose linguistic information was collected})}$$

Counting the denominator: The 29.64 million people in Canada in 2001 were studied in relation to mother tongue and other variables.
Counting the numerator: The census classified 7.14 million people whose mother tongue was French.

Therefore, P(mother tongue is French) = (7.14 million)/(29.64 million) = 0.241

Example 6.8

 *(Based on the **Coin_Flips** file)*

Sasha, who is majoring in statistics, had some time on his hands. He decided to flip a coin 500 times, and recorded the full sequence of results (heads and tails). He then used this data to calculate empirical probabilities. The full sequence of heads and tails can viewed in the file saved as **Coin_Flips.** (The coin flips recorded in the file were simulated on the computer. See "In Brief 6.2" on re-creating the computer simulation.)

With reference to the sequence of outcomes saved in the **Coin_Flips** file, answer the following questions:

1. Suppose Sasha had stopped after 1 coin flip. Based on that limited data set, what is the empirical probability of tossing a head?
2. Suppose, instead, that Sasha had stopped after 20 coin flips. Based on the first 20 flips, what is the empirical probability of tossing heads?
3. What if he had stopped after 40 coin flips? Based on the first 40 flips, what is the empirical probability of tossing heads?
4. The original case recorded that Sasha continued until reaching 500 coin flips. Based on the full set of data available, what is the empirical probability of tossing heads?
5. Based on classical probability, what is the probability of tossing heads? How do the two approaches compare in this example, and can you reach any conclusions?

Solution

1. After just one flip, the data are insufficient for calculating a meaningful empirical probability. The one flip will either be a head (so P(head) = 1/1 = 1.00) or a tail (so P(head) = 0/1 = 0.00). By chance, the first flip was a head.
2. After the first 20 flips, we find that 12 happened to be heads. By empirical probability at that point, P(heads) = 12/20 = 0.600.
3. After the first 40 flips, we find that 21 happened to be heads. By empirical probability, P(heads) = 21/40 = 0.525.
4. Following the last flip (the 500th), we find that 253 flips were heads. Using all of the data available, we calculate P(heads) = 253/500 = 0.506.

5. The probabilities for flipping a coin would typically be decided using classical probability. On a given throw, there are two possible outcomes (heads or tails), and exactly one outcome counts as the event "heads was tossed." Therefore $P(\text{heads}) = 1/2 = 0.500$. In theory, that value represents the proportion of times heads would appear *whenever* fair coins are being tossed.

In practice, what this example shows is that in the *short run* (i.e., if the number of experiments is relatively limited), the actual proportion of times that the event occurs may not be equal to the theoretical proportion, as given by the classical probability. (This becomes an issue only when both a classical and an empirical probability can be calculated.) As an indicator of what might happen in practice, classical probabilities do work well for predicting the *long-run* proportion of cases (i.e., over many experiments) in which the event will likely occur; this is called the **Law of Large Numbers.** This pattern is illustrated in Figure 6.4. Using the data in the file **Coin_Flips,** the empirical probability for $P(\text{heads})$ is recalculated after each new coin toss. We see that the empirical value takes a while to "settle" close to 0.500, the value predicted by the classical method.

Sections 6.1, 6.2, and 6.4

Figure 6.4
Empirical Probability after Many Experiments

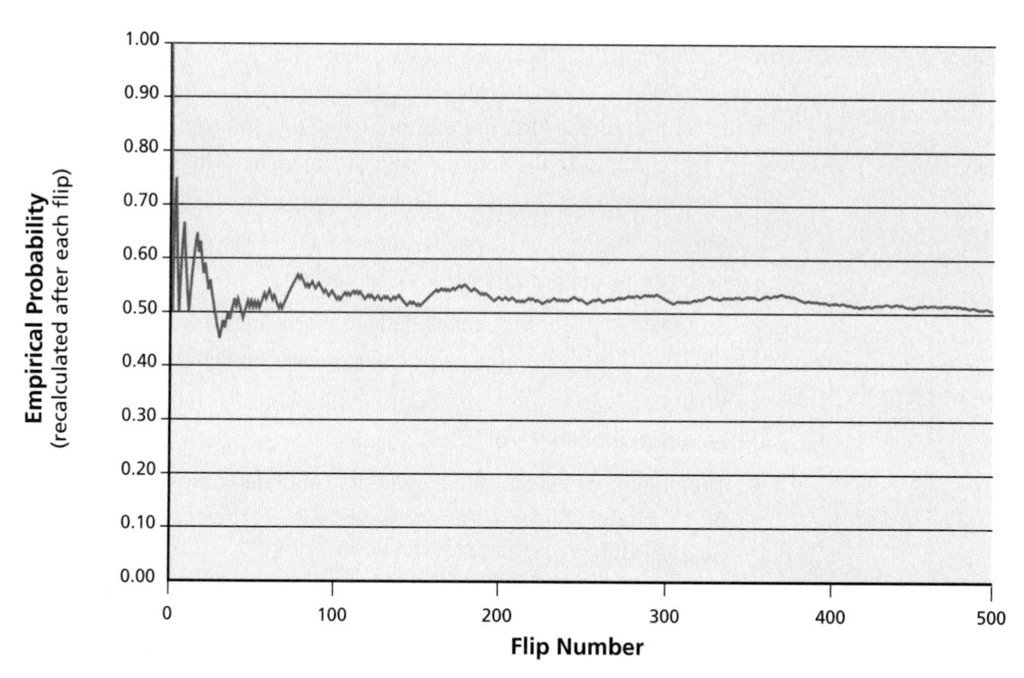

Some Errors to Avoid

On the surface, the calculations for empirical probability can be quite simple: Find the proportions of times that an event has occurred (or construct a relative frequency distribution, as discussed in Section 3.1) and then interpret the relative frequencies or proportions as probabilities. If not used carefully, however, this approach can lead to error.

1. **Too few experiments:** As demonstrated in Example 6.8, if we calculate the empirical probability after too few experiments, the results can be misleading. The

short-run proportion of times that heads occurs, compared to other outcomes, may not be representative of the long-term probability. If you try a new cold medicine and it appears to work, this is not sufficient evidence that it will always work for you. And the fact that one or two scoops of Toronto lakefront water are relatively free of *E. coli* bacteria might not truly indicate that it is safe to swim there. Recall the distinction that we made in Chapter 2 between sampling error (which is unavoidable if working from a sample) versus nonsampling error. Ensuring a sufficiently large sample is one way to lessen the possible impact of sampling error.

2. **Improperly conducted experiments:** These errors can lead to the nonsampling errors discussed in Chapter 2. If our sample is biased due to such problems as addressing the wrong population, or selecting a nonrepresentative sample, or using an unreliable test instrument, then the relative frequencies obtained in the sample will distort conclusions about the probabilities.
3. **Waiting for the "law of averages":** While there *is* a Law of Large Numbers (see Example 6.8), what many think of as a "law of averages" is *not* a real law. Suppose you toss a run of seven heads—it can happen. What is the probability of the next flip being tails? It is 0.50, *just like it is for every toss.* It is probable that, in the long run, streaks of tails and heads will balance out (that is the Law of Large Numbers); but in the short run—and particularly in the very next throw—all bets are off. There is no law, or even a small tendency, that says it's time for a tail.

   Compare a batter on a hitless streak: Even if his "normal" batting average is a whopping 0.400, there is no law that says he has to snap out of his hitless streak on any particular bat. (A complication for batters is that they have emotions and aches and pains, and that can *alter* their "true" batting abilities for stretches of time.) Expecting a law of averages is, unfortunately, a common downfall among compulsive gamblers. If the chance of winning on a roulette wheel is rather small, then it is small *every* time you play, even if you play a lot.

Using Computer Simulations for Probabilities

Computer simulations are used in many fields to help understand processes and make predictions. They can be used as part of weather forecasting, or for predicting the spread of a disease outbreak, or to model economies or the formation of galaxies. The basic principle of simulations is really simple. Start from a model of key features of the process, and then let the process unfold.

Increasingly, this approach is also being applied for statistical and probability calculations. Suppose we do not know the classical probability for tossing heads with a fair coin. A computer simulation, such as the one used to create the **Coin_Flips** file, could provide an estimate. By instructing the computer to randomly generate a number that has only one of two possible values, we built a model for what an actual coin flip accomplishes. (We ignored other physical aspects of the coin flips, since we do not think that these are relevant to the probabilities.) Then we replicated the flips many times. By applying empirical probability, the distribution of the many simulated outcomes is interpreted to approximate the real probabilities for tosses of real coins.

In "In Brief 6.2," we cannot demonstrate generically how to construct every possible computer simulation: The researcher must determine the key elements of a process in order to determine the probabilities of the outcomes. Then the researcher must devise a way to repeat that process many times, and then calculate the empirical probabilities from the results. The model used to create the **Coin_Flips** file is demonstrated in "In Brief 6.2."

• *In Brief 6.2 Computer Simulation for a Probability*

*(Based on the files **Coin_Flips** and **Five_Hundred_Cases**)*

The objective is to use a computer simulation to estimate the probability of flipping heads with a fair coin.

The computer software models flipping a fair coin many times, and records the results. Empirical probabilities are then calculated from those results.

Excel

The only column absolutely needed for this simulation is column B, labelled "Random Outputs" in the illustration. The formula **=RANDBETWEEN(0,1)** is replicated 500 times. This simulates many repetitions of a process in which every outcome is randomly either one of two possible values. Using the formulas shown, the ones and zeros are converted into "Heads" and "Tails" in column C to give a more realistic appearance. The formula in column D keeps a running total of the cumulative number of heads (i.e., the number of times "1" is recorded). The formula in column E recalculates the empirical probability for each row. Only the final calculation is really necessary to meet the objective of this simulation.

	A	B	C	D	E
1	Flip_Num	Random_Outputs	Simulated Results	Cumulative Number of Heads	Empirical Probability (So Far)
2	1	1	Heads	1	1.0000
3	2	0	Tails	1	0.5000
4	3	1	Heads	2	0.6667
5	4	1	Heads	3	0.7500
6	5	0	Tails	3	0.6000
7	6	0	Tails	3	0.5000
8	7	1	Heads	4	0.5714
9	8	1	Heads	5	0.6250
10	9	1	Heads	6	0.6667
	⋮	⋮	⋮	⋮	⋮
493	492	0	Tails	250	0.5081
494	493	1	Heads	251	0.5091
495	494	0	Tails	251	0.5081
496	495	1	Heads	252	0.5091
497	496	1	Heads	253	0.5101
498	497	0	Tails	253	0.5091
499	498	0	Tails	253	0.5080
500	499	0	Tails	253	0.5070
501	500	0	Tails	253	0.5060
		=RANDBETWEEN(0,1)	=IF(B501=1,"Heads","Tails")	=B501+D500	=D501/A501

Minitab

Open a new worksheet and begin with the menu sequence *Calc / Random Data / Integer.* Specify in the resulting dialogue box a large number of rows (e.g., 500) to generate, and indicate to which column to send the data (C1) (see the screen capture). Input where shown that the randomly generated integers should vary between a minimum of 0 and a maximum of 1. Click *OK.* A possible result of the procedure is shown in the table "Tally for Discrete Variables: C1" that appears at the lower left of the illustration. (Since the zeros and ones are random, their counts are different from the outputs shown for the other software programs.)

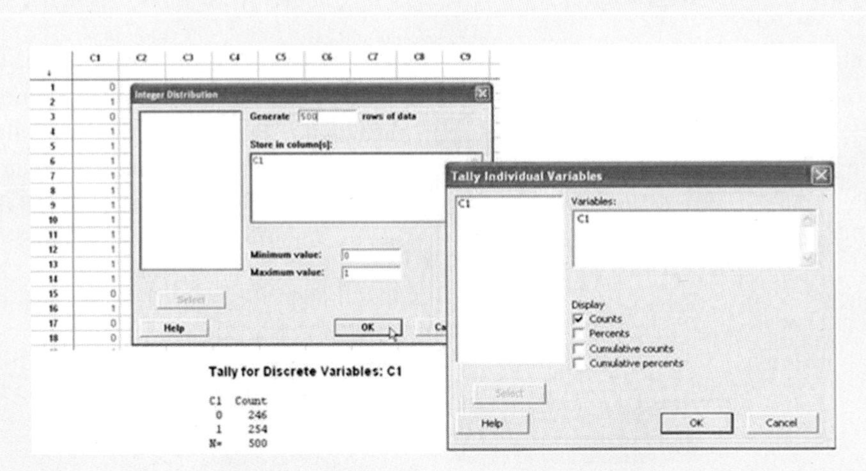

Once the simulated data have been generated, you can use the commands *Stat / Tables / Tally Individual Variables* to call up the second dialogue box in the screen capture to tally the counts for the specified column. This gives the number of heads out of 500 flips that you can use for estimating the probability.

SPSS

Open the SPSS file **Five_Hundred_Cases** on the CD, which contains the numbers 1 to 500 in the left column. (*This step enables a simple parallel with the other software approaches.*) Next use the menu sequence *Transform / Compute* to open the dialogue box shown on the left in the screen capture. Input where shown a name for the output column (e.g., "Outcomes") and input the numerical expression **Rnd(RV.uniform(0,1))** where indicated. This will cause many random values (in this case, 500) to be generated next to the numbers brought into column 1. (The function generates a decimal number between 0 and 1, and then rounds the result to one of those integers.) Click *OK*. A possible output is

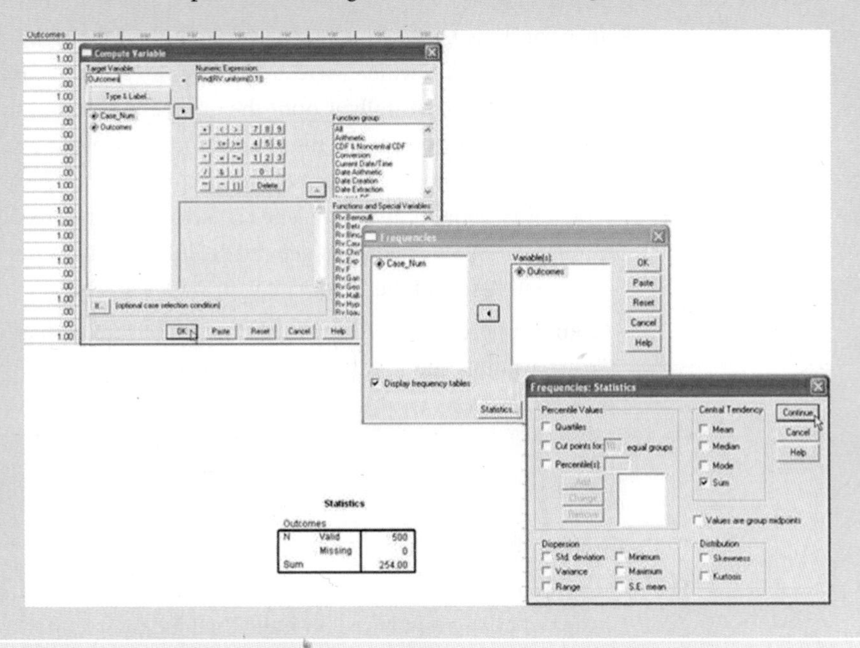

shown in the illustration. (Since the numbers are random, the display is different from the outputs shown for the other software programs.)

Once the simulated data have been generated, you can use the menu sequence *Analyze / Descriptive Statistics / Frequencies* and then, in the resulting "Frequencies" dialogue box, select the variable Outcomes for analysis. Click on the *Statistics* button and then, in the "Frequencies: Statistics" dialogue box, select *Sum;* click on *Continue* and then *OK.* This gives the number of heads out of 500 flips that you can use for estimating the probability.

• Hands On 6.1 *Design Your Own Simulations*

Suppose that you have four 52-card decks of cards, each fully shuffled. Use a computer simulation to estimate each of the two probabilities shown below. These problems actually have a classical solution, but for this approach, you would need a very large table of outcomes or else some advanced methods that are discussed in the next chapter. In the first problem below, you are shown a possible solution but are asked to create your own equivalent on the computer. For the second problem, you are asked to start from scratch.

1. Suppose you draw one card randomly from each of the four shuffled card decks. *Estimate* the probability that *at least three* of the selected cards are hearts.

   Figure 6.5 illustrates a possible type of simulation: Four separate columns of 500 numbers are randomly generated, with each number having a one-quarter probability of being a heart (represented by "1"), and a three-quarters chance of being another suit. Each *row* represents a possible experiment of drawing the four cards. The column at the right shows a "1" for those rows only in which there are at least three 1's. The final probability estimate follows naturally: Count the number of ones in the "AtLeastThreeHearts" column (equals the number of times the event occurred) divided by 500 (equals the number of experiments conducted). This is the estimated probability. (For comparison: The classical solution is 0.0508.) Using Excel, Minitab, or SPSS, attempt to generate columns that do the same basic job as illustrated here and then estimate your own value for the probability.

2. Again, suppose you draw one card randomly from each of the four decks. Create a simulation to estimate the probability that all four of the cards you draw are jacks.

Subjective Probability

The third approach to probability is called **subjective probability.** Here, the probability value is assigned to an event based on individual judgment, after taking experience and the available evidence into account. You make an educated guess.

Figure 6.5

Estimated Probability: Drawing at Least Three Hearts from Four Decks

	(1="heart")	(1="heart")	(1="heart")	(1="heart")	(1="at least three hearts")
Experiment_Num	Deck1Card	Deck2Card	Deck3Card	Deck4Card	AtLeastThreeHearts
1	0	1	1	1	1
2	0	0	1	1	0
3	0	0	0	0	0
4	0	0	0	0	0
5	0	0	0	0	0
6	0	0	0	0	0
7	0	0	1	0	0
8	0	0	0	0	0
9	1	0	0	1	0
⋮	⋮	⋮	⋮	⋮	⋮
493	0	0	0	1	0
494	0	0	0	1	0
495	0	1	0	0	0
496	1	0	1	0	0
497	0	1	0	1	0
498	0	0	0	0	0
499	0	0	1	1	0
500	1	1	0	1	1

Total number of experiments (out of 500 experiments)
for which at least three cards were hearts: 24
Estimated: *P*(at least 3 hearts) = 24/500 0.048

This approach is unavoidable when there is too little information for using another approach or when the event is a one-time-only occurrence. When a company is deciding whether to buy a new building, it does not have the luxury to repeat that same purchase hundreds of times to see how frequently the move is profitable. It buys or fails to buy on this one occasion, and then faces the consequences. Similar companies may have made similar decisions under similar circumstances—but none of the variables are exactly the same from decision to decision.

On a cloudy day, your decision whether to carry an umbrella may include elements of subjective probability. You assess the chance that when you leave work it will be raining. You may take advantage of empirical evidence that is available, by checking the *Weather Network* for example. But we all know that weather forecasters' predictions are not perfect and may not apply to your work location. You could also factor in the empirical probability of a bus arriving within five minutes of your leaving work. Yet, ultimately, you are left to take your own best guess as to whether you will need the umbrella.

Once the probability value is assigned, it is meant to be interpreted much like any other probability. But because these values are a bit fuzzy, be aware that if you lined up the assigned subjective probabilities for all possible outcomes, the sum of the sample space may not turn out be 1.0, which technically violates a condition for probabilities.

6.2 Conditional Probability

As we will discover, the concept of **conditional probability** is critically important in inferential statistics. It also can carry a lot of weight in court. For example, suppose that DNA discovered at a crime scene closely matches the DNA of a suspect. The defence may argue that the match is just a coincidence. The prosecutor will claim: "*If* [i.e., on the condition that] the DNA found at the scene did not come from the suspect, the probability that it would match the suspect's DNA so closely is one divided by several thousand trillions." The implied conclusion is not that the DNA did not match the suspect's or that the match is at all surprising; it is that the *conditional probability* of such a match happening by mere coincidence is too small to seriously consider.

Referring to Table 6.3, we can compare the calculations for **absolute probabilities** (i.e., the type discussed in Section 6.1) with the calculations for conditional probabilities. Table 6.3 is based on data from a survey conducted in 2005 by SES Research. Responding to public discussion about whether private use of marijuana should be decriminalized, the surveyors asked a follow-up question: If decriminalized, what should be the legal minimum age for use of the substance? The responses were broken down by the ages of the respondents.

Table 6.3 **Marijuana Decriminalization Opinion Survey, 2005**

	Respondent's Age Group					
Minimum Age Suggested	**18 to 29**	**30 to 39**	**40 to 49**	**50 to 59**	**60 plus**	**Totals**
12 years of age and older	0	0	1	1	4	6
16 years of age and older	21	13	7	6	5	52
18 years of age and older	159	192	172	124	129	776
No age given	21	22	25	27	47	142
Unsure	8	3	1	4	8	24
Totals	209	230	206	162	193	1000

Source: SES Research Inc. SES Research Poll. Found at: http://www.sesresearch.com/news/press_releases/PR%20February%2028%202005.pdf

We have learned how to calculate absolute probabilities from such a table. For example, what is the probability that a randomly selected respondent was at least 60 years old? Of a total of 1,000 respondents, 193 met this criterion, so the probability is 193/1000 = 0.193. Or what is the probability that a randomly selected respondent thought that the minimum age for marijuana use (if it becomes applicable) should be 18 or older? Of the 1,000 respondents, 776 gave this answer, so the probability is 0.776.

Note that we found the numbers needed for these calculations on the *margins* of the table, which is where we look to find the totals for how many cases meet the criteria for any particular event (e.g., the event of the respondent being a certain age, or expressing a certain opinion). Because the absolute probabilities can be calculated in this way from the marginal values, they are sometimes referred to as **marginal probabilities.**

But sometimes we have a particular interest in one of the groups, as defined by one of the variables. Perhaps a youth worker would like to know how the younger respondents have answered the question about minimum age. *Given that* a randomly selected respondent is between 18–29 years old, what is the probability that he or she thought that the minimum age should be 18 or older? The basic formula (Formula 6.2) still applies, but we have to be careful—not all the 1,000 original respondents are relevant to this calculation. In this case, the relevant column in the table shows that only 209 people were sampled from the 18–29 age group specifically. Of these, 159 met the condition of believing that the minimum age should be 18. Thus, the conditional probability is 159/209 = 0.761.

In symbols, the conditional probability of event *A, given that B occurs,* is expressed as $P(A \mid B)$. As shown in the above example, attention is restricted to only those cases where *B* occurs or applies. Counting just those cases, the expression asks how many cases also represent the occurrence or presence of event *A*.

Example 6.9

Use the data in Table 6.3 to answer the following questions:

1. What percentage of respondents aged at least 60 agreed that if the use of marijuana is decriminalized, the legal minimum age for use of the substance should be 18?
2. If a randomly selected respondent thought that the minimum age should be 18, what was the probability that the respondent was at least 60 years old?
3. What is the probability that a respondent was at least 60 years old if we know that the respondent was unsure about what minimum legal age would be appropriate?
4. Suppose you randomly select an individual from the 50–59 age group. What is the probability that this respondent was unsure about what minimum legal age would be appropriate?

Solution

All of these problems are solved using conditional probabilities. We begin by reexpressing each problem in symbols to clarify what is the condition and what is the desired probability.

1. Although this question is worded in terms of percentages, the solution can be interpreted as an empirical probability:

$P(\text{believes minimum age should be 18} \mid \text{respondent 60+ years old}) = 129/193 = 0.668$

2. $P(\text{respondent 60+ years old} \mid \text{believes minimum age should be 18})$
$= 129/776 = 0.166$
3. $P(\text{respondent 60+ years old} \mid \text{unsure}) = 8/24 = 0.333$
4. $P(\text{unsure} \mid \text{respondent 50–59 years old}) = 4/162 = 0.025$

• Hands On 6.2 Conditional Probabilities with Simulations

In "Hands On 6.1," you used computer simulations to estimate probabilities for specific events. There are other occasions when the possible inputs to a modelled system may vary, and you want to forecast the probabilities of outcomes *if* one or another possible input occurs. This is, in effect, a type of "what-if" analysis. The "what-ifs" in the model comprise the condition, and you can then use a computer simulation to estimate the conditional probability of a particular event occurring as a result. Reconstruct and run the simulations in "Hands On 6.1" to answer the following questions.

1. Suppose you draw one card randomly from each of four shuffled card decks.
   a) *Estimate* the probability that *at least three* of the cards selected subsequently will be hearts *if* the first selected card has turned out to be a heart. (Hint: Begin as in "Hands On 6.1" but (compare Figure 6.5) add a new column to the right called "AtLeastThree_If." This column takes the value 1 only for rows where Deck1Card = 1 (a heart) and AtLeastThreeHearts = 1. Otherwise, the value is zero. The conditional probability will be based on (denominator) the number of rows for which Deck1Card = 1, and (numerator) the number of cases where AtLeastThree_If = 1.)
   b) *Estimate* the probability that *at least three* of the cards selected subsequently will be hearts *if* the *second* selected card has turned out *not* to be a heart.
2. Again, suppose you draw one card randomly from each of the four decks.
   a) *Estimate* the probability that *all four* of the cards selected subsequently will be jacks *if* the first selected card has turned out to be a jack.
   b) *Estimate* the probability that *all four* of the cards selected subsequently will *not* be jacks *if* the first selected card has turned out *not* to be a jack.

6.3 Combining Probabilities

Section 6.1 showed how to calculate the probability of a single event, A, occurring—i.e., $P(A)$. Given sufficient time and information, all probability calculations in this chapter could be reduced back to the two basic formulas introduced in that first section: Formula 6.1 (for classical probability) or Formula 6.2 (for empirical probability). However, some common applications of probability are cumbersome for directly calculating from the basic formulas; some of these can be calculated more simply if we introduce some additional rules.

First, we will need some terminology:

1. **Compound events:** Every experiment results in one outcome, from all those possible in the sample space. But depending on the context, there is some flexibility in naming the "event" that has occurred. Suppose you draw the five of hearts from a deck. By standard conventions, that unique outcome could be classified as an "event" in several possible ways: You have drawn a heart, a five, a red card, an odd-numbered card, and so on. If you are able to name the event with a single description of this sort and to compute directly the number of outcomes that "count as" the event—for use in Formula 6.1 or Formula 6.2—then the event can be called a **simple event.**

   Sometimes, it is easier to describe an "event" and to calculate its probabilities in terms of other events by combining the descriptors, rather than by using a single name. An event that is defined in terms of *other* simple events is a **compound event.** For example, what is the probability of drawing a red five card?

Viewed as a combination of the events "a red card" *and* "a five card," we are describing a compound event. Similarly, the event of drawing a card that is red *or* is a five is a compound event.

2. **Joint events: Joint events** (also called **conjunctive events**) are compound events that are said to occur only if *all* of the named simple events occur. For example, drawing a card that is red *and* a five has occurred only if the event "draw a red card" *and* the event "draw a five" have occurred. In notation, if A and B are the names of two events, $P(A \text{ and } B)$ is the probability of the joint event, based on A and B.

3. **Disjunctive events: Disjunctive events** are compound events that are said to occur *if at least one* of the named simple events occurs. For example, drawing a card that is red *or* a five has occurred only if the event "draw a red card" *or* the event "draw a five" (*or both*) have occurred. In notation, if A and B are the names of two events, $P(A \text{ or } B)$ is the probability of the disjunctive event, based on A or B.

4. **Independent events:** Two events are **independent events** if the probability that one of the events will occur is not affected by the other event's occurrence or nonoccurrence, and vice versa. For example, if you buy two packs of candies, and randomly choose one candy from each bag, the events A "the candy from bag 1 is mauve" and B "the candy from bag 2 is mauve" are independent. This can be expressed in terms of conditional probabilities: Events A and B are independent if (a) $P(A \mid B) = P(A)$ and (b) $P(B \mid A) = P(B)$.

5. **Dependent events:** Two events are **dependent events** if the probability that one of the events will occur *is* affected by the occurrence or nonoccurrence of the other event. For example, suppose a weak football team, "Team 1," has only a small probability of winning each game. If every player on the opposing team suddenly contracts food poisoning, this could *change* the probability of Team 1's winning. This can be expressed in terms of conditional probabilities: Events A and B are dependent if one or both of (a) $P(A \mid B) \neq P(A)$, or (b) $P(B \mid A) \neq P(B)$.

6. **Mutually exclusive events:** Two events are **mutually exclusive** if there is no possible outcome that could count as occurrence of both of the events. For example, if you pick one candy from a single bag of candies, two possible events are A "the selected candy is mauve" and B "the selected candy is green"; it is *not possible* to pick a candy that could be classified as both A and B. Therefore, the two events are mutually exclusive; the probability of their joint occurrence is zero.

   Note that if two possible events A and B are mutually exclusive events, they cannot also be independent. If they are mutually exclusive, then the independence condition $P(A \mid B) = P(A)$ is not satisfied, since $P(A \mid B) = 0$ for mutually exclusive events, but $P(A) > 0$ if A is possible. Similarly, the independence condition $P(B \mid A) = P(B)$ is not satisfied, since $P(B \mid A) = 0$ for mutually exclusive events, but $P(B) > 0$ if B is possible.

7. **Complement of an event:** The compound event that is said to occur when a named simple event *does not* occur is called the **complement** of the named event. For example, if you draw a card that *does not* count as the event "drawing a five," then what has occurred is the complement of the "drawing a five" event; this is the same as the event "not drawing a five." In notation, if A is the name of an event,

then $P(\overline{A})$ [or $P(\text{not } A)$] is the probability of the complement of event A. Note that the events A and $\overline{A}$ are mutually exclusive.

Based on concepts 4 and 7 above, these additional relationships apply. If two events A and B are independent, then each of the following event pairs are also independent: A and $\overline{B}$, $\overline{A}$ and B, and $\overline{A}$ and $\overline{B}$.

Some Rules for Calculating Compound Probabilities

For probabilities that involve compound events, the rules described in this section can often simplify the calculations, as alternatives to using the basic rules in Formulas 6.1 or 6.2.

Rules for Complementary Events

Formula 6.3 Complementary Event Rule 1

$$P(A) + P(\overline{A}) = 1$$

where A represents any event, and $\overline{A}$ is the complement of that event. Combined, the mutually exclusive events A and $\overline{A}$ represent every possible outcome in the sample space, so the sum of their two probabilities adds to 1.

Formulas 6.4 and 6.5 are algebraically equivalent to Formula 6.3.

Formula 6.4 Complementary Event Rule 2

$$P(A) = 1 - P(\overline{A})$$

Formula 6.5 Complementary Event Rule 3

$$P(\overline{A}) = 1 - P(A)$$

It is sometimes easier to calculate the probability of the complement and then use one of these rules, rather than solve directly for the desired probability.

Example 6.10

As noted in Example 6.7, the 2001 federal census counted 7.14 million Canadians whose mother tongue is French, out of a total Canadian population of 29.64 million people.[2] If you randomly select one Canadian this year (assuming that the 2001 data are representative), what is the probability that his or her mother tongue is *not* French?

Solution

We are given no direct count of any individuals other than those for whom their mother tongue is French, but if we find P(mother tongue is French), then we can use Formula 6.5 to calculate $P\,(\overline{\text{mother tongue is French}})$. Therefore:

$$P(\overline{\text{mother tongue is French}}) = 1 - P(\text{mother tongue is French})$$
$$= 1 - 7.14/29.64 = 1 - 0.241$$
$$= 0.759$$

Example 6.11

Refer to the data in Table 6.3. If you randomly select one individual who responded to the survey, what is the probability that the selected person is younger than 60?

Solution

Given how respondents' ages have been grouped, respondents who are younger than 60 could be in any of the age categories—*except* the group "60 plus." We could use Formula 6.2, but we would have to add the counts for all people who were in age groups other than "60 plus." It might be faster to just find $P(\text{60 plus})$ and then use Formula 6.5 to calculate $P(\overline{\text{60 plus}})$. Therefore:

$$P(\overline{\text{60 plus}}) = 1 - P(\text{60 plus})$$
$$= 1 - 193/1000 = 1 - 0.193$$
$$= 0.807$$

Rules for Joint Probability

Formulas 6.6 and 6.7 represent the two rules that are generally given for joint probability, depending on whether the joint events are independent events or dependent events, as defined earlier in this section.

Formula 6.6 Multiplication Rule for Independent *Events A and B*

$$P(A \text{ and } B) = P(A) \cdot P(B)$$

Formula 6.7 Multiplication Rule for Dependent *Events A and B*

$$P(A \text{ and } B) = P(A) \cdot P(B|A)$$

Apply Formula 6.6 to find joint probabilities of independent events. For example, what is the probability of throwing heads on the tosses of two fair coins? The events are independent, therefore:

$$P(\text{coin 1 heads } \textit{and} \text{ coin 2 heads})$$
$$= P(\text{coin 1 heads}) \times P(\text{coin 2 heads})$$
$$= 0.5 \times 0.5 = 0.25$$

Note that Formula 6.6 is extendible in the following fashion for joint probabilities that involve three or more events:

$$P(\text{coin 1 heads } \textit{and} \text{ coin 2 heads } \textit{and} \text{ coin 3 heads})$$
$$= P(\text{coin 1 heads}) \times P(\text{coin 2 heads}) \times P(\text{coin 1 heads})$$
$$= 0.5 \times 0.5 \times 0.5 = 0.125$$

Example 6.12

Use the multiplication rule to find the following probabilities.

1. If you throw two standard six-sided dice, what is the probability that both land on a 2?

2. If you throw four standard six-sided dice, what is the probability that all four land on a 2?
3. If you throw two standard six-sided dice, what is the probability that both land on even numbers?

Solution

1. This is solved by simply applying Formula 6.6.

$$P(\text{die 1 lands on a 2 } and \text{ die 2 lands on a 2})$$
$$= P(\text{die 1 lands on a 2}) \times P(\text{die 2 lands on a 2})$$
$$= 1/6 \times 1/6 = 0.028$$

2. This is solved by using an extended version of Formula 6.6:

$$P(\text{die 1 on a 2 } and \text{ die 2 on a 2 } and \text{ die 3 on a 2 } and \text{ die 4 on a 2})$$
$$= P(\text{die 1 on a 2}) \times P(\text{die 2 on a 2}) \times P(\text{die 3 on a 2}) \times P(\text{die 4 on a 2})$$
$$= 1/6 \times 1/6 \times 1/6 \times 1/6$$
$$= 0.000772$$

3. This example reminds us that at any stage of a probability calculation, we can fall back on the basic rules (Formulas 6.1 or 6.2). Prior to applying Formula 6.6, we find the probability for a *single* die landing on an even number. (Three of the six possible numbers—2, 4, and 6—are "even," so $P(\text{even}) = 3/6 = 0.5$.)

$$P(\text{die 1 even } and \text{ die 2 even})$$
$$= P(\text{die 1 even}) \times P(\text{die 2 even})$$
$$= 0.5 \times 0.5 = 0.25$$

If events are *dependent,* then the occurrence of one event impacts on the probabilities that other events in the sample space will occur. To find the joint probabilities for such cases, apply Formula 6.7. Consider, for example, the data on hurricane strength (measured in "Category" numbers) presented in Table 6.4. What is the probability that if you randomly select two of the hurricanes that occurred in Massachusetts, they will *both* be Category 3 hurricanes?

The probability that the first hurricane you select is Category 3 is easy to calculate, since two of the six hurricanes recorded in Massachusetts were Category 3:

$$P(\text{hurr 1 is Cat 3}) = 2/6 = 0.3333$$

But when you randomly select a second hurricane, the numerator and denominator for calculating probabilities have changed. The number of hurricanes left to choose from (the denominator) is reduced by one and so, too, is the number of remaining Category 3 hurricanes (the numerator), given that one Category 3 has already been selected. Therefore, for the second, randomly selected hurricane:

$$P(\text{hurr 2 is Cat 3} \mid \text{hurr 1 is Cat 3}) = 1/5 = 0.200$$

We are now ready to calculate the joint probability using Formula 6.7:

$$P(\text{hurr 1 is Cat 3 } and \text{ hurr 2 is Cat 3})$$
$$= P(\text{hurr 1 is Cat 3}) \times P(\text{hurr 2 is Cat 3} \mid \text{hurr 1 is Cat 3})$$
$$= 0.333 \times 0.200 = 0.067$$

Table 6.4 **U.S. Hurricanes, 1900–2000**

	Hurricane Strength Category Number					
States Impacted*	**1**	**2**	**3**	**4**	**5**	**Totals**
Alabama	5	2	5	0	0	12
Connecticut	2	3	3	0	0	8
Florida	19	17	17	6	1	60
Georgia	1	4	0	0	0	5
Louisiana	9	5	8	3	1	26
Maine	5	0	0	0	0	5
Maryland	0	1	0	0	0	1
Massachusetts	2	2	2	0	0	6
Mississippi	1	2	5	0	1	9
North Carolina	10	6	10	1	0	27
New Hampshire	1	1	0	0	0	2
New Jersey	1	0	0	0	0	1
New York	3	1	5	0	0	9
Rhode Island	0	2	3	0	0	5
South Carolina	6	4	2	2	0	14
Texas	12	9	10	6	0	37
Virginia	2	1	1	0	0	4

*A single hurricane can have an impact on more than one state, so the rows of this table cannot be interpreted as mutually exclusive events.

Source: U.S. Department of Commerce, National Oceanic and Atmospheric Administration, Atlantic Oceanographic and Meteorological Laboratory, Hurricane Research Division. Based on data in *NOAA Technical Memorandum NWS TPC-1:* "The Deadliest, Costliest, and Most Intense United States Hurricanes from 1900 to 2000." Found at: http://www.aoml.noaa.gov/hrd/Landsea/deadly/Table9.htm

This case is relatively simple, so you could alternatively use a table of outcomes approach, as in Section 6.1. Figure 6.6 displays a single box for each outcome that is possible to occur. The large Xs, in effect, depict the conditional aspect of the probability: If a specific hurricane was selected first, it is not available to be chosen second. Thirty outcomes

Figure 6.6
Table of Outcomes for Hurricane Selection

First Selected Hurricane

Second Selected Hurricane	1a	1b	2a	2b	3a	3b
1a	✗					
1b		✗				
2a			✗			
2b				✗		
3a					✗	✓
3b					✓	✗

✗ = **These outcomes are not possible (don't count them)**
✓ = **Both selected hurricanes were Category 3**

Source: U.S. Department of Commerce, National Oceanic and Atmospheric Administration, Atlantic Oceanographic and Meteorological Laboratory, Hurricane Research Division. Based on data in *NOAA Technical Memorandum NWS TPC-1:* "The Deadliest, Costliest, and Most Intense United States Hurricanes from 1900 to 2000." Found at: http://www.aoml.noaa.gov/hrd/Landsea/deadly/Table9.htm

are possible, of which two count as the event "both hurricanes are Category 3," so the joint probability is $2/30 = 0.067$. It is clear that this method would become too cumbersome if, for instance, you were calculating the joint probability of selecting two Category 3 hurricanes among the 60 hurricanes recorded in Florida.

Example 6.13

Use Formula 6.7 to find the following probabilities.

1. If you draw two cards randomly from a standard deck, without replacement (i.e., you do not return the first card to the deck before drawing the second card), what is the probability that both cards are jacks?
2. If you draw three cards randomly from a standard deck, without replacement, what is the probability that all three cards are jacks?
3. From a bowl containing one each of eight colours of candies, select two candies randomly, without replacement. What is the probability that the first candy is *not* green, and the second candy *is* green?

Solution

1. The probability that the first card drawn is a jack is 4/52. Given that a jack was drawn, of the 51 remaining cards, only three jacks are left for the second draw. Therefore:

$$P(\text{card 1 is a jack } and \text{ card 2 is a jack})$$
$$= P(\text{card 1 is a jack}) \times P(\text{card 2 is a jack} \mid \text{card 1 is a jack})$$
$$= 4/52 \times 3/51 = 0.0045$$

2. This is solved by using an extended version of Formula 6.7.

$$P(\text{card 1 is a jack } and \text{ card 2 is a jack } and \text{ card 3 is a jack})$$
$$= P(\text{card 1 is a jack}) \times P(\text{card 2 is a jack} \mid \text{card 1 is a jack})$$
$$\times P(\text{card 3 is a jack} \mid \text{card 1 and card 2 are } both \text{ jacks})$$
$$= 4/52 \times 3/51 \times 2/50 = 0.00018$$

3. There are eight candies of different colours in the bowl. $P(\text{candy 1 is green}) = 1/8 = 0.125$. By the complement rule: $P(\text{candy 1 is not green}) = 1 - 1/8 = 7/8$. After the first candy is selected, seven candies will be left in the bowl. If the first candy was not green, then one of the remaining seven candies must be the green one. In other words: $P(\text{candy 2 is green} \mid \text{candy 1 is not green}) = 1/7$. Applying the formula:

$$P(\text{candy 1 is not green and candy 2 is green})$$
$$= P(\text{candy 1 is not green}) \times P(\text{candy 2 is green} \mid \text{candy 1 is not green})$$
$$= 7/8 \times 1/7 = 1/8 = 0.125$$

In Formula 6.7, the joint probability is defined in terms of a conditional probability. If the joint probability is already known, the equation can be solved to find the conditional probability:

$$P(A \mid B) = \frac{P(A \text{ and } B)}{P(B)}$$

We will apply this information in Section 6.4.

Probability Trees

Probability trees are diagrams designed to help you visualize the conditional relationships among probabilities, and to guide calculations. Figure 6.7 displays the probabilities relating to hurricane strengths in the example given earlier.

The chart is read from left to right, and each solid dot represents a "probability point" or "chance point." If one of these points is reached, then one or more outcomes (or subsequent probability points), shown to its right, will occur—with probabilities as indicated on the lines extending to the right. When you construct the chart, you must determine the probabilities to write at each point; this may be done by classical, empirical, or subjective methods, as suits the application. Each probability shown on a line is *conditional* on everything having occurred to lead to that point.

Figure 6.7
Probability Tree for Hurricane Selection

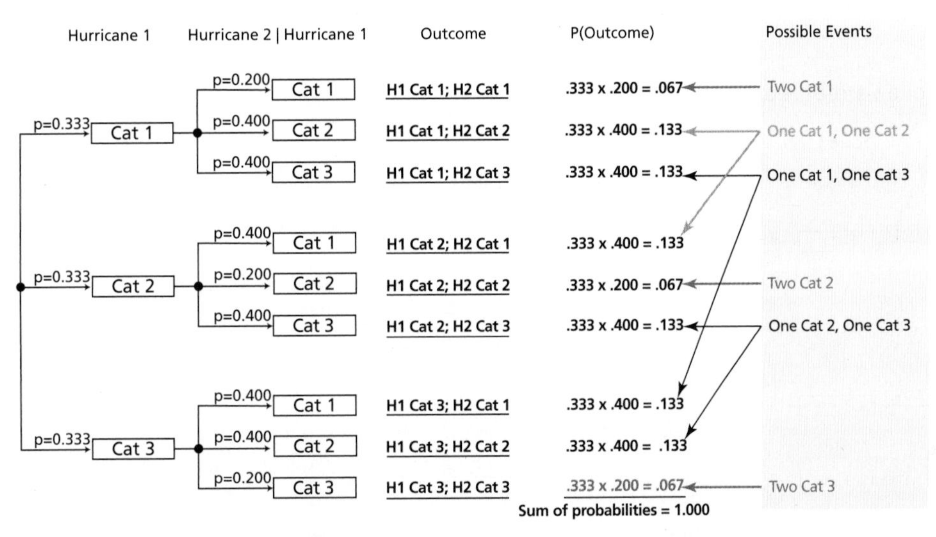

Source: U.S. Department of Commerce, National Oceanic and Atmospheric Administration, Atlantic Oceanographic and Meteorological Laboratory, Hurricane Research Division. Based on data in *NOAA Technical Memorandum NWS TPC-1:* "The Deadliest, Costliest, and Most Intense United States Hurricanes from 1900 to 2000." Found at: http://www.aoml.noaa.gov/hrd/Landsea/deadly/Table9.htm

Each possible sequence from the left to the right of the chart leads to a possible outcome. The probabilities of each outcome are calculated consistently with Formula 6.7: All probabilities on the path leading to the outcome are multiplied. The left-most probability on that path is an absolute probability for the first branch. Every subsequent probability is conditional on that next point being reached.

Applying Figure 6.7 to the hurricane example: There is one outcome that counts as the event "both hurricanes are Category 3." The left-most probability on that path [*P*(hurr 1 is Cat 3)] is given as 0.333, since two of the six recorded hurricanes were in that category. The path to the right has a branch toward the box showing that hurricane 2 is a Category 3. The (conditional) probability of following that branch is 0.200 because (based on Table 6.4) after the first hurricane selected is a Category 3, there are only five remaining hurricanes, and only one of these could be another Category 3. The product of the probabilities on the path, 0.333 and 0.200, is 0.067, as previously calculated.

Example 6.14 If you draw three cards randomly from a standard deck, without replacement, what is the probability that the first card drawn is a face card and that the *next two* cards drawn are *not* face cards? (Jacks, queens, and kings are face cards.) Use a probability tree to illustrate the necessary calculations.

Solution Figure 6.8 illustrates a probability tree for this problem. Unlike the hurricane example in Figure 6.7, the probabilities here are classical; that is, they are based on our knowledge of card decks, rather than on empirical information about frequencies. As we move to the right of the tree, the probabilities are conditional on the occurrence of every outcome to the left of a given probability point. The event has been defined as the case of first drawing a face card ($P = 12/52$); then drawing two successive non-face cards (conditional probabilities 40/51 and then 39/50, respectively). The joint probability for the three events is shown—clearly corresponding to the formula version: $P(\text{face} \rightarrow \text{not-face} \rightarrow \text{not-face}) = (12/52) \times (40/51) \times (39/50) = 0.141$.

Figure 6.8
Probability Tree for a Card-Based Event

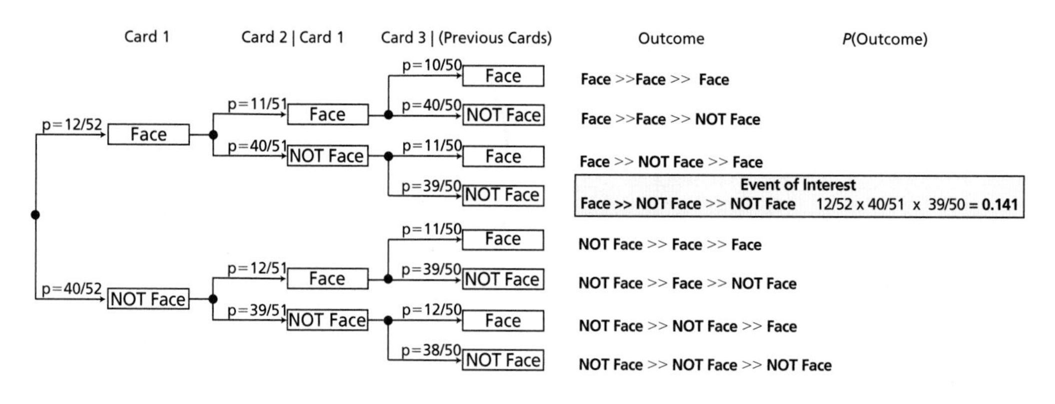

Rules for Disjunctive Probabilities

We defined disjunctive events as compound events, such that "*A* or *B*" occurs if *either* of the events *A* or *B* (or both) occurs individually. By extension, for any number of events A, B, *C*, . . ., "*A* or *B* or *C* or . . ." occurs if *any of* the individual events *A*, *B*, *C*, or . . . occurs. Depending on whether the component events are mutually exclusive, one of the two basic calculation rules shown in Formulas 6.8 and 6.9 can be applied.

Formula 6.8 Addition Rule for Mutually Exclusive *Events A and* B

$$P(A \text{ or } B) = P(A) + P(B)$$

Formula 6.9 Addition Rule for Not Mutually Exclusive *Events A and* B

$$P(A \text{ or } B) = P(A) + P(B) - P(A \text{ and } B)$$

The need for two separate rules is illustrated in Figure 6.9. Suppose you throw two standard dice in turn. The probability *P*(first die thrown was a 4 *or* a 6) can be calculated by the basic rule for classical probability (Formula 6.1), with the aid of the table of outcomes in Figure 6.9(a). Of the 36 outcomes possible, six count as the event "the first die thrown was a 4," and *another six* count as the event "the first die thrown was a 6." Because these two event descriptions do not overlap (i.e., the events are mutually exclusive), the number of outcomes that count as the disjunctive event "first die thrown was 4 *or* a 6" is simply the sum of outcomes that count as the individual events that are compounded. Therefore:

$$P(\text{first die thrown was a 4 } or \text{ a 6}) = (6 + 6)/36 = 6/36 + 6/36 = 12/36 = 0.333$$

This simple principle of addition can be used to find the probability for any disjunctive event if the probabilities of the individual events are known (or obtainable) and the events are mutually exclusive.

Figure 6.9
Tables of Outcomes for Disjunctive Events

a)

Number depicted on the first die thrown

Number depicted on the second die thrown	1	2	3	4	5	6
1				✓		✓
2				✓		✓
3				✓		✓
4				✓		✓
5				✓		✓
6				✓		✓

First die thrown was a 4 ✓ OR a 6 ✓

b)

Number depicted on the first die thrown

Number depicted on the second die thrown	1	2	3	4	5	6
1				✓		
2				✓		
3				✓		
4				✓		
5				✓		
6	✓	✓	✓	✓✓	✓	✓

First die thrown was a 4 ✓
OR Second die thrown was a 6 ✓

Formula 6.8 does not apply if the compounded events are not mutually exclusive. Again, suppose you throw two standard dice. With the aid of the table of outcomes in Figure 6.9(b), you see that 36 outcomes are possible. What is numerator for the probability, *P*(first die thrown was a 4 *or* the second die thrown was a 6)? If you simply add the number of ways in which the first event could occur (the green check marks) and the number of ways in which the second event could occur (the red check marks), you would double-count the box for the outcome "the first die thrown was a 4 *and* the second die thrown was a 6." This occurs because the two event descriptions *do* overlap (i.e., the events are *not* mutually exclusive). They overlap precisely at the outcome that represents "the first event occurs *and* the second event occurs." Formula 6.9, therefore, subtracts that overlapping probability from the simple sum of probabilities for the compounded events:

$$P(\text{first die thrown was a 4 } or \text{ the second die thrown was a 6})$$
$$= (6 + 6 - 1)/36 = 6/36 + 6/36 - 1/36 = 11/36 = 0.306$$

Example 6.15

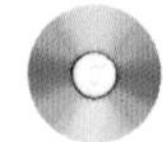 *(Based on the **Players_Toronto_Blue_Jays** file)*

In Example 3.7 in Chapter 3, we examined some data from the file **Players_Toronto_Blue_Jays.** A selection of the data is summarized below. If you randomly select one of the players, what is the probability that:

1. He bats only right or bats only left?
2. He bats only right or throws left?

Table 6.5 Contingency Table for Left/Right Preferences in Batting and Throwing

	Throws		
	Left	Right	Total
Bats Both*	0	2	2
Only Left	3	6	9
Only Right	0	15	15
Total	3	23	26

*"Both" refers to a switch hitter who is able to bat from either side of the plate.

Source: USA Today. 2005. Reprinted with permission. Found at: http://asp.usatoday.com/sports/baseball/salaries/teamdetail.aspx?team=14&year=2005

Solution

Problem 1 can be solved with Formula 6.8 because the events "batting only right" or "batting only left" are mutually exclusive. From Table 6.5, we can easily find the marginal probabilities $P(\text{bats only right}) = 15/26$ and $P(\text{bats only left}) = 9/26$; thus, by the formula:

$$P(\text{bats only left or bats only right}) = 15/26 + 9/26 = 24/26 = 0.923$$

Problem 2 requires Formula 6.9 because the events "batting only right" and "throwing left" are not mutually exclusive. One can (theoretically) do both. Using Table 6.5, we can find the marginal probabilities $P(\text{bats only right}) = 15/26$ and $P(\text{throws left}) = 3/26$. We can also calculate $P(\text{bats only right } and \text{ throws left}) = 0/26$. Therefore, by the formula:

$$P(\text{bats only left or throws right}) = 15/26 + 3/26 - 0/26 = 18/26 = 0.692$$

(Even though the combination of batting only left and throwing right is theoretically possible, the table shows that, empirically, no one has done it.)

A Hint for Solving Probability Problems with Contingency Tables

At several points in this chapter, we have used contingency tables to help solve probability problems. In Example 6.5, we used Table 6.1 to find both marginal probabilities [e.g., *P*(draw a red card)] and what we identified subsequently as joint probabilities [e.g.,

P(draw a card that is a black card *and* a face card)]. Then, based on Table 6.3, in Example 6.9) we calculated conditional probabilities [e.g., *P*(respondent is unsure | respondent is 50–59 years old)]. Finally, we have just used Table 6.5 in Example 6.15 to calculate a disjunctive probability. In every case, a contingency table helps to summarize the frequencies of various events that may be relevant for a calculation of probability.

Often you will be presented with a word problem or a real-life situation for which you need to calculate probabilities when you will *not* be provided with a ready-made contingency table. At times like this, if you can *create* a contingency table from the data, the actual probability calculations may be relatively simple.

First, when encountering a probability problem, ask yourself if a contingency table applies. The requirements are (1) data that have been, or could be, grouped into relative frequencies, based on mutually exclusive classes; (2) two interacting variables (such as a player's batting versus hitting preferences); and (3) sufficient available data to make the table.

If these requirements are met, you can begin to draw the contingency table. List names for the possible values of one variable at the top, and then list names for the other variables at the left. On the right-hand side and at the bottom of the table, provide the marginal totals (i.e., the frequencies for each value of one variable, ignoring the other variable). In the *internal* cells of the table, give the frequencies for specific combinations of variable values. For example, where the column and row intersect for "batting preference is only right" and "throwing preference is left," fill in the number of cases (if any) that meet *both* conditions.

When you build a table in this way, it is typical that some of its cells will initially be blank. You may not be presented, directly, with all of the marginal and internal frequencies. However, you can often fill in missing values by using these simple rules:

- All values in a given row should add to the marginal total on the right.
- All the values in a given column should add to the marginal total at the bottom.
- The sum of all column (or row) marginal totals should equal the total number of sampled cases.

Once the table is complete, you can calculate any desired probabilities by using the appropriate formulas.

Example 6.16

In a class of 42 students, 6 scored an A. Of all of the students 10 are left-handed. Four of the left-handed students scored an A. Based on these data, find the following probabilities.

What is the probability that a randomly selected student is left-handed?

What is the probability that a randomly selected student is left-handed *and did not* score an A?

What is the probability that a randomly selected student is left-handed *or did* score an A?

What is the probability that *if* you select a left-handed student, the student you select did *not* score an A?

Solution

A contingency table can help you to calculate probabilities. The conditions are met for constructing such a table: We are given some frequencies in relation to two intersecting variables—student hand preferences and their grades (see Table 6.6).

Table 6.6 Contingency Table for Students' Marks

		Students' Hand Preference		
		Left-Handed	Other	Total
Students' Marks	A	4	2	6
	Not A	6	30	36
	Total	10	32	42

Cells contain *directly provided* information
Cells completed by inference

Note how only the data shown in black font in Table 6.6 are provided directly in the problem. All of the remainder have been inferred. It does not matter which variable is depicted in the rows and which in the columns. Make sure that the row and column data add to their respective marginal totals, and that the marginal totals for the rows (or columns) add to the grand total for sample size; then we can fill in all of the blanks in the table (indicated in colour in the table). Using this new table, we find:

1. P(left-handed) $= 10/42 = 0.238$
2. P(left-handed *and* not A) $= 6/42 = 0.143$
3. P(left-handed *or* A) $= 10/42 + 6/42 - 4/42 = 12/42 = 0.286$
4. P(not A | left-handed) $= 6/10 = 0.600$

• Hands On 6.3 *Matching Birthdays*

A favourite trick enjoyed by many teachers of mathematics or statistics is this: Ask each student in the class to write down the month and day of their birth. If the class is large enough, the chances are very good that two students will have exactly the same birthday. To novices, this can seem quite surprising.

A traditional way to demonstrate why the trick works is based on conditional probabilities. There are 365 days in a year (most years). The first birthday selected could be any day ($P = 365/365$). The probability that a randomly selected second student's birthday is *different* from the first birthday is (364 days left to choose from)/(365 days in the year). The probability that a third selected birthday matches neither of the first two is (363 days left to choose from)/(365 days in the year); and so on for all the students in the class.

For all birthdays to be different would require that each new selection be different from the preceding. So, applying an extended version of Formula 6.7, multiply *all* of the conditional probabilities obtained in the previous paragraph. Try it for a class of 40. You will find it is very unlikely that no two of the birthdays in the class are a match.

As a "Hands On" exercise, try this in a simulated class of 40. A possible technique might be to generate 40 columns of random numbers in parallel. Generate the column numbers so as to represent a student's possible birthday. (A random integer between 1 and 365 could be used as an approximation.) One row of 40 such random numbers represents one possible set of birthdays in a particular class. Add an extra column that will be set to a "1" if a match is found among the 40 numbers, and a "0" otherwise. If you create several hundred or a thousand such rows, you can estimate the proportion of cases in which there are, or are not, matched birthdays. See for yourself how rare it is to have no matches in a class of 40.

6.4 Bayes' Rule

Bayes' Rule (or **Bayes' Theorem**) was first developed in the 17th century by an English clergyman, the Reverend Thomas Bayes. Its original intent was to permit the *revision* of our initial probability assignments (called the **prior probabilities**) based on new information. In terms introduced in this chapter, prior probabilities are marginal probabilities, which are originally assigned based on classical, empirical, or subjective considerations. With new information about relevant conditions, we can later develop and refine conditional probabilities.

Suppose you have two bags of 100 marbles each, "BlueBag" and "Not BlueBag." BlueBag contains 70% blue marbles and 30% marbles of other colours. Not BlueBag has only 20% blue marbles and 80% marbles of other colours. The bags look identical from the outside.

If you randomly select one bag, the marginal probability that you will be selecting a marble from BlueBag is 50%. 100 of the 200 marbles are in that bag; but you have no idea which bag you have reached into. If by chance you grabbed BlueBlag, you have an increased probability of picking a blue marble, compared to the (conditional) probability had you not picked BlueBag. At the outset, you do not know which case confronts you.

Things change, however, if you reach into a selected bag and pick your first marble. Intuitively: If the selected marble was blue, you are more likely to have chosen from the predominantly blue-marble bag, rather than the other bag. So the **posterior probability** of selecting BlueBag is different from its probability before you had this new information.

The formula for Bayes' Rule can be a bit daunting to new students of statistics. Figure 6.10 is based on a Bayes' Rule template that is available in Excel on the CD you received with your textbook. (See file **Template_Bayes.**)

The initial information in the problem can be found in the yellow-highlighted areas of Figure 6.10. We knew that half of the 200 marbles were grouped in each bag; thus the initial 50% probability of picking any particular bag. Based on the verbal description, we could say that the probability of picking a blue marble, *given that* BlueBag is picked, is 70%. We also have information that $P(\text{pick a non-blue marble} \mid \text{pick Not BlueBag}) = 80\%$.

We now have sufficient information to complete the contingency table. If 70% of the 100 marbles in BlueBag are blue, this means that $0.70 \times 100 = 70$ is the frequency for the cell corresponding to "BlueMarbles in BlueBag." Similarly, we can see that $0.80 \times 100 = 80$ is the value to be written in the cell for "Not BlueMarble in Not BlueBag." The remainder of the table can be filled in, based on the information on contingency tables that has already been provided.

Returning to the problem, what is the newly available conditional probability for "BlueBag was picked," given that you have a picked a blue marble? Simply use the table

Figure 6.10
Contingency Table for the Marble Problem

EVENT NAMES: *{Input Below}*	
A:	*BlueBag*
B:	*BlueMarble*

A \ B	BlueMarble	Not BlueMarble	TOTALS
BlueBag	70	30	100
Not BlueBag	20	80	100
TOTALS	90	110	200

Input these conditional probabilities:		i.e.,
P(BlueMarble|BlueBag) =	0.7	70/100
P(notBlueMarble|notBlueBag) =	0.8	80/100
Bayesian Result:		
P(BlueBag|BlueMarble) =	0.777777778	70/90

to find the answer, much as we did for problems on conditional probability in Section 6.2. We see that the total number of blue marbles is 90. Of these, 70 are found in BlueBag. Therefore, $P(\text{BlueBag was picked} \mid \text{picked a blue marble}) = 70/90 = 0.778$.

Compare our steps using the table with the three versions of Bayes' Rule that are presented in Formulas 6.10–6.12.

Formula 6.10 The Core Idea

$$P(A \mid B) = \frac{P(B \mid A) \cdot P(A)}{P(B)}$$

Formula 6.11 Basic Bayes' Rule

$$P(A \mid B) = \frac{P(B \mid A) \cdot P(A)}{P(B \mid A) \cdot P(A) + P(B \mid \bar{A}) \cdot P(\bar{A})}$$

Formula 6.12 Extended Formal Version

$$P(A_i \mid B) = \frac{P(B \mid A_i) \cdot P(A_i)}{P(B \mid A_1) \cdot P(A_1) + P(B \mid A_2) \cdot P(A_2) + \ldots + P(B \mid A_k) \cdot P(A_k)}$$

Figure 6.11 reexpresses the quantities in the table as relative frequencies (i.e., relative to the total sample size). These can then be interpreted as empirical probabilities.

Formula 6.10 is simpler than formulas usually given for Bayes' Rule but it captures its essence. In the table approach, we find a value for the upper left cell, $P(A \text{ and } B) = P(A) \times P(B \mid A) = 0.5 \times 0.7 = 0.35$ exactly as the formula indicates: In the word problem, $P(B \mid A)$ is the probability of drawing a blue marble, given that you've selected the blue bag; this value was input near the lower right of Figure 6.11. We multiply this value times the marginal probability for event A (selecting the blue bag) to get the value for the upper left cell. Just as the formula divides the resulting value by $P(B)$ (i.e., the marginal probability of selecting a blue marble), the table procedure does the same. We divide the upper left cell value (0.35) by the column total (0.45) that represents $P(B)$. This gives us

Figure 6.11
Contingency Table for the Marble Problem: Relative Frequencies

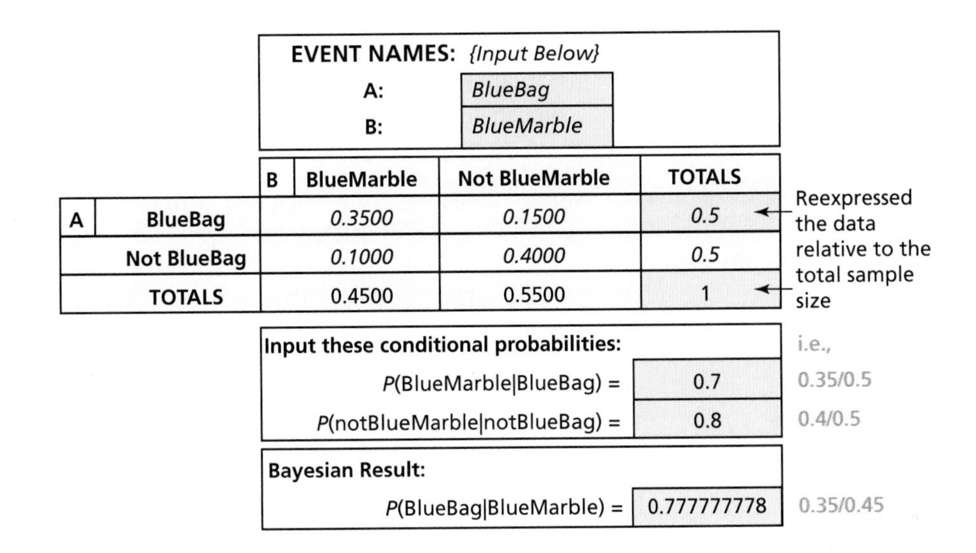

EVENT NAMES: *{Input Below}*

A:	*BlueBag*
B:	*BlueMarble*

A \ B	BlueMarble	Not BlueMarble	TOTALS
BlueBag	*0.3500*	*0.1500*	*0.5*
Not BlueBag	*0.1000*	*0.4000*	*0.5*
TOTALS	0.4500	0.5500	1

Input these conditional probabilities:		
P(BlueMarble|BlueBag) =	0.7	0.35/0.5
P(notBlueMarble|notBlueBag) =	0.8	0.4/0.5

Bayesian Result:		
P(BlueBag|BlueMarble) =	0.777777778	0.35/0.45

$P(A \mid B) = 0.778$. Depending on a problem's complexity, you could solve it either by manipulating a table or by using the formula.

Formula 6.11 reexpresses the denominator of the core formula as the sum of $P(B$ and $A)$ and $P(B$ and $\overline{A})$—which is equivalent to $P(B)$. Formula 6.12 further generalizes the same principle, and can handle cases where the A variable has more than two possible values. Applied to the previous example, A_1 corresponds to Bag colour = Blue; A_2 corresponds to Bag colour = Not blue. Had Not blue been subdivided into Green, Red, Brown, etc., the formula could represent these as values A_2, A_3, A_4, and so on up to the last category A_k. A_i in the formula represents the event whose conditional probability is of interest. Although the extended denominator of Formula 6.12 looks complicated, it represents what is obvious from the contingency diagram: The total value for $P(B)$, shown at the bottom of the column, is obtained by adding probabilities for *all* the possible joint events that make up that column. In the example, the $P(B)$ total = 0.45 is the sum of the probabilities 0.35 and 0.15 that are listed above the total.

Example 6.17

Bayes' Rule has special interest for those who must assess the strength of evidence, such as in legal or scientific investigations. Suppose an eyewitness observes a person in a Queen sweatshirt defacing public property in a high school building. He calls Security, who immediately block all of the exits. The principal calls all 60 people on the site to an assembly. Three of the people are wearing Queen sweatshirts. If one of these three is randomly selected, what is the probability that he or she is innocent of the crime?

Solution

There are only three people with Queen sweatshirts and (we are assuming) one of these did the damage; two are innocent. Given the 59 innocent people on the property, $P(\text{Queen\_sweatshirt} \mid \text{innocent}) = 2/59 = 0.0339$. It looks bad for the suspect, one would think, to have to explain the seemingly unlikely circumstances. But the real question is not the probability of wearing this or that clothing; it is the *probability of being innocent, given what one is wearing* (i.e., we need to calculate $P(\text{innocent} \mid \text{Queen\_sweatshirt})$).

We can use Formula 6.10, letting A be "the person is innocent" and B "the person is wearing a Queen sweatshirt." Note that 59 of the 60 people are innocent, so the prior probability is $P(A) = 0.9833$. Three of the 60 are wearing the sweatshirts, so $P(B) = 3/60 = 0.05$. By the simplified formula:

$$P(A \mid B) = \frac{P(B \mid A) \times P(A)}{P(B)} = \frac{0.0339 \times 0.9833}{0.05} = 0.667$$

In other words, the suspect's clothing is not quite incriminating as it first appeared. Without additional evidence, we find it is more likely that the selected person is innocent than that he or she is guilty.

Note that if you draw the contingency table, it might seem that $P(A \mid B)$ can be just read off the table without even needing Bayes' Rule. If you noticed that fact, you've discovered the secret to mastering Formula 6.12 without panicking: The formula is simply accomplishing algebraically what you could otherwise glean by constructing and manipulating a contingency table and using the basic rules of probability.

Example 6.18

A serious matter for Olympic athletes is the outcome of required testing for the presence of banned substances, such as steroids, in their blood or urine. A given type of test might have a 1% chance of showing a false positive—that is, of indicating the presence of a banned substance in an athlete when the substance is not really present. If there are hundreds of athletes at an event, several false positives could easily result. (We are ignoring the precautions of multiple sampling, repeat testing, and so on.)

Suppose that for a certain event, the probability that a randomly selected athlete is clear of banned substances is 98.5%. Each athlete must submit to a test, which has a 1% probability of a false positive. Historically, 2% of all athletes who take the test show a positive result for the presence of banned substances. If a certain athlete does test positive for banned substances, what is the probability that he or she is in fact clear of the banned substances?

Solution

We have sufficient information to use the core version of Bayes' Rule in Formula 6.10. Let A = "the athlete is clear of banned substances" and B = "the test result is positive for banned substances." The prior probability $P(A)$ that the athlete is clear is given as 0.985. The given probability of a false positive (= 0.01) can be interpreted as $P(B \mid A)$; i.e., the probability that the test result is positive, given that the athlete is in fact clear of banned substances. The marginal probability $P(B)$ that an athlete will test positive is given in the problem as 0.02. Using the formula:

$$P(A \mid B) = \frac{P(B \mid A) \times P(A)}{P(B)} = \frac{0.01 \times 0.985}{0.02} = 0.493$$

In other words, even an athlete who tests positive has almost a 50% chance of, in fact, being clear of the banned substances.

• Hands On 6.4 *Some Bayesian (and Other) Problems in Probability*

As you may have noticed, the computer has not had as much to offer in this chapter as in most of the other chapters in this book. The basic rules for probability (see Formulas 6.1 and 6.2) are actually quite simple. The practical challenge lies not so much in calculations that are complex as in how to interpret word problems and data, and how to find ways to count the cases that meet various conditions. Software programs (especially Excel) can be quite useful, on a case-by-case basis, in organizing the data and solving certain functions. The file **Template_Bayes.xls,** provided on the text CD, is useful to illustrate the concepts, but is not intended as a calculator for sophisticated problems in Bayesian statistics.

The Internet can provide a wealth of example data and problems for those who might enjoy practising on a range of problems, with or without some help from the computer. Just two examples of such websites, which can lead you to many others, are:

- http://www.benbest.com/science/theodds.html (by Ben Best)
- http://www.ds.unifi.it/VL/VL_EN/index.html (*Virtual Laboratories in Probability and Statistics,* by Kyle Siegrist, on the Università degli Studi di Firenze [University of Florence] website)

You are encouraged to look at media stories about probability and assess their credibility for yourself.

Chapter Summary

6.1.1 Describe the core concepts of probability.

Probability theory provides a method to analyze uncertainty and compare the risks of different choices. The probability of an **event** is the expected proportion of times that the event would occur if the random experiment that produces the event was repeated a large number of times. The term **random experiment** refers to a planned process of observation to obtain a value for a random variable. Each distinct observation that results from an experiment is called an **outcome,** and for a given experiment, the set of all possible outcomes is called the **sample space.** Combinations of outcomes can be grouped or classified into a lesser number of events that the researcher considers to be of interest.

The following **conditions of probability** will always apply: (1) The assigned value of probability is always a number between 0 and $+1$, with 0 representing an impossible event and $+1$ signifying a certain event; (2) the sum of probabilities for all of the possible outcomes always equals 1.

6.1.2 Explain and apply the three basic approaches to probability and explain some errors to avoid.

To calculate specific values for probabilities, there are three basic approaches or **types of probabilities:**

1. For **classical** (or **a priori**) **probability,** the probability of an event is determined from previously known definitions and properties of things, and no actual observations (experiments) are required. (See Formula 6.1.) The method depends on accurate, theoretical counts of possible outcomes, and some useful counting techniques include calculating **combinations** and **permutations,** and constructing **tables of outcomes** or **contingency tables.**
2. For **empirical** (or **relative frequency** or **experimental**) **probability,** we learn from experience how often a particular event has occurred in the past, relative to all the other alternatives, and take this as an estimate of the future proportion. (See Formula 6.2.)
3. For **subjective probability,** the probability value is assigned to an event based on individual judgment, after taking experience and the available evidence into account.

Some common errors to avoid are relying on too few experiments, or on improperly conducted experiments, or depending on the nonexistent law of averages.

6.2 Explain and apply the concepts of conditional probability.

The **absolute probability** of event A [in symbols: $P(A)$] is calculated without reference to other possible events. The **conditional probability** of "event A, given event B" [in symbols: $P(A \mid B)$] indicates that the occurrence of B may have impacted on the probability of A. Absolute probabilities are also called **marginal probabilities,** based on their relationship to numbers in the margins of contingency tables.

6.3 Define key terminology related to compound probabilities and apply the basic rules for compound probabilities.

Every experiment results in one outcome, from the sample space, but there is flexibility in naming the event that has occurred. If you can name the event without reference to other defined events, this is a **simple event.** An event that is defined in terms of *other* simple events is a **compound event.**

Joint (or **conjunctive**) **events** are compound events that occur only if *all* of the named simple events occur. In notation, $P(A \text{ and } B)$ is the probability of the joint event, based on the simple events A and B. **Disjunctive events** are compound events that occur *if at least one* of the named simple events occurs. In notation, $P(A \text{ or } B)$ is the probability of the disjunctive event, based on A or B.

Two events are **independent events** if the probability that one of the events will occur is not affected by the other event's occurrence or nonoccurrence, and vice versa. Two events are **dependent events** if the probability that one of the events will occur *is* affected by the occurrence or nonoccurrence of the other event. Two events are **mutually exclusive** if there is no possible outcome that could count as both of the events' occurring.

The compound event that occurs when a named simple event *does not* occur is called the **complement** of the named event. In notation, $P(\overline{A})$ [or $P(\text{not } A)$] is the probability of the complement of event A.

If we know the probabilities for the simple events that are combined into a compound event, we can apply some rules for calculating the compound probabilities.

For complementary events, the basic rule is given in Formula 6.3. Two rules for joint probability are given in Formula 6.6 (for independent events) and in Formula 6.7 (for dependent events). For two rules for disjunctive probability, see Formula 6.8 (for mutually exclusive events) and Formula 6.9 (for not mutually exclusive events).

Probability trees are diagrams that can help you to visualize the conditional relationships among probabilities and can guide calculations. Contingency tables can also be useful, especially for solving word problems in probability.

6.4 Discuss the applications of Bayes' Rule and apply the rule to calculate posterior probabilities.

Bayes' Rule (or **Bayes' Theorem**) was developed to facilitate *revision* of initial probability assignments (called the **prior probabilities**) based on new information, to develop revised **posterior probability** estimates. The core idea is presented in Formula 6.10, with more extended versions shown in Formulas 6.11 and 6.12.

• Key Terms

• Exercises

Basic Concepts 6.1

1. Explain the meaning of each of the following terms:
 a) Experiment
 b) Sample space
 c) Outcomes
 d) Events

2. Name the three types of probability and describe the key features of each type. List two examples of each type of probability.

3. Consider the following experiment: Roll two dice and observe whether the sum of the dots on the dice add to 7.
 a) Describe the sample space.
 b) List four possible outcomes of the experiment.
 c) Describe the event.
 d) What is the probability that the event of interest occurs?

4. Consider the following experiment: Observe the gender of all 200 participants in the next Terry Fox run in your area.
 a) Describe the sample space.
 b) List four possible outcomes.
 c) What is the probability of the event "more than half of the participants are female"?

5. Each morning, 500 drivers turn right at a certain stop sign to enter their workplace's front gate. One morning, you watch all of the cars to observe the number of cars that do not stop fully at the stop sign.
 a) Describe the sample space.
 b) List four possible outcomes.

6. 250 students attended a concert and were surveyed on whether they enjoyed it. The results were:

Response	Frequency
Yes, enjoyed the concert	162
No, did not enjoy the concert	67
Don't know *or* No response	21

a) What is the probability that a selected student clearly enjoyed the concert?
b) What is the probability that a selected student stated a clear opinion about whether she or he enjoyed the concert?

7. On your way to a dental appointment, you take along your iPod. You expect you will *probably* have time to listen to it while sitting in the waiting room. What kind of probability are you using?

8. What is the probability of obtaining each of the following events from a fully shuffled standard deck of 52 playing cards? (Each part of the exercise represents a new experiment.)
a) Ace of hearts.
b) Any spade.
c) Any one of a 10, a jack, a queen, a king, or an ace.

9. A child has a bowl containing 8 marbles—4 red, 2 white, and 2 blue.
a) If the child randomly selects one marble from the bowl, what is the probability that the marble is red?
b) Suppose that the child randomly selects one marble from the bowl and then returns it to the bowl after noting its colour; then the child randomly selects a second marble. What is the probability that the child selected one blue and one white marble—in either order?

Applications 6.1

10. A survey was conducted by the Asia Pacific Foundation of Canada. Each respondent checked off the names of countries with which he or she would like to do business. Respondents could check off more than one.

Country	Positive Responses
Japan	33
Mainland China	68
Hong Kong	27
South Korea	24
Taiwan	24
Singapore	23
Other Southeast Asian country	37
India, Pakistan, and/or Sri Lanka	39
Australia	25
Other	12
Total Positive Responses	**312**

Source: Asia Pacific Foundation of Canada. Based on data in *2006 Asian Investment Intentions Survey.* Retrieved from http://www.asiapacific.ca/analysis/pubs/pdfs/invest_survey/invest_intentions2006.pdf

a) Every positive response was put into a hat. One response was drawn, and whoever marked that response won a free ticket to the indicated country. What is the probability that the winning response was "Mainland China"?
b) From the viewpoint of "what countries did a particular respondent choose," are the listed choices mutually exclusive? Do we have sufficient information to find the probability that half of the respondents chose Mainland China? Explain.

11. In the periodic table of chemical elements, six elements belong to the family (i.e., group) "alkaline earth." The atomic numbers of these six elements are, from low to high: 4, 12, 20, 38, 56, and 88, respectively.[3] A chemist randomly picks two of the elements.
a) What is the probability that the sum of the two elements' atomic numbers is exactly 58?
b) What is the probability that the sum of their atomic numbers is less than 60?

12. In the Northern Hemisphere, the north-pointing pointer of a magnetic compass is attracted to the north magnetic pole of the earth.[4] If, mistakenly, the label "North" was painted on the wrong side of the compass pointer, the compass would appear to be pointing south. Suppose an exhausted compass maker randomly picks one side on each of three compass pointers to label as "North." What is the probability that all three compass pointers will be correctly labelled?

13. In the summer of 2003, SES Canada Research Inc. conducted a Canada Day survey of 500 randomly selected beer drinkers. Each was asked what one Canadian item they'd most like to have if they were stranded on a desert island. (For the set of results, see the file **Desert_Island.**)
a) Assuming that the respondents are typical of *all* beer drinkers, find the probability that a randomly selected beer drinker would select "Canadian beer" as his or her preferred item. (Hint: You can use the computer first to construct a frequency distribution.)
b) Could the proportions in this data set be used to find probabilities for the choices of nondrinkers? Explain why or why not.

14. In 2003, the Ontario Chamber of Commerce surveyed employers on their satisfaction with employees of varying education levels. Partial results are found in the file **College_Graduates.** Each record (i.e., row) shows one employer's satisfaction level with its college-grad employees, and also indicates the approximate size of the employer's organization in terms of numbers of employees. Suppose you randomly select one employer who took the survey. (Hint: For the following questions, you can first use the computer to construct a contingency table of the data. See "In Brief 3.5(b).")
a) What is the probability that the selected employer has less than 51 employees?

b) What is the probability that the employer describes his or her level of satisfaction as "Good"?
c) What is the probability that you selected an employer who has over 250 employees and who did not answer the question?

15. The file **Religion** shows the stated religions of Canadians, according to the 2001 Canadian census. Suppose that one Canadian was randomly selected in 2001. What is the probability that the selected person was:
a) Living in Manitoba?
b) Living in a province located west of Ontario? (If you need a map, the Government of Canada website has one.)
c) Identified as being Catholic?
d) Identified as a Buddhist living in the Yukon?

16. Those responsible for security on the networks of large institutions have a serious concern about the consumption of system bandwidth (i.e., resources) by spam and other unwanted code. Concordia University published these weekly data on e-mails that have arrived at the university.

Identified E-Mail Type	Number: July 4–10/05	Number: March 1–7/04
Spam	983,677	509,606
Viruses	3,632	16,908
Acceptable for delivery	369,214	590,928

Source: Concordia University, Instructional and Information Technology Services. Based on data in *Concordia University E-Mail Statistics.* Retrieved from http://iits.concordia.ca/statistics/email

Suppose that one e-mail was randomly selected from the e-mails analyzed in the table. What is the probability that the email selected was:
a) Received in 2004?
b) Spam?
c) A virus received in 2005?
d) *Not* acceptable for delivery on campus?

Basic Concepts 6.2

17. The university library has found that some students do not return books on time. The library classifies the students based on their faculty of study. The following table records some of the relative frequencies.

Book Return Status	Engineering	Business and Humanities	Health Sciences	Total
On time	0.07			0.87
Late		0.06		
Total	**0.10**	**0.55**		

a) Complete the contingency table by filling in the missing probabilities.
b) What is the probability that a randomly selected returned book is late?
c) If a returned book is from a Health Sciences student, what is the probability that it was returned late?
d) If a book was returned on time, what is the probability that it was signed out to a Health Sciences student?

18. 250 parents bought tickets for a peewee hockey tournament. Four seating options were available. The frequency distribution for the ticket options sold is shown below.

Ticket Price	Frequency
$10.00	45
$15.00	75
$20.00	60
$25.00	70
Total	**250**

a) What is the probability that a randomly selected ticket costs $25?
b) A parent paid more than $15 for her ticket. What is the probability that she paid $25 for the ticket?
c) Another parent paid less than $20 for his ticket. What is the probability that he paid $15 for the ticket?

19. 250 students who attended a concert were surveyed on whether they enjoyed it. The results are shown below. Selecting only from those who voiced an opinion, what is the probability that the selected person enjoyed the concert?

Response	Frequency
Yes, enjoyed the concert.	162
No, did not enjoy the concert.	67
Don't know *or* No response.	21

20. A child has a bowl containing 8 marbles—4 red, 2 white, and 2 blue.
a) Suppose one red marble is taken from the bowl and *not* replaced, and then another marble is selected randomly from the remaining 7 marbles. What is the probability that the randomly selected marble is red?
b) Three red marbles are taken from the original set of marbles and not replaced, and then another marble is selected randomly from the remaining marbles. What is the probability that the randomly selected marble is red?

21. A school's administrators are concerned about absenteeism at their school and the effect it has on students' grades. The following data were compiled, based on a sample of students.

Student's Grade	Absent Infrequently	Medium Absenteeism	Absent Frequently
Low	1	9	8
Medium	6	35	7
High	8	9	2

a) What is the probability that a randomly selected student will have a low grade?
b) What is the probability that a student will have a low grade, given that he or she is frequently absent?

c) In general, what is the probability that a randomly selected student will be frequently absent?
d) Given that a student has a high grade, what is the probability that that student is absent frequently?

Applications 6.2

22. A data set on coal production in the United States was compiled by an energy planner based in Alabama. (See the file **Coal_Production.**) The number of coal mines in each state that were active in 2004 are recorded. Underground mines (Variable: **Mines_U04**) and surface mines (Variable: **Mines_S04**) are distinguished.
    a) If any active coal mine in 2004 (surface *or* underground) is randomly selected, what is the probability that the mine was located in Alabama?
    b) If any active *surface* coal mine in 2004 is randomly selected, what is the probability that the mine was located in Alabama?
    c) If any active coal mine in 2004 (surface *or* underground) is randomly selected, what is the probability that the mine was an underground mine?
    d) If any active coal mine in Alabama in 2004 (surface *or* underground) is randomly selected, what is the probability that the mine was an underground mine?

23. A chemistry teacher describes some facts about the periodic table of chemical elements. (i) There are six elements in the family alkaline earth. Their atomic numbers are 4, 12, 20, 38, 56, and 88. (ii) The smallest atomic number for any element in the family transition metals is 21.[5]
    a) If one of the alkaline earth elements is randomly selected, what is the probability that its atomic number is smaller than any atomic number for a transition metal?
    b) A chemist randomly selects one alkaline earth element whose atomic number is smaller than the atomic number of any transition metal. What is the probability that the selected element has an atomic number of 12?

24. The file **Religion** shows the stated religions of Canadians according to the 2001 Canadian census.
    a) If one person is randomly selected from Manitoba, what is the probability that his or her survey response was "No Religion"?
    b) For which of the three territories is the probability greatest that a randomly selected person from the territory responded "No Religion" in the census?
    c) Given that a person identified his or her religion as an "Eastern Religion" (not in any other category), what is the probability that the person was residing in Alberta?
    d) Given that a person's religion is identified as "Other" in the results, it is most likely that the person resided in what province?

25. In the summer of 2003, SES Canada Research Inc. conducted a Canada Day survey of 500 randomly selected beer drinkers. Each was asked what one Canadian item they'd most like to have if they were stranded on a desert island. (For the set of results, see the file **Desert_Island.**)
    a) For a random individual who did *not* choose Canadian beer what is the probability that the individual chose maple syrup?
    b) For a random individual who chose a liquid beverage, what is the probability that the individual chose screech? (Screech is a strong liquor bottled in Newfoundland.)

26. To prepare for an appearance on a game show, a student is brushing up on trivial facts—including a list of the world's tallest towers. (See the file **Tallest_Towers.**)
    a) If any one tower on the list is randomly selected, what is the probability that the tower is in China?
    b) If one of the three tallest towers on the list is randomly selected, what is the probability that the tower is in China?
    c) If any one of the towers listed as being in China is selected, what is the probability that the tower is one of the three tallest towers on the entire list?

27. Concerned about network security, Concordia University published these weekly data on e-mails that arrived at the university.

Identified E-Mail Type	Number: July 4–10/05	Number: March 1–7/04
Spam	983,677	509,606
Viruses	3,632	16,908
Acceptable for delivery	369,214	590,928

Source: Concordia University. Found at: http://iits.concordia.ca/statistics/email

    a) If an e-mail received during one of the indicated one-week periods is randomly selected, what is the probability that the e-mail was spam?
    b) During which of the two time periods was it more likely that a received e-mail was spam?
    c) If an e-mail received during one of those time periods was a virus, was it more likely to have come in 2004 or 2005?

Basic Concepts 6.3

28. Explain the meaning of each of the following terms:
    a) Compound events
    b) Joint events
    c) Disjunctive events
    d) Complement (of an event)
    e) Dependent events
    f) Independent events

29. Are the two events A and B dependent or independent? Explain your answers.

a) $P(A) = 0.20$; $P(B) = 0.70$; $P(A \text{ and } B) = 0.14$

b) $P(A) = 0.80$; $P(B) = 0.50$; $P(A \text{ and } B) = 0.36$

30. 81% of the students who enroll in a certain tough course complete it successfully. What is the probability that a randomly selected student who enrolls in the course will not complete it successfully? (Assume that the past probabilities will hold in the future.)

31. Every year, 5% of the students earn a prestigious scholarship. What is the probability that a randomly selected student will *not* win the scholarship?

32. To test for pregnancy, two different tests are available. Test 1 gives a correct result 85% of the time. Test 2 gives a correct result 90% of the time. If both tests are taken, let A be the event "Test 1's result is correct" and B be the event "Test 2's result is correct."

a) Find $P(A \text{ and } B)$.

b) Find $P(A \text{ or } B)$. (Hint: Are the events mutually exclusive?)

c) What is the probability that neither test result is correct?

d) Are the events independent or dependent? Explain.

33. Randomly select two cards, without replacement, from a standard deck of 52 playing cards. What is the probability that:

a) Both cards are aces?

b) Both cards are face cards (i.e., jacks, queens, and/or kings)?

c) The first card is an ace or the second card is an ace (or both)?

d) Neither card is an ace?

34. Randomly select two cards from a standard deck of 52 playing cards, *with* replacement (i.e., return the first card and reshuffle before drawing the second card). What is the probability that

a) Both cards are aces?

b) Both cards are face cards (i.e., jacks, queens, and/or kings)?

c) The first card is an ace or the second card is an ace (or both)?

d) Neither card is an ace?

35. In a survey on pollution, respondents were asked whether they believed poor air quality was a serious problem in their own community. These are the results.

Gender	Yes	No
Male	60	40
Female	86	16

a) Given that a respondent was male, what was the probability that he answered "No"?

b) What is the probability that a randomly selected respondent was a male who answered "No"?

c) What is the probability that a respondent was a female and she answered "Yes"?

d) What is the probability that the respondent was a male who answered "No" or a female who answered "Yes"?

36. A parent group is promoting healthy eating at the local school. On "Healthy Treat Day," students have a choice of healthy snacks at a very reduced cost. A parent wanted to compile a chart of students' preferences. This is what she has found so far.

Grades	Fruit	Yogurt	Totals
Grades 1–3	50		100
Grades 4–6			
Totals		135	220

a) Fill in the blanks in the chart.

b) What is the probability that a randomly selected student who participated was in Grades 4–6?

c) What is the probability that a randomly selected participant was in Grades 1–3 and chose yogurt?

d) What is the probability that a participant was in Grades 1–3 or chose yogurt?

e) Given that a participant chose yogurt, what is the probability that the person was in Grades 4–6?

37. What is the probability of rolling a pair of sixes with two dice, given that you just rolled a pair of sixes with the dice?

Applications 6.3

38. The file **Religion** shows the religions (if any) recorded for all Canadians during the 2001 census. If you randomly select one person from the entire census population, what is the probability that:

a) The person lives in New Brunswick and responded "No religion"?

b) The person lives in New Brunswick or responded "No religion"?

c) The person gave some response *other* than "No religion"?

d) The person's religion was recorded as *one* of Catholic, Protestant, Muslim, or Jewish?

39. Two scouts for a baseball team are discussing players who have made it to Major League Baseball's "Hitting Milestones" list. (See the file **Baseball_hitting**—ignoring the records with missing data for "Team.") Answer the following questions, which involve the team name and the type of memorable event. (Hint: First, use the computer to construct a contingency table.)

a) If one milestone event is randomly selected, what is the probability that it is one of the home runs scored by Boston?

b) If two of the milestone events are randomly selected, what is the probability that both are home runs scored by Boston?

c) If one milestone event is randomly selected, is it more likely to be one of the RBIs scored by Houston or one of the bases on balls (BB) earned by Philadelphia? Explain.
d) What is the probability that a randomly selected milestone event was earned by a player for either Houston or Baltimore?
e) If a milestone event is randomly selected, what is the probability that either it is for an RBI or it is for some event for a Detroit player?

40. In the survey described in Exercise 25, each of 500 randomly selected beer drinkers were asked what one Canadian item they'd most like to have if they were stranded on a desert island. (See the file **Desert_Island.**) If two of the beer drinkers in the sample are selected, what is the probability of each of the following?
a) They both chose Canadian Beer.
b) At least one chose melons.
c) One chose melons and the other chose lobster.
d) Neither chose fiddleheads.

41. One estimate for the probability that a given person will be struck by lightening in a specific year is 0.00000032.[6] (If you climb trees or mountains during thunderstorms, or else you never leave a cave, your personal values may differ.) If two normally living people in the world are randomly selected, and monitored for a year, what is the probability of the following?
a) At least one is struck by lightning.
b) Both are struck by lightning.
c) Neither is struck by lightning.

42. To prepare for an appearance on a game show, a student is brushing up on trivial facts—including a list of the world's tallest towers. (See the file **Tallest_Towers.**) If two of the listed towers are randomly selected, what is the probability of each of the following?
a) The sum of their heights equals at least 1,000 metres.
b) The sum of their heights is no more than 670 metres.
c) Both towers were built in the 21st century.
d) Both selected towers are located in Russia.
e) The first tower (ignore the second tower) is located in Malaysia or Kuwait.

43. Concerned about network security, Concordia University published these weekly data on e-mails that have arrived at the university.

Identified E-Mail Type	Number: July 4–10/05	Number: March 1–7/04
Spam	983,677	509,606
Viruses	3,632	16,908
Acceptable for delivery	369,214	590,928

Source: Concordia University, Instructional and Information Technology Services. Based on data in *Concordia University E-Mail Statistics*. Retrieved from http://iits.concordia.ca/statistics/email

What is the probability that a randomly selected e-mail during the recorded periods was:
a) Received in 2005 and was spam?
b) Received in 2005 or was spam?
c) A virus received in 2005?
d) A virus received in 2004?

44. A taxonomist is studying the 106 species of the pine tree. (See the file **Pine**—excluding data for which chromosome numbers are not listed.)
a) What is the probability that a randomly selected species from the list:
i) Is from the genus *Pinus*?
ii) Is from the genus *Pinus and* has 12 chromosomes?
iii) Has 13 or 33 chromosomes?
iv) Does not have 12 chromosomes?
b) What is the probability that two randomly selected species from the list both have more than 12 chromosomes?

Basic Concepts 6.4

45. Given $P(A) = 0.81$, $P(B) = 0.14$, and $P(B \mid A) = 0.05$, calculate $P(A \mid B)$.

46. Paul Francis coaches the Moosonee Old-Timers Baseball Team. Concerned with his players' injuries, he has compiled five years of data on the number of injuries sustained by players by position.

	Infielder	Outfielder	Pitcher	Catcher
Number of players having held that position	24	18	20	6
Number of players injured in that position	10	5	12	2

Given that a randomly selected player is injured, what is the probability that the player is:
a) An infielder?
b) An outfielder?
c) A pitcher?
d) A catcher?

Solve the problems first with Bayes' Rule and then confirm using a contingency table.

47. A total of 65% of Daeglo lamps are made on the day shift, and the remainder are made on the night shift. The quality control manager has determined that there is a 0.10 probability that a lamp constructed during the day will be defective; the probability is 0.35 that a lamp constructed at night will be defective. Given that a customer has just returned a defective lamp, what is the probability that it was constructed during the night shift?

48. The following information is provided.

Given:

$P(H) = 0.4$ $P(G) = 0.4$ $P(F) = 0.2$
$P(S \mid H) = 0.6$ $P(S \mid G) = 0.4$ $P(S \mid F) = 0.2$

Find:

a) $P(H \mid S)$
b) $P(G \mid S)$
c) $P(F \mid S)$

49. A would-be cat breeder has lost track of his cats' activity. He has determined the following probabilities:

P(the Siamese cat is the father of the litter) = 0.35

P(a kitten with a curly tail is born | the Siamese cat is the father of the litter) = 0.03

P(the Ocicat is the father of the litter) = 0.60

P(a kitten with a curly tail is born | the Ocicat is the father of the litter) = 0.07

P(the Devon Rex cat is the father of the litter) = 0.05

P(a kitten with a curly tail is born | the Devon Rex cat is the father of the litter) = 0.04

Find:

a) P(the Siamese cat is the father of the litter | a kitten with a curly tail is born)
b) P(the Ocicat is the father of the litter | a kitten with a curly tail is born)
c) P(the Devon Rex cat is the father of the litter | a kitten with a curly tail is born)

50. Joyce Green is very concerned about the environment and regularly buys a newspaper every day to look for articles on that subject. On any particular day, her probability of buying *The Globe and Mail* is 0.6, of buying *The Toronto Sun* is 0.3, and of buying the *Toronto Star* is 0.1. She has noticed that *The Globe and Mail* has an article on the environment 40% of the time, *The Toronto Sun* has an article on the environment 30% of the time, and the *Toronto Star*, 70% of the time. Given that today Joyce has read a newspaper article on the environment, what is the probability that today she bought:

a) *The Globe and Mail*?
b) *The Toronto Sun*?
c) *The Toronto Star*?

51. A furniture store is willing to sell products on credit. When a payment is overdue, the store contacts the customer by telephone 50% of the time, by letter 30% of the time, and by a personal visit 20% of the time. The store found that the probabilities of collecting the moneys due were 0.60, 0.30, and 0.80, respectively, depending on the type of follow-up that was used. Given that a customer has paid up an overdue account, what is the probability that the customer:

a) Was contacted by telephone?
b) Received a reminder letter?
c) Received a personal visit?

52. A school's administrators are concerned about absenteeism at their school and its effect on students' grades. These data were compiled.

Student's Grade	Absent Infrequently	Medium Absenteeism	Absent Frequently
Low	1	9	8
Medium	6	35	7
High	8	9	2

If a randomly selected student has a low grade, what is the probability of each of the following? Use Bayes' Rule to solve these problems.

a) The student was absent infrequently.
b) The student had medium absenteeism.
c) The student was absent frequently.

53. The following information is provided:

Given:

$P(A) = 0.3$ $P(B) = 0.4$
$P(C) = 0.2$ $P(D) = 0.1$
$P(X \mid A) = 0.4$ $P(X \mid B) = 0.2$
$P(X \mid C) = 0.1$ $P(X \mid D) = 0.5$

Find:

a) $P(A \mid X)$
b) $P(B \mid X)$
c) $P(C \mid X)$
d) $P(D \mid X)$

Applications 6.4

54. Concerned about network security, Concordia University published these weekly data on e-mails that have arrived at the university.

Identified E-Mail Type	Number: July 4–10/05	Number: March 1–7/04
Spam	983,677	509,606
Viruses	3,632	16,908
Acceptable for delivery	369,214	590,928

Source: Concordia University. Found at: http://iits.concordia.ca/statistics/email

If a randomly selected e-mail from these periods was acceptable for delivery, what is the probability that it was received in which year? Use Bayes' Rule to solve these problems.

a) 2004
b) 2005

55. A data set on coal production in the United States was compiled by an energy planner. (See the file **Coal_Production.**) The numbers of coal mines in each state that were active in 2004 are recorded. Underground mines (Variable: Mines_U04) and surface mines (Variable: Mines_S04) are distinguished.

a) If any active coal mine in 2004 (surface or underground) is randomly selected, what is the probability that the mine was an underground mine?
b) If any active coal mine in 2004 is randomly selected, what is the probability that the mine was located in Kentucky?
c) What is the probability that a 2004 mine is underground, given that the mine is in Kentucky?
d) Applying Bayes' Rule to answers (a)–(c), what is the probability that a 2004 mine is in Kentucky, given that it is an underground mine?

56. The file **Religion** shows the stated religions of Canadians according to the 2001 Canadian census.
a) If any one Canadian is randomly selected, what are the following probabilities?
  i) *P*(Catholic)
  ii) *P*(Protestant)
  iii) *P*(Muslim)
  iv) *P*(Jewish)
b) What are the following conditional probabilities?
  i) *P*(New Brunswick | Catholic)
  ii) *P*(New Brunswick | Protestant)
  iii) *P*(New Brunswick | Muslim)
  iv) *P*(New Brunswick | Jewish)
c) Apply Bayes' Rule to the preceding information to find:
  i) *P*(Catholic | New Brunswick)
  ii) *P*(Protestant | New Brunswick)
  iii) *P*(Muslim | New Brunswick)
  iv) *P*(Jewish | New Brunswick)

57. A naturalist observed that the brood of a Gray Jay usually begins with three young—but two of the young (66.7%) are banished from the family's territory by June. If a young Gray Jay is banished, its probability of surviving until autumn is only 20%.[7] Suppose that, in general, 43% of all young Gray Jays survive until autumn. If the naturalist photographs a young Gray Jay at the end of autumn, what is the probability that the bird had been banished from its family's territory earlier in the year?

58. Preparing for an appearance on a trivia game show, a student is studying a list of the world's tallest towers. (See the file **Tallest_Towers.**)
a) If any one tower on the list is randomly selected, what is the probability that the tower is over 400 metres tall?
b) If any one tower on the list is randomly selected, what is the probability that the tower is located in Russia?
c) If one of the towers listed as being in Russia is selected, what is the probability that the tower is over 400 metres tall?
d) Apply Bayes' Rule to the above information to find:

P(the tower is located in Russia | the tower is over 400 metres tall)

59. To entertain a guest, a mayor has downloaded information on Australian wines from the website of the PEI Liquor Control Commission. (See the file **Australian_Wines.**)
a) If any wine on the list is randomly selected, what is the probability that it retails for more than $14.00?
b) If any wine on the list is randomly selected, what is the probability that its level of sweetness is 1 on the "SC" scale?
c) If one of the wines costing over $14.00 is selected, what is the probability that its level of sweetness is 1 on the "SC" scale?
d) Apply Bayes' Rule to the above information to find:

P(the wine costs over $14.00 | the wine's sweetness is 1 on the scale).

60. The two baseball scouts in Exercise 39 are continuing to discuss some "Hitting Milestones." (See the file **Baseball_hitting;** use the contingency table that you constructed for the previous problem). If a milestone event relates to getting a base on balls (BB), what is the probability that the player's team is one of the following?
a) Philadelphia
b) Baltimore
c) Houston

Solve the problems first with Bayes' Rule and then confirm using the contingency table.

Review Applications

For Exercises 61–66, use data from the file **Brilliants.**

61. A dealer in fine diamonds is studying the data published by Canada Diamonds to learn the factors that contribute to good prices for these jewels. Begin by constructing a contingency table to relate the diamonds' coded colours with their carat weights.

62. Based on the table constructed in Exercise 61:
a) What is the probability that a randomly selected diamond will weigh more than two carats?
b) What is the probability that a randomly selected diamond's colour is category "G"?

63. If one of the analyzed diamonds is randomly selected, what are the following probabilities?
a) It weighs more than two carats and has a colour of category "G."
b) It weighs more than two carats or has a colour of category "G."
c) It weighs more than 1 carat.
d) It is faint yellow (i.e., colour categories K–M).

64. With regard to the analyzed diamonds, what are the following probabilities?
a) A diamond weighs up to 0.5 carats, given that its colour is category "G."
b) A diamond's colour is category "G," given that it weighs up to 0.5 carats.

65. If two of the analyzed diamonds are randomly selected, what are the following probabilities?
a) Both have a colour that is category "G."
b) Neither has a colour that is category "G."
c) At least one has a colour that is category "G."

66. For the following questions, presume that the proportions in the table apply, in general, to the whole population of sale-quality diamonds.
a) If any two sale-quality diamonds are randomly selected, what is the probability that both have a colour that is category "G"?
b) Does your answer to (a) differ from your answer to Exercise 65(a)? If so, explain why.
c) If any three sale-quality diamonds are randomly selected, what is the probability that all three have a colour that is category "G"?
d) If any two sale-quality diamonds are randomly selected, what is the probability that one has a colour that is category "G" and one has a colour that is category "H"?

For Exercises 67–70, use data from the file **Grace_Alumni_World.**

67. Globe-trotting graduate of Grace University hopes to meet fellow alumni during her travels. She has found on the university website a list that shows the number of grads known to have moved to various countries. If she randomly meets one of the alumni listed, what are the following probabilities?
a) The alumnus is living in Canada.
b) The alumnus is living in Cote d'Ivorie.
c) The alumnus is living in Canada or Cote d'Ivorie.
d) The alumnus is living on the African continent.

68. If one of the alumni living in Africa is selected, what is the probability that:
a) The alumnus is living in Kenya?
b) The alumnus is living in Nigeria?

69. If two alumni are randomly selected from the full list, what is the probability that:
a) Both live in Africa?
b) One (in either order) lives in Africa and the other lives in Europe?

70. Suppose that the travelling graduate makes two separate trips around the world. Each time, she independently selects one listed alumnus to visit. What is the probability that both visits will be with a Canadian?

For Exercises 71–75, use data from the file **National_Parks.**

71. Natural Resources Canada publishes a list of all Canada's national parks. A tourist is studying the list to plan her summer activities. If one of the parks is randomly selected, what is the probability of each of the following?
a) Its area is over 1,000 square kilometres.
b) It is located in British Columbia.
c) Its area is over 1,000 square kilometres and it is located in British Columbia.
d) Its area is less than 30 square kilometres and it is located in Ontario.
e) Its area is less than 30 square kilometres or it is located in Ontario.

72. Given that a park is located in British Columbia, what is the probability that its area is over 1,000 square kilometres?

73. Using applicable answers from Exercises 71–72, apply Bayes' Rule to find:

$$P(\text{the park is in British Columbia} \mid \text{the park's area is over 1,000 square km}).$$

74. If two of the parks are randomly selected, what are each of the following probabilities?
a) Both are in Saskatchewan.
b) Neither is in Saskatchewan.
c) Exactly one is in Saskatchewan.
d) The areas of the two parks add up to at least 80,000 square kilometres.

75. The tourist will visit a Canadian national park during each of the next three years. His plan is to, each year, make an independent random selection of the park to visit. What is the probability that:
a) He visits an Ontario park each year?
b) He never visits an Ontario park?

For Exercises 76–79, use data from the file **Canadian-Health.**

76. A public health nurse is studying recent results from the *Canadian Community Health Survey.* She is alarmed by the percentages of adults having problems with stress, obesity, and other ailments.
a) For each province and territory, what is the probability that a randomly selected resident is an adult?
b) What is the probability that a randomly selected Canadian is either an adult or living in one of the three territories?
c) What is the probability that a randomly selected Canadian is an adult living in one of the three territories?

77. If any two Canadians are randomly selected, what are the following probabilities?
a) Both are adults.
b) Neither is an adult.

c) Exactly one is an adult.
d) Both live in Alberta.
e) Both live in Alberta or Manitoba.

78. What is the probability that a randomly selected adult Canadian lives in Alberta and:
a) Suffers from a lot of stress?
b) Is obese?
c) Suffers from stress and is obese?
d) Suffers from stress or is obese?
e) Are the variables "is obese" and "suffers from stress" necessarily independent? If not, how might this affect our calculations for (c) and (d)?

79. What is the probability that a randomly selected adult Canadian:
a) Lives in Ontario?
b) Suffers from stress? (Hint: Estimate the numbers of adults suffering from stress in each province and then compare their total to the overall population of adults.)
c) Suffers from stress, given that the person lives in Ontario? (Hint: This value is included in the original data set.)
d) Apply Bayes' Rule to the above information to find:

$$P(\text{the adult lives in Ontario} \mid \text{the adult is suffering from stress})$$

For Exercises 80–83, use data from the file **Canadian_Equity_Funds.**

80. Two considerations when choosing an equity fund for investment are the funds' fee policies and also the funds' ratings by various criteria. Using the data in the file **Canadian_Equity_Funds,** create a contingency table with columns for the included funds' ratings (omit funds where this field is empty) and rows for the load-fee policies of the funds.

81. For funds included in the table (see Exercise 80), what is the probability that a randomly selected fund:
a) Charges no sales fee?
b) Charges no sales fee or is front-end loaded (i.e., has this fee option)?
c) Has a rating of 4?
d) Has a rating between 2 to 4, inclusive?
e) Has a rating of 4 or charges no sales fee?

82. What is the probability that two randomly selected funds:
a) Both charge no sales fee?
b) Both are not front-end loaded?

83. What is the probability that:
a) A fund charges no sales fee, given that it has a rating of 5?
b) A fund has a rating of 5, given that it charges no sales fee?
c) A fund is front-end loaded if it has a rating less than 4?

For Exercises 84–87, use data from the file **Major_Bridges.**

84. A student of architecture is planning a tour to the sites of six major bridges to see each of the main types of bridges. The data file **Major_Bridges** lists her selection of sites to visit. What is the probability that a randomly selected site on her tour:
a) Is in Canada?
b) Is in Canada or South Korea?
c) Is a segmental concrete bridge?
d) Is a steel arch bridge and located in the United States?
e) Is a steel arch bridge or located in the United States?
f) Has a span of over 1,000 metres?
g) Does not span as much as 100 metres?

85. What is the probability that a bridge on her tour:
a) Is in Canada, given that it is a steel arch bridge?
b) Is a steel arch bridge, given that it is located in Canada?
c) Is a cable-stayed bridge, given that it is over 600 metres long?
d) Is in Canada, given that it is less than 200 metres long?

86. What is the probability that two randomly selected bridges on her tour:
a) Are both in Canada?
b) Are both not in Canada?
c) Have spans that, if added, add to less than 200 metres?
d) Have spans that, if added, add to more than 2,100 metres?
e) Are both steel truss bridges, given that both are located in the United States?
f) Are both located in the United States, given that both are steel truss bridges?

87. What is the probability that a randomly selected bridge on her tour:
a) Was built before 1930?
b) Was a simple truss bridge?
c) Given that it was a simple truss bridge, that it was built before 1930?
d) Apply Bayes' Rule to the above information to find:

$$P(\text{bridge was a simple truss type} \mid \text{bridge was built before 1930})$$

For Exercises 88–92, use data from the file **Kansas_Wells.**

88. A Kansas hydrologist, concerned about the water supply, is looking at historical data for Atchison County. The file **Kansas_Wells** gives data from a random sample of historical well readings. The right two columns indicate the distances from the well-water surface to the ground above it, at two times in a given year. What is the probability that from April/May to July/August, for a well in the sample:

a) The distance from the water level to the ground level increased?
b) The distance from the water level to the ground level decreased?

89. What is the probability that the well depth of a given sample is:

a) Greater than 50 feet?
b) Less than 20 feet?
c) Between 22 to 60 feet?

90. What is the probability that a well reading in the sample:

a) Was taken in 1968?
b) Was taken in 1968 from a test well?
c) Was taken in 1968 from a well used for withdrawal of water?
d) Was from an unused well or from a well whose depth was less than 15 feet?
e) Was from a domestic-use well, given that its depth was less than 20 feet?
f) Showed a depth less than 20 feet, given that the well was for domestic use?

91. What is the probability that two randomly selected wells from the file:

a) Are both used for domestic water supply?
b) Are both used for public water supply?
c) Have depths that, if added, add to less than 25 feet?
d) Have depths that, if added, add to more than 100 feet?

92. What is the probability that a randomly selected well in the file:

a) Had a depth greater than 50 feet?
b) Was for the public water supply?
c) Had a depth greater than 50 feet, given that it was for the public water supply?
d) Apply Bayes' Rule to the above information to find:

$$P(\text{the well was for public water supply} \mid \text{the well had a depth greater than 50 feet})$$

Discrete Probability Distributions

Chapter 7

• Chapter Outline

7.1 The Discrete Probability Distribution

7.2 The Binomial Distribution

7.3 The Hypergeometric Probability Distribution

7.4 Poisson Probability Distribution

• Learning Objectives

7.1 Explain the concept of a discrete probability distribution and calculate summary statistics for such distributions.

7.2 List the criteria for a binomial probability distribution, calculate probabilities and cumulative probabilities based on the distribution, and calculate summary statistics for such distributions.

7.3 List the criteria for a hypergeometric probability distribution, calculate probabilities and cumulative probabilities based on the distribution, and calculate summary statistics for such distributions.

7.4 List the criteria for a Poisson probability distribution, and calculate probabilities and cumulative probabilities based on the distribution.

7.1 The Discrete Probability Distribution

Building on the probability theory covered in Chapter 6, this chapter introduces the concept of a probability distribution. For the development of probability models that can be applied for forecasting and decision making in a wide variety of fields, the concept of probability distributions is fundamental. For example, a manufacturer wants to estimate the number of defects he can expect in a new batch of parts. A biologist may be estimating the number of microbes with mutations that can be collected within a given time. An engineer wants to know the number of concrete beams expected to fail due to flexure in the next year. Probability distributions can provide the information needed to answer such questions.

You will recall from Chapter 1 that a *random variable* is a variable whose numeric value is determined by the result of a random experiment or trial. *Discrete random variables*, we saw, can assume only a limited number of values in an experiment. Examples of discrete random variables are the number of children in a family, the number of reported safety violations during a shift, or the number of printer malfunctions in a day. In each case, the "experiment" is to count the actual number of children, violations, or malfunctions. The observed result is the value of the variable. Examples of discrete random variables can occur at any of the four levels of measurement discussed in Chapter 1: *number of children* (ratio level), *brand of lawnmower* (nominal level), *one-to-five-star rating* (ordinal level), or *thermostat temperature* (in integer degrees Celsius: interval level).

When tossing two coins, only one of three events will occur: None, one, or both of the two coins will be a head. Table 7.1 shows the specific sequences of coin-toss results that can be interpreted as the events "0 heads," "1 head," or "2 heads."

Table 7.1 Possible Outcomes from a Toss of Two Coins

Outcome Sequence		
First Coin	**Second Coin**	**Event**
T	T	0 Heads
T	H	1 Head
H	T	1 Head
H	H	2 Heads

By the basic rule of probability [$P(x)$ = (how many outcomes are interpreted as the event)/(number of outcomes that are possible)], a probability can be calculated for each of the possible head-count events, as shown in Table 7.2. Interpreting x as the variable

Table 7.2 Probability Distribution for the Number of Heads in Two Tosses of a Coin

Value of Variable (Number of Heads: x)	Probability of x
0	$1/4 = 0.25$
1	$2/4 = 0.5$
2	$1/4 = 0.25$

"numbers of heads thrown in two coin tosses," the table shows a distribution of *all* possible values of the variable, together with the probabilities that each of these possible values will occur. This is the **probability distribution**.

A graphical representation of a probability distribution is shown in Figure 7.1. For a discrete random variable, the figure is similar to a relative frequency histogram, showing on the y-axis the probability of occurrence for each value of the variable on the x-axis.

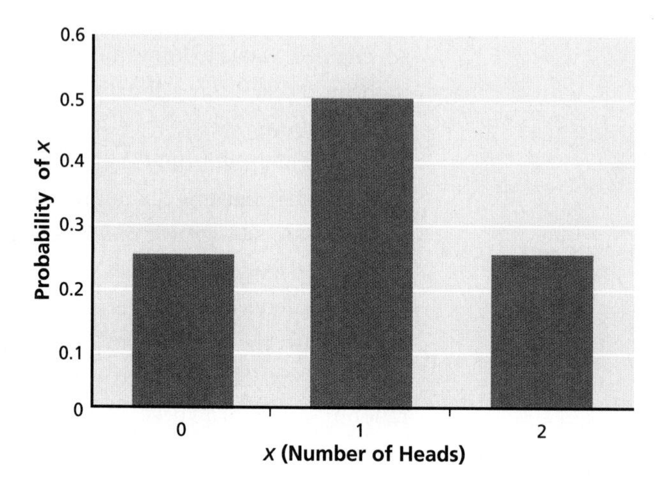

Figure 7.1
Probability Distribution for the Number of Heads in Two Tosses of a Coin

The sum of all probabilities in a probability distribution is 1.00, since all of the possible outcomes from the experiment and their corresponding probabilities are considered. These values are *collectively exhaustive,* meaning no outcome could occur that is not included as one of the listed events. The events listed in the probability distribution are also *mutually exclusive,* meaning that any particular outcome (such as "getting the sequence of tosses: tail then head") can be classified as *only one* of the events listed in the distribution (e.g., as "1 head" as opposed to any of the other listed choices).

Seeing Statistics
Section 6.4.2

The coin-toss example is based on classical probability. The probabilities for each event can be calculated without conducting an experiment. Probability distributions can also be derived from relative frequency probabilities, which are based on observed data and then grouped into frequency distributions. An example of a probability distribution based on relative frequencies is shown in Table 7.3. In Figure 7.2, the same distribution is displayed in a chart format.

Table 7.3 **Families by Size in Canada: 2001 Sample**

x (Number of Persons in the Family)	Frequency	% Frequency	Probability
2	15,206	41.2%	0.412
3	6,990	19.0%	0.190
4	7,583	20.6%	0.206
5	4,602	12.5%	0.125
6	1,798	4.9%	0.049
7+	686	1.9%	0.019
Total	**36,865**	**100.0%**	**1.000**

Source: Statistics Canada. Census families and economic families on unincorporated farms by size of family, 2001. Found at: http://www.statcan.ca/english/freepub/95F0303XIE/tables/html/agpop09.htm. Accessed 10 Aug. 2007.

Figure 7.2
Probability Distribution for the Number of Persons in a Family in Canada, 2001

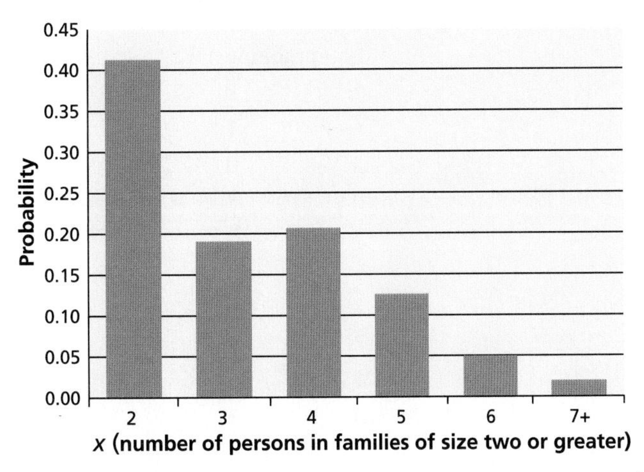

Source: Statistics Canada. Census families and economic families on unincorporated farms by size of family, 2001. Found at: http://www.statcan.ca/english/freepub/95F0303XIE/tables/html/ agpop09.htm. Accessed 10 Aug. 2007.

The information in Table 7.3 is based on the 2001 census of Canada. In reality, the census also includes sampling, since about 20% of households receive a long form to fill out. The table approximates the frequency distribution for the variable "persons in the sampled family" (considering only families of at least two members), taken from the long-form sample. In the right-hand column, the proportions for values of the variable "persons in the sampled family" have been converted into probabilities. For example, since 19.0% of families are composed of three persons, the probability that a randomly selected family will be composed of three persons is 0.190.

The probability distribution in Figure 7.2 is based on empirical probabilities. For discrete probability distributions that are based on classical probabilities, there are a variety of possible types. Table 7.4 (on the next page) and Figure 7.3 illustrate data consistent with three of the types of discrete probability distribution. The **equiprobable distribution** means that each value in the sample space has the same probability of occurring as any other. The other two distributions (binomial and Poisson) will be discussed later in this chapter, in Sections 7.2 and 7.4.

Figure 7.3
Graphical Displays for Three Discrete Probability Distributions

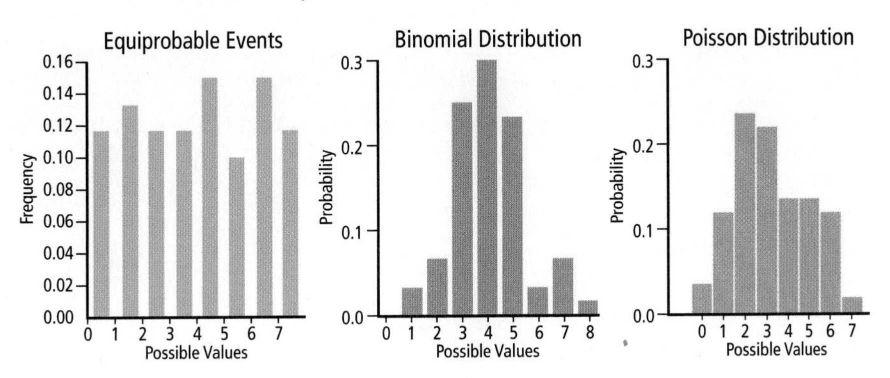

Table 7.4 **Sample Data Generated from Three Distributions**

Equiprobable		Binomial		Poisson	
2	7	5	2	5	4
4	1	2	4	4	5
3	0	4	4	3	2
0	7	3	5	2	6
3	6	5	3	2	2
1	5	4	4	3	3
2	6	3	2	1	1
7	4	5	2	5	2
1	1	3	3	3	6
0	1	8	7	1	6
5	4	5	3	2	4
1	2	4	5	6	3
1	4	5	4	3	5
7	2	5	3	5	1
4	3	3	4	6	3
2	4	3	5	2	6
4	3	3	6	5	4
5	4	6	3	5	4
7	3	3	4	3	5
0	6	1	4	4	4
1	6	7	4	2	3
4	0	3	4	4	0
6	5	4	3	1	1
3	0	4	4	7	2
3	0	7	4	2	6
5	6	1	5	2	2
6	6	4	5	3	2
7	5	4	5	0	3
2	7	7	5	3	3
2	6	5	3	2	1

• *In Brief 7.1(a) Generating a Discrete Probability Distribution by Computer*

 *(Based on the* ***Three_Distributions*** *file)*

You can use the computer to generate a random number sequence that is based on a specific probability distribution. That is, the probabilities determining what numbers are being generated are consistent with the classical probabilities that define the distribution.

Excel

Start with the menu sequence *Tools / Data Analysis / Random Number Generation.* For the equiprobable discrete distribution, the easiest approach is to select *Discrete* in the *Distribution* box (see the screen capture). Input 1 as the number of variables and specify how many random numbers to generate (60 are specified in the screen capture). In the *Parameters* section, supply the address for a table (such as the one shown in the spreadsheet behind the dialogue box in the illustration). The table (which must be prepared first) should display the possible values of the variable on the left (in this case, integers from 0 to 7), along with their corresponding probabilities (in this case, the probabilities are all the same value 1/8 = 0.0125 for each). Specify an output range and then click *OK.* A sample output is shown in column D of the spreadsheet behind the dialogue box in the illustration.

Minitab

Start with the menu sequence *Calc / Random Data.* Specify the desired distribution and then specify details of the desired distribution and the number of data points. For the equiprobable discrete distribution, the easiest approach is to initially select the *Discrete* distribution. In the resulting dialogue box (shown in the screen capture), from the box displaying available columns, select where indicated the columns for the possible values and for their corresponding probabilities. (Prepare these two columns prior to using the program. In this case, the probabilities are all the same value 1/8 = 0.0125 for each). Specify an output column and then click *OK.* A sample output is shown in column C10 of the spreadsheet to the right of the dialogue box in the illustration.

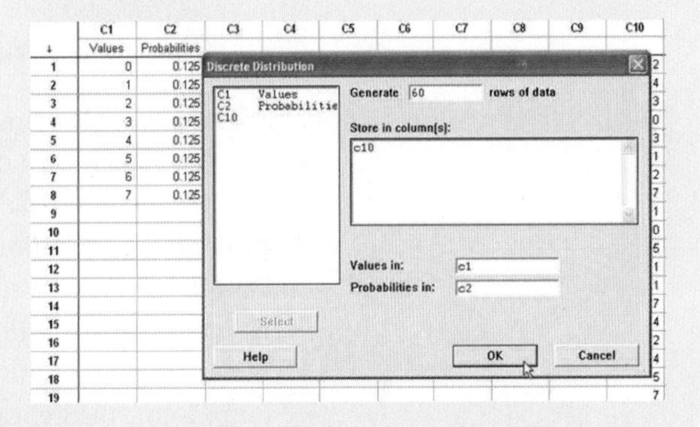

SPSS

Populate a spreadsheet column with 1.00 (or another number) in each cell to set up the desired number of output rows (see the spreadsheet behind the dialogue box in the illustration). Use the menu sequence *Transform / Compute;* then specify a column for outputting the results. Input a random value (RV) formula as shown in the screen capture to generate random numbers from the specified distribution. Because there is no *Discrete* option in this dialogue box, an adjustment has to be made to the formula so that only integers are output, as in Figure 7.3, The RV formula for a uniform distribution is embedded in the formula **Trunc( )**, and the limits for the uniform distribution are shown as 0 to 7.9999999. The result will be randomly generated numbers anywhere in that range that are truncated (i.e., reduced to integers from 0 to 7). A sample output column is shown in the spreadsheet behind the dialogue box in the illustration.

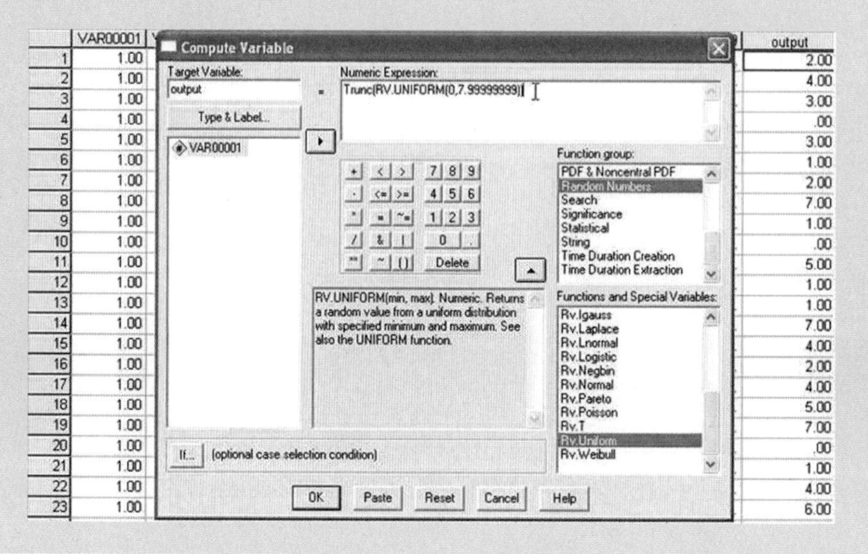

• Hands On 7.1(a) *Theoretical and Actual Distributions*

We have described how some common discrete probability distributions can be simulated easily on a computer. Note that due to chance, the *actual* values that are generated by a particular software program may not line up exactly with the theoretical distribution pattern. For example, although the first graph in Figure 7.3 was based on an equiprobable distribution, the actual frequencies for different values were a bit *un*equal. In this exercise, you will get a sense of how much variation of this sort there can be when you are comparing the computer outputs to a theoretical model.

Using Excel, Minitab, or SPSS, create an output similar to Table 7.4 for the equiprobable distribution. (Refer back to "In Brief 7.1(a)" for more specific instructions.) Once you have created the data, you can use the techniques described in Chapter 3 to create relative frequency distributions and bar charts for the data, to be interpreted as the probability distributions.

Try regenerating the same set of equiprobable random numbers several times and in each case, regenerate the tables and charts. Notice how the probability distributions generated each time are similar to those shown in Figure 7.3, but the exact patterns vary from experiment to experiment.

If you are feeling adventurous, you can try this also for the binomial and Poisson distributions, which are depicted in Figure 7.3 as well. The inputs and selections required by the dialogue boxes will become clearer in the next sections. And, although the outputs will not be integers, the general idea can still be illustrated: A single model of classical probability distribution will produce a certain amount of variation when it runs in practice.

As illustrated in "Hands On 7.1(a)," a discrete probability distribution is, in effect, the relative frequency distribution for a discrete random variable. By calculating summary measures such as the *mean* and *standard deviation,* some important characteristics of the probability distribution can be represented numerically.

Mean or Expected Value of a Discrete Probability Distribution

The mean of a probability distribution is also referred to as the **expected value** of the distribution. Consider the example of tossing one coin twice. How many heads can we expect to throw? Put another way: Suppose we repeat hundreds of times an experiment of tossing one coin twice. What would be the mean number of heads thrown for all those two-toss experiments? In this context, the terms *expected value* of the distribution and its *mean* are identical.

The notation for the expected value of a random variable X is $E(X)$, and the mean of the random variable x is written as μ. Therefore,

$$E(X) = \mu$$

The calculation for the mean of a discrete probability distribution is a special case of the calculation for weighted mean: Here, each possible value x of the random value is weighted by its probability of occurring $P(x)$. Multiply each x times its probability, and add all the products. The sum of the weights (i.e., of the probabilities) is 1.

Formula 7.1 Expected Value of a Discrete Probability Distribution

$$E(X) = \sum [xP(x)]$$

where $E(X)$ = Expected value of the random variable
x = A value of the random variable x
$P(x)$ = Probability of occurrence of the value x

The formula has the effect that, if there is an outlier, it will not necessarily skew the mean by much—provided the probability of the outlier occurring is very small. As the probability of a value x increases, its potential impact on the mean increases.

Example 7.1

Refer to Table 7.2, which displays all possible values for x (numbers of heads) and corresponding probabilities, $P(x)$, if a fair coin is tossed twice. Find the mean or expected value for the number of heads that would be tossed.

Solution We can apply Formula 7.1:

$$\begin{aligned} E(X) &= \sum[xP(x)] \\ &= (0 \times 0.25) + (1 \times 0.50) + (2 \times 0.25) \\ &= 0 + 0.5 + 0.5 \\ &= 1.0 \end{aligned}$$

The mean or expected value of the number of heads in two tosses is 1.

Knowledge of expected values can be very useful for planning. For example, if a manufacturer knows the expected value for numbers of work incidents requiring first aid in a month, the manufacturer can estimate the need for first-aid supplies and for the time lost completing incident reports. If a farmer knows the expected value of crop yield per hectare planted, the farmer can estimate the revenue potential from planting an entire field.

Expected values are less useful for one-shot experiments. The expected value of buying a lottery ticket is almost always negative (over a lifetime of playing, you will probably have lost money). If you buy a ticket anyway, you are accepting the long-term risk. Your focus is on great things that *might* happen—*this one time.*

Example 7.2

Based on the U.S. 2000 census, the probability distribution for numbers of households in different size categories (counting related people only) is shown in Table 7.5. For households in the United States during that census year, what is the mean value for the household size?

Table 7.5 Household Sizes in the United States, 2000

Household Size (Related Persons)	Probability
1	0.255
2	0.331
3	0.164
4	0.146
5	0.067
6	0.023
11*	0.014
Total	**1.000**

* The reported size category is "7 or more," but the average size of households in that category is 11.

Source: Information Please® Database, © 2005 Pearson Education, Inc. All rights reserved. Found at: http://www.infoplease.com/ipa/A0884238.html

Solution The question can be restated in probabilities: Suppose we repeat many times the experiment of randomly selecting a single family from the U.S. population in 2000 and noting its size. For all of the family sizes selected, what is the expected value of the sizes? Applying Formula 7.1, we obtain

$$\begin{aligned} E(X) &= \sum[xP(x)] \\ &= (1 \times 0.255) + (2 \times 0.331) + (3 \times 0.164) + (4 \times 0.146) \\ &\quad + (5 \times 0.067) + (6 \times 0.023) + (11 \times 0.014) \\ &= 2.62 \text{ persons per household} \end{aligned}$$

Statistical software packages can provide varying degrees of assistance in solving problems based on expected value.

• *In Brief 7.1(b)* *Calculating Expected Value with the Computer*

Excel

Input the x and $P(x)$ data in a column format (see the illustration). At the top of a third column, input an Excel formula to multiply that row's values in the x and $P(x)$ columns. Copy the formula down the remainder of the third column. Then use an Excel formula to find the sum of the [x times $P(x)$] values in this new column.

	A	B	C	D
1	H_Size	Prob		
2	1	0.255	=A2*B2	=SUM(C2:C8)
3	2	0.331	=A3*B3	
4	3	0.164	=A4*B4	
5	4	0.146	=A5*B5	
6	5	0.067	=A6*B6	
7	6	0.023	=A7*B7	
8	11	0.014	=A8*B8	

Minitab

Input the x and $P(x)$ data in the first two columns of the spreadsheet, as shown in the illustration. Use the menu sequence *Calc / Calculator* and input the formula **C1*C2** in the *Expression* section of the resulting dialogue box. Specify that the result is to be stored in C3. (The results are illustrated behind the screen capture in the illustration.) Next use the menu sequence *Calc / ColumnStatistics.* Select *Sum* and enter "C3" as the input variable. Click *OK.* The final output is shown at the lower right of the illustration.

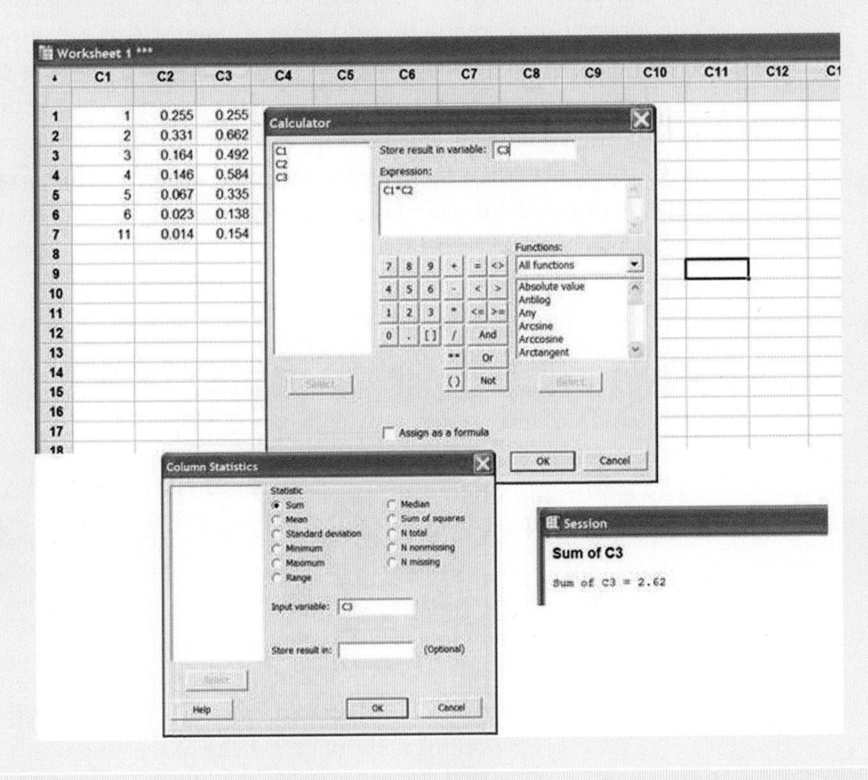

SPSS

In SPSS, the procedure we used for finding weighted averages (see Section 4.3) can be used here to find the expected value. Input the x and $P(x)$ data in columns but, for the probabilities, adjust the probability decimal points so that all values are whole numbers (see the illustration). (For Example 7.2, we multiplied each $P(x)$ by 1,000.) To interpret the probability data as weights, use the menu sequence *Data / Weight Cases.* Then select a technique for generating the mean (see Chapter 4). The displayed result will be the probability-weighted mean.

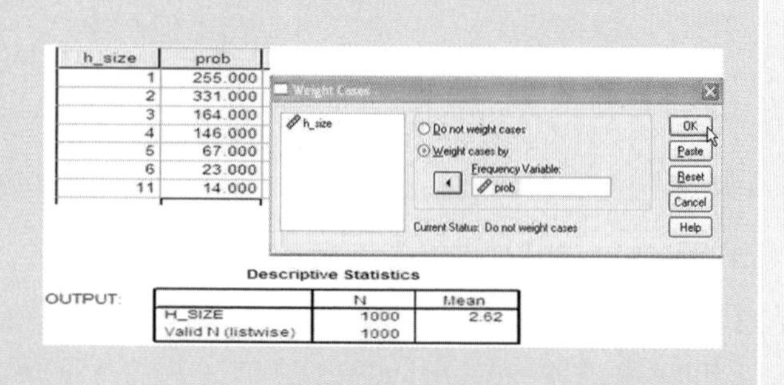

Variance and Standard Deviation of a Discrete Probability Distribution

In Chapter 5, we examined the concept of dispersion, or variation, in the data and introduced two commonly used measures for dispersion: *variance* and *standard deviation.* Both measures supplement the information provided by the mean. The latter tells where the data are centred, but does not indicate the degree of dispersion of actual values around that centre point.

The mean for a probability distribution is subject to the same limitation: It indicates only the data's centre. To get a sense of how the data are spread, the variance and standard deviation of the distribution can be calculated.

The calculations for variance of a discrete probability distribution are a special case of calculations for the weighted variance. Each squared difference between an x-value and the distribution mean μ is weighted by the probability $P(x)$ of that x-value occurring. Multiply each squared difference times its probability, and add all of the products. Similar to the calculation for expected value, the sum of the weights (i.e., of the probabilities) is 1.

Formula 7.2 Variance of a Discrete Probability Distribution

$$\sigma^2 = \sum [(x - \mu)^2 \times P(x)]$$

Formula 7.3 Standard Deviation of a Discrete Probability Distribution

$$\sigma = \sqrt{\sigma^2}$$

where σ^2 = Variance of the distribution

σ = Standard deviation of the distribution

x = A possible value of the random variable

μ = Mean, or expected value, of the random variable

$P(x)$ = Probability of occurrence of the value x

Example 7.3

In Example 7.1, we found the expected value for the probability distribution based on numbers of heads thrown in two tosses of a coin. Calculate the variance and standard deviation for that distribution.

Solution

Using Formula 7.2:

$$\begin{aligned}\sigma^2 &= \sum[(x - \mu)^2 \times P(x)] \\ &= [(0 - 1)^2 \times (0.25)] + [(1 - 1)^2 \times (0.50)] + [(2 - 1)^2 \times (0.25)] \\ &= [1 \times (0.25)] + [0 \times (0.50)] + [1 \times (0.25)] \\ &= 0.5\end{aligned}$$

Using Formula 7.3:

$$\begin{aligned}\sigma &= \sqrt{\sigma^2} \\ &= \sqrt{0.5} \\ &= 0.71\end{aligned}$$

The variance of the probability distribution is 0.5, and the standard deviation is 0.71. Since the standard deviation is almost as large as the mean (which equals 1), there is proportionally a great deal of variation in this probability distribution.

Example 7.4

In Example 7.2, we found that the mean, or expected value, of household size in the United States was 2.62 persons per household. Using the data given in Table 7.5, calculate the variance and standard deviation for the variable: persons per household.

Solution

As when calculating the mean, statistical packages can offer some assistance in performing the calculation. (The final answers are displayed in the Minitab screen capture in "In Brief 7.1(c).") The computer-based procedures are straightforwardly based on automating the steps in Formulas 7.2 and 7.3.

• *In Brief 7.1(c) Calculating Variance and Standard Deviation of a Probability Distribution with the Computer*

Excel

Open the worksheet that you created for calculating the mean. At the top of a new column, use Excel formulas to perform this calculation on the x and $P(x)$ values in the top row: $(x - \mu)^2 \times P(x)$. (In the illustration, note the use of an absolute address for the mean, which was calculated previously.) Copy

	A	B	C	D	E
1			Calculations for Mean	Calculations for the Variance	
2	H_Size	Prob	x(P(x))	(x-mean)^2	[(x-mean)^2](P(x))
3	1	0.255	=A3*B3	=(A3-$C$11)^2	=D3*B3
4	2	0.331	=A4*B4	=(A4-$C$11)^2	=D4*B4
5	3	0.164	=A5*B5	=(A5-$C$11)^2	=D5*B5
6	4	0.146	=A6*B6	=(A6-$C$11)^2	=D6*B6
7	5	0.067	=A7*B7	=(A7-$C$11)^2	=D7*B7
8	6	0.023	=A8*B8	=(A8-$C$11)^2	=D8*B8
9	11	0.014	=A9*B9	=(A9-$C$11)^2	=D9*B9
10					
11		mean:	=SUM(C3:C9)	variance:	=SUM(E3:E9)
12				stardard deviation:	=SQRT(E11)

this formula down the remainder of the new column. Then use Excel formulas to (1) find the sum of the added column (which will be the variance) and (2) take the square root of the sum (which will be the standard deviation).

Minitab

Open the worksheet that was created for calculating the mean. To create a column containing the equivalent of $(x - \mu)^2$ for each x-value in column C1 (assuming the mean is stored in C4), use the menu sequence *Calc / Calculator;* input the formula **(C1–C4)**2** in the *Expression* box, and signify that the result is to be stored in column C5. To weight the squared differences times their corresponding probabilities, rerun *Calc / Calculator* but this time, input **C5*C2** in the *Expression* box and store the result in C6. Find the sum of these products (which is the variance) by using the sequence *Calc / Column Statistics;* then select *Sum* and specify the input column as C6. The output, which is the variance, is shown in the illustration. Take the square root of this value to find the standard deviation.

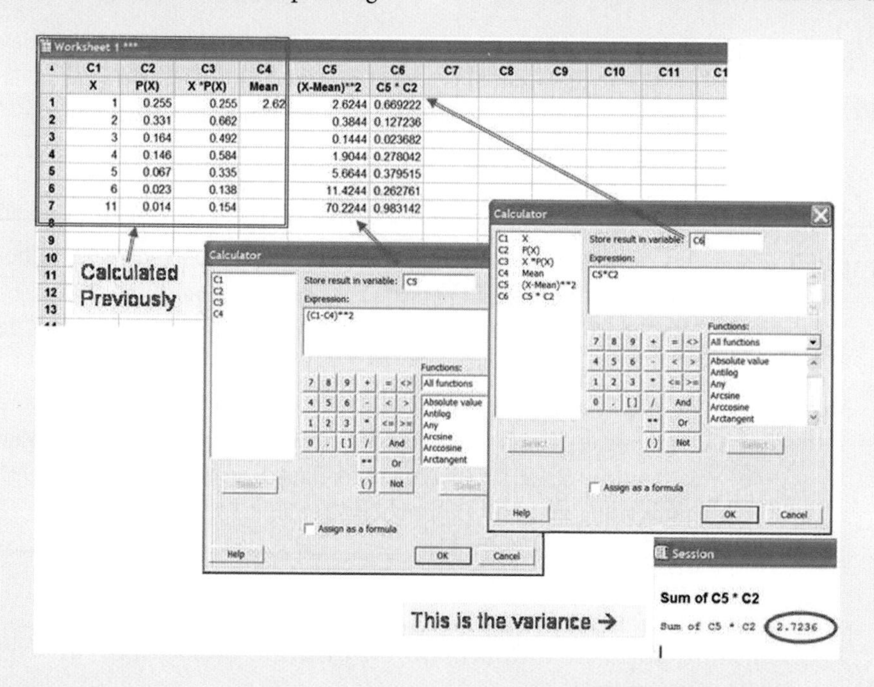

SPSS

If the worksheet that was created for calculating the expected value is still open, then assigning the probability values as weights is probably still in effect. (If not, rerun the sequence for *Data / Weight Cases,* as detailed in "In Brief 7.1(b).") Then select an SPSS technique (see Chapter 5, Section 5.1) for generating the variance and standard deviation. These values will automatically be weighted by the probabilities.

Descriptive Statistics

	N	Mean	Std. Deviation	Variance
H_SIZE	1000	2.62	1.651	2.726
Valid N (listwise)	1000			

• Hands On 7.1(b) Expected Values and the Mean

(Based on the ***Powerball*** *file)*

The key element of the expected value is that, ultimately, it is a mean, so the expected value is subject to all of the strengths and weaknesses of the mean. The calculation of the expected value considers all of the data—including extreme values, such as U.S. household sizes of 11 (see Table 7.5) or lottery returns of $42 million (see Table 7.6). In any case, it represents a *central value* of anticipated outcomes over a great many new experiments or observations. How accurate is this central value as a predictor for the very next observation? Or even as a predictor for the mean of the next 5, 500, or 1,500 observations?

Table 7.6 **Probability Distribution for Prize Values of the Powerball Lottery, 2004**

Cash Value*	x Return Value**	Published "Odds"	$P(x)$ Probability	Calculations for the Mean $xP(x)$	Calculations for the Standard Deviation $(x - mean)^2$	$(x - mean)^2P(x)$
$42,541,557	$42,541,556	1 in 120,526,770.00	0.0000000083			
$100,000	$99,999	1 in 2,939,677.32	0.0000003402			
$5,000	$4,999	1 in 502,194.88	0.0000019913			
$100	$99	1 in 12,248.66	0.0000816416			
$100	$99	1 in 10,685.00	0.0000935891			
$7	$6	1 in 260.61	0.0038371513			
$7	$6	1 in 696.85	0.0014350291			
$4	$3	1 in 123.88	0.0080723281			
$3	$2	1 in 70.39	0.0142065634			
$0	−$1	1 in 1.03	0.9722713577	Mean =	Variance =	
					Standard Deviation =	

* The actual cash value for the grand prize varies, depending in part on whether there was more than one winning ticket. The value listed for the grand prize in the table was the mean value per winning ticket for the 20 winning tickets prior to June 2004

** Return value = Cash value of prize (if any) minus cost of ticket ($1)

NOTE: Apparently, repeated prize values (e.g., $100) are awarded by different criteria.

Source: Multi-State Lottery Association MUSL. Based on data in Probability Distribution for Prize Values of the Powerball Lottery. Found at: http://www.powerball.com/powerball/pb_prizes.asp

In this exercise, you can see for yourself. Table 7.6 and Figure 7.4 are based on Excel templates for the simulation, but Minitab and SPSS versions are also available (see the file named **Powerball**). The context is the U.S. multi-state lottery game "Powerball."

Winning in the Powerball lottery is based on matching user-selected numbers with randomly selected numbered balls—five white balls and one red Powerball (http://www.powerball.com/powerball/pb_howtoplay.asp). Table 7.6 shows the probabilities in 2004 for winning the available prizes, although the exact values may now differ. A player's actual return is the cash value less the

Figure 7.4
Simulation of Multiple Plays of the Powerball Lottery

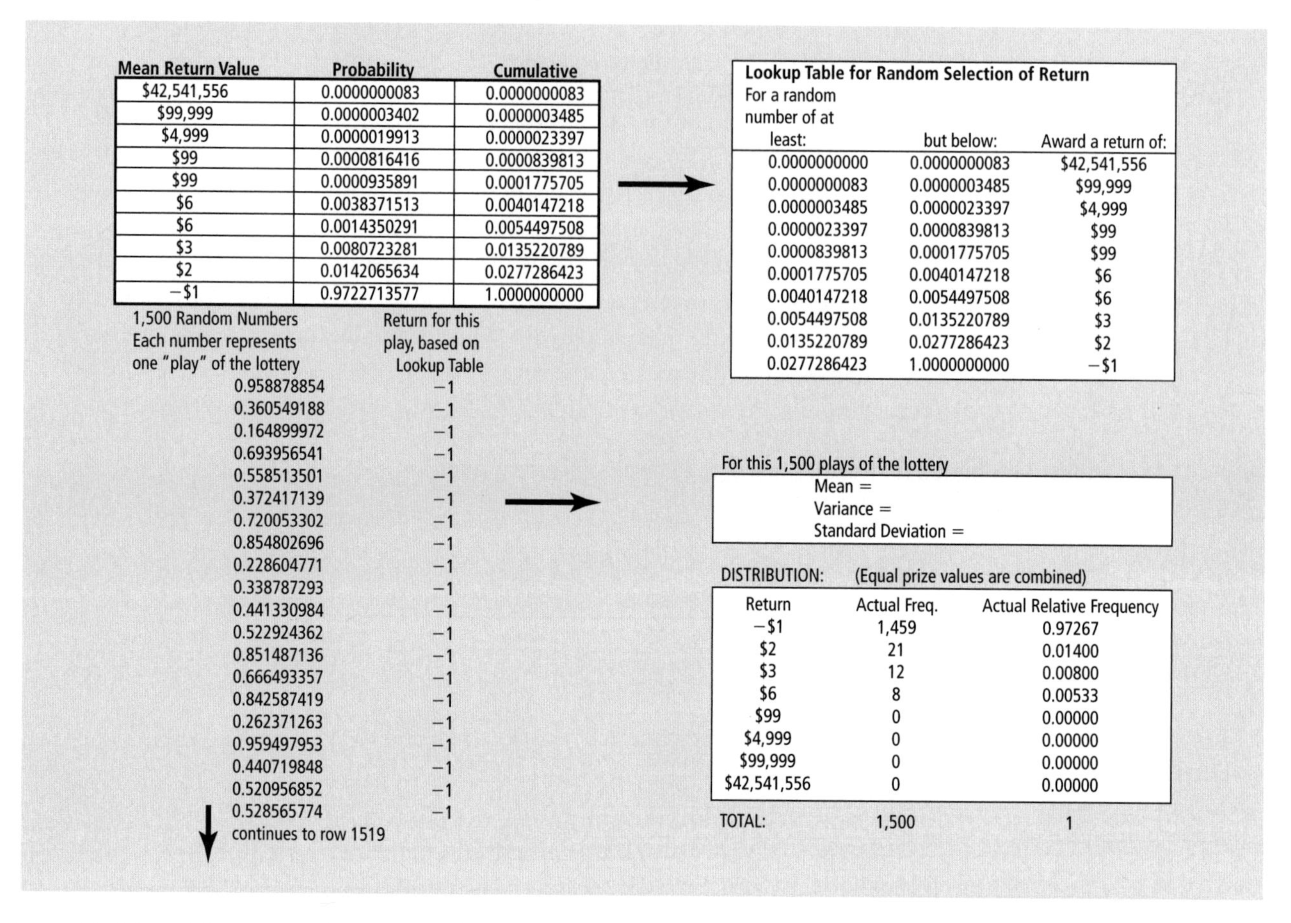

Mean Return Value	Probability	Cumulative
$42,541,556	0.0000000083	0.0000000083
$99,999	0.0000003402	0.0000003485
$4,999	0.0000019913	0.0000023397
$99	0.0000816416	0.0000839813
$99	0.0000935891	0.0001775705
$6	0.0038371513	0.0040147218
$6	0.0014350291	0.0054497508
$3	0.0080723281	0.0135220789
$2	0.0142065634	0.0277286423
−$1	0.9722713577	1.0000000000

Lookup Table for Random Selection of Return

For a random number of at least:	but below:	Award a return of:
0.0000000000	0.0000000083	$42,541,556
0.0000000083	0.0000003485	$99,999
0.0000003485	0.0000023397	$4,999
0.0000023397	0.0000839813	$99
0.0000839813	0.0001775705	$99
0.0001775705	0.0040147218	$6
0.0040147218	0.0054497508	$6
0.0054497508	0.0135220789	$3
0.0135220789	0.0277286423	$2
0.0277286423	1.0000000000	−$1

1,500 Random Numbers Each number represents one "play" of the lottery	Return for this play, based on Lookup Table
0.958878854	−1
0.360549188	−1
0.164899972	−1
0.693956541	−1
0.558513501	−1
0.372417139	−1
0.720053302	−1
0.854802696	−1
0.228604771	−1
0.338787293	−1
0.441330984	−1
0.522924362	−1
0.851487136	−1
0.666493357	−1
0.842587419	−1
0.262371263	−1
0.959497953	−1
0.440719848	−1
0.520956852	−1
0.528565774	−1

continues to row 1519

For this 1,500 plays of the lottery
Mean =
Variance =
Standard Deviation =

DISTRIBUTION: (Equal prize values are combined)

Return	Actual Freq.	Actual Relative Frequency
−$1	1,459	0.97267
$2	21	0.01400
$3	12	0.00800
$6	8	0.00533
$99	0	0.00000
$4,999	0	0.00000
$99,999	0	0.00000
$42,541,556	0	0.00000
TOTAL:	1,500	1

Source: Multi-State Lottery Association. Based on data in *Powerball: Prizes and Odds* (June 2004). Retrieved from http://www.powerball.com/powerball/pb_prizes.asp

cost of the ticket ($1). The highest return value can vary, based on the number of tickets purchased for the draw and whether multiple winners share the available jackpot. This simulation uses, for simplicity, a single average-based jackpot value.

1. Complete the blank columns in Table 7.6. Then, where indicated, calculate the expected value and standard deviation of the probability distribution for winnings. The CD **Powerball** file contains all of the necessary data.
2. Use your software to simulate 1,500 successive replays of the lottery. Figure 7.4 shows the principles of the simulation, which have been automated as much as possible in the **Powerball** files. (Figure 7.4 shows the Excel approach. The exact procedures for running this simulation differ in Excel, Minitab, and SPSS; see "In Brief 7.1(d).") Based on the probability distribution function (in the upper left of the figure), a lookup table for generating simulated outcomes can be created (upper right of the figure). For each

replay, a random number from 0 to 1 is generated, and the simulated outcome is based on the range in which the random number falls (e.g., from 0.0000000000 to 0.0000000083); the probabilities of falling in each range correspond to probabilities in the probability distribution. The lower left of Figure 7.4 shows some of the 1,500 random numbers that have been generated, and the corresponding simulated lottery returns for each of those "replays." The actual distribution of those returns is displayed at the lower right.

After you run the simulations, what is the outcome of the first replay? Did you win? What is the mean of the first five replays? Of the total 1,500 replays? Do any of these means reach the "expected value" predicted theoretically? Generate the absolute and relative frequency distributions for the 1,500 plays. Generate another 1,500 plays, then another, and then another; and repeat the above steps each time. Each time, use the methods from previous chapters to calculate the mean and standard deviation of the updated column of returns and to generate a new frequency distribution table. How often do your frequency distributions actually reach returns of $4,999 or more? Describe and explain any differences between the theoretical probability distributions and the actual relative frequency distributions from your replays.

• *In Brief 7.1(d) Simulating Replays of the Powerball Lottery*

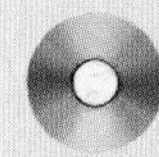

*(Based on the **Powerball** file)*

These instructions are specifically for running the "Hands On 7.1(b)" exercise.

Excel

Open the file **Powerball.xls**; click on tab at the bottom to select the worksheet "Hands-On 7.1b Template," which looks similar to Figure 7.4. Press the F9 key. Each repetition of this keystroke regenerates the column of 1,500 random numbers on the lower left. Each random number is converted, in the next column to the right, into the simulated payoff for *one* replay. On the lower right, spaces are provided for displaying the mean, variance, standard deviation, and frequency distribution of one simulated set of 1,500 replays. (Hint: For finding the frequency distribution in Excel, use the numbers in the range E27:E34 as the "bins.") Press the F9 key again to start over.

Minitab

Open the Minitab project named **Powerball**; within that project, open the worksheet named **Powerball_2**. The worksheet includes a column (C6) for the 1,500 random numbers and another column (C7) for the simulated returns. Use the menu sequence *Calc / Random Data / Uniform* to regenerate the random numbers. Fill column C7 with recodes of the C6 values, to represent the payoffs of each random replay; use the sequence *Manip / Code / Numeric to Numeric* (see the resulting dialogue box in the screen capture). Specify to code the data from column C6 into a new column "Simulated Returns." As shown in the figure, list the ranges of original values and their corresponding coded values, i.e., the values of the prizes, based on the lookup table. Click OK.

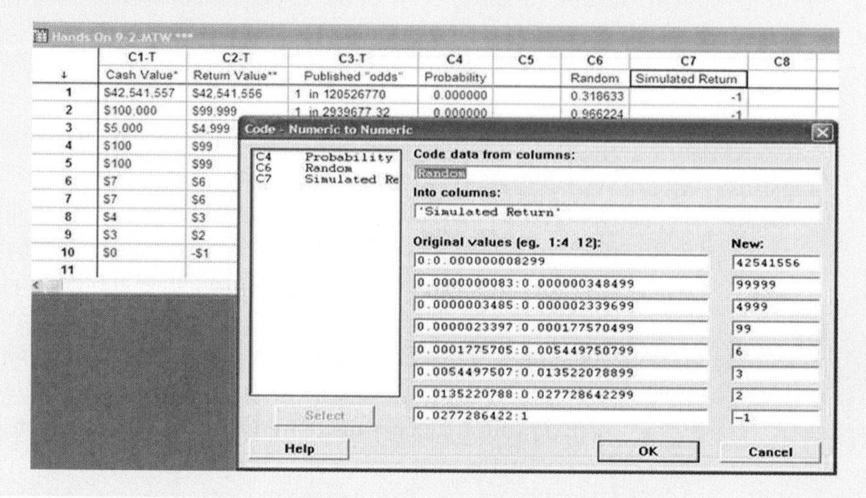

SPSS

Open the data file for **Powerball.** The worksheet includes a column for the 1,500 random numbers ("random") and a column for the simulated returns ("sim_rtrn"). To regenerate the random numbers, use the menu sequence *Transform / Compute;* then input *random* as the target variable and **rv.uniform(0,1)** as the numeric expression for the function. To re-code these random values into simulated returns, use the menu sequence *Transform / Recode / Into Different Variables.* The ranges illustrated in the lookup table (see the upper right of Figure 7.4) must be input into the SPSS dialogue box in the manner shown in the

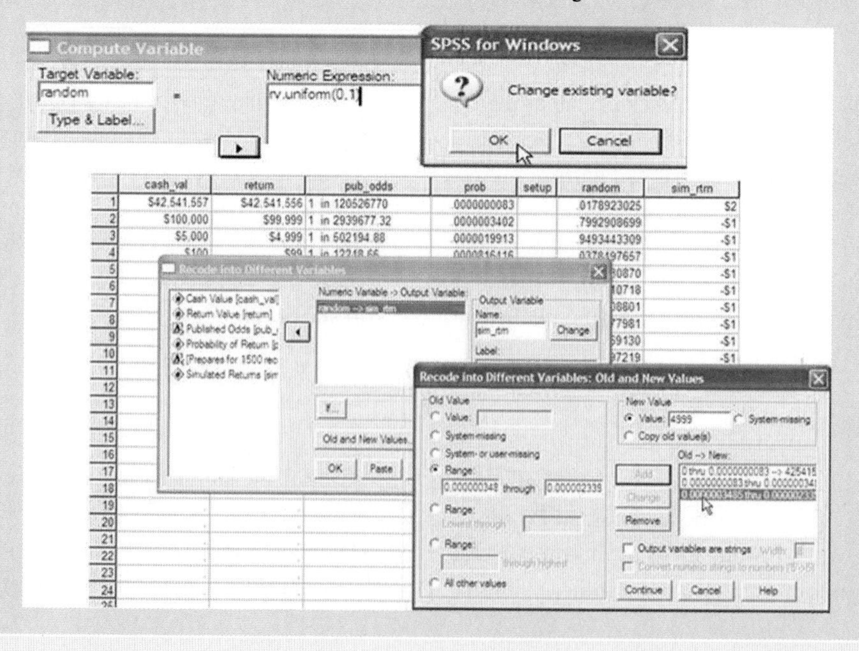

screen capture. (Note: The upper bound "*below* 0.0000000083" must be coded explicitly as "0.00000000829"—and so on for all of the upper bounds.) SPSS will remember these conversion codes until you change them *or close the program,* so generating and re-coding the second and third sets of simulated replays and so on will be easier if you leave SPSS running until all are completed.

7.2 The Binomial Distribution

The **binomial distribution** (more formally, the **binomial probability distribution**) is a special case of discrete probability distribution. The probabilities are classical probabilities, so for a given problem, we can identify all possible outcomes in advance and list the theoretical probabilities for each outcome. Many situations in business, social science, and other applications can be modelled by this particular distribution. It also provides a foundation for a number of inferential statistics procedures that will be used in upcoming chapters.

The word *binomial* is derived from *bi* for "two" and *nomial* for "numbers," suggesting a reference to trials (such as coin flips) that have only two possible values or numbers that could result. When you toss a coin, either a head turns up or a tail turns up. Traditionally, one possible outcome of a binomial experiment is denoted a *success* and the other a *failure.* These are used as technical terms: A "success" denotes the occurrence of an outcome in which we are interested, such as tossing a head, and a "failure" denotes obtaining the opposite outcome. The terminology can sometimes lead to odd descriptions, as in a quality control experiment, for example. If the event of interest is determining whether a product is defective, then a "success" may be obtaining a defective product and a "failure" obtaining a nondefective product.

A binomial experiment can also be called a **trial** (sometimes called a **Bernoulli trial**), since in this context, a trial is a two-possible-outcome random experiment of the sort described above. These trials can be combined into a sequence of identical trials. For instance, you might toss a coin twice. Each individual flip is one trial and can have only one of two possible outcomes (heads or tails). However, each unique *sequence* of outcomes for *all of the trials combined* may define *one* possible outcome for the overall experiment. In the coin-flip case, there are four possible outcomes, overall: (1) T $\rightarrow$ T, (2) T $\rightarrow$ H, (3) H $\rightarrow$ T, and (4) H $\rightarrow$ H. If the *order* of trial outcomes is immaterial, and each "head" result is counted as a success, then the possible results of the above experiment are (1) 0 heads; (2) 1 head; or (3) 2 heads.

A **binomial experiment** is an experiment, such as the two coin flips, that consists of a sequence of identical and independent Bernoulli trials. A variable that counts the number of successes (regardless of exact sequence) among the trial results in the binomial experiment is called a **binomial random variable.**

We now have the elements for a more formal definition. A binomial probability distribution is a discrete probability distribution for all of the possible values of a binomial random variable. The possible values of the variable are the numbers of successes that result from trials in a binomial experiment. These conditions must always be met:

1. **The experiment consists of n identical trials.** Examples: 15 tosses of a coin, or 50 microbes sampled from a pond. The outcome of one experiment is interpreted as a count of successes for all of the trials that comprise the experiment. For example, "eight (of the 15) tosses were heads"; or "four (of the 50) microbes were mutated."
2. **There are only two, mutually exclusive, results possible for each trial.** For example, head vs. tail in each coin toss, or mutated vs. nonmutated for each microbe.
3. **All of the trials are independent.** The outcome of one trial is not affected by the outcome of any other trial or trials.
4. **The probabilities of success are constant for each trial.** For example, the probability of getting a tail is the same each time you toss the coin.

In applied situations, the researcher is responsible for ensuring that the preceding conditions are actually met. For example, if the binomial distribution is being applied to test for defective parts, the probability of getting a defective part must remain the same throughout the entire experiment. If there is evidence that the rate of defective parts is changing during the course of sampling, then the binomial distribution cannot be applied.

Similarly, to meet the condition that each trial is *independent,* the probability of getting one defective part cannot depend on the results of previous checks for defective parts. An example of this condition being violated is if parts tend to be damaged in pairs, so that if one part is damaged, the neighbouring part has an increased probability of also being damaged.

Calculating the Probabilities

Calculating probabilities for a binomial random variable follows the same bottom-line principle as any other calculation of a classical a priori probability. For an event A:

$$P(A) = \frac{\text{The number of ways } A \textit{ can } \text{occur}}{\text{The number of different } \textit{outcomes possible} \text{ in the sample space}}$$

The challenge is to determine what numbers are required for the numerator and denominator.

Before giving an example, we present some standard notation. In calculating probabilities for a binomial random variable:

X	Name of a binomial variable for the number of successes
x	Specific number of successes observed in a binomial experiment (Example: "4 mutated microbes")
n	Fixed number of trials (Example: "50 sampled microbes")
p	Probability of success in each trial (Example: "0.05")
$(1 - p)$	Probability of failure in each trial (Example: $1 - 0.05 = 0.95$)
$P(X = x)$ or $P(x)$	Probability of getting exactly x successes among n trials

In the simplest cases, no new special formula is required.

Example 7.5 If a fair coin is tossed two times, what is the probability of tossing exactly one tail? What is the probability of tossing exactly two tails?

Solution The experiment consists of two trials (coin tosses), with two possible results for each trial (head or tail). A table of possible outcomes for this experiment was provided in Table 7.1. Two of the displayed outcomes can be interpreted as the event "toss exactly 1 tail"; namely, the outcomes "tail, then head" and "head, then tail." The total number of outcomes possible is four (based on 2 tosses × 2 possible outcomes per toss). Since all of the listed outcomes are equally likely, $P(X = 1) = 2/4 = 0.5$.

Of the four possible outcomes, only one outcome ("tail, then tail") counts as the event "exactly two tails." Therefore, $P(X = 2) = 1/4 = 0.25$.

Example 7.6 Construct the complete probability distribution for the number of tails thrown, if a fair coin is tossed two times.

Solution The probability distribution displays the probability for each possible outcome. We have already calculated $P(X = 1)$ and $P(X = 2)$. Only one other outcome is possible: ("head, then head"). In this last case, the number of tails is 0. Therefore, $P(X = 0) = 1/4 = 0.25$. The probability distribution can be summarized as

For $n = 2$; $p = 0.5$

x	$P(x)$
0	$1/4 = 0.25$
1	$2/4 = 0.50$
2	$1/4 = 0.25$

As problems become more complex (for example with a large number n of trials, or small values for the probability of success for each trial), calculating binomial probabilities can become more complicated. Directly calculating a numerator and denominator for the basic probability formula becomes difficult. Instead, you can always use the following general formula to find $P(x)$ for any given x (number of successes), n (number of trials), and p (probability of success for each trial):

Formula 7.4 Binomial Probability of an Outcome

$$P(x) = {}_nC_x p^x (1 - p)^{(n-x)}$$

or

$$P(x) = \frac{n!}{x!\,(n - x)!} p^x (1 - p)^{(n-x)}$$

where $0 \leq X \leq n$
$0 < p < 1$

The left-hand factor in the formula, ${}_nC_x$ (pronounced "n choose x"), is the formula for calculating combinations. It is equivalent to $\frac{n!}{x!\,(n - x)!}$, which is shown in the expanded second version of the formula. There are ${}_nC_x$ ways that x successes can occur among

n trials. Each one of these distinct combinations has a probability of occurring equal to $p^x(1-p)^{(n-x)}$. In short, the probability of obtaining x successes equals (the number of combinations that include exactly x successes) times (the probability of getting *any one* of these combinations having x successes).

The factorial symbol "!" represents an operator such that $k! = 1$, if $k = 1$, and for all other positive integers, $k! = k \times (k-1) \times (k-2) \times \ldots \times 1$. That is, for $k > 1$, $k!$ equals the product of k with all other positive integers smaller than itself.

Example 7.7

Microbes in a certain colony exposed to radiation have a 1/6 probability of being mutated.

1. If three microbes from the colony are sampled, what is the probability that exactly two of the microbes will be mutated?
2. What is the probability that none of the microbes will be mutated? Solve each problem manually *and* also solve question 1 with the computer.

Solution

The sampling of each microbe is a trial, so $n = 3$. The probability of "success," p, for each trial is given as $1/6 = 0.1667$.

1. First, calculate the probability that the number of successes x equals 2, using the formula:

$$P(2) = \frac{3!}{2!(3-2)!} 0.1667^2(1 - 0.1667)^{(3-2)}$$
$$= 0.069$$

The same solution appears in the computer displays for "In Brief 7.2(a)."

2. Next, use the formula for where the number of successes x equals 0:

$$P(0) = \frac{3!}{0!(3-0)!} 0.1667^0(1 - 0.1667)^{(3-0)}$$
$$= 0.579$$

Section 6.5

(Note that, by definition: $0! = 1$; also, for any number k, $k^0 = 1$.)

• *In Brief 7.2(a) Computing Values for Binomial Probabilities*

Excel

To calculate $P(x)$, if x, p, and n are known: Choose any cell in the worksheet and input a formula in the format **=BINOMDIST(*x*,*n*,*p*,0)** (see the illustration). The required fourth input, zero (**0**), indicates that the desired probability is for exactly $P(x)$; i.e., it is *not* for a cumulative probability.

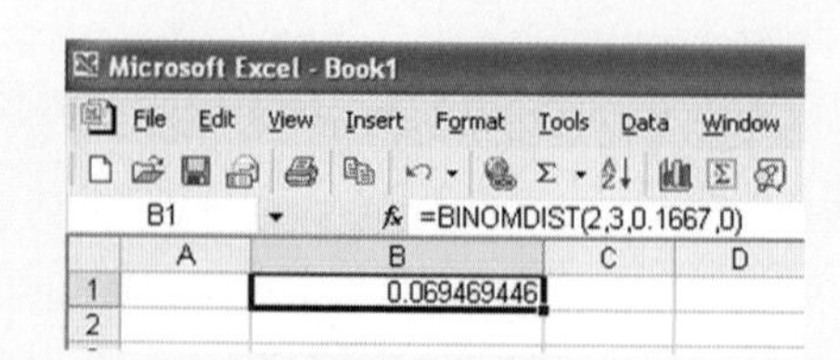

Minitab

To calculate $P(x)$, if x, p, and n are known: Input the value for x into a worksheet cell (usually, at the top of an empty column). Use the sequence *Calc / Probability Distributions / Binomial* to bring up the calculation screen and input the values for n and p where indicated, as shown in the screen capture. Select *Probability* in the dialogue box and then specify the column that contains x and the column to hold the solution. (Alternatively, you can input the x-value in the *Input constant* area; in that case, the output will not be added to the worksheet page.)

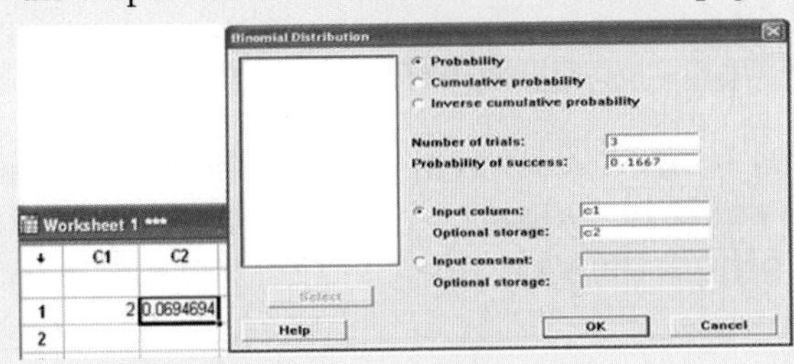

SPSS

To calculate $P(x)$, if x, p, and n are known: Input the value for x into a worksheet cell (usually, at the top of an empty column). Use the menu sequence *Transform / Compute* to call up the "Compute Variable" dialogue box. In the *Target Variable* window, input the name of the column ("Target Variable") to hold the solution. In the *Numeric Expression* window, input a formula of the form **pdf.binom(VAR00001,*n*,*p*)**, replacing **VAR00001** with the name of the column holding x. (If needed, adjust the displayed number of decimal places for the solution variable.) Click *OK*.

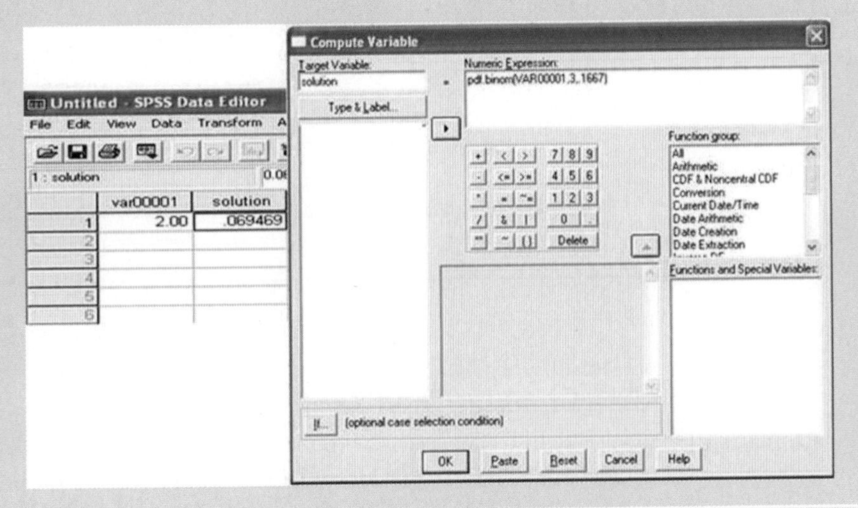

Example 7.8 In Example 7.7, the probabilities for obtaining exactly two or exactly zero successes were calculated, given three trials, and $p = 0.1667$. Construct the *full* binomial probability distribution for the possible numbers of successes.

Solution If three microbes are selected, the number mutated could be one, two, three, or none at all (i.e., zero). In Example 7.7, you calculated the probabilities for $P(0)$ and $P(2)$. To construct the probability distribution, we need to calculate only the remaining probabilities, for $P(1)$ and $P(3)$, and then organize the results into a table.

Minitab and SPSS can combine some steps if you first line up the possible x-values into one column and then proceed as described above. Figure 7.5 displays the SPSS solution; the approach in Minitab is similar.

Figure 7.5
Computing the Full Binomial Probability Distribution

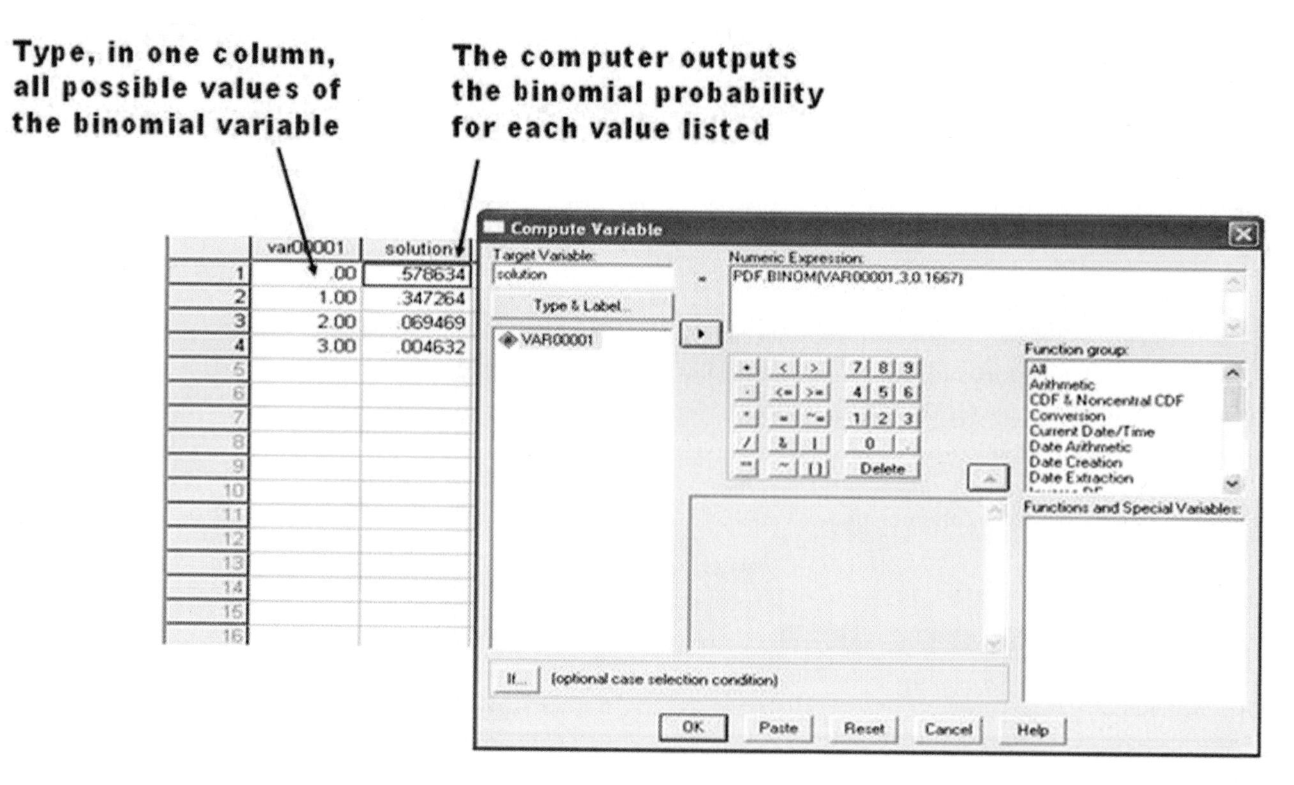

Finding Cumulative Probabilities

Often you might be concerned with the binomial probability for a *range* of possible X-values, above or below a particular cutoff. For example, what is the probability of selecting *fewer than three* mutants in the sample of three microbes? Or what is the probability of getting *at least 5* questions right on a 10-question multiple-choice test if all answers are selected randomly?

As always, if the problem is simple and you have time, you could apply the core rules of probability from Chapter 6. The complete binomial probability distribution for the number of mutants included in a sample of three microbes was presented in the solution for Example 7.8. To select "fewer than three" mutants is equivalent to the compound event "select 0 mutants *or* 1 mutant *or* 2 mutants." Applying the addition rule, the cumulative probability is $P(0) + P(1) + P(2)$. The numbers to add are presented clearly in the second column of Figure 7.5: $P(X < 3) = P(0) + P(1) + P(2) = 0.578634 + 0.347264 + 0.069469 \approx 0.995$.

If you are using a computer, there is no need to construct the full table for this problem. Statistical software generally has a function that will find $P(X \leq x)$ for any value of x that you specify. In the above example, you would request to find $P(X \leq 2)$, which is equivalent to $P(X < 3)$.

• *In Brief 7.2(b) Computing Cumulative Probabilities for Binomial Distributions*

Excel

Choose any empty cell in the worksheet and input a formula in the format =**BINOMDIST(*x*,*n*,*p*,1)** (see the illustration). The required fourth input must be one (**1**), to indicate that the desired output is the *cumulative* probability: $P(X \leq x)$ for a specified value of x. Input in the positions indicated: the upper bound x, the number of trials n, and the probability of success p for each trial.

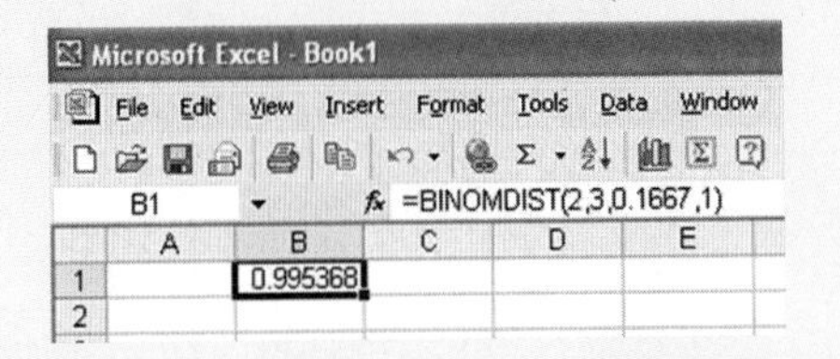

Minitab

To calculate $P(X \leq x)$ for a specified value of x, choose an empty cell in the worksheet (usually at the top of an empty column) and input the value for x into the cell. Use the menu sequence *Calc / Probability Distributions / Binomial* to bring up the"Binomial Distribution" dialogue box (see the screen capture) and input where indicated the values for the number of trials n and the probability of success p for each trial. Specify the input column that contains x and then the column to hold the solution. In the dialogue box, select *Cumulative Probability* for the output.

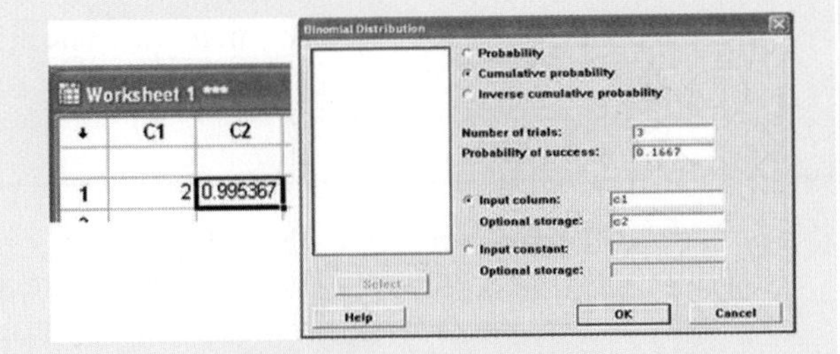

SPSS

To calculate $P(X \leq x)$ for a specified value of x, choose an empty cell in the worksheet (usually at the top of an empty column) and input the value for x into the cell. Use the menu sequence *Transform / Compute* to bring up the "Compute Variable" dialogue box shown in the screen capture. In the *Numeric Expression* window, input a formula of the form **cdf.binom(var00001,*n*,*p*)**, replacing **var00001** with the name of the column holding x (**cdf** stands for "cumulative distribution function"). In the formula positions indicated, input the number of trials n and the probability of success p for each trial. Specify in the *Target Variable* window the name of the column ("Target Variable") to hold the solution. (If needed, adjust the displayed number of decimal places for the solution variable.)

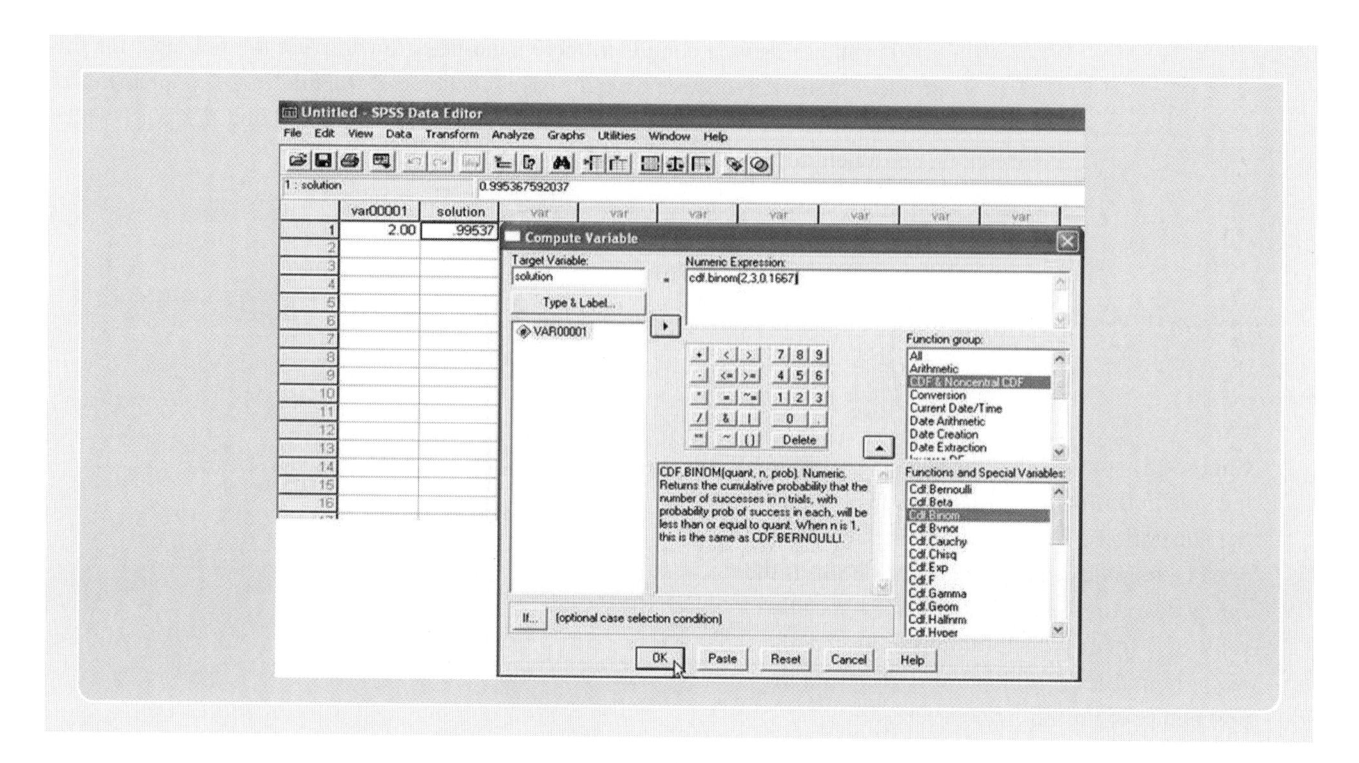

All three software versions handle this one type of case directly: "find the probability of obtaining a number of successes *less than or equal to a specified value x.*" Not all real-world problems regarding cumulative probability will necessarily take this form. Some flexible critical thinking may be required. The problem must be adjusted, or "translated," into an equivalent expression that *can* take advantage of the computer's power. Often there is more than one approach that could solve the problem; your solution depends on common sense, plus a basic fluency in English and probability.

Example 7.9

According to an anti-spam report published in 2004, 80% of mobile text messages that are received in Japan are unsolicited.[1] Suppose that in 2004, the variable "spam or not spam" was binomially distributed, and consider a "success" as the receipt of a *non*-spam message. If a person in Japan received seven mobile text messages in 2004, what is the probability that:

1. At most 4 were non-spam?
2. More than 4 were non-spam?
3. At least 4 were non-spam?
4. 4 or more were non-spam?
5. At least 4 were spam?
6. Between 2 and 4, inclusive, were non-spam?

Solution

The report states that the probability that year of receiving a mobile text message that was spam was equal to 80%. For our purpose, each receipt of a mobile text message is one "trial." Each trial's probability of a success (i.e., being non-spam) is $1 - 0.80 = 0.20$. Therefore:

1. P(at most 4 were non-spam) is equivalent to $P(X \leq 4)$. The computer can calculate this directly by the methods shown above. Use $x = 4$; $n = 7$; $p = 0.20$. The solution is 0.995.
2. P(more than 4 were non-spam) is equivalent to $P(X > 4)$. The computer cannot handle this directly, but the complement of $P(X > 4)$ is $P(X \leq 4)$. (Think of an integer number line: What is not "above four" is "equal to or below four.") The computer *can* handle $P(X \leq 4) = 0.995$. (We did this in solution 1.) The solution, therefore, is the complement of $P(X \leq 4) = 1 - 0.995 = 0.005$.
3. P(at least 4 were non-spam) is equivalent to $P(X \geq 4)$. Again, the computer cannot handle this directly, but the complement of $P(X \geq 4)$ is $P(X \leq 3)$. (On the *integer* number line: What is not "four or more" must be "three or less.") The computer *can* handle $P(X \leq 3) = 0.967$. The solution, therefore, is the complement of $P(X \leq 3) = 1 - 0.967 = 0.033$.
4. P(4 or more were non-spam) is equivalent to P(at least 4 were non-spam). Therefore (see solution 3), the solution is 0.033.
5. Successes were defined as getting non-spam. Out of 7 messages, if 4 were spam, then $7 - 4 = 3$ were *non*-spam. If *at least* 4 were spam, then *at most* 3 were non-spam. Now we have something the computer can handle: $P(X \leq 3) = 0.967$.
6. To plan a solution to $P(2 \leq X \leq 4)$, consider the number line:

   0---1---2---3---4---5---6---7

   The computer cannot solve the problem at once, but it can find the probability for the boxed region: $P(X \leq 4)$. That area includes a bit extra on the lower end: $P(X \leq 1)$. But if we subtract the "unwanted" probability at the left $P(X \leq 1)$ from $P(X \leq 4)$, the remainder is the probability we require:

   $$P(2 \leq X \leq 4) = P(X \leq 4) - P(X \leq 1) = 0.995 - 0.577 = 0.419$$

Summary Measures of the Binomial Distribution

In Section 7.1, we showed the calculations for the expected value (mean) and variance of a probability distribution. Those formulas can be adapted for use with the binomial distribution, in particular.

Recall that the general formula for expected value of a probability distribution is

$$E(X) = \mu = \sum [xP(x)]$$

where x represents the possible values of the random variable and $P(x)$ represents the probability that that value will occur. This formula could be applied, for example, to find the number of mutated microbes expected to be included if three microbes are randomly selected from an irradiated colony.

The left two columns of Figure 7.5 (reproduced below) comprise the binomial probability distribution for the microbe example.

X	*P(X)*
0	0.578634
1	0.347264
2	0.069469
3	0.004632

where X is the variable for the number of mutants found among the three selected microbes. By the formula:

$$E(X) = \mu = (0 \times 0.578634) + (1 \times 0.347264) + (2 \times 0.069469) + (3 \times 0.004632)$$
$$= 0.500$$

For binomial probability distributions, the ordered nature of the x-values (always 0, 1, 2, 3, . . .) permits a simpler formula to be applied, as well: Formula 7.5.

Formula 7.5 Expected Value of a Binomial Probability Distribution

$$E(X) = \mu = np$$

For the microbes example:

$$E(X) = \mu = 3 \times 0.1667 = 0.500$$

Of course it is meaningless to count half of a mutated microbe among the sample of three selected. This fractional answer gives a sense that if the experiment is repeated many times, the numbers of mutants counted each time will tend to be close to zero or one.

Calculating the variance gives an indication of how far the actual x-values would tend to vary from this mean from experiment to experiment. It would be possible to use the general formula that applies for any discrete probability distribution. (See Formula 7.2.) However, because of the restrictions on x-values for the binomial distribution, a more simplified formula can again be developed: Formula 7.6.

Formula 7.6 Variance of a Binomial Probability Distribution

$$\sigma^2 = np(1 - p)$$

Applied to the ongoing example of the microbes:

$$\sigma^2 = 3 \times 0.1667 \times (1 - 0.1667)$$
$$= 0.4167$$

Therefore, the standard deviation is

$$\sqrt{\sigma^2} = \sqrt{0.4167}$$
$$= 0.646$$

Example 7.10

In a recent year, 31.5% of applicants for Natural Sciences and Engineering Research Council of Canada (NSERC) postgraduate fellowships in life sciences/biology were successful.[2] Twenty applicants for the fellowship that year were randomly selected.

1. How many of these would you expect to have been successful?
2. Suppose that the actual number of successes for the 20 applicants was 4. Would that result be surprising? Justify your answer.
3. Is it possible that the distribution of successes is *not* binomially distributed?

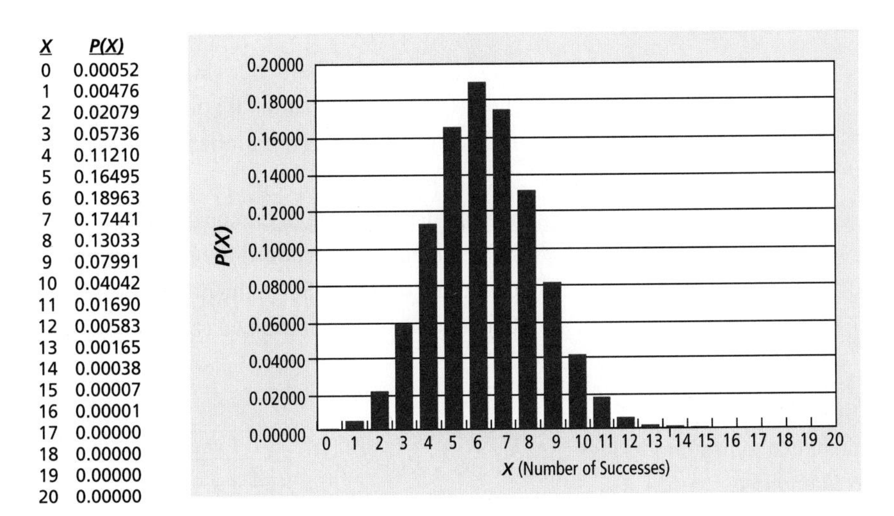

X	P(X)
0	0.00052
1	0.00476
2	0.02079
3	0.05736
4	0.11210
5	0.16495
6	0.18963
7	0.17441
8	0.13033
9	0.07991
10	0.04042
11	0.01690
12	0.00583
13	0.00165
14	0.00038
15	0.00007
16	0.00001
17	0.00000
18	0.00000
19	0.00000
20	0.00000

Figure 7.6
Probability Distribution for the Numbers of Successful Applications (out of 20)

Based on what you have just seen, answer these questions:

1. Is the binomial probability distribution *always* symmetrical?
2. What factor or factors tend to increase or decrease the symmetry of the distribution?
3. For a given number of trials, what happens to the distribution shape as the p approaches 0.0? As it approaches 1.0?

7.3 The Hypergeometric Probability Distribution

A condition for using the binomial probability distribution is that the probability of success remains the same regardless of the number of trials. This condition applies if you throw a coin or die two times and observe the outcomes. But it does not apply if you draw two cards from a single deck—*unless you sample with replacement.* To **sample with replacement** means that after drawing one item, you return it to the sample space (e.g., you return a drawn card into the deck, and reshuffle). Otherwise, as we learned in Section 6.2, the *conditional probabilities* will change for each new selection; consequently, the outcomes will not follow a binomial distribution.

An experiment that consists of randomly drawing a sample of size n from a population of size N, *without replacement,* and counting the number of items that have a certain property (i.e., are "successes") is a **hypergeometric experiment.** The variable that counts the number of successes (regardless of exact sequence) among the sampled individuals in the hypergeometric experiment is called a **hypergeometric random variable.**

Calculating the Probabilities

The underlying principle for calculating probabilities with this distribution is the same as for every calculation involving classical probabilities. In the simplest cases, no new special formula is required.

Example 7.11

You have combined the four kings and four queens of a standard deck of cards to make a deck of just eight cards. If you randomly draw two cards from the new deck, without replacement, what is the probability that both of those cards are diamonds?

Solution

We can use the table of outcomes in Figure 7.7. For the first draw, any of eight cards can be selected. The second draw is conditional on the first, since the first card selected cannot be redrawn. Since the order of selection does not matter, we can identify 28 distinct combinations of the two cards that could be selected.

Figure 7.7
Table of Outcomes for Picking Two Diamonds

First Selected Card

Second Selected Card	Q♣	Q♦	Q♥	Q♠	K♣	K♦	K♥	K♠
Q♣	×	Q♦Q♣	Q♥Q♣	Q♠Q♣	K♣Q♣	K♦Q♣	K♥Q♣	K♠Q♣
Q♦		×	Q♥Q♦	Q♠Q♦	K♣Q♦	K♦Q♦	K♥Q♦	K♠Q♦
Q♥		⇒	×	Q♠Q♥	K♣Q♥	K♦Q♥	K♥Q♥	K♠Q♥
Q♠				×	K♣Q♠	K♦Q♠	K♥Q♠	K♠Q♠
K♣			⇒		×	K♦K♣	K♥K♣	K♠K♣
K♦						×	K♥K♦	K♠K♦
K♥					⇒		×	K♠K♥
K♠								×

× = These outcomes are not possible (don't count them).
⇒ = Because order does not matter, each combination on the lower left is equivalent to an outcome at the upper right.
■ = Both of the selected cards were diamonds.

Of these 28 possible outcomes (the denominator), only one outcome (K♦Q♦) counts as the event "both of the cards are diamonds." Therefore, P(both are diamonds) = 1/28 = 0.0357.

For more complex problems, Formula 7.7 can be applied.

Formula 7.7 Hypergeometric Probabilities

$$P(x) = \frac{\dfrac{r!}{x!\,(r-x)!} \times \dfrac{(N-r)!}{(n-x)! \times [(N-r)-(n-x)]!}}{\dfrac{N!}{n!\,(N-n)!}}$$

where x = Specific number of successes (Example: "draw *2* diamonds")
n = Number of elements in the sample (Example: "draw *2* cards in total")
N = Number of elements in the population (Example: "a special deck with *8* cards")
r = Number of elements in the population with the characteristic of interest. (Example: "*2* diamonds are originally available in the deck")

The three ratios that appear in the formula are all combination formulas. The ratio in the denominator gives the total number of distinct samples of size n that could be drawn from the specified population. (This returns the same value that we could obtain using a table, as in Example 7.11.) The numerator overall finds the number of possible samples in which there are specifically x successes. (Again, we found this answer previously using a table). Formula 7.7 could have been used to solve Example 7.11 by inputting the appropriate values for x, N, n, and r.

Example 7.12

The Webbwood Kitchen Warehouse receives faucets from its supplier in batches of 80. Westwood has discovered that often some of these faucets need a minor adjustment before they can be sold. If 10% or more of the faucets require this adjustment, then carrying the product is not profitable. Therefore, the company has begun to randomly sample 6 faucets from every new shipment of 80. If any sampled faucets need the adjustment, the entire shipment is rejected.

1. In cases where a shipment does contain 10% of faucets requiring adjustment, what is the probability that exactly one of the faucets in the random sample will require adjustment? Calculate either using the formula or the computer, using the procedures in "In Brief 7.3(a)."
2. In the above case, what is the probability that *one or more* (i.e., "any") of the sampled faucets will require adjustment?
3. Is this method of sampling likely to reject shipments that are unprofitable, due to the need for adjustments? Explain.

Solution

This application of hypergeometric distribution is called **acceptance sampling**. Based on sampling, shipments are rejected that are likely to contain an unacceptable level of defective products.

1. Regardless of whether using the formula or the computer, we first have to determine the values for x, N, n, and r. The number of successes x for which we want the probability is given in the question as 1. The total number N of faucets in the shipment is 80. The sample size n is 6 faucets. If 10% of the shipment require adjustment, that means that the number of cases in the population having the property of interest (i.e., the number of "successes" in the population) $r = 0.10 \times 80 = 8$. By formula, $P(1)$ works out to 0.3725. The computer solution is visible in the computer displays in "In Brief 7.3(a)."

• *In Brief 7.3(a) Computing Values for Hypergeometric Probabilities*

Excel

Choose any empty cell in the worksheet and input a formula in the format **=HYPGEOMDIST(1,6,8,80)** as shown in the illustration. The output is the hypergeometric probability $P(x)$, where x is the number of successes specified in the first input. The second input is n, the sample size; the third input is r, the number in the population with the property of interest; and the final input is the population size N.

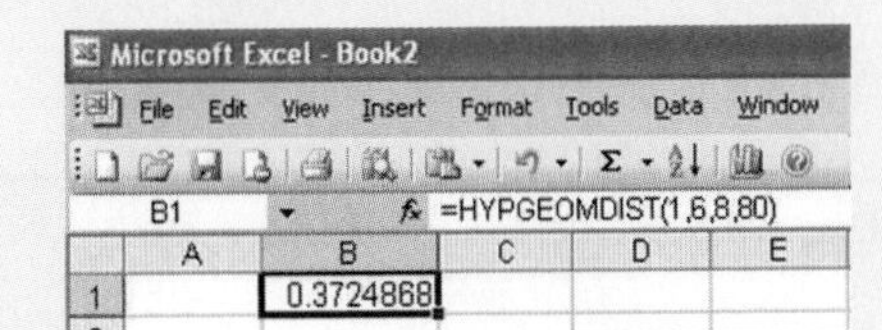

Minitab

To calculate $P(x)$, begin with the menu sequence *Calc / Probability Distributions / Hypergeometric* and in the resulting dialogue box (see the screen capture), input where indicated the values for the population size N, the number of successes in the whole population r (which Minitab refers to as "M"), and the sample size n. Select *Input Constant* and input the specified number of successes x. In the dialogue box, select *Probability*. The output of the procedure is also shown in the illustration.

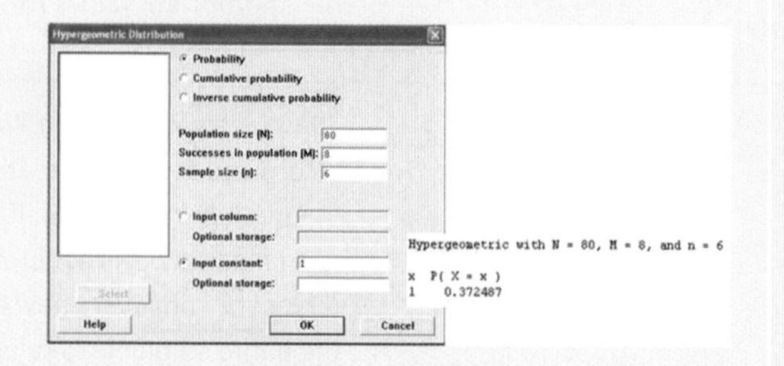

SPSS

To calculate $P(x)$ for a specified number of successes x, choose an empty cell in the worksheet (usually at the top of an empty column) and input the value for x into the cell. Use the menu sequence *Transform / Compute* to bring up the "Compute Variable" dialogue box (see the screen capture). In the *Numeric Expression* box, input a formula of the form **PDF.HYPER(VAR00001,*N*,*n*,*r*)**, replacing **VAR00001** with the name of the column holding x. Input in the formula positions indicated: the population size N, the sample size n, and the number of successes in the population r. In the *Target Variable* box, input the name of the column to hold the solution. (If needed, adjust the displayed number of decimal places for the solution variable.)

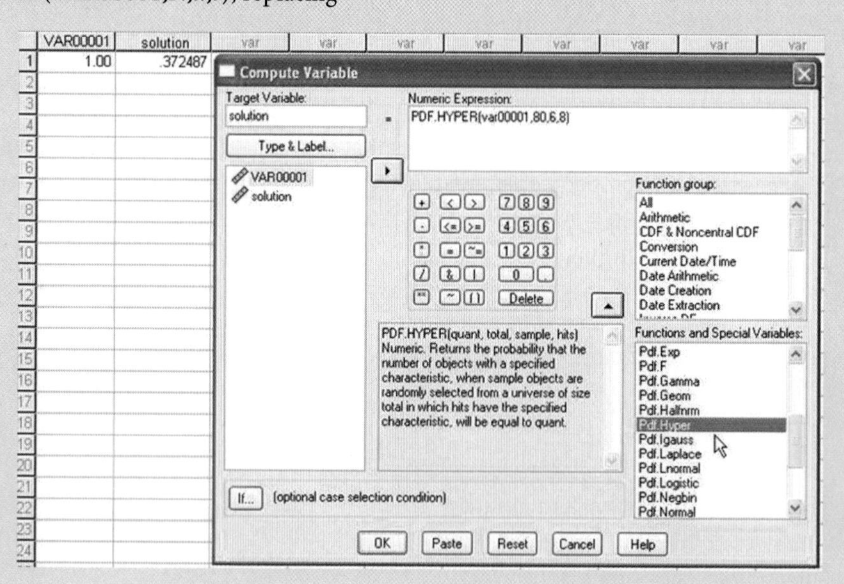

2. The second question in Example 7.12 requires the cumulative probability for all numbers of successes greater than or equal to 1. Since the values of X are discrete, $P(X \geq 1)$ is the complement of $P(X = 0)$, which can easily be found by the

computer. $P(X = 0) = 0.5199$; therefore, $P(X \geq 1) = 1 - 0.5199 \approx 0.48$. This is the probability that if 10% of the shipped items are defective, at least one of the sampled items will be defective.

• In Brief 7.3(b) Computing Cumulative Probabilities for a Hypergeometric Distribution

Excel

Excel does not directly offer a cumulative hypergeometric probability. However, you can easily calculate $P(X \leq x)$ in this way: Use formulas to create a column for the hypergeometric probabilities $P(0)$, $P(1)$, $P(2)$, and so on, up to $P(x)$. The sum of these values is the desired probability.

Minitab

To find $P(X \leq x)$ for a specific number of successes x, proceed as discussed in "In Brief 7.3(a)." In the dialogue box, click on *Cumulative Probability*, and for x, input the upper bound of values for which you want the cumulative probability.

SPSS

To find $P(X \leq x)$ for a specific number of successes x, proceed as discussed in "In Brief 7.3(a)." In the *Numeric Expression* area of the dialogue box, replace **PDF** in the formula **PDF.HYPER(VAR00001,*N*,*n*,*r*)** with **CDF** (for cumulative distribution function). For the x position of the input, input the upper bound of values for which you want the cumulative probability.

3. The acceptance sampling method is not foolproof. If 10% of the faucets in the shipment need adjustment, we found in solution 2 that there is less than a 50% probability that the acceptance sample will include a defective faucet, which would lead to rejecting the shipment. So half of the time, shipments with defective faucets would *not* be stopped.

Summary Measures of the Hypergeometric Distribution

Similar to the other types of discrete probability distribution, the mean, variance, and standard deviation can be calculated for the hypergeometric distribution, as shown in Formulas 7.8 and 7.9.

Formula 7.8 Expected Value for a Hypergeometric Distribution

$$E(X) = \frac{nr}{N}$$

Observe the similarity to the expected value formula for the binomial distribution ($E(X) = np$). For this case, the probability factor p is replaced with ratio r/N, which is the proportion of elements in the population that have the characteristic of interest.

Formula 7.9 Variance for a Hypergeometric Distribution

$$\sigma^2 = \frac{r(N-r) \cdot n(N-n)}{N^2(N-1)}$$

Formula 7.10 Standard Deviation for a Hypergeometric Distribution

$$\sigma = \sqrt{\sigma^2} = \sqrt{\frac{r(N-r) \cdot n(N-n)}{N^2(N-1)}}$$

where n = Number of elements in the sample
N = Number of elements in the population
r = Number of elements in the population with the characteristic of interest

---

Example 7.13

An employee at Casino Rama has been dealing cards to customers for many years.

1. For all of the occasions that she has dealt 10 cards from a newly shuffled deck, estimate the mean for the number of hearts she has dealt each time.
2. What is the estimated standard deviation for the number of hearts she has dealt each time?
3. If none of the 10 cards dealt on one specific occasion are hearts, would this be unusual? Explain.

Solution

1. The mean of the distribution and the expected value are equivalent. Over the long run, the actual mean is expected to be close to this calculated value. Applying Formula 7.8:

$$E(X) = \mu = \frac{nr}{N}$$
$$= \frac{10 \times 13}{52}$$
$$= 2.5 \text{ hearts}$$

2. Using Formulas 7.9 and 7.10 to find the standard deviation:

$$\sigma^2 = \frac{[r(N-r) \times n(N-n)]}{[N^2(N-1)]}$$
$$= \frac{[13(52-13) \times 10(52-10)]}{[52^2(52-1)]}$$
$$= 212{,}940/137{,}904$$
$$= 1.5441$$
$$\sigma = \sqrt{\sigma^2} = \sqrt{1.5441} \approx 1.24 \text{ hearts}$$

3. Although we are not working with a normal distribution, the empirical rule can still provide some reasonable approximations, as long as the sample size is reasonably large and the probability of success (or of failure) is not too small. By the empirical rule, we would expect most values to fall within ± 2 standard deviations of the mean, which in this case is the range [2.5 – 2(1.24)] to [2.5 + 2(1.24)], or roughly from 0 to 5. Since drawing zero hearts falls within this range, it would not appear to be a rare occurrence. (For a more accurate assessment, we could apply Formula 7.7, which shows $P(0 \text{ Hearts}) = 0.040$.)

The Hypergeometric Distribution and Survey-Based Probabilities

Firms that conduct surveys and public opinion polls often report percentages of "successes" in some population. For example, how many respondents support the current prime minister or governor general? If probabilities are calculated based on these results, the binomial distribution is often used for the calculations. Technically speaking, the hypergeometric distribution should be applied because once someone has filled out a survey or answered a questionnaire, that person is removed from consideration when the next individual is approached. As we know from preceding sections, this "sampling without replacement" alters the conditional probabilities that successive respondents will be successes, and this violates the requirements for the binomial distribution.

For example, according to a study published by the Adoption Council of Canada, 45% of adult Canadians are "very favourable" toward adoption.[3] Suppose that there is some community in which, according to a local sample, only 35% of the adults are "very favourable" toward adoption. Compared to the national trend, is the local statistical result unusual? If it is, this might suggest that the community, in fact, does *not* follow the general trend. (This train of thought will be discussed further in Part 5 in the section on hypothesis testing.)

First, we will try the binomial distribution to answer the question. If the national finding applies to all communities, then we expect 45% of all sampled adults to have the property "very favourable toward adoption." So, $p = 0.45$. Suppose the sample size n for the local study is given as 360, and 126 of these (35% of the sample) were successes. If we ignore that the sampling is without replacement, we can use the formulas and procedures of Section 7.2 to find the *cumulative* binomial probability of getting *no more than* 126 successes out 360 sampled, if indeed the probability of success for each trial is 0.45. (We use the cumulative probability because the chance of getting *any one particular* number of successes—124 or 125 or 126 or . . .—is relatively low. At issue is whether getting a value no higher than 126 is rare.) The calculated probability is 0.0000733, a quite unusual result if the national p applies everywhere.

Can we apply, instead, the more formally correct hypergeometric distribution? Thanks to the computer, the greater complexity of calculations for that method need not concern us. But a more significant barrier exists: To use the hypergeometric distribution, we need to know the precise size N of the full population. This may not have been published in the original survey, and perhaps it could not be known if the population is infinite. The original poll results, depended on the national sample size and not on any particular value for N. Suppose that N was 20 million, so r (number of successes for the population) was 9 million (45% of 20 million). Given a local sample of size n equal to

360, you can calculate by the techniques you have learned in this section that $P(X \leq 126) = 0.0000846$—which is very close to the result from the binomial probability distribution. As long as the sample size is less than 5% of the (estimated) population size, the binomial approximation will generally be acceptable in such cases.

• Hands On 7.3 *Comparing Binomial and Hypergeometric Probabilities*

As we have seen, there are times when probabilities that formally should be modelled using the hypergeometric distribution are modelled in practice with the binomial distribution. In these cases, the effect of sampling *without* replacement is relatively small and therefore is ignored. In this exercise, you can observe the consequences of this strategy.

Review "In Brief 7.2(b)" and "In Brief 7.3(b)" for calculating the cumulative binomial and the cumulative hypergeometric probabilities, respectively. Set up columns with the labels and interpretations shown in the following table. Be sure to allow quite a few decimal places in the last four columns.

Column Heading	Interpretation	Values Created by . . .	Calculations Based On . . .
*N*pop	Population size	Manual input	
*n*samp	Sample size	Manual input	
p	Proportion of successes in population	Manual input	
r	Number of successes in population	Calculate with the software	$r = p \times N$pop
X	Number of successes in the sample	(Artificial, calculated inputs)	[Use: 0.8 × ($p \times n$samp)]
Binom	Cumulative, binomial probability of obtaining up to *X* successes in the sample, given the other values	Calculate with the software	(Use cumulative binomial distribution calculation, given: *X*, *n*, and *p*)
Hyper	Cumulative, hypergeometric probability of obtaining up to *X* successes in the sample, given the other values	Calculate with the software	(Use cumulative hypergeometric distribution calculation, given: *X*, *N*pop, *n*, and *r*)
ArithDiff	Arithmetic difference between the calculated probabilities, based on the two distributions	Calculate with the software	Hyper – Binom
%Diff	Percentage difference between the calculated probabilities, based on the two distributions	Calculate with the software	(Hyper – Binom)/Hyper

For the cells requiring manual input, input values following the guidelines shown in Figure 7.8. The objective is to experiment with a wide range of cases—with sample sizes ranging from small to large compared to the population size, and the probabilities of success for each trial ranging from small (near zero) to large (near 1)—in various combinations. In every case, we are comparing the probabilities of obtaining *up to or including x* successes, using the two different distributions. For consistency of comparison, *x* is set arbitrarily as 80% of the expected value of successes for the particular sample size.

Figure 7.8
Guidelines for Completing the Comparison

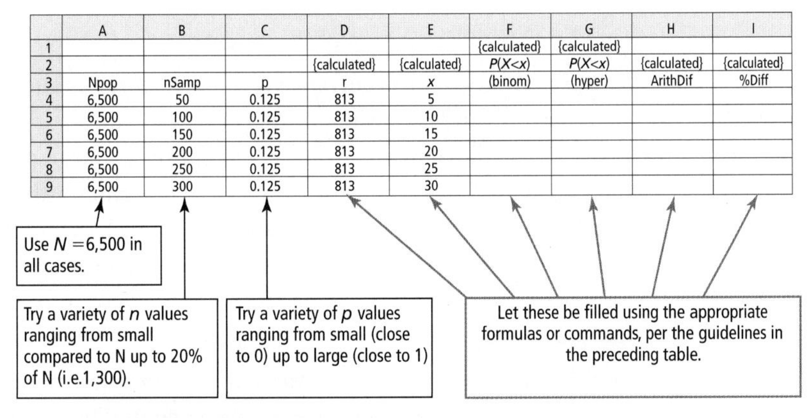

	A	B	C	D	E	F	G	H	I
1						{calculated}	{calculated}		
2				{calculated}	{calculated}	P(X<x)	P(X<x)	{calculated}	{calculated}
3	Npop	nSamp	p	r	x	(binom)	(hyper)	ArithDif	%Diff
4	6,500	50	0.125	813	5				
5	6,500	100	0.125	813	10				
6	6,500	150	0.125	813	15				
7	6,500	200	0.125	813	20				
8	6,500	250	0.125	813	25				
9	6,500	300	0.125	813	30				

After you have created a number of rows in the worksheet, you can see how large a difference in outputs can get by the two methods, both as an absolute difference and as a percentage difference. Under what conditions, if any, does the absolute difference become large? Under what conditions, if any, does the *percentage* difference become large? In some cases, do you think the guidelines for using the binomial distribution as a substitute for the hypergeometric distribution are reasonable? Explain.

7.4 The Poisson Probability Distribution

When we calculated probabilities for the binomial distribution, we were interested in how many successes occurred over a given number of trials. Our calculations depended on sample size and expected probability of success for each trial. Sometimes these notions of distinct trials, and of outcomes of success versus failure, do not seem to apply. Instead, we are interested in the number of *occurrences* over some stretch of time or space of a particular object or circumstance. For example, how many patients will arrive at a hospital emergency department in Ontario over the next minute or hour? Or how many wolves will be living in a particular area of Algonquin Park this winter? How many Wal-Mart customers will use checkout counter 5 in a day? How many car accidents per month will occur in the Selkirk district in Manitoba?

Given certain conditions, the probability distribution we would use for these examples is the **Poisson probability distribution,** for which we do not need to know a sample size or an expected probability. The required input is the mean number of occurrences of the outcome (e.g., number of wolves or customers at counter 5) in a comparable period of time or unit of space. For the model to be valid, the following conditions must always apply:

1. **All occurrences are independent.** For example, whether a wolf resides in a particular section of the park should be independent of whether there are one or more wolves already in that section. Wolves usually cluster in packs or families, so that could violate this condition of the model.
2. **The probability of an occurrence is constant throughout all subintervals of the space or time under consideration.** For this condition, you do not need to formally specify the number of subintervals or their units, but some reasonable subdivision is implicit when using the model. For example, the road system in Selkirk can be subdivided into metres. To apply the model, the probabilities of an accident within any particular metre in the road system should be the same as within any other. If the region has some high-risk curves or intersections, this would violate the model.
3. **Within any one subinterval of the space or time under consideration, the probability of having two or more occurrences should be inconsequential.** For example, even if two accidents occur in the vicinity of each other on a single stretch of highway, their exact locations are likely to be different by at least one metre.

Formula 7.11 is used for calculating a probability using the Poisson distribution. As observed for all of the probability distributions, the researcher has a responsibility to ensure that all of the necessary conditions are met or, if there are possible exceptions, he or she must at least take them into account.

Formula 7.11 Poisson Probabilities

$$P(x) = \frac{\mu^x e^{-\mu}}{x!}$$

where μ = Mean or expected number of occurrences per interval (alternatively, this parameter can be represented with the Greek letter lambda: λ)

x = Number of occurrences of a particular outcome

e = A constant, known as *Euler's constant,* which = 2.71828

Example 7.14

Based on data published by the Canadian Association of Emergency Physicians, the average number of clients who arrive per minute at the doors of an Ontario hospital emergency department is approximately 9.7.[4]

1. Suppose that the average rate applies at any time of day. At any randomly selected minute, what is the probability that (a) no patients will arrive at an emergency department in Ontario? (b) Eight patients will arrive? Or (c) Sixteen patients will arrive? Which of these would be considered a rare number of occurrences?
2. Do you think that the conditions for applying the Poisson distribution are met in the above example? Can you envision any exceptions that could possibly impact on the Poisson-based results?

Solution

1. We can use Formula 7.11 to answer the first set of questions. The mean (μ) was given in the problem.

a)

$$P(0) = \frac{(9.7^0 \times e^{-9.7})}{0!} = \frac{1 \times 0.0000613}{1} = 0.0000613$$

A minute in which no new patients arrived would be very rare.

b)

$$P(8) = \frac{(9.7^8 \times e^{-9.7})}{8\,!} = \frac{78374335.9 \times 0.0000613}{40320} = 0.119$$

A minute in which eight new patients arrived would not be unusual.

c)

$$P(16) = \frac{(9.7^{16} \times e^{-9.7})}{16!} = \frac{6142536534626850 \times 0.0000613}{20922789888000} = 0.0180$$

A minute in which 16 new patients arrived would be relatively uncommon but, given the number of minutes available each day, this event would not appear to be rare.

2. To assume that the mean arrival rate was the same during all parts of the day was probably not realistic. But *within* a given minute (the interval used for the formula) the rate is not likely to change by much during the subintervals. Individual arrivals may not be independent if, for example, there have been car accidents involving several people, or if groups of people are arriving at the hospital with a contagious flu or other disease.

• *In Brief 7.4(a) Computing Values for Poisson Probabilities*

Excel

To calculate $P(x)$ if x and μ are known: Choose any cell in the worksheet and input a formula in the format =**POISSON(x,μ,0)** (see the illustration). The required third input, zero (**0**), indicates that the desired probability is for exactly $P(x)$; i.e., it is *not* for a cumulative probability.

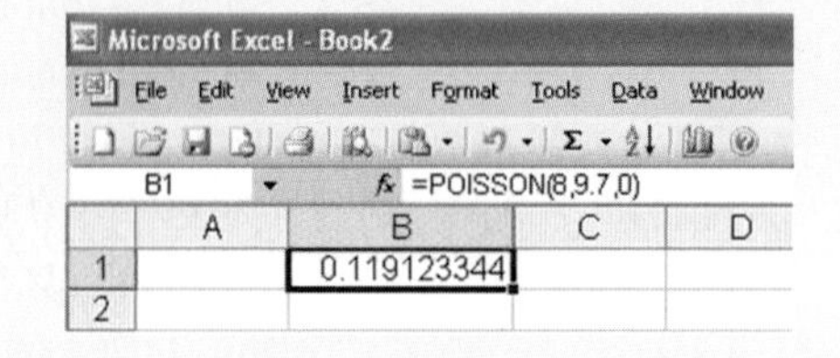

Minitab

To calculate $P(x)$ if x and μ are known: Input the value for x into a worksheet cell (usually, at the top of an empty column). Use the sequence *Calc / Probability Distributions / Poisson* to bring up the dialogue box shown in the screen capture and input the value for μ where indicated. Specify the column that contains x and the column to hold the solution. Select the *Probability* option.

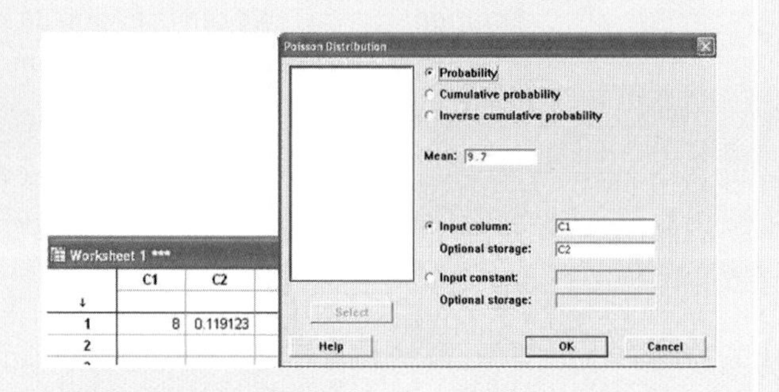

SPSS

To calculate $P(x)$ if x and μ are known: Input the value for x into a worksheet cell (usually, at the top of an empty column). Use the menu sequence *Transform / Compute* to call up the dialogue box shown in the screen capture. In the *Target Variable* window, input the name of the column to hold the solution. In the *Numeric Expression* window, input a formula of the form **pdf.poisson(VAR00001,μ)**, replacing **VAR00001** with the name of the column holding x. (If needed, adjust the displayed number of decimal places for the solution variable.) Click *OK*.

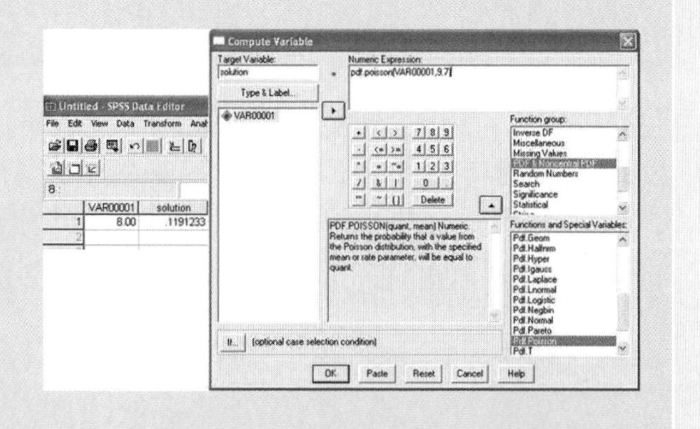

Example 7.15 The Friends of Algonquin Park have observed that, on average, 2.7 wolves can be found in every 100-km^2 sector of the park during the winter.[5] Presume that this pattern follows a Poisson distribution.

1. In a randomly selected 100-km^2 area of the park, what are the probabilities of finding 0, 3, or 10 wolves, respectively? Which, if any, of these findings would appear to be rare?

2. What is the probability of finding *at most 5* wolves in a given area? *More than* 5?
3. In a randomly selected 500-km^2 area of the park, what is the probability of finding *exactly 10* wolves?

Solution

1. Whether you use the formula or the software (see the procedures in "In Brief 7.4(a)"), start by determining the values for the mean ($\mu = 2.7$) and for x (given as 0, 3, or 10, respectively). The probabilities for each x-value are calculated to be as shown below.

	X	Prob_X
1	.00	.0672055
2	3.00	.2204677
3	10.00	.0003813

It would not be unexpected to find 0 or 3 wolves in a given sector of the park, but finding 10 wolves would be quite unusual.

2. The second question asks for cumulative probabilities. This is done most easily with the help of a software program that can find the Poisson probability $P(X \leq x)$ for any specified x-value. Using the procedures shown in "In Brief 7.4(b)," we find that $P(X \leq 5)$ (i.e., "at most 5") is 0.943. The probability that more than 5 wolves are found is the complement of the probability that at most 5 wolves are found. Therefore, $P(X > 5) = 1 - 0.943 = 0.057$.

• *In Brief 7.4(b) Computing Cumulative Probabilities for a Poisson Distribution*

Excel

To find $P(X \leq x)$ for a specific number of occurrences x, proceed as discussed in "In Brief 7.4(a)." Modify the formula by making the third input a "1" instead of a "0." The result is now the desired cumulative probability.

Minitab

To find $P(X \leq x)$ for a specific number of occurrences x, proceed as discussed in "In Brief 7.4(a)." In the dialogue box, select *Cumulative Probability,* and for x, input the upper bound of values for which you want the cumulative probability.

SPSS

To find $P(X \leq x)$ for a specific number of occurrences x, proceed as discussed in "In Brief 7.4(a)." In the *Numeric Expression* area of the dialogue box, replace **pdf** in the expression **pdf.poisson (VAR00001,μ)** with **cdf.** For the x position of the input, input the upper bound of values for which you want the cumulative probability.

3. The third part of Example 7.15 requires a change of value for μ in the calculations. If there are 2.7 wolves per 100 km^2, then for each area five times larger (500 km^2) we would expect $5 \times 2.7 = 13.5$ wolves (13.5 per 500 km^2). We can calculate:

$$P(10) = \frac{(13.5^{10})(2.71828^{-13.5})}{10\,!} = 0.076$$

Note that the mean for a Poisson distribution is generally given as the input for other calculations, so a separate formula for the mean is not needed here. Interestingly, the variance of a Poisson distribution is also equal to its mean. Thus, one strategy when checking to see if a set of raw data follows the Poisson distribution is to see if the mean and variance of the data set are roughly equal.

• Hands On 7.4 *Comparing the Poisson and Binomial Distributions*

The Poisson distribution is really a special application of the binomial distribution. (Formally, it is "the continuous limit" of the binomial distribution.[6]) Consider Example 7.14, on the daily number of arrivals at Ontario hospital emergency departments. We provided a mean number of *occurrences,* 9.7, but we never described anything as a *success,* as in a Bernoulli trial. However, we *did* say that the Poisson interval in this case (a single day) must be dividable into many small subintervals, all with the same constant probability of a particular event occurring.

The key to relating the two distributions is to think of each small subinterval as one "trial." An occurrence during any subinterval corresponds to a success. If the interval does not have an occurrence, this corresponds to a failure. The total number of occurrences is analogous to the total number of successes in the binomial distribution. We can try this out for ourselves. We will use the binomial distribution to estimate the probabilities for the Poisson distribution in the following example.

Suppose customers for a Calgary Stampeders football game are arriving at the ticket booth at a Poisson-distributed rate. The mean number of arrivals every minute is 5. What is the distribution of probabilities that, at a given minute, 0, 1, 2, 3, etc., up to 10 customers will arrive at the booth? We will describe a simulation for estimating the missing values in the following table.

X (Number of Arrivals)	*P(X)*
0	
1	
2	
3	
4	
5	
6	
7	
8	
9	
10	

Adapting the procedures discussed in "Hands-On 7.1(a)," generate a long set of random numbers based on the *binomial* distribution. To generate the binomially distributed random numbers, the computer will ask you to input the number of trials n and the probability of success for each trial p. As described earlier in this feature, we can divide the given time unit of minutes into subunits of, say, one-thousandths of minutes—interpreting each subunit as a "trial" and an occurrence during a subunit as a "success." So, input 1,000 as the value of n when creating the random numbers. As for p, recall that for a binomial distribution, $E(X) = np$. The mean number of arrivals given above, 5, is the expected value of the distribution. We can now solve for $p = E(X)/n = 5/1{,}000 = 0.005$; use this as the input for p when generating the random numbers.

You have now created a long set of random binomially distributed numbers. Use the procedures provided in Chapter 3 to create a relative frequency distribution of all of the individual numbers (e.g., the zeros, ones, twos, and other numbers in the list). By the Law of Large Numbers, we can interpret relative frequencies as estimates for the probabilities of obtaining those numbers under the given conditions. So the table on page 310 can be completed.

Now generate a more conventional Poisson distribution table using classical probabilities and compare this with the table you have just completed. That is, use the methods described in "In Brief 7.4(a)" to find the Poisson probability for each of $X = 0, X = 1, \ldots, X = 10$, given a known mean of 5. How do these probabilities compare to those estimated in the table on page 310?

For another simulation attempt, try, in turn, "slicing" the one-minute time interval into smaller subunits and using a longer set of generated random numbers. Do either or both of these strategies improve the estimates, compared to the true Poisson probabilities? If so, why do you think that is?

Chapter Summary

7.1 Explain the concept of a discrete probability distribution and calculate summary statistics for such distributions.

The distribution of all possible values of a variable, together with the probabilities that each of these values will occur, is a **probability distribution**. A special case is the **equiprobable distribution,** where each value in the sample space has the same probability of occurring as any other.

If an experiment produces outcomes with probabilities consistent with the probability distribution and the experiment is repeated many times, then the mean (or **expected value**) of the discrete probability distribution is the mean expected from all of the outcomes generated by the repeated experiment. For calculations, use Formula 7.1 to find the expected value for a discrete probability distribution, and Formulas 7.2 and 7.3 to find the variance and standard deviation, respectively, of the distribution.

7.2 List the criteria for a binomial probability distribution, calculate probabilities and cumulative probabilities based on the distribution, and calculate summary statistics for such distributions.

A special type of discrete probability distribution is the **binomial probability distribution** (or **binomial distribution**). An underlying concept is the **trial** (or **Bernoulli trial**), which is a random experiment for which only two, mutually exclusive outcomes are possible. (One of the two outcomes is identified as a success.) A **binomial experiment** is a random experiment, such as two coin flips in a row, consisting of a sequence of identical and independent Bernoulli trials. A variable that counts the number of successes (regardless of exact sequence) among the trial results in the binomial experiment is called a **binomial random variable.**

A binomial distribution meets these four criteria: (1) The experiment consists of n identical trials. (2) There are only two mutually exclusive results possible for each trial. (3) All of the trials are independent. (4) The probabilities of success are constant for each trial.

Formula 7.4 can be used to calculate the probabilities for specific values of a binomial random variable. Cumulative probabilities for the variable are sums of individual value probabilities, within specified ranges of the sample space. For summary statistics, the mean or expected value of the distribution can be calculated with Formula 7.5 and the variance with Formula 7.6. (The standard deviation is the square root of the variance.)

7.3 List the criteria for a hypergeometric probability distribution, calculate probabilities and cumulative probabilities based on the distribution, and calculate summary statistics for such distributions.

To **sample with replacement** means that after drawing one item, you return it to the sample space. An experiment that consists of randomly drawing a sample of size n from a population of size N, *without replacement,* and counting the number of items that have a certain property (i.e., are successes) is a **hypergeometric experiment**. The variable that counts the number of successes (regardless of exact sequence) among the sampled individuals in the hypergeometric experiment is called a **hypergeometric random variable.**

Formula 7.7 can be used to calculate probabilities for specific values of a hypergeometric random variable. For summary statistics use Formulas 7.8, 7.9, and 7.10, respectively, to find the mean, variance, and standard deviation of a hypergeometric probability distribution.

7.4 List the criteria for a Poisson probability distribution, and calculate probabilities and cumulative probabilities based on the distribution.

When we are not interested in the success or failure of trials, but rather in the number of *occurrences* of an object or circumstance over some stretch of time or space, the appropriate distribution model may be the **Poisson probability distribution**. Neither sample size nor expected probability are required as inputs; the one input needed is the mean number of occurrences of the outcome (e.g., number of customers at a teller's window) over a specified period of time or unit of space.

For the model to be valid, these conditions must apply: (1) All occurrences are independent. (2) The probability of an occurrence is constant throughout all subintervals of the space or time under consideration. (3) Within any one subinterval of the space or time under consideration, the probability of having two or more occurrences should be inconsequential. Formula 7.11 can be used to calculate probabilities for specific values of a Poisson random variable.

Key Terms

acceptance sampling, p. 299
binomial experiment, p. 285
binomial probability distribution (*or* binomial distribution), p. 285
binomial random variable, p. 285
equiprobable distribution, p. 271
expected value, p. 275
hypergeometric experiment, p. 297
hypergeometric random variable, p. 297
Poisson probability distribution, p. 306
probability distribution, p. 270
sample with replacement, p. 297
trial (*or* Bernoulli trial), p. 285

Exercises

Basic Concepts 7.1

1. Consider the following probability distribution.

x	−3	0	3	6
$P(x)$	0.3	0.2	0.2	0.3

a) What discrete values can the random variable assume?
b) What is the probability that $x > 0$?
c) Is this probability distribution collectively exhaustive?

2. Consider the following probability distribution.

x	1	2	3	4	5
$P(x)$	0.45	0.25	0.15	0.10	0.05

a) Find $P(2)$.
b) Find $P(2 \text{ or } 5)$.
c) Find $P(X > 2)$.

3. A six-sided die is thrown two times. What is the probability distribution for the number of fives that will be thrown?

4. A publisher is reviewing the potential market for a new textbook. Based on data such as user surveys and projected course enrolments, the marketing department has prepared the following table.

Number of Copies Sold	Probability
15,000	0.35
20,000	0.40
25,000	0.20
30,000	0.05

a) What is the probability that at least 25,000 copies will be sold?
b) What is the probability that at most 25,000 copies will be sold?
c) What is the expected number of copies that will be sold?
d) If each book sells for $110, what is the expected gross revenue from the sales?

5. A convenience store owner has decided to carry three flavours of the new product "Gator Gum." The distributors of Gator Gum claim that these are the probabilities for every customer who passes by the introductory display.

X	No Gator Gum Purchased (Revenue = $0)	Cherry Gator Gum Purchased (Revenue = $1.50)
$P(x)$	0.81	0.06

X	Lime Gator Gum Purchased (Revenue = $1.50)	Blue Gator Gum Purchased (Revenue = $1.75)
$P(x)$	0.01	0.12

a) What is the probability that a customer will purchase some flavour of Gator Gum?
b) What is the expected revenue from each customer who passes the display?
c) What is the standard deviation for the revenue expected from each customer?
d) If 450 customers are expected to pass the display, what is the expected revenue from sales of Gator Gum?

6. Highway regulations specify a maximum load-weight limit of 32,500 kg for any truck trailer. A typical load of a DayStar Lines trailer is carried in three containers. A probability distribution for possible loads in any one container is shown below.

Contents	Granite	Limestone
Load weight (x)	12,000 kg	9,000 kg
$P(x)$	0.05	0.30

Contents	Hardwood	Talcum Powder
Load weight (x)	8,000 kg	6,000 kg
$P(x)$	0.50	0.15

a) What is the expected value for the weight of any one container?
b) What is the variance and standard deviation for a container's weight?
c) What is the expected weight of a three-container load for the truck trailer?
d) If three containers are combined on a truck trailer, what combination of loads for the three containers would exceed the maximum allowed weight on the highway? For the combinations that exceed that allowed weight, construct a probability distribution.

7. The 250 students who attended a concert were surveyed on whether they enjoyed it. The results were as follows.

Response	Frequency
Yes, enjoyed the concert	162
No, did not enjoy the concert	67
Don't know or no response	21

a) For the possible survey response of a randomly selected student, construct a discrete probability distribution.
b) Is it possible to calculate an expected value for this distribution? Why or why not?

8. A child has a bowl of Canadian coins that contains four 1-cent coins, two 5-cent coins, two 25-cent coins, and three $1.00 coins.
a) If one coin is selected randomly, show the probability distribution for that coin's denomination.
b) What is the standard deviation for the denomination of the coin that is drawn?
c) Suppose that the child draws a random coin 10 times. Between each draw, she stirs the coin back into the bowl. What is the expected combined value of the 10 coins that have been drawn?

Applications 7.1

9. In the periodic table of chemical elements, six elements belong to the family "alkaline earth." The atomic numbers of these six elements are, from low to high: 4, 12, 20, 38, 56, and 88, respectively.[7] A chemist randomly picks one of the elements.
a) For the atomic number that will apply to the selected element, give the probability distribution.
b) What is the expected value for the atomic number that will apply to the selected element?

10. A regulator is investigating the number of complaints made by consumers regarding oversales (i.e., double selling) of seats by different U.S. airlines. (See the file **Airline_Complaints**, in the column "Ov_Sale.")
a) If any one airline is randomly selected, what is the probability distribution for the numbers of oversales-related complaints made against that airline?
b) For a randomly selected airline, what is the expected value of the number of oversales-related complaints made against that airline? What is the variance and standard deviation?
c) Suppose it costs the regulator $4,500 to investigate each complaint. How much does it cost the regulator, on average, per airline to investigate complaints for oversales?

11. Concordia University published the following weekly data on e-mails that arrived at the university in one week during July 2005.[8]

Identified E-Mail Type	Quantities
Spam	983,677
Viruses	3,632
Acceptable for delivery	369,214

a) What is the probability distribution for the *type* of a randomly received piece of e-mail?
b) Suppose that auditors determine that each piece of spam received costs the university $0.005 extra to process. Each virus received costs the university $0.12 extra to deal with. (Other e-mail has no extra cost.) For every 200,000 pieces of e-mail received, what is estimated extra cost due to viruses or spam?

12. An investor is studying the ratings that have been assigned to Canadian equity funds by various experts. (In the file **Canadian_Equity_Funds,** ignore funds without ratings.)
a) If a fund is randomly selected from that list, what is the expected value of its rating?
b) What is the standard deviation for the assigned ratings?

13. A telemarketing company is planning a phone campaign to a number of Asian cities. The table below shows the per-minute call rates to the cities of interest. (Based on data in the file **Phone_Rates.**) Calls will be placed randomly. The fourth column in the table shows the relative frequency of calls that will be made to phone numbers in each city.

Country	City	Rate (CDN$/Min.)	Percentage of Randomized Calls
China	Beijing	0.064	0.037
China	Chengdu	0.064	0.044
China	Guangzhou	0.064	0.068
China	Shanghai	0.064	0.045
China	Shenzen	0.064	0.017
Japan	Kanto	0.084	0.035

Country	City	Rate (CDN$/Min.)	Percentage of Randomized Calls
Japan	Okinawa	0.084	0.051
Japan	Osaka	0.084	0.026
Japan	Tokyo	0.084	0.015
South Korea	Seoul	0.077	0.076
Malaysia	Kuala Lumpur	0.103	0.020
Russia	Irkutsk	0.108	0.100
Russia	Ulan-Ude	0.108	0.098
Russia	Vladivostok	0.108	0.089
Thailand	Bangkok	0.120	0.096
Vietnam	Hanoi	0.341	0.082
Vietnam	Ho Chi Minh	0.341	0.101

Source: T-ONE Corporation Inc. Found at: http://www.t-one.ca/bbldrates.xls

If each random phone call will be billed for one minute, then the above table represents the probability distribution for the cost of the next phone call.

a) What is the probability that the next phone call will cost $0.084?
b) Find $P(\text{cost} > \$0.084)$.
c) What is the expected value for the cost of the next call?
d) The telemarketer plans to make 200 calls per hour. What is the expected cost per hour to make those calls?

14. Typically, the clutch size (i.e., number of eggs in the nest) for the species Wood Duck is from 10 to 14 eggs.[9] Supposing that any one of these numbers is equally likely:
a) What is the expected value for the number of eggs in a clutch?
b) What is the variance and standard deviation?
c) If there are 30 Wood Duck clutches in a wildlife reserve, what is the expected number of eggs in the reserve?

Basic Concepts 7.2

15. List the conditions that must be present in order to use the binomial distribution.

16. The Cold River Fundraisers estimate that the probability that a homeowner will buy one of their chocolate bars is 0.12. If 15 households are approached, what is the probability of selling:
a) Exactly 3 chocolate bars?
b) No more than 3 chocolate bars?
c) Up to, at most, 3 chocolate bars?
d) Less than 6 chocolate bars?
e) More than 2 chocolate bars?

17. Sheila Jackson is a sportswriter for a national newspaper. Each Saturday, she writes her predictions for the next week's Major League Baseball games. The success rate of her predictions has been 72%. Presuming that her success rate is a constant, and independent of which teams are playing, what is the probability that she will select the winners for:
a) All of the next 15 games?
b) Exactly 5 of the next 15 games?
c) More than 7 of the next 15 games?
d) At least 7 of the next 15 games?
e) Between 3 to 7, inclusive, of the next 15 games?

18. If there are generally 45 games a week, how many winners do you expect Sheila Jackson to predict correctly each week? (See Exercise 17.) What is the standard deviation for her number of correct weekly predictions?

19. Surveys have found that 65% of all students approve of their university's new logo. If you randomly select 10 of the students, what is the probability that:
a) All will approve of the logo?
b) None will approve of the new logo?
c) Between 2 and 4, inclusively, of the students will approve of the logo?

20. On any day, the probability that your Internet provider will be out of service is 0.06. Over the next 30 days, what is the probability that your provider will be out of service for:
a) Exactly 5 days?
b) More than 2 days?
c) At least 2 days?
d) No more than 5 days?

21. In a 30-day-period, how many days without service do you expect from your Internet provider? (See Exercise 20.) What is the standard deviation for the number of days out of service?

22. A child has a bowl containing 8 marbles—4 red, 2 white, and 2 blue. Suppose that 10 times in a row, she draws one marble and then returns it to the bowl for the next draw. What is the probability that:
a) A red marble was selected for 5 of the draws?
b) A blue marble was selected for at least 3 of the draws?
c) For 5 or more draws, the selected marble was white or blue?

23. Find the mean and standard deviation for the following binomial distributions:
a) $n = 20; p = 0.3$
b) $n = 120; p = 0.6$
c) $n = 100; p = 0.5$
d) $n = 250; p = 0.35$
e) $n = 56; p = 0.75$

24. For a binomial distribution with $n = 12; p = 0.30$, determine:

a) $P(X = 8)$
b) $P(X > 6)$
c) $P(X < 9)$
d) The mean of the distribution.
e) The standard deviation of the distribution.

Applications 7.2

25. The probability that twins will be delivered from a full-term pregnancy is about 0.003.[10] This statistic is independent of genetics and world location.
    a) If a city hospital's maternity ward has 200 deliveries in a year, how many pairs of twins are there likely to be in one year?
    b) At that rate, how many pairs of twins are likely after 10 years?
    c) In a given year, what is the probability of having more than one pair of twins delivered at that hospital?
    d) It appears that older mothers are more likely to have twins than younger mothers. Explain the impact of this fact on the above use of the binomial model.

26. As part of a study on humpback whales, an observation post was established on the roof of the Hyatt Hotel in Maui, Hawaii. On any given day when parasail boats were not nearby, the probability of spotting a pod (i.e., group) of whales from this viewpoint was 0.80.[11] If the researchers' records are examined for 10 randomly selected days when there were no parasail boats nearby:
    a) On how many days do you expect that they observed a pod of humpbacks?
    b) What is the probability that they observed a pod of humpbacks on all 10 of the days? On none of the days? On at least 7 of the days?

27. According to one estimate, only 13% of Canadian start-up businesses survive their first year.[12]
    a) If 300 newly started businesses are randomly selected, what is the expected number that will survive to a second year?
    b) A business consultant provides start-up services to 20 new businesses; 9 of these businesses survive to a second year. Is this number of successes unusually high? Explain.
    c) By definition, the clients in (b) are showing a willingness to accept advice (and have the money to be able to afford it) that may or may not be shared by all who start a business. Explain the possible impact of this on the binomial-based conclusion in (b).

28. Hospitals are concerned about the rates of patient infection by the drug-resistant bacteria methicillin-resistant *staphylococcus aureus* (MRSA). The Canada-wide rate of MRSA infection in hospitals is approximately 5 infections per 1,000 admissions.[13] If this rate was constant for all hospitals and types of patients:
    a) What would be the expected number of infections among a random sample of 400 patients? What would be the standard deviation?
    b) If 1,000 patients are admitted, what is the probability that none become will become infected with MRSA? That at most 10 will become infected? That at least six will become infected?
    c) Is it realistic to assume, for a binomial model, that the infection rate is "constant for all hospitals and types of patients"? Explain. How could you adjust the calculations for (a) and (b) if the assumption of a constant rate does not hold?

29. City officials are concerned about the risk of dog bites in their community. They cite research from Pittsburgh that estimates an annual risk to humans of dog bites of 0.00589.[14]
    a) What assumptions are necessary to calculate probabilities with the above data, using the binomial model? To what extent do you think these conditions apply in Pittsburgh? Explain.
    b) Assuming the binomial model does apply and that the population of the city is 150,000, how many dog bites would you expect annually in the city?
    c) Based on the same assumptions, what is the probability that from 10 to 15 people will be bitten in the city in a year?

30. A survey conducted by the Asia Pacific Foundation of Canada[15] indicated that 26% of respondents planned to invest in China during the next year. Suppose that the results are representative of a large growth-oriented business community of 11,000 companies.
    a) How many of these companies do you think are planning to invest in China during the next year?
    b) What is the probability that no more than 500 of the companies are planning to invest in China in the next year?

31. It is reported that 9% of Nova Scotians aged 12 and over have experienced major symptoms of depression.[16]
    a) If the above rate is constant for all Nova Scotians of that age group, how many students in a Nova Scotia high school with 230 students could be expected to have been depressed? What would be the standard deviation?
    b) In the same high school, what is the probability that none of the students have ever been depressed? That 20 or more have been depressed? That at least 40 have been depressed?
    c) Is it realistic to assume, for the binomial model, that the proportion of Nova Scotians aged 12 and over who have experienced depression is constant—regardless of gender, backgrounds, work conditions, and so on? If

not, how could you adjust the calculations for (a) and (b) if the assumption of a constant rate does not hold?

32. Astronomers now believe that if a star has about three times more metal than our sun, it has a 20% chance of harbouring planets.[17] Hoping to become famous, an amateur is examining 100 stars of the above-mentioned description.
   a) How many of those stars could be expected to harbour planets?
   b) What is the probability that the number of planets turns out to be less than 10? To be more than 25?

Basic Concepts 7.3

33. Explain how the hypergeometric distribution differs from the binomial distribution.

34. For a population $N = 7$, a sample $n = 4$ is taken. The random variable X has a hypergeometric probability distribution. The number of successes in the population is $r = 2$. Compute $P(x)$ for $x = 0$, 1, and 2, respectively.

35. For a population $N = 8$, a sample $n = 5$ is taken. The random variable X has a hypergeometric probability distribution; and the number of successes in the population is $r = 3$. Compute:
   a) $P(X = 1)$
   b) $P(X \text{ is at least } 2)$
   c) $E(X)$
   d) The variance and standard deviation for X.

36. Of the 24 new bikes just produced by Sykle Motors, 8 have defective brake lights. Unknowingly, Oshawa Motorcycles has taken delivery of 6 of the new bikes. What is the probability that:
   a) All six of the bikes have the defective part?
   b) None of the bikes have the defective part?
   c) More than two of the bikes have the defective part?

37. In the previous problem, how many defective bikes would you *expect* Oshawa Motorcycles to receive in its order of 6 bikes?

38. Suppose that you select, without replacement, 3 cards from a standard deck of 52 cards. What is the probability that:
   a) All 3 cards are red?
   b) None of the 3 cards is red?
   c) Exactly 2 of the 3 cards are red?

39. In a survey on pollution, respondents were asked whether they believed poor air quality was a serious problem in their own town or city. These are the results.

Gender	Yes	No
Male	60	40
Female	86	16

   a) Randomly select 10 of the males, without replacement. What is the probability that all 10 selected males answered "Yes"?
   b) Randomly select 10 of the females, without replacement. What is the probability that at least half of the selected females answered "No"?

40. Of the 37 birds eating your newly sown grass seed, 21 are Canada geese. A tourist passing by takes pictures of 8 of the birds, selected randomly. What is the probability that:
   a) All 8 of the birds in the photographs are Canada geese?
   b) None of the birds in the photographs is a Canada goose?
   c) No more than 4 of the birds in the photographs are Canada geese?

41. In the previous problem, how many birds in the photographs would you *expect* to be Canada geese?

Applications 7.3

42. At one point in 2006, the Toronto Blue Jays' roster included 12 active pitchers. Five of these pitchers throw left-handed.[18] For a practice session, the pitching coach randomly picks six of the pitchers. What is the probability that:
   a) Three or more of the selected pitchers throw left-handed?
   b) None of the selected pitchers throws left-handed?
   c) One or two of the selected pitchers throws left-handed?

43. In the previous problem, about how many left-handed pitchers would you expect to have among the six that were selected?

44. Biologists estimate that only 30 Vancouver Island marmots (VIMs) are alive in the wild.[19] Suppose that 14 of these animals have been caught, tagged, and released back into the wild. If 10 of the 30 living VIMs are now randomly photographed to check for tags, what is the probability that:
   a) None of the photographed animals has been tagged?
   b) More than 2 of the photographed animals have been tagged?
   c) All of the photographed animals have been tagged?

45. In the previous problem, what is the expected number of photographs in which a VIM was tagged? What is the standard deviation?

46. In a survey, 500 beer drinkers were asked what one Canadian item they'd most like to have if they were stranded on a desert island. (See the file **Desert_Island.**) Suppose that the 500 respondents were the full population at a large family reunion.
   a) If a success was choosing "maple syrup," what was r for the population?

b) Suppose you randomly select 50 of the respondents:
   i) What is the probability that more than half of your sample selected maple syrup in the original survey?
   ii) What are $E(x)$ and the standard deviation for the numbers in your sample who chose maple syrup?

47. To prepare for an appearance on a game show, a student in China is brushing up on trivial facts, including a list of the world's tallest towers. (See the file **Tallest_Towers.**) If six of the listed towers are randomly selected, what is the probability that:
   a) All six towers are located in China?
   b) None of the towers are located in China?
   c) Exactly half of the towers are located in China?
   d) At least 3 of the towers are located in China?

48. In the previous problem, out of the six towers that are selected from the full list, what is the expected number of towers that are located in China? What is the standard deviation?

Basic Concepts 7.4

49. If X has a Poisson probability distribution in each example below, find:
   a) $P(X = 4)$, given $\mu = 3.0$
   b) $P(X = 3)$, given $\mu = 2.0$
   c) $P(X = 7)$, given $\mu = 4.0$

50. In a case where x has a Poisson probability distribution and $\mu = 4.0$, calculate:
   a) $P(0)$
   b) $P(2)$
   c) $P(6)$

51. In a case where X has a Poisson probability distribution and $\mu = 5.0$, calculate:
   a) $P(X < 2)$
   b) $P(X > 6)$
   c) $P(3 \leq X \leq 6)$

52. A representative on the company joint health and safety committee estimates that the average number of accidents per month in the warehouse is 3.5. Calculate:
   a) The probability that exactly 2 accidents will occur in the warehouse next month.
   b) The probability that at least 4 accidents will occur in the warehouse next month.
   c) The probability that the number of accidents that occur next month will be 5 at the most.

53. Barbara's chicken pies contain an average of five pieces of chicken. (The number of pieces per pie has a Poisson distribution.) What is the probability that in a randomly selected chicken pie:
   a) There are less than 5 pieces of chicken?
   b) There are exactly 7 pieces of chicken?

54. Paul Francis, the coach of the Moosonee Old-Timers baseball team, has compiled data on injuries suffered by his players. It appears that, on average, his players suffer (collectively) 5.8 injuries a year, and the annual number of injuries is Poisson-distributed. What is the probability that, next year, his players will suffer:
   a) Exactly 6 injuries?
   b) No more than 8 injuries?
   c) At least 4 injuries?
   d) No injuries at all?

55. The quality manager has determined that for every day of production of DaeGlo lamps, 6 of the lamps will be defective. Daily numbers of defective lamps generally follow a Poisson distribution. What is the probability that during the next day's production:
   a) No lamps will be defective?
   b) Exactly 6 lamps will be defective?
   c) More than 8 lamps will be defective?
   d) At least 5 lamps will be defective?

56. A breeder of Devon Rex cats has determined that, each year, an average of five of the breeder's kittens are born with a curly tail. There is a Poisson distribution for this number. What is the probability that, next year:
   a) None of the new kittens will have a curly tail?
   b) At least 2 new kittens will have a curly tail?
   c) Fewer than 6 new kittens will have a curly tail?

57. Joyce Green is very concerned about the environment and regularly buys a newspaper a day to look for articles on that subject. She has noticed that, every week, her favourite paper averages 5 articles on the environment. (This weekly number has a Poisson distribution.) Given that Joyce has read this newspaper every day this week, what is the probability that she found:
   a) No articles on the environment?
   b) Exactly 5 articles on the environment?
   c) Between 4 and 8 articles, inclusive, on the environment?

Applications 7.4

58. A criminologist is studying the annual rates of Canadian deaths due to firearms. According to a 2002 report by Statistics Canada, the average number of firearm-related deaths (including suicides) per 100,000 persons is 4.9.[20] Presuming that this average is constant from year to year and that the number is Poisson-distributed, what is the probability that next year:
   a) 4.2 persons per 100,000 will die due to firearms?
   b) No one will die due to firearms?
   c) More than 6 persons per 100,000 will die due to firearms?

d) Do you think that, in practice, the rate is constant from year to year? If not, how could that affect the above calculations?

59. When Boeing established its Enterprise Help Desk in 2004, it was anticipated that the average number of incoming calls per minute would be 2.[21] Numbers of calls to a help desk are often Poisson-distributed. If the average call rate is in fact 2 per minute, find these probabilities for the numbers of calls in the next minute:
   a) *P*(0 calls)
   b) *P*(5 calls)
   c) *P*(between 2 and 4 calls, inclusive)
   d) *P*(at least 2 calls)

60. A company that plans to expand its Internet-based services is looking for a suitable Internet backbone service provider. According to the literature, large provider Global Crossing currently handles 8.8 billion page-views per month.[22] This translates into 1.26 million hits per minute—which is 21 hits per millisecond. Presuming this rate is constant throughout the day and that the hits are Poisson-distributed, what is the probability that:
   a) There will be no hits in the next millisecond?
   b) There will be at least 20 hits in the next millisecond?
   c) There will be more than 25 hits in the next millisecond?
   d) There will be no more than 15 hits in the next millisecond?
   e) Do you think it is likely that the rate of hits per millisecond is the same for all parts of the day? Explain.

61. For foresters in northern Minnesota, the spread of the parasitic species dwarf mistletoe is a serious concern. Numbers of infected spruce trees per acre have increased dramatically since 1962, reaching 173 per acre as of 1998.[23] If the number of infections per acre is Poisson-distributed, what was the probability that year that the number of infected spruce trees in a randomly selected acre was:
   a) Less than 150?
   b) More than 180?
   c) Between 150 and 180?

62. In order to maintain adequate service, providers of personal communication services need to estimate the present and future call traffic on their systems. According to a simulation teletraffic study for the San Francisco Bay area, the average number of call arrivals on the system, at about 9 a.m. on weekdays, is approximately 30 per minute.[24] Call arrivals are modelled with the Poisson distribution. During a random minute selected around 9 a.m. on a weekday, what is the probability that there will be:
   a) At least 25 call arrivals?
   b) At most 40 call arrivals?
   c) Exactly 30 call arrivals?

63. Access to health care is a serious concern for all Canadians. Capacity must be able to meet demand. To better understand the demand, a six-year study of a hospital emergency department (ED) determined that the average arrival rate of patients, per hour, at the ED is 6.46.[25] During the next hour, what is the probability that the number of new arrivals at the ED will be:
   a) No patients?
   b) Fewer than 4 patients?
   c) More than 8 patients?
   d) Do you think that the arrival rate is constant for all parts of the day? If not, how would you recommend adjusting the calculations? Explain.

64. One challenge faced by facilities for aquaculture (i.e., fish farms) is the hunting of the fish by birds. For example, in the northeastern United States, the Department of Agriculture determined that ospreys are catching, on average, two fish per hour from aquaculture facilities.[26] For aquaculture facilities in that region, what is the probability that during the next hour the number of fish caught by ospreys will be (presuming that the numbers are Poisson-distributed):
   a) None?
   b) Between 1 and 3, inclusive?
   c) More than 3?

Review Applications

65. During the campaign for the Canadian federal election in 2004, there were at least 4 candidates competing in each riding, and often more.
   a) If the numbers of candidates competing for individual ridings in the country is considered a random variable, use the file **Early Returns** to construct a probability distribution for that variable in Manitoba, only.
   b) Calculate the mean and standard deviation for the distribution in (a).
   c) For the next election, the town hall staff in Provencher, Manitoba, needs to determine whether its stage is large enough for conducting an all-candidate's debate. A statistics grad on staff suggests basing a solution on the probability distribution for numbers of candidates competing in ridings in Manitoba, as constructed in (a). If the stage in the Provencher town hall can fit at most 6 candidates, then (based on above probability distribution) what is the probability that the stage will not have enough room for all the candidates who compete in this Manitoba riding in the next election?
   d) If you were planning for this event, would you recommend booking the town hall? Explain.

66. A social researcher is tracking trends of infanticide in the United States. These are her findings for 2004:

Age (0 = less than 1)	Number of Victims
0	201
1	174
2	89
3	60
4	54
Total	**578**

Source: U.S. Department of Justice, Office of Justice Programs, Bureau of Justice Statistics. Based on data in *Homicide trends in the U.S.: Infanticide.* Found at: http://www.ojp.gov/bjs/homicide/tables/kidsagetab.htm. Accessed Sept. 4, 2007.

a) Construct the probability distribution for the age of a victim of infanticide in that year.
b) Find the mean and standard deviation of the distribution.
c) Create a graph of the probability distribution.
d) Does it appear that a child becomes less likely to face this risk as he or she grows older? Are there any other variables we would need to know before we could confidently answer the preceding question? Explain.

67. Among the players acquired by the Toronto Blue Jays during the 2004 first-year player draft was outfielder Cory Patton. Cory's on-base percentage of 0.428 meant that he made it on base—by hitting, walking, or being a hit batsman—over 42% of his times at bat.[27] Suppose that next season Cory maintains this on-base percentage of 0.428 and that the conditions for a binomial distribution apply.
a) If Cory is at bat 450 times during the season, how many times would you expect him to reach first base?
b) What is the standard deviation of the probability distribution for 450 at-bats?
c) If he is at bat 6 times during a long game, what is the probability that he will make it on base at least 4 times? At most 2 times?
d) By the eleventh inning of a tie game, Cory has batted 5 times and reached base only once. Given the assumptions provided in the question, what is the probability that Cory will reach base on his sixth at bat? Explain your answer.
e) The assumption of a binomial distribution for the on-base variable could be questioned. Name at least three factors that could change the probabilities of Cory's being successful for different at-bats (i.e., for different trials).

68. According to data posted by the National Eye Institute, 3.74% of all North Dakotans over the age of 40 exhibit some degree of vision impairment or blindness.[28] A health researcher randomly selects 36 adults from North Dakota who are over the age of 40.
a) What is the probability that at least 4 of the persons selected exhibit some degree of vision impairment or blindness? Fewer than 3 of them? Between 2 and 4 of them?
b) If 5 persons in the sample exhibit the symptoms, would this result be unusual? Explain.

69. Suppose that the experiment described in Exercise 68 is repeated many times. For each sample of 36 adults, what is the expected number of individuals with vision impairment or blindness who will be selected? What is the standard deviation of the distribution?

70. A globe-trotting graduate of Grace University hopes to meet fellow alumni during her travels. (See the file **Grace_Alumni_World.**) She will randomly pick alumni who have moved to other countries and (if they agree) fly to visit them. Using the data in the file, construct the probability distribution for the *continent* of residence of next-selected alumnus. What is the probability that the next selected alumnus resides in Africa?

For Exercises 71–73, use data from the file **Airline_Complaints.**

71. A public relations specialist has been hired to address concerns about the number of consumer complaints made against airline companies. The data in the file **Airline_Complaints** represent three months' worth of complaints. Estimate the average number of complaints per workday, for each category of complaint (e.g., for the variable Ov_Sale = oversales). [Hint: Add all of the complaints for that category and divide by 66 (the approximate number of workdays in three months).] The actual number of complaints received per workday is likely Poisson-distributed. What is the probability that, during the next work-day:
a) More than 6 complaints about oversales are received?
b) No complaint about oversales is received?
c) More than 6 complaints about discrimination are received?
d) No complaint about discrimination is received?

72. For the same data set, create a probability distribution for the number of complaints any individual airline will receive in a three-month period for discrimination. (See the column "Discrim.")

73. Based on your answer to Exercise 72, what is the expected value of the number of discrimination-related complaints made against an airline? What is the variance and standard deviation?

For Exercises 74–76, use data from the file **Grotto_Geyser.**

74. A naturalist observed the numbers of eruptions per day of the Grotto Geyser over an extended period. Distinguishing days that have (a) 1 to 2 eruptions from days that have (b) 3 to 4 eruptions, she concluded that the distribution of category (a) days is roughly binomial, with a probability of 0.46. On this assumption, what is the probability that in the next 10 days, the number of days having 1 to 2 eruptions will be 5? More than 6? Less than 3?

75. Based on the preceding exercise, what is the expected number of days, out of 10, that will have from 1 to 2 eruptions that day? What is the standard deviation?

76. Overall, the **Grotto_Geyser** data set includes 83 distinct days of observations. How many of those days had from 1 to 2 eruptions? If you randomly select 15 of the days in the data set, what is the probability that:
a) 7 of those days had from 1–2 eruptions?
b) Fewer than 9 of those days had from 1–2 eruptions?
c) At least 5 of those days had from 1–2 eruptions?

Exercises 77–81 are based on data reported by the Canadian Hospitals Injury Reporting and Prevention Program (CHIRPP).[29]

77. A public health nurse is concerned about children's injuries due to in-line skating. She has found that over half (0.568) of injuries due to this activity are sustained by youths 10–14 years old. If this rate is constant, and 100 records for in-line skating injuries are randomly selected, what is the probability that:
a) Over half of the recorded injuries were suffered by youths 10–14 years old?
b) More than two-thirds of the injuries were suffered by youths 10–14 years old?

78. The nurse in Exercise 77 also examined 962 records of in-line skating injuries and observed that for 486 of the cases, the accident resulted from a loss of control with no identified cause. Suppose that she randomly assigns 400 of the original injury records to a colleague to reexamine. What is the probability that, in the 400 reexamined records, the accident resulted from a loss of control with no identified cause according to:
a) At least half of the records?
b) Less than half of the records?
c) At least 370 of the 400 records?
d) What is the expected value for the number of cases in the 400 reexamined records for which the accident resulted from a loss of control with no identified cause?

79. The risk of in-line skating injuries to youths 10–14 years old was further emphasized by the finding that, for every 1,000 injuries of all types recorded in the CHIRPP database for youths of that age group, an average of 20.1 of the injuries are due to in-line skating. If this rate is constant, then what is the probability that, among the next 1,000 records of injuries for youths 10–14 years old:
a) At least 20 of the injuries will be due to in-line skating?
b) None of the injuries will be due to in-line skating?
c) Between 14 and 17, inclusive, of the injuries will be due to in-line skating?

80. Of those whose in-line skating injuries were serious enough for admission to hospital, and where their safety equipment status was known, only 44.8% were wearing safety helmets. If this proportion holds true in the population, what is the probability that, of the next 12 hospital admissions due to in-line skating injuries:
a) All the victims were wearing their helmets?
b) More than half of the victims were wearing their helmets?
c) None of the victims were wearing their helmets?

81. For the next 12 hospital admissions due to in-line skating injuries (see Exercise 80), what is the expected value for the number of victims who were wearing their helmets? What is the standard deviation?

Continuous Probability Distributions

Chapter 8

Chapter Outline

Learning Objectives

8.1 Describe the concepts of a probability density curve, a probability density function, and a cumulative distribution function; apply these concepts to calculate areas under the normal curve and to find an *x*-value that bounds an area of known size under the normal curve.

8.2 Apply a variety of methods to assess whether a continuous probability distribution is reasonably approximate to a normal distribution.

8.3 Apply the concepts of a probability density function and a cumulative distribution function to calculate areas within a uniform distribution and also to calculate summary statistics for a uniform distribution.

8.4 Apply the concepts of a probability density function and a cumulative distribution function to calculate areas within an exponential distribution and also to calculate summary statistics for an exponential distribution.

8.1 The Normal Probability Distribution

For many people, the normal (bell shaped) distribution is perhaps the most familiar probability distribution. It can be applied to model the distributions of many real-world variables, such as the lengths, heights, weights, and other measured values of natural or manufactured objects. For example, many common objects today, from jewellery to cases for musical instruments to medical inhalers, are made of thermoplastics. A manufacturer has learned that the stress value at which thermoplastic composite objects deform may centre on the value 66.2 megapascals (MPas), with a standard deviation of 9.2 MPa. If a product is designed to these specifications and a sample of 500 outputs is tested, typical results may appear similar to Figure 8.1. Most values are clustered near the mean, and the frequencies drop off after two or more standard deviations from the mean. (Compare the discussion of the empirical rule, in Section 5.2.)

Figure 8.1
Distribution of Yield Stress Values for 500 Tested Objects

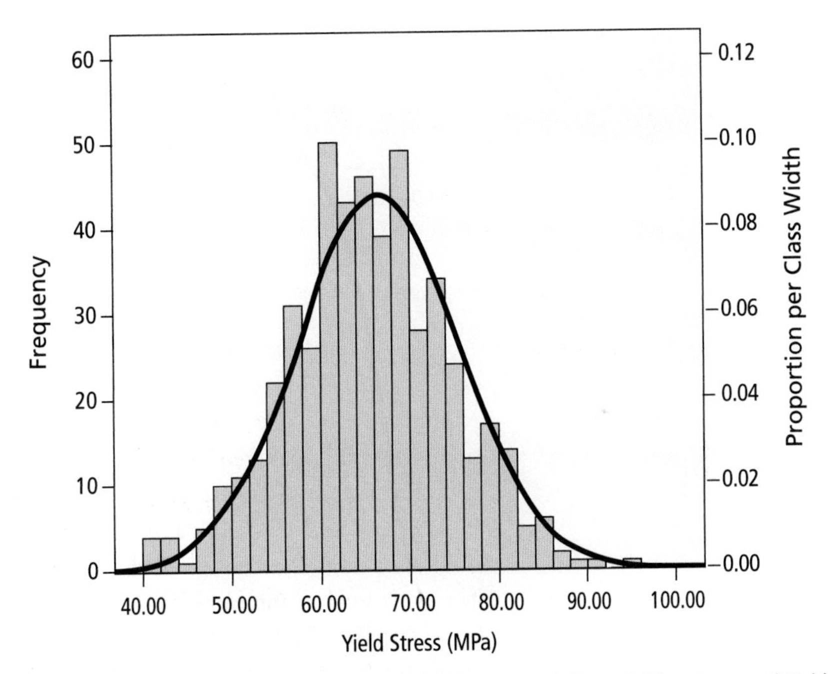

Source: David Kazmer and David S. Roe, *Exploiting Melt Compressibility to Achieve Improved Weld Line Strengths*. Pg. 12. Found at: http://kazmer.uml.edu/Staff/Archive/1998JPRCP_ Weld_Strength.pdf

Seeing Statistics
Section 7.1

The frequencies in this figure can be converted into relative frequencies or proportions. Each bar is drawn over a range of possible outcomes (a class) and a bar's length represents the proportion of outcomes that occur somewhere in that class. If we think of a class width as one unit, then the area under each bar represents the proportion of the total frequency that occurred in that unit of class width. In short, we have a density scale histogram, as defined in Chapter 3: The *area* encompassed by each bar represents, out of the total number of observations, the percentage that fall within the class boundaries for that bar. These percentages can be interpreted as probabilities, and the total area under the curve (like the sum of probabilities for a sample space) equals 1.

A density scale histogram, however, cannot fully model the probability distribution for a continuous variable. The probabilities may not be equal for all values of the random variable within a given class width; yet we cannot possibly represent each unique outcome on the x-axis to show its exact frequency or probability of occurrence. If we tried, we would get a rather useless graph, such as Figure 8.2.

Figure 8.2
Distribution of Yield Stress Values—Displaying Each Unique Value

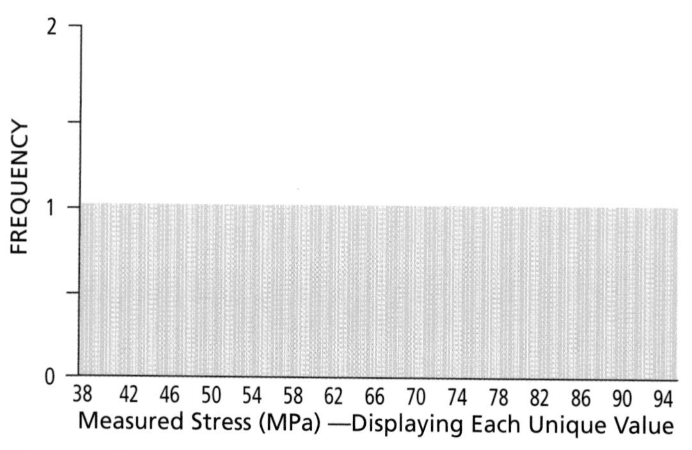

Source: David Kazmer and David S. Roe, *Exploiting Melt Compressibility to Achieve Improved Weld Line Strengths*. Pg. 12. Found at: http://kazmer.uml.edu/Staff/Archive/1998JPRCP_ Weld_Strength.pdf

There are at least two problems with this figure: (1) There are so many distinct x-values, that it is impossible to distinguish them all on the horizontal axis and (2), unless we round off the x-values to perhaps the nearest unit, there will likely be no exact repetitions of values, so all frequencies will be either 0 or 1.

In Figure 8.1, we avoided these problems by clustering the continuous x-values, as we did when creating histograms in Chapter 3. This meaningfully displays *ranges* of x-values that are large enough to show relative frequencies. Now suppose that an infinite population had this same basic distribution. There would then be enough data that we could make all class widths for the histogram very small and still obtain a similar-looking histogram.

The bell-shaped curve that is superimposed on Figure 8.1 is a mathematical construct, representing a limit for the theoretical process of redrawing the histogram—with ever smaller class widths, down to a class width of virtually 0. At a class width of exactly zero, the proportion of the area under the curve cannot be defined. Yet the curve in the figure, commonly known as the **normal curve** or **bell curve,** is such that for any class whose width is greater than zero, the *area* within the curve lying above a class range for the variable does approximate, like a density scale histogram, the probability of getting an outcome within that particular range of x-values.

Insofar as it models a probability distribution for a continuous variable, the normal curve is one example of a **probability density curve.** Other examples will be introduced later in the chapter. Mathematically, the ordinate value of the curve is determined for each x-value in the sample space by a **probability density function.** As mentioned above, there is no interpretation for the height of the curve over any specific point, since the class width would be zero. But for a range of x with a width > 0, the *area under the curve* represents the probability of randomly drawing a value in that range from the population. Formula 8.1 presents the probability density function for the normal curve.

Formula 8.1 Probability Density Function for the Normal Curve

$$f(x) = \frac{1}{\sigma \sqrt{2\pi}} e^{\frac{-(x-\mu)^2}{2\sigma^2}}$$

where μ = Population mean
σ = Population standard deviation
π = The constant 3.14159
e = The constant 2.71828

Formula 8.1 tells us that for a normally distributed population, we need to know only its mean and standard deviation to define the entire curve. Everything else in the formula is a constant. Using the curve to model the probability density, we can calculate the proportions (probabilities) of data falling into any particular range of values.

When we speak of "the" normal distribution, we are really talking about a class of distributions that is infinitely large. There are as many different normal distributions as there are combinations of different μ- and σ-values for populations. Figure 8.3 compares the distributions of weights for adult human males and females. Since males have the larger mean weight, the curve for that population (in red) is centred on a larger value on the x-axis. Since the standard deviation of weights is larger among females, the curve for that population (in blue) appears slightly wider. Note that even a single normal curve can have a wide variety of applications. The blue curve could represent the distribution of times, in minutes, that students required to write a test, or the distribution of observed widths of galaxies in arc-minutes.

Figure 8.3
Two Normal Distributions of Weights

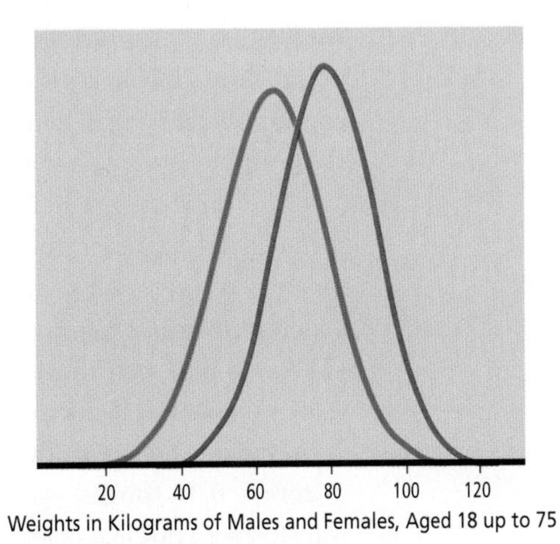

Source: U.S. EPA (1997) *Exposure Factors Handbook Volume 1: General Factors*. August 1997, U.S. Environmental Protection Agency, National Center for Environmental Assessment, Washington, D.C.

Seeing Statistics
Section 7.2

Regardless of what the data represent, the mathematics of the density curve allows us to calculate the probabilities based on any normal distribution, provided we know the mean and standard deviation of the population. We hinted at this fact in Section 5.2 when we introduced the empirical rule, which is based on the normal distribution. For example, we learned that for any normally distributed population, about 68% of the data fall within ± 1 standard deviation of the mean. If we know the values of the mean and standard deviation, we can calculate the actual numbers that correspond to the limits "–1 standard deviation" and "+1 standard deviation" from the mean.

Example 8.1

For the population of adult human males from age 18 to less than 75, the mean weight $\mu = 78.1$ kg and the standard deviation for the weights $\sigma = 13.5$ kg. What is, approximately, the probability that a randomly selected adult male will weigh between 64.6 kg and 91.6 kg?

Solution

For this population, a weight of 64.6 kg is one standard deviation below the mean, since the mean (78.1) minus one standard deviation (13.5) equals 64.6. The weight of 91.6 kg is one standard deviation above the mean, since the mean (78.1) plus one standard deviation (13.5) equals 91.6. Therefore, for a normal distribution, the probability of a value falling between those bounds is approximately 68%.

Based on the mathematical function given in Formula 8.1, we can *extend* the line of thinking that is applied in the empirical rule for just the simple cases. For instance, we can know the probability of a value falling between 2.1 standard deviations below the mean up to 1.7 standard deviations above the mean, or any other combination. And, knowing μ and σ, we can calculate the real-world values that define those class limits.

Probability Calculations Based on the Standard Normal Curve

Notice that the empirical rule is not limited to any particular values for μ and σ of a population. This makes the rule general, but to apply it in Example 8.1, we first had to convert actual weight values into numbers, signifying how many standard deviations those numbers were away from the mean.

The same step is needed to use normal distribution tables that commonly are included in statistics textbooks. It is not possible to create a distinct table for every combination of values for μ and σ for a possible normal distribution. Instead, users must convert the *actual* values of the data into *standard* values, called **z-values** or **z-scores.** (These z-values were introduced in Chapter 5.) A "z-score $= +1$" corresponds to a value exactly one standard deviation above the mean; a "z-score $= -1.5$" corresponds to a value 1.5 standard deviations below the mean, and so forth. The mean itself is defined as z-score $= 0$. In converting all values of a normal distribution into z-values, we in effect transform an original distribution into a **standard normal distribution,** for which the mean $= 0$ and the standard deviation $= 1$. Traditional z tables, like the empirical rule, are designed to take values in this format as their inputs. If needed, you can later convert the results back to real-world units.

For convenience, the formula to calculate a z-score is restated here as Formula 8.2. (Compare Formulas 5.7 and 5.8.)

Formula 8.2 Calculating a z-Score

$$z = \frac{x - \mu}{\sigma}$$

where μ = Population mean
σ = Population standard deviation
x = Actual value of a member of the population, to be converted to a z-score

Technically, these calculated z-values are unitless numbers, because the units in the formula (e.g., kg or arc-minutes) are the same in both the numerator and denominator, and so cancel each other out. Informally, you might think of "standard deviations" as the units for z. For example, knowing that the weight of an adult male has a z-value $= +1.3$ tells us that his weight is "+1.3 standard deviations" from the mean for that population.

Example 8.2

We have seen that for adult human males aged 18 to less than 75, the mean weight μ is 78.1 kg and the standard deviation for the weights σ is 13.5 kg.

1. Find the z-scores corresponding to each of the following weights among adult males:
   a) 105.1 kg
   b) 51.1 kg
   c) 60 kg
   d) 78.1 kg
2. Approximately, what is the probability that a randomly selected adult human male in the age group of 18 years to less than 75 years will have a weight between 51.1 kg and 105.1 kg?

Solution

1. In each case, apply Formula 8.2:

   a) $z = \frac{105.1 - 78.1}{13.5} = +2$

   b) $z = \frac{51.1 - 78.1}{13.5} = -2$

   c) $z = \frac{60 - 78.1}{13.5} = -1.3407$

   d) $z = \frac{78.1 - 78.1}{13.5} = 0$

(Observe that for the mean of the distribution (78.1), $z = 0$.)

2. The empirical rule for normally distributed populations can be rephrased in terms of z-values. We found in solution 1 that for this population, 51.1 kg corresponds to $z = -2$ and 105.1 kg corresponds to $z = +2$. Applying the empirical rule to this information, the probability that a randomly selected individual will fall between $z = -2$ and $z = +2$ is approximately 95%.

Employing the calculus of the normal curve, we are not limited, in the manner of Example 8.2 (2), to finding probabilities for only the simple cases where $z = \pm 2$ or other whole numbers. With techniques based on the **cumulative distribution function** (CDF) for the normal curve (see Figure 8.4), we can solve problems such those in Example 8.3.

Figure 8.4
Cumulative Probability Based on z

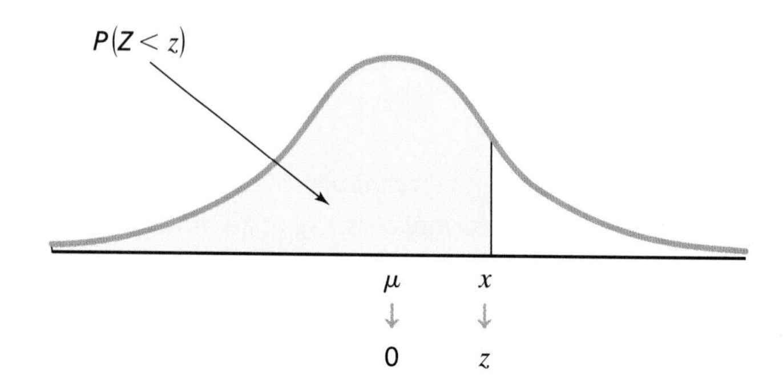

Example 8.3

Use the z-values that were calculated in Example 8.2 (1). What is the probability that a randomly selected adult human male in the age group of 18 years to less than 75 years will have a weight less than:

1. 105.1 kg?
2. 51.1 kg?
3. 60 kg?
4. 78.1 kg?

Solution

For all of the solutions, refer to the model displayed in Figure 8.4. Given the mean and standard deviation for the normal distribution, we can convert any specific x-value in the population to a corresponding z-value. With statistical software (or a z table) based on the CDF, you can then calculate or find the probability that a randomly selected member of the population will have a z-value *smaller* than the z-value that is specified. The answer is found by integration—accumulating all areas (representing probabilities) for z that fall below the specified upper boundary. The procedure is analogous to finding a cumulative probability $P(X \leq x)$ for a discrete probability distribution, except that (1) we are looking up z instead of x, (2) the possible values in the indicated range are not limited to discrete values, and (3) the area under the curve over the *exact* upper z-value in the range is zero, so the area corresponding to $P(Z \leq z)$ is the same as for $P(Z < z)$.

Inputting the values for μ, σ, and z into one of the procedures described in "In Brief 8.1(a)," we obtain the solutions: (1) 0.9772, (2) 0.0228, (3) 0.0900, and (4) 0.5000.

Note that, as expected, one-half of the curve (0.5000) is below the mean for the distribution (see solution 4). Also, we expected about 95% of the distribution to fall between $z = -2$ (i.e., 51.1) and $z = +2$ (i.e., 105.1). More accurately, we can now conclude that 95.44% of the curve lies in that range, because the portion of the curve below $z = +2$ is 97.72%. But observe that 2.28% of the curve *also* falls below $z = -2$; this leaves in the range between $z = -2$ and $z = +2$: 97.72% − 2.28% = 95.44%.

• *In Brief 8.1(a)* *Computing a Cumulative Probability for a Normal Distribution*

All three software packages can calculate directly the cumulative probability for a normal distribution, that is, find the area under the curve (representing a probability) to the left of a specified value. If you know μ and σ for the normal distribution and input an x-value for the upper bound of the range, the software will take care of converting that value into a z-score as part of its calculations.

Excel

To calculate $P(X < x)$ for a normal distribution if μ, σ, and value x are known: Choose any cell in the worksheet and input a formula in the format **=NORMDIST(x,μ,σ,1)**. The required fourth input (**1**) indicates that the desired output is the cumulative probability.

Alternatively, to calculate $P(Z < z)$ for a normal distribution if z has already been determined: Choose any cell in the worksheet and input a formula in the format **=NORMSDIST**(z). Both approaches are shown in the illustration.

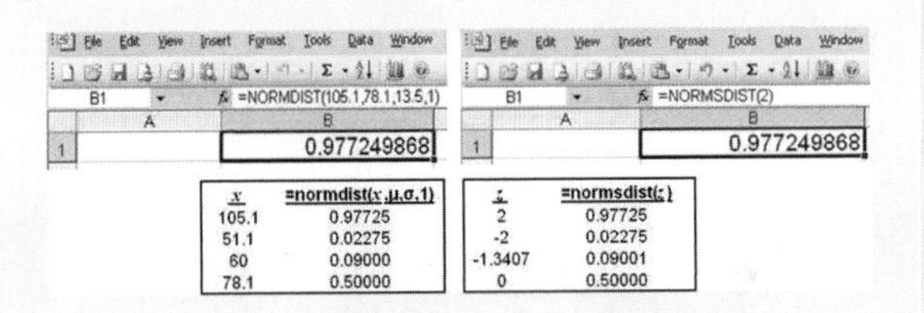

Minitab

To calculate $P(X < x)$ for a normal distribution if μ, σ, and value x are known: Input the value for x into a worksheet cell (usually, at the top of an empty column). Use the sequence *Calc / Probability Distributions / Normal* to bring up the "Normal Distribution" dialogue box and input the values for μ and σ where indicated in the screen capture. Specify the column that contains x and the column to hold the solution. Select *Cumulative probability* and then click *OK*. (Alternatively, you can input the x-value in the *Input constant* area; in that case, the output will not be added to the worksheet page.)

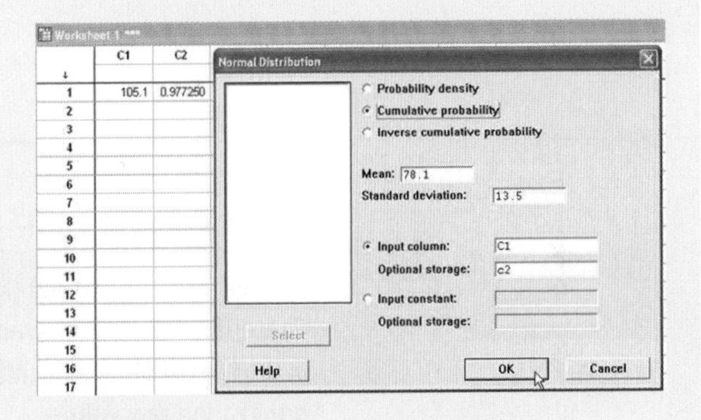

Although Minitab does not explicitly accept a z-value as the input for these calculations, this can be accomplished by using the same dialogue box as just described. Simply define the mean of the normal distribution as 0 and the standard deviation as 1. In relation to those standard values, any x-value that you input is, in effect, also the z-value.

SPSS

To calculate $P(X < x)$ for a normal distribution if μ, σ, and value x are known: Input the value for x into a worksheet cell (usually, at the top of an empty column). Use the menu sequence *Transform / Compute* to call up the "Compute Variable" dialogue box as shown in the left-hand screen capture. In the *Target Variable* window, input the name of the column to hold the solution. Input a formula in the *Numeric Expression* window of the form **CDF.NORMAL(VAR00001,μ,σ)**, replacing

VAR00001 with the name of the column holding x. (If needed, adjust the displayed number of decimal places for the solution variable.) Click *OK*.

Alternatively, calculate $P(Z < z)$ for a normal distribution if z has already been determined: Input the value for z into a worksheet cell (usually, at the top of an empty column). Proceed as described in the previous paragraph but, as shown in the right-hand screen capture, input a formula in the format **CDFNORM(z),** where **z** is the name of the column holding z.

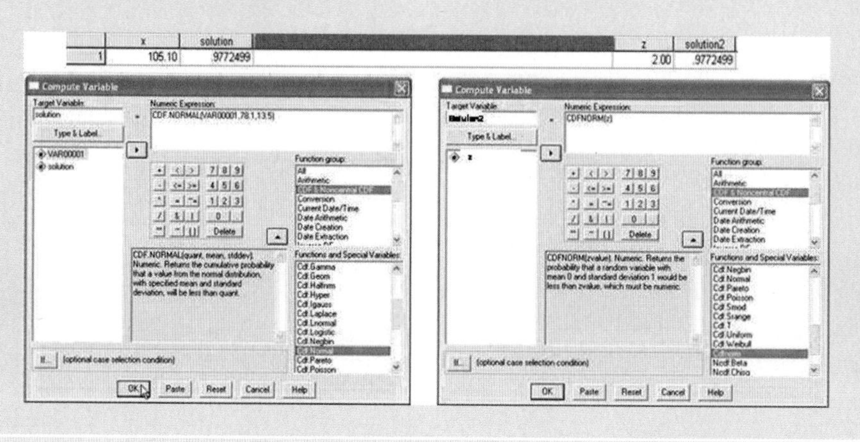

Example 8.4

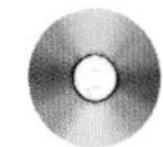
*(Based on the **US State Education** file)*

The performance of public education in the United States is monitored by the National Center for Public Policy and Higher Education. Partial results for 2002 have been saved in the file **US State Education.** To facilitate comparisons, the values displayed are indexes rather than raw values.

The variable Math_12 is based on the percentage of high school students from each state who take advanced math in their senior year. The distribution of this variable is very close to normal. Answer the following question based on the mathematics of the normal distribution.

If a U.S. state is selected randomly, what is the probability that the index value for the variable Math_12 is:

1. Below 75?
2. Below 100?
3. Above 75?
4. Above 100?
5. Between 75 and 100?
6. Between $z = -1$ and $z = +1$?

Solution

The solutions to (1) and (2) can be calculated directly with the computer, provided you know μ and σ for the population. (Review "In Brief 8.1(a).") If you are starting from raw data, the computer also can help to calculate the mean and standard deviation for the population (or estimate these from a sample) using the procedures given in Part 2 of this text. In the present case, we have the data from *all* U.S. states, so we calculate $\mu = 75.860$ and $\sigma = 17.988$. Inputting these values into the computer, together with the specified upper bound for the cumulative distribution, gives these results for (1) and (2):

1. $P(X < 75) = 0.48$
2. $P(X < 100) = 0.91$

Solutions to (3) and (4) require an extra step. As shown in Figure 8.5, the probability $P(Z > z)$ for any given z-value is the complement of $P(Z < z)$ for that z-value. (On a continuous scale, "Z exactly equals z" has zero probability.) The same principle applies if we start calculations from μ and σ; the standard z model underlies the subsequent calculations. Therefore:

3. $P(X > 75) = 1 - 0.48 = 0.52$
4. $P(X > 100) = 1 - 0.91 = 0.09$

As illustrated in this solution, you are encouraged to sketch the relationship between the probability returned by the CDF function (built into the software program or used in a table) and the probability that is required by the problem.

Figure 8.5
Greater Than Cumulative Probability Based on z

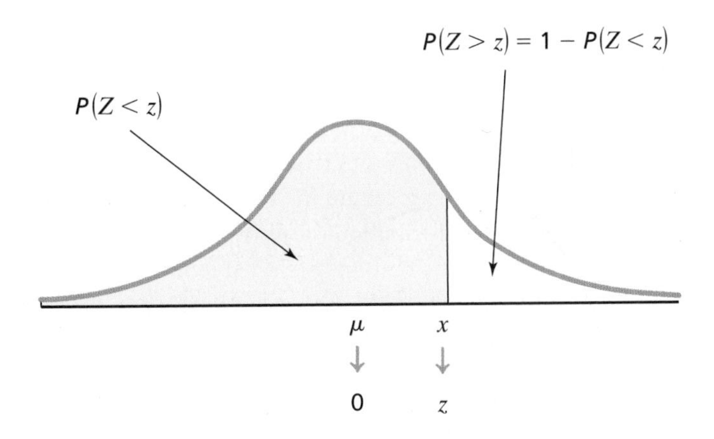

The solution to (5) also requires some additional manipulation. Statistical software can calculate the probability that if a member of the normally distributed population is randomly selected, its corresponding z-value will be *less than* a specified upper limit. In symbols, the software calculates $P(Z < z)$, for a z-value that is specified. Suppose we specify two z-values, z_1 and z_2, with z_2 being the larger. What is the probability that for a randomly selected individual, the corresponding z-value will lie between those limits?

Figure 8.6 illustrates that such problems can be solved in three steps: (1) First find $P(Z < z_2)$, for the larger z-value. (2) Then find $P(Z < z_1)$, for the smaller z-value. (3) Take the difference, which is the probability of selecting a value between those two limits. If we start calculations from μ and σ, rather than from z, the same principle of calculation applies. Therefore:

5. $P(75 < X < 100) = P(X < 100) - P(X < 75) = 0.91 - 0.48 = 0.43$
6. $P(-1 < Z < +1) = P(Z < +1) - P(Z < -1) = 0.8413 - 0.1587 \approx 0.68$
   (Compare the empirical rule.)

Figure 8.6
Between Cumulative Probability Based on z

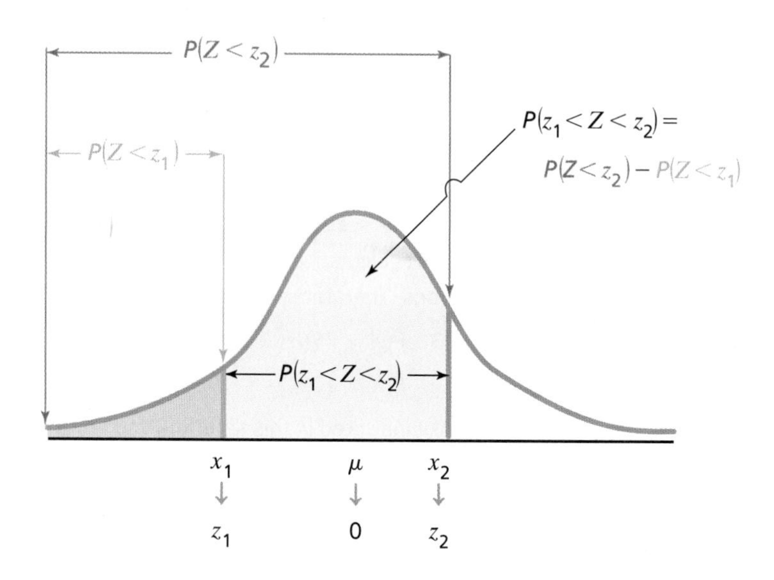

Section 7.3

Finding an x-Value That Corresponds to a Probability

The basic procedure in "In Brief 8.1(a)" was to start from a known x-value within a normal distribution (or from the corresponding z-value) and then calculate the probability that a randomly selected individual will have a z- or x-value less than that upper bound. In Example 8.4, we learned how to extend this technique to find cumulative probabilities for values greater than or between specified limits. This section presents the inverse operation: If the less than cumulative probability is the known starting point, how can we calculate the upper limit x in the normal distribution?

A manual solution would, again, be based on the mathematics of the standard normal curve. The upper bound of the region having the specified cumulative probability would be determined as a z-value. By solving Formula 8.2 for x, we can find the real-world value that corresponds to z. The software procedures are provided in Example 8.5.

Example 8.5

In its promotional material, Itron reports that for 40G gas meter modules that were supplied to Atlanta Gas Light, the estimated mean battery life is 20.30 years, with a standard deviation of 4.80 years.[1] Suppose that, based on this information, an Itron executive is developing a warranty for the company's batteries: Batteries that fail before x-years will be replaced at no charge. What value of x should be used in this promise?

Supplemental information is that the executive wants a large enough x-value that at least some batteries will fail within the warranty period; otherwise, the promise would look rather empty. Of course, she does not want to replace too many batteries free of charge. The executive plans for a warranty such that 5% of the batteries will fail within the warranty period. She believes that the population of battery lives is normally distributed.

Mathematically, the problem can now be framed like this: For a normally distributed population, with the specified mean and standard deviation, what is the value of *x* such that the cumulative probability of selecting a value less than *x* is 0.05? If it is desirable to choose a whole number as the *x*-value, what would be the impact of rounding *x* up to a whole number?

Solution

The basic model for the solution is displayed as Figure 8.7. (You are encouraged to start with such a sketch before proceeding with calculations.) This figure is similar to Figure 8.4, except that the inverse calculations are used. The cumulative probability $P(Z < z)$ is supplied as an input: 0.05, in this example.

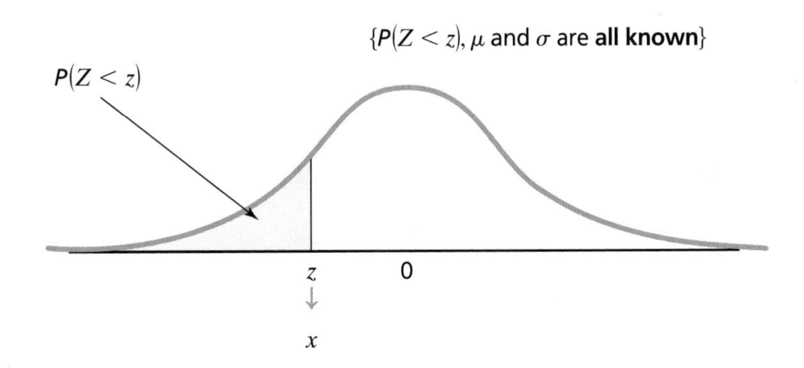

Figure 8.7
x Given the Less Than Cumulative Probability

Software programs (see "In Brief 8.1(b)") can find the *z*-value at the upper limit of that range. Given μ and σ as additional inputs, the software program can then solve for the real-world value of *x*. In "In Brief 8.1(b),"you will see that the precise *x*-value calculates to 12.4047 years.

In this example, the executive would probably prefer a whole-number value for the limit of the warranty period. Reusing techniques from the last section, she can estimate about how many batteries would fail during the warranty period if the warranty was rounded up to 13 years. In that case, 6.4% of the batteries would fail during the warranty period, somewhat increasing the cost of the warranty program from the goal of 5%.

• *In Brief 8.1(b) Finding x Given the Less Than Cumulative Probability*

All three software packages can directly calculate the upper bound of *x* within a normal distribution for a given less than cumulative probability. That is, they can find the limit, below which the area under the curve (representing a probability) has a given magnitude. If you know μ and σ for the normal distribution and input the less than cumulative probability, the software finds (internally) the upper bound of the range as a *z*-value and converts the result into an *x*-value for display. Excel will also allow you to output the *z*-score directly.

Excel

To find x for a normal distribution if μ, σ, and the probability $P(Z < z)$ are known: Choose any cell in the worksheet and input a formula in the format **=NORMINV**(P(Z < z),μ,σ) (left side of the illustration). Note that once x is determined: $P(Z < z) = P(X < x)$.

Alternatively, choose any cell in the worksheet, and input a formula in the format **=NORMSINV**(P(Z < z)) (right-hand side of the illustration). The output is a z-value that can be converted manually to x by solving Formula 8.2 for x: $x = \mu + z\sigma$.

Minitab

To find x for a normal distribution if μ, σ, and the probability $P(Z < z)$ are known: Input the value for $P(Z < z)$ into a worksheet cell (usually, at the top of an empty column). Use the sequence *Calc / Probability Distributions / Normal* to bring up the dialogue box shown in the screen capture and input the values for μ and σ where indicated. Specify the column that contains the probability and the column to hold the solution. Select *Inverse cumulative probability* and then click *OK*. (Alternatively, you can input the x-value into the *Input constant* area; in that case, the output will not be added to the worksheet page.) Note that once x is determined: $P(Z < z) = P(X < x)$.

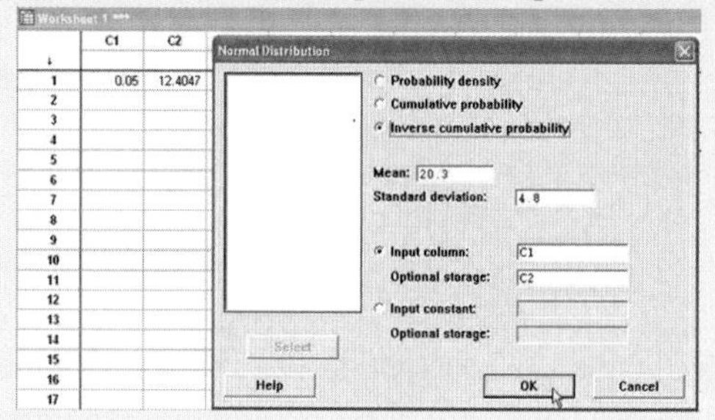

Although Minitab does not explicitly output a z-value for this calculation, this can be accomplished by using the same dialogue box as above. Simply define the mean of the standard normal distribution as 0 and the standard deviation as 1. In relation to those standard values, the x-value that is output is, in effect, also a z-value.

SPSS

To find x for a normal distribution if μ, σ, and the probability $P(Z < z)$ are known: Input the value for $P(Z < z)$ into a worksheet cell (usually, at the top of an empty column). Use the menu sequence *Transform / Compute* to call up the dialogue box shown in the screen capture. In the *Target Variable* window, input the name of the column to hold the solution. In the *Numeric Expression* window, input a formula of the form **IDF.NORMAL**(x,μ,σ), replacing ***x*** with the name of the column holding the value x. (If necessary, adjust the displayed number of decimal places for the solution variable.) Click *OK*. Note that once x is determined: $P(Z < z) = P(X < x)$.

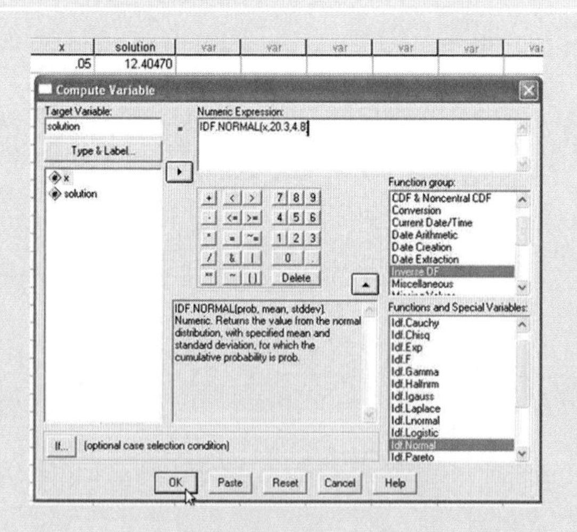

Although SPSS does not explicitly output a z-value for this calculation, this can be accomplished by using the same dialogue box. Simply input the mean of the standard normal distribution as 0 and the standard deviation as 1. In relation to those standard values, the x-value that is output is, in effect, also a z-value.

Example 8.6

In Example 8.5, we reported a mean battery life of 20.30 years, with a standard deviation of 4.80 years, for modules of certain gas meters.

1. A supplier recommends that each module be replaced automatically after a certain number of years—even if it is still functioning. At what number of years x should the modules be replaced automatically in order to meet an expectation that 10% of the modules would still be functional when replaced?
2. The supplier proposes advertising a "standard working life" for its batteries. The company defines this concept as the range of battery lives that applies to 80% of its batteries, centred around the mean. What are the years x_1 and x_2 that define this range of time? Is this advertising plan a good idea?

Solution

1. Compared to the warranty problem in Example 8.5, the solution to the first problem in this example requires an extra step. As shown in Figure 8.8, if the greater than cumulative probability is given, then the less than cumulative probability can be calculated as the complement. (The probabilities can be expressed either in terms of a limiting x-value or as an equivalent limiting z-score. In either case, the area over *exactly* the limiting x- or z-value is zero.) In the current example, the less than cumulative probability can be calculated as $P(Z < z) = 1 - 0.10 = 0.90$. Applying the procedures given in "In Brief 8.1(b)" for probability $= 0.90$, $\mu = 20.3$ years, and $\sigma = 4.8$ years, we estimate that $x = 26.45$. If the supplier replaced all units aged about 26.50 years, 10% of them would still be functional when replaced.

Figure 8.8
x Given the More Than Cumulative Probability

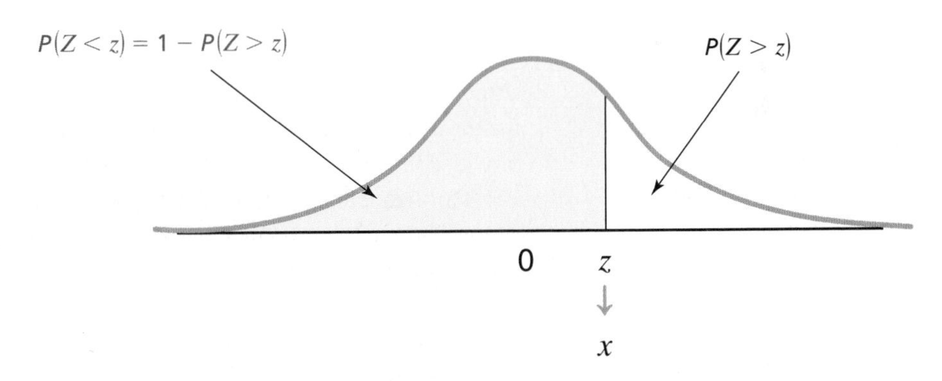

2. The solution to the second problem also requires additional manipulation. Consider year x_1 the lower bound of the "standard battery life" range, and year x_2 as the upper bound. If 80% of the data are included in that range, 20% is left outside the boundaries. We are told that the range is centred at the mean, so half (10%) must be below boundary x_1 and half (10%) above boundary x_2.

   Applying the procedures in "In Brief 8.1(b)" for probability $= 0.10$, $\mu = 20.3$ years, and $\sigma = 4.8$ years, we estimate $x_1 = 14.15$. We already found in solution 1 that the value for x_2 (less than cumulative probability is 0.90) $= 26.45$. Therefore, the standard battery life can be estimated at 14 to 26.50 years. Because the range is so wide, the supplier might choose not to publicize this variance in its product quality.

• Hands On 8.1 Are This Chapter's Procedures Robust?

*(Based on the files **Early_Returns, Meteors_N_Amer, Purchasing Power_World, Star_Distances**)*

Many techniques in this book are designed to work with data that follow a certain distribution. For example, procedures described in this section are intended to apply if the underlying data follow a normal probability distribution. That is why in Example 8.5, it said: "She believes that the population . . . is normally distributed."

We say that a statistical method is **robust** if it reaches reasonably accurate conclusions about the data—even if its assumptions about the data set do not fully apply. Therefore, we would say that the procedures in Section 8.1 of this book are reasonably robust if their calculated results for cumulative probabilities mirrored the actual cumulative probabilities in a data set—even if the data are not that normally distributed.

In "Hands On 3.4," we already experimented graphically with some actual distributions and compared them with the normal distribution. You may recall producing histograms and stemplots for the following data sets: The winning margin column (W_Margin) in the file **Early_Returns;** the approximate age in millions of years column (Av_Age) in the file **Meteors_N_Amer;** the purchasing power per capita column (GDP_CAD) in the file **Purchasing Power_World;** and the distances column (Distance) in the file **Star_Distances.**

Now take the analysis of those data sets one step further: For each data set construct a cumulative relative frequency distribution table. Observe the cumulative relative frequencies reached at the end of each class. For example, the "Winning Margins" table may have a class that ends at 400, showing a less than cumulative relative frequency of about 3.9%. With your software, you can find the mean and standard deviation for that full data set and use the procedures provided in "In Brief 8.1(a)" to estimate the cumulative relative frequency below $x = 400$—realizing that the procedure assumes a normal distribution. How close are the computer's estimated cumulative percentages from the actual values in your tables? Repeat this comparison for each of the displayed cumulative relative frequencies for each of the four data sets.

Based on this small sample, would you say that the techniques introduced in this chapter appear to be robust in terms of making reasonable predictions for less than normally distributed data? Explain and justify your conclusion.

8.2 Assessing the Normality of a Distribution

The model of the normal curve provides a powerful tool for solving problems in statistics. The model helped us in Section 8.1 to estimate cumulative probabilities given a specified upper limit for a variable x or, given a cumulative probability, to find that upper limit. In "Hands On 3.4" and "Hands On 8.1," we alluded to a practical obligation that follows from having this powerful tool: We must be sure when we use the model that it really applies to the data set in front of us. In this section, we examine ways to check whether the condition of normality is met. Fortunately, with software programs, it is no longer difficult nor very time-consuming to make this check a regular practice.

If you have already compiled summary statistics for a data set, you can inspect to see if some well-known properties of the normal curve appear applicable for your data: Are the mean and median quite close relative to the range of the data? Are the large majority of values (about 95%) within two standard deviations of the mean? Do virtually all of the values fall within three standard deviations of the mean? (Allow for the occasional more extreme value as a possible outlier.)

Example 8.7(a)

*(Based on the **Vehicle_Ratings** file)*

The data file **Vehicle_Ratings** is based on a survey conducted in 2002 by the Canadian Automobile Association. Seven of the included variables, and their interpretations, are as shown in the table below. Calculate some basic summary statistics for each variable and then use these data to assess which of the variables appears to have a normal distribution.

Variable Names	**Variable Descriptions**
Rep_Cost	Average repair cost (CDN$)
Fuel_Cst	Average fuel cost per week (CDN$)
Ins_Cost	Average insurance cost (CDN$)
Satisfac	Customer satisfaction with the vehicle (%)
Sat_Sale	Customer satisfaction with the dealer: Sales Department (%)
Sat_Serv	Customer satisfaction with the dealer: Service Department (%)
Repeat	Repeat purchase (%)

Solution

The first step is to calculate basic summary statistics for each variable. The yellow-highlighted statistics in Figure 8.9 could easily be determined by the methods given in Chapters 4 and 5. Interpret the standard deviation s for each sampled variable as an estimate for σ in the full population. Then the green-highlighted statistics can be calculated by simply adding or subtracting 2σ or 3σ from the mean, as indicated. Perhaps the easiest way to

Figure 8.9
Summary Statistics Compiled to Assess Normality

Variable Names:	Rep_Cost	Fuel_Cst	Ins_Cost	Satisfac	Sat_Sale	Sat_Serv	Repeat
n:	31	31	31	31	31	31	31
Means:	412.77	31.25	1,058.03	85.16	71.32	73.94	88.65
Medians:	412.00	30.18	1,037.00	89.00	71.00	77.00	91.00
Ranges:	564.00	38.75	296.00	30.00	30.00	33.00	26.00
Standard deviations:	134.38	7.81	80.94	7.52	7.28	8.82	6.53
Mean -2σ	144.01	15.64	896.14	70.11	56.76	56.30	75.58
Mean $+2\sigma$	681.54	46.86	1,219.92	100.21	85.89	91.57	101.71
Mean -3σ	9.62	7.83	815.20	62.59	49.48	47.49	69.04
Mean $+3\sigma$	815.92	54.67	1,300.87	107.73	93.17	100.38	108.25
Frequency below -2σ	0	0	0	1	1	0	2
Frequency beyond $+2\sigma$	1	1	1	0	0	0	0
Frequency below -3σ	0	0	0	0	0	0	0
Frequency beyond $+3\sigma$	0	1	0	0	0	0	0

Source: Canadian Automobile Association. *Vehicle Ownership Survey Results, 2004.* Found at: http://www.caa.ca/e/automotive/vos-results.shtml

complete the table for each variable is to sort the variable values from low to high and manually count the numbers of cases below or above the limits specified. (Formulas or macros could also be written for this step.)

Unfortunately, for assessing normality, not all these results are easy to interpret without additional information. The Rep_Cost variable appears closest to the conditions for being normal: The mean and median are less than 1 unit apart, compared to the much larger standard deviation of 134.38 and range of 564. There are no outliers beyond $\pm 3\sigma$ from the mean; and only 1 case out of 31 (3.2%) has a value beyond $+2\sigma$ from the mean.

The variable Fuel_Cost gives some evidence of skewing: The mean is larger than the median and there are two large values, one greater than the mean by 2σ and the other by more than 3σ, which is 3.2% of the cases. The variable Repeat may also be somewhat skewed. Its mean and median are farther apart in comparison to the size of the standard deviation, and two of the 31 cases (6.5%) are both more than 2σ below the mean. Yet for such small samples, these types of discrepancies do not present a clear picture. The other variables appear to be relatively symmetrical, based on the differences between the mean and median. Most have only one case (3.2%) beyond 2σ from the mean.

With readily available software programs, a second approach for assessing the normality of a distribution has become easy to implement: Creating charts such as histograms, stemplots, and boxplots for a distribution can be highly useful to supplement the analysis from the first approach. You have already learned to create these charts in Sections 3.2, 3.4, and 5.4. For the histograms, you do not need (except in Excel) to formally design the class boundaries; it is sufficient to work with the default values generated by the software.

Example 8.7(b)

For each of the seven variables discussed in Example 8.7(a), create a histogram, stemplot, and boxplot. Compare and contrast what these figures tell us about the normality of these variables' distributions, in relation to our previous conclusions.

Solution

Using methods from Chapters 3 and 5, create charts similar to those in Figure 8.10. For the histograms, the apparent distributions may vary slightly based on the class widths and class boundaries that you select. Figure 8.10 uses the default boundaries generated by SPSS. Your software may include also an option to overlay a normal curve on the histogram to enhance visually assessing the normality.

The histogram for the Rep_Cost variable is informative; we see that the distribution is not quite as symmetrical as we thought in our solution to Example 8.7(a). The variable Sat_Sale looks reasonably normal, consistent with our earlier solution. As expected from the summary statistics, the variable Repeat is quite skewed to the left. Interestingly, the variable Fuel_Cost (which we had thought earlier might be skewed) turns out to be rather normally distributed—*if* we interpret the one largest value as an outlier from the main distribution.

The charts for Ins_Cost, Satisfac, and Sat_Serv are also quite revealing. Despite the ambiguity of our earlier analysis, based on the descriptive measures, we find that each of these distributions, in its own way, varies noticeably for the normal distribution. One distribution (Satisfac) is skewed to the left, and the other distributions appear to be bimodal.

Figure 8.10

Charts Created to Assess Normality

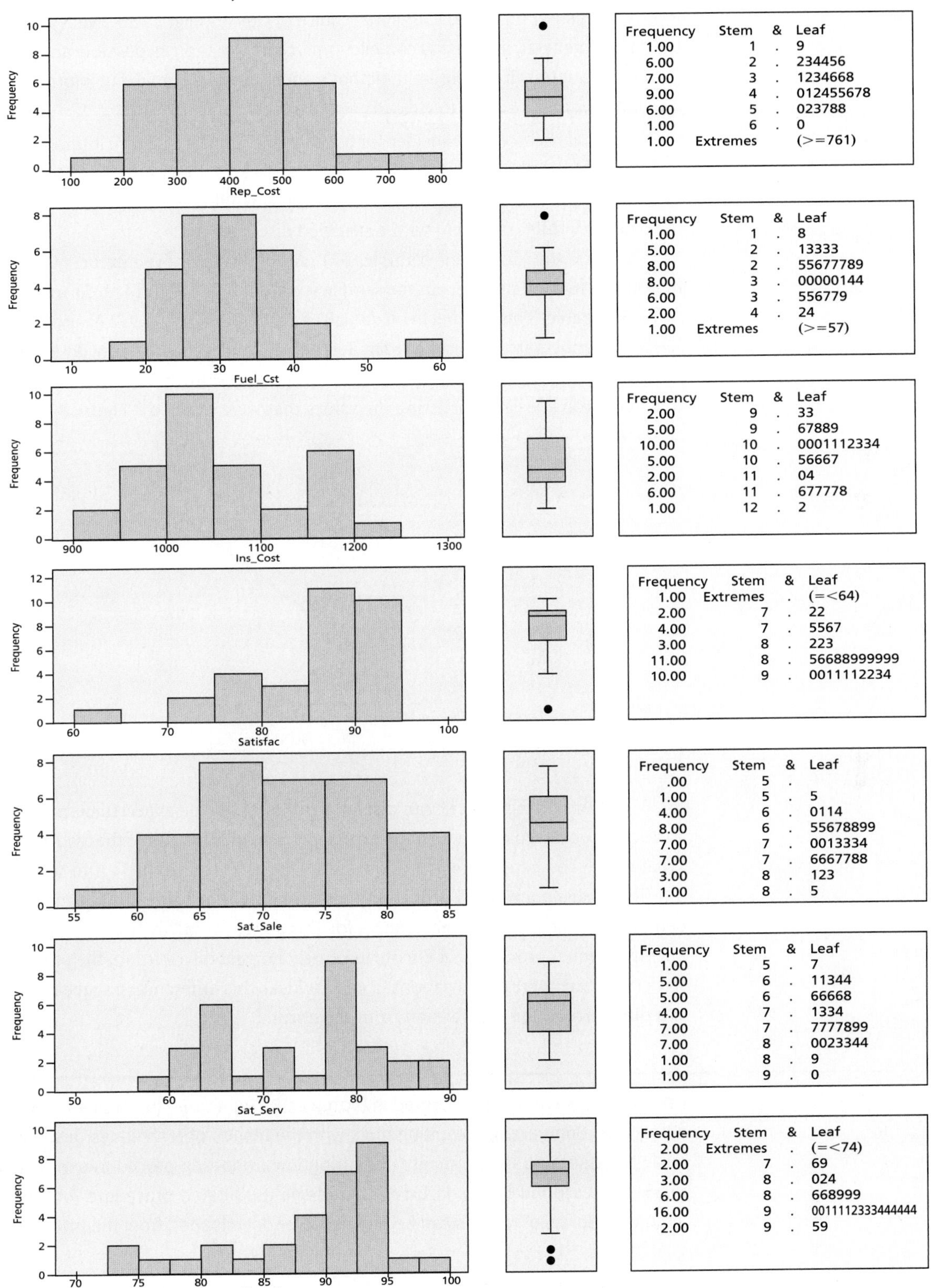

Source: Canadian Automobile Association. *Vehicle Ownership Survey Results, 2004.* Found at: http://www.caa.ca/e/automotive/vos-results.shtml

It is clear from these charts that looking *only* at the summary measures to assess normality is generally insufficient to get a reasonable picture of the data. On the other hand, if you do a histogram check, your initial choice of classes might also give a false image of the data. In general, it is best to compile as much information as possible and explore the data more carefully if the different methods appear to give conflicting information.

The final method we consider for assessing normality of a distribution is based on the **normal probability plot**. In general, a probability plot compares to the actual distribution of a set of n-values, the distribution you would *expect* the n-values to follow *if* the data were fully consistent with a specified distribution.

Consider this simplified example. You randomly select five numbers from a distribution in which all numbers in the sample space (of values from 1 to 5) are expected to be equiprobable. The numbers you actually select are 2.4, 1.1, 4.9, 2.2, and 4.8. Sort the selected numbers in increasing order: 1.1, 2.2, 2.4, 4.8, and 4.9. How do these ordered numbers compare to the values you would expect to draw if all numbers in the range were equiprobable (after ordering the values that were selected)? Figure 8.11 shows the comparison.

Figure 8.11
Simple Example of a Probability Plot

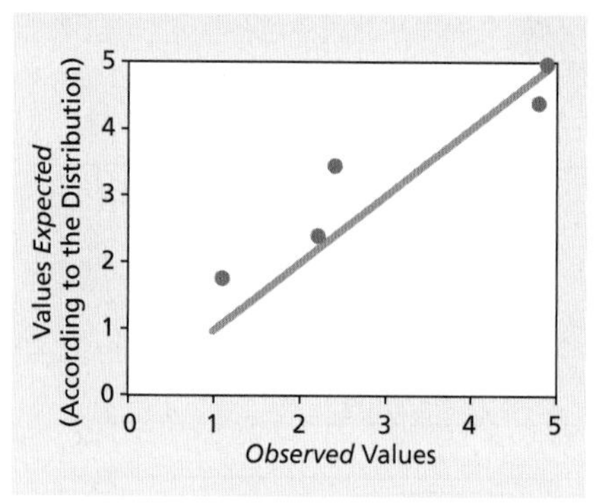

Variations of Figure 8.11 can display, for the y-axis, the expected cumulative probabilities for observed x-values, or the expected values of quantiles of the distribution (e.g., What would you expect the first of 5 values to be? The second of 5? And so on.)

To construct a normal probability plot, we apply the same basic principles as just described. Some computer procedures for this are given in "In Brief 8.2." (We will not attempt manual procedures.) For an absolutely normal distribution, the plots would be straight. Where there are deviations from that ideal, this difference in shape is interpreted as a difference of the distribution from the normal.

Example 8.7(c)

For each of the variables discussed in Examples 8.7(a) and 8.7(b), create a normal probability plot. Compare the resulting plots with the results of techniques described previously for assessing the normality of distributions. (Plotting procedures in Minitab and SPSS are described below. In Excel, there is no automated procedure for these plots, although add-ins or macros can be purchased or downloaded from the Internet.)

Solution

The outputs in Figure 8.12 are based on the Minitab version of the procedure. Minitab's approach uses the expected cumulative probabilities that correspond to each value in the actual sample. It also displays confidence intervals around the line of expected values. (Confidence intervals, which express an estimate as a *range* of values, are discussed in detail in Chapter 10.)

If viewed together with the charts in Figure 8.10, the charts in Figure 8.12 can be very informative. The actual value for Rep_Cost differs most from the expected normal at around 250. In the histogram, this is where the bar is taller than might have been

Figure 8.12

Probability Plots Created to Assess Normality

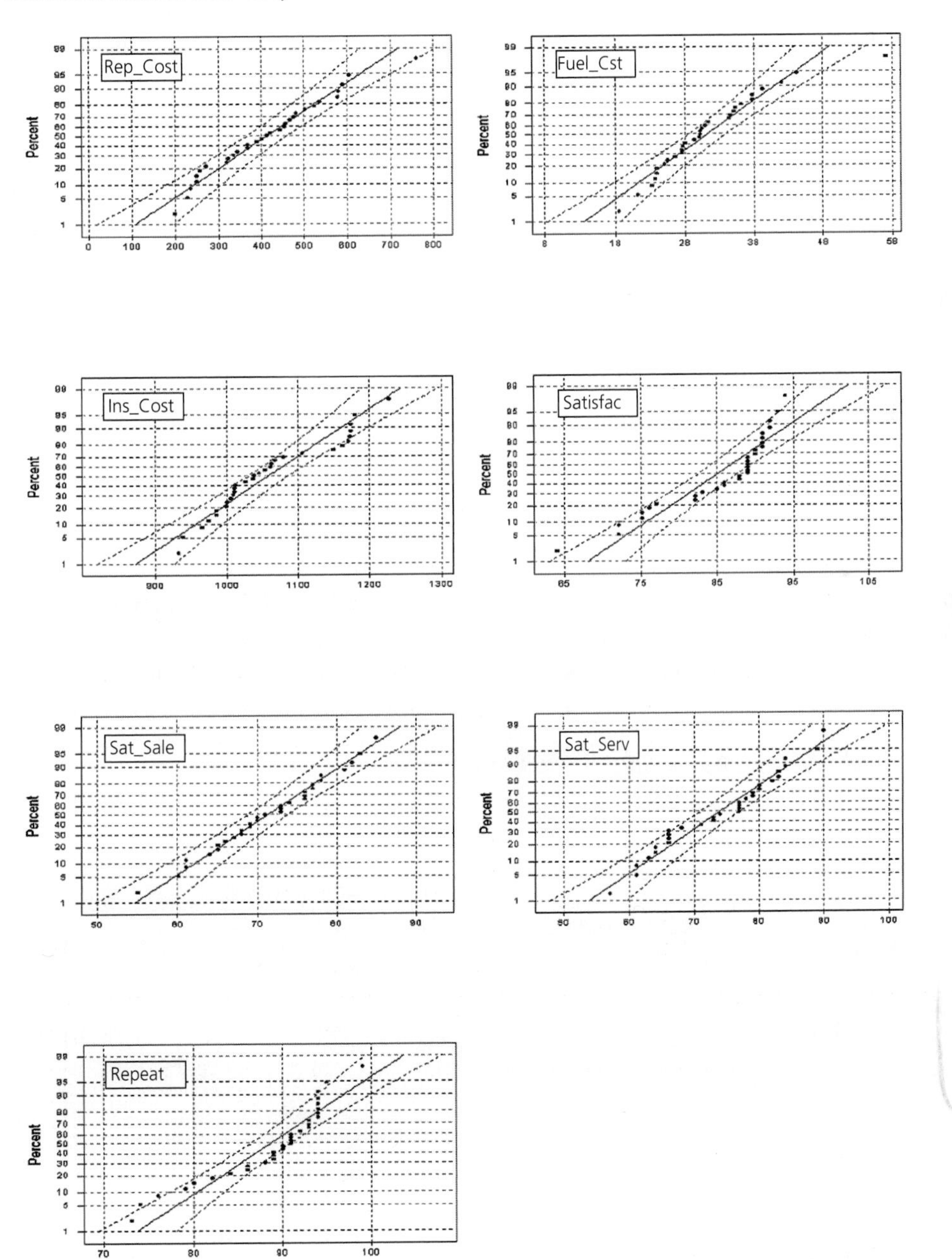

Source: Canadian Automobile Association. *Vehicle Ownership Survey Results, 2004.* Found at: http://www.caa.ca/e/automotive/vos-results.shtml

expected. Notice how the outlier that appears in the histogram and boxplot for Fuel_Cost shows up in the probability plot (at the upper right) as a point very far from the line showing expected values.

Noteworthy about the plot for Ins_Cost is that, at about 1,150, some points fall outside the confidence intervals. This implies that the difference in the actual values from the expected values in that area is greater than might be explained by simple chance variation of the data. On the histogram, we see that the second modal class falls in that same area, which caused us to question the normality in Example 8.7(b), but then we did not have a way to measure whether the amount of deviation was noteworthy. The plots for Satisfac, Sat_Serv, and Repeat also have some points outside their respective confidence bands, which confirms the appearance of some non-normality in the histograms. In contrast, the small discrepancies in Sat_Sale are not greater than might be expected, if the underlying population was known to be normal.

• *In Brief 8.2 Creating a Normal Probability Plot*

Excel

Excel does not have a specific function for this procedure, but add-ins or macros for the purpose can be purchased or developed using built-in functions.

Minitab

Begin with the menu sequence *Graph / Probability Plot.* In the first dialogue box (not shown in the illustration), a default option of *Single* is highlighted; click *OK.* In the next dialogue box (the first screen capture), select the name of the variable to be analyzed and then select *Distribution.* In the next dialogue box (the second screen capture) ensure that the name displayed in the *Distribution* box is "Normal." Click *OK* twice to exit both dialogue boxes.

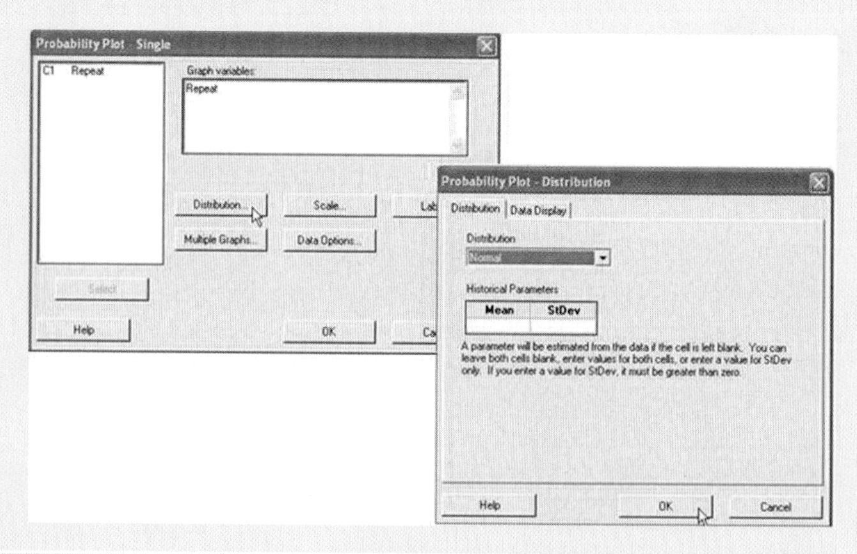

SPSS

To generate a plot in the format expected normal value versus corresponding observed value, use the menu sequence *Graphs / Q-Q* to access the dialogue box shown in the screen capture. Select the variable to be analyzed. The output shown in the graph on the left in the illustration is somewhat different from the version in Figure 8.12, whose dependent variable is the expected cumulative percent.

SPSS does not offer a confidence interval feature, which in Example 8.7(c) provided a sense of how much variation from the expected value is particularly noteworthy. The graph on the right in the illustration, however, serves a similar purpose. The plot is redrawn to show, for each actual observed value, the standard deviation of that value from the expected value. Compare the extreme points on that plot to the extreme points in Figure 8.12.

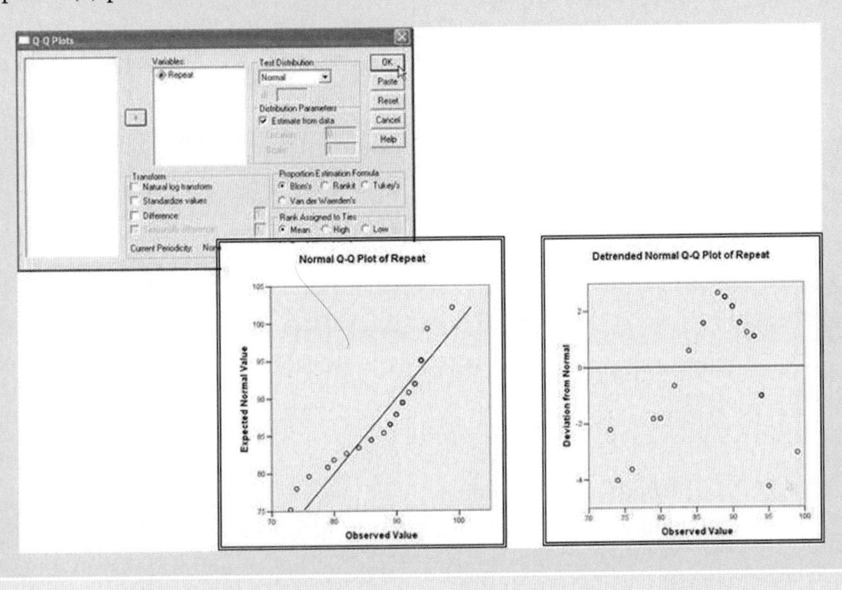

Seeing Statistics

Section 7.5

• Hands On 8.2 *Anticipating the Limits of Robustness of This Chapter's Procedures*

(Based on the ***Vehicle_Ratings*** *file)*

In "Hands On 8.1," we explored the relationship between actual cumulative frequencies and the predictions for those cumulative frequencies, based on assuming some distributions were normally distributed. In effect, we were testing the robustness of the normal-based model.

In this section, we have examined some techniques for evaluating the assumption that a given distribution is normal. By using these techniques, it would be useful to gain a hands-on sense of how non-normal the distribution has to *appear* for the normal-based calculations to start being unacceptably wide of the mark.

Open the file **Vehicle_Ratings,** and consider the seven variables that were introduced in Example 8.7(a). In Figures 8.9, 8.10, and 8.12, we created models to assess the relative degrees to which these distributions are normal.

Now use the procedure followed in "Hands On 8.1" to analyze those data sets: For each data set, construct a cumulative relative frequency distribution table. (Use the same class boundaries as in the histograms for Figure 8.10.) Observe the cumulative relative frequencies reached at the end of each class. Use software to find the mean and standard deviation for each full data set, and use the procedures in "In Brief 8.1(a)" to *estimate* the cumulative relative frequencies as of the end of each class. Repeat this for all seven data sets. How close are the computer's estimated cumulative percentages from the actual values in your tables?

In particular, which of the estimated cumulative probabilities, for which of the data sets, seems farthest removed from the corresponding actual cumulative probabilities in the sample? Reexamine all of the patterns in Figure 8.10. Can you identify features in the individual histograms that could tell you in advance, by visual inspection, when the estimates were likely to be too large or too small, and to what degree? Explain.

8.3 The Uniform Probability Distribution

Another familiar distribution is the **uniform probability distribution.** Suppose you insert a coin into a vending machine to obtain approximately 150 mL of coffee. It is unlikely that you will get *exactly* 150 mL of refreshment. Depending on the precision of the dispenser's timing mechanism and the size of its cups, you might, with equal probability, obtain *any* amount between, for example, 135 mL and 165 mL of coffee.

The above example is for a *continuous* uniform distribution, which is the focus of this chapter. As a comparison, we will first review the *discrete* version of the distribution, which might help to clarify the model.

A common example of a discrete uniform probability distribution is the result of spinning a roulette wheel. There are a finite number of values possible, and any one of these is equally likely. Another example is depicted in Figure 8.13. The Quebec Banco lottery is based on selecting 20 numbers out of a possible 70 numbers (the integers 1, 2, 3, . . . , up to 70). Within one game, the probabilities do not remain constant, due to drawing without replacement (see Section 6.3). But, over the course of many complete games, a roughly equal number of ones, twos, threes, and every other number will have been drawn, as we see in the figure.

The frequencies in Figure 8.13 can be converted into relative frequencies and interpreted as probabilities. Each bar represents a possible outcome and its length corresponds to the probability of its occurrence. Before a game begins, all numbers have an equal probability of being among the 20 numbers chosen.

Suppose that in a new version of Banco, you draw numbers that range from 1.000 to 70.000 *in increments of 0.001*. It would now be impossible to draw a discrete distribution for all of the probabilities. The individual values on the x-axis could not be distinguished, and any particular number, down to that precision, would have very little chance

Figure 8.13
Distribution of Numbers Drawn in the Quebec Banco Lottery, Sept. 1999–Nov. 2005

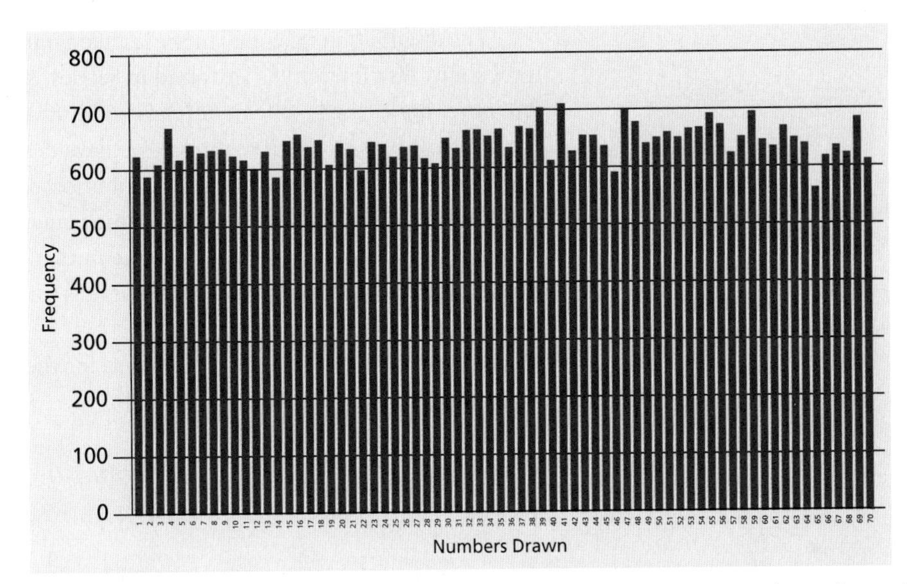

Source: Used by permission of Lottery Canada.com. Found at: http://www.lotterycanada.com/lottery/?job=frequency_chart&lottery_name=qc_banco&order=0&years=0

of occurring. Instead, as in Figure 8.14, we could use a histogram to display small *ranges* of x-values and frequencies at which actual values fall within those ranges.

Figure 8.14
Distribution of Numbers Drawn If the Scale Was Nearly Continuous

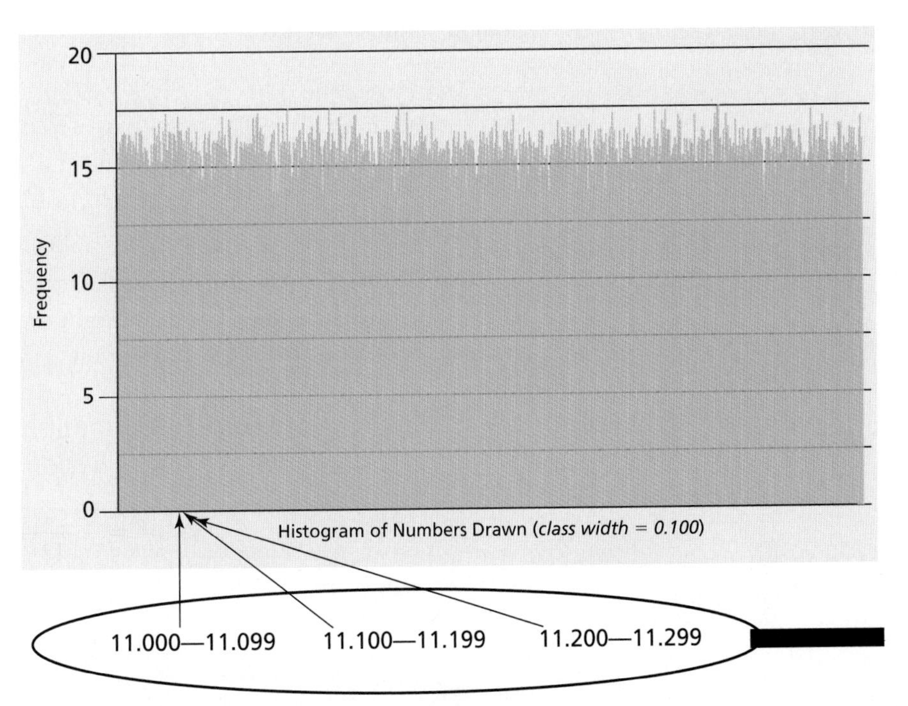

Source: Used by permission of Lottery Canada.com. Found at: http://www.lotterycanada.com/lottery/?job=frequency_chart&lottery_name=qc_banco&order=0&years=0

Next, suppose the population was infinite, and that the balls could have numbers *anywhere* in a continuous range between 1 and 70. There would be enough data to make all class widths for the histogram infinitesimally small and still obtain a similar-looking (rectangular) frequency distribution.

The distribution imagined in the last paragraph is, in effect, the continuous uniform probability distribution. As we found in Section 8.1, the class widths would approach—but never quite reach—a *limit* of (class width = zero). Converting this histogram (with very narrow bars) to a curve format, we have described a probability density curve, in this case for a uniform probability distribution of a continuous variable. As in Section 8.1, the *area under the curve* for any x-value range of width > 0 can be interpreted as the probability of randomly drawing a value in that range from the population. Again, by convention, the area under the entire curve is interpreted as the full sample space, and has a numerical value equal to 1.00.

The probability distribution function for the continuous uniform probability distribution is as shown in Formula 8.3.

Formula 8.3 Probability Distribution Function for the Continuous Uniform Probability Distribution

$$f(x) = \frac{1}{b - a}$$

where x = Any value within the sample space for the distribution
b = Upper bound of the distribution
a = Lower bound of the distribution

The expected value (mean), variance, and standard deviation of this distribution are given in Formulas 8.4–8.6.

Formula 8.4 Expected Value for a Continuous Uniform Probability Distribution

$$E(X) = \mu = \frac{b + a}{2}$$

Note that the formula simply finds the mean of the upper and lower values of the range—since the curve has a constant height over the range. Graphically, this value is always in the exact centre of the distribution.

Formula 8.5 Variance of a Continuous Uniform Probability Distribution

$$\sigma^2 = \frac{(b - a)^2}{12}$$

Formula 8.6 Standard Deviation of a Continuous Uniform Probability Distribution

$$\sigma = \sqrt{\sigma^2} = \sqrt{\frac{(b - a)^2}{12}}$$

Example 8.8

According to a Niagara Falls website (http://www.iaw.com/~falls), the Horseshoe Falls are currently receding at a rate of 0.3 to 0.4 inches (approximately 0.76 to 1.02 cm) per year.[2] Presume that the actual amount of recession in any given year has an equal chance of falling anywhere within that range.

1. What is the expected value for how far the Horseshoe Falls will recede in a year?
2. What is the standard deviation for that variable?

Solution

To solve these problems, we can use Formulas 8.4–8.6. The lower and upper bounds of the sample space (a and b) are 0.3 and 0.4, respectively.

1. $E(X) = \frac{b + a}{2} = \frac{0.4 + 0.3}{2} = 0.35$ inches

2. $\sigma^2 = \frac{(0.4 - 0.3)^2}{12} = 0.000833333$

   $\sigma = \sqrt{\sigma^2} = \sqrt{0.000833333} = 0.029$ inches

Probability Calculations Based on the Continuous Uniform Probability Distribution

The models for performing these calculations are similar to those we used for the normal distribution. In the two parts of Figure 8.15, probabilities are represented by the proportions of the area under the curve that fall below or above certain, specified values. (The area of the curve above [$X =$ *exactly* x] is zero, and so not represented in the probability expressions.)

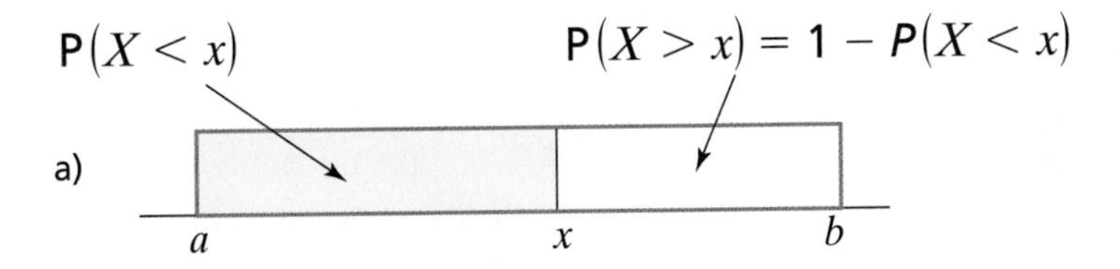

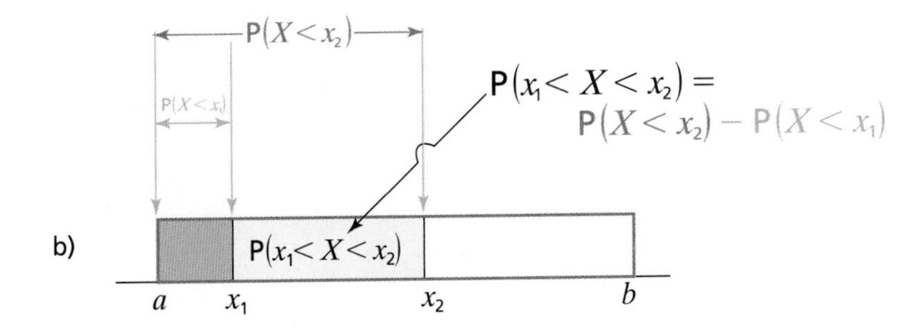

Figure 8.15
Cumulative Probabilities Within a Uniform Distribution

To calculate cumulative probabilities, the mathematics is easier than for the normal distribution, because the area under the curve is just a rectangle (i.e., the height is constant). In Figure 8.15(a), we know the bounds of the uniform distribution, a and b, and specify a value x within those bounds. The proportion of values in the distribution that are less than x is simply the ratio of values between a and x, versus the whole range of values between a and b. Formula 8.7 is the expression of this.

Formula 8.7 Less Than Cumulative Probabilities for a Uniform Distribution

$$P(X < x) = \frac{x - a}{b - a}$$

Other types of cumulative probability can also be derived from the above procedure. The probability that $P(X > x)$ is found as the complement of $P(X < x)$. For two values from the distribution x_1 and x_2 (x_2 being the larger), the probability that a selected individual falls between those values is $P(X < x_2) - P(X < x_1)$.

Example 8.9

Annually, the Horseshoe Falls in Niagara, Ontario recede between 0.3 and 0.4 inches (approximately 0.76 to 1.02 cm).[3] Presuming a uniform distribution of this variable, what is the probability that in a randomly selected year, the Horseshoe Falls will recede:

1. Less than 0.35 inches (0.89 cm)?
2. Less than 0.38 inches (0.97 cm)?
3. More than 0.38 inches (0.89 cm)?
4. Between 0.35 and 0.38 inches (0.89 and 0.97 cm)?

Solution

Steps for the manual solutions are shown below. (For the portions that can be automated, see also "In Brief 8.3(a).") For this problem, *a* and *b* are 0.3 and 0.4 inches, respectively. Parts (1) and (2) can be solved directly with Formula 8.7. Parts (3) and (4) need to be derived.

1. $P(X < 0.35) = \frac{0.35 - 0.3}{0.40 - 0.3} = \frac{0.05}{0.10} = 0.50$
2. $P(X < 0.38) = \frac{0.38 - 0.3}{0.40 - 0.3} = \frac{0.08}{0.10} = 0.80$
3. $P(X > 0.38) = 1 - P(X < 0.38) = 1 - 0.80 = 0.20$
4. $P(0.35 < X < 0.38) = P(X < 0.38) - P(X < 0.35) = 0.80 - 0.50 = 0.30$

• *In Brief 8.3(a) Computing a Less Than Cumulative Probability for a Uniform Distribution*

Excel

For Excel, you must use the manual approach and formula, as described in Example 8.9.

Minitab

To calculate $P(X < x)$ for a continuous uniform distribution if *a*, *b*, and value *x* are known: Input the value for *x* into a worksheet cell (usually, at the top of an empty column). Use the sequence *Calc / Probability Distributions / Uniform* to bring up the "Uniform Distribution" dialogue box and then input the values for a and b where indicated (see the screen capture). Specify the column that contains the probability value x and the column to hold the solution. Select *Cumulative probability* and then click *OK*. (Alternatively, you can input the probability value x in the *Input constant* area; in that case, the output will not be added to the worksheet page.)

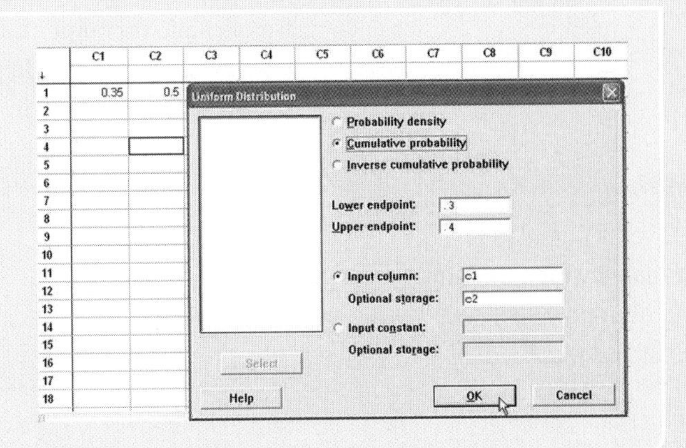

SPSS

To calculate $P(X < x)$ for a continuous uniform distribution if a, b, and value x are known: Input the value for x into a worksheet cell (usually, at the top of an empty column). Use the menu sequence *Transform / Compute* to call up the dialogue box shown in the screen capture. In the *Target Variable* window, input the name of the column to hold the solution. Input a formula into the *Numeric Expression* window of the form **CDF.UNIFORM(VAR00001,*a*,*b*)**, replacing **VAR00001** with the name of the column holding x. (If necessary, adjust the displayed number of decimal places for the solution variable.) Click *OK*.

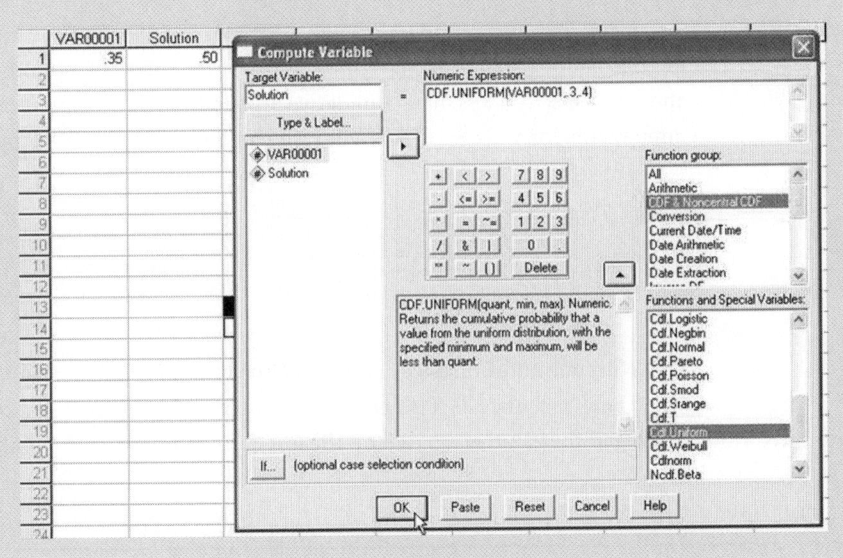

Finding an x-Value That Corresponds to a Probability in a Uniform Distribution

We have seen how to start from a known x-value within a uniform distribution and calculate the probability that the value of a randomly selected individual will be less than x as an upper limit. In Example 8.9, we also illustrated how to find cumulative probabilities for values greater than or between specified limits. This section presents the inverse operation: If the less than cumulative probability is the known starting point, how can we calculate the upper limit x for the uniform distribution? The problem is presented visually in Figure 8.16. In the figure, the area of the curve above $[X = \textit{exactly } x]$ is zero, and so not represented in the probability expressions.

Figure 8.16
x Given a Cumulative Probability for a Uniform Distribution

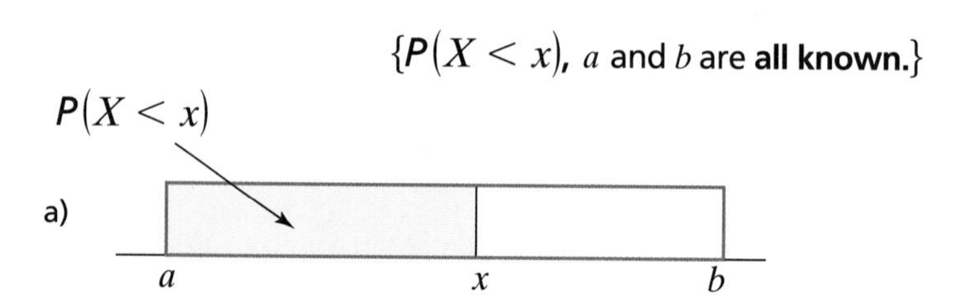

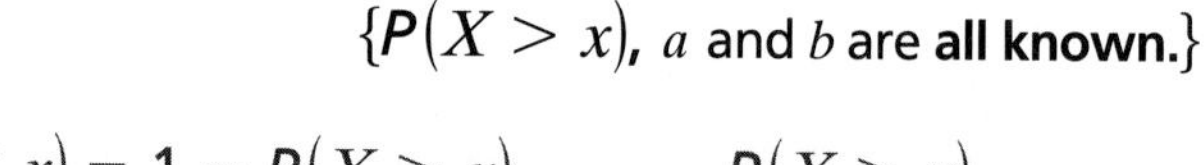

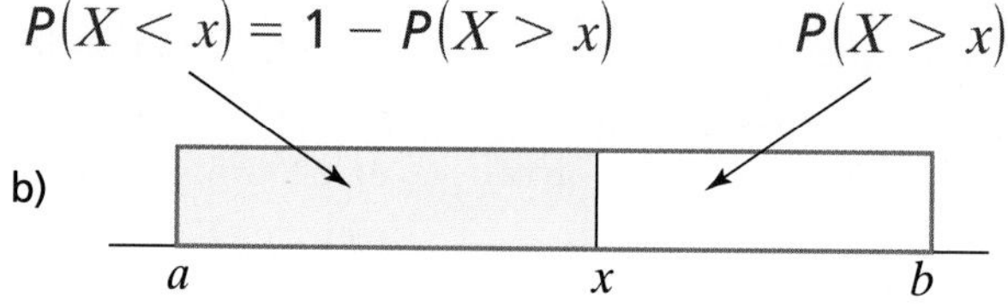

Again, because of the rectangular shape of the distribution and its constant height, the calculations are rather straightforward. If you know the bounds of the distribution a and b and specify a less than cumulative probability $P(X < x)$, you can calculate the upper limit x with Formula 8.8. (This can be derived from Formula 8.7 by solving for x.)

Formula 8.8 Upper Bound of an Area Corresponding to a Cumulative Probability in a Uniform Distribution

$$x = a + \{[P(X < x)] \times (b - a)\}$$

With one additional step, you can also calculate a *lower* limit x if you are given $P(X > x)$: First, calculate $P(X < x)$, as the complement of $P(X > x)$; i.e., use $1 - P(X > x)$. Then find the value of x based on the less than cumulative probability that you have just found.

Example 8.10

The daily water discharge of Little River near Grand Falls, New Brunswick, varies seasonally. However, the quantity discharged from early August to early September each year appears to be uniformly distributed, ranging anywhere from 2.0 to 3.5 cubic metres per second.[4]

A commercial enterprise has requested permission to periodically siphon off water from that location between those dates, on days when the discharge rate is near its highest. The authorities are willing to let this occur on about 13% of the days (i.e., about four days) during that time period each year. What should they set as the minimum daily discharge rate at which siphoning is allowed to occur, intending that this value will be exceeded only 13% of the time?

Solution Steps for the manual solutions are shown below. (For the portions that can be automated, see also "In Brief 8.3(b).") The problem describes a uniformly distributed set of data ranging from 2.0 to 3.5 m^3/sec. The minimum rate of discharge at which water can be siphoned off is the value x such that $P(X > x) = 0.13$.

In order to use Formula 8.8 (or a software program), we must first find the less than cumulative probability:

$$P(X < x) = 1 - P(X > x) = 1 - 0.13 = 0.87$$

Given $a = 2.0$ and $b = 3.5$:

$$x = 2.0 + (0.87)(3.5 - 2.0) = 2.0 + 1.305 = 3.305$$

That is, if the authorities allow siphoning only on days that the discharge rate exceeds 3.305 m^3/sec, then siphoning would be expected to occur on only about 13% of the designated days.

• *In Brief 8.3(b) Finding x Given the Less Than Cumulative Probability for a Uniform Distribution*

Excel

For Excel, you must use a manual approach, as described in Example 8.10.

Minitab

Within a uniform distribution whose range is from a to b, to find the upper limit x for which the cumulative probability $P(X < x)$ is known: Input the value for $P(X < x)$ into a worksheet cell (usually, at the top of an empty column). Use the sequence *Calc / Probability Distributions / Uniform* to bring up the dialogue box shown in the screen capture and then input the values for a and b where indicated. Specify the column that contains the probability value $P(X < x)$ and the column to hold the solution. Select *Inverse cumulative probability* and then click *OK*. (Alternatively, you can input the $P(X < x)$ value in the *Input constant* area; in that case, the output will not be added to the worksheet page.)

SPSS

Within a uniform distribution whose range is from a to b, to find the upper limit x for which the cumulative probability $P(X < x)$ is known: Input the value for $P(X < x)$ into a worksheet cell (usually, at the top of an empty column). Use the menu sequence *Transform / Compute* to call up the dialogue box shown in the screen capture. In the *Target Variable* window, input the name of the column to hold the solution.

Input a formula into the *Numeric Expression* window of the form **IDF.UNIFORM(VAR0001,*a*,*b*)**, replacing **VAR0001** with the name of the column holding the probability value $P(X < x)$. (If necessary, adjust the displayed number of decimal places for the solution variable.) Click *OK*.

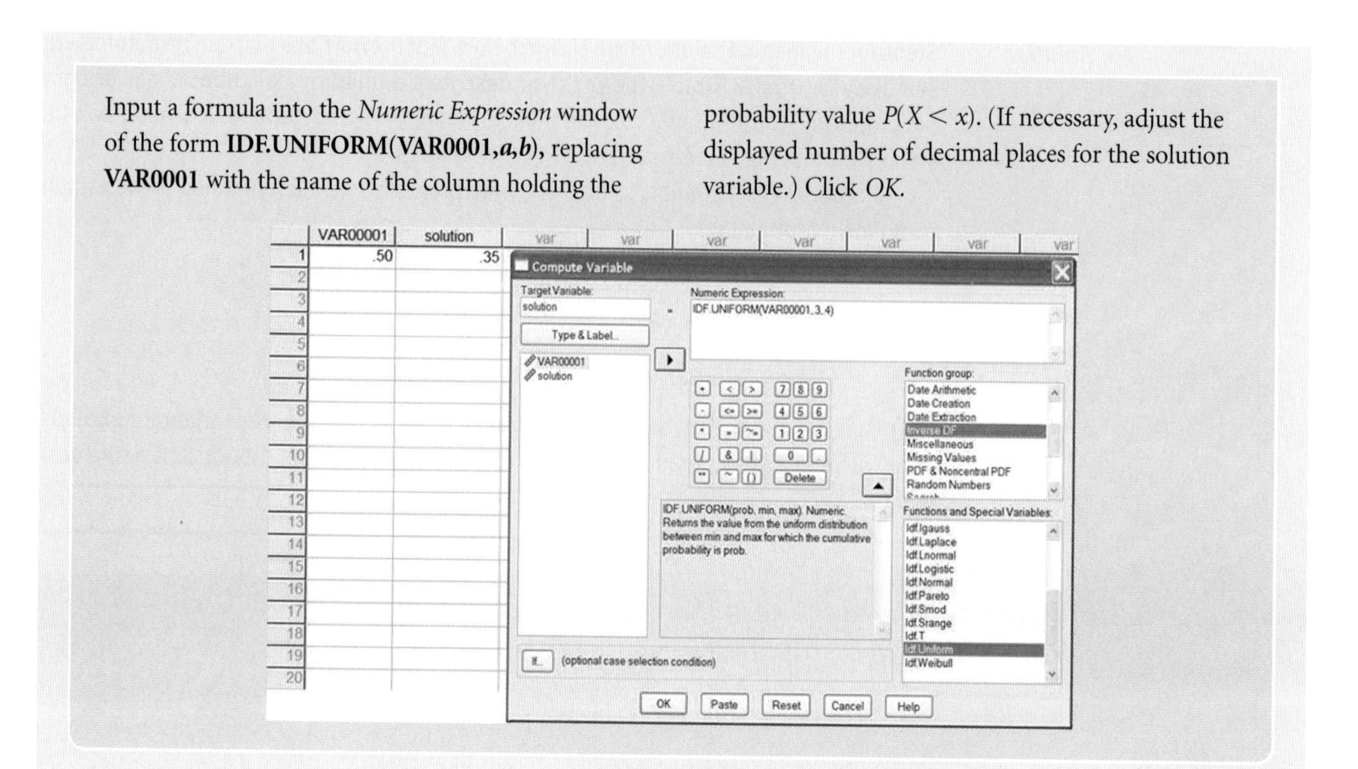

• Hands On 8.3 The Uniform Distribution as a Model for Uncertainty

(*Based on the files* ***Early_Returns, Meteors_N_Amer, Purchasing Power_World, Star_Distances,*** *and* ***Vehicle_Ratings.***)

Except for artificial contexts like the outcomes of lotteries or shuffles of cards, a true uniform distribution is not encountered very often. When we need to make decisions in contexts of uncertainty (such as what to wear, when it *might* rain, or how to handle a propane tank when it might explode), it would be useful if we could know the real underlying probability distributions for possible outcomes. This information is not always available.

In decision theory, a uniform distribution is often used to assign probabilities when there is nothing else to go on. Suppose you are deciding what team to favour in the office betting pool for the next Grey Cup. If you shrug your shoulders and say, "It's a tossup," you are implying a situation similar to that of flipping a coin—a context with equally probable outcomes. Do you have positive evidence that the probabilities in this case are equal? In reality, perhaps, one team's quarterback has just caught the flu, so his team's chances of winning have been reduced. But lacking this information, you assume equal probabilities by default. How prudent is this strategy?

In both "Hands On 8.1" and "Hands On 8.2," we explored the relationship between actual cumulative frequencies and the predictions for those cumulative frequencies, based on assuming some distributions were normally distributed. We have looked so far at 11 variables: the column "W_Margin" in the file **Early_Returns;** column "Av_Age" in the file **Meteors_N_Amer;** column "GDP_CAD" in the file **Purchasing Power_World;** column "Distance" in the file **Star_Distances;** and seven columns ("Rep_Cost," "Fuel_Cost," "Ins_Cost," "Sat-isfac," "Sat_Sale," "Sat_Serv," and "Repeat") in the file **Vehicle_Ratings.**

If you have not done so for the previous exercises, construct a cumulative relative frequency distribution table for each of these 11 variables in turn. Observe the relative frequencies reached by the upper limit (x) of each class. Now, using the procedures you have learned in this section, estimate the less than cumulative probabilities that would be expected for each of these x-values—based on a presumption that the outcomes follow a uniform probability distribution.

For each upper limit of each class of each variable, compare the actual cumulative frequency with the estimates (1) assuming a normal distribution and (2) assuming a uniform distribution. How do these all compare? Does one of the assumptions (1) or (2) always yield the closer estimate of the two? If so, which? If not, can you detect a pattern of when assumption (1) seems to yield the best results and when (2) seems better suited? Based on your experience with these data, would you recommend a strategy of assuming that a distribution is uniform until gaining more information? Explain.

8.4 The Exponential Probability Distribution

When the random variable relates to *time,* the distribution of possible values often follows the **exponential probability distribution.** For example, how long does it take for a randomly selected customer to be served in a restaurant or at a grocery store checkout? Or how many hours of continuous use can a randomly selected AAA battery provide? Another variation of a time variable is the length of time between successive random events. Examples include the lengths of time between patients arriving at a hospital emergency department or the lengths of time between breakdowns of a machine.

Formula 8.9 represents the exponential probability distribution.

Formula 8.9 Exponential Probability Distribution

$$f(x) = \lambda e^{-\lambda x}$$

where x = Any value within the sample space for the distribution

λ = The mean or expected number of occurrences of an event within a given time interval (this is the *same parameter* as λ (or μ) encountered for the Poisson distribution)

e = The constant 2.71828

Since variable x for this distribution usually represents a time period, some books denote this variable with the letter t. When using this model, the time units should be consistent. For example, if the mean number of occurrences λ is given as 5 *per minute,* then the units for values of x should be in minutes as well.

The graphical equivalent to Formula 8.9 is shown in Figure 8.17, which defines the probability density curve for the distribution. As described in Sections 8.1 and 8.3, a probability density curve cannot assign a probability for one exact value of x. For any *range* of x-values, however, the probability of drawing a value within that range equals the area under the curve in that range.

Figure 8.17
The Exponential Probability Distribution

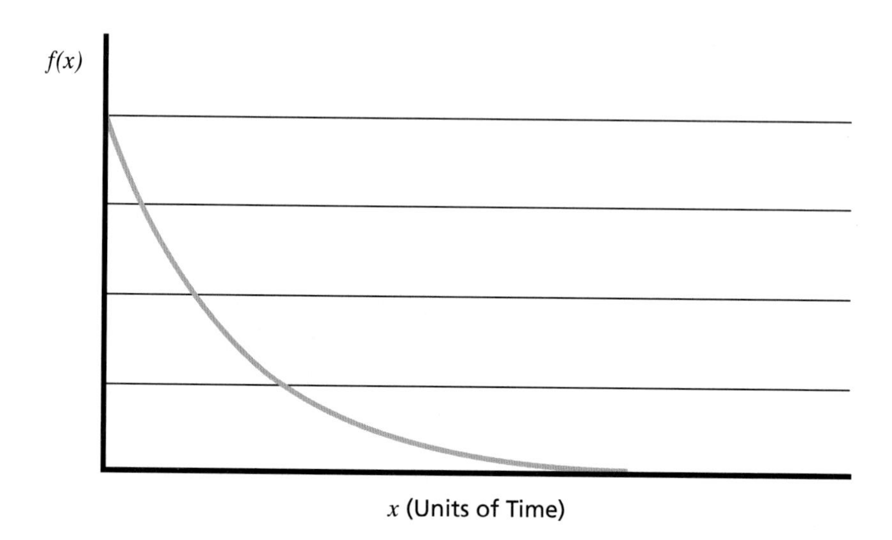

One important feature of this model is that, like time, there can be only *positive* values. No negative values for $f(x)$ are defined. Also, the distribution is *skewed to the right.* If you have waited in lines at banks or grocery stores, you may recognize the pattern: Every so often, there is a wait time that greatly exceeds the expected average.

The expected value (mean), variance, and standard deviation of this distribution are given in Formulas 8.10, 8.11, and 8.12.

Formula 8.10 *Expected Value for an Exponential Probability Distribution*

$$E(X) = \mu = \frac{1}{\lambda}$$

Formula 8.11 *Variance for an Exponential Probability Distribution*

$$\sigma^2 = \frac{1}{\lambda^2}$$

Formula 8.12 *Standard Deviation for an Exponential Probability Distribution*

$$\sigma = \sqrt{\sigma^2} = \sqrt{\frac{1}{\lambda^2}} = \frac{1}{\lambda}$$

Perhaps you observed this interesting feature of the exponential distribution: The mean and standard deviation have identical values. The exponential distribution is also closely associated with the Poisson discrete probability distribution (see Section 7.4). If the *number of events* that occur in a time period can be modelled by the Poisson distribution, then the *time between occurrences* of the events can be modelled with the exponential distribution.

Example 8.11

According to one study, the mean rate at which passengers arrive at the Vancouver Airport security checkpoint between 9:10 and 9:30 a.m. is 15 passengers per minute.[5] Presume that the number of arrivals per minute follows a Poisson distribution.

1. What is the expected value for the *time between arrivals,* in minutes?
2. What is the standard deviation for that variable?

Solution

To solve these problems, we can use Formulas 8.10–8.12. The mean rate of arrivals per minute λ is given in the problem as 15.

1. $E(X) = \frac{1}{\lambda} = \frac{1}{15} = 0.0667$ minutes per arrival
2. $\sigma^2 = \frac{1}{\lambda^2} = \frac{1}{15^2} = 0.00444$

   $\sigma = \sqrt{\sigma^2} = \sqrt{0.00444} = 0.0667$ minutes per arrival

Probability Calculations Based on the Exponential Probability Distribution

We use similar models for performing these calculations as were used in Sections 8.1 and 8.3. Applying a cumulative distribution function, we can determine the proportions of the area under the curve that fall below certain values. From this, we can also find the proportions of the curve lying above, or between, certain values (refer to Figure 8.18).

Formula 8.13 shows the cumulative distribution function.

Formula 8.13 Less Than Cumulative Probabilities for an Exponential Probability Distribution

$$P(X < x) = 1 - e^{-\lambda x}$$

where

x = Any positive value of time within the sample space

λ = The mean or expected number of occurrences of an event within a given time interval (this is the same parameter as λ (or μ) encountered for the Poisson distribution)

e = The constant 2.71828

If we know the expected number of occurrences per unit of time (λ), then the probability that a randomly selected time between occurrences is less than a specified x-value $[P(X < x)]$ can be determined from the formula. Other cumulative probabilities can be derived. The probability $P(X > x)$ is found as the complement of $P(X < x)$. For two values

Figure 8.18
Cumulative Probabilities Within an Exponential Distribution

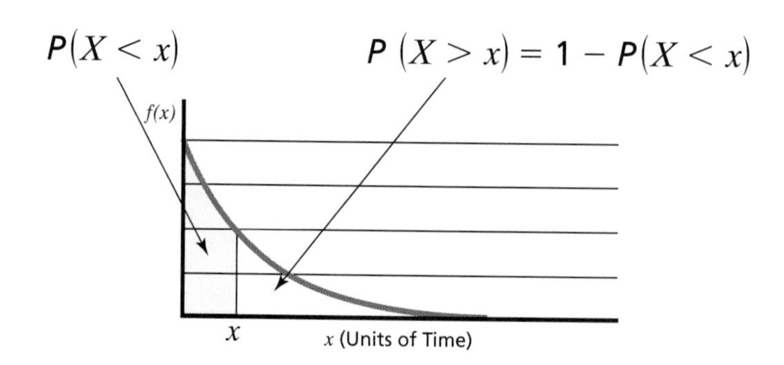

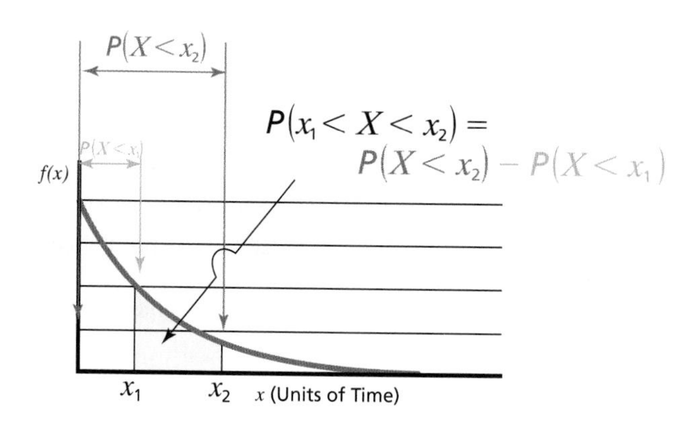

from the distribution x_1 and x_2 (x_2 being the larger), the probability that a selected time period falls between those values is $P(X < x_2) - P(X < x_1)$.

Example 8.12

Continuing from Example 8.11, λ for passenger arrivals at the Vancouver Airport security checkpoint at a certain time of day is 15 passengers per minute.

1. If arrivals follow a Poisson distribution at that time of day, what is the probability that a randomly selected time between arrivals is:
   a) Less than 0.05 minutes?
   b) Less than 0.07 minutes?
   c) More than 0.07 minutes?
   d) Between 0.05 and 0.07 minutes?
2. If the time between arrivals is more than 0.07 minutes, what is the probability that this time period is actually more than 0.09 minutes?

Solution

Steps for the manual solutions are shown below. (For the portions that can be automated, see "In Brief 8.4.") The problem specifies that $\lambda = 15$. Problems 1(a) and 1(b) can be solved directly with Formula 8.13. The remaining solutions need to be derived.

1. a) $P(X < 0.05) = 1 - e^{(-15 \times 0.05)} = 0.5276$
   b) $P(X < 0.07) = 1 - e^{(-15 \times 0.07)} = 0.6501$
   c) $P(X > 0.07) = 1 - P(X < 0.07) = 1 - 0.6501 = 0.3499$
   d) $P(0.05 < X < 0.07) = P(X < 0.07) - P(X < 0.05) = 0.6501 - 0.5276 = 0.1225$

2. This problem illustrates what is called the "memoryless" property of the distribution. We've already granted that $X > 0.07$ minutes, so $P(X > 0.09 \text{ minutes} \mid X > 0.07 \text{ minutes})$ is adding $(0.09 - 0.07) = 0.02$ minutes extra. The solution just requires finding $P(X > 0.02) = 1 - P(X < 0.02) = 1 - 0.2592 = 0.7408$.

   In symbols, the memoryless property says: $P(X > (x + t) \mid X > x) = P(X > t)$, where t is some extra increment of time by which $x + t$ exceeds x.

• *In Brief 8.4 Computing a Less Than Cumulative Probability for an Exponential Distribution*

Excel

To calculate $P(X < x)$ for an exponential distribution if values for x and λ are known: Choose any cell in the worksheet and input a formula in the format: **=EXPONDIST(*x*,λ,1)** as shown in the illustration. The required third input, **1**, signifies that the desired output is the cumulative probability.

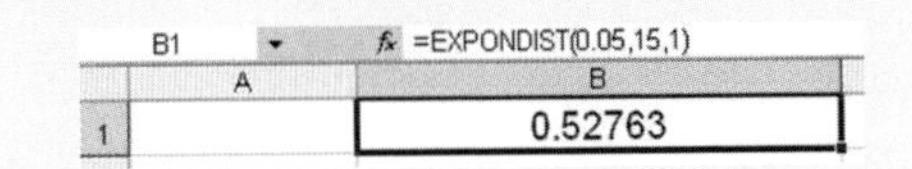

Minitab

To calculate $P(X < x)$ for an exponential distribution if values for x and λ are known: First use Formula 8.10 to calculate the mean of the distribution, given the known value for λ. Then input the value for x into a worksheet cell (usually, at the top of an empty column). Use the sequence *Calc / Probability Distributions / Exponential* to bring up the dialogue box shown in the screen capture and then input the value for the mean where indicated. Specify the column that contains x and the column to hold the solution. Select *Cumulative probability* and then click *OK*. (Alternatively, you can input the x-value in the *Input constant* area; in that case, the output will not be added to the worksheet page.)

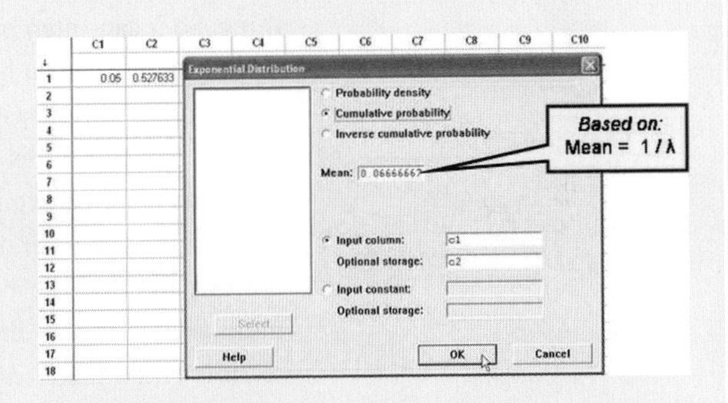

Note that Minitab can also calculate the value x if the probability $P(X < x)$ is known. Use the same dialogue box as above but select *Inverse cumulative probability*. In the input column, specify the probability rather than the x-value.

SPSS

To calculate $P(X < x)$ for an exponential distribution if values for x and λ are known: Input the value for x into a worksheet cell (usually, at the top of an empty column). Use the menu sequence *Transform / Compute* to call up the dialogue box shown in the screen capture. In the *Target Variable* window, input the name of the column to hold the solution. Input a formula into the *Numeric Expression* window of the form **CDF.EXP(*x*,λ)**, replacing ***x*** with the name of the column holding the *x*-value. Note that SPSS refers to the λ term as the "scale parameter." (If necessary, adjust the displayed number of decimal places for the solution variable.) Click *OK*.

Note that SPSS can also calculate the value for x if the probability $P(X < x)$ is known. Proceed as above, but the formula to input into the *Numeric Expression* window should have the form **IDF.EXP(*prob*,λ)**, replacing ***prob*** with the name of the column holding the value for $P(X < x)$.

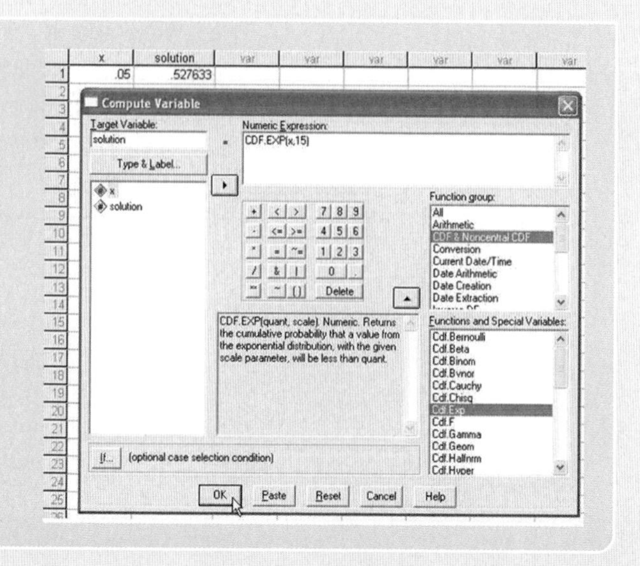

• Hands On 8.4 *Finding Some Exponential Distributions*

Exponential distributions are often associated with durations of time; for example, how long something will last or how long it takes to accomplish something. So it can often be difficult to actually observe all of these distributions at the same time. If you have time (and the weather permits), you might sit at a street corner and measure the time that elapses between passing cars at a certain time of day. Input this data into your computer and compare the frequency distribution with the exponential. For the same result, you could count (instead of measure) the *number* of cars passing through one of the intersections per minute, and find the average (λ). If you obtain a Poisson distribution for this number of cars, the times that elapsed between cars passing through the intersection would then have an exponential distribution.

Here is an experiment you can try at your desk: Conduct a two-word search with an Internet search engine. For the first word, write in any topic, such as "birds," "business," or "friction." For the second word, write the URL extension that corresponds to a remote country, such as ".dk" for Denmark or ".au" for Australia. (We want a place whose Internet pages load a bit slowly from North America.) The search engine will probably return a long list of related sites.

Next, devise a sampling plan to randomly select 100 of the sites that are listed. (Do not enter these sites yet; just plan your sequence for entering the sites in the next step.) Also, in preparation, locate a stopwatch (or a watch with clear, readable numbers and a second hand) and a piece of paper.

Proceed as follows for each of the 100 sites in turn: Note the time and click on a site's link in the search engine's output. Observe and record how many seconds elapse before *all* text and photos are downloaded and your browser displays "Done." If "Done" is never displayed or is taking a long time to appear, discard the site from the sample. Also, if a site's contents seem to be objectionable or if a site starts feeding graphics nonstop, you might want to stop the download and delete the site from your sample. Once you have followed this procedure for each site, input all of the recorded download times into your computer and compare the distribution with the exponential. Does the model seem to apply?

What if you could download the *same* site 100 times and record its download time each repetition? (This would not work as intended if your computer stores the pages locally when first downloaded.) Would you expect that downloading repeatedly from the same remote site would provide a better example of the exponential distribution than the first version approach? Explain why or why not?

Chapter Summary

8.1 Describe the concepts of a probability density curve, a probability density function, and a cumulative distribution function; apply these concepts to calculate areas under the normal curve and to find an *x*-value that bounds an area of known size under the normal curve.

A probability distribution for a continuous quantitative variable can be modelled with a graph called a **probability density curve.** The graph looks similar to a smoothed frequency polygon for a data set (see Chapter 3). The familiar **bell curve** or **normal curve** is one possible type of probability density curve. The formula that determines the ordinate value of the curve for each *x*-value in the sample space is a **probability density function.** For the normal curve, this function is seen in Formula 8.1. For any range of *x*-values greater than zero, the area under the curve that lies above that range represents the proportion of the sample space that falls within that range of values. This proportion, in turn, represents the probability that a randomly selected individual will have an *x*-value in that range.

To calculate probabilities based on the normal curve, the actual data values must be converted into *standardized* values called ***z*-values** or ***z*-scores.** If we converted all values of a normal distribution into *z*-values (using Formula 8.2), we would transform the original distribution into a **standard normal distribution,** for which the mean $= 0$ and the standard deviation $= 1$. The **cumulative distribution function** (CDF) for the normal curve allows us to find the areas under the curve (i.e., probabilities) corresponding to specific values of the variable. Inversely, we can find the areas under the curve that are bounded by particular values of *x*.

We say a statistical method is **robust** if it reaches reasonably accurate conclusions about the data—even if its assumptions about the data set do not fully apply.

8.2 Apply a variety of methods to assess whether a continuous probability distribution is reasonably approximate to a normal distribution.

If we use the powerful model for the normal curve, we should check whether the condition of normality is actually met by using the following procedure: (1) Examine summary statistics for the data set to see if well-known properties of the normal curve appear to apply: Are the mean and median relatively close? Are about 95% of the data within two standard deviations of the mean? Do virtually all of the values fall within three standard deviations of the mean?

(2) Create charts, such as histograms, stemplots, and boxplots, to compare the symmetry, skewing, outliers, and visual aspects with a normal distribution model. (3) Construct a **normal probability plot,** which compares the actual distribution of all data values to the distribution you would *expect* the values follow *if* the data were normally distributed.

8.3 Apply the concepts of a probability density function and a cumulative distribution function to calculate areas within a uniform distribution and also to calculate summary statistics for a uniform distribution.

A **uniform probability distribution** is a continuous probability distribution whose probability density curve appears rectangular, since the probability density function (see Formula 8.3) returns the same ordinal value for any value of x. The cumulative distribution function for this distribution (see Formula 8.7) allows us to find the areas under the curve (i.e., probabilities) corresponding to specified ranges of values of the variable. Inversely, we can find the areas under the curve that are bounded by particular values of x. Using Formulas 8.4, 8.5, and 8.6, respectively, we can calculate the mean, variance, and standard deviation of a uniform probability distribution.

8.4 Apply the concepts of a probability density function and a cumulative distribution function to calculate areas within an exponential distribution and also to calculate summary statistics for an exponential distribution.

When the random variable relates to *time,* the distribution of possible values often follows the **exponential probability distribution.** For the probability density function for this distribution, see Formula 8.9. For probability calculations, use the cumulative distribution function, Formula 8.13. Inversely, we can find the areas under the curve (i.e., probabilities) that are bounded by particular values of x. Using Formulas 8.10, 8.11, and 8.12, respectively, we can calculate the mean, variance, and standard deviation of an exponential probability distribution.

• Key Terms

bell curve (*or* normal curve), p. 324
cumulative distribution function, p. 328
exponential probability distribution, p. 353
normal probability plot, p. 340
probability density curve, p. 324
probability density function, p. 324
robust (*statistical method*), p. 336
standard normal distribution, p. 326
uniform probability distribution, p. 344
z-values (*or* z-scores), p. 326

Exercises

Basic Concepts 8.1

1. In a certain northern community, snow covers the ground for 240 days a year, on average. The standard deviation for annual days of snow cover is 20 days. If this distribution of days of snow cover is approximately normal, then *without* using the computer, estimate the probability that on a randomly selected year, snow covers the ground:
   a) Between 220 and 260 days.
   b) Between 200 and 280 days.
   c) For fewer than 180 days or more than 300 days.

2. In the preceding example, what are the *z*-scores for the following durations of annual snow cover in the community?
   a) 220 days.
   b) 280 days.
   c) 245 days.
   d) 254.5 days.
   e) 240 days.

3. The mean height of flowering dandelions in your local park is 7.6 cm. The standard deviation is 1.8 cm. Find the *z*-scores for the following heights of flowering dandelions in the park:
   a) 8.1 cm.
   b) 7.9 cm.
   c) 7.2 cm.
   d) 7.45 cm.

4. Presume that the heights of flowering dandelions are normally distributed. Based on the parameters given in Exercise 3, what is the probability that if you randomly select a flowering dandelion, its height will be following?
   a) Less than 8.1 cm.
   b) More than 7.9 cm.
   c) Less than 7.2 cm.
   d) Between 7.2 and 8.1 cm.

5. According to the product label, every piece of Gator Gum contains 15 grams of sucrose sweetener. The actual amounts of sweetener in Gator Gum pieces have a normal distribution. The mean amount of sucrose per piece is 15.03 grams, and $\sigma = 0.79$ grams. If the following amounts of sucrose were found in selected pieces of Gator Gum, determine their *z*-scores:
   a) 13.99 grams
   b) 14.47 grams
   c) 14.65 grams
   d) 16.23 grams
   e) 15.03 grams

6. If a piece of Gator Gum is randomly selected (see Exercise 5), what is the probability that the amount of sucrose is:
   a) Less than 13.99 grams?
   b) Less than 15.03 grams?
   c) More than 15.03 grams?
   d) Between 13.99 grams and 15.03 grams?

7. An outdoor vendor is selling a very large number of used DVDs. The marked prices of the DVDs are normally distributed, with $\mu = \$9.79$ and $\sigma = \$1.49$.
   a) What is the probability that one randomly selected DVD costs less than $9.00?
   b) What is the probability that two randomly selected DVD *both* cost less than $9.00? (Hint: Because there is such a large population, you can ignore that the sampling is without replacement.)

8. A recipe that comes with a home bread-making machine says that the resulting bread will weigh 1.4 kg. A consumer group confirmed that 1.4 kg is the mean weight of the bread, but the standard deviation is 0.43 kg. The weights are normally distributed.
   a) If a customer follows the recipe, what is the probability that the resulting bread will weigh between 1.5 and 1.6 kg?
   b) The lowest 10% of bread weights that are produced are below what weight value?
   c) The top 15% of bread weights that are produced are above what weight value?
   d) What is the interquartile range for the bread weights that are produced? (Hint: What percentage of cases is below the lower bound? What percentage of cases is above the upper bound?)

9. An office thermostat is set to 22°C. The actual temperature varies around that setting with a normal distribution: $\mu = 22.2$°C and $\sigma = 1.4$°C.
   a) At any given time, what is the probability that the temperature is above 23°C?
   b) At any given time, what is the probability that the temperature is between 22°C and 23°C?
   c) The lowest 10% of temperatures that occur in the office fall below what temperature value?
   d) The temperature in the office is between what two values 75% of the time? (Identify a range that is symmetrical around the mean.)

Applications 8.1

10. The footcandle is a unit of measurement for light intensity—the amount of illumination on a one square-foot

surface placed one foot from a specified type of candle. An experimenter has found that, by this measurement, candles of a certain brand provide a mean of 0.94 footcandles of illumination on a one square-foot surface placed one foot away.[6] The distribution of this measurement is normal: $\sigma = 0.09$. If one candle of this type is selected randomly and placed one foot from the one square-foot surface, what is the probability that the measure of illumination is:

a) At least 1.04 footcandles?
b) Between 1.00 and 1.05 footcandles?
c) At most 0.93 footcandles?

11. The experimenter in Exercise 10 is especially interested in the candles that appear to provide an unusually high amount of illumination—namely, the top 5% of candles in the distribution of illumination measurements.
    a) What is the lower bound, in footcandles, for the illumination provided by one of the candles in the top 5%?
    b) For the illumination emitted by a candle in the bottom 5%, what is the highest value that it could have in footcandles?

12. Provincial governments in Canada are seriously concerned with the escalation of health care costs due to various causes, including the need for hospital stays. A researcher has found that for senior citizens who are admitted to hospital due to a motor vehicle-related incident, the mean hospital stay is 11.6 days.[7] Suppose that the distribution of hospital-stay lengths is normal, with a standard deviation of 3.1 days.
    a) If a senior citizen is admitted to hospital due to a typical motor vehicle-related cause, what is the probability that the individual will stay in hospital less than 5 days?
    b) What is the probability that the hospital stay will be more than 13 days?
    c) What is the probability that the hospital stay will be between 5 and 13 days?
    d) Hospital records may round the number of hospital-stay durations to whole numbers of days. Will that cause your answer to (c) to be an underestimate, an overestimate, or neither? Explain.
    e) In reality, the distribution of lengths of hospital stays is positively skewed, rather than normally distributed. Will that cause your answer to (c) to be an underestimate, an overestimate, or neither? Explain.

13. A criminologist is studying the lengths of jail sentences given in Canada. She found that, in 2001/2002 the mean sentence given was for 4.10 months: $\sigma = 9.75$ months. The median sentence was for about 1 month.[8]
    a) If the distribution of sentence lengths was normal, what would be the probability that the next sentence given is between 5.0 and 8.0 months?
    b) If the distribution of sentence lengths was normal, what would be the sentence length corresponding to the 90th percentile?
    c) Comparing the values for the population mean and median, is there reason to suspect that the distribution is *not* normal? If so, in what direction does the distribution appear to be skewed? Explain your answer.
    d) Given your answer to (c), do you expect that your answer to (b) is an underestimate, an overestimate, or neither? Explain.

14. The criminologist in Exercise 13 is also concerned about the increasing trial delays faced by those who have been charged. She found that, in 2001–2002, the mean elapsed time from first to last court appearance for a person charged with robbery was 218 days.[9]
    a) If the distribution of trial delays was normal, with $\sigma =$ 109 days, what would be the probability that the trial delay for a randomly selected robbery suspect would be at least 200 days?
    b) At most 300 days?
    c) Between 200 and 300 days?
    d) What is the lower bound, in days of delay, for the longest 5% of trial delays?

15. If a cancer patient is correctly diagnosed, the effectiveness of the treatment may depend on the accuracy of the dose delivery system to reach the diseased tissues. This, in turn, requires a reliable dose measuring tool (a dosimeter), in order to check the physical distribution pattern of the dose in the patient's body. To help ensure this reliability of measurement for a certain treatment, scientists compared the measured dose distribution patterns in the body with the calculation-based expected patterns.

    The results for one of these experiments were as follows: The mean difference between measured and calculated doses, for all of the observed body locations, was –0.15 Gy (Gy is a unit of radiation dose for cancer treatment). Standard deviation was 4.59 Gy. The distribution of differences was apparently normal.[10]
    a) Estimate the boundaries of the interquartile range for the differences between the measured and calculated doses.
    b) If one pair of measured/calculated doses is randomly selected, what is the probability that the difference is no more than +6.2 Gy?
    c) At least –6.2 Gy?
    d) Between –6.2 and +6.2 Gy?

16. A movie fan wants to buy a subwoofer for his home theatre. (A subwoofer is a speaker that specializes in very low frequencies—such as the deep rumble sound of approaching enemy spacecraft.) He has narrowed his search to the model Velodyne SPL-1000 subwoofer. According to his search in July 2006, these could be purchased online at various prices: $\mu =$ US\$1,175, $\sigma =$ US\$176.[11] If these prices are normally distributed, and one price is chosen at random, what is the probability that the price is:

a) More than US$1,200?
b) Less than US$1,200?
c) Between US$1,100 and US$1,150?

Basic Concepts 8.2

17. Describe three checks you can make, using numeric descriptive statistics, to assess whether a data set is normally distributed. Explain why it is not always sufficient to rely on only these numeric checks.

18. Explain how histograms, boxplots, and stemplots could be used to help assess whether a data set is normally distributed.

19. Describe the concept of a normal probability plot and explain how you would interpret its results.

20. Relying on the data from the following samples, assess whether the underlying populations appear to be normally distributed. List the checks that you use, and describe their results.

Data set 1:	10	14	9	17	16	5	13	15	
Data set 2:	9	4	6	8	9	5	7	3	8
Data set 3:	20	24	29	25	26	27	22	20	24

21. Ten bottles of a certain soft drink were sampled; each claims to hold 750 mL of liquid. From the following sample of 10 bottles, determine whether the distribution of volumes is approximately normal. List the checks that you use and describe their results.

749.5	748.0	749.7	748.6	750.1
751.3	750.5	749.7	750.1	750.8

22. The numbers of houses built each year by Shepitka Construction are variable. In each of the last 13 years, respectively, the numbers of houses shown below have been built. Assess whether the distribution appears to be normal and justify your results.

262	265	267	212	291	314	
220	196	281	244	233	285	298

23. Relying on the data from the following samples, assess whether the underlying populations appear to be normally distributed. List the checks that you use and describe their results.

Data set 1:	1884	1588	2350	1584	2087
	1922	2047	2361	1801	1896
	1835	1982	1959	2328	2256
Data set 2:	45.1	51.3	44.9	36.6	53
	42.4	85.1	57	46	47
	50	62.3	51.5	44.8	42.1
Data set 3:	16	21	15	16	16
	20	17	16	18	27
	18	23	18	14	20

24. Do any data sets in Exercise 23 contain an outlier? If so, for any sets that contain an outlier, reassess whether the data set—*excluding* the outlier—is approximately normal. Justify your conclusion.

25. A biologist is observing the numbers of insects being carried by ants each day into a very large anthill. Assess whether the distribution of his results, shown below, appears to be normal and justify your results.

2,369	2,279	2,368	2,318	2,343	2,311
2,041	2,166	2,098	2,124	2,224	2,215
2,064	2,362	2,129	2,052		

Applications 8.2

26. A Canadian specialist in waterfalls is studying the vertical drops of selected falls. Her sample data for the heights of these drops is recorded in the file **Waterfalls.**
a) Describe the shape of the distribution.
b) Are the conditions met to apply the probability calculations of Section 8.1? Explain.

27. To plan for a manufacturing process, a chemist has compiled a list of gases, including their solubilities in water at 293° Kelvin under standardized test conditions. A sample of these solubilities is stored in the file **Solubilities_of_Gases.**
a) Describe the shape of the distribution.
b) Are the conditions met to apply the probability calculations of Section 8.1? Explain.

28. A public relations officer for Gamblers Anonymous wants to calculate, and communicate, the probability that a casino in Alberta will have fewer than 200 slot machines. (See the file **Casinos_Alberta.**)
a) Is the full data set normally distributed?
b) The highest number of slot machines in a casino appears to be an outlier. Is the data set—*excluding* the outlier—approximately normally distributed?
c) Applying the methods of Section 8.1 on the data set—*excluding* the outlier—estimate the probability that the Alberta casino will have fewer than 200 slot machines. Devise a sentence that the public relations officer can use that (a) includes the above probability estimate, but (b) accounts honestly for the exclusion of the outlier from the calculation.

29. A company specializing in exhibiting gemstones has discovered the website for the United States Faceters Guild (USFG). The manager wonders how much exposure the website would provide if her company posted some announcements on the site. The USFG publishes openly the data on its site's message history.[12] The manager of the exhibition company randomly selected 17 monthly summaries from about six years of records. The numbers of messages posted on the website for the selected months were:

114	373	352	299	495	362	366	377	
193	334	343	317	802	360	618	257	535

a) Is the full data set normally distributed?
b) Are there any apparent outliers? Is the data set—*excluding* any outliers—approximately normally distributed?
c) Applying the methods of Section 8.1—*excluding* any outliers—estimate the probability that on a given month the number of posted messages will be above 250.
d) For the website to be a viable tool for advertising the exhibition company, the probability of at least 250 messages per month must be at least 80%. Based on your answer to (c) *and* your awareness of any outliers, do you feel that the website meets that criterion? Explain.

30. An economist is studying the distribution of egg prices in China from the years 2000 to 2004. The data below, selected randomly from the file **Eggs_China,** are in U.S. dollars.

4.94	5.31	5.61	4.97	5.99	6.13
5.49	5.14	5.94	5.42	5.48	5.73
5.95	5.63				

Do the egg prices in that period appear to have been normally distributed? Justify your answer.

31. Next summer, the general manager of an NHL hockey team plans to sign a contract with a free agent. To prepare, he is studying some of the salary data (in millions of U.S. dollars) published early in the free agent trading period of 2006. (See the file **Free_Agents.**)
a) Are the selected annualized salaries normally distributed? Justify your answer.
b) Suppose you used the methods of Section 8.1 to estimate the probability that a randomly selected player in the data set earned less than $1 million per year. Given the actual distribution of the data, do you think that the normal-based probability estimate would be, in this case, an overestimate or an underestimate? Explain.

32. A new graduate would like to move before seeking a full-time job. To compare costs in different cities, he compiled the table shown below of average dwelling rental costs in 28 cities across Canada.[13] (See the file **Housing_Facts**)
a) Are the costs normally distributed? Justify your answer.
b) Are the conditions met to enable application of the probability calculations in Section 8.1? Explain.
c) In (a) you checked the distribution of these *average* rental costs by city. Supposed you listed the rental costs of every *individual* rental unit across Canada. Do you think that the distribution of these individual costs would have the same shape as the distribution of the average costs? Explain.

575	673	483	439	538	446	419	529
573	914	709	799	1027	740	680	722
683	738	620	657	605	568	558	783
654	645	919	751				

Basic Concepts 8.3

33. A random variable X is uniformly distributed over the interval between 15 and 30, inclusive.
a) Find the mean and standard deviation of X.
b) What proportion of values of X are greater than 25?
c) If one member of the population is randomly selected, what is the probability that its value is between 18 and 25, inclusive?

34. According to a regular commuter, travel times by train between Ottawa and Oshawa are uniformly distributed, lasting between 3.69 and 4.11 hours.
a) What is the average travel time for a train trip between Ottawa and Oshawa?
b) What is the standard deviation for the travel time?
c) What proportion of all trips between those cities takes less than four hours?
d) What is the probability that a randomly selected trip takes more than 4.05 hours?

35. The number of marshmallows in each box of a cereal is uniformly distributed, with a mean of 34. The maximum number of marshmallows in a box is 42.
a) Is this a discrete or continuous uniform distribution?
b) What is the minimum number of marshmallows in a box?
c) Sketch a graph of this probability distribution.
d) What is the probability of getting fewer than 32 marshmallows in a box of this cereal?
e) What is the probability of getting between 32 and 40 marshmallows, inclusive, in a box of this cereal?

36. A manager is concerned about an employee's Internet surfing on company time. Using spyware on the employee's computer, the manager finds that the number of minutes per day spent by the employee Internet surfing during work hours is uniformly distributed, ranging from 26 to 54 minutes. Find the probability that on the next work day, the employee's time spent Internet surfing during work hours will be:
a) Less than 30 minutes.
b) More than 50 minutes.
c) Between 30 and 50 minutes.

37. In the previous problem, what is the mean amount of time spent Internet surfing each day by the employee? What is the standard deviation?

38. Another concern of the manager in Exercise 36 is the employee's time spent walking to and from the water cooler. Observation has determined that the employee takes an average of 7.2 minutes to go to and from the cooler. The distribution of times is uniform, and the maximum time is 9.4 minutes. Find the probability that the next time the employee walks to the cooler, the time taken will be:
a) Less than 7 minutes.

b) More than 8 minutes.
c) Between 7 and 8 minutes.

39. The waiting times at South Mall Muffins during rush hour are uniformly distributed, with a range from 6 to 16 minutes. When the next customer arrives at the store, what is the probability that his or her wait will be:
a) Less than 10 minutes?
b) Not more than 8 minutes?
c) At least 12 minutes?

40. Refer to Exercise 39.
a) What are the mean and standard deviation of the waiting times?
b) For those customers whose waits are in the longest 10% of waiting times, how much time, at minimum, are they having to wait?
c) For lucky customers whose waits are in the shortest 15% of waiting times, how long, at most, do they have to wait?

41. Applicants to Filbert's Flying School must pass an aptitude test before they can be accepted. Applicants' scores on the test are uniformly distributed, with a mean score of 65 and a range of ± 20 marks.
a) What is the probability that a randomly selected applicant will receive a score of between 40 and 55?
b) What score corresponds to the first quartile of scores?
c) What score corresponds to the 95th percentile of scores?

Applications 8.3

42. Jacob has been training for the next international marathon to be held in Seoul, South Korea. In all of his practice runs, his times to complete the distance have been uniformly distributed, within a range from 2.1144 to 2.1331 hours. What is the estimated probability that Jacob will complete the Seoul marathon in:
a) Less than 2.13 hours?
b) No more than 2.12 hours?
c) Between 2.12 and 2.13 hours?

43. Sometimes, when a variable's distribution is irregular, there may be *portion* of the distribution that is, functionally, roughly uniform. For example, the runner in Exercise 42 might be interested to know that international data show that for marathon runners whose run times fell in the range 2.1144 to 2.1331 hours, their distribution of times is comparatively flat (i.e., uniform).
a) For this set of runners, *use the uniform model* to estimate their mean time for completing the marathon, as well as the standard deviation.
b) For all of the runners in this category who will compete at Seoul, estimate the upper bound for the completion time for the slowest 20% of these competitors.

44. For all of the Canadian equity funds that are ranked by Globefund.com, there is a wide distribution of fund assets. An analyst discovered, however, that for funds with assets in the range of \$225 million to \$975 million, the distribution of the funds' assets is quite uniform.[14] Based on that finding:
a) Estimate the mean and standard deviation for the assets of funds in that range.
b) Find these probabilities:
i) *P*(assets < \$500 million | assets are in the range \$225 million to \$975 million)
ii) *P*(assets > \$700 million | assets are in the range \$225 million to \$975 million)
iii) *P*(assets are between \$500 million and \$700 million | assets are in the range \$225 million to \$975 million)

45. The great majority of known chemical elements were discovered at some time after 1750. However, some very important elements, such as silver, iron, and tin, were discovered between –3250 (i.e., 3250 B.C.E.) and –750. The distribution of element discovery dates in that range is approximately uniform.[15]
a) Estimate the "average year" for the chemical element discoveries of that era.
b) Find these probabilities:
i) *P*(element was discovered prior to 1500 B.C.E. | element was discovered some time in the range 3250 B.C.E. to 750 B.C.E.)
ii) *P*(element was discovered between 2500 B.C.E. to 1500 B.C.E. | element was discovered some time in the range 3250 B.C.E. to 750 B.C.E.)

46. From a cross-section viewpoint, it is not unusual for a rain shower to dump 1 mm of water uniformly across a stretch of earth 100 metres long.[16] Suppose that before such a rain shower, a meteorologist places on the ground a long ruler that has been labelled from 0 to 100 at intervals of one metre, from left to right. What proportion of the rainfall will land:
a) To the left of the 10-metre mark?
b) To the right of the 50-metre mark?
c) Between the 20- and 60-metre marks?

47. Would your answers to the previous exercise change if the rain shower actually dumps 1.2 mm of water uniformly across the distance? Explain.

48. A health and safety inspector in Nova Scotia is examining the load on a heavy-duty scaffold. The scaffold surface measures 1 metre by 8 metres (i.e., 8 square metres). Regulations require that the scaffold be capable of holding a uniformly distributed load of 366 kilograms per square metre.[17] The observed load on the scaffold (see Figure 8.19) is $\frac{86 \text{ blocks} \times 34 \text{ kg/block}}{8 \text{ m}^2} = 365.5$ kg per m^2. But is the load *uniformly distributed?* If not, it may exceed the designed capability of the structure.

a) Refer to Figure 8.19, which depicts a physical placement of a set of blocks on a scaffold. Assuming a uniform distribution of the blocks, what proportion of the blocks should be to the left of position 10? What proportion should be to the right of position 10?

b) Compare your answers to (a) to the actual numbers of blocks on the left and right halves of the scaffold. Given this answer, plus any other information in the problem, do you think this load is safe? Explain.

Basic Concepts 8.4

49. If X has an exponential probability distribution with $\lambda = 3.0$, find:

a) $P(X > 2)$

b) $P(X > 1)$

c) $P(X < 1.5)$

50. Based on Exercise 49, what is the expected value for X? What is the standard deviation?

51. In a case where X has an exponential probability distribution and $\lambda = 1.5$, calculate:

a) $P(X > 1)$

b) $P(X < 1)$

c) $P(0.5 < X < 2)$

d) $E(X)$

52. The shelf life of baked goods is an important consideration for both the consumers and the bakers of the goods. The shelf life is the mean time until a type of good spoils. If the shelf life of bran muffins is two days:

a) What is the numerical value of λ?

b) Interpret, in nontechnical language, the meaning of λ in this example.

c) What proportion of bran muffins will be spoiled 2.5 days after they are baked?

d) What proportion of bran muffins can still be sold (hopefully at a discount) three days after they are baked?

53. The time X between car arrivals at the Wet 'n' Wooly Car Wash on a hot clear day is exponentially distributed, with a mean time of one minute.

a) What is the value of λ?

b) Find these probabilities:

i) $P(X < 0.5 \text{ minutes})$

ii) $P(X > 1 \text{ minute})$

iii) $P(0.4 \text{ minutes} < X < 1.4 \text{ minutes})$

54. In Exercise 54 in Chapter 7, we learned that the annual number of injuries sustained by players on the Moosonee Old-Timers baseball team is Poisson-distributed. It follows that the times between the players' injuries are exponentially distributed. If, on average, the players suffer 5.8 injuries a year, calculate the following:

a) Expected average, and standard deviation, for the time between injuries, in years.

b) $P(\text{time between injuries} < 1.4 \text{ months})$

c) $P(\text{time between injuries} > 73 \text{ days})$

d) $P(65 \text{ days} < \text{time between injuries} < 73 \text{ days})$

(Note: One year $\approx$ 365 days = 12 *approximately equal-length* months.)

55. Northern Subs is examining the need to hire an extra lunch-hour employee. The time between customer arrivals

Figure 8.19
Blocks on a Scaffolding

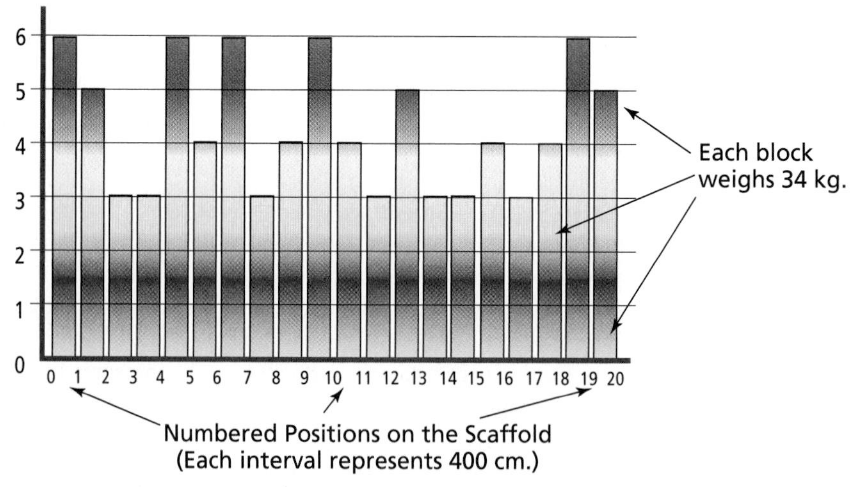

Source: Province of Nova Scotia. *Fall Protection and Scaffolding Regulations Made Under Section 82 of the Occupational Health and Safety Act,* S.N.S. 1996, c. 7 O.I.C. 96-14 (January 3, 1996), Nova Scotia. Reg. 2/96. Found at: http://www.gov.ns.ca/just/regulations/regs/ohs296f.htm

has been found to be exponentially distributed with a mean of two minutes.

a) What is the value of λ?
b) For the time X between arrivals, find:
   i) $P(X < 1 \text{ minute})$
   ii) $P(X < 0.5 \text{ minutes})$
   iii) $P(0.5 \text{ minutes} < X < 1 \text{ minute})$

56. For a breeder of Devon Rex cats, an average of five kittens with a curly tails are born each year. There is a Poisson distribution for this number. Calculate:

a) Expected mean time (in years) between births of curly-tailed cats.
b) Standard deviation of the time between births of curly-tailed cats, in *months*. (See note for Exercise 54.)
c) P(time between births of curly-tailed cats $<$ 2 months)
d) P(time between births of curly-tailed cats is between 2.0 and 4.0 months)

57. Joyce Green, who is very concerned about the environment, has noticed that every week, her favourite paper averages five articles on the environment. (This weekly number has a Poisson distribution.) Calculate:

a) Expected mean time (in weeks) between published articles on the environment.
b) Standard deviation of the time (in days) between published articles on the environment.
c) P(time between published articles on the environment $>$ 1.0 day)
d) P(time between published articles on the environment is between 1.0 and 3.0 days)

Applications 8.4

58. A reliability expert for compressors and turbines in a refinery environment has found that for rotary compressors that are combined with gas turbines, a typical failure rate for the compressor part is 83 months between failures.[18] Presume the distribution of months between failures is exponential.

a) What is the value of λ in the units "failures/month"?
b) In what proportion of cases is the time between failures less than 60 months?
c) In what proportion of cases, is the time between failures between 60 and 100 months?

59. When Boeing established its Enterprise Help Desk in 2004, it anticipated that the average number of incoming calls per minute would be two.[19] Times between calls to a help desk are often exponentially distributed.

a) What is the expected average time between incoming calls?
b) What is the anticipated standard deviation for times between incoming calls?
c) If the estimate of λ given in the problem is correct, calculate:
   i) P(time between calls $<$ 0.25 minutes)
   ii) P(time between calls $>$ 45 seconds)
   iii) P(15 seconds $<$ time between calls $<$ 45 seconds)

60. A company looking for an Internet backbone service provider is reading literature about the large provider Global Crossing. That firm handles 1.26 million hits per minute—which is 21 hits per millisecond.[20] Presume that this rate is constant throughout the day and that the hits are Poisson-distributed.

a) What is the expected average time, in milliseconds, between hits?
b) What is the standard deviation for times between hits, in milliseconds?
c) Based on the assumptions in the question, calculate:
   i) P(time between hits $<$ 0.05 milliseconds)
   ii) P(time between hits $>$ 0.08 milliseconds)
   iii) P(0.05 milliseconds $<$ time between hits $<$ 0.08 milliseconds)

61. X for an exponential variable does not necessarily have to be time. According to a Transport Canada report for the year 2000, the accident rate for Canadian rail services was 13.3 accidents per million train miles.[21] Using this data, find:

a) The average number of train miles between accidents.
b) The standard deviation for the number of train miles between accidents.
c) P(the number of train miles between accidents $<$ 60,000)
d) P(the number of train miles between accidents $>$ 80,000)
e) P(the number of train miles between accidents is between 70,000 and 80,000)

62. According to a simulation study of teletraffic for the San Francisco Bay area, the average number of call arrivals on the system, at about 9 a.m. on weekdays is approximately 30 per minute.[22] Times between calls can be modelled with the exponential distribution. For a time period near 9 a.m. on a weekday, what is the probability that the time between calls is:

a) Less than 1.5 seconds?
b) Less than 3 seconds?
c) More than 2 seconds?

63. An IT manager is planning to buy a number of laptop computers and is preparing a long-term budget. She anticipates an eventual cost to replace the rechargeable batteries that provide power backup for the RAM. Her research finds that these batteries have an average life span of about 2.5 years and, presumably, the distribution of life spans is exponential.[23]

a) What is the value of λ (i.e., the expected number of battery failures per year per laptop)?
b) If 20 laptops are bought for a department, how many battery failures could be expected within:
   i) One year?
   ii) Two years?
   iii) Three years?

64. According to a six-year study of a hospital emergency department (ED), the average arrival rate of patients, per hour, to the ED is 6.46.[24] The distribution of times between arrivals is exponential. Calculate:
a) The average time between arrivals at the ED.
b) The standard deviation for time between arrivals at the ED.
c) P(the time between arrivals $>$ 0.16 hours)
d) P(the time between arrivals $<$ 8 minutes)
e) Do you think that the time between arrivals is constant for all parts of the day? If not, how would you recommend adjusting the calculations for (c) for times near 3:30 a.m.?

Review Applications

For Exercises 65–67, use data from the file **Early_Returns**, "Tot_Vote" column.

65. To prepare for the next federal election, planners for Elections Canada in the province of Quebec are studying the distribution of ballots used per riding in the 2004 election. Hired as their consultant, you have been directed to the data for Quebec in the **Early_Returns** data set (in the "Tot_Vote" column).
a) Presumably, each vote cast required one ballot. For the Quebec ridings in the 2004 election, calculate the mean and standard deviation for the numbers of ballots required per riding.
b) Presume that the parameters in (a) will be similar in the next election, and that the vote numbers per riding will be normally distributed. On these assumptions, estimate the proportion of ridings in Quebec that will require the following numbers of ballots:
   i) Fewer than 43,000 ballots.
   ii) More than 49,000 ballots.
   iii) Between 43,000 and 49,000 ballots.

66. To save paper, one election planner had this idea: At the beginning of Election Day, give each Quebec riding just enough ballots so that only 20% of the ridings would run out of ballots by the end of the day. (Ridings that run out could order additional ballots for immediate delivery using a hotline.) If this planner's idea is followed, and given the assumptions in Exercise 65, how many ballots should be given to each Quebec riding at the beginning of Election Day?

67. Both of the preceding questions assumed a normal distribution for the vote-numbers data.
a) Use the raw data for the 2004 election to assess whether, in that election, the numbers of votes cast in Quebec ridings were (i) very close to normally distributed, (ii) very far from normally distributed, or (iii) somewhat normally distributed. Justify your conclusions. If your conclusion was (iii), describe how the distribution does, or does not, conform to the normal model.
b) In light of your answer to Exercise 67(a), explain how your solutions for Exercises 65 and 66, respectively, might be overestimates or underestimates. Or would they be neither?

For Exercises 68 and 69, use data from the file **Grotto_Geyser**.

68. A naturalist observed the durations of eruptions of the Grotto Geyser over an extended period. She found that the average duration (i.e., mean time for a geyser to cease erupting) is about 212 minutes. The distribution of the durations is approximately exponential. Based on the exponential model, what is the probability that the next eruption lasts:
a) Less than 200 minutes?
b) At least 250 minutes?
c) Between 180 and 230 minutes?

69. An impatient tourist wants to limit his time at the geyser site before moving on to another attraction. When the next geyser begins, he decides to stay at the site just long enough to have a 12% chance of missing the last part of the eruption when he leaves (i.e., if it's a long one). For that plan, how long should he stay at the geyser site?

For Exercises 70–73, use data from the file **Meteors_N_Amer** ("Latitude" and "Longitude" columns).

70. A tour operator is planning a new bus route that will take tourists to the sites of meteor craters in North America. As the tour company's consultant, you are reviewing data for the latitudes and longitudes of all of these sites.
a) Calculate the mean and standard deviation for (i) all of the sites' latitudes and (ii) all of the sites' longitudes.
b) Assuming that latitudes and longitudes of the sites are both normally distributed, estimate the proportion of meteor sites with a:
   i) Latitude below 45° north.
   ii) Longitude between 100.0° west and 110.0° west.

c) Based on your answers to (b), what is the probability that a randomly selected meteor site in North America is located at a latitude below 45° north *and* at a longitude between 100.0° west and 110.0° west?

71. To save on fuel, the tour operator plans to exclude the most northern 15% of meteor sites from the new tour route. Based on Exercise 70, what would be the upper bound of northern latitudes for sites included on the tour?

72. The two preceding questions assumed normal distributions for the latitudes and longitudes of the meteor sites.

a) Use the raw data on your CD to assess whether, in fact, the longitudes are (i) very close to normally distributed, (ii) very far from normally distributed, or (iii) somewhat normally distributed. Justify your conclusions. If your conclusion was (iii), describe how the distribution does, or does not, conform to the normal model.

b) In light of your answer to (a), explain how your answer to Exercise 70(b)(ii) might be an overestimate or an underestimate or does not need to be revised.

73. Repeat Exercise 72(a) but this time, assess the normality assumption for the *latitudes* of the meteor sites.

For Exercises 74–76, use data from the file **Harp_Times**.

74. A music fan has discovered a site for a Canadian musician's album of Celtic music. To see how many songs will fit on her MP3 player, the fan is examining the songs' track lengths. (See the column "DecMin" in the data set.)

a) Calculate the mean and standard deviation for the songs' lengths, in decimal minutes.

b) Assuming that the distribution of lengths is normally distributed, estimate the proportion of songs that:

i) Last longer than 4.0 decimal minutes.

ii) Last between 3.0 and 4.0 decimal minutes.

75. The music fan in Exercise 74 prefers not to include short pieces of music in her collection, so she decides to omit the shortest 20% of pieces on the Celtic CD. What is the shortest length of song on the CD that she would record?

76. The two preceding questions assumed normal distributions for the lengths of songs.

a) Use the raw data on your CD to assess whether, in fact, the song lengths are (i) very close to normally distributed, (ii) very far from normally distributed, or (iii) somewhat normally distributed. Justify your conclusions. If your conclusion was (iii), describe how the distribution does, or does not, conform to the normal model.

b) In light of your answer to (a), explain how your answer to Exercise 74(b)(i) might be an overestimate or an underestimate, or does not need to be revised.

For Exercises 77–79, use data from the file **Low_Fee_Accounts**.

77. To direct his clients to the best deals, a financial analyst is reviewing the fees required by low-fee bank accounts. A sample of the options available is found in the data set file (see the "Fees" column). The analyst has a special interest in accounts for which the fees are in the range of $2.00 to $3.50. Examine the data set to assess whether the set of fees *in that particular range* has an approximately *uniform* distribution? Explain your conclusion.

78. Assuming that fees in the range $2.00 to $3.50 *do* have an approximately uniform distribution, estimate the mean and standard deviation for the fees in that range.

79. On the above assumption, calculate the following probabilities:

a) $P(\text{fee} < \$3.00 \mid \text{fee is in the range } \$2.00 \text{ to } \$3.50)$

b) $P(\text{fee} > \$2.45 \mid \text{fee is in the range } \$2.00 \text{ to } \$3.50)$

c) $P(\$2.45 < \text{fee} < \$3.00 \mid \text{fee is in the range } \$2.00 \text{ to } \$3.50)$

80. In a report by a City of Saskatoon traffic engineer, the distributions of car speeds in the vicinity of public schools was studied to determine the effect of a speed zone change. While speeds were generally close to normally distributed near each school, a school on Kingsmere Boulevard was an interesting exception: Following the zone change, most drivers travelled under 41 km/hour. But for those travelling between 45 to 55 km/hour, the distribution of speeds was nearly uniform.[25] On that basis, estimate these probabilities:

a) $P(\text{speed} < 52 \text{ km/hr} \mid \text{speed is in the range 45 to 55 km/hour})$

b) $P(\text{speed} < 47.5 \text{ km/hr} \mid \text{speed is in the range 45 to 55 km/hour})$

c) $P(\text{speed is between 47.5 and 50 km/hr} \mid \text{speed is in the range 45 to 55 km/hour})$

81. In 2005, Toronto set a record by issuing 26 heat alerts over the summer.[26]

a) What is the value of λ in the units "heat alert arrivals" per month? (Estimate a summer as three equal-length months.)

b) If the times between alerts were exponentially distributed, what was the mean and standard deviation for the time (in months) between alerts?

c) What proportion of times between alerts were:
   i) Less than 3 days?
   ii) Less than 1.5 days?
   (Note: Estimate a summer month as equal to 30.3 days.)

82. Residents in Florida are concerned with a perceived increase in storm activity during hurricane season. But what are the historical patterns, for comparison? The *South Florida Sun-Sentinel* published a summary of hurricane and storm data for 1944 to 1996. From its data, it appears that the average arrival rate of "named storms" during the season was 3.4 per month.[27]

a) For that era, what were the mean and standard deviation (in months) for the times between named storms in south Florida?
b) If the times between storms are exponentially distributed, what was the probability during that era that the time between two named storms was:
   i) Less than 0.3 months?
   ii) At most 0.4 months?
   iii) Between 0.35 and 0.45 months?

PART 4

Samples and Estimates

Chapter

9

The Distributions of Sample Statistics

• Chapter Outline

• Learning Objectives

9.1 Explain and apply the concepts of a sampling distribution; calculate the mean and standard deviation of a sampling distribution of the sample mean under specified conditions; and describe what is meant by the term *central limit theorem.*

9.2 Apply the central limit theorem to calculate probabilities associated with the sampling distribution of the sample mean under specified conditions.

9.3 Calculate the mean and standard deviation for the sampling distribution of the sample proportion and, using calculations based on binomial distributions, find probabilities associated with the sampling distribution of the sample proportion.

9.4 Explain what is meant by *bootstrap methods* and how in general they can be used to approximate sampling distributions and their associated probabilities.

9.1 Sampling Distribution of the Sample Mean

In Chapter 1, we distinguished inferential statistics from descriptive statistics. In either approach, we can begin with techniques described in Parts 1 to 3. For a given data set, we can find measures of location and variation, construct charts and tables, and assess the distribution of the data. For descriptive statistics, our interest stops there—with displaying and analyzing the properties of that data set. But inferential statistics goes a step further; its goal is to *infer from* the observations of the sample some characteristics of the *population.*

This chapter prepares the way for the inferential step. We explore some key relationships between the sample and the population that it comes from. Based on that foundation, the next chapter will show how to estimate a population parameter when related sample statistics are known.

The core concept for this chapter is called, traditionally, the **sampling distribution.** The sampling distribution is a special case of probability distribution, wherein the variable of interest is itself a sample statistic whose value is determined by a process of random sampling. If the variable of interest is the sample mean, its distribution is called the **sampling distribution of the sample mean.** Suppose there is a population from which we plan to sample. From the sample, we will calculate a certain statistic, such as the mean $\bar{x}$, which corresponds to a population parameter (μ) whose value we hope to infer.

What is the relationship between the set of values in the population and the possible sets of values that we could randomly select in our sample? The distribution of the sample mean (i.e., its sampling distribution) expresses that relationship in terms of probabilities. Namely, given (1) the population we are drawing from and (2) the size n of the sample we plan to draw, the sampling distribution is defined as the probability distribution for all of the values of the sample statistic that could be obtained from a sample of that size.

For a finite population, one could always (given enough time and paper) list all possible samples that could occur, find the mean (or other sample statistics) from each sample, and construct for that statistic a frequency distribution table. If you interpret this relative frequency distribution for the statistic as a probability distribution, you have just described its sampling distribution. For a population that is *infinite,* the sampling distribution, like a probability distribution for a continuous variable, can determine probabilities only for ranges of values for the statistic, rather than for any exact value.

Examples 9.1 and 9.2 illustrate, informally, the relationship between a population and a sampling distribution that is derived from the population. The distributions illustrated are not the complete sampling distributions, as defined above, so these examples are not intended to convey precise methods for generating these distributions.

Example 9.1

*(Based on the file **Oil_Reserves;** variable = **end2004**)*

The proven oil reserves at the end of 2004 for all significant oil-producing countries (in billions of barrels) are displayed on the left-hand side of Figure 9.1. The mean reserve quantity for this population is 24.652 billion barrels. Suppose we did not have access to the full population of data, but had to rely on taking one sample of three countries' reserve data. Can we determine in advance how the sample mean we calculate might compare to the real population mean if the population data are accessible? What if our sample consisted of six countries' data instead of three? What is the effect of this change relative to what has been found for the smaller sample size?

Figure 9.1

Partial Sampling Distribution of the Sample Mean

POPULATION: Proven Oil Reserves of the Main Oil-Producing Countries of the World (in billions of barrels)

Actual Population Mean = 24.652

Country	Reserves	Country	Reserves
Saudi Arabia	262.7	Indonesia	4.7
Iran	132.5	United Kingdom	4.5
Iraq	115	Malaysia	4.3
Kuwait	99	Australia	4
United Arab Emirates	97.8	Egypt	3.6
Venezuela	77.2	Syria	3.2
Russian Federation	72.3	Vietnam	3
Kazakhstan	39.6	Yemen	2.9
Libya	39.1	Argentina	2.7
Nigeria	35.3	Gabon	2.3
USA	29.4	Congo - Brazzaville	1.8
China	17.1	Colombia	1.5
Canada	16.8	Denmark	1.3
Qatar	15.2	Equatorial Guinea	1.3
Mexico	14.8	Brunei	1.1
Algeria	11.8	Trinidad & Tobago	1
Brazil	11.2	Peru	0.9
Norway	9.7	Chad	0.9
Angola	8.8	Italy	0.7
Azerbaijan	7	Tunisia	0.6
Sudan	6.3	Uzbekistan	0.6
Oman	5.6	Turkmenistan	0.5
India	5.6	Thailand	0.5
Ecuador	5.1	Romania	0.5

Seventy-Five Possible SAMPLES of Size n = 3
Mean of the 75 Sample Means = 24.761

			Sample Means:				Sample Means:
11.8	3.2	0.5	5.167	2.3	29.4	99	43.567
39.6	99	11.8	50.133	77.2	4.5	0.9	27.533
72.3	0.6	3	25.300	2.3	16.8	16.8	11.967
99	3	1.3	34.433	0.5	2.9	99	34.133
0.5	11.2	72.3	28.000	0.6	3.2	17.1	6.967
16.8	1.3	132.5	50.200	0.6	4.5	9.7	4.933
1.3	3.2	0.6	1.700	2.7	2.3	72.3	25.767
0.5	39.1	17.1	18.900	0.9	39.6	1.3	13.933
262.7	97.8	1	120.500	11.2	7	5.6	7.933
0.6	4.5	0.5	1.867	17.1	115	4.7	45.600
0.6	3.6	14.8	6.333	8.8	4.5	0.9	4.733
1.8	5.6	0.7	2.700	6.3	1.3	14.8	7.467
2.7	1.3	1.5	1.833	2.7	3.2	11.8	5.900
1	0.5	5.6	2.367	1.3	2.9	72.3	25.500
0.6	115	3.6	39.733	8.8	1.5	1.1	3.800
0.9	115	4	39.967	262.7	0.7	4.7	89.367
0.5	132.5	4.7	45.900	5.1	5.1	4.7	4.967
16.8	262.7	5.6	95.033	0.6	1.1	72.3	24.667
3	0.6	132.5	45.367	4	0.6	1.5	2.033
3.2	1.5	1.3	2.000	14.8	5.6	35.3	18.567
1	3.2	7	3.733	4.7	16.8	3.6	8.367
4.7	0.5	1.3	2.167	5.1	262.7	1.3	89.700
1	17.1	39.1	19.067	16.8	1.8	4.7	7.767
15.2	2.3	11.2	9.567	0.5	35.3	5.6	13.800
72.3	7	2.7	27.333	2.9	11.2	97.8	37.300
0.6	2.3	72.3	25.067	0.5	35.3	262.7	99.500
8.8	2.9	4.5	5.400	77.2	2.9	4.7	28.267
115	11.8	1.8	42.867	0.5	3	4.7	2.733
5.1	1.8	72.3	26.400	3.2	5.1	9.7	6.000
4	39.1	0.5	14.533	11.8	4.3	4	6.700
8.8	0.6	1.3	3.567	1.5	1.3	1.3	1.367
0.5	11.2	3	4.900	11.2	6.3	3.2	6.900
99	3.2	11.2	37.800	3.6	14.8	29.4	15.933
0.5	2.9	0.9	1.433	1.8	0.5	262.7	88.333
262.7	2.9	17.1	94.233	5.1	0.5	0.6	2.067
15.2	29.4	1.3	15.300	17.1	29.4	11.8	19.433
4.5	1.1	11.8	5.800	0.6	3.2	39.1	14.300
8.8	4.3	115	42.700				

Source: BP Statistical Review of World Energy June 2006. "Energy: proved oil reserves (in 1000 million barrels)." Found at: http://www.geohive.com/charts/charts.php?xml=en_oilres&xsl=en_res

Informal Solution

Over 17,000 distinct samples of size $n = 3$ could be drawn from a population of $N = 48$. (To confirm, calculate the number of combinations.) By chance, a sample could happen to be the three smallest values of the population (0.5, 0.5, and 0.5). In that case, our estimate of μ would be 0.5, significantly below the true population value (24.652). On the other hand, a random sample could include just the highest values of the population, leading to a large overestimate of μ. Fortunately, the most extreme sample combinations are unlikely to be selected. The table on the right-hand side of Figure 9.1 shows 75 randomly selected samples of size $n = 3$. As expected, there is lot of variation in the means of individual samples, but note that taken together, the mean of the 75 sample means is very close to the actual population mean. This tells us that, for all the variability of these small samples' means, they do tend at least to converge toward, or be centred on, the correct value.

Figure 9.2 shows that the significant skewing of the original population is echoed in the distribution of the 75 random samples. If we choose a larger sample size, such as $n = 6$, observe that the distribution of sample means becomes somewhat less skewed and so more closely centred on the true population value for μ. In other words, if we take one sample of six observations, we are less likely than with the smaller sample size to have a large error in the estimate of the mean. And if there is an error, it is not as likely to be *biased* in one direction (in this case, toward being too small rather than too large).

Figure 9.2
Partial Sampling Distribution of the Sample Mean (n = 3 and n = 6)

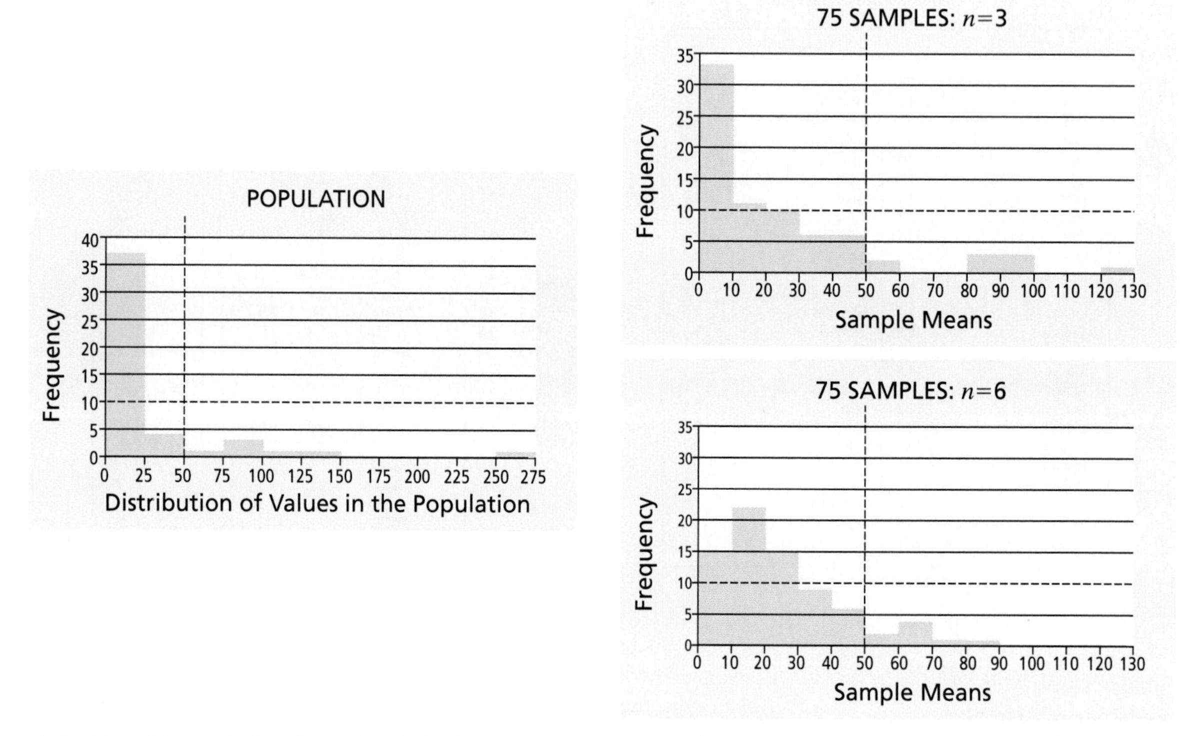

Source: BP Statistical Review of World Energy June 2006. "Energy: proved oil reserves (in 1000 million barrels)." Found at: http://www.geohive.com/charts/charts.php?xml=en_oilres&xsl=en_res

Example 9.1 models some important aspects of the sampling distribution of the sample mean. However, some formulas introduced in this chapter assume sampling *with replacement.* That is, every successive item that is selected in the random sample should have the same probabilities of falling within certain ranges of values. This requirement is not met when the population is finite and relatively small, as in Example 9.1. After each observation in the sample, the sample space decreased, which changes the conditional probabilities for the next selections. (See Chapter 6.) On the other hand, the population in Example 9.2 is so large that the conditional probabilities are virtually unchanged after each item is selected.

Example 9.2

*(Based on the **WindSpeed_Alta** file)*

The Government of Alberta has monitored wind speeds in certain areas on an hourly basis. The file **WindSpeed_Alta** records every hourly wind speed observation (43,824 records in all) at a Peace River station over a five-year period. (Values are in metres per second.) With such a large population, absolute probabilities and the conditional probabilities when sampling without replacement are, in most cases, quite close. Also, if we expect wind speed patterns to remain similar into the future, we could sample from the past distribution *with replacement* to get a sense of how the sample statistic tends to be distributed and how well the statistic value has tended to approximate the corresponding population parameter. Based on simulations to approximate the sampling distributions of the sample mean for the wind speeds population, using sample sizes of 5, 15, 30, and 100, respectively, compare the shapes of the sampling distributions with the shape of the original distribution of raw data.

Solution The original distribution of wind speeds is considerably skewed to the right. The population mean is approximately 2.67 metres per second. Figure 9.3 (middle left) shows output from a computer simulation in which 1,000 independent samples of size $n = 5$ were selected from the population distribution, using sampling with replacement. (Note: For this example, consider the simulation results as a given; to try such simulations on your own, see "Hands On 9.1.") Means were calculated for each of the 1,000 samples and their distribution is shown in Figure 9.3. The same process was repeated for samples of $n = 15$, 30, and 100, respectively.

Observe that as sample size is increased, the distribution of sample means tends to cluster more tightly and more symmetrically around the true population mean. This suggests that probabilities increase with larger samples, that the mean obtained from the sample will be in the vicinity of the population mean, and that the sample mean becomes a **non-biased indicator** of the population mean. That is, the value obtained from the sample has an equal probability of being larger *or* smaller than the population parameter to be estimated.

For an infinite population, the sampling distribution of the sample means is generally represented with a probability density curve (see Chapter 8) rather than with a discrete histogram such as in Figure 9.3. Areas under the curve can then be used to estimate the probabilities of a sample mean falling within a certain range. Important features of the sampling distribution of the sample means have been approximated reasonably well in simulation-based Figures 9.1 and 9.3. For example, the means of the sample distributions are consistent with Formula 9.1.

Formula 9.1 Mean for the Sampling Distribution of the Sample Mean

$$\mu_{\bar{x}} = \mu$$

where $\mu_{\bar{x}}$ = Mean of the sampling distribution of the sample mean, for a given sample size

μ = Mean of the full population from which the samples are drawn

Figure 9.3
Partial Sampling Distribution of the Sample Mean (Sampling with Replacement)

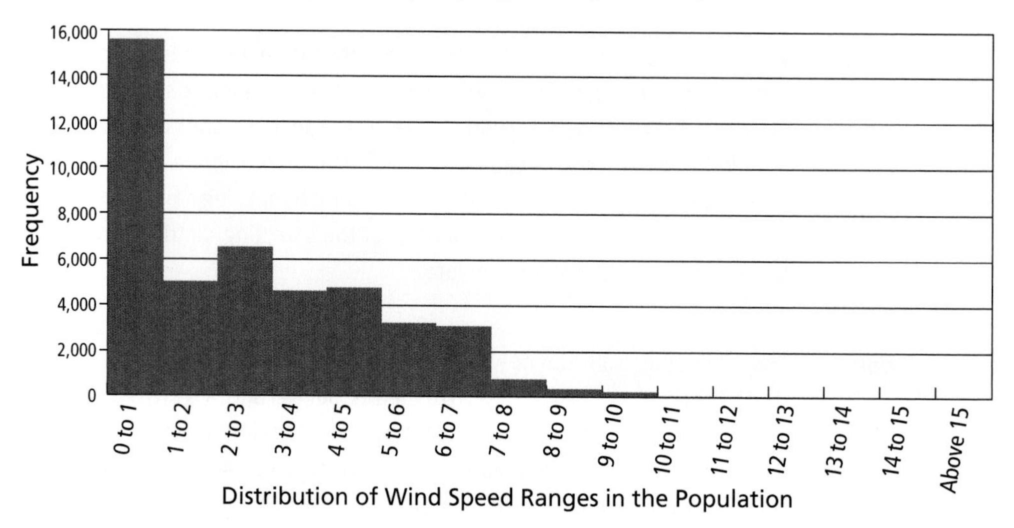

Distribution of Means of Wind Speed Ranges in Samples

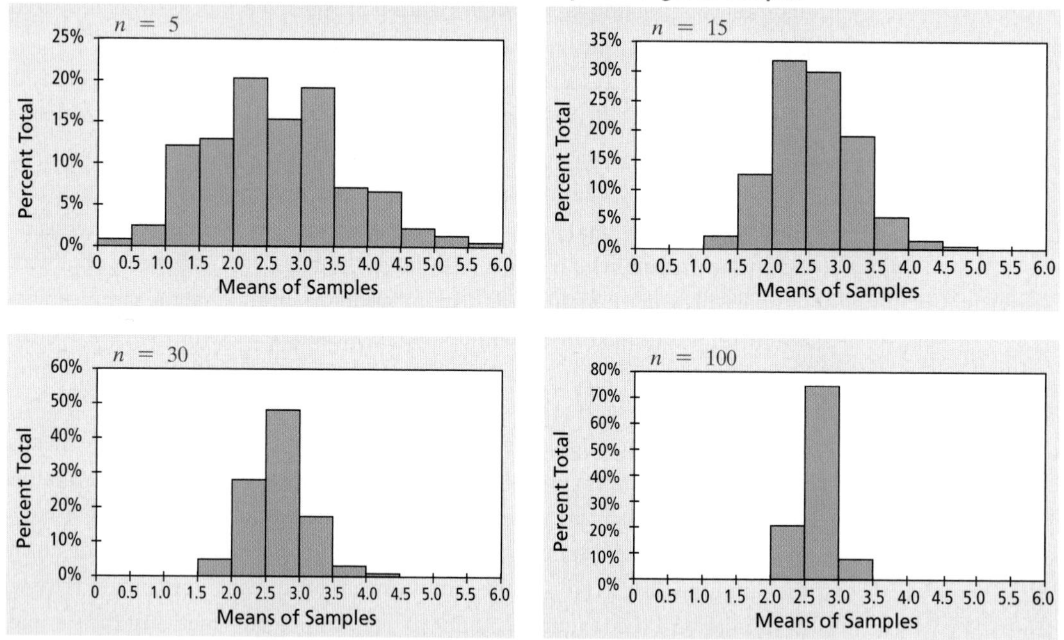

Source: Alberta Environment. Table 2.1 "Wind speed distribution summary for each AENV Administration Region." Found at: http://www3.gov.ab.ca/env/air/pubs/compmet.pdf. Source Document: *Comparison of Meteorology Elements in the Alberta Environment, Regional Screening Dispersion Modelling Data Sets*. Prepared by Conor Pacific Environmental Technologies, Edmonton, Alberta, Project No. 9316234, July 1999. Pub. T/461 ISBN: 0-7785-0643-6.

Figures 9.2 and 9.3 also illustrate that as the sample size increases, the sampling distribution of the sampling means increasingly becomes more symmetrical—tending toward the normal distribution. The centre of the distribution is its mean, which we have seen is equal to the mean of the population. The standard deviation can be calculated by Formulas 9.2 and 9.3.

Formula 9.2 Variance for the Sampling Distribution of the Sample Mean

$$\sigma_{\bar{x}}^2 = \frac{\sum_{i=1}^{N} (\bar{x}_i - \mu)^2}{N}$$

Formula 9.3 Standard Deviation for the Sampling Distribution of the Sample Mean

$$\sigma_{\bar{x}} = \sqrt{\sigma_{\bar{x}}^2} = \sqrt{\frac{\sum_{i=1}^{N} (\bar{x}_i - \mu)^2}{N}}$$

where $\sigma_{\bar{x}}^2$ = Variance of the sampling distribution of the sample means

$\sigma_{\bar{x}}$ = Standard deviation of the sampling distribution of the sample means (also called the **standard error** of the sampling distribution)

$\bar{x}_i$ = For each value of i, the mean of the ith possible sample of size n from the population

μ = Mean of the sampling distribution, which is also the mean of the population

N = Number of sample means

In other words, we can calculate the standard deviation on the same basic model as in Formulas 5.1–5.3. In this case, the population comprises all of the possible sample means, and μ is the mean of the entire sampling distribution. The variance is based on the squared deviations of all individual means from the population mean.

Provided all of the n-values in a sample are selected independently, then the formula for standard deviation of the sampling distribution can be simplified, as shown in Formula 9.4.

Formula 9.4 Standard Deviation for the Sampling Distribution of the Sample Mean (Simplified Formula)

$$\sigma_{\bar{x}} = \frac{\sigma}{\sqrt{n}}$$

where $\sigma_{\bar{x}}$ = Standard deviation of the sampling distribution of the sample means

σ = Standard deviation of the population from which the samples are drawn

n = Sample size for each individual sample in the sampling distribution

Example 9.3

*(Based on the **WindSpeed_Alta** file)*

The hourly wind speeds over a Peace River, Alberta, weather station, were recorded for five years (see file **WindSpeed_Alta**). Over that extensive period, the 43,824 hourly records showed that the mean wind speed was 2.67 metres per second and the standard deviation was 2.21 metres per second. For this population, what are the estimated mean and standard deviation for a sampling distribution of the sample means if (1) the sample size $n = 30$ and (2) the sample size $n = 100$?

Solution

The values for μ and σ are provided in the problem. Therefore:

1. $\mu_{\bar{x}} = \mu = 2.67$ metres per second

$$\sigma_{\bar{x}} = \frac{\sigma}{\sqrt{n}} = \frac{2.21}{\sqrt{30}} = 0.40 \text{ metres per second}$$

2. $\mu_{\bar{x}} = \mu = 2.67$ metres per second

$$\sigma_{\bar{x}} = \frac{\sigma}{\sqrt{n}} = \frac{2.21}{\sqrt{100}} = 0.22 \text{ metres per second}$$

Example 9.4

(Based on the ***Oil_Reserves*** file)

Data for the proven oil reserves (in billions of barrels) for significant oil-producing countries can be found in the file **Oil_Reserves.** For this population, what is the standard deviation for the distribution of sample means, if (1) the sample size $n = 15$ and (2) the sample size $n = 30$?

Solution

Formula 9.4 was used to find the standard error for Example 9.3, but the formula presupposes that sampling is done *with replacement.* If not, then as discussed in Chapter 6, the selections of each element in the sample would no longer be independent; once an item was selected and removed from the sample space, probabilities for the next selection would change. Because the population in Example 9.3 was very large, even though we actually sampled without replacement (if a weather record was selected, it was not considered available to be selected again), the impact of ignoring a correction factor was negligible. This type of sampling situation is rather common.

This example, however, illustrates that *sometimes* the error can be noticeable if we sample without replacement and do not make some adjustment. The adjustment for those cases is to include in the standard error estimate a **finite population correction factor,** as shown in Formula 9.5. A rule of thumb is to switch from Formula 9.4 to Formula 9.5 if the sample size n equals more than 5% of the population size N. For smaller ratios of sample size to population size (or if the population size is infinite), the correction factor equals approximately one, so in practice, the correction factor has little impact and can be left out of calculations.

Formula 9.5 Standard Deviation for the Sampling Distribution of the Sample Means Adjusted by a Finite Population Correction Factor

$$\sigma_{\bar{x}} = \frac{\sigma}{\sqrt{n}}\underbrace{\sqrt{\frac{N-n}{N-1}}}_{\text{Finite population correction factor}}$$

where $\sigma_{\bar{x}}$ = Standard deviation of the sampling distribution of sample means
σ = Standard deviation of the population from which the samples are drawn
n = Sample size for each individual sample in the sampling distribution
N = Number of items in the population from which the samples are drawn

For both parts of this example, the sample sizes *are* large compared to N for the finite population (n is 31% of N and 63% of N, respectively). Therefore, for this problem, we include the correction factor in the solutions:

1. $\sigma_{\bar{x}} = \frac{\sigma}{\sqrt{n}}\sqrt{\frac{N-n}{N-1}} = \frac{47.47}{\sqrt{15}}\sqrt{\frac{48-15}{48-1}} = 10.27$ billions of barrels

2. $\sigma_{\bar{x}} = \frac{\sigma}{\sqrt{n}}\sqrt{\frac{N-n}{N-1}} = \frac{47.47}{\sqrt{30}}\sqrt{\frac{48-30}{48-1}} = 5.36$ billions of barrels

The Central Limit Theorem

Figure 9.3 illustrated, experimentally, a relationship among (1) a population of values, (2) a sample size n, and (3) the distribution of possible sample means (sample size = n) for that population. The sample mean can be viewed as a random variable, and under certain conditions, its distribution appears to be approximately normal—with values for the distribution mean and standard deviation as determined by formulas (e.g., Formulas 9.1 and 9.4). This is clearly observed in Figure 9.3 and is formally affirmed as a principle in the **central limit theorem.** (For rigorous mathematical proof, see http://mathworld.wolfram.com/CentralLimitTheorem.html.) The theorem can be summarized in this form:

When the sample size n is sufficiently large and sampling is done with replacement:

1. The sampling distribution of the sample means is approximately normal, regardless of the distribution of the underlying population.
2. This approach toward the normal distribution becomes closer as the sample size n increases.

The mean and standard deviation of the distribution are as given in Formulas 9.1 and 9.4, respectively.

The size needed for n to be "sufficiently large" can vary, based on the distribution of the underlying population. If the population is itself normally distributed, the distribution of the sample mean becomes normal—even given a small value of n. For a population that is, for example, very skewed or bimodal, n needs to be larger for the sampling distribution to become normal. Generally, the theorem applies by at least $n \geq 30$.

• Hands On 9.1 Confirming the Central Limit Theorem

In "In Brief 7.1," we explored how to generate a sequence of random numbers based on a specific probability distribution. By experimenting with your software, you can easily discover related similar options or menu sequences for generating data from other types of distributions, including the uniform, normal, and exponential distributions.

As a first step for this exercise, randomly generate three long columns of data, each representing a large population. Let the first data set be uniformly distributed; the second set, normally distributed; and the third set, exponentially distributed. Using the graphical procedures provided in Chapter 3, confirm the distribution patterns of these three data sets. Also use the computer to find the mean and standard deviation of each of the populations.

Procedures for taking a random sample from a given column of data were presented in "In Brief 2.2." Using those techniques, perform the following sequence (save your outputs for future use):

1. Take a random sample of size $n = 5$ from the uniformly distributed population.
2. Find and record the mean of that sample.
3. Repeat steps 1 and 2 until you have recorded the means for 40 independently drawn samples of size $n = 5$ from the uniformly distributed population.
4. Find and record the mean of the 40 recorded sample means computed in step 3.
5. Find and record the standard deviation of the 40 recorded sample means computed in step 3.
6. Graphically display the frequency distribution for the 40 recorded sample means computed in step 3.

This completes one simulation of one sampling distribution for one population.

7. Repeat steps 1 through 6, except substitute "$n = 10$" for "$n = 5$" wherever it occurs.
8. Repeat steps 1 through 6, except substitute "$n = 30$" for "$n = 5$" wherever it occurs.

This completes one set of three simulations of a sampling distribution (each using a different sample size) from a common type of population.

9. Repeat steps 1 through 8, except substitute "normally distributed" for "uniformly distributed" wherever it occurs.
10. Repeat steps 1 through 8, except substitute "exponentially distributed" for "uniformly distributed" wherever it occurs.

When these steps are completed you will have simulated three sets of three simulations each of sampling distributions of the mean, using three different sample sizes from three types of population. In each case, compare the means and standard deviations of the sampling distributions with the original population parameters. How well do Formulas 9.1 and 9.4 appear to apply? Also compare the shapes of the sampling distributions with the shape of the population. Based on this experiment, are you comfortable accepting that the central limit theorem appears to be accurate? Does the rule of thumb seem accurate that the theorem applies, regardless of the source population, once n is at least equal to 30?

9.2 The Central Limit Theorem and Probability

Section 9.1 has shown how a sample statistic (the mean, in particular) and the corresponding population parameter are related. This is important, because generally we do not know the population mean in advance. (If we did, why would we sample?) In practice, we have no idea whether a particular sample has a mean that is really too small or too large, compared to the population mean. However, if certain conditions are met, we *can* know the probabilities that the sample mean will fall on one side or the other of the population, and by how much. For a large enough sample, and if the population sigma is known (or reasonably assumed from, for example, past data or published standards), then, thanks to the central limit theorem, we can derive the probabilities we seek from the model of the normal curve.

The calculations based on the normal curve that were introduced in Section 8.1 can now be applied to the sampling distribution. All that changes is the population being modelled by the probability density curve. Instead of modelling a population of raw data values, the population modelled here is the set of means of all possible samples of a given size from the population of raw data (i.e., the sampling distribution of the sample means). As shown in "In Brief 9.1," the calculations and procedures *within* the model are the same as in Chapter 8.

• *In Brief 9.1 Computing a Cumulative Probability for a Normally Distributed Sampling Distribution*

Excel, Minitab, and SPSS

For each of these packages, review the corresponding procedures in "In Brief 8.1(a)." You are shown how to calculate $P(X < x)$ for a normal distribution if μ,σ, and the value x are known. The underlying calculations are based on the model of Figure 9.4, wherein the inputs are converted internally to z-values and the less than cumulative probability is calculated from the probability density curve. The only significant difference between Figure 8.4 and Figure 9.4 is that the x term now represents one individual mean $\bar{x}$ in a distribution determined by the μ and σ *for a sampling distribution* (i.e., $\mu_{\bar{x}}$ and $\sigma_{\bar{x}}$) instead of a particular raw datum in a distribution determined by μ and σ for a raw data population. *Your software will not know the difference.*

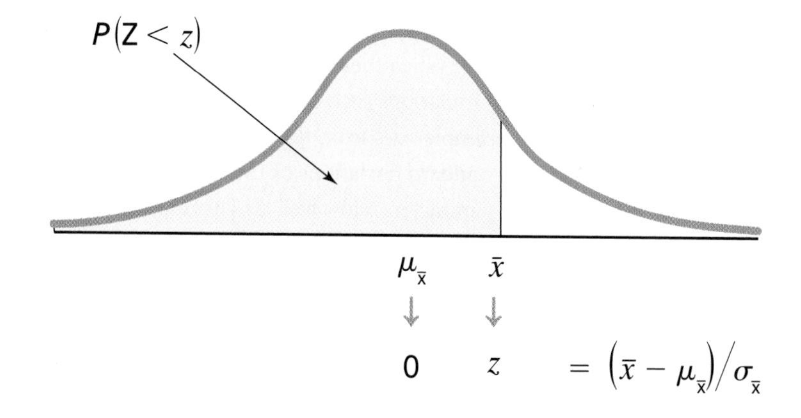

Figure 9.4
Cumulative Probability for the Normally Distributed Sampling Distribution Based on z

In short, use the computer procedures exactly as you did in Chapter 8. Ensure that your input for σ is really the value $\sigma_{\bar{x}}$—which means you may have to manually calculate $\sigma_{\bar{x}} = \frac{\sigma}{\sqrt{n}}$ *prior* to inputting a value. Also ensure that the x term that you input is really a particular mean ($\bar{x}$) within the sampling distribution.

Note that for a normally distributed sampling distribution, you can also apply the procedures given in "In Brief 8.1(b)." Using your software, you have learned how to determine an x-value within a normal distribution if $P(X < x)$, μ, and σ are known. Now you can apply the same procedures to find an $\bar{x}$ within a normally distributed sampling distribution if the other inputs are interpreted as applying to that distribution. Again, be sure to calculate $\sigma_{\bar{x}} = \frac{\sigma}{\sqrt{n}}$ prior to inputting a value for the σ term.

Example 9.5

The mean amount of pressure that causes a composite thermoplastic object to deform is 66.2 megapascals (MPas). The standard deviation is 9.2 MPa.

1. If pressure is applied to a single randomly selected composite thermoplastic object, what is the probability that it will deform at a pressure less than 60 MPa?
2. If a random sample of 30 composite thermoplastic objects is taken from the population, what is the probability that the *mean* pressure at which the objects deform will be: (a) less than 60 MPa? (b) Less than 70 MPa? (c) Greater than 70 MPa? (d) Between 60 and 70 MPa?
3. How might our calculations for (1) and (2) be affected if the underlying population is not normally distributed?
4. If the random sample taken from the population has only 5 records, how might this affect our calculations for (2)?

Solution

1. The values for $\mu = 66.2$ and for $\sigma = 9.2$ are given in the problem. We are not told if the population is normally distributed. (If it is not, see solution 3 below.) Assuming that the distribution *is* normal, we can calculate $P(X < 60) \approx 0.25$ by using procedures from "In Brief 8.1(a)."

2. In contrast to solution 1, these calculations will be based on the sampling distribution. Because $n \geq 30$, whether the population is normally distributed is not a concern. The central limit theorem suggests that the sampling distribution will approach normality in any case. We can, therefore, adapt the computer procedures from "In Brief 8.1(a)" for the case of the sampling distribution.
   (a) Our estimate for $\mu_{\bar{x}}$ equals the population value for μ, given as 66.2. Input this value as the μ term in the computer procedure. The estimate for $\sigma_{\bar{x}}$ equals $\frac{\sigma}{\sqrt{n}} = \frac{9.2}{\sqrt{30}} = 1.67968$. Input this as the σ term in the procedure. Based on these modifications, the computer output for the cumulative probability can be interpreted as the solution for $P(\bar{x} < 60)$. The probability is 0.00011.
   (b) Apply the computer procedure as above, but for $P(\bar{x} < 70)$, replace the x term in the procedure with the mean value 70 to obtain a probability equal to 0.98816.

      The solutions for (c) and (d) require additional steps, similar to those required for Example 8.4, solutions 4 and 5.
   (c) On a continuous scale, $P(\bar{x} > 70)$ is the complement of $P(\bar{x} < 70)$. Therefore, $P(\bar{x} > 70) = 1 - 0.98816 = 0.01184$.
   (d) $P(60 < \bar{x} < 70) = P(\bar{x} < 70) - P(\bar{x} < 60) = 0.98816 - 0.00011 = 0.98805$
3. The calculations in solution 1 could vary widely from the mark if the population is highly skewed or irregular. Solution 2 is relatively unaffected because we are applying the sampling distribution, and the sample size is considered large.
4. If the original population is reasonably normally distributed, then according to the central limit theorem, the sampling distribution of the sample means is also reasonably normal, even for small samples. So the answers for (2) would still apply. However, if the source population is highly non-normal, then at $n = 5$, the samples are too small to apply the normal model with confidence.

• Hands On 9.2 *Confirming Probability Calculations Based on the Central Limit Theorem*

In "Hands On 9.1," we generated three populations of different distributions—uniform, normal, and exponential. For each population, we simulated the generation of partial sampling distributions for $n = 5$, 10, and 30, respectively. If you have saved these data, you can reuse them for this exercise.

According to the central limit theorem, five of the sampling distributions that we generated earlier should be approximately normal: All three distributions (i.e., $n = 5$, 10, and 30) for the normal population should be roughly normal, as well as the distributions for $n = 30$ for the uniform and exponential populations. For each of these five cases, in turn, perform the following sequence:

1. Determine, by formula from values in the simulated population, the values in the sampling distribution that fall at (a) the mean of the distribution, (b) one standard error below the mean, and (c) one standard error above the mean. In theory, based on the mathematics of the normal curve, 50% of the values in the sampling distribution should fall below value (a), about 16% should fall below value (b), and about 84% below value (c).
2. For the sampling distribution of the sample means under consideration, construct a cumulative percentage frequency table. Identify the approximate x-values in the table at which the cumulative frequencies 16%, 50%, and 84% are reached.
3. Do the distributions in the table appear to roughly match the expected distributions according to the central limit theorem?

Once you have done this for all five cases, ask whether the fit is equally good in all cases or better for some distributions than others. If the latter, can you explain why this might be? Note that the simulated sampling distributions are not complete (each contains only 40 of many possible samples). Does this appear to have been a factor in your observed results?

9.3 The Distribution of the Sample Proportion

In Chapters 7 and 8, we distinguished the concepts of discrete versus continuous probability distributions. In this chapter, we have taken a continuous distribution (i.e., the normal distribution) to a next step, and introduced concepts of the sampling distribution and related probability calculations. Are there also sampling distributions for discrete variables?

For discrete variables that are numerical, the concepts of Sections 9.1 and 9.2 can be applied as approximations. If the population is large, then for samples of n at least 30 (but less than 5% of population size), the distribution of sample means will tend to approach normality. The corresponding calculations from the central limit theorem can then be applied. For example, Figure 9.5(a) displays the distribution of a discrete variable that has only five possible values. A computer was used to take many independent, random samples of size $n = 30$, each based on sampling with replacement, which simulates an infinite population. The distribution of the many sample means approximates the true sampling distribution—which is clearly seen in Figure 9.5(b) to approach the normal distribution.

Many textbooks include a section on the **sampling distribution of the sample proportion,** which usually follows the section on the central limit theorem. Proportions are usually associated with *categorical* (nonnumerical) variables. For example, what proportion of voters prefers the Green Party to the alternative parties? For such variables, we cannot blend the many individual answers (e.g., "Prefer Green," "Prefer Liberal," "Prefer Conservative," "Prefer Other") into one "central" value that is something like a mean. Without having a mean, the central limit theorem cannot directly apply.

Figure 9.5
Distribution of a Discrete Quantitative Variable and Its Sampling Distribution of the Sample Means

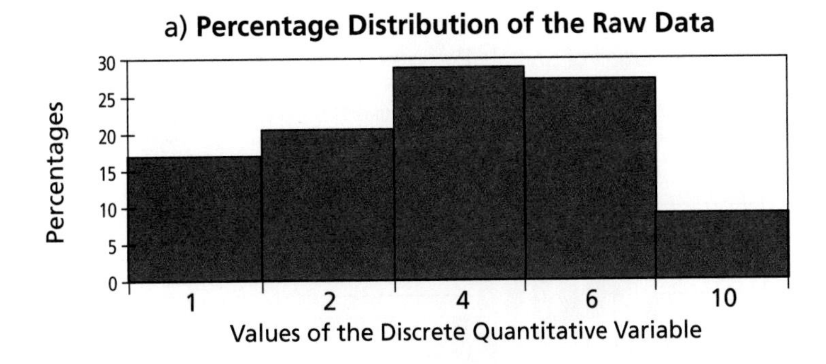

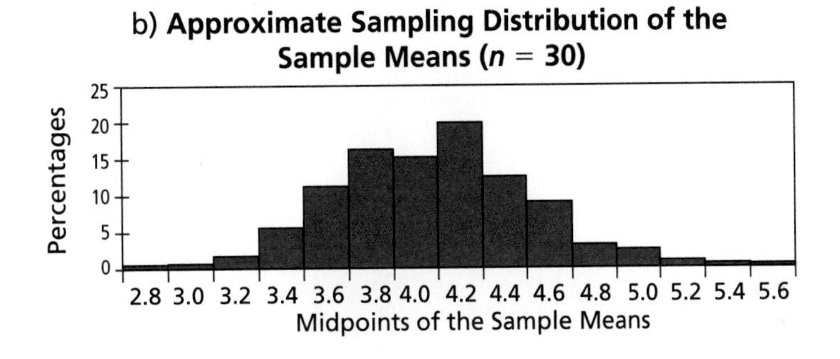

Yet under certain conditions, the normal curve and its mathematics *can* be applied as an approximation for a distribution of proportions. This is illustrated in Figure 9.6. If the distribution of a variable meets the conditions for a binomial distribution (see Section 7.2), then the distribution of sample proportions approaches the normal distribution as the sample size increases. You will recall that for the binomial distribution, the variable X can be interpreted as the number of "successes" that result from n identical trials of a binomial experiment. If we interpret n as the sample size, then x/n is the proportion of the sample that meets the condition to be a success. The outcome of each trial—which corresponds to a characteristic of one sampled individual—can have only one of two mutually exclusive values. The probability of success p for each trial is presumed to remain constant.

As indicated, for large enough samples, the distribution of proportions is similar to the normal curve. In those cases, the methods of Section 9.2 can be used to estimate probabilities associated with that distribution. The mean and standard deviation of the sampling distribution can be derived from the mean and standard deviation of the binomial and then applied (manually or with the computer) to calculate cumulative probabilities.

Example 9.6

In Example 7.10, we noted that 31.5% of applications for funding to the Natural Sciences and Engineering Research Council of Canada were successful in a particular year.[1]

1. Suppose that for every possible sample of 20 applicants that could be selected from the population, the proportion of successes p_s for that sample is recorded. This is the sampling distribution of the sample proportions for $n = 20$. (a) What is the expected mean for that distribution? (b) What is the standard deviation?
2. Using the values for the mean and standard deviation exactly as calculated for (1), apply the methods of Section 8.1 (on probability calculations for the normal curve)

Figure 9.6

Comparison of the Distribution of Sample Proportions and the Classical Binomial Probability Distribution

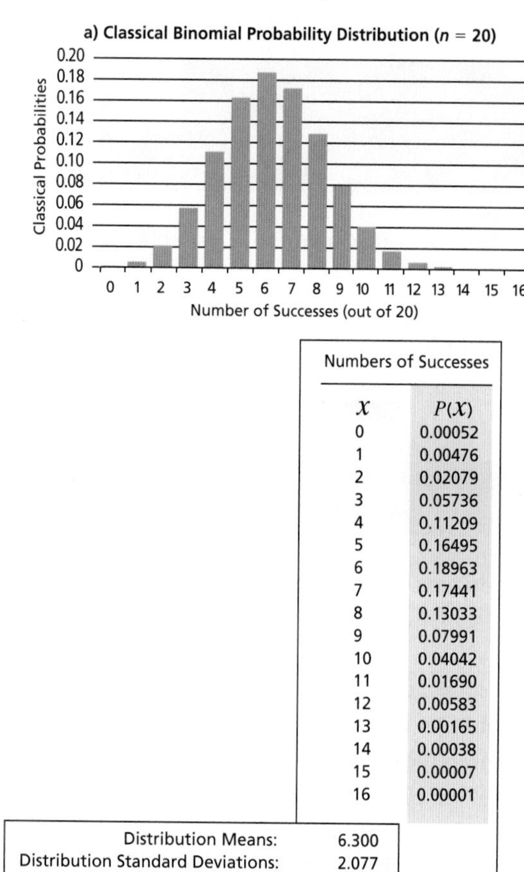

Numbers of Successes

X	$P(X)$
0	0.00052
1	0.00476
2	0.02079
3	0.05736
4	0.11209
5	0.16495
6	0.18963
7	0.17441
8	0.13033
9	0.07991
10	0.04042
11	0.01690
12	0.00583
13	0.00165
14	0.00038
15	0.00007
16	0.00001
Distribution Means:	6.300
Distribution Standard Deviations:	2.077

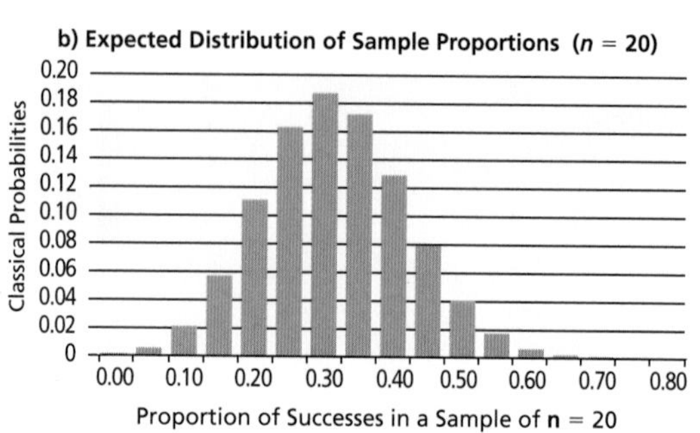

Proportions

X	$\frac{X}{n}$	$P\left(\frac{X}{n}\right)$	Cumulative Probability
0	0.00000	0.00052	0.00052
1	0.05000	0.00476	0.00528
2	0.10000	0.02079	0.02606
3	0.15000	0.05736	0.08342
4	0.20000	0.11209	0.19551
5	0.25000	0.16495	0.36046
6	0.30000	0.18963	0.55010
7	0.35000	0.17441	0.72451
8	0.40000	0.13033	0.85483
9	0.45000	0.07991	0.93474
10	0.50000	0.04042	0.97516
11	0.55000	0.01690	0.99206
12	0.60000	0.00583	0.99789
13	0.65000	0.00165	0.99954
14	0.70000	0.00038	0.99992
15	0.75000	0.00007	0.99999
16	0.80000	0.00001	1.00000
	0.315		
	0.104		

to answer these questions: If one sample of 20 applicants is selected, what is the *approximate* probability that the proportion of successes is (a) less than 0.25? (b) Less than or equal to 0.25?

3. Compare your answers to (2) with the cumulative probabilities displayed in Figure 9.6(b). Do you see a need for adjustments if the normal model for questions about the sampling distribution of a proportion is applied? Explain.

Solution

1. The first two solutions can be derived from the solutions given for Example 7.10. We previously calculated a mean for the distribution of X, the possible numbers of successes. The solution was $\mu_{(X)} = n \times p = 20 \times 0.315 = 6.30$. We later calculated $\sigma_{(X)} = \sqrt{\sigma^2_{(X)}} = \sqrt{20 \times 0.315 \times (1 - 0.315)} = 2.077$. These values can be found at the lower left of Figure 9.6. Because each value for p_s is directly equivalent to an X-value divided by n, the mean and standard deviation of the sampling distribution can easily be derived:

a) $\mu_{(p_s)} = \dfrac{\mu_{(X)}}{n} = \dfrac{6.30}{20} = 0.315$

b) $\sigma_{(p_s)} = \dfrac{\sigma_{(X)}}{n} = \dfrac{2.077}{20} = 0.104$

2. Applying the methods described in "In Brief 8.1(a)" for estimating a cumulative probability that a selected value will fall within a certain range, we could input the values from solution 1 of 0.315 for μ, of 0.104 for σ, and of the specified range boundary (0.25) for *x:*

   a) $P_{\text{normal}}(X < 0.25) \cong 0.27$

   b) $P_{\text{normal}}(X \leq 0.25) \cong P_{\text{normal}}(X < 0.25) \approx 0.27$

3. The precise cumulative probabilities up to the end of each proportion value are shown at the lower right of Figure 9.6. Because the number of successes is a discrete variable, the possible exact values for the proportions are also discrete. So whereas in the normal model, $P_{\text{normal}}(X < 0.25) \cong P_{\text{normal}}(X \leq 0.25) \cong 0.27$, we see in the cumulative probability column that $P\left(\frac{X}{n} < 0.25\right) = P\left(\frac{X}{n} \leq 0.20\right) \cong 0.20$, but $P\left(\frac{X}{n} \leq 0.25\right) \cong 0.36$. The problem is that the normal model is *continuous,* whereas the true binomial is not. We could try a normal estimate for $P(X < 0.25)$ that acknowledges the discrete jump to the next-lower proportion value, such as $P_{\text{normal}}(X \leq 0.20) \cong 0.13$ but now we have underestimated the true value. Clearly, some additional refinements would be needed to use the normal model for this application.

   Solutions 2 and 3 allude to what is traditionally called the *normal approximation of the binomial.* It is useful to have seen the relationship between the two concepts. However, if you are using statistical software, it is unnecessary in practice to use the normal model for finding proportion-related probabilities. As Figure 9.6 makes evident, the distribution of sampling proportions is a direct variation of the binomial distribution, which was discussed in Chapter 7.

Example 9.7

As given in Example 9.6, the proportion of successes among applicants for Natural Sciences and Engineering Research Council of Canada funding is 31.5%.[2] One sample of 20 applicants has been randomly selected. What is the probability that the proportion of successes is (1) exactly 25%? (2) Less than or equal to 25%? (3) Below 25%?

Solution

1. Unlike the value of a continuous variable, a proportion *can* have a unique, exactly definable value. If using the binomial procedures, there is no need for the adjustments that the normal model requires. To use your software, directly input the known values for n and population p in the appropriate dialogue box. Where you are given a proportion of successes (e.g., 25%), you can derive the corresponding number of successes X, using (proportion) times (n). Then input the X-value (in this case, $0.25 \times 20 = 5$) where expected. For this solution, do *not* request a cumulative probability:

$$P\left(\frac{X}{n} = 0.25\right)_{(\text{for } n=20)} = P(X = 5) = 0.165$$

2. In your software package, input the same values as for (1) but this time request a cumulative probability:

$$P\left(\frac{X}{n} \leq 0.25\right)_{(\text{for } n=20)} = P(X \leq 5) = 0.360.$$

3. For the cumulative less than probability, this model acknowledges the discrete nature of the data. If the X-value input for solution 2 was 5, then the next lower bound for the probability range to be input for this solution is the next smaller $X = 4$:

$$P\left(\frac{X}{n} < 0.25\right)_{(\text{for } n=20)} = P(X < 5)$$

$$P(X < 5)_{(\text{for this discrete variable})} = P(X \leq 4) = 0.196$$

• Hands On 9.3 ***Comparing the Two Models for the Sampling Distribution of the Sample Proportions***

At the end of this chapter, there are exercises to calculate cumulative probabilities based on the sampling distribution of sample proportions. It is expected that you will use the binomial model for the solutions, as illustrated in Example 9.7. That model is more precise and, with computers, no more difficult to calculate (perhaps easier) than using the normal model.

As a comparison, choose at least five complete end-of-chapter problems that call for probability calculations related to the sample distribution of the sample proportion. After finding the solutions with the binomial model, solve the problems again using the approach illustrated in Example 9.6. That is, find the distribution means and standard deviations and then use the normal model for your estimates.

How would you handle a case where a problem asks for the probability that a proportion is exactly some value? Can you figure out how to make adjustments to solve this when using the continuous normal curve as your model?

9.4 Additional Notes on Sampling Distributions

As you will discover, the concept and application of a sampling distribution will play an important role in virtually all of the remaining sections of this textbook. If we did not have a model for how the characteristics of a sample might reflect the features of a population, there would be no way to use the sample to make inferences about the population. That would be like receiving a gift-wrapped box and trying to guess its contents. Through experience, you know the likely size and weight of the gift, based on the features of the box. Outliers are possible—like receiving a small jewel in a large container—but generally your guesses about the size of the gift based on the size of the box will not be far off.

We have looked in particular at two models of sampling distribution: (1) If certain conditions about sample size, parameter of interest, and so on, are met, then we can model the sampling distribution by the normal curve. (2) Under other conditions, if we are studying the proportions of a binomial variable, we can model the sampling distribution with the binomial distribution. The advantage of these models is that the procedures for calculating probabilities based on these curves are well established. Once we learn how to "map" key information from a problem into the mathematical model, we can easily generate a solution.

Where the conditions for the above models do not apply, there is a variety of alternative models that can be considered. For example, if n is too small to apply the normal curve, then perhaps the t-distribution is the best model for the sampling distribution. For sampling from a population of variances, there is an F-distribution; and for tests on contingency tables, there is a chi-square distribution; and so on. These distributions will be addressed later in the book.

There are other sampling distributions for sampling from more than one population. Suppose you are comparing the mean incomes of university graduates with the mean incomes of those without this qualification. This would call for a sampling distribution of the difference between two means. Or you might want to compare the proportion of college graduates who are millionaires with the proportion of nongraduates who are millionaires. This comparison requires a sampling distribution of the difference between two proportions. As in the case of the F- and chi-square distributions, these additional variations will be introduced later in the book as they are needed.

Using Bootstrap Methods to Approximate Sampling Distributions

The strength of the methods referred to in the preceding section lies in the well-documented mathematics for applying them. A common weakness is that each model makes assumptions about the distributions of the raw data for which it applies. As far as the data vary from that mould, the calculations based on a model may not apply or may be less accurate than realized or acknowledged.

Here is one example: If you attempted the "Hands On 8.2" exercise on checking a distribution for normality, you may have realized it is not that easy to find a perfectly normal distribution. Perhaps you plan to work with a sample of $n = 6$ from a population taken to be normal and, based on the central limit theorem, you plan to assume that the sampling distribution is normal. But how normal is that population distribution, really? Is it normal enough to proceed without further questions?

To avoid these problems, you might consider using what are sometimes called **bootstrap methods.** These methods are not as widely applied as the alternatives and are often less easy for a nonspecialist to apply using conventional software. The techniques will not be covered formally in this textbook, but these methods are increasingly being mentioned in articles and other research, so you are strongly encouraged to learn more about them through other readings or courses.

The word *bootstrap* relates to the expression "pick yourself up by your own bootstraps," which basically means that you have to call on all of your available resources if you want to succeed. In this case, the data set in front of you is used to provide its own basis for estimating a sampling distribution. The sampling distribution is not picked out, like the normal curve, from a standard library of distributions that have proven useful in the past.

For an example, refer back to Figure 9.5. The construction of Figure 9.5(b) was derived by a bootstrap-type simulation from the population depicted in Figure 9.5(a). The goal was to illustrate that the distribution of random samples of size $n = 30$ that could be drawn from that population would be approximately normal. The figure does not back up this claim mathematically. Instead, repeated samples (a thousand of them) were taken independently from the source data. Each sample was selected randomly, using sampling with replacement from the full set of original data. (There were 120 numbers: 20 ones, 24 twos, 34 fours, 32 sixes, and 10 tens.) By this method, every item sampled had the identical distribution of probabilities for which value might be drawn. Figure 9.5(b) does not literally show *every* possible sample that could be drawn from the population, but because (1) every possible sample had an equal chance of being selected and (2) a great many samples were selected as examples, we can be quite confident that the full sampling distribution looks something like the one that is shown.

Once the bootstrapped output is generated, the probabilities of drawing samples of given sample sizes with statistics of certain values can now be calculated using cumulative probabilities from the tables or figures that are output. In Figure 9.5(b) for example, we can see simply by looking at it that, from this population, there is a less than 1% chance that a sample of $n = 30$ would have sample mean less than about 2.9.

Another advantage of the bootstrap method is that we can estimate a sampling distribution for *any* statistic that can be calculated. We are not limited to statistics like the mean or proportion for which formal models have been developed for the sampling statistic.

Example 9.8

Following a string of severe hurricanes in recent years, such as Ivan, Katrina, and Rita, some people wonder if hurricanes in general are becoming more severe. Figure 9.7 gives one picture of the old patterns: It displays the relative severities of all hurricanes that landed in North Carolina during the 20th century.

Figure 9.7
Hurricane Landings in North Carolina, 1900–2000

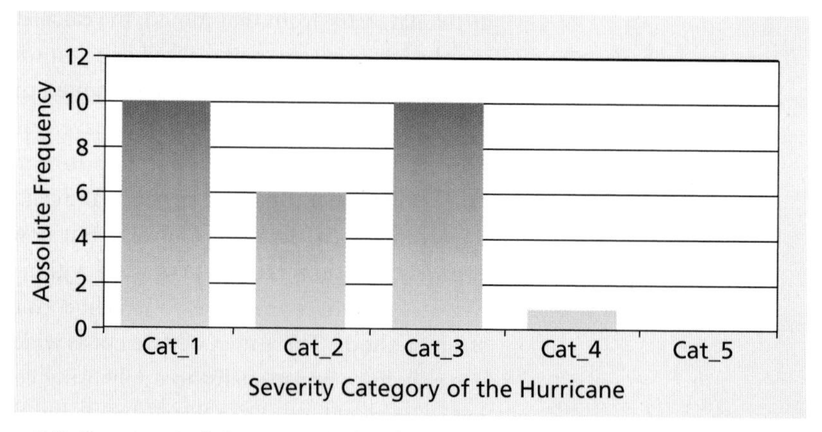

Source: U.S. Department of Commerce, National Oceanic and Atmospheric Administration, Atlantic Oceanographic and Meteorological Laboratory, Hurricane Research Division. Based on data in *NOAA Technical Memorandum NWS TPC-1:* "The Deadliest, Costliest, and Most Intense United States Hurricanes from 1900 to 2000." Found at: http://www.aoml.noaa.gov/hrd/Landsea/deadly/Table9.htm

If the pattern is *not* changing, then the probability distribution for the severities of hurricanes in North Carolina in the 20th century may continue to be the probability distribution for the 21st century. If that assumption holds, and if five of the North Carolina hurricanes of the 21st century are randomly selected, display the sampling distributions for the following:

1. The ratio of Category 3 to Category 2 hurricanes.
2. The absolute number of hurricanes that will be a Category 3 or greater in a sample.

Solution

1. We are asked for the sampling distribution of a sample statistic that has not been modelled in the preceding sections—the ratio between the counts of two possible values of the variable within any one sample. The data for this new variable, the ratio, is quantitative, but it is not clear what its distribution would be nor how that distribution would relate to the original distribution of hurricane strengths.

   With bootstrapping, we avoid the need to develop a complex mathematical model to solve the problem. Imagine a bowl filled 27 marbles corresponding to the $N = 27$ hurricanes in the specified population. Let 10 of the marbles be red, 6 green, 10 blue, and 1 yellow to correspond to the relative numbers of Category 1, 2, 3, and 4 hurricanes in the population. Randomly sample five marbles *with replacement.* Count the blue (Category 3) and green (Category 2) marbles and record the ratio. That gives the ratio statistic for one possible sample. If you repeat the procedure many times, the relative frequencies of the ratios will approximate the sampling distribution for that statistic. (See Figure 9.8(a).)

Figure 9.8
Bootstrapped Sampling Distributions Based on North Carolina Hurricanes

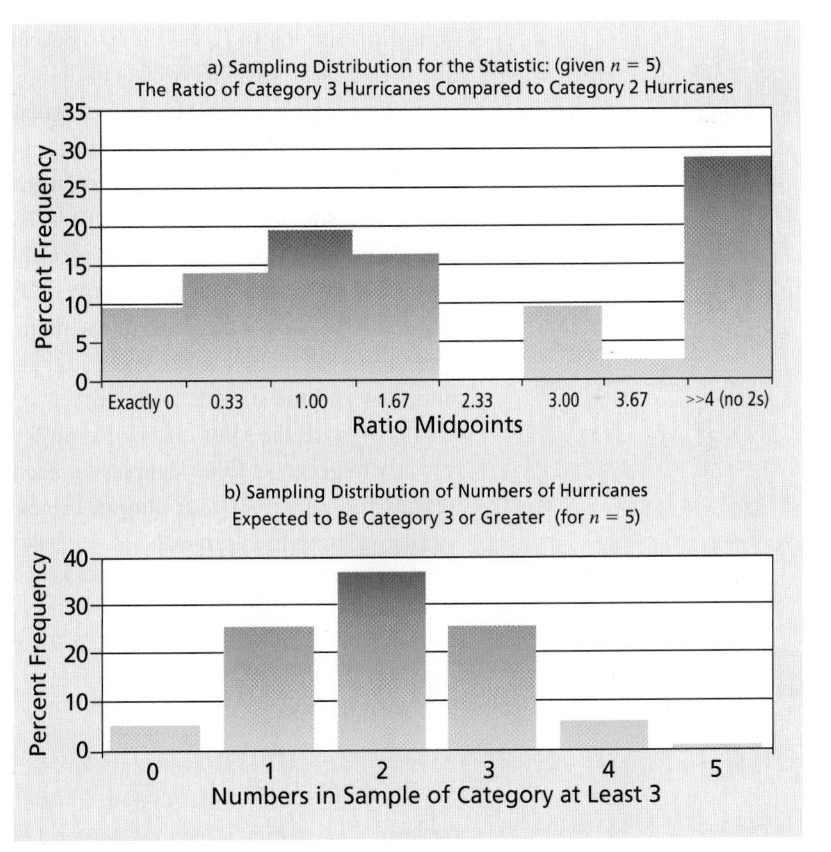

Source: U.S. Department of Commerce, National Oceanic and Atmospheric Administration, Atlantic Oceanographic and Meteorological Laboratory, Hurricane Research Division. Based on data in *NOAA Technical Memorandum NWS TPC-1:* "The Deadliest, Costliest, and Most Intense United States Hurricanes from 1900 to 2000." Found at: http://www.aoml.noaa.gov/hrd/Landsea/deadly/Table9.htm

2. This problem can be solved using bootstrapping, but note that with a minor adjustment, it can also be solved conventionally using the binomial distribution. For the latter, note that any outcome of Category 3 or Category 4 would count as "Category 3 or greater"; any outcome of Category 1 or Category 2 would count as

"Not Category 3 or greater." Since all of the probabilities are supposed to be constant, we now have the conditions to use the binomial: There is a binomial variable "Being Category 3 or greater" (or not), and there are fixed probabilities for samples of fixed size.

On the other hand, we could reuse the coloured marbles in solution 1 for a resample solution: Again, randomly sample *with replacement* five marbles. Count the blue (Category 3) and yellow (Category 4) marbles, and record the sum. That gives the desired statistic for the sample. Repeat the procedure many times. The relative frequencies of the sums approximate the sampling distribution for that statistic. (See Figure 9.8(b).)

• Hands On 9.4 *Design Resampling Routines on the Computer*

In previous exercises, you have generated sequences of random numbers of various distributions and used these to develop simulations. Bootstrapping employs the same basic tools. The trick is to design ways to select with replacement from an original column of data, whose frequency distribution will be taken as the probability distribution for the subsequent sampling. Then figure out a way to take a first random sample (with replacement) from that initial column. Record the desired statistic for that one sample. Now you can take *many* independent samples of the same sort for the source column and record the statistic of interest. If you know how to use macros or do programming in SPSS or Minitab, this is a good chance to use them. Otherwise, it may take more time and ingenuity. (For example, you could generate one new column per resample and then find the statistic for each.)

Try it with the solutions for Example 9.8. Using the software available to you, try to generate 1,000 samples for each of the two problems, based on resampling, and see if your outputs can be graphed as probability distributions similar to those in Figure 9.8.

Chapter Summary

9.1 Explain and apply the concepts of a sampling distribution; calculate the mean and standard deviation of a sampling distribution of the sample mean under specified conditions; and describe what is meant by the term *central limit theorem*.

The **sampling distribution** is a special case of a probability distribution, wherein the variable of interest is itself a sample statistic (such as the mean) whose value is determined by a process of random sampling. If every possible sample of a given size was taken from a population and the mean (or other statistic) was calculated for every one of the samples, the sampling distribution is the distribution of that statistic as taken from all the samples. For the sampling distribution of the sample means as sample size is increased, the distribution of the sample means tends to cluster more tightly and symmetrically around the true population mean; therefore, with larger samples, the mean obtained from the sample is increasingly likely to be in the vicinity of the population mean. For large enough samples, the sample mean is a **non-biased indicator** of the population mean; the sample statistic has an equal probability of being larger *or* smaller than the population parameter to be estimated. The mean and the standard deviation of the sampling distribution of the sample mean can be found, under specified conditions, with Formulas 9.1 through 9.5. Formula 9.5 includes an adjustment (the **finite population correction factor**) for cases of sampling without replacement if the sample size is relatively large compared to the population.

The **central limit theorem** can be summarized in this form: When the sample size n is sufficiently large and sampling is done with replacement, then (1) the sampling distribution of the sample means is approximately normal, regardless of the distribution of the underlying population and (2) this approach toward the normal distribution becomes closer as the sample size n increases.

9.2 Apply the central limit theorem to calculate probabilities associated with the sampling distribution of the sample mean under specified conditions.

When the conditions for the central limit theorem are satisfied, it is possible to calculate the probability that a sample of a given size drawn from a population of specified parameters will fall within certain regions of the distribution: First estimate the mean and standard deviation of the sampling distribution and then apply the procedures from Chapter 8 to find probabilities associated with the normal distribution.

9.3 Calculate the mean and standard deviation for the sampling distribution of the sample proportion and, using calculations based on binomial distributions, find probabilities associated with the sampling distribution of the sample proportion.

The distribution of sample proportions can be approximated (for large enough samples) by the normal distribution; for such samples, we can estimate the mean and standard deviation of the distribution by the normal model. More precisely, the **sampling distribution of the sample proportion** follows the binomial distribution. The latter gives the exact probability for obtaining each of the possible numbers of successes X in a sample of n trials. But the number of successes X divided by sample size n is the *proportion* of successes in the sample. So, if we know the proportion and the sample size, we can calculate X—and thus calculate the true probabilities associated with the sampling distribution by using the binomial model.

9.4 Explain what is meant by *bootstrap methods* and how in general they can be used to approximate sampling distributions and their associated probabilities.

Bootstrap methods for approximating sampling distributions are not as widely applied as alternative techniques, which tend to be based on well-studied models such as the normal and binomial distribution. The sample data themselves are used to provide the basis for estimating a sampling distribution. Following repeated random sampling, with replacement from the sample data, the obtained frequency distribution of the sample statistic (calculated from each of the many samples) is used as the estimate for sampling distribution, from which probabilities can be calculated.

Key Terms

bootstrap methods, p. 391
central limit theorem, p. 381
finite population correction factor, p. 380
non-biased indicator, p. 377
sampling distribution, p. 374
sampling distribution of the sample mean, p. 374
sampling distribution of the sample proportion, p. 386
standard error, p. 379

Exercises

Basic Concepts 9.1

1. Does the term *error* in the expression *standard error of the mean* imply that a mistake has been made? Explain how the term should be interpreted.
2. Does it follow from the central limit theorem that if you take two large samples from the same population, the sample means will be identical? Explain your answer.
3. When estimating the standard error of the mean, for which of the following cases should the population correction factor be employed? For which cases is it not necessary to use the factor?
   a) Taking a sample $n = 200$ from the population of voters in Alberta (provincial population in 2001 = 2,907,882).[3]
   b) Taking a sample $n = 200$ from the population of voters in Alix, Alberta (village population in 2001 = 825).[4]
   c) Taking a sample of 50 countries from the 192 member-countries of the United Nations.[5]
   d) Taking a sample of 50 communities from a list of 3,427 communities in Canada.[6]
4. Suppose you take all possible samples of a given sample size n from a population; find the means for all of those samples, and find the interquartile range for these sample means. If you then repeat this sequence using a larger sample size, what happens to the size of the interquartile range? Explain.
5. For a given population, if you take all possible samples of a given sample size from the population, the mean of all the sample means will equal what parameter of the population?
6. Explain how, for large samples, knowing the central limit theorem simplifies the calculation for the standard deviation of the sampling distribution of the sample means.
7. The population mean for all steel rods produced by Sutton Steel Supplies is 300 cm, with a standard deviation of 25 cm.
   a) Estimate the mean of the sampling distribution of the sample means for $n = 30$.
   b) Estimate the mean of the sampling distribution of the sample means for $n = 90$.
   c) Estimate the standard deviation for the sampling distribution of the sample means for $n = 30$.
   d) Estimate the standard deviation for the sampling distribution of the sample means, for $n = 90$.
8. Suppose that Sutton Steel Supplies is a new company and has produced a total of only 1,200 rods, with the parameters described in Exercise 7. Given this extra information, recalculate the solutions to Exercises 7(c) and 7(d).
9. The Langston family is considering a switch to a new long-distance telephone provider. A check of the family's records (for hundreds of calls) shows that, on average, their long-distance calls last 11.4 minutes, with a standard deviation of 4.1 minutes. If 35 of their records are randomly sampled:
   a) Estimate the mean for the sampling distribution of the sample means.
   b) Estimate the standard deviation for the sampling distribution of the sample means.

Applications 9.1

10. A footcandle measures the amount of illumination on a one-square-foot surface placed one foot from a specified

type of candle.[7] By this measure, candles of a certain brand provide a mean of 0.94 footcandles of illumination on a surface one foot square placed one foot away. The distribution of this measurement is normal, $\sigma = 0.09$. If 50 candles of this type are selected randomly and placed one foot from the one-foot-square surface, what is:

a) The mean for the sampling distribution of the footcandle measurements?
b) The standard deviation of the sampling distribution of the footcandle measurements?

11. Suppose that, in fact, only 500 candles with the parameters described in Exercise 10 have ever been produced. Given this extra information, recalculate the solution to Exercise 10(b).

12. For each of the following sampling situations calculate the population correction factor.

a) Taking a sample $n = 200$ from the population of voters in Alberta (provincial population in 2001 = 2,907,882).[8]
b) Taking a sample $n = 200$ from the population of voters in Alix, Alberta (village population in 2001 = 825).[9]
c) Taking a sample of 50 countries from the 192 member-countries of the United Nations.[10]
d) Taking a sample of 50 communities from a list of 3,427 communities in Canada.[11]
e) Compare the above solutions to your answers for Exercise 3. Complete the following sentence and explain your answer: "When the value of the population correction factor is close to ______, it is usually not necessary, in practice, to calculate the population correction factor."

13. A criminologist is studying the lengths of jail sentences given in Canada. She found that, in 2001/2002 the mean sentence given was for 4.10 months, $\sigma = 9.75$ months.[12]

a) If samples are taken of size $n = 30$, what is the mean of the sampling distribution of the sample means for the lengths of sentences?
b) If samples are taken of size $n = 30$, what is the approximate variance of the sampling distribution of the sample means for the lengths of sentence? Estimate the standard deviation.
c) If the distribution of sentence lengths is *not normal*, will the sampling distribution of sample means—for samples of size $n = 5$—be normally distributed? In that case, do the statements that introduced these exercises provide enough information to find the variance of the sampling distribution? Explain your answers.

14. The criminologist in the previous exercise also found that, in 2001/2002, the mean elapsed time from first to last court appearance for a person charged with robbery was 218 days.[13] Assume that the distribution of trial delays is normal, with $\sigma = 109$ days, and a sample is taken of $n = 20$.

a) Estimate the mean for the sampling distribution of the sample mean.
b) Estimate the standard deviation for the sampling distribution of the sample mean.

15. Suppose that the parameters given in Exercise 14 for trial delays involving robbery had been based on a population of only 198 cases.

a) Based on this new information, what is the value of the population correction factor that should have been used when solving Exercise 14(b).
b) Recalculate the solution to Exercise 14(b), using the correction factor.

16. As described in Exercise 16 in Chapter 8, a movie fan is planning to buy a model of subwoofer for his home theatre and has found that many versions are available online. Available prices are distributed normally with $\mu =$ US\$1,175, $\sigma =$ US\$176.[14] If these prices represent hundreds of online options, are normally distributed, and a sample of size $n = 40$ is taken:

a) Estimate the mean of the sampling distribution of sample means.
b) Estimate the standard deviation of the sampling distribution of sample means.
c) Recalculate (b) if the sample size is reduced to $n = 20$.

Basic Concepts 9.2

17. Explain how, for large samples, the central limit theorem provides a model to calculate probabilities for the sample means.

18. For each of the following cases, indicate whether the central limit theorem provides a model to reliably calculate probabilities for the sample means.

a) The population is not normally distributed, and sample sizes are >50.
b) The population is not normally distributed, and sample sizes are <30.
c) The population *is* normally distributed, and sample sizes are <30.

19. The population mean for all steel rods produced by Sutton Steel Supplies is 300 cm, with a standard deviation of 25 cm. If a random sample of 45 steel rods is taken, what is the probability that the sample mean is:

a) Less than 294 cm?
b) At least 303 cm?
c) Between 294 cm and 303 cm?

20. When solving Exercise 19, did it matter whether the lengths of rods produced by Sutton Steel were normally distributed? Explain.

21. The Langston family makes many long-distance telephone calls. On average, the family's long-distance calls last 11.4 minutes, with a standard deviation of 4.1 minutes. If data for 35 of their calls are randomly sampled, what is the probability that the sample mean is:
    a) No more than 12.3 minutes?
    b) At most 13.1 minutes?
    c) Between 10.9 minutes and 11.1 minutes?
22. Increase the sample size in Exercise 21 to 50 calls. Then redo parts (a), (b), and (c) of that exercise.
23. A bank teller spends 3.5 minutes per customer, on average. The standard deviation is 0.5 minutes. For a random sample of 40 of his customers, what is the probability that the mean time spent per customer in the sample is less than 3.4 minutes?
24. Dorrie's Drugs sells an average of 340 packages of decongestants per week. The distribution is normal and the standard deviation is 39 packages. In a random sample of seven weeks, what is the probability that:
    a) Average sales are more than 350 packages a week?
    b) Average sales are less than 300 packages per week?
    c) Average sales are between 300 and 350 packages per week?
25. In Exercise 24, why was it necessary to mention that the distribution of sales was normal?

Applications 9.2

26. A criminologist is studying the lengths of jail sentences given in Canada. She found that, in 2001/2002 the mean sentence given was for 4.10 months, $\sigma = 9.75$ months.[15] If a random sample is taken of size $n = 35$, what is the probability that the mean sentence length in the sample is
    a) More than 4.44 months?
    b) Between 4.00 and 4.44 months?
27. The criminologist in the previous exercise also found that, in 2001/2002, the mean elapsed time from first to last court appearance for a person charged with robbery was 218 days.[16] Assume that the distribution of trial delays is normal, with $\sigma = 109$ days, and a sample is taken of $n = 20$.
    a) Estimate the probability that the mean trial delay in the sample is between 196 and 248 days.
    b) When calculating an answer to (a), why was it *not* a concern that the sample size was small?
28. Suppose that the parameters given in Exercise 27 for trial delays involving robbery had been based on a population of only 198 cases.
    a) Use the population correction factor to revise your estimate for the standard error in the problem.
    b) Recalculate the solution to Exercise 27(a) using the revised estimate for the standard error.
29. A movie fan is planning to buy a certain model of subwoofer for his home theatre and has found that it is available online, with μ = US\$1,175, σ = US\$176.[17] Assume that there are hundreds of online offerings and that the prices are normally distributed. If a sample of size $n = 40$ is taken:
    a) Estimate the probability that the sample's mean price (in US\$) is between \$1,141 and \$1,212.
    b) Recalculate (a) if the sample size is increased to $n = 60$.
    c) Suppose we learn that the population of prices is not normally distributed. Would this significantly impact the solutions to (a) and/or (b)? Explain.
30. Saskatchewan Highways and Transportation publishes analysis results on the composition of asphalt to assist in quality control for that product. In 2004, tests found that the bulk density of coarse aggregate in the brand Marshal Mix averaged 2.667 g/cm^3, with a standard deviation of 0.013 g/cm^3.[18]
    a) If a new sample of this product, made up of 45 observations, is randomly selected from across the province, what is the probability that the sample's mean for bulk density will be between 2.663 g/cm^3 and 2.668 g/cm^3?
    b) Recalculate your solution to (a) if the sample size is changed to $n = 55$.
31. Saskatchewan Highways and Transportation also tests asphalt for water absorption of fine aggregate. In 2004, the mean for this variable in Marshal Mix was 1.297%, with a standard deviation of 0.30%.[19]
    a) If a new sample of this product, made up of 45 observations, is randomly selected from across the province, what is the probability that the sample's mean for water absorption will be between 1.20% and 1.30%?
    b) The solution for (a) assumes that the 2004 mean for water absorption, given in the problem, can be interpreted as the true population parameter. In reality, the quoted mean was itself a *sample* result obtained from various labs. Use the central limit theory to explain why interpreting the 2004 mean value as a population parameter is not too risky *provided* the original sample size was very large.
32. In 2002, the Tolly Group published a comparative evaluation of the audio product OCT6100 Series Echo Canceller. For repeated measurements of perceptual evaluation of speech quality (PESQ), the OCT6100 obtained a mean score of 3.43 (0 is poor and 5 is excellent). The standard deviation for the measurements was 0.26.[20] If these statistics are close to the actual population parameters and a new sample of 40 observations is randomly selected, what is the probability that the sample's mean for PESQ will be less than 3.32?

Basic Concepts 9.3

33. For a particular population, the proportion of cats that are female is 0.55. A sample $n = 30$ is randomly selected to see what proportion in the sample is female. Use the normal distribution to approximate the distribution of the proportion.
a) What is the mean of the distribution?
b) What is the standard deviation of the distribution? (Hint: First estimate the standard deviation for the number of successes [i.e., females] in the sample using Formula 7.6. Convert the number of successes to a proportion by dividing by n.)
c) Based on the normal distribution, estimate the probability that the proportion of females in the sample is:
i) Greater than 0.75.
ii) Less than 0.40.

34. Recalculate the solution to Exercise 33(c) but use computer procedures based on the binomial distribution, not the normal distribution. Compare your solutions to those in Exercise 33(c) and explain any differences.

35. Brake King Car Repairs has determined from past experience that 9% of their repair jobs need redoing. To confirm this, 70 of their old repair records are randomly selected as a sample. Using the normal distribution to approximate the distribution of the proportion:
a) What is the mean of the distribution?
b) What is the standard deviation of the distribution?
c) What is the probability that more than 9% of the repairs in the sample needed to be redone?

36. Recalculate the solution to Exercise 35(c) but use computer procedures based on the binomial distribution. Compare your solutions to those in Exercise 35(c) and explain any differences.

Unless otherwise instructed by your instructor, use the computer software procedures based on the binomial distribution for the following problems.

37. In general, 5% of students at your institution take bus route "C." In a random sample of 100 students, what is the probability that, on that occasion, less than 4% take bus route "C"?

38. After years of observing, you determine it rains in your neighbourhood only 82% of the days that the Weather Channel predicts it will rain. In the coming year, the channel predicts rain on 55 days. Based on your research, what is the probability that it will rain at least 85% of the days that rain is predicted?

39. Government auditors suspect that of all the business expenses reported by Y. Lee Consultants, 15% are criminally inflated. If that is so, and a sample of 60 of the company's reported expenses is selected, what is the probability that:
a) More than 10% have been criminally inflated?
b) Less than 20% have been criminally inflated?
c) Between 10% and 20% have been criminally inflated?

40. In past elections, multiple candidates competed for Sherry Brewster's riding, so she figures she can win the seat in the next election with just 35% of the votes cast. Her backers say that, in fact, 37% of the voters in the riding support her.
a) If her backers are correct, what is the probability that in a sample of 40 voters in the riding who are willing to share their opinions, more than 40% will say they intend to vote for Sherry Brewster?
b) In these circumstances, what is the probability that less than 35% will say they intend to vote for Sherry Brewster?

41. Of all those who enroll for the first time for lessons at Sarah's Swim School, 12% eventually attain a Level 10 certificate.
a) If 50 people enroll in the school this year for the first time, what is the probability that:
i) At least 5 of them will eventually attain Level 10?
ii) No more than 15% of them will eventually attain Level 10?
iii) Between 10% and 25%, inclusive, of them will eventually attain Level 10?
b) Although the above-mentioned proportion equalled 0.12 *in the past*, will this proportion apply necessarily to the (nonrandomly selected) set of all students enrolling *this* year? Suppose that 30 of this year's enrollees were in fact transfers from a Level 8 swimming program that just closed down. Would this cause your answer to (a)(i) to be an underestimate or overestimate, or make no difference? Explain.

Applications 9.3

42. As part of a study on humpback whales, an observation post was established on the roof of the Hyatt Hotel in Maui, Hawaii. On 80% of the observation days when parasail boats were not nearby, a pod (i.e., group) of whales could be observed.[21] For 40 randomly selected observation days when no parasail boats were nearby, what is the probability that a pod of humpbacks was observed on:
a) At least 75% of the days?
b) More than 90% of the days?
c) Between 75% and 90% of the days, inclusive?

43. According to one estimate, only 13% of Canadian start-up businesses survive their first year.[22]
a) If 300 newly started businesses are randomly selected, what is the probability that at least 25% of them will survive to a second year?
b) A business consultant provides start-up services to 20 new businesses. If we assume that the consultant has

no impact on the probabilities of success for the businesses, what is the probability that at least 9/20 of these businesses will survive to a second year?

c) If in fact 9/20 of the clients in (b) *do* succeed, is this fact consistent with the assumption made in (b) (i.e., that $p = 0.13$ for the clients *as well as* for non-clients)? Explain.

44. A variety of methods have been tried by cities for having residents sort their recyclables. Blue boxes versus blue bags are two examples. According to a study conducted in Northumberland County, Ontario, 69.4% of residents said blue bags were their method of choice.[23] If that is the true population proportion and a new random sample is taken of 60 Northumberland County residents, what is the probability that the proportion preferring blue bags will be between 70% and 80%, inclusive?

45. A survey conducted by the Asia Pacific Foundation of Canada found that 26% of respondents planned to invest in China during the next year.[24] Suppose the results are representative of a large growth-oriented business community. If 1,000 of those companies are randomly selected, what is the probability that:

a) At least 30% plan to invest in China during the next year?

b) Between 20% and 30%, inclusive, plan to invest in China during the next year?

46. It is reported that 9% of Nova Scotians aged 12 and over have experienced major symptoms of depression.[25] *If this rate was constant for all Nova Scotians of that age group:*

a) What would be the probability that more than 10% of students in a Nova Scotia high school of 230 students have been depressed?

b) What is the probability that in the above high school, 16% or more have been depressed?

c) Suppose that due to high unemployment and drug problems in the area of the above high school, the school has a higher proportion of students that have been depressed than in some other areas. Knowing this, should the estimate in (b) be increased, decreased, or left the same? Explain.

47. An economist is concerned with the decreasing percentage of working Canadians who are covered by registered pension plans (RPPs). In 1979, 45.9% of the labour force had this coverage. By 2003, only 39.3% of the labour force had this coverage.[26]

a) If a random sample of 1,000 workers was selected in 1979, what is the probability that at most 39.3% of them had RPPs?

b) If a random sample of 1,000 workers was selected in 2003, what is the probability that at least 45.9% of them had RPPs?

c) Based on your answers to (a) and (b), do the percentages reported in 1979 and 2003 appear to be significantly different, as opposed to just being variations due to sampling error? Explain. (Note: The concept of "significance" is introduced more formally in Chapter 11.)

48. If a star has about three times more metal than our sun, the probability of it harbouring planets is 20%.[27] Hoping to become famous, an amateur is examining 100 stars of the above-mentioned description. What is the probability that the proportion of these stars having planets is less than 10%? More than 25%?

Basic Concepts 9.4

For Exercises 49–51, refer to this statement: For any type of quantitative data, there is an accepted formal model (such as the normal or uniform distribution) that can be used to model the sampling distribution of a desired statistic.

49. Is the above statement true or false?

50. Explain in general terms how, when there is an accepted mathematical model for a sampling distribution, you can use the model to calculate probabilities for the value range of the sample statistic.

51. If the statement above is false, describe a situation in which the statement does not apply.

52. Briefly describe what is meant by a bootstrap model of a sampling distribution and how this model could be used to calculate probabilities for the value range of the sample statistic.

53. Explain the relationship between the bootstrap model for a sampling distribution and the concept of a relative frequency distribution histogram.

Review Applications

54. An environmental concern related to cigarette smoke is the emission of the toxic gas formaldehyde. Health Canada reports that under standard test conditions, the mean formaldehyde content in direct cigarette smoke is 43.4 μg (micrograms) per cigarette. The standard deviation is 32.8 μg.[28] Suppose that 50 cigarettes of various brands are randomly selected and subjected to the test for formaldehyde content in smoke.

a) What is the mean for the sampling distribution for the formaldehyde content?

b) What is the standard deviation of the sampling distribution for the formaldehyde content?

55. Based on the data in the previous exercise, if a random sample of 40 cigarettes is tested, what is the probability that the mean formaldehyde content in the smoke of each is:

a) More than 55.0 μg?

b) No more than 35.0 μg?
c) Between 35.0 and 55.0 μg?

56. The calculated mean for formaldehyde emissions (See Exercise 54) was influenced by one brand whose smoke contains 128 μg of formaldehyde per cigarette.[29] Given this extra information, explain whether the answer to Exercise 55(a) is likely to be an overestimate, and underestimate, or neither.

57. A hang-gliding enthusiast is studying possible locations for his next outing. He has heard of possibilities in North Cyprus, at the locations listed below.[30] He plans to conduct surveys of individual adult residents in each location. (He estimates that, in each location, about 67% of the population are adults.)

Gazimağusa (Famagusta): Population 27,637
Girne (Kyrenia): Population 14,205
Güzelyurt: Population 12,865
Lefke: Population 6,490
Yeni Iskele: Population 2,814

a) A sample of $n = 250$ will be taken from each location, and the researcher intends to calculate a standard error for each sample. For which locations should he use the population correction factor? Explain.
b) For each location where it was recommended to use the population correction factor in (a), calculate the correction factor.

Exercises 58–62 are based on the following table of laboratory attenuation values for foam earplugs, under specified conditions, according to U.S. Occupational Safety and Health Administration testing.

Laboratory Attenuation Values for Foam Earplugs

Test Frequency (Hz)	125	250	500	1,000
Mean attenuation (decibels)	17.9	19.0	21.0	24.7
Standard deviation (decibels)	7.3	6.3	7.3	6.4

Test Frequency (Hz)	2,000	4,000	8,000
Mean attenuation (decibels)	29.9	35.6	34.6
Standard deviation (decibels)	5.3	5.0	5.4

Source: Safe@Work[SM], Inc. Based on National Institute for Occupational Safety and Health data in *New Developments in Hearing Protection Labeling.* Found at: http://www.safe-at-work.com/Reference/OSHA-HP-Labeling.html

58. For workers in loud environments, ear plugs are recommended to attenuate (i.e., reduce) the effective exposure to the noise. Notice that for foam earplugs, the level of protection varies, based on the frequency of the sound (e.g., a deep rumble versus a high-pitched sound). Assume that the parameters in the table represent the whole population of earplugs of a given type. An inspector randomly selects 50 such earplugs and tests them for attenuation at the frequency 125 Hz.

a) For the level of attenuation at this frequency, what is the mean of the sampling distribution of the sample means?
b) What is the variance of the sampling distribution of the sample means? What is the standard deviation?
c) What is the probability that the mean attenuation for that particular sample of 50 earplugs is less than 16.5 decibels (dB)? Between 16.5 dB and 20.0 dB? More than 20.2 dB?

59. Repeat Exercises 58(a) and (b) but for the frequency 250 Hz.

60. If an inspector randomly selects 50 foam earplugs and tests them for attenuation at the frequency 250 Hz, what is the probability that the mean attenuation for the sample is less than 17.0 dB? Between 17.0 dB and 21.4 dB? More than 21.4 dB?

61. Repeat Exercises 58(a) and (b) but for the frequencies (a) 500 Hz, (b) 1,000 Hz, (c) 2,000 Hz, and (d) 4,000 Hz.

62. What is the probability that the mean attenuation at 8,000 Hz for a sample of 50 earplugs is between 32.0 dB and 36.0 dB?

63. Commonly accepted values for the mean and standard deviation of normal eye pressure are 15.5 mmHg (millimetres of mercury) and 2.5 mmHg, respectively.[31] A technician selects 60 individuals randomly to measure their eye pressure.

a) Estimate the mean for the sampling distribution of the sample mean of the eye pressures.
b) Estimate the standard deviation for the sampling distribution of the sample mean of the eye pressures.
c) The distribution of eye pressures in the human population is not normally distributed; rather, it is positively skewed. Knowing this fact, would your estimates in a) and b) need to be significantly revised. Explain why or why not.

64. Suppose that the eye pressure parameters in Exercise 63 were not for the general population, but only for a population of 150 anthropologists. The population values are normally distributed. Based on this new information, recalculate the solution to Exercise 63(b).

65. Based on the data in Exercise 63 for the sample of $n = 60$, what is the probability that the mean eye pressure is:

a) More than 16 mmHg?
b) More than 16.5 mmHg?
c) Between 16 mmHg and 16.5 mmHg?

Exercises 66–69 are based on the following table, which shows the percentages of captured frogs in a study that were "recaptures"—i.e., they had been captured and marked at that site at a previous time.

Percentages of Frogs in Study That Were Recaptures

	Breeding Habitats	Non-Breeding Habitats
Juvenile	26%	12%
Mature male	66%	9%
Mature female	56%	53%

Source: David S. Pilliod, Charles R. Peterson, and Peter I. Ritson. *Seasonal migration of Columbia spotted frogs (Rana luteiventris) among complementary resources in a high mountain basin.* Found at: http://article.pubs.nrc-cnrc.gc.ca/ppv/RPViewDoc?_handler_=HandleInitialGet&journal=cjz&volume=80&calyLang=eng&articleFile=z02-175.pdf

66. As part of a study on the seasonal migration of Columbia spotted frogs in a high mountain basin, the frogs were captured and released at various sites and times. Many of the captured frogs were found to be recaptures. The percentages of recaptures among captured frogs, over several years, are shown in the table.

Suppose that 40 records of captures are randomly selected from the historical data, for each of these three categories, respectively:

- Juveniles captured in non-breeding habitats.
- Mature males captured in breeding habitats.
- Mature females captured in breeding habitats.

a) Use the normal model for the sampling distribution of the sample proportions to estimate the mean of the sampling distribution for each of the three categories.
b) Use the normal model for the sampling distribution of the sample proportions to estimate the standard deviation of the sampling distribution for each of the three categories.

67. Refer to the data in the table "Percentages of Frogs in Study That Were Recaptures" and use binomial-based calculations for this exercise: If a sample of 50 records is randomly selected from the historical data for juveniles captured in breeding habitats, what is the probability that:
a) At least 25% are recaptures?
b) No more than 30% are recaptures?
c) Between 20% and 35%, inclusive, are recaptures?

68. If a sample of 50 records is randomly selected from the historical data for mature males captured in non-breeding habitats, what is the probability that:
a) At most 10% are recaptures?
b) At least 11% are recaptures?
c) Between 8% and 12%, inclusive, are recaptures?

69. If a sample of 45 records is randomly selected from the historical data for mature females captured in non-breeding habitats, what is the probability that:
a) No more than 56% are recaptures?
b) At least 50% are recaptures?
c) Between 52% and 54%, inclusive, are recaptures?

Exercises 70–73 are based on the following table, which shows some results from a survey of commercial drivers in Alberta.

Commercial Driver Fatigue Management Study: Selected Results

Driver reports needing at least 7 hours sleep per night.	60%
Driver reports regularly getting less than [his or her] perceived ideal amount of sleep per night.	71%
Driver reports making several mistakes or mental errors per year due to fatigue.	84%

Source: Canadian Sleep Institute. Based on data in *North American Pilot Fatigue Management Program (FMP) for Commercial Motor Carriers (Alberta Results)*. Found at: http://www.csisleep.com/pdf/TH%20AMERICAN%20PILOT%20FATIGUE%20MANAGEMENT%20PROGRAM%20FOR%20COMMERCIAL%20MOTOR%20CARRIERS%20.pdf

70. Driver fatigue is a known hazard for drivers, both commercial and private. The prevalence of this risk is illustrated in the table above, which is derived from a study of commercial motor carriers in Alberta. For this problem assume similar proportions apply for *all* commercial drivers across Canada and suppose that a random sample of 50 commercial drivers is selected.
a) Use the normal model for the sampling distribution of the sample proportions to estimate the mean of the sampling distributions for:
 i) Drivers who report getting less than his or her perceived ideal amount of sleep.
 ii) Drivers who report making several mistakes per year due to fatigue.
b) Use the normal model for the sampling distribution of the sample proportions to estimate the standard deviations for each of the above two sampling distributions.

71. Refer to the data in the commercial driver fatigue management study table and use binomial-based calculations for the following exercise: A sample of 75 drivers for commercial motor carriers is randomly selected and the drivers are asked how much sleep they need per night. What is the probability that:
a) No less than 55% report needing at least 7 hours sleep?

b) No more than 67% report needing at least 7 hours sleep?
c) Between 55% and 67%, inclusive, report needing at least 7 hours sleep?

72. A sample of 65 drivers for commercial motor carriers is randomly selected and asked whether they regularly get less than their perceived ideal amount of sleep. What is the probability that:
a) More than 75% report getting less than their ideal amount of sleep?
b) More than 80% report getting less than their ideal amount of sleep?
c) Between 75% and 80%, inclusive, report getting less than their ideal amount of sleep?

73. A sample of 75 drivers for commercial motor carriers is randomly selected and asked whether they have made mistakes or mental errors due to fatigue in the past year. What is the probability that between 80% and 85%, inclusive, report making such errors?

Chapter 10

Estimates and Confidence Intervals

• Chapter Outline

• Learning Objectives

10.1 Distinguish among point, interval, and confidence interval estimates, explain the concepts of confidence levels and confidence limits, and also describe some basic desirable properties of estimators.

10.2 From sample data, calculate the margin of error and construct confidence intervals for the mean in cases where population σ is known, for varying levels of confidence.

10.3 From sample data, calculate the margin of error and construct confidence intervals for the mean in cases where population σ is not known, for varying levels of confidence, and explain the concepts of the *t*-distribution and the associated degrees of freedom (*df*).

10.4 From sample data, construct approximate confidence intervals for the proportion based on the normal approximation, and calculate values for sampling distribution mean and standard deviation to be used for the estimate.

10.5 Describe the tradeoff for constructing confidence intervals between increasing the precision and increasing the level of confidence.

10.1 Point Estimates and Interval Estimates

Based on the information in a sample, we can make inferences about one or more characteristics of a population. The most direct inference we can make is an **estimation,** which is a process of obtaining a single numerical value, or a set of values, intended to approximate the actual value of a parameter in the population. For example, a scientist may inquire into the mean daytime surface temperature of the moon. Yet the population of all daytime temperatures, on all places on the moon, is infinite. By sampling many specific temperatures at many specific locations, the scientist can *estimate* a value for that parameter—107°C.[1]

The estimate just illustrated is a **point estimate.** A point estimate is a single value intended to approximate the true corresponding value of a population parameter, such as a mean, median, variance, or proportion. Is the true mean surface temperature of the moon, at this very moment (in daytime locations), *exactly* 107°C? Probably not, but that one figure is presented as a best guess for the approximate mean value. We presume that the actual value is close, if not identical, to the estimate.

An **interval estimate** provides an alternative to a point estimate. The interval estimate gives a *range* of values within which the true value of the population characteristic is expected to lie. For example, a biologist may ask at what mean surface temperature of sand do tiger beetles (*C. theatina*) engage in foraging activities. The population of such temperatures is highly variable, so naming just one value may not give the best estimate. Instead, he might say: When foraging occurs, the current mean temperature is somewhere between 22.5° and 51.5°C.[2]

The interval estimate employed in statistics is called a **confidence interval.** A confidence interval does not just specify the range of possible values for the parameter, it also indicates a degree of confidence or likelihood that the true value does, in fact, lie in that range. The probabilities for this assessment are based on the expected sampling distribution of the studied population. (This is discussed in more detail in the next section.)

Typical results for a published news survey can illustrate these approaches for estimation. For example, the lead paragraph in a *Straight Goods* online magazine article (December 14, 2005) reported simply that "tracking has the Liberals at 38 percent" support.[3] That was the point estimate of support as of that date. Later, the article refers to a "margin of error" of "± 2.9 percent, 19 times out of 20." In other words, the interval estimate for Liberal support was the range from (point estimate *minus* 2.9% ≈ 35%) on the low end, up to (point estimate *plus* 2.9% ≈ 41%) on the high end. The "19 times out of 20" (which is 95%) indicates the **degree of confidence** or **confidence level** that the true level of support was really within the interval estimate; it could happen (1 chance in 20) that the true value is *not* in that interval after all. The lower and upper boundaries of the interval estimate are called the **confidence limits.** (See Figure 10.1.)

The degree of confidence is *not* a probability with respect to a population parameter. On the day of sampling, the percentage of the population who supported the Liberals was some value—independently of our estimate for it. A 95% level of confidence means that when a parameter is estimated by the techniques and sample size that have been employed, it will accurately estimate the confidence interval 95% of the time.

Seeing Statistics
Refer to Section 8.4.3

Figure 10.1
Point Estimate versus Confidence Interval

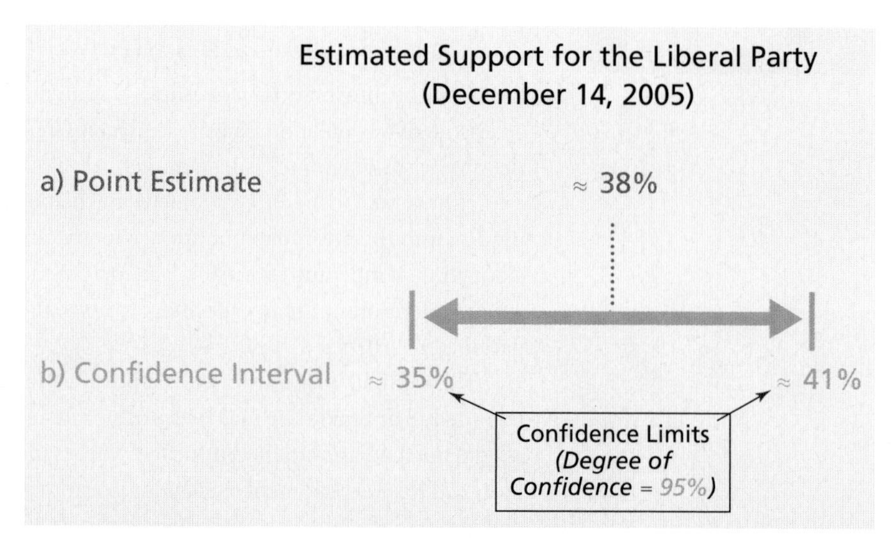

Source: Straight Goods. Based on data in "NDP Losing Ground." Found at: http://www.straightgoods.ca/Election2006/ViewNews.cfm?Ref=3. Accessed Dec. 14, 2005.

Desirable Properties of Estimators

An **estimator** is a measured or calculated value based on a sample, used to estimate a population parameter. Generally, the best estimator for a parameter is the corresponding sample statistic. For example, to estimate μ (for a population), find $\overline{x}$ (for a sample from the population); or to estimate σ, find s from a sample. This is not always possible. For example, if a salary survey includes an open-ended category "greater than \$100,000," you cannot calculate an exact sample mean. An alternative estimator would be required.

If possible, a good estimator should include these properties:

1. **Unbiasedness:** An estimator for a population parameter is **unbiased** if the mean of its sampling distribution is equal to the true parameter. We have seen that for the population mean, this requirement is met by the sample mean: The mean of the sampling distribution of the means equals the population mean. Although a particular estimate can be in error, the condition of unbiasedness ensures that the true parameter is at least at the centre of the range of possible estimates.
2. **Efficiency:** Depending on the parameter to be estimated, there may be more than one statistic that could be used as the estimator. For example, the population mean for a normally distributed population could be estimated by the sample mean, the sample median, or even by $\dfrac{\text{Sample minimum} + \text{Sample maximum}}{2}$.
   When comparing unbiased estimators for a population parameter, the most **efficient** estimator is the most precise, in that its sample statistics are distributed most closely around the population parameter (i.e., its sampling distribution has the least variance).

   Figure 10.2 illustrates, with simulations, the distributions of sample means and of sample medians taken from a population with parameters $\mu = 0$ and $\sigma = 1$. The sample mean is the more efficient estimator because, as shown, its variance and standard deviation are smaller than for the medians.

Figure 10.2
Comparative Efficiencies of the Mean and Median as Estimators of the Population Mean

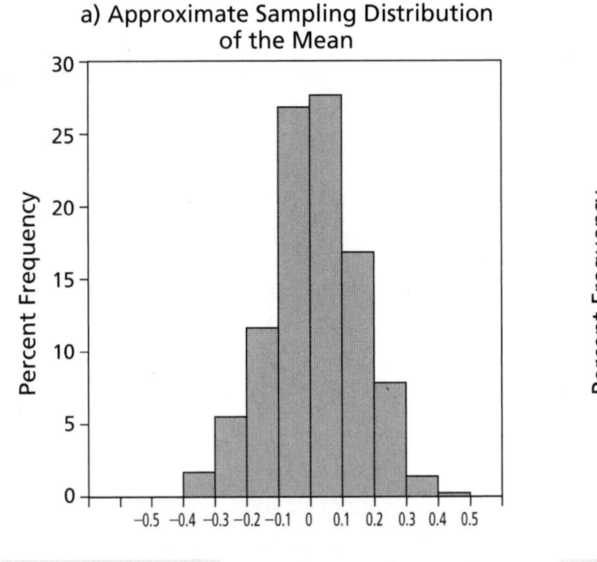

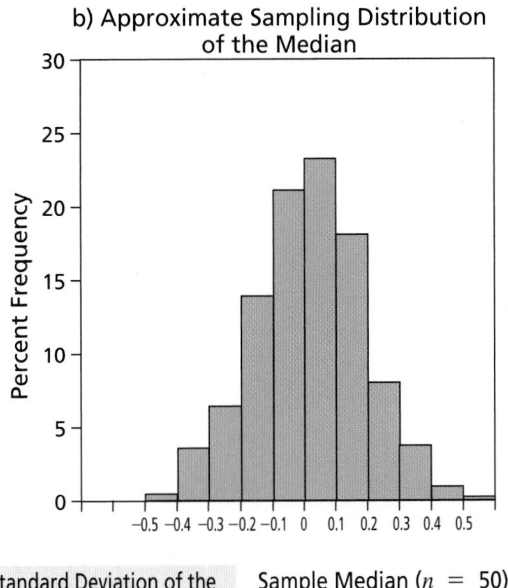

3. **Consistency:** A **consistent** estimator becomes more efficient as the sample size increases. That is, with larger samples, the values of the statistic are more likely to be closer to the parameter being estimated. The sample mean is a consistent estimator. If the sample size is expanded to equal the population size, the variance around the estimate would decrease to zero.

• *Hands On 10.1* *Comparing Estimators*

In Chapters 4 and 5, we introduced parameters that measure the centre and spread of a population distribution. There are a number of ways to use sample statistics as estimators for these parameters. Table 10.1 provides some examples:

Population Parameter	Estimator
Mean (μ)	Sample mean ($\bar{x}$)
Mean (μ) (*for a normally distributed population*)	Sample median
Mean (μ) (*for a normally distributed population*)	(Sample minimum + Sample maximum)/2
Standard deviation (σ)	Sample standard deviation (s)
Standard deviation (σ) (*for a normally distributed population*)	(Sample range)/6
Standard deviation (σ) (*for normally distributed population*)	(Interquartile range)/1.33

Use simulations to test how well the proposed estimators for these parameters meet the conditions for good estimators, namely: unbiasedness, efficiency, and consistency. A possible procedure is as follows:

1. First, generate a long column of normally distributed random numbers on your computer. This can represent the population data. Calculate the true value of μ for this population and then repeatedly generate samples of a given size from this population. For each sample, record the values for $\overline{x}$ and median and $\frac{\text{Sample maximum} + \text{Sample minimum}}{2}$. After many repetitions, construct a frequency distribution for each of the three estimators; then repeat the entire three-part sequence several times, choosing different sample sizes for each pass.

   Compare all resulting distributions with the conditions for being a good estimator. Based on your findings, which estimator for μ do you think is most reliable, assuming a normally distributed population? For what reasons might you choose to use the other estimators instead?

2. Repeat the procedures given in (1), except the comparison is now with the true value of σ (rather than μ) for the population. For each sample that you generate, now record the values for s and (Range/6) and (IQR/1.33). Again, after generating many samples, for each of several sample sizes, compare all of the resulting sampling distributions with the conditions for being a good estimator. Based on your findings, which estimator for σ do you think is most reliable, assuming a normally distributed population? For what reasons might you choose to use the other estimators instead?

10.2 Confidence Interval for Population Mean (σ Known)

Each of the next three sections describes how to construct a confidence interval estimate under certain conditions. In this section, we learn how to construct the confidence interval for the mean in cases where σ for the population is known. The method presumes that the sampling distribution of the sample means is normally distributed. According to the central limit theorem, this presumption is realistic if (1) the population itself is normally distributed or (2) the sample size is large (greater than 30).

Some textbooks discourage using the model presented in this section, believing that its assumptions are unrealistic in practice. If a population mean needs to be estimated, are you likely to know the population standard deviation in advance? Also, the normal approximation of the sampling distribution is never *entirely* accurate, even for $n > 30$. From this viewpoint, the methods that are presented in Section 10.3 can be used to estimate the confidence interval of the mean instead.

Yet the basic principles for estimating a confidence interval from a sample statistic are essentially the same in more advanced models as they are in this section. As long as a probability density curve can be used to model a sampling distribution for a parameter, the principles shown here can be adapted to estimate the confidence interval.

The upper curve in Figure 10.3 represents a sampling distribution for the sample means that corresponds to a particular sample size and population. This section's model assumes that the sampling distribution is normal. Therefore, we can use techniques from Section 8.1 to calculate the probability that a randomly selected sample mean will fall within, or outside of, the highlighted range under the curve. For example, if the upper and lower bounds of the highlighted area fall at 1.96 standard deviations above and below the mean (i.e., $z = \pm 1.96$), there is 95% chance that the selected mean will fall within that region. Or if the bounds of the highlighted area fall at 1.0 standard deviation above and below the mean (i.e., $z = \pm 1.0$), the probability that the selected, sample mean will fall within that region is about 68%.

Suppose that the displayed model is based on $z = \pm 1.96$. We would expect that 95% of the sample means would fall within the highlighted bounds. This is illustrated in Figure 10.3. By chance, $\bar{x}_3$ is one exception that happens to fall outside the indicated range.

Figure 10.3
Sampling Distribution of the Sample Mean in Relation to Confidence Intervals Based on Selected Samples

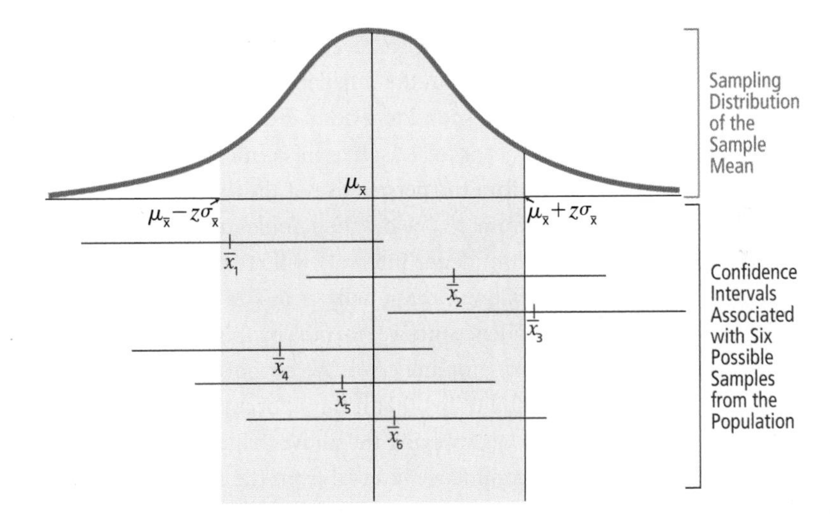

However, in many real applications, the upper part of Figure 10.3 is not available. We have only the data from our sample. We do not know if our sample mean is like $\bar{x}_4$, which falls within the expected region, or like $\bar{x}_3$, which does not. Yet based on the model's assumptions, we *do* know how far a sample mean would have to deviate from the population mean to be outside the expected region.

In Figure 10.4, the *x*-axis for the sampling distribution has been standardized. The mean of the distribution falls at *z*-value = 0. The *z*-values that define the upper and lower bounds of where we expect most samples to fall are called the **critical values.** By convention, the probability that a randomly selected sample mean will fall beyond that expected

Figure 10.4
Standardized Sampling Distribution of the Sample Mean, and Probabilities Associated with Confidence Intervals

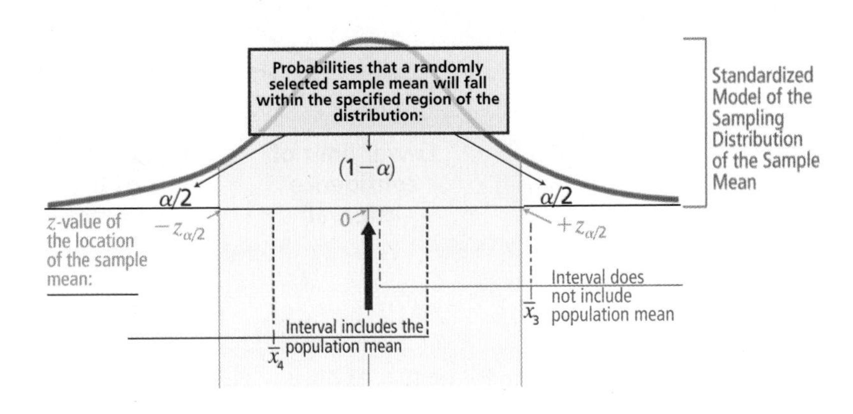

range is designated as α (e.g., a 5% chance of error). But there are two ways to fall outside the expected range: The sample mean might be less than the lower critical value or the mean might be greater than the upper critical value. The probability of either case occurring is $\alpha/2$; so the total probability of falling beyond the range is $\alpha/2 \times 2 = \alpha$.

If the event "a randomly selected sample mean falls beyond the critical values" is symbolized by A, the complement of that event is $\overline{A}$ ("the selected mean does not fall beyond the critical values"). Since $P(A)$ was defined as α, we can use Formula 6.4 for complementary events to determine that $P(\overline{A})$ (which is the probability that the mean of a random sample falls *between* the critical values) is equal to $(1 - \alpha)$. In other words, the value $(1 - \alpha)$ is the degree of confidence or level of confidence that the interval based on the sample mean has successfully estimated the population mean. This means that the sample did not (like $\overline{x}_3$) happen to fall beyond the expected range. Typical values used for confidence levels are 90%, 95%, or 99%.

From the calculations for a normal curve, we can find the critical values $\pm z_{\alpha/2}$ that correspond to a desired degree of confidence. Since most (i.e., $(1 - \alpha)$%) of the sample means fall less than this critical distance from the population mean, we can use this fact from the perspective of the sample: $(1 - \alpha)$% of the time, if you extend a line reaching from $z_{\alpha/2}$ below the sample mean to $z_{\alpha/2}$ above the sample mean, this line (which is the confidence interval) will cross the true population value. This is the case for $\overline{x}_4$ in the figure. Occasionally, as in $\overline{x}_3$, a sample mean is far from the population mean, so the confidence interval formed by this procedure will fail to include the value of μ. This is a case of sampling error. We cannot avoid sampling error, but now we can quantify the probability (α) that it might occur when we estimate the confidence interval.

Applying the above principles, Figure 10.5 expresses a procedure for calculating a confidence interval from the sample mean. The centre of the confidence interval is the point estimate, which is equal to the sample mean itself. The lower and upper bounds of the interval are found by subtracting from or adding to the point estimate, respectively, a **margin of error** (E). The margin of error is a distance from the mean expressed in the original units of the raw data. For example, satellite data have been used to suggest that the mean increase in global temperature per decade since 1979 has been +0.126°C, with a margin of error of ±0.05°C.[4] The confidence interval for this mean increase, therefore, is 0.126°C ± 0.05°C. This can be expressed as $0.076°C < \mu < 0.176°C$.

Figure 10.5
The Confidence Interval and Margin of Error

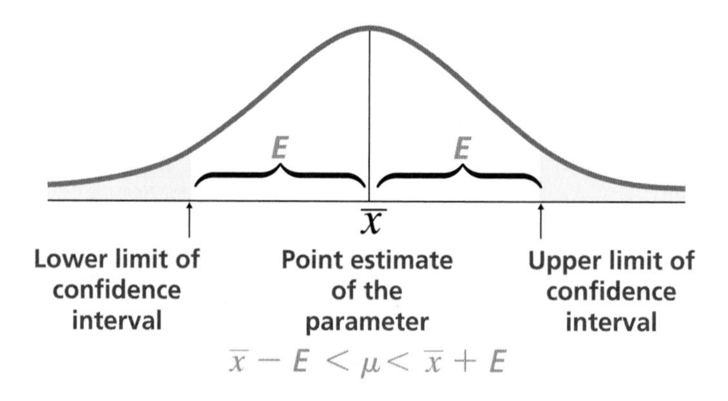

A complete formula for calculating the margin of error is shown in the inner boxes in Figure 10.6. As discussed earlier, the confidence interval is defined in terms of the critical values that correspond to the desired degree of confidence $(1 - \alpha)$. These are denoted as $\pm z_{\alpha/2}$ in the figure. In order to convert the standardized z-values back to the units of the data, we multiply the critical values times the standard error (i.e., times the standard deviation for the sampling distribution). Calculations are illustrated in Examples 10.1 and 10.2.

Figure 10.6
Formula to Calculate the Confidence Interval for μ (σ known)

$$\bar{X} - Z_{\alpha/2}\frac{\sigma}{\sqrt{n}} < \mu < \bar{X} + Z_{\alpha/2}\frac{\sigma}{\sqrt{n}}$$

Lower Limit **Upper Limit**

$$\bar{x} - E < \mu < \bar{x} + E$$

Example 10.1

A critical factor in the survival of victims buried by avalanches is the time that it takes to rescue them. According to a sample reported by the Colorado Avalanche Information Center, the mean burial time for victims who were recovered or saved by professionals was 18.3 minutes ($n = 32$).[5] Presume that the burial times for all victims eventually reached by professional searchers are normally distributed and that the standard deviation is 14.7 minutes. Based on the 32 sampled burial times:

1. Construct a 95% confidence interval for the population mean.
2. Construct a 98% confidence interval for the population mean.
3. How do you interpret the difference between the two solutions?

Solution

Confidence intervals for the population mean can be calculated using the formula in Figure 10.6. The sample mean of 18.3 minutes is the point estimate of the population mean and provides the centre point for both the requested confidence interval estimates. The standard error is defined as $\frac{\sigma}{\sqrt{n}} = \frac{14.7}{\sqrt{32}} = 2.5986$.

1. We have seen that for a 95% confidence interval, $z_{\alpha/2} = 1.96$. Accordingly, the margin of error $(E) = \pm(1.96 \times 2.5986) \approx \pm 5.1$. The confidence interval is

   $$(18.3 - 5.1) < \mu < (18.3 + 5.1)$$

   Therefore: $13.2 < \mu < 23.4$, in minutes

2. Most of us have not memorized the critical value that corresponds to a 98% confidence interval, so it has to be calculated. (Compare "Hands On 10.2.") For this, Figure 10.7 is analogous to Figure 8.7 in Example 8.5. There we learned to calculate a cutoff value within the range of possible values, for which the probability of selecting a smaller value than the cutoff is known. Figure 10.7 applies this to the standardized model (wherein $\mu = 0$ and $\sigma = 1$), so the cutoff is the desired critical z-value. Using the inverse distribution function on your software (see Chapter 8), you can find that for $\mu = 0$, $\sigma = 1$, and $P(Z < z_{\alpha/2}) = 0.01$, the value of $z_{\alpha/2}$ is -2.326. Since the normal distribution is symmetrical, the upper critical value is $+2.326$.

Figure 10.7
Calculating Critical Value for a Specified Degree of Confidence

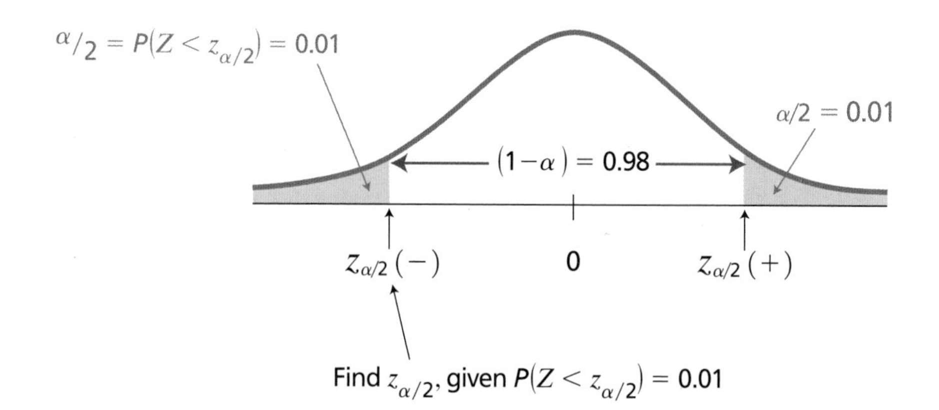

Now that we know the critical value and earlier calculated the standard error, we can find the margin of error (E) $= \pm(2.326 \times 2.5986) \approx \pm 6.0$. The confidence interval is

$$(18.3 - 6.0) < \mu < (18.3 + 6.0)$$

Therefore: $12.3 < \mu < 24.3$, in minutes

3. Comparing solutions 1 and 2 illustrates the tradeoff between the degree of confidence and the relative tightness of the interval estimate. It is less likely that the wider estimate ($12.3 < \mu < 24.3$) is mistaken (α is only 2%). But even if correct, we cannot tell if the true mean time to rescue is closer to 12.3 or to almost double that time value, 24.3. If we can accept a somewhat larger chance of error (α = 5%), we can be more precise in estimating the mean time for the population.

When interpreting the degree of confidence, it is important to realize that it does *not* represent a "probability that the population parameter falls in a certain interval." The population parameter has whatever value it has—regardless of our efforts to make an estimate. That probability is certain. What is uncertain is whether our interval estimate, in *this* case, happens to really contain the population parameter. When our degree of confidence is at a particular level, say 95%, we are expecting that 95% of the interval estimates constructed by using the methods in this textbook will correctly capture the true parameter. The remaining 5% (α) is the chance that, on a particular occasion, the constructed interval estimate will fail to capture the parameter. Unfortunately, for a given sample, we do not know which case applies.

In Example 10.1, our inputs were data summaries (e.g., values of $\bar{x}$ and n) rather than raw data. For Example 10.2, the raw data are available. As we will see, computers can play a role in solving such problems.

Example 10.2

 *(Based on the **Federal_Political_Polls** file)*

From mid-December 2003 to mid-January 2006, popular support for the federal New Democratic Party was quite stable. Interpreting the poll results during that period as scores, we see in Figure 10.8 that the scores are normally distributed, with a standard deviation of only 2.079. The normal pattern is so pronounced that the population σ probably has about the same value. Construct the 95% confidence interval for the true mean of the support level for the party during this period.

Figure 10.8
Distribution of Scores for NDP Support, December 2003 to January 2006

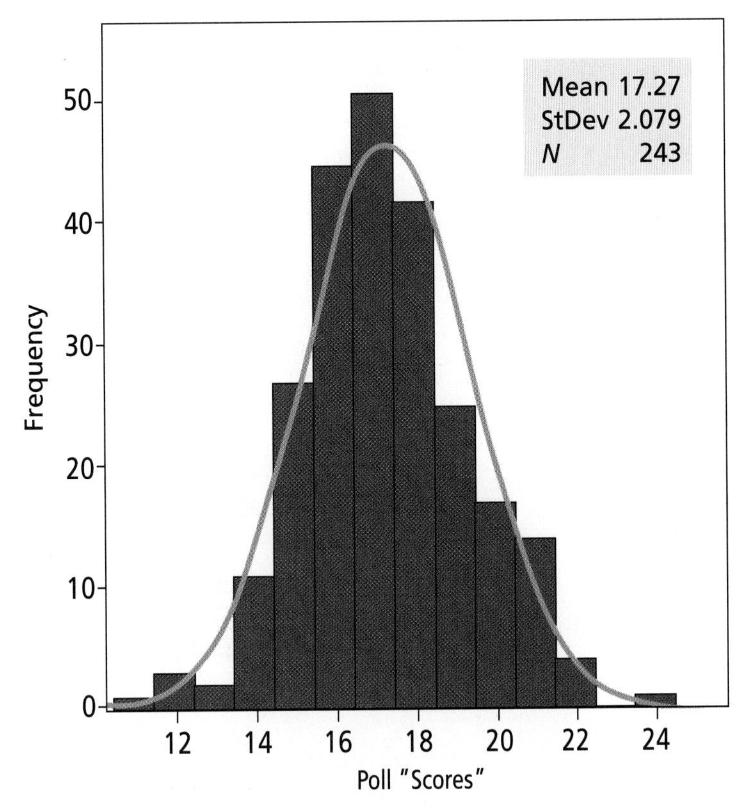

Source: Politics Canada. Based on data in published political poll results. Found at: http://www.canadawebpages.com/pc-polls.asp. Accessed Jan. 18, 2006.

Solution At a minimum, the computer can help to calculate $\bar{x}$ and n for this large sample. In the problem, a value of σ was pre-estimated and assumed as a given. (To avoid making that assumption, see Section 10.3.) Knowing $\bar{x}$, n, and σ, we could calculate the confidence interval as in Example 10.1. Some software can take further advantage of the available raw data. The confidence interval is displayed in the figures for "In Brief 10.1."

• *In Brief 10.1 Finding the Confidence Interval for Population Mean (σ Known)*

*(Based on the **Federal_Political_Polls** file)*

Excel

To use the model of this section (where σ is known) requires a combination of Excel formulas and manual calculation. In a blank cell, key in a formula of this format: **=CONFIDENCE(alpha,stdev,size)** (see the illustration). For ***alpha,*** input a value that is equal to 1 minus the desired degree of confidence. For ***stdev,*** input the known (or presumed) value of σ. ***Size*** is the sample size, n. The output of this procedure is actually the margin of error, *not* the confidence interval.

For the confidence interval, we also need to know the sample mean. This can be found with the formula **=AVERAGE(data_range)** (replacing ***data_range*** with the worksheet address for the block of raw data). We can now add and subtract the margin of error from $\bar{x}$ to find the upper and lower bounds of the confidence interval, respectively.

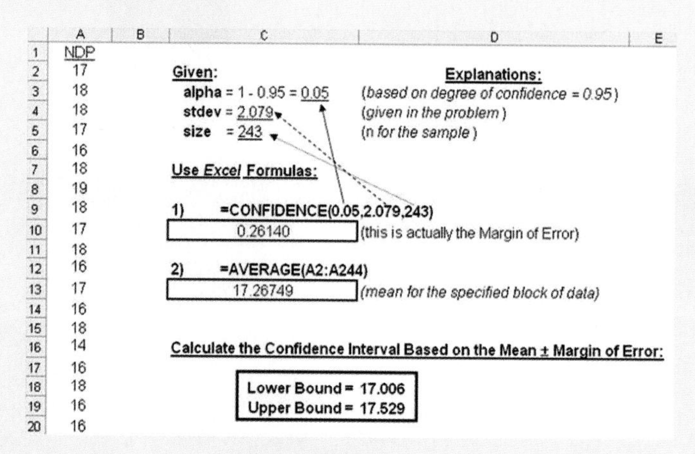

Minitab

Minitab will accept either raw data or summary data as inputs. Begin with the menu sequence *Stat / Basic Statistics / 1-Sample Z.* Starting from raw data (as shown in the illustration), select *Samples in columns* and then select the name of the column that contains the data. If instead you select *Summarized data,* input the values for the sample size and mean where indicated.

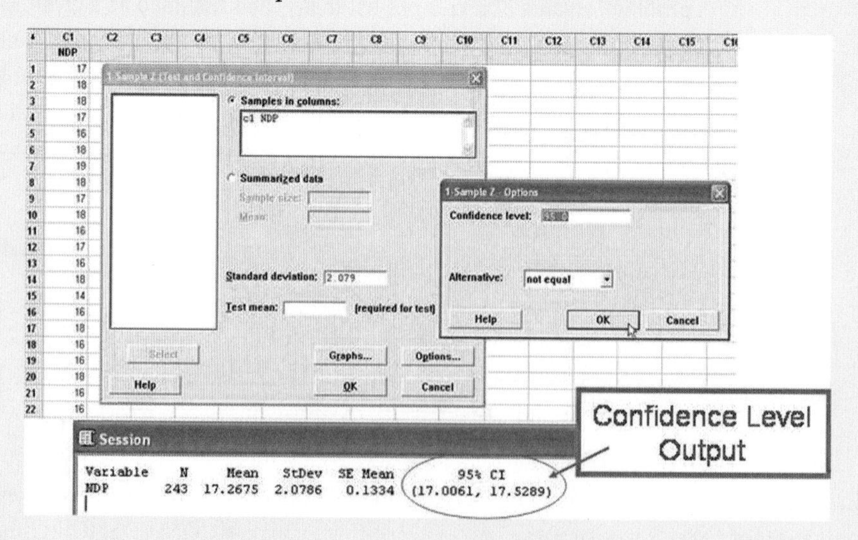

In either case, a predetermined value for σ must be input. Then click on *Options* and specify the confidence level in the next dialogue box. Click *OK* and then *OK* again to proceed. The result is shown in the illustration.

SPSS

Strictly speaking, SPSS does not provide automation for this model, but it can assist in finding the sample size and mean, which are part of the calculation. However, its procedure for finding the confidence interval of the mean takes the standard deviation value s from the raw data; there is no provision for inputting a σ-value separately. (The method is based on t-values, which will be explained in the next section.)

Observe in the illustration that, for the large sample described in this problem, the results based on using t-values are identical, to four significant digits, to the solution based on z-values. See "In Brief 10.2" for more details on the general method and on the keystrokes required in SPSS.

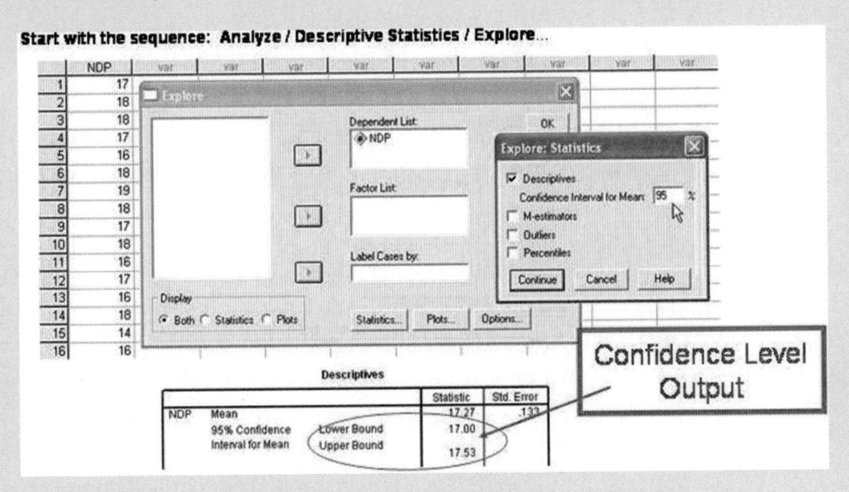

• Hands On 10.2 *Constructing a Table of Critical Values*

As you explore the examples and exercises in this chapter, you will notice that certain levels of confidence (such as 95%) tend to recur. So when constructing confidence intervals, the corresponding critical values (such as $z_c = \pm 1.960$) reappear as well. As a convenience, construct a table to show the critical values corresponding to various levels of confidence and fill in the blanks shown in Table 10.2.

Table 10.2 Table of Critical z-Values

Confidence Level	$\alpha/2$	Critical z-Values
99%	0.005	
98%	0.010	± 2.326
97%	0.015	
96%	0.020	
95%	0.025	± 1.960
94%	0.030	
93%	0.035	
92%	0.040	
91%	0.045	
90%	0.050	

10.3 Confidence Interval for Population mean (σ unknown)

Constructing a confidence interval for the mean with the techniques discussed in the previous section is not always feasible. That method presumed that we know σ for the overall population but, in many sampling situations, this value is unknown. At best, we have an s-value from a pilot study or a previous survey. We still need to estimate σ itself from a sample value for s.

In the previous section, we tried to get around this problem with reference to the central limit theorem: For a large sample size, (1) s will approach σ, in any case, and (2) the sampling distribution of the sample means will approach the normal. But sometimes it is impractical or not cost-effective to obtain a large sample. For example, it is very expensive to crash-test a car, so a manufacturer would be hesitant to crash-test 30 cars for each new experiment. As an alternative, we introduce in this section the ***t*-distribution** (also called **Student's *t*-distribution**) as the probability distribution for the mean associated with sampling. Based on this distribution, we then construct a confidence interval with the same fundamental steps as before.

Student's *t*-Distribution

The t-distribution was first developed by an employee of the Guinness Brewery in Ireland, William Sealy Gossett. He published under the pseudonym "Student"—thus giving the distribution its name.

In this model, the sampling distribution of the sample means is said to follow a t-distribution, rather than a normal distribution. Recall that for a normal distribution of sample means, each possible value of the sample mean for given sample size can be converted to its equivalent z-value. The normal curve is the probability distribution of all of these possible z-values. The alternative approach converts all of the means in the sampling distribution to their equivalent t-values. The probability distribution of these t-values, for a given sample size, follows the t-distribution. The formula for a t-value is shown in Formula 10.1.

Formula 10.1 Formula for a t-Value

$$t = \frac{\bar{x} - \mu}{\frac{s}{\sqrt{n}}}$$

You can see from Formula 10.1 that t is calculated similarly to z in the previous section. Both start by taking the difference between a value of one sample mean and the mean of the population. (As for z, the mean of the sampling distribution $\mu_{\bar{x}}$ is presumed to approximate the population mean μ.) Before, we standardized the difference found in the numerator in terms of how many standard errors $\sigma_{\bar{x}}$ (presuming we *know* that quantity) are included in that numerator value; here, we divide the numerator by the *approximate* standard deviation of the sampling distribution, as given by Formula 10.2. Observe that in the absence of knowing population σ, this formula relies on s (sample standard deviation) as an estimator for σ.

Formula 10.2 Sample-Based Estimate for Standard Deviation of the Sampling Distribution of the Sample Mean

$$s_{\bar{x}} = \frac{s}{\sqrt{n}}$$

Once we have a model for the sampling distribution for the sample means, corresponding to a particular sample size and population, we can use formulas to calculate the probability that a randomly selected sample mean will fall within, or outside of, any particular range under the curve. This general procedure is similar to the approach discussed in the previous section, but the specific calculations are different due to the non-normal distribution. (See Figure 10.9.)

Figure 10.9
Comparison of the Normal and t-Distributions

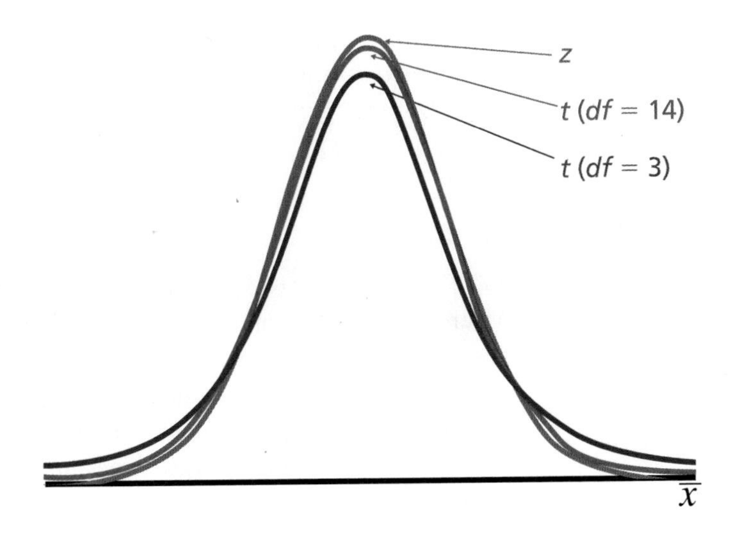

As shown in Figure 10.9, there is actually a family of t-distributions, rather than just one distribution. For a sample size that is infinite, the t-curve is identical to the z-distribution. For smaller samples, with smaller than infinite **degrees of freedom** (***df***), the curve becomes relatively flatter than the normal distribution, with proportionally more area under the tails.

Degrees of freedom is a technical term that relates the sample size to the nature of the sample statistic that you wish to calculate. When sampling for the mean of a single population, the degrees of freedom equal the sample size minus one (i.e., $n - 1$). The degrees of freedom are determined by the number of observations that could freely change in value without changing the sample statistic (e.g., the mean that has been calculated for the sample).

This concept can explained with reference to Figure 10.10. For each sample illustrated, the mean = 25 is specified. Three of the four observations in each sample (which is $n - 1$ observations) could take any value whatsoever without being inconsistent with the mean. But only one number is possible for the fourth observation, given the mean that was specified for the full sample. In effect, if we are given information about a sample and its mean, one piece of that information is redundant. As Figure 10.9 showed, the disadvantage of having reduced information is a wider sampling distribution; if interval estimates are created, they will be wider (i.e., less precise) than if the degrees of freedom are larger.

Figure 10.10
$n - 1$ Degrees of Freedom

Samples of 4, Each Having a Mean = 25

Sample Numbers:	1	2	3	4	5	6	7	8	9	10	11	12
	−852	−578	154	690	813	344	−141	480	−656	800	470	921
	−569	141	384	653	160	427	−543	−996	−135	−246	−222	−113
	−947	−353	−639	140	207	835	−118	−170	346	−700	183	382
	2468	890	201	−1383	−1080	−1506	902	786	545	246	−331	−1090
Means:	25	25	25	25	25	25	25	25	25	25	25	25

If mean = 25, there is only one possible value for this number.

These values can be any randomly selected numbers.

Constructing the Confidence Interval with the t-Distribution

Figure 10.11 has the same structure as Figure 10.4. All values for a sampling distribution have been converted to corresponding t-values (instead of z-values), for a particular value of n. The mean of the distribution falls at t-value = 0. As before, we can specify critical values—now in terms of t-values—to define the upper and lower bounds of where we expect most samples to fall. These values are set based on the mathematics of the curve, for the value α (chance of error) that corresponds to the desired degree of confidence

$(1 - \alpha)$. The probability value α has been split into $\alpha/2$, which equals both (1) the chance that the random sample mean could fall above the upper critical value and (2) the chance that it could fall below the lower critical value.

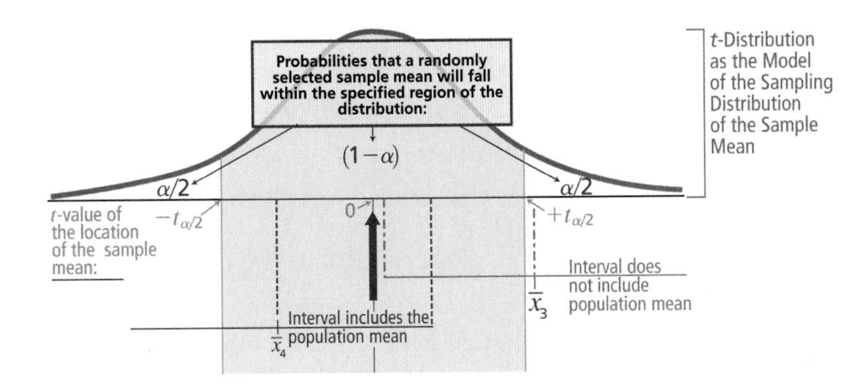

Figure 10.11
t-Distribution Model for the Sampling Distribution of the Sample Mean, and the Probabilities Associated with Confidence Intervals

Figure 10.11 is drawn from the perspective of the sampling distribution of the mean. From this, we learn the positive and negative *t*-values (the critical values) that correspond to the desired degree of confidence. In Figure 10.12, we apply this information from the perspective of the sample. As in the previous section, the centre of the confidence interval for the population mean is the point estimate, which is equal to the sample mean. We again find the bounds of the confidence interval by subtracting from or adding to, respectively, a margin of error to the point estimate. The margin of error is based on the critical *t*-value times the estimated standard deviation for the sampling distribution. It is expressed in the original units for the raw data.

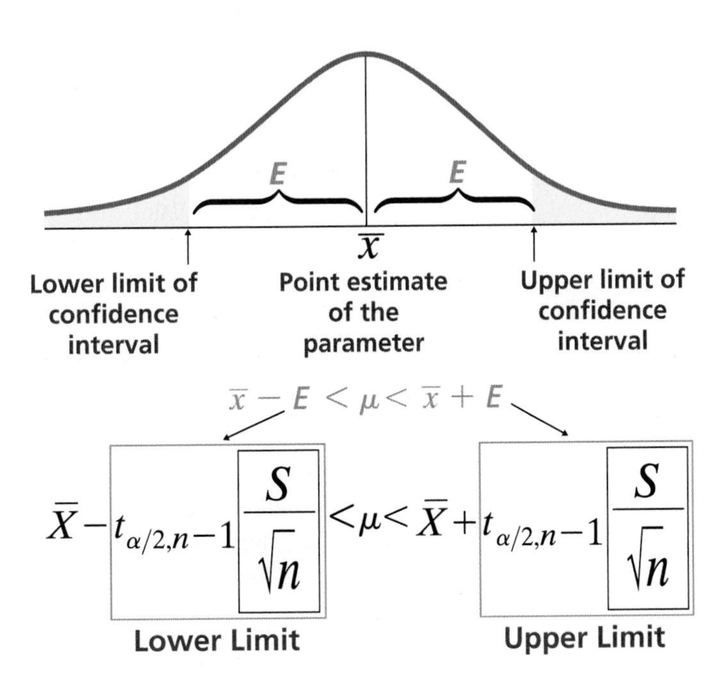

Figure 10.12
Calculating a Confidence Interval for μ (σ unknown)

Compare Examples 10.3 and 10.4 to their counterparts in the previous section.

Example 10.3 A critical factor in the survival of victims buried by avalanches is the time that it takes to rescue them. We learned in Example 10.1 that the mean burial time for victims who were recovered or saved by professionals was 18.3 minutes ($n = 32$).[6] Presume that the burial times for all such victims are normally distributed, but *we do not know* the population standard deviation. For the given sample, $s = 14.7$ minutes. Based on the 32 sampled burial times:

1. Construct a 95% confidence interval for the population mean.
2. Construct a 98% confidence interval for the population mean.
3. How do these solutions compare with solutions 1 and 2 in Example 10.1?

Solution The assumption of population normality is particularly important when using the *t*-distribution, particularly if n is small. Figure 10.12 includes a formula for constructing the confidence intervals for the population mean. The point estimate for the mean, 18.3 minutes, provides the centre point for both the confidence intervals requested in the problem. We estimate the standard deviation for the sampling distribution as

$$\frac{s}{\sqrt{n}} = \frac{14.7}{\sqrt{32}} = 2.5986$$

1. To solve this problem, we next have to find the critical values ($\pm t_{\alpha/2,n-1}$). The $n - 1$ term in the notation reminds us that the *t*-value is relative to the degrees of freedom for the sample size. For this problem, $\alpha/2 = (1 - 0.95)/2 = 0.025$, and $n - 1 = 31$. We can use the computer to determine the *t*-value that corresponds to these inputs. (See "In Brief 10.2" for details.) The computer shows that $t_{0.025,31} = -2.04$. Since the *t*-distribution is fairly symmetrical, the upper critical value is $+2.04$. Accordingly, the margin of error (E) $= \pm(2.04 \text{ x } 2.5986) \approx \pm 5.3$. The confidence interval is

$$(18.3 - 5.3) < \mu < (18.3 + 5.3)$$

   Therefore: $13.0 < \mu < 23.6$, in minutes

2. Due to the change in confidence level, we recalculate $\alpha/2 = (1 - 0.98)/2 = 0.01$. $n - 1$ is unchanged. Using the procedures in "In Brief 10.2," we find that the magnitude of $t_{0.01,31} = 2.45$ (i.e., its sign is -2.45, but for calculating the margin of error (E), we are interested in both the positive *and* negative critical values.) The margin of error (E) is therefore $\pm(2.45 \times 2.5986) \approx \pm 6.4$. The confidence interval is

$$(18.3 - 6.4) < \mu < (18.3 + 6.4)$$

   Therefore: $11.9 < \mu < 24.7$, in minutes

3. Comparing these estimates with the confidence intervals computed in Exercise 10.1, we find that for a given level of confidence, the confidence interval will be wider (less precise) for the *t*-based model compared to the *z*-based model. This reflects the flatter shape of the *t*-distribution compared to the normal distribution.

The *t*-based approach to finding the confidence interval requires that both $\bar{x}$ *and* s be calculated from the sample. You will often have the raw data available when you are making the calculations. Example 10.4 is based on that assumption.

Example 10.4

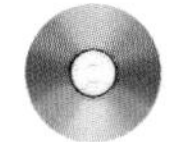

*(Based on the **World_Apple_Production** file)*

For the 10 years from 1992 to 2001, apple production in Canada was relatively stable. Interpret these past data (given in thousands of metric tonnes) as samples of yearly production numbers, in general, for Canada. On that basis, construct the 90% confidence interval for the population mean of annual apple production in Canada.

Solution

The sample size is small, so we can use the *t*-distribution for the estimate—*provided that* the underlying population is approximately normally distributed. (See Section 8.2 for methods to make this assessment.) A quick check of Figure 10.13 shows that, based on the sample, the population does appear to be reasonably close to normal, although the model is not exact.

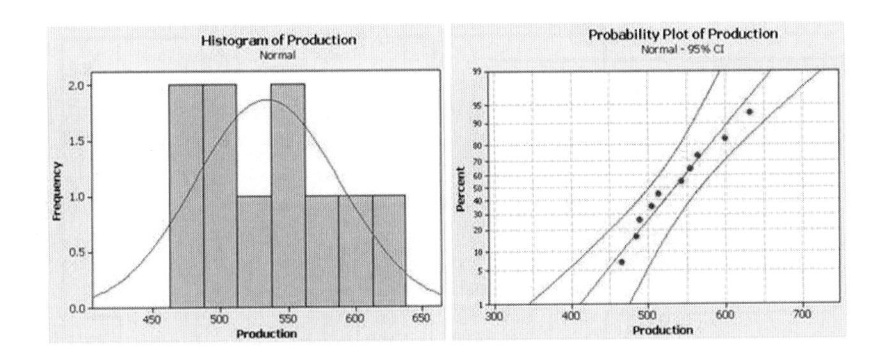

Figure 10.13
Checks of Normal Distribution for the Population

At this point, the computer can calculate for the sample $\overline{x} = 534.7$, $s = 53.5$, and $n = 10$. The critical *t*-value would be 1.833. We could now use the procedures in Example 10.3 to compute the 90% confidence interval as $503.6 < \mu < 565.8$. Alternatively, given access to raw data, your software may be able to automate some of the procedures, as discussed in "In Brief 10.2." Note that because the distribution is not fully normal (see Figure 10.13), the calculated boundaries of the confidence interval should not be considered exact.

• *In Brief 10.2 Finding the Confidence Interval for Population Mean (σ Unknown)*

*(Based on the **World_Apple_Production** file)*

Excel

If you start from summary values for $\overline{x}$ and s and have a specified α, then Excel can help with finding the critical *t*-value. (See step 3 in the illustration). Note the inputs for the Excel formula **=TINV(α,df)**. The first input is the full value of **α**, *not* **α/2**. The second input is the degrees of freedom (df). The remaining steps could be done manually, or with a calculator, or with Excel formulas. If $\bar{x}$ and s are not provided, then you can also use the formulas shown in the first two steps to obtain these from the data.

If you have access to raw data, there is an easier procedure. Begin with the menu sequence *Tools / Data Analysis / Descriptive Statistics,* and then press *OK.* In the "Descriptive Statistics" dialogue box (shown on the right-hand side of the illustration), specify the Excel range that contains the data. Select both *Summary statistics* and *Confidence Level for Mean* and input the requested value for $(1 - \alpha) \times 100$. Specify an output range and then click *OK.* The output of this procedure shown at the bottom of the illustration is actually the margin of error, although it is labelled as the confidence interval in the dialogue box. You must still add and subtract this value from the sample mean to find the confidence interval boundaries.

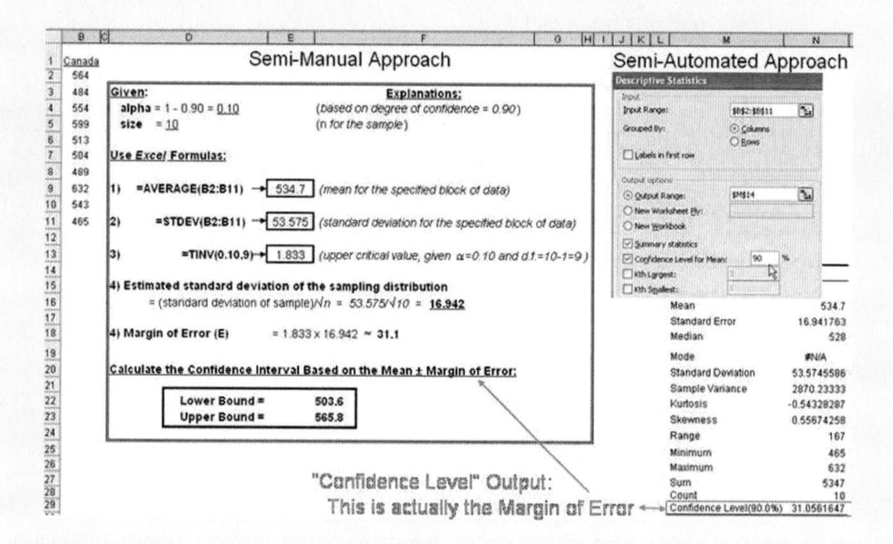

Minitab

Minitab will accept either raw data or summary data as inputs. Begin with the menu sequence *Stat / Basic Statistics / 1-Sample t.* If starting from raw data (as shown in the illustration), select *Samples in columns* and then select the name of the column that contains the data. If instead you select *Summarized data,* then input the values for the sample size, sample mean, and sample standard deviation where indicated.

Now click on *Options* and in the resulting dialogue box, specify the confidence level (e.g., 90). Click *OK* and then *OK* again to proceed. The result is shown in the illustration.

Alternatively, you can use Minitab to find the critical t-value and then proceed semimanually to calculate the margin of error and the bounds of the confidence interval. Start with the menu sequence *Calc / Probability Distributions / t.* Then, in the dialogue box (not shown here), select *Inverse Cumulative Probability* and input the degrees of freedom where indicated. Then select *Input Constant* and input the value of $\alpha/2$, which is the proportion of t-distribution to the left of the (negative) critical t-value; the output (if we ignore its negative sign) gives the magnitude of the critical t-value.

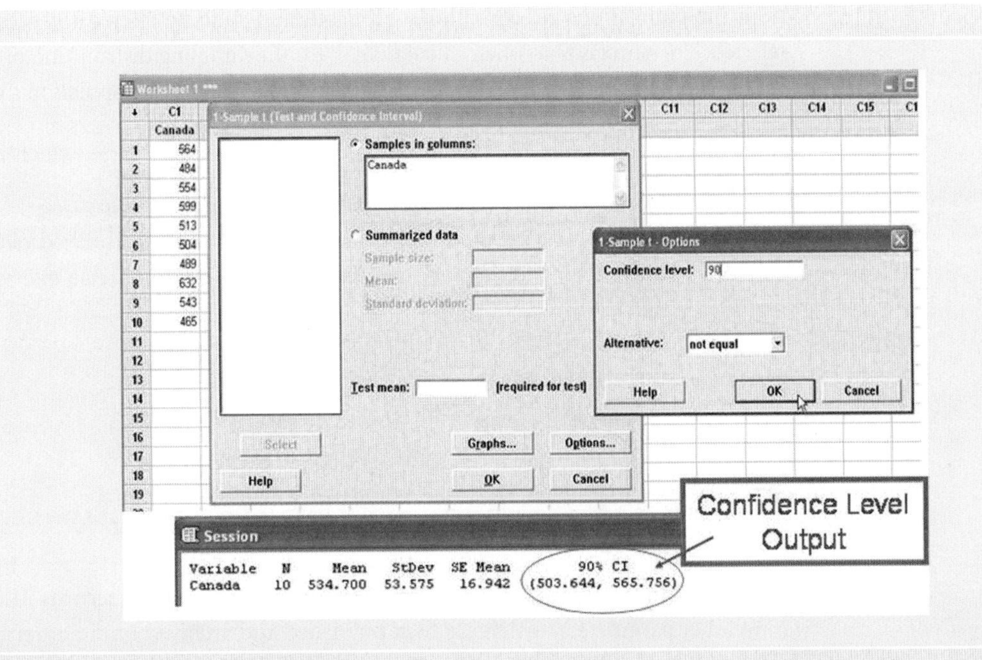

SPSS

The SPSS illustration in "In Brief 10.1" depicted the t-based method of calculation for the confidence interval; SPSS does not provide automation for the z-based model. Starting from a column of raw data, use the menu sequence *Analyze / Descriptive Statistics / Explore.* Select the variable to be analyzed. Click on *Statistics,* and in the resulting dialogue box, select *Descriptives.* Specify, where asked, the desired confidence interval for the mean. Click *Continue* and then *OK.* See "In Brief 10.1" for an illustration of the dialogue boxes. The output that corresponds to Example 10.4 is shown in the illustration.

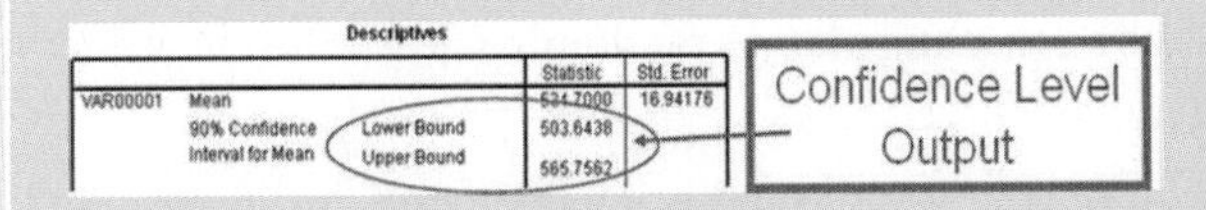

Alternatively, you can use SPSS to find the critical t-value and then proceed semimanually to calculate the margin of error and the bounds of the confidence interval. Ensure that there is at least one data value in the worksheet, then use the menu sequence *Transform / Compute.* Then, in the resulting dialogue box, input in the *Target Variable* area the name of the column where the answer will appear. In *Numeric Expression,* input a formula of the form **idf.t(a,b)**, replacing **a** with the value of $\alpha/2$ and **b** with the degrees of freedom. (Note: An alternative to inputting the formula directly is to select *Inverse DF* (for "Inverse Distribution Function") from the *Numeric Expression* area of the dialogue box, and then scroll and find/select **idf.t** in the box at the lower right; then replace the **a** and **b** as described above.) Click *OK.*

You may recall that in Chapter 9, Example 9.4, the formula for the standard error of the sampling distribution of the sample mean (σ known) had to be adjusted when the population size N was small, compared to the sample size n. If sampling without replacement from the population, the conditional probabilities would be altered for each new item selected. The same is true when calculating $s_{\bar{x}}$ for the sampling distribution of the sample mean (σ unknown): If the sample size is greater than 5% of the population size, then the value of $s_{\bar{x}}$ per Formula 10.2 should be multiplied times $\sqrt{\frac{N-n}{N-1}}$, which is the *finite population correction factor.* This multiplier $\cong 1$ for large populations, so we can generally ignore it. If you are using the computer, you can multiply the displayed value for standard error by the correction factor and then construct the confidence interval with the modified value.

• Hands On 10.3 *Testing the Confidence Interval for the Mean*

Every 15 minutes, the University of Waterloo Weather Station records 21 key weather parameters, which are then published and archived on the Internet. The complete population of readings for barometric pressure in 2005 consisted of over 35,000 records that were highly normal in distribution. The few large outliers did not materially affect the mean. The population μ and σ are 101.727 kP (kilopascals) and 5.724 kP, respectively.[7]

Since we happen to know the population mean, we are in a good position to test out the assumptions for procedures to build confidence intervals. The basic plan for this exercise is to independently draw samples of (1) size $n = 35$ and (2) size $n = 6$, and construct from them confidence intervals for the mean. How often do these interval estimates include, or not include, the true population mean?

To draw one sample from the population, go to the University of Waterloo Weather Station's website at http://weather.uwaterloo.ca/data.html. Scroll down the page to the option "Choose Yearly Data Requirements." Specify the year 2005, then select *Barometric Pressure,* and then click on *Request Data.* After a short pause, the entire population of values will appear. Now follow these steps:

Step 1

a) Design and execute a sampling plan to randomly select 35 of the values from the population.

b) Based on the mean of your sample and using the methods discussed in Section 10.2, construct the 90% confidence interval. (Use the true σ-value, which is known.)

c) Observe and record whether this sample-based confidence interval includes the true population value for μ (which is known).

Repeat steps (a)–(c) many times, ideally at least 50. (The more times, the better.) For what percentage of the samples did the 90% confidence interval *not* include the true population mean? What percentage of the samples would you have expected, in theory, to not include the population mean? How do the previous two answers compare?

Step 2

Repeat the full sequence of step 1 with these modifications: Use the methods discussed in Section 10.3 and, when constructing confidence intervals for the mean from each sample, find and use the standard deviation from the sample without making any assumptions about population σ. How do these recomputed interval estimates compare in width to the confidence intervals calculated earlier? And how often do these interval estimates include, or not include, the true population mean?

Step 3

Repeat the full sequence of step 2, with this modification: Each time you independently draw a sample, use a size $n = 6$, instead of $n = 35$. How do these recomputed interval estimates compare in width to the confidence intervals calculated earlier? And how often do these interval estimates include, or not include, the true population mean?

10.4 Confidence Interval for the Proportion

In Chapter 9, we observed that the binomial distribution, although discrete, can look similar in shape to a normal distribution under certain conditions (see Figure 9.6). Accordingly, in Section 9.3 we noted the concept of the normal approximation for the binomial, wherein the normal curve can be used to approximate the sampling distribution of the sample proportion—where the "proportion" is the proportion of trials in an experiment that are successes. With the computer, however, the proportion-based problems of Section 9.3 could be solved directly, instead, using the binomial model to avoid the limitations of the normal approximation.

On the other hand, for finding the confidence interval for a proportion, in practice, it is easier—and generally quite accurate—to retain the traditional approach, using a normal approximation of the binomial, so this is the approach introduced here. To validly use the model, both $n \times p$ and $n \times (1 - p)$ must equal at least 5 (some statisticians prefer 10). Where this condition is not met, then a binomial-based alternative is required.

Recall that for a binomial distribution, $\mu = n \times p$ and $\sigma = \sqrt{[n \times p \times (1 - p)]}$. These parameters refer to the numbers of successes, X, in experiments involving n trials. Now we are interested in the *proportions* of successes (X/n) in those trials. So we adapt the above equations from this new perspective:

$$\mu_{\mathrm{p}} = \frac{\mu_{\mathrm{X}}}{n} = p$$

$$\sigma_{\mathrm{p}} = \frac{\sqrt{[n \times p \times (1 - p)]}}{n}$$

$$= \sqrt{\frac{[n \times p \times (1 - p)]}{n^2}}$$

$$= \sqrt{\frac{[p \times (1 - p)]}{n}}$$

Once we have a μ and σ for the normal approximation, we construct the confidence interval as we did in Section 10.2. In effect, we construct the confidence interval estimate for the mean (μ_p), as in a case where the standard deviation (σ_p) is known. In Example 10.5, we compare solutions by this method with the solutions using the true binomial. For more details on the latter approach, see the "In Brief 10.3."

Example 10.5

The Pacific Salmon Commission forecasts how many Fraser River sockeye salmon are expected to divert through the Johnstone Strait during a season. In 2005, the expected diversion rate was 50%.[8]

1. Suppose that during the season, a random sample of 10 sockeye salmon are tagged and observed. Five of the salmon divert through the Johnstone Strait. Use the normal approximation of the binomial to construct the 95% confidence interval for the proportions of salmon that are diverting through the strait.
2. Construct the confidence interval if 100 salmon are observed and 50 of these divert through the strait.
3. Repeat (1) and (2), but use the true binomial calculations. Compare the solutions.

Solution

1. The necessary calculations are shown in the upper left of Figure 10.14, and $n \times p$ meets the conditions for using the normal approximation of the binomial (i.e., $10 \times 0.50 = 5$, thus $= \geq 5$). The proportion of successes in the sample ($5/10 = 0.5$, which is the point estimate) provides the centre of the confidence interval. σ_p is approximated by $\sqrt{\frac{0.50 \times 0.50}{10}} = 0.158$

Figure 10.14
Calculating a Confidence Interval for a Proportion

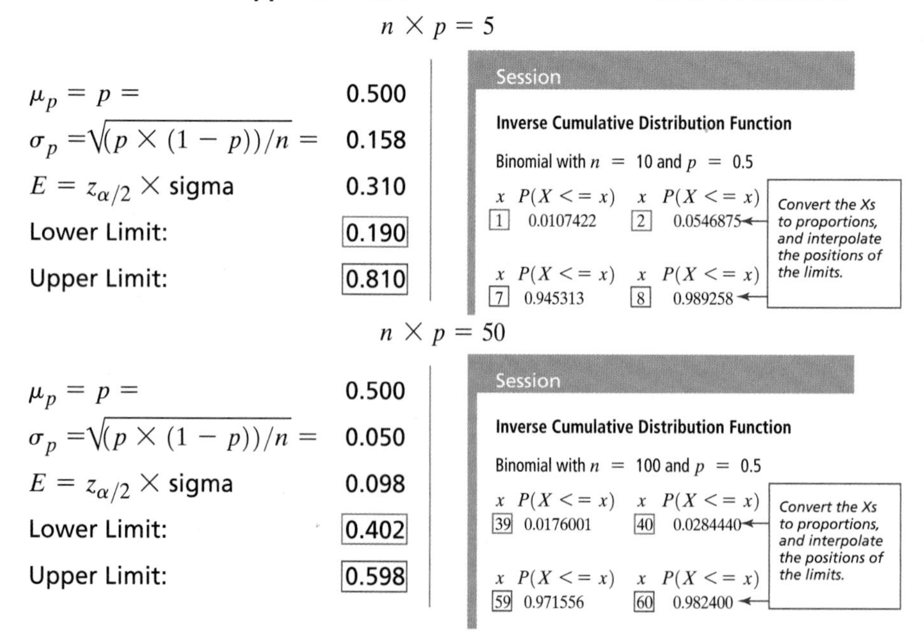

We now follow the basic model from Section 10.2. Find the critical *z*-value ($z_{\alpha/2}$) that corresponds to the degree of confidence: For a confidence level of 95%, $\alpha = 1 - 0.95 = 0.05$, and $\alpha/2 = 0.025$. We calculate $z_{0.025} \approx -1.96$. (Note: The convention in this book for the *calculated* critical value is to display the sign that follows if the subscript for *z* (or *t*) stands for the *less than* cumulative probability below that particular *z*-value (or *t*-value); this corresponds to how you "look up" a *z*-value (or a *t*-value) on the computer for a given probability. But traditionally, the value for $z_{0.025}$, for example, is often given as positive, either viewing the answer as the magnitude or as based on the area to the *right* of the positive *z*-value or *t*-value.) In any case, using the magnitude of 1.96, the margin of error *E* equals $z_{\alpha/2}$ times sigma, which is 1.96 x 0.158 = 0.310. Subtracting *E* from the point estimate gives the lower bound of the confidence interval, and adding *E* to the point estimate gives the upper bound. Based on the small sample of 10, our estimate of the proportion of salmon that were diverted is given by this very large interval:

$$0.19 < p < 0.81$$

2. When the sample size is increased to 100, $n \times p$ still meets the conditions for using the normal approximation of the binomial (i.e., $100 \times 0.50 = 50$, thus ≥ 5). The point estimate becomes $50/100 = 0.50$, providing the centre of the confidence interval. (See the lower left of Figure 10.14.) σ_p changes to

$$\sqrt{\frac{0.50 \times 0.50}{100}} = 0.050$$

The increased sample size does not affect the critical value: $z_{0.025} \approx -1.96$. But using the magnitude of 1.96, the margin of error *E* is now 1.96 x 0.050 = 0.098. Adding and subtracting *E* from the point estimate gives the upper and lower bounds of the confidence interval, respectively. With this larger sample, our estimate for the true proportion of salmon that are diverting to the strait is given by this approximate range:

$$0.40 < p < 0.60$$

3. In the right-hand portion of Figure 10.14, we see the binomial solutions to the above problem. The estimates for the larger sample (40/100 for the lower bound and 60/100 for the upper bound) are indistinguishable from those for solution 2 when rounded to two decimal places.

For the smaller sample, at the top right-hand side of the figure, the extra details of the binomial solution, compared to solution 1, do not really help in constructing a confidence interval and *are not a recommended procedure.* We see that the lower bound of the confidence interval (for $\alpha/2 = 0.025$) is somewhere *between* 1 success in 10 (where $P(X \leq 1) = P(p \leq 1/10) = 0.0107$) and 2 successes in 10 (where $P(X \leq 2) = P(p \leq 2/10) = 0.0547$). It would be possible to interpolate between those two values to estimate a p_c, for which $P(p \leq p_c)$ equals exactly $\alpha/2$. But this adds another layer of approximation, so we lose the advantage of not using the traditional normal approximation. The normal approximation should generally be retained for this application as long as the minimum size conditions are met.

• *In Brief 10.3* *Finding the Confidence Interval for the Proportion*

Excel

For applying the normal approximation to the binomial, Excel can help in finding the critical value. (Use the function **=normsinv(**α**/2)** to find the magnitude of $z_{\alpha/2}$.) The remaining steps are performed semimanually, similar to the left-hand side of Figure 10.14. We *cannot* use the "=Confidence()" formula that we applied in Section 10.2 because σ for the proportion differs from the σ presupposed by that formula.

For the binomial approach, see the steps and calculations in the illustration.

Given:		Explanations:
Sample Size *n*	100	
Confidence Level	0.95	
X	50	*Observed number of successes*
Calculate:		
α	0.050	*1 − (confidence level)*
CumProb(Left)	0.025	$P(p<P_{c(left)}) = \alpha/2$
CumProb(Right)	0.975	$P(p<P_{c(right)}) = 1 - \alpha/2$
Sample Proportion *p*	0.500	*X/n (i.e., estimated from the sample)*

Use this *Excel* Formula (twice):
=CRITBINOM(n,p,CumProb)

1) **=CRITBINOM(100,0.5,0.025)** = *40*

Interpretation:
40 is the first discrete number of successes at which the cumulative probability to its left is $\alpha/2$.
Therefore, the critical proportion (left) is below 0.40 *(i.e., 40/n = 40/100)*

2) **=CRITBINOM(100,0.5,0.975)** = *60*

Interpretation:
60 is the first discrete number of successes at which the cumulative probability to its left is $(1 - \alpha/2)$.
Therefore, the critical proportion (right) is below 0.60 *(i.e., 60/n = 60/100)*

Therefore, the confidence interval for the proportion ≈

Lower Bound = "A little" below 0.40
Upper Bound = "A little" below 0.60

Minitab

For applying the normal approximation to the binomial, see "In Brief 10.1." Select *Summarized data* and then input, respectively, the sample proportion as the mean and the calculated value of ($\sigma_p \times \sqrt{n}$) where the standard deviation is expected. Proceed as before; the displayed confidence interval will be for the proportion. (Note: Minitab will divide the standard deviation you input by $\sqrt{n}$ to obtain the standard error, so we input (σ_p times $\sqrt{n}$) as an adjustment—because σ_p is *already* the standard error of the model.)

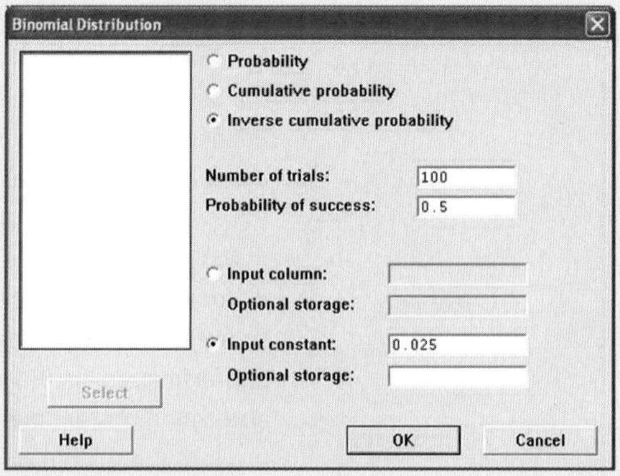

Binomial Distribution
○ Probability
○ Cumulative probability
◉ Inverse cumulative probability
Number of trials: 100
Probability of success: 0.5
○ Input column:
Optional storage:
◉ Input constant: 0.025
Optional storage:
Select
Help OK Cancel

For the binomial approach, begin with the menu sequence *Calc / Probability Distributions / Binomial.* In the resulting dialogue box shown in the screen capture, select *Inverse cumulative probability* and specify the number of trials n and the probability of success p for each trial. Select *Input constant* and input the less than cumulative probability reached at or before the discrete boundary of the confidence interval. Click *OK.* (The first constant will be the value for $\alpha/2$.) Repeat, but input the value for $(1 - \alpha/2)$ as the constant. The right-hand side of Figure 10.14 illustrates the output of these procedures.

SPSS

SPSS does not offer automation explicitly for this model. However, it can provide, if needed, a "look-up" for the z-value corresponding to the confidence interval.

10.5 A Preliminary Note on Sample Sizes

To see the importance of sample size for constructing confidence intervals, compare the solutions for Example 10.5, parts (1) and (2). The goal was to estimate a proportion, and confidence intervals were constructed, based on samples of $n = 10$ and $n = 100$, respectively. In the first case, the interval $(0.19 < p < 0.81)$ was so wide that, in practice, it would probably be unhelpful. With the larger sample, the interval $(0.40 < p < 0.60)$ can provide at least a ballpark sense of where the population parameter may lie.

In general, all things being equal, a larger sample can provide a narrower estimate for the confidence interval. At some point, increasing sample size may be too expensive or impractical. A tradeoff is to reduce the confidence level. To be 95% confident that the population parameter fell within the confidence interval, for solution 1 in Example 10.5, we had to construct a wide interval. But if we could accept only 90% confidence, then the interval could be narrower. If you are designing a sample and hope to keep sample size to a minimum, you must balance these considerations: To increase confidence that the parameter lies in the constructed interval, you will have to widen the boundaries of the interval, or vice versa.

When sample size is discussed further in Chapter 11, we will look at an additional consideration related to hypothesis tests: the power we want for the test.

Chapter Summary

10.1 Distinguish among point, interval, and confidence interval estimates, explain the concepts of confidence levels and confidence limits, and also describe some basic desirable properties of estimators.

Estimation is a process of obtaining a single numerical value, or a set of values, intended to approximate the actual value of a parameter in the population. A **point estimate** is a single value intended to approximate the true corresponding value of a population parameter. An **interval estimate** gives a *range* of values within which the true value of the population characteristic is expected to lie. A **confidence interval** does not just specify the range of possible values for the parameter, it also indicates a **degree of confidence** or **confidence level** that the computed confidence interval really does contain the true value of the parameter. The lower and upper boundaries of the confidence interval estimate are called the **confidence limits.**

An **estimator** is a measured or calculated value based on a sample, used to estimate a population parameter. Generally, the best estimator for a parameter is the corresponding sample statistic. If possible, a good estimator should be **unbiased, efficient,** and **consistent.**

10.2 From sample data, calculate the margin of error and construct confidence intervals for the mean in cases where population σ is known, for varying levels of confidence.

To construct a confidence interval for the mean (or other statistic) we need a model for the sampling distribution. For the confidence interval for the mean—if the population σ is known, we can use the normal distribution as the model for the sampling distribution of the sample means. When the conditions for the central limit theorem are satisfied, it is then possible to define the upper and lower bounds of where we expect most samples to fall, based on **critical values** for the z-statistic, which in turn are calculated based on the specified degree or level of confidence. More specifically, the lower and upper bounds of the confidence interval are found by subtracting or adding, respectively, a **margin of error** to the point estimate for the mean. The margin of error is a distance from the mean, based on the product of the critical value and the standard error for the sampling distribution.

10.3 From sample data, calculate the margin of error and construct confidence intervals for the mean in cases where population σ is not known, for varying levels of confidence, and explain the concepts of the t-distribution and the associated degrees of freedom (df).

For the confidence interval for the mean if the population σ is *not* known, then (provided the underlying population is normal or the sample is large) we can use the ***t*-distribution** (or **Student's *t*-distribution**) as the model for the sampling distribution of the sample means. (See Formulas 10.1 and 10.2 for calculating t-values.) There is actually a family of t-distributions, and you must use the distribution that is applicable for a given **degrees of freedom (*df*)**. Degrees of freedom are determined by the number of observations that could freely change in value without changing the sample statistic (e.g., the mean that has been calculated for the sample). For estimating the mean of a single population with the t-distribution, the degrees of freedom equals the sample size minus one (i.e., $n - 1$).

10.4 From sample data, construct approximate confidence intervals for the proportion based on the normal approximation, and calculate values for sampling distribution mean and standard deviation to be used for the estimate.

Although the sampling distribution for the sample proportion is actually binomial, it is generally easier—and reliable—to use the normal approximation to find the confidence interval for the estimate. Use the formulas below to find the mean and standard error of the normal model, where p is the observed value of the proportion:

$$\mu_{\mathrm{p}} = \frac{\mu_{\mathrm{X}}}{n} = p$$

$$\sigma_{\mathrm{p}} = \sqrt{\frac{[p \times (1 - p)]}{n}}$$

Centre the confidence interval around the point estimate for the proportion and find the margin of error as in Section 10.2, but using σ_{p} as the standard error for the sampling distribution.

10.5 Describe the tradeoff for constructing confidence intervals between increasing the precision and increasing the level of confidence.

When constructing confidence intervals, you need be aware of the tradeoffs among increasing the precision, increasing the level of confidence, and increasing costs (due to larger sample sizes). (Sample sizes will be discussed in more detail in Chapter 11.)

• Key Terms

confidence interval, p. 405
confidence limits, p. 405
critical values, p. 409
degree of confidence (*or* confidence level), p. 405
degrees of freedom (*df*), p. 418
estimation, p. 405
estimator, p. 406
interval estimate, p. 405
margin of error, p. 410
point estimate, p. 405
***t*-distribution (*or* Student's *t*-distribution), p. 416**
unbiased, efficient, and consistent estimators, p. 406–407

• Exercises

Basic Concepts 10.1

1. Describe the difference between a point estimate for a parameter and an interval estimate for the parameter. Give two examples of each type of estimate.
2. Explain the relationship between an interval estimate for a parameter and a confidence interval for the parameter.
3. For each of the following cases, identify whether it is (i) a point estimate, (ii) a confidence interval, or (iii) an interval estimate—but *not* a confidence interval.
   a) Tomorrow's mean temperature is predicted to be 27°C.
   b) Tomorrow's mean temperature is expected to be between 19°C–23°C.

c) Support for expanding the recycling program is somewhere between 65–70%.
d) According to a daily poll, 67% of homeowners favour the blue box program. (The poll is expected to be accurate within ±3%, 19 times out of 20.)
e) Based on my research, I am 95% sure that the average student age at my university is between 19.2 and 20.5.

4. Explain the relationship between an estimator for a population parameter and the value of the population parameter.
5. Explain what it means for an estimator to be unbiased, and give an example.
6. Explain what it means for an estimator to be efficient, and give an example.
7. Is the value of $\frac{\text{Sample maximum} + \text{Sample minimum}}{2}$ a consistent estimator for the population mean? (Hint: Using examples, consider whether increasing the sample size always tends to make the estimate more reliable.)
8. Is the value of $\frac{\text{Sample maximum} + \text{Sample minimum}}{2}$ a consistent estimator for the population median? (See hint in Exercise 7.)
9. Is the value of $\frac{\text{Sample maximum} - \text{Sample minimum}}{6}$ a consistent estimator for the population standard deviation? (See hint in Exercise 7.)

Applications 10.1

For each of the following quotations, indicate whether the quotation is providing (a) a point estimate, (b) a confidence interval, or (c) an interval estimate—but *not* a confidence interval.

10. "The population estimate [for harp seals] for 2000 is 5.2 million seals."[9]
11. "There is a 95% chance that [the harp seal population] is in the range of 4.0 to 6.4 million." [10]
12. "Total pup production for harp seals in the Northwest Atlantic in 1999 . . . [was in the] range 800,000 to 1,200,000 pups. . . ."[11]
13. "The 'estimated wait time' for customs processing (non-commercial travellers) at the St-Bernard-de-Lacolle border crossing between Lacolle, Quebec, and Champlain, New York, is '1 hour 30 minutes.'" [12]
14. "The 'estimated wait time' for Customs processing (non-commercial travellers) at the North Portal border crossing between North Portal, Saskatchewan, and Portal, North Dakota, is 'No Delay.'" [13]
15. "For sections of Grade 12 calculus taught in British Columbia, the curriculum component 'Functions, Graphs, and Limits' comprises between '10–15' percent of the instructional time."[14]
16. "An estimated 95% of the visitors [to the Trade Show] register online. . . ."[15]

Basic Concepts 10.2

17. Toward preparing a budget, the Langston family randomly sampled 35 long-distance telephone calls that they had made and recorded the durations of these calls. For the sample, the mean duration of the calls was 11.4 minutes. For 95% confidence, the margin of error was found to be 1.36 minutes. Estimate the confidence interval for the mean duration of the family's long-distance calls.
18. For a random sample of 40 customers, a bank teller spent 3.5 minutes per customer, on average. Suppose the standard deviation of the teller's time spent per customer has been previously determined to be 0.5 minutes. What critical z-value can be used for constructing a 95% confidence interval for mean time per customer? What is the standard error for the mean time per customer?
19. For Exercise 18, what would the critical z-value be if:
a) The confidence level had been 90%?
b) The degree of confidence had been 98%?
c) The sample size had been 50?
20. For Exercise 18, find (i) the standard error and (ii) the margin of error if:
a) The confidence level had been 90% ($n = 40$).
b) The degree of confidence had been 98% ($n = 40$).
c) The confidence level had been 90% and the sample size was equal to 50.
21. For Exercise 18, find the confidence interval if:
a) The confidence level had been 90% ($n = 40$).
b) The degree of confidence had been 98% ($n = 40$).
c) The confidence level had been 90% and the sample size was equal to 50.
22. To construct a 92% confidence interval of the mean rod length, a random sample of 45 steel rods is taken from the output produced by Sutton Steel Supplies. The mean rod length for the sample is 300 cm. Suppose that σ is known to equal 25 cm. Find the:
a) Standard error for the sample mean.
b) Critical value for constructing the confidence interval.
c) Margin of error for constructing the confidence interval.
d) Lower and upper boundaries of the confidence interval.
23. When solving Exercise 22, do we need to know the exact distribution of the population of rod lengths? Explain.
24. In a random sample of seven weeks, Dorrie's Drugs sold an average of 340 packages of decongestants per week. The distribution of this variable is normal, and the standard

deviation is 39 packages. Estimate the 95% confidence interval for the mean number of packages that Dorrie's sells per week.

25. In Exercise 24, why was it necessary to mention that the distribution of sales was normal?

Applications 10.2

26. A criminologist is studying the lengths of jail sentences given in Canada. She found that in 2001/2002, the standard deviation for length of sentence was 9.75 months. She randomly selects 40 of the sentences and finds that their mean length is 4.05 months.[16] Find the:
    a) Standard error for the sample mean.
    b) Critical value for constructing the 94% confidence interval.
    c) Margin of error for constructing the 94% confidence interval.
    d) Lower and upper boundaries of the 94% confidence interval.

27. The Yukon government is concerned about the losses of wildland to forest fires each year. Numbers of hectares burned each year in the Yukon since 1950 are recorded in the file **Yukon_Wildfires.**
    a) Presume that these data are representative of annual data that could be expected into the future, and that σ for the distribution of annual hectares burned is approximately 271,000. On those assumptions, estimate the 95% confidence interval for the mean number of hectares burned every year in the Yukon.
    b) In (a) we assumed that the nonrandom historical data are representative of the population of *all* years' fire losses (including the future) in the Yukon. Explain some problems with that assumption.
    c) In (a), the estimate for σ was actually based on the sample value s. For using a z-based critical value, is it valid to estimate σ in this way? Explain.

28. A movie fan is planning to buy a certain model of subwoofer for his home theatre and has found that it is available online. Experience shows that for the online prices, σ equals US$176. The movie fan has sampled 38 prices, in particular, for online offerings; the mean price for the sample is US$1,125.[17]
    a) Construct the 90% confidence interval for the mean of online prices.
    b) What is the critical z-value, and what is the margin of error for the online prices?
    c) Recalculate (a) and (b) for a sample size of $n = 60$.

29. Over 50 countries participate in the Program for International Student Assessment (PISA). A sample of 8 countries' composite student scores in reading, based on PISA, can be found in the file **Pisa_Results_International.** Presume that the population of national scores for reading is normally distributed, with a population standard deviation of 19.
    a) For constructing the 96% confidence interval for the mean national score in student reading, what is the critical z-value?
    b) What is the margin of error?
    c) What are the lower and upper bounds of the confidence interval?

30. Saskatchewan Highways and Transportation publishes analysis results on the composition of asphalt to assist in quality control of that product. Over a period of time, the standard deviation for the bulk density of coarse aggregate in Marshal Mix has been 0.013 g/cm^3. An inspector samples 45 batches of coarse aggregate produced by the supplier Aggie Asphalt, and finds that the sample mean for bulk density is 2.709 g/cm^3.[18]
    a) Construct the 95% confidence interval for the mean bulk density of coarse aggregate produced by Aggie Asphalt.
    b) For completing (a), where did you get a value for σ? Is it possible that the σ-value you used does not apply for the output of this particular supplier? Explain why or why not.

31. Saskatchewan Highways and Transportation (see Exercise 30) also tests asphalt for water absorption of fine aggregate. Over a period of time, the standard deviation for this variable has been 0.30%. An inspector now samples 35 batches of fine aggregate produced by Aggie Asphalt. The sample mean is 1.312%.[19] Construct the 92% confidence interval for the mean water absorption of fine aggregate produced by Aggie Asphalt.

32. In 2002, the Tolly Group published a comparative evaluation of the audio product OCT6100 Series Echo Canceller. Over repeated testing for perceptual evaluation of speech quality (PESQ), the OCT6100 obtained a mean score of 3.43. (0 is poor and 5 is excellent). The standard deviation for the measurements was 0.26.[20] Suppose that a competing evaluation company (Golly Group) sampled 50 OCT6100s that same year and obtained a sample mean for PESQ of 3.30.
    a) Based on the Golly Group sample and using the sigma value reported by the Tolly Group, construct a 95% confidence interval for PESQ.
    b) Does the confidence interval in (a) contain the sample mean originally reported by the Tolly Group? What might your answer suggest about the populations sampled by the two evaluation groups? Explain.

Basic Concepts 10.3

33. Toward preparing a budget, the Langston family randomly sampled 35 long-distance telephone calls made by family members. The sample mean for call duration was 11.4 minutes and the standard deviation was 4.1 minutes.

a) For the family's long-distance calls, construct the 95% confidence interval for the mean duration.
b) If n had been 40, instead of 35, would the interval estimate be wider or narrower than in (a)? Explain.

34. For a random sample of 40 customers, a bank teller spent 3.5 minutes per customer, on average, with a standard deviation of 0.5 minutes per customer.
a) What critical t-value can be used for constructing a 95% confidence interval for mean time per customer?
b) What are the boundaries of the 95% confidence interval?
c) In Exercise 18, 0.5 minutes was given as σ for the time per customer, rather than as the *sample* s for time per customer. Which approach generates a wider confidence interval? Why?

35. A camper is unhappy with the motorboat sounds on the campground's lake. To prepare a complaint to the campground owner, the camper randomly samples 15 quarter-hour blocks of time throughout a day. The mean duration of motorboat sounds per quarter-hour is 9.3 minutes, with a standard deviation of 3.4 minutes.
a) Based on these data, construct a 94% confidence interval for the mean duration of motorboat sounds per quarter-hour.
b) To calculate (a), what assumption was needed about the distribution of the variable: sound durations per quarter-hour?
c) For (a), what is the critical t-value? What is the margin of error?

36. A bored outfielder in a sandlot baseball game observes a yellow plant growing in clusters on the field. She mentally divides the outfield into 30 equal sections and randomly samples 9 of the sections. For the sample, the mean number of plants per cluster is 11 and the standard deviation is 3. The distribution appears to be normal.
a) What is standard error for the mean number of yellow plants per cluster on the outfield? (Hint: Remember to use the finite population correction factor.)
b) Based on your answer to (a), construct the 95% confidence interval for the mean number of yellow plants per cluster.

37. A tire dealer is conducting research using vehicles parked at a large mall. He randomly selects 40 of the tires on parked vehicles and measures the minimum tread depth on the selected tires. The table below shows the results (in cm).

0.43	0.32	0.42	0.41	0.39	0.32	0.41	0.30
0.35	0.35	0.28	0.39	0.36	0.37	0.29	0.35
0.35	0.32	0.29	0.29	0.36	0.35	0.30	0.29
0.42	0.36	0.30	0.33	0.31	0.40	0.41	0.42
0.37	0.29	0.33	0.40	0.33	0.41	0.37	0.39

For all of the tires on parked vehicles on the lot, construct a 95% confidence interval for the mean of the minimum tread depths.

38. To construct a 92% confidence interval of the mean rod length, the table below records a sample of 15 steel rods taken from the output produced by Sutton Steel Supplies. The rod lengths are normally distributed. The results are in centimetres.

294.1	345.0	267.1	294.7	298.6	281.3	308.0
308.6	266.0	300.8	313.3	318.0	266.3	332.3
310.6						

Find the:
a) Standard error for the sample mean.
b) Critical value for constructing the confidence interval.
c) Margin of error for constructing the confidence interval.
d) Lower and upper boundaries of the confidence interval.

39. The number of packages of decongestants sold at Dorrie's Drugs for each of seven randomly selected weeks is shown below.

369 342 251 291 329 359 314

a) Assuming that the distribution of weekly sales numbers is normal, construct the 90% confidence interval for the mean number of packages that Dorrie's sells per week.
b) Suppose that, in fact, the distribution of weekly sales numbers is highly skewed to the right. Is the confidence interval calculated in (a) likely to be too small or too large? Explain.

40. The tube pressure in a generating station is monitored each hour. A random sample of 10 of the hourly pressure readings, in pounds per square inch (psi), is shown below.

11,167	11,598	11,481	10,007	10,790	10,871
11,970	11,461	9,784	11,008		

Assuming that the pressure readings are normally distributed, construct a 95% confidence interval for the mean pressure in the tube.

41. Suppose that all 10 pressure readings in Exercise 40 were selected from 24 hourly readings taken in the course of a single day. Construct a 95% confidence interval for the mean pressure in the tube for that one day.

Applications 10.3

42. A sample of 8 countries' composite student scores in math, based on the Program for International Student Assessment (PISA), can be found in the file **Pisa_Results_International.** Presume that the population of national scores in math is normally distributed.
a) For constructing the 96% confidence interval for the mean national math score, what is the critical value?
b) What is the margin of error?
c) What are the lower and upper bounds of the confidence interval?

43. In the year that the national PISA scores were sampled (see Exercise 42), only 41 countries participated in the project. Utilizing this additional information, answer again all parts of the previous exercise.

44. Saskatchewan Highways and Transportation publishes analysis results on the composition of asphalt to assist in quality control of that product. An inspector samples 45 batches of coarse aggregate produced by the supplier Aggie Asphalt. The sample mean for bulk density is 2.709 g/cm^3, and the sample standard deviation is 0.013 g/cm^3.[21]
   a) Construct the 95% confidence interval for the mean bulk density of coarse aggregate produced by Aggie Asphalt.
   b) Compare your answer to (a) with your answer to Exercise 30(a). Explain why the answers are not exactly the same.

45. An inspector for Saskatchewan Highways and Transportation is testing asphalt for water absorption of fine aggregate. In a sample of 40 batches of fine aggregate produced by Aggie Asphalt, the sample mean is 1.299% and the standard deviation is 0.30%.[22] Construct the 95% confidence interval for water absorption of fine aggregate produced by Aggie Asphalt.

46. In research conducted by the Promocycle Foundation in Quebec, the 2001 sport model Honda CBR929RR bicycle was tested for its ability to stop, based on hard braking from an initial speed of 100 km/h. For 214 such tests, the mean stopping distance was 41.67 metres.[23] Supposing the standard deviation for stopping distance is 7.5 metres, construct a 95% confidence interval for the mean stopping distance of this bicycle under test conditions. What is the margin of error?

47. The data in the file **Marathons_Selection** are a random unsorted selection from the best recorded completion times for the men's marathon as of June 27, 2006. Take as a sample the first 100 decimal completion times ("DecTime") in the file. Based on that data, construct the 95% confidence interval for the mean marathon completion time for all top marathoners in the original list.

48. In the source data from which the **Marathons_Selection** data are taken, the total number of best completion times on record is 1,131. Utilizing this additional information, solve Exercise 47 again.

Where appropriate, do the following exercises using the normal approximation for the binomial. If instead your instructor requests that you use the binomial method, follow the procedures in "In Brief 10.3." Due to the different approximations of the two methods, your final solutions may vary slightly from the author's solutions.

Basic Concepts 10.4

49. Suppose you plan to construct confidence intervals for the proportion. As a preliminary step, given the sample sizes n, numbers of successes X, and degrees of confidence that are shown below, find for each case: (i) the point estimate for the proportion and (ii) the margin of error.

a) $n = 2{,}000$	$X = 450$	90%
b) $n = 2{,}000$	$X = 450$	95%
c) $n = 694$	$X = 552$	95%
d) $n = 1{,}870$	$X = 500$	96%
e) $n = 4{,}850$	$X = 4{,}049$	90%

50. For each of the cases shown in Exercise 49, construct the confidence interval estimate.

51. Of 75 cats of a certain breed that are sampled, 7% have kinked tails.
   a) Construct the 95% confidence interval for the population proportion.
   b) What is the margin of error for estimating the population proportion?

52. A quality consultant for Brake King Car Repairs is concerned about how many repairs need to be redone. She randomly selects 67 of their old repair records and finds that 6 of these repairs needed to be redone.
   a) What is the point estimate for the population proportion of repairs that need to be redone?
   b) What is the margin of error for estimating the population proportion with 90% confidence?
   c) Construct the 90% confidence interval for the population proportion.

53. In a random sample of 200 students at your institution, 22 take bus route "B."
   a) What is the point estimate for the proportion of students, overall, who take bus route "B"?
   b) Construct the 99% confidence interval for the population proportion.
   c) What is the margin of error for estimating the population proportion?

54. In a random sample of 55 days for which the Weather Channel predicted rain, it actually rained on only 44 of those days.
   a) Construct a 95% confidence interval for the proportion of days that it actually rains when the Weather Channel has predicted rain.
   b) Suppose an advertisement claims that the channel is accurate 85% of the time when it predicts rain. Is the advertised claim consistent with your findings in (a)? Explain.

55. Government auditors randomly selected 60 expense reports submitted by Y. Lee Consultants with its tax return. 15% of the reports were found to be criminally inflated.

Construct a 95% confidence interval for the percentage of reports submitted by Y. Lee Consultants that are criminally inflated.

56. A poll predicted that candidate Sherry Brewster would win 35% of the votes cast in her riding in the next election. The margin of error for 95% confidence was reported to be ±2%. In the actual election, Brewster won 1,267 of the 3,454 votes cast. Were the published poll results (a) correct or (b) mistaken? Explain.

57. Of 50 students who enroll for beginner's lessons at Sarah's Swim School, 6 eventually attain a Level 10 certificate. If this pattern is representative for all of her students:
    a) Construct a 92% confidence interval for the proportion of Sarah's students who go on to attain a Level 10 certificate.
    b) What is the margin of error?

Applications 10.4

58. A large bank polled 2,000 customers who have access to the Internet. Of these, 1,210 did their banking online.[24]
    a) What is the point estimate for the proportion of the bank's Internet-linked customers who do their banking online?
    b) Construct a 95% confidence interval for the population proportion.
    c) A similar survey among Internet-linked bank customers in general was conducted by Statistics Canada. In its survey, the proportion that banked online was 58%. Explain two reasons why the bank's proportion for the variable may be different from your answer to (a).

59. As part of a study on humpback whales, an observation post was established on the roof of the Hyatt Hotel in Maui, Hawaii. On 40 randomly selected observation days when no parasail boats were nearby, a pod (i.e., group) of whales could be observed on 32 of those days.[25]
    a) Considering only days when no parasail boats are nearby, what is the point estimate for the proportion of days when a pod of whales is observable from the roof of the Hyatt Hotel?
    b) Construct a 93% confidence interval for the proportion.
    c) Assuming no parasail boats are nearby, would it be unusual to observe a pod of whales on 9 of the next 10 days from the top of the Hyatt Hotel? Explain.

60. In a survey of 1,000 adult Torontonians, 500 reported being very concerned about acid rain as an environmental issue.[26]
    a) Give the point estimate for the proportion of adult Torontonians who have this concern.
    b) Construct a 95% confidence interval for the proportion.

61. In the same survey of 1,000 adult Torontonians (see Exercise 60), 750 said they were very concerned about air pollution as an environmental issue, and 730 said they were very concerned about the quality of drinking water.[27]
    a) Construct a 95% confidence interval for the proportion of adult Torontonians who are very concerned about air pollution.
    b) Construct a 95% confidence interval for the proportion of adult Torontonians who are very concerned about the quality of drinking water.
    c) Based on (a) and (b), is it possible that, in the population, *more* adult Torontonians are concerned about the quality of drinking water than about air pollution? Explain.

62. An economist randomly sampled 1,000 Canadians in the labour force and found that only 393 of them were covered by registered pension plans (RPP).[28]
    a) Give the point estimate for the proportion of Canadians in the labour force who were covered by RPPs.
    b) Construct the 98% confidence interval for the proportion.

63. In a study that sampled about 2,200 male adults, the U.S. Centers for Disease Control found that about 1,560 of the men were overweight.[29]
    a) Give the point estimate for the proportion of American adult males who were overweight.
    b) Construct the 95% confidence interval for the proportion.

64. In the same study (see Exercise 63), it was found that about 1,365 of 2,200 adult American females surveyed were overweight.[30]
    a) Construct the 95% confidence interval for the proportion of adult American females who were overweight.
    b) Based on your answers to Exercises 63(b) and 64(a), is it (i) reasonably likely, (ii) possible but not very likely, or (iii) virtually impossible that the proportions of adult American males and adult American females who were overweight are really about the same? Explain.

There are no separate exercises for Section 10.5.

Review Applications

65. Health Canada reports that under standard test conditions, the standard deviation for formaldehyde content in direct cigarette smoke is 32.8 μg (micrograms).[31] Suppose that 50 cigarettes of various brands are randomly selected and subjected to the test for formaldehyde content in smoke. The sample mean is 45.1 μg per cigarette. You plan to use a normal model to construct the 95% confidence interval for the mean formaldehyde content in the population.
    a) What is the critical z-value?
    b) What is the standard error for the mean?
    c) What is the margin of error?
    d) What are the lower and upper bounds of the confidence interval?

66. Recalculate your solutions to Exercise 65, but assume that the sample size is $n = 20$. What additional assumption must be made in order for your calculations to be valid? Explain.

67. The standard deviation for the calcium content in the Rideau River can be estimated at approximately 3 mg/L.[32] Suppose that in 100 samples taken randomly from the river, the mean calcium content is 18.5 mg/L. Construct a 95% confidence interval for the mean calcium content in the Rideau River.

68. In a follow-up study (see Exercise 67) the Rideau River was independently sampled from two different locations—Smiths Falls to Burritts Rapids ($n = 50$), and Burritts Rapids to Long Island ($n = 65$). In these samples, the mean calcium contents were 18.2 mg/L and 19.1 mg/L, respectively.[33] Use the original standard deviation estimate (3 mg/L).
   a) Construct 95% confidence intervals for the mean calcium contents in each of these two sections of the Rideau River.
   b) Based on your answers to (a), is it (i) reasonably likely, (ii) possible but not very likely, or (iii) very unlikely that the mean calcium contents are really about the same in both of these parts of the Rideau River? Explain.

Exercises 69 and 70 are based on the following table, which shows the standard deviations of laboratory attenuation values of foam earplugs, under specified conditions, according to U.S. Occupational Safety and Health Administration testing.

Test Frequency (Hz)	125	250	500	1,000
Standard Deviation (decibels)	7.3	6.3	7.3	6.4

Test Frequency (Hz)	2,000	4,000	8,000
Standard Deviation (decibels)	5.3	5.0	5.4

Source: Safe@Work[SM], Inc. Based on National Institute for Occupational Safety and Health data in *New Developments in Hearing Protection Labeling*. Found at: http://www.safe-at-work.com/Reference/OSHA-HP-Labeling.html

69. For workers in loud environments, the level of protection provided by foam earplugs—and the variance in the levels of protection—varies, based on the frequency of the sound. Assume that the parameters in the above table represent the whole population of earplugs of a given type. An inspector randomly selects 100 such earplugs and tests them for attenuation (sound reduction) at the frequency 125 Hz. The mean attenuation of sound for the sample is 15.8 decibels (dB).
   a) Construct a 95% confidence interval for the mean attenuation provided by the earplugs at this frequency.
   b) Suppose that a government department's standard for earplugs at this frequency is a mean attenuation of 17.9 dB. Based on your answers to (a), is it (i) reasonably likely, (ii) possible but not very likely, or (iii) very unlikely that the government department's standard is being met? Explain.

70. a) Repeat Exercise 69(a), but for a frequency equal to 250 Hz and a sample mean attenuation = 19.4 dB.
   b) Repeat Exercise 69(b), based on your answer to 70(a), presuming that the government department's standard at this frequency is for a mean attenuation of 19.0 dB.

71. The standard deviation of "normal" eye pressure is about 2.5 mmHg (millimetres of mercury).[34] A technician randomly selects 40 individuals to measure their eye pressure and obtains a sample mean of 15.7 mmHg. He plans to construct the 95% confidence interval for the population mean.
   a) What is the critical z-value?
   b) What is the standard error for the mean?
   c) What is the margin of error?
   d) What are the lower and upper bounds of the confidence interval?

72. Recalculate your solutions to Exercise 71, but assume that the sample size is $n = 60$.

73. A ski enthusiast wants to know (with 95% confidence) how much snowfall occurs each January or February near Iqaluit in Nunavut. Drawing from many years of records for those months, she obtained the following random sample of snowfall amounts (in centimetres).[35]

   19.2 22.6 76.2 28.8 13.0 8.2

   a) What is the critical t-value?
   b) What is the standard error for the mean?
   c) What is the margin of error?
   d) What are the lower and upper bounds of the confidence interval?

74. Recalculate your solutions to Exercise 73, but use the following larger sample (in centimetres).[36]

   45.7 17.0 76.5 47.8 11.8 44.7 36.8 8.4
   12.3 27.8 27.2 3.6

75. A new investor in Latin America and the Caribbean is examining the size of those markets. Obtaining data from the Economic Commission for Latin America and the Caribbean, he has selected this sample of populations (in 1,000s) from countries in the region.[37]

228	700	147	322	4	103	427	347
12	915	100	151	947	266	321	79
152	80	74	22	442	215	369	577
39	875	121	597	113	592		

For constructing a 95% confidence interval of the mean population in the region:
   a) What is the critical t-value?
   b) What is the standard error for the mean?
   c) What is the margin of error?
   d) What are the lower and upper bounds of the 95% confidence interval?

76. Recalculate your solutions to Exercise 75, but adjust for the fact that there are only 40 countries in the region from which the investor sampled.

77. With the exception of one outlier, the annual numbers of first marriages in Sweden by males, 50 years or older, are quite normally distributed. A marriage broker has randomly selected the following sample of annual numbers for such marriages. Construct a 90% confidence interval for the mean number of such marriages each year in Sweden.[38]

342 333 330 298 385 458 406

78. Suppose the interest of the marriage broker (see Exercise 77) is limited to an historical period of only 17 years. Recalculate the mean for the annual number of such marriages over that 17-year period.

79. A new resident of Quebec is exploring the popular lottery game Banco. Each play, some combination of 20 possible numbers is drawn. Examining records for draws over several years, she counted the number of times that any particular number was drawn as part of the winning combination. She randomly sampled six of the values for that variable.[39]

612 638 650 620 655 668

Construct a 95% confidence interval for the mean number of times that any particular number was drawn as part of the winning combination over the period being studied.

Exercises 80–83 are based on the following table, which shows the numbers of frogs captured in different aspects of a study, and how many of these were "recaptures"(i.e., they had been captured and marked at that site at a previous time and then released). Assume that, in all cases, the recapture percentages would be consistent if future similar studies were conducted.

Frogs in Study That Were Recaptures

	Breeding Habitats		Non-Breeding Habitats	
	n	Recaptures	*n*	Recaptures
Juvenile	177	46	69	8
Mature male	615	406	32	3
Mature female	339	190	219	116

Source: David S. Pilliod, Charles R. Peterson, and Peter I. Ritson. *Seasonal migration of Columbia spotted frogs (Rana luteiventris) among complementary resources in a high mountain basin.* Found at: http://article.pubs.nrc-cnrc.gc.ca/ppv/RPViewDoc?_handler_=HandleInitialGet&journal=cjz&volume=80&calyLang=eng&articleFile=z02-175.pdf

80. As part of a study on the seasonal migration of Columbia spotted frogs in a high mountain basin, the frogs were captured and released at various sites and times. As shown in the table, many of the captured frogs were found to be "recaptures." Construct a 95% confidence interval for the proportion of juveniles caught in breeding habitats that turn out to be recaptures.

81. Construct a 95% confidence interval for the proportion of juveniles caught in non-breeding habitats that turn out to be recaptures. Compare your solution for this exercise with the solution for Exercise 80. Allowing for sampling error, do you think it is likely that the true population proportions for the two cases are about the same? Why or why not?

82. From the table "Frogs in Study That Were Recaptures," calculate the *total* number of frogs caught in breeding habitats and the total number of these frogs that were recaptures. Construct a 95% confidence interval for the proportion of all frogs caught in breeding habitats that are recaptures.

83. Construct a 95% confidence interval for the proportion of all frogs caught in non-breeding habitats that were recaptures. Compare the solutions for Exercises 82 and 83. Allowing for sampling error, do you think it is likely that the true population proportions for the two cases are about the same? Why or why not?

Exercises 84–86 are based on the following table.

Commercial Driver Fatigue Management Study: Selected Results

Driver reports needing at least 7 hours sleep per night.	60%
Driver reports regularly getting less than [his or her] perceived ideal amount of sleep per night.	71%
Driver reports making several mistakes or mental errors per year due to fatigue.	84%

Source: Canadian Sleep Institute. Based on data in *North American Pilot Fatigue Management Program (FMP) for Commercial Motor Carriers (Alberta Results)*. Found at: http://www.csisleep.com/pdf/TH%20AMERICAN%20PILOT%20FATIGUE%20MANAGEMENT%20PROGRAM%20FOR%20COMMERCIAL%20MOTOR%20CARRIERS%20.pdf

84. Driver fatigue is a known hazard for drivers, both commercial and private. The above table is derived from a study of drivers of commercial motor carriers in Alberta. Suppose that the proportions in the table are based on the responses of 75 randomly selected commercial drivers in that province.

a) How many of the sampled drivers reported needing at least 7 hours of sleep?

b) What is the margin of error for estimating the population proportion for this variable (with 95% confidence)?

85. a) In the commercial driver fatigue management study, how many of the sampled drivers reported making several mistakes or mental errors per year due to fatigue?

b) Construct a 92% confidence interval for the true population proportion.

86. For the commercial driver fatigue management study, construct a 98% confidence interval for the true population of Albertan commercial drivers who (if asked) would report making several mistakes or mental errors per year due to fatigue.

Exercises 87–89 are based on data in the file **SmallCaps**, from a report on the annual performance of Canada's top small-capitalization companies.

87. An investor in small-capitalization, publicly traded Canadian companies is examining the data in *The Globe and Mail*'s "Report on Business" list of the top 200 companies in this category. She randomly samples 22 of those companies for a closer look. What percentage of the sample decreased in revenue (coded "0" in the file) compared to the previous year? If the sample is representative of the population, estimate (with degree of confidence 95%) the proportion in the population of these firms that decreased in revenue.

88. The values for assets in the sample are coded "0" for firms whose assets decreased in the past year. Projecting to the population, construct a 95% confidence interval for the population proportion of firms that decreased in assets over the year.

89. The analyst is starting to doubt whether the rich always do get richer—at least, by the measure of increasing assets year to year. Presume that the top 11 rows in the data file are representative of the higher-ranked half of companies in the "Report on Business" list and the bottom 11 rows in the data file are representative of the lower-ranked companies in the list.

a) Based on the top 11 values for assets in the sample, construct a confidence interval (95%) for the proportion among the higher-ranked companies that increased in assets over the past year.

b) Based on the bottom 11 values for assets in the sample, construct a confidence interval (95%) for the proportion among the lower-ranked companies that increased in assets over the past year.

c) Based on your answers to (a) and (b), do you think it is likely that the true proportion of the higher-ranked companies in the data set that increased in assets is higher than the true proportion of the lower-ranked companies in the data set that increased in assets? Explain your answer.

PART 5

Tests for Statistical Significance

Chapter

One-Sample Tests of Significance

11

• Chapter Outline

• Learning Objectives

11.1 Describe and apply the core concepts, terminology, and steps for implementing hypothesis tests by either a critical value or *p*-value approach.

11.2 Conduct a *z*-test for the population mean and a *t*-test for the population mean, and determine which of these tests is suitable for a given problem.

11.3 Describe some key reasons why the sample size—for hypothesis testing *or* for constructing confidence intervals—may need to be increased, and calculate minimum needed sample sizes for constructing one-sample confidence intervals.

11.4 Conduct a hypothesis test for the population proportion based on the binomial distribution.

11.1 The Hypothesis-Testing Process

In inferential statistics, we make a generalization about a population parameter on the basis of data collected in a sample. Chapter 10 addressed one of two categories of inferential statistics—the use of sample data to make an *estimate* of a population parameter. There, we presumed at the outset that the population parameter was unknown. In this chapter, on the other hand, we address the second category of inferential statistics—the use of sample data to *test a hypothesis* about a population parameter. In this case, some hypothesis has been made in advance about the value of the population parameter. Through a process of hypothesis testing, we determine whether to reject or not to reject this initial hypothesis.

In general language, a **hypothesis** is a tentative proposal or statement about a fact. For example, to some people, a phenomenon of global warming is a good hypothesis to explain the receding of the ice cap in the Canadian Arctic. Others do not accept that this event is explained by that hypothesis. This suggests that a hypothesis does not stand on its own, but needs to be backed up, if possible, by some proven standard of evidence.

In this book, we distinguish between a **claim** and a hypothesis. People claim or assert all sorts of things for all kinds of reasons. At a restaurant, you may claim that food is too salty or undercooked. One newspaper writer may claim that global warming exists; another may claim that it does not exist. A claim, in this sense, is just part of normal discourse. However, in this book, we use the term *hypothesis* in a technical sense, because hypotheses play a very specific role in an important statistical method of reasoning. In this sense, a hypothesis is a statement or claim made specifically in order to *test* that statement or claim by the methods to be described below.

To get a sense of how the testing method works, consider the following example. Someone claims that the mean age of trees in the Statistical Forest is 15 years. If he puts that claim to the test, the hypothesis looks like this: $\mu = 15$. Thirty trees in the forest are randomly selected, and the sampled ages are as shown in Figure 11.1. As it happens, the mean age of the *sampled* trees is only 7.2 years. Knowing that the mean age of the sampled trees is so far removed from 15 years, is it likely that, as hypothesized, the population mean is really 15 years?

We cannot realistically expect that the mean of a randomly selected sample will be exactly equal to the population mean. But, taking sample size into consideration, it may be unlikely (i.e., statistically improbable) that a sample mean would be so far removed from the hypothesized population mean as in the tree-age example. In that case, we would reject the hypothesis that was put forward. Otherwise, the hypothesis stands.

Taking the steps to evaluate a formally stated hypothesis in the manner just described is called **hypothesis testing.** The exact procedures can vary, based on a number of factors. For example, one population parameter (e.g., the mean, median, or variance) may be the focus of the hypothesis. Or a comparison of parameters for more than one population (e.g., the difference between two population means) may be the focus. But whatever the variations, all hypothesis tests are based on a common set of core concepts and follow a very similar sequence of main steps.

The Core Concepts for Hypothesis Tests

Null and Alternative Hypotheses

In order to conduct a hypothesis test, we need to reexpress our claim or belief formally as a hypothesis. Actually, there are *two* hypotheses required for hypothesis testing: the **null**

Figure 11.1
Sampled Ages of Trees

Photo credit: Shutterstock/Valenta

hypothesis (represented as H_0) and an **alternative hypothesis** (represented as H_1 or sometimes by H_a). H_0 and H_1 are logical opposites, meaning that if the null hypothesis is rejected, then the alternative hypothesis has effectively been accepted or supported. For the example about the mean age of trees in a forest, we can represent the two hypotheses as

Null hypothesis:	$H_0: \mu = 15$
Alternative hypothesis:	$H_1: \mu \neq 15$

In the above, why did we assign these two hypotheses in this order? That is, why could H_1 not be $\mu = 15$? The simplest answer is that:

- The null hypothesis must always include the concept of equality. That is, the logical operator must be $=$ or $\leq$ or $\geq$.
- The operator for the alternative hypothesis must always express the logical opposite of the operator used for the null hypothesis. That is:
  - If the null expresses $=$, the alternative expresses $\neq$.
  - If the null expresses $\leq$, the alternative expresses $>$.
  - If the null expresses $\geq$, the alternative expresses $<$.

The above rules are needed because each hypothesis, H_0 and H_1, plays a distinct role in the hypothesis-testing process. The null hypothesis is the direct focus of the hypothesis test: It asserts that a parameter equals some value. The results of sampling are either consistent with that hypothesis or else so widely inconsistent with it that we reject the null hypothesis.

Compare the principle of "innocent until proven guilty": H_0 is like the presumption of innocence in a trial. It is the hypothesis that we assume to be true, and we reject the null hypothesis only if we have sufficient evidence against it. If we do reject H_0, then H_1 (as its logical opposite) is the position supported by the outcome of the test. Whatever our decision, it should be based on the evidence provided by the data. The method that investigates what the data tell us on the matter is the hypothesis test.

Some books encourage (or require) that if you make a claim, it should take the structure of an alternative hypothesis (i.e., use an inequality operator). This strategy is useful if you are planning to conduct a hypothesis test and, if successful, to publish a statement that says the test supports your claim. If instead your claim is based on equality, which has the structure of the null hypothesis, it is more difficult to make an impressive headline. Even if H_0 has not been rejected, you cannot say that it has been supported, but only that "there is not sufficient evidence to reject the claim." Nonetheless, there is no rule about how to make a claim. What is important is to know what structure of claim is suitable to be an H_0 and what structure takes the form of H_1.

The Critical Region of the Hypothesized Sampling Distribution

In the tree-age example, we reject H_0 (that the mean age is 15 years) if the mean of the sample (7.2 years) is very far from the proposed population parameter. How far is "very far"? Far enough to be statistically improbable, if indeed H_0 is true. This probability assessment can be calculated, based on the sampling distribution of the sampling statistic.

Chapter 9 describes how to calculate the sampling distribution if we know the relevant population parameters. In the tree-age example, the parameter μ is not actually *known*, but for the sake of the test, we have hypothesized its value. Suppose, for the simplest case, that σ for the population is also known; then, we can calculate the mean and standard error for the sampling distribution as in Section 9.1. On the basis of this normal sampling distribution, the probability of a random sample's mean reaching any particular distance (in standard errors) from the population mean can be calculated. The **critical region** refers to that part (or parts) of the sampling distribution where the sample mean *could* (with a very small chance) occur, but with a probability too low to be considered likely in practice. We will reject the null hypothesis if the sample statistic falls within that critical region.

The concept of the critical region is illustrated in Figure 11.2. Each of the curves represents the sampling distribution of the sample mean, based on particular values for the sample size and for σ, and centred around the "equal" value for μ given in H_0. In the tree-age example, the H_0 operator is the = sign, so the critical region consists of two parts, as shown in curve 3 of the figure. The test is considered be "two-tail" because if either a very large or a very small value is obtained for the sample mean, this would fall in the critical region; H_0 would be rejected. Curves 1 and 2 in the figure correspond to one-tail tests, meaning that the critical region for rejecting H_0 occurs on only one side, or tail, of the distribution.

If someone had claimed that the mean tree age is "at least 15 years," then the first curve would apply. A sample mean falls in the critical region only if its value is much smaller than 15. Lastly, in curve two, the null hypothesis is that the mean age is at most 15 years. To fall in the critical region, a sample mean has to be much larger than 15.

Figure 11.2
Critical Region of a Sampling Distribution

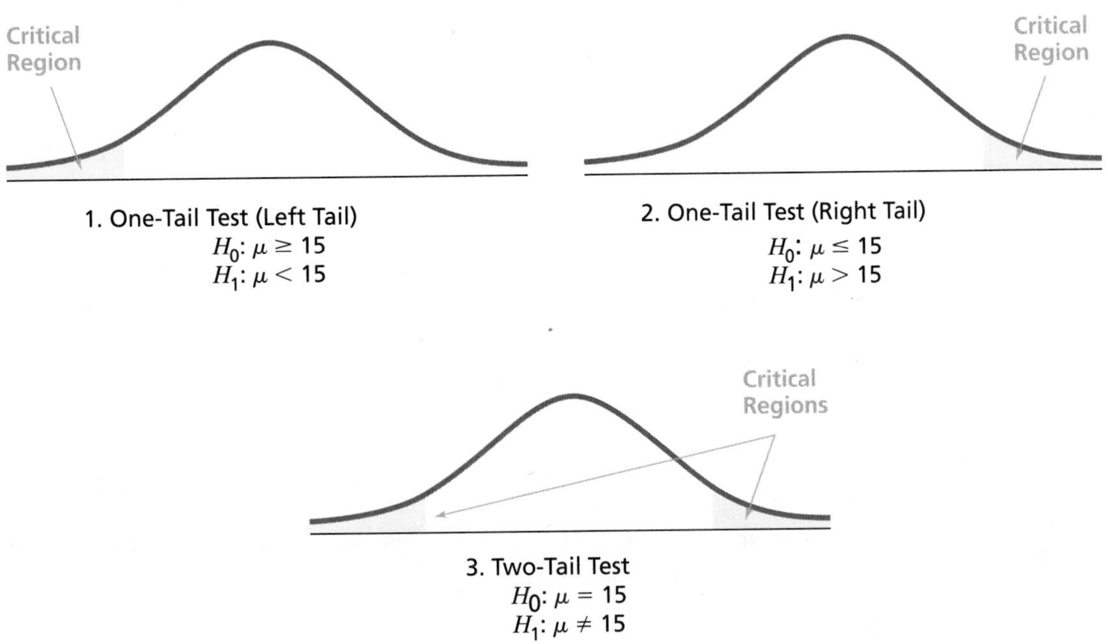

The Level of Significance, Critical Value(s), and Test Statistic

In our example, the sample mean was equal to 7.2 years, which was 7.8 years less than the hypothesized population mean. By itself, this information is not sufficient to tell us if the sample mean is small enough to belong in the critical region—that is, small enough to reject the null hypothesis. In the traditional approach to hypothesis testing, three additional details are required. (Later in this chapter, an alternative approach is introduced.)

First we need to know the **level of significance** of the hypothesis test. This value, symbolized by α, quantifies how unlikely (expressed as a probability) a sample statistic needs to be to trigger the rejection of the null hypothesis. Formally, α gives the probability of committing a **Type I error** (mistakenly rejecting a correct null hypothesis) when you conduct the hypothesis test (the various types of possible errors are discussed in detail later in this chapter). If H_0 happens to be true, there is a small chance that the sample mean will fall in the critical region. (See Figure 11.3.) If by chance this occurs, then when you correctly follow the test procedure and reject the null hypothesis, you will unwittingly make an error. The probability α is the probability of making this error.

Typical values for α are 0.01 and 0.05, and sometimes 0.10. The closer α is set to zero, the smaller the chance of making a Type I error. However, if α is very small, it is also more difficult to reject a null hypothesis, since the critical region becomes very small as well; in practical terms, this can be a disadvantage, as discussed later. In any case, it is important that the researcher set the significance level *before* undertaking the hypothesis test. You should be prepared to state and defend your reasons for choosing the level that you do. If someone conducts a test and *then* resets α later to ensure that H_0 is rejected (or not rejected), this compromises the objectivity of the testing method.

As shown in Figure 11.3, the boundaries that mark the beginning of the critical region are called the **critical values.** (For a one-tail test, there is only one critical value,

Figure 11.3
The Critical Region and Level of Significance

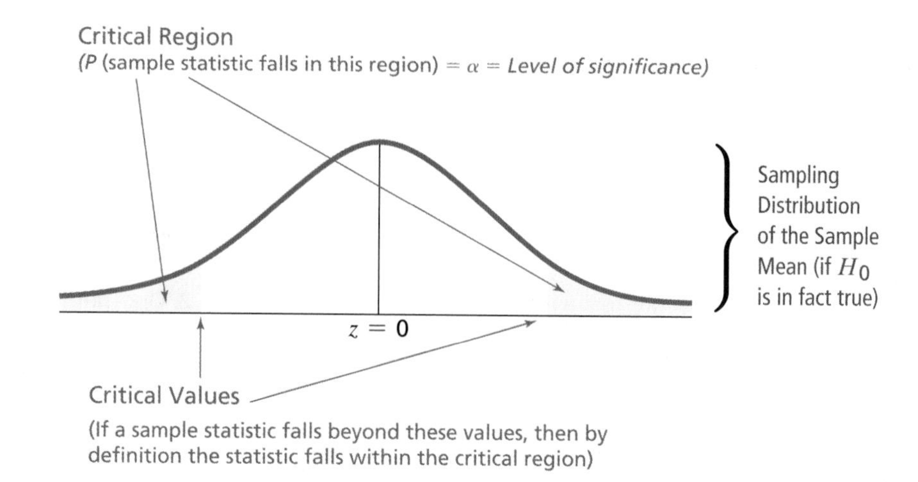

on the tail that has the critical region.) These values are given in standardized units, depending on the applicable model for the sampling distribution. For a large-sample test of the mean with a known σ, we mark the distances from the mean in z scores. If the population is normally distributed but σ is unknown, then the t-distribution is appropriate to model the sampling distribution (see Chapter 10); in that case, the critical values would be given as t scores. Similarly for other sampling distributions that might apply (F, χ^2, etc.), the critical values are relative to the appropriate curve.

The **test statistic** is used to determine whether the sampling result falls in the critical region. The sample mean (or the sample variance or other **sample statistic**) is converted into the appropriate standardized score for the relevant sampling distribution. This test statistic can be directly compared with the critical values to see if the sample statistic falls in the critical region.

Example 11.1

Biologist A claims that the mean age of trees in the Statistical Forest is *not* 15 years. The significance level of the test is 0.05. Thirty trees are sampled, and the sample mean of their ages is 7.2 years. Based on past data on this forest, σ for the population of tree ages is 4.58 years.

1. Write out H_0 and H_1 for testing the biologist's claim. Determine the critical values for the test and the test statistic. Is the test statistic in the critical region?
2. Biologist B claims that the mean age of trees in the forest is equal to 8 years. Retaining all other values from the original problem, redo the solutions for (1) in relation to B's claim.
3. Biologist C claims that the mean age of trees in the forest is *less than* 15 years, and takes a new sample of 30 trees; the sample mean is 13.2 years. For a significance level equal to 0.05 and σ equal to 4.58 years, redo the solutions for (1).

Solution

1. The operator for Biologist A's claim is "not equal" ($\mu \neq 15$). Since the operator does not contain equality, the claim is the alternative hypothesis in the testing process and its opposite is the null hypothesis:

$$H_0\!: \mu = 15 \quad H_1\!: \mu \neq 15$$

Because the sample is large and σ is known, we can apply the normal distribution to determine the critical values. The significance level 0.05 gives the proportion of the distribution that will be included in the critical region. The test is two-tailed (i.e., whether the mean is, or is not, equal to 15), so the critical region is divided into two parts—containing $0.05/2 = 0.025$ of the curve on each side. Using methods from previous chapters, we can calculate the value of $z_{critical}$ on each tail; it is the *z*-value such that the probability of selecting a sample whose mean falls beyond the critical value on that tail equals $\alpha/2$.

If the sample mean is 7.2, then the test statistic *z* equals:

$$\frac{7.2 - \mu}{\frac{\sigma}{\sqrt{n}}} = \frac{7.2 - 15}{\frac{4.58}{\sqrt{30}}} = -9.328$$

This test statistic falls in the critical region, since the value lies beyond the left critical value of -1.96.

2. The operator for Biologist B's claim is "equals" ($\mu = 8$). Since the operator *does* contain equality, this claim becomes the null hypothesis and its opposite becomes the alternative hypothesis:

$$H_0: \mu = 8 \quad H_1: \mu \neq 8$$

α is unchanged from solution 1 and the test is two-tailed, so the test has the same structure as in Figure 11.4(a). But, this time, because the sample mean is 7.2 and the null hypothesis for μ is 8, the test statistic *z* equals:

$$\frac{7.2 - \mu}{\frac{\sigma}{\sqrt{n}}} = \frac{7.2 - 8}{\frac{4.58}{\sqrt{30}}} = -0.957$$

This test statistic does not fall in the critical region, since the value does not lie beyond either the left critical value of -1.96 or the right critical value of $+1.96$.

3. The operator for Biologist C's claim is "less than" ($\mu < 15$). The operator does not contain equality, so it becomes the alternative hypothesis; its opposite becomes the null hypothesis:

$$H_0: \mu \geq 15 \quad H_1: \mu < 15$$

α is unchanged from solutions 1 and 2, but this test is one-tailed, so the sampling distribution under the null hypothesis is as shown in Figure 11.4(b). The full alpha (i.e., the probability of a Type I error) lies on the left side of the distribution. For a sample mean equal to 13.2 and the null hypothesis μ equal to 15, the test statistic *z* equals:

$$\frac{13.2 - \mu}{\frac{\sigma}{\sqrt{n}}} = \frac{13.2 - 15}{\frac{4.58}{\sqrt{30}}} = -2.153$$

This test statistic falls in the critical region, since the value is beyond the one critical value of -1.64.

Figure 11.4

Critical Values for the Hypothesis Tests

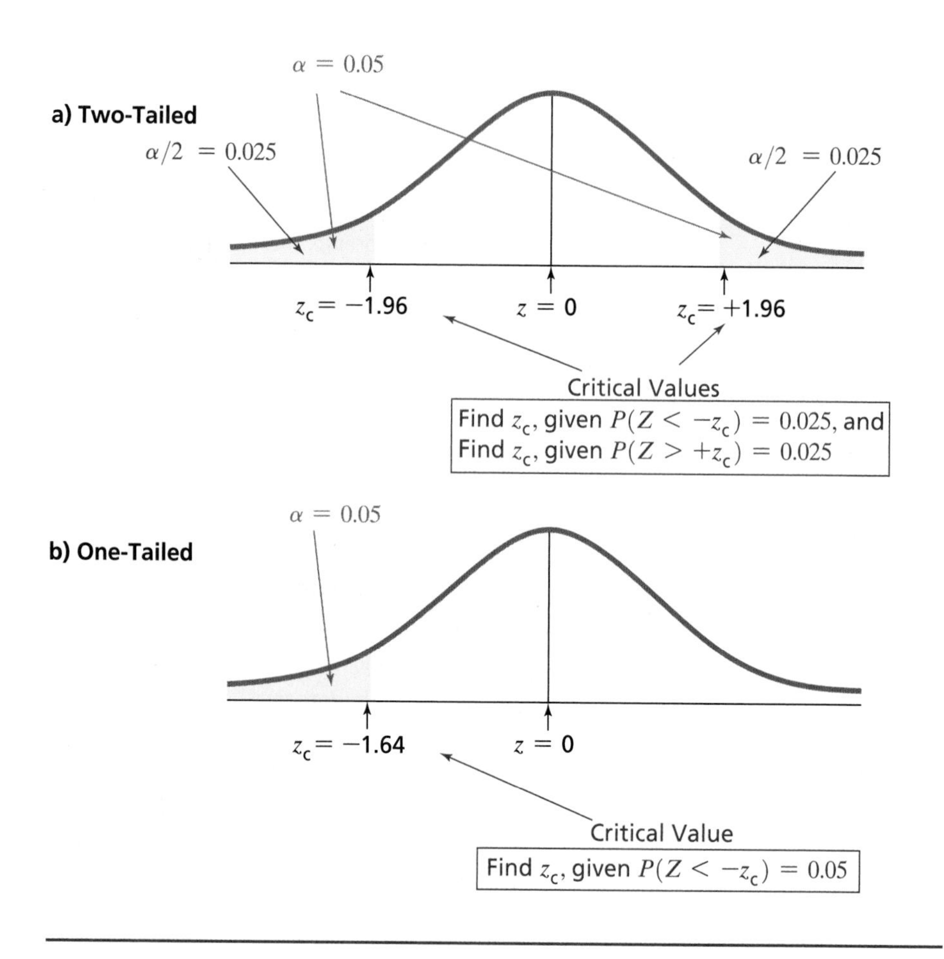

The Two Types of Errors

Since we rely on sample data to test the hypothesis, we cannot be absolutely sure that the conclusions are correct. Had we chosen a different sample, we might have reached a different conclusion. Even if we flawlessly follow the testing procedures, the reliance on sample data opens the door to possible error.

Four possible scenarios can occur when a hypothesis test has been conducted:

1. The null hypothesis (H_0) is in fact true, and we *did not reject* H_0.
2. The null hypothesis (H_0) is in fact false, and we *rejected* H_0.
3. The null hypothesis (H_0) is in fact true, but we mistakenly *rejected* H_0. (As mentioned previously, this is called a *Type I error.*)
4. The null hypothesis (H_0) is in fact false, but we mistakenly *failed to reject* H_0. (This is called a **Type II error.**)

The first two scenarios describe successful results: The hypothesis test has done its job. The last two scenarios signify that an error has occurred. Unfortunately, we can never be sure when this has happened.

The classic example to illustrate these two types of errors can be drawn from our legal system. An accused person is presumed innocent until proven guilty. In hypothesis-testing terms, "presumed innocent" is the null hypothesis. The alternative hypothesis is "not innocent" (i.e., guilty).

H_0: Accused is innocent.

H_1: Accused is guilty.

If a court declares an innocent person to be guilty, then a serious Type I error has occurred. If the accused is guilty but declared not guilty by the court, then a serious Type II error has occurred. Table 11.1 relates this example to the possible scenarios for the outcome of a hypothesis test.

Table 11.1 **Result of the Hypothesis Test Compared to the Actual State**

	Do Not Reject H_0 (Declare Innocent)	Reject H_0 (Declare Guilty)
H_0 Is True (Is Innocent)	Decision is correct. (Probability $= 1 - \alpha$)	Type I error (Probability $= \alpha$)
H_0 Is False (Is Guilty)	Type II error (Probability $= \beta$)	Decision is correct. (Probability $= 1 - \beta =$ Power*)

* The power of a test measures its ability to detect a difference of the actual population parameter from its presumed value in the null hypothesis.

This table is worth considering when making a choice for significance level α. How much risk of Type I error can you accept when you make your decision? You might wish that you could choose none or very little, but in that case the risk of Type II error would become very large. You would lose the practical ability to ever make distinctions or judgments with this tool.

Generally, we choose a smaller α when the consequences of making a Type I error are more serious. For example, when we test an athlete for steroids or other banned drugs prior to an Olympic event, the consequences of having a false positive could be very damaging to the athlete's career. But even then, we do have other options besides making α impractically small. For example, one positive result on a steroids test, with $\alpha = 0.01$, need not be taken as conclusive; we could order that a second independent test be made. Then the probability that *both* tests would be false positives, due only to chance, would be the more acceptable $0.01 \times 0.01 = 0.0001$.

Sections 8.03, 8.1, and 8.7

The Steps in Hypothesis Testing

This section introduces a systematic approach for hypothesis testing. With minor variations, the method can be applied to a wide variety of hypothesis-testing situations. The exact number of steps involved is counted differently by almost every author. What is important is to follow the basic sequence, knowing that steps can be grouped or can be applied with some flexibility.

Concrete examples of the procedure in action will be presented in the next section of this chapter, as well as in subsequent chapters. The model presented below focuses on a critical value approach. An alternative p-value approach can also be used, and this adaptation of the model is described in step 7.

Step 1: Write Down the Claim in Symbols

This step is often omitted in lists of hypothesis-testing procedures, but it can make all the difference in performing the second step correctly. What exactly is the statement that someone is claiming or wanting to test? As mentioned earlier, real people make claims of all sorts. Before deciding whether their claim is an H_0 or an H_1, begin with a more basic question: What is the claim itself? If possible, reframe the claim mathematically.

Example A.1: Biologist A claims that the mean age of trees in the Statistical Forest is *not* 15 years. This can be denoted as:

$$\mu \neq 15$$

When abstracted in this format, we do not care if the original word claim was about trees, bees, or bears. The essence is a claim about a population parameter—the mean—and its relationship—inequality—to a particular value.

Example B.1: A claim is made that the mean life span of black bears is *at least* 15 years. This could be denoted as:

$$\mu \geq 15$$

Again, the statistical interest is on the parameter (the mean), the operator ("at least," which means "equal to or greater than"), and the constant (15).

Step 2: State the Null and Alternative Hypotheses

If you have taken the time for step 1, then step 2 can follow almost automatically. Does the operator in the step 1 expression signify equality? If it does, that same expression becomes the null hypothesis. If it does not, it becomes the alternative hypothesis. Either way, there is one hypothesis remaining to be filled in. Do this by using the mathematical opposite of the expression generated in step 1.

Example A.2: In Example A.1, we represented the claim in symbols as $\mu \neq 15$. Since the operator does not contain equality, this expression becomes the alternative hypothesis, H_1. Therefore, the null hypothesis, H_0, must be the opposite: $\mu = 15$.

$$H_0: \mu = 15$$

$$H_1: \mu \neq 15$$

Although, following convention, we list H_0 first, observe that in this example we actually determined H_1 first.

Example B.2: The claim in Example B.1 was represented in symbols as $\mu \neq 15$. This operator does contain equality, so the expression becomes the null hypothesis, H_0. Therefore, the alternative hypothesis, H_1, must be the opposite: $\mu < 15$.

$$H_0: \mu \geq 15$$

$$H_1: \mu < 15$$

Again, we list H_0 first, by convention; in this case, we also determined H_0 first.

Step 3: Choose and List the Level of Significance

We have seen that the significance level (α) is the probability of committing a Type I error. The advantages and disadvantages of choosing higher or lower values have been discussed. Make your decision, and list the value of α in this step. The choice of α can often depend on circumstances, so no one significance level is inherently better than another. However, it is important that once you make your decision, you continue the testing process based on that specific value of α. This principle of keeping α constant is called *fixed-level testing.* [A variant of this approach that may apply when using p-values is provided in step 7(b).]

Examples A.3/B.3: In textbook examples such as these, you will often be told the significance level. Suppose the level of 0.05 is specified. Simply list the value in this step:

$$\alpha = 0.05$$

If you are conducting primary research, then setting the alpha value is part of the process. Also, if you plan to publish the results, many journals prefer that $\alpha = 0.05$.

Step 4: Specify the Test and the Sample Size(s)

Hypothesis tests can be specified with names like "*z*-test for the mean" or "*t*-test for the mean difference (two dependent samples)." The second part of these names indicates the parameter to be tested. The key to performing the hypothesis test is the sampling distribution (z, t, χ^2, etc.) that is chosen. The selected distribution models the variance that could be expected for the sample statistic, among all possible samples of a given size. The relative position of the statistic taken from the test sample, compared to the posited sampling distribution—presuming H_0 is true—provides the ultimate basis for accepting or rejecting the null hypothesis.

In common textbook problems, the value of sample size n is often given, or it is implicit in the number of data items in the problem. If you are conducting primary research, then determining the value of n will be a required step. (Section 11.3 provides some guidelines.)

Example A.4: In Example 11.1, we learned that the sample size is 30 and that population σ was known to be 4.58. Under those conditions, the normal curve could be selected as the sampling distribution (this will be discussed further in Section 11.2). Formally:

Use the z-test for the mean, $n = 30$.

Example B.4: Suppose that for practical reasons, the sample size for this case is restricted to 20. Population σ is not given. Presume that the standard deviation for the sample is 2.6, and that the life spans of black bears are normally distributed. Under those conditions, a t-distribution is the appropriate sampling distribution (more information on this topic will be provided in Section 11.2). Formally:

Use the t-test for the mean ($df = n - 1 = 20 - 1 = 19$).

Step 5: Display the Null Distribution

The **null distribution** is short for "the expected sampling distribution of the test statistic if the null hypothesis were actually true." Many errors of testing can be avoided by drawing a simple sketch of the null distribution; neither artistic skills nor drawing to scale are required. This sketch will provide the opportunity to decide if the test is one-tail (right-tail or left-tail) or two-tail. Depict the critical region or regions in the sketch and mark the value of α (or of $\alpha/2$ for two-tail) in the critical region(s). Specify the distribution (z or t, etc.) that you are using. If you are using the critical value approach, indicate the appropriate location(s) and value(s) in the sketch.

The resulting diagram depicts the essence of the entire test. If the equal part of the null hypothesis is true, then you have illustrated the sampling distribution from which you would expect to pick your sample. You have determined that if the sample statistic falls in the highlighted areas, you will reject H_0 as being too unlikely to be true. For a critical value approach, you have also specified the values of the test statistic that mark the beginning of that critical region.

Example A.5: All of the elements required for this step were illustrated previously in Figure 11.4(a). Observe how for a critical value approach, you calculate specific numbers for the critical values and display these on the horizontal axis in the figure.

Example B.5: Step 5 for the bear life-span case is shown in Figure 11.5. The figure clearly indicates that the null distribution is a *t*-distribution, but for our sketch, it can look more or less like a normal curve. Again, observe how, for a critical value approach, the specific critical value has been calculated and displayed. The test for this case is one-tail, because of the $<$ operator in H_1.

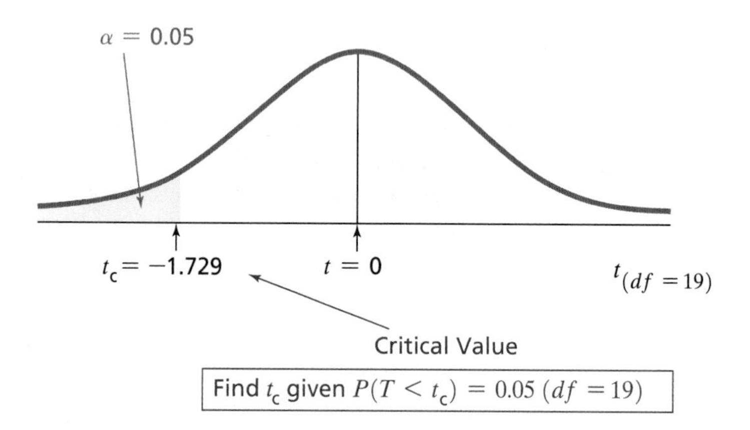

Figure 11.5
Step 5 for Example B.5

Step 6: Collect the Data

In textbook examples, the raw data or even summary statistics are often provided at the outset. If you are conducting primary research, you may require many days or weeks to obtain (and clean) the data that are necessary.

Examples A.6 / B.6: In both cases, the necessary summary statistics are provided in the text. For Example B.6, presume that the mean life span of the bears in the sample is 14.6.

Step 7: Calculate the Test Statistic or p-Value/Compare It to Step 5/ Make the Statistical Decision

7(a) Critical Value Approach: This traditional approach may be easiest when you are provided with summary statistics (e.g., the sample mean and standard deviation) at the outset, or if you are working with a *z*-based model. From the data in your sample, you find the relevant sample statistic (e.g., the mean) and convert it into a test statistic; that is, convert the sample statistic into the units (z, t, χ^2, etc.) of the standardized sampling distribution.

The resulting value can now be compared to the critical value or values depicted in step 5. Draw an arrow from your calculated test statistic to the appropriate portion of Figure 11.5 (e.g., to the corresponding critical region or to the middle area, if applicable).

If the test statistic falls in the critical region, then the decision is to reject H_0; otherwise do not reject H_0.

Example A.7(a): Example 11.1 provided the information that the sample mean is 7.2 and the population standard deviation is 4.58. Steps 5 through 7 are illustrated in Figure 11.6.

Figure 11.6
Steps 5–7 for Example A.7(a)

Step 5:

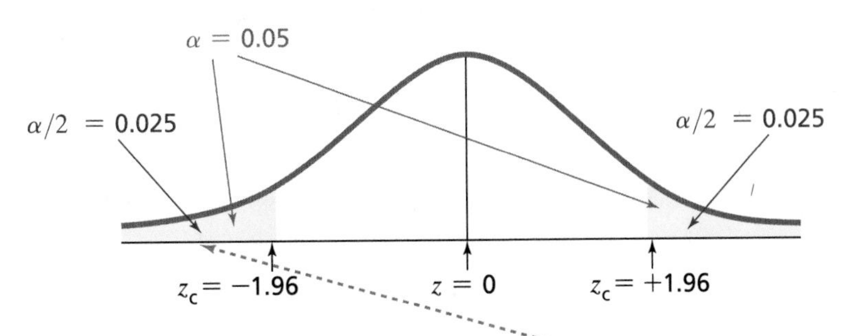

Step 6: Collect sample data.

Step 7:
- Calculate test statistic: $z = (7.2 - \mu)/(\sigma/\sqrt{n}) = (7.2 - 15)/(4.58/\sqrt{30}) = -9.328$
- Compare the test statistic to the regions in step 5 (see red arrow).
- Because the test statistic falls in a critical region, reject H_0.

Example B.7(a): In the bear example, the sample mean was found to be 14.6 years and the sample standard deviation was 2.6 years. Steps 5 through 7 are illustrated in Figure 11.7.

Figure 11.7
Steps 5–7 for Example B.7(a)

Step 5:

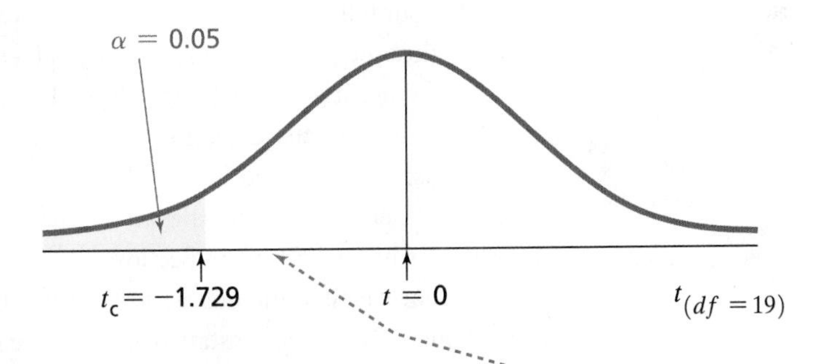

Step 6: Collect sample data.

Step 7:
- Calculate test statistic: $t = (14.6 - \mu)/(s/\sqrt{n}) = (14.6 - 15)/(2.6/\sqrt{20}) = -0.688$
- Compare the test statistic to the regions in step 5 (see red arrow).
- Because the test statistic does not fall in a critical region, do not reject H_0.

7(b) **p-*value Approach*:** This approach can be a practical alternative when software to help generate *p*-values is available. The ***p*-value** is a conditional probability (for this concept, see Chapter 6). In step 5, you drew the sampling distribution that would be expected if the population parameter had the value specified in the null hypothesis. *If* that sampling distribution applied, what is the probability of obtaining a sample statistic at least as far from the hypothesized parameter as has occurred in your sample? That conditional probability is the p-value.

In contrast to the critical value approach, the p-value approach does not rely on the visual concept of a rejection (or critical) region that is bounded by a critical value; but for fixed-level testing, the p-value approach is functionally equivalent to the critical value approach: Rather than using α to calculate a critical value as a cutoff point for rejecting H_0, we use instead a *decision rule* that is based on α, namely:

If:	**Then the statistical decision is:**
p-value $\leq \alpha$	**Reject H_0**
p-value $> \alpha$	**Do not reject H_0**

Recall that the critical value bounds the rejection region, which encompasses, by definition, a proportion α of the area (representing probability) under the null distribution. If a test statistic falls beyond the critical value, this means that (1) it is somewhere in the rejection region (so reject H_0) and (2) the proportion of area under the curve beyond that test statistic must be somewhat less than the full α. The decision rule follows from the above observations; that is, reject the null hypothesis only if the p-value (graphically, the proportion of area under the curve that is beyond the test statistic) is less than or equal to α.

It should be noted that in published research, a variant of fixed-level testing has evolved that has more than one reference value for deciding significance. A single publication might assert that one finding is "significant at the 0.001 level," a second finding is "significant at the 0.01 level," and a third finding is "significant at the 0.05 level." The idea is to claim significance at the smallest level p_i (e.g., 0.001, or 0.01, or 0.05) for which the p-value $< p_i$. This approach conveys whether the p-value is just barely small enough to reject H_0 (e.g., at the 0.05 level) or whether it is so small that a Type I error is highly unlikely (e.g., at the 0.001 level). In the latter case, the statistic may be viewed as providing stronger evidence against the null hypothesis.

When interpreting the latter approach to p-values, be sure to remember the lessons of Bayes' Rule in Section 6.4. A very low p-value may appear to be strong evidence against the null hypothesis when the following probability is small: P(obtaining a sample statistic such as we did $|\, H_0$ is true). But this value does *not* directly tell us a probability that may be of more interest to us: $P(H_0$ is true $|$ we obtained a sample statistic such as we did), and this latter probability may sometimes be larger than we think.

Example A.7(b): Refer back to Figure 11.6, step 7. For the p-value, you similarly need to calculate the test statistic:

$$z = \frac{7.2 - \mu}{\frac{\sigma}{\sqrt{n}}} = \frac{7.2 - 15}{\frac{4.58}{\sqrt{30}}} = -9.328$$

What is the probability of obtaining such a low z score under the null hypothesis? This can be calculated as $P(Z \leq -9.328) \cong 0.000$. (For calculations based on the normal curve, see "In Brief 8.1.") Because this is a two-tail test, multiply the above result times 2 (i.e., p-value $= 0.000 \times 2 = 0.000$). This is the probability of obtaining a result at least as far from the null mean—in either direction—as we observed in our sample. Put another way: This p-value is the probability that by chance alone we could have obtained evidence against the null hypothesis as strong as (or stronger than) the evidence provided in the current sample.

Since the p-value is less than the reference value of $\alpha = 0.05$, the statistical decision is to reject the null hypothesis.

Example B.7(b): Refer back to Figure 11.7, step 7. For the p-value, you similarly need to calculate the test statistic:

$$t = \frac{14.6 - \mu}{\frac{s}{\sqrt{n}}} = \frac{14.6 - 15}{\frac{2.6}{\sqrt{20}}} = -0.688$$

What is the probability of obtaining a t-value at least this small under the null hypothesis? This can be calculated as $P(T \leq -0.688) \cong 0.250$. (For the calculations based on the t-distribution, see "In Brief 11.1" later in this chapter.) This is the probability of obtaining a result at least as far from the null mean—in just one direction—as we in fact observed. (If the test was two-tail, multiply the calculated result times 2.)

Since the p-value is greater than the reference value of $\alpha = 0.05$, the statistical decision is to not reject the null hypothesis.

Step 8: Clearly Express Your Conclusion—In Words

In step 7, the last component was to state the **statistical decision**—to either reject or fail to reject H_0. For people who have an interest in your findings, that form of conclusion may not be intelligible. For this step, relate your statistical decision back to the original claim that was being tested. Clearly state your findings as they relate to that claim. If the original claim could be translated into H_1, it may have been supported or not supported by the evidence. If the claim was translatable to H_0 (i.e., contained equality), then there may or may not have been sufficient evidence to reject that claim. Use common sense to express the statistical result in a way that an interested party could understand.

Example A.8: Biologist A claimed that the mean age of trees in the Statistical Forest was not 15 years. This became H_1 in step 2 of the process. In step 7, we rejected H_0, in favour of H_1 (which was, essentially, the claim). Therefore, we conclude:

The evidence supports the claim that the mean age of trees in the Statistical Forest is not 15 years.

Example B.8: The claim was that the mean life span of black bears is at least 15 years. This claim became H_0, as formulated in step 2 of the process. By step 7, we had failed to reject the claim as expressed in H_0. Therefore, we conclude:

Based on the available sample, there is not sufficient evidence to reject a claim that the mean life span of black bears is at least 15 years.

Section 8.2.1

More examples and details about the process, including specific tests that can be conducted, will be provided in the sections and chapters that follow.

• *In Brief 11.1 Determining the* p*-Value Given a Test Statistic* t

The null distribution for the hypothesis test can be viewed as a probability density curve. The *p*-value for the test is the probability that corresponds to the area under the curve that is at least as far from the null parameter as the test statistic—in the direction of the *one or two* tails of the test. If the null distribution is normal, then use procedures from "In Brief 8.1(a)" to perform the calculations (using the test statistic *z*). The instructions below apply when the test statistic is a known *t*-value, and the degrees of freedom are also known.

Excel

In any empty cell of a spreadsheet, input a formula of this format: **=TDIST(*t,df,tails*)**, where ***t*** is the *magnitude* of a previously calculated test statistic, ***df*** is the degrees of freedom for the particular test, and ***tails*** is the number of tails for the hypothesis test. (Note: The **TDIST** function will not accept a negative value, so disregard the sign of the calculated *t* statistic and just input its (positive) magnitude in the function.)

Minitab

Use the menu sequence *Calc / Probability Distributions / t.* In the resulting dialogue box shown in the screen capture, select *Cumulative probability* and fill in the box for the degrees of freedom. Select *Input constant* and input the magnitude of the test statistic *as a negative number.* (The Minitab procedure finds the cumulative probability to the *left* of the *t*-value that is input but the distribution is symmetrical, so the answer is the same as the cumulative probability to the right of a positive value for *t.*) Click *OK.* For a two-tail test, multiply the displayed result times 2.

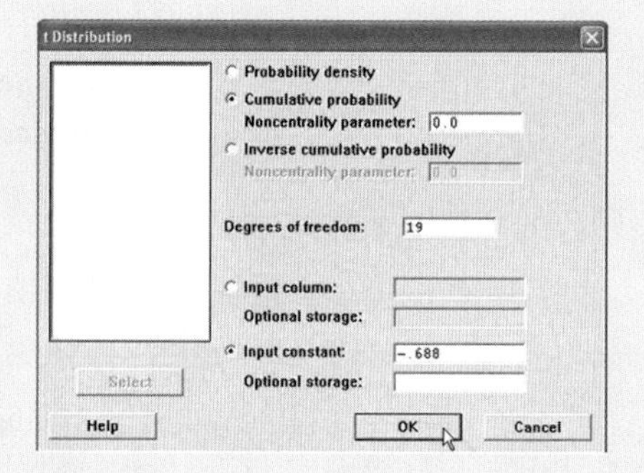

SPSS

Ensure there is at least one value in the worksheet, then use the menu sequence *Transform / Compute.* In the resulting dialogue box, input in the *Target Variable* area a name for the column in which the solution should appear. In the *Numeric Expression* box, input a formula of this format: **cdf.t(*t,df*)**, where ***t*** is the magnitude of the test statistic *input as a negative number,* and ***df*** is the degrees of freedom for the particular test. (The SPSS procedure finds the cumulative probability to the *left* of the *t*-value that is input but the distribution is symmetrical, so the answer is the same as the cumulative probability to the right of a positive value for *t.*) Click *OK.* If necessary, use *Variable View* to display more decimal places in the solution. For a two-tail test, multiply the displayed result times 2.

11.2 One-Sample Tests for Population Mean

This section is one of many throughout this text that will apply the hypothesis-testing procedures described in the previous section. Possible applications of the testing procedures can be distinguished based on three main factors:

1. Is the test based on one sample or on more than one?
2. What is the parameter of interest? (Examples: the mean, the median, a difference in population means, a population proportion.)
3. What is the model employed for the sampling distribution? (Examples: The normal curve, a t-distribution, an F-distribution, a chi-square distribution.)

In this section, hypothesis tests have these three criteria: (1) They each draw from a single sample; (2) the parameter of interest is the population mean; and (3) the model for the sampling distribution is either the normal curve (i.e., a z-test) or a t-distribution. The series of examples A.1–A.8 in the previous section was based on the normal model. The series B.1–B.8 assumed a t-distribution. The determination of which model to use depends on whether the population standard deviation σ is known prior to the test.

z-Test for the Population Mean (σ known)

It is not always clear how you would know the population standard deviation prior to drawing a sample from that population. In some cases, the value may have been estimated from a pilot study or based on previous years' data. In any case, this model is the traditional starting point when introducing hypothesis testing. In its favour, the model displays very clearly the essential principles of the eight-step hypothesis-testing procedure.

The z-based model requires that the sampling distribution of sample means be normally distributed. This requires, in turn, that either the population itself is normally distributed or else that the sample size is large (at least 30), so that the central limit theorem applies.

Example 11.2

A farmer who lives in Enka, North Carolina, claims that the standard forecasts for the date of the first frost in her region have become inaccurate. Based on 30 years of data ending in 1980, the North Carolina Cooperative Extension Service reported that, for Enka, the mean date number for the first frost is 293 (October 20); the standard deviation has been 10 days.[1] Presume that this distribution of first-frost dates is normal, and that the farmer has collected data on 15 randomly selected first frosts—from years subsequent to 1980. The mean date number for that sample of first frosts is 297 (October 24). Test the farmer's claim at the 0.05 level of significance.

Solution

We follow the eight-step procedure described in the previous section.

Step 1: The farmer disputes that the recorded mean date number of 293 remains accurate. In symbols, the farmer's claim can be formulated as:

$$\mu \neq 293$$

Step 2: Since the claim of step 1 does not contain equality, it can be interpreted as the alternative hypothesis. The opposite of that hypothesis becomes H_0.

$$H_0: \mu = 293$$
$$H_1: \mu \neq 293$$

Step 3: The level of significance is given in the problem:

$$\alpha = 0.05$$

Step 4: The sample size is also given in the problem. The choice of test can be determined, given that the farmer is testing for the mean based on one sample from a population that is normally distributed, with a known value of σ. Therefore, she can:

Use the z-test for the mean, $n = 15$.

(Caution: The farmer is said to know σ for a period *prior to 1980*. Does she necessarily "know" if this value still applies?)

Step 5: A good depiction of the test model is already available in Figure 11.4(a). It includes the null distribution (i.e., a curve of the sampling distribution of the sample means as it would be if H_0 was true), an indication of α, and the critical values that mark the bounds of the critical regions. If necessary, computer software can help to calculate the critical values for a given number of tails and the value of alpha.

Steps 6/7: Typically, the data necessary for a textbook exercise will be provided. In the problem statement we learn that

$$\bar{x} = 297$$

We can calculate the test statistic:

$$z = \frac{\bar{x} - \mu}{\frac{\sigma}{\sqrt{n}}}$$

$$= \frac{297 - 293}{\frac{10}{\sqrt{15}}}$$

$$= 1.55$$

For the critical value approach, a line can be drawn from the calculated test statistic ($z = 1.55$) to a middle-right portion of Figure 11.4(a)—showing that z falls short of the critical value ($z = 1.96$) required to reject H_0. Therefore, the statistical decision is:

Do not reject H_0.

The same decision would be reached if we used the p-value approach. The two-tail p-value that corresponds to the test statistic $z = 1.55$ can be calculated as $2 \times P(Z \geq =1.55) \mid H_0 \text{ is true}) = 0.12$, which is greater than $\alpha = 0.05$.

Step 8: Relating the statistical decision back to the original claim (which corresponded to H_1), we conclude:

> The evidence does not support the claim that the mean date of the first frost is different from the accepted forecast model, which puts the mean first frost date at day 293 (October 20).

Sections 8.2.2–8.2.5, 8.3

t-Test for the Population Mean (σ unknown)

If the population σ is not known, then the z-test for the mean should not be used. Instead, we have to rely on s, the sample standard deviation, to estimate σ. As we saw in Chapter 10, the more appropriate model for the sampling distribution of the sample means, when σ is unknown, is the t-distribution. As with the z-test, the t-based model also requires that either the population is normally distributed or else that the sample size is sufficiently large (at least 30).

In Example 11.3, we are provided with only summary data. In Example 11.4, raw data are available.

Example 11.3

The farmer in Example 11.2 was convinced by a statistician that she should not have used a z-test. Recall her claim that the reported mean date number for first frosts (293) in her region is inaccurate. We said the distribution of these dates is normal and that the farmer collected a sample of 15 more recent dates for first frosts—obtaining a sample mean date of 297 (October 24). Following the advice of the statistician, the farmer will not use the 25-year-old standard deviation number (10 days) as a current estimate of σ; instead she will utilize the s-value from her own sample, which is 7 days. Again, test the farmer's claim at the 0.05 level.

Solution

Step 1: As in the previous example, her claim can be formulated as

$$\mu \neq 293$$

Step 2: The hypotheses can now be interpreted as

$$H_0: \mu = 293$$

$$H_1: \mu \neq 293$$

Step 3: The level of significance is unchanged:

$$\alpha = 0.05$$

Step 4: The sample size is also unchanged. Given that the farmer is testing for the mean based on one sample from a population that is normally distributed, with an *unknown* value of σ:

Use the t-test for the mean, $n = 15$ ($df = 14$).

Steps 5–7: Figure 11.8 shows the critical value approach. The same decision would be reached if we used the p-value approach: The two-tail p-value that corresponds to the test statistic $t = 2.21$ can be calculated (see "In Brief 11.1") as $2 \times P(T_{df=14} \geq 2.21 \mid H_0 \text{ is true}) = 0.044$, which is less than $\alpha = 0.05$. Reject H_0.

Step 8: Relating the statistical decision back to the original claim, we can conclude:

The evidence supports the farmer's claim that the accepted forecast model, based on a mean first frost date of 293 (October 20), is inaccurate.

Figure 11.8
Steps 5–7 for Example 11.3

Step 5:

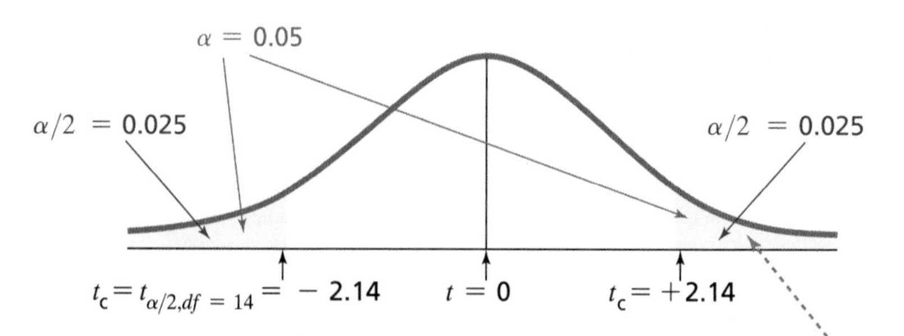

Step 6: Collect sample data.

Step 7:

- Calculate test statistic: $t = (\bar{x} - \mu)/(s/\sqrt{n}) = (297 - 293)/(7/\sqrt{15}) = +2.21$
- Compare the test statistic to the regions in step 5 (see red arrow).
- Because the test statistic falls in a critical region, reject H_0.

Example 11.4

As part of reports on water use in Canada, Environment Canada publishes updated lists of water-using communities with populations greater than 1,000. In 1996, the mean population of all listed communities was 19,368. Suppose that 30 of those communities were randomly selected in 1999 for an audit of their water usage. The following were the populations in 1999 of the selected communities.[2]

1,345	6,972	35,000	6,749	11,174	2,117	6,972	12,000	24,285
38,981	67,000	22,500	1,109	5,500	3,492	1,068	136,976	6,148
44,882	322,252	4,700	46,000	5,322	971	6,500	2,195	2,107
5,450	85,000	24,525						

Based on this sample, test the claim (using $\alpha = 0.01$) that in 1999 the mean population of the listed communities was greater than their mean population in 1996.

Solution

Again, we follow the eight-step procedure but with the variation of using the *p*-value approach only.

Step 1: The recorded mean population in 1996 was 19,368. It is claimed that for 1999, the population value was larger than that value:

$$\mu > 19{,}368$$

Step 2: Since the claim of step 1 does not contain equality, it can be interpreted as the alternative hypothesis. The opposite of that hypothesis becomes H_0.

$$H_0: \mu \leq 19{,}368$$
$$H_1: \mu > 19{,}368$$

Step 3: The level of significance is given in the problem:

$$\alpha = 0.01$$

Step 4: The sample size is also given in the problem. We are testing for the mean of the population based on one sample, in a case where σ is not known. The population is not normally distributed, as can be demonstrated with a probability plot or other method as described in Section 8.2. Nonetheless, because the sample size is reasonably large, normality of the population is not required. Therefore:

Use the t-test for the mean, $n = 30$ $(df = 29)$.

Step 5: The test model is depicted in Figure 11.9. It is not necessary to specify the exact value for t_c, since we will be using the p-value approach, which does not utilize the critical value. The figure reminds us what test to use, where the critical region lies, and what is the value of α.

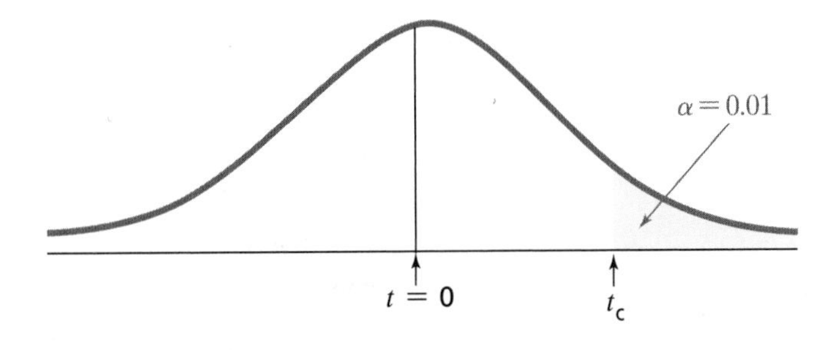

Figure 11.9
Depiction of Model for Step 5, Example 11.4

Steps 6/7: The data are provided in the problem. Semimanually, we could compute $\bar{x}$ and s from the sample data, determine the test statistic, and find the corresponding p-value with the procedures from "In Brief 11.1." But if using software and if raw data are available, much of this can be automated: Apply to the data the procedures discussed in "In Brief 11.2" on page 465; then the output will include the test statistic (t-value $= +1.04$) and the corresponding p-value $= P(T_{df=29} \geq +1.04 \mid H_0 \text{ is true}) =$

p-value $= 0.153$

That is, if the null hypothesis accurately described the population, it would not be unusual for a sample statistic for the mean to be as much above the population mean as happened in this sample (this would occur about 15% of the time). Since this p-value is not less than α, the statistical decision is

Do not reject H_0.

Step 8: Relating the statistical decision to the original claim, we conclude:

The evidence does not support a conclusion that the mean population of listed communities was greater in 1999 than the mean recorded for 1996 (19,368).

Seeing Statistics
Sections 9.2 and 9.3

Example 11.5

In recent decades, the mean weight of adult human males, aged 18 to less than 75, has been 78.1 kg, with a standard deviation of 13.5 kg.[3] In a study of whether weights are changing, a researcher samples 40 males in that age group and obtains the sample statistics $\bar{x} = 82.3$ kg and $s = 15.7$ kg. At a significance level of 0.05, can the researcher conclude that the mean weight has increased?

Solution

Step 1: Though phrased as a question, the implied claim is that the mean weight is greater than 78.1 kg (i.e., the previous value):

$$\mu > 78.1$$

Step 2: The claim of Step 1 does not contain equality, so it becomes H_1.

$$H_0: \mu \leq 78.1$$

$$H_1: \mu > 78.1$$

Step 3: The level of significance is given in the problem:

$$\alpha = 0.05$$

Step 4: The sample size $n = 40$ is given in the problem. We are testing for the mean of the population based on one sample. A previous value for σ is known, but the premise of the research is that things could be changing. If the value of σ has possibly changed, then its new value is unknown. We are not told the distribution of the population, but the sample size is large. Given these considerations:

Use the t-test for the mean, $n = 40$ ($df = 39$).

Step 5: The test model is depicted in Figure 11.10.

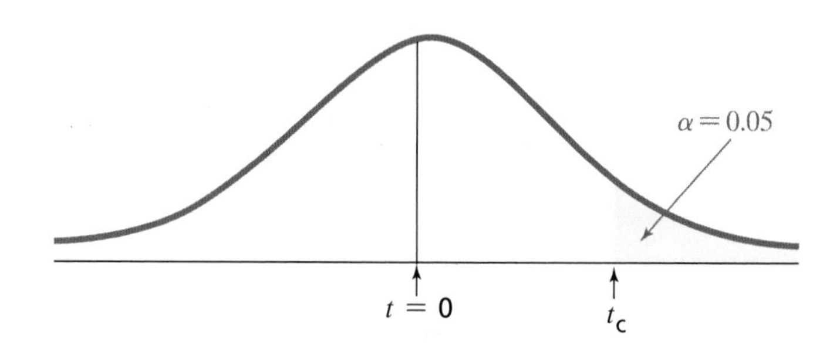

Figure 11.10
Depiction of Model for Step 5, Example 11.5

Steps 6/7: Summary data are provided in the problem, but not raw data. We can find the test statistic:

$$t = \frac{82.3 - \mu}{\frac{s}{\sqrt{n}}} = \frac{82.3 - 78.1}{\frac{15.7}{\sqrt{40}}} = +1.69$$

Applying the methods provided in "In Brief 11.1," we find the corresponding p-value for the test statistic:

$$p\text{-value} \cong 0.050$$

Since this p-value is less than or equal to α, the statistical decision is

Reject H_0.

Step 8: Relating the statistical decision to the original claim, we conclude:

The evidence supports the conclusion that the mean weight of males aged 18 but less than 75 years has increased from its previous value of 78.1 kg.

• *In Brief 11.2* *Computer Assistance for One-Sample* t-*Test for the Mean*

(Based on the **Water_Use** *file)*

The computer does not automate a hypothesis test in the same sense that it automates finding a mean or a variance. It remains our job to construct the test hypotheses and to identify an appropriate statistical test. We must also know how to utilize the computer output. In the "In Brief" sections for hypothesis testing, we will focus generally on how to generate values that can be applied within the eight-step testing procedure.

For this "In Brief" exercise, a sample of 30 values in the "Pop_1999" column in the **Water_Use** file were selected, as shown below.

1,345	6,972	35,000	6,749	11,174
2,117	6,972	12,000	24,285	38,981
67,000	22,500	1,109	5,500	3,492
1,068	136,976	6,148	44,882	322,252
4,700	46,000	5,322	971	6,500
2,195	2,107	5,450	85,000	24,525

Excel

The writers of Excel opted not to include this specific t-test, but the program can be "tricked" to obtain the p-value that is needed for the one-sample t-test for the mean. As shown in the illustration, you first create a column of zeros next to the full column of data. Then use the menu sequence *Tools / Data Analysis / t-Test: Paired Two Sample for Means,* and click *OK.* In the resulting dialogue box (see the screen capture), input the actual data range as the *Variable 1 Range;* input the range of zeros as the *Variable 2 Range.* Next, input the test value for μ in the box *Hypothesized Mean Difference.* Optionally, to learn the critical t-value for the test, specify the test alpha where indicated. Finally, specify the upper left cell address for the output range, and click *OK.*

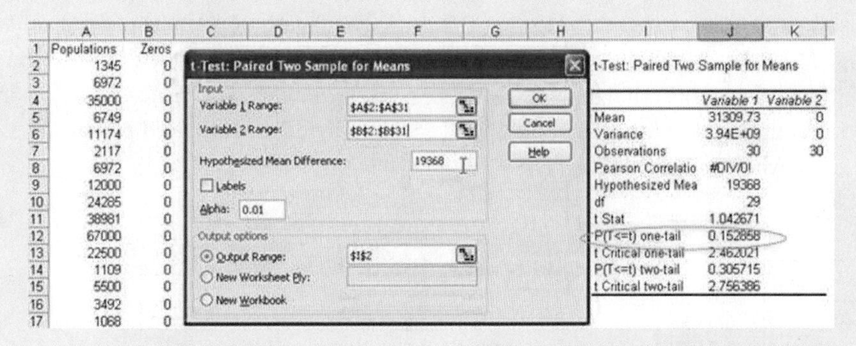

The illustration also shows the software output and highlights the one-tail p-value that corresponds to Example 11.4. Had the test been two-tailed, the two-tail p-value would be read instead.

Minitab

With the raw data in one column, use the menu sequence *Stat / Basic Statistics / 1-Sample t.* In the resulting dialogue box (see the screen capture), select the variable for the sample and specify the test mean where indicated. (Note: Minitab gives the option of starting from summary data instead of raw data; if applicable, click in the box for that option.) Now select *Options* and, in the next dialogue box, use the drop-down menu to select the operator of the alternative hypothesis. (If needed, also change the default for degree of confidence.) Click *OK* twice. The illustration also shows the software output.

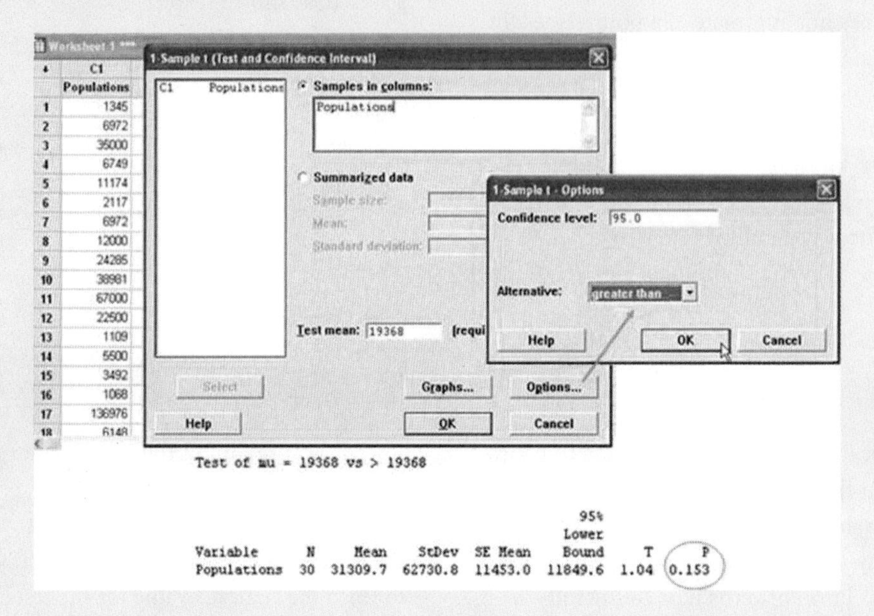

SPSS

With the raw data in one column, use the menu sequence *Analyze / Compare Means / 1-Sample T Test*. In the resulting dialogue box, shown in the screen capture, select the variable for the sample and specify the test mean where indicated. Click *OK*.

The illustration also shows the software output and highlights the two-tail p-value that is automatically generated. For a one-tail test, as in Example 11.4, divide the displayed p-value by 2.

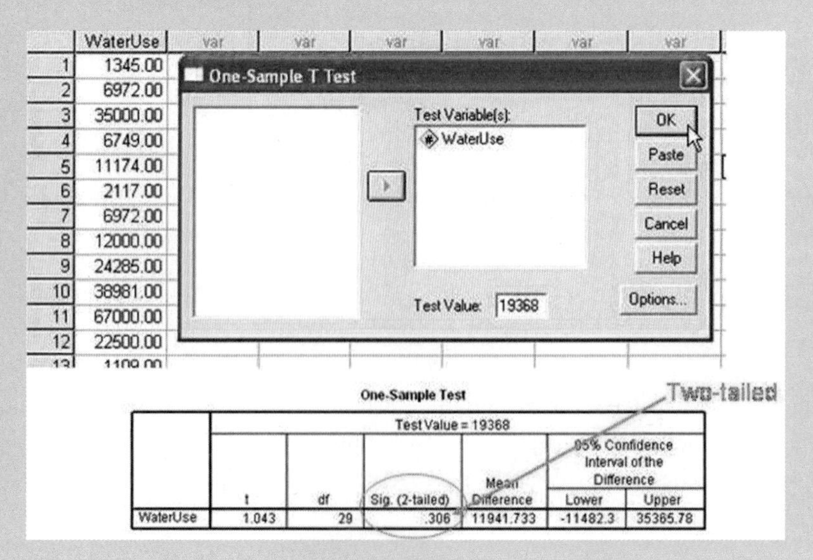

• Hands On 11.1 Setting the Effective Alpha

When we set a value for α during a hypothesis test, we accept a probability $= \alpha$ that the null hypothesis will be rejected in error—that is, the null hypothesis will be rejected, even though it is true. How do we obtain this probability? We suppose that we know the correct model for the sampling distribution of the sample statistic; we then derive the probabilities from the mathematical properties of that model. This "Hands On" illustrates what can happen if our model of the sampling distribution is not quite accurate.

Step 1: Use two columns of your spreadsheet or data view screen to represent the values of two complete populations. Label the first column "Population 1" and the second column "Population 2."

- **Population 1:** Using the procedures from "In Brief 5.2," generate 600 normally distributed numbers based on a mean of (approximately) 1,000 and a standard deviation of 50. Find and record the true mean for this population.
- **Population 2:** Copy and paste the values of population 1 into a new column. Replace the top four values with the numbers 2,000, 3,000, 4,000, and 5,000, respectively. Find and record the true mean for population 2.

Question: Compared to population 1, how do you think the change made in population 2 would impact the sample distribution of the sample mean for samples with $n = 30$? Explain your reasons.

Step 2: Using the procedures provided by "In Brief 2.2," create and store a random sample of 30 numbers from population 1. Repeat this process 49 times in order to collect 50 independent samples of $n = 30$ from the population.

Step 3: Suppose each of the above samples was collected to test whether the mean of the source population was larger than a certain value μ. For μ, use the true mean of population 1 that you found in step 1. (Your test result will be erroneous if you do reject the null hypothesis.) Use α equal to 0.10 for the test. For each of the 50 samples in turn, find the t-statistic for its sample mean in comparison with the true mean. (Do not presume to know σ.) Compare this t-value with t_c (given a one-tail $\alpha = 0.10$ and $df = 29$).

Question: For what proportion of the samples was $t > t_c$? (This is the proportion of Type I errors that would have been made.) Is it about 10%? If you repeated steps 2 and 3, would you expect to obtain about the same proportion? Would you obtain exactly the same proportion?

Step 4: Repeat steps 2 and 3 as above, but this time use population 2 as the source population for the sampling. Use the mean μ of *this* population as the basis for calculating each sample's t-statistic and then comparing its t-value with t_c (given a one-tail $\alpha = 0.10$ and $df = 29$).

Question: In this revised case, for what proportion of the samples was $t > t_c$? Is it about 10%? If not, why not? What are the implications for making a Type I error in this type of case? For making a Type II error? Explain.

11.3 Determining a Sample Size

If we always had timely, affordable, and reliable access to the full set of population data, we would not need to sample. Directly knowing the population parameter, the methods discussed in Chapter 10 (on finding confidence intervals) and in this chapter (on testing a claim about a parameter) would not be necessary. But as we learned in Section 1.2, this is not realistic. There may be time or cost constraints in collecting the data; the measurement process may be destructive; the population may be infinite—even stretching into the future; or, in practice, the accuracy of a sample may be greater than that of a census.

Given that sampling is often a necessity, the issue of appropriate sample size must be raised. Some issues in this section are more applicable to confidence interval construction (Chapter 10), and others apply more to this chapter's topics.

If you work with secondary data, including the problem sets in this textbook, the sample sizes are generally given, although sometimes you might sample just a subset from a larger data set that is provided. For primary research, however, determining the sample size is part of the job. Generally, this means determining the *minimum* sample size that meets certain criteria. If your time and resources allow, you may opt to go for more than the minimum size, which will serve to reduce the standard error of your sample.

General Considerations for Determining Sample Size

Several considerations should guide the selection of sample size.

1. **Acceptable level of risk for Type I error:** Recall that Type I error occurs when a hypothesis test mistakenly results in a rejection of a true null hypothesis. The risk of this happening is represented by α. The complement of α can be applied to confidence intervals; that is, $1 - \alpha$ gives the *confidence* that the constructed interval will include (rather than mistakenly exclude) the true population parameter. In practice, this α risk cannot be reduced to zero. But other factors being equal, *increasing the sample size reduces the α risk.*
2. **Variability of the data** (σ)**:** If a data set is highly variable, then randomly selected individual items may be far from the true population parameter in one direction or the other. In a small sample, selected values that are large may not be evenly balanced by smaller values, or vice versa. This could pull the sample statistic farther away from the true population value. Keeping all other factors equal, you must *increase the sample size as variance increases.*
3. **For constructing confidence intervals—Acceptable magnitude of the error term** (E)**:** In Chapter 10, we learned that confidence intervals are centred on a point estimate for a parameter and include a range of plus or minus distances (E) from

the point estimate. For example, media reports of public opinion polls often include a footnote that the actual proportion could be about ± 3% from the published proportion. Likely, the pollsters decided *before* their research that this approximate error magnitude would be acceptable. Other factors being equal, *increasing the sample size decreases the error magnitude E* for the confidence interval.

4(a) **For hypothesis testing—Acceptable power for the test:** The **power of a test** measures its ability to detect a difference of the actual population parameter from its presumed value in the null hypothesis. The power of a hypothesis test is, mathematically, $1 - \beta$; i.e., it is the complement of β for the test. Recall that β is the probability of a Type II error, which occurs when a hypothesis test mistakenly fails to a reject a false null hypothesis. Thus, the power of a test is the probability that when the null hypothesis *should* be rejected (because it is false), it *will* be rejected. The relationship between the test α and its β is illustrated in Figure 11.11.

Figure 11.11
The Relationships among α, β, and Power

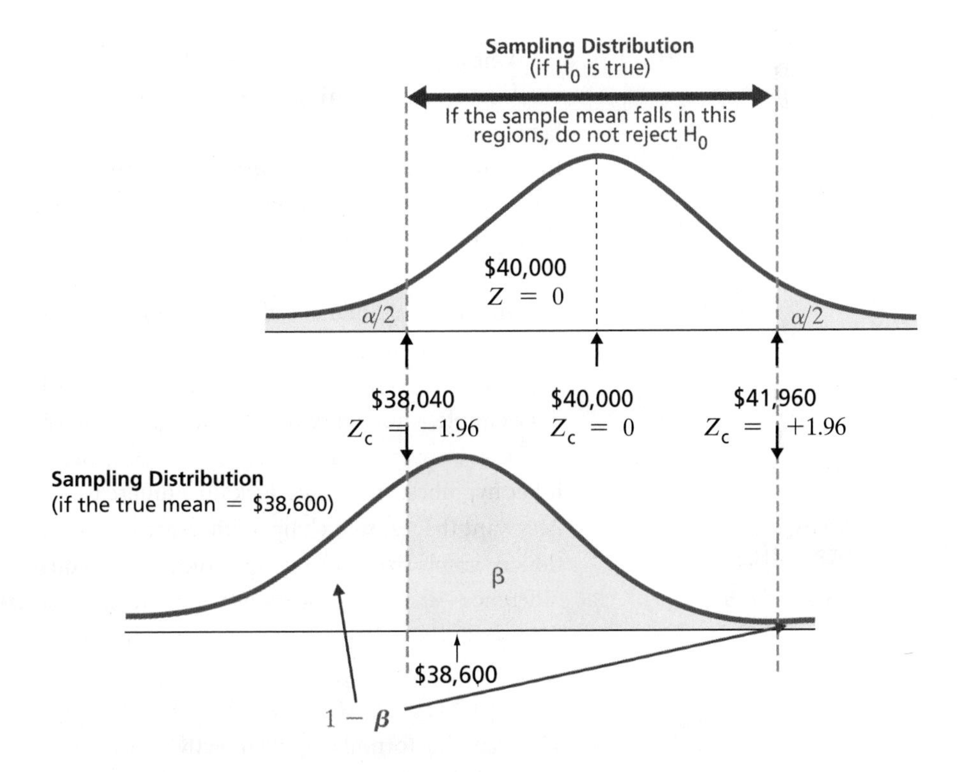

In this figure, we see that H_0: μ = \$40,000 and $\alpha = 0.05$. The critical values depend on α and also on sample size. Suppose the true mean of the population is \$38,600. When we take a sample, it will not actually be taken from the sampling distribution depicted in the upper part of the figure; instead, we take the sample from an alternative sampling distribution—the distribution shown in the lower half of the figure—but we are not aware of the location of the lower distribution. The area labelled β represents the large probability that the random sample will have a mean that falls within the *non*critical region, *as originally specified in the test model.* In that case, we would not reject the null mean and therefore would make a Type II error. If a test has a large power, the area 1 −

β in the lower part of the figure (i.e., the probability of correctly rejecting the false null hypothesis) is relatively large.

Recall that increasing sample size decreases the size of the standard error for the sampling distribution, which in turn decreases the range of values in the nonrejection region for the test. The latter is the range susceptible to the **β risk**—the risk of making a Type II error. Thus, if all other factors are kept equal, *increasing the sample size reduces the β risk, and so increases the power.*

4(b) **For hypothesis testing—Minimum difference to be detected:** Unlike the α for a given hypothesis test, the β (and therefore, the power) is not a unique value. It depends on the (usually unknown) distance between the true location of the population parameter from its assumed value according to H_0. Figure 11.11 illustrates one possible case, but the true population mean might really be $37,200, or $38,700, or $38,999. In each case, we would need to revise Figure 11.11 to reflect the revised position of the true mean relative to the hypothesized mean. As a result, the proportion of the lower curve representing β would vary with each possible scenario.

One approach to the problem of multiple β-values for a single test is to construct a chart called an **operating characteristic curve.** The x-axis displays possible distances between the actual and hypothesized parameters; the y-axis displays, for each distance, the corresponding value of β. The complement of the operating characteristic curve is called the **power curve,** which shows, for any distance between the true and hypothesized parameters, the corresponding probability $(1 - \beta)$ of correctly rejecting the null hypothesis.

Sections 8.5 and 8.6

To select a sample size, the entire power curve does not have to be constructed. We generally have in view some *minimum difference* or *minimum effect* that we hope the hypothesis test can detect. Use this distance as the basis for calculating the β and power. For example, a human resources manager claims that women in a certain occupation are paid less, on average, than the $40,000 that men are generally paid. She is not concerned if her hypothesis test could pick up a difference of one or two hundred dollars, but she *does* want the test to pick up a difference of, say, one thousand dollars or more. In relation to sample size: If all other factors are kept equal—including the specified minimum difference—*increasing the sample size reduces the β risk, and so increases the power.*

Sample Sizes for Constructing Confidence Intervals

The following formulas apply directly to confidence intervals (see Chapter 10), but confidence intervals can also be used in a modified procedure for hypothesis testing: If the confidence interval that is based on the sample fails to include the parameter assumed in H_0, this is evidence for rejecting the null hypothesis. So the formulas given in this section can be applied to either constructing a confidence interval or conducting a hypothesis test.

Sample Size for Determining the Mean

The procedure to construct a confidence interval for the mean is given in Section 10.2. The interval is centred on the point estimate of the mean $\bar{x}$, and extends a distance E in either direction. (See Figures 10.5 and 10.6.) This error distance around the point estimate reflects the acceptable level of α, as well as the standard error of the sampling distribution. We saw that for the confidence interval for the mean, the value $E = z_{\alpha/2} \times \frac{\sigma}{\sqrt{n}}$. Solving this equation for n gives us Formula 11.1 to use in finding the minimal sample size.

Formula 11.1 Sample Size Determination to Construct a Confidence Interval for the Mean

$$n = \frac{z^2_{\alpha/2}\,\sigma^2}{E^2}$$

Observe that this formula reflects three of the general considerations for sample size listed in the previous section:

1. The term $z_{\alpha/2}$ represents the critical value for z that is used to construct the confidence interval for an intended value of α (i.e., for a given level of confidence $1 - \alpha$). The formula shows that in order to decrease α risk, which requires a larger value for $z_{\alpha/2}$ in the numerator, the sample size must be increased as well.
2. The variance in the population is denoted by σ^2. The formula shows that as σ^2 increases in the numerator, the sample size must be increased also to keep all the other factors constant.
3. When determining a sample size to find the mean, the researcher must decide on the acceptable magnitude of the error term E in the confidence interval. Since E appears in the denominator of Formula 11.1, it follows that to reduce the acceptable size of E, the sample size must be increased, and vice versa.

   (Note: If the confidence interval is constructed in the context of a hypothesis test, then the E term of the interval is related to the power of the test. The mean assumed in H_0 is more likely to fall outside the confidence interval constructed from the sample (so that H_0 is rejected) if the interval (based on E) is smaller. To obtain a smaller interval, and thus more power, the formula requires a larger sample size.)

Example 11.6 Determine the sample size required to estimate the mean family income in Windsor, Ontario, to within $1,000 at a 95% level of confidence. Assume that the population standard deviation is $5,000.

Solution The allowable error for the solution is $1,000. The critical z-value corresponding to a 95% level of confidence (i.e., to $\alpha = 0.05$) is 1.96. The standard deviation is given as $5,000. Using Formula 11.1:

$$n = \frac{z^2_{\alpha/2} \times \sigma^2}{E^2}$$

$$= \frac{1.96^2 \times 5{,}000^2}{1{,}000^2}$$

$$= 96.04$$

Always round up to the next highest integer. If you round down, your sample size would be slightly lower than the required minimum. Therefore, use a sample size of 97 families.

Sample Size for Determining the Proportion

Section 10.4 gives procedures to construct a confidence interval for the proportion. A common approach uses the normal approximation for the binomial. The confidence interval is centred on the point estimate for the proportion p_s and extends a distance E in either direction. This error distance E reflects the acceptable level of α, as well as the standard error of the sampling distribution—assuming that the distribution is approximately normal. For the confidence interval for the proportion, the value

$$E = z_{\alpha/2} \times \sqrt{\frac{p \times (1 - p)}{n}}$$

Solving this equation for n gives us Formula 11.2 (where p is a preliminary estimate of the proportion) to find the minimal sample size.

Formula 11.2 Sample Size Determination to Construct a Confidence Interval for the Proportion

$$n = \frac{z^2_{\alpha/2}p(1 - p)}{E^2}$$

Similar to the sample-size formula for the mean, this formula addresses three of the general considerations for sample size discussed earlier:

1. As before, the term $z_{\alpha/2}$ represents the critical value for z that is used to construct the confidence interval for an intended value of α. To decrease α risk, the value for $z_{\alpha/2}$ in the numerator must be increased, requiring an increase in sample size.
2. The expression in the numerator $p \times (1 - p)$ is derived from the standard error calculation for the normal approximation of the binomial, so it plays the role of the variance in the sample-size calculations. In this context p is a *preliminary estimate* of the population proportion. Either p *or* $(1 - p)$ will have a value between 0.00 and 0.50; as that particular value increases, the formula will yield a larger sample size if all other factors remain constant. If p cannot be estimated, then in the formula use $p = 0.50$; this will maximize the sample size compared to any other possible value for p.
3. As in the case of the mean, when determining a sample size to find the proportion, the acceptable magnitude of E in the confidence interval must be decided ahead of time. Since E appears in the denominator of Formula 11.2, it follows that to reduce the acceptable size of E, the sample size must be increased, and vice versa.

Example 11.7 Ontario Provincial Police officers are concerned about the number of traffic violations involving elderly drivers. How many traffic violations must be surveyed in order to determine the proportion of violations that involve elderly drivers? Use a 90% level of confidence and plan to determine the proportion within $\pm 4\%$. Suppose that in other provinces, the proportion of violations involving elderly drivers has been 0.19.

Solution

The allowable error for the solution is 0.04. The critical z-value corresponding to a 90% level of confidence (i.e., to $\alpha = 0.10$) is 1.645. An initial estimate for population p might be based on the other provinces' data: 0.19. Using the formula,

$$n = \frac{z_{\alpha/2}^2 \times p \times (1 - p)}{E^2}$$

$$= \frac{1.645^2 \times 0.19 \times 0.81}{0.04^2}$$

$$= 260.29$$

Always round up to the next highest integer. Therefore, inspect a sample of 261 records for traffic violations.

Power-Based Calculations for Sample Size

In "General Considerations for Determining Sample Size," parts 4(a) and 4(b) applied particularly to hypothesis testing. We saw the role that test power and the minimum detectable difference could play in determining the value of n. As illustrated in Figure 11.11: As the distance between the true mean and the assumed (null hypothesis) mean becomes smaller, β grows larger and so power grows smaller. If you want a small distance between the means to be detectable, then the sample size must increase to compensate. Some computer packages can handle the related calculations. Example 11.8 will be solved using Minitab because Excel and the core modules of SPSS do not offer this feature.

Example 11.8

A breeding association is preparing to test a claim that cats in a certain genetic line have a longer average life span than is typical for the species. The association plans to use a t-test for the mean, using α equal to 0.04. Suppose that life spans for cats in general are normally distributed, with a mean of 16.5 years and a standard deviation of about 1.7 years. What minimum sample size is required to ensure that a minimum difference of 1.5 years could be detected with a power of (1) 70%, (2) 80%, and (3) 90%? Interpret the results.

Solution

H_1 for the intended one-tail t-test is $\mu > 16.5$. If we input the test specifications into the computer, we find that the required sample sizes are (1) 9, (2) 11, and (3) 14 (see the screen capture in "In Brief 11.3"). These are the sample sizes needed to have the specified degrees of confidence that a difference of means equal to $+1.5$ years would be detected as significant by the test.

• *In Brief 11.3 Power-Based Determinations of Sample Size*

Of the software programs used in this textbook, only Minitab offers this feature.

Minitab

Minitab can determine necessary sample sizes for a variety of tests. The options all work similarly to the one-sample *t*-test option described here. In this case, start with the menu sequence *Stat / Power and Sample Size / 1-Sample t.* Input where indicated the desired minimum difference and the desired power. Note that more than one choice can be input; each solution will be listed separately. (It is also possible to input the sample size as given and calculate the resulting power or minimum difference.) Click on *Options.* You can now specify the significance level, as well as whether the alternative hypothesis is two-tail, left-tail, or right-tail. Click *OK* twice.

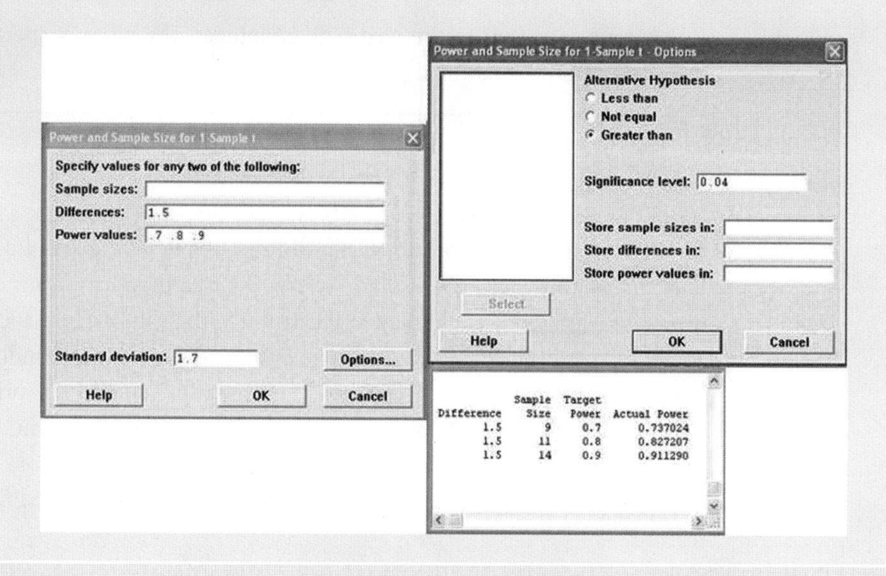

11.4 Test for a Single Proportion

Hypothesis tests can also be performed using proportions. If the data are categorical and can be slotted into two categories, then the formally appropriate model for the sampling distribution is the binomial distribution. This model presumes that if H_0 is true, then the probabilities of success will remain the same as the proportion of successes in H_0, assuming a random selection from an infinite population or sampling with replacement from a finite population. All of the observations should be independent.

When researchers had to rely on printed binomial tables, there were restrictions on the values of p and n that could be used, so testing for the proportion has traditionally been taught using the normal approximation for the binomial. (In this text, we referred to that model in Sections 9.3 and 10.4.) Provided the sample is large enough (i.e., $n \times p \geq 5$, and $n \times (1 - p) \geq 5$), the normal model works reasonably well—*if* the assumed $p \approx 0.5$ and/or the test is one-tailed. (Some texts recommend $n \times p \geq 10$ and $n \times (1 - p) \geq 10$). Using a software program, we can avoid these restrictions by using the binomial model (p-value approach). In Example 11.9, the p-value is taken from computer outputs as shown in "In Brief 11.4" starting on page 477.

Example 11.9

Disagreeing with a paper by G. C. de Strobel, an astronomer claims that fewer than 11% of stars that have extra-solar planets (ESPs) also have low metallicity.[4] Suppose that in the next five years, 200 ESP stars are discovered and that 18 of these have low metallicity. Test the astronomer's claim at the 0.05 level of significance.

Solution We follow the eight-step hypothesis-testing procedure described in Section 11.1.

Step 1: The astronomer claims that the proportion of "successes" (stars with low metallicity) in the population of ESP stars is less than 11%. In symbols:

$$p < 0.11$$

Step 2: Since the claim of step 1 does not entail equality, we interpret it as the alternative hypothesis. The opposite of that hypothesis becomes H_0.

$$H_0: p \geq 0.11$$

$$H_1: p < 0.11$$

Step 3: The level of significance is given in the problem:

$$\alpha = 0.05$$

Step 4: The sample size (200) is given. The test is for a claim about a single proportion. This can be described as:

Use a binomial test for a single proportion, $n = 200$.

Step 5: A formal depiction of the test model would show a histogram of the binomial distribution, shown on the left in Figure 11.12. A quick sketch, shown on the right in the figure, would be sufficient for displaying the basic elements of the test (e.g., it is left-tailed and *approximately* normal in appearance (although the tails are *not*

Figure 11.12
Depiction of Model for Step 5, Example 11.9

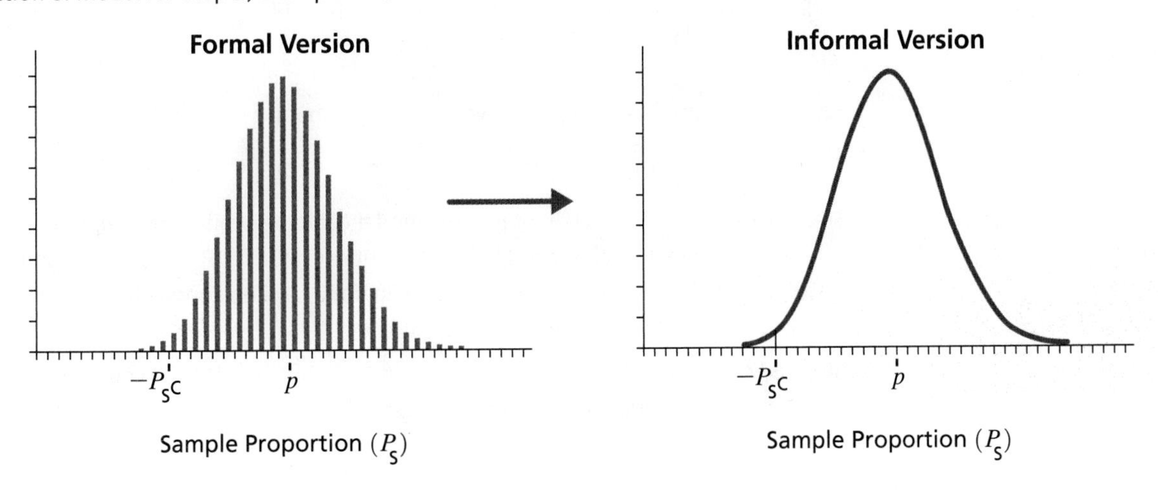

really symmetrical when $p \neq 0.5$)). Although a critical value for the sample proportion is displayed, it is not necessary to calculate the exact value when using the p-value approach.

Steps 6/7: The required data are provided in the problem. With the computer (see "In Brief 11.4" starting on page 477), we can calculate the p-value:

$$p\text{-value} = 0.218$$

The p-value of 0.218 is not less than α (0.05). In other words, it would not be usual, if H_0 was true, to draw a sample proportion that is less than the assumed value by the magnitude $(18/200 - 0.11)$. Therefore, the statistical decision is:

Do not reject H_0.

Step 8: Relating the statistical decision back to the original claim, we conclude:

The evidence does not support the claim that the proportion of ESP stars with low metallicity is less than 11%.

Example 11.10

In 20 years, the technology to detect stars that have extra-solar planets (ESPs) will have improved considerably. Suppose at that time that a new sample is taken of 200 ESP stars to test a claim that the proportion of ESP stars with low metallicity is *equal to* 13%. Of the 200 stars sampled, 36 of the stars have low metallicity. Test the claim at the 0.04 level of significance.

Solution

Step 1: The revised claim is that the proportion of "successes" (ESP stars with low metallicity) is equal to 13%. In symbols:

$$p = 0.13$$

Step 2: Since the claim of step 1 entails equality, we interpret it as the null hypothesis. The opposite of that hypothesis becomes H_1.

$$H_0: p = 0.13$$

$$H_1: p \neq 0.13$$

Step 3: The level of significance is given in the problem:

$$\alpha = 0.04$$

Step 4: The sample size (200) is given. For a test about a single proportion:

Use a binomial test for a single proportion, $n = 200$.

Step 5: An informal depiction of the test model could be similar to the right-hand portion of Figure 11.12. However, a second critical value for P_s should appear on the right tail, to indicate a two-tail test. Again, for a p-value approach, it is not necessary to calculate its exact location.

Steps 6/7: The required data are provided in the problem. With the computer (see "In Brief 11.4" starting on page 477) we can calculate the p-value:

$$p\text{-value} = 0.045$$

The p-value of 0.045 is not less than α (0.04). In other words, it would not be considered usual, if H_0 was true, to draw a sample proportion as distant as (36/200 − 0.13)—*in either direction*—from the assumed value. Therefore, the statistical decision is:

Do not reject H_0.

Step 8: Relating the statistical decision back to the original claim, we conclude:

There is not sufficient evidence to reject the claim that the proportion of ESP stars with low metallicity is equal to 13%.

If we take another look at the data underlying Example 11.10, we will find that, depending on where the reference value for α falls, the normal approximation for the binomial might lead us to an opposite statistical decision than using the true binomial model. The solution for Example 11.10 uses a true binomial two-tail test. For this procedure, when $p \neq 0.5$, we must find a separate p-value for each tail and then add the left and right p-values; i.e., the distribution is not symmetrical. (For more details, see "In Brief 11.4".)

The normal approximation, on the other hand, is symmetrical, so the two-tail p-value is simply 2 times the one-tail p-value. If Example 11.10 was done again with the normal approximation, the one-tail p-value (if based on z) would be 0.0178; multiplying times two, the calculated p-value would be $0.0178 \times 2 = 0.036$, which is less than α. As long as you report the p-value and the method used, readers can perhaps make their own judgment about such borderline cases.

• *In Brief 11.4 Computer Assistance for a One-Sample Binomial Test for the Proportion*

Excel

Although Excel does not offer an explicit test for the proportion, you can easily generate the p-value on which to base the statistical decision for the binomial test. For this we apply Excel functions that compute values of cumulative binomial probabilities; these functions were introduced originally in "In Brief 7.2(b)."

Three possible cases are described in the illustration on the next page. All models presume that we know the sample size n, the test statistic for the proportion p_s, and the population proportion p as assumed in H_0. The number of successes (X) in the sample can be derived by $p_s \times n$.

Generating the* p*-Value in Excel

Case 1: Find the p-value for a left-tail test.
Knowing X, n, and p, use the function shown in the illustration to find the probability of randomly obtaining a number of successes equal to or less than X. This is the desired p-value.

Case 2: Find the p-value for a right-tail test.
Knowing X, n, and p, use an expression in the format shown in the illustration for Case 2. The first term in the function **BINOMDIST** is $X - 1$. This function

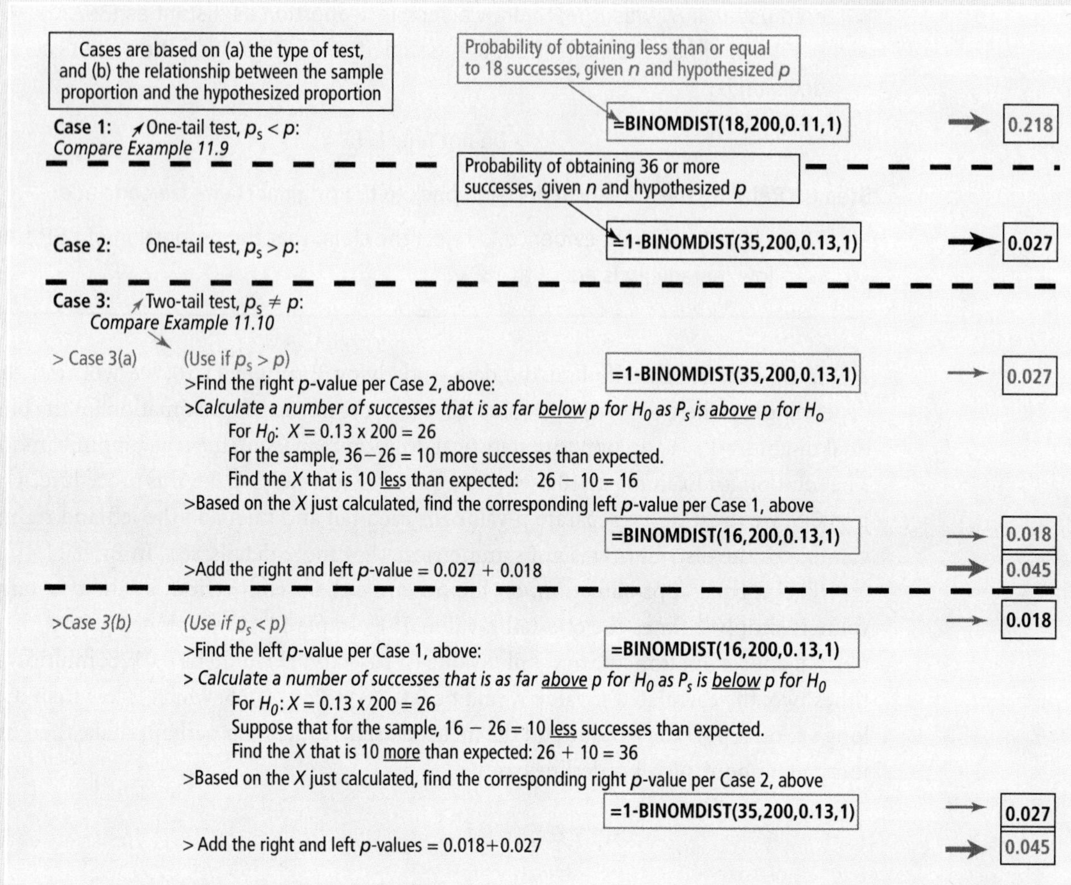

finds the probability of randomly obtaining a number of successes equal to or less than $(X - 1)$. Taking the complement of this value (i.e., subtracting this probability from 1) gives the probability of randomly obtaining a number of successes *equal to or greater than X*. This is the desired *p*-value.

Case 3: Find the *p*-value for a two-tail test. The probabilities on the two tails are not symmetrical. We can use the methods of Cases 1 and 2 to find left *and* right *p*-values, respectively, and then *add* the two values. The illustration shows what to use as the *X*-values for the calculations: On the tail where p_s was found, you calculate the one-tail *p*-value normally. Then, as shown, find an X_2 on the other tail that is equidistant from *p* compared to *X*—and use *this second X*-value to find the second *p*-value. The desired *p*-value is the sum of these two intermediate values.

Minitab

Minitab can generate the *p*-value on which to base the statistical decision for the binomial test. Use the menu sequence *Stat / Basic Statistics / 1 Proportion.* Input (as shown in the screen capture) the summary data for the number of trials and the number of events (i.e., successes). Alternatively, if raw qualitative data are available, select *Samples in columns* and specify the column containing the data. Then select *Options* and, in the resulting dialogue box, input the test proportion as specified in H_0. Your choice in the *Alternative* window will indicate a two-tail, right-tail, or left-tail test, as appropriate. Click *OK* twice.

Results may differ from other programs due to Minitab's internal procedures.

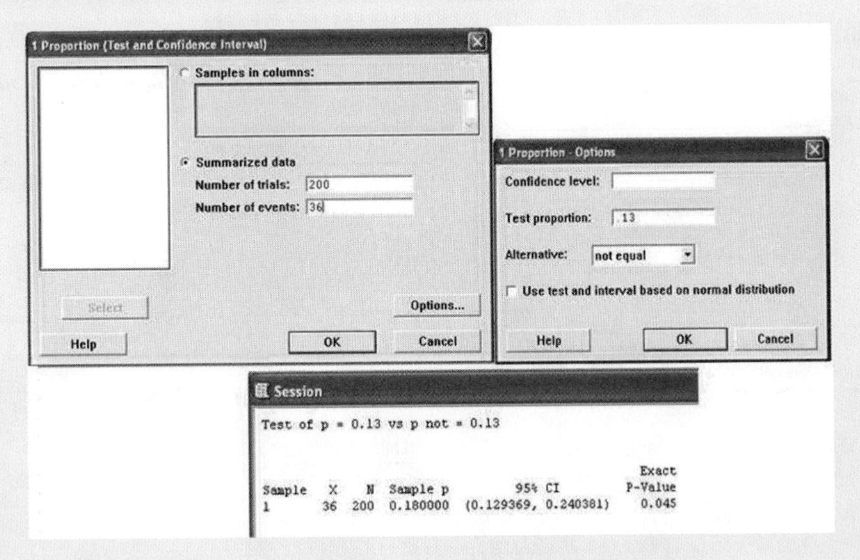

SPSS

SPSS provides a function (not shown here) explicitly for testing the binomial; this can be used for a hypothesis test for the proportion. (Begin with the menu sequence *Analyze / Nonparametric Tests / Binomial.* Specify the variable being tested, then click on *Options,* and then select *Exact Test.*) However, this function is quite restrictive in its requirements: All raw data must be available, not merely the counts of X and n; the categorical data must be coded into numbers (e.g., zeros and ones); and the test must be one-tail only—except for the case of testing H_0: $p = 0.5$, which is the only value where the tails are symmetrical. Described below is a more general procedure, for which any choice of one or two tails can be used for any hypothesized value of the proportion. For this, we apply *SPSS* functions that compute values of cumulative binomial probabilities; these functions were originally introduced in "In Brief 7.2(b)."

The boxed formulas shown in the illustration on the next page represent the formulas to input into the *Transform / Compute* dialogue box in SPSS.

Generating the p*-Value in SPSS*

Three possible cases are described in the illustration. All models presume that we know the sample size n, the test statistic for the proportion p_s, and the population proportion p as assumed in H_0. The number of successes (X) in the sample can be derived by $p_s \times n$.

Case 1: Find the *p*-value for a left-tail test. Knowing X, n, and p, use the function shown in the illustration to find the probability of randomly obtaining a number of successes equal to or less than X. This is the desired p-value.

Case 2: Find the *p*-value for a right-tail test. Knowing X, n, and p, use an expression in

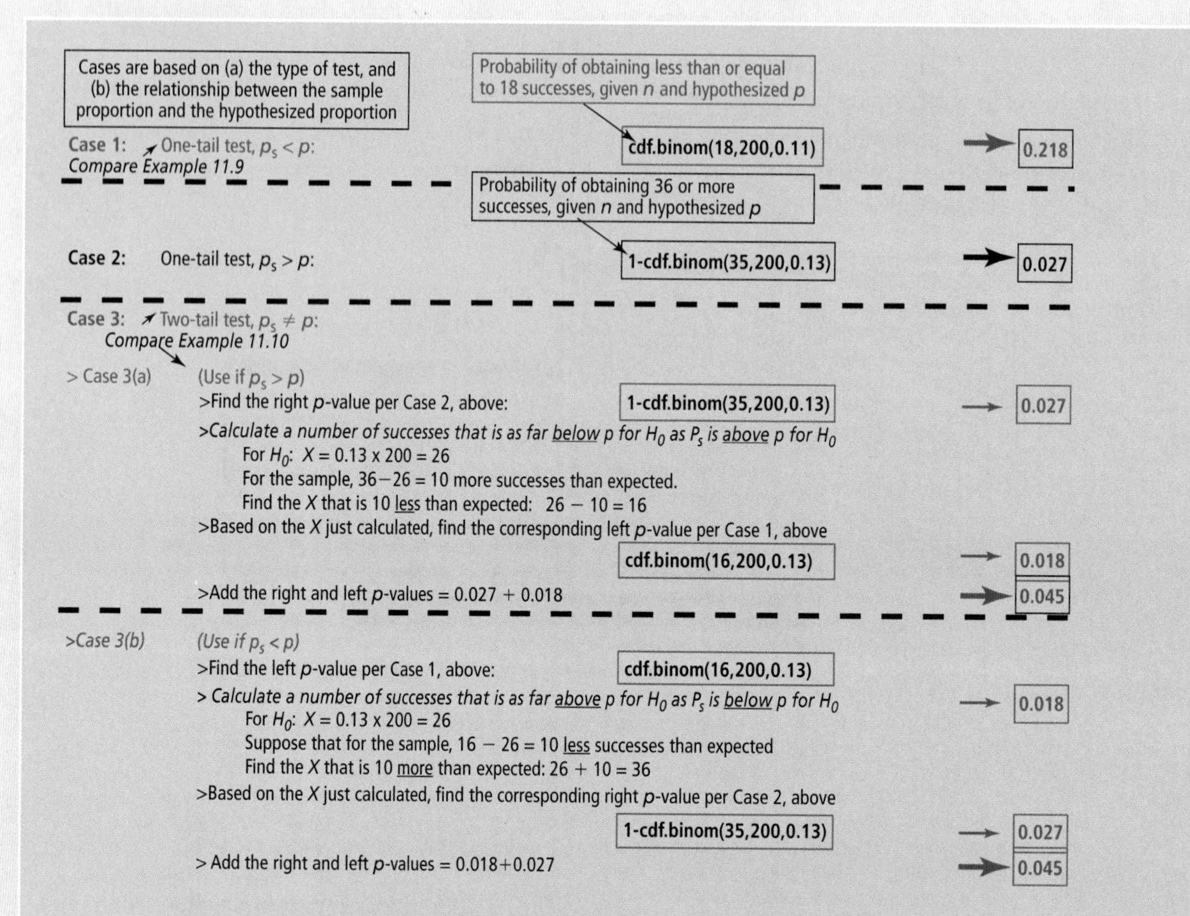

the format shown in the illustration for Case 2. The first term in the function **CDF.BINOM** is $X - 1$. This function finds the probability of randomly obtaining a number of successes equal to or less than $(X - 1)$. Taking the complement of this value (i.e., subtracting this probability from 1) gives the probability of randomly obtaining a number of successes *equal to or greater than X*. This is the desired p-value.

Case 3: Find the p-value for a two-tail test. The probabilities on the two tails are not symmetrical. We can use the methods in Cases 1 and 2 to find left *and* right p-values, respectively, and then *add* the two values. The illustration shows what to use as the X-values for the calculations: On the tail where p_s was found, you calculate the one-tail p-value normally. Then, as shown, find an X_2 on the other tail that is equidistant from p compared to X, and use this second X-value to find the second p-value. The desired p-value is the sum of these two intermediate values.

• Hands On 11.2 *To Use—Or Not to Use—the Normal Approximation of the Binomial?*

If in solving Example 11.10 we had used the normal approximation for the binomial (which is a conventional test for the proportion), we would have obtained a p-value = 0.036, instead of 0.045 using the binomial model. This would have reversed our statistical decision for the test, given that $\alpha = 0.04$. Yet the p-values differed by only 0.009. Is this problem important? How far apart can the p-values become when comparing the two models and under what conditions does this distance make a difference?

To answer these questions, try creating a series of examples for testing the proportion. Factors to vary include (1) one or two tails, (2) the H_0 assumption for p, (3) the sample size n, and (4) the number of successes (X) in the sample. For each trial case, find and compare the p-value by these methods:

1. Use the normal approximation for the binomial by finding

$$z = \frac{p_s - p}{\sqrt{\dfrac{p \times (1 - p)}{n}}}$$

Then the p-value (one-tail) is the probability of obtaining a z of that magnitude or greater if, per the null, the true $z = 0$. (See "In Brief 8.1(a).") For a two-tail p-value, use 2 times the one-tail p-value.

2. Find the binomial-based p-value by one of the methods described in "In Brief 11.4." (Note: When using the formula in Excel or SPSS to find the "other tail's" p-value, restrict if necessary the value of X to 0 for the minimum and to the original X for the maximum.)

Can you detect a pattern of where the p-values obtained by the different methods diverge? Especially consider cases where the values are on either side of a typical α-value (e.g., 0.01 or 0.05). A rule of thumb is to not use the normal approximation of the binomial if either $(n \times p)$ or $[n \times (1 - p)]$ is less than 5. If you used this rule, would this avoid most of the discrepancies in the p-values that you discovered in your experiment?

Chapter Summary

11.1 Describe and apply the core concepts, terminology, and steps for implementing hypothesis tests by either a critical value or *p*-value approach.

People make many **claims** or assertions about facts. In this text, we use the term **hypothesis** in a technical sense, to mean a statement or claim *made specifically in order to test that statement or claim,* using the methods described in this section. Taking steps to formally evaluate a hypothesis is called **hypothesis testing.** There are many variations, but hypothesis tests are all based on a common set of core concepts and follow similar steps. The core concepts include:

1. The **null hypothesis** (H_0) and **alternative hypothesis** (H_1 or H_a), which are logically opposed. The null hypothesis always includes the concept of equality (i.e., the logical operator: $=$ or $\leq$ or $\geq$).
2. The **critical region,** which is a portion of the hypothesized sampling distribution for which the probability that the sample statistic will occur there is low, so we reject the null hypothesis if the sample statistic does fall in the critical region.
3. a) The **level of significance** of the hypothesis test (α)**,** which expresses how small the probability is that a sample statistic will fall in the critical region if the null hypothesis is true.

   b) A **Type I error,** which occurs if H_0 happens to be true but the null hypothesis is rejected.

   c) The **critical value** or **values,** which are the boundaries (given in standardized units, depending on the model used for the sampling distribution) that mark the beginning of the critical region.

   d) The **sample statistic,** which is converted into the appropriate standardized score—the **test statistic**—for the sampling distribution and used to determine whether the sample result falls in the critical region.
4. A **Type II error,** which occurs if the null hypothesis (H_0) is in fact false but we mistakenly *failed to reject* H_0.

Eight steps are identified for the hypothesis-testing process: (1) Write down the claim, preferably in symbols; (2) state the null and alternative hypotheses; (3) choose and list the level of significance; (4) specify the test and the sample size or sizes; (5) display the null distribution, which is the expected sampling distribution of the test statistic if the null hypothesis is true; (6) collect the data; (7) calculate the test statistic or ***p*-value,** compare it to step 5, and make the **statistical decision** in relation to the hypothesis test. (The *p*-value is the conditional probability of obtaining a sample statistic at least as far from the expected parameter as has occurred, if in fact the sampling distribution based on H_0 applied); (8) clearly express the conclusion—in words.

11.2 Conduct a *z*-test for the population mean and a *t*-test for the population mean, and determine which of these tests is suitable for a given problem.

Applications of the above procedures can be distinguished based on three main factors: Is the test based on one sample or more than one? What is the parameter of interest? And what is the model for the sampling distribution? The tests in this section are both one-sample tests for the population mean. If the population is normally distributed, or the sample size is large, *and* if the population value for σ is known, then the sampling distribution can be presumed to be normal, so a *z*-test for the mean can be conducted. Under the same conditions, but if the population value for σ is *not* known, then the sampling distribution to use is the *t*-distribution with degrees of freedom $= n - 1$.

11.3 Describe some key reasons why the sample size—for hypothesis testing *or* for constructing confidence intervals—may need to be increased, and calculate minimum needed sample sizes for constructing one-sample confidence intervals.

Counter to pressures such as time and cost in favour of small samples, there are several reasons you may require an increased sample size: (1) to reduce the risk α for Type I error in hypothesis testing; (2) if the variability of the data (σ) is large; (3) when constructing confidence intervals, to lower the magnitude for the error term (E) considered acceptable; and (4) when hypothesis testing, to reduce the **β risk** for Type II error, and to increase the **power of a test** to detect a difference of the population parameter from its presumed value in H_0.

The power of a hypothesis test is, mathematically, $1 - \beta$. There are actually many possible values of β, depending on the actual (but unknown) value for the population parameter. An **operating characteristic curve** can display, over a range of possible distances between the actual and hypothesized parameters, the values of β that correspond to each distance. The complement of the operating characteristic curve is called the **power curve,** which shows, for any distance between the true and hypothesized parameters, the corresponding probability $(1 - \beta)$ of correctly rejecting the null hypothesis.

11.4 Conduct a hypothesis test for the population proportion based on the binomial distribution.

For a hypothesis test on the sample proportion, the binomial distribution is the formally appropriate model to use for the sampling distribution. When the hypothesized p-value is other than 0.5, the distribution is not symmetrical, so (unlike cases for the mean) the two-tail p-value does not simply equal two times the one-tail p-value.

• Key Terms

alternative hypothesis (H_1 or H_a), p. 445
β risk, p. 470
claim, p. 444
critical region, p. 446
critical value (*or* values), p. 447
hypothesis, p. 444
hypothesis testing, p. 444
level of significance (α), p. 447
null distribution, p. 453
null hypothesis (H_0), p. 445
operating characteristic curve, p. 470
power curve, p. 470
power of a test, p. 469
p-value, p. 455
sample statistic, p. 448
statistical decision (for the hypothesis test), p. 457
test statistic, p. 448
Type I error, p. 447
Type II error, p. 450

• Exercises

Basic Concepts 11.1

1. What is a null hypothesis?
2. What is an alternative hypothesis?
3. Is every "claim" the same as an alternative hypothesis? Why or why not?

4. Which of the following expressions has the form of a null hypothesis?
   a) $\mu \neq 29$
   b) $\sigma = 5.46$
   c) $y = 2x + 4$
   d) $\sigma > 5.46$
   e) $\mu_{Winnipeg} \leq \mu_{Vancouver}$
5. Which of the following expressions has the form of an alternative hypothesis?
   a) $\mu \neq 29$
   b) $\sigma = 5.46$
   c) $y = 2x + 4$
   d) $\sigma > 5.46$
   e) $\mu_{Winnipeg} \leq \mu_{Vancouver}$
6. A student born in British Columbia who is attending university in Manitoba complains that the mean annual amount of snowfall in Winnipeg is greater than the mean annual amount of snowfall in Vancouver.
   a) Formally represent the implied claim in symbols.
   b) What is the null hypothesis (in symbols)?
   c) What is the alternative hypothesis (in symbols)?
7. Explain the difference between Type I and Type II errors. Why are we concerned with Type I errors? Why are we concerned with Type II errors?
8. Define what is meant by *level of significance.*
9. Define the term *critical region* and explain the role of this concept in a hypothesis test.
10. What is the relationship between a critical region and a critical value?
11. What is the relationship between a sample statistic and a test statistic?
12. Differentiate between one-tailed and two-tailed hypothesis tests.
13. For some hypothesis test, $z_c = \pm 2.58$, and the test statistic is found to be $z = 2.71$. What is the correct statistical decision?
14. For another hypothesis test, $\alpha = 0.05$ and the p-value is found to be 0.077. What is the correct statistical decision?
15. What are some advantages of using the p-value approach to hypothesis testing? Under what circumstances might using a critical value approach be more practical? Give an example.
16. Farmer A claims that a larger proportion of his cows yield top-grade milk than do his neighbour's. His neighbour, Farmer B, is putting the claim to a hypothesis test.
   a) Express Farmer A's claim in symbols.
   b) What is the null hypothesis for testing A's claim?
   c) What is the alternative hypothesis?
17. Suppose that the significance level is 0.05 for the hypothesis test in Exercise 16. After sampling, it turns out that the p-value is 0.062.
   a) What is the statistical decision?
   b) What conclusion can Farmer B express to Farmer A, *in clear words?*
18. Based on exit polls, J. J. Jackson believes that he attained more than half of today's votes for Ward 7 Trustee.
   a) Formally express J. J. Jackson's claim.
   b) If the claim is put to a test, what is the null hypothesis?
   c) What is the alternative hypothesis?
19. Given the test hypotheses in Exercise 18, describe the possible Type I and Type II errors.
20. In spite of attempted improvements, the manager at Red Deer Driving Schools believes that there is still an average of four near-accidents per outing with new driving students.
   a) Formally express the manager's claim.
   b) If the claim is put to a test, what is the null hypothesis?
   c) What is the alternative hypothesis?
21. Would a null hypothesis be accepted or rejected if the level of significance was 0.01 (one-tailed test) and the p-value was 0.008? Explain.
22. Would your answer to Exercise 21 change if the test was a two-tailed test? Explain.
23. What test hypotheses would you use to back up a claim that a coin is biased to land on heads (i.e., it is more likely to land on a head) when tossed?
24. Is it possible to get a large difference between the sample statistic and the hypothesized population parameter, even when the null hypothesis is true? Explain.
25. If the null hypothesis is not rejected:
   a) Has the null hypothesis been proved true?
   b) Has the null hypothesis been supported?

   Explain your answers.
26. If the null hypothesis *is* rejected:
   a) Has the alternative hypothesis been proved true?
   b) Has the alternative hypothesis been supported?

   Explain your answers.

For applied problems, see the Applications exercises in the following sections.

Basic Concepts 11.2

27. Under what conditions is the normal distribution used for hypothesis testing?
28. Is it possible for the null hypothesis to be rejected at the 5% level of significance, but not rejected at the 1% level of significance?

a) If so, explain how this apparent contradiction could occur.

b) Explain how, even in such a case, the statistician can still make a statistical decision.

29. In each of the following questions, suppose you are conducting a *z*-test for the mean (σ known), using a critical value approach. For each case, sketch and label a model of the test—as recommended in step 5 of the hypothesis-testing process. That is, show a curve of the sampling distribution of the sample means as it would be if H_0 was true, and show/label the critical value or values and critical region(s).

a) Right-tailed test, $\alpha = 0.01$.
b) Two-tailed test, $\alpha = 0.05$.
c) Left-tailed test, $\alpha = 0.005$.
d) Two-tailed test, $\alpha = 0.10$.
e) Right-tailed test, $\alpha = 0.02$.
f) Two-tailed test, $\alpha = 0.02$.

30. In each of the following questions, suppose you are conducting a *t*-test for the mean (σ unknown), using a critical value approach. For each case, sketch and label a model of the test—as recommended in step 5 of the hypothesis-testing process. Show/label the critical value or values, and critical region(s).

a) Right-tailed test, $\alpha = 0.01$, $n = 15$.
b) Two-tailed test, $\alpha = 0.05$, $n = 10$.
c) Left-tailed test, $\alpha = 0.005$, $n = 22$.
d) Two-tailed test, $\alpha = 0.10$, $n = 18$.
e) Right-tailed test, $\alpha = 0.02$, $n = 9$.
f) Two-tailed test, $\alpha = 0.02$, $n = 12$.
g) Two-tailed test, $\alpha = 0.02$, $df = 12$.

31. In each of the following questions, suppose you are conducting a *z*-test for the mean (σ known), using a critical value approach. For each case, what is the value of the test statistic?

a) $H_0: \mu = 50$ $\quad \sigma = 5$ $\quad \bar{x} = 51.75$ $\quad n = 35$
b) $H_0: \mu = 22$ $\quad \sigma = 1.5$ $\quad \bar{x} = 20.85$ $\quad n = 15$
c) $H_0: \mu \leq 33$ $\quad \sigma = 6$ $\quad \bar{x} = 35.73$ $\quad n = 65$
d) $H_1: \mu < 70$ $\quad \sigma = 8$ $\quad \bar{x} = 66.9$ $\quad n = 30$
e) $H_1: \mu > 15$ $\quad \sigma = 2.5$ $\quad \bar{x} = 14$ $\quad n = 25$

32. Suppose each hypothesis test in Exercise 31 is at the 0.05 level of significance. What should be the statistical decision for each case? Which results, if any, would be different if the tests were at the 0.01 level of significance? Determine the *p*-values for each of the above tests.

33. In each of the following questions, suppose you are conducting a *t*-test for the mean (σ unknown), using a critical value approach. For each case, what is the value of the test statistic?

a) $H_0: \mu = 50$ $\quad s = 5.57$ $\quad \bar{x} = 52.60$ $\quad n = 35$
b) $H_0: \mu = 22$ $\quad s = 1.38$ $\quad \bar{x} = 20.71$ $\quad n = 15$
(normally distributed population)
c) $H_0: \mu \leq 30$ $\quad s = 3.46$ $\quad \bar{x} = 30.90$ $\quad n = 65$
d) $H_1: \mu < 75$ $\quad s = 6.48$ $\quad \bar{x} = 71.96$ $\quad n = 30$
e) $H_1: \mu > 14.2$ $\quad s = 3.19$ $\quad \bar{x} = 14.35$ $\quad n = 40$

34. Suppose each hypothesis test in Exercise 33 is at the 0.05 level of significance. What should be the statistical decision for each case? Which results, if any, would be different if the tests were at the 0.01 level of significance?

35. Determine the *p*-values for each of the tests described in Exercise 33 and then redo Exercise 34 using this *p*-value information. Do the statistical decisions ever differ from the critical value approach?

Applications 11.2

As part of the solution for all hypothesis tests, formally show the eight-step sequence illustrated in the text. Unless otherwise specified by your instructor, use your own judgment on whether to use the critical value or *p*-value approach.

36. A criminologist found that the mean length of a jail sentence given in Canada in 2001/2002 was 4.10 months, $\sigma =$ 9.75 months.[5] If in 2008 the variance has not changed, but a random sample of 35 sentences has a mean sentence length of 3.97 months, test a claim that mean sentence length in 2008 is less than the previous value. Use a 0.05 level of significance.

37. A movie fan named Elwy is planning to buy a certain model of subwoofer for his home theatre and has found it is available online, with $\mu =$ US\$1,175, $\sigma =$ US\$176.[6] Assume that there are hundreds of online offerings and that the prices are normally distributed. To test a claim by a new website that its online prices for subwoofers are lower than the online average, Elwy samples 10 related products from the new website's catalogue. The sample mean price is US\$1,152. Test the website's claim at the 0.02 level of significance.

38. Based on published test results by Saskatchewan Highways and Transportation, the bulk density of coarse aggregate in Marshal Mix (asphalt) in 2004 had a mean value of 2.667 g/cm^3. Standard deviation is 0.013 g/cm^3.[7] Based on a new random sample of 45 observations for which the mean bulk density is 2.661 g/cm^3, test a claim that the mean bulk density is different from its earlier value. Use the 0.01 level of significance.

39. In Exercise 38, an old (2004) standard deviation was interpreted as the population standard deviation ($\sigma = 0.013$ g/cm^3) for the hypothesis test. Carry out the hypothesis test again using the sample value for *s* ($s = 0.019$ g/cm^3). Is the conclusion different from Exercise 38? If so, which answer do you believe is more reliable, and why?

40. According to an evaluation performed by the Tolly Group, the audio product OCT6100 Series Echo Canceller obtained a mean score of 3.43 on measurements of perceptual evaluation of speech quality (PESQ). (0 is poor and 5 is excellent).[8] Suppose this mean was close to the true population parameter for the year of testing. If a new sample of 40 observations is randomly selected and the PESQ scores have a mean of 3.55 and a standard deviation of 0.26, test a claim that the product is performing better than in the past. Use a 0.05 level of significance.

41. The list of mutual funds in the file **Canadian_Equity_ Funds** is a random sample from a large list of funds published in 2005 by Globefund.com. Test the claim that for the full population of Globefund listings at that time, the mean asset value was $130 million. Use the 0.05 level of significance. (Disregard funds for which no asset value is given in the file.)

42. From the sample of funds described in Exercise 41, use the 0.05 level of significance to test the claim that for the full population:
    a) The mean net asset value per share (NAVPS) was at least $42.
    b) The mean NAVPS was at most $20.
    c) Explain how the answers to the two preceding questions can *both* be correct.

43. From the sample of funds described in Exercise 41, use the 0.05 level of significance to test the claim that for the full population:
    a) The mean NAVPS was greater than $20.
    b) The mean NAVPS was greater than $16.

Basic Concepts 11.3

44. Last year you conducted a hypothesis test on a certain topic. This year, you plan to conduct a follow-up test. For each case below, tell whether the sample needs to be larger than last year or can be smaller than last year *if all other factors remain equal.*
    a) You want a reduced probability of a Type I error.
    b) You want a reduced probability of a Type II error.
    c) There is now less variability in the data than there used to be.
    d) You want the test to have more power than last year.
    e) You want to be able to detect a smaller difference of the true mean, relative to the hypothesized mean, compared to last year's test.

45. Last year you estimated a confidence interval for the mean for a certain variable. This year, you plan to reestimate the mean. For each case below, tell whether the sample needs to be larger than last year or can be smaller than last year *if all other factors remain equal.*
    a) You want to increase the level of confidence in the estimate.
    b) In the estimation process, you want to increase the value of E.
    c) There is now more variability in the data than there used to be.
    d) You want to decrease the margin of error.

46. Last year you estimated a confidence interval for the proportion for a certain variable. This year, you plan to reestimate the proportion. For each case below, tell whether the sample needs to be larger than last year or can be smaller than last year *if all other factors remain equal.*
    a) You are willing to accept a lower level of confidence in the estimate.
    b) In the estimation process, you want to increase the value of α.
    c) Whereas last year you had no prior estimate for population p, this year your preliminary estimate is that $p = 0.25$.
    d) You are willing to accept an increase in the margin of error.

47. Explain this passage from the text: "If all other factors are kept equal, *increasing the sample size reduces the β risk.*"

48. If *β risk* decreases, what happens to the power of the test? Explain.

49. Explain the meaning of the terms *operating characteristic curve* and *power curve.*

50. What sample size is required to estimate the proportion of first-year full-time students at postsecondary institutions who buy all of the textbooks that are assigned? The estimate is to be at the 95% level of confidence and accurate within ± 0.06. (No prior estimate is available for this proportion.)

51. Recalculate your answer to Exercise 50 if a preliminary estimate for the proportion is 85%.

52. What sample size is required to estimate the mean cost of textbooks for first-year full-time students at postsecondary institutions across Canada? The estimate is to be at the 90% level of confidence and accurate within $\pm\$65$. Past studies suggest that σ for the cost of first-year textbooks is $210.

Applications 11.3

53. Determine the sample size required to estimate the grizzly bear population density in British Columbia to within ± 2 bears per 1,000 square kilometres at a 96% level of confidence. For bear population densities across the province, the standard deviation is about 7 bears per 1,000 square kilometres.[9]

54. Determine the sample size required to estimate the percentage of grizzly bears in British Columbia that are female to within $\pm 4\%$, at a 99% level of confidence.

55. Determine the sample size required to estimate the percentage of grizzly bears that reside primarily in the Upper

Skeena-Nass and Tweedsmuir areas of British Columbia to within ±4% at a 95% level of confidence. (Note that, in the past, roughly 6% of British Columbia bears have lived in one of those two regions.)

56. Textbook publishers in the United States find that about 60% of faculty who order textbooks prefer to use the most current editions.[10] A publisher wants to confirm if the proportion is about the same in Canada. How big a sample is needed to estimate the proportion to within ±7% at a 95% level of confidence?

57. According to one estimate, only 13% of Canadian start-up businesses survive their first year.[11] If you want to draft a new estimate, to be accurate within ±5% at the 95% level of confidence, how big a sample should be used?

58. A published study examines the prevalence of various physical and emotional symptoms in patients who are undergoing maintenance hemodialysis for kidney disease. To estimate the mean number of distinct symptoms experienced by such patients, how large a sample size is required? The desired margin of error is ±0.25 symptoms and the intended confidence level is 95%. A previous study suggests that σ is about 1.8.[12]

59. Also examined in the study described in Exercise 58 was the percentage of patients who suffered from dry skin, in particular. In order to estimate this percentage to within ±6.5% at a confidence level of 94%, how big a sample is needed? Assume that there has been no prior estimate for this proportion.

Basic Concepts 11.4

60. For each of the following cases, determine whether the null hypothesis would be accepted or rejected at the 0.05 level of significance; p represents the population proportion and x is the number of successes in the sample.
    a) $H_1: p < 0.4$ $n = 15$ $x = 5$
    b) $H_1: p < 0.85$ $n = 20$ $x = 13$
    c) $H_1: p < 0.3$ $n = 20$ $x = 4$

61. Redo each of the tests in Exercise 60 at the 0.01 level of significance. Do any of the statistical decisions change? If so, which ones? Explain how to interpret this change from one version of the test to another.

62. For each of the following cases, determine whether the null hypothesis would be accepted or rejected at the 0.05 level of significance.
    a) $H_0: p \leq 0.8$ $n = 15$ $x = 13$
    b) $H_0: p \leq 0.2$ $n = 10$ $x = 5$
    c) $H_0: p \leq 0.8$ $n = 20$ $x = 19$

63. Redo each of the tests in Exercise 62 at the 0.01 level of significance. Do any of the statistical decisions change? If so, which ones?

64. For each of the following cases, determine whether the null hypothesis would be accepted or rejected (a) at the 0.05 level of significance and (b) at the 0.01 level of significance.
    a) $H_1: p \neq 0.4$ $n = 15$ $x = 5$
    b) $H_1: p \neq 0.7$ $n = 20$ $x = 16$
    c) $H_1: p \neq 0.3$ $n = 20$ $x = 5$
    d) $H_0: p = 0.8$ $n = 15$ $x = 9$
    e) $H_0: p = 0.2$ $n = 10$ $x = 5$
    f) $H_0: p = 0.8$ $n = 20$ $x = 17$

65. The tests for proportion described in this chapter presume that the distribution of sample proportions under the null hypothesis follow the binomial distribution. Explain why this is a valid model, provided that certain conditions are met.

66. List the conditions required for the binomial distribution to be a valid model for the sampling distribution of the sample proportion. Describe a situation where you might want to test the proportion, but the conditions for using the binomial model are *not* met.

67. A company makes team jackets for sports teams. In the past, 10% of all jackets needed some repair before being distributed to customers. This year, however, the manager claims that quality has improved: Only five jackets out of a random sample of 64 jackets needed some repair. Test the manager's claim ($\alpha = 0.02$).

68. Galluci's Home Heating believes that 90% of its customers are satisfied with the products and service they have received from the company. A survey of 150 customers found that 130 of them were satisfied. Test whether the data are consistent with the company's belief about customer satisfaction ($\alpha = 0.05$).

Applications 11.4

As part of the solution for all hypothesis tests, formally show the eight-step sequence illustrated in the text.

69. According to experts on the seasonal migration of Columbia spotted frogs in Idaho, about 8.5% of adult males of the species living in a certain region migrate from their breeding ponds to a separate summer habitat. To confirm this, a new sample is taken of 86 adult male frogs of the species in the region. About 7% of the observed frogs were found to migrate to a summer habitat.[13] At the 0.05 level of significance, test whether the experts' claim is accurate.

70. In Alberta, a way for organizations to raise funds for charitable or religious purposes is to obtain a licence for a casino gaming event. A variety of card and roulette-based games are permitted, with much of the revenue going to the good cause. A gaming official claims that about 10% of these casino gaming licences are granted to support the

arts.[14] An auditor is testing the claim, at $\alpha = 0.05$, and selects a random sample of 100 licences granted in the past year. She finds that 8% of these licences were to support the arts. What conclusion should be reached?

71. In spite of published articles to the contrary, a snowmobile fan believes that less than a quarter of all avalanche victims in Canada were riding snowmobiles at the time of the tragedy. A concerned relative is testing the fan's claim at the 0.05 level of significance. Based on selected past data, 33 of 142 avalanche victims were riding snowmobiles at the time.[15] What conclusion can the relative report to the snowmobiler?

72. In the month before Canada Day, 2003, SES Canada Research asked a random sample of Canadians how they planned to spend the holiday. Responses to the survey are in the file **Celebration_Canada_Day.** Test the claim that a third of the population planned to spend the day with friends and family. (Note: Because more than two types of responses are recorded in the file, you might have to convert all responses that are not "With Friends/Family" responses into "Other".) ($\alpha = 0.05$).

73. Referring to the pre-Canada Day survey in Exercise 72, test a claim that less than a quarter of the population was "Unsure" how they would spend the holiday. First use ($\alpha = 0.10$) and then retest using ($\alpha = 0.05$). Are the two test results the same? If not, explain any apparent difference.

74. Referring to the pre-Canada Day survey in Exercise 72, test a claim that 10% of the population *either* intended to be "Working" or intended to do "Nothing" during the holiday ($\alpha = 0.05$). (Hint: Let the binomial category for a "success" include giving *either* of those two answers.)

75. Referring to the pre-Canada Day survey in Exercise 72, test a claim that at most, half of the population did not intend to be with friends and/or family during the holiday ($\alpha = 0.01$).
    a) Base the test on an assumption that every answer *other* than "With Friends/Family" counts as an intention not to be with friends and/or family.
    b) Do you think that the assumption in (a) is realistic? If not, do you have enough information to test the claim in this question? Explain your answers. (Hint: Consider whether everyone who answered "Watching Fireworks" or "Road Trip" planned to do this activity *by themselves.*)

Review Applications

As part of the solution for all hypothesis tests, formally show the eight-step sequence illustrated in the text.

76. Health Canada reports that under standard test conditions, the mean formaldehyde content in direct cigarette smoke is 43.4 μg (micrograms) per cigarette. The standard deviation is 32.8 μg.[16] Suppose that 50 cigarettes of various "light" brands are randomly selected by a cigarette manufacturer, and the mean level of formaldehyde content in each cigarette's smoke is 41.2 μg. Test a claim by the manufacturer that the mean formaldehyde content in direct cigarette smoke from the "light" brands is less than the mean content in the population of regular cigarettes. Use $\alpha = 0.05$.

77. Repeat Exercise 76, but instead of assuming that the recorded value for σ applies also to the "light" brands, use in your calculations the sample standard deviation $s = 34.1$ μg. Does this affect the conclusion in this case?

78. In response to the manufacturer's claim in Exercise 76, Health Canada wants to establish a confidence interval for the mean formaldehyde content in direct cigarette smoke from "light" brands. They presume that the standard deviation is approximately equal to the value for regular brands—32.8 μg. The goal for the estimate is to have a margin of error of ± 4 μg and a confidence level of 95%. Find the sample size needed.

Exercises 79–82 are based on the following table of laboratory attenuation (noise reduction) values for foam earplugs under specified conditions, according to U.S. Occupational Safety and Health Administration testing.

Test Frequency (Hz)	125	250	500	1,000
Mean Attenuation (decibels)	17.9	19.0	21.0	24.7

Test Frequency (Hz)	2,000	4,000	8,000
Mean Attenuation (decibels)	29.9	35.6	34.6

Source: Safe@Work[SM], Inc. Based on National Institute for Occupational Safety and Health data in *New Developments in Hearing Protection Labeling.* Found at: http://www.safe-at-work.com/Reference/OSHA-HP-Labeling.html

79. For foam earplugs, the level of protection varies, based on the frequency of the sound (e.g., a deep rumble versus a high-pitched sound). Suppose that the means in the above table represent standards that any new brand of earplugs must satisfy. An inspector randomly selects 50 earplugs from the new brand EasyEar and assesses their mean attenuation at the frequency 125 Hz. The sample standard deviation is 7.2 dB and the sample mean is 16.4 dB. Test the claim that the EasyEar product provides less mean attenuation than is required ($\alpha = 0.01$).

80. Repeat Exercise 79 for the frequency 250 Hz. The sample standard deviation is 6.3 dB and the sample mean is 17.9 dB.

81. Repeat Exercise 79 for the frequency 1,000 Hz. The sample standard deviation is 6.2 dB and the sample mean is 21.3 dB.

82. Repeat Exercise 79 for the frequency 8,000 Hz—but this time the claim is that the product provides *more* attenua-

tion than the requirement. The sample standard deviation is 5.5 dB and the sample mean is 35.7 dB.

83. Suppose the U.S. Occupational Safety and Health Administration also wants to assess attenuation by earplugs at a frequency of 6,000 Hz. For the currently available products, it wants to estimate the mean attenuation to within ± 3 decibels, with 95% confidence. The experts estimate that the population $\sigma = 5.2$ dB. How large a sample should be taken?

84. A company that makes a type of earplug has been having a problem with quality control: Some (but not all) of their products have been found deficient in providing attenuation for frequencies above 4,000 Hz. As a first step, company managers want to estimate what proportion of the earplugs are deficient in this respect. How large a sample should be taken to estimate the percentage of defective products to within $\pm 4\%$, with 96% confidence? (A pilot study suggests that the proportion of defective products is about 5.5%.)

85. Commonly accepted values for the mean and standard deviation of "normal" eye pressure are 15.5 mmHg (millimetres of mercury) and 2.5 mmHg, respectively. In a study of 60 randomly selected patients at risk of glaucoma, their mean eye pressure was 16.2 mmHg.[17] Test a claim that this population has a higher than "normal" mean eye pressure ($\alpha = 0.05$).

86. Suppose that for the sample of patients in Exercise 85 the standard deviation for their eye-pressure parameters was 2.9 mmHg. Using this additional information from the sample, recalculate the solution to Exercise 85.

Exercises 87–90 are based on data in the file **Water_Use**.

87. Environment Canada publishes a summary of water use by every Canadian municipality with a population of over 1,000. For analysis, take a systematic sample that includes (a) the 12th record in the data file and (b) every 30th record thereafter. On the basis of that sample, test a claim that the mean domestic water use in all municipalities of this size is greater than 6,000 cubic metres per day. Use the 0.04 level of significance.

88. On the basis of the sample described in the previous exercise, test a claim that the mean commercial/institutional water use in municipalities with populations over 1,000 is equal to 2,500 cubic metres per day. Use the 0.04 level of significance.

89. On the basis of the sample described in Exercise 87, test a claim that the mean latitude for Canadian municipalities with populations over 1,000 is *not* equal to 47° N. Use the 0.04 level of significance.

90. On the basis of the sample described in Exercise 87, test a claim that over three-quarters of Canadian municipalities with populations greater than 1,000 disinfect their water using chlorine. Use the 0.01 level of significance.

91. Suppose a researcher wants to estimate domestic water use for Canadian municipalities having a population from 200–1,000. She plans for the estimate to have a margin of error of ± 100 cubic metres per day and a 95% confidence level. Based on past years data, σ is estimated to be 200. How many municipalities should be included in the sample?

92. The researcher in Exercise 91 also wants to estimate the proportion of municipalities with populations under 1,000 that use chlorine to disinfect the town's water supply. How large a sample should be taken to estimate this percentage to within $\pm 6\%$, with 95% confidence? (The researcher has no initial estimate for this proportion.)

Two-Sample Tests of Significance

Chapter 12

Chapter Outline

Learning Objectives

12.1 Describe the distinction between independent and dependent samples, and explain some key concepts and terminology relating to collecting data for making inferences for two samples.

12.2 Conduct a *z*-test or a *t*-test for the difference in population means for independent samples—knowing which test is appropriate—and construct confidence intervals for the difference in population means for independent samples.

12.3 Conduct a *t*-test for the difference in population means for data taken from two dependent samples, and construct confidence intervals for the difference in population means for two dependent samples.

12.4 Conduct a *z*-test for the difference in population proportions, and construct confidence intervals for the difference in population proportions.

12.1 Inferences for Two Samples

Each of the inferences in the previous two chapters was based on one sample. From the sample data, we learned to either estimate the value of a parameter or test a hypothesis about the value of a parameter. In this chapter, each technique begins with *two* samples. In general, we compare sample statistics from the two samples (such as for means or proportions) and make inferences about the corresponding population parameters. For example, we can estimate the difference between these parameters in the populations, or test whether the populations have similar values for the characteristic. In fact, we might conclude that both samples were drawn from the same population.

As in previous chapters, all of the procedures and tests described here assume that the samples have been randomly drawn. Two categories of these procedures can be distinguished: (1) Procedures that assume independent samples and (2) procedures that assume dependent samples.

Independent Samples

Samples are **independent** if the selection of elements for one sample is not influenced by the selection of elements for the other. That is, each sample is chosen independently of the others. Suppose a company is planning to relocate in British Columbia, in either Kamloops or Kelowna. By randomly sampling workers from each community, the company hopes to estimate the mean wage rate in the two competing areas. Since, presumably, the wages of selected workers in Kamloops have no bearing on the wages of selected workers in Kelowna, the two samples are independent.

There are two main approaches for selecting two samples that are independent of each other. Figure 12.1 illustrates the model in which the compared populations are predetermined on some basis prior to sampling. Perhaps we are testing the relative proportions of Ontarians and Quebeckers for whom French is the first language. Persons from each province could be sampled independently, and data from the two samples could then be compared.

Figure 12.1
Comparing Sample Statistics to Make Inferences about Population Parameters

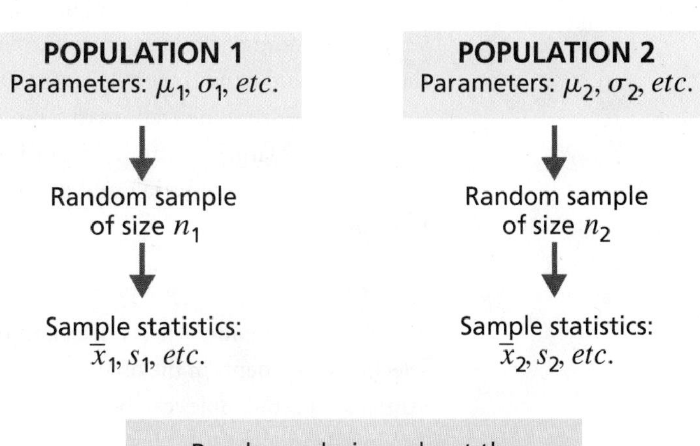

A second approach is to select the two independent samples as part of a **controlled experiment.** This method is often used in a laboratory setting. An original sample is taken from a single specified population; the distinction into two groups is then introduced by the experimenter. To the elements or participants in one group (called the **experimental group** or **treatment group**), an extra experimental procedure (treatment) is applied. To the elements or participants in the other group (called the **control group**), the extra procedure is not applied. The statistics associated with each group are then compared. (See Figure 12.2.)

Figure 12.2
Comparing Sample Statistics for the Control and Experimental Groups to Detect Differences in Results

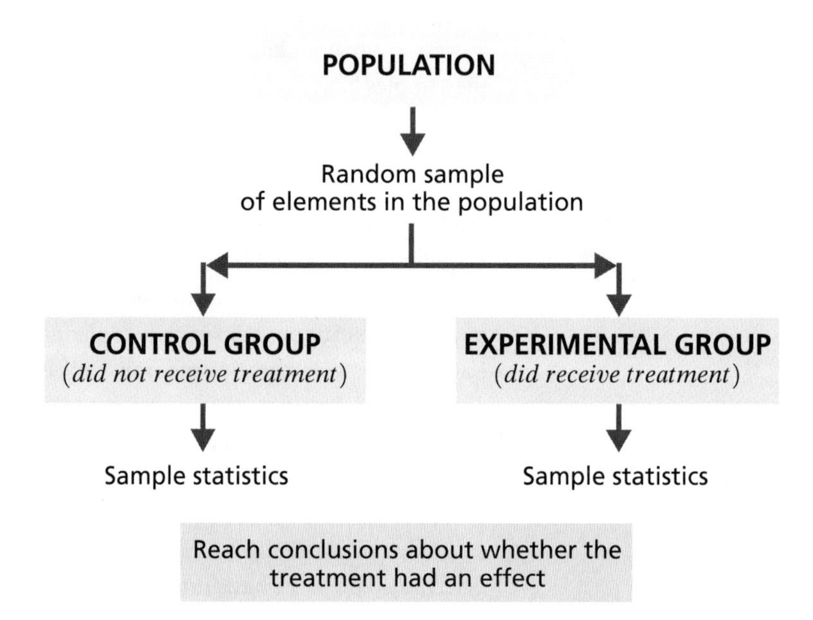

An example of a procedure of the second type is a **clinical trial,** in which new drugs or medical procedures are tested using human patients. The group receiving the new drug or treatment is the experimental group. If the statistics are different for the treatment and nontreatment groups, this *may* suggest that the treatment has had a successful impact.

Before reaching conclusions, however, it is very important to be aware of other, possibly confounding factors: During sampling, were there preexisting differences between the control and experimental groups? Did the two groups experience totally identical interactions during the trial, apart from the experimental treatment? Was the sampling variation very large? To minimize misleading results from such factors, clinical testing often requires repeated trials and other safeguards, not illustrated in the simple model in Figure 12.2.

Dependent Samples

Samples are **dependent** if the selection of elements in one sample is influenced by the selection of elements in the other sample. These are also called **related samples** or **paired samples.** Paired samples can be very useful for controlling unwanted variation in the data due to **extraneous variables.**

Suppose you are testing the effectiveness of a weight-loss program and are trying to use a model similar to Figure 12.1. You could sample participants *before* they start the program (population 1) and then sample participants *after* they take the program (population 2); then you might compare their mean weights. The problem occurs if, by chance, the "after" sample includes some of the heaviest people in the program or if the "before" sample includes some of the lightest people. In that case, the initial weight of participants is an extraneous variable that is being ignored, which could create misleading results.

One method of obtaining a paired sample is the **before/after method,** also known as the **pre-test/post-test method** or the **repeated-measures method.** Applied to the weight-loss example, we could take one sample of participants and weigh each participant *twice*—once before the program and again after the program. If some members of the population are initially heavier or lighter than the norm, then those individuals appear in either *both* parts of the sample (before and after) or in neither. Their weights in the "after" group are compared with their own weights in the "before." The before/after approach to dependent samples is illustrated in Figure 12.3.

Figure 12.3
The Before/After Approach to Dependent Samples

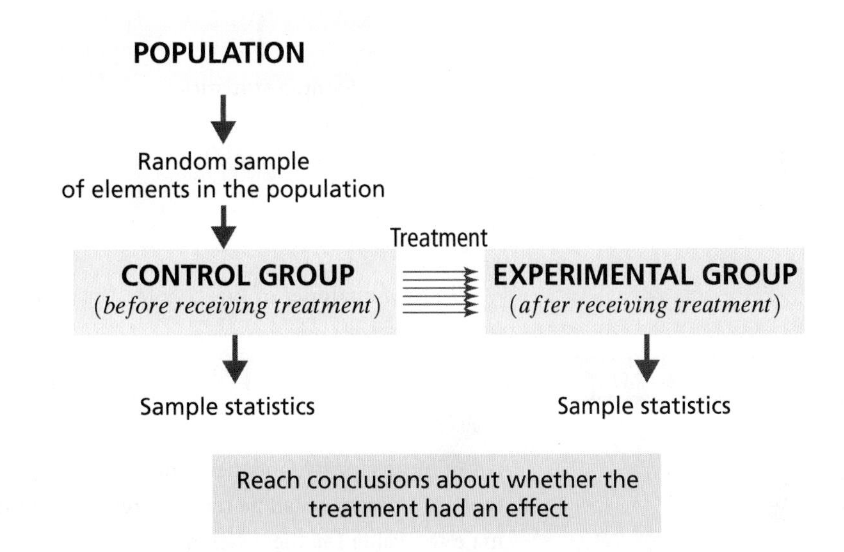

Another approach to using dependent samples is the **matched-pairs approach.** Again, the objective is to control for extraneous factors, but unlike the previous method, each element is measured or assessed only once.

Suppose we want to compare two brands of tires for the amount of wear exhibited after 5,000 km of driving. If independent samples are used, we might compare the mean amounts of wear for samples from each of the two brands. But those results may be influenced by various extraneous factors: For example, where were the tires placed on the vehicles? On what types of vehicles? Under what road conditions? In the matched-pair approach, we divide the sample into pairs of individuals, one from each population (e.g., from each brand of tire). Place one pair, for example, on the back wheels of a vehicle and another on the front wheels; one pair is to be driven on paved roads and another on country lanes, and so forth. In that way, we compare the different populations' performances, given common values for the extraneous variables.

Figure 12.4 illustrates the basic matched-pair method for dependent samples. The figure presumes a controlled experiment in which one member of each pair is given a certain treatment (such as a life-extending spray for tires) and the other is not. Variations of the model can include dual treatments (e.g., a different brand of tire spray for each tire) or even drawing the paired elements from different populations (e.g., from different brands of tires).

Figure 12.4
Matched-Pair Approach to Dependent Samples

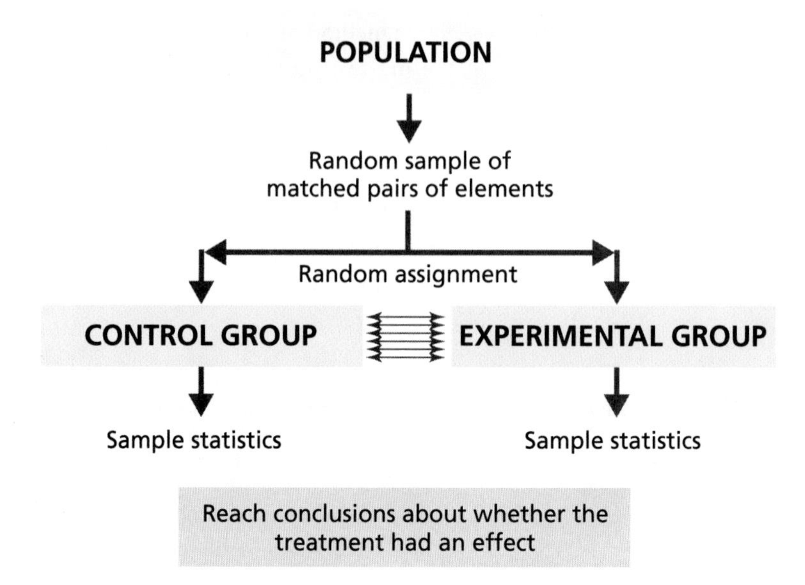

In the remainder of this chapter, the basic concepts provided in this section apply.

12.2 Differences between Population Means: Two Independent Samples

Given means for two independent samples, as defined in the previous section, two main types of inferences can be made in relation to the underlying populations. We can estimate the value for the difference, if any, between the means of the populations, or we can test a hypothesis about the value of the difference in mean for the two populations. The models in this section presume that the samples are drawn randomly from independent, normally distributed populations.

Whether estimating or hypothesis-testing, we may know the population variances at the outset (or at least have a working value for them). This allows us to use the normal curve for modelling the sampling distribution. If the population variances are not known, then the t-distribution is used.

Figure 12.5 is intended as a simplified model of the sampling distribution that applies when making inferences in this section. Unlike the similar figure in Chapter 10 (Figure 10.4), the distribution depicted here is not the distribution of all possible values of sample means for a given sample size. In this section, we take *two* samples and each sample has its own mean. The variable that is distributed is the *difference in values* between the means of the two samples. The set of all possible values for this variable, if all possible samples of sizes n_1 and n_2 were taken, is the sampling distribution that is the basis for inferences.

In the first type of two-sample inference mentioned in the first paragraph of this section, we presume to know both population variances and test for whether there is a significant difference between the population means. The null hypothesis is "no difference" between the means. Figure 12.5 provides the core model for this case. The curve is the distribution for all possible differences in means that might be obtained from the

Figure 12.5
Standardized Sampling Distribution of the Difference in Sample Means, and Probabilities Associated with the Confidence Intervals

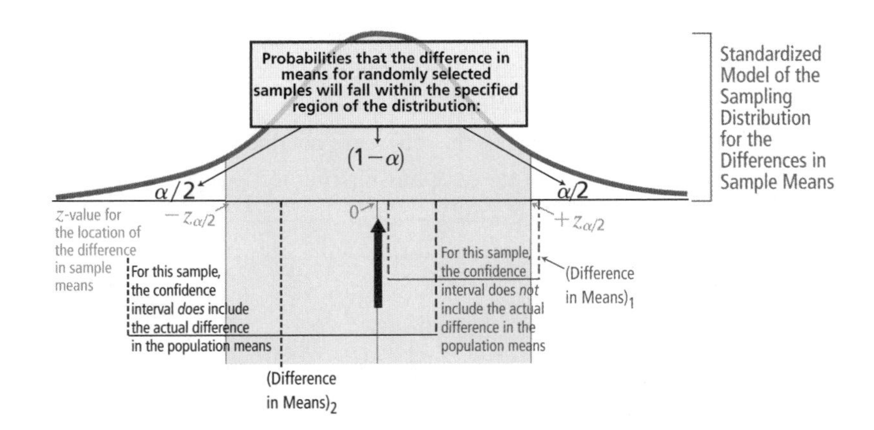

samples. For an actual pair of samples, the difference in their means can be located on the horizontal axis in terms of *z*-values. That is, using the standard deviation of the distribution, how many standard deviations above or below zero is the actual difference in sample means? It is unlikely that the *z*-value of the difference in sample means will have a very large magnitude.

Alternatively, we can make an *estimate of the difference* between population means, based on the Figure 12.5 model for the sampling distribution. Centre the estimate on the obtained difference between the two sample means. Create a confidence interval around the central value, similar to the procedures in Chapter 10. There is $1 - \alpha$ probability that the calculated interval will include the true value for the difference in population means.

With minor changes, the model in Figure 12.5 can be adapted for other types of inferences as well. For example, the hypothesis might be that the two population means are not identical, but are some specified distance apart: 5 units, for instance. In that case, the centre of the distribution, at $z = 0$, represents samples with means exactly five units distant. Other locations on the horizontal axis are measured as some $\pm$ *z*-value above or below that exact distance value of 5 units.

If we do not know the variances for the two populations, then the correct model for the sampling distribution is the *t*-distribution. The basic concepts in Figure 12.5 still apply, but the distribution shape is slightly flatter and the measures of location for the differences are specified in *t*-values rather than in *z*-values.

z-Test for the Difference in Population Means (Known Variances)

The *z*-test is used traditionally to introduce the concept of hypothesis testing for difference in means. Figure 12.5 illustrates the paradigm; however, the approach is not applicable if the population variances are unknown. The model also requires that the sampling distribution for the differences in sample means be normally distributed. This requires, in turn, that either both underlying populations are normally distributed or else that both sample sizes are large (at least 30) in order that the central limit theorem applies.

The test statistic (a *z*-value) locates the obtained difference in sample means with respect to the sampling distribution for the parameter. As in previous chapters, the *z*-value is based on this general calculation:

$$z = \frac{\text{Point estimate for the parameter} - \text{Hypothesized value for the parameter}}{\text{Standard error of the point estimate}}$$

The point estimate in this case is simply the difference in sample means. The hypothesized value of the difference in means is often taken as zero but, depending on the problem, another value may be used (this will be discussed in Example 12.4). To normalize the difference obtained in the numerator, we divide it by the standard deviation of the sampling distribution—also called the *standard error of the estimate.* In this case, the standard error is based on combining the variances of the two separate populations, divided by their sample sizes. This gives the denominator of Formula 12.1.

Formula 12.1 z-Value Calculation for the Difference in Sample Means

$$z = \frac{(\bar{X}_1 - \bar{X}_2) - (\mu_1 - \mu_2)}{\sqrt{\dfrac{\sigma_1^2}{n_1} + \dfrac{\sigma_2^2}{n_2}}}$$

Example 12.1

(Based on the ***Playoff_Stats*** *file)*

Midway through the playoff series in 2006 between the Edmonton Oilers and the San Jose Sharks, a hockey fan examined player statistics in hopes of predicting the series winner. The data in the file **Playoff_Stats** were updated on the morning of May 9, 2006. Interpret these preliminary stats as a sample of what the players will accomplish by the end of the playoffs.

The sports fan argues that the Sharks' players have, on average, a better "+/–" performance than the Oilers' players. (The +/– record tells how many more or fewer goals are scored *for* a player's team than *against* the player's team, when that player is on the ice.) Assuming that the +/– records, although discrete data, are approximately normally distributed on each team, with a standard deviation of 2.0, test the sports fan's claim at the 0.05 level of significance.

Solution

We follow the eight-step procedure for hypothesis testing provided in Section 11.1.

Step 1: The fan argues that the mean value for +/– is larger for the Sharks than for the Oilers. In symbols, his claim can be formulated as:

$$\mu_1 > \mu_2$$

But be sure to specify which team is represented by the subscript 1 and which team is represented by the subscript 2; alternatively, use a notation such as:

$$\mu_{\text{Sharks}} > \mu_{\text{Oilers}}$$

Step 2: Since the claim of step 1 does not contain equality, it can be interpreted as the alternative hypothesis. The opposite of that hypothesis becomes H_0.

$$H_0\!: \mu_1 \leq \mu_2$$

$$H_1\!: \mu_1 > \mu_2$$

An alternative approach for this step is to show the following expressions, which are mathematically equivalent to the preceding ones:

$$H_0: \mu_1 - \mu_2 \leq 0$$

$$H_1: \mu_1 - \mu_2 > 0$$

When comparing these expressions with the word problem, the first approach may be easier to interpret when the hypothesized difference is zero. However, the second approach is more clearly related to the test procedure: The sketch to be drawn in step 5 will show the expected distribution of *differences* in means, should the equality assumption of H_0 be true.

Step 3: The level of significance is given in the problem:

$$\alpha = 0.05$$

Step 4: The sample sizes are found in the data. We will test for difference in means between two populations with known variances. "Assuming" the distributions are normal, we can:

Use the *z*-test for difference in population means (known variances), $n_1 = 20$, $n_2 = 22$.

It is good practice, however, to briefly check our assumptions about the nature of the distribution. Seeing in Figure 12.6 that the normal assumption is approximate at best, we should approach any conclusions based on *z* with some care.

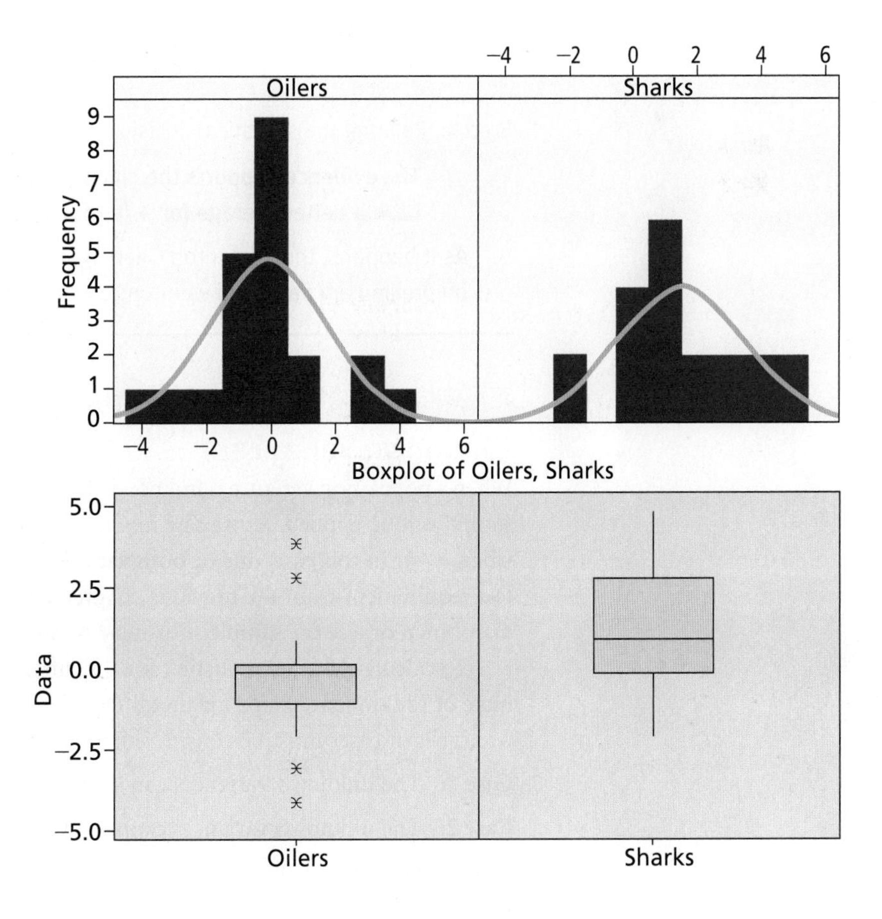

Figure 12.6
Quick Check of the Distribution Assumptions

Steps 5–7: For the critical value approach, these steps are illustrated in Figure 12.7. The same decision would be reached if we used the *p*-value: The one-tail *p*-value that corresponds to the test statistic $z = 2.575$ can be calculated as $P[Z = 2.575 \mid$ (the sampling distribution is as shown in step 5)$] = 0.005$, which is less than $\alpha = 0.05$. Reject H_0.

Figure 12.7
Steps 5–7, Example 12.1

Step 5:

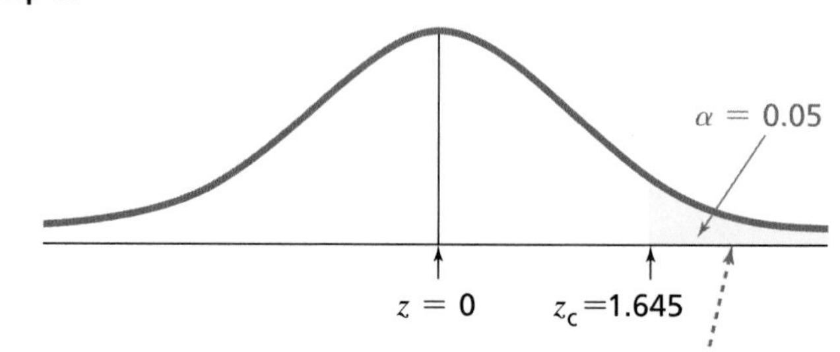

Step 6: Use the sample data in the CD file to calculate: $\bar{x}_1 = 1.5$ and $\bar{x}_2 = -0.0909$

Step 7:

- Calculate test statistic, given:
  - Sample means: $\bar{x}_1 = 1.5$ and $\bar{x}_2 = -0.0909$
  - Hypothesized: $\mu_1 - \mu_2 = 0$
  - $\sigma_1 = \sigma_2 = 2.0$

$$z = \frac{(1.5 - (-0.0909)) - (0)}{\sqrt{\frac{2^2}{20} + \frac{2^2}{22}}} = 2.575$$

- Compare the test statistic to the critical region in step 5 (see red arrow).
- Because the test statistic falls in a critical region, reject H_0.

Step 8: Relating the statistical decision back to the original claim, we conclude:

The evidence supports the claim of the sports fan that the Sharks players have a better average for +/− than the Oilers players.

As it happens, the Edmonton Oilers ignored these test results, which were based on preliminary data, and went on to win the playoff series against San Jose.

t-Test for the Difference in Population Means (Unknown Variances)

When a population variance is unknown, then the *z*-based test for the difference in means should not be applied. As we saw in Chapters 10 and 11, the more appropriate model when σ (or in this case, one or both variances) is unknown is the *t*-test. Extending the test requirement from the previous chapter, either both populations must be normally distributed or else the sample sizes must be sufficiently large (at least 30).

To calculate the test statistic *t* for the difference in sample means, we need an estimate of the standard error for the difference in means. To estimate the standard error, two types of cases must be considered:

Case 1: The unknown variances can be reasonably assumed to be equal.

Case 2: The unknown variances cannot be reasonably assumed to be equal (i.e. they appear to be unequal).

For **Case 1,** the estimate of the standard error becomes the denominator of the formula for the test statistic in Formula 12.2(a). Observe that this formula is very similar to Formula 12.1, except that σ is unknown, so it must be estimated. The best guess for the value of σ^2 in this case is called the *pooled variance,* as calculated in Formula 12.2(b). The latter can be seen as the weighted average of the variances from the two samples, weighted by their individual degrees of freedom.

Formula 12.2 t-Value Calculation for the Difference in Sample Means (Case 1)

a) $$t = \frac{(\bar{X}_1 - \bar{X}_2) - (\mu_1 - \mu_2)}{\sqrt{\frac{s_p^2}{n_1} + \frac{s_p^2}{n_2}}}$$

where

b) $$s_p^2 = \frac{(n_1 - 1)s_1^2 + (n_2 - 1)s_2^2}{(n_1 - 1) + (n_2 - 1)}$$

The degrees of freedom for this test statistic are equivalent to the denominator of Formula 12.2(b): $df = n_1 + n_2 - 2$

Case 2 is more difficult, because the sample variances—which themselves only approximate the population variances—are not assumed to be equal. In one approach, called *Welch's t-test,* the formula to find the test statistic for Case 2 is relatively simple (see Formula 12.3), but using it requires an estimate for the overall degrees of freedom. A simple estimate is provided by $df =$ the smaller of $(n_1 - 1, n_2 - 1)$. Or your software may provide an alternative estimate (but its complex formula is not shown here).

Formula 12.3 Welch's t-Test Calculation for the Test Statistic (Case 2)

$$t = \frac{(\bar{X}_1 - \bar{X}_2) - (\mu_1 - \mu_2)}{\sqrt{\frac{s_1^2}{n_1} + \frac{s_2^2}{n_2}}}$$

where s_1 and s_2 are the individual variances of the two samples.

Not all authors and textbooks agree on how or when to apply the two cases. Some believe that, in practice, the formula for Case 1 is sufficient to handle most examples. (You are encouraged to try "Hands On 12.1" on page 514 and draw your own conclusion.)

One common approach is to *test* the Case 1 assumption (that both populations' variances are equal) as a null hypothesis. Then switch from Case 1 to Case 2 only if that null hypothesis is rejected. To conduct the test for significant difference in variances, a standard *F*-test is sometimes utilized, but this test is highly sensitive to violations in the normal distribution requirement, so its results may be biased. If you rely on software for the testing, then different packages (including Excel, Minitab, and SPSS) may not give the same results because each uses different algorithms.

Another approach to deciding between the cases is to graph the distributions and visually inspect them to assess whether their variances are roughly equal. We illustrate some of the above approaches in Examples 12.2–12.4. Based on the software you have available, the nature of the data, plus the guidelines of your instructor, you can assess when and how to apply the case that is most appropriate.

Example 12.2

(*Based on the files* ***Calgary_Flames*** *and* ***Calgary_Stampeders***)

A colleague of the sports fan in Example 12.1 claims that, on average, the Calgary Flames' hockey players are younger than the Calgary Stampeders' football players. For a test, she located data for each team's players in a particular year (Flames' players in 2003, Stampeders' players in 2006). For the test, she assumes that the teams' rosters on those years were representative of the age distributions of players on the team for *any* year in general. Estimate ages from birth years, using (2003 – birth year) for Flames' players and (2006 – birth year) for Stampeders. Test the colleague's claim at the 0.05 level of significance.

Solution

Again, we follow the eight-step procedure for hypothesis testing, although in practice there will be some overlap of steps in this case.

Step 1: The claim, in words, is that the mean age of the Flames is less than the mean age of the Stampeders. If we arbitrarily specify that the Flames are team 1 and the Stampeders are team 2, then the claim in symbols is:

$$\mu_1 < \mu_2$$

Step 2: The claim in step 1 does not contain equality, so it can be interpreted as the alternative hypothesis. The opposite of that hypothesis becomes H_0.

$$H_0: \mu_1 \geq \mu_2$$

$$H_1: \mu_1 < \mu_2$$

Step 3: The level of significance is given in the problem:

$$\alpha = 0.05$$

Step 4(a): We know that we are testing for the difference in means in two populations, independently sampled. Variances are not known. Sample sizes (determined from the data) are reasonably large (although n_1 is not quite 30), and if we checked, we would see that the distributions for the populations are more or less similar to the normal (see Figure 12.8, upper left). We can therefore *partly* complete step 4:

Use the *t*-test for difference in population means, two independent samples (____________), $n_1 = 27$, $n_2 = 61$.

We have not yet decided, however, whether to approach this problem as one of equal or unequal variances; i.e., the blank shown in the above test description still needs to be filled in.

Step 4(b): To fill in the blank left in (a), we have to start collecting and manipulating the data—which overlaps procedures we have identified as step 6 or 7 in the formal simplified process. Figure 12.8 shows some *possible* manipulations you might perform to determine whether Case 1 or Case 2 applies.

Figure 12.8
Possible Analyses for Deciding Case 1 vs. Case 2

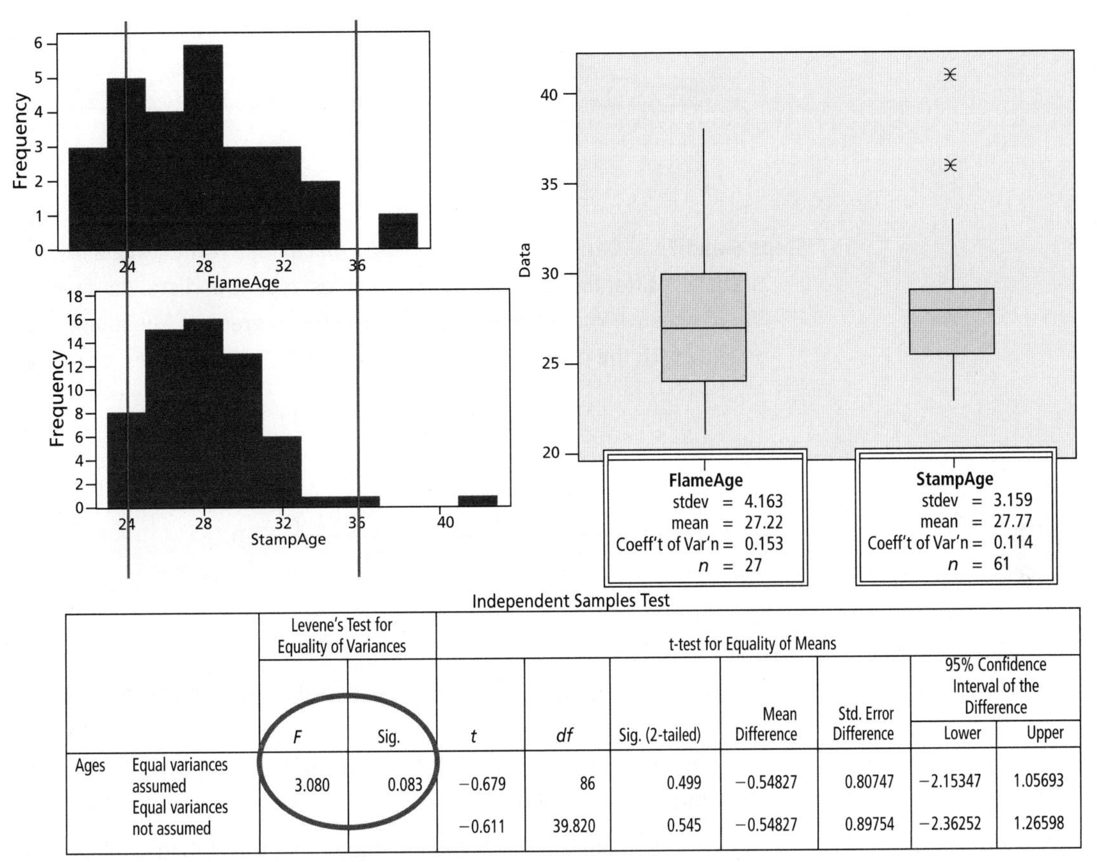

Independent Samples Test

		Levene's Test for Equality of Variances		t-test for Equality of Means					95% Confidence Interval of the Difference	
		F	Sig.	t	df	Sig. (2-tailed)	Mean Difference	Std. Error Difference	Lower	Upper
Ages	Equal variances assumed	3.080	0.083	−0.679	86	0.499	−0.54827	0.80747	−2.15347	1.05693
	Equal variances not assumed			−0.611	39.820	0.545	−0.54827	0.89754	−2.36252	1.26598

On the one hand, the boxplots appear to show non-identical variances. Yet, including all of data in the calculations, the samples' standard deviations differ by only about one unit (4.163 versus 3.159). When this difference in standard deviations is compared to the means, it appears rather small.

The figure also shows (in the oval near the bottom) another approach for deciding the cases: testing formally whether a null hypothesis of "no difference in variance" should be rejected. The software used for Figure 12.8 generated a test statistic for such a test ($F = 3.08$) and has calculated the corresponding p-value ("Sig") equal to 0.083. Presuming $\alpha = 0.05$ for this preliminary test of the null hypothesis of "equal variances," the p-value $> \alpha$, and so does not support rejecting the equal variances (Case 1) hypothesis.

In light of these considerations, it does seem reasonable to select "equal variances" to complete the test description for step 4 of the process:

Use the *t*-test for difference in population means, two independent samples (equal variances), $n_1 = 27$, $n_2 = 61$.

Step 5: Figure 12.9 depicts the null distribution (i.e., the sampling distribution of differences in sample means under the assumption of "no difference"). For the test identified in step 4, the null distribution is a t-distribution with degrees of freedom equal to $n_1 + n_2 - 2$.

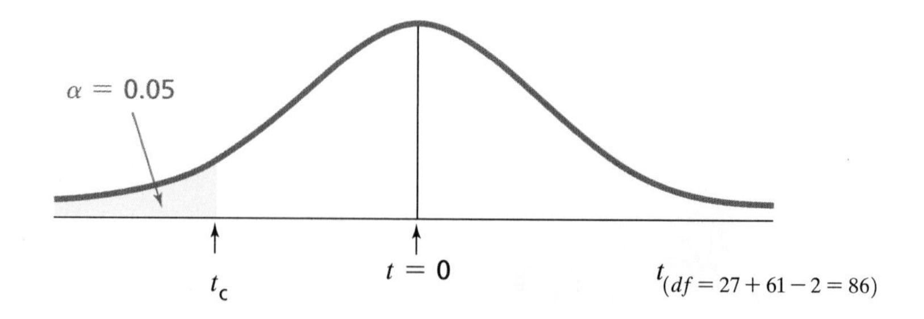

Figure 12.9
Step 5 for the Hypothesis Test, Example 12.2

Steps 6 and 7: Data have been provided. We can use Formula 12.2 to find the test statistic *t*, after first calculating the samples' means and standard deviations. Then, once we know the test statistic and the degrees of freedom, we can find the *p*-value with the procedures from "In Brief 11.1."

$$s_p^2 = \frac{(27 - 1) \times 4.163^2 + (61 - 1) \times 3.159^2}{(27 - 1) + (61 - 1)} = 12.2018$$

$$t = \frac{(27.22 - 27.77) - (0 - 0)}{\sqrt{\frac{12.2018}{27} + \frac{12.2018}{61}}} = -0.681$$

The degrees of freedom = 27 + 61 − 2 = 86; therefore:

p-value = $P[T_{df=86} \leq -0.681 \mid$ (the sampling distribution is as shown in step 5)]
= 0.25

(The *p*-value can also be obtained using the computer. See "In Brief 12.1(a)" starting on page 504.)

Since this *p*-value is not less than or equal to α, the statistical decision is:

Do not reject H_0.

Step 8: Relating the statistical decision to the original claim, we conclude:

The evidence does not support the claim that the mean age of Calgary Flames players is less than the mean age of Calgary Stampeders players.

Example 12.3

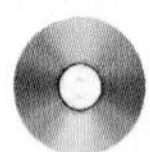

*(Based on the **Enplaned_Passengers** file)*

A company that produces packages of peanuts for airline passengers is studying the potential market for its product. The company has collected some past data on numbers of passengers who have travelled on various U.S. airlines and believes that these numbers are representative of more current years. At the 0.05 level of significance, assume the population distributions are normal and test a claim that for "major" airlines the mean number of passengers who are transported is the same as the mean for "national" airlines.

Solution

Step 1: If we designate the major airlines as 1 and the national airlines as 2, then the claim in symbols is:

$$\mu_1 = \mu_2$$

Step 2: The claim in step 1 contains equality, so it can be interpreted as H_0.

$$H_0: \mu_1 = \mu_2$$
$$H_1: \mu_1 \neq \mu_2$$

Step 3: $\alpha = 0.05$

Step 4(a): We are testing for the difference in means in two populations, independently sampled. Variances are not known. Sample sizes are small, but the problem assumes that the distributions are normal. (In practice, a visual check of the data (see Figure 12.10) might discourage acceptance of this normal assumption, so a nonparametric technique from Chapter 13 might be preferable.) We can *partly* complete step 4:

Use the *t*-test for difference in population means, two independent samples (____________), $n_1 = 13$, $n_2 = 12$.

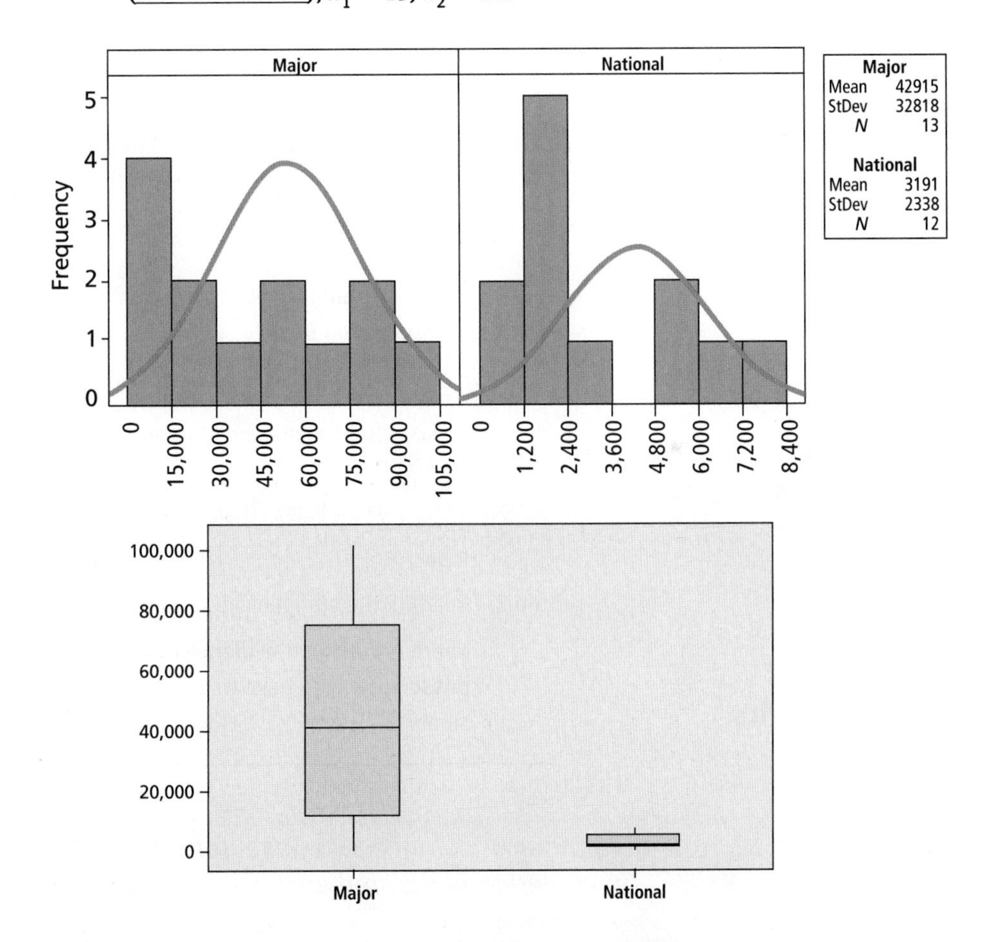

Figure 12.10
Possible Analyses for Deciding Case 1 vs. Case 2

Step 4(b): To fill in the blank left in (a) we examine the data to assess whether Case 1 or Case 2 applies. Based on comparing the boxplots in Figure 12.10, the sample variances could not be more different—so there is no need for a more formal assessment. Thus, select "unequal variances" to complete the test description for step 4 of the process:

Use the *t*-test for difference in population means, two independent samples (unequal variances), $n_1 = 13$, $n_2 = 12$.

Step 5: Figure 12.11 depicts the null distribution, which is a t-distribution for the selected test. (For the case of unequal variances, we have not presented here the formula to estimate the degrees of freedom. The value shown in Figure 12.11 has been estimated by the Welch-Satterwaite approximation, which is used in conjunction with Welch's t-test for unequal variances.)

Figure 12.11
Step 5 for the Hypothesis Test, Example 12.3

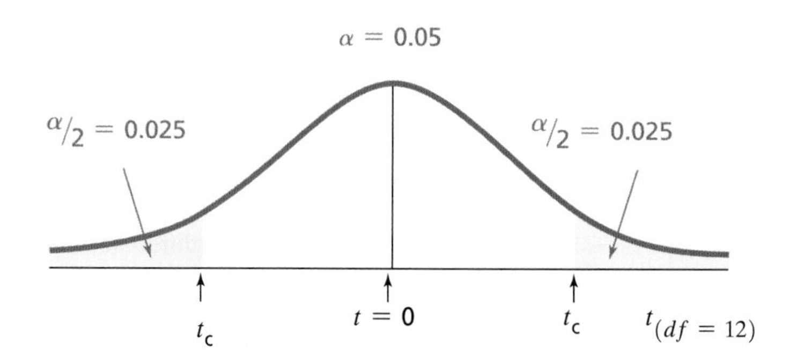

Steps 6 and 7: Using the provided data for a Case 2 problem, it is easiest to use the computer (see "In Brief 12.1(a)") to at least calculate the test statistic t (t-value = 4.35) and the degrees of freedom ($df \cong 12$). We can then find the p-value with the procedures in "In Brief 11.1":

p-value (2-tailed) = $2 \times P[T_{df=12} \geq 4.35) \mid$ (the sampling distribution is as shown in step 5)]

$= 0.001$

The p-value can also be obtained directly using the computer. See "In Brief 12.1(a)."

Since this p-value is less than α, the statistical decision is:

Reject H_0.

Step 8: Relating the statistical decision to the original claim, we conclude:

There is sufficient evidence to reject the claim that the mean numbers of passengers who have travelled on "major" versus "national" airways is the same.

Sections 10.0–10.3

• *In Brief 12.1(a)* *Computer Assistance for t-Tests for Difference in Means: Two Independent Samples*

*(Based on the files **Calgary_Flames** and **Calgary_Stampeders**)*

Any of the three software packages can contribute to testing for (1) difference in means, independent samples—for equal *or* unequal variances. For deciding on the case, each package also provides a test for (2) a null hypothesis of no difference in variance. (Remember when calculating columns for players' ages to use (2003 – birth year) for Flames' players and (2006 – birth year) for Stampeders' ages.)

Excel

1. In Excel, the options for a *t*-test presume a prior decision about the equality or unequality of the variances. The data for each of the samples must be in columns on the worksheet. Use the menu sequence *Tools / Data Analysis / t-Test: Two Sample Assuming Equal Variances* or *Tools / Data Analysis / t-Test: Two Sample Assuming Unequal Variances*—depending on your prior decision. Click *OK*. In the resulting dialogue box (which has a similar layout for both cases), input the ranges for the variable 1 data and variable 2 data, respectively (see the screen capture). Next, input the value of the mean difference in the null hypothesis (which was 0 in Example 12.2). Finally, specify the upper left cell address for the output range, and click *OK*.

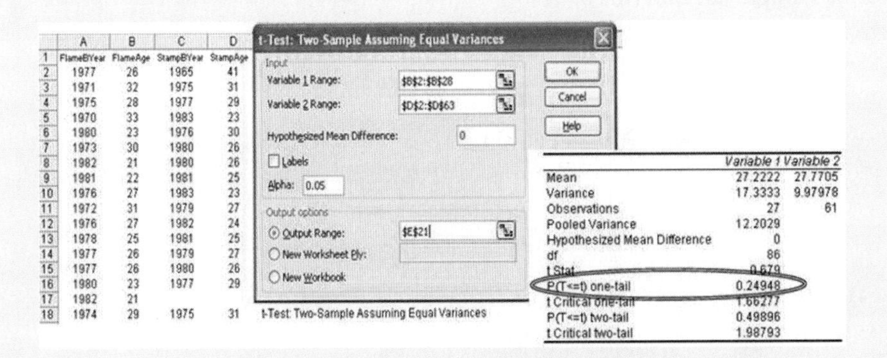

Also shown in the illustration is the software output, and circled is the one-tail *p*-value that corresponds to Example 12.2. If the test had been two-tailed, the two-tail *p*-value would be read instead.

2. Your instructor may require that the decision in (1) on equality of variance be based on a prior hypothesis test, setting the null hypothesis as "variances are equal." In that case, use the menu sequence *Tools / Data Analysis / F-Test: Two Sample for Variances* and then click *OK*. In the resulting dialogue box (see the second screen capture), input the ranges for the two variables—but *always input the data set with the larger variance as "Variable 1."* Specify the upper left-hand cell address for the output range and then click *OK*.

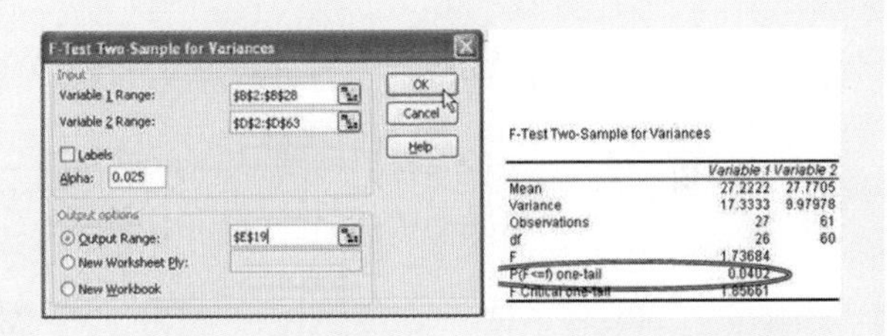

Because the *F*-distribution is not symmetrical, the input rule above ensures that the outputs (critical value and *p*-value) always refer to the upper tail. Since under the assumption of equal variance, either sample could by chance have had the larger variance, the two-tailed *p*-value is *twice* the value shown in the output. If the two-tailed *p*-value is not less than or equal to 0.05, choose the *Tools / Data Analysis / t-Test: Two Sample Assuming Equal Variances* menu sequence to proceed.

Minitab

1. In Minitab, a prior decision is assumed about the equality or unequality of the variances. To begin, use the menu sequence *Stat / Basic Statistics / 2-sample t.* Depending on your prior decision, check or leave blank the option *Assume equal variances.* Minitab is flexible about the input format. The illustration shows the sample data in two separate columns. Alternative options are to use summary data for the sample statistics (the *Summarized data* option) or to use a single column of data with group labels in a separate column (*Samples in one column* option). Then click on the *Options* button and, in *Test difference* in the resulting dialogue box, input the test value for the mean difference as given in the null hypothesis (which was 0 in Example 12.2). Then specify the operator ($\neq$, $<$, or $>$) in the *Alternative* window. Click *OK* twice. The illustration also shows the software output, including the p-value for the test.

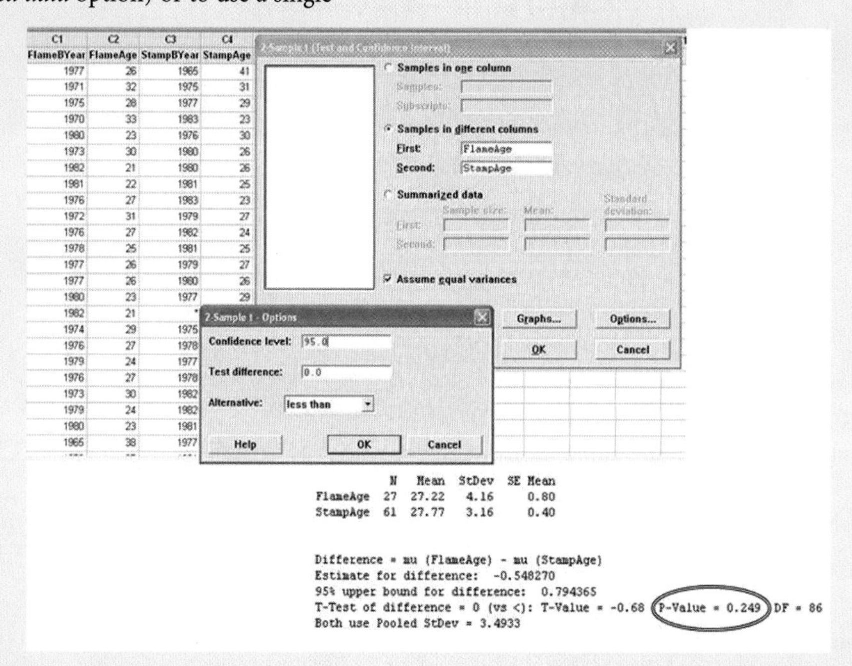

2. Your instructor may require that the decision in procedure 1 on equality of variance be based on a prior hypothesis test, setting the null hypothesis as "variances are equal." In that case, use the menu sequence *Stat / Basic Statistics / 2 Variances.* As in procedure 1, Minitab is flexible about the input format; the illustration shows the option of the sample data in separate columns. Click *OK.* The displayed output shows two alternative p-values: The F-test p-value is most reliable if the samples are very normally distributed; if the distributions are

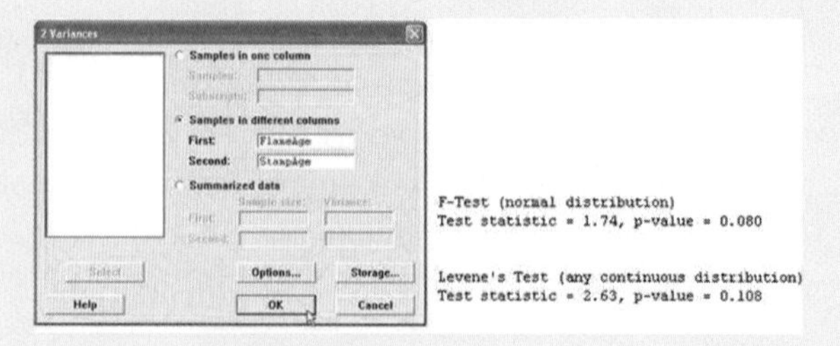

not that normal, the second Levene's Test (based on deviations from the median) may be preferable. If the p-value from the selected test is not less than or equal to 0.05, choose the *Assume equal variances* option in the first dialogue box.

SPSS

1. Arrange the data for both samples in a single column, using a separate column to label which values come from which samples. Then use the menu sequence *Analyze / Compare Means / Independent-Samples T Test.* In the resulting dialogue box, shown in the screen capture, select the variable to be tested and then specify the grouping variable—that is, the label column that distinguishes the two samples. Select the *Define Groups* option and, in the next dialogue box, input the *exact* labels (including capitalization) used in the grouping variable. (Note: To avoid problems, it is often a good prior step to transform/recode the grouping labels into numerical ones and twos, etc. In fact, sometimes SPSS will accept as a grouping variable *only* values that are numbers.) Select *Continue* and then *OK.*

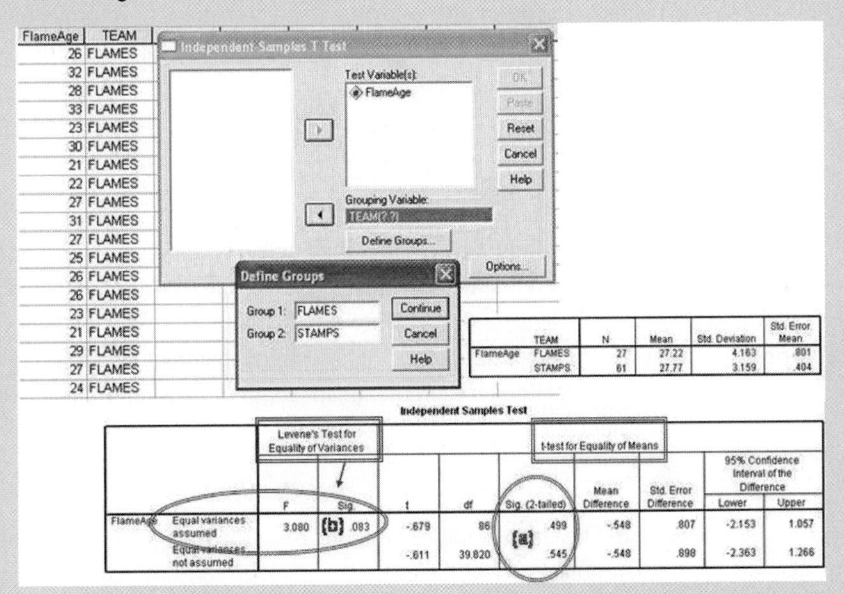

	TEAM	N	Mean	Std. Deviation	Std. Error Mean
FlameAge	FLAMES	27	27.22	4.163	.801
	STAMPS	61	27.77	3.159	.404

Independent Samples Test

		Levene's Test for Equality of Variances		t-test for Equality of Means						
									95% Confidence Interval of the Difference	
		F	Sig.	t	df	Sig. (2-tailed)	Mean Difference	Std. Error Difference	Lower	Upper
FlameAge	Equal variances assumed	3.080	(b) .083	-.679	86	.499 (a)	-.548	.807	-2.153	1.057
	Equal variances not assumed			-.611	39.820	.545	-.548	.898	-2.363	1.266

Note that in SPSS, the test procedure always assumes that in the null hypothesis the difference in means is zero. If the claim is for a different value, you need to create a new column by subtracting the claimed difference in means from every original value of x_1. Then test for a null hypothesis of zero difference between means between the new column of adjusted x_1 values and the original x_2 values.

The p-value results for the test are shown in the area labelled "(a)" in the illustration. Two-tailed results are always displayed, so for a one-tailed test (such as in Example 12.2), divide the displayed value by 2. Use the upper p-value if equal variances are assumed; otherwise, use the lower p-value.

2. Your instructor may require that the decision in procedure 1 on equality of variance be based on a prior hypothesis test, setting the null hypothesis as "variances are equal." Refer to the output displayed in the area labelled "(b)" in the illustration. The p-value corresponds to a test whose null hypothesis is that the two variances are equal. If the p-value is not less than or equal to 0.05, choose the "Equal Variance" portion in the t-test display area (a) in the illustration; otherwise, use the lower figure in area (a).

Example 12.4

*(Based on the **Notable_Tunnels** file)*

An engineer was discussing with a statistician the construction of long tunnels. (Subway tunnels were excluded.) The engineer claimed that, on average, a railroad tunnel that's considered to be long is 6 km *longer* than a vehicular tunnel that is considered to be long. (This fact may be of interest to track manufacturers.) To prove her point, she randomly sampled from a published list of noteworthy tunnels. The lengths of tunnels in her sample (in kilometres) were as follows.

Railroad	50	13.5	18.5	16.3	14.6	13.5	14.6	15.2	14.9	14.3
Vehicular	11.2	11.4	11.3	12.9	10.9	11	14	16.4	8.5	24.5

The statistician objected that the 50-km Chunnel (which is the combined length of the three railroad tunnels under the English Channel) is an outlier. Omit the outlier from the sample and test the engineer's claim at the 0.01 level of significance.

Solution

Step 1: The engineer claims that the mean length of long railroad tunnels (population 1) is 6 km longer than the mean length of long vehicular tunnels (population 2). In symbols:

$$\mu_1 - \mu_2 = 6$$

Step 2: The claim in step 1 contains equality, so it can be interpreted as the null hypothesis.

$$H_0: \mu_1 - \mu_2 = 6$$
$$H_1: \mu_1 - \mu_2 \neq 6$$

Step 3: The level of significance is given in the problem:

$$\alpha = 0.01$$

Step 4(a): We are testing for the difference in means in two populations, independently sampled. Variances are not known. Sample sizes are small, but if the outlier is omitted, the distributions are skewed but vaguely normal. (See Figure 12.12.) We can *partly* complete step 4:

Use the t-test for difference in population means, two independent samples (__________), $n_1 = 9$, $n_2 = 10$.

Figure 12.12
Check of the Distribution Shapes

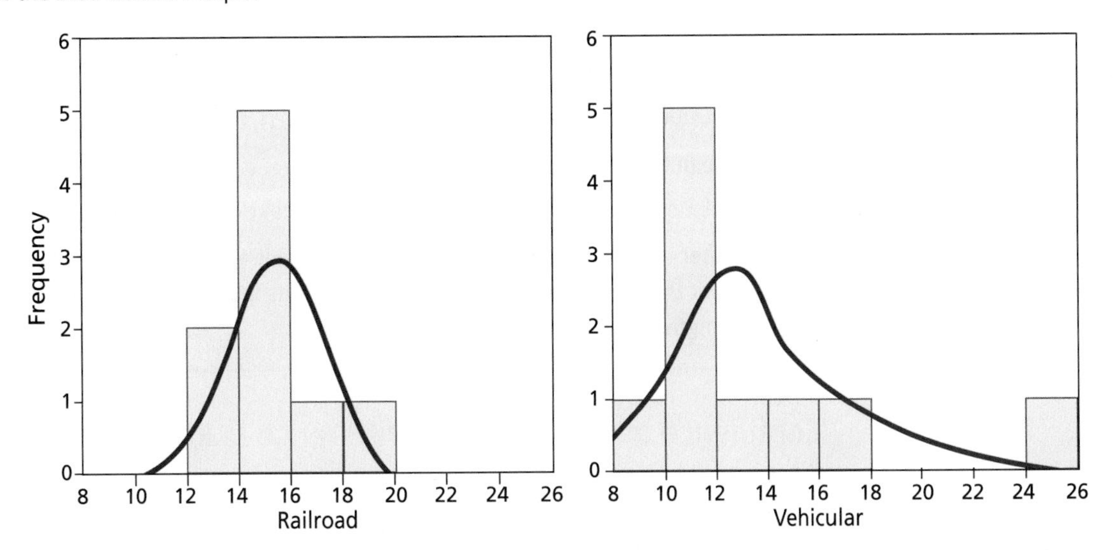

Step 4(b): To fill in the blank left in (a) we examine the data to assess whether Case 1 or Case 2 applies. Based on comparing the histograms in Figure 12.12, the sample variances are somewhat different. And if you formally test for equal variances on the computer, you might obtain a small p-value (based on an F-test), which also suggests a difference in variances. On the other hand, software that uses Levene's Test would return a p-value > 0.05—suggesting *no difference* in variances.

Fortunately (as commonly happens), the results for the t-test in step 7 will be the same whichever case you select for step 4. Completing the test description in (a):

Use the t-test for difference in population means, two independent samples (equal variances), $n_1 = 9$, $n_2 = 10$. (Or unequal variances. It's your call on this one.)

Step 5: The sampling distribution under the null hypothesis is depicted in Figure 12.13. Formula 12.2(a) specifies that t equals 0 if (the actual difference in sample means) minus (the presumed difference in population means) is zero; that is, if the difference in sample means *equals* the claimed population difference—in this case, 6. We will use a p-value approach.

Figure 12.13
Step 5 for the Hypothesis Test, Example 12.4

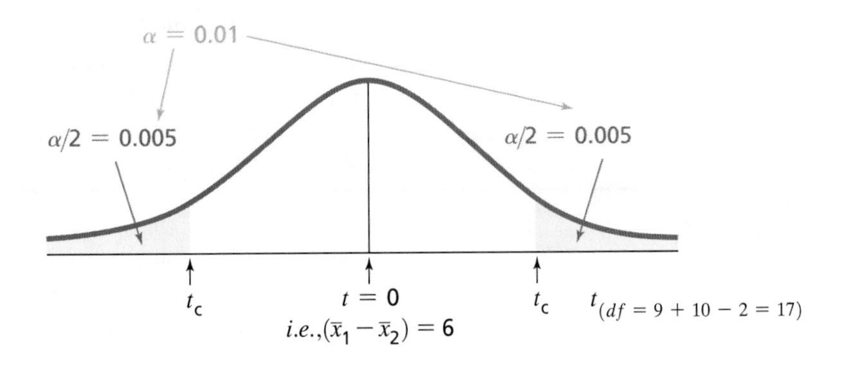

Steps 6 and 7: We can perform the test with the computer, using the procedures in "In Brief 12.1(a)." For Case 1, the displayed *p*-value is:

$$p\text{-value} \approx 0.017$$

(The *p*-value differs, but only slightly, if you assumed unequal variances.) Since this *p*-value is not less than or equal to α (= 0.01), the statistical decision is:

Do not reject H_0.

Step 8: Relating the statistical decision to the original claim, we conclude:

After removing outliers, there is not sufficient evidence to reject the claim that the mean length of long railway tunnels is 6 km longer than the mean length of long vehicular tunnels.

Confidence Intervals for the Difference in Population Means

When we collect two independent samples, we may not be interested in testing a hypothesis. The goal might be to estimate the difference between the two corresponding population means. The precise difference between the means of the two samples that we collect is the point estimate for the difference. Following the same principles as introduced in Chapter 10, we can extend the point estimate into an interval estimate—that is, a *range* of values within which we are confident (to the level $1 - \alpha$) that the true difference in population means lies. This is the confidence interval for the difference in population means.

Review the basic formulas for confidence intervals in Chapter 10. You will find:

For known population variances, the confidence interval is:

(Point estimate) ± [$z_{\alpha/2}$ × (Standard error of the estimate)]

For unknown population variances, the confidence interval is:

(Point estimate) ± [$t_{\alpha/2}$ × (Standard error of the estimate)]

In either case, under the null hypothesis, the standard error is the standard deviation of the sampling distribution for the parameter. $z_{\alpha/2}$ or $t_{\alpha/2}$ is the magnitude of the critical value for the sample statistic that determines the boundary of the confidence interval for a given level of confidence $(1 - \alpha)$. In Chapter 10, these estimates were applied for a one-population parameter, but here we apply them to the difference in population means.

For Known Population Variances

The sampling distribution can be modelled with the normal curve *for known population variances.* The critical *z*-value can be determined from the degree of confidence, as discussed in previous chapters. The calculation for standard error of the estimate for this model was introduced in Formula 12.1 (the denominator).

Example 12.5

The Fraser Institute is an economic think tank whose research is intended to help inform and influence the making of public policy. In one study, it compared prescription drug prices (in U.S. dollars) in various states and provinces. The institute found, for example, that a 40-mg prescription of the cholesterol-control drug Lipitor cost, on average, \$34.82 in Ontario (standard deviation, \$1.96), but only \$32.36 in Manitoba (standard deviation, \$1.60).

(Results are based on 41 pharmacists from Ontario and 36 from Manitoba who responded to a survey.)[1] Assume that (1) the prices were normally distributed and (2) the standard deviations can be interpreted as the population's sigma (for example, perhaps the standard deviations have been consistent over several studies). Construct a 95% confidence interval for the difference in mean price of Lipitor in the two provinces. (Use U.S. dollars.)

Solution

Let Ontario prices be population 1 and Manitoba prices be population 2. The point estimate for $\mu_1 - \mu_2$ is the difference in sample means: \$34.82 – \$32.36 = \$2.46. Since the population variances are known, the sampling distribution is the normal distribution. We know from previous chapters that the critical z-value for 95% confidence is 1.96.

The standard error of the estimate is:

$$\sigma_{(\bar{x}_1 - \bar{x}_2)} = \sqrt{\frac{\sigma_1^2}{n_1} + \frac{\sigma_2^2}{n_2}}$$

$$= \sqrt{\frac{1.96^2}{41} + \frac{1.60^2}{36}}$$

$$= 0.4060$$

The resulting confidence interval is:

Lower limit = (Point estimate) – [$z_{\alpha/2}$ × (Standard error of the estimate)]
= 2.46 – (1.96 × 0.4060)
= 1.66
Upper limit = (Point estimate) + [$z_{\alpha/2}$ × (Standard error of the estimate)]
= 2.46 + (1.96 × 0.4060)
= 3.26

The confidence interval for the difference in prices between Lipitor sold in Ontario and Lipitor sold in Manitoba is from \$1.66 to \$3.26 (in US\$).

For Unknown Population Variances

In the case of unknown variances, the z-distribution cannot be used to model the sampling distribution for the difference in means. Provided that the populations have a normal distribution or that the samples are large, we can use the t-distribution instead. A critical t-value can be determined from the degree of confidence, as discussed in previous chapters. The calculation for standard error of the estimate for this model (for the case of equal population variances) was introduced in Formula 12.2(a) (the denominator) Depending on your software, the calculations can be automated to varying degrees. The solution to Example 12.6 assumes minimal automation, but see "In Brief 12.1(b)" for information about what your software provides.

Example 12.6

*(Based on the files **Calgary_Flames** and **Calgary_Stampeders**)*
In Example 12.2, the solution failed to establish that Calgary Flames players are, on average, younger than Calgary Stampeders players. We could stop there—or we might ask: "What *is* the mean difference players' ages for the teams?" To answer this question, construct a confidence interval, estimating a 95% level of confidence. (Recall that in

Example 12.2, the players' years of birth were converted into ages, based on the years that were sampled.)

Solution Let the Flames players' ages be population 1 and the Stampeders' ages be population 2. Whatever your software, the intermediate figures highlighted in Figure 12.14 can readily be calculated from the data.

Figure 12.14
Key Data for Estimating the Difference in Means: Unknown Variance

t-Test: Two-Sample Assuming Equal Variances

	Variable 1	*Variable 2*
Mean	27.2222	27.7705
Variance	17.3333	9.97978
Observations	27	61
Pooled Variance	12.2029	
Hypothesized Mean Difference	0	
df	86	
t Stat	−0.679	
P(*T*<=*t*) one-tail	0.24948	
t Critical one-tail	1.66277	
P(*T*<=*t*) two-tail	0.49896	
t Critical two-tail	1.98793	

The point estimate for $\mu_1 - \mu_2$ is the difference in sample means, which we see is 27.2222 – 27.7705 = –0.5483. Since the population variances are unknown, the sampling distribution is the *t*-distribution. We could calculate the critical *t*-value for confidence level 95% as was done in Chapter 10, but notice that the software (using the steps provided in "In Brief 12.1(b)") can determine this for us: $t_{\text{critical(2-tail)}} = 1.9879$.

Assuming equal variances, the standard error of the estimate is:

$$s_{(\bar{x}_1-\bar{x}_2)} = \sqrt{\frac{s_p^2}{n_1} + \frac{s_p^2}{n_2}}$$

where s_p^2 is the pooled variance from the two samples.

If you have some time, you can calculate the pooled variance manually with Formula 12.2(b), but notice that the software has already told us that pooled variance = 12.203. Continuing the calculation:

$$s_{(\bar{x}_1-\bar{x}_2)} = \sqrt{\frac{12.203}{27} + \frac{12.203}{61}} = 0.807$$

The resulting confidence interval is:

Lower limit = (Point estimate) – [$t_{\alpha/2}$ × (Standard error of the estimate)]
= –0.5483 – (1.9879 × 0.807)
= –2.15
Upper limit = (Point estimate) + [$t_{\alpha/2}$ × (Standard error of the estimate)]
= –0.5483 + (1.9879 × 0.807)
= 1.06

The confidence interval for the difference in ages between Flames players and Stampeders players is from –2.15 to +1.06.

• *In Brief 12.1(b) Computer Assistance for Estimating Confidence Intervals for Difference in Means: Two Independent Samples*

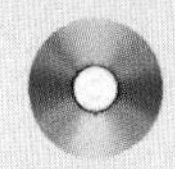

(Based on the files ***Calgary_Flames*** *and* ***Calgary_Stampeders****)*

Excel

Excel does not directly automate this calculation, but it *can* create an output similar to Figure 12.14. We saw at the end of Example 12.6 how to complete the calculation from that point forward. Figure 12.14 is generated as if you requested a *t*-test for two independent samples, assuming equal variances. (You could also assume unequal variances, but the standard error calculation would change. The impact is usually small on the results for the confidence interval.)

For the *t*-test instructions, see the first part of the Excel instructions in "In Brief 12.1(a)." In this context, always choose zero (0) as the value for mean difference in the null hypothesis, and always specify a value for alpha [i.e., 1 – (the confidence level)]. From the output of the procedure, interpret $t_{\text{critical(2-tail)}}$ as the desired critical value for the interval.

Minitab

Minitab's output includes a confidence interval whenever you initiate a *t*-test for two independent samples. When deciding whether to assume equal or unequal variances, you can use the procedures given in the second part of the Minitab instructions in "In Brief 12.1(a)."

For the *t*-test instructions, see the first part of the Minitab instructions in "In Brief 12.1(a)." In this context, always select *Options* in the first dialogue box (see the screen capture) and then (1) choose zero (i.e., 0.0) as the value for *Test difference,* (2) select "not equal" as the operator for the alternative hypothesis, and (3) input the desired level of confidence.

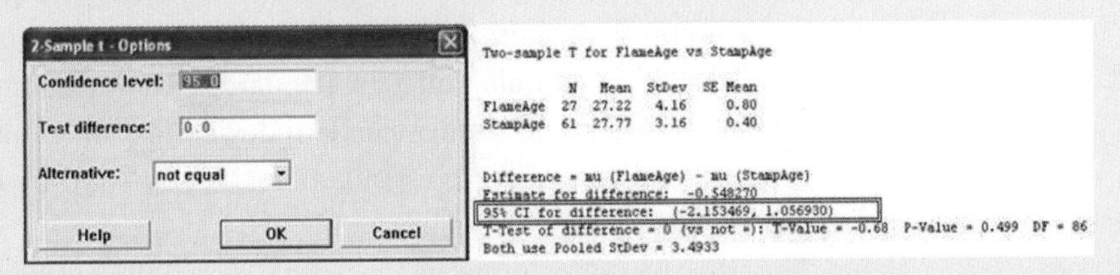

SPSS

SPSS's output includes a confidence interval whenever you initiate a *t*-test for two independent samples. When deciding whether to assume equal or unequal variances, you can use the procedures given in the second part of the SPSS instructions in "In Brief 12.1(a)."

For the *t*-test instructions, see the first part of the SPSS instructions in "In Brief 12.1(a)." In this context, always select *Options* in the first dialogue box, and then input the desired level of confidence (see the screen capture).

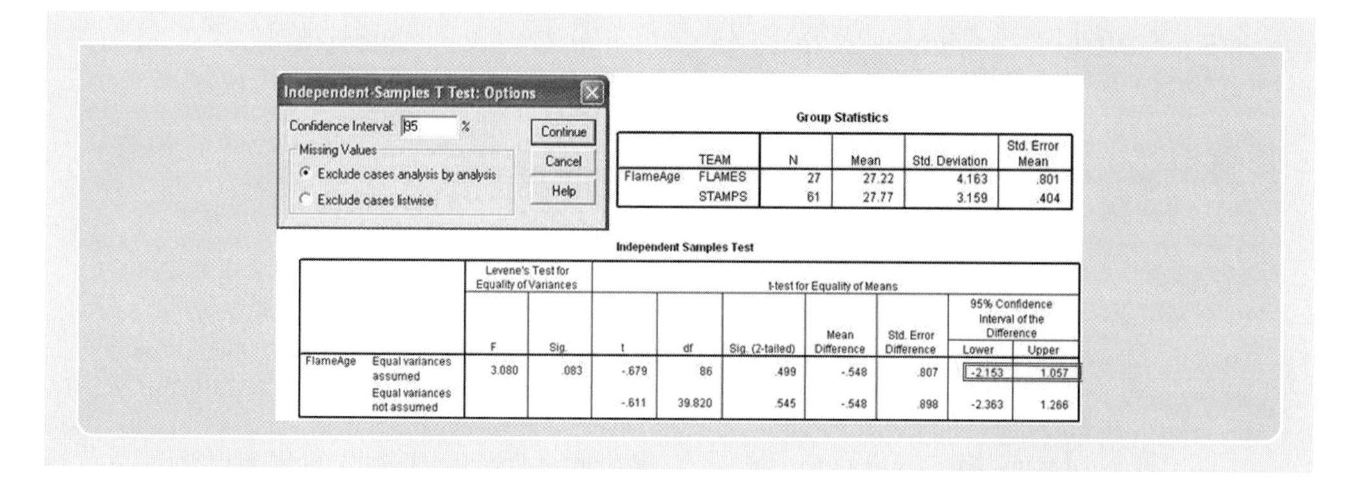

Group Statistics

	TEAM	N	Mean	Std. Deviation	Std. Error Mean
FlameAge	FLAMES	27	27.22	4.163	.801
	STAMPS	61	27.77	3.159	.404

Independent Samples Test

		Levene's Test for Equality of Variances		t-test for Equality of Means					95% Confidence Interval of the Difference	
		F	Sig.	t	df	Sig. (2-tailed)	Mean Difference	Std. Error Difference	Lower	Upper
FlameAge	Equal variances assumed	3.080	.083	-.679	86	.499	-.548	.807	-2.153	1.057
	Equal variances not assumed			-.611	39.820	.545	-.548	.898	-2.363	1.266

Sections 10.4–10.8

• Hands On 12.1 *Impacts of Assuming Equal or Unequal Variances*

When sampling for a difference in means when σ is unknown, we found that it makes a mathematical difference whether we assume equal or unequal variances for the populations. This applies whether estimating a confidence interval or testing a hypothesis. Some may argue that in practice, especially if the data themselves are a bit noisy, the distinction between the two methods is not too critical. Thanks to the computer, we can now experiment directly on this.

On a spreadsheet, generate many pairs of data columns. Using procedures from "In Brief 5.2," cause each column to be generated from a normal distribution—each column having a mean and standard deviation of your choice. Choose the means and standard deviations using the following plan: Have some pairs of columns with similar means and similar standard deviations; some pairs with similar means but quite different standard deviations; some pairs with quite different means and similar standard deviations; and others with quite different means and quite different standard deviations—and try many variations in between.

For every pair of columns you generate, record the means and standard deviations you selected. Next, for every pair of columns, use the computer to perform two *t*-tests for difference in means. In the first test, assume equal variances for the data columns; in the second test, assume unequal variances. For each test, record the *p*-values for the one-tail test and the *p*-values for the two-tail test. You could organize your results to look something like Table 12.1.

After accumulating data, try to find cases in which the *p*-values are significant according to the equal variances assumption but not according to the unequal assumption (or vice versa). Especially look for cases that straddle the usual values for alpha (such as 0.01 or 0.05). How many such cases can you find (or create, if you try)? Interpret your results.

Table 12.1 Comparative Effects on *p*-Values of Selecting Case 1 versus Case 2 for Tests of Unequal Means

	......................Trial Numbers......................															
	...	24		...	29		...	35		36		...	47		...
sample means sample variances		0.1735 0.8062	1.2452 1.9016		−0.2972 0.7228	1.8734 0.9999		0.0373 1.1457	2.0078 10.1680	0.4540 1.0782	3.2689 21.3750		−0.0276 1.2100	−1.1442 17.8070	
		Sample 1	Sample 2		Sample 1	Sample 2		Sample 1	Sample 2	Sample 1	Sample 2		Sample 1	Sample 2	
P(one-tail) {Assume = Variance}		0.0030			0.0000			0.0063		0.0057			0.1297		
P(one-tail) {Assume **UN** = Variance}		0.0032			0.0000			0.0077		0.0074			0.1323		
P(two-tail) {Assume = Variance}		0.0060			0.0000			0.0126		0.0115			0.2594		
P(two-tail) {Assume **UN** = Variance}		0.0064			0.0000			0.0153		0.0148			0.2645		

12.3 Differences between Population Means: Two Dependent Samples

In Section 12.1, we distinguished between independent samples and dependent samples. Samples are dependent if selected pairs of elements, one from each sample, are in some way connected. For example, each pair of individuals may in fact be the same person—before and then after some experimental intervention. Or you may pair items for a price comparison between stores, selecting one item of each type (pears, apples, etc.) from each store. As mentioned earlier, we select dependent samples to control, or eliminate, the effect of extraneous variables on the result of the experiment. This can also reduce the variability associated with random sampling.

We can test for the difference in means of dependent samples with the dependent samples *t*-test. (We will not cover the corresponding *z*-test; for a realistic sample, it would be unlikely that a population value for σ would be known.) The following assumptions must apply for the *t*-test to be valid:

- *Random selection of the pairs:* For example, if pricing corresponding goods (such as apples and pears) at competing stores, randomly select one apple, one pear, and so on from each store to form the matched pairs. For a controlled experiment, such as evaluating a diet program, randomly select the participants and then determine randomly which member of each pair is assigned the experimental treatment.
- There is a normal distribution of the population of differences between paired values, or a large sample of paired values.
- The measurement level of the observations is at least at the interval level.

The sampling statistic is not interpreted as the difference in sample means for two groups; it is based on the mean of the value differences for all of the paired values. In Table 12.2, for instance, the right-hand column compares the prices between two U.S. regions for similar grocery items. The mean of the column of differences is shown at the bottom right. This mean difference can be represented with the symbol $\bar{d}$.

Table 12.2 Grocery Price Comparisons for Two U.S. Regions

Grocery Item	Northeast Urban	South Urban	Differences
Flour, white, all purpose, per lb. (453.6g)	0.359	0.304	0.055
Spaghetti and macaroni, per lb. (453.6g)	1.031	0.901	0.130
Ground beef, lean and extra lean, per lb. (453.6g)	3.017	2.970	0.047
Bacon, sliced, per lb. (453.6g)	3.846	3.906	−0.060
Ham, boneless, excluding canned, per lb. (453.6g)	3.100	3.313	−0.213
Chicken, fresh, whole, per lb. (453.6g)	1.086	0.943	0.143
Eggs, Grade A, large, per doz.	1.756	1.167	0.589
Apple, Red Delicious,per lb. (453.6g)	1.085	1.113	−0.028
Bananas, per lb. (453.6g)	0.529	0.441	0.088
Potatoes, white, per lb. (453.6g)	0.481	0.533	−0.052
Mean Difference			**0.070**

Source: U.S. Bureau of Labor Statistics. Based on data in the 2004 *U.S. Consumer Price Index.* Found at: http://data.bls.gov/PDQ/outside.jsp?survey=ap

For the two-tailed hypothesis test, the null and alternative hypotheses are:

$$H_0\text{: } \mu_D = 0$$

$$H_1\text{: } \mu_D \neq 0$$

where μ_D is the population mean of the differences in pairs of values. The null hypothesis states that, apart from sampling error, there is *no* difference, on average, between paired values from the two groups. Applied to Table 12.2, this would mean that the prices are generally the same in the two regions, regardless of the item selected. The alternative hypothesis states that there is a difference in prices between the regions.

Formulas 12.4, 12.5, and 12.6 show the mean, variance, and standard deviation of the differences, respectively. In structure, these formulas are identical to Formulas 4.2, 5.4, and 5.6. This is because the column of differences (d_i) is just a special case of a set (x_i) of sample numbers; their mean, variance, and standard deviation can be calculated in the usual way.

Formula 12.4 Mean Difference in Pairs of Values

$$\bar{d} = \frac{\sum_{i=1}^{n} d_i}{n}$$

Formula 12.5 Variance in Pairs of Values

$$s_d^2 = \frac{\sum_{i=1}^{n} (d_i - \bar{d})^2}{(n - 1)}$$

Formula 12.6 Standard Deviation in Pairs of Values

$$s_d = \sqrt{s_d^2} = \sqrt{\frac{\sum_{i=1}^{n} (d_i - \bar{d})^2}{(n - 1)}}$$

Under the null hypothesis we expect the mean of the differences ($\bar{d}$) to be approximately zero. The sampling distribution of all of the possible sample differences will be centred around that value. Given the test assumptions, we can represent the sampling distribution as having a *t*-distribution with $(n - 1)$ degrees of freedom; n is the number of *pairs* of values. The standard error for the distribution is, again, a special case of the standard error for *any* *t*-distributed sampling distribution (see Formula 12.7), and the *t*-statistic is calculated analogously to Formula 10.1 (see Formula 12.8).

Formula 12.7 Standard Error of the Differences

$$s_{\bar{d}} = \frac{s_d}{\sqrt{n}}$$

Formula 12.8 t-Value for a Specific Value Difference

$$t = \frac{\bar{d} - \mu_D}{\frac{s_d}{\sqrt{n}}}$$

To perform the statistical test, either compare the *t*-value that corresponds to the sample's $\bar{d}$ to the critical *t*-value (given the test α and appropriate degrees of freedom), or else compare the *p*-value that corresponds to the test statistic to the test value for α.

Example 12.7

*(Based on the **Price_Comparison** file)*

An entrepreneur is considering a move from an urban centre in the U.S. northeast to an urban centre in the U.S. south. One concern is whether food prices may be higher in the new community. Refer to the sample data in Table 12.2. Test whether there is a difference in grocery prices between the regions at the 5% level of significance. Assume that the population of price differences is normally distributed.

Solution

Step 1: We are asked to test a possible claim that grocery prices are different in the two regions. The data have been collected as a paired sample, so the claim relates to the mean of the differences for the paired values. The claim in symbols is:

$$\mu_D \neq 0$$

Step 2: Since the claim does not contain equality, it can be interpreted as the alternative hypothesis. The opposite of that hypothesis is H_0.

$$H_0: \mu_D = 0$$

$$H_1: \mu_D \neq 0$$

Step 3: The level of significance is given in the problem:

$$\alpha = 0.05$$

Step 4: We are testing for the mean difference in paired samples from the two populations. Variances are not known. The population of differences is presumed to be normally distributed. (The histogram and probability plot in Figure 12.15 are roughly consistent with the assumption.) The sample size is the number of data *pairs*. Therefore, we select this test:

Use the *t*-test for the mean difference, two dependent samples, $n = 10$.

Figure 12.15
The Population of Price Differences

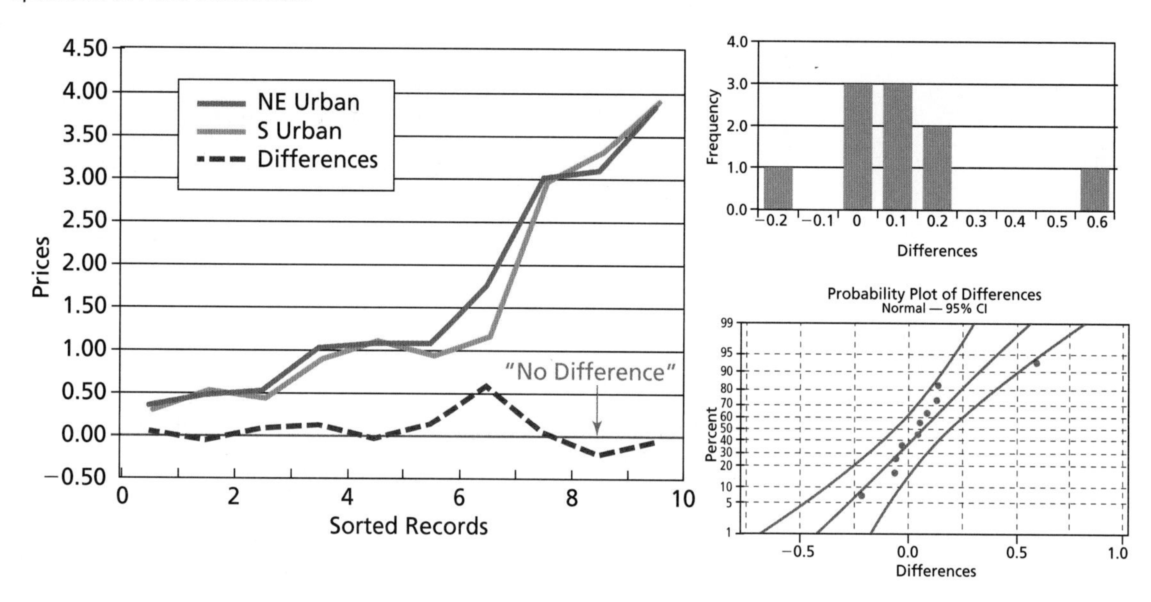

Step 5: Figure 12.16 shows the null distribution for the mean difference. The *t*-value is zero when the mean of the sample differences is zero. The degrees of freedom for the *t*-distribution equals (the number of data *pairs*, i.e., *n*) minus 1. The test is two-tailed.

Figure 12.16
Step 5 for the Hypothesis Test, Example 12.7

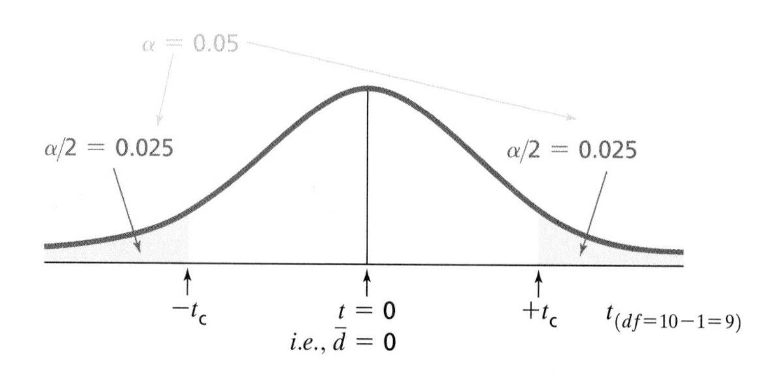

Steps 6 and 7: With the computer, create the column of differences using the provided data, and find the mean and standard deviation of the difference column. These values are $\bar{d} = 0.0699$ and $s_d = 0.2110$, respectively. Find the test statistic t.

$$t = \frac{\bar{d} - \mu_D}{\frac{s_d}{\sqrt{n}}} = \frac{0.0699 - 0.0}{\frac{0.2110}{\sqrt{10}}}$$

$$= 1.048$$

Knowing the t-value and the degrees of freedom, apply the procedures from "In Brief 11.1" to calculate the p-value (two-tail) $= 2 \times P(T \geq 1.048 \mid$ (the sampling distribution is as shown in step 5)):

p-value = 0.32, which is greater than $\alpha = 0.05$

Therefore, the statistical decision is:

Do not reject H_0.

Step 8: Relating the statistical decision to the original claim, we conclude:

You cannot conclude based on the evidence that grocery prices are different in the two regions.

Example 12.8

*(Based on the **Kansas_Wells** file)*

A hydrologist is studying the drainage patterns of wells in Atchison County, Kansas, during the middle third of the last century. She claims that, for any well, the water level decreased between April/May and July/August in any part of the county. Using data in the file **Kansas_Wells**, test the hydrologist's claim at the 0.01 level of significance. In the file, note that water levels are measured in feet *from the surface to the water.* Assume that the population of water level differences is normally distributed.

Solution

Step 1: The claim—in conventional language—is that the water level decreases. How does this translate into symbols? If the water level decreases in July/August, then the recorded distance from the water level to the surface increases, compared to April/May. So if we calculate the pair-differences as (April/May distances to the surface) minus the *larger* (July/August distances to the surface), we expect *negative* results. Therefore, the claim translates into:

$$\mu_D < 0$$

Step 2: The claim, which does not contain equality, becomes the alternative hypothesis. The opposite of that hypothesis is H_0.

$$H_0: \mu_D \geq 0$$
$$H_1: \mu_D < 0$$

Step 3: The level of significance is given in the problem:

$$\alpha = 0.01$$

Step 4: We are testing for the mean difference in paired samples from the two populations based on before/after measurements. Variances are not known, and the differences are presumed to be normally distributed. (The assumption could be checked with a probability plot or histogram.) The sample size is the number of data *pairs*.

Use the *t*-test for the mean difference, two dependent samples, $n = 7$.

Step 5: Figure 12.17 shows the null distribution for the mean difference. The *t*-value is zero when the mean of the sample differences is zero. This is a left-tail test.

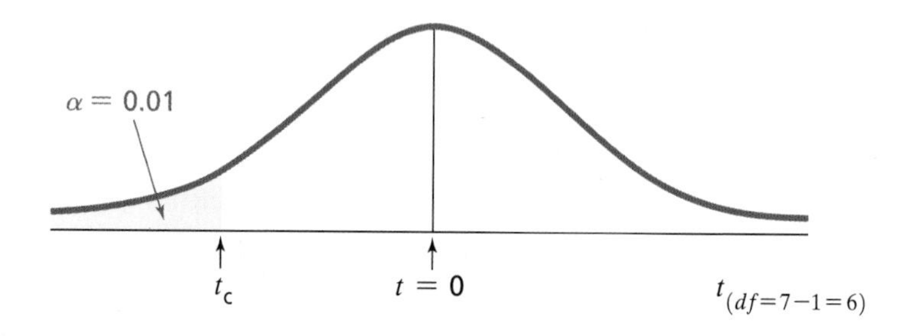

Figure 12.17
Step 5 for the Hypothesis Test, Example 12.8

Steps 6 and 7: As in Example 12.7, you can create the column of differences using the provided data and find the mean and standard deviation of the difference column. These values are $\bar{d} = -0.1814$, and $s_d = 0.3259$, respectively. Find the test statistic *t*.

$$t = \frac{\bar{d} - \mu_D}{\frac{s_d}{\sqrt{n}}} = \frac{-0.1814 - 0.0}{\frac{0.3259}{\sqrt{7}}}$$
$$= -1.47$$

Knowing the *t*-value and the degrees of freedom, apply the procedures from "In Brief 11.1" to calculate the one-tail *p*-value = $P(T_{df=6} \le -1.47 \mid$ (the sampling distribution is as shown in step 5)). You can also obtain this *p*-value more directly, using the procedures in "In Brief 12.2(a)."

***p*-value = 0.096, which is greater than $\alpha = 0.01$**

Therefore, the statistical decision is:

Do not reject H_0.

Step 8: Relating the statistical decision to the original claim, we conclude:

The evidence does not support the claim that the water levels generally decreased, between the April/May and the July/August reporting periods in Atchison County, Kansas, during the period of study.

• *In Brief 12.2(a) Computer Assistance for* t*-Tests for Mean Difference: Two Dependent Samples*

*(Based on the **Kansas_Wells** file)*

Excel

Begin with the menu sequence *Tools / Data Analysis / t-Test: Paired Two Sample for Means.* In the resulting dialogue box (see the screen capture), input the ranges for the *Variable 1* data (e.g., the "before" data) and the *Variable 2* data (e.g., the "after" data), respectively. Excel will take the pair-differences in that order. It is optional to input a non-zero value for the hypothesized mean difference, per the null hypothesis. (In Example 12.8, we left that value at zero). Finally, in *Output Range,* specify the upper left cell address for the output range and then click *OK.*

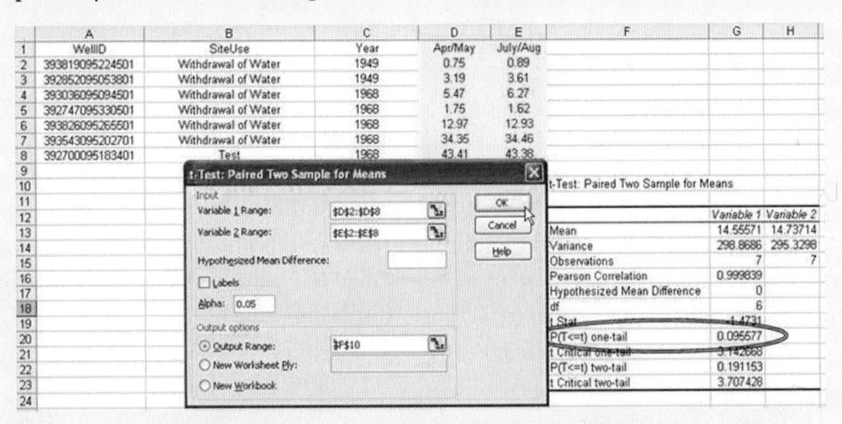

The illustration also shows the software output, and highlights the one-tail *p*-value that corresponds to Example 12.8. If the test had been two-tailed, the two-tail *p*-value would be read instead.

Minitab

In Minitab, begin with the menu sequence *Stat / Basic Statistics / Paired t.* The input format is flexible. The spreadsheet in the illustration shows the sample data in two separate columns. Alternatively, you could input the summary statistics for the data, if available. For the column approach, Minitab will

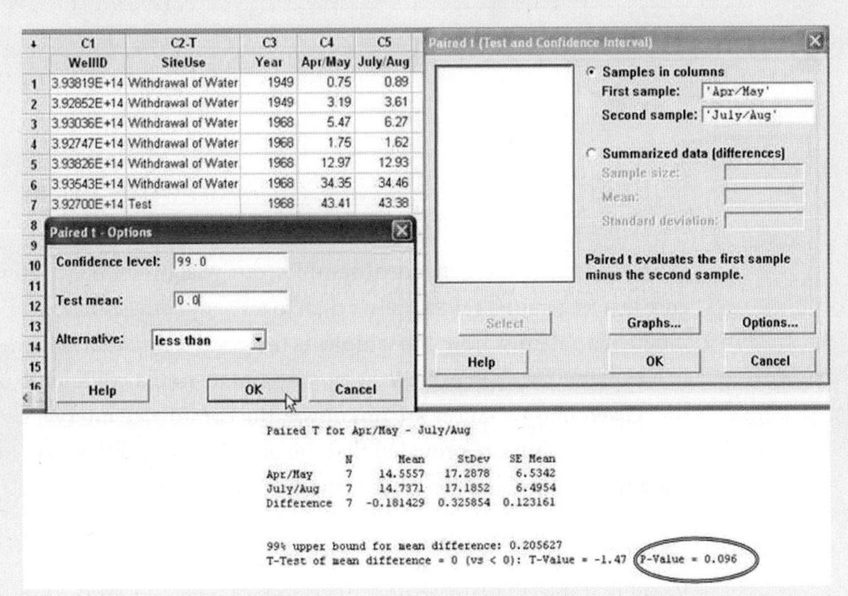

calculate the differences in the first and second order that you specify. Now click on the *Options* button and, in the resulting dialogue box, input in the *Test mean* window the test value for the mean difference (if any) assumed in the null hypothesis (this was 0 in Example 12.8). Then specify the operator ($\neq$, $<$, or $>$) in the *Alternative* window. (It is optional at this point to also input a confidence level equal to $1 - \alpha$.) Click *OK* twice. The illustration also shows the software output, including the *p*-value for the test.

SPSS

In SPSS, begin with the menu sequence *Analyze / Compare Means / Paired-Samples T Test.* In the resulting dialogue box (see the screen capture), select both variables to be compared. Simply select the variables from the upper left box one at a time and they will both appear in *Current Selection* on the lower left; then click on the arrow button to select the pair. SPSS will take the differences from left to right, in the order in which the variables appear in the worksheet. Click *OK.*

The *p*-value results for the test are shown at the right of the output, circled in the illustration. Two-tailed results are always displayed, so for a one-tailed test (such as Example 12.8), divide the displayed value by 2.

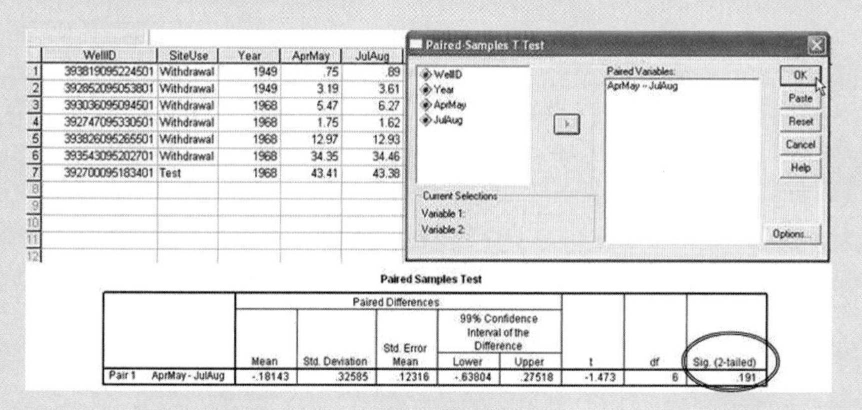

Confidence Intervals for the Mean Difference: Two Dependent Samples

When collecting two dependent samples, our goal might be to estimate the mean difference between paired values, rather than to test a hypothesis. The precise value for the mean difference obtained from the samples is the point estimate. Using the principles described in Chapter 10, we can extend the point estimate into a confidence interval estimate.

Given that variances are unknown, the confidence interval can be estimated based on the *t*-distribution, provided that the population of differences between paired values has a normal distribution or that the samples are large:

$$\textbf{(Point estimate)} \pm [\mathbf{t}_{\alpha/2} \times \textbf{(Standard error of the estimate)}]$$

Recall that the standard error is the standard deviation of the sampling distribution for the parameter, under the null hypothesis. $t_{\alpha/2}$ is the critical value for the sample statistic that determines the boundary of the confidence interval for a given level of confidence $(1 - \alpha)$. Apply this general model to estimating the mean difference.

The critical *t*-value can be determined from the degree of confidence, as discussed in previous chapters. The calculation for standard error of the estimate is given in Formula 12.7 on page 517. Depending on your software, the calculations can be automated to varying degrees. The solution to Example 12.9 assumes minimal automation, but see "In Brief 12.2(b)" to determine what your software provides.

Example 12.9

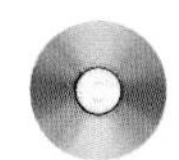

*(Based on the **Price_Comparison** file)*

For Example 12.7, the solution failed to establish a difference in grocery prices between cities in the northeastern United States and cities in the southern United States. It may still be useful to construct a confidence interval to estimate how far apart the two sets of prices may be. Construct this estimate at the 95% confidence level.

Solution

Let the values from the northeastern cities be population 1 and the values from the southern cities population 2. Whatever your software, the intermediate figures highlighted in Figure 12.18 can readily be calculated from the data:

Figure 12.18
Key Data for Estimating the Mean Difference: Two Dependent Samples

t-Test: Paired Two Samples for Means

	Variable 1	*Variable 2*
Mean	1.629	1.5591
Variance	1.567866	1.735386
Observations	10	10
Pearson Correlation	0.987788	
Hypothesized Mean Difference	0	
df	9	
t Stat	1.047394	
P(*T*<=*t*) one-tail	0.161115	
t Critical one-tail	1.833113	
P(*T*<=*t*) two-tail	0.32223	
t Critical two-tail	2.262157	

The point estimate μ_D for the mean difference is mathematically the same number as $\mu_1 - \mu_2$, the difference in sample means. In this case, the point estimate equals 1.629 − 1.559 = 0.070. The differences have a *t*-distribution with $df = 9$, so using procedures from "In Brief 10.2," we can find the critical *t*-value for confidence level 95%: $t_{\text{critical(2-tail)}} = 2.262$.

The standard error of the estimate is $\frac{s_d}{\sqrt{n}}$. If you create a column of differences, as in Table 12.2, s_d is the standard deviation of that column (0.21104). Therefore, the standard error is $\frac{0.21104}{\sqrt{10}} = 0.06674$. The resulting confidence interval is

Lower limit = (Point estimate) − [$t_{\alpha/2}$ × (Standard error of the estimate)]
$= 0.070 - (2.262 \times 0.06674)$
= −0.081

Upper limit = (Point estimate) + [$t_{\alpha/2}$ × (Standard error of the estimate)]
= 0.070 + (2.262 × 0.06674)
= 0.221

We have 95% confidence that the mean difference in grocery prices between the two regions is in the range from –0.081 to +0.221.

• *In Brief 12.2(b) Computer Assistance for Estimating Confidence Intervals for Mean Difference: Two Independent Samples*

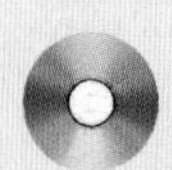

(*Based on the* ***Price_Comparison*** *file*)

Excel

Excel does not directly automate this calculation, but it *can* create an output similar to Figure 12.18, which can provide the means and critical *t*-value. You will still need to construct a column of differences (see Table 12.2) and then use the Excel formula (**=stdev( )**) to find its standard deviation. Then proceed as described in the solution for Example 12.9.

Minitab

Minitab's output includes a confidence interval whenever you initiate a paired sample *t*-test. For the *t*-test instructions, see "In Brief 12.2(a)." In this context, always click on the *Options* button in the first dialogue box and then, as shown in the screen capture, (1) choose zero (i.e., 0.0) as the value for the test difference to be inserted into the *Test mean* window, (2) select *not equal* as the operator in the *Alternative* window, and (3) input the desired level of confidence.

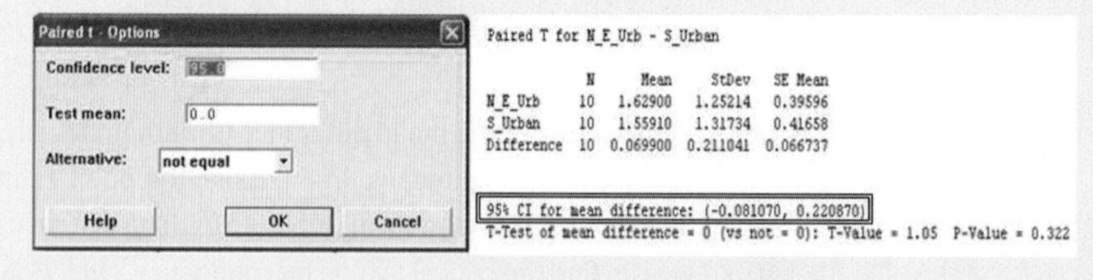

SPSS

SPSS's output includes a confidence interval whenever you initiate a paired sample *t*-test. For the *t*-test instructions, see "In Brief 12.2(a)." In this context, always click on the *Options* button in the first dialogue box and then input the desired level of confidence (see the screen capture).

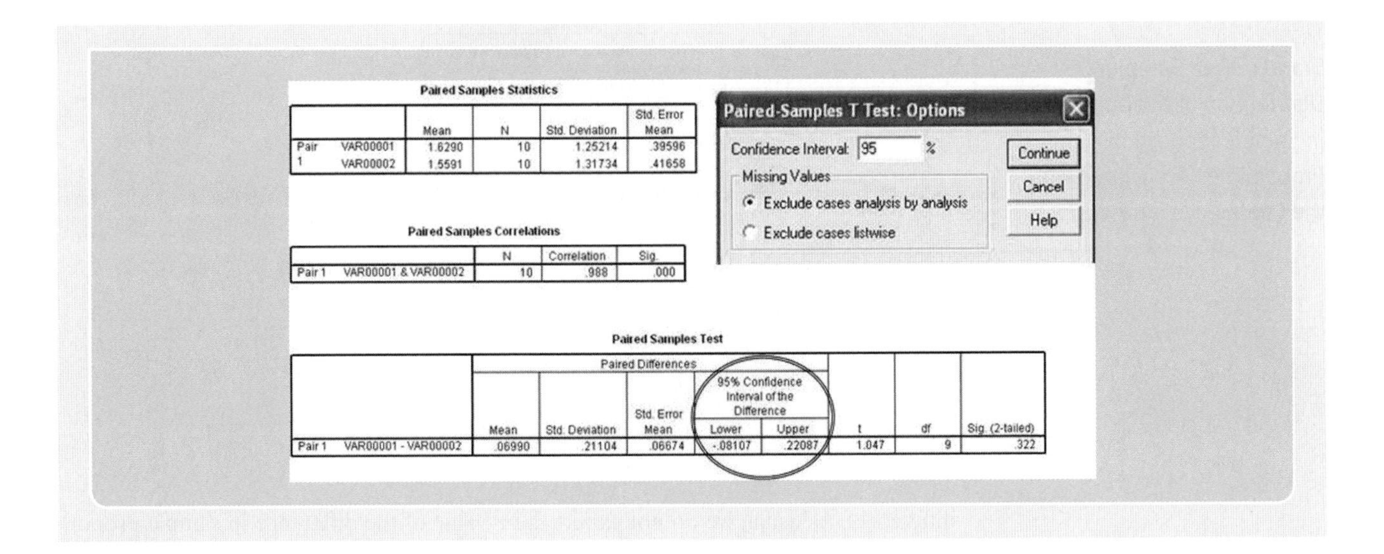

Paired Samples Statistics

		Mean	N	Std. Deviation	Std. Error Mean
Pair 1	VAR00001	1.6290	10	1.25214	.39596
	VAR00002	1.5591	10	1.31734	.41658

Paired Samples Correlations

		N	Correlation	Sig.
Pair 1	VAR00001 & VAR00002	10	.988	.000

Paired Samples Test

		Paired Differences					t	df	Sig. (2-tailed)
					95% Confidence Interval of the Difference				
		Mean	Std. Deviation	Std. Error Mean	Lower	Upper			
Pair 1	VAR00001 - VAR00002	.06990	.21104	.06674	-.08107	.22087	1.047	9	.322

12.4 Differences between Two Population Proportions

We can also use a two-sample hypothesis test to compare population proportions. In this section, we consider binomially distributed data; that is, the raw data are categorical—with two categories, so that for each element collected, there is a fixed probability that one particular value will be drawn (i.e., be a "success"). The presumption is that all observations in each sample are collected independently, and that random selection is from infinite populations or else with replacement from finite populations.

We used similar assumptions when introducing a one-sample hypothesis test for the proportion in Section 11.4. There, we could use the binomial model directly for the sampling distribution of the proportions—and bypass the traditional z-based model for approximating the distribution. For testing a *difference* in proportions, however, even computers use the traditional normal approximation for the sampling distribution. (The difference in proportions is *not* a binomial variable.) The normal approximation works reasonably well, provided that for the *smaller* of the two sample sizes: $n_{\text{smaller}} \times \bar{p} \geq 5$, and $n_{\text{smaller}} \times (1 - \bar{p}) \geq 5$, for both samples. (Some statisticians more stringently prefer that both of these products should equal at least 10.) $\bar{p}$ is the "pooled" estimate of the proportion, based on the two samples combined. (An alternative approach, the chi-square distribution, will be introduced in Section 15.4.)

When making inferences in this section for differences in proportion, the sampling distribution is presumed to be as shown in Figure 12.19. As in the similar figure for difference in means (Figure 12.5), the variable of interest is *not* taken from one sample individually; there are two samples, and each sample has its own proportion of successes. The variable that is distributed is *the difference in values* of the proportions for the two samples. The set of all possible values for this variable, if all possible samples of size n_1 and n_2 were taken, is the sampling distribution used for the inferences.

For the hypothesis test, a common null hypothesis is "no difference" between the sample proportions. For an actual pair of samples, the difference in their two proportions can be located on the horizontal axis of Figure 12.19 in terms of z-values. In this context, z specifies the number of (plus or minus) standard deviations between the observed difference in sample proportions and a difference of zero (i.e., the case of *no*

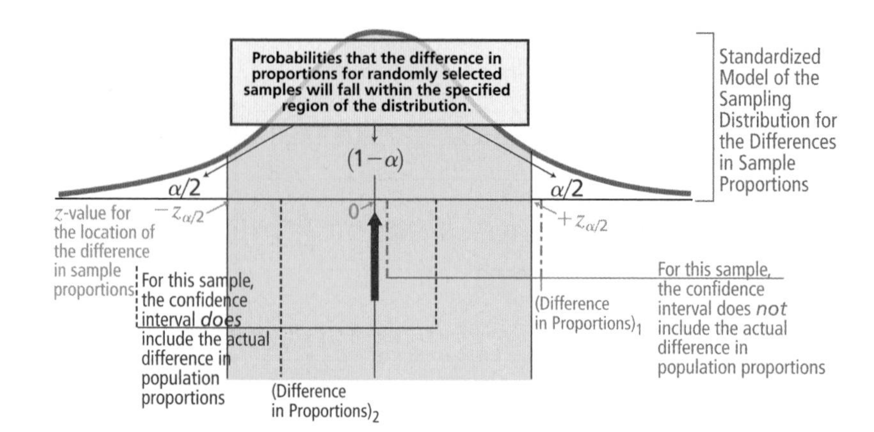

Figure 12.19
Standardized Sampling Distribution of the Difference in Sample Proportions, and Probabilities Associated with the Confidence Intervals

difference). As usual, we do not expect the z-value of the difference in sample proportions to have a very large magnitude.

We can also use the model to make an estimate of the difference between the population proportions. Centre the estimate on the difference between the sample proportions. Create a confidence interval around the central value, similar to the procedures in Chapter 10. There is $1 - \alpha$ probability that the calculated interval will include the true value for the difference in population proportions.

We can also adapt the model in Figure 12.19 to handle other types of inferences. For example, the hypothesis might be that the two populations' proportions are not identical, but rather are some specified distance apart, for instance, 0.1. In that case, the centre of the distribution, at $z = 0$, represents samples with proportions differing by exactly 0.1. Other locations on the horizontal axis are measured in z-units above or below that exact distance.

z-Test for the Difference in Population Proportions

By now, the general procedures for a z-based hypothesis test should be familiar. The model assumptions were discussed above. After taking the two samples, find the proportions of successes in each, find their difference, and then convert that distance to a z-value in relation to the sampling distribution that is modelled.

In this application, the formula for finding the test statistic, z, is shown in Formula 12.9. Observe that, in the numerator, the obtained difference in sample proportions is compared to the hypothesized difference (which is commonly zero). The standard error term, in the denominator, requires a preliminary calculation of the pooled estimate for the proportion, as shown.

Formula 12.9 z-Value for the Difference in Sample Proportions

a) $$z = \frac{(p_{s_1} - p_{s_2}) - (p_1 - p_2)}{\sqrt{\dfrac{\bar{p}(1-\bar{p})}{n_1} + \dfrac{\bar{p}(1-\bar{p})}{n_2}}}$$

where p_{s_1} and p_{s_2} are the two *sample* proportions
p_1 and p_2 are the two *population* proportions
$\bar{p}$ is the pooled estimate of the proportion based on:

b) $\overline{p} = \frac{x_1 + x_2}{n_1 + n_2}$

x_1 and x_2 are the numbers of successes in each of the samples

Example 12.10

*(Based on the **Metallic_Minerals** file)*

The lustre of a mineral refers to how brightly it reflects external light. The lustres of metallic minerals tend to fall into two classifications: metallic and submetallic. Find the lustres of known metallic minerals in the file **Metallic_Minerals.** A specialist claims that the proportion of minerals with submetallic lustres is different, based on the density of the mineral. Divide the data set into minerals with lower density and with higher density (use a cutoff of density < 5 and density ≥ 5). Assuming that the data set is representative also of minerals yet to be discovered, test the specialist's claim at the 0.01 level of significance.

Solution

Step 1: The specialist claims that the proportions of the minerals with submetallic lustres are different for the two density-based groups. Let the subscript "1" refer to the group with densities less than 5. In symbols, the claim can be formulated as:

$$p_1 \neq p_2$$

Step 2: The claim in step 1 can be interpreted as the alternative hypothesis, since it does not contain equality. The opposite hypothesis becomes H_0.

$$H_0: p_1 = p_2$$
$$H_1: p_1 \neq p_2$$

These hypotheses could also be written as:

$$H_0: p_1 - p_2 = 0$$
$$H_1: p_1 - p_2 \neq 0$$

Figure 12.20 in step 5 relates more clearly to this version of the hypotheses. The distribution is for the *differences* in sample proportions, provided that the equality assumption of H_0 is true.

Step 3: The level of significance is given in the problem:

$$\alpha = 0.01$$

Step 4: The sample sizes can be taken from the data. We choose a test for the difference in proportions for independently sampled populations; the lustre variable appears to be binomial.

First check the criteria for using the normal approximation:

- $\overline{p} = \frac{x_1 + x_2}{n_1 + n_2} = \frac{13 + 8}{29 + 71} = \frac{21}{100} = 0.21$
- $n_{smaller} \times \overline{p} = 29 \times 0.21 = 6.09 \geq 5$
- $n_{smaller} \times (1 - \overline{p}) = 29 \times 0.79 = 22.91 \geq 5$

The weaker criterion for using this model is satisfied. (If you use the stricter criterion that the products should equal at least 10, then a nonparametric method that will be discussed in Chapter 13 might be preferred.)

Use the z-test for difference in population proportions, $n_1 = 29$, $n_2 = 71$.

Steps 5–7: The same decision as that shown in Figure 12.20 would be reached if we used the p-value approach. The (two-tail) p-value that corresponds to the test statistic $z = 3.73$ [i.e., $2 \times P(Z) \geq 3.73$ | (the sampling distribution is as shown in step 5)] can be calculated as 0.0002, which is less than $\alpha = 0.01$. Reject H_0.

Figure 12.20
Steps 5–7 of the Hypothesis Test, Example 12.10

Step 5:

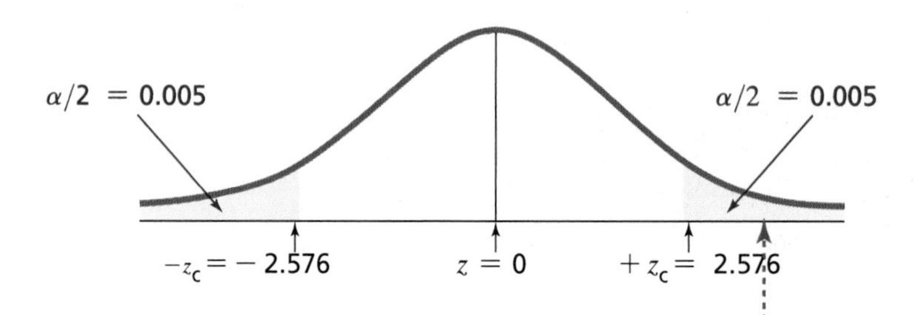

Step 6: Use the sample data in the CD file to calculate: $p_{s_1} = 0.448$ and $p_{s_2} = 0.113$.

Step 7:

- Calculate test statistic given:
  - Difference in sample proportions ≈ 0.335
  - Hypothesized: $p_1 - p_2 = 0$
  - $\bar{p} = 0.21$

$$z = \frac{(0.335) - (0.0)}{\sqrt{\frac{(0.21)(0.79)}{29} + \frac{(0.21)(0.79)}{71}}} = \mathbf{3.73}$$

- Compare the test statistic to the critical region in step 5 (see red arrow).
- Because the test statistic falls in a critical region, reject H_0.

Step 8: Relating the statistical decision back to the original claim, we conclude:

The evidence supports the claim of the specialist that the proportions of minerals having submetallic lustres are different, depending on the density of minerals.

• *In Brief 12.3 Computer Assistance for z-Tests for Difference in Proportions*

*(Based on the **Metallic_Minerals** file)*

Excel

Excel does not offer any options that explicitly test for difference in proportions. However, its functions can help to automate some calculations, such as finding the critical z-value(s) for the test. In Example 12.10, an Excel *Sort* procedure provides a simple way to separate the data for the low-density and high-density minerals, in order to next find the n and p_s for each group. For counting the number of successes in each group, the illustration shows how the Excel function **=COUNTIF()** could be used. The first input in the function is the range for one of the sample sets of the variable; the second input identifies the "success" value to look for, and count, in that range.

Range of all the data in Sample 1

=COUNTIF(F2:F30,"Sub Metallic") ⇒ 13

Range of all the data in Sample 2

=COUNTIF(F31:F101,"Sub Metallic") ⇒ 8

Minitab

In Minitab, begin with the sequence *Stat / Basic Statistics / 2 proportions.* The program is flexible about the input format. The worksheet in the illustration shows the sample data in one column (as they are in the CD data set). Alternative options are to use summary data for the sample statistics (then select *Summarized data* in the dialogue box) or to have each sample's data in its own column (then select *Samples in different columns* in the dialogue box). The one-column approach requires that there be another, "Subscript" column to specify which data row belongs to which sample. In the worksheet in

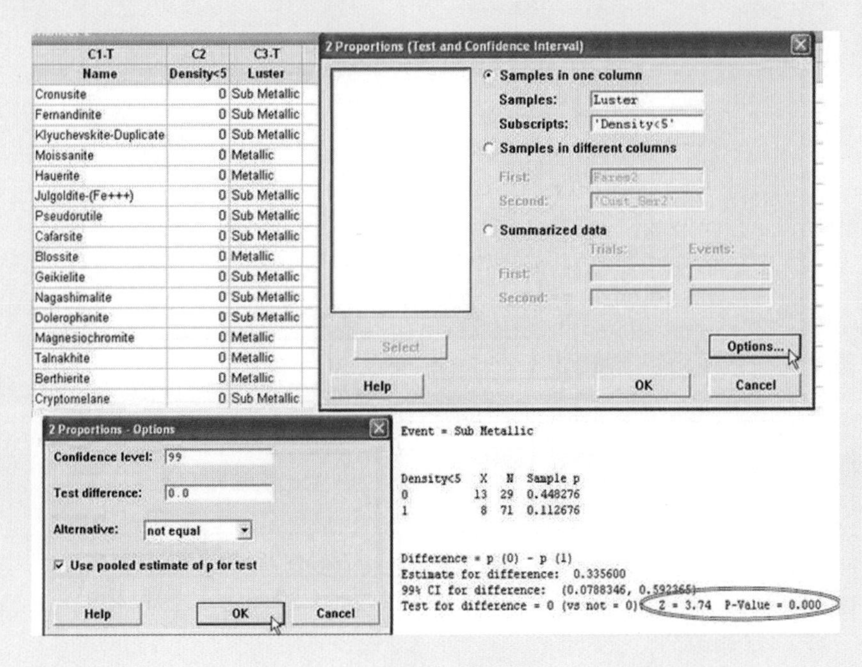

the illustration, this has been done by sorting the data from low to high density and adding a "Subscript" column that shows "0" for the low-density samples and "1" for the high-density samples.

In the first dialogue box (shown in the screen capture), click on the *Options* button. In the next dialogue box, input into the *Test difference* window the difference in proportions as given in the null hypothesis (which was 0 in Example 12.10). Then specify the operator ($\neq$, $<$, or $>$) in the *Alternative* window. For *Confidence level,* input the value of $1 - \alpha$. Following the method in this text, specify a pooled estimate for the proportion. Click *OK* twice. The illustration shows the software output. Note that you can use the z-value or the p-value, depending on how you set up the test.

SPSS

SPSS does not offer options that explicitly test for difference in proportions. However, its functions can help to automate some calculations, such as finding the critical z-value(s) for the test. In Example 12.10, an SPSS *Sort* procedure provides a simple way to separate the data for the low-density and high-density minerals; then a data column can be added to differentiate between those two groups (for example, coding the lower-density minerals with "0" and the higher-density minerals with "1").

A convenient way to find n and p_s for each group is to then use the *Crosstabs* function: Begin with the sequence *Analyze / Descriptive Statistics / Crosstabs.* In the first dialogue box (see the screen capture), select the variable that distinguishes the two samples as the *Rows,* and the variable with the data as the *Columns.* Click *Cells* and in the next dialogue box, select *Observed, Row,* and *Round cell counts,* and then click on *Continue* to find percentages for the rows. The output, also shown in the illustration, includes summary data that are useful for the solution steps.

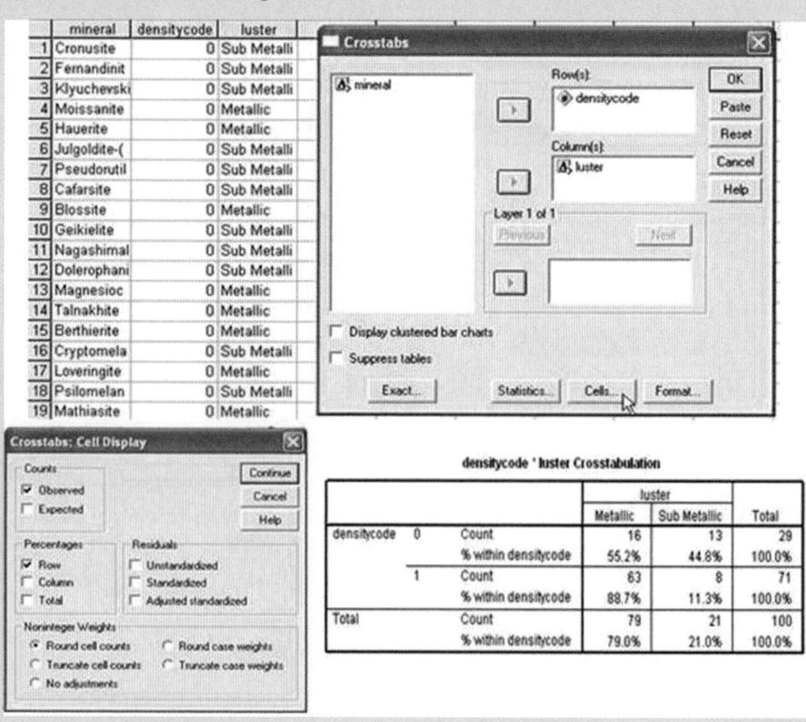

	mineral	densitycode	luster
1	Cronusite	0	Sub Metalli
2	Fernandinit	0	Sub Metalli
3	Klyuchevski	0	Sub Metalli
4	Moissanite	0	Metallic
5	Hauerite	0	Metallic
6	Julgoldite-(	0	Sub Metalli
7	Pseudorutil	0	Sub Metalli
8	Cafarsite	0	Sub Metalli
9	Blossite	0	Metallic
10	Geikielite	0	Sub Metalli
11	Nagashimal	0	Sub Metalli
12	Dolerophani	0	Sub Metalli
13	Magnesioc	0	Metallic
14	Talnakhite	0	Metallic
15	Berthierite	0	Metallic
16	Cryptomela	0	Sub Metalli
17	Loveringite	0	Metallic
18	Psilomelan	0	Sub Metalli
19	Mathiasite	0	Metallic

densitycode * luster Crosstabulation

			luster		
			Metallic	Sub Metallic	Total
densitycode	0	Count	16	13	29
		% within densitycode	55.2%	44.8%	100.0%
	1	Count	63	8	71
		% within densitycode	88.7%	11.3%	100.0%
Total		Count	79	21	100
		% within densitycode	79.0%	21.0%	100.0%

Confidence Intervals for the Difference in Population Proportions

When we compare two proportions, our goal might not be to test a hypothesis. Perhaps we want to estimate the size of the difference between the two proportions. The point estimate for the difference is the difference in proportions for the two samples. Following the principles discussed in Chapter 10, we can construct a confidence interval around the point estimate so that we have confidence (to the level $1 - \alpha$) that the true difference in population proportions lies in that interval.

In this section, the estimate is based on the normal approximation of the sampling distribution. Therefore, the confidence interval is:

$$\textbf{(Point estimate)} \pm [z_{\alpha/2} \times \textbf{(Standard error of the estimate)}]$$

As in previous sections, the boundary of the confidence interval for a given level of confidence $(1 - \alpha)$ is a function of the critical value for the sample statistic $(z_{\alpha/2})$. The critical z-value itself can be determined from the desired degree of confidence. The calculation for standard error of the estimate for this model was introduced in Formula 12.9(a) (the denominator), on page 526.

Example 12.11

The conclusion of the hypothesis test in Example 12.10 was that, as claimed by the specialist, the proportions of minerals having submetallic lustres are different, depending on the density of the minerals. We might ask: What *is* the magnitude of difference between those proportions for the two density-based groupings? Find the confidence interval for this difference at the 99% level of confidence.

Solution

As before, let the low-density minerals (with density < 5) be denoted as group 1. The point estimate for $p_1 - p_2$ is the difference in sample proportions: $0.448 - 0.113 = 0.335$. The sampling distribution is approximately normal, so we find the critical z-value corresponding to 99% confidence, which is 2.576.

The pooled proportion of the estimate is:

$$\bar{p} = \frac{x_1 + x_2}{n_1 + n_2} = \frac{13 + 8}{29 + 71} = \frac{21}{100} = 0.21$$

Therefore, the standard error of the estimate is:

$$\sigma_{(p_{s1} - p_{s2})} = \sqrt{\frac{\bar{p} \times (1 - \bar{p})}{n_1} + \frac{\bar{p} \times (1 - \bar{p})}{n_2}}$$

$$= \sqrt{\frac{0.21 \times 0.79}{29} + \frac{0.21 \times 0.79}{71}}$$

$$= 0.0898$$

The resulting confidence interval is:

Lower limit = (Point estimate) – [$z_{\alpha/2}$ × (Standard error of the estimate)]

= 0.335 – (2.576 × 0.0898)

= 0.104

$$\text{Upper limit} = (\text{Point estimate}) + [z_{\alpha/2} \times (\text{Standard error of the estimate})]$$
$$= 0.335 + (2.576 \times 0.0898)$$
$$= 0.566$$

The confidence interval for the difference in the proportions for submetallic lustre minerals in low-density versus higher density metallic minerals is approximately from 0.10 to 0.57.

• Hands On 12.2 Testing the Significance of Comparative Poll Results

Among the most common statistical displays in the media are comparative proportions. For example, in a press release by the Environics Research Group on April 4, 2006, it was announced that public support for the Liberal Party leader dropped from 45% in fall 2005 to 35% in March 2006.[2] Was this drop significant or could the difference be merely statistical error? In theory, we now have the tools to test such claims for ourselves.

At least the press release provided the sample size—for the 2006 survey: 2,035. If the press release is picked up by various media, such as the *Toronto Star* or a regional newspaper, the publishers may or may not think to tell us the sample sizes. In fact, even the press release did not publish the sample size (n_1) for the fall 2005 sample. To get a sense of whether the drop is significant, we may have to make some guesses. For example, could the survey of 2005 have had a comparable sample size to the 2006 survey, of around 2000 people? As a critical reader of statistics, be warned: Your conclusions about significance are only as good as your guesses about unpublished sample sizes. Also, be wary of informal Internet surveys, the sample selection of which may be biased.

Your hands-on exercise is to locate recent comparative poll results that have been published in Canadian media. These are not hard to find using Internet search engines. For each case, assess the validity of the reported data collection methods and see whether you are given information about *both* sample sizes. If these first conclusions are positive, use the methods discussed in this section to assess the significance of the claimed findings. (Focus especially on cases where the apparent differences in proportion are not large.) Often—short of writing to the publisher—you will have to make guesses about sample sizes. For example, an article may tell you the overall size of a study, but then report proportion differences for men versus women. Ask yourself: Is it reasonable to take a guess (e.g., half of the n are men and half are women)? With some care, use your guess to test the significance of the claim.

Repeat this process for at least five polls.

Chapter Summary

12.1 Describe the distinction between independent and dependent samples, and explain some key concepts and terminology relating to collecting data for making inferences for two samples.

Samples are **independent** if the selection of elements for one sample is not influenced by the selection of elements for the other. By one approach, sampling is taken from two pre-identified populations. For a **controlled experiment,** such as a **clinical trial,** a sample is taken from a single population and then a distinction is introduced by the experimenter—between an **experimental group** (or **treatment group**), to which an experimental procedure (or treatment) is applied, versus a **control group,** to which the treatment is not applied.

Samples are **dependent** (also called **related samples** or **paired samples**) if the selection of elements in one sample is influenced by the selection of elements in the other sample. Paired samples can help to control unwanted variation due to **extraneous variables.** One method of obtaining a paired sample is the **before/after method** (also called the **pre-test/post-test method,** or **repeated-measures method**), in which the same individuals are sampled both before and after an intervention or experience and results are compared. Alternatively, in the **matched-pairs approach,** items are selected in pairs, one each from a predetermined set of categories (e.g., selecting from each of two grocery stores, one each of an apple, pear, fig, etc., to compare prices).

12.2 Conduct a *z*-test or a *t*-test for the difference in population means for independent samples—knowing which test is appropriate—and construct confidence intervals for the difference in population means for independent samples.

To test for a difference in population means when the populations are normally distributed, or when the samples are large, *and* you know the values of the population variances, you can use the *z*-test for the difference in population means. Use Formula 12.1 to calculate the test statistic z. Under the same conditions, but if the population variances are *not* known, use the *t*-test for the difference in population means. For this test, distinguish Case 1, where the two unknown variances can be reasonably assumed equal, and Case 2, where variances appear to be unequal. The formulas for the test statistic t differ somewhat for the two cases (compare Formulas 12.2 and 12.3), as do estimates for degrees of freedom.

To construct confidence intervals for the difference in population means, use these basic models:

- For known population variances, the confidence interval is:
  (Point estimate) ± [$z_{\alpha/2}$ × (Standard error of the estimate)]
- For unknown population variances, the confidence interval is:
  (Point estimate) ± [$t_{\alpha/2}$ × (Standard error of the estimate)]

12.3 Conduct a *t*-test for the difference in population means for data taken from two dependent samples, and construct confidence intervals for the difference in population means for two dependent samples.

For the *t*-test to be valid, the data pairs must be randomly selected and the population of differences between paired values must be normally distributed (or else the sample of paired differences must be large). First, align all of the paired values and take their differences. The mean and standard deviation for the paired differences can be found with Formulas 12.4 through 12.6. Then base the *t*-test on a hypothesis about the value of the mean difference, using Formulas 12.7 and 12.8 to calculate, respectively, the standard error of the differences and the test statistic *t*. To construct a confidence interval for the mean difference, use this basic model:

$$\textbf{(Point estimate)} \pm [t_{\alpha/2} \times \textbf{(Standard error of the estimate)}]$$

12.4 Conduct a *z*-test for the difference in population proportions, and construct confidence intervals for the difference in population proportions.

To test for the difference in population proportions, the normal distribution can be used to estimate the sampling distribution for the difference in sample proportions. (Unlike for the single population proportion, this sampling distribution is not binomial.) Certain minimal sample size requirements must be met. Then use a *z*-test to perform the hypothesis test, and calculate the test statistic *z* using Formula 12.9. To construct confidence intervals for the difference in population proportions, use this basic model:

$$\textbf{(Point estimate)} \pm [z_{\alpha/2} \times \textbf{(Standard error of the estimate)}]$$

• Key Terms

before/after method (also called **pre-test/post-test method** or **repeated-measures method**), **p. 493**
Case 1 versus **Case 2** (for **comparing means of two independent samples**) **p. 499**
clinical trial, p. 492
control group, p. 492
controlled experiment, p. 492
dependent samples, p. 492
experimental group (or **treatment group**), **p. 492**
extraneous variables, p. 492
independent samples, p. 491
matched-pairs approach, p. 493
related samples or **paired samples, p. 492**

• Exercises

Basic Concepts 12.1

1. Explain the difference between independent samples and dependent samples.
2. Does matched-pairs sampling involve independent or dependent samples? Explain.
3. Explain the distinction between repeated-measures sampling and matched pairs sampling.
4. Which of the following examples of sampling constitute a randomized control study?

a) The heights of randomly selected Albertans are compared with the heights of randomly selected New Brunswickers.
b) One hundred randomly selected Albertans participate in a trial of a new mouthwash. Fifty are given the new mouthwash; the other fifty (who are not told who they are) are given the old version. Compare the two groups' satisfaction with the mouthwash they were given.
c) One of two groups in an experiment is the "treatment group."
d) Statistics Canada compares the incomes for families headed by single fathers versus families headed by single mothers.
e) An experiment is a clinical trial for a new prescription drug.

5. Consider the following statement: "In a flawed study to assess the effectiveness of an archery course, the mean accuracy of a sample of people who started the course is compared with the mean accuracy of an independent sample of people who finished the course. But the study ignores the possibility that course participants may differ in their initial skills and experience in archery." Would such differences in initial skills be an extraneous variable? Why or why not?
6. Explain how a before/after study could be used to minimize the impact of the concern in Exercise 5.
7. Identify at least one extraneous variable that could affect the following plan for a study: To test how much house prices have changed since last year, you compare a random sample from all house prices last year with a random sample from all house prices this year.
8. Explain a type of dependent sampling that could be used to minimize the impact of the extraneous variable in Exercise 7.
9. What type of basic study design is the following: Two individuals are randomly selected from each university in Canada. One individual from each pair receives a free one-person pass to attend three of his or her university's hockey games; the other individual receives a free *two*-person pass to attend three of his or her university's hockey games. Test if the individuals who receive the one-person passes are less likely to bring a friend when they attend the three games.

For applied problems, see the Applications exercises in the following sections.

Basic Concepts 12.2

10. What is the difference between Case 1 and Case 2 when conducting a t-test for the difference in means?
11. Test whether the following two means are equal at the 5% level of significance, using these data:
$\bar{x}_1 = 36 \quad \sigma_1 = 5.8 \quad n_1 = 15$
$\bar{x}_2 = 41 \quad \sigma_2 = 6.3 \quad n_2 = 15$
Assume that the samples are independent.
a) What is the null hypothesis?
b) What is the statistical decision?
c) What is your conclusion (in words) based on the test?
12. The following samples were taken from two independent populations. Test at the 5% level of significance whether the first population has a larger mean than the second population.
$\bar{x}_1 = 2904 \quad \sigma_1 = 148 \quad n_1 = 16$
$\bar{x}_2 = 2825 \quad \sigma_2 = 95 \quad n_2 = 13$
a) What is the null hypothesis?
b) What is the test statistic?
c) What is the p-value for the test?
d) What is the statistical decision?
e) What is your conclusion (in words) based on the test?
13. The following samples were taken from two independent normally distributed populations. Population sigmas are not known but are presumed to be about equal. Test at the 5% level of significance whether the first population has a larger mean than the second population.
$\bar{x}_1 = 2904 \quad s_1 = 148 \quad n_1 = 16$
$\bar{x}_2 = 2825 \quad s_2 = 95 \quad n_2 = 13$
a) What is the null hypothesis?
b) What is the test statistic?
c) What is the p-value for the test?
d) What is the statistical decision?
e) What is your conclusion (in words) based on the test?
14. The following samples were taken from two independent normally distributed populations. Population sigmas are not known but are presumed to be about equal. Test at the 5% level of significance whether the difference in population means equals 50.
$\bar{x}_1 = 2904 \quad s_1 = 148 \quad n_1 = 16$
$\bar{x}_2 = 2825 \quad s_2 = 95 \quad n_2 = 13$
a) What is the null hypothesis?
b) What is the test statistic?
c) What is the p-value for the test?
d) What is the statistical decision?
e) What is your conclusion (in words) based on the test?
15. Test whether the following two means are equal at the 1% level of significance, using these data:
$\bar{x}_1 = 30.6 \quad \sigma_1 = 5.1 \quad n_1 = 30$
$\bar{x}_2 = 34 \quad \sigma_2 = 5.48 \quad n_2 = 38$
Assume that the samples are independent.
a) What is the null hypothesis?
b) What is the statistical decision?
c) What is your conclusion (in words) based on the test?

16. Redo Exercise 15 but test whether the first population has a *smaller* mean than the second population. Does your answer differ from the previous answer? If so, explain why.

17. Values from two independent samples from normally distributed populations are displayed below. Test whether the two population means are equal at the 1% level of significance.

Sample 1	8.1	4.6	10.1	13.8	13.5
Sample 2	7.6	−1.8	8.3	8.5	9.9
Sample 1	15.4	1.3	8.4	13.1	5.3
Sample 2	12.8	−0.1	6.1	8.0	4.4

a) What is the null hypothesis?
b) What is the *p*-value if Case 1 is presumed?
c) What is the *p*-value if Case 2 is presumed?
d) After making a judgment about which case applies, what is the statistical decision for the test?
e) What is your conclusion (in words) based on the test?

18. Redo Exercise 17 but test whether the first population has a *larger* mean than the second population.

19. Redo Exercise 17 but test whether the two population means differ by 3.0.

20. Construct the 95% confidence interval estimate for difference in population means $(\mu_1 - \mu_2)$ for the following data:
$\bar{x}_1 = 28$ $\sigma_1 = 6.24$ $n_1 = 25$
$\bar{x}_2 = 25$ $\sigma_2 = 5.63$ $n_2 = 35$
Assume the samples are independent.

21. Construct the 95% confidence interval estimate for difference in population means $(\mu_1 - \mu_2)$ for the following data:
$\bar{x}_1 = 234$ $s_1 = 21.3$ $n_1 = 35$
$\bar{x}_2 = 257$ $s_2 = 18.9$ $n_2 = 45$
Assume the samples are independent, and that the population variances are approximately equal.

22. Repeat Exercise 20 using a 99% confidence level estimate.

23. Repeat Exercise 21 using a 98% confidence level estimate.

24. Values from two independent samples from normally distributed populations are displayed below. Construct the 95% confidence interval estimate for difference in population means $(\mu_1 - \mu_2)$.

Sample 1	8.1	4.6	10.1	13.8	13.5
Sample 2	7.6	−1.8	8.3	8.5	9.9
Sample 1	15.4	1.3	8.4	13.1	5.3
Sample 2	12.8	−0.1	6.1	8.0	4.4

25. Repeat Exercise 24 using a 99% confidence level estimate.

Applications 12.2

As part of the solution for any hypothesis tests, formally show the eight-step sequence as illustrated in the text. Unless otherwise specified by the question or your instructor, use your own judgment on whether to use (1) the critical value or *p*-value approach and (2) the Case 1 or Case 2 assumption regarding population variances.

26. According to a salary survey of industrial hygiene workers in British Columbia and the Yukon, the following were the comparative salaries for workers of two different education levels in 1999[3]:
$\bar{x}_{\text{master's}} = \$64{,}300$ $s_{\text{master's}} \cong \$33{,}800$ $n_{\text{master's}} = 31$
$\bar{x}_{\text{bachelor's}} = \$59{,}600$ $s_{\text{bachelor's}} \cong \$22{,}600$ $n_{\text{bachelor's}} = 26$
a) Test a claim that the mean salaries are equal for workers in this field with master's and bachelor's degrees. Use the 5% level of significance. Assume that both distributions are roughly normal and that the variances are equal.
b) Would it be reasonable to apply your same conclusion about the workers' salaries to workers who live in another province or who work in a different career field? Explain.

27. Using the data in Exercise 26, construct the 95% confidence interval estimate for difference in population means $(\mu_1 - \mu_2)$. Assume that the population variances are approximately equal.

28. In the survey described in Exercise 26, salaries were also compared based on the levels of experience of the industrial hygiene workers. For the test below, group 1 represents 0–5 years' experience; group 2 represents 11–20 years' experience.
$\bar{x}_1 = \$55{,}500$ $s_1 \approx \$15{,}000$ $n_1 = 10$
$\bar{x}_2 = \$65{,}000$ $s_2 \approx \$33{,}800$ $n_2 = 23$
Test a claim that the mean salary for workers in this field who have 11–20 years of experience is greater than the mean salary for workers in the field who have only 0–5 years of experience. Use the 5% level of significance. Assume that both distributions are roughly normal and variances are equal.

29. Using the data in Exercise 28, construct the 95% confidence interval estimate for difference in population means $(\mu_1 - \mu_2)$. Assume that the population variances are approximately equal.

30. Research has been conducted for the Public Health Agency of Canada on the durations of major depressive episodes among Canadians. The mean episode duration for young women aged 12–24 (group 1) was found to be 5.4 months, and for women aged 45–64 (group 2) the mean duration was 8.7 months.[4]
a) Suppose the above conclusions were based on samples of 100 (for group 1) and 120 (for group 2), and that based on past experience, the variances were estimated at $\sigma_1 = 2.5$ months and $\sigma_2 = 4.4$ months. On these assumptions and using the 0.05 level of significance,

test a claim that the duration of depressive incidents is longer for women aged 45–64 than for women aged 12–24.

b) The researchers for the study found that distributions of episode durations were very skewed to the right, rather than normally distributed. Does this information have much effect on the reliability of the conclusions in (a)? Why or why not?

31. From the study on depressive episodes described in Exercise 30, another finding was that the mean duration among women at least 65 years old (group 3) was 5.7 months, compared to the previously cited mean duration for women aged 12–24 (group 1), which was 5.4 months.[5] Suppose the above conclusions were based on samples of 100 (for group 1) and 90 (for group 3) and that variances were estimated at $\sigma_1 = 2.5$ months and $\sigma_3 = 2.7$ months. On these assumptions and using the 0.01 level of significance, test a claim that the duration of depressive incidents is longer for women aged 65 or older than for women aged 12–24.

32. Using the data in Exercise 30(a), construct the 99% confidence interval estimate for difference in population means $(\mu_1 - \mu_2)$.

33. Using the data in Exercise 31, construct the 95% confidence interval estimate for difference in population means $(\mu_1 - \mu_3)$.

34. A concerned Yukon citizen claims that, on average, more of the territory's forests have been destroyed yearly by fire since 1981 than was previously the case. To put this view to the test, a friend compiled the data stored in the file **Yukon_Wildfires.** She interprets the annual data for hectares lost by fire from 1950 to 1980 as a representative sample of pre-1981 patterns of annual fire loss, and the data from 1981 to 2004 as representing the more current pattern.

a) Based on the model described, use the eight-step procedure to test the Yukon resident's claim at the 5% level of significance.

b) Does the final conclusion differ depending on whether you use Case 1 or Case 2?

c) Using the procedures provided in Section 8.2, assess whether the **Yukon_Wildfires** data set for hectares lost meets the condition of normality, which is formally required to use the *t*-test. In this particular example, do you think your findings about normality impact the final conclusion of the test? Explain.

35. The Yukon resident of Exercise 34 also claims that since 1981, there are more cases per year in which lightning causes a forest fire than was previously the case. Again, her friend uses the data in the **Yukon_Wildfires** file to test the claim, comparing lightning-caused fires per year up to 1980 with the numbers of such fires per year from 1981 to 2004.

a) Use the eight-step procedure to test the Yukon resident's claim at the 5% level of significance.

b) Does the final conclusion differ depending on whether you use Case 1 or Case 2?

c) Using the procedures in Section 8.2, assess whether the wildfires data set for numbers of lightning-caused fires meets the condition of normality, which is formally required to use the *t*-test. In this particular example, do you think your findings about normality impact the final conclusion of the test? Explain.

36. Use the eight-step procedure to test at the 1% level of significance the claim that, during the period from 1981 onward, an average of *60 more* lightning-caused forest fires have occurred per year than was previously the case. (Compare Exercise 35.)

37. Using the data in Exercise 35, construct the 99% confidence interval estimate for difference in population means $(\mu_1 - \mu_2)$.

Basic Concepts 12.3

38. What is the difference between a test for equality of two sample means when the samples are independent and a test for equality of two sample means given dependent samples?

39. If samples are dependent, is it still possible to perform the test for equality of sample means using the same techniques as for independent samples? If yes, explain what difference is made by which test you use in such a case.

40. Describe how a paired sample can be selected.

41. Paired sample data from normally distributed populations are displayed below. Test at the 5% level of significance a claim that there is no difference in the means.

Group A	161	192	219	91	160	132	57	87
Group B	212	200	220	87	158	143	86	91

a) What is the null hypothesis?
b) What is the test statistic?
c) What is the *p*-value?
d) What is the statistical decision?
e) What is your conclusion (in words) based on the test?

42. Use the data presented in Exercise 41 but test the claim that the mean for group A is smaller than the mean for group B. Test at the 0.05 level of significance.

a) What is the null hypothesis?
b) What is the test statistic?
c) What is the *p*-value?
d) What is the statistical decision?
e) What is your conclusion (in words) based on the test?

43. A brand of toothpaste introduced a new formula. For a period, both the original and new versions were available at a number of supermarkets. Test a claim, at the 0.02

significance level, that prices are generally higher for the new version of the toothpaste.

Supermarket	A	B	C	D
Original formula	2.31	2.32	2.33	2.19
New formula	2.42	2.33	2.43	2.37

Supermarket	E	F	G
Original formula	2.29	2.18	2.27
New formula	2.42	2.31	2.36

a) What is the statistical decision?
b) What is your conclusion (in words) based on the test?

44. A random sample of 10 cats was selected to test their preferences for two recipes of cat food. Two bowls were placed in front of each cat, with each containing a sample of one of the recipes. Trained observers scored each cat's preferences based on a combination of the cat's behaviour and measurement of the amounts of food eaten from each of the bowls. Test a claim, at the 0.01 level of significance, that the cats had generally equal preferences for the two recipes.

Cat ID	A	B	C	D	E
Recipe A	9.4	4.9	9.3	6.8	7.5
Recipe B	8.4	4.7	7.3	6.0	8.3

Cat ID	F	G	H	I	J
Recipe A	6.2	7.5	6.7	8.6	7.2
Recipe B	5.4	8.0	5.5	6.9	6.1

45. Construct a 95% confidence interval for the data in Exercise 41.

46. Construct a 95% confidence interval for the data in Exercise 43.

47. Construct a 99% confidence interval for the data in Exercise 44.

Applications 12.3

As part of the solution for any hypothesis tests, formally show the eight-step sequence illustrated in the text. Unless otherwise specified by the question or your instructor, use your own judgment on whether to use the critical value or p-value approach.

48. Generally, the prices of groceries, like other commodities—keeping quality and other variables constant—change over time due simply to inflation. Suppose that each month, an inspector makes one purchase from each of several grocery categories (such as bread, beef, chicken, and so on), using a standardized quantity and the same criteria for each purchase. Interpret the data in the **Prices_for_groceries** file as a record (in U.S. dollars) of the inspector's monthly purchase costs.

a) Test a claim ($\alpha = 0.03$) that the mean price for the commodities purchased by the inspector increased between March and April 2005.
b) Test a claim ($\alpha = 0.03$) that the mean price for the commodities purchased by the inspector increased between February and March 2005.
c) When you use the methods provided in Chapter 8, do the data appear to satisfy the test condition of a normal distribution? If they do not, explain how this might impact on your conclusions in (a) and (b)?

49. Based on the data set described in Exercise 48, construct a 94% confidence interval for the mean difference in purchase prices between prices in March 2005 and those in April 2005.

50. Based on the data set described in Exercise 48, construct a 95% confidence interval for the mean difference in purchase prices between prices in January 2000 and those in April 2005.

51. A school guidance counsellor collected data to see whether, on average, there is a salary difference for Canadians based on whether they have a college diploma/degree or a trades education. (See data in the file **Earnings_Working.**) The counsellor claims that there is *no* difference in salaries based on this criterion.

a) Test the guidance counsellor's claim, using $\alpha = 0.05$. Presume, for this exercise, that the data set **Earnings_Working** was constructed by randomly selecting one person of each education level from each province and territory, and that the numbers shown are salaries (in Canadian dollars) of those selected individuals.
b) Construct a 95% confidence interval for the mean difference in salaries based on a college diploma/degree versus a trades education.

52. In reality, the numbers in the **Earnings_Working** data set were not selected as described in the exercise. Instead, they are *averages* for the salaries of individuals in the various combinations of education level and location. Explain how this deviation from the test conditions for the t-test could impact the answers to the previous exercise.

53. Recently a dairy farmer made this optimistic claim: "After years of decline, the average number of dairy cows being kept in each province is now increasing." She based her claim on data provided by the Canadian Dairy Information Centre (see file **Dairy_Cattle**).

a) Use the provincial data in the file (omit the national total) for the years 2004 and 2005 to test the farmer's claim. ($\alpha = 0.05$)
b) In reality, the values given in the data set are totals (i.e., populations) for each year/province combination. If (a) is interpreted as a test for a trend that may continue, this assumes that data for 2004–2005 is a representative sample of paired data to come in future years. Is this a realistic assumption? How could it be justified?

54. Referring to data in the file **Meat_Consumption**, a poultry farmer claimed that unlike chicken consumption, "people's annual consumption of chicken did not change between 1993 and 2003."
    a) Assuming the non-chicken data in the file are based on sampling, test the farmer's claim. ($\alpha = 0.05$)
    b) Construct the 95% confidence interval for the change in meat consumption—of *all* types—from 1993 to 2003.

Basic Concepts 12.4

55. For modelling the sampling distribution of the difference between two sample proportions, we use the normal approximation. What conditions are required to validly use the *z*-test for a difference in proportion?

56. A university's Student Association sent a questionnaire to selected first- and fourth-year students. The association wants to test a claim that the proportion of fourth-year students who buy at least one secondhand textbook is larger than the proportion of first-year students who do this. In the study, 300 first-year students responded; 105 of these said they buy at least one textbook secondhand. Of the 130 fourth-year students who responded, 98 said they buy at least one textbook secondhand.
    a) Write the null and alternative hypotheses for the test.
    b) Calculate the pooled proportion of first- and fourth-year students, combined, who buy at least one secondhand textbook.
    c) Are the conditions met for using a *z*-test to test the claim? Explain why or why not.

57. Conduct a hypothesis test for the claim described in Exercise 56. (Use $\alpha = 0.01$.) What is the statistical decision? State, in words, the conclusion of the test.

58. Construct a 98% confidence interval for the difference in proportion of the two groups of students who buy at least one secondhand textbook (based on data in Exercise 56).

59. In the east end of a large park, a random sample of $n_1 =$ 50 trees was selected; in the west end, a random sample of $n_2 = 60$ trees was selected. On inspection, six trees in the first sample are infested with tent caterpillars; eleven trees in the west-end sample are infested. Comparing the two ends of the park, find the 95% confidence interval for the difference in proportions of trees that are infested.

60. Based on the data in Exercise 59, test a claim that a larger proportion of trees in the west end, compared to the east end, is infected with the caterpillars.

61. A company makes team jackets for sports teams. Demand is so great that two shifts (day and evening) are used to produce the product. For quality control, 150 jackets are sampled from the output of each of the two shifts. Flaws were found in 10% of the jackets in the evening-shift sample. Only 5% of the jackets sampled from the day shift output were flawed. At a 0.02 level of significance, test a claim that, proportionally, more flawed jackets are produced during the evening shift than during the day shift.

62. Based on the jacket samples described in Exercise 61, construct the 95% confidence interval for the difference in proportions of jackets that are flawed during the two shifts.

63. According to a survey of two universities, 22% of the engineering students in university A are women, while 17% of the engineering students in university B are women. Sample sizes were $n_A = 73$, $n_B = 65$. Test a claim that university A has recruited a higher proportion of female students for its engineering program than university B has. ($\alpha = 0.05$).

Applications 12.4

As part of the solution for all hypothesis tests, formally show the eight-step sequence illustrated in the text.

64. Experts on the seasonal migration of Columbia spotted frogs in Idaho are sampling from two groups of adult male frogs of a certain species who live in nearby regions of the state. From the first group ($n = 150$), 14 of the frogs migrated from their breeding ponds to a separate summer habitat. In a second sample ($n = 100$), 7 of the observed frogs migrated to a summer habitat.[6]
    a) At the 0.05 level of significance, test if the proportions of frogs who migrate to a summer habitat are the same for the two neighbouring populations.
    b) For the difference in proportions of frogs that migrate, construct the 95% confidence interval for the two groups.

65. For the safety of passengers and drivers, MADD (Mothers Against Drunk Driving) Canada recommends a lowering of the legal limit for the allowed blood alcohol level of drivers. Under the current limit, a 200-lb. (about 91-kg) man could, under typical conditions, consume up to six standard drinks within two hours of driving before exceeding the limit. (This boundary decreases given a person who weighs less, or if the drinks are not evenly spaced, and so on.) Under MADD's recommendation, the 200-lb. man could drink only four or fewer drinks over the two hours before driving. To test the response of the public, SES Research surveyed 926 people who have never had more than four drinks within two hours of driving, and 30 other people who have had, at least once, more than four drinks within two hours of driving. In the first group, 73 thought that the reduced alcohol limits would be too strict; in the second group, 9 thought that the reduced limits would be too strict.[7] At the 0.05 level of significance, test whether, for the two groups, the same proportions of respondents think that the reduced limits would be too strict.

66. To dissuade their daughter from snowmobile riding in the hills, her protective parents use selected historical data to warn her of the dangers. The data show that 33 of 142 avalanche victims were riding snowmobiles at the time of their deaths. Their daughter (a college stats student) responds with a sample that she prefers, based on 24 avalanche victims: Only 5 of these were riding snowmobiles at the time.[8] At the 0.04 level of significance, test the claim that there is no difference in the proportions for the two samples described.

67. Based on the data in Exercise 66, construct a 95% confidence interval for the difference in proportions for the two samples.

68. Because of our aging population, governments are concerned about the financial security of the next generations who retire. Many in the current work force will have some retirement income due to registered pension plans (RPPs) established by their employers or unions. But some worry that the proportion of workers covered by RPPs is declining. In a labour force survey in 1999, Statistics Canada found that 41% of paid workers had RPPs; in a 2004 survey, 39% of paid workers had RPPs. (These are large surveys, with sample sizes of workers equal to about $n = 100{,}000$).[9] On this basis, test the claim that the proportion of paid workers covered by RPPs is declining. (Use $\alpha = 0.01$.)

69. Based on the data in Exercise 68, construct a 99% confidence interval for the difference in proportions of workers who have RPPs in the two samples.

70. Continuing Exercise 68, test whether the proportion of workers covered by RPPs in 1989 (42,800 out of a similar sample size) was any different from the proportion covered in 1999. (Use $\alpha = 0.01$.)

Review Applications

As part of the solution for any hypothesis tests, formally show the eight-step sequence illustrated in the text. Unless otherwise specified by the question or your instructor, use your own judgment on whether to use (1) the critical value or *p*-value approach and (2) the Case 1 or Case 2 assumption regarding population variances.

71. A health lobby group is comparing different brands of cigarettes for their respective mean levels of formaldehyde content in direct cigarette smoke. For a sample of 50 cigarettes of brand A, the mean formaldehyde content in the smoke is 41.5 μg (micrograms) per cigarette. In the smoke for a sample of 45 cigarettes of brand B, the mean formaldehyde content is 44.1 μg (micrograms) per cigarette. Health Canada reports that under standard test conditions, the standard deviation for this variable is 32.8 μg per cigarette.[10] Test whether the smoke from brand B cigarettes has a higher mean formaldehyde content than the smoke from brand A cigarettes. Use $\alpha = 0.05$.

72. Based on the data in Exercise 71, construct a 96% confidence interval for the difference in the population means for the formaldehyde content of smoke from these two brands of cigarettes.

73. Redo Exercise 71 but instead of using the Health Canada value for σ, use the sample values for s. (Presume that $s_{\text{brand A}} = 29.9$ μg per cigarette and $s_{\text{brand B}} = 36.4$ μg per cigarette.) Do these changes affect what test to use? (If so, explain.) What is the new conclusion?

74. Redo Exercise 72, but use the sample values for s given in Exercise 73.

Exercises 75–79 are based on data in the file **Water_ Use.**

75. Many Canadian communities add chlorine to their water supply as a disinfectant. A researcher claims that comparing communities that do or do not use chlorine, both groups consume the same mean quantity of water per day in the domestic sector. Test this claim using data in the **Water_Use** file, interpreting the one-year's data as a sample to represent longer-term patterns. (The water-use data are in the column "WatUseDom." Sort or filter the data to find separate means for the rows where (1) the Disinf_Chlorine variable = *Y* and (2) the Disinf_Chlorine variable is left blank.) For σ find and use the standard deviation of the domestic water use for *all* of the communities, regardless of chlorine use. Use the 0.05 level of significance.

76. Redo Exercise 75 but test the revised claim that, for communities using chlorine, the mean domestic water use is 800 cubic metres per day greater than for communities that do not use chlorine.

77. Based on the data in Exercise 75, construct a 95% confidence interval for the difference in the mean domestic water consumption for communities that do, versus those that do not, use chlorine as a disinfectant.

78. Redo Exercise 75 but instead of using the indicated estimate for σ, find and use the values for s for each of the two groups—those using and those not using chlorine. What test should you use? What is the new conclusion?

79. Redo Exercise 77 but use the sample values for s that are described in Exercise 78.

Exercises 80–82 are based on data in the file **Canadian_Equity_Funds.**

80. Two investors are debating whether it is worth paying a sales fee to buy into a mutual fund. Based on the file's sample data for three-month returns, investor Smith claims that the mean return is not any better for funds charging

a fee than for funds not charging a fee. (The relevant data are in the column "Ret_3Mth." Sort or filter the data to separate the two samples (1) where the LoadFees variable = *N* and (2) where the LoadFees variable equals any other value.) Test the claim at the 0.02 level of significance.

a) What is the null hypothesis for the test?

b) What is the statistical decision?

c) For the benefit of investor Smith, state your conclusion in words.

81. Continuing the debate about paying fees, Smith also claims that the proportion of no-fee funds rated at least 3 by GlobeFund was actually better than the proportion of other funds that achieved that rating. Test Smith's claim at the 3% level of significance. (First, depending on your software, you may need to re-code the variable values or calculate the relevant proportions manually. The GlobeFund ratings are in the column "Ratings" and the no-fee funds are those with an *N* in the "Loadfees" column. Ignore the records for funds with no rating.)

82. As the above investors explored the data, investor Haines noted that, in general, the sampled funds had worse returns on investments for one month (column: "Ret_1Mth") than for three months (column: "Ret_3Mth").

a) Interpreting the one-month returns and the three-month returns as two variables, are these variables paired or independent?

b) Use the appropriate method to test the investor's claim. ($\alpha = 0.05$.)

c) Construct a 95% confidence interval for the mean difference for the two variables.

Exercises 83–87 are based on data in the file **CanadianHealth**.

83. Health officials are worried about the impacts of stress and obesity on Canadian adults. The file **CanadianHealth** includes columns for the proportions of adults who suffer from obesity and stress, respectively, in each province and territory. Presume that these proportions were calculated based on samples of $n = 1{,}000$ taken from each province or territory. Given this data, test whether the same proportions of British Columbians and Albertans suffer from obesity. (Use $\alpha = 0.05$.)

84. Using the same data set as above, test a claim that the proportion of adults from Saskatchewan who are obese is larger than the proportion of adults who are obese in British Columbia. ($\alpha = 0.01$).

85. Construct a 95% confidence interval for the difference in proportions of adult Quebeckers and New Brunswickers who suffer from stress.

86. Construct a 98% confidence interval for the difference in proportions of adult Ontarians and residents of the Northwest Territories who suffer from stress.

87. Interpret the columns of proportions in the data file as "scores." For example, Ontario has a worse (i.e., higher) score for stress (0.231) than it does for obesity (0.151). In this model, test a claim that, in general, for all provinces and territories, the two scores for obesity and stress have about the same mean. Test at the 5% level of significance.

Chapter 13

Nonparametric Tests of Significance

• Chapter Outline

13.1 One-Sample Tests about the Median

13.2 Two-Sample Tests about the Median

13.3 Additional Notes about Nonparametric Tests

• Learning Objectives

13.1 Explain the difference between parametric and nonparametric (or distribution-free) hypothesis tests; convert interval or ratio data into rank values; and conduct a sign test or a Wilcoxon signed-ranks test for the median—and recognize when each would be preferable.

13.2 Conduct a Wilcoxon rank-sum test for two independent samples or a Wilcoxon signed-ranks test for two dependent samples.

13.3 Describe some conditions under which using nonparametric procedures is an advantage and when it is a disadvantage, compared to using parametric methods for hypothesis tests, and explain what is meant by a resampling approach to hypothesis tests.

13.1 One-Sample Tests about the Median

All of the significance tests that we have looked at have required certain assumptions if they are to be used validly. Most have required that the population be normally distributed. Based on the central limit theorem, that assumption is less important for large samples, but if the distribution is very non-normal even a large sample of $n \approx 30$ may not be large enough to compensate. Also, the measurement level for data used for a normal-based approach should ideally be at least at the interval level. Proceed with caution if the raw data are actually ordinal, as in published rankings for universities or on 1–5 scale evaluations of textbooks.

Nonparametric hypothesis tests do not require such preconceptions about the distribution of a population or variable. Another name (perhaps more appropriate) for these tests is **distribution-free tests.** The hypothesis tests described in this chapter are just an introduction to some of the available tests of this type. Another example will be presented in Chapter 15.

Some of the methods in this chapter rely on the concept of **ranks,** or **rank-ordering,** of data. They also use the median, rather than mean, as the relevant measure for location of the distribution.

Assigning Ranks to Data

As shown in Section 1.4, data at the ordinal level convey information only in relation to the ordering of values based on some criteria. If you have ever rated a professor or textbook on a scale of 1–5, you have used an ordinal scale. As the values move upward from 1 to 2 and beyond, we know only that some property (e.g., clarity of lectures or readability of the text) has increased—but the data do not specify by exactly how much. The differences between the assigned numbers (or ranks) can vary among the people who fill in the survey, or even for the same person; for example, perhaps you require more improvement to move your course assessment from 3 to 4 than merely to move it from 1 to 2.

If your raw data are ordinal, you can run the methods of this chapter. If not, a preliminary step may be needed to transform raw interval or ratio-level data into their corresponding ranks. In effect, you convert the raw data into an ordinal format.

First, sort your raw data into ascending or descending order. (You can use the computer: See "In Brief 13.1(a)," following Examples 13.1 and 13.2.) In the simplest case, assign the consecutive, integer values 1, 2, 3, . . . , respectively, to the increasing (or decreasing) values of the raw data. (See Example 13.1.) If there are ties in the raw data values, the procedure is slightly modified (this will be discussed in Example 13.2).

Example 13.1

*(Based on the **World_Apple_Production** file)*

During the summer, you are working for Agriculture and Agri-Food Canada. The department has asked you to help prepare data for an analysis of Canadian and world apple production. Find the raw production data for Canada for 10 years in the file **World_Apple_Production,** and convert these raw production data into ranks.

Solution

With the help of a computer, the data are sorted (see the top row of the following table). Ranks are assigned consecutively, from low to high. There are no ties in the raw data.

Raw data, sorted	465	484	489	504	513	543	554	564	599	632
Assigned ranks	1	2	3	4	5	6	7	8	9	10

Example 13.2

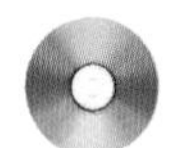

*(Based on the **World_Apple_Production** file)*

Continuing your work for Agriculture and Agri-Food Canada, you have been asked to prepare production data for other countries. Again using data in the file **World_Apple_Production,** convert raw production data for Turkey into ranks.

Solution

With the help of a computer, the data are sorted (see the top row of the table below). Preliminary ranks are assigned consecutively, from low to high. However, there are ties in the raw data—two instances of 2,100 and two of 2,450. It is not sensible to rank one of the instances of 2,100 higher than the other. A common solution for tied raw values is to assign the mean of their preliminary ranks to each of them. Thus, the assigned ranks for each of the two instances of 2,100 are the mean of the preliminary rank of each, 3 and 4: $\frac{3+4}{2} = 3.5$. The assigned ranks for both of the two instances of 2,450 are $\frac{7+8}{2} =$ 7.5. (This method for handling tied ranks is typical but, if you are using software, your package may handle ties somewhat differently.)

			Raw Values Are Tied				Raw Values Are Tied			
Raw data, sorted:	2,080	2,095	2,100	2,100	2,200	2,400	2,450	2,450	2,500	2,550
Preliminary ranks:	1	2	3	4	5	6	7	8	9	10
Assigned ranks:	1	2	3.5	3.5	5	6	7.5	7.5	9	10

• *In Brief 13.1(a) Assigning Ranks Using the Computer*

*(Based on the **World_Apple_Production** file)*

Excel

Excel can automate this step but does not quite follow the conventions described in Examples 13.1 and 13.2. First, the program assigns the lowest rank number (1) to the largest raw value, not to the smallest. This reversal does not impact any other procedures in this chapter, *provided you remember how the rank numbers have been assigned.* More importantly, Excel does not handle ties among raw values by averaging their preliminary ranks. Instead, it assigns to all of the tied values the value of the lowest preliminary rank for any of them. This modification results in slightly altered values if you use the ranks for other procedures in this chapter.

The steps are to input the raw values into a single column (or row) in Excel. (In the illustration, the variable to be ranked is in column "N.") Then use the menu sequence *Tools / Data Analysis,* and choose the *Rank and Percentile* dialogue box. In that dialogue box, as shown in the screen capture, specify the raw data's location and, where asked for

Output Range, input the cell location for what will become the upper left cell for the procedure's output. Then click *OK.* The output is also depicted in the illustration.

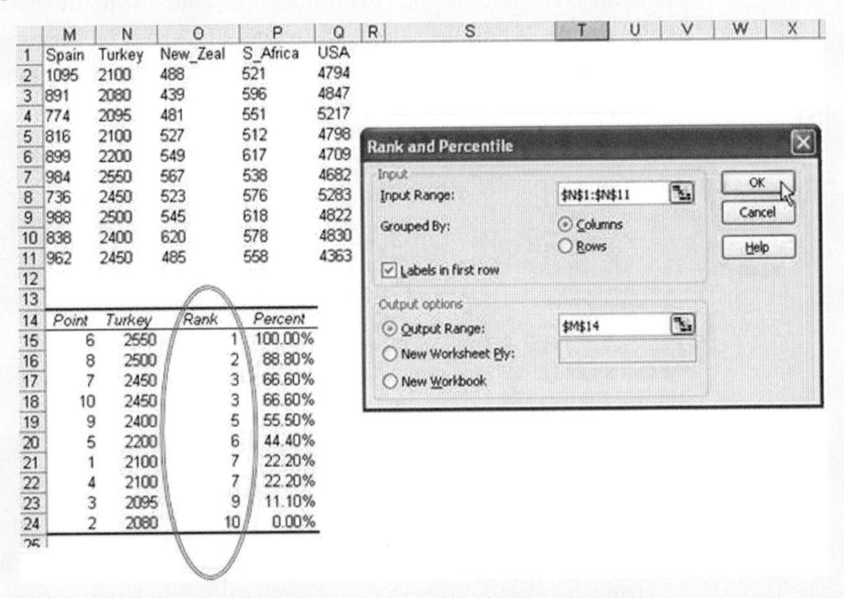

Minitab

Input the raw values into a single column in Minitab. (In the illustration, the variable to be ranked is "Turkey," in the second column.) Then use the menu sequence *Data / Rank.* In the dialogue box shown in the screen capture, select the variable for which the values are to be ranked. In the *Store ranks in* window, name the worksheet column that will display the assigned ranks; then click *OK.* The output is also depicted in the illustration.

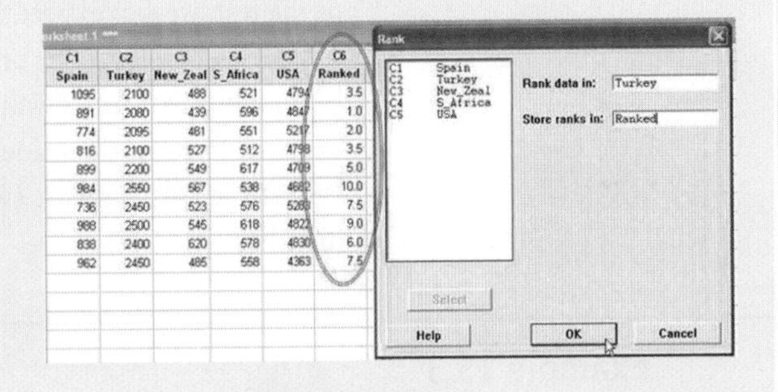

SPSS

Input the raw values into a single column in SPSS. (In the illustration, the variable to be ranked is "Turkey," in the second column.) Then use the menu sequence *Transform / Rank Cases.* In the dialogue box shown in the screen capture, select the variable for which the values are to be ranked. By default, rank "1" will be assigned to the lowest raw value, and the means-based procedure will assign the ranks for ties. Note that you have options in the dialogue box to change the defaults. When satisfied, click *OK.* The output is also depicted in the illustration.

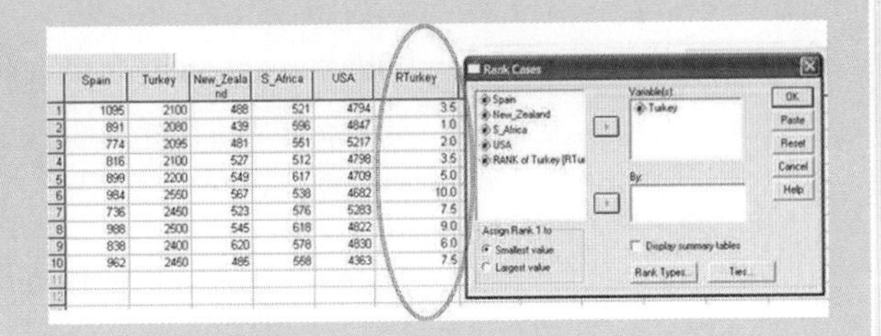

A Hypothesis Test about the Median: The Sign Test

The concept for the sign test is quite simple and requires relatively few assumptions. Consider the sample of data in Figure 13.1. Intuitively, it seems obvious that if H_0 is correct about the median value, then that value should be close to the centre of the sorted values in the sample. If not, then the median hypothesis may be wrong—so reject H_0. But we have to allow for sampling error, so how far must the H_0 median be from its expected centre position in the sample before we reject H_0?

Figure 13.1
Basic Concept of the Sign Test

(Hypothesized Median = 65)

Values below (−) or above (+) the hypothesized median	62.18	62.5	63.21	63.42	63.51	63.59	64.48	65.19	65.22	65.87
	−	−	−	−	−	−	−	+	+	+

Source: Agriculture and Agri-Food Canada. Statistics Canada. Table 18. Consumption of Fruits in Canada (kg per person) 1992-2001. Found at: http://atn-riae.agr.ca/applecanada/production-e.htm. Accessed Aug. 10, 2007.

The assumptions for validly using the sign test are:

- The raw data are at least at the ordinal level.
- Observations are independent and are taken from a randomly drawn sample.
- Relatively few values *exactly* equal the median. (This will be the case if the underlying distribution is continuous or else the population is very large.)

Assign to each sampled value either a + sign or a – sign, depending on whether the value falls numerically above or below the hypothesized median. Under the assumptions of the test, we can expect the plus and minus distribution to be binomial: There are two categories of values (+ or –) and each randomly sampled value represents a trial. The probability that any given trial is a plus (or a minus) is $p = 0.50$, since the median is *expected* to be in the middle of the sorted values. Therefore, the basis for assigning a *p*-value to the test statistic (which is the number of + signs in the sample) is the binomial distribution.

Example 13.3

(Based on the ***Fruit_Consumption*** *file)*

In discussions with Agriculture and Agri-Food Canada, a spokesperson for the farming industry claims that, in general, the median level of annual Canadian consumption of fruit has been 65 kg per person. As evidence, she presents the column labelled "Total" in the CD data file **Fruit_Consumption.** This lists 10 years of annual fruit consumption data for all fruits combined. Test the claim at the 0.05 level of significance.

Solution

We can apply the formal model for hypothesis testing.

Step 1: The formal claim:

$$\textbf{Median} = \textbf{65}$$

Step 2: The null and alternative hypotheses:

$$H_0\textbf{: Median} = \textbf{65}$$

$$H_1\textbf{: Median} \neq \textbf{65}$$

Step 3: The level of significance:

$$\alpha = 0.05$$

Step 4: Specify the test and sample size:

Use the sign test for the median, $n = 10$.

Alternatively, we could write:

Use a binomial test for the number of plus signs, $n = 10$.

Step 5: Depict the test model (see Figure 13.2). Exact artistry is not required. The purpose is to clarify (a) that the test statistic is the number of plus signs (successes), (b) that the expected distribution of this statistic is binomial, and (c) that the planned test is two-tailed. (We will use the p-value approach.)

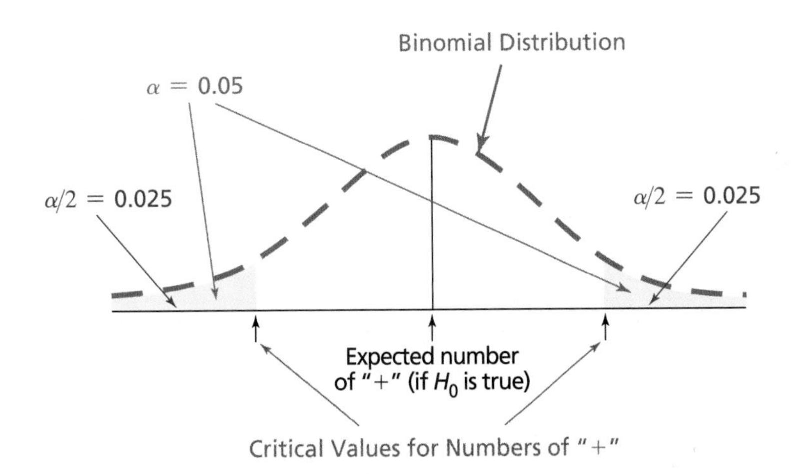

Figure 13.2
Depiction of Model for Step 5, Example 13.3

Step 6: Collect the data. (Data are provided in the problem.) There were three successes (plus signs).

Step 7: Calculate the p-value and make the statistical decision. On the binomial model, the number of trials $= n = 10$; the probability of a success (a plus sign) each trial $= p = 0.5$. The left-tail p-value is the cumulative binomial probability of obtaining 3 or fewer successes in 10 trials. Applying methods from Section 7.2: p-value (left tail) $= P(X \leq 3 \mid \text{the null distribution is the true distribution}) = 0.172$.

For a two-tail test, we double the one-tail p-value. This applies because outcomes of $(n - 3)$ successes or greater, on the right tail, would be just as rare as outcomes of three successes or less on the left tail. Therefore:

$$p\text{-value} = 2 \times 0.17 = 0.344$$

Given that $\alpha = 0.05$:

Do not reject the null hypothesis, because the p-value is not less than or equal to α.

Step 8: Clearly express your conclusion, in words:

There is not sufficient evidence, based on the sign test, to reject the claim that the median Canadian consumption of fruit is 65 kg per person per year.

• In Brief 13.1(b) The Sign Test on the Computer

(*Based on the* **Fruit_Consumption** *file*)

Excel

Although Excel does not offer this particular test, it can help to perform it:

1. By using Excel to sort the sample data, you can more easily count the number of cases that are above (+) and below (–) the hypothesized mean.
2. Use Excel to calculate the binomial probabilities, as shown in "In Brief 7.2(a)" or "In Brief 7.2(b)."

Minitab

If the raw data are in a Minitab column, the program can directly perform a sign test. (The illustration is based on using the data in the "Total" column.) Use the menu sequence *Stat / Nonparametrics / 1-Sample Sign*. In the resulting dialogue box, shown in the screen capture, select the variable for the sample. Select the option *Test median* and input the median value as hypothesized in H_0. In the *Alternative* window, input the operator for the alternative hypothesis. Click *OK*. The output is also shown in the illustration.

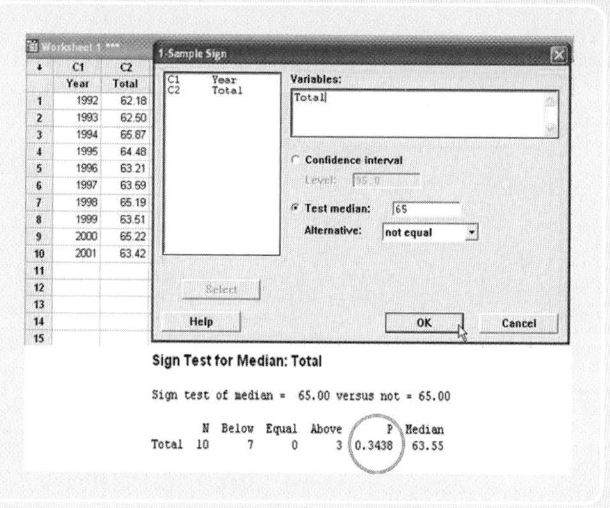

SPSS

This procedure is a special case of the SPSS binomial test. (The illustration is based on using the data in the "Total" column.) The sample data should be in one column. Begin with the menu sequence *Analyze / Nonparametric Tests / Binomial.* In the resulting dialogue box (see the screen capture), select the appropriate data column and then input 0.50 as the *Test Proportion* where indicated. Click the *Exact* button and then, in the next dialogue box, select *Exact* and optionally retain the default time limit of 5 minutes (it would take a *very* large sample to require more than 5 minutes for the exact calculation). Then click on *Continue,* which will return you to the first dialogue box (shown at the top centre of the illustration). On the lower left of this dialogue box, select *Cut point* and input the value of the hypothesized median according to H_0. (SPSS will count the data values below and above this cutoff.) Click *OK*. The output is also depicted in the illustration.

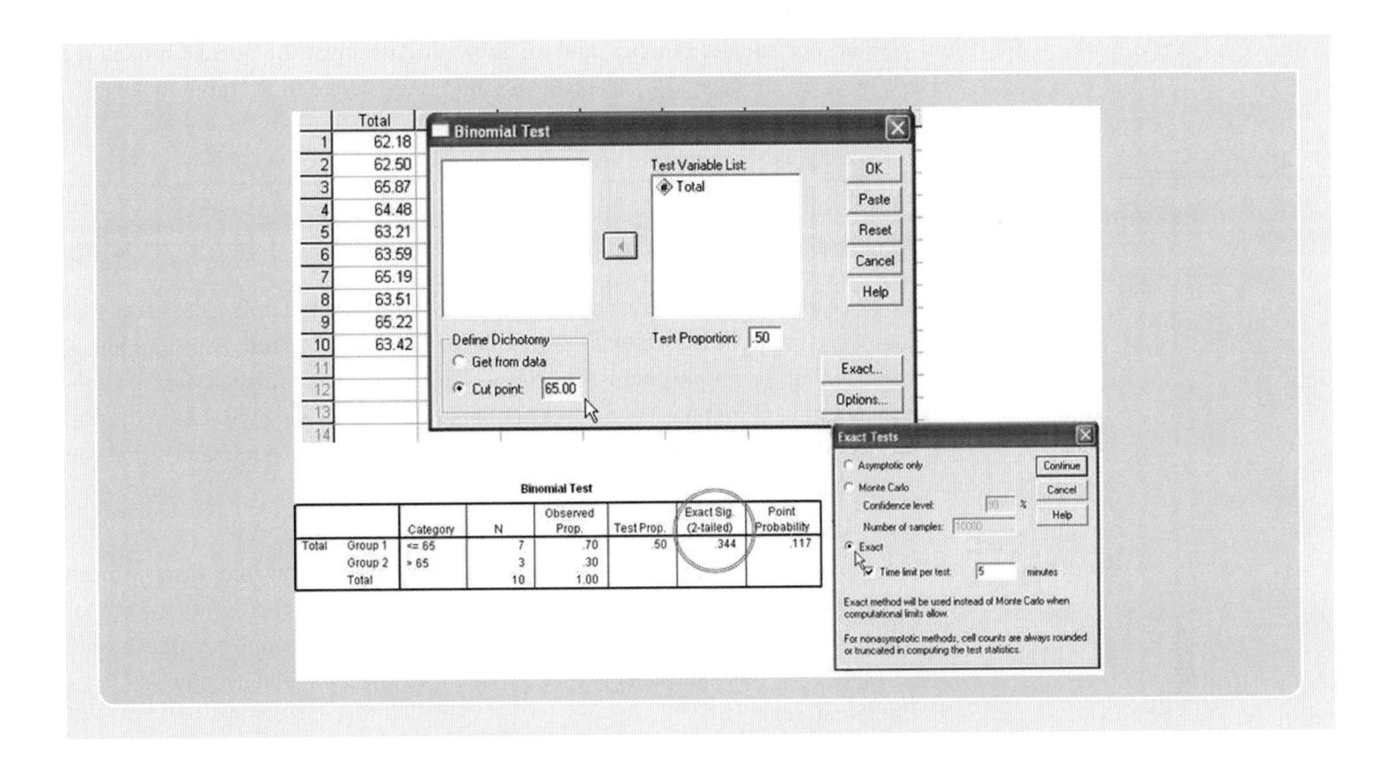

A Hypothesis Test about the Median: The Wilcoxon Signed-Ranks Test

The Wilcoxon signed-ranks test is a variation of a two-sample test that will be introduced in the next section. Instead of testing a hypothesis about a population mean, as in Chapter 11, the null hypothesis for this test centres on the population median. The test is not altogether distribution-free—and has more requirements than the sign test—but it does remove the parametric requirement that the population be normally distributed.

The assumptions for validly using the Wilcoxon signed-ranks test are:

- The raw data are at least at the ordinal level.
- Data are taken from a randomly drawn sample.
- The underlying distribution is continuous (or else the population is very large).
- The underlying distribution is essentially symmetrical.

A normally distributed population would meet the above requirements but, in that case, we could use the parametric approach. However, a population with a uniform or bimodal distribution, for example, could be a case of a continuous, symmetrical, but non-normal distribution. The reason for requiring a near-continuous distribution (or a very large n) is to avoid obtaining too many ties (in proportion to the number of data elements) when the data are converted into ranks.

One final condition for the model is not absolute but depends on the software you are using [see "In Brief 13.1(c)" on page 552]:

- The sample size must be sufficiently large.

In theory, the Wilcoxon test can handle samples of any size. For software that uses a normal approximation of the signed-ranks distribution, the requirement of a large sample size is reintroduced (ideally $n \geq 30$, but no smaller than about 10 or 15). (For a chart to find p-values for smaller samples, and for more discussion of the normal model, see "Hands On 13.1" on page 554.) The basic method is illustrated in Example 13.4.

Example 13.4

*(Based on the **Fruit_Consumption** file)*

Use the Wilcoxon signed-ranks test as an alternative for solving Example 13.3. Recall that the data were found in the column "Total" in the CD data file **Fruit_Consumption.** A spokesperson for the farming industry claims that, in general, the median level of Canadian consumption of fruit has been 65 kg per person per year. Using the data in the file, test the claim at the 0.05 level of significance. Compare the solution to the decision you made in Example 13.3.

Solution

A full solution would include all steps of the formal hypothesis-testing process. The focus here is on how the p-value is determined. Recall that the p-value represents the probability of obtaining the sample results under the assumptions of the null hypothesis, and if the p-value is less than or equal to α (0.05), then we reject the null hypothesis.

The formal claim is:

$$\textbf{Median} = 65$$

The test hypotheses are, therefore:

$$H_0\text{: Median} = 65$$

$$H_1\text{: Median} \neq 65$$

In Figure 13.3(a), the sorted sample data are visually compared to the hypothesized median. We can see just by looking at the figure that the median is not exactly in the middle location as expected. This alone suggests a reduced probability that the sample is consistent with the null hypothesis—but it does not account for *how far* each data value is from the hypothesized median. If the data had comparable spreads above and below the hypothesized median, in terms of some farther points and some closer points in either direction, the hypothesis might stand. If not, the p-value will be low and the null hypothesis will be rejected.

Figure 13.3(b) shows how the p-value will be assigned. In the third column, the difference is taken between each sampled value and the hypothesized median. Keeping in mind the signs of the differences for values above or below the proposed median, take the absolute value of each difference to focus on the magnitude of difference. Using the procedures described earlier in this chapter, convert the absolute difference values into ranked values. This produces the fifth column in the figure.

Next, add up all of the ranks for cases in which the difference was originally positive and add separately all of the ranks for cases in which the difference was originally negative. The test statistic T is the smaller of the resulting sums. Under the null hypothesis, the sum of the ranks in *either* direction from the median should be about the same—in this example, equal to about 27, being half of the total of all rank differences (ignoring direction). The one-tail p-value is this conditional probability: P(obtaining a test statistic less than or equal to the sample's value for $T \mid H_0$ is true).

Figure 13.3

Basic Concept of the One-Sample Wilcoxon Signed-Ranks Test for Median

(Hypothesized Median = 65)

a)

Sample
62.18
62.50
63.21
63.42
63.51
63.59
64.48
65.19
65.22
65.87

b)

	Sample	Difference	Absolute Difference	Ranked Absolute Difference	+ Difference	− Difference
65	62.18	2.82	2.82	10	10	
65	62.50	2.50	2.5	9	9	
65	63.21	1.79	1.79	8	8	
65	63.42	1.58	1.58	7	7	
65	63.51	1.49	1.49	6	6	
65	63.59	1.41	1.41	5	5	
65	64.48	0.52	0.52	3	3	
65	65.19	−0.19	0.19	1		1
65	65.22	−0.22	0.22	2		2
65	65.87	−0.87	0.87	4		4
				Sums:	48	7

Test statistic T = The smaller of these two sums:

H_0: Median is 65

Test statistic $T = \text{Min}[\Sigma(+\text{Differences}), \Sigma(-\text{Differences})]$

p-value (one-tail) $= P(T \leq T \mid H_0 \text{ is true})$

Source: Agriculture and Agri-Food Canada. Statistics Canada. Table 18. Consumption of Fruits in Canada (kg per person) 1992-2001. Found at: http://atn-riae.agr.ca/applecanada/production-e.htm. Accessed Aug. 10, 2007.

The exact value for this probability could be calculated (in principle) by the core rule of classical probability (see Formula 6.1): The denominator is the number of combinations of ranked absolute differences that could have appeared below the line for the hypothesized median in Figure 13.3 (for the given sample size); the numerator is the number of those combinations for which the sum (T) of those ranks would equal T (i.e., 7) or less.

Your computer software might either calculate the above conditional probability exactly or it might make an estimate based on a normal approximation for the T-distribution. (If your software uses the normal approximation for T [see "In Brief 13.1(c)" for details], the computer output may be less reliable for small values of n. For a discussion of small sample cases, see "Hands On 13.1.")

Applying these principles to the present example, we find that p-value (one-tail)$_{n=10}$ $= P(T \leq 7 \mid \text{Median} = 65) \cong 0.02$. Since the test was two-tail, and given the presumption of a symmetrical distribution, p-value (two-tail) $= 2 \times p$-value (one-tail) $\cong 0.04$. This calculated p-value is less than alpha, so we reject the null hypothesis and conclude:

There is sufficient evidence to reject the claim that the median level of consumption is 65 kg per person per year.

Note that in the solution to Example 13.3, we did *not* reject the same claim, using the same data, that we did reject here. The sign test uses less of the available information than does the Wilcoxon test and so has less power to reject possibly false null hypotheses. The Wilcoxon signed-ranks test would be preferable—provided its extra assumption of a symmetrical population does apply.

• In Brief 13.1(c) Wilcoxon Signed-Ranks Tests on the Computer

(*Based on the* **Fruit_Consumption** *file*)

The following notes on software apply to both of the following closely related procedures:

1. Wilcoxon signed-ranks test for hypotheses about the median
2. Wilcoxon signed-ranks test for two dependent samples

The Wilcoxon signed-ranks test for hypotheses about the median was discussed in Example 13.4; the Wilcoxon signed-ranks test for two dependent samples will be introduced in the next section.

Excel

Excel does not include the Wilcoxon formulas or procedures. For students using this software, you may download the file **Wilcoxon_Template** from this book's CD to use as a specialized calculator. Click on the spreadsheet tab *Workspace1*. If inputs remain from a previous use, click on *Clear All Inputs*. If you are testing one sample for the median, use the first column to copy the hypothesized value of the median (see the illustration) and input the list of sample values into the second column. If, instead, you are comparing dependent samples, input the data, pair by pair, into the two left columns. As shown, the differences and absolute differences will be calculated automatically, as will the value of *n*.

	A	B	C	D	E	F	G	H	I	J	K
1	Clear All Inputs										
2	How many values (pairs) have been input:				Manually	*Notes: (a) If there is *no difference*, type in 0 for ranked difference.					
3			n = 10		Fill in this column*			*n* will be adjusted.			
4					*after sorting* columns A to D	(b) Adjust rank numbers, as needed, if successive raw differences are tied					
5	Input Values Manually.....		Computer Calculated			Computer Calculated		Computer Calculated			
6	Sample 1 (or Hypothesized Median)	Sample 2	Difference	Absolute Difference	Ranked Absolute Difference	+ Difference	- Difference	Sum + Difference	Sum - Difference		
7	65	65.19	-0.19	0.19	1	0	1	48	7		
8	65	65.22	-0.22	0.22	2	0	2				
9	65	64.48	0.52	0.52	3	3	0	z-Approximation for the			
10	65	65.87	-0.87	0.87	4	0	4	sum of - differences			
11	65	63.59	1.41	1.41	5	5	0	(Sample1 or Median) - Sample 2			
12	65	63.51	1.49	1.49	6	6	0		2.03859		
13	65	63.42	1.58	1.58	7	7	0				
14	65	63.21	1.79	1.79	8	8	0				
15	65	62.50	2.5	2.5	9	9	0	Approximate 1-tailed p-value:			
16	65	62.18	2.82	2.82	10	10	0		0.0207		
17											

The next step is to manually input the ranks corresponding to the absolute differences. If there are many pairs, it may be easiest to *Data / Sort* the left four columns of the data, based on ascending absolute differences; then the ranks can easily be determined by eye. Wherever the absolute difference is zero, input its rank as 0.0.

Excel automates the next steps, first summing the plus and minus difference ranks. Using an accepted normal approximation for the distribution of the negative sums, the program finds a *z*-value corresponding to the test statistic. From this, a *p*-value is determined (this can be doubled for the two-tail *p*-value), which can applied back in your hypothesis test.

Minitab

For the one sample test around the median, input the raw values into a single column in Minitab. (See column "C2" in the illustration.) Use the menu sequence *Stat / Nonparametrics / 1-Sample Wilcoxon*. In the resulting dialogue box shown in the screen capture, select the variable for the sample. Select the option *Test median* and input the median value as hypothesized in H_0. In the *Alternative* window, input the operator for the alternative hypothesis. Click *OK*. The output is shown in the illustration.

(Slight changes required in the procedure for the paired two-sample test are discussed in the next section.)

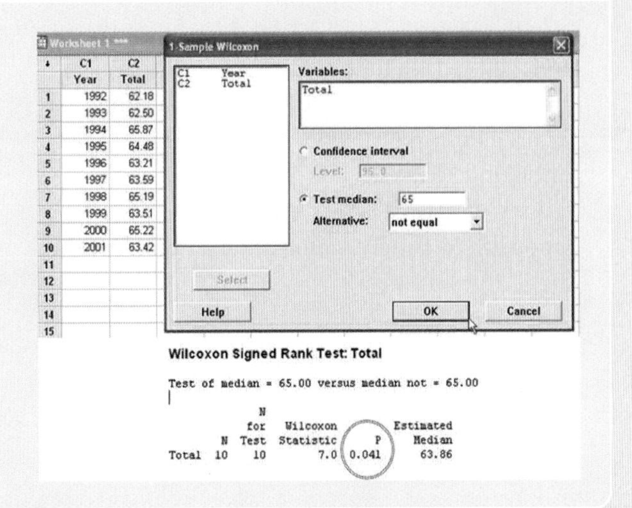

C1 Year	C2 Total
1992	62.18
1993	62.50
1994	65.87
1995	64.48
1996	63.21
1997	63.59
1998	65.19
1999	63.51
2000	65.22
2001	63.42

	N	N for Test	Wilcoxon Statistic	P	Estimated Median
Total	10	10	7.0	0.041	63.86

SPSS

If you are testing one sample for the median, use the first column to copy the hypothesized value of the median, as shown in the illustration; input the list of sample values in the second column. If, instead, you are comparing dependent samples, input the data, pair by pair, into the two left columns. In either case, next use the menu sequence *Analyze / Nonparametric Tests / 2 Related Samples.* In the resulting dialogue box (first screen capture) select the pair of columns to be analyzed. Then select *Wilcoxon* and then click on the *Exact* button.

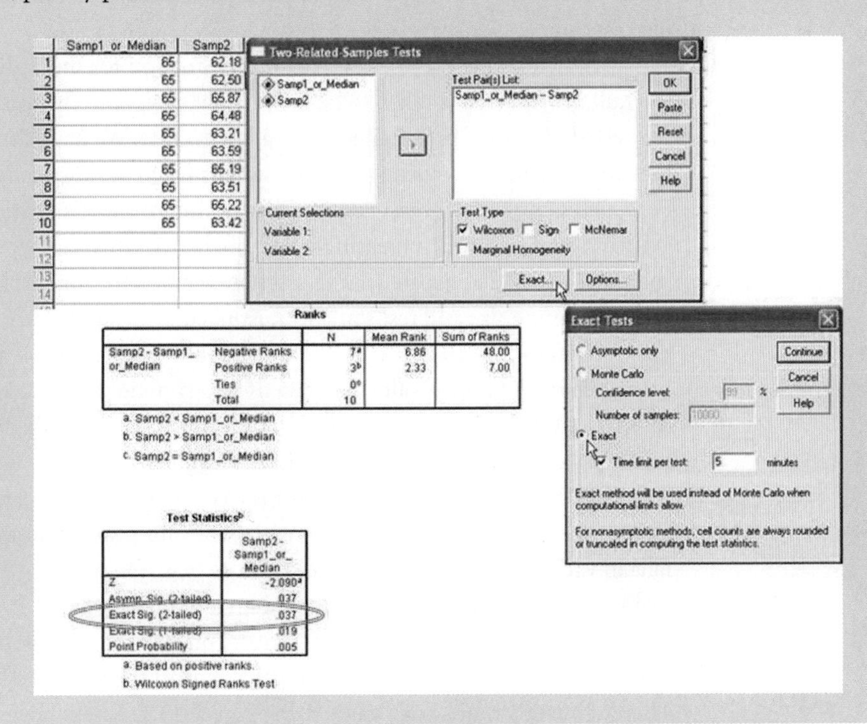

	Sampl_or_Median	Samp2
1	65	62.18
2	65	62.50
3	65	65.87
4	65	64.48
5	65	63.21
6	65	63.59
7	65	65.19
8	65	63.51
9	65	65.22
10	65	63.42

Ranks

		N	Mean Rank	Sum of Ranks
Samp2 - Sampl_or_Median	Negative Ranks	7[a]	6.86	48.00
	Positive Ranks	3[b]	2.33	7.00
	Ties	0[c]		
	Total	10		

a. Samp2 < Sampl_or_Median
b. Samp2 > Sampl_or_Median
c. Samp2 = Sampl_or_Median

Test Statistics[b]

	Samp2 - Sampl_or_Median
Z	-2.090[a]
Asymp. Sig. (2-tailed)	.037
Exact Sig. (2-tailed)	.037
Exact Sig. (1-tailed)	.019
Point Probability	.005

a. Based on positive ranks.
b. Wilcoxon Signed Ranks Test

In the next dialogue box, if the data set is not too large, you can ask for an exact test by selecting *Exact*. In effect, the exact test uses permutations to calculate exact probabilities of obtaining the sample test statistic, given the null hypothesis. For large data sets, it may be faster to choose the *Monte Carlo* option. (SPSS switches automatically from the *Exact* test to a *Monte Carlo* simulation if the time limit for the *Exact* test—shown as 5 minutes in the illustration—is exceeded. This is quite unlikely to happen.) The *Asymptotic only* choice uses a normal approximation of the test statistic distribution. Click *Continue* and then *OK*. The output is also shown in the illustration.

• Hands On 13.1 Small Sample Estimates for p-Values for the Wilcoxon Test

The general form we have been using for calculating the *p*-value (one-tail) for the Wilcoxon signed-ranks test is similar to the calculations used for many other tests: $P(\mathbf{T} \leq T \mid$ the null distribution reflects the true distribution), where the **T** on the left is the name of the test statistic and the *T* on the right is a particular value for that statistic for the given sample. For a parametric test, this conditional probability can be calculated based on some theoretical model for the null distribution.

If your software uses the normal approximation for the *T*-distribution, then the null distribution is taken to be the normal curve and the test statistic *T* is converted into an approximately equivalent *z*-value. We have already learned how to calculate the *p*-value that corresponds when we know a test statistic *z*. For small sample sizes, however, the distribution of the *T* statistic for the Wilcoxon test becomes increasingly *non*-normal. In that case, you could use a chart like the one in Figure 13.4 to estimate the *p*-value for specific combinations of sample size and test statistic *T*. Applied to Example 13.4, look for the intersection of $n = 10$ and $T = 7$ in the graph in Figure 13.4; it appears just below the red line for p-value $= 0.025$. If we estimate the *p*-value (one-tail) for $n = 10$ and $T = 7$ as a little less than 0.025 (say, 0.02), then the two-tail *p*-value $= 0.04$, which is close to the answer in the original problem.

This "Hands On" exercise calls for the construction of a series of data sets of sizes ranging from $n = 5$ to $n = 15$, with data values such that the H_0 median falls in various positions within the data sets (e.g., for $n = 10$, try having only two values of data less than the H_0 median and then try with three values less, and then with four values less, and so on). For each case, use your software to find the one-tail *p*-value with (if you get a choice) the normal approximation. Then estimate a more exact value using Figure 13.4. Assess how well the normal approximation works as *n* decreases from about 15 to about 5.

Figure 13.4
p-Value Estimates for Wilcoxon Signed-Ranks Test: Small-Sample Cases

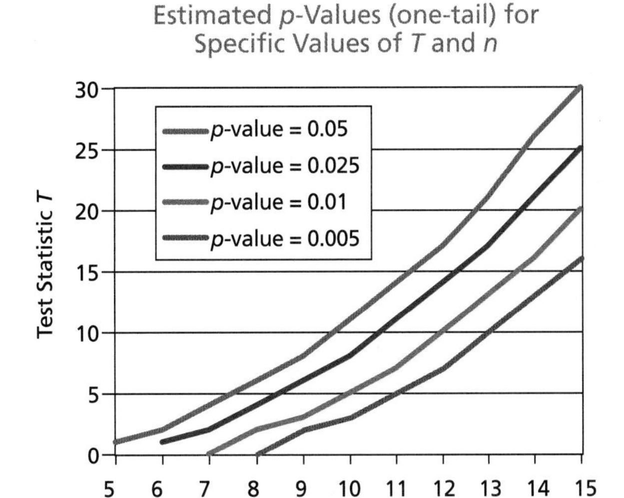

Source: Based on data from *Some rapid approximate statistical procedures.* Frank Wilcoxon and Roberta A. Wilcox. (Pearl River, NY: Lederle Laboratories, ©1949, 1964).

13.2 Two-Sample Tests about the Median

The tests described in this section provide alternatives to the means-based two-sample tests that were introduced in Sections 12.2 and 12.3. The alternatives can be useful when the assumptions needed for the parametric tests do not apply.

Wilcoxon Rank-Sum Test: Two Independent Samples

The Wilcoxon rank-sum test can be used to compare the medians of two independent samples to test for their equality. It is equivalent to another test called the *Mann-Whitney U test* in the sense that, given the same data, it would give the same test results. The general assumptions for using the Wilcoxon rank-sum test are:

- The raw data are at least at the ordinal level.
- Each sample value is drawn randomly.
- The samples are independent.
- The two sample distributions are similar to one another (but no particular shape is required).

An additional condition for the model is not absolute, but depends on the software you are using (see "In Brief 13.2(a)" beginning on page 557):

- The sample size must be sufficiently large (at least 10 observations per sample).

Like the signed-ranks Wilcoxon test, the Wilcoxon rank-sum test can in theory handle samples of any size. Larger samples are needed, however, if your software relies on a normal approximation of the rank-sum distribution. The basic test method is illustrated in Example 13.5.

Example 13.5

*(Based on the **Brilliants** file)*

A young man is preparing to buy a diamond on eBay for his fiancée. To get a sense of diamond prices and the related variables, he compiled a set of diamond sales data from the

Canada Diamonds website. He noticed to his surprise that sometimes a diamond of less than a half-carat can cost more than a diamond of between 0.5 and 1 carat. He decides to test a claim that there is no difference in median cost for diamonds in either of those two weight categories. Test his claim at the 0.01 level of significance, applying the Wilcoxon rank-sum test.

Solution Rather than show formally all steps of the hypothesis test, the focus here is on the basic concept for determining the p-value. The formal claim can be modelled as:

$$\textbf{Median}_{0.5\text{–}1.0} = \textbf{Median}_{0.0\text{–}0.5}$$

The test hypotheses are, therefore,

$$H_0\text{: }\textbf{Median}_{0.5\text{–}1.0} = \textbf{Median}_{0.0\text{–}0.5}$$

$$H_1\text{: }\textbf{Median}_{0.5\text{–}1.0} \neq \textbf{Median}_{0.0\text{–}0.5}$$

As Figure 13.5 illustrates, the next step (if you are working manually) is to *sort all values*—for both samples combined. That done, use the methods discussed in the previous section to assign ranks to all of the sorted values. If both populations have roughly the same median, then (1) the median of the two samples combined would be about equal to the two individual medians and (2) (assuming comparable shapes of distribution) the spreads of lower and higher values around the median would be similar. Under the null hypothesis, two samples of the same size should each have about the same share of low values (now assigned low ranks), medium values (now assigned medium ranks), and high values (now assigned high ranks), so if you added these ranks for each sample, the rank sums should be approximately equal. If the samples are of different sizes, then the rank sums also depend on the different numbers of ranks to be added for each sample.

Figure 13.5 conveys how the p-value will be assigned. The expected rank sum for the sample of higher-carat diamonds, assuming both samples had the same median, has been calculated. (For samples of size n_1 and n_2, the expected rank sum for the sample of size $n_1 = \dfrac{n_1 \times (n_1 + n_2 + 1)}{2}$. In the figure, the expected rank sum for sample 1 equals $\dfrac{28 \times (28 + 66 + 1)}{2} = 1{,}330$.) The *actual* rank sum for the same sample has also been calculated. Denote the sample's actual rank sum as W. Under the null hypothesis, W should approximately equal the expected rank sum for the sample. The right-tail p-value is the following conditional probability: p-value (one-tail) $= P(W \geq 2{,}096 \mid H_0 \text{ is true})$. What is the probability of obtaining such a large W-value, which indicates that the sample contains an unexpectedly large number of large rank values? The one-tail p-value times two gives the two-tail p-value (compare "In Brief 13.2(a)"):

$$p\textbf{-value} \approx \textbf{2} \times \textbf{0.00} = \textbf{0.00}$$

Since this value is less than alpha:

Reject the null hypothesis.

Our conclusion is:

There is sufficient evidence to reject the claim that diamonds with lower and higher carats have the same median price.

Figure 13.5

Basic Concept of the Two-Sample Wilcoxon Rank-Sum Test

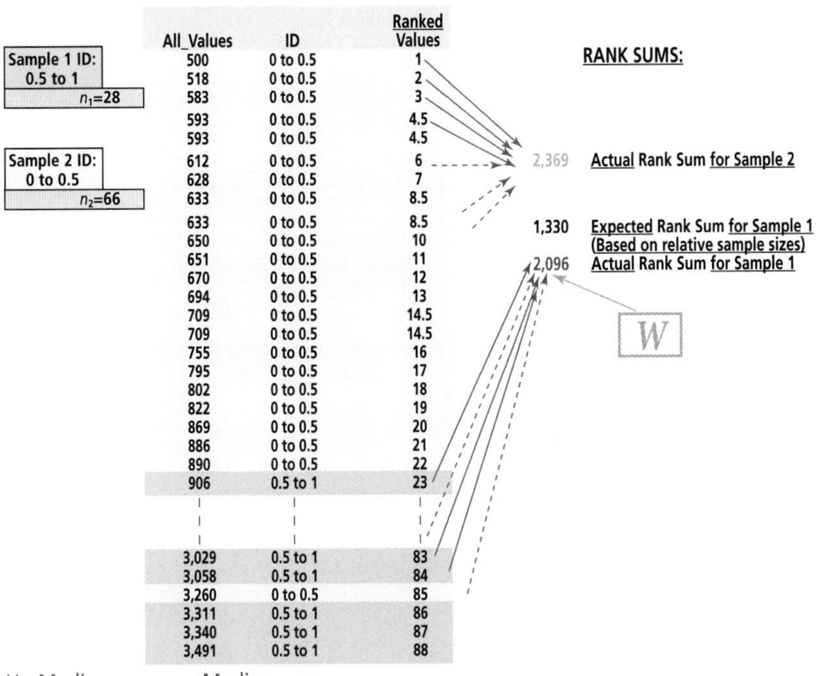

H_0: $\text{Median}_{0.5-1.0} = \text{Median}_{0.0-0.5}$

One-Tail p-Value $= P(\,(W - \text{Expected Rank Sum}) \geq 766 \mid \text{Median}_{0.5-1.0} = \text{Median}_{0.0-0.5})$

Source: Canada Diamonds. Found at: http://www.canadadiamonds.com/round_brilliant_canadian_diamonds.php?page=650&limit=10&customer=CAD

• *In Brief 13.2(a) Wilcoxon Rank-Sum Test*

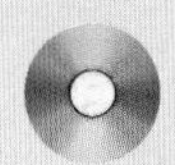

(Based on the ***Brilliants*** *file)*

Excel

Excel does not include the Wilcoxon formulas or procedures. For students using this software, you may download the file **Wilcoxon_Template** from this book's CD to use as a specialized calculator. Click on the spreadsheet tab for *Workspace2.* If inputs remain from a previous use, click on *Clear All.* Use the first column, as shown in the illustration, to input all of the values from both samples. Use the second column to identify from which sample each data value was taken. Be sure to input the two sample identifiers at the top, where indicated, and to use consistent spelling when inputting these labels. The computer will automatically count the sample sizes.

The next step is to manually input the ascending ranks for all data values—regardless of their source. Assuming that you will input the data one sample at a time, it may be easiest to *Data / Sort* the left two columns of the data (in order of ascending "All_Values") after they are input. Then the ranks can be easily determined by eye.

The program automates the remaining steps. Using an accepted normal approximation for the distribution of the ranked sums, it finds a z-value corresponding to the test statistic. From this, a p-value is determined (this can be doubled for the two-tail p-value), which can be applied back in your hypothesis test.

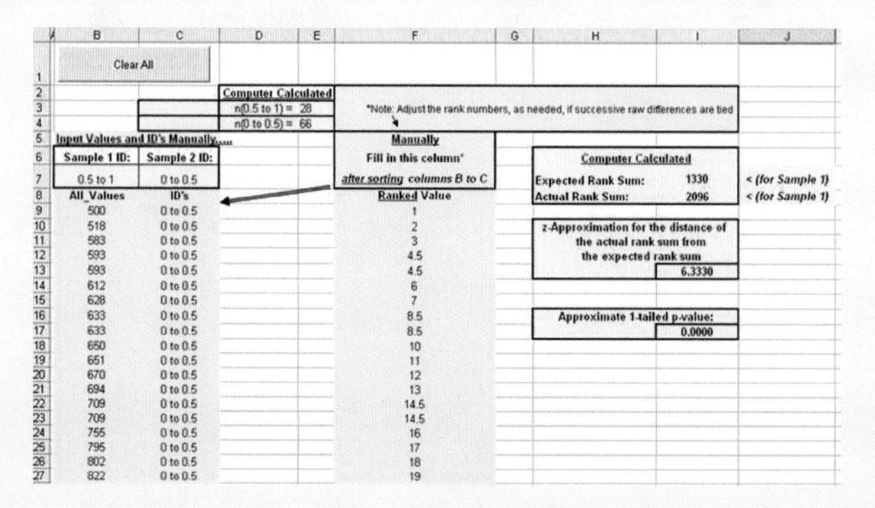

Minitab

As shown in the spreadsheet behind the screen capture, input the raw values for each sample into its own separate column. Use the menu sequence *Stat / Nonparametrics / Mann-Whitney.* In the resulting

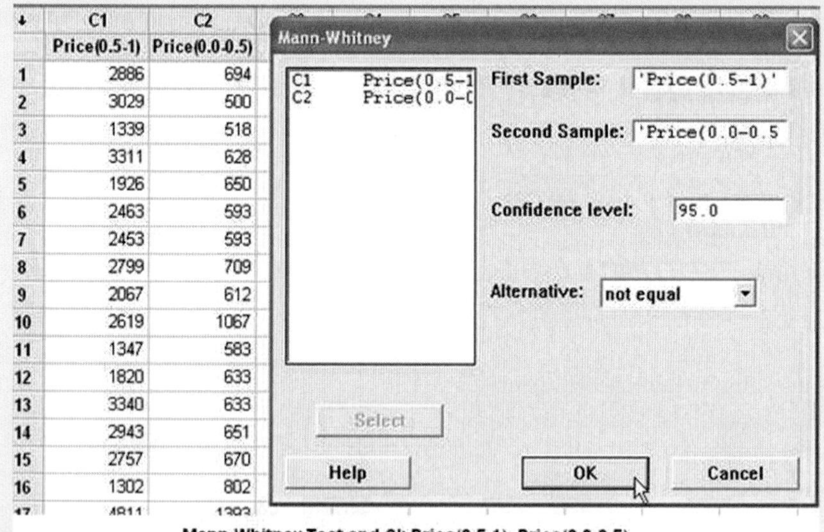

dialogue box, select the columns representing the first and second samples, respectively. The software will calculate values for the expected rank sum and *W*, with reference to the first sample identified. In the appropriate box, input the operator for the alternative hypothesis. Click *OK*. The output is also shown in the illustration.

SPSS

Use the first column, as shown in the spreadsheet behind the screen capture, to input all of the values from both samples. In the second column, identify from which sample each data value was taken, using a *numerical code*. Now begin with the menu sequence *Analyze / Nonparametric Tests / 2 Independent Samples*. In the resulting dialogue box shown in the illustration, select the column containing the sample data. Then specify the *Grouping Variable*, that is, the coded column that identifies the two samples. Once you input that variable's name, click on *Define Groups*.; in the next dialogue box (*Two Independent Samples: Define*), input the two identifying codes in the appropriate boxes and then click *Continue*.

Back in the first dialogue box, select *Mann-Whitney U* as the test type. Then click on *Exact*. If the data set is not too large, in the resulting dialogue box you can ask for an exact test by selecting *Exact*. In effect, this uses permutations to calculate exact probabilities of obtaining the sample test statistic, given the null hypothesis. For large data sets, it may be faster to use the *Monte Carlo* option. (The *Asymptotic only* choice uses a normal approximation of the test statistic distribution.) Click *Continue* and then *OK*. The output is also depicted in the illustration. (SPSS chooses which sample to use for the *W* calculation.)

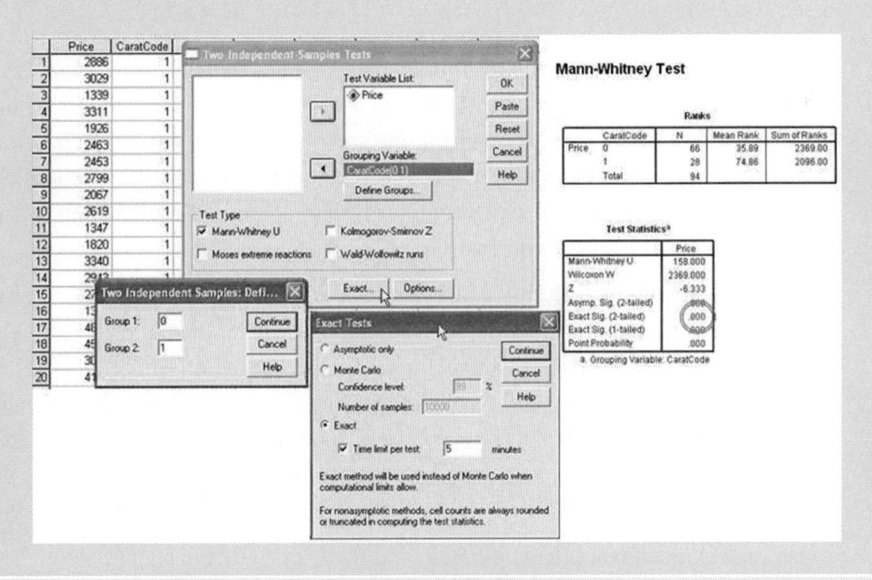

	CaratCode	N	Mean Rank	Sum of Ranks
Price	0	66	35.89	2369.00
	1	28	74.86	2096.00
	Total	94		

	Price
Mann-Whitney U	158.000
Wilcoxon W	2369.000
Z	-6.333
Asymp. Sig. (2-tailed)	.000
Exact Sig. (2-tailed)	.000
Exact Sig. (1-tailed)	.000
Point Probability	.000

Example 13.6

(*Based on the* **Brilliants** *file*)

The diamond shopper in Example 13.5 has talked to his accountant and has decided to look only at diamonds of no more than 0.5 carats. From that subset of the data set **Brilliants,** he has concluded that the median price is higher for diamonds that have "ideal" symmetry than for all of the other low-carat diamonds, collectively. Again using the Wilcoxon rank-sum method, test his claim at the 0.02 level of significance.

Solution Apply the formal model for hypothesis testing.

Step 1: The formal claim:

$$\textbf{Median}_{\text{IdealSymmetry}} > \textbf{Median}_{\text{Other Symmetry}}$$

Step 2: The null and alternative hypotheses:

$$H_0\text{: }\textbf{Median}_{\text{IdealSymmetry}} \leq \textbf{Median}_{\text{Other Symmetry}}$$

$$H_1\text{: }\textbf{Median}_{\text{IdealSymmetry}} > \textbf{Median}_{\text{Other Symmetry}}$$

Step 3: The level of significance:

$$\alpha = 0.02$$

Step 4: Specify the test and sample size:

Use the Wilcoxon rank-sum test for two independent samples,

$$n_1 = 13; n_2 = 53$$

Alternatively, we could write:

Use the Mann-Whitney U test for two independent samples,

$$n_1 = 13; n_2 = 53$$

Step 5: Depict the test model (see Figure 13.6). Precision artistry is not required. The purpose is to clarify (a) that the test statistic is the difference between *W* (the actual sum of ranks for one sample) and the expected sum of ranks (for the same sample) and (b) that the planned test is one-tailed. (Whether left- or right-tailed depends on which sample is used for calculating *W*.) Using a *p*-value approach, the critical values do not need to be calculated.

Step 6: Collect the data. A data set is provided, but we need to extract the relevant data. Consider only the diamond prices for diamonds up to 0.5 carats and create a column that distinguishes diamonds based on "ideal" versus non-ideal symmetry. You might want to explore features on your software (not covered here) on how to automatically "unstack" or pick out relevant data from a stored data set.

Step 7: Calculate the *p*-value and make the statistical decision. The test statistic *W* can be calculated: 569, and the expected rank sum is $\frac{13 \times (13 + 53 + 1)}{2} =$ 435.5. Your software can find the *p*-value (see "In Brief 13.2(a)"):

$$p\text{-value} \approx 0.015$$

The exact value may vary with your software. (Obtaining a negative *z*-value for the test statistic indicates that the software has calculated *W* for the non-ideal, rather than for the ideal, sample.)

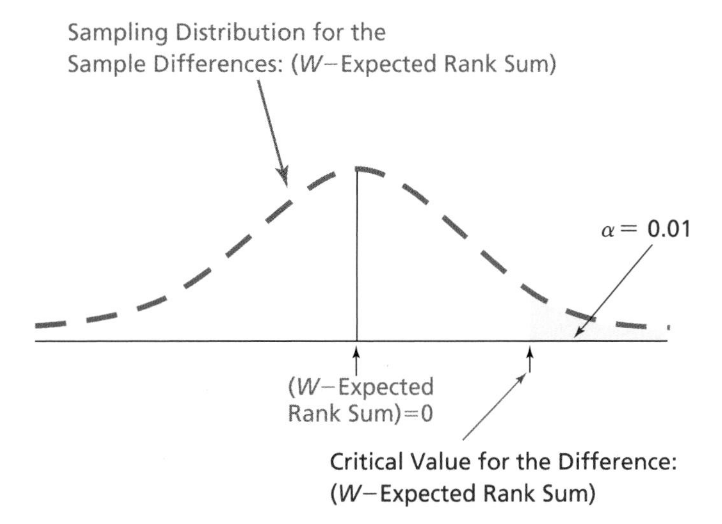

Figure 13.6
Depiction of Model for Step 5, Example 13.6

Because the p-value is less than $\alpha = 0.02$:

Reject the null hypothesis.

Step 8: Clearly express your conclusion, in words:

The evidence supports the claim that, for low-carat diamonds, the median price is higher for diamonds with ideal, rather than non-ideal, symmetry.

Wilcoxon Signed-Ranks Test: Two Dependent Samples

In Section 12.1, we distinguished between independent and dependent samples. Recall that samples are dependent if selected pairs of elements, one from each sample, are in some way connected. In Section 12.3, we used a parametric test, a t-test, to test for the difference in means of dependent samples. The Wilcoxon signed-ranks test provides an alternative if the assumptions for the t-test are not met sufficiently. The assumptions required for the Wilcoxon test are:

- The pairs are randomly selected. (For example, if pricing goods at competing stores, randomly select one corresponding item from each store to form the matched pairs. For a controlled experiment, such as evaluating a diet program, randomly select the participants and then determine randomly which member of each pair is assigned the experimental treatment.)
- The measurement level of the observations is at least at the ordinal level.
- The population of differences (as calculated from all the paired values) is approximately symmetrical.

Procedurally, this test is quite similar to the one-sample Wilcoxon signed-ranks test for the median, covered in previous section and in "In Brief 13.1(c)." We line up all of the paired values from the dependent samples and take their differences. Then we test for whether the median of this one column of differences is equal to zero. Under the null hypothesis (that the median of differences does equal zero), it follows that there is no systematic difference between the paired values of the two samples.

Example 13.7

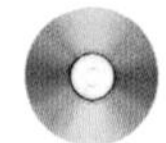

*(Based on the **School_Scores_Ontario** file)*

Every year, Grade 3 students across Ontario take a series of standardized tests to help evaluate their schools and curricula. Like many other boards, the Durham District School Board has posted the test results from all of the district schools. After studying the results (found in the file **School_Scores_Ontario**), a board trustee claims that the median student reading score for all schools is different from the median student writing score for all schools. Presuming that the data from this trustee's board is representative of a wider population, test the trustee's claim at the 0.05 level of significance.

Solution

Step 1: The formal claim:

$$\textbf{Median}_{\textbf{reading}} \neq \textbf{Median}_{\textbf{writing}}$$

Step 2: The null and alternative hypotheses:

$$H_0\textbf{: Median}_{\textbf{reading}} = \textbf{Median}_{\textbf{writing}}$$

$$H_1\textbf{: Median}_{\textbf{reading}} \neq \textbf{Median}_{\textbf{writing}}$$

Step 3: The level of significance:

$$\alpha = 0.05$$

Step 4: Specify the test and sample size. The data are paired: For each school, its reading score and its writing score can be directly compared. The number of data pairs is n (i.e., the number of schools in the sample). (Exclude the two schools for which the data in the file are incomplete.)

Use the Wilcoxon signed-ranks test for two dependent samples, $n = 30$.

Step 5: Depict the test model (see Figure 13.7). Create a sketch to illustrate (a) that the test statistic is the sum of all ranked negative (or positive—depending on software) differences between paired values and (b) that the planned test is two-tailed. (We will use a p-value approach.)

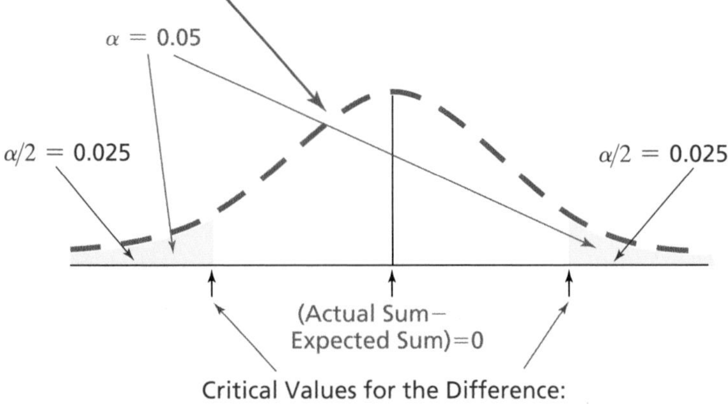

Figure 13.7
Depiction of Model for Step 5, Example 13.7

Step 6: Collect the data. (Data are provided in the problem.)

Step 7: Calculate the p-value and make the statistical decision. Applying methods discussed in the previous section you can obtain:

$$\textbf{Test statistic } T = 91.5$$

$$p\textbf{-value} \approx 2 \times 0.029 = 0.058$$

(The exact value may vary with your software.)

Because the p-value is greater than $\alpha = 0.05$:

Do not reject the null hypothesis.

Step 8: Clearly express your conclusion, in words:

The evidence does not support the claim that the schools' median scores for reading and for writing are different.

• *In Brief 13.2(b)* *Wilcoxon Signed-Ranks Test: Two Dependent Samples*

(Based on the ***School_Scores_Ontario*** *file)*

Excel

See "In Brief 13.1(c)" for the description of the file **Wilcoxon_Template** on this book's CD, in the spreadsheet tab *Workspace1*. Those same procedures can also be used for the case of two dependent samples. The only difference is that, instead of using the left column of the template to copy the H_0 value of the median, you now input the sample 1 data in that column; input each corresponding sample 2 value in the second column. As before, a one-tail p-value is output; for a two-tail test, multiply this value times two.

Minitab

See the procedures described in "In Brief 13.1(c)"; these can also be used for the case of two dependent samples. The only difference is the need for a preparatory step.

Input the paired data into two columns in Minitab. Create a third column, which is composed of the differences between each pair of values. The screen capture shows the dialogue box that can automate that step: Begin with the menu sequence *Calc / Calculator.* In the dialogue box, name the variable to hold the differences and input the mathematical expression to calculate the difference from the appropriate columns. Click *OK.* At this point, proceed as you did in "In Brief 13.1(c)" but use the new difference variable as the input for the test. (For the test, specify the test median as 0.0.)

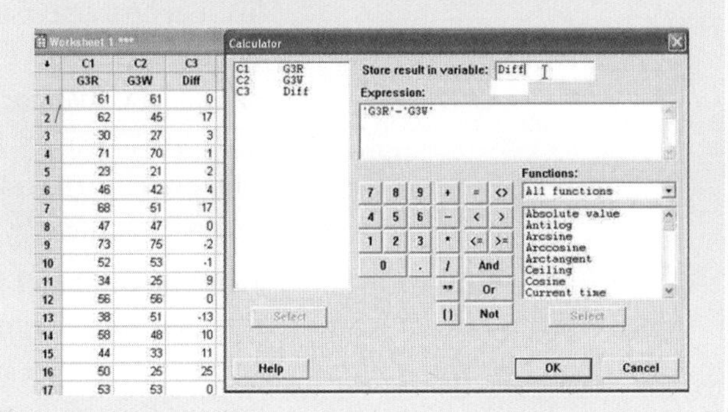

SPSS

See the procedures described in "In Brief 13.1(c)"; these can also be used for the case of two dependent samples. Instead of using the first column to copy the hypothesized value of the median, you now input the data, pair by pair, into the two left columns. In either case, proceed as described in "In Brief 13.1(c)."

Chapter 14

13.3 Additional Notes about Nonparametric Tests

Advantages and Disadvantages

When is using nonparametric tests more advantageous than using parametric tests? Every test proceeds by certain assumptions and has limits as to what it can tell us. To the extent that a test's assumptions are not met, the test results might not be reliable, so perhaps another type of test should be considered. This section takes that approach to the question.

When Using Nonparametric Procedures Is an Advantage

1. **When the usually more rigid conditions for using a parametric test are not satisfied.** A one-sample *t*-test for the mean, for example, presumes that the population is normally distributed, or else that the sample is large. If a sample is relatively small and the population is not normally distributed, the *t*-test results may be suspect. However, the conditions for using a nonparametric test (such as a Wilcoxon signed-ranks test for the median) may still be met, allowing us to reach a valid conclusion.
2. **When the data's level of measurement is not sufficient for the parametric test.** Strictly speaking, you cannot take a mean of ordinal data, for example, because the distances between successive values are not clear. This suggests choosing a test, like a nonparametric one, that is not based on the mean and so can handle the ordinal data. (Note that, in practice, some researchers *do* at times take the means of ordinal values; for instance, if the data are questionnaire results of the "Rank from 1 to 5" variety. In that case, some may choose parametric procedures, but with caution. The tradeoff in this case is added restrictions on how to interpret results meaningfully.)
3. **When no computer is available.** For pen and calculator methods, nonparametric formulas are sometimes easier to work with than their parametric counterparts. Obviously, with the many available choices of software programs, this consideration has become much less important.

When Using Nonparametric Procedures Is a Disadvantage

1. **When you could validly use an alternative that preserved more information.** In the tests discussed in this chapter, for example, we converted raw data into ranks—or even, in one case, just to positive or negative signs. The precise magnitudes of raw data are not accounted for in the nonparametric tests' conclusions. If it would be valid (based on test conditions and assumptions) to use some alternative parametric test, then you could say that data are lost if you choose the nonparametric procedure.
2. **When you could validly use an alternative that was more efficient.** Of two tests, the more **efficient testing procedure** is the one that, for a given sample size, has the more power. Recall that power (which is calculated as $1 - \beta$) is the probability that the test will correctly reject a false null hypothesis. In other words, if the conditions apply for validly using a parametric test, then the parametric test can use a smaller sample size without increasing the risk of a Type II error. The key question, of course, is whether those conditions apply.

The Resampling Approach to Hypothesis Tests

In this chapter, we have seen that nonparametric methods can be useful when the assumptions for a parametric test are not satisfied. It is likewise useful to have another alternative if the nonparametric test's assumptions are not met. For example, the Wilcoxon rank-sum test for two independent samples does not require a normality assumption, but it still works best if the two distributions are at least similar. The sign test on the other hand, has fewer distribution assumptions—but at the price of losing more data (i.e., only the signs above and below the hypothesized median are utilized).

A **resampling test** can be used to make fuller use of the data without imposing any distribution assumptions from outside the data set. Although this text does not formally cover this class of procedures, it is useful to be aware of what they can do. Students are encouraged to pursue this topic further.

Resampling takes the sample data itself as the best evidence for what the population distribution might be. With the power of the computer, we can take repeated samples of a given size, with replacement, from the original set of data and record a statistic of interest (such as a median) from each sample. The result of this repeated sample-taking is a simulation of a sampling distribution, which theoretically would be based on every possible sample of the given size. We can then discover by inspection the percentage of resampled cases in which the test statistic falls within a certain specified range. This is, in effect, the p-value for obtaining that statistic from a randomly selected sample.

The principles of the test are illustrated in Example 13.8.

Example 13.8

(Based on the ***Brilliants*** *file)*

In Example 13.6, a diamond shopper reviewed price data for diamonds up to 0.5 carats. Based on a Wilcoxon rank-sum test, he apparently confirmed a claim that the median price for ideally symmetrical diamonds of that weight is different from the median price for less than ideally symmetrical diamonds of that weight. However, his fiancée, who is taking statistics, disputed his use of the Wilcoxon test for these data. She created the graphs in Figure 13.8 to show that the price distributions for the two compared groups are not at all similar. Therefore, she recommends that a resampling test should be tried instead. Perform that test at the 0.02 level of significance and compare the results with Example 13.6.

Figure 13.8
Comparative Price Distributions for Ideally and Non-Ideally Symmetrical Diamonds

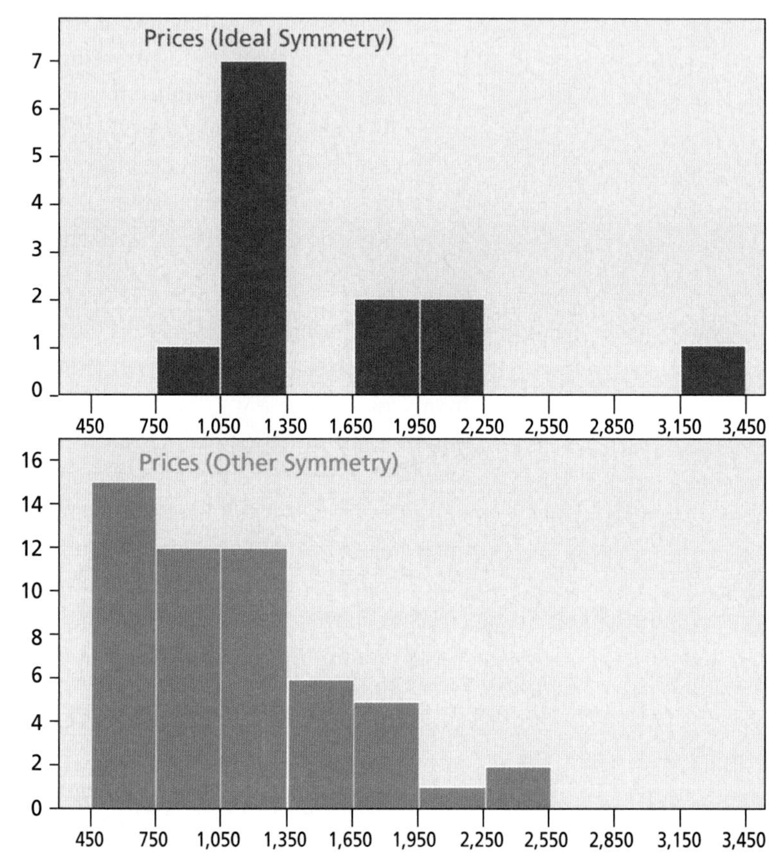

Source: Canada Diamonds. Found at: http://www.canadadiamonds.com/round_brilliant_canadian_diamonds.php?page=650&limit=10&customer=CAD

Solution

Step 1: The formal claim:

$$\textbf{Median}_{\textbf{IdealSym}} > \textbf{Median}_{\textbf{Non-IdealSym}}$$

Step 2: The null and alternative hypotheses:

$$H_0\textbf{: Median}_{\textbf{IdealSym}} \leq \textbf{Median}_{\textbf{Non-IdealSym}}$$

$$H_1\textbf{: Median}_{\textbf{IdealSym}} > \textbf{Median}_{\textbf{Non-IdealSym}}$$

Step 3: The level of significance:

$$\alpha = 0.02$$

Step 4: Specify the test and sample size. Because the distributions do not meet the requirement for either the parametric *t*-test or the nonparametric Wilcoxon rank-sum test:

Use a resampling-based test, $n_{\text{IdealSym}} = 13$, $n_{\text{Non-IdealSym}} = 53$

Step 5: We usually depict the test model in this step, but for a resampling test, there is no a priori model for the sampling distribution and so nothing to draw until we reach step 7. In short, leave step 5 blank for a resampling test.

Step 6: Collect the data. (Data are provided in the problem.)

Step 7: Calculate the *p*-value and make the statistical decision.

For the two samples:

$$\text{SampleMedian}_{\text{IdealSym}} - \text{SampleMedian}_{\text{Non-IdealSym}} = \$1{,}131 - \$1{,}035 = \$96.$$

The desired *p*-value is actually this conditional probability:

$$p\text{-value} = P(\text{SampleMedian}_{\text{IdealSym}} - \text{SampleMedian}_{\text{Non-IdealSym}} \geq \$96 \mid H_0 \text{ is true})$$

To determine the above conditional probability we presume no initial knowledge of the population distributions, but we *do* have all of the data from our two samples. Under the equal aspect of the null hypothesis, our best guess for the population distribution is modelled by the frequency distribution for all data combined. This would mean that, under H_0, both groups of diamonds have the same median and any apparent differences in their price distributions are due to sampling error.

To calculate the conditional probability (*p* value), we need to estimate (by simulation) the sampling distribution for the differences in the two sample medians under H_0. To do this, draw with replacement two independent samples of the same sizes as those that appear in the data. Find their medians and calculate the difference. Repeat this step thousands of times. The distribution of all of the calculated differences is an approximation of the expected sampling distribution of the differences under the null hypothesis.

This estimated sampling distribution, based on 5,000 resamples, is shown in Figure 13.9. (In this example, the distribution is approximately normal, but this will not always be the case.) We see that, conditional on H_0, it is not rare to obtain a difference in sample medians of at least \$96. In the simulation, this occurred 1,168 times. Therefore:

$$p\text{-value} \approx 1{,}168/5{,}000 = 0.23$$

(The exact value may vary each time you run the simulation.)

Because the *p*-value is greater than $\alpha = 0.02$:

Do not reject the null hypothesis.

Step 8: Clearly express your conclusion in words:

The evidence does not support the claim that, for low-carat diamonds, the median price is higher for diamonds with ideal, rather than non-ideal, symmetry.

This result directly contradicts the solution in Example 13.6. As always, we should ask critically which model better conforms to the real nature and distribution of the data. It is clear in Figure 13.8 that the Wilcoxon requirement of "similar" distributions has not been met, so the conclusion in Example 13.8 may be more reliable in this instance.

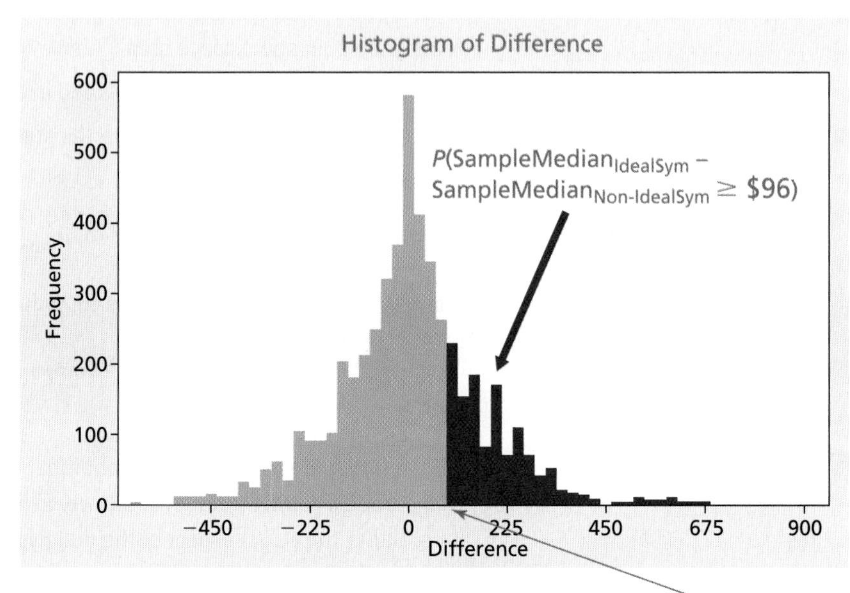

Figure 13.9
Resampling Distribution of the Sample Differences in the Median

• Hands On 13.2 *Resampling Based Confidence Intervals*

Although this textbook does not formally cover resampling, we have used simulations a number of times in these "Hands On" exercises. Resampling is a special case of a simulation: You are simulating the drawing of random samples from a population that has the frequency distribution of the original data. We record a statistic of interest for each generated sample, and these statistics are compiled into a simulated sampling distribution. This distribution can be the basis for constructing a confidence interval or for hypothesis testing.

To obtain data for this exercise, think of four types of objects that can be bought on eBay. For each of the four types of objects, do the following: Log on to eBay, starting on the page http://hub.ebay.ca/buy. From here, you can do a search for the category and specific name of the object that you would like to consider. For example on a particular day, a search for "DVD monster" in the category "DVDs & Movies" returned a list of 9,778 such objects, each showing the current bid price. Draft and follow a random sampling plan to select a reasonable number of the listings for your object of interest, and record the latest bid price for each selected object. Repeat this sampling process for each of the four types of objects that you are considering. Then, for each type of object, use resampling to construct a 95% confidence interval for the median bid price for that type of object.

In other words, for a given object, interpret your sample data as your best guess of the distribution of prices for that type of object. Find a way to simulate new samples for the prices of that object type. (Hint: Select each element for each resample in such a way that it has an equal probability of being any one of the values in the original sample. Repeat this process to create many such samples of the desired size.) Once all samples are ready, find the median price in each sample; construct a frequency distribution of all of the median prices. Find the bounds so that only 2.5% of the sample medians are below the lower bound and only 2.5% of the sample medians are above the upper bound. These are the bounds of a 95% confidence interval. Do the same for four types of objects.

Chapter Summary

13.1 Explain the difference between parametric and nonparametric (or distribution-free) hypothesis tests; convert interval or ratio data into rank values; and conduct a sign test or a Wilcoxon signed-ranks test for the median—and recognize when each would be preferable.

Hypothesis tests known as **nonparametric** (or **distribution-free**) require fewer assumptions about the distribution of the population data than do parametric tests. Some of the methods in this chapter rely on the concept of **ranks, or rank-ordering,** of data.

For data that are randomly selected from a single sample at, at least, the ordinal level and with relatively few values that *exactly* equal the median, we can perform a sign test based on the relative numbers of data values falling above and below the hypothesized median. An alternative that retains more information about how far the data items are from the median on either side is the Wilcoxon signed-ranks test, which adds a requirement that the underlying distribution be essentially symmetrical.

13.2 Conduct a Wilcoxon rank-sum test for two independent samples or a Wilcoxon signed-ranks test for two dependent samples.

The Wilcoxon rank-sum test can be used to test a hypothesis about the difference between medians for two independent random samples, provided the data are, at least, at the ordinal level and that the two sample distributions are roughly similar to one another. For two dependent samples, the Wilcoxon signed-ranks test can be used, as long as the population of differences (as calculated from all the paired values) is approximately symmetrical.

13.3 Describe some conditions under which using nonparametric procedures is an advantage and when it is a disadvantage, compared to using parametric methods for hypothesis tests, and explain what is meant by a resampling approach to hypothesis tests.

There are advantages to using nonparametric procedures for hypothesis testing when (1) the usually more rigid conditions for using a parametric test are not satisfied, (2) the data's level of measurement is not sufficient for the parametric test or (3) a computer is not available. Using nonparametric procedures is a disadvantage when (1) you could validly use an alternative that preserved more information or (2) you could validly use an alternative that is a more **efficient testing procedure** (i.e., one that has more power for the given sample size).

As an alternative, a **resampling** test can make fuller use of the data without imposing any distribution assumptions from outside the data set. Resampling takes the sample data itself as the best evidence for what the population distribution might be.

Key Terms

efficient testing procedures, p. 565
nonparametric (*or* distribution-free) hypothesis tests, p. 543
ranks (*or* rank-ordering) of data, p. 543
resampling tests, p. 565

Exercises

Basic Concepts 13.1

1. Assign ranks to the numbers in each of the following sets:
   a) 1946 2408 2206 1999 2058
   b) 34 65 96 41 89 59
   44 32 58 47 22
   c) 65 59 46 71 63 41
   41 41 59 41 59
   d) 1974 2058 2111 2206 2058 1999 2058
2. What are the assumptions required to validly use the sign test for the median?
3. When you are conducting the sign test for the median, what is the value you calculate for use as the test statistic?
4. Explain why the sign test can be associated with the binomial distribution.
5. What are the assumptions required to validly use the Wilcoxon signed-ranks test for the median?
6. When you are conducting the Wilcoxon signed-ranks test for the median, what is the value you calculate for use as the test statistic?
7. The Wilcoxon signed-ranks test is used with which level of measurement? Is it possible to use this test with higher levels of measurement? Explain.
8. The baggage handlers at an airport believe that the median weight of luggage is 100 kg. A sample of 15 pieces of luggage was selected with the following weights (in kg):
   98 104 106 90 102 107 103 99
   95 101 96 97 103 91 111
   Test the baggage handlers' claim at the 0.05 level of significance, using the sign test.
   a) State the null and alternative hypothesis.
   b) What is the test statistic?
   c) What is the p-value for the test?
   d) State the statistical decision and then state your conclusion in words to the baggage handlers.

9. Redo Exercise 8 but use the Wilcoxon signed-ranks test instead of the sign test.

Applications 13.1

As part of the solution for any hypothesis test, formally show the eight-step sequence, as illustrated in the text.

10. Federal and provincial governments in Canada often specify a minimum-wage level in approved government-tendered contracts. To ensure the minimum-wage levels are reasonable, government officials may survey the wage rates for similar contracts elsewhere. A key official believes that the median wage rate for part-time construction millwrights on contract is $15 per hour. A sample of 12 workers in this category was selected, and their wage rates were recorded as follows:

$15.50	$14.46	$17.00	$13.40
$17.25	$18.50	$13.50	$14.37
$16.79	$17.50	$16.20	$18.45

Source: Human Resources Development Canada. *Ontario Wage Survey,* for *Construction Millwrights and Industrial Mechanics Excluding Textile Industries.* Found at: http://onestep.on.ca/ows/Ows99_Eng/7311_ON_e.html#Full

Test the official's belief about the median wage rate at the 0.05 level of significance using the Wilcoxon signed-ranks test.

11. The list of mutual funds in the file **Canadian_Equity_Funds** is a random sample from a large list of funds published in 2005 by GlobeFund.com. Using the Wilcoxon signed-ranks test, test the claim that for the full population of GlobeFund listings at that time, the median asset value was $40 million. Use the 0.05 level of significance. (Disregard funds for which no asset value is given in the file.)

12. Redo Exercise 11, but use the sign test for the hypothesis test.

13. From the sample of funds described in Exercise 11, use the 0.05 level of significance to test the claim that for the full population, the median net asset value per share (NAVPS) was less than $16.

14. a) Redo Exercise 13, but use the sign test for the hypothesis test.
    b) Try to explain why Exercises 13 and 14 give opposite results. (It might help to use the computer to generate a quick histogram of the variable distribution.)

15. While preparing a report for the provincial Labour Standards Division in Nova Scotia, an official realized that different jurisdictions in Canada have different thresholds for how many hours employees may work per week before they must be paid time-and-a-half. The official believes that the median threshold is 42 hours per week. Using the following sample from different jurisdictions,[1] test the official's view at the 0.02 level of significance.

44	40	40	44	48	40	40	40
44	48	40					

16. Redo Exercise 15 but use the sign test for the hypothesis test.

Basic Concepts 13.2

17. What are some conditions under which a parametric test for comparing two samples using a t-test may not be appropriate?

18. Under what conditions is it appropriate to use the Wilcoxon rank-sum test for two independent samples?

19. In general, what do the null and alternative hypotheses look like for the Wilcoxon rank-sum test?

20. When you are conducting the Wilcoxon rank-sum test for two independent samples, what is the value you calculate for use as the test statistic?

21. For the following sets of numbers, independently selected, determine whether there is a difference in the locations of the distributions. That is (depending on the test used), are the means or medians of the sets the same or different? Use a 0.01 level of significance.

Set 1	64	79	80	64	64	74	89	58	60	59
Set 2	54	68	82	54	71	68	80	63	79	

    a) Use the Wilcoxon rank-sum test.
    b) Use the t-test for independent samples (assuming equal variances).

22. Under what conditions is it appropriate to use the Wilcoxon signed-ranks test for two dependent samples?

23. In general, what do the null and alternative hypotheses look like for the Wilcoxon signed-ranks test for two dependent samples?

24. When you are conducting the Wilcoxon signed-ranks test for two dependent samples, what is the value you calculate for use as the test statistic?

25. For the following two dependent sets of numbers, determine whether there is a difference in the locations of the distributions. That is (depending on the test used), are the means or medians of the sets the same or different? Use a 0.01 level of significance.

Set 1	64	79	80	64	64	74	89	58	60	59
Set 2	54	68	82	54	71	68	80	63	79	63

    a) Use the Wilcoxon signed-ranks test.
    b) Use the t-test for dependent samples.

Applications 13.2

As part of the solution for any hypothesis test, formally show the eight-step sequence, as illustrated in the text.

26. A concerned Yukon citizen claims that, based on the median, more of the territory's forests are being destroyed yearly by fire since 1981 than was previously the case. Use the data stored in the file **Yukon_Wildfires** and presume that the annual data for hectares lost by fire from 1950 to 1980 is representative of previous patterns of annual fire loss, while the data from 1981 to 2004 is representative of a more current pattern.
    a) Based on the model described, test the Yukon citizen's claim using the Wilcoxon rank-sum test, at the 5% level of significance.
    b) Inspecting the data, the value for 2004 appears to be an outlier. Does this fact pose a concern for using the Wilcoxon test? Explain why or why not.
27. The Yukon resident in Exercise 26 also claims that, since 1981, there are more cases per year in which lightning causes a forest fire than was previously the case. Use the data in **Yukon_Wildfires** to test the claim at the 5% significance level—comparing the median number for lightning-caused fires per year up to 1980 with the median number of such fires per year from 1981 to 2004.
28. An additional claim of the Yukon resident (compare Exercise 26) is that, since 1981, there are more cases per year in which humans play a role in causing forest fires than there were previously. Use the data in **Yukon_Wildfires** to test the claim at the 5% significance level—comparing the median number for human-caused fires per year up to 1980 with the median number of such fires per year from 1981 to 2004.
29. A parent of young children is considering alternative types of education. With a limited budget, she is preparing to place an order from a book list on this topic. Selected prices (chosen independently) for three different categories of such books are recorded in the file **Education_Books.**
    a) Test a claim that the median price for books on education theory is less than the median price for books on home schooling ($\alpha = 0.05$).
    b) Use a t-test for a claim that the *mean* price for books on education theory is less than the mean price for books on home schooling ($\alpha = 0.05$).
    c) Compare your answers for (a) and (b). Do they give similar answers regarding the measures of locations? If not, explain why that might be. (It may help to use the computer to sketch the two distributions.)
30. The parent in Exercise 29 is continuing to review the book list whose prices are sampled in the file **Education_Books.** Test a claim that the median price for books on education theory is the same as the median price for books on alternative methods of education ($\alpha = 0.05$).
31. To monitor inflation in the prices of groceries, an inspector each month makes one purchase (selecting standardized quantities and criteria) from each of several grocery categories (such as bread, beef, chicken, and so on). Interpret the data in the **Prices_for_Groceries** file as a record (in U.S. dollars) of the inspector's monthly purchase costs.
    a) Test a claim ($\alpha = 0.03$) that the median price for the commodities purchased by the inspector increased between March and April 2005.
    b) Test a claim ($\alpha = 0.03$) that the median price for the commodities purchased by the inspector increased between February and March 2005.
32. Recently a dairy farmer made this optimistic claim: "After years of decline, the median number of dairy cows being kept in each province is now increasing." She based her claim on data provided by the Canadian Dairy Information Centre (see file **Dairy_Cattle**). Use the provincial data in the file (omit the national total) for the years 2004 and 2005 to test the farmer's claim ($\alpha = 0.05$).
33. Referring to data in the file **Meat_Consumption,** a poultry farmer claimed that unlike chicken consumption, "People's annual consumption of non-chicken meat did not change between 1993 and 2003." Assuming the non-chicken data in the file are based on sampling, compare medians to test the farmer's claim ($\alpha = 0.05$).

Basic Concepts 13.3

34. Describe and explain the conditions under which using nonparametric procedures is an advantage over using parametric procedures for a similar test.
35. Describe conditions under which using nonparametric procedures is *not* an advantage, compared to using parametric procedures for a similar test.
36. Briefly explain what is meant by a resampling test. What are the conditions when you **might use one?**

There are no Applied questions for this section.

Review Applications

As part of the solution for all hypothesis tests, formally show the eight-step sequence, as illustrated in the text. If a specific test is not specified, use an appropriate test from among those introduced in this chapter.

Exercises 37–40 are based on the data in the file **Animal_Weights.**

37. The owners of a small petting zoo are compiling statistics about their animals. These data are collected in the file

Animal_Weights. Suppose they own one specimen for each species shown in the file. Test, using the Wilcoxon signed-ranks test, a claim that for the domestic animals in the collection, the median weight is 500 grams ($\alpha = 0.05$).

38. Test, using the signed-ranks test for the median, a claim that for the livestock animals in the collection, the median weight is 40 kg ($\alpha = 0.05$).

39. Do the livestock animals consume, on average, a greater amount of water per kilogram of body weight than do the domestic animals? Perform the test at the 0.01 level of significance, based on the Wilcoxon ranks-sum test for the medians.

40. Idly examining the data while on the phone one day, the petting zoo owners' son concluded that, in general (if the units are ignored), each domestic animal consumes about the same magnitudes of water and food per kilogram of body weight. Test this claim at the 0.05 level of significance, using the Wilcoxon signed-ranks test.

Exercises 41–46 are based on data in the file **Water_ Use.**

41. Environment Canada publishes a summary of water use by every Canadian municipality with a population of over 1,000. For analysis, take a systematic sample that includes (a) the 12th record in the data file and (b) every 30th record thereafter. On the basis of that sample, test a claim that the median domestic water use in all municipalities of this size is greater than 1,500 cubic metres per day. Use the 0.04 level of significance.

42. On the basis of the sample described in the previous exercise, test a claim that the median commercial/institutional water use in municipalities with populations over 1,000 is equal to 400 cubic metres per day. Use the 0.04 level of significance.

43. On the basis of the sample described in Exercise 41, test a claim that the median longitude for Canadian municipalities with populations over 1,000 is *not* equal to 72°W. Use the 0.05 level of significance.

44. For the sample described in Exercise 41, compare the populations of municipalities of two groups: those that disinfect their water using chlorine and those that do not. Test a claim that the median populations for the municipalities of both groups are the same. Use the 0.02 level of significance.

45. For all of the municipalities included in the sample (see Exercise 41), compare their populations in 1999 with their populations in 1996. Test a claim, using medians, that in general the municipalities' populations have increased from 1997 to 1999. Use the 0.05 level of significance.

46. For the sample described in Exercise 41, compare the populations of municipalities of two groups: those that fluoridate their water and those that do not. Test a claim that the median populations for the municipalities of both groups are the same. Use the 0.02 level of significance.

47. Health officials are worried about the impacts of stress and obesity on Canadian adults. The file **CanadianHealth** includes columns for the proportions of adults who suffer from obesity and stress, respectively, in each province and territory. Interpret the proportion data as *scores* for the two variables for the corresponding regions. Test whether, on average, the obesity scores are the same as the stress scores for the various regions. Use $\alpha = 0.05$.

Exercises 48–50 are based on data in the file **Carriers_ revenue_tons.**

48. The U.S. Internal Revenue Service is examining the "revenue tons" transported by different airlines across the country. (See the file: **Carriers_revenue_tons.**) One official believes that, in general, major airlines service more revenue tons in a year than national airlines do. Test the claim at the 0.05 level of significance.

49. Test the claim that regional airlines (which includes large and medium airlines) tend to service fewer revenue tons than national airlines do ($\alpha = 0.05$).

50. For a variety of reasons, an airline may not carry exactly the size of load that was scheduled. Nonetheless, an airline spokesperson says, on average, the amount of revenue tons transported and the amount scheduled for any given airline tend to be about equal. Test this claim at the 0.02 level of significance.

Chapter 14

Analysis of Variance (ANOVA)

• Chapter Outline

14.1	One-Way ANOVA
14.2	Randomized Block Design
14.3	Two-Factor ANOVA

• Learning Objectives

14.1 Construct a one-way ANOVA table and apply this model, where applicable, for a hypothesis test.

14.2 Construct an ANOVA table and apply the model for a hypothesis test for randomized block design and one-factor repeated measures.

14.3 Construct a two-factor ANOVA table and apply this model, where applicable, to tests for interactions of factors and for main effects.

14.1 One-Way ANOVA

This section extends the type of analysis that was presented in Section 12.2. There, we compared sample means for two independent samples to test for a difference in means of the populations that correspond. With analysis of variance (ANOVA) techniques, we can test for a difference among more than two means. If a difference in means is confirmed, ANOVA does not actually specify *which* of the means is different from the others. Depending on your software, the latter question can be addressed with a follow-up test.

The model for **one-way ANOVA** (also called **one-factor ANOVA**) presumes that in the data set there are two variables: (1) the dependent variable, and (2) a **grouping** (or **factor**) **variable.** The latter provides the basis for dividing the data of variable (1) into separate samples or groups—also called *levels.* The concept is illustrated in Figure 14.1.

In part (a) of Figure 14.1, we see a sample of mutual funds. For each fund, values are shown for (1) its one-year financial return (i.e., the dependent variable) and (2) the fee option for the fund, which is the grouping variable. (The options displayed are "N,"

Figure 14.1
Example of Data Divided into Groups

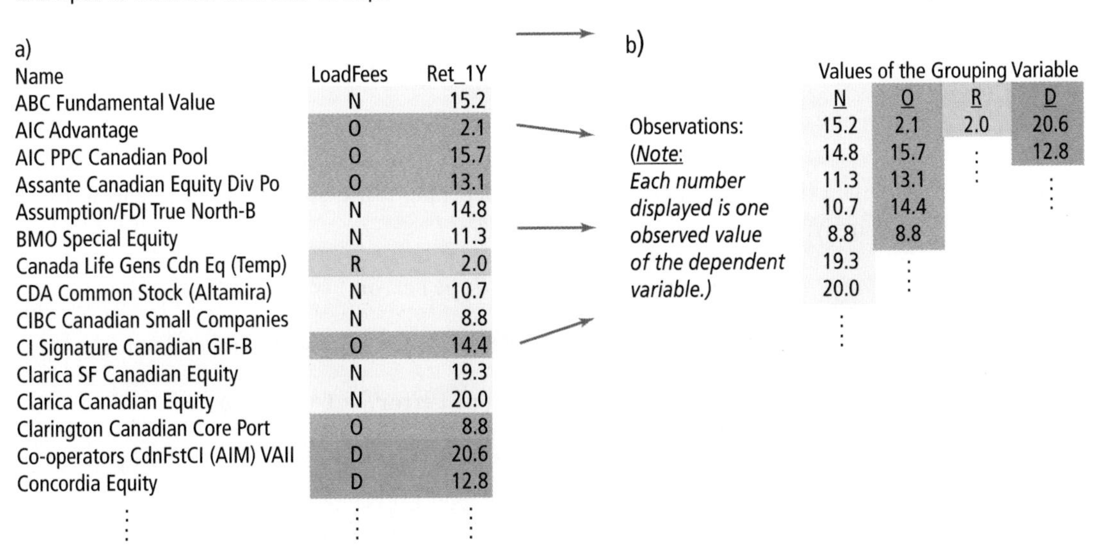

a)

Name	LoadFees	Ret_1Y
ABC Fundamental Value	N	15.2
AIC Advantage	O	2.1
AIC PPC Canadian Pool	O	15.7
Assante Canadian Equity Div Po	O	13.1
Assumption/FDI True North-B	N	14.8
BMO Special Equity	N	11.3
Canada Life Gens Cdn Eq (Temp)	R	2.0
CDA Common Stock (Altamira)	N	10.7
CIBC Canadian Small Companies	N	8.8
CI Signature Canadian GIF-B	O	14.4
Clarica SF Canadian Equity	N	19.3
Clarica Canadian Equity	N	20.0
Clarington Canadian Core Port	O	8.8
Co-operators CdnFstCl (AIM) VAII	D	20.6
Concordia Equity	D	12.8
⋮	⋮	⋮

b)

Observations: (Note: Each number displayed is one observed value of the dependent variable.)

Values of the Grouping Variable			
N	O	R	D
15.2	2.1	2.0	20.6
14.8	15.7	⋮	12.8
11.3	13.1		⋮
10.7	14.4		
8.8	8.8		
19.3	⋮		
20.0			
⋮			

Source: The Globe and Mail, "Report on Business." Based on data in "Asset Class: Canadian Equity." Found at: GlobeFund.com at http://www.globefund.com/static/romf/generic/tabceq.pdf. Reprinted with permission from *The Globe and Mail.*

which stands for "no fee;" "O" "optional fee;" "R" "redemption fee;" or "D" "deferred load.") In part (b) of the figure, we see how the sample values for the return can be subdivided, based on the values of the grouping variable (i.e., based on the factor "fee option").

If the data are collected initially in the format shown in Figure 14.1(b), it might appear that the columns "N," "O," "R," and "D" represent four different variables, instead of values for a single grouping variable. We know this is not the case, however, because the dependent variable, Ret_1Yr, does not have its own separate column in that figure. Instead, all values for the one-year return (the dependent variable) are displayed as numbers under the headings for the various factor values.

For test purposes, all numerical values associated with a single group are interpreted as a sample from a separate population. In Figure 14.1(b), four samples from four populations are shown. The null hypothesis is that all of the population means are equal; the alternative hypothesis is that at least one of the population means is different from the others. In general form:

$$H_0: \mu_1 = \mu_2 = \mu_3 = \ldots = \mu_k$$

H_1: **At least one of the population means is different.**

ANOVA is not the only method we could use to test the above hypothesis. You could separately test every pair of means in H_0 (e.g., μ_1 versus μ_2; μ_1 versus μ_3; μ_3 versus μ_k, etc.) for equality. If equality is rejected for any one of these pairs, then reject H_0. Two compelling reasons for *not* using this approach are:

1. Depending on the number of groups, the number of pairs of means to test separately would become quite large. For three groups, there are three possible pairs of means, which would be manageable. But for 10 groups, for instance, 45 pairs of means would have to be tested for equality. This would be very impractical.
2. If tests for mean-pairs are combined in this fashion, the risk of making a Type I error grows significantly. For any one test:

   P(there has not been a Type I error) = $(1 - \alpha)$

   For example, if $\alpha = 0.05$, the probability is 0.95.

   If we combine n hypothesis tests and want *all* of them to *not* be Type I errors, then (if individual test results are independent) the multiplication rule would apply:

   P(n results are all not Type I errors) = $(1 - \alpha)^n$

   which means $(1 - \alpha) \times (1 - \alpha) \times \ldots \times (1 - \alpha)$, for all n repetitions. Thus:

   P(there is at least one Type I error) = $1 - (1 - \alpha)^n$

   For the grouped data illustrated in Figure 14.1, it can be shown that six pairs of means would have to be tested for the four groups. The probability that no Type I error occurs in all six tests could equal $0.95^6 = 0.735$. Thus, the probability that there *will* be a Type I error is large: $1 - 0.735 = 0.265$. (Note: This answer is given to illustrate the general problem; the exact probability of a Type I error could vary if not all individual test results are independent.)

Assumptions for the ANOVA Model

For the use of one-way ANOVA to be appropriate, a number of conditions must be met:

1. All of the samples must be random and must be drawn independently from one another.
2. Each sample should be drawn from a population with an approximately normal distribution. If distributions depart only moderately from the normal, the test is reasonably robust.

3. The variances of all of the populations must be nearly equal. (Note: The different samples do not need to be of the same size. However, if the sample sizes are not equal, the test becomes more sensitive to this requirement that the samples have equal variances.)
4. All data should be at the interval or ratio level of measurement.

The test used for ANOVA is the *F*-test, which is used for comparing variances. The idea is to compare the variance *between or among* the different groups with the variance *within* the groups. Since the model assumes that all of the groups' variances are equal, the variance among groups should not differ from the variance within groups if all the population means are also equal. In other words, all groups would in effect be samples from the same population. The same sampling error would account for *all* of the variance—whether within samples, viewed individually, or among samples, if the data elements are grouped. On the other hand, if the data groups have unequal means, this difference adds to the variance among the groups, but it has no impact on the sampling error within the groups. Thus, a high *F*-ratio will suggest a significant difference among the group means.

In the model illustrated in Figure 14.2, three groups have the same variance. Although their means vary slightly, the variance among the groups (2.00^2) is equal to the variance within the groups ($\approx 2.00^2$). This accords with a null hypothesis that, in fact, all three groups have the same mean and that their apparent differences reflect the variance that they share in common. If the figure included a fourth group that had the same variance but had a mean far to the right of the others, this would increase the variance among groups compared to the unchanged variance within groups. In that case, the *F*-ratio would become large, leading to a rejection of the null hypothesis of a common mean for all groups.

Figure 14.2
Model for the Null Hypothesis for One-Way ANOVA

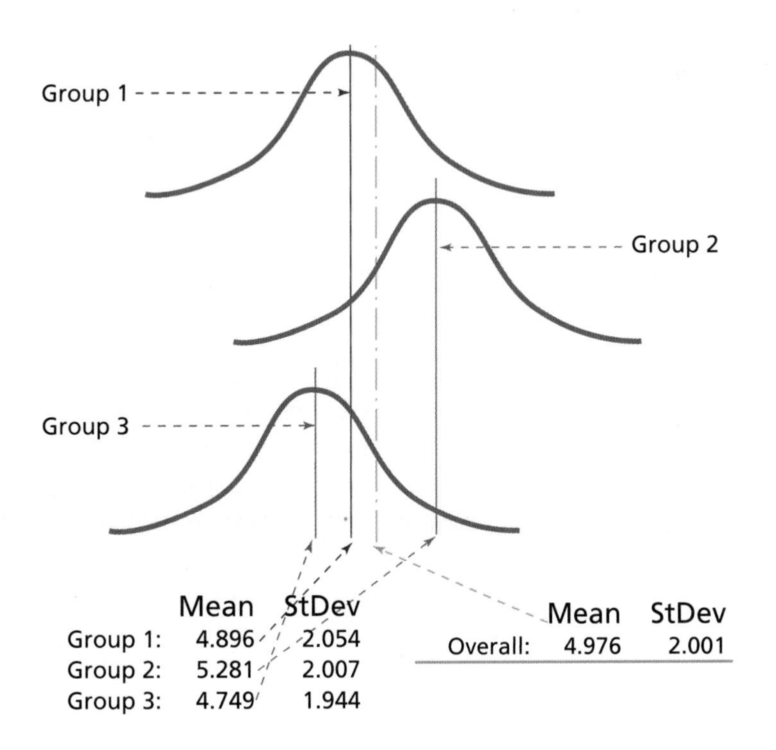

Constructing the ANOVA Table

The **ANOVA table** shows the basic elements that lead to the final calculation for the test statistic (the *F*-value) when conducting an ANOVA hypothesis test. For a critical value-based test, the test statistic can be compared to the critical *F*-value; otherwise a *p*-value can be determined that corresponds to the test statistic *F* for the given degrees of freedom.

The basic structure of an ANOVA table is shown in Figure 14.3. Note that *two* different variances are compared: (1) the variance "between groups," which is a measure of dissimilarity between the groups and (2) the variance "within groups," which is a pooled measure of the variance or noise that the groups all share in common. The test statistic *F* is the ratio calculated between the variances (1) and (2). A large *F*-value suggests that the variance between groups is larger than could be attributed to sampling error, just based on the pooled sample variance.

Figure 14.3
Basic Structure of an ANOVA Table

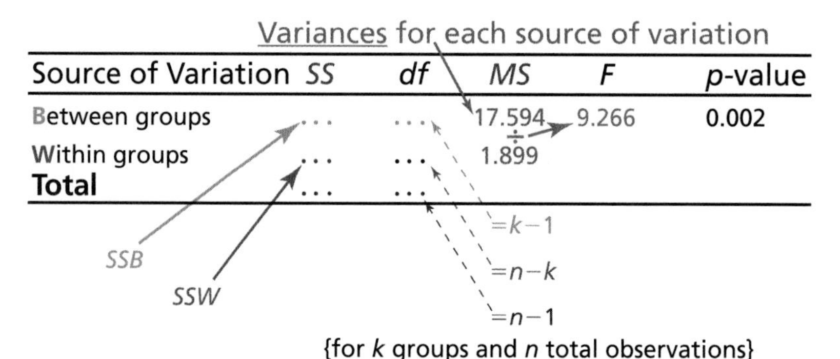

The calculations for variance that are needed to complete the table are, not surprisingly, related to the variance formula introduced in Chapter 5 (see Formula 5.4). The basic structure is:

$$\text{Variance} = \frac{(\text{Sum of the squared differences from the mean})}{(\text{Degrees of freedom})}$$

which can be expressed as

$$\textbf{Variance} = SS/df$$

where SS = Sum of the squared differences from the mean
df = Degrees of freedom

As indicated in Figure 14.3, the full table has columns to display for each row the corresponding values for *SS* and *df*, from which the variance can be directly calculated.

Sum of Squares between Groups (SSB)

The **sum of squares between groups** can be denoted as *SSB*. Other names for this same value are **sum of squares among groups** (*SSA*), *sum of squares model, sum of squares treatment,* and *sum of squares explained.* Formula 14.1 is used to calculate *SSB*.

Formula 14.1 Sum of Squares between Groups (SSB)

$$SSB = \sum_{i=1}^{k} n_i(\bar{x}_i - \bar{\bar{x}})^2$$

where k = Number of groups

Compare Formula 14.1 with Figure 14.4. A mean can be calculated for each group of raw values. These means are the values for $\bar{x}_1$, $\bar{x}_2$, up to $\bar{x}_k$. The differences that are squared are the differences, respectively, between each sample mean and the "grand" mean, $\bar{\bar{x}}$, which is calculated as the mean of all group means. Each group mean represents more than one element of raw data. So, before summing the squared differences, each group's value of $(\bar{x}_i - \bar{\bar{x}})^2$ is multiplied (or weighted) times the sample size n_i for that group.

Figure 14.4
Calculating the Sum of Squares between Groups *(SSB)*

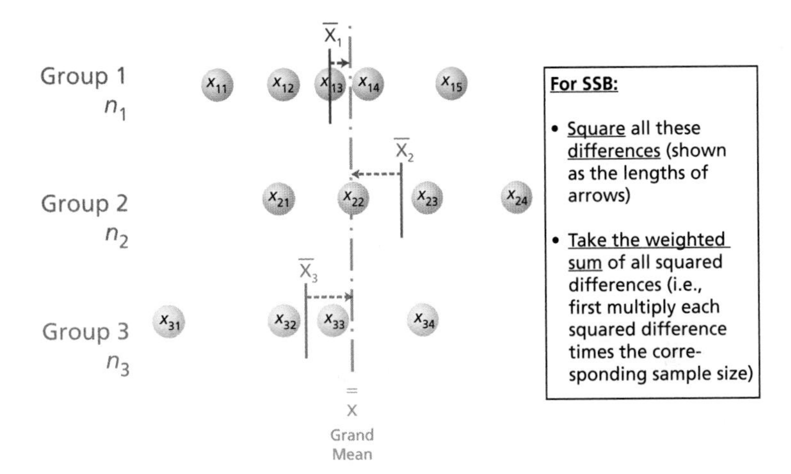

We calculate the variance for between groups as *SSB* divided by the between-groups degrees of freedom ($k - 1$ is the number of groups minus one). (To review the concept of degrees of freedom, see Chapter 10, Figures 10.9 and 10.10.)

Degrees of freedom between groups = $k - 1$

Note that in ANOVA tables, the column for displaying variances is generally labelled "*MS*" for *mean of the squared differences.*

Sum of Squares within Groups (SSW)

The **sum of squares within groups** (i.e., the estimate of common variation) can be denoted as *SSW*. Other names for this value are **sum of squares error** (***SSE***) and *sum of squares unexplained.* This reflects the amount of variance in the data set that can be expected even if the groups are not actually distinct. Formula 14.2 is used to calculate *SSW*.

Formula 14.2 Sum of Squares within Groups (SSW)

$$SSW = \sum_{j=1}^{k}\sum_{i=1}^{n_j}(x_{ij} - \bar{x}_j)^2$$

where $\bar{x}_j$ = Sample mean of group j
x_{ij} = The ith observation in group j
k = Number of groups
n_j = Number of observations in group j

Compare Formula 14.2 with Figure 14.5. There are $k = 3$ groups of raw values. For each group of raw values, a mean $\bar{x}_j$ can be determined. For each of the k groups, in turn, find the squared differences between individual raw values in the group and the particular group's mean, and then add all of the squared differences. This will result in the k sums that are represented by $\sum (x_{ij} - \bar{x}_j)^2$. Next add the k sums.

Figure 14.5
Calculating the Sum of Squares within Groups *(SSW)*

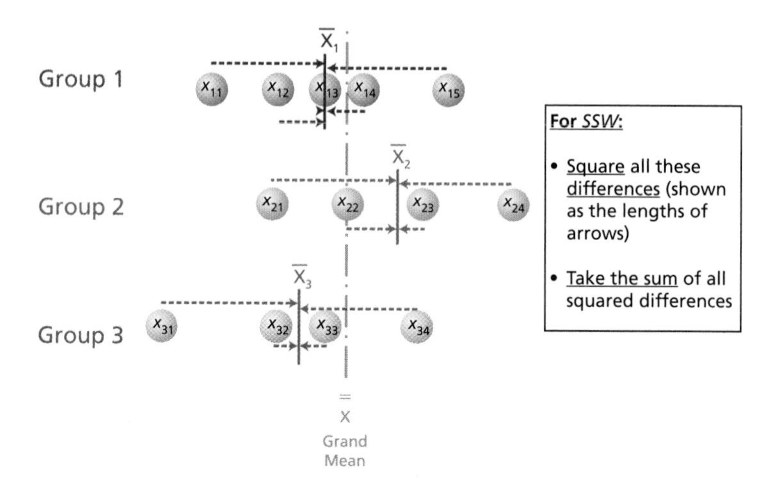

Calculate the within-group variance as *SSW* divided by the within-group degrees of freedom ($n - k$ is the overall sample size minus the number of groups).

Degrees of freedom within groups = $n - k$

The total sum of squares can now be calculated by $SSB + SSW$; $df = n - 1$. For an alternative approach, see Figure 14.6.

Figure 14.6
Calculating the Sum of Squares Total *(SST)*

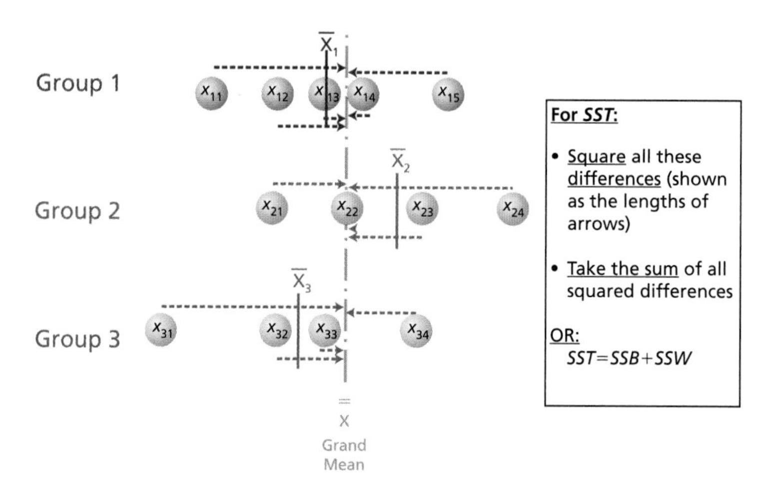

Example 14.1

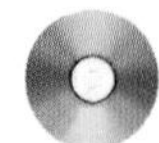

*(Based on the **Free_Agents** file)*

A bargaining representative for a free agent NHL hockey player compared samples from three groups of contracts recently obtained by NHL free agents. Groups were based on the *durations* of the contracts: Group 1—contracts for 1–2 years; Group 2—contracts for 3–4 years; Group 3—contracts for 5–7 years. Table 14.1 shows the selected values of contracts in millions of U.S. dollars (selected randomly from the file **Free_Agents**).

Table 14.1 **Selected Values of NHL Free Agent Contracts (in millions of US$)**

Group 1	Group 2	Group 3
1.800	2.500	1.800
1.250	2.000	6.000
1.250	3.700	4.500
1.400	2.625	7.500
2.100	3.500	2.800
1.400	2.800	6.500
0.463	4.850	2.660
	3.667	

Based on these data, the representative tested a claim that the mean annual salaries for free agents are not the same for the three groups. What would be the *p*-value for making the statistical decision?

Solution The test hypotheses are:

$$H_0: \mu_1 = \mu_2 = \mu_3$$

H_1: **At least one of the population means is different.**

The number of distinct observations is $n = 22$. For three groups, $k = 3$. To solve the problem, we can use formulas to fill in the structure of the ANOVA table. (Compare Figure 14.3 on page 578.) The degrees of freedom (for the *df* column) are:

$$df_{\text{between}} = k - 1 = 3 - 1 = 2$$

$$df_{\text{within}} = n - k = 22 - 3 = 19$$

$$df_{\text{total}} = n - 1 = 22 - 1 = 21$$

The means for each group and the grand mean (i.e., the mean of the means) have been calculated and are displayed in Table 14.2.

Next, using Formula 14.1:

$$SSB = \sum_{i=1}^{k} n_i(\bar{x}_i - \bar{\bar{x}})^2$$

$$= n_1(\bar{x}_1 - \bar{\bar{x}})^2 + n_2(\bar{x}_2 - \bar{\bar{x}})^2 + n_3(\bar{x}_3 - \bar{\bar{x}})^2$$

$$= 7(1.3804 - 3.0409)^2 + 8(3.2052 - 3.0409)^2 + 7(4.5371 - 3.0409)^2$$

$$= 35.187$$

This is the value to place in the ANOVA table in the position for *SSB*. The corresponding value to place in the variance column (under "*MS*") is:

$$SSB/df_{\text{between}} = 35.187/2 = 17.594$$

Table 14.2 also shows most of the calculations needed to find *SSW*. For each group, the squared differences are shown for each data value, compared to its group mean, and

Table 14.2 **Calculations to Determine *SSW***

(Note: *SSW* is the sum of the totals in the bottom row.)

				3.0409		← Grand Mean
1.3804		3.2052		4.5371		← Group Means
Group 1	$(x_{i1}$ − Group 1 Mean$)^2$	Group 2	$(x_{i2}$ − Group 2 Mean$)^2$	Group 3	$(x_{i3}$ − Group 3 Mean$)^2$	
1.800	0.1761	**2.500**	0.4973	**1.800**	7.4920	
1.250	0.0170	**2.000**	1.4525	**6.000**	2.1400	
1.250	0.0170	**3.700**	0.2448	**4.500**	0.0014	
1.400	0.0004	**2.625**	0.3366	**7.500**	8.7785	
2.100	0.5179	**3.500**	0.0869	**2.800**	3.0177	
1.400	0.0004	**2.800**	0.1642	**6.500**	3.8528	
0.463	0.8425	**4.850**	2.7053	**2.660**	3.5237	
		3.667	0.2129			
	1.5712		**5.7007**		**28.8059**	← **Sums of Squared Differences within Groups**

the sum of the squared differences for each group is displayed. The sum of these sums for all groups is *SSW*.

$$SSW = \sum_{j=1}^{k}\sum_{i=1}^{n_j}(x_{ij} - \bar{x}_j)^2$$

$$= 1.5712 + 5.7007 + 28.8059$$

$$= 36.078$$

Place this value in the ANOVA table in the position for *SSW*. The corresponding value to enter in the variance column (under "*MS*") is $SSW/df_{\mathbf{within}} = 36.078/19 = 1.899$. To complete the ANOVA table, as shown in Figure 14.7, calculate the value for *F* and the corresponding *p*-value, and input these values where shown.

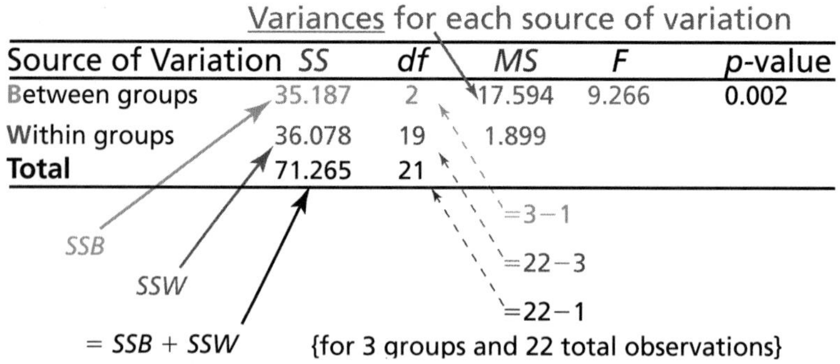

Figure 14.7
The Completed ANOVA Table

Variances for each source of variation

Source of Variation	*SS*	*df*	*MS*	*F*	*p*-value
Between groups	35.187	2	17.594	9.266	0.002
Within groups	36.078	19	1.899		
Total	71.265	21			

Source: The Globe and Mail (July 3, 2006). Based on data in a table of NHL free agent trade results, p. S1.

The *F*-value is a ratio of the variances—mean of the squared difference *between* groups (*MSB*) versus mean of the squared difference *within* groups (*MSW*). In symbols:

F-value = MSB/MSW = 17.594/1.899 ≅ 9.265

For a ratio of variances, the test distribution that can be used is the *F*-distribution—which, if compared to the normal distribution, appears skewed to the right. Although we will test an apparently *two-tailed* claim ("equal" or "not equal"), a visual representation of the test (see Figure 14.8 on page 587) appears right-tailed. This occurs because, as mentioned, the statistical test is actually for the ratio of two variances. If all group means are equal, then so are the between and within variances, so the ratio is approximately one. If the means are unequal, then the numerator (the variance for between groups) will always be greater than the denominator (the variance for within groups), so the ratio is always positive. The *F* model will always, therefore, be one-tailed for the ANOVA test.

As always, the *p*-value is a conditional probability, reflecting the relative position of the test statistic (*F*) in the null distribution (which here is the *F*-distribution):

$$p\text{-value} = P[F \geq \text{test statistic } F \mid H_0 \text{ is true (with specified degrees of freedom)}]$$

For the current example:

$$p\text{-value} = P(F \geq 9.265 \mid H_0 \text{ is true, } df_{\text{between}} = 2, df_{\text{within}} = 19) = 0.002$$

(See "In Brief 14.1(a)" for additional details.)

• *In Brief 14.1(a) One-Way ANOVA and the* F*-Distribution*

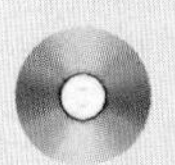

(*Based on the* ***Free_Agents*** *file*)

Outputs may differ slightly from Example 14.1, where only a subset of the data was used.

Excel

Given data in the format depicted in the spreadsheet behind the screen capture, Excel can automatically generate the full ANOVA table. Input one separate data column per group, all starting on the same row. Depending on your source data, this may require some initial cutting and pasting.

	A	B	C
1	DurGroup1	DurGroup2	DurGroup3
2	0.950	1.877	4.500
3	1.200	2.625	7.500
4	6.000	2.800	6.000
5	2.100	3.700	6.500
6	1.250	2.100	1.800
7	0.463	4.850	2.800
8	1.100	5.000	2.660
9	1.400	2.000	
10	0.700	4.000	
11	0.475	3.500	
12	1.250	2.500	
13	1.800	2.500	
14	3.100	3.667	
15	1.500	5.000	
16	3.500	2.200	
17	1.400	4.500	
18		3.025	

Anova: Single Factor
Input
Input Range: $A$1:$C$18
Grouped By: Columns / Rows
Labels in first row
Alpha: 0.05
Output options
Output Range: $E$15
New Worksheet Ply:
New Workbook
OK
Cancel
Help

Anova: Single Factor

SUMMARY

Groups	Count	Sum	Average	Variance
DurGroup1	16	28.1875	1.7617188	1.9752389
DurGroup2	17	55.843333	3.284902	1.1793873
DurGroup3	7	31.76	4.5371429	4.8009905

ANOVA

Source of Variation	SS	df	MS	F	P-value	F crit
Between Groups	42.010052	2	21.005026	10.053538	0.0003258	3.2519238
Within Groups	77.304724	37	2.0893169			
Total	119.31478	39				

When the data are ready, use the menu sequence *Tools / Data Analysis / Anova: Single Factor;* click *OK*. In the dialogue box shown in the screen capture, specify the address range containing the data. (Be sure the range includes the bottom values of the longest data set). Select *Labels in first row* if group headings are included in the first row of the range. Optionally, select *Alpha* to obtain the critical *F*-value. Finally, specify the output range and then click *OK*. An example output is included in the illustration.

If alternatively you obtain the test statistic *F* by using formulas, find the *p*-value using the cumulative distribution function for the *F*-distribution: In any blank cell of the spreadsheet, input a formula of this format: **=FDIST(F,df1,df2)** where **F** is the *F*-value, **df1** is the degrees of freedom between (i.e., for the numerator) and **df2** is the degrees of freedom within (i.e., for the denominator). This function returns a value = (1 − the less than cumulative probability corresponding to *F*) = (the area under the curve to the right of *F*), which equals the *p*-value.

Minitab

Given data in either of the two formats shown in the illustration, Minitab can automatically generate the full ANOVA table. In the first case (see the worksheet behind the first dialogue box), all values of the dependent variable are aligned in a single column. A second column denotes which factors correspond to each data value (i.e., to what group each data value belongs). Use the menu sequence *Stat / ANOVA / One Way;* click *OK*. In the first dialogue box shown in the illustration, highlight and select the *Response* variable (i.e., the dependent variable) and then highlight and select the *Factor* variable.

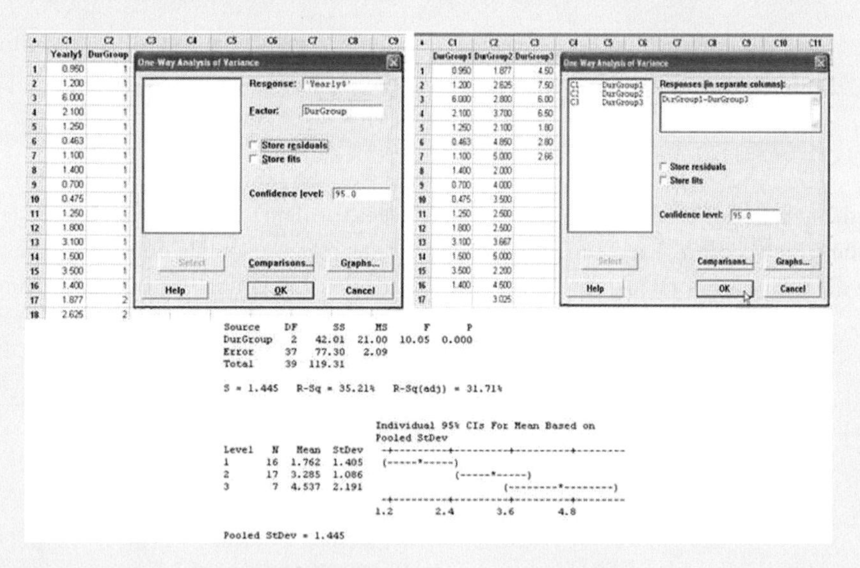

For a second approach (see the worksheet behind the second dialogue box), the data are stored using one separate data column per group. In this case, use the menu sequence *Stat / ANOVA / One Way (Unstacked);* click *OK*. In the second dialogue box, use control-clicks to highlight the names of all of the data columns at once, then choose *Select.*

For either of these two approaches, you have the option of specifying the confidence level. Then click *OK*. An example output is included in the illustration.

If alternatively you obtain the test statistic *F* by using formulas, find the *p*-value by using the menu sequence *Calc / Probability Distributions / F*. In the resulting dialogue box (not shown here), input the values for *Numerator degrees of freedom* (i.e., "between"), and for *Denominator degrees of freedom* (i.e., "within"); then click where indicated to input the constant and input the *F*-value. Click *OK*. The output will be a less than cumulative probability; subtract this result from 1 to obtain the conditional probability that $F \geq$ the test statistic, which is the *p*-value.

SPSS

Given data in the format in the spreadsheet to the left of the first dialogue box in the screen capture, SPSS can automatically generate the full ANOVA table. The second column denotes which factors correspond to each data value (i.e., to what group each data value belongs). The factors must have numerical labels ("1," "2," etc.), rather than text ones (and be sure that the variable is set to *Numeric* in the *Variable View*).

When the data are ready, use the menu sequence *Analyze / Compare Means / One-Way ANOVA;* click *OK*. In the first dialogue box, highlight and select, in turn, the names of the *Dependent* variable, and of the *Factor* variable. To also generate a display of the group means, click the *Options* button in the first dialogue box and then select *Descriptive* in the second one. Click *Continue* and then *OK*. An example output is also shown in the illustration.

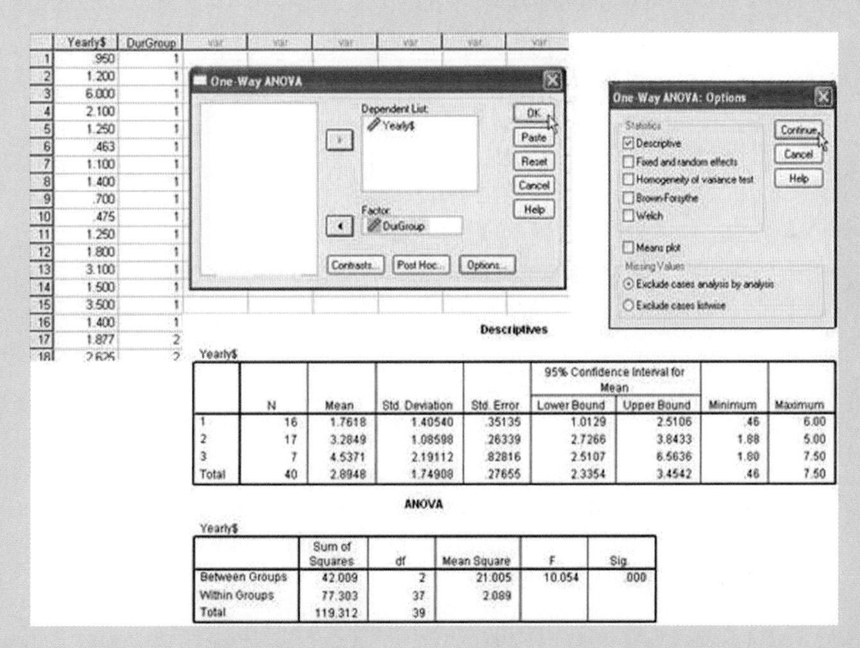

	Yearly$	DurGroup
1	.950	1
2	1.200	1
3	6.000	1
4	2.100	1
5	1.250	1
6	.463	1
7	1.100	1
8	1.400	1
9	.700	1
10	.475	1
11	1.250	1
12	1.800	1
13	3.100	1
14	1.500	1
15	3.500	1
16	1.400	1
17	1.877	2

Descriptives

Yearly$

	N	Mean	Std. Deviation	Std. Error	95% Confidence Interval for Mean		Minimum	Maximum
					Lower Bound	Upper Bound		
1	16	1.7618	1.40540	.35135	1.0129	2.5106	.46	6.00
2	17	3.2849	1.08598	.26339	2.7266	3.8433	1.88	5.00
3	7	4.5371	2.19112	.82816	2.5107	6.5636	1.80	7.50
Total	40	2.8948	1.74908	.27655	2.3354	3.4542	.46	7.50

ANOVA

Yearly$

	Sum of Squares	df	Mean Square	F	Sig.
Between Groups	42.009	2	21.005	10.054	.000
Within Groups	77.303	37	2.089		
Total	119.312	39			

If alternatively you obtain the test statistic *F* by using formulas, find the *p*-value by using the cumulative distribution function for the *F*-distribution: Place at least one value in the worksheet area and use the menu sequence *Transform / Compute*. In the resulting dialogue box (not shown here), specify a name for the target variable (i.e., column) to display the answer; then in the area for *Numeric Expression*, input a formula of this format: **1-CDF.F(F,df1,df2)** where **F** is the *F*-value, **df1** is the degrees of freedom between (i.e., for the numerator), and **df2** is the degrees of freedom within (i.e., for the denominator). Note that the CDF result (which is less than cumulative probability) is subtracted from 1 to obtain the *p*-value, which is the conditional probability that $F \geq$ the test statistic.

In Example 14.2, we use the ANOVA procedures in the context of testing a hypothesis.

Example 14.2

*(Based on the **Canadian_Lakes** file)*

While planning for a fishing contest, a promoter was wondering whether Manitoba, Ontario, or Quebec had the most suitable large lakes for hosting the event. From a list of principal lakes compiled by Statistics Canada, she downloaded a random sample (provided in the **Canadian_Lakes** file). The promoter concluded that there is no difference among the provinces Manitoba, Ontario, or Quebec in relation to the mean size of their principal lakes. Presuming that the lake size distributions within the provinces are normal, test the promoter's claim at the 0.05 level of significance.

Solution

We follow the eight-step procedure for hypothesis testing.

Step 1: The promoter claims that the mean sizes of principal lakes are the same for the provinces of interest. The claim can be expressed informally as:

All of the means are equal.

(An alternative option for this step is to use formal symbols (e.g., $\mu_1 = \mu_2 = \mu_3$) and list what provinces are coded by the subscripts.)

Step 2: Since the claim in step 1 contains equality, it represents the null hypothesis. The opposite of that hypothesis becomes H_1.

H_0: All of the means are equal.

H_1: Not all of the means are equal.

(If using formal symbols, then H_1 can be represented as:

$$H_1\text{: Not } (\mu_1 = \mu_2 = \mu_3)$$

Step 3: The level of significance is given in the problem:

$$\alpha = 0.05$$

Step 4: The sample sizes can be taken from the data. The choice of test can be determined, given that the promoter is testing for the differences in means for more than two normally distributed groups, with one factor variable (province). Therefore, use this test:

Use the one-way ANOVA test, $n_{\text{Man.}} = 4$, $n_{\text{Ont.}} = 3$, $n_{\text{Que.}} = 6$.

Step 5: The depiction of the test model can be informal (see Figure 14.8). We plan to use the *p*-value approach, so the precise critical value for *F* does not need to be calculated.

Step 6: In this example, the raw data are provided on disc, but they have to be copied and pasted for the analysis. First, select only the records for Manitoba, Ontario, and Quebec; then put the relevant data into the input format appropriate for your software.

Step 7: Whether by formulas or automated, the following data for the one-way ANOVA table can be generated:

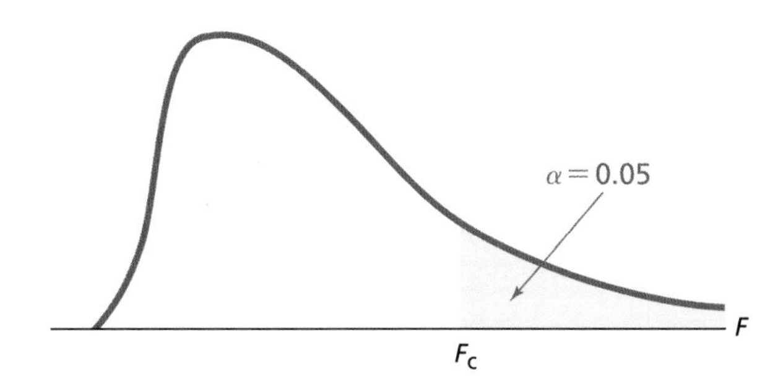

Figure 14.8
Depiction of Model for Step 5, Example 14.2

Source	DF	SS	MS	F	p
Province	2	2,699,460	1,349,730	0.81	**0.473**
Error	10	16,735,427	1,673,543		
Total	12	19,434,888			

If *F* is found by formulas, then calculate:

$$p\text{-value} = P(F \geq 0.81 \mid H_0 \text{ is true}, df_{\text{between}} = 2, df_{\text{within}} = 10) = 0.473$$

Since this *p*-value is not less than or equal to α, the statistical decision is:

Do not reject H_0.

Step 8: Relating the statistical decision back to the original claim, we conclude:

There is not sufficient evidence to reject the claim that the mean lake areas are about the same in each of the three provinces: Manitoba, Ontario, and Quebec.

One caution is that the samples are quite small, which limits the power of the test to detect any difference in means that may exist. Also, the population of lake sizes may not be normally distributed—especially if any of the Great Lakes are included. (It would be useful to check for normality in step 4 before committing to the *F*-test, but note that for small samples, checks such as the normal probability plot may be difficult to assess.)

A Note on Follow-Up Tests

If we reject the null hypothesis in an ANOVA test, we conclude that not all population means are the same. The ANOVA test itself does not specify *which* mean or means are different from the others. To take this next step, a variety of procedures, called **multiple-range tests** or **multiple-comparison tests,** have been developed. Among the options that might be available on your software are the Scheffe, Tukey, Duncan, or Dunnett tests, among others.

One quaintly named test of this sort is the Tukey Honestly Significant Difference (or Tukey HSD) test. The procedure outputs a table in which the mean differences for all possible group pairs are displayed and assessed for significance. These data, combined with a graphic portrayal of all groups' distributions, can facilitate a judgment of which groups' means are different from the others. If applied to the data of Example 14.1, the test output would appear like the upper part of Figure 14.9. Observe that the group 1 mean is significantly different from both other means (more literally: the differences in means between group 1 and group 2, and between group 1 and group 3, are both significantly different from zero); whereas the group 2 and group 3 means are not significantly different from each other. This *may* suggest that of the three groups, group 1 is the "odd man out."

Figure 14.9
Example of a Tukey HSD Multiple-Comparison Test

(I) Group	(J) Group	Mean Difference (I–J)	Std. Error	Sig.	95% Confidence Interval Lower Bound	Upper Bound
1.00	2.00	−1.82482*	0.71317	0.048	−3.6366	−0.0131
	3.00	−3.15671*	0.73656	0.001	−5.0279	−1.2855
2.00	1.00	1.82482*	0.71317	0.048	0.0131	3.6366
	3.00	−1.33189	0.71317	0.175	−3.1437	0.4799
3.00	1.00	3.15671*	0.73656	0.001	1.2855	5.0279
	2.00	1.33189	0.71317	0.175	−0.4799	3.1437

* The mean difference is significant at the 0.05 level.

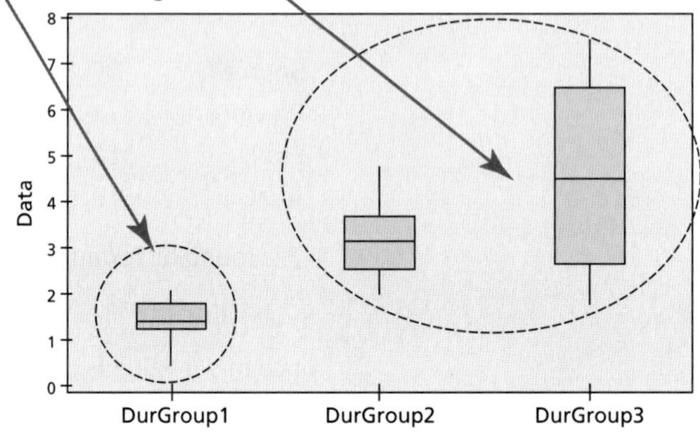

Source: "Free-Agent Frenzy: Who Went Where for What?" *The Globe and Mail*, July 3, 2006, page S1. Reprinted with permission from *The Globe and Mail*.

One caution mentioned earlier in the chapter was the heightened risk of Type I error if multiple paired tests are conducted. Tukey's and some of the other tests (but not all of them) attempt to adjust their *p*-value calculations to compensate for multiple testing. Also, such tests require the assumption of equal variances for all of the tested groups. As shown in Figure 14.9, the precaution of making boxplots can give a good sense of whether the equal variance assumption is plausible.

• *In Brief 14.1(b) Follow-Up Tests for ANOVA*

Excel

No follow-up tests are provided.

Minitab

When you are in the ANOVA dialogue box (see "In Brief 14.1(a)"), click the *Comparisons* button. For explanation of the various options, click *Help*.

SPSS

When you are in the ANOVA dialogue box (see "In Brief 14.1(a)"), click the *Post Hoc* button. For explanation of the various options, click *Help* and then click *Show Details.*

• Hands On 14.1 Mean Search Results

For many who use Internet search engines, it can be difficult to determine which of the many search results will lead to the information that is required. The author once used the word-pair **moderate mistrust** (not placed in quotes) for a search and Ask.com returned 28,400 results; Yahoo! returned 324,000 results; and Google returned 586,000 results. On the face of it, these search engines do not return the same number of results for the same queries.

Conduct a study using this basic outline:

1. Repeat the following procedure for each of three (or more) different search engines of your choice, in turn:
   - Independently generate 20 or more pairs of randomly and independently selected words from the dictionary. Do not intentionally use the same words for the different engines.
   - Conduct a search for each word-pair in turn and record the number of returns for each.
   - For cases where no matches are found, record the number of returns as zero (0).
2. Following the first step, you should now have a data set for which the dependent variable is the number of returns and the factor is the search engine used. At the 95% level of significance, test the claim that, for the different search engines, the mean numbers of search returns for the random word-pairs are not all the same. Base the test on the one-way ANOVA procedures.

What is your finding? Do you think that the required conditions for the test are sufficiently met to rely on this method? If the returns do appear to be different, which engine appears to generate the largest mean number of returns?

14.2 Randomized Block Design

The simplest of the ANOVA procedures was described in the previous section. The same basic model can be extended for more complex cases—for example, where two or more factors may impact the response. Preliminary calculations are modified for the particular context, but the ANOVA hypothesis test always hinges on the same basic *F*-test for the ratio of between-group to within-group variances.

The relationship between the one-way ANOVA just presented and the **randomized block design** in this section is analogous to the relationship between a *t*-test for difference in population means and a matched-pair *t*-test for the mean difference. In the former, the means of two *independent* samples are compared; in the latter, the samples are *dependent*. By linking related elements in the two samples, we control for some of the variation among individuals. And, just as dependent samples for the *t*-test can involve the *same* individuals (as in a before/after test), there is a variation of the block design for **one-factor repeated measures** that can involve repeated measures on the same individuals.

In randomized block design, the **blocks** are the intentionally connected observations. For example, a production manager may suspect that productivity in the operation is not equal for each of the three shifts (morning, afternoon, and night). With one-way ANOVA, she could randomly select work periods from each of the three shifts and test for the equality of mean production output per shift. But suppose that the company has several production plants and that (independent of shift) some sites are more productive than the others. If a high- or low-productivity plant happened to be overrepresented in the sample, this could bias the test results.

The above problem is addressed in the block approach, illustrated in Table 14.3. By eye, we can see that for all treatments (i.e., all shifts), plant 2 produces less output than the others. However, we sample in such a way that each block (i.e., each plant) is subjected to the same treatments and, inversely, each treatment is applied across all blocks. In that way, if the blocking factor does influence the response variable, it is not over- or underrepresented in the sample.

Table 14.3 Example of a Randomized Block Design

Response Variable = Units Output for the Shift		Production Plant		
		1	2	3
Work Shift	Morning	39	27	33
	Afternoon	30	22	25
	Night	34	21	27

For test purposes, the numeric values associated with each treatment are interpreted as a sample from a separate population. As for one-way ANOVA, the null hypothesis is that all of the population means are equal; the alternative hypothesis is that at least one of the population means is different from the others. Semiformally:

H_0: All of the treatment means are equal.

H_1: At least one of the treatment means is different.

Valid use of the model requires these conditions:

1. The observations for each treatment are independent of the observations for the other treatments. The observations are, ideally, taken from a randomized experiment; use caution if applying the method to secondary data.
2. All of the potentially observed values for each block/treatment combination are normally distributed.
3. The variances of the distributions [see (2)] are approximately equal.
4. All data should be at the interval- or ratio-level of measurement.

Constructing the ANOVA Table

Just as in Section 14.1, the test used for this ANOVA is the F-test, and the basic principles and formulas are the same. But because we have introduced a new source of variation (the blocks), some calculations for degrees of freedom are revised. The basic ANOVA table structure for randomized block design is shown in Figure 14.10.

For testing the equality of means for the treatment groups, we compare the variances (in the "MS" column) for (1) "between treatments" and (2) the "error" that mea-

Figure 14.10
Basic Structure for an ANOVA Table for a Randomized Block Design

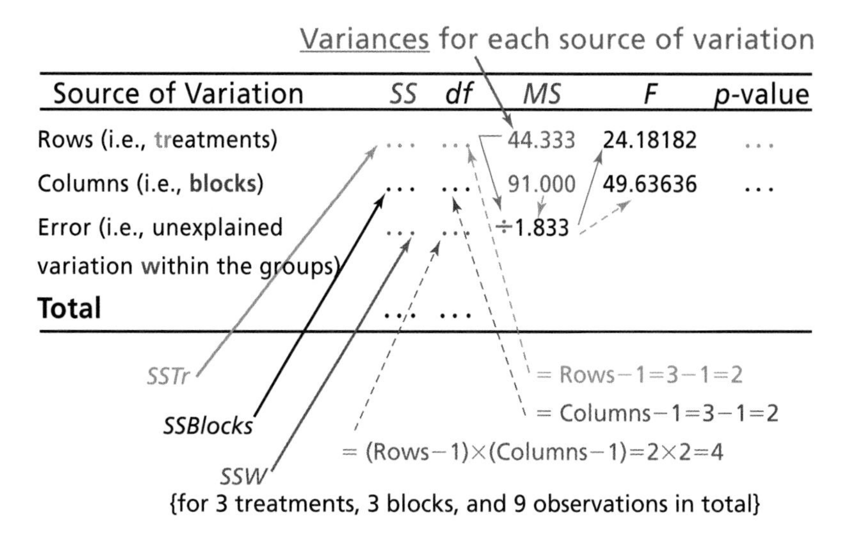

sures the unexplained sample variance. As before, this ratio is the *F*-statistic, and a large *F* suggests that the variance between groups is larger than could be attributed to sampling error; hence, the *p*-value would be small. Also as in a one-factor ANOVA, the needed calculations for variance to complete the table are related to the basic structure of the variance formula:

$$\text{Variance} = \frac{(\text{Sum of the squared differences from the mean})}{(\text{Degrees of freedom})} = \frac{SS}{df}$$

In the completed table (Figure 14.11, on page 593), the necessary values for *SS* and *df* will be displayed; from these the required variances can be calculated and from these the *F*-value.

Sum of Squares between Treatments (SSTr) *and between Blocks* (SSBlocks)

The **sum of squares between treatments** can be denoted as ***SSTr***. It could also be called SSB, as in one-way ANOVA. (See Formula 14.3.)

Formula 14.3 Sum of Squares between Treatments (SSTr)

$$SSTr = \sum_{i=1}^{k} b(\bar{x}_i - \bar{\bar{x}})^2 = b\sum_{i=1}^{k} (\bar{x}_i - \bar{\bar{x}})^2$$

where k = Number of treatment groups
b = Constant number of blocks

Formula 14.3 is really a special case of Formula 14.1, which finds *SSB*. In the case of randomized blocks, each treatment group has one observation per block—so the number of observations per group (the n_i term in Formula 14.1) is here simply equal to the constant number of blocks (b). To calculate the variance for between treatments, divide *SSTr* by the between-treatments degrees of freedom:

Degrees of freedom between treatments equals $k - 1$

where
$k - 1$ = Number of treatment groups minus one

To complete the second row of the ANOVA table for randomized block design, the **sum of squares blocks** (*SSBlocks*) must be calculated. This is a measure for the amount of variation explained by distinguishing groups of data based on the blocks to which treatments were applied, and can be calculated analogously to *SSTr*, but from the viewpoint of the blocks (i.e., column data instead of row data). We could adapt Formula 14.3, with *k* becoming the number of blocks and *b* being replaced with the constant number of treatments. Degrees of freedom for the blocks is the number of blocks minus one.

Sum of Squares within Groups (SSW)

As for one-way ANOVA, the sum of squares within groups (*SSW*) or "Error" is an estimate of common variation for the sample. *SSW*, in this case, equals the total sum of squares minus the combined sum of squares for both treatments and blocks. Its degrees of freedom is based on (Number of treatments − 1) × (Number of blocks − 1). Again, the corresponding variance term MS_{error} is the sum of squares for error divided by the degrees of freedom for error; this is the within-group variance.

F-values can be calculated for both the treatments and the blocks. The former is the ratio between the between-group (i.e., between treatment) variance and the within-group (i.e., Error) variance; the *F*-value for blocks compares the block variance with the error. In each case, the *p*-value is based on the probability of obtaining an *F* of at least that magnitude, given the null hypothesis and the numbers of treatments and blocks.

Example 14.3

Having compiled Table 14.3 on page 590, a production manager claims that the mean number of units output per shift are not the same. (1) Manually determine the values of *SS* and *MS* for the treatment groups. (Interpret the three production plants as the blocks.) (2) What would be the *p*-value for making the statistical decision?

Solution

1. The means for each treatment can be calculated as $\bar{x}_{morning} = 33.0000$, $\bar{x}_{afternoon} = 25.6667$, and $\bar{x}_{night} = 27.3333$; the mean of the means (i.e., the grand mean) = 28.6667. From Formula 14.3:

$$SSTr = b\sum_{i=1}^{k}(\bar{x}_i - \bar{\bar{x}})^2$$

$$= b \times [(\bar{x}_1 - \bar{\bar{x}})^2 + (\bar{x}_2 - \bar{\bar{x}})^2 + (\bar{x}_3 - \bar{\bar{x}})^2]$$

$$= 3 \times [(33.0000 - 28.6667)^2 + (25.6667 - 28.6667)^2 + (27.3333 - 28.6667)^2]$$

$$= 88.666$$

   Degrees of freedom for the treatments = (3 treatments) – 1 = 2. Therefore, the variance (*MSTr*) = 88.666/2 = 44.333.

2. Using the procedures provided in "In Brief 14.2," we can next complete the remainder of the ANOVA table, as shown in Figure 14.11. Confirm from the table that $MS_{error} = SS_{error}/df_{error} = 7.333/4 = 1.833$, and that the *F*-value for treatments $= MS_{treatments}/MS_{error} = 44.333/1.833 = 24.182$.

$$p\text{-value} = P(F \geq 24.182 \mid H_0 \text{ is true}, df_{numerator} = 2 \text{ and } df_{denominator} = 4) = 0.006$$

Figure 14.11
Completed ANOVA Table for a Randomized Block Design

	Source of Variation	SS	df	MS	F	p-value
(i.e., treatments) →	Rows	88.667	2	44.333	24.182	0.006
(i.e., blocks) →	Columns	182.000	2	91.000	49.636	0.002
(i.e., unexplained variation within the groups) →	Error	7.333	4	1.833		
	Total	278.000	8			

$df_{\text{treatments}} = \text{Rows} - 1$

$df_{\text{blocks}} = \text{Columns} - 1$

$df_{\text{error}} = (\text{Rows} - 1) \times (\text{Columns} - 1)$

$df_{\text{total}} = n - 1$

• *In Brief 14.2 Two-Way ANOVA*

Unfortunately, software packages do not all agree on the terminology for the procedures just described. They also bundle the procedures in different ways with other features that they offer. The focus in the following is strictly on how to tweak the packages to accomplish the procedures in this section.

Excel

The data must be in a table format, with rows for the treatments and columns for the blocks, as shown in the spreadsheet behind the screen capture.

When the data are ready, use the menu sequence *Tools / Data Analysis / Anova: Two Factor without Replication;* click *OK*. (The terminology refers to the

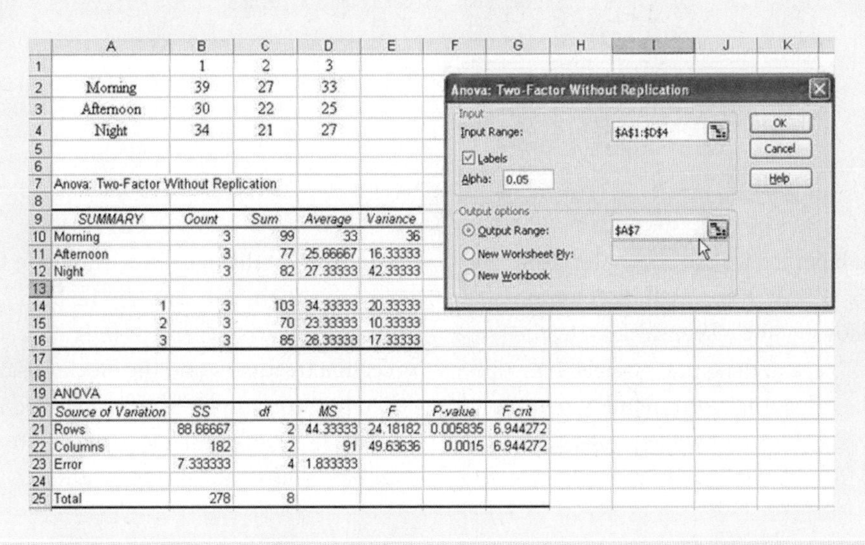

	A	B	C	D	E	F	G
1		1	2	3			
2	Morning	39	27	33			
3	Afternoon	30	22	25			
4	Night	34	21	27			
5							
6							
7	Anova: Two-Factor Without Replication						
8							
9	SUMMARY	Count	Sum	Average	Variance		
10	Morning	3	99	33	36		
11	Afternoon	3	77	25.66667	16.33333		
12	Night	3	82	27.33333	42.33333		
13							
14	1	3	103	34.33333	20.33333		
15	2	3	70	23.33333	10.33333		
16	3	3	85	28.33333	17.33333		
17							
18							
19	ANOVA						
20	Source of Variation	SS	df	MS	F	P-value	F crit
21	Rows	88.66667	2	44.33333	24.18182	0.005835	6.944272
22	Columns	182	2	91	49.63636	0.0015	6.944272
23	Error	7.333333	4	1.833333			
24							
25	Total	278	8				

fact that for each combination of row and column variable, there is only one data element.) In the resulting dialogue box, specify the address range containing the data. You have the option to include headings in the address range; in that case, select *Labels*. Specify the upper left-hand cell for the output range and then click *OK*. An example output is also shown in the illustration.

Minitab

The data must be in the format shown behind the screen capture. Namely, use a separate column for each of the following: the response variable, the treatment variable, and the block variable. Line up, in the rows, the values that correspond. Use the menu sequence *Stat / ANOVA / Two Way;* click *OK*.

In the dialogue box, highlight and select the names of each of each of the following: the response variable (i.e., the dependent variable), the row factor (i.e., treatment) variable, and the column factor (i.e., block) variable. Click *OK*. An example output is also included in the illustration.

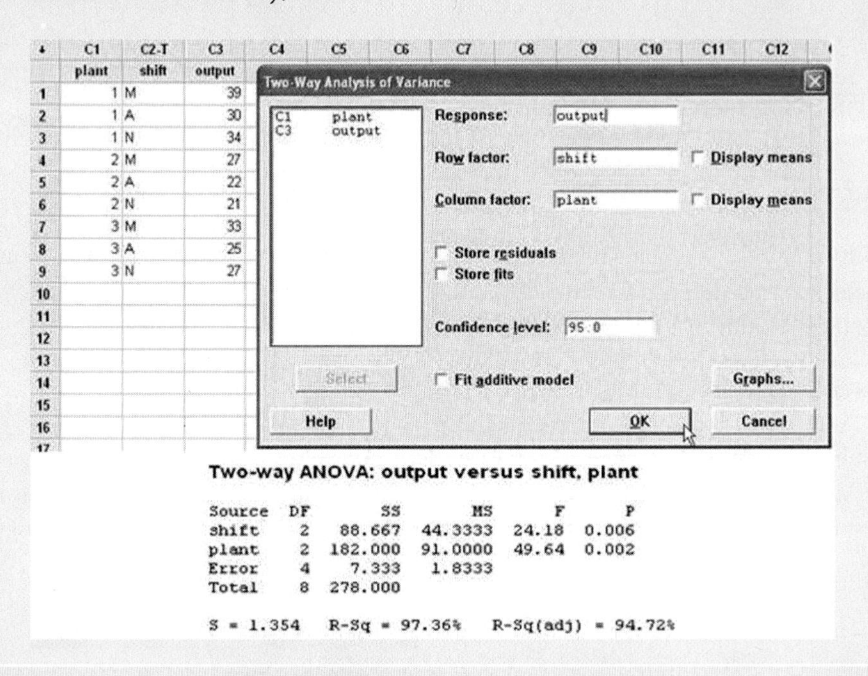

SPSS

The data must be in the format shown behind the screen capture. Namely, use a separate column for each of the following: the response variable, the treatment variable, and the block variable. Line up, in the rows, the values that correspond. When the data are ready, use the menu sequence *Analyze / General Linear Model / Univariate;* click *OK*.

In the first dialogue box (shown in the illustration), highlight and select, in turn, the names of the dependent variable, and of the two factor variables (i.e., the treatment and the block variables). Note that, in the example, "shift" is selected as *Fixed Factor,* since *all* of the company's shifts appear in the data set. "Plant" would be a *Random Factor* if,

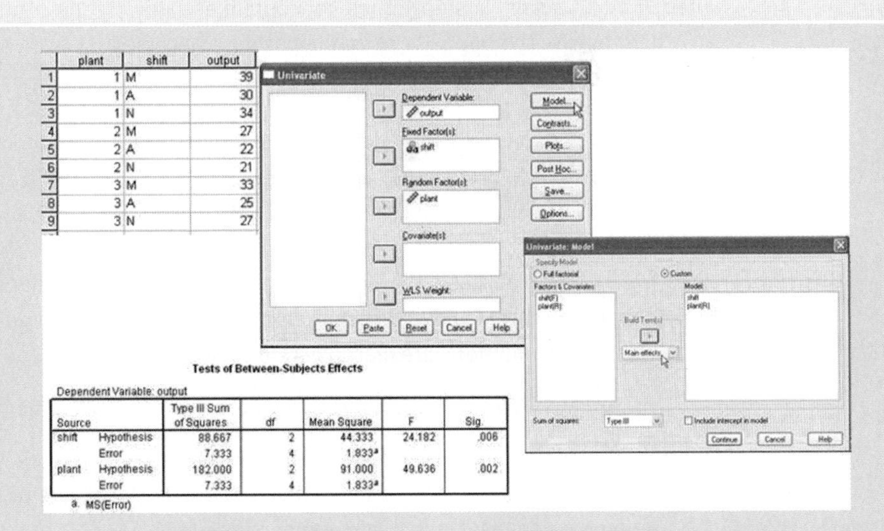

Tests of Between-Subjects Effects

Dependent Variable: output

Source		Type III Sum of Squares	df	Mean Square	F	Sig.
shift	Hypothesis	88.667	2	44.333	24.182	.006
	Error	7.333	4	1.833[a]		
plant	Hypothesis	182.000	2	91.000	49.636	.002
	Error	7.333	4	1.833[a]		

a. MS(Error)

for example, the company has additional plants, and only a random selection of the plants is included in the sample.

Now click the *Model* button and, in the resulting dialogue box (the second screen capture in the illustration), select *Custom.* To move the names of the two factor variables into the *Model* box, highlight their names and then use the drop-down menu in the centre of the dialogue box to choose *Main effects;* then click the arrow under *Build Term(s).* (If necessary, click on the box to deselect the option *Include intercept in model.*) Click on *Continue* and then *OK.* An example of the relevant output is also included in the illustration.

Example 14.4

In the context of studying soil quality, scientists J. Chaudhury, U. Mandal, and others tested plant yields in response to a variety of growth treatments.[1] The response variable shown in the table below is an indexed measure of yield as described in their paper:

Sustainable Yield Index for Crops

Treatment	Jute	Rice	Wheat
1	0.479	0.379	0.400
2	0.528	0.433	0.478
3	0.641	0.439	0.529
4	0.662	0.538	0.537

Suppose another colleague is studying whether all of these treatments have the same effect on crop yields. Because different crop types respond differently to the same treatments, a block design uses the several crop types as the blocks. Test a claim that the effect on crop yield is *not* the same for all four types of treatment. Use the 0.05 level of significance.

Solution

Step 1: The claim can be expressed informally as:

Not all of the treatment means are equal.

Step 2: As the claim in step 1 does not contain equality, it represents the alternative hypothesis. The opposite of that hypothesis becomes H_0.

H_0: All the treatment means are equal.

H_1: Not all the treatment means are equal.

Step 3: The level of significance is specified in the problem:

$$\alpha = 0.05$$

Step 4: The sample sizes can be taken from the data. The description of the experiment is consistent with a randomized block design (provided normality and equal-variance assumptions are met). Therefore, use this test:

Use the ANOVA test, for randomized block design, *4 treatments*, *3 blocks*

Step 5: The depiction of the test model can be informal (see Figure 14.12), since we plan to use the computer to find a *p*-value. Due to the nature of the *F*-test for comparing the variances, the model appears one-tailed.

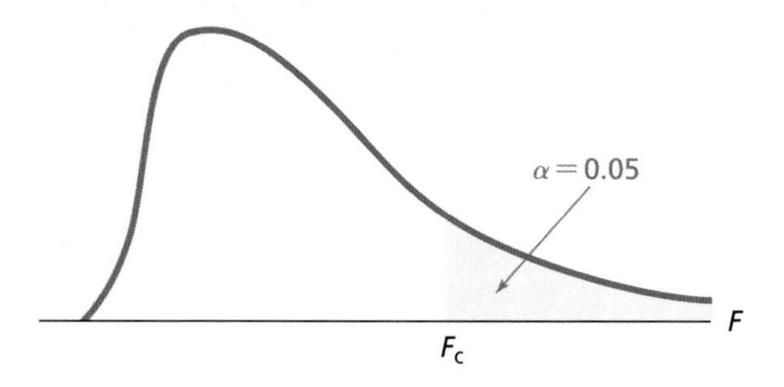

Figure 14.12
Depiction of Model for Step 5, Example 14.4

Step 6: The data are provided in the question. Depending on your software, they may need to be copied and pasted for computer input.

Step 7: Using the computer as discussed in "In Brief 14.2," perform the ANOVA test to obtain an output such as that shown in Table 14.4.

Table 14.4 **Two-Way ANOVA Table**

Source of Variation	*SS*	*df*	*MS*	*F*	*p*-value
Rows	0.043290917	3	0.01443031	16.94748	0.002469
Columns	0.035785167	2	0.01789258	21.0137	0.00195
Error	0.005108833	6	0.00085147		
Total	0.084184917	11			

$$p\text{-value (for treatment)} = P(F \geq 16.95 \mid H_0 \text{ is true}, df_{\text{numerator}} = 3, df_{\text{denominator}} = 6) = 0.0025$$

Since this *p*-value is less than α, the statistical decision is:

Do reject H_0.

Step 8: Relating the statistical decision back to the original claim, we conclude:

The evidence supports the claim that the mean crop yields are different for the different growth treatments.

One-Factor Repeated Measures

As indicated earlier, the model of one-factor repeated measures is closely related to the randomized block design. In the latter, the treatments are applied to blocks or groups of related or similar individuals. In one-factor repeated measures, *all* treatments are applied to all of *the same* individuals in the study. For example, in a study on three varieties of poison ivy, each of five guinea pigs may be exposed to all three varieties to see if all types produce the same level of irritation. By using the same individuals (assuming this study is ethically approved), we minimize variation due to different, individual sensitivities.

This design also has some disadvantages. For example, is there an extraneous effect of subjecting the same individuals to all of the repeated measurements? Also if one individual is unusually sensitive, or misrepresentative, this could have a large impact on the results.

Example 14.5

Using their computers, baseball announcers provide timely conditional statistics for almost every new context in the game. They can tell you the current hitter's batting average early in the game, late in the game, against lefties, against righties, and so on. A certain coach disputes whether the statistics are meaningful for making decisions. For the 2006 season, up to late August, he compiled the following batting-average statistics for a selection of Toronto Blue Jays players.[2]

	Overbay	Catalanotto	Zaun	McDonald
Bases empty	0.312	0.257	0.252	0.212
Runner on 1st base	0.323	0.415	0.368	0.283
Runner on 2nd base	0.276	0.361	0.167	0.200
Runner on 3rd base	0.231	0.222	0.500	0.333

Source: Toronto Blue Jays website. Found at: http://toronto.bluejays.mlb.com/NASApp/mlb/stats/sortable_player_stats.jsp?c_id=tor. Accessed August 30, 2006.

Test a claim that the mean batting averages do not differ based on the hitting contexts (i.e., the data about base runners). Use the 0.05 level of significance. How could the coach use this information?

Solution

Step 1: The claim can be expressed informally as:

All of the treatment means are equal.

Step 2: As the claim in step 1 does contain equality, it represents the null hypothesis. The opposite of that hypothesis becomes H_1.

H_0: All of the treatment means are equal.

H_1: Not all of the treatment means are equal.

Step 3: The level of significance is specified in the problem:

$$\alpha = 0.05$$

Step 4: The treatments in the experiment are the different hitting contexts, and all treatments were applied to each of the four selected batters. This experimental design is consistent with one-way repeated measures procedures.

Use the ANOVA test, for one-way repeated measures, 4 treatments, 4 repetitions.

Step 5: Depict the test model informally (see Figure 14.13); we plan to use the *p*-value approach.

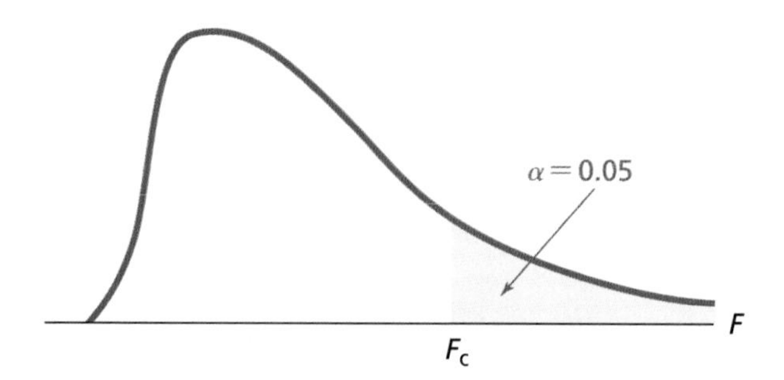

Figure 14.13
Depiction of Model for Step 5, Example 14.5

Step 6: The data are provided in the question. Depending on your software, they may need to be copied and pasted for computer input.

Step 7: Using the same computer procedures as for the randomized block design (see "In Brief 14.2"), generate an ANOVA table similar to Table 14.5.

Table 14.5 **ANOVA Table for One-Way Repeated Measures**

Source of Variation	*SS*	*df*	*MS*	*F*	*p*-value
Rows	0.026872	3	0.008957	1.048698	0.417339
Columns	0.010402	3	0.003467	0.405933	0.752485
Error	0.076871	9	0.008541		
Total	0.114144	15			

The ***p*-value** for the treatments (rows), i.e., the base running contexts is:

$$P(F \geq 1.0487 \mid H_0 \text{ is true}, df_{\text{numerator}} = 3, df_{\text{denominator}} = 9) = 0.417$$

Since this *p*-value is not less than or equal to α, the statistical decision is:

Do not reject H_0.

Step 8: Relating the statistical decision back to the original claim, we conclude:

There is insufficient evidence to reject the claim that the mean batting averages are the same regardless of the listed batting contexts.

Using this information, the coach could, for example, resist the temptation to change the scheduled batter, based solely on whether there are runners on a particular base.

14.3 Two-Factor ANOVA

In one-way ANOVA, only one variable is used as a grouping variable or factor. With **two-factor ANOVA** (also called **two-way ANOVA),** it is possible to examine the effects on a response variable of two factors. Each factor is a variable that has a number of possible values, called **levels.**

In two-factor analysis, we can pose and test a number of hypotheses. We generally test first the null hypothesis that the two factors do not interact. The effects of each factor, individually, on the response variable are called the **main effects.** If the two factors do *not* interact, we can proceed to test the significance of the main effects. If the F-value for an interaction or a main effect is relatively large, we say that the interaction or single factor accounts for or explains much of the variance among values in the dependent variable.

Some of the total variance, as mentioned, might be explained by the **interaction effects** of the two factors. For example, compare the response variable "the amount of steam rising from an electric kettle after two minutes" with two factors: the power setting of the kettle (from zero to full power, in discrete steps) and the quantity of water in the kettle (from zero to full volume, in discrete steps). The power setting alone may not explain the amount of steam after two minutes; there may be no water in the kettle or too much water to heat quickly. But the amount of water in the kettle may also not explain, by itself, the amount of steam after two minutes—since the power may be off or set to very low. So, in this case, the variance in the observed amount of steam can be explained largely by the *interaction* of the "power setting" factor *and* the "water level" factor. Any portion of variance still left unexplained is the error; this is the total error minus *all* factor-explained variance—both the main effects and the interactions.

The structure of the ANOVA table is quite similar to the previous sections. As in Section 14.1, the test basis is the F-test, but having introduced new sources of variation (a second factor and the interaction between factors), the calculations for degrees of freedom are revised. For each row in the table, the variance ("MS") = SS/df for that row. Formulas to find the degrees of freedom are given in Figure 14.14. An F-value can be calculated for each factor and also for the factor interaction. In each case, the F-value for a factor (or interaction) equals the ratio: $MS_{\text{(for the factor or interaction)}}/MS_{\text{error}}$. Once the F-value is known, a p-value can be calculated for F, given the degrees of freedom in the numerator and the denominator.

Figure 14.14
Basic Structure of a Two-Factor ANOVA Table

Variances for each source of variation

Source	df	SS	MS	F	p
Factor 1 (rows)	= Rows − 1	. . .	49.5	0.48	0.627
Factor 2 (columns)	= Columns − 1	. . .	261.6	2.54	0.129
Interaction of the factors	= (Rows − 1) × (Columns − 1)	. . .	3.3	0.03	0.969
Error	$= df_{total} - \Sigma(df_1, df_2, df_{interaction})$	. . .	103.1		
Total	$= n - 1$				

Source: Chicken Farmers of Canada. *Data Handbook 2004*. Found at: http://www.chicken.ca/app/DocRepository/1/Data_Handbook/2004_Data_Booklet.pdf

Valid use of the model requires these conditions:

1. The observations for each treatment are independent of the observations for the other treatments. The observations are, ideally, taken from a randomized experiment; use caution if applying the method to secondary data.
2. All of the potentially observed values for each combination of factor values are normally distributed.
3. The variances of the distributions [see (2)] are approximately equal.
4. All data should be at the interval or ratio level of measurement.

Example 14.6

A distributor of broiler chickens for supermarkets has been comparing prices for this product across Canada. In the table below, prices are compared by national region (east and west) and also by an indicator of local product supply (i.e., high, medium, and low annual production numbers in the specific region of sale). The numbers in the cells are simulated samples of individual transaction prices for broiler chickens in cents per kilogram.

	High Production Level				Medium Production Level				Low Production Level			
West	132.3	122.7	112.3	90.1	119.3	117.5	116.6	116.9	119.6	125.2	117.4	120.1
East	118.0	120.6	127.9	121.7	102.6	130.5	140.1	124.8	123.4	125.3	129.8	124.7

The distributor is wondering whether the variance in transaction prices can be explained by the interaction between national region of sale and the local production level, or if not, by either of those individual variables. Using an ANOVA-based test, provide some answers for the distributor.

Solution

Step 1: The distributor's open question can be restated in terms of the following claims:

a) **The interaction of the two factors does not account for a significant amount of the variance among the means.**

(Note: For convenience, we will write out the three tests in parallel. In practice, you would test claim 1 for the interaction first and proceed to the other tests only if claim 1 is not rejected; i.e., if there is not an interaction.)

b) **All means, for groupings based on national regions, are equal.**

c) **All means, for groupings based on local production levels, are equal.**

Step 2: All of the claims in step 1 contain equality, so they represent the null hypotheses.

a) **For the *interaction of factors:*** **H_0: The impact on total variance is not significant.**

H_1: The impact on total variance is significant.

b) **For the *regions* factor:** **H_0: All of the group means are equal.**

H_1: Not all of the group means are equal.

c) **For the *production levels:*** **H_0: All of the group means are equal.**

H_1: Not all of the group means are equal.

Step 3: The level of significance is not specified in the problem. We can use:

$$\alpha = 0.05$$

Step 4: The description of the problem is consistent with using a two-factor ANOVA test (provided the normality and equal-variance assumptions are met):

Use the two-factor ANOVA test

Step 5: For each test, in turn, the same model will apply (see Figure 14.15). We will use the p-value approach.

Figure 14.15
Depiction of Model for Step 5, Exercise 14.6

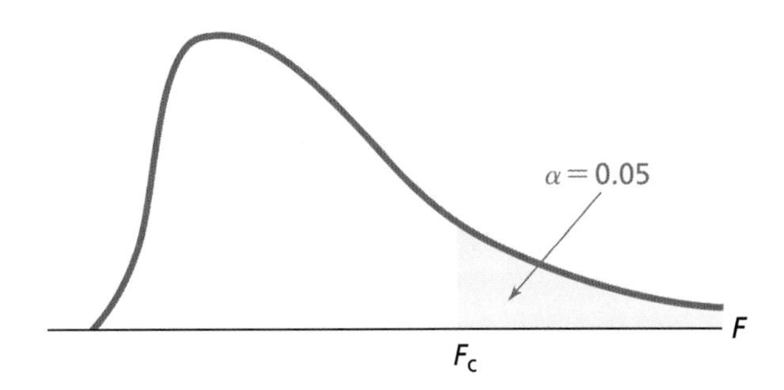

Step 6: The data are provided in the question. Your software will determine how they need to be copied and pasted for computer input. A single ANOVA table will be output by the computer, containing data relevant to all of the three tests.

Step 7: Using the computer as discussed in "In Brief 14.3," perform the two-factor ANOVA test and output a table similar to Table 14.6.

Table 14.6 **Two-Factor ANOVA Table**

Source of Variation	*SS*	*df*	*MS*	*F*	*p*-value
Rows (regions)	261.6201	1	261.6201	2.539372	0.128447
Columns (prod'n levels)	98.99495	2	49.49748	0.480439	0.626219
Interaction	6.54574	2	3.27287	0.031768	0.968786
Error ("within")	1854.46	18	103.0255		
Total	2221.62	23			

a) For the interaction of factors:

$p\text{-value} = P(F \geq 0.032 \mid H_0 \text{ is true}, df_{\text{numerator}} = 2, df_{\text{denominator}} = 18) = 0.969$

Since this *p*-value is not less than or equal to α, the statistical decision is:

Do not reject H_0.

(Since there appears to be no interaction, you can now proceed to test the main effects.)

b) For the *national regions* factor:

$p\text{-value} = P(F \geq 2.539 \mid H_0 \text{ is true}, df_{\text{numerator}} = 1, df_{\text{denominator}} = 18) = 0.128$

Since this *p*-value is not less than or equal to α, the statistical decision is:

Do not reject H_0.

c) For the local production levels factor:

$p\text{-value} = P(F \geq 0.480 \mid H_0 \text{ is true}, df_{\text{numerator}} = 2, df_{\text{denominator}} = 18) = 0.626$

Since this *p*-value is not less than or equal to α, the statistical decision is:

Do not reject H_0.

Step 8: Relating the statistical decisions back to the original, multipart claim, we conclude:

> **There is insufficient evidence to reject claims that variance in the transaction prices for broiler chickens cannot be explained by either the interaction of region of sale and local production volume, or by either of those individual factors.**

• *In Brief 14.3 Two-Factor ANOVA*

Once again, the software packages do not agree on terminology for these procedures, and they bundle the procedures in different ways with their other features.

The focus below is strictly on how to adapt the packages to accomplish the procedures in this section.

Excel

The data must be in a table format, with rows for one factor and columns for the other, and include labels as shown in the spreadsheet behind the screen capture. Notice how the multiple data values for each factor combination are input in sequence down the appropriate column. Row labels appear only on the rows for the first data value of each given factor combination. The software can handle only balanced designs with a constant number of data values per factor combination.

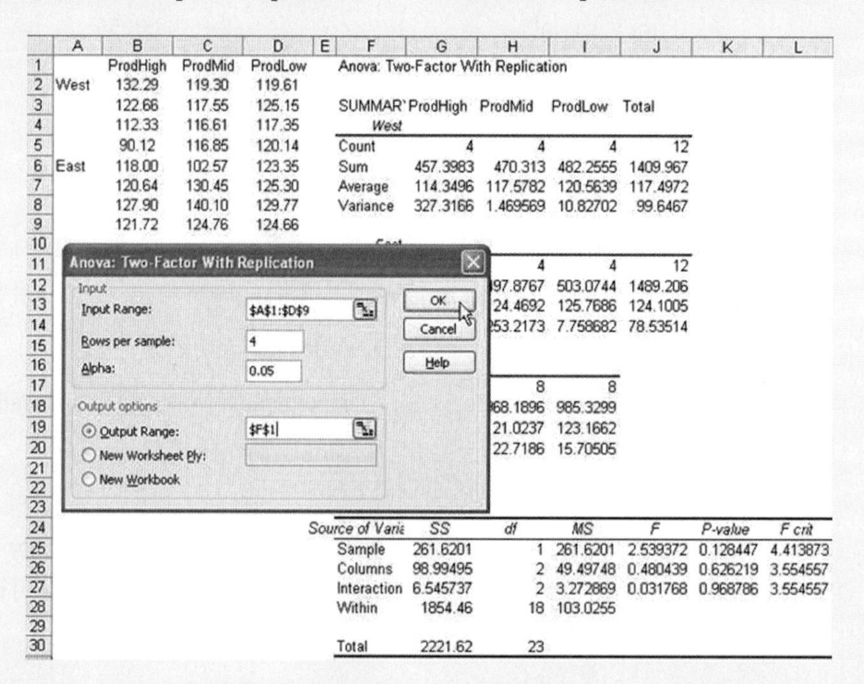

When the data are ready, use the menu sequence *Tools / Data Analysis / Anova: Two-Factor With Replication;* click *OK*. (The terminology refers to the fact that more than one data element exists for each combination of row and column variable.) In the resulting dialogue box (see the screen capture), specify the address range containing the data, including labels. In *Rows per sample*, input the number of data values for each factor combination. Select *Output Range* and input the cell address for what will become the upper left cell for the displayed output, and then click *OK*. An example output is also included in the illustration.

Minitab

The data must be in the format depicted in the three columns at the left in the illustration: Use a separate column for each of the two factor variables and for the response variable and line up, in the rows, the values that correspond. The software can handle only balanced designs with a constant number of data values per factor combination.

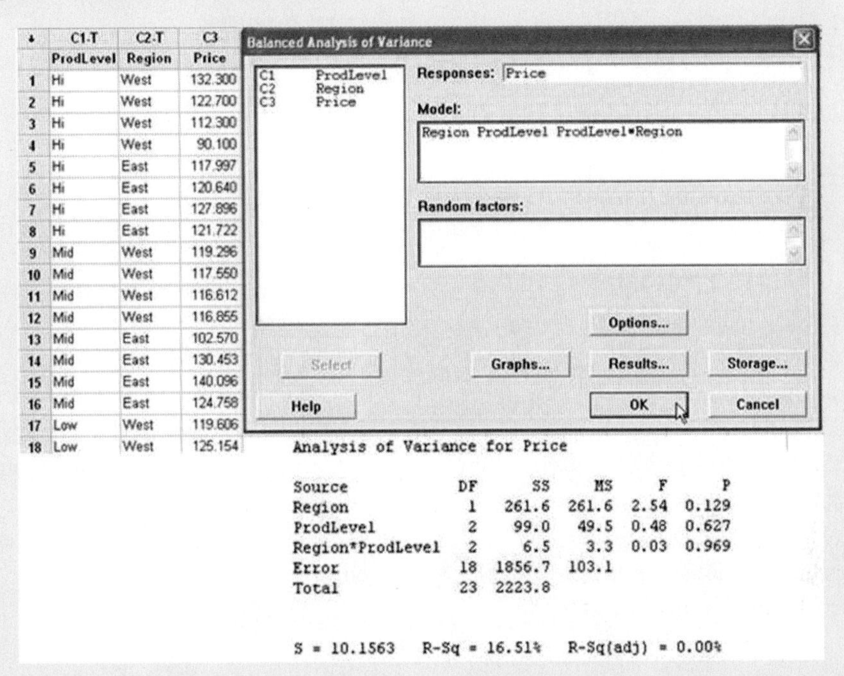

Use the menu sequence *Stat / ANOVA / Balanced ANOVA;* click *OK*. In the resulting dialogue box (see the screen capture), highlight and select the name of the response variable (i.e., the dependent variable). Then highlight and select into the *Model* area each of the two factor variables' names (for example: "Region" and "ProdLevel"). For interactions, select into the *Model* area the names of the individual factors to be combined, and then link those names with an asterisk symbol (in the example: "ProdLevel*Region.") Click *OK*. An example output is also included in the illustration.

An alternative approach in Minitab (not shown here) is to begin with the menu sequence *Stat / ANOVA / Two-way*. Minitab will see that there are replications for each factor combination and will, by default, produce a two-way ANOVA table that includes the interaction component.

SPSS

The data must be in the format depicted in the spreadsheet behind the screen capture: Use a separate column for each of the two factor variables and for the response variable and line up, in the rows,

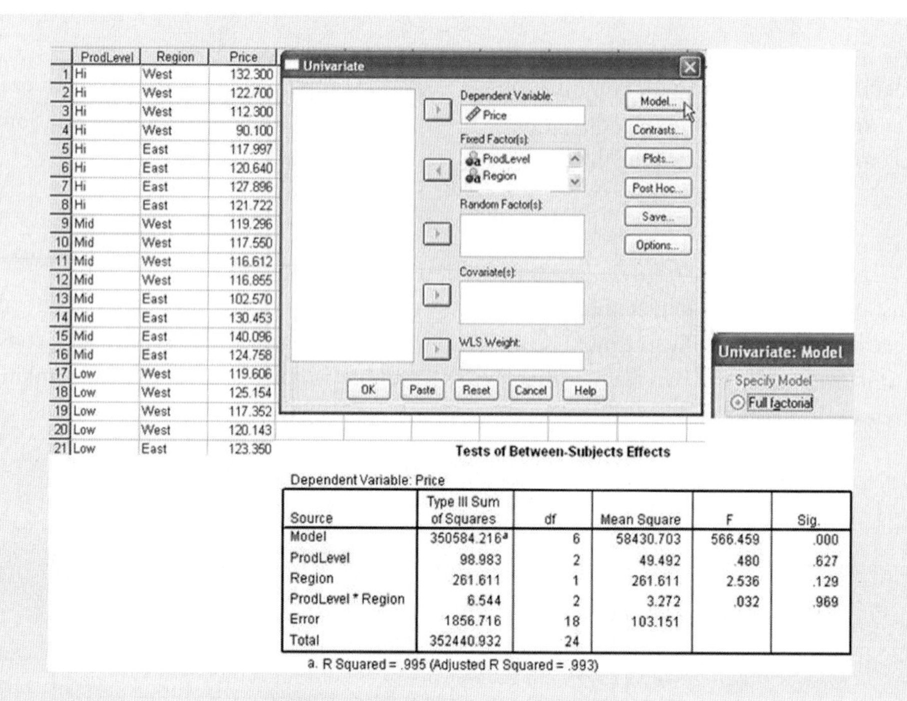

	ProdLevel	Region	Price
1	Hi	West	132.300
2	Hi	West	122.700
3	Hi	West	112.300
4	Hi	West	90.100
5	Hi	East	117.997
6	Hi	East	120.640
7	Hi	East	127.896
8	Hi	East	121.722
9	Mid	West	119.296
10	Mid	West	117.550
11	Mid	West	116.612
12	Mid	West	116.855
13	Mid	East	102.570
14	Mid	East	130.453
15	Mid	East	140.096
16	Mid	East	124.758
17	Low	West	119.606
18	Low	West	125.154
19	Low	West	117.352
20	Low	West	120.143
21	Low	East	123.350

Tests of Between-Subjects Effects

Dependent Variable: Price

Source	Type III Sum of Squares	df	Mean Square	F	Sig.
Model	350584.216[a]	6	58430.703	566.459	.000
ProdLevel	98.983	2	49.492	.480	.627
Region	261.611	1	261.611	2.536	.129
ProdLevel * Region	6.544	2	3.272	.032	.969
Error	1856.716	18	103.151		
Total	352440.932	24			

a. R Squared = .995 (Adjusted R Squared = .993)

the values that correspond. When the data are ready, use the menu sequence *Analyze / General Linear Model / Univariate;* click *OK.*

In the resulting dialogue box, highlight and select, in turn, the dependent variable and then the two factor variables. Now click the *Model* button. In the "Univariate Model" dialogue box (partially shown on the right of the illustration), select *Full Factorial;* then click *Continue* and then *OK.* An example of the relevant output is also included in the illustration.

(Note: If you get error messages, it may be because SPSS is remembering options selected earlier in your work session. Be sure to open a new data view for every new problem, rather than just deleting columns and rows from a view that is open.)

• Hands On 14.2 *Mean Search Results: Continued*

In "Hands On 14.1," you explored whether different search engines return different numbers of matches for randomly selected word pairs. We did not examine how they compare for general searches versus searches restricted to Canada. Use ANOVA to expand your analysis.

Conduct a study using this basic outline:

1. For each of these three different search engines: (1) http://www.google.ca; (2) http://ca.yahoo.com; (3) http://sympatico.msn.ca.
    - Independently generate 20 pairs of randomly and independently selected words from the dictionary. Do not intentionally use the same words for the different engines.

- Conduct a search for each of the first 10 word-pairs without restricting country of search. Record the numbers of returns. For cases where no matches are found, record the number of returns as zero (0).
- Conduct a search for each of the next 10 word-pairs, restricting the country of search to Canada. (Each of the search engines has a box to click for this option.) Record the numbers of returns. For cases where no matches are found, record the number of returns as zero (0).

2. Following step 1, you should have a data set that could be represented like this:

Scope of Search	Google	Yahoo	Sympatico-MSN
Unrestricted search	10 observations	10 observations	10 observations
Canada only	10 observations	10 observations	10 observations

At the 95% level of significance, test claim 1, below. If claim 1 is *not* supported, then proceed to test claim 2.

1. Test that there is an interaction between the factors "scope of search" and "search engine."

2. Test that there is a difference in number of returns when the search is restricted to Canada.

Chapter Summary

14.1 Construct a one-way ANOVA table and apply this model, where applicable, for a hypothesis test.

The model for **one-way ANOVA** (also called **one-factor ANOVA**) presumes that in the data set there are two variables—(1) the dependent variable, and (2) a **grouping** (or **factor**) **variable,** on the basis of which the data of variable (1) are divided into separate samples or groups (called **levels**). The ANOVA model requires that the samples be drawn randomly and independently from populations that are roughly normally distributed, and that the variances of all of the populations are nearly equal. For all models in this chapter, data should be measured at the interval or ratio level.

The **ANOVA table** shows the basic elements that lead to the final calculation for the test statistic (the F-value) when conducting an ANOVA hypothesis test. (For the basic structure, see **Figure 14.3**.) The variance calculations used to complete the table follow the same basic formula as for all variance:

$$\text{Variance} = \frac{\text{(Sum of the squared differences from the mean)}}{\text{(Degrees of freedom)}}$$

which can be expressed as:

$$\textbf{Variance} = SS/df$$

The ratio of the variance between groups and the common variance (within groups) is the *F*-ratio. The *p*-value is the conditional probability that $F_{\text{(for specified degrees of freedom)}} =$ test statistic *F*, given that H_0 is true.

To calculate the numerators for the two compared variances, we need to calculate, respectively, the **sum of squares between groups** (***SSB***) (also called **sum of squares among groups** (***SSA***) or *sum of squares explained*) and the **sum of squares within groups** (***SSW***) (also called **sum of squares error** (***SSE***) and *sum of squares unexplained*). These sums can be calculated by Formulas 14.1 and 14.2, respectively. If the ANOVA test's null hypothesis is rejected, we conclude that not all population means are the same. To then determine *which* mean or means are different from the others, we can use procedures called **multiple-range tests** or **multiple-comparison tests.**

14.2 Construct an ANOVA table and apply the model for a hypothesis test for randomized block design and one-factor repeated measures.

In **randomized block design,** we link related elements in the two samples to control for some of the variation among individuals. The **blocks** are the intentionally connected observations. The model expects that the observations for each treatment are independent of the observations for the other treatments, and that the values for each block/treatment combination are normally distributed, with approximately equal variances.

The ANOVA table and related calculations are similar to those for one-way ANOVA, except we have introduced the blocks, which appear in the table as an extra row and impact on calculations for degrees of freedom and for unexplained variance. (For the basic structure, see Figure 14.10.) To complete the table, we calculate the **sum of squares between treatments** (***SSTr***) (corresponding to *SSB*) (see Formula 14.3) and the **sum of squares blocks** (***SSBlocks***), as well as the unexplained variance (*SSW*)—which, compared to one-way ANOVA, is reduced since some variance is now explained by *SSBlocks*. The ANOVA table for **one-factor repeated measures** is similar, except (analogous to a before/after *t*-test) the successive treatments are applied to the same individuals.

14.3 Construct a two-factor ANOVA table and apply this model, where applicable, to tests for interactions of factors and for main effects.

With **two-factor ANOVA** (also called **two-way ANOVA**) we examine the effects on a response variable of two factors, each having several possible values or **levels.** Some of the total variance might be explained by the **interaction effects** of the two factors; failing that, we can test for significant effects of each factor individually (called the **main effects**) on the response variable. If an interaction or single factor accounts for or explains much of the variance among values in the dependent variable, then the *F*-value for the interaction effect or main effect will be relatively large. The test requires that the observations for each treatment be independent of the observations for the other treatments, and that values for each combination of factor levels are approximately normally distributed, with approximately equal variances. (For the basic structure of the table, see Figure 14.14.)

Key Terms

ANOVA table, p. 578
blocks, p. 590
grouping (*or* factor) variable, p. 575
interaction effects, p. 599
levels, p. 598
main effects, p. 599
multiple-range tests *or* multiple-comparison tests, p. 587
one-factor repeated measures, p. 589
one-way ANOVA (*or* one-factor ANOVA), p. 575
randomized block design, p. 589
sum of squares between treatments (*SSTr*), p. 591
sum of squares between groups (*SSB*) *or* sum of squares among groups (*SSA*), p. 578
sum of squares blocks (*SSBlocks*), p. 592
sum of squares within groups (*SSW*) *or* sum of squares error (*SSE*), p. 579
two-factor ANOVA (or two-way ANOVA), p. 598

Exercises

Basic Concepts 14.1

1. What is the purpose of analysis of variance tests?
2. What is the general form of the null hypothesis for an analysis of variance test?
3. When testing for the possible difference between three or more means, why not just use a series of *t*-tests applied to each possible pair of the means to detect a difference in the means?
4. What conditions must be met to validly conduct a one-way ANOVA test?
5. For each of the following one-way ANOVA tests, state the degrees of freedom (df) for the between sums of squares (SSB) and for the within sums of squares (SSW).
   a) 5 groups, 10 observations per group
   b) 3 groups, 6 observations per group
   c) 4 groups, 12 observations per group
   d) 6 groups, 8 observations per group
   e) 4 groups, 9 observations per group
6. For each case below, calculate the mean for sum of squares (MS) between:
   a) $SSB = 93$; $df = 3$
   b) $SSB = 78$; $df = 5$
   c) $SSB = 135$; number of groups $= 5$
   d) $SSB = 105$; number of groups $= 4$
7. For each case below, calculate the mean for sum of squares (MS) within:
   a) $SSW = 21$; $df = 10$
   b) $SSW = 28$; $df = 25$
   c) $SSW = 52$; number of groups $= 5$; $n = 35$
   d) $SSW = 14$; number of groups $= 4$; $n = 42$
8. For the following ANOVA table, fill in the missing numbers (where applicable). Then state whether the null hypothesis would be rejected at the 0.05 level of significance.

Source of Variation	*SS*	*df*	*MS*	*F*	*p*
Between groups	22	3			
Within groups	126	25			
Total					

9. For the following ANOVA table, fill in the missing numbers (where applicable). Then state whether the null hypothesis would be rejected at the 0.01 level of significance.

Source of Variation	*SS*	*df*	*MS*	*F*	*p*
Between groups	51.67				
Within groups		25			
Total	82.17	27			

10. Which of the following terms are used to distinguish the different values of the grouping variable: *dependent variable; factors; levels*?
11. Explain why there can be a need to conduct a follow-up test, once an ANOVA test has been run.

Applications 14.1

Unless otherwise specified by your instructor, formally show the eight-step sequence as part of the solution for any hypothesis tests, and display the completed ANOVA table as part of step 7.

12. According to an Australian wine taster, the prices of wine depend in part on the sweetness of the wine. To test her theory, she examined the prices and sweetness of wines carried in various liquor stores. (See the file **Australian_Wines.**) For each brand of wine in the file, the sweetness

count (from 0 to 2, on a standard scale) is given, along with the price of a 750-mL bottle. The wine taster expects that the mean prices will be different for the different sweetness categories of the wine. Test her theory at the 0.05 level of significance.

13. A university alumni association tracks the whereabouts of its university's graduates. The numbers of alumni in various non-U.S. countries are recorded in the file **Grace_Alumni_World.** A student remarked that the mean numbers of Grace alumni per country listed in each of Asia, Africa, and Europe are the same. Test this claim at the 0.05 level of significance.

14. A second student at Grace University became aware of the study described in Exercise 13. She claimed that if Canada was excluded from consideration as being an outlier, then the mean numbers of Grace alumni per country listed would also be the same in each of Asia, Africa, Europe, and the Americas. Test this claim at the 0.05 level of significance.

15. According to a diamond appraiser, if you consider only the different levels of a diamond's clarity, ranging from "imperfect" to "very, very slight incursions," the mean prices of diamonds are the same at all of the levels. To convince a skeptical colleague, he compiled the data set **Brilliants.** Recode the clarity data, as necessary, to create these levels for diamonds: (1) the clarity code begins with "V," (2) the clarity code begins with "S," and (3) the clarity code begins with "I." Test that the mean prices for diamonds in all three categories are the same, at the 0.10 level of significance.

16. According to the appraiser in Exercise 15, the mean prices of diamonds do vary if the factor levels are based on the carat-weights of the diamonds. To test this claim, re-code the carat data, as necessary, to create these levels: (1) diamonds with carats < 0.50, (2) diamonds with carats between 0.50 and 0.99, and (3) diamonds with carats ≥ 1.00.
    a) Test that the mean prices differ for diamonds in these categories. Use the 0.01 level of significance.
    b) If there is a difference, which level or levels appear to have the highest mean prices? How can you tell?

17. An entrepreneur is deciding where to best locate her new business. To help decide, she collected the data in the file **Canadian_Cities_for_Business.** Test the claim that the mean living costs are the same for all of the provinces listed. (Exclude the record that lists "Ontario/Quebec" as the province.) (Use $\alpha = 0.05$).

18. Another relevant factor for the entrepreneur's decision in Exercise 17 is the crime level in the different provinces. Again using the data in the file **Canadian_Cities_for_Business,** test the claim that the mean levels of crime are not the same for all of the provinces listed. (Exclude the record that lists "Ontario/Quebec" as the province.) (Use $\alpha = 0.05$). If this proves to be the case, which province or provinces appear to have higher mean levels of crime than the others?

Basic Concepts 14.2

19. When and why would you use a randomized block design to test a hypothesis, instead of using one-way ANOVA?

20. In what way is the randomized block design analogous to the t-test for difference in population means? Explain.

21. What are the blocks in the randomized block design?

22. What is the name of the variation of the block design that involves taking repeated measures on the same individuals?

23. To validly use the method of randomized block design, what conditions must be satisfied?

24. A writer for a car magazine is testing a new brand of tire. In particular, she is testing whether the average tire wear, over a period of testing, is different for the tires, based on the load that the tire is carrying. Wear may also vary with the test device used, since each device uses a different rotation speed. Therefore, the writer uses a randomized block design, with the blocks based on the test device. In the table below, fill in the missing numbers and then state whether the null hypothesis would be rejected at the 0.05 level of significance.

Source of Variation	*SS*	*df*	*MS*	*F*	*p*-Value	F_{crit}
Loads	2,000.080	3				4.757063
Test device	3,629.405					5.143253
Error		6				
Total	7,570.180	11				

25. The writer in Exercise 24 conducted a similar test for a different brand of tire. For a test of whether mean wear is different for different loads (again using a test device as the blocks), find the p-value for a test in difference in means. Presuming $\alpha = 0.05$, state the conclusion of the test in words.

		Test Device Categories		
		1	2	3
Load Categories	1	15.26	24.66	47.97
	2	6.69	8.45	13.39
	3	2.17	2.00	13.06
	4	1.12	2.26	2.89

26. A geneticist is exploring ways to increase the pollen production of various plants. She is testing whether certain manipulations of a gene have an effect on the mean pollen yields (in mg). Using a randomized block design, the blocks are based on three species of plant used for each treatment. In the table below, fill in any missing numbers and then

state whether the null hypothesis would be rejected at the 0.01 level of significance.

Source of Variation	*SS*	*df*	*MS*	*F*	*p*-Value	F_{crit}
Treatment	4.245127					6.944272
Plant Species	1.229549					6.944272
Error		4				
Total	5.769808	8				

27. The geneticist in Exercise 26 conducted a similar test in a different country. For a test of whether mean pollen yield (in mg) is different based on treatment (again using plant species as the blocks), find the *p*-value for a test for difference in means. Presuming $\alpha = 0.03$, state the conclusion of the test in words.

		1	2	3
Treatment	1	2.38	2.61	2.19
Categories	2	2.06	2.27	1.87
	3	3.31	3.75	3.24

Applications 14.2

Unless otherwise specified by your instructor, formally show the eight-step sequence as part of the solution for any hypothesis tests, and display the completed ANOVA table as part of step 7.

28. In order to plan efficiently for affordable health care, government officials try to anticipate the categories of service that will be required. A health official has randomly sampled discharge records from selected short-stay hospitals in four regions of the United States—reviewing several days of records for each hospital. She is now studying the relative numbers of discharges in relation to patients' diagnoses, as listed in the table below.[3]

Number of Discharges in the Sampled Records Related to the Diagnosis

Diagnosis	East	Midwest
Neoplasms	8,228	7,916
Coronary atherosclerosis	8,524	9,305
Diseases of the genitourinary system	13,242	14,853

Diagnosis	South	West
Neoplasms	13,226	5,803
Coronary atherosclerosis	12,026	4,496
Diseases of the genitourinary system	25,194	10,612

Test a claim that the mean numbers of discharges are not the same for the three diagnoses. To control for the different regions, use the regions as blocks ($\alpha = 0.05$).

29. An experiment was conducted in Cserszegtomaj, Hungary, to assess the impacts on grapevine yield of different rootstock-scion combinations for the plants. In the table below, the yields are shown (in kilograms per vine) for the different types of rootstock treatments.[4] The blocks represent the different scions (i.e., graft cuttings) used for the grape production.

Rootstock	Hungarian Riesling	Vinitor	Italian Riesling
Georgikon 28	4.250	3.375	3.175
Berl. X Rip. T. 8B	4.475	3.055	3.282
Berl. X Rip. T.K. 5BB	5.200	3.215	3.725
Berl. X Rip. T. 5C	4.375	3.237	3.227
Ruggeri 140	4.195	3.020	4.177

At the 5% level of significance, test the claim that there is no difference in the mean yields for different types of rootstocks.

30. As part of the experiment described in Exercise 29, the sugar content of the produced grape juice was also assessed. The table below shows the sugar content (in grams per litre) for juice resulting from the different rootstock treatments.[5] Again, the blocks represent the different scions used for production.

Rootstock	Hungarian Riesling	Vinitor	Italian Riesling
Georgikon 28	179.50	190.25	204.25
Berl. X Rip. T. 8B	190.25	182.75	193.25
Berl. X Rip. T.K. 5BB	189.50	189.75	200.00
Berl. X Rip. T. 5C	173.00	192.00	199.75
Ruggeri 140	185.00	198.00	190.00

At the 5% level of significance, test the claim that there *is* a difference in the mean sugar levels for juice produced from the different types of rootstocks.

31. In another aspect of the study described in Exercise 29, the potassium content of the grape leaves at harvest was assessed. The table shows the potassium content (in percent/dry weight) for the harvested leaves, for different rootstock treatments.[6] The different scions used for production are shown as the blocks.

Rootstock	Hungarian Riesling	Vinitor	Italian Riesling
Georgikon 28	3.00	3.08	3.24
Berl. X Rip. T. 8B	3.11	3.09	3.02
Berl. X Rip. T.K. 5BB	3.23	3.39	3.52
Berl. X Rip. T. 5C	3.43	3.28	3.16
Ruggeri 140	3.06	3.71	3.24
Fercal	3.06	3.00	3.33

At the 5% level of significance, test the claim that there *is* a difference in the mean potassium level for the harvested leaves for the different types of rootstocks.

32. In the course of research to assess the resistance of barley to spot blotch, a biologist manipulated a variety of barley genotypes and then measured the resistance of the resulting plants to the condition.[7] For each treatment, the corresponding measurements were taken separately from detached leaves, from seedlings, and from adult plants. Suppose one measurement was taken for each combination of treatment and block.

Measurement of Resistance Levels

		Measurement Location		
Genotype Treatment	Genotype Origin	Detached Leaf	Seedling	Adult Plant
1	Australia	6.44	6.35	6.18
2	Pakistan	5.44	4.60	5.10
3	Syria	5.36	5.82	4.30
4	Syria	5.16	5.58	6.00
5	U.S.A.	5.14	4.28	4.42
6	England	4.70	4.66	5.00
7	Germany	4.02	4.02	4.34
8	Syria	3.98	2.98	3.88
9	Ethiopia	3.38	3.06	2.96
10	Germany	3.06	2.83	2.18

Test the claim that the mean resistances resulting from the treatments are not the same. Use the 0.05 level of significance.

33. Repeat Exercise 32 but test only the claim with respect to genotypes that originated in Syria.

34. Repeat Exercise 32 but test only the claim with respect to genotypes that originated in England or Germany.

Basic Concepts 14.3

35. Explain when and why you would use two-factor ANOVA to test a hypothesis.

36. You can use the ANOVA model to test a hypothesis about the main effects of the factors, or about their interaction.
    a) Explain what is meant by the *main effects.*
    b) Explain what is meant by an *interaction of the factors.*

37. To validly use the method of two-factor ANOVA, what conditions must be satisfied?

38. A writer for a car magazine is comparing the durability of three new brands of tires under different load conditions. For each load category, she samples three tires each from each brand and makes a measurement of durability. In the table below, fill in any missing numbers.

Source	*df*	*SS*	*MS*	*F*	*p*
Load	3	3,299.74			
Brand	2	28.24			
Load*Brand	6	73.94			
Error	24				
Total		3,425			

39. Based on the completed table for Exercise 38, is there a significant interaction between the variables Load and Brand in relation to the durability of the tires? Explain how you know.

40. Based on the completed table for Exercise 38, are there significant main effects in relation to the durability of the tires? If so, explain what they are and justify your answer.

41. To verify the results shown in Exercise 38, a competing car magazine conducted its own test for the same variables. These were the results:

	Brand A	Brand B	Brand C
Load Level 1	29.29	24.12	30.89
	29.70	24.34	29.96
	30.14	24.25	30.74
Load Level 2	9.51	4.03	11.33
	8.85	3.39	10.75
	9.37	4.26	9.69
Load Level 3	5.74	3.93	5.63
	5.80	4.05	7.07
	4.92	3.33	3.37
Load Level 4	2.09	5.04	0.48
	2.18	5.17	1.55
	2.07	5.04	1.64

Find the *p*-value for a test that brand and load have an interaction effect on the mean durability of tires. Presuming $\alpha = 0.01$, state the conclusion of the test in words.

42. Use the data in Exercise 41 but suppose there is no interaction effect between the variables (i.e., disregard your previous answer). Find the *p*-value for a test that tires bearing different levels of load do not all have the same mean durability. Presuming $\alpha = 0.01$, state the conclusion of the test in words.

43. Use the data in Exercise 41 but suppose there is no interaction effect between the variables (i.e., disregard your previous answers). Find the *p*-value for a test that tires of different brands do not all have the same mean durability. Presuming $\alpha = 0.01$, state the conclusion of the test in words.

Applications 14.3

Unless otherwise specified by your instructor, formally show the eight-step sequence as part of the solution for any hypothesis tests, and display the completed ANOVA table as part of step 7.

44. A conservationist is examining data for water use in municipalities across Canada. The table below is based on sampling from the full set of **Water_Use** data in the CD file. The displayed values are magnitudes of water use by each of three nondomestic sectors, in cubic metres per day. A second factor is a size code for each sampled community.

	Sector of Water Use		
Size Code	Commercial	Industrial	Other
2	100	30	20
2	80	50	30
2	40	40	45
2	126	180	135
2	136	136	68
3	360	60	120
3	43	174	43
3	233	36	90
3	125	63	124
3	320	240	160
4	626	2,878	2,127
4	605	1,210	303
4	1,155	2,310	385
4	2,640	1,760	880
4	494	380	76
5	6,764	4,509	2,255
5	13,714	457	3,657
5	10,001	10,001	5,455
5	5,159	4,299	14,188
5	7,407	16,047	3,703
6	61,920	59,340	12,900
6	394,479	285,658	108,822
6	198,990	252,054	490,842
6	108,000	18,000	36,000
6	44,888	151,497	72,943

Formally test a claim that means for water use are affected by the interaction of use sector and community size code. Use $\alpha = 0.01$.

45. Based on the data in Exercise 44:

a) Formally test a claim that means for water use are the same for all of the listed sectors. Use $\alpha = 0.01$. If the means do vary, which use sectors appear to have the greater water use?

b) Formally test a claim that means for water use vary for the different sizes of community. Use $\alpha = 0.01$. If the means do vary, which community size or sizes appear to have the greater water use?

46. Do the data in Exercise 44 meet the ideal test condition that "the variances of the distributions (within each combination of factor values) are approximately equal"? If not, how might this deviation from the model affect your conclusions to Exercises 44 and 45?

47. An engineering student is conducting a study of the world's major bridges. He is particularly interested in the effects of *when* the bridges were built (variable = "era") and the *type* of bridge on the response variable (the span of the bridge in metres). The table below is extracted from the data in the file **Major_Bridges.**

Type	Pre-1970	1970+
Cantilever	376	510
	360	280
	326	267
Box and/or plate	229	300
	137	192
	146	140
Suspension	1,006	1,385
	704	1,210
	655	1,074

Based on the table, formally test a claim that the span of a major bridge is affected by the interaction of its era and type. Use $\alpha = 0.05$.

48. a) Re-do Exercise 47 but for $\alpha = 0.01$.

b) Formally test a claim that the mean spans of major bridges are different for groups based on the era of the bridges' construction. Use $\alpha = 0.01$.

49. Several drivers in the Vancouver area noticed that gas prices varied in the communities in the surrounding areas. They wondered if any community appeared to have consistently lower prices. One of the drivers, a stats student, began collecting randomized data from the website http://gastips.com. The results are shown in the table below. The row headings list communities, the column headings list different brands of gasoline, and the values are gas prices at selected stations in cents per litre.[8]

	Chevron	Esso	Shell
Langley	93.9	101.9	104.2
	101.9	93.9	104.9
	101.9	101.9	101.9
Surrey & Area	101.9	103.2	101.9
	104.9	101.9	105.2
	101.9	105.2	105.5
Abbotsford	92.5	89.0	92.5
	92.5	88.9	88.9
	98.5	92.5	87.9

a) Formally test a claim that mean gas prices are not affected by the interaction of community and brand of gas. Use $\alpha = 0.05$.

b) Formally test a claim that the gas prices are no different for groups based on the brand of gas. Use $\alpha = 0.05$.

50. Based on the data in Exercise 48:

a) Formally test a claim that the mean gas prices are not the same for the three sampled communities. Use $\alpha = 0.05$.

b) If they are not the same, which community appears, by eye, to offer the lower prices?

Review Applications

Unless otherwise specified by your instructor, formally show the eight-step sequence as part of the solution for any hypothesis tests, and display the completed ANOVA table as part of step 7.

51. A geology student has compiled a sample of waterfalls within, or on the border of, Canada. (See file **Waterfalls.**) She claims that the mean vertical drops of the waterfalls are not the same in all listed provinces and territories.
    a) Test this claim at the 0.05 level of significance.
    b) If her claim is supported, what region or regions appear to have larger means for vertical drops of their waterfalls compared to other regions?

Exercises 52–54 are based on data in the file **Water Mineral_Content.**

52. A buyer for a large chain of supermarkets is deciding what brands of mineral waters to carry. A consideration is the mineral content of the various options available, and also the countries of origin. In his research, the buyer has compiled the large related data set, stored in the file **Water_Mineral_Content.** Test a claim that the mean contents of phosphorus ("P") in mineral water products are the same for all of the countries listed. Use $\alpha = 0.05$.

53. Repeat Exercise 52 but test whether the mean content of calcium ("Ca") is the same in all of the listed countries' mineral water products.

54. Repeat Exercise 52, except test whether the mean contents of bicarbonate ("HCO3") are *not* the same in the all listed countries' mineral water products.

Exercises 55–57 are based on data in the file **Calgary_Flames.**

55. A sports fan has compiled statistics for the Calgary Flames for the 2003–2004 season. She is looking for factors that impact on various variables. For example, she claims that heavier players tend to be assessed more time in the penalty box. (See the variable Penalties in Minutes—"PIM"—in the data set. For player weights, create a new variable that codes weights *less than* 90 kg as "1," weights from 90 kg up to 100 kg, inclusive, as "2," and weights *larger than* 100 kg as "3.") Based on these data, test whether indeed the mean time spent in the penalty box is different for the different weight categories. Use $\alpha = 0.05$.

56. Using the data file for Exercise 55, test whether the mean number of points earned by players in the season was different based on players' positions ("POS"). Use $\alpha = 0.05$.

57. Using the data file as modified in Exercise 55, test whether the mean number of points earned by players in the season was the same for all of the coded weight categories. Use $\alpha = 0.05$.

58. Two real estate agents were having a debate about whether having extra bathrooms in a house necessarily increases the price of the house. One of the agents claims that the mean prices of houses do not differ based on number of bathrooms. To prove his point, he randomly sampled houses on the market that had different numbers of bathrooms. He controlled for location and used three different locations as blocks. The list of prices of the sampled houses is shown in the table below.[9]

	Location Blocks		
Bathrooms	Greater Sudbury	Kingston	Windsor
1	179,700	202,500	157,899
2	69,200	219,900	189,800
3	529,900	172,500	243,900
4	359,000	329,000	749,999

Test the agent's claim at the 0.05 level of significance.

59. For a variety of reasons, airlines do not always transport the same value in revenue tons of goods and passengers as they may have scheduled. The table below is based on data from the file **Carriers_revenue_tons** and shows sample data for the performance rates (actual revenue tons divided by scheduled revenue tons) for a number of air carriers. The performance rates have been converted into categories, based on relative size. For each block (group) of airlines, one airline has been randomly selected for each of the three performance categories. For the selected airlines, values of revenue tons carried ("All_RevP") are shown in thousands of dollars.

Group	All_RevP	All_RevS	PerformRate	PerformCat
Major	271.898	275.149	0.988	0
Major	58.921	50.210	1.173	1
Major	120.142	6.540	18.370	2
National	46.754	48.111	0.972	0
National	43.906	30.589	1.435	1
National	0.955	0.025	38.200	2
Regional	3.005	38.265	0.079	0
Regional	4.570	4.142	1.103	1
Regional	4.464	2.554	1.748	2

Test a claim that mean values for revenue tons carried are about the same regardless of the performance category of the airline. Use the 0.05 level of significance.

60. The governors of several states (Alaska, Arizona, Arkansas, California, Colorado, Connecticut, and Delaware) are comparing their states' records of educational progress, using seven different index measures. (These data are included in the file **US State Education.**) The governor of Colorado is bragging that the mean index scores for the seven states are

not the same—and that Colorado, in particular, has a different (higher) mean than the rest.

a) If a randomized block design is used to test the first part of the governor's claim, which variable (the states or the named index measures) would be the blocks?
b) Test the first part of the governor's claim at the 0.02 level of significance.
c) Do the data appear to support the second part of the governor's claim? Explain.

61. A nutritionist is studying the characteristics of the food items listed in the file **Composition_of_Food.** He claims that if the foods are divided into the categories shown in the table, the mean calories of the foods within different groups are not the same. Test his claim at the 0.05 significance level.

62. The nutritionist also claims that if the foods are grouped by category, the within-group means (a) for protein, (b) for carbs, (c) for fat, and (d) for fibre, respectively, are not the same. Test each of these four claims separately, using $\alpha = 0.05$.

63. After collecting the data shown below (based on the file **Metallic_Minerals**), a student geologist concluded that there is no interaction between the variables main colour and hardness code in relation to density. Test this conclusion at the 0.05 level of significance.

Density	Main Colour	Hardness Code	Density (Continued)	Main Colour (Continued)	Hardness Code (Continued)
5.16	black	1	6.60	grey	1
6.25	black	1	5.56	grey	1
5.73	black	1	4.30	grey	1
5.95	black	1	6.70	grey	1
6.20	black	2	3.46	grey	2
5.10	black	2	14.30	grey	2
4.78	black	2	4.90	grey	2
7.30	black	2	4.95	grey	2
4.75	black	3	5.05	grey	3
4.64	black	3	6.50	grey	3
5.30	black	3	7.15	grey	3
4.69	black	3	5.20	grey	3

64. Using the data in Exercise 63, test whether the mean densities of minerals grouped by hardness codes are different ($\alpha = 0.05$).

65. Based on the data in Exercise 63, test whether mean densities of minerals grouped by main colour are equal.

66. The ministers of education of several countries (Finland, Japan, Canada, Australia, Sweden, Ireland, Germany, and the United States) are comparing their countries' records of educational progress, using three index measures. (See the file **Pisa_Results_International.**)

a) Using a randomized block design, test the claim that the mean of the index measures are basically the same for all of the countries. Test at the 0.05 level of significance.
b) The scores listed in the data set are not really individual values of a student's performance. They are summary statistics for a whole country's performance on a particular measure. Is this fact consistent with what the test model expects? If not, what type of inaccuracy might be caused when performing the test in (a)?

67. Politicians often complain that there is bias in media political reporting. The media generally deny this, and claim, for example, that reports of changing political popularity are simply based on the objective results of polls. But could there be some bias in the poll results themselves?

The following table is based on data randomly extracted from the file **Federal_Political_Polls.** The original file collects poll results taken over three years on voters' party preferences, obtained by a variety of polling organizations. Interpret the data values as scores of popularity for the named parties.

Organization	Liberal	Conservative	NDP	Other/None
Compas	35	30	19	16
Compas	38	29	18	15
Compas	39	31	17	13
Compas	30	34	18	18
Decima	33	30	20	17
Decima	32	29	20	19
Decima	37	28	18	17
Decima	37	25	20	18
Ekos	29	35	18	18
Ekos	39	29	17	15
Ekos	30	37	18	15
Ekos	37	27	17	19
Environics	36	30	19	15
Environics	33	33	18	16
Environics	39	28	21	12
Environics	27	33	24	16
Ipsos-Reid	31	32	17	20
Ipsos-Reid	32	31	16	21
Ipsos-Reid	33	31	17	19
Ipsos-Reid	38	26	17	19
Leger	39	27	16	18
Leger	40	24	15	21
Leger	36	28	17	19
Leger	31	34	18	17
Pollara	37	34	17	12
Pollara	36	31	16	17
Pollara	36	30	19	15
Pollara	38	30	17	15
SES	33	34	18	15
SES	34	31	21	14

(continued)

Organization	Liberal	Conservative	NDP	Other/None
SES	31	35	16	18
SES	37	34	17	12
Strategic	29	37	15	19
Strategic	28	38	16	18
Strategic	33	29	17	21
Strategic	34	30	17	19

a) Test a claim that the mean preference scores do show an interaction effect between polling organization and party preference.

b) Are the results of (a) *consistent with* a concern that some organizations may exhibit bias in their collection or reporting? Explain.

c) Do the results of (a) *prove* that some organizations are exhibiting bias in their collection or reporting? If not, explain other factors that could be involved.

68. Several parents were comparing their experiences in buying children's books online. They noticed that the book listings are not sorted in an obvious order, such as alphabetically by title or author. They wondered if perhaps the order partly reflected the prices. By random selection from related pages on a bookseller's site, the data shown below were collected. The rows give the target audience ages for the selected books, the columns refer to the order of presentation of the books (i.e., in the first, second, or third screen of the displayed list), and the values in the cells are the prices of the randomly selected books.[10]

	Screen 1	Screen 2	Screen 3
Baby to age 3	6.50	10.87	10.44
	5.50	8.99	6.50
	8.99	10.35	8.15
Ages 4–8	5.95	9.89	1.50
	5.99	1.25	10.91
	8.99	10.35	8.99
Ages 9–12	25.83	1.50	10.80
	10.87	14.56	22.05
	14.40	14.56	11.68

a) Test whether the mean prices are different based on an interaction of target audience age for the book and the screen on which the book list is displayed ($\alpha = 0.05$).

b) Test whether the mean prices are different depending on which screen of the book list is displayed ($\alpha = 0.05$).

69. Using the data in Exercise 68, test a claim that the mean prices of children's books, when grouped by the target audience ages of the books, are not the same ($\alpha = 0.05$).

70. A student noticed that, when looking up phrases in search engines on the Internet, widely different numbers of hits (matches) may be returned. She experimented to see what factors might influence the mean number of hits, varying the number of random words in the search phrases and/or the search engines that she used. The following were the results (the numbers represent thousands of hits).[11]

	One Word	Two Words	Three Words
Google	184,000	936	180
	85,700	1,160	1,270
	2,680	348	187
Ask	42,600	116	7.8
	25,090	252	214
	729	27.7	4.14
AltaVista	145,000	727	80.1
	105,000	1,570	1,310
	2,080	194	62.3

Test whether the mean number of hits is affected by the interaction of the search engine and the number of words (up to three) in the search phrase ($\alpha = 0.05$).

71. Using the data in Exercise 70, test whether:

a) The mean number of hits is the same regardless of the number of words (up to three) in the search phrase ($\alpha = 0.05$).

b) The mean number of hits is different for the different search engines ($\alpha = 0.05$).

PART 6

Measures and Tests for Association

Measuring and Testing Associations for Bivariate Data

Chapter 15

• Chapter Outline

• Learning Objectives

15.1 Identify a linear correlation and calculate the sample coefficient of correlation r, assess the strength and significance of the correlation, and describe some cautions when interpreting the correlation results.

15.2 Calculate for suitable—or transformed—data the Spearman rank correlation coefficient r_s, assess the strength and significance of the correlation and describe some cautions required when interpreting the correlation results.

15.3 Describe and distinguish the concepts of tests for homogeneity, independence, and goodness of fit, and apply these tests where appropriate.

15.1 Linear Correlation for Continuous Bivariate Data

An association of variables can be expressed statistically in terms of the variables' conditional probabilities: If $P(A|B)$ is different for different values or ranges of B and/or $P(B|A)$ is different for different values or ranges of A, then there is an **association between two variables** A and B. Associations can be found among variables of all different types and measurement levels. For example, the genders of university students and their chosen degree programs (e.g., nursing, engineering) are two *categorical* variables that may be historically associated. (Compare Section 15.3.) On the other hand, the students' "1 to 5" ratings of their courses and their "1 to 5" ratings of their course textbooks may be two associated *ordinal* variables. (Compare Section 15.2.)

The initial focus of the first section of this chapter is on *continuous* bivariate data. We will learn to recognize and test whether these data are associated in the manner called a *linear correlation.* If two continuous variables are shown to have an association of this type, a natural next step is to build a model (called a *regression model*) that shows the nature of the association and provides a tool for predicting or estimating the value of one variable from the value of the other, as discussed in Section 16.1. (Note: Depending on the design of your course, it is optional to skip from the completion of Section 15.1 directly to Chapter 16, and then return later to this chapter to consider other types of data.)

To identify an association between two continuous variables, a helpful first step is to graph their relationship. Recall how we created *scatter diagrams* in Section 3.5 to illustrate associations between variables that are measured at the integer or ratio level. Figure 15.1 is an example of a scatter diagram. Each point represents one pair of data corresponding to one day of trading for Bank of Montreal (BMO) stock; the coordinates of each point, relative to the horizontal and vertical axes, represent the corresponding date (horizontal axis) and closing price (vertical axis) at the end of one trading day. The figure

Figure 15.1
Association between a Stock's Prices and Time (January 2000–September 2006)

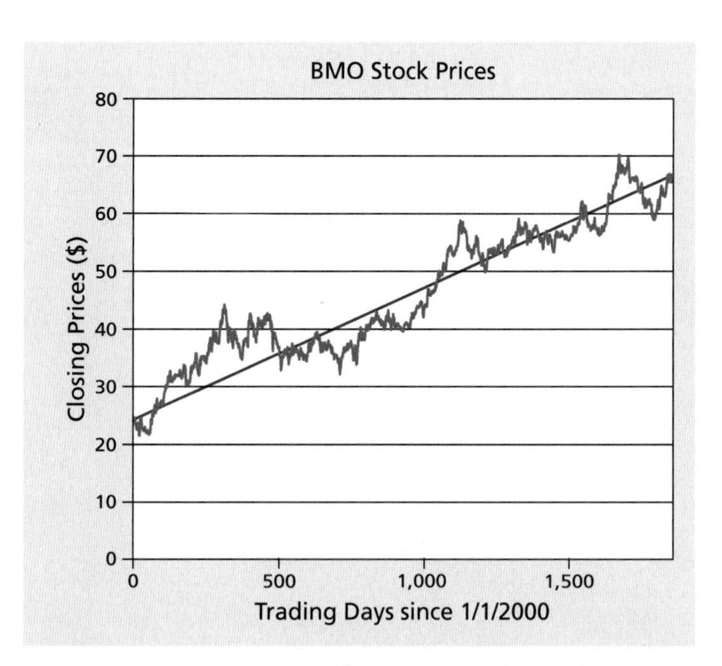

Source: Bell Globemedia. Found at: http://www2.bmo.com/content/0,1089,divId-3_langId-1_navCode-4644,00.html

suggests that dates closer to January 2000 are associated with lower trading prices for BMO stock than are more recent dates.

By eye, the pattern of points in Figure 15.1 appears to more or less follow a straight line (i.e., be linear; compare the wavy blue line to the straight line added for comparison in the figure). When the association of two variables follows a linear pattern in a scatter diagram, we say there is a **linear correlation** between the two variables.

Strength and Sign of the Correlation

A scatter diagram provides a useful first impression of whether two variables are correlated but cannot provide an objective measure for the strength (if any) of the relationship. Depending on the scale used for the drawing and on an observer's subjective factors, the same variations from linear may appear significant to one observer and not to another. We calculate a **sample coefficient of correlation (r)** to generate a more objective measure of the strength of linear correlation.

A sample coefficient of correlation r has both a sign (+ or −) and a magnitude. A negative sign means that there is an **inverse correlation** between the values of the two associated variables. That is, if the value of one variable increases (compared to other sampled items), the corresponding value of the second variable (compared to the other items) decreases. In a **positive correlation,** if the value of one variable increases, then the corresponding value of the other variable increases as well.

Possible values for r range from -1, meaning a perfectly negative correlation, up to $+1$, which is a perfectly positive correlation. Sample coefficient of correlation r-values are applied to several example correlations in Figure 15.2. In part (a), we see that for the 2004 federal election, the number of candidates competing in particular ridings had no apparent relationship to the number of polls that were counted in those ridings, as of the press deadline for *The Globe and Mail* on election night. Therefore, r is very low. In part (b) we see (as we might expect) that the number of polls counted in the ridings as of election night is very closely and positively related to the total number of polls available to be counted in the ridings. So, r is quite close to the maximum value of $+1$.

Part (c) of Figure 15.2 illustrates a moderately strong correlation of $r \approx +0.75$. This is the relationship between the total number of votes for a winning candidate versus the margin of victory for the candidate (i.e., how many more votes the winner obtained than did the candidate with the second-highest vote count). On the other hand, part (d) of the figure shows a moderately strong but *negative* correlation between the numbers of votes for winning candidates and the losing margins of the corresponding second-highest scoring candidates.

There are no precise cutoffs for concluding that an r-value is strong, moderate, weak, or any other degree of strength, although an approximate guide is provided in Figure 15.3, which is roughly based on the percentage of variance explained, for different values of r. Compare a given r-value to a number on the horizontal axis of the figure and then read off a corresponding descriptive term (e.g., "Moderate") that generally would be appropriate. [The vertical axis in Figure 15.3 will be explained more fully in Section 16.1, but in general terms, it shows for any given value of r how much of the variation of one variable is explained by the other with which it is correlated. For example, if $r \cong 1.0$, as in Figure 15.2 (b), then changes in the number of polls counted in various ridings can be fully explained by noting the corresponding changes in the total number of polls in those ridings.]

Figure 15.2
Possible Measures of Association

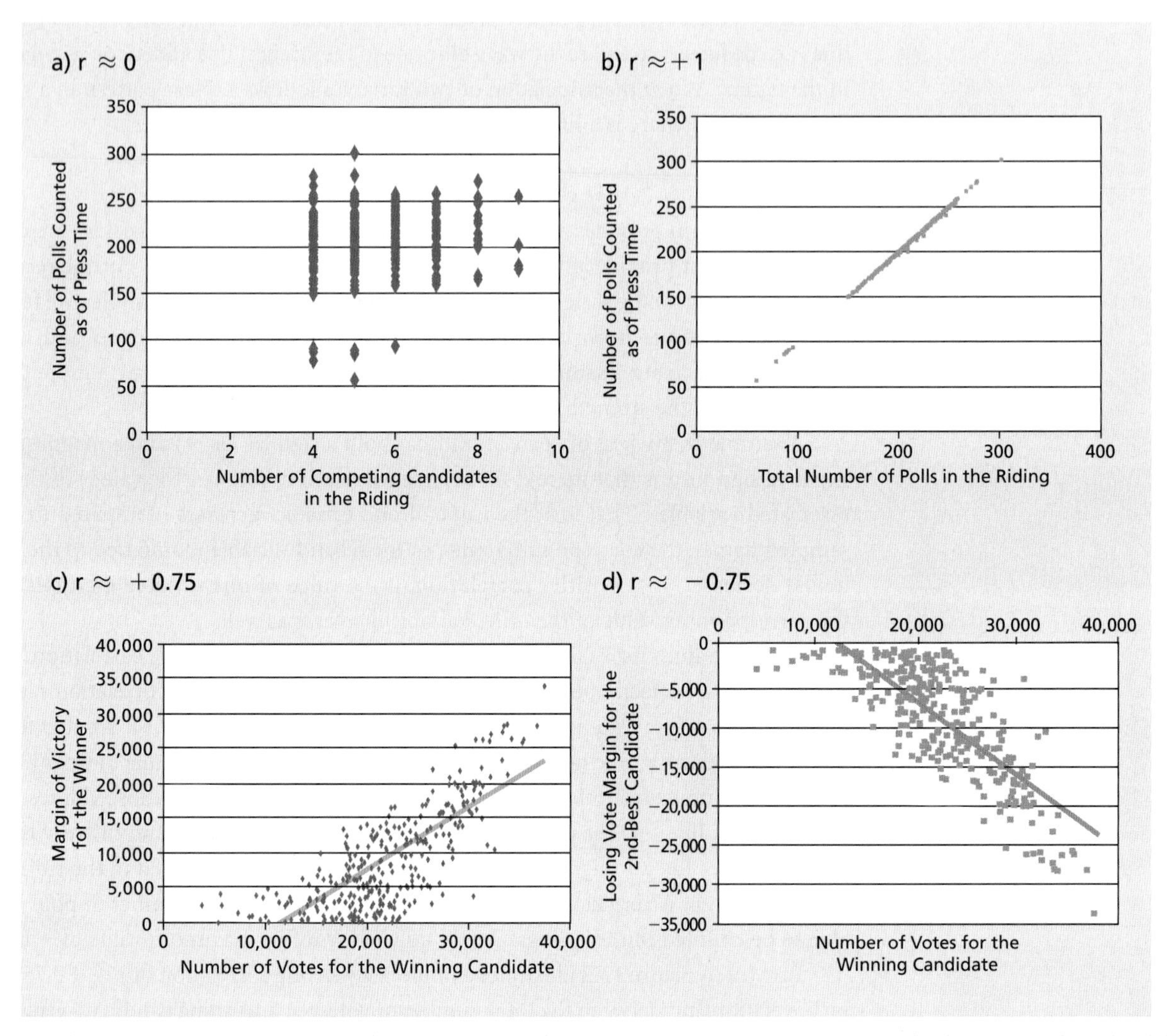

Source: "Federal Election Results, 2004." *The Globe and Mail.* Wednesday, June 30, 2004. Pages A12–A13. Reprinted with permission from *The Globe and Mail.*

In order to validly use the model presented in this section, the following conditions for the data set must apply:

- The data must be quantitative and, ideally, at least at the interval level of measurement. (For ordinal data, see Section 15.2.)
- The set of paired data is a random sample.
- The population from which the sample is drawn has a **bivariate normal distribution.** In terms of the scatter diagram, this means that the deviations of actual *y*-values (i.e., graphed on the vertical axis) from the predicted *y*-values (as estimated by a regression line) are normally distributed for all parts of the line.

To illustrate this concept, compare the scatter diagrams in Figure 15.4. Part (a) compares, by riding, the actual winning margins in the 2004 federal election (on the *y*-axis) to the winning margins in those ridings in the previous election (on the *x*-axis). It does not satisfy the condition of a bivariate normal distribution very well. For example, for

Figure 15.3
Strengths of Correlation Based on r

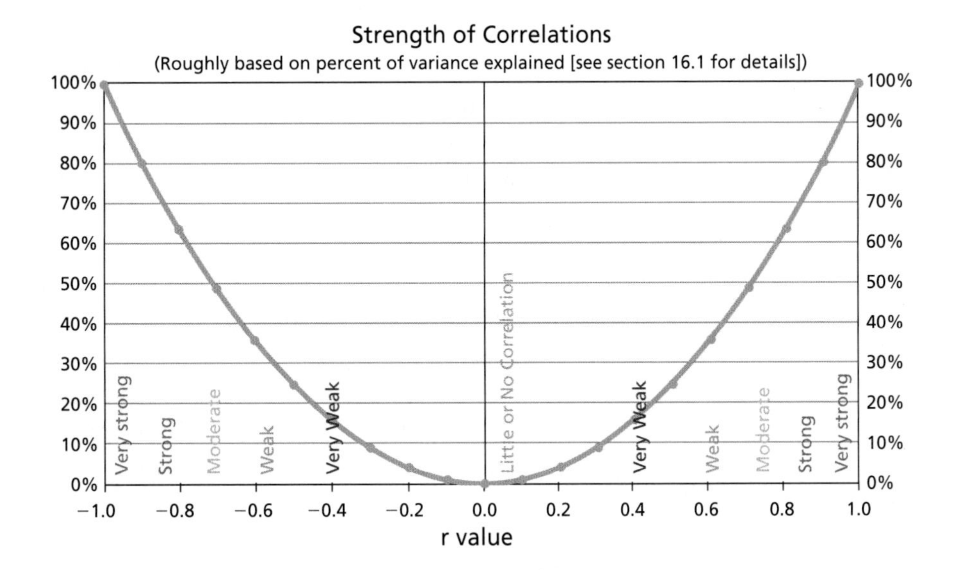

x-values of approximately 30,000, observe that *all* of the variation is above (i.e., greater than) the linear estimate. For x-values of approximately 10,000, the centring is more reasonable—but, unlike the normal distribution, there are proportionally more y-values farther from the centre of the estimate than closer to the centre. Compare the actual

Figure 15.4
Assumption of a Bivariate Normal Distribution

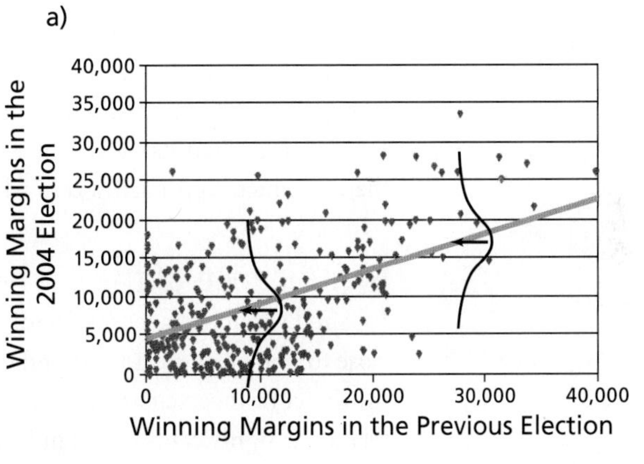

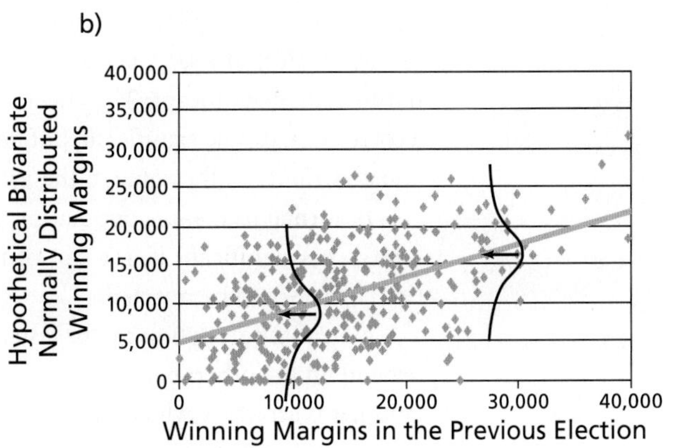

Source: "Federal Election Results, 2004." *The Globe and Mail.* Wednesday, June 30, 2004. Pages A12–A13. Reprinted with permission from *The Globe and Mail.*

distribution in part (a) with the hypothetical distribution shown in part (b): In the latter, the y-values are truly distributed normally above and below the estimate for essentially all portions of the line. This is the condition expected by the model.

Formula 15.1 is used to find r. For convenience, the numerator can be abbreviated as *SSXY* (with *SS* standing for "sum of squares," as in ANOVA) and, similarly, the two summations in the denominator (within the square root symbols) can be abbreviated *SSX* and *SSY*, respectively. The form of the denominator will look familiar, as it is based on variances for the individual x- and y-values in the data pairs. The numerator may look less familiar; but it is based on the covariance of how the two values in each pair vary together. In the case of a perfect correlation, *all* of the variation from the x- and y-values individually (denominator) is reflected in the covariances of x- and y-values moving up or down in tandem (numerator). Therefore, the ratio r approaches ± 1 (depending on whether the relationship is direct or inverse).

Formula 15.1 Value of the Correlation Coefficient r

$$r = \frac{\sum_{i=1}^{n} (x_i - \bar{x})(y_i - \bar{y})}{\sqrt{\sum_{i=1}^{n} (x_i - \bar{x})^2} \times \sqrt{\sum_{i=1}^{n} (y_i - \bar{y})^2}}$$

On the other hand, if there is no correlation, then the numerator tends toward zero: There will be cases where one of the products being summed in the numerator is positive (e.g., for x- and y-values that are both larger than their respective means), but these cases will cancelled out, when taking the sum, by cases where the product is negative (e.g., an x is larger than its mean but y is smaller, or vice versa). Therefore, the sum of all of these products approaches zero when there is little correlation.

Significance of the Correlation

If paired variables meet the conditions for assessing linear correlation, then a value of r close to *plus or minus* 1.0 can indicate a *strong* correlation, and the sign of r can indicate the direct or inverse nature of the correlation. But, in itself, r does not tell us if the correlation is **significant**—r is significant if the evidence suggests that the appearance of correlation in the sample represents a true population parameter and not just a random sampling effect. If a sample is small, a correlation that seems to be strong could still turn out to be nonsignificant: The effect could be due merely to chance. On the other hand, a correlation can be significant, but not strong, if there are other, extraneous factors that also contribute to the value of the y variable.

In formal terms, the population parameter that corresponds to r is ρ (pronounced "rho"). r is significant if in a test of null hypothesis "$\rho = 0$" (i.e., "there is no correlation"), H_0 would be rejected.

Note that correlation can be tested without yet committing to which variable's values depend on the other; perhaps they both depend in reality on a third factor. For a correlation, it is arbitrary which of the two variables we visualize or test as the y variable that appears in Formula 15.1.

To test the significance of r, an ANOVA test based on the F-statistic can be used, analogous to the ANOVA tests introduced in Chapter 14. If a correlation is significant, the correlation accounts for much of the variation in y-values from the y mean (i.e., the variance "between") compared to the residual variance of y-values due simply to sampling error. As in Chapter 14, the test statistic F is the ratio of the variance-between and the variance-error, and if the F-value is unusually high, then the p-value [i.e., $P(F \geq$ the test statistic $F \mid H_0$ is true)] is low.

Example 15.1

While studying for his master's degree in resource and environmental management, a student devised a sampling plan to ensure that both large and small cities had adequate representation in the sample. This is part of a table that appeared in his thesis.[1]

City	Population	*n* (Actual)
Vancouver	1,829,854	130
Victoria	288,346	21
Calgary	879,277	69
Edmonton	782,101	66
Winnipeg	626,685	48
Saskatoon	196,816	16
Toronto	4,366,508	257
Ottawa–Hull	827,854	58
Hamilton	618,820	36
Kitchener	387,319	23
London	337,318	24
St. Catharines–Niagara	299,935	21
Windsor	263,204	17
Oshawa	234,779	17
Montreal	3,215,665	216
Quebec	635,184	43
Halifax	276,221	62
St. John's	122,709	30

It can be shown that the sample sizes are not strictly proportional to city populations. Nonetheless, if this sampling approach was followed in many studies, (1) does there *appear to be* a linear correlation between the cities' sample sizes and their population sizes? (2) If so, would the correlation be strong, and how would we know? (Include any necessary formula-based calculations.) (3) Is the correlation, if any, significant? How would we know? (At this stage, refer to the computer output described in "In Brief 15.1.")

Solution

1. The first question is asking, "Do the data conform reasonably to the model of a linear correlation?" A good start is to construct a scatter diagram such as Figure 15.5, showing the relationship between the two variables "population" and "sample size." Given the relative lack of samples on the upper right-hand part of the line, it is hard to be sure—but by eye, it does seem reasonable to allow that the paired data are approximately normally distributed around a line of predicted y-values. So, there *appears* to be a linear correlation.

Figure 15.5
Scatter Diagram to Assess If Linear Correlation Applies

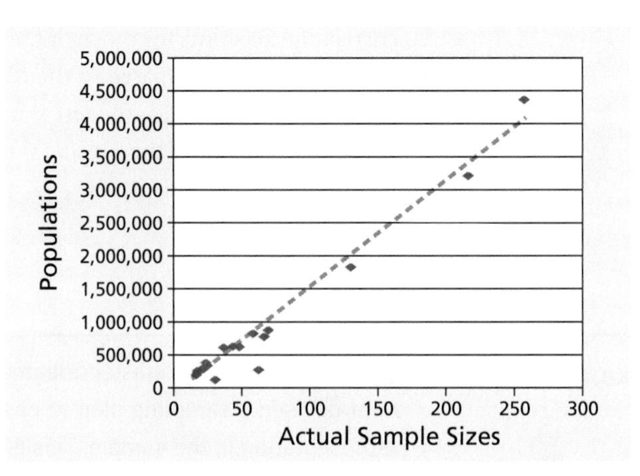

Source: Matthew Horne. *Incorporating Preferences for Personal Urban Transportation Technologies into a Hybrid Energy-Economy Model.* Simon Fraser University, School of Resource and Environmental Management.

2. The measure for strength of correlation is the value of *r*. Preliminaries for using the following formula are displayed in Figure 15.6.

$$r = \frac{SSXY}{\sqrt{SSX} \times \sqrt{SSY}}$$

$$= \frac{1{,}318{,}543{,}635.22}{\sqrt{22{,}136{,}457{,}661{,}658.30} \times \sqrt{81{,}115.78}}$$

$$= 0.984$$

Being very close to 1.0, this *r*-value signifies a very strong correlation.

Figure 15.6
Preliminary Calculations for *r*

x Population	*y* *n* (Actual)		$(X_i - \bar{X})$	$(Y_i - \bar{Y})$	$(X_i - \bar{X}) \times (Y_i - \bar{Y})$	$(X_i - \bar{X})^2$	$(Y_i - \bar{Y})^2$
1,829,854	130		930487.61	65.89	61308794.82	865807194431.26	4341.35
288,346	21		−611020.39	−43.11	26341767.88	373345915637.93	1858.57
879,277	69		−20089.39	4.89	−98214.79	403583545.93	23.90
782,101	66		−117265.39	1.89	−221501.29	13751171431.26	3.57
626,685	48		−272681.39	−16.11	4393200.15	74355139846.37	259.57
196,816	16		−702550.39	−48.11	33800479.82	493577048927.93	2314.68
4,366,508	257		3467141.61	192.89	668773092.99	12021070951498.20	37206.12
827,854	58		−71512.39	−6.11	437020.15	5114021764.60	37.35
618,820	36		−280546.39	−28.11	7886470.71	78706276318.60	790.23
387,319	23		−512047.39	−41.11	21050837.10	262192528467.93	1690.12
337,318	24		−562048.39	−40.11	22544385.38	315898391452.60	1608.90
299,935	21		−599431.39	−43.11	25842153.21	359317989985.26	1858.57
263,204	17		−636162.39	−47.11	29970316.99	404702585036.82	2219.46
234,779	17		−664587.39	−47.11	31309450.32	441676397470.15	2219.46
3,215,665	216		2316298.61	151.89	351820022.38	5365239255835.26	23070.23
635,184	43		−264182.39	−21.11	5577183.77	69792334599.04	445.68
276,221	62		−623145.39	−2.11	1315529.15	388310175693.49	4.46
122,709	30		−776657.39	−34.11	26492646.49	603196699715.71	1163.57
16,188,595.00	1,154.00	← Sums →			1318543635.22	22136457661658.30	81115.78
18.00	18.00	← *n*			*SSXY*	*SSX*	*SSY*
899,366.39	64.11	← Means					

Source: Matthew Horne. *Incorporating Preferences for Personal Urban Transportation Technologies into a Hybrid Energy-Economy Model.* Simon Fraser University, School of Resource and Environmental Management.

3. For this section we will rely on software to provide the p-value ≈ 0.000. (Additional details are provided in Section 16.1.) The implied hypothesis test is for H_0: $\rho = 0$. (α is unstated, so generally a significance level of 0.05 is presumed.) Since the p-value is less than α, reject H_0 in favour of the implied alternative hypothesis: H_1: $\rho \neq 0$. In other words, the evidence suggests that there *is* a significant correlation.

• *In Brief 15.1 Measuring Correlation*

On the computer you may be able to measure correlation and test its significance using the correlation procedure. If instead your software requires regression steps to find significance, you will be directed to skip also to "In Brief 16.1."

Excel

Excel can find *r* directly. The paired data can be aligned in two parallel columns, as shown in the spreadsheet behind the screen capture, *or* in two parallel rows. Begin with the menu sequence *Tools / Data Analysis / Correlation;* then click *OK.* In the resulting dialogue box, specify the address range containing the paired data. Check *Labels in first row* as indicated, if the variable names (labels) are included at the start of the data range. Select whether the data are aligned by columns or rows. Finally, specify the upper left-hand cell for the output range, then click *OK.*

	A	B	C
1	City	Population	n (Actual)
2	Vancouver	1,829,854	130
3	Victoria	288,346	21
4	Calgary	879,277	69
5	Edmonton	782,101	66
6	Winnipeg	626,685	48
7	Saskatoon	196,816	16
8	Toronto	4,366,508	257
9	Ottawa-Hull	827,854	58
10	Hamilton	618,820	36
11	Kitchener	387,319	23
12	London	337,318	24
13	St. Catharines-Niagara	299,935	21
14	Windsor	263,204	17
15	Oshawa	234,779	17
16	Montréal	3,215,665	216
17	Québec	635,184	43
18	Halifax	276,221	62
19	St. John's	122,709	30

Correlation
Input
Input Range: $B$1:$C$19
Grouped By: Columns / Rows
Labels in first row
Output options
Output Range: $E$19
New Worksheet Ply:
New Workbook
OK
Cancel
Help

	Population	n (Actual)
Population	1	
n (Actual)	0.983983	1

An example output is included in the illustration. *r* is displayed in a table or grid format because you can actually input several columns (or rows) of variables at once, and the output displays *r* for each variable pair. If you wish to also output the *p*-value for the correlation, use instead the procedures provided in "In Brief 16.1."

Minitab

The paired data should be aligned in two parallel columns, as shown in the spreadsheet behind the screen capture. Begin with the menu sequence *Stat / Basic Statistics / Correlation;* click *OK.* In the dialogue box, control-click to highlight and then select the names of both variables to be assessed for correlation. Check the box for *Display p-values* and then click *OK.* An example output is included in the illustration.

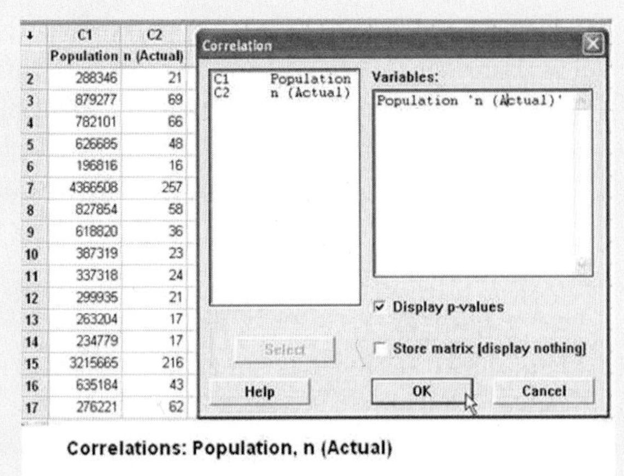

SPSS

The paired data should be aligned in two parallel columns, as shown in the spreadsheet behind the screen capture. Begin with the menu sequence *Analyze / Correlate / Bivariate;* then click OK. In the dialogue box, control-click to highlight the names of both variables to be assessed for correlation and then click the arrow to select them. Click *OK.*

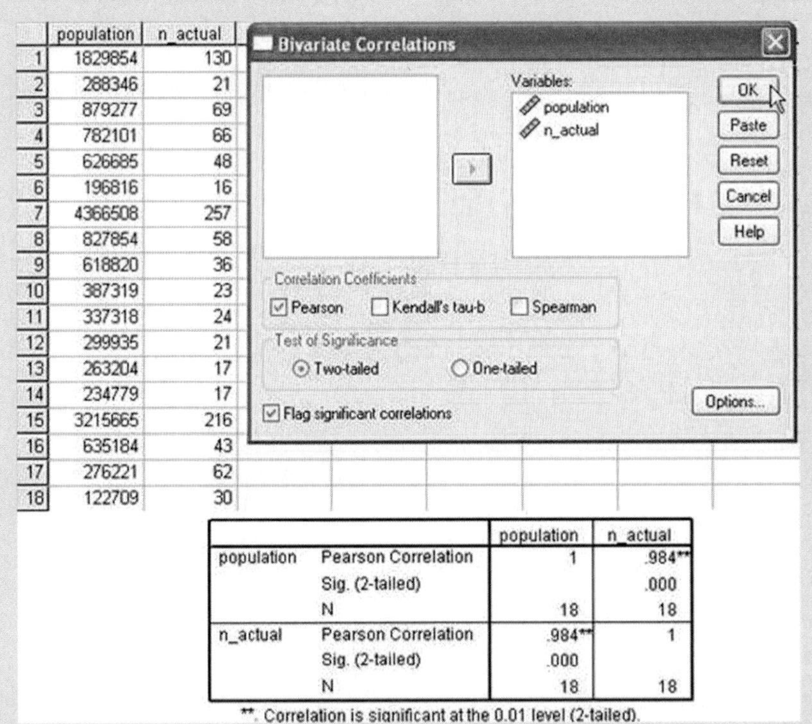

	population	n_actual
1	1829854	130
2	288346	21
3	879277	69
4	782101	66
5	626685	48
6	196816	16
7	4366508	257
8	827854	58
9	618820	36
10	387319	23
11	337318	24
12	299935	21
13	263204	17
14	234779	17
15	3215665	216
16	635184	43
17	276221	62
18	122709	30

		population	n_actual
population	Pearson Correlation	1	.984**
	Sig. (2-tailed)		.000
	N	18	18
n_actual	Pearson Correlation	.984**	1
	Sig. (2-tailed)	.000	
	N	18	18

**. Correlation is significant at the 0.01 level (2-tailed).

An example output is included in the illustration. The *r*- and *p*-values are displayed in a table or grid format because you can actually input several columns of variables at once, and the output displays *r*- and *p*-values for each variable pair.

Cautions When Interpreting Correlation Results

Of course, the output of a correlation analysis becomes less reliable, the farther the distribution is from a true bivariate normal distribution. But even if all conditions for using the measure are satisfied, we still must be careful how we interpret the results.

- Beware of concluding that because a linear correlation between two variables is significant that there is therefore a *causal* relationship between them.

  Suppose, for example, an instructor finds that her students' marks and their levels of class attendance are significantly correlated. Does it follow that attending class more frequently is a cause for getting better marks? Maybe, but not necessarily. Perhaps a third factor, students' motivation, is the causal factor; better attendance and better marks may both follow from this motivation.
- Be careful when drawing conclusions from data that are based on rates or averages. In particular, avoid drawing conclusions that are more specific than the data you started with.

  Suppose you have collected data for the average annual rainfall and the average annual temperatures of various randomly selected cities. And suppose you do find a correlation. This may entitle you to say that if a certain city has a high average temperature, then this tells you something about its likely average rainfall. However, if you learn that a city has a high temperature *this week,* then knowing that there is a correlation between the annual averages tells you nothing about the likelihood of rain this week.

  Similarly, for rates: If you find a correlation between annual unemployment rates and annual usage rates for social services, you cannot draw reliable conclusions about unemployed *individuals* and their particular uses of the services.
- If for two paired variables $r \approx 0$ (i.e., there is no linear correlation), it does not necessarily follow that there is no relationship at all between the variables.

  Try throwing a ball into the air. Call the time of your throw t and the height of the ball above the ground h. The clear relationship between these two variables is shown in Figure 15.7. But the relationship is not linear, so it will not be measured by the value of r.

Figure 15.7
A Nonlinear Correlation

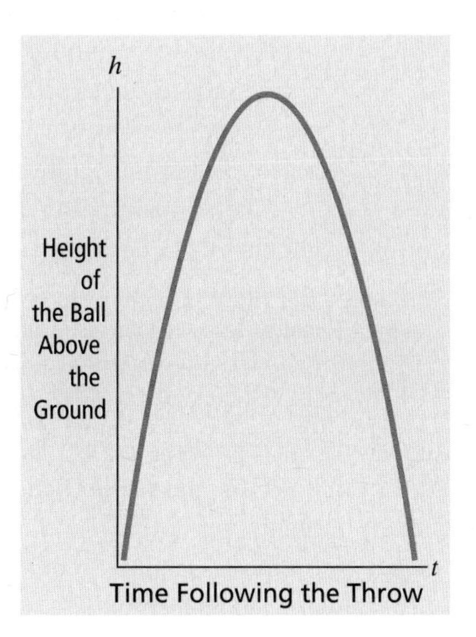

• Hands On 15.1 *Measuring and Interpreting Time-Based Correlations*

One of the easiest types of paired data to obtain from the Internet is time series, where the annual values of one variable (such as unemployment rates, numbers of building permits, numbers of employees in a career field) are given for each year over a period of several years. Good sources of such data include Statistics Canada and Industry Canada. (For more information on time series, see Chapter 17.)

Using the Internet, obtain five sets of such time series paired data of the type mentioned above. In each case, determine if there is a significant correlation of the other variable with time, and if so, note the strength of the correlation.

Then, for each case where a correlation is found, question what it means for a variable to be correlated with time. Is this a causal relationship? If so, which variable causes which? Or is the causation due to some external variable? If so, speculate how this might occur. Obviously, the causal variable does not "cause" the years to progress as they do. So in what sense does the causal variable explain the time series?

NOTE: To proceed directly to building a regression model for a linear correlation that is significant, skip now to Chapter 16.

15.2 Rank-Order Correlation for Ordinal Data

The coefficient of correlation r that we calculated in the previous section is technically known as the *Pearson product-moment coefficient of correlation.* To properly interpret that coefficient, data should be measured at least at the interval scale. If data are at the ordinal (i.e., rank-order) scale, the necessary conditions for r are not satisfied. Also, even data measured at the interval or ratio levels may not meet the other requirements for using r, perhaps because their distributions are skewed rather than normal.

In Chapter 13, we addressed similar issues and introduced nonparametric tests of significance. In Section 13.1, we saw that interval or ratio data can be converted into ranks, if necessary, to prepare for a nonparametric test. Although the finding of a correlation coefficient is not, by itself, a hypothesis test, we have seen that this measure can be subjected to a test for significance.

The nonparametric correlation measure introduced in this section is named in honour of its developer, psychologist Charles Spearman. The measure is called the **Spearman (or Spearman's) rank correlation coefficient,** and is symbolized by r_s. Ironically, despite its revised name and symbol, the calculations for r_s are really no different from those for r. What matters is whether the procedure is applied to raw data at the interval or ratio level (for r), or whether it is applied to ordinal or ranked data (for r_s). The impact is on how the results are interpreted and applied.

(You may possibly encounter elsewhere a different formula for r_s—including a distinctive term: $6\Sigma d^2$—but this is just a shortcut for calculations from the days prior to computers, and not required here. That shortcut formula is based on the properties of ordered integers and ideally requires that there be no tied ranks.)

When calculating r_s, the direction of ranking does not matter, as long as both of the variables are ranked on the same basis. For example, "1" can represent the highest or lowest raw data value, or the best or worst ordinal category. Similar to the case for r, r_s calculates to $+1.0$ if the paired variables' rankings are perfectly and directly correlated. That is, for any element in the sample, the ranks for both compared variables are always identical. On the other hand, if there is no correlation whatsoever between the paired ranks, r_s calculates to 0.0. Finally, if the variables are perfectly inversely correlated, such that when the ranks for one variable get higher, the ranks for the other get lower, the r_s calculates to -1.0.

In short, r_s measures the degree of association between the ranks, but *not* the degree of association between the raw values of the variables if these were originally interval or ratio values.

Example 15.2

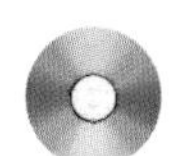

(Based on the ***Canadian_Equity_Funds*** *file)*

A corporate investor claims that there is a correlation between the total asset value of a corporation and the net asset value per share (NAVPS) of the corporation stock. He has hired you as a consultant to test his claim. You have collected the sample data in the file **Canadian_Equity_Funds** (see the columns "Assets" and "NAVPS"). (Note: Ignore records that do not have data for *both* of these variables.)

1. Construct a scatter diagram to compare the values of the assets and NAVPS variables. Describe any appearance of correlation.
2. Use the Pearson correlation coefficient r to test the investor's claim.
3. Use the Spearman rank correlation coefficient r_s to test the investor's claim.
4. Based on all of your results in (1) through (3), what would you conclude in your reply to the investor's claim?

Solution

1. The scatter diagram is shown in Figure 15.8. (The Asset variable is placed on the x-axis and the NAVPS variable is situated on the y-axis.) Although there is a definite visual pattern of the data points that suggests some type of correlation, the pattern is clearly not linear.
2. Using the procedures provided in Section 15.1, we can determine that $r = -0.055$ and the p-value $= 0.691$. (Your software may show only the magnitude of r; but its negative sign can be inferred from the scatter diagram.) The implied null hypothesis for the test is no significant *linear* correlation. Presuming a typical $\alpha = 0.05$ for the test, the p-value is far from being less than or equal to α, so do not reject the null hypothesis. Based on linear correlation, the investor's claim is not supported.
3. Using the procedures discussed in "In Brief 15.2," we can determine that $r_s = -0.454$ and the p-value $= 0.001$. (Your software may show only the magnitude of r_s, but its negative sign can be inferred from the scatter diagram.) The implied null hypothesis for the test is no significant *rank correlation*. Presuming a typical $\alpha = 0.05$ for the test, the p-value is very much less than α, so we *do* reject the null hypothesis. Based on the rank correlation coefficient, the investor's claim *is* supported.

Figure 15.8
Scatter Diagram for Corporate Assets versus NAVPS

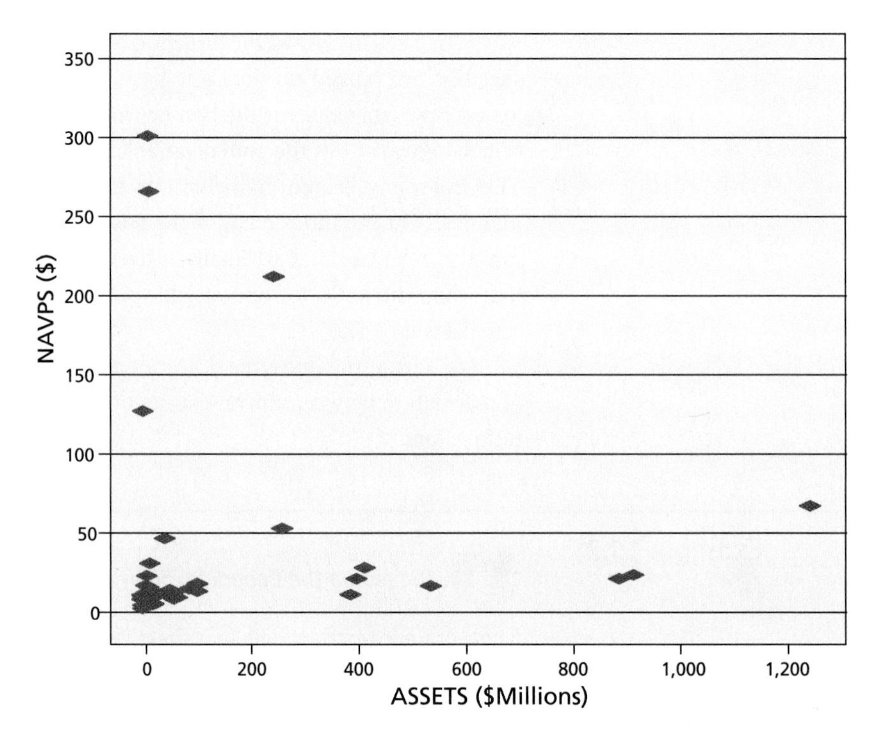

Source: Canadian Equities published on the Globefund.com website, sponsored by *The Globe and Mail.* Found at: http://www.globefund.com/static/romf/generic/tabceq.pdf. Reprinted with permission from *The Globe and Mail.*

4. We can first inform the investor that there is no *linear* correlation between the raw data values for total assets and net asset value per share. Therefore, the investor would be ill-advised to try predicting a specific value for assets based on a specific value for NAVPS, or vice versa. On the other hand, there is a highly significant (inverse) correlation between the ordered ranks of the two variable values. This means it is reasonable to infer *in general* that as the value of total assets increases, the net asset value per share decreases, and vice versa. But we cannot be sure by what exact amount.

• *In Brief 15.2 Measuring Rank Correlation*

Excel

Excel can find r_s directly—provided the data are ordinal numbers or have been converted into ordered-rank numbers. If that is the case, proceed exactly as discussed in "In Brief 15.1," using the ordered numbers as the inputs. If, instead, the data are measured at the interval or ratio level, first use the procedures provided in "In Brief 13.1(a)" to convert them into rank number form. Then proceed as described in "In Brief 15.1," using the newly created rank numbers as the inputs.

If you also want to output the *p*-value for the rank correlation, use instead the procedures given in "In Brief 16.1." Again, if necessary, first convert the raw data into ranked numbers and then use these ranked numbers as the regression inputs.

Minitab

Minitab can find r_s directly—provided the data are ordinal numbers or have been converted into ordered-rank numbers. If that is the case, proceed exactly as discussed in "In Brief 15.1," using the ordered numbers as the inputs. If, instead, the data are measured at the interval or ratio level, first use the procedures provided in "In Brief 13.1(a)" to convert them into rank-number form. Then proceed as described in "In Brief 15.1," using the newly created rank numbers as the inputs.

SPSS

SPSS can find r_s directly in two different ways:

1. If the data are ordinal numbers or have already been converted into ordered-rank numbers, then proceed exactly as discussed in "In Brief 15.1," using the ordered numbers as the inputs.
2. If, instead, the data are measured at the interval or ratio level, begin as in "In Brief 15.1" but add this step: When you are in the dialogue box for correlation, near the middle, check the option to use the Spearman correlation coefficient. The software will then rank the data automatically. The relevant output will appear under the heading "Nonparametric Correlations."

Cautions When Interpreting Rank Correlation Results

A key concern was mentioned earlier: Be sure not to interpret the results as applying at the level of raw numerical measurements. A significant rank correlation tells only how the relative ranks of the variables are correlated.

Also, the Spearman approach should be applied only if the relationship between the paired ranks is roughly **monotonic.** This means that, for the most part, if you steadily increase the value or rank for one variable, the other variable consistently increases (or decreases) in value or rank. Note that in Figure 15.8, the relationship is reasonably, but not perfectly, monotonic: Although NAVPS tends to decrease as assets increase, this is not the case, for example, between the two right-most data points in the figure.

15.3 Contingency Table and Goodness of Fit for Categorical Data

In this chapter, we have explored tools to measure correlations between variables at the ordinal, interval, and ratio levels, and how to test for significance of the correlations. In this section, we measure the association between categorical variables. Discrete quantitative variables can also be compared, if the ordering of specific numerical values is not considered.

This section also introduces **goodness of fit**—a measure of how well a set of data correlates to an expected distribution pattern. Although goodness of fit is not a measure of association (since we analyze just one variable), the analysis shares some procedures with the methods to be introduced next.

Tests for Homogeneity and Independence

For each of years 2001/2002 and 2002/2003, the U.K. Health and Safety Executive recorded dangerous incidents that occurred in the nation's mines. A subset of the organization's data is shown in the following table (assume that these data are also representative of unreported incidents and "near-misses" over those same years).

Incident Type	2001/2002 Number of Incidents	2002/2003 Number of Incidents
Fire underground	9	10
Fan/ventilation problem	19	14
Locomotive incident	35	20

Does it appear from the table that both years' samples have about the same proportions for each of the three incident types, respectively? If so, we could say that in relation to these characteristics, the two populations (i.e., the two years) are **homogeneous.** If instead the proportions of these incidents are different for the different samples, this suggests an association or relationship between the years sampled and the proportions of incident types observed. Essentially, this describes a **test for homogeneity.**

A **test for independence** involves similar calculations, but the question addressed is different. Suppose we take *one* sample of Canadians and find the relative proportions of values for each of two different variables: mother tongue (Anglophone, Francophone, Other) and province of birth (Quebec, Ontario, New Brunswick, Other). If in the sample the proportional distributions of mother tongues are not the same for all of the listed provinces of birth, this suggests that the two variables are related (i.e., not independent). That is, the value of one variable appears to have an impact on the value of the other.

Once collected, the data for either of these tests can be modelled with a contingency table. The table shown below is based on the data used originally in Example 12.10, which in turn were derived from the file **Metallic_Minerals.** There, we tested a claim that the proportion of minerals with submetallic lustres is different for the different mineral densities. If so, we could now infer that the lustre variable is related to the density variable; they are not independent.

	Low Density	High Density	Total
Metallic lustre	16	63	79
Submetallic lustre	13	8	21
Total	29	71	100

This time we will use a **chi-square** (in symbols: χ^2) **distribution** test. The name refers to the expected sampling distribution of the values in the cells of the contingency table under the null hypothesis of independence, provided that the test conditions are satisfied. Since the raw data are at the nominal level, we are really looking at the distributions of counts (frequencies) within the cells of the contingency table, and the χ^2 distribution is an approximation of that distribution.

These are the minimum test conditions required to validly use the χ^2-distribution as described in this section:

- The sample data must be selected randomly. For a test of homogeneity, each sample should be selected independently.
- The frequency expected in each cell of the contingency table should be at least 1. It is preferable if all or most cell values have an expected frequency of at least 5.

In Example 12.10 in Section 12.4, we used the z-test for the differences between proportions. For a 2×2 contingency table (i.e., a table with only two rows and two columns), both the z and chi-square approximations will yield the same test conclusions for the equality of proportions for a given value of α. If the table has more than two rows or columns, then the z model cannot be used.

Before we present the details of the model, recall how we used contingency tables to calculate conditional probabilities in Section 6.2. Based on the table for minerals shown earlier, you should be able to calculate each of the following: (1) P(submetallic), i.e., the marginal probability; (2) P(submetallic | low density); and (3) P(submetallic | high density). The null hypothesis is in effect, that (1) = (2) = (3). That is, that density has no impact on the conditional probabilities. In the chi-square test, if the test statistic χ^2 is very large, this means that the conditional probabilities (2) and (3) are far from being equal to each other or to the marginal probability for submetallic lustre. The **degrees of freedom for the test are based on (Number of rows minus 1) × (Number of columns minus 1).** The p-value for the test can be determined as $P[\chi^2 \geq$ the test statistic $\chi^2 \mid H_0$ is true (for the given degrees of freedom)].

Key mathematical details of the method are described in the context of Example 15.3. The related computer procedures are described in "In Brief 15.3(a)" beginning on page 636.

Example 15.3

*(Based on the **Aircraft_Delay** file)*

A travel broker is comparing two airlines on a number of factors. One factor is the proportion of the airlines' flights that have been cancelled or not on time over a given period. She collected the data in the file **Aircraft_Delay** and compiled the following table. (The category "Other" includes flights delayed for a variety of causes, plus flights that have been diverted.)

	Atlantic Southeast	JetBlue	Total
On time	14,564	5,624	20,188
Cancelled	1,360	10	1,370
Other	6,670	2,201	8,871
Total	22,594	7,835	30,429

At the 0.05 level of significance, test the claim that both sampled airlines experienced the same relative proportions of flights that are on time, cancelled, or have "Other" status, respectively. What are the test statistic, degrees of freedom, and the p-value for the test?

Solution

Step 1: The broker claims that the proportions of flights that are on time, cancelled, or "Other," respectively, are the same for both of the airlines. The claim can be expressed semiformally as:

The respective proportions of cases in each category are the same for both populations.

An alternative to this step is to use formal symbols, such as:

$$\text{Proportion}_{(\text{Atlantic,ontime})} = \text{Proportion}_{(\text{JetBlue,ontime})}$$

$$\text{Proportion}_{(\text{Atlantic,cancelled})} = \text{Proportion}_{(\text{JetBlue,cancelled})}$$

$$\text{Proportion}_{(\text{Atlantic,other})} = \text{Proportion}_{(\text{JetBlue,other})}$$

Step 2: Since the claim in step 1 contains equality, it represents the null hypothesis. The opposite of that hypothesis becomes H_1.

H_0: The respective proportions of cases in each category are the same for both populations.

H_1: The respective proportions of cases in each category are not the same for both populations.

Step 3: The level of significance is given in the problem:

$$\alpha = 0.05$$

Step 4: The choice of test can be determined, given that the broker is testing for the homogeneity of proportions of occurrence of the different values of a categorical variable for two sampled groups (the two airlines). Using chi-square, the degrees of freedom is (Number of rows − 1) × (Number of columns − 1) = (3 − 1) × (2 − 1) = 2. Therefore, the test is:

Use chi-square test for homogeneity, $df = 2$.

Step 5: The depiction of the test model (see Figure 15.9) can be informal. Using the computer, we plan to use the *p*-value approach, so the precise critical value for chi-square does not need to be calculated.

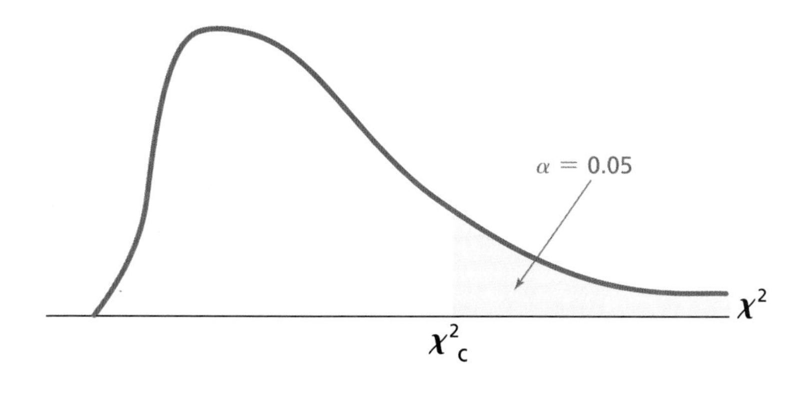

Figure 15.9
Depiction of Model for Step 5, Example 15.3

Note that, similar to an *F*-test, Figure 15.9 appears one-tailed, although the wording of the null hypothesis suggests two-tailed ("equal" versus "not equal"). As its name implies, the test statistic chi-square is based on squaring distances of actual from expected values, and the squares of both positive and negative distances become positive numbers.

Step 6: In this example, the raw data were pre-grouped into a contingency table format. For your software's input procedures, see "In Brief 15.3(a)."

Step 7: Use the computer as discussed in "In Brief 15.3(a)" to perform the analysis. The key automated steps of the procedure are illustrated in Figure 15.10.

The upper left-hand table in the figure shows the observed frequencies, represented by *O*, for each row–column combination. (**Observed frequencies** are the actual numbers of cases that fall into each cell of a contingency table or of a frequency distribution—regardless of what numbers were expected.) In other words, the upper left-hand table is the contingency table for the observed data. The upper

Figure 15.10
Basic Procedures for Chi-Square

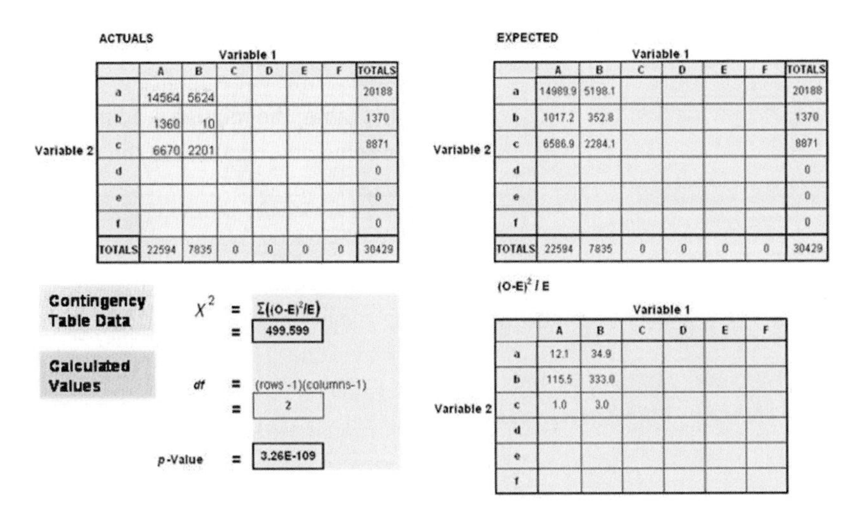

ACTUALS

Variable 2 \ Variable 1	A	B	C	D	E	F	TOTALS
a	14564	5624					20188
b	1360	10					1370
c	6670	2201					8871
d							0
e							0
f							0
TOTALS	22594	7835	0	0	0	0	30429

EXPECTED

Variable 2 \ Variable 1	A	B	C	D	E	F	TOTALS
a	14989.9	5198.1					20188
b	1017.2	352.8					1370
c	6586.9	2284.1					8871
d							0
e							0
f							0
TOTALS	22594	7835	0	0	0	0	30429

Contingency Table Data

Calculated Values

$\chi^2 = \Sigma((O-E)^2/E) = 499.599$

$df = (\text{rows}-1)(\text{columns}-1) = 2$

p-Value $= 3.26\text{E-}109$

$(O-E)^2 / E$

Variable 2 \ Variable 1	A	B	C	D	E	F
a	12.1	34.9				
b	115.5	333.0				
c	1.0	3.0				
d						
e						
f						

Source: U.S. Department of Transportation. Based on data in *Air Travel Consumer Report,* April 2005. Found at: http://airconsumer.ost.dot.gov/reports/2005/0504atcr.pdf

right-hand table records for each cell the corresponding expected frequency (represented by E), under the null hypothesis of equal proportions. (In a chi-square test for two variables, **expected frequencies** are the frequencies that would be expected in each cell of the contingency table if the variables were independent.)

Formally, any cell's expected value is determined by the following formula, where "Row" is the row containing that cell and "Column" is the column containing that cell:

$$\text{Expected frequency} = \frac{\text{Row total}}{\text{Grand total}} \times \text{Column total}$$

In other words, we expect that the *overall* proportion of cases in a given row [i.e., (Row total) / (Grand total)] will be the same for each and every column, so that this proportion times a particular column total gives the expected value for that row–column combination.

The table on the lower right-hand side of Figure 15.10 measures the deviations of each observed value from its corresponding expected value. The deviations are squared so that plus and minus deviations do not cancel each other out. The squared differences are made relative to the expected frequencies (E); otherwise, for example, a difference of 20 when 1,000 is the expected frequency, would appear greater than a difference of 10 when a count of only 50 is expected.

Chi-square is the sum of all difference measures in the lower right-hand table. The computer can determine the degrees of freedom and finds the probability of obtaining at least that high a chi-square if in fact H_0 is true. This probability is the p-value.

In this example:

Test statistic $\chi^2 = 499.6$

p-value $= P(\chi^2 \geq 499.6 \mid H_0$ is true, $df = 2) \approx 0.000$

Since this p-value is less than α, the statistical decision is:

Do reject H_0.

Step 8: Relating the statistical decision back to the original claim, we conclude:

There is sufficient evidence to reject the claim that for both sampled airlines in the table, the relative proportions of flights that are on time, cancelled, or had some other status are the same.

• *In Brief 15.3(a) Chi-Square Tests for Contingency Tables*

(Based on the ***Aircraft_Delay*** *file)*

Excel

Although Excel does not provide a related wizard, it does provide two useful functions.

1. If you manually set up the two tables for observed and expected frequencies, as in Figure 15.10, you can use the function =**chitest(*range1,range2*)**, where ***range1*** is the data range holding the observed data and ***range2*** is the data range holding the expected data. The output of the function is the *p*-value for the chi-square test.
2. If you already know the chi-square test statistic and the degrees of freedom, then you can use the function =**chidist(*chsq,df*)**, where ***chsq*** is the known chi-square value and ***df*** is the degrees of freedom. Again, the output is the *p*-value for the test.

Alternatively, a template has been created for this textbook, which looks like Figure 15.10. Open the text CD file **Template Chi-Square,** and click on the spreadsheet tab "Contingency Tables." Input the observed frequencies in the yellow area, starting from the upper left. All other calculations will be automated. If the raw data are not yet compiled into a contingency table, see "In Brief 3.5(b)."

Minitab

Minitab can accept the data either as (1) a contingency table or as (2) raw data in columnar form. Both cases are explained in the text below and illustrated behind the relevant screen captures. If you already know the chi-square test statistic and the degrees of freedom, then see item (3) below to find the *p*-value.

1. In this case [see part (a) of the illustration], input the data from the contingency table without totals, as shown in the upper left-hand side of the illustration. Use the menu sequence *Stat / Tables / Chi-square test (table in worksheet).* In the dialogue box, control-click to highlight and then select the headings of the columns containing the data. Then click *OK.* An example line of the output is included in the illustration.
2. In this case [see part (b) of the illustration], input the data in columnar form (each row representing the data from a single observation) as shown. (Due to the large number of data observations, the example is simulated.) Use the menu sequence *Stat / Tables / Cross Tabulation and Chi-square.* In the dialogue box,

highlight and select the headings, in turn, for the variable to become the rows of the table and the variable to become the columns. Under *Display,* select *Counts.* Then click on the button *Chi-Square* and, in the next dialogue box, choose *Chi-Square analysis.* Click *OK* and then *OK* again. An example line of the output is included in the illustration. The contingency table will also be displayed.

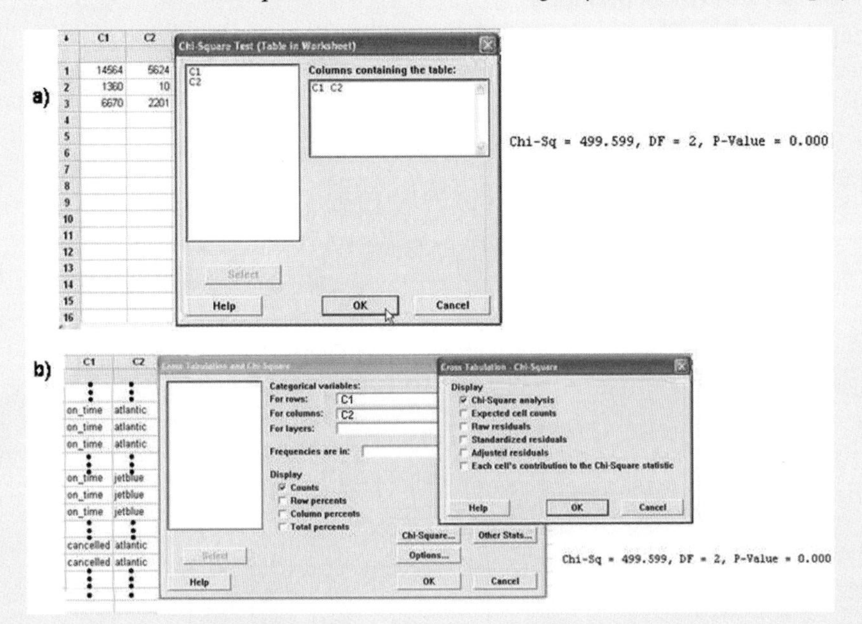

3. You may have predetermined the test statistic χ^2 and the degrees of freedom by using formulas. To find the p-value, use the menu sequence *Calc / Probability Distributions / Chi-square.* In the resulting dialogue box (not shown here), input the value for the degrees of freedom and then click *Input constant* and input the χ^2-value. Click *OK.* The output will be a less than cumulative probability; subtract this result from 1 to obtain the p-value, which is the conditional probability that $F \geq$ the test statistic.

SPSS

SPSS can accept the data either as (1) grouped (i.e., similar to a contingency table) or as (2) raw data in columnar form. Both cases are explained in the text below and illustrated with the relevant screen captures. If you already know the chi-square test statistic and the degrees of freedom, then see item (3) below to find the p-value.

1. In this case [see part (a) of the illustration], input the grouped data in the format shown in the upper left corner of the illustration. (Use the left two columns to input—once only—each possible combination of the different variables. To the right of each combination, input the frequency for that combination.) Then use the menu sequence *Data / Weight Cases* and, in the resulting dialogue box (not shown here), click *Weight Cases by* and then highlight and select the variable containing the frequencies. Click *OK.*

(Warning: Once you have found the desired *p*-value, below, rerun the *Weight Cases* routine and reset before you perform any subsequent calculations.)

2. In this case [see part (b) of the illustration], input the full set of raw data in columnar form (each row representing the data from a single observation), as shown. (Due to the large number of data observations, the example is simulated.)

For either case (1) or (2): At this point, use the menu sequence *Analyze / Descriptive Statistics / Cross Tabs.* In the dialogue box, highlight and select the headings, in turn, for the variable to become the rows of the table and the variable to become the columns. Click the button *Statistics,* and choose *Chi-square* in the next resulting dialogue box. Click *Continue* and then *OK.* An example portion of the output is included in the illustration.

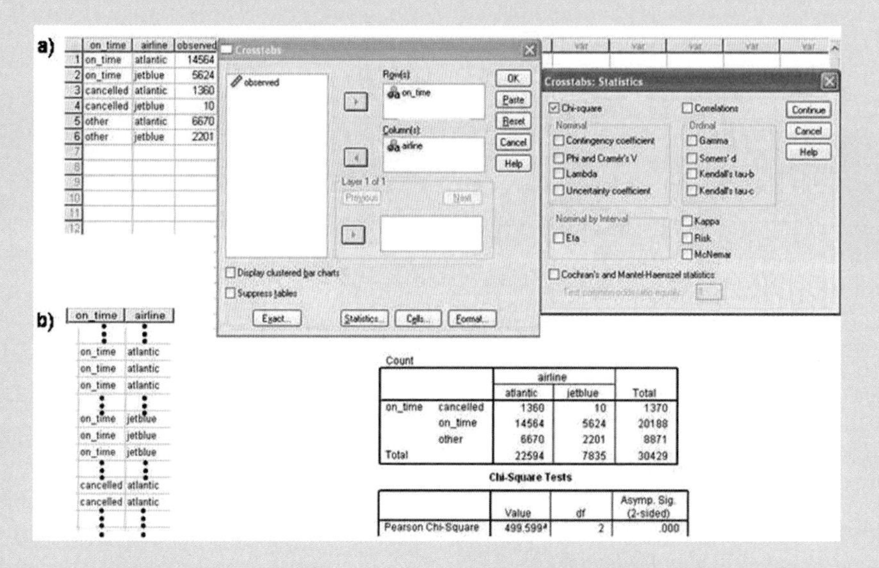

3. You may have predetermined the test statistic χ^2 and the degrees of freedom by using formulas. Find the *p*-value by using the cumulative distribution function for χ^2. Place at least one value in the worksheet area and use the menu sequence *Transform / Compute.* In the resulting dialogue box (not shown here), specify a name for the target variable (i.e., column) to display the answer; then in the area for *Numeric Expression,* input a formula of this format: **1-CDF.CHISQ(Chisq,df)** where **Chisq** is the chi-square value, and **df** is the degrees of freedom. Note that the CDF result (which is less than cumulative probability) is subtracted from 1 to obtain the *p*-value, which is the conditional probability that $\chi^2 \geq$ the test statistic.

In Example 15.4, we apply the "In Brief 15.3(a)" methods to a test for independence.

Example 15.4

*(Based on the **SmallCaps** file)*

An investment banker has been comparing the performance of top small-capitalization companies, with a focus on changes in their revenues and accumulated assets. He compiled the sample data that are stored in the file **SmallCaps.** Based on this research, he claims that from one year to the next, whether a firm's revenue increases or decreases is independent on whether its assets increase or decrease. Test this claim at a 0.04 level of significance.

Solution

Step 1: The claim can be expressed, in words, as:

Revenue increases and asset increases are not related.

What the procedures actually test is the same type of claim as for the test of homogeneity (compare Example 15.3):

The respective proportions of cases increasing or decreasing in revenue are the same for both asset groups.

Step 2: From the second version of the claim in step 1, we see that it contains equality. Therefore, that claim becomes the null hypothesis and the opposite claim becomes H_1.

H_0: The variables are independent (i.e., the respective proportions are equal).

H_1: The variables are related (i.e., the respective proportions are not equal).

Step 3: The level of significance is given in the problem:

$$\alpha = 0.04$$

Step 4: We select a test for independence (i.e., for an equality of proportions or conditional probabilities for the values of one variable, given the values of the other). Each variable has two levels or categories: "0" for a decrease and "1" for an increase. Using chi-square, the degrees of freedom is $(2 - 1) \times (2 - 1) = 1$. Therefore, the test is:

Use the chi-square test for independence, $df = 1$.

Step 5: We can informally depict the test model (see Figure 15.11). Using the computer we plan to use the *p*-value approach, so the chi-square critical value does not need to be calculated.

Step 6: The data are stored in a columnar format. Use the input procedures given in "In Brief 15.3(a)." (You might also check at this point regarding the condition that every cell in the contingency table should have an expected value of at least 5. Alternatively, depending on your software, cells where the condition is not met will be flagged in the output for step 7.)

Step 7: Using the procedures in "In Brief 15.3(a)," perform the analysis, find the test statistic, and calculate the *p*-value. Our results are:

Test statistic $\chi^2 = 1.383$
p-value $= P(\chi^2 \geq 1.383 \mid H_0$ is true, $df = 1) = 0.240$

Figure 15.11
Depiction of Model for Step 5, Example 15.4

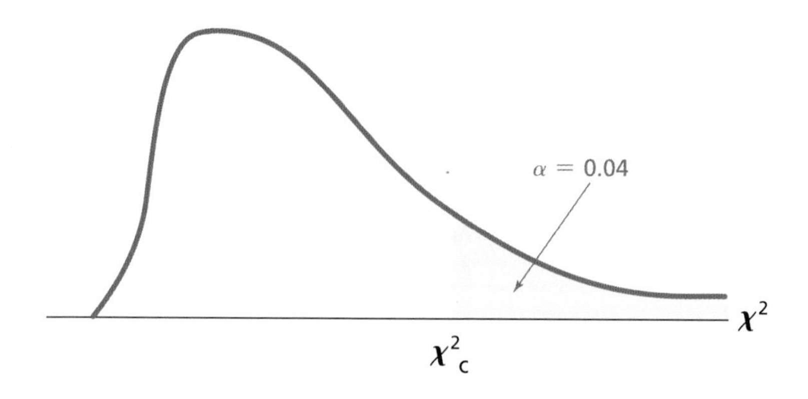

Since this p-value is not less than or equal to α, the statistical decision is:

Do not reject H_0.

Step 8: Relating the statistical decision back to the original claim, we conclude:

There is not sufficient evidence to reject the claim that for top small-capitalization companies, the variables for revenue increase and asset increase are independent.

Tests for Goodness of Fit

We have discussed the chi-square distribution for testing the association between two categorical or discrete variables. In this section, we introduce "one-way" chi-square tests for goodness of fit. With this type of test, we can assess the correspondence of an observed frequency distribution with our expectation of what that distribution might be. For example, if we expect that all of the values of a variable are equally likely, but in practice the frequencies are not exactly equal, then we can test for the significance of that deviation. Or if we plan to conduct a certain test for which the Poisson or binomial distribution is a required condition, we can test for goodness of fit of the sample data with the required distribution.

The basic formula for the test statistic χ^2 is the same as for the previously described tests. As shown in Example 15.5, you can set the observed frequencies into an array and compare each value to an expected frequency based on a presumed distribution model. (In a goodness of fit test, expected frequencies are the frequencies expected in the frequency distribution if the null hypothesis was true.) χ^2 is based on the squared deviations of the actual from the expected frequencies and if χ^2 is very large (compared to the critical value), this means that the observed frequencies do not fit the expected pattern. The **degrees of freedom for the test are based on (number of categories or discrete values of the variable) minus 1.** The p-value for the test can be determined as $P[\chi^2 \geq \text{the test statistic } \chi^2 \mid H_0 \text{ is true (for the given degrees of freedom)}]$.

Example 15.5

*(Based on the **National_Parks** file)*

An environmental group is arguing with a politician about whether the numbers of national parks in different provinces have been fairly allocated. The politician claims that, at least for the non-Atlantic provinces, the number of national parks per province is consistent with the distribution of provincial populations. To back this up, she compiled the data shown below (based on data in the file **National_Parks**). At the 0.05 level of significance, test the politician's claim.

Province	Population	Proportional Distribution of Population	Number of National Parks
British Columbia	4,146,600	0.142088	8
Alberta	3,153,700	0.108065	5
Saskatchewan	994,800	0.034088	2
Manitoba	1,162,800	0.039845	2
Ontario	12,238,300	0.419358	6
Quebec	7,487,200	0.256557	4
Total	**29,183,400**	1	27

(Note: Technically, the parks data are for a population, not a sample, but more parks are likely to be designated in the future. The following solution assumes that the current parks distribution is a *sample* of the long-term distribution that politicians are developing over time.)

Solution

Step 1: The politician claims that the frequency distribution of the parks is approximately the same as the frequency distribution of the populations. The claim can be expressed as:

For every individual category, its observed and expected proportions are equal.

Step 2: Since the claim in step 1 contains equality, it represents the null hypothesis. The opposite of that hypothesis becomes H_1.

H_0: For every individual category, its observed and expected proportions are equal.

H_1: The observed and expected proportions are not equal for every individual category.

Step 3: The level of significance is given in the problem:

$$\alpha = 0.05$$

Step 4: We are testing whether the proportions in the observed distribution match the proportions in an expected distribution. Using chi-square, the degrees of freedom is (Number of categories − 1) = (6 − 1) = 5. Therefore, the test is:

Use the chi-square test for goodness of fit, $df = 5$.

Step 5: We can depict the test model informally (see Figure 15.12). We plan to use the p-value approach and do not need to calculate the chi-square critical value.

Step 6: In this example, the raw data for the observed frequency distribution is given and the population distribution provides the basis for the expected distribution. For your software's input procedures, see "In Brief 15.3(b)."

Step 7: Use the computer procedures discussed in "In Brief 15.3(b)" to perform the analysis, find the test statistic, and calculate the p-value. Key automated steps of the procedure are illustrated in Figure 15.13.

Figure 15.12
Depiction of Model for Step 5, Example 15.5

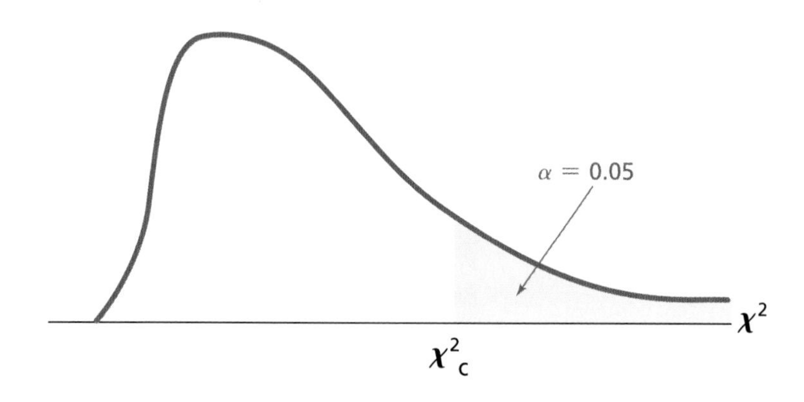

Figure 15.13
Basic Procedures for Goodness of Fit Test

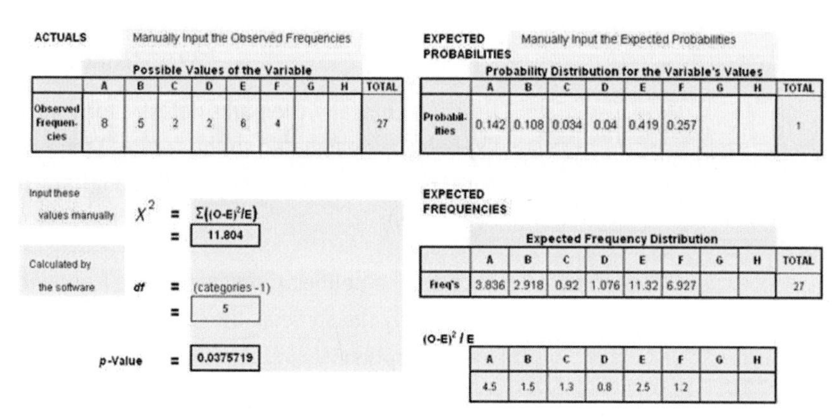

ACTUALS — Manually Input the Observed Frequencies

Possible Values of the Variable

	A	B	C	D	E	F	G	H	TOTAL
Observed Frequencies	8	5	2	2	6	4			27

EXPECTED PROBABILITIES — Manually Input the Expected Probabilities

Probability Distribution for the Variable's Values

	A	B	C	D	E	F	G	H	TOTAL
Probabilities	0.142	0.108	0.034	0.04	0.419	0.257			1

Input these values manually

X^2 = Σ((O-E)²/E)
= 11.804

Calculated by the software

df = (categories - 1)
= 5

p-Value = 0.0375719

EXPECTED FREQUENCIES

Expected Frequency Distribution

	A	B	C	D	E	F	G	H	TOTAL
Freq's	3.836	2.918	0.92	1.076	11.32	6.927			27

(O-E)² / E

A	B	C	D	E	F	G	H
4.5	1.5	1.3	0.8	2.5	1.2		

Source: Original map data provided by *The Atlas of Canada.* Found at: http://atlas.gc.ca/site/english/learningresources/facts/parks.html. Produced under licence from Her Majesty the Queen in Right of Canada, with permission of Natural Resources Canada.

The upper left-hand table shows the actual ("Observed") frequencies of parks for each category of the variable province. The upper right-hand table records for each category (i.e., each province) the corresponding *expected proportion* of cases for that category. One more table is needed—the middle table at the right of the figure: The expected proportions of parks for each province are converted into expected frequencies for each province. For any category (province): Expected frequency = (Expected proportion) × (Total number of actual observations). The table on the lower right of the figure measures the deviations of each observed frequency from the corresponding expected frequency. Observe that this formula is the same as for the previous chi-square tests.

Chi-square is again the sum of all difference measures in the lower right-hand table. The computer determines the degrees of freedom and then it finds the probability of obtaining at least that high a chi-square if in fact the respective proportions are equal. This probability is the *p*-value.

In the example:

Test statistic $\chi^2 = 11.804$
p-value $= P(\chi^2 \geq 11.804 \mid H_0 \text{ is true}, df = 5) = 0.038$

Since this *p*-value is less than α, the statistical decision is:

Do reject H_0.

Step 8: Relating the statistical decision back to the original claim, we conclude:

There is evidence to reject the claim that distribution of national parks in the non-Atlantic provinces is essentially the same as the relative distribution of populations in the provinces.

• *In Brief 15.3(b) Chi-Square Tests for Goodness of Fit Tests*

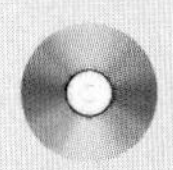

(Example data based on the ***National_Parks*** *file; population data based on CanadaLegal.info)*

Excel

Excel does not offer a related function but it does have two functions that could be utilized for a semi-automated approach: (1) **=chitest(*range1,range2*)** and (2) **=chidist(*chsq,df*)**. For details on these functions, see "In Brief 15.3(a)."

Alternatively, a template that looks like Figure 15.13 has been created for this textbook. Open the text CD file **Template Chi-Square** and click on the spreadsheet tab for *Goodness of Fit.* Input the observed frequencies in the yellow area on the upper left, starting from the left cell. Input the expected *proportions* in the yellow area on the upper right, also entering from left to right. All other calculations will be automated.

Minitab

(*Minitab versions earlier than 14.20 may not include this option.*) Input first one column with the expected frequencies (or the expected proportions) and then another column with the observed frequencies. Then use the menu sequence *Stat / Tables / Chi-square goodness of fit test (one variable)*. On the upper right of the dialogue box (see the screen capture), specify *Observed counts* and select the variable for the observed data. If you have a column of expected frequencies, click the radio button on the lower right, and specify the column; if you have a column of expected proportions, click on *Proportions specified by historical counts* and specify the column. Click *OK.* The partial output is displayed in the illustration.

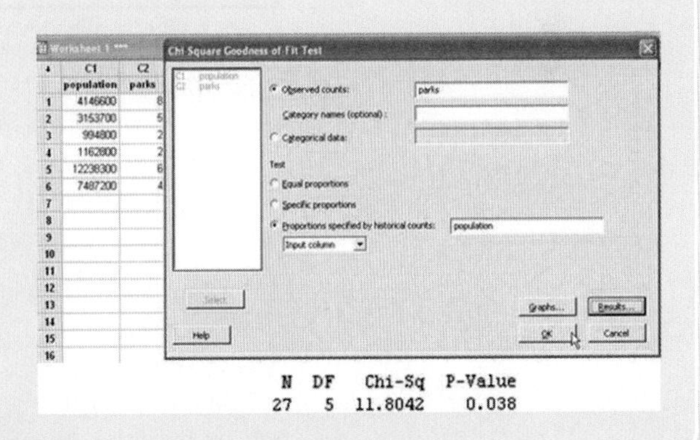

SPSS

For this procedure, SPSS requires two columns of initial input: (1) a sequence of identifiers for the categories and (2) the corresponding observed frequencies (see the spreadsheet behind the screen capture). Next, use the menu sequence *Data / Weight Cases* and in the resulting dialogue box (not shown), click on *Weight Cases by* and then highlight and select the variable containing the frequencies. Click *OK.*

Now use the menu sequence *Analyze / Nonparametric tests / Chi-square.* In the dialogue box (see the screen capture), highlight and select the heading for just the categories column—*not* for the frequencies column. If the null hypothesis is for equal proportions for all categories, select that option in the *Expected Values* area. Otherwise, select *Values* in the *Expected Values* area and then input all of the categories' proportions *without showing decimals,* one at a time, clicking *Add* after each entry. Then click *OK.* (Your results may vary slightly, based on the number of decimal places you include for these proportions.) An example portion of the output is included in the illustration.

(Warning: Remember to rerun the *Weight Cases* routine and reset before performing any subsequent calculations.)

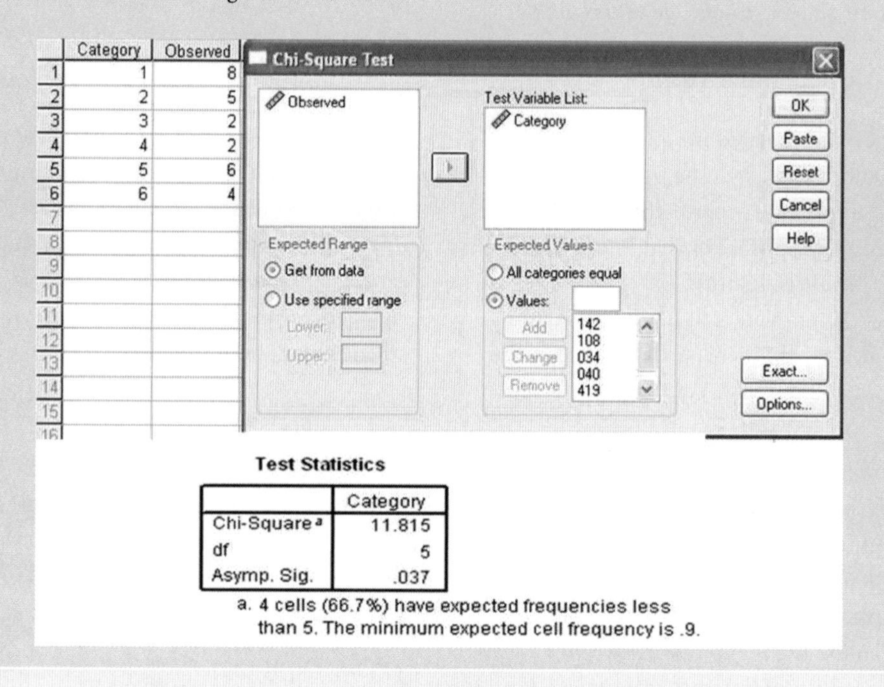

	Category
Chi-Square[a]	11.815
df	5
Asymp. Sig.	.037

a. 4 cells (66.7%) have expected frequencies less than 5. The minimum expected cell frequency is .9.

• Hands On 15.2 Testing Goodness of Fit to a Presumed Uniform Distribution

One commonly used null hypothesis for a goodness of fit test is that all categories have the same expected frequency. In other words, the presumed distribution is uniform. The expected proportions in these cases are easy to calculate: Each cell's expected proportion of the total is $1/k$, where k is the number of distinct categories.

Try the following application for such a test. Search online for an L. L. Bean catalogue (or a similar catalogue, such as Sears). You will notice that for any given product category, there are a finite number of alternative models or styles. For example, at the time of writing, the footwear section of the L. L. Bean catalogue listed 20 different models under "Mocs and Clogs," 17 models under

"Handsewn Loafers and Moccasins," 15 variations under "Oxfords and Chukka's," and so on.

You could sample additional footwear product categories and record the number of different types or models in each product category. Test the hypothesis that the distribution of the number of types per product category is roughly uniform.

Then do the same for several other sections of the catalogue, in turn. These are two additional suggestions:

1. Look in the girls' outerwear clothing section. Product categories include "Jackets and Parkas" and "Vests," among others. Is the distribution of numbers of models for each product category uniform?
2. Try the backpack section. Product categories include "Youth Back Packs," "Teen Back Packs," and others. Again, is the distribution of number of models for each product category uniform?

Find some other product lines to check in the same way.

Chapter Summary

15.1 Identify a linear correlation and calculate the sample coefficient of correlation *r*, assess the strength and significance of the correlation, and describe some cautions when interpreting the correlation results.

When there is an **association** between two quantitative variables that follows a linear pattern in a scatter diagram, we say there is a **linear correlation** between the variables. For a measure of the strength of the linear correlation, we calculate a **sample coefficient of correlation** (***r***) (see Formula 15.1). r has both a sign ($+$ or $-$) and a magnitude. In a **positive correlation,** if the value of one variable increases, then the corresponding value of the other variable increases as well. If the sign is negative, this means that there is an **inverse correlation** between the values of the two associated variables.

For the most effective use of this model, the set of paired data should be a random sample and the population should have a **bivariate normal distribution;** i.e., in the scatter diagram, the deviations of actual y-values from the predicted y-values (as estimated by a regression line) should be normally distributed for all parts of the line. The data should be measured at the interval or ratio level. The correlation is **significant** if there is sufficient evidence that the apparent correlation in the sample represents a true population parameter and not just a random sampling effect. The population parameter that corresponds to r is ρ (pronounced "rho"); r is significant if in a test of the null hypothesis $\rho = 0$ (i.e., "there is no correlation"), H_0 would be rejected.

When a significant correlation is found, beware of concluding that there is necessarily a *causal* relationship between the variables. Also, for data involving rates or averages, avoid drawing conclusions that are more specific than the data you started with. Further, note that if there is no linear correlation between variables, there may still be an association between them.

15.2 Calculate for suitable—or transformed—data the Spearman rank correlation coefficient r_s, assess the strength and significance of the correlation and describe some cautions required when interpreting the correlation results.

If associated data are ordinal or do not meet other requirements for calculating *r*, we can assess the correlation by calculating the **Spearman (or Spearman's) rank correlation coefficient, r_s.** If the data are not ordinal, first convert the data into rank numbers; then find r_s and assess its strength and significance as you would *r*. But keep in mind that you are testing the correlation of *ranks*—not the correlation of the original numbers. A requirement for the Spearman approach is that the relationship between the paired ranks should be roughly **monotonic.** For the most part, if you steadily increase the value or rank of one variable, the other variable consistently increases (or decreases) in value or rank.

15.3 Describe and distinguish the concepts of tests for homogeneity, independence, and goodness of fit, and apply these tests where appropriate.

Suppose you sample a group of men and a group of women and observe the proportion of each group that owns (1) a car, (2) a truck, or (3) no vehicle. If each sampled group has about the same proportion of individuals in each of the named categories, we could say that in relation to these categories, the two gender populations are **homogeneous.** If instead the respective proportions are different for the different samples, this suggests a relationship between the groups sampled and the proportions found in the different categories. Determining which case applies is a **test for homogeneity.**

A **test for independence** involves similar calculations but addresses a different question. We take *one* sample (e.g., of Canadians) and find the proportions in different combinations of two categories (e.g., mother tongue and province of birth). If the proportional distributions of one variable are not the same for all of the listed categories of the other, this suggests that the two variables are related (i.e., not independent).

The basis for both of the above tests is the **chi-square (χ^2) distribution,** by which we assess the extent to which the **observed** and **expected** frequencies differ for each combination of values for the two categorical variables. For these tests, the sample data must be selected randomly, and for a test of homogeneity, each sample should be selected independently. It is preferable if all or most cell values have an expected frequency of at least 5.

A related chi-square test, although it applies to just a single categorical variable, is the test for **goodness of fit.** We assess the correspondence of an observed frequency distribution with our expectation of what that distribution might be.

• Key Terms

association between two variables, p. 618
bivariate normal distribution, p. 620
chi-square (χ^2) distribution, p. 632
expected frequencies, p. 635
goodness of fit, p. 631
homogeneous, p. 632
inverse correlation, p. 619
linear correlation, p. 619
monotonic, p. 631
observed frequencies, p. 634

Exercises

Basic Concepts 15.1

1. Explain, using conditional probabilities, what it means to say that there is an association between two variables.
2. What is the symbol for the sample coefficient of correlation?
3. What is the range of possible values for the sample coefficient of correlation? What is the interpretation of a negative value for this coefficient?
4. Give examples of values for the sample coefficient of correlation that would suggest a strong correlation between two variables. Give examples of values that would suggest a weak correlation.
5. List the conditions that are required in order to validly use the model of correlation as presented in this chapter.
6. Explain the difference between assessing the *strength* of a correlation and assessing the *significance* of a presumed correlation. What procedures would you use to assess each of these values?
7. Explain the difference between the symbols r and ρ and the role that each plays in assessing correlation.
8. List some cautions to keep in mind when interpreting the results of a correlation analysis between two variables.
9. The following data refer to the shelf space allocated to each product at a small grocery store and the corresponding sales of those products.

Shelf Space (cm)	Sales ($)
30	250
40	299
50	350
60	361
70	375

a) Calculate r.
b) Does the correlation appear to be strong? Explain the basis for your answer.
c) Is the correlation significant? Explain the basis for your answer.
d) Repeat (a) through (c) after adding the following additional row to the data table:

Shelf Space (cm)	Sales ($)
54	359

Applications 15.1

10. An official of a dental association observed that the mean ages of workers in various dental-related fields tend to vary over time. She wondered if, in general, the mean ages of workers in dental-related occupations are correlated. She found a data set for worker ages in health occupations (see the file **Age_Health_Occupation**). The ages are displayed for each of a series of years; presume that these data are representative for past and future years.
    a) Create a scatter diagram to compare the ages, for any given year, of dental assistants and dental hygienists. Based on the diagram, do you expect that there is a linear correlation? If so, how well do the conditions seem to apply for calculating a correlation coefficient? Explain.
    b) What is the apparent strength of the correlation between the ages of workers in these two groups? Justify your answer.
    c) Is the apparent correlation (if any) between the two groups' ages significant? Justify your answer.
11. Repeat all parts of Exercise 10 but this time compare ages for the occupations dental assistant and dental technician.
12. Repeat all parts of Exercise 10 but this time compare ages for the occupations dental assistant and denturist.
13. The cost of health care is a major concern that all political parties have recognized. A researcher has compiled data to compare expenditures across Canada on each of several health-related categories, such as drugs, hospitals, and physicians. (See the file **Health_Expenditures.**)
    a) Create a scatter diagram to compare expenditures for any given year for hospital care versus drugs. Does the diagram suggest a linear correlation between these variables? How well do the conditions seem to apply for calculating a correlation coefficient? Explain.
    b) What is the apparent strength of the correlation between these two categories of expenditures? Justify your answer.
    c) Is the apparent correlation (if any) between these areas of expenditure significant? Justify your answer.
14. Repeat all parts of Exercise 13 but this time compare expenditures for the categories drugs and physicians.
15. Review your results from Exercises 13 and 14 in connection with these facts: The data are not a random sample,

but rather a time-ordered series of results for several years in a row. Over that time, *all* health-related costs—of whatever category—increased. Would this effect influence whether we found correlations between particular pairs of variables? Explain.

16. A baseball player has just become a free agent and is studying recent data on contracts negotiated by other free agents. (See the file **Free_Agents.**)
    a) Create a scatter diagram to compare the durations (in years) of recent free-agent contracts and the annual value of those contracts ("Yearly$"). How well does the diagram suggest a linear correlation between these variables? Explain.
    b) What is the apparent strength of the correlation between the length and yearly value of the free agent contracts? Justify your answer.
    c) Is the apparent correlation (if any) between these two variables significant? Justify your answer.
    d) If there is a significant correlation between the length and yearly value of the free agent contracts, which of the following statements do you think is most plausible?
        i) If a free agent negotiates a longer contract, this will tend to cause the negotiated yearly salary to be greater.
        ii) If a free agent negotiates a greater yearly salary, this will tend to cause the negotiated contract duration to be longer.
        iii) There is more likely a third unlisted factor that, if present, tends to cause both higher salaries and longer contract durations.

    Explain your answer.

Basic Concepts 15.2

17. Explain the distinction between the types of correlation coefficient: r and r_s.
18. The Spearman rank correlation coefficient measures the degree of association between variables that are at what level of measurement?
19. Is the following statement true or false? Explain your answer. "If the raw data for a pair of related variables are measured at the ratio level, then there is no valid way to compare the variables by calculating an r_s coefficient."
20. Is the following statement true or false? Explain your answer. "If the raw data for a pair of related variables are measured at the ratio level, then there would never be a good reason to compare the variables using r_s instead of using r."
21. Explain what it means that the relationship between ranks that can be compared with a Spearman rank correlation coefficient must be roughly monotonic.
22. Give examples of values for r_s that would generally indicate a strong relationship between the ranks for the two compared variables.
23. How can you assess whether, for a given set of data, the computed value for r_s is significant?
24. Ten provincial parks were ranked by experienced campers on these two characteristics: cleanliness of the washrooms and the uniqueness of the nature trail. These were the results:

	Parks				
	A	B	C	D	E
Cleanliness rank	5	9	3	1	4
Nature trail rank	3	9	6	2	5
	F	G	H	I	J
Cleanliness rank	8	6	10	2	7
Nature trail rank	10	7	8	1	4

    a) Calculate r_s between these rankings.
    b) Is the Spearman correlation significant? Is it strong?

Applications 15.2

Exercises 25–28 are based on data in the file **Environmental_Sustainability.**

25. The Yale Center for Environmental Law and Policy, in association with three other research groups, has published an international *Environmental Sustainability Index.* This index documents the perceived ability of the world's nations to protect and sustain their individual and collective environments. "Sustainability" is assessed for each of a number of variables—including air quality, water quality, water quantity, and reduction of stresses on ecosystems.

    Data from 50 randomly selected countries, of 146 nations in the report, have been stored in the file **Enviromental_Sustainability.** For each variable, there are two columns in the data file: The first lists each country's score on that variable; the second column lists the relative ranks of the score data for that variable (in relation to all countries tested by the researchers).
    a) Calculate the Spearman rank correlation coefficient between the ranked-data columns for water quantity ("RWatQuan") and water quality ("RWatQual").
    b) Based on your answer to (a), does the relationship between the ranked scores for sustaining water quality and sustaining water quantity appear to be strong or weak?
    c) Is the relationship between these ranked variables significant?
26. Using the data set specified in the previous exercise, find r_s for the correlation between countries' sustainability *ranks*

for water quality ("RWatQual") and for reducing waste/consumption pressures ("RWaste"). Describe the strength and significance of any apparent correlation.

27. Using the data set specified in Exercise 25:
    a) Calculate the Spearman rank correlation coefficient between the ranked values for air quality ("RAirQual") and for reduction of ecological stresses ("REco_Str"). Describe the strength and significance of any apparent correlation.
    b) Calculate the Pearson correlation coefficient r for the *unranked* column scores for air quality ("AirQual") and for reduction of ecological stresses ("Eco_Str"). Describe the strength and significance of any apparent correlation.
    c) Is there a difference in the strengths of correlation in the two cases above—the first comparing ranked values and the second comparing unranked values for the same underlying data? If so, describe the difference and explain why this might occur.

28. Using the data set specified in Exercise 25:
    a) Calculate the Spearman rank correlation coefficient between ranked values for air quality ("RAirQual") and for population ("RPopul"). Describe the strength and significance of any apparent correlation.
    b) Calculate the Pearson correlation coefficient r for the unranked column scores for air quality ("AirQual") and for population ("Popul"). Describe the strength and significance of any apparent correlation.
    c) Is there a difference in the strengths of correlation in the two cases above—the first comparing ranked values and the second comparing unranked values for the same underlying data? If so, describe the difference and explain why this might occur.

Exercises 29–31 are based on data in the file **Tech100_Companies.**

29. An investor is examining the returns for companies identified as the "Top 100" technology companies by *Canadian Business* magazine. (See the file **Tech100_Companies.**)
    a) Calculate the Spearman rank correlation coefficient between the ranked-data columns for total sales in 2004 ("Rank2004") and total sales in the previous year ("Rank2003"). (Omit records for companies that did not make the "Top 100" list in both the years.)
    b) Based on your answer to (a), for companies that made the "Top 100" list in both 2003 and 2004, is there a significant correlation between their rankings in each of those two years? Is the correlation, if any, strong?

30. Using the data set specified in the previous exercise, find r_s for the correlation between the listed companies' sales ranks in 2004 ("Rank2004") and their profit ranks in 2004 ("Rank_pro"). Describe the strength and significance of the correlation, if any.

31. In the data set specified in Exercise 29, the variable "Change" shows, for each listed company, the percentage positive or negative change in the company's sales dollars from 2003 to 2004. *Rank* these "Change" values for all of the listed companies and create a new column "RChange" that records each company's respective rank. [For example, the company Sirit Inc. had the highest change (+341%) in sales, so the value for "RChange" next to Sirit is 1, the highest rank.]
    a) Calculate r_s for the correlation, if any, between the ranks for change in sales versus (ranked) total sales. Describe the strength and significance of any correlation.
    b) In preparing for (a), why was it necessary to create the new column "RChange"? That is, why not directly compare the original change values in "Change" with the ranked values in "Rank2004"?

Basic Concepts 15.3

32. What is meant by a test for independence?
33. Explain the difference between a test for independence and a test for homogeneity.
34. For a chi-square test of independence based on contingency tables, how are the degrees of freedom calculated?
35. Explain what is actually measured by the test statistic chi-square (χ^2).
36. List the minimum test conditions that are required to validly use the χ^2-distribution as described in this chapter.
37. What is the name of a test that uses χ^2 for assessing the correspondence of an observed frequency distribution with some particular distribution that is expected to apply for the data?
38. For the type of test described in Exercise 37, how is the degrees of freedom calculated?
39. Diane, of Dianne's Cheesecake Shop, believes that the gender of a customer is related to the type of cheesecake that is purchased. She collected the following data on 100 customers.

	Type of Cheesecake		
Gender	Plain	Cherry	Total
Male	30	15	45
Female	20	35	55
Total	50	50	100

    a) Develop a table of expected frequencies for this problem.
    b) Calculate the sample χ^2-value.
    c) What is the value of df?
    d) If Diane's claim is formally tested with a χ^2-test, what would be the resulting p-value and statistical decision? What is the conclusion, in words?

40. The owner of the cheesecake shop (see Exercise 39) also believes that she knows the distribution of sales quantities for the different subvarieties of cherry cheesecake. Her expectations about the distribution are given in the top row of the following table; actual observed frequencies are shown in the second row:

	Variations of Cherry Cheesecake					
	A	B	C	D	E	Total
Expected probabilities	0.20	0.20	0.25	0.20	0.15	1.00
Observed frequencies	10	12	15	10	8	55

a) Develop a table of expected frequencies for this problem.
b) Calculate the sample χ^2-value and degrees of freedom.
c) If Diane's claim is formally tested with a χ^2-test, what would be the resulting p-value and statistical decision? What is the conclusion, in words?

Applications 15.3

As part of all test solutions for this section, show the calculated test statistic, the degrees of freedom, and the p-value.

41. Concerned about network security, Concordia University published these data on e-mails that have arrived at the university.

Identified E-Mail Type	Quantities in July 2005	Quantities in March 2004
Spam	983,677	509,606
Viruses	3,632	16,908
Acceptable for delivery	369,214	590,928

The quantities in the centre and right-hand columns represent, in each case, e-mails received over a period of 168 hours. Assume that the hours sampled were selected randomly over the course of the specified month. At the 5% level of significance, were the same proportions of different e-mail types received in July 2005 as in March 2004?

Exercises 42–43 are based on data in the file **Religion.**

42. The file **Religion** shows the stated religions of Canadians across various regions of the country according to a recent census. Interpret these data as representative of Canadian's religious-choice distributions into the future and focus on (a) the four Atlantic Provinces (Newfoundland-Labrador, Prince Edward Island, Nova Scotia, and New Brunswick) and (b) the four columns for subdivisions of the Christian religion. Test a claim that for stated Christians who live in the Atlantic Provinces, the province they live in is unrelated to the subdivision of Christianity with which they identify ($\alpha = 0.05$).

43. Repeat Exercise 42, but focus instead on residents of Ontario and Quebec. Test a claim that for Christians living in one of those two provinces, their stated branch of Christianity is independent of which province they live in.

44. A sociologist is studying attitudes of Canadians toward potentially decriminalizing marijuana. One survey question was: If marijuana was decriminalized, what should be the minimum allowable age for possession? A portion of the results are shown in this table. Values in the cells are the frequencies of the response.

		Age Group of the Respondent		
		18 to 29	30 to 39	40 to 49
Suggested Minimum Age for Possession	16 years of age and older	21	13	7
	18 years of age and older	159	192	172
	No age given	21	22	25

		50 to 59	60 plus
Suggested Minimum Age for Possession	16 years of age and older	6	5
	18 years of age and older	124	129
	No age given	27	47

Test a claim that in the populations of people aged 50–59 and 60-plus, respectively, the proportions who prefer the various minimum-age options are the same. Use the 0.05 level of significance.

45. Based on the data set for Exercise 44, test a claim that in the populations of people aged 18–29 and 40–49, respectively, the proportions who prefer the various minimum-age options are *not* the same. Use the 0.05 level of significance.

46. Based on the full data set for Exercise 44, test a claim that the variables age of respondent and preference for the minimum age are not independent. Use the 0.01 level of significance.

47. Of all the bicycles produced in the world, about 85% are produced in just five countries. Suppose an expert claims that the international distribution of bicycles built by these top producers always follows the pattern shown below (in the bottom row of table). At the 0.05 level of significance,

test a claim that the recorded production numbers in 1998 follow the predicted distribution.

Distribution	China	India	Japan
Actual (1998, in millions)	23.1	10.5	5.9
Actual (1999, in millions)	42.7	11.0	5.6
Expected distribution	57%	14%	9%

Distribution	Taiwan	U.S.A.
Actual (1998, in millions)	10.1	2.5
Actual (1999, in millions)	8.3	1.7
Expected distribution	12%	8%

48. Using the data from Exercise 47, test a claim that the recorded production numbers of bicycles in 1999 follow the predicted distribution.

Review Applications

49. According to data collected by the International Bicycle Fund, the following are the comparative production numbers, worldwide, for bicycles versus automobiles.[2]

Year	1990	1991	1992	1993
Bikes (millions)	92	99	102	102
Autos (millions)	36	35	36	34

Year	1994	1995	1996	1997
Bikes (millions)	105	106	98	92
Autos (millions)	35	35	37	39

Year	1998	1999	2000
Bikes (millions)	76	96	104
Autos (millions)	38	40	41

a) Find the linear correlation coefficient r for the comparative numbers of bikes and autos produced annually. Is there a strong correlation?

b) Is the correlation, if any, significant? How do you know?

50. Manually or with the computer, convert the data in Exercise 49 into ranks and then compute the Spearman rank correlation coefficient for the two variables. Is there a significant correlation between the ranked values?

For Exercises 51–54, use data from the file **Canadian_Equity_Funds.** For all calculations, omit records that lack data for one of the variables being analyzed.

51. An investor is comparing the performance of several Canadian equity funds and has compiled related data on several variables for analysis. (See file **Canadian_Equity_Funds.**)

a) Find the linear correlation coefficient r for these two pairs of fund variables:

i) Assets ↔ MER (i.e., management expense ratio: % ratio between a fund's management fees and expenses vs. the fund's assets)

ii) MER ↔ NAVPS (i.e., net asset value per share as of the end of the previous month, in CDN$)

b) Which, if any, of the above two correlations is strong?

c) Which, if any, of the above two correlations is significant?

52. Manually or with the computer convert all values for the variables Assets, NAVPS, and MER to their respective ranks. Then repeat all three parts of Exercise 51 but this time measure and test for the Spearman rank correlation coefficient r_s, instead of for the Pearson coefficient r. Are all of the conclusions about the strength and significance of the correlations the same? If not, what is it about the data that might explain the difference or differences?

53. As part of her research, the funds investor constructed this contingency table, to compare the numbers of funds having either no fees or optional fees, and those belonging to groups based on a ranking by *The Globe and Mail* "Report on Business" GlobeFund.com.

	Ratings 1–2 Rate Group 1	Ratings 3–5 Rate Group 2	Unrated Rate Group 3
No fees	6	11	3
Optional fees	5	17	7

Test the claim, at the 0.05 level of significance, that there is no relationship between the variable for fees and the variable "Rategroup" for these funds. As part of your solution, show the test statistic, the degrees of freedom, and the p-value.

54. The investor also compared the actual distribution of fee options listed in the file with an expected distribution based on an article she read on fee options.

	Deferred Load ("D")	Front-End Loaded ("F")	No Sales Fee ("N")
Frequency in sample	5	3	20
Expected distribution	0.10	0.05	0.30

	Optional Fee ("O")	Redemption Fee ("R")
Frequency in sample	29	3
Expected distribution	0.40	0.15

Test the claim, at the 0.05 level of significance, that the sampled funds in the file follow the same distribution regarding fee options as the investor read in the article. As part of your solution, show the test statistic, the degrees of freedom, and the p-value.

For Exercises 55–59, use data from the file **Weather_Data_Canada.**

55. After compiling the data in the file **Weather_Data_Canada,** a meteorology student began to explore the relationships between different weather variables. The file displays, for various cities, a number of their weather norms for each of two months of the year.

a) Find the linear correlation coefficient r between the variables rainfall and snowfall.

b) How strong is the correlation, if any, between a city's rainfall norm for a month and its snowfall norm for the same month? Is there a significant correlation between those variables?

56. For many of the recorded months in the above file there is rainfall data, but snowfall never occurs during those months. Compare again the variables rainfall and snowfall but omit the months for which snowfall is zero.

a) Find the linear correlation coefficient r between the specified variables.

b) How strong is the correlation, if any, between the specified variables? Is there a significant correlation?

57. In the above file, the variable Ex_MaxT displays the maximum-ever temperature recorded (in degrees Celsius) for the indicated city and month of the year. The variable Ex_MinT displays the minimum-ever temperature recorded (in degrees Celsius) for the indicated city and month of the year.

a) Find the linear correlation coefficient r between the specified variables.

b) How strong is the correlation, if any, between these variables? Is there a significant correlation?

c) Bearing in mind the meaning and aggregate nature of the variables compared, how would you explain in ordinary language to someone else the correlation (if any) that you found in the previous steps?

58. Manually or with the computer, convert all values for the variables rainfall and snowfall to their respective ranks. Then repeat all parts of Exercise 55 but this time, measure and test for the Spearman rank correlation coefficient r_s, instead of for the Pearson coefficient r. Are all of the conclusions about the strength and significance of the correlation the same? If not, what is it about the data that might explain the difference or differences?

59. Manually or with the computer, convert to ranks all of the values for, respectively, the variables Ex_MaxT (for maximum-ever temperature recorded for the indicated city and month of the year) and Ex_MinT (the minimum-ever temperature recorded in that city/month combination). Then repeat all parts of Exercise 57 but this time measure and test for the Spearman rank correlation coefficient r_s instead of for the Pearson coefficient r. Are all of the conclusions about the strength and significance of the correlation the same? If not, what is it about the data that might explain the difference or differences?

For Exercises 60–63, use data from the file **Mishandled_Baggage.**

60. Every month the U.S. Department of Transportation receives thousands of reports about mishandled baggage, affecting every registered airline. Data for February 2005 are found in the file **Mishandled_Baggage.** One might expect that airlines that carry more passengers are likely to generate more complaints about baggage. Test this claim informally, using the Spearman rank correlation coefficient between enplaned passengers and numbers of baggage reports as the test statistic. Is there a significant correlation between these two variables? Justify your conclusion.

61. Suppose that sample data were collected for the five best-ranked airlines, in terms of the above complaints, and the data were as shown below.

Airline	Number of Reports Filed	Number of Passengers Sampled
Airline A	7	2,602
Airline B	8	3,000
Airline C	8	2,602
Airline D	9	3,000
Airline E	9	3,000

Test the claim, at the 0.05 level of significance, that the proportion of reports filed in relation to passengers sampled is not the same for all of these airlines. What is the statistical decision? What is your conclusion in ordinary language? As part of your solution, show the test statistic, the degrees of freedom, and the p-value.

62. Suppose that sample data were collected for the five worst-ranked airlines, in terms of these complaints about baggage, and the data were as shown below.

Airline	Number of Reports Filed	Number of Passengers Sampled
Airline F	24	3,079
Airline G	32	2,954
Airline H	37	3,041
Airline I	46	3,462
Airline J	56	2,903

Test the claim, at the 0.05 level of significance, that the proportion of reports filed in relation to passengers sampled is not the same for all of these airlines. What is the statistical decision? What is your conclusion in ordinary language? As part of your solution, show the test statistic, the degrees of freedom, and the p-value.

63. The sample data in Exercise 62 on the number of reports filed by each of the five worst-ranked airlines in terms of baggage complaints were compared with an expert's expected distribution for these numbers, given her knowledge of the industry and the sample sizes.

Airline	Number of Reports in the Sample	Expected Distribution of Reports
Airline F	24	0.15
Airline G	32	0.15
Airline H	37	0.20
Airline I	46	0.20
Airline J	56	0.30

Test the claim, at the 0.05 level of significance, that the number of reports in the sample followed the expected distribution. As part of your solution, show the test statistic, the degrees of freedom, and the *p*-value.

Chapter

16

Simple and Multiple Linear Regression

• Chapter Outline

• Learning Objectives

16.1 Construct and interpret a simple linear regression equation; assess the strength, significance, and validity of the model; and use the model to make inferences.

16.2 Construct and interpret a multiple linear regression equation; assess the strength, significance, and validity of the model; and use the model to make inferences.

16.3 Describe and apply some rules of thumb for developing a good regression model and recognize some key hazards to avoid in the process.

16.4 Construct and interpret a multiple linear regression equation that includes dummy variables to represent qualitative data.

16.1 Simple Linear Regression and Inferences from the Regression Model

In Chapter 15, we learned to identify and measure associations between two variables for a variety of data types. For continuous bivariate data (Section 15.1), we explored using scatter diagrams to assess whether a possible correlation appears to be linear. If so, we calculated r to measure the strength of the correlation and tested whether the value of r was significant. If, following these steps, we establish that there is a strong and significant linear correlation between two paired variables, we may find it useful to create a model for the relationship between the variables. **Regression analysis** is a powerful tool to create such a model (in the form of an equation) with which we can estimate or predict the value of one of the variables based on its linear relationship with the other.

When applying the scatter diagrams in Section 15.1, we borrowed the graph-based notations of an x-variable (graphed on the horizontal axis) and a y-variable (graphed on the vertical axis) in order to refer easily to the two paired variables. But it did not matter which variable was denoted x or y for just testing correlation. This is not the case for regression.

For regression, the y-variable should represent the **dependent** (or **response**) **variable** and the x-variable is the **independent** (or **predictor**) **variable.** Ask yourself this question: If there *is* a correlation between the paired variables, which variable's values do I intend to estimate or predict, based on the other variable's values? The variable to be predicted is the dependent variable; the variable used to make those predictions is the independent variable.

In practice, which variable plays which role depends on the problem you need to solve. Suppose the number of potholes repaired in regional roads is correlated to the size of the region's road budget. A town planner may inspect the number of potholes in the spring to estimate what road budget will be needed; "size of budget" is the dependent variable. On the other hand, a tire store manager may examine the region's road budget to predict how many potholes will be repaired; the unrepaired ones could cause tire damage and perhaps lead to increased tire sales. For him, "potholes repaired" is the dependent variable.

Once x and y have been identified, the **regression equation** and **regression line** can be developed and interpreted in relation to the scatter diagram. In Figure 15.1 in Chapter 15, we saw the apparent scatter of (x, y) value pairs (for the dates and trading values of Bank of Montreal stock) around a single line, which suggested a correlation of the variables. Once a correlation is confirmed, a linear model can be developed, as illustrated in Figure 16.1. The regression line in the figure is known as the **line of best fit;** this line shows, for any value of x found in the population, the estimated centre of an interval in which the corresponding y-value is expected to lie. The regression equation expresses the same relationships mathematically that the regression line expresses graphically. In the following paragraphs, we will look first at the general form and interpretation of the regression equation and then at how the equation can be generated.

Many readers of this book may be familiar, from some prior course, with the following equation for a straight line:

$$y = mx + b$$

The constant term b is called the **y-intercept.** Notice that if the black line in Figure 16.2 is extended to cross the vertical (y) axis, it crosses that axis at the value $y = b \approx 27$. In other words, b gives the value of y when $x = 0$.

Figure 16.1

Regression Equation and Regression Line (January 2000–September 2006)

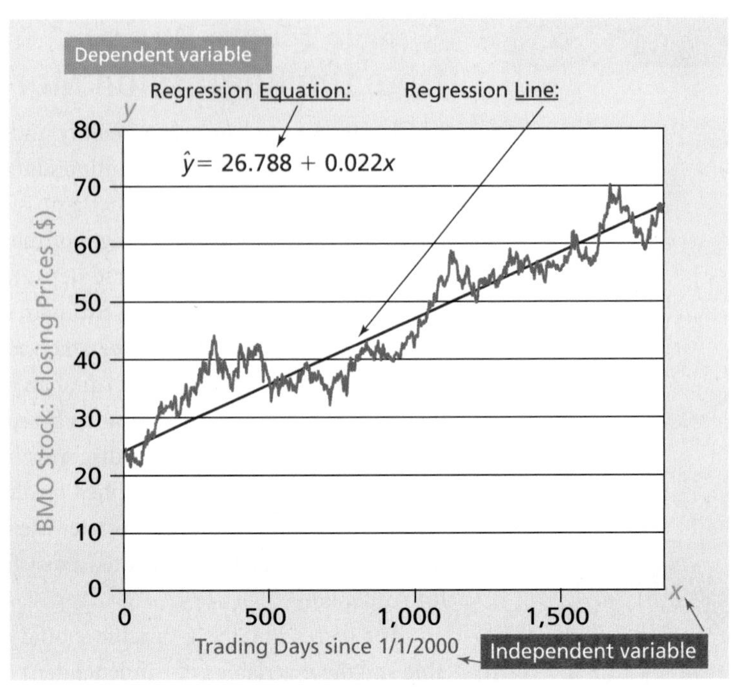

Source: Bank of Montreal. BMO stock closing prices from January 2000 to September 2006. Found at: http://www2.bmo.com/content/0,1089,divId-3_langId-1_navCode-4644,00.html. Accessed September 6, 2006. Used with permission of Bell Globemedia.

Figure 16.2

Equation for a Straight Line

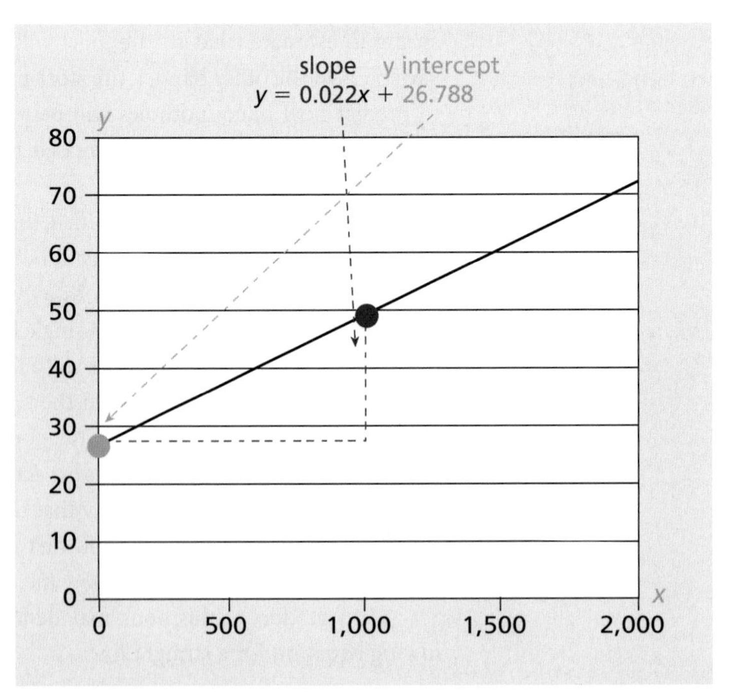

Source: Bank of Montreal. BMO stock closing prices from January 2000 to September 2006. Found at: http://www2.bmo.com/content/0,1089,divId-3_langId-1_navCode-4644,00.html. Accessed September 6, 2006. Used with permission of Bell Globemedia.

In $y = mx + b$, the **x-coefficient** (shown as m) gives the **slope of the line.** The slope is a constant ratio that applies for all pairs of points on the line—namely, the ratio between the difference in y-values for the two points and the corresponding difference in x-values for the same points. The slope is sometimes described as the ratio of rise (change in vertical distance on the y-axis) over run (change in horizontal distance on the x-axis) between any pair of points.

Observe in Figure 16.2 two illustrated points with the approximate coordinates (0, 27) and (1,000, 49). The slope equals (change in y) / (change in x) = (49 − 27)/(1,000 − 0) = 0.022. The slope can also be looked at in this way: For every one unit of increase in the value of x, expect the corresponding change in y-value to equal m, which is the slope.

The standard form of the regression equation that will be used in this book takes a slightly different format:

$$\hat{y} = b_0 + b_1 x$$

where

x can assume any value within the range of possible or plausible values for the independent variable.

$\hat{y}$ is the *expected* corresponding value for the dependent variable, given a specified value x for the independent variable.

b_0 and b_1 define, respectively, the y-intercept and slope of the line of regression.

Compared to $y = mx + b$, the different symbols and order of terms on the right of the equation are not mathematically significant. But on the left of the equation, the use of $\hat{y}$ (pronounced "y cap") instead of y does make a difference: The equation that starts "$y =$" mathematically defines a line; for each value of x, the equation tells exactly what is the corresponding value of y. But compare the equation in Figure 16.1. Almost *none* of the actual values of the dependent variable—i.e., the actual y-values—fall exactly on the regression line. In the context of regression, the equation does *not* tell us the real value of y; it tells us only the predicted or expected value of y (which is what $\hat{y}$ denotes), insofar as the regression models the world accurately.

The switch to symbols b_0 (instead of b) for the constant (i.e., y-intercept), and b_1 (instead of m) for the x-coefficient, makes the notation more flexible for use in the next section of this chapter, where we introduce regressions with more than one independent variable. For example, the b_1 notation can easily be adapted for cases with three independent variables:

$$\hat{y} = b_0 + b_1 x_1 + b_2 x_2 + b_3 x_3$$

Note there is a way to express a regression equation that does not use the $\hat{y}$ symbol. Instead, we could display:

$$y = \beta_0 + \beta_1 x + \varepsilon$$

where the β (pronounced "beta") notations indicate that β_0 and β_1 are the population parameters for the constant and slope, corresponding to the sample estimates b_0 and b_1. The Greek letter epsilon (ε) represents the error in calculating y that is not explained by the regression; i.e., it is the difference between "real life" and what the model predicts "real life" to be. By definition, the value of ε remains unknown to us, so the equation is a reminder that our prediction is ultimately just an estimate.

Finding the Regression Equation

If we knew what line, graphically, best fit all of the points in the data set, we easily could find the regression equation for the line: Simply determine the slope and y-intercept for the line and write the equation.

But how can we visualize the regression line if there are several independent variables? And even for simple regression, with only one independent variable, to choose the best-fitting line by eye can be very difficult and unreliable. For example, which is the best-fitting line in Figure 16.3?

Figure 16.3
The Line of Best Fit

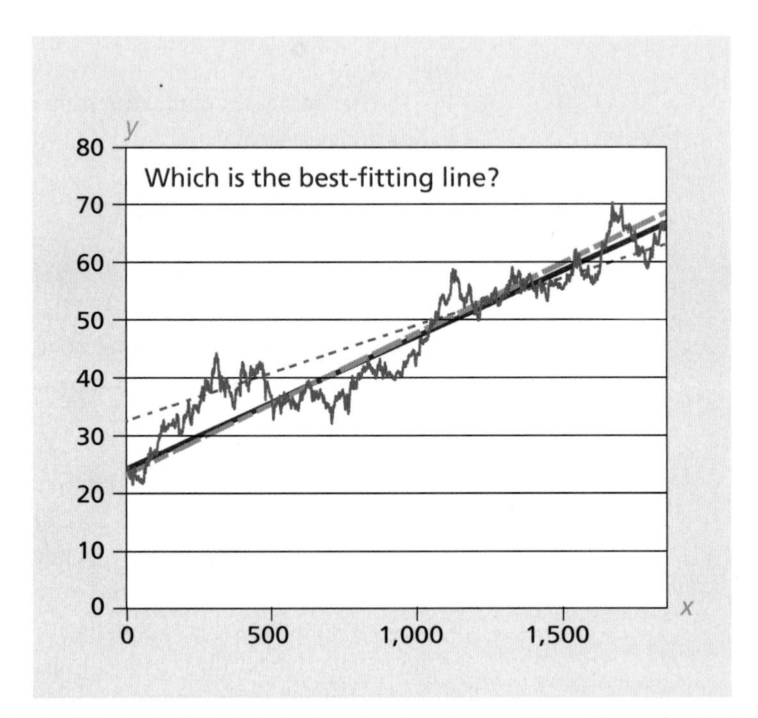

Source: Bank of Montreal. BMO stock closing prices from January 2000 to September 2006. Found at: http://www2.bmo.com/content/0,1089,divId-3_langId-1_navCode-4644,00.html. Accessed September 6, 2006. Used with permission of Bell Globemedia.

In practice, regression proceeds in the opposite order: Using mathematical procedures, we find the regression equation first. Then, if desired, the equation can be graphed—often displayed in the same illustration as the scatter diagram, as in Figure 16.1.

The basic procedure to find the regression equation is called **ordinary least squares,** which extends the now-familiar notion (from variance) of comparing actual with expected values of a variable by taking the squared distances between those values. In this case, there are *many* expected values for y; namely, all of the $\hat{y}$-values that form the regression line. Each value of $\hat{y}$ is the expected value of the dependent variable for one x-value in the range of the independent variable. The distance between some actual value of y and the corresponding expected value ($\hat{y}$), for a given x-value, is called a **residual.** (See Figure 16.4.) Essentially, each residual is an error left unexplained by the regression. The least squares method, in summary, produces the best-fitting line *by minimizing the sum of all of the squared residuals between actual and expected y values.*

Two definitional formulas are used for this technique (see Formulas 16.1 and 16.2). Observe that the numerator for the b_1 formula is the same as for r (refer back to For-

Figure 16.4
Regression Line and Residuals

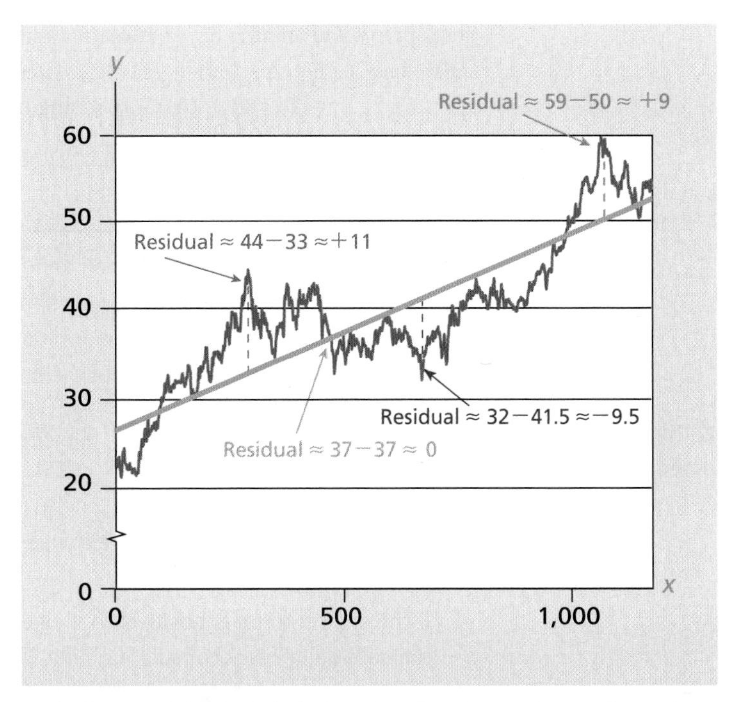

Source: Bank of Montreal. BMO stock closing prices from January 2000 to September 2006. Found at: http://www2.bmo.com/content/0,1089,divId-3_langId-1_navCode-4644,00.html. Accessed September 6, 2006. Used with permission of Bell Globemedia.

mula 15.1). If there is no covariance between the variables, then the correlation coefficient r and the slope b_1 will both be zero (0).

Formula 16.1 Slope of the Regression Line

$$b_1 = \frac{SSXY}{SSX} = \frac{\sum_{i=1}^{n}(x_i - \bar{x})(y_i - \bar{y})}{\sum_{i=1}^{n}(x_i - \bar{x})^2}$$

Formula 16.2 y-Intercept of the Regression Line

$$b_0 = \bar{y} - b_1\bar{x}$$

The centre point of the regression line is at the mean for both the independent and dependent variables. If the line passed through the origin (0,0), then the slope b_1 would be equal to the rise of the line (from the origin to the y-mean) divided by the run of the line (from the origin to the corresponding x-mean), so the slope b_1 would equal $\frac{\bar{y}}{\bar{x}}$. Solving for $\bar{y}$ in this case: $\bar{y}$ would equal b_1 times $\bar{x}$. If in fact $\bar{y}$ does not equal b_1 times $\bar{x}$, then the difference between them is b_0. (See Formula 16.2.) In other words, b_0 can be viewed as the constant distance between every $\hat{y}$ value on the regression line and the corresponding term $(x \times b_1)$.

Once the values of b_0 and b_1 are determined, replace those symbols with their values within the generic regression equation. For example, in the relationship between BMO stock prices and trading days between January 2000 and September 2006, b_0 can be calculated as 26.788; the slope b_1 is 0.022. Therefore, the regression equation for this relationship is: $\hat{y} = 26.788 + (0.022)x$. Using the specific variable names:

$$\textbf{Predicted price} = 26.788 + (0.022)(\text{Number of days since January 1, 2000})$$

Example 16.1

A petting zoo keeps a variety of small animals. A veterinarian has tracked a sample of these animals and has compiled the data shown in the table below, which compares each animal's weight with its daily consumption of water.[1] Her goal is to estimate the amount of water consumption if the weight of the animal is known.

Animal weight (in g)	85	100	400	500	700	1,000	2,500	2,800	3,000	3,000	5,000
Water consumed per day (in mL)	5	20	50	50	50	100	150	130	300	300	470

1. Assess whether it is appropriate to use linear regression to solve the veterinarian's problem.
2. If the answer to (a) is positive, then use a regression equation to estimate an animal's daily water consumption if it weighs about 1,500 grams.

Solution

1. To first check whether regression applies, create a scatter diagram (see Figure 16.5). There does seem to be a linear correlation between the variables. (For regression, consumption would be the *y*-variable, since that is the variable to be estimated by the model.)

Figure 16.5
Scatter Diagram for Water Consumption versus Weight

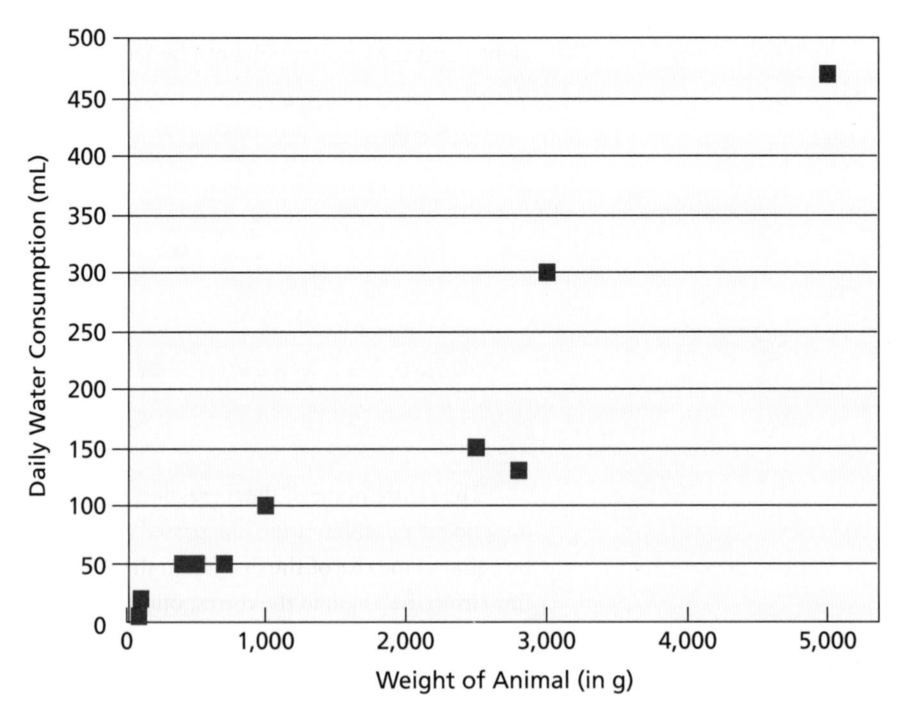

Source: Water Management Branch; Environment and Resource Division; Ministry of Environment, Land and Parks; Government of British Columbia. http://wlapwww.gov.bc.ca/wat/wq/reference/foodandwater.html#table1.

By computer or by formula (see Figure 16.6), we can find an r value:

$$r = \frac{SSXY}{\sqrt{SSX} \times \sqrt{SSY}} = \frac{2{,}252{,}050}{\sqrt{25{,}894{,}750} \times \sqrt{218{,}168}} = 0.947$$

Figure 16.6

Useful Calculations for r and for the Regression Equation

In grams x Weight(g)	In mL y WatPerDay		$(X_i - \bar{X})$	$(Y_i - \bar{Y})$	$(X_i - \bar{X}) \times (Y_i - \bar{Y})$	$(X_i - \bar{X})^2$	$(Y_i - \bar{Y})^2$
85	5		−1650.00	−142.73	235500.00	2722500.00	20371.07
100	20		−1635.00	−127.73	208834.09	2673225.00	16314.26
400	50		−1335.00	−97.73	130465.91	1782225.00	9550.62
500	50		−1235.00	−97.73	120693.18	1525225.00	9550.62
700	50		−1035.00	−97.73	101147.73	1071225.00	9550.62
1,000	100		−735.00	−47.73	35079.55	540225.00	2277.89
2,500	150		765.00	2.27	1738.64	585225.00	5.17
2,800	130		1065.00	−17.73	−18879.55	1134225.00	314.26
3,000	300		1265.00	152.27	192625.00	1600225.00	23186.98
3,000	300		1265.00	152.27	192625.00	1600225.00	23186.98
5,000	470		3265.00	322.27	1052220.45	10660225.00	103859.71
19,085.00	1,625.00	← Sums →			**2252050.00**	**25894750.00**	**218168.18**
11.00	11.00	← n			***SSXY***	***SSX***	***SSY***
1,735.00	147.73	← Means					

Source: Water Management Branch; Environment and Resource Division; Ministry of Environment, Land and Parks; Government of British Columbia. http://wlapwww.gov.bc.ca/wat/wq/reference/foodandwater.html#table1. Copyright © Province of British Columbia. All rights reserved. Reprinted with permission of the Province of British Columbia. www.ipp.gov.bc.ca.

The correlation is clearly very strong, and it can also be shown to be significant (see Section 15.1): p-value = 0.000. Therefore, the regression model might be very useful for the estimation.

2. First, we can use the formulas to find the slope and intercept for the regression equation:

$$b_1 = \frac{SSXY}{SSX} = \frac{2{,}252{,}050}{25{,}894{,}750} = 0.087$$

$$b_0 = \bar{y} - b_1\bar{x} = 147.73 - 0.087 \times 1{,}735 = -3.22$$

Thus, the regression equation is $\hat{y} = -3.22 + (0.087)x$, where x is the animal's weight in grams. For an animal that weighs 1,500 grams, the estimated water consumption is $-3.22 + (0.087)(1{,}500) \cong 127$ mL. (A more accurate expression would be: If there were many animals that weighed 1,500 grams, the mean for water consumption for all of these animals, collectively, would be expected to be 127 mL.)

Example 16.2

*(Based on the **Brilliants** file)*

A jeweller has previously determined that the prices of diamonds are significantly correlated with their weights, measured in carats. Her sample data are stored in the file **Brilliants.**

1. Find the regression equation that relates the prices and the weights of diamonds.
2. Use the regression equation to estimate the price of a diamond that weighs 0.75 carats.
3. Are there any reasons for using caution when applying the regression equation for (2)? Explain.

Solution

1. The variable to be estimated (the price variable) is the dependent variable y. Using formulas or the computer procedures in "In Brief 16.1" starting on page 667, we can find the values for b_0 and b_1. These are $b_0 = -3{,}153.2$ and $b_1 = 11{,}887$. Therefore, the regression equation is $\hat{y} = -3{,}153.2 + 11{,}887x$, which can be written and simplified as, for example:

   Predicted price = −3,153 + 11,890 (carats)

2. By substituting the x-value 0.75 carats in the solution 1 equation, the corresponding y-value can be calculated:

   Predicted price = −3,153 + 11,890(0.75) ≈ $5,760

3. The same conditions for validly assessing correlation apply, by extension, to using the regression equation. Unfortunately, as shown in Figure 16.7 on the next page, the distribution of y-residuals is not consistent for the whole range of values for x. (Note: We will discuss the analysis of residuals more fully in the next section.) This non-consistent distribution seems particularly bad at precisely $x = 0.75$, where we hoped to make our prediction. So, taking the scatter diagram into account, the regression results must be considered as approximate, at best. For example, we might conclude:

   The price is likely to be closer to $5,000 than to any value above $10,000.

Additional Cautions When Interpreting Regression Results

As noted in Example 16.2, solution 3, all of the cautions and conditions for interpreting correlation apply also to regression. The terminology can be restated for the regression context. For example, all of the y-values must be normally distributed around the line of regression and their variances around that line should be constant for all portions of the line. Also, all of the errors must be independent from one another (i.e., the residuals for particular values of x should have no impact on the size or magnitude of residuals for any other values of x). Of course, it is assumed that all observations have been correctly measured and that any outliers have been properly addressed.

Another type of caution arises from interpreting the regression equation as a model for causation. Sometimes, *other* independent variables not included in the model are having some unknown impact on both the x- and y-variables together. This can sometimes distort the value or sign of the x-coefficient in the model, so that the real impact of the independent variable is fully or partly masked. It requires appropriate knowledge of the real-world application, as well as of statistics, to help avoid this problem.

Finally, attempts to **extrapolate** from the data used for the regression should be made with caution. Suppose a doctor has compiled a regression between the heights and ages of normally developing young children. Once there is a regression equation, it would be possible—mathematically—to input the age "25" as the independent variable and read

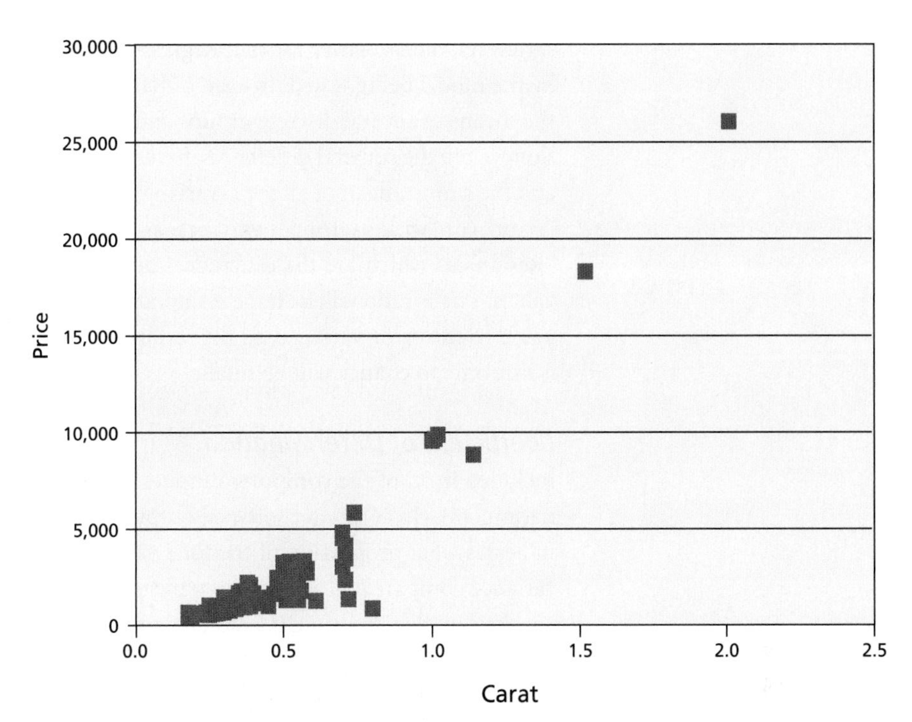

Figure 16.7
Scatter Diagram for Price versus Weight of Diamonds

Source: Canada Diamonds. Found at: http://www.canadadiamonds.com/round_brilliant_canadian_diamonds.php?page=650&limit=10&customer=CAD

off some large estimate for height. But, clearly, the data for basing a regression between age and heights of children does not extend into a person's twenties, and estimates in that region should be avoided. That said, all regression-based *predictions* do extrapolate somewhat past the time periods (the x-variables) for which data were collected. This is unavoidable, but be careful about predicting too far into the future.

Other Important Concepts in Regression

Significance and the ANOVA Table

Statistical significance for a regression is a positive answer to the question: Do the data support that there is a linear dependency of the response variable on the predictor variable? For regression (as for correlation), significance can be determined by use of a p-value that is displayed on computer outputs. (For details about your software, see "In Brief 16.1" starting on page 667.) For correlation, we interpreted significance as the rejection of the null hypothesis that there is no correlation between the two paired variables (i.e., H_0: $\rho = 0$). The p-value for a simple regression is the same number as for the correlation, but we generally obtain its value through an ANOVA test. (For more on ANOVA tests, see Chapter 14.)

For the data in Example 16.2, for instance, the following ANOVA table could be generated:

	df	*SS*	*MS*	*F*	Significance *F*
Regression	1	1,007,453,452	1,007,453,452	594.06587	0.000
Residual	97	164,498,568.4	1,695,861.53		
Total	98	1,171,952,020			

As always, the *MS* entry for the "Regression" row signifies the variance that is explained by the model being tested. Instead of basing that value on squared differences between the means of several discrete groups and the grand mean, as in Chapter 14, for regression we use the squared differences, for each x, between the y predicted by the model ($\hat{y}$) and the simple mean of all the y-variables ($\bar{y}$). The second row in the table measures the "error" (or "within-group") variance, as in Chapter 14; here, this is the variance due to "Residuals," which are the differences between actual and predicted y-values for all x-values. The F-ratio will be large if the variance explained by the regression is large compared to the error variance, so the resulting probability (p-value) that the large F-ratio is due only to chance will be small.

Coefficient of Determination

Included in all of the computer outputs is a measure called the **coefficient of determination (r^2)**. (In computer software output, look for the label "R Square" or "R-Sq.") This indicates what proportion of the total variance of the dependent variable (which is the variance compared to simply the y-mean), is explained by, or attributable to, the variance due to the regression. If a correlation is perfect, like the relationship between Fahrenheit temperatures and their Celsius equivalents, then $r^2 = 1$, which means that the regression between the two scales totally explains the variance of y-values from the mean. Lesser values of r^2 indicate that some portion $(1 - r^2)$ of the variation in y-values is due to non-modelled causes and/or is due to chance.

If you have constructed the ANOVA table, the calculation of r^2 is very straightforward, as shown in Formula 16.3.

Formula 16.3 Coefficient of Determination

$$r^2 = \frac{SSregression}{SSTotal}$$

where *SSregression* is the sum of squared differences between the model-predicted y-values (i.e., $\hat{y}$) and $\bar{y}$.
SSTotal is the sum of squared differences between all the actual y-values and $\bar{y}$.

This ratio is the proportion of variance explained by the regression. For the example given above: $r^2 = SSregression/SSTotal = 1{,}007{,}453{,}451.6/1{,}171{,}952{,}020.0 = 0.85964$.

Standard Error and Confidence Interval of the Estimate

We know that not all of the scatter points fall on the estimated regression line. Due to the influence of non-modelled factors and/or chance, there is some deviation between the actual y-values and the values of the dependent variable predicted by the regression.

Think of the regression line as specifying, in a sense, the centre of the dispersed y-values. In Chapter 5, we learned how to measure dispersion from a mean in the units of standard deviation. The **standard error of the estimate** is like a standard deviation of all actual y-values, compared to predicted y-values. However, unlike a true standard deviation, there is not a single central $\hat{y}$-value to which all y-values are compared. Instead, deviations are taken between each specific y and the $\hat{y}$ that corresponds for its particular value of x. (See Formula 16.4.)

Formula 16.4 Standard Error of the Estimate

$$s_y = \sqrt{\frac{\sum_{i=1}^{n} (y_i - \hat{y}_i)^2}{n - 2}}$$

Given the modelling assumption that the errors (residuals) are normally distributed, we can use the empirical rule, combined with the standard error, to generate an approximate confidence interval around an estimate for $\hat{y}$. In the diamond price example, we can calculate (or learn from the computer printouts) that the standard error is approximately $1,300. By the empirical rule, we therefore expect that, 95% of the time, predictions based on the regression will be accurate to within about ± 2 standard errors (i.e., $\pm(2 \times 1{,}300) = \pm\$2{,}600$).

Be aware, however, that the above estimate applies only to the regression *as a whole*, and an estimate for a specific $\hat{y}$ gives only a predicted *mean* for the values of Y that might correspond to a value of X. But if you have in mind *one particular* value of the independent variable and you would like to know the range of likely y-values that might occur, then the interval estimate you require is called a **prediction interval.** For any given x-value, the prediction interval is wider than the confidence interval for the estimate. (For the prediction interval formula, see "Hands On 16.1" on page 669.) In practice, most data sets do not meet the ideal conditions for the regression and they are even less likely to be normally distributed around any specific value of x—so exact precision of the prediction interval is elusive. For practical purposes, it is often sufficient to combine your point estimate with a ballpark error band based on the standard error.

Example 16.3

In Example 15.1, we examined a table from a master's student's research plan, which compared the populations of Canadian cities with the numbers of individuals (n) actually sampled from each city. You were asked to assess the significance and strength of the correlation between those two variables. Using that same data, answer these additional questions:

1. Provide the regression equation that relates the sample size taken from a city to the population of that city.
2. If the master's student sampled from an additional city that is not shown in the table, estimate the sample size that he would take for that city if the city's population is 850,000.
3. What percentage of the variation among the sizes of samples is accounted for by the regression between sample size and city population?
4. Your estimate for question 2 is a point estimate. In general, 95% of the predictions using this regression model tend to fall within $\pm$ what amount from the point estimate?

Solution

1. Note that for correlation (compare Figure 15.5), we did not care which variable was graphed on the y-axis. For this regression problem, the goal is to predict the sample size based on population. Therefore, sample size is the dependent variable y.

   Applying formulas or procedures from "In Brief 16.1" and using sample size as the y-variable, we can obtain an output similar to the right-hand side of Figure 16.8. The information we need is circled at the bottom of the figure.

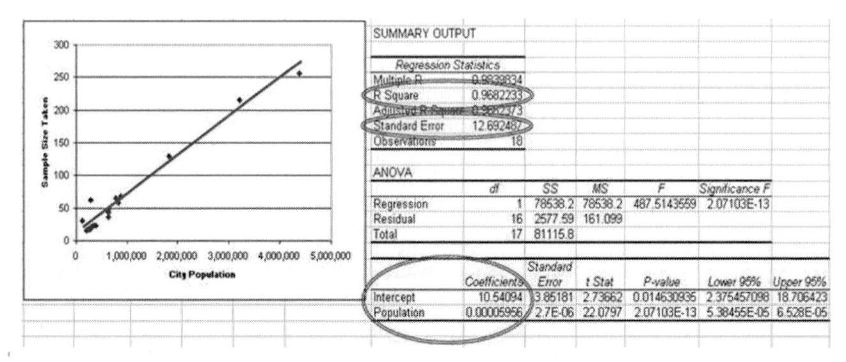

SUMMARY OUTPUT

Regression Statistics	
Multiple R	0.9839834
R Square	0.9682233
Adjusted R Square	0.9662373
Standard Error	12.692487
Observations	18

ANOVA

	df	SS	MS	F	Significance F
Regression	1	78538.2	78538.2	487.5143559	2.07103E-13
Residual	16	2577.59	161.099		
Total	17	81115.8			

	Coefficients	Standard Error	t Stat	P-value	Lower 95%	Upper 95%
Intercept	10.54094	3.85181	2.73662	0.014630935	2.375457098	18.706423
Population	0.00005956	2.7E-06	22.0797	2.07103E-13	5.38455E-05	6.528E-05

Figure 16.8
Regression for Sample Size versus Population Size

Source: Matthew Horne. *Incorporating Preferences for Personal Urban Transportation Technologies into a Hybrid Energy-Economy Model.* Simon Fraser University, School of Resource and Environmental Management.

We see that the y-intercept (b_0) equals 10.54 and that the x-coefficient (b_1) equals 0.00005956. (On your software display, this value may appear in scientific notation as 5.956E-05.) Therefore, the regression equation is:

$$\hat{y} = \mathbf{10.54 + 0.00005956}x$$

You could also write:

Predicted Sample Size = 10.54 + 0.00005956(Population Size)

2. For this question, we utilize the regression equation as constructed in solution 1.

   Predicted Sample Size = 10.54 + 0.00005956(850000)
   Predicted Sample Size = 61.16 $\approx$ 61

3. The percentage of the variation among the sizes of samples accounted for by the regression is given by the value for r^2. According to the upper circled area in Figure 16.8, r^2 equals 0.9862233. (We could also find this value by using *SSregression/SSTotal.*) Therefore, about 98.6% of the variation in sample size can be attributed to the regression.

4. The basis for solving this question is the standard error of the regression, which is also circled in Figure 16.8. For the regression as a whole, this represents a standard deviation for how far actual y-values are varying from the predicted values. Therefore, using the empirical rule, we expect about 95% of the predictions from the regression to fall within ± 2 standard errors (i.e., $2 \times 12.6925 \approx 25$) from the estimate. In other words, 95% of the time, the actual sample size will be within ± 25 from the predicted size. Judging from the scatter diagram in Figure 16.8, this estimate seems quite reasonable.

• *In Brief 16.1 Simple Linear Regression*

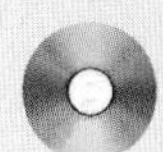

*(Based on the **Brilliants** file)*

The outputs from regression software provide more than just the b_0 and b_1 terms for determining the equation. For example, also included are outputs for r, the p-value, standard error, and the full ANOVA table. Other features, such as possible residual analysis, will be discussed in subsequent sections. Minitab and SPSS also have options for estimating prediction intervals.

Excel

The paired data must be recorded in columns, such that each (x, y) data pair is recorded on the same row. Begin with the menu sequence *Tools / Data Analysis / Regression;* then click *OK.* In the dialogue box (see the screen capture), specify the separate address ranges for the dependent (y) and independent (x) variables, respectively. Select *Labels* if the variable names (labels) are included at the start of the data range. Then, specify the upper left cell for the output range and then click *OK.*

	A	B
1	Carat	Price
2	2.01	26047
3	1.01	9655
4	1.02	9874
5	1.14	8851
6	1.52	18304
7	0.51	2886
8	0.51	3029
9	0.51	1339
10	0.52	3311
11	0.52	1926
12	0.53	2463
13	0.53	2453
14	0.54	2799
15	0.54	2067
16	0.55	2619
17	0.55	1347
18	0.56	1820
19	0.57	3340
20	0.58	2943
21	0.58	2757
22	0.61	1302
23	0.70	4811
24	0.70	4538
25	0.70	3058
26	0.71	4138
27	0.71	3491
28	0.71	2382
29	0.72	1382
30	0.74	5847

Regression

Input
Input Y Range: $B$1:$B$100
Input X Range: $A$1:$A$100
Labels
Constant is Zero
Confidence Level: 95 %
Output options
Output Range: $D$13
New Worksheet Ply:
New Workbook
Residuals
Residuals
Residual Plots
Standardized Residuals
Line Fit Plots
Normal Probability
Normal Probability Plots
OK
Cancel
Help

SUMMARY OUTPUT

Regression Statistics	
Multiple R	0.92717
R Square	0.85964
Adjusted R Square	0.85819
Standard Error	1302.25
Observations	99

ANOVA

	df	*SS*	*MS*	*F*	*Significance F*
Regression	1	1007453452	1007453452	594.0658678	3.81329E-43
Residual	97	164498568	1695861.53		
Total	98	1171952020			

	Coefficient	*Standard Error*	*t Stat*	*P-value*	*Lower 95%*	*Upper 95%*
Intercept	-3153.17	262.803903	-11.998181	7.09249E-21	-3674.761754	-2631.575781
Carat	11887.2	487.712312	24.3734665	3.81329E-43	10919.26578	12855.2136

An example output is included in the illustration. The desired x-coefficient (b_1) appears at the bottom left of the output. The y-intercept (b_0) appears immediately above b_1. If you did not assess for linear correlation separately, you can do that now: In the spreadsheet behind the screen capture, the magnitude of r appears in the "Regression Statistics" section (with the label "Multiple R"); for the + or − sign for r, use the same sign as for the x-coefficient. The p-value for the regression equals the p-value for the correlation—displayed below the heading *Significance F* in the ANOVA table. (Note that the extremely small p-value is displayed in scientific notation: 3.81329E-43 is equivalent to $3.81329/10^{43} \cong 0.000$.)

Minitab

The paired data must be recorded in columns, such that each (x, y) data pair is recorded on the same row. Begin with the menu sequence *Stat / Regression / Regression.* In the dialogue box (see the screen capture), control-click to highlight and then select the name of the *Response* (y) variable and then do the same for the *Predictor* (x) variable. Click *OK.*

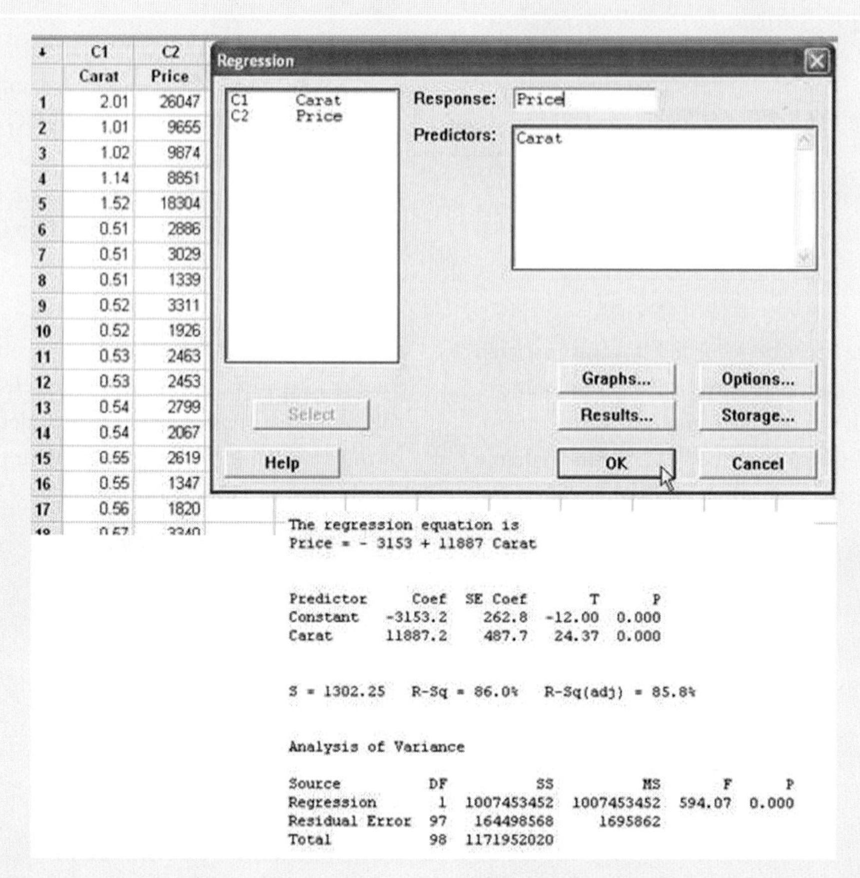

An example output is included in the illustration. The desired x-coefficient (b_1) appears at the left of the output in the table below the regression equation. The constant (i.e., y-intercept, b_0) appears immediately above b_1. The full equation is provided on the second line of the output.

If you did not assess the p-value for linear correlation separately, you can do that now: The magnitude of r can be calculated as the square root of r^2 (labelled "R-Sq" in the third segment of the output); for the + or − sign for r, use the same sign as for the x-coefficient. The p-value for the regression equals the p-value for the correlation—displayed below the heading "P" in the "Analysis of Variance" table.

Not shown in the illustration is an additional option to generate a prediction interval for a specific x-value. When in the Regression dialogue box, click on *Options;* then, in the resulting dialogue box, select *Fit Intercept.* Under *Prediction Intervals for New Observations,* input the x-value of interest, and below this specify the degree of confidence. Click *OK.*

SPSS

The paired data must be recorded in columns, such that each (x, y) data pair is recorded on the same row. Begin with the menu sequence *Analyze / Regression / Linear.* In the dialogue box (see the screen capture), highlight and select, in turn, the names of the dependent (y) variable, then the independent (x) variable. Click *OK.*

An example output is included in the illustration. The desired x-coefficient (b_1) appears at the bottom left of the "Coefficients" table. The y-intercept (b_0) appears immediately above b_1. If you did not assess for linear correlation separately, you can do that now: The magnitude of r appears in the "Model Summary" table (under the label "R"); for the + or

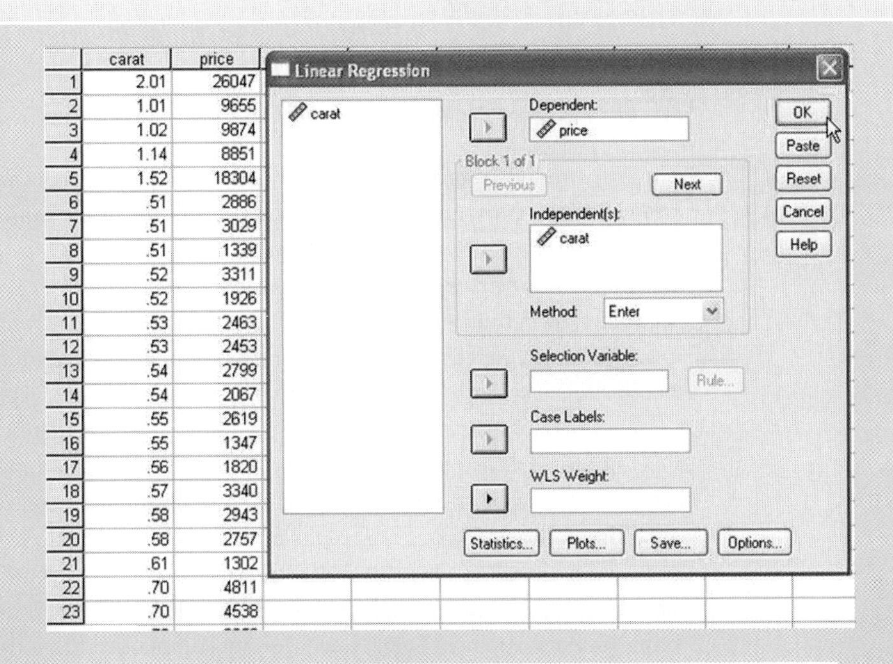

	carat	price
1	2.01	26047
2	1.01	9655
3	1.02	9874
4	1.14	8851
5	1.52	18304
6	.51	2886
7	.51	3029
8	.51	1339
9	.52	3311
10	.52	1926
11	.53	2463
12	.53	2453
13	.54	2799
14	.54	2067
15	.55	2619
16	.55	1347
17	.56	1820
18	.57	3340
19	.58	2943
20	.58	2757
21	.61	1302
22	.70	4811
23	.70	4538

Model Summary

Model	R	R Square	Adjusted R Square	Std. Error of the Estimate
1	.927[a]	.860	.858	1302.252

a. Predictors: (Constant), carat

ANOVA[b]

Model		Sum of Squares	df	Mean Square	F	Sig.
1	Regression	1.0E+009	1	1007453452	594.066	.000[a]
	Residual	1.6E+008	97	1695861.530		
	Total	1.2E+009	98			

a. Predictors: (Constant), carat

b. Dependent Variable: price

Coefficients[a]

Model		Unstandardized Coefficients		Standardized Coefficients	t	Sig.
		B	Std. Error	Beta		
1	(Constant)	-3153.169	262.804		-11.998	.000
	carat	11887.240	487.712	.927	24.373	.000

a. Dependent Variable: price

− sign for r, use the same sign as for the x-coefficient. The p-value for the regression equals the p-value for the correlation—displayed below the heading "Sig" in the ANOVA table.

Not shown in the illustration is an additional option to generate a prediction interval for a specific x-value. When in the Regression dialogue box, click on Save and then, in the resulting dialogue box, select *Unstandardized* under *Predicted Values,* and select *Individual* under *Prediction Intervals,* and then indicate the desired level of confidence. Click on *Continue* and then *OK.* As a result, three new columns will be added to the data view with your original data: "PRE_1," "LICI_1," and "UICI_1." These columns list, respectively, the point estimate and the lower and upper bounds of the prediction intervals *corresponding to each data element, individually, in the table.*

• Hands On 16.1 Interpreting an Outlier in Linear Regression

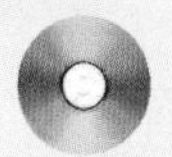

(*Based on the* **FloridaVote2000** *file*)

The data and basis for the case in this feature were generously provided by Jim Stallard of the University of Calgary. Also, for a copy of the confusing "butterfly" ballot design that may have created an outlier, go to http://www.asktog.com/columns/042ButterflyBallot.html

An outlier, by definition, does not lie within the range of expected values for a variable. In fact, we generally take a second look at an outlier, on the premise that in some way it may not be a member of the same "natural" population as the other data. This is especially true if the other data, considered by themselves, appear to form a clear pattern (e.g., a normal distribution)—to which the outlier is a clear exception. The same principle can be applied to regression.

Consider the controversial results in the U.S. presidential election of 2000. The close national results in the election hinged on the results in Florida, and many people in Florida complained of irregularities in voting procedures. In one example, many voters in Palm Beach County complained that—due to a confusing ballot design—they voted *accidentally* for candidate Pat Buchanan when they intended to vote for Al Gore. If their claim is true, Buchanan's votes would have been artificially inflated. Can regression help us test their claim?

1. Download the file **FloridaVote2000**, which compares the votes received in each county by candidates Buchanan and Bush, respectively. Create a scatter diagram (using Buchanan's votes as the response variable) and assess whether there appears to be, in general, a linear correlation between the votes received by those candidates. Does any data point appear to be an outlier? Without peeking at the county names, can you guess which data point in your scatter diagram represents Palm Beach County?
2. Your conclusions by eye in question 1 may not be reliable, so construct a regression equation with Buchanan's votes as the response variable and Bush's votes as the independent variable. Test whether the model is significant at the 0.05 level. How much of the variance in the response variable is explained by the regression?
3. Test whether the data point for Palm Beach County is unusually far from the model-predicted value for Buchanan's votes, given the number of Bush's votes. [Suggestions: If your software can generate prediction intervals, see whether the Palm Beach data point lies within the expected interval. If you need a formula, the prediction interval for a given x-value x_k equals $\hat{y} \pm$ (the margin of error), as shown below.]

$$\text{Margin of error} = t_{(\alpha/2,n-2)} \times S_y \times \sqrt{1 + \frac{1}{n} + \frac{(x_k - \bar{x})^2}{\sum_{i=1}^{n}(x_i - \bar{x})^2}}$$

4. Create a column showing the residuals for each x-value in the data set. Now create a boxplot for the set of residuals. Is the residual for Palm Beach

County an outlier? If that residual was removed, would the set of residuals be close to normal?

5. Remove the data pair for Palm Beach County and redo step (2). What is the coefficient of determination for the data—not including Palm Beach County?

What do you think? Do you think something unusual may have happened in Palm Beach County in that election? Explain.

16.2 Basic Concepts of Multiple Linear Progression

Multiple linear regression extends the techniques that were introduced in the first section of this chapter. Instead of only one independent variable (x) being used to explain the variation in the dependent variable (y) relative to the y-mean, two or more independent variables are used. As far as including new independent variables in the model allows us to explain more of the variation in y, this increases the coefficient of determination r^2 for the regression and, under certain conditions, improves the ability of the regression to accurately predict the values of the dependent variable. This concept is illustrated in Figure 16.9.

Figure 16.9
Variables Associated with the Best Cities Score

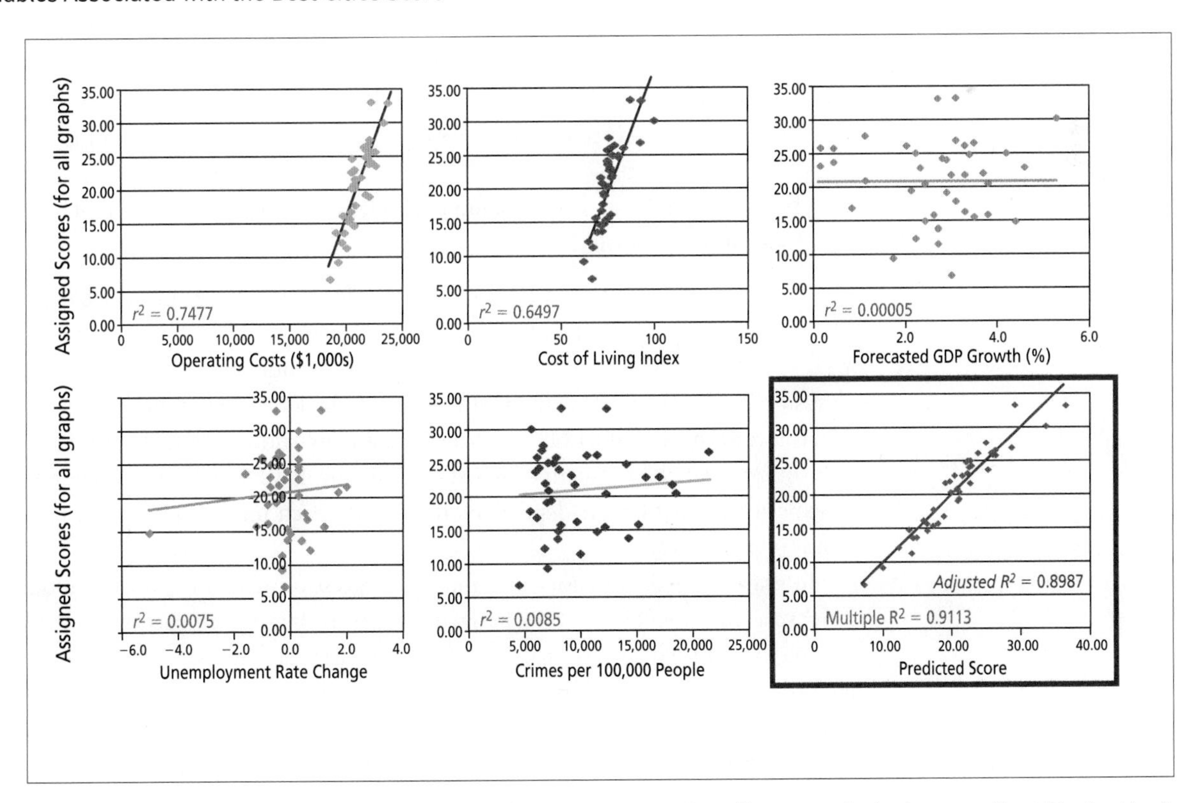

Source: Canadian Business Online. "Best Cities for Business in Canada." Found at: http://www.canadianbusiness.com/bestcities/ranking.jsp

At least five factors contribute to the "Best Cities for Business" scores that are assigned by *Canadian Business Online.* (Lower scores are preferred.) If each factor is compared individually to the assigned scores, we see that operating costs have the biggest impact on the scores: $r^2 = 0.7477$. This means that about 75% of the variation in the scores can be attributed to the differences in businesses' operating costs in various cities. The other factors, individually, explain less of the variation of the dependent variable, but if we use multiple regression to include all of the five factors as independent variables, then the coefficient of determination $R^2 = 0.9113$. (R^2 for multiple linear regression is analogous to r^2 for simple linear regression.) This means that the regression-based predictions, utilizing all of the variables, can explain 91.13% of the variation in the y-values—which enables more precise estimations.

The equation for the multiple regression can be interpreted as a model for how the independent variables $x_1, x_2, \ldots, x_n$ relate to the dependent variable y. The coefficient for each x, individually, shows the impact of a unit increase in that variable on the value of y—if all other variables remain constant. This effect of just one independent variable on the response is called the *pure* or *partial effect* specified by the multivariate regression model.

The general form of a multiple regression equation is:

$$\hat{y} = b_0 + b_1x_1 + b_2x_2 + \ldots + b_kx_k$$

where

$x_1, x_2, \ldots, x_k$ is the set of all independent variables, each of which can assume any value within the range of possible or plausible values for that particular variable.

$\hat{y}$ is the *expected* value for the dependent variable, given a specified combination of values for the independent variables.

b_0 defines the y-intercept (or constant) for the regression model. This represents the mean value expected for the dependent variable whenever all of the x_i-values equal zero.

$b_1, b_2, \ldots, b_k$ is the set of **regression coefficients** for each of the different x-variables, respectively.

Each regression coefficient b_i represents the slope of a line, as in simple regression. But in this case, it is the slope of the line for the equation $\hat{y} = b_0 + b_ix_i$, when (1) x_i is one specific independent variable, (2) b_i is the corresponding coefficient as given by the regression, and (3) the values of all other independent variables happen to be zero (so the terms for other variables have no effect and are not shown in this equation). As long as the other x-variables remain constant (they do not have to be zero), the slope b_i tells the expected impact on the dependent variable for each unit of increase in the corresponding x_i.

To graph the full equation for a multiple regression would require a multidimensional space. If there are two independent variables, then (given specialized software), we could depict their relationship in a three-dimensional diagram such as Figure 16.10. But given more than two independent variables, we cannot visually depict the regression model.

As also mentioned for simple regression, there is another way to express the regression equation that does not use the $\hat{y}$ symbol. The alternative is:

$$y = \beta_0 + \beta_1x_1 + \beta_2x_2 + \ldots + \beta_kx_k + \varepsilon$$

Figure 16.10
Response Surface Diagram (Simulated)

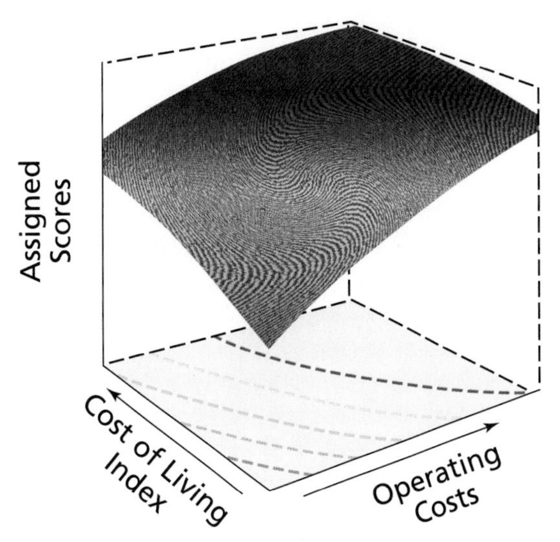

Source: Stat-Ease Software. Figure adapted from *Response Surface Methods for Process Optimization* (Earn 2.4 CEU's)*. Found at: http://www.statease.com/clas_rsm.html

where

β is the population parameter for the relationship in each case.

ε (epsilon) represents the unexplained "error" between an actual y-value and the corresponding output of the regression equation.

Finding and Assessing the Regression Equation

For simple regression, we initially assessed the strength and significance of a correlation between variables (in Section 15.1) before we constructed a regression model for that relationship in the first section of this chapter. If the correlation is weak or not significant, there may be little value in continuing with the regression. Also, at times, a researcher may explore for interrelationships between variables, in various combinations, without proposing a firm model of their relationships.

For multiple regression, we could use the same sequence, but in practice, the basic elements for constructing and assessing the regression equation are provided in the same computer outputs. As before, the calculations for the linear regression are based on ordinary least squares. While the computer output does display a value for a correlation coefficient (i.e., "multiple R"), the strength of a multiple regression is generally assessed by the coefficient of determination, R^2. This tells the proportion of total variation in y that can be explained by the regression equation. In Figure 16.9, for example, we saw that significantly more of the variation in cities' desirability scores could be explained by the regression of the five factors than by any one of the factors individually.

It might appear from the last paragraph that multiple R^2 provides a basis for recommending one regression model in favour of another—the model that explains more of the variation would seem to be preferable. However, for this use, R^2 has a limitation. As more independent variables are added to the model, degrees of freedom are lost in the calculations. This reduces the effective information content of the source data—perhaps more than is compensated for by the higher R^2 of the model. Therefore, **adjusted R^2** (see Figure 16.11) is recommended as an alternative comparative measure that takes into account the costs of reduced degrees of freedom as new variables are added. As shown in Figure 16.9, even after adjustment, a good regression model with multiple variables can still compare favourably with models using fewer variables.

Figure 16.11
Adjusted R^2

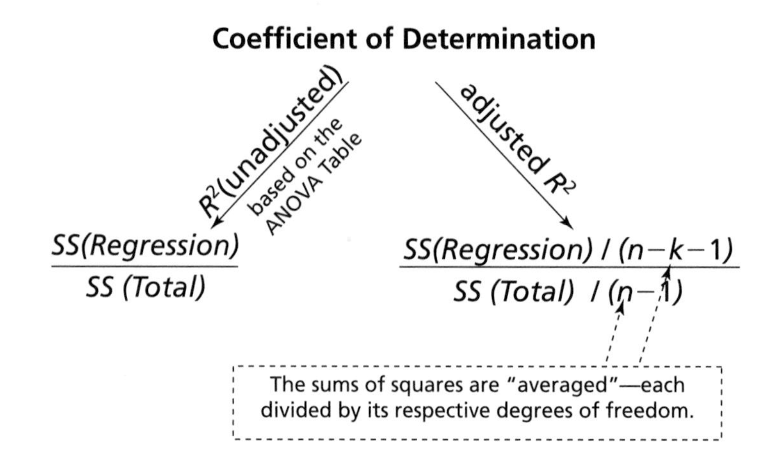

Similar to simple regression, we can test whether the regression is significant with reference to the p-value, which the computer generally outputs in connection with an ANOVA table. The larger the ratio of variation explained by the regression compared to the total variation (this is also the R^2 ratio), the larger the F-value for an implied test of a null hypothesis that $\rho^2 = 0$. Following the usual method: The regression is deemed significant if the p-value is not more than a predetermined value of α (e.g., 0.05), where the p-value $= P(F \geq$ the obtained test statistic $F \mid H_0$ is true, at the given the degrees of freedom).

The computer output for the regression will include estimated values for the constant (b_0) and for all of the x-coefficients ($b_1 - b_k$). Once these are determined, replace those symbols with their values within the generic regression equation. This is illustrated in Example 16.4. Note that in the solutions for parts (5) and (6) in the example, we also introduce the important concepts of t-tests for the individual coefficients and of **residual plots.**

Example 16.4

*(Based on the **Canadian_Cities_for_Business** file)*

The online version of *Canadian Business* posts the results of a survey of the best Canadian cities in which to operate a business. Each listed city obtains a net score, such that the lowest score is the best. An urban planner is studying some of that data, which are found in the file **Canadian_Cities_for_Business.** However, he is not sure how the journal determines the score for each city. Undaunted, he decides that using regression, he can estimate how the net score for a city is related to these five variables:

- Annual operating costs, in $1,000s.
- Cost of living index.
- Forecasted growth in GDP (in %).
- Change in unemployment rate.
- Number of crimes reported per 100,000 population.

1. Find the regression equation that relates a city's net score to the five variables listed above.
2. Is the regression significant?
3. How strong is the apparent relationship?
4. Use the regression equation to estimate the net score for a city that has these characteristics: (a) annual operating costs = $21 million; (b) cost of living index =

75.0; (c) forecasted growth in GDP = 0.8%; (d) unemployment rate has increased by 0.6; (e) reported crimes last year were 6,200 per 100,000 population.

5. Does every input variable appear to make a significant contribution to the net score? Explain.
6. In general, does the data set appear to meet the criteria for applying a linear regression? Explain.

Solution First input the data into a program for linear regression. (For software particulars, see "In Brief 16.2" starting on page 677.) Most of the above questions can be answered directly, based on the computer output. A relatively standard output screen is shown in Figure 16.12.

Figure 16.12
Output from a Regression for the Best Cities Score

Regression Statistics	
Multiple R	0.954632147
R Square	0.911322536
Adjusted R Square	0.898654327
Standard Error	1.953232384
Observations	41

ANOVA

	df	*SS*	*MS*	*F*	Significance *F*
Regression	5	1372.254689	274.4509379	71.937756	2.06743E-17
Residual	35	133.5290861	3.815116746		
Total	40	1505.783776			

	Coefficients	*Standard Error*	*t Stat*	*P-value*	*Lower 95%*	*Upper 95%*
Intercept	−78.19360169	6.189048309	−12.63418829	1.34E-14	−90.75803765	−65.62916573
Operate	0.002990622	0.000435286	6.870477184	5.62E-08	0.002106944	0.003874299
Liv_Cst	0.462019727	0.078456926	5.888832882	1.08E-06	0.3027437	0.621295753
GDP_fore	−1.205442564	0.316780359	−3.805294515	0.000546	−1.848540879	−0.562344249
Unemploy	0.909638899	0.292315513	3.111839292	0.00369	0.316206862	1.503070935
Crime	0.00033997	8.20439E-05	4.143751435	0.000206	0.000173412	0.000506528

Source: Canadian Business Online. "Best Cities for Business in Canada." Found at: http://www.canadianbusiness.com/bestcities/ranking.jsp

1. The estimated constant and *x*-coefficients are shown in the box at bottom of the figure on the left-hand side. These are $b_0 = -78.19360$; $b_1 = 0.00299$; $b_2 = 0.46202$; $b_3 = -1.20544$; $b_4 = 0.90964$; $b_5 = 0.00034$. Rounding each coefficient to three significant digits, the resulting equation is:

$$\hat{y} = -78.2 + 0.00299x_1 + 0.462x_2 - 1.21x_3 + 0.910x_4 + 0.000340x_5$$

where

x_1 = Annual operating costs (in $1,000s)

x_2 = Cost of living index

x_3 = Forecasted growth in GDP

x_4 = Change in unemployment rate

x_5 = Reported crimes per 100,000 population

2. The regression is highly significant. The p-value (in the box at the far right of the ANOVA table) is virtually equal to zero:

$$p\text{-value} = P(F \geq 71.938 \mid H_0 \text{ is true}, df_1 = 5, df_2 = 35) \cong 0.000$$

The value shown in the figure, in scientific notation, is equivalent to $2.07/10^{17}$.

3. The regression is also very strong. Focusing on the multiple R^2, this indicator equals 0.911. This tells us that over 90% of the variation in the net scores is explainable by the regression.

4. By substituting into the regression equation that was developed above (see solution 1) the x-values that are specified in part 4 of the problem, the corresponding y-value can be estimated. (For operating costs, note that units are in \$1,000s, thus \$21 million = 21,000.) Therefore:

$$\hat{y} = -78.2 + 0.00299(21000) + 0.462(75.0) - 1.21(0.8) + 0.910(0.6) + 0.000340(6200)$$

$$\hat{y} \approx 20.9$$

This is the estimated score for the city described.

5. We can answer the question about the contribution of individual variables affirmatively by performing (in effect) separate t-tests on each individual coefficient. In the bottom section of Figure 16.12, observe to the right of each coefficient a t-statistic and a corresponding p-value. For each row, the implied null hypothesis is that—keeping all other coefficients (including the y-intercept term) constant—the true population value for this row's coefficient is equal to zero (i.e., for the variable x_i in the population, H_0 is that the coefficient $\beta_i = 0$). If the t-statistic is large, accounting for degrees of freedom, then the p-value is small. We would therefore reject the null hypothesis and conclude that this coefficient does make a significant, non-zero contribution to the y-estimate.

   In this example, the p-values are very small for every coefficient (in most cases, significant at the 0.01 level), including for the y-intercept value.

6. This important question asks us to consider the validity of relying on the regression model, and some of these issues will be discussed further in the next section. Some of the same standards apply as for correlation and simple linear regression. For example: The distribution of the residuals (i.e., distances between actual y-values and their corresponding predicted values on the line of regression) should ideally be normal and the distributions of residuals across all portions of the line should have about the same variance. The errors must be independent from one another; i.e., the residuals for particular values of x should have no impact on the size or magnitude of residuals for any other values of x.

   For simple linear regression, we could visually inspect a scatter diagram to get a sense of whether the data are consistent with the above rules. We could ask: Do the points appear to follow the line of regression with approximately equal variance for any x-value, and are most points clustered relatively close to the line? In multiple dimensions, we cannot inspect the scatter diagram directly. Instead, we make various plots of the residuals (i.e., distances between actual and predicted y-values), as the x- and/or expected y-values vary. Examples of such diagrams are shown in Figure 16.13.

Figure 16.13
Plot of Residuals versus the y-Estimate

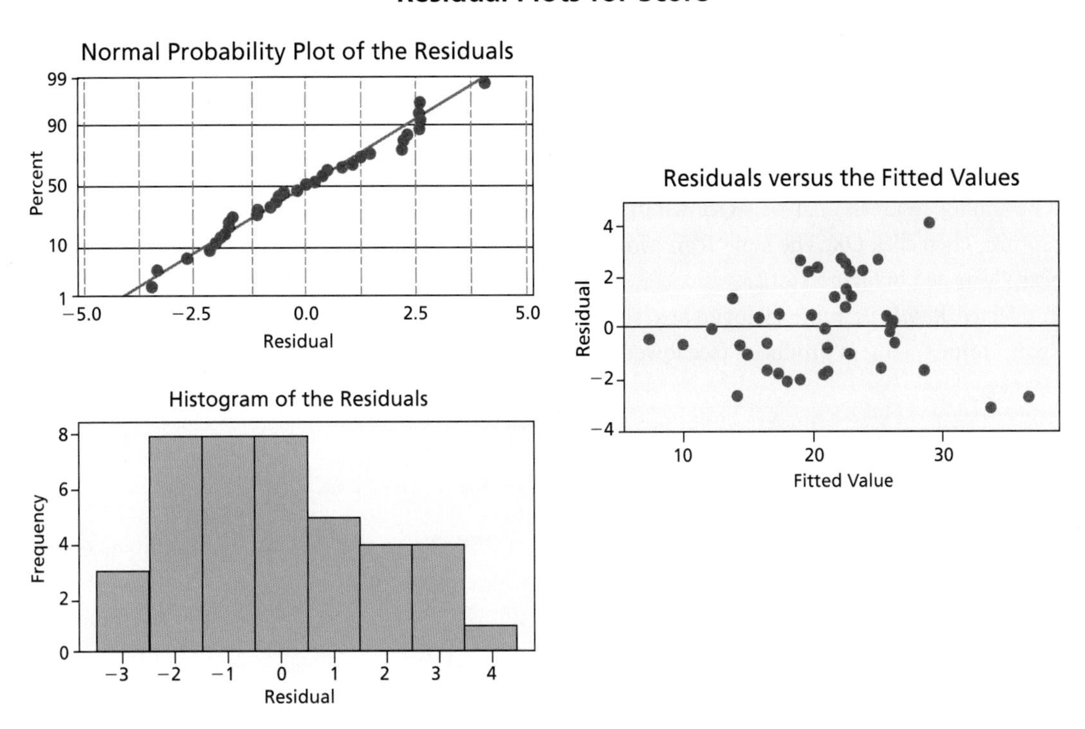

Source: Canadian Business Online. "Best Cities for Business in Canada." Found at: http://www.canadianbusiness.com/bestcities/ranking.jsp

From the histogram of residuals and the normal probability plot, it appears that, for the most part, the errors do tend to be normally distributed. From the graph on the right in Figure 16.13, however, there does appear to be a marked increase in variance for the largest values of the predicted (fitted) y-value. Use caution if making estimates in that area.

• *In Brief 16.2 Multiple Linear Regression*

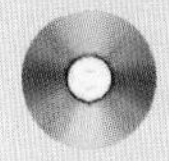

(Based on the ***Canadian_Cities_For_Business*** *file)*

Excel

The associated data must be recorded in columns, such that for each given case or observation, all of the related x_1–x_k data, as well as the y data, are recorded on the same row. All of the x-variables must be stored in a contiguous block of columns; there is no restriction on the y column location. Proceed with the menu sequence *Tools / Data Analysis / Regression;* then click *OK*. In the resulting dialogue box (see the screen capture), specify the separate address ranges for the dependent (y) and independent (x) variables, respectively. Be sure that the address range for the x-variables extends from the upper left-most value of the data block to the lower right-most value.

Select *Labels* if the variable names (labels) are included at the start of the data range. Then, specify the upper left cell for the output range and, if desired, select an option for residuals analysis, as shown in the screen capture. Then click *OK.* The main regression output was illustrated in Figure 16.12.

If you selected *Residuals* in the dialogue box, then a "Residuals Output" table is produced (see lower right-hand corner of the illustration). For every observation (row) in the data set, the predicted y-value is shown, together with the residual for that estimate. These two columns can be graphed using a scatter diagram (see the "Residuals vs. Predicted Values" graph in the illustration). If you selected *Residual Plots,* you will get several graphs—one per independent variable x_i—showing the residuals that correspond to each individual value for x_i (see the illustration).

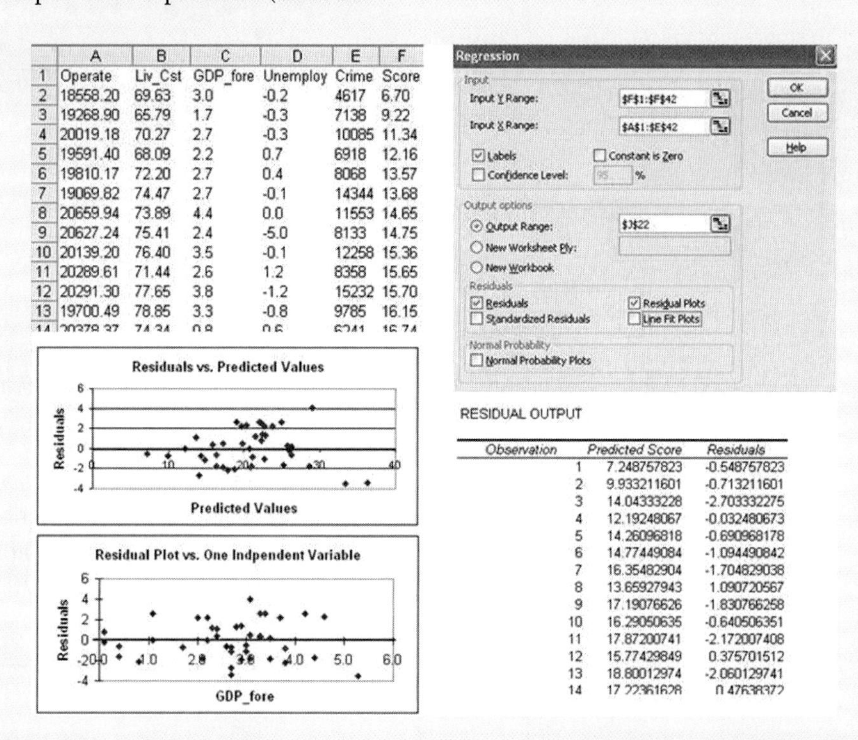

Minitab

The associated data must be recorded in columns, such that for each given case or observation, all of the related x_1–x_k data, as well as the y data, are recorded on the same row. Begin with the menu sequence *Stat / Regression / Regression;* then click *OK.* In the dialogue box (see the screen capture), highlight and select the column name for the response (dependent y) variable and then all of the names for the predictor (independent x) variables. (The columns for the x-variables do not need to be contiguous.) To choose residual analysis, select *Graphs.* Click *OK* when finished. The main regression output is shown at the bottom left of the illustration.

If you selected *Graphs,* then use the *Regression* dialogue box shown in the illustration to select some residuals options. The histogram plot in the illustration or the normal probability plot (see Figure 16.13) show if, collectively, the residuals are normally distributed. The plot for "Residuals versus the Fitted Values" in Figure 16.13 graphs the predicted y-value for each observation versus the residual for that particular estimate. Also shown in the illustration here is the plot for residuals versus each value of a particular variable [GDP_fore(cast) was selected].

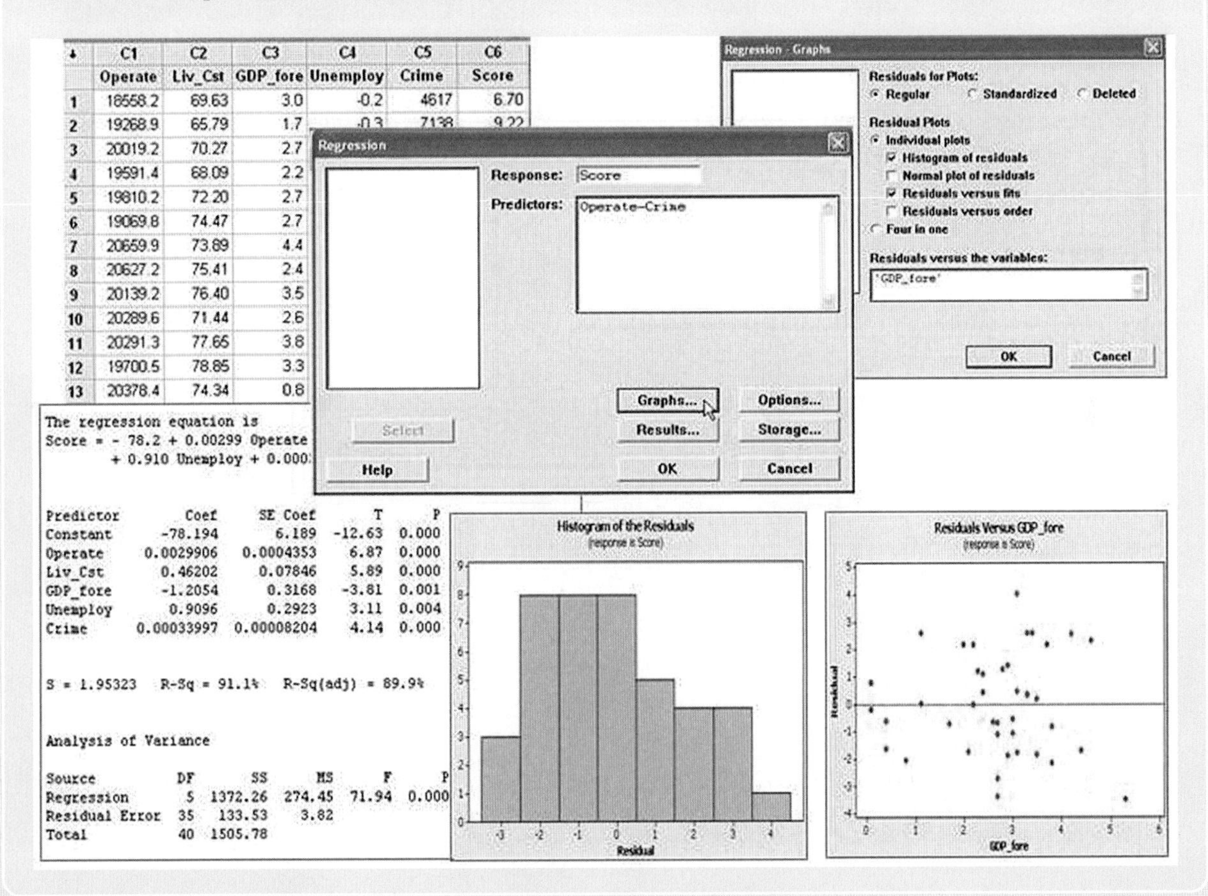

SPSS

The associated data must be recorded in columns, such that for each given case or observation, all of the related x_1–x_k data, as well as the data for y, are recorded in the same row. Begin with the menu sequence *Analyze / Regression / Linear.* In the dialogue box (see the first screen capture), highlight and select the variable name for the dependent (y) variable and then all of the names for the independent (x) variables. (The columns for the x-variables do not need to be contiguous.) To choose residual analysis, click on the *Plots* button. Click *OK* when finished. The main regression output is shown at the bottom left of the illustration.

If you selected *Plots,* the second dialogue box shown in the illustration allows you to select some residuals options. The histogram plot for standardized residuals (bottom centre in the illustration) shows whether, collectively, the residuals are normally distributed. This dialogue box also gives you the choice of graphing all standardized residual values versus the corresponding standardized predicted values for each observation (in the right centre of the illustration). Illustrated under this graph is one of the requested partial plots—in this case, for the standardized residuals versus each value of a particular x-variable, GDP_fore(cast).

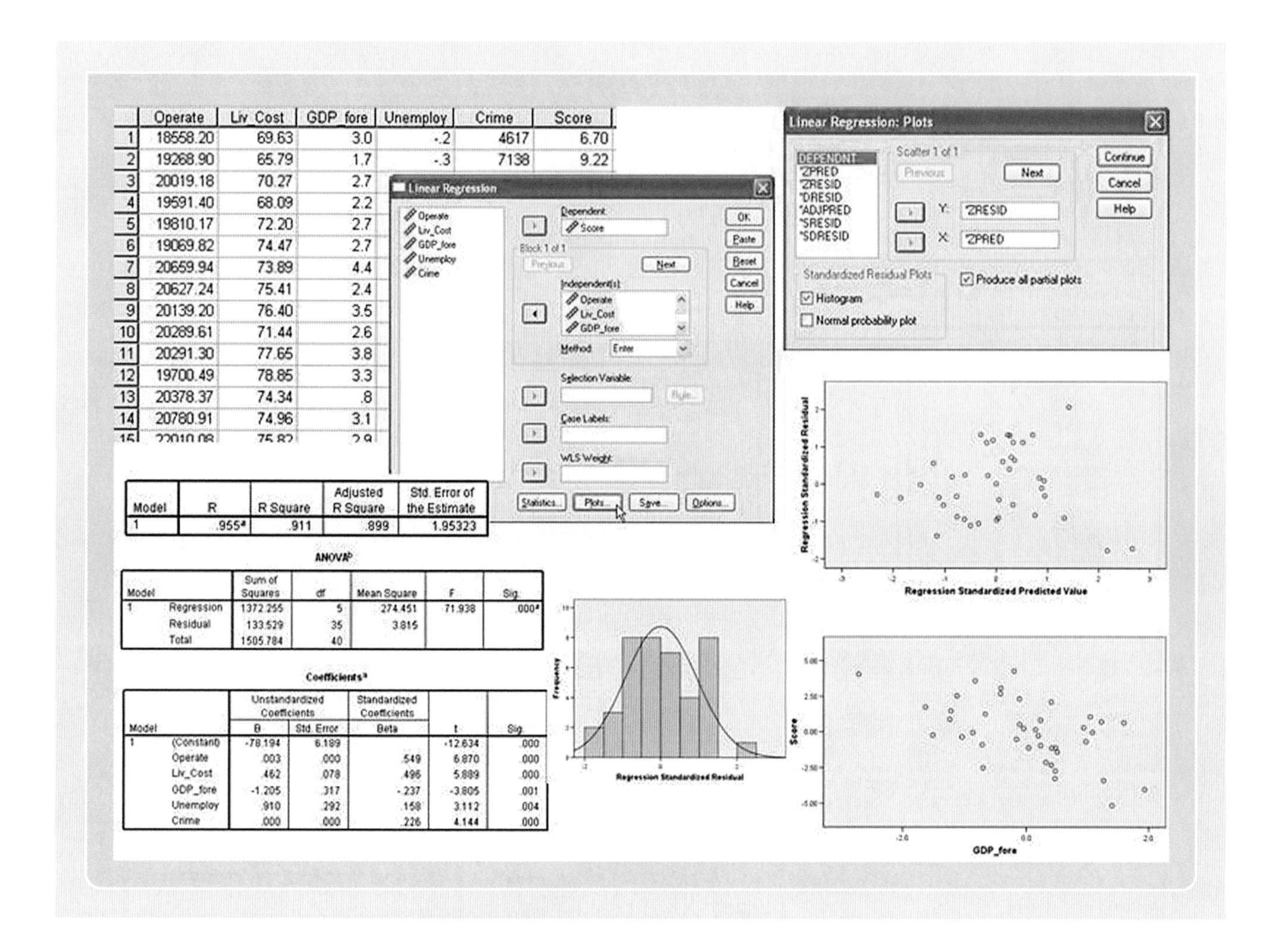

16.3 Developing and Refining the Model

Performing multiple linear regression is deceptively easy when using the computer. As long as a data set contains several numerical columns, the software can provide an equation of some sort or other. But it takes more care and experience to tease out a regression equation that is meaningful and useful as a model for how the variables interact.

Although there are techniques for model-building that appear to be relatively systematic (for example, **stepwise regression**), these may introduce some problems of their own, such as problems related to repeated hypothesis testing. In this section we introduce guidelines for building a model that are more informal. Some of the points overlap, and the steps do not need to be performed in any particular sequence.

Obviously, do not use a regression at all if the basic model assumptions do not apply or if the regression tests are not significant. Similarly, omit any variable whose coefficient is clearly not significant according to its *t*-test. The decisions are less clear-cut in the rules of thumb given below. The goal is to select from among several models, *all* of which may be technically valid, a candidate that is particularly useful and informative.

Rules of Thumb for Developing the "Best" Regression Model

The following are guidelines for developing a regression model.

1. **Keep things simple, including:**
   - Keep the result easy to interpret.
   - Keep the input data easy to collect.

   For someone who is familiar with the basic variables in the data set, the regression equation should be easy to interpret. In Example 16.4, for instance, the urban planner would naturally understand the concepts of cost of living indexes, GDP growth, crime rates, and so on. On finding the equation described in the solution, it would make sense to her that, for instance, increased crime rates contribute to a city's having less desirable (i.e., higher) location score, while increased economic activity in a city (positive GDP growth) contributes to its having a more desirable (i.e., lower) score.

   Suppose the planner was considering a model that would have added another variable, such as the number of visits to households in the city by travelling salespeople. Even if this variable potentially had a significant coefficient in the equation, it is not clear how or why this relates to a city's desirability. In any case, it is certainly not clear how you would obtain such data to use in the equation; the other variables are more readily available. For both of these reasons, if the regression is strong without this extra variable, it may be a good idea to omit it from the model.

2. **Keep the number of independent variables relatively small.** This guideline follows partly from the first one: It is generally easier to collect data for fewer variables than for a greater number, and it is easier to interpret an equation that is less complex. Another consideration is that having more variables requires a larger sample size to obtain a given level of significance for the regression. (Some statisticians suggest having at least four times as many observations as there are independent variables.) Moreover, adding to the number of variables increases the likelihood of encountering collinearity, which is discussed in guideline 4.

3. **But don't omit highly influential variables.** Imagine trying to predict whether people will buy a luxury car but failing to include any factor that represents income or personal wealth. This gap could inflate the apparent effect of some **proxy variable,** such as the brand of shoes or tie being worn when shopping. The proxy variable may stand in for the missing variable of wealth and may sometimes have an unexpected value for its coefficient. Omitting the key relevant variable might make it impossible to find a significant model.

4. **Among the variables included, minimize the collinearity (or multicollinearity).** If there is **collinearity** (or **multicollinearity**) among the variables in the model, then they are not all totally independent from one another. If variables are correlated, this suggests that given a value for one of them, the value of the other provides information that is not altogether new. The value of one variable is derivable from the values of the other variables.

   On the one hand, this type of case runs counter to the second guideline: Why include an extra variable that does not offer new information? Perhaps more serious is the fact that collinearity (like proxy effects) can greatly increase the error in the coefficients of the model. Also, collinearity can sometimes have the effect that neither of the correlated x-variables' coefficients appear to be significant in the model—even though if just one of the variables was included, its contribution *would* be significant. Despite this, it is not always realistic to find variables that have no collinearity at all. In Example 16.4, a depressed economy (with a falling GDP) may in fact be correlated with the crime rate, but neither can predict the other with certainty. But where the x-variable correlations are very strong, you might consider whether one of the related variables could be omitted.

5. **Maximize explanatory benefits—but balance these against the costs.** Ideally, the best model would account for the most variation in the dependent variable (i.e., have the highest R^2, compared to possible alternatives). But as noted in the previous section, for models that have different numbers of independent variables, it is misleading to compare R^2 values directly. So, all things being equal, we might hope to maximize the *adjusted* R^2 value when selecting a model.

   It is a benefit if a model meets this condition, but there are generally costs in terms of being able to satisfy the other rules of thumb. If you can markedly increase adjusted R^2 without great cost in adding new variables, or in increasing the difficulty of data collection and so on, then you are possibly on the right track. On the other hand, if adding some extra variable or extra complexity results in little gain in R^2, then it may be time to simplify the model.

Additional Cautions for Regression

Naturally, all of the warnings and conditions given in the first section of this chapter apply also for multiple regression. Residual analysis becomes especially important for multiple regression because you can no longer judge normality and heteroscedacity assumptions (equal variance for the whole line) by eye from a scatter diagram. Also, the more variables, the greater the risk that some of these will exhibit collinearity. Because of this, it is useful to run correlation analyses among the variables to discover pairs that may be closely linked.

For regressions in which time is an independent variable, there is an additional risk of **autocorrelation** for some variables. Instead of each observation for the variable being independent, the values of observations may be correlated with the values of previous observations. For example, if you are tracking weekly sales of swimsuits, this year's high-value observations in August will be correlated with a similar surge of values last August, and this year's cluster of low-value observations (in January) will be correlated with a similar pattern last January. In short, residuals from the regression are not independent.

Finally, be careful to identify any **influential observations** that, if included in the model, can have a major (and often misleading) effect on the slope or slopes of the model. Consider the two outliers in Figure 16.14, which compares the annual numbers of formal complaints about airlines for baggage issues versus refund issues. Two airlines had a particularly bad year for baggage complaints. If those outliers are included in the model, all baggage complaint estimates rise sharply for the entire regression. The effects are easy to spot in the scatter diagram but would be less obvious for multivariate cases that cannot be graphed as simply.

Figure 16.14
Example of Influential Values

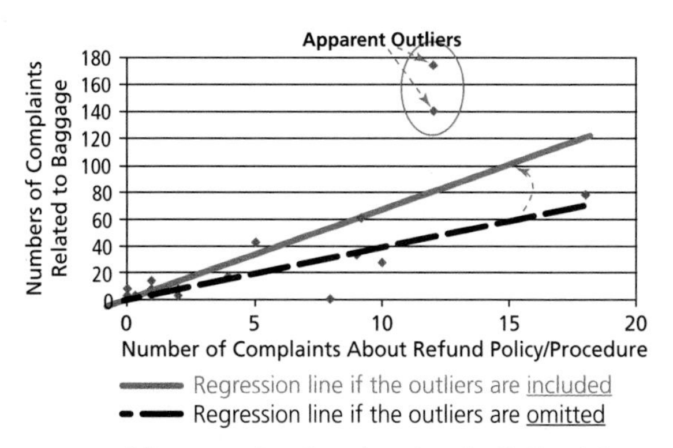

Source: U.S. Department of Transportation. Based on data in *Air Travel Consumer Report.* Found at: http://airconsumer.ost.dot.gov/reports/2005/0505atcr.pdf

---

Example 16.5

(*Based on the* ***Toronto_Listings*** *file*)

An accomplished musician has been offered a position in Toronto. Concerned about house prices in the city, he is trying to assess what factors are determining the list prices of the city's houses. Sampling from data published by the *Virtual Agent Network* (at http://on.virtual-agent.com), he compiled the data file **Toronto_Listings.** Drawing from the data and variables in that file, construct the "best" multiple regression model for predicting the list price of a house in Toronto. (Assume that the sample is representative of houses and prices in Toronto for an indefinite block of time.)

Solution

There is rarely just one "best" model, but here we try to find as good a model as possible. The nature of the variables is shown in Figure 16.15. Besides the dependent variable price, we have only the following quantitative variables to choose from: bedrooms, bathrooms, kitchens, and parking spots (i.e., the number of these), plus the dwelling's age and square footage. ("PC2" is not a true number; it represents the second character of the postal code.) In the next section, we learn to use "dummy variables" to handle promising nonquantitative variables, but for this section, we will consider only quantitative variables.

Figure 16.15
Toronto Listings Data

	Price	Type	Bedrooms	Bathrooms	Kitchens	Family_Room	Age	Sq. Feet	Garage	Parking	PC2	PC3	Basement
1	939 000	Detached	4	2	1	Y	60	2,000	Y	4	6	C	
2	379,000	Detached	4	2	1	Y	.	.	N	3			F
3	279,900	Semi-detached	4	3	3	N	.	.	N	4			A
4	639,000	Detached	7	4	3	Y	.	.	Y	5	3	H	A
5	365,000	Detached	5	2	1	N	.	.	Y	5	1	L	
6	2195 000	Detached	5	4	1	Y			Y	4			F

Source: Virtual Agent Network. Based on data found in a search of the *Virtual Agent Network* for Toronto real estate listings as of October 13, 2006. Found at: http://on.virtual-agent.com/toronto-real-estate.html

The variables age and square feet can be rejected by the first guideline: The many blanks show that these variables are hard to collect values for, at least by the musician in our example. Figure 16.16 explores some of the characteristics of the data and their possible intercorrelations. Take note that the numbers of kitchens and parking spots are not normally distributed. We can still try to use those variables because it is the normality of the *residuals* distribution that matters, but bear in mind that their distributions may impact on the distribution of the response variable. Also note the collinearity of the bedrooms variable with all of the other candidate variables. This is a reason to omit that variable from further consideration (see rule of thumb guideline 4).

Figure 16.16
A Preliminary Look at the Data

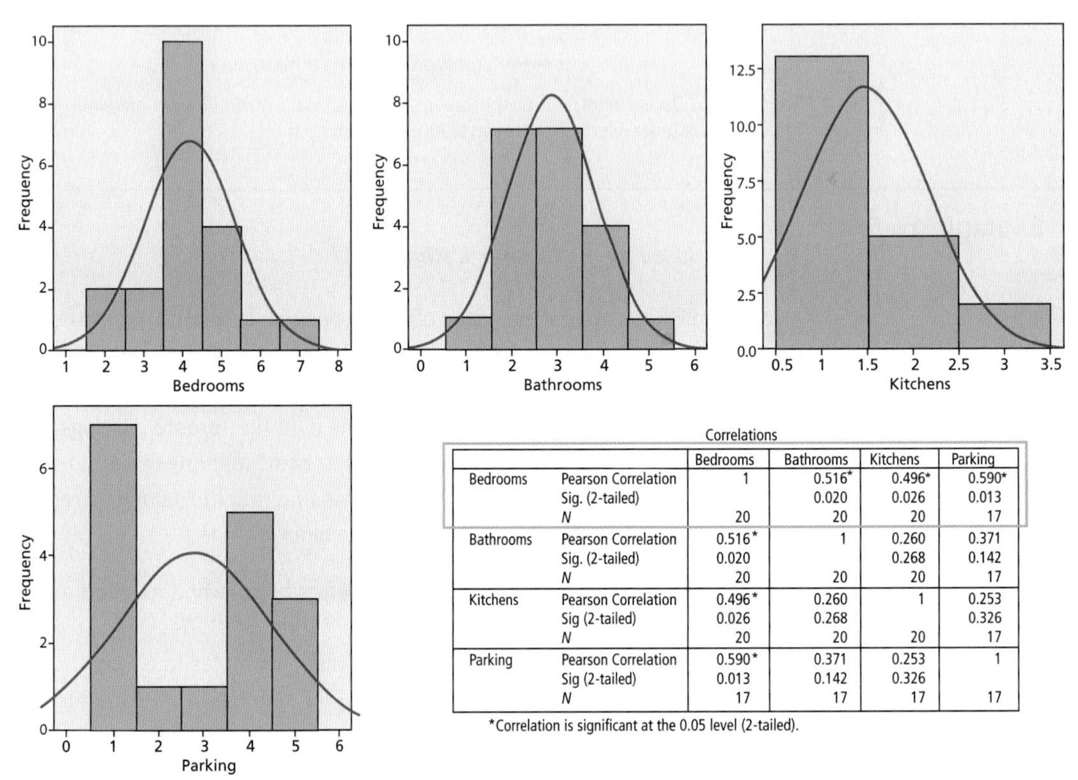

Correlations

		Bedrooms	Bathrooms	Kitchens	Parking
Bedrooms	Pearson Correlation	1	0.516*	0.496*	0.590*
	Sig. (2-tailed)		0.020	0.026	0.013
	N	20	20	20	17
Bathrooms	Pearson Correlation	0.516*	1	0.260	0.371
	Sig. (2-tailed)	0.020		0.268	0.142
	N	20	20	20	17
Kitchens	Pearson Correlation	0.496*	0.260	1	0.253
	Sig (2-tailed)	0.026	0.268		0.326
	N	20	20	20	17
Parking	Pearson Correlation	0.590*	0.371	0.253	1
	Sig (2-tailed)	0.013	0.142	0.326	
	N	17	17	17	17

*Correlation is significant at the 0.05 level (2-tailed).

Source: Virtual Agent Network. Based on data found in a search of the *Virtual Agent Network* for Toronto real estate listings as of October 13, 2006. Found at: http://on.virtual-agent.com/toronto-real-estate.html

Using the most promising variables from the above preliminaries, we can run a multiple regression with the remaining variables. (Other approaches are possible. Some prefer to add the possible predictor variables one at time, starting with the most plausible, and then run a partial *F*-test after each new variable is attempted. Stop adding variables when the improvement in the *F*-value is not worth the cost of adding the extra variable.) For a less formal approach, you can run the regression with the variables that remain under consideration, and the software will generate something similar to Figure 16.17. Although the regression as a whole is significant (p-value $= 0.006$), we see from the *t*-tests on the coefficients that one of the variables, parking, is apparently not making a significant contribution. The p-value for its coefficient equals 0.175. This suggests that the regression should be tried again.

Outputs from a second regression pass are shown in Figure 16.18. The regression on variables bathrooms and kitchens explains about two-thirds of the variation in prices ($R^2 = 0.611$). The relationship is also highly significant (p-value ≈ 0.000). The

Figure 16.17

First-Pass Output

SUMMARY OUTPUT

Regression Statistics	
Multiple R	0.778675
R Square	0.606334
Adjusted R Square	0.515488
Standard Error	374700.1
Observations	17

ANOVA

	df	*SS*	*MS*	*F*	Significance *F*
Regression	3	2.81122E+12	9.37E+11	6.674307	0.005757249
Residual	13	1.8252E+12	1.4E+11		
Total	16	4.63642E+12			

	Coefficients	Standard Error	*t*-Stat	*p*-value	Lower 95%	Upper 95%
Intercept	−335906	316777.7987	−1.06038	0.308274	−1020262.824	348450.8276
Bathrooms	436041.6	123808.3944	3.521907	0.003753	168569.8337	703513.382
Kitchens	−358927	142755.6142	−2.51427	0.025882	−667331.5447	−50522.03685
Parking	87012.69	60616.6484	1.435459	0.174778	−43941.61785	217966.9962

Source: Virtual Agent Network. Based on data found in a search of the *Virtual Agent Network* for Toronto real estate listings as of October 13, 2006. Found at: http://on.virtual-agent.com/toronto-real-estate.html

Figure 16.18

Second-Pass Output

Model Summary[b]

Model	R	R Square	Adjusted R Square	Std. Error of the Estimate
1	0.782[a]	0.611	0.565	376281.724

a. Predictors: (Constant), Kitchens, Bathrooms

b. Dependent Variable: Price

ANOVA[b]

Model		Sum of Squares	*df*	Mean Square	*F*	Sig
1	Regression	3.8E+012	2	1.892E+012	13.360	0.000[a]
	Residual	2.4E+012	17	1.416E+011		
	Total	6.2E+012	19			

a. Predictors: (Constant), Kitchens, Bathrooms

b. Dependent Variable: Price

Coefficients[a]

Model		Unstandardized Coefficients		Standardized Coefficients		
		B	Std. Error	Beta	*t*	Sig
1	(Constant)	−165794	289688.1		−0.572	0.575
	Bathrooms	448887.5	90477.203	0.777	4.961	0.000
	Kitchens	−350545	130256.6	−0.422	−2.691	0.015

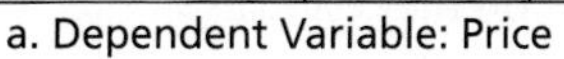

a. Dependent Variable: Price

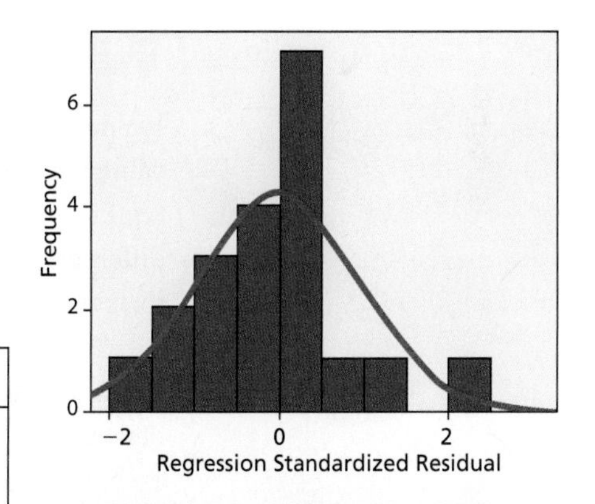

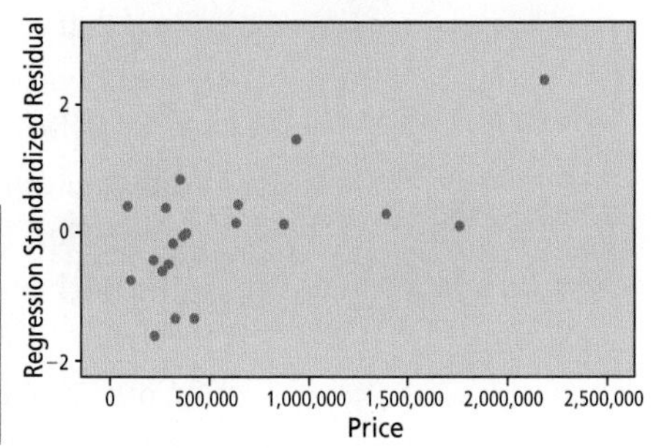

Source: Virtual Agent Network. Based on data found in a search of the *Virtual Agent Network* for Toronto real estate listings as of October 13, 2006. Found at: http://on.virtual-agent.com/toronto-real-estate.html

distribution of residuals looks relatively reasonable, as shown by the histogram of the residuals and by the scatter of residuals around the actual y-values.

The x-variable's coefficients are also highly significant (the p-values equal 0.000 and 0.015, respectively). It is unexpected, perhaps, that adding kitchens to a house is predicted to *reduce* the price of the house (Kitchens coefficient = *Negative* 350,545). This may indicate that the kitchens variable is a proxy for some variable we have overlooked. For example, we can speculate (and test in the next section) that the richest homes may not include basement apartments; extra kitchens are often associated with basement apartments. Therefore, the following model is probably about the best we can get from the given set of quantitative variables:

$$\hat{y} = -165{,}794 + (448{,}887.5)\text{Bathrooms} - (350{,}545)\text{Kitchens}$$

Despite its strengths, the model does have problems. The standard error is shown as \$376,282. This means that, at best, the model can distinguish the most expensive homes from something a bit less—but has no useful precision for a finer estimate. Also, the y-intercept does not test as significant, so applying that constant for a particular calculation could well increase the error of the estimate. If the musician wants to build a better model, we recommend that (1) he obtain a larger sample size and (2) look for some additional explanatory variables.

• Hands On 16.2 *Influential Observations in Multiple Regression*

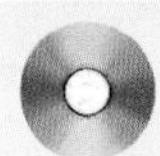

*(Based on the **FloridaVote2000** file)*

We noted that in regression a data point is called an influential observation if its value for the response variable is so far removed from $\hat{y}$ that it pulls the entire regression line toward itself (since regression is trying to minimize the squared differences from $\hat{y}$). This is analogous to an outlier for univariate data that distorts the position of the data set's mean.

To see this effect for linear regression, open the file for **FloridaVote2000**. If you attempted "Hands On 16.1," you drew scatter diagrams for the relationship between Buchanan votes (y) and Bush votes (x) in Florida, including and then excluding the data for Palm Beach County. Now add to each of those diagrams a line that corresponds to the regression, and compare the impact on the regression of including or excluding the outlier.

A direct comparison like the above is not possible for multiple regression, but you *can* visually compare the effects of an outlier on the residuals. Using the same Florida votes file, run a regression program for the relationship between the Buchanan votes (y) and *all* of the other variables in the data set as the independent variables. Repeat this with, and then without, the data for Palm Beach County. In each case, generate residual plots and assess the normality of the residuals. What is the apparent impact, if any, on the residuals of including the Palm Beach County data? Explain why this effect (if any) is occurring.

16.4 Including Qualitative Variables in the Model

Multiple regression opens the door to including qualitative variables in the regression model. In Example 16.5, our model had a very large standard error. Perhaps the equation could be improved if we considered some of the qualitative variables in the data set, such as type of dwelling, the presence of a garage, or the inclusion of a separate apartment in the dwelling.

The trick to including this information is to use **dummy variables.** By convention, a dummy variable usually takes a default value of "0" if some qualitative condition is not met; the value of the dummy variable is "1" if that condition *does* apply for an observation. For example, a revised data set for the house listings could include a dummy variable "Garage"; its value is 0 for houses that lack a garage and 1 for houses that have a garage.

Sometimes, the original qualitative variable has more than two possible values; let us call the number of possible values or categories k. In that case, we encode the original qualitative information using up to $k - 1$ dummy variables in this way: Choose one of qualitative values as the default case. There is no special dummy variable for the default case. Then create a separate dummy variable for each of the other possible values (or, if desired, for clusters of other possible values). If, for a given observation, all of these dummy values = 0, then the default case is presumed to be applicable for that observation. Otherwise, there must be one dummy variable that = 1 for that observation, and this conveys the relevant information about the case.

For example, in the **Toronto_Listings** data set, there is a variable called "basement." The variable takes $k = 4$ possible values: "F" (for "finished"), "A" (for a basement apartment), "N" (for no basement), and blank (presumed unfinished). In Example 16.6, we will use $k - 1 = 3$ dummy variables to analyze this qualitative data. The default case used is a dwelling presumed to have an unfinished basement. The dummy variables are finished basement, basement apartment, and no basement. Any dwelling that falls into one of those three categories shows the value 1 for the corresponding dummy variable; otherwise, the dummy variable shows as 0. For any dwelling where none of these three categories applies, then (as per the default case), we presume instead that there is an unfinished basement.

When a regression analysis includes a dummy variable and the coefficient of the variable is significant, it tells us to do the following: Whenever the qualitative variable takes the value modelled by the dummy variable, then add (in full) the dummy variable's coefficient to the predicted output. Otherwise, that particular variable contributes *nothing* to the predicted value. This process is illustrated in Figure 16.19.

Suppose that the new owner of Sally's Salad Bar cannot find the old pricing information for purchases (by weight) of egg salad and for side orders of garlic toast. But some relevant data are found on the electronic cash register, as shown at the left of Figure 16.19. The owner will use regression to estimate the previous owner's price scheme for these items. In the data table, note the default case is that garlic toast is *not* ordered. A value of 1 for the dummy variable means that garlic toast *is* ordered.

Perform the regression calculations like any other multiple regression. (See "In Brief 16.2.") As shown in the top right-hand side of the figure, the significant coefficient for the garlic toast variable is \$1.50. When Garlic toast = 1, then an extra 1 × \$1.50 will be added to the predicted charge. When Garlic toast = 0, then 0 × \$1.50 (i.e., nothing) will be added to the predicted charge.

Figure 16.19

The Value of a Dummy Variable

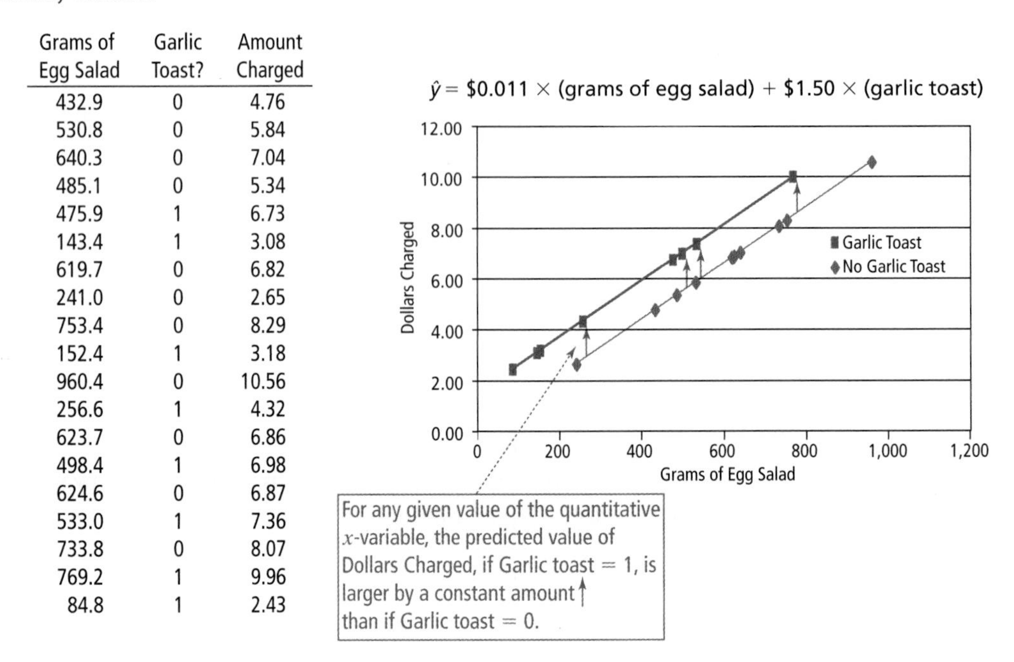

Grams of Egg Salad	Garlic Toast?	Amount Charged
432.9	0	4.76
530.8	0	5.84
640.3	0	7.04
485.1	0	5.34
475.9	1	6.73
143.4	1	3.08
619.7	0	6.82
241.0	0	2.65
753.4	0	8.29
152.4	1	3.18
960.4	0	10.56
256.6	1	4.32
623.7	0	6.86
498.4	1	6.98
624.6	0	6.87
533.0	1	7.36
733.8	0	8.07
769.2	1	9.96
84.8	1	2.43

The graph portion of Figure 16.19 shows the best interpretation of how the dummy variable has its effect. For quantitative-only variables, we would draw a single regression line for the scatter diagram, but due to the either/or effect of the qualitative variable, the scatter of the points does not follow a normal distribution around a single line. Instead, we envision *two* regression lines for the quantitative variable (or variables). The blue line in the figure connects cases for which Garlic toast = 0. The red line connects cases for which Garlic toast = 1. The vertical distance between the lines, above any given x-value in this two-dimensional example, is equal to the coefficient of the related dummy variable. In the figure, the blue line predicts the charge for any given purchase weight of egg salad, but if garlic toast is ordered, then the red line adds $1.50 to that prediction.

A critical *assumption* of the model is that the slopes of the non-dummy variables are in no way affected by the values taken by the qualitative variable. In other words, the quantitative variables must be independent of the dummy variables. A case of violating this requirement would be if Sally's Salad Bar offered a discount on the price of egg salad for those who also buy garlic toast. The slopes of the two lines in the figure would then be unequal and there would no longer be one constant distance between them that symbolizes the impact of the dummy variable.

Example 16.6

*(Based on the **Toronto_Listings** file)*

In Example 16.5, a musician used regression to estimate house prices in Toronto, based on the values of some quantitative variables. He is not pleased with the error size of the model and hopes to produce a better model by including some qualitative variables. Use suitable dummy variables to resume the search for the "best" model for estimating house prices in Toronto.

1. What is the equation for the regression model?
2. How does the model compare with the solution to Example 16.5?
3. Use the model to estimate the price of a house with the following characteristics.[2]

Type	Bedrooms	Bathrooms	Kitchens	Family Room	Garage	PC2	Use of Basement
Detached	5	3	2	Y	Y	6	Apartment

Solution To begin, we carry over the quantitative results from the previous analysis (see the solution for Example 16.5). Only the variables for numbers of bathrooms and kitchens were found to be helpful.

Next, to add qualitative variables, they must first be coded as dummy variables. This requires some judgment. For example, the majority of houses in the original sample are detached; there are only a few cases each of the other types, such as condo or semidetached. To maintain reasonable sample sizes for the subgroups, we show only one related dummy variable in Figure 16.20. The default case is a detached house; any other alternative is coded by NotDetached = 1.

Figure 16.20
Revised Data Set with Dummy Variables

1	Price	Not Detached	Bathrooms	Kitchens	Family room	Garage	PC2=6	Apartment	No Basement	Finished Bsmt
2	939,000	0	2	1	1	1	1	0	0	0
3	379,000	0	2	1	1	0	0	0	0	1
4	279,900	1	3	3	0	0	0	1	0	0
5	639,000	0	4	3	1	1	0	1	0	0
6	365,000	0	2	1	0	1	0	0	0	0
7	2,195,000	0	4	1	1	1	0	0	0	1
8	329,900	0	3	1	0	1	0	0	0	0
9	224,900	1	3	1	0	1	0	0	0	1
10	91,900	1	1	1	0	1	0	0	1	0
11	1,395,000	0	4	1	1	1	0	0	0	1
12	219,999	1	2	1	0	1	1	0	1	0
13	647,000	0	3	2	1	0	0	0	0	1
14	318,999	0	2	1	0	1	0	0	0	1
15	429,000	0	4	2	0	0	1	1	0	0
16	879,000	0	3	1	1	0	0	0	0	0
17	1,769,000	0	5	1	1	1	0	0	0	0
18	289,900	0	3	2	0	1	0	0	0	1
19	349,900	1	2	2	0	0	0	0	0	1
20	259,999	0	3	2	0	0	1	0	0	1
21	103,500	1	2	1	0	1	0	0	1	0

Source: Virtual Agent Network. Based on data found in a search of the *Virtual Agent Network* for Toronto real estate listings as of October 13, 2006. Found at: http://on.virtual-agent.com/toronto-real-estate.html

Family room and garage are both easily coded: Code Y as 1 and N as 0 for each variable. Also, viewing the original data, it appears that for a cluster of the observations, the second character of the postal code was 6. In the revised data set illustrated in Figure 16.20, the default case for the dummy variable PC2 = 6 is that the second character of the postal code is *not* 6. Code the variable as 1 if the second character *is* 6. The original variable PC3 (third character of the postal code) has been omitted from further analysis; there are too many subcategories, with few cases of each. The codings of the three dummy variables for apartment, no basement, and finished basement were discussed earlier in this section.

Figure 16.21 shows all correlations between variables in the revised data set and, in the upper left, Figure 16.22 shows a preliminary analysis using these variables. First, all variables in Figure 16.20 (except price) are considered as potential independent variables. We see at the left of Figure 16.22 that only two variables yield significant coefficients: bathrooms (quantitative) and family room (qualitative). (It was possible that other coefficients may have been significant were it not for collinearities with the variables kitchens and no basement; however, rerunning the analysis without those variables had no significant effect.) Also note that by adding new information the oddly negative coefficient for numbers of kitchens no longer appears as significant as it did in Example 16.5.

Figure 16.21
Correlations between the Variables in Figure 16.20

	Nondetached	Bathrooms	Kitchens	FamilyRoom	Garage	PC2=6	Apartment	NoBasement
Bathrooms	−0.465 0.039							
Kitchens	0.049 0.838	0.260 0.268						
FamilyRoom	−0.535 0.015	0.445 0.049	−0.092 0.701					
Garage	0.023 0.924	−0.005 0.982	−0.447 0.048	−0.043 0.858				
PC2=6	−0.055 0.819	−0.052 0.828	0.037 0.876	−0.153 0.519	−0.157 0.508			
Apartment	0.031 0.898	0.356 0.123	0.764 0.000	−0.057 0.811	−0.279 0.234	0.140 0.556		
NoBasement	0.642 0.002	−0.516 0.020	−0.283 0.227	−0.343 0.139	0.308 0.186	0.140 0.556	−0.176 0.457	
FinishedBsmt	−0.154 0.518	0.037 0.878	−0.008 0.975	0.082 0.731	−0.179 0.450	−0.201 0.395	−0.380 0.098	−0.380 0.098

Source: Virtual Agent Network. Based on data found in a search of the *Virtual Agent Network* for Toronto real estate listings as of October 13, 2006. Found at: http://on.virtual-agent.com/toronto-real-estate.html

1. At the bottom of Figure 16.22, we have rerun the regression using only bathrooms and family room as the independent variables. Again, both are highly significant. From this we can get the equation:

$$\hat{y} = -341{,}958 + (245{,}546)\text{Bathrooms} + (618{,}489)\text{Family Room}$$

2. This model compares favourably with the solution model for Example 16.5. For both models, the regression p-value ≈ 0.000; i.e., highly significant. Both require only two input variables, using readily available data. But if we compare the adjusted R-squares for the models, the previous value was 0.565, which has now improved to 0.646.

   The standard error for the estimate has improved—from \$376,282 to \$339,556—but this is still too large for the model to be very useful, especially if you are looking for a house costing less than \$300,000. Similarly, although the large error on the y-intercept has improved—from \$289,688 to \$240,942, the coefficient remains nonsignificant.

   As in the previous example, more sample data are needed, with possibly extra variables, to make a more practical model. Most likely, a variable for location

Figure 16.22
Outputs With Dummy Variable

Second Pass

Regression Statistics	
Multiple R	0.89191799
R Square	0.7955177
Adjusted R Square	0.611483631
Standard Error	355782.6676
Observations	20

ANOVA					
	df	*SS*	*MS*	*F*	Significance *F*
Regression	9	4.92452E+12	5.47E+11	4.3226654	0.015974787
Residual	10	1.26581E+12	1.27E+11		
Total	19	6.19033E+12			

	Coefficients	Standard Error	*t* Stat	P-value	Lower 95%	Upper 95%
Intercept	−406623.5058	473577.3139	−0.85862	0.4106581	−1461819.514	648572.5022
NotDetached	194368.488	281680.5401	0.690032	0.5058723	−433254.8646	821991.8407
Bathrooms	315237.5653	116722.7315	2.700738	0.0222851	55163.11351	575312.0171
Kitchens*	−180025.4163	221359.0022	−0.81327	0.4349972	−673244.0071	313193.1746
FamilyRoom	595018.9137	212471.2132	2.800468	0.0187795	121603.5507	1068434.277
Garage	157053.3907	210412.4287	0.746407	0.4726041	−311774.7146	625881.496
PC2=6	73310.8266	223894.3082	0.327435	0.7500905	−425556.7781	572178.4313
Apartment	−159796.9967	459285.365	−0.34793	0.7351088	−1183148.558	863554.5649
NoBasement*	−176139.5078	389682.1062	−0.45201	0.6609091	−1044405.345	692126.3293
FinishedBsmt	26230.31714	224152.7825	0.11702	0.9091612	−473213.2042	525673.8385

Third Pass

Predictor	Coef	SE Coef	T	P
Constant	−341958	240942	−1.42	0.174
Bathrooms	245546	88041	2.79	0.013
FamilyRoom	618489	173075	3.57	0.002

S = 339556 R-Sq = 68.30% R-Sq(adj) = 64.60%

Analysis of Variance

Source	*df*	*SS*	*MS*	*F*	*P*
Regression	2	4.23026E+12	2.11513E+12	18.34	0.000
Residual Error	17	1.96007E+12	1.15298E+11		
Total	19	6.19033E+12			

Price	Bathrooms	FamilyRoom
939,000	2	1
379,000	2	1
279,900	3	0
639,000	4	1
365,000	2	0
2,195,000	4	1
329,900	3	0
224,900	3	0
91,900	1	0
1,395,000	4	1
219,999	2	0
647,000	3	1
318,999	2	0
429,000	4	0
879,000	3	1
1,769,000	5	1
289,900	3	0
349,900	2	0
259,999	3	0
103,500	2	0

* Results are similar if, to reduce collinearity, the variables Kitchens and NoBasement are removed.

Source: Virtual Agent Network. Based on data found in a search of the *Virtual Agent Network* for Toronto real estate listings as of October 13, 2006. Found at: http://on.virtual-agent.com/toronto-real-estate.html

of the property or for recent selling prices of neighbouring houses would be very helpful.

3. Based on the model:

$$\hat{y} = -341{,}958 + (245{,}546)(3) + (618{,}489)(1) = \$1{,}013{,}169$$

But bear in mind that, roughly speaking, this estimate might be off by as much as two standard errors: ±$680,000!

• In Brief 16.3 Coding the Dummy Variables

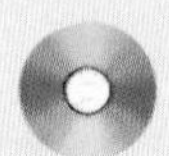

(*Based on the* ***Toronto_Listing*** *file*)

Regardless of whether any of the variables are dummy variables, the computer procedures for regression are unchanged from those in "In Brief 16.2." Where the computer can help is in the process of coding the dummy variables prior to running the regression.

Excel

In the illustration, columns A and F contain data for qualitative variables (basement and type of dwelling). A common use of a dummy variable is to take the value 1 if the qualitative variable has one particular value; otherwise, take the value 0. For example, the variable apartment has the value 1 when, and only when, Basement = A.

	A	B	C	D	E	F
1	Basement	Apartment	NoBasement	FinishedBsmt	Type	NotDetached
2	←	=IF(A2="A",1,0)	=IF(A2="N",1,0)	=IF(A2="F",1,0)	Detached	=IF(OR(E2="Semi-Detached",E2="Attached",E2="Condo"),1,0)
3	F	0	0	1	Detached	0
4	A	1	0	0	Semi-Detached	1
5	A	1	0	0	Detached	0
6		0	0	0	Detached	0
7	F	0	0	1	Detached	0
8		0	0	0	Detached	0
9	F	0	0	1	Attached	1
10	N	0	1	0	Condo	1
11	F	0	0	1	Detached	0
12	N	0	1	0	Condo	1
13	F	0	0	1	Detached	0
14	F	0	0	1	Detached	0
15	A	1	0	0	Detached	0
16		0	0	0	Detached	0
17		0	0	0	Detached	0
18	F	0	0	1	Detached	0
19	F	0	0	1	Semi-Detached	1
20	F	0	0	1	Detached	0
21	N	0	1	0	Condo	1

A simple Excel function can be used to automate the coding, as illustrated at the top of the "Apartment" column. The format is **=if(A2="A",1,0)**, where **A2** is the cell address for the corresponding value of the original qualitative variable. If that value equals the specified value (e.g., A), then apply the code **1**; otherwise, apply the code **0.** Copy the formula down the column. A similar formula is shown for the other dummy variables NoBasement and FinishedBsmt.

Sometimes in the coding process, more than one value of the qualitative variable are given the same code for the dummy variable. In the figure, the variable NotDetached is set to 1 if the original qualitative variable Type has any of these three values: Attached, Semi-Detached, or Condo. Again, use the **=if** function in Excel as shown above but in this case, use an **OR( )** function for the condition; in the parentheses, list each of the conditions, separated by commas, as shown in the illustration. If any of the **OR** conditions apply, the dummy is set to **1;** otherwise, it is set to **0.**

Minitab

In the illustration, columns C1-T and C5-T contain data for qualitative variables (basement and type of dwelling). Prepare columns for the intended dummy variables. Apply the menu sequence given below for each dummy variable separately. (In the illustration, the results of the process are displayed.)

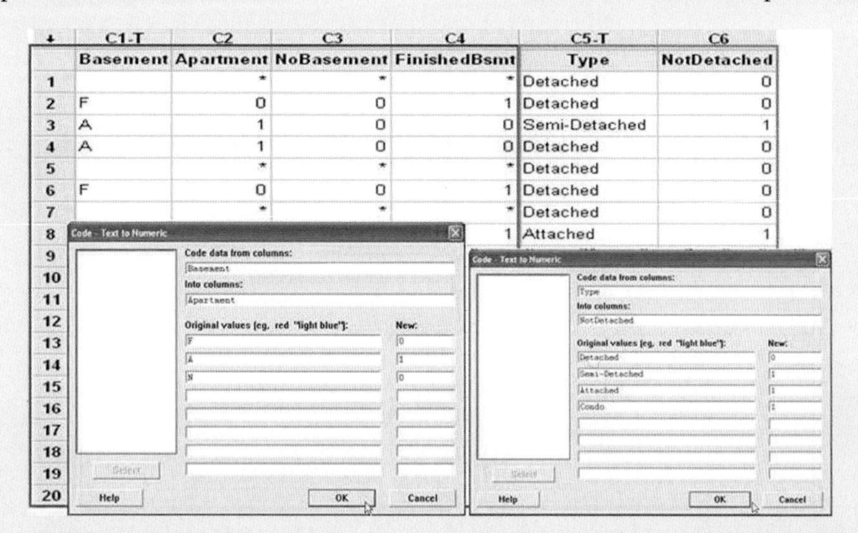

Use the menu sequence *Data / Code / Text to Numeric*. In the dialogue box, highlight and select the name of a variable whose data need to be encoded. Then highlight and select the name of the column into which the coded value should be placed. Below that, use the space provided to create a table, showing on the left each possible value to be found in the original data column; to its right, show what code should be placed in the corresponding row for the dummy variable. For example, in the left-hand screen capture, the table says to encode an "A" as 1 and any other value as 0. (Note that Minitab will not handle cases where the original value was blank; these will have to be filled in manually.)

Sometimes in the coding process, more than one value of the qualitative variable is encoded by a single dummy variable. In the right-hand screen capture, the variable NotDetached is set to 1 if the original qualitative variable Type has any of these three values: Attached, Semi-Detached, or Condo. The basic procedure is unchanged from the above.

SPSS

In the illustration, the column "Basement" contains data for the corresponding qualitative variables. Plan your columns for the intended dummy variables. Apply the menu sequence below for each dummy variable separately. (In the illustration, the results of the process are displayed.)

Use the menu sequence *Transform / Recode / Into Different Variables*. In the first dialogue box (the top screen capture in the illustration), highlight and select (by pressing the arrow button) the name of a variable whose data need to be encoded. Then, in the *Output Variable* area, input the name of the column into which the coded values should be placed and then click on the *Old and New Values* button. This calls up a second dialogue box, as illustrated in the lower part of the illustration.

For the next steps, there is some flexibility. In the *Old Value* area of the dialogue box at the bottom

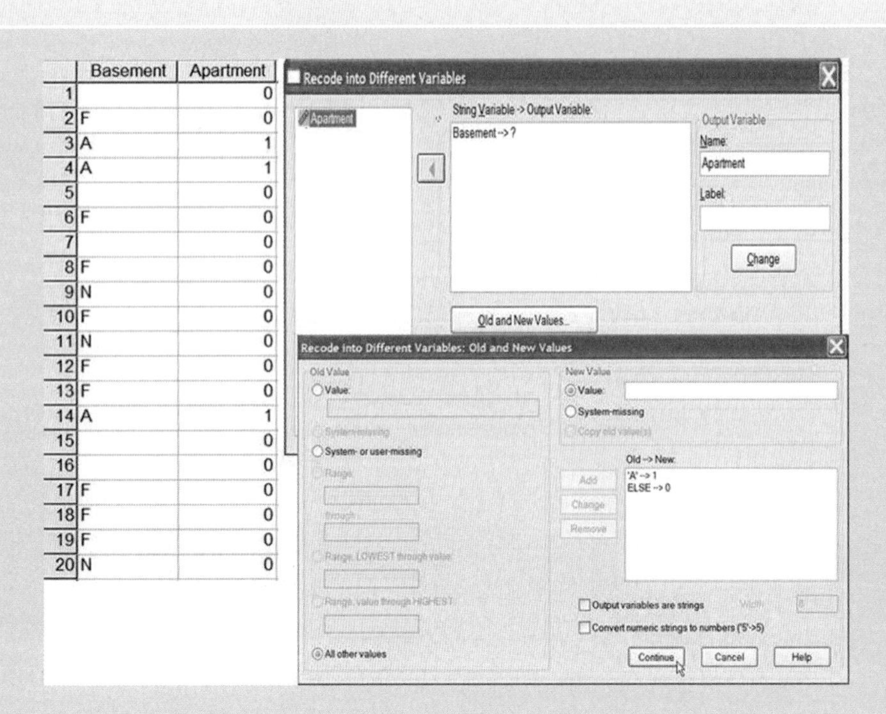

left, *Value* was selected (not shown here) and the raw data value "A" was input; then in the *New Value* area, *Value* was selected again and the code value "1" was entered; then the *Add* button was clicked. Notice how each new instruction is added automatically to a list in the *Old* → *New* box.

Additional combinations of *Old* → *New* values could be added to that list in similar fashion. Finally, *All other values* was selected in the *Old Value* area, and the coded value "0" was input as the *Value* in the *New Value* area; then *Add* was clicked.

When the list of coding choices is complete, click on *Continue* and then *OK*. Follow a similar process for each new dummy variable to be encoded.

• Hands On 16.3 *Predicting World Temperatures*

Construct a good multiple regression equation that could be used to predict the current temperature of any city in the world. For independent variables, choose from data that are readily available on the Internet. (One very good site is http://www.wunderground.com.) A partial list of possible independent variables (use your imagination to identify more) include latitude, current local time, elevation, dew point, and wind speed, plus categorical variables such as cloud cover conditions.

First examine the kinds of data that your Internet weather service makes available and then set up a spreadsheet. Each row will represent one randomly selected city in the world; then fill in the column entries in that row based on the current time, latitude, etc., of the particular city. Try to collect the data for at least 50 cities from any continent. Include at least one qualitative variable, recoded into a suitable number of dummy variables.

See if you can construct an equation that is strong and significant and does not require a great number of independent variables to be used.

Chapter Summary

16.1 Construct and interpret a simple linear regression equation; assess the strength, significance, and validity of the model; and use the model to make inferences.

Simple **regression analysis** is a powerful tool to create a model (in the form of an equation) with which we can estimate or predict the value of one variable based on its linear relationship with another. The y-variable (if the relationship is graphed) represents the **dependent** (or **response**) **variable;** the x-variable (graphed on the horizontal axis) is the **independent** (or **predictor**) **variable.** The **regression equation** and **regression line** are both models of the relationship between the variables. The regression line is known as the **line of best fit,** which shows, for any value of x found in the population, the estimated centre of an interval in which the corresponding y-value is expected to lie.

The general form of the simple linear regression equation is $\hat{y} = b_0 + b_1x$. The constant term b_0 is called the ***y*-intercept,** and the term b_1 is the ***x*-coefficient,** which gives the **slope** of the line. The slope is a constant ratio that applies for all pairs of points on the line—namely, the ratio between the difference in y-values for the two points and the corresponding difference in x-values for the same points. $\hat{y}$ (pronounced "y cap") is the expected or predicted value of the dependent variable for a specified value of x.

The procedure to find the regression equation is called **ordinary least squares.** The distance between some actual value of y and the corresponding, expected value ($\hat{y}$) for a given x value is called a **residual.** Each residual is an error left unexplained by the regression. To obtain the best fitting line, minimize the sum of all of the squared residuals between actual and expected y-values. Use Formulas 16.1 and 16.2 to find the slope and intercept of the regression line, respectively.

Be careful of attempts to **extrapolate** from the data—using x-values that fall beyond the original range of data. A p-value can be obtained from the ANOVA table that is output by regression software (based on an F-test) to determine significance. The **coefficient of determination (r^2)** indicates what proportion of the total variance of the dependent variable (which is the variance compared to simply the y-mean) is explained by, or attributable to, the variance due to the regression. If a correlation is perfect, then $r^2 = 1$. To find r^2, use Formula 16.3. For a measure of the precision of the estimate, use the **standard error of the estimate** derived from Formula 16.4. If you have in mind one particular value of the independent variable, the interval estimate for the range of likely y-values that might occur is called the **prediction interval;** for any given x-value, the prediction interval is wider than the **confidence interval for the estimate.**

16.2 Construct and interpret a multiple linear regression equation; assess the strength, significance, and validity of the model; and use the model to make inferences.

The general form of a multiple regression equation is:

$$\hat{y} = b_0 + b_1x_1 + b_2x_2 + \ldots + b_kx_k$$

where

b_0 defines the y-intercept (or constant) for the regression model

$b_1, b_2, \ldots, b_k$ is the set of regression coefficients for each of the different x-variables.

Each regression coefficient b_i represents the **slope of the line** for an equation $\hat{y} = b_0 + b_ix_i$, in which (1) x_i is one specific independent variable, (2) b_i is the corresponding coefficient as given by the regression, and (3) the values of all other independent variables are presumed to equal zero. As long as the other x-variables remain constant, the slope b_i tells the expected impact on the dependent variable for each unit of increase in the corresponding x_i.

For comparing multiple regression models with different numbers of variables, it can be misleading to use the relative sizes of multiple R^2 as a measure of model superiority, because adding variables to a model causes degrees of freedom to be lost in the calculations. Instead, **adjusted R^2** is recommended for a comparative measure between models. For assessing the compliance of the data with regression assumptions, the use of **residual plots** is encouraged.

16.3 Describe and apply some rules of thumb for developing a good regression model and recognize some key hazards to avoid in the process.

Although there are techniques for model-building that appear relatively systematic (for example, **stepwise regression**), these may also introduce some problems related to repeated hypothesis testing. This section introduces some rules of thumb for developing the "best" (or at least a good) regression model. Aim, as much as possible, to (1) keep things simple (e.g., for data collection and output interpretation); (2) keep the number of independent variables relatively small; but (3) don't omit highly influential variables (to avoid the problems of **proxy variables**); (4) minimize **collinearity** (or **multicollinearity**) among the variables that are included; and (5) maximize explanatory benefits—but balance these against the costs.

For regressions in which time is an independent variable, there is a risk of **autocorrelation,** such that observed values may be correlated with the values of previous observations. Another caution is to watch for **influential observations,** which are outliers that, if included in the model, can have a major (and often misleading) effect on the slope or slopes of the model.

16.4 Construct and interpret a multiple linear regression equation that includes dummy variables to represent qualitative data.

Multiple regression opens the door to also include qualitative variables in the regression model by using **dummy variables.** A dummy variable usually takes a default value of 0 if some qualitative condition is not met; the value of the dummy variable is 1 if that condition *does* apply for an observation.

Key Terms

adjusted R^2, p. 673
autocorrelation, p. 682
coefficient of determination (r^2 or R^2), p. 664
collinearity (*or* multicollinearity), p. 681
dependent (*or* response) variable, p. 655
dummy variables, p. 687
extrapolate, p. 662
independent (*or* predictor) variable, p. 655
influential observations, p. 682
line of best fit, p. 655
ordinary least squares, p. 658
prediction interval, p. 665
proxy variables, p. 681
regression analysis, p. 655
regression coefficients, p. 672
regression equation, p. 655
regression line, p. 655
residual, p. 658
residual plots, p. 674
slope (of the regression line), p. 657
standard error of the estimate, p. 664
stepwise regression, p. 680
x-coefficient, p. 657
y-intercept, p. 655

Exercises

Basic Concepts 16.1

1. Explain the difference between the independent variable and the dependent variable in a regression analysis.
2. Name an equivalent pair of terms for the independent variable and the dependent variable, respectively, in a regression analysis.
3. Describe the difference between a regression equation and a regression line.
4. List the conditions that are required to be met (at least approximately) in order to validly use the model of regression as presented in this chapter.
5. In each of the following cases, regression will be used as a method to solve a problem. Indicate which variable is the dependent variable for each case.
   a) Given a regression between each year's average winter temperature and the corresponding size of annual tree-ring growth (for a certain species), measure this year's average winter temperature and predict the size of this year's tree-ring growth for the species.
   b) Given a regression between each year's average winter temperature and the corresponding size of annual tree-ring growth (for a certain species), measure the amount of tree-ring growth in 1893 for that species to estimate the average winter temperature that year.
   c) Given a regression between annual numbers of housing starts and the average unemployment for a year, forecast the number of housing starts next year based on an estimate for next year's unemployment.
6. Given the equation $\hat{y} = 15 + 25x$:
   a) What is the value of the regression coefficient x?
   b) What is the value of the y-intercept?
   c) What is the value of the slope?
   d) Can this equation predict the exact value of the dependent variable if the value of x is known? Why or why not?
   e) Is the width of the prediction interval the same for all values of x?
   f) Explain the difference between the y in the equation $y = 15 + 25x$ and the $\hat{y}$ in the equation given above.
7. Explain the meaning and role of each of these expressions in regression analysis:
   a) Line of best fit
   b) Ordinary least squares
   c) Residuals
   d) Extrapolation
   e) Coefficient of determination
   f) r^2
   g) Standard error of the estimate
   h) Prediction interval
   i) Significance of the regression
8. What are the risks of extrapolating based on a regression equation?

9. In Exercise 9 of the previous chapter, we examined data for the shelf space allocated to products at a small grocery store and the corresponding sales of those products.

Shelf Space (cm)	Sales ($)
30	250
40	299
50	350
60	361
70	375

The goal of the present analysis is to predict the sales for a new item by knowing the amount of shelf space allotted.

a) Construct the regression equation for the data.

b) How much of the variation in items' sales is explained by the regression? On what did you base your answer?

c) If an item is allotted 45 cm of shelf space, estimate the sales for the item.

d) Fill in the dollar value in the blank: Ninety-five percent of the time, estimates based on the above regression will be accurate within ±$________. Explain how you reached this conclusion.

e) Repeat (a) through (d) after adding this additional row to the data table:

Shelf Space (cm)	Sales ($)
54	359

Applications 16.1

10. The dental association official in Chapter 15, Exercise 10, plans to extend her analysis of mean ages of workers in various dental-related fields. Her intent is to use regression to forecast the mean ages of different groups in future years. (See the **Age_Health_Occupation** file.)

a) Construct a regression equation for forecasting the mean ages of dental hygienists for a given year based on the mean ages of dental assistants for that year.

b) Is the regression significant? Justify your claim.

c) What is the standard error of the regression?

d) Use the regression to estimate the mean age of dental hygienists in a year when the mean age of dental assistants is 32.2.

e) In a certain year, the mean age of dental assistants is 33.5 and the mean age of dental hygienists is 34.1. In relation to the above regression equation, what is the *y*-residual for the year?

f) The general population is known to be aging, on average. Can you use the above regression to estimate the mean age of dental hygienists in a future year when the mean age of dental assistants is 46.9? Why or why not?

11. a) Repeat the first two parts of Exercise 10 but this time the goal is to forecast the mean ages of dental technicians for particular years, given the corresponding mean ages of dental assistants.

b) For estimating the mean age of dental technicians in a year when the mean age of dental assistants is 32.2, why would it be preferable to simply use the mean of the *y*-values than to use the above regression equation for the estimate?

12. Refer back to the analysis of health-care related costs in Chapter 15, Exercise 13. By using regression, a researcher hopes to estimate some of these costs, based on knowing other costs. (In the file **Health_Expenditures,** costs are in CDN$100,000.)

a) Construct a regression equation for forecasting a year's drug expenditures, based on knowing the year's expenditures on hospital care.

b) Is the regression significant? Justify your claim.

c) What is the standard error of the regression?

d) Use the regression to estimate expenditures for drugs in a year when expenditures for hospital care are $2,532,100,000.

e) In a certain year, the expenditure on hospital care is $3,325,300,000 and expenditure on drugs is $1,777,700,000. In relation to the above regression equation, what is the *y*-residual for the year?

f) All health-related costs appear to be increasing. Can you use the above regression to estimate the drug expenditures for a year when hospital-care expenditures reach $45,000? (Units are in CDN$100,000.) Why or why not?

13. a) Repeat (a) through (d) of Exercise 12 but replace "drug expenditures" (or equivalent expression) with "physician expenditures" where appropriate.

b) In a certain year, the expenditure on hospital care is $33,253 and expenditure on physicians is $14,321. (Units are in CDN$100,000.) In relation to the above regression equation, what is the *y*-residual for the year?

Exercises 14–16 are based on data in the file **Water_ Mineral_Content.**

14. A chemist is researching mineral contents for various commercially available brands of mineral water. She has compiled for her research the data file **Water_Mineral_ Content.** In particular she is interested in the potassium content of the different waters. (Units for this problem are all in mg/L.)

a) Construct a regression equation for estimating the level of potassium (K) in a water product, based on its level of sodium (Na).

b) What percentage of the variance in the level of potassium is explained by its regression with sodium? Explain the basis for your answer.
c) In general (95% of the time), the regression-based estimates for the level of potassium will be correct within + or − what amount? Explain the basis for your answer.
d) Comparing two brands of mineral water, the second brand has 20 more mg/L of sodium than the first brand. How much more or less potassium would you expect in the second brand compared to the first?

15. The chemist in Exercise 14 next wants to construct a regression equation for estimating the level of potassium (K) in a water product, based on its level of calcium (Ca).
a) Construct the regression equation just described.
b) Is the regression significant? Explain the basis for your answer.
c) Is the regression strong? Explain the basis for your answer.
d) Bearing in mind your answers to (a) through (c), use either the regression *or a more suitable estimate* to estimate the level of potassium, given that the calcium content is 90 mg/L. What basis did you use for the estimate?

16. The chemist in Exercise 14 is continuing her analysis.
a) Construct a regression equation for estimating the level of potassium (K) in a water product, based on its level of chloride (abbreviated Cl).
b) Is the regression significant?
c) Is the regression strong?
d) Bearing in mind your answers to (a) through (c), use either the regression *or a more suitable estimate* to estimate the level of potassium, given that the chloride content is 110 mg/L. What basis did you use for the estimate?

Basic Concepts 16.2

17. Explain the difference between simple linear regression and multiple linear regression.

18. In the context of multiple linear regression, what is the interpretation of the coefficient of determination R^2?

19. In multiple linear regression, what is the difference between R^2 and adjusted R^2, and when it is recommended to use the adjusted R^2?

20. When evaluating a model based on multiple linear regression, why it is useful to examine a histogram of the residuals and/or a normal probability plot? What can each of those tools tell us about the regression model?

21. In the context of multiple linear regression, how is the slope of a single variable interpreted?

22. The p-value for a multiple linear regression is based on what statistical test?

23. Given a displayed value for the coefficient of a variable in a multiple linear regression, what statistical test can be performed to assess whether the value of that coefficient is significantly different from zero?

24. According to the ANOVA table output from a regression program, the F-value for the ratio of the mean sum of squares between the regression and the residuals is 90.117, with degrees of freedom equal 6 for the numerator and 33 for the denominator. What is the p-value for the regression?

25. In Exercise 24, if $\alpha = 0.01$, should the regression be considered significant?

Applications 16.2

26. The official in Exercise 10 is developing a model to forecast the mean ages of dental assistants by means of multiple regression.
a) Construct a regression equation for forecasting the mean ages of dental assistants, for a given year, based on the mean ages of both dental hygienists and dental technicians. (See the **Age_Health_Occupation** file.)
b) Is the regression significant at the 0.05 level? Justify your claim.
c) What is the standard error of the regression?
d) Use the regression to estimate the mean age of dental assistants in a year when the mean age of dental hygienists is 35.5 and of dental technicians is 40.5.
e) In a certain year, the mean ages of dental hygienists, dental technicians, and dental assistants are 35.5, 40.5, and 38.8, respectively. In relation to the above regression equation, what is the y-residual for the year?

27. The official in Exercise 26 hopes to improve the regression model by adding another independent variable. (See the **Age_Health_Occupation** file.)
a) Construct a regression equation for forecasting the mean ages of dental assistants for a given year, based on the mean ages of dental hygienists, dental technicians, and denturists.
b) Is the regression significant at the 0.05 level? Justify your claim.
c) Did R^2 increase, decrease, or stay the same, compared to the regression in Exercise 26?
d) Did adjusted R^2 increase, decrease, or stay the same, compared to the regression in Exercise 26? What does that answer tell us about the advantages, if any, of adding the extra independent variable, compared to Exercise 26?

28. In order to improve the model for estimating drug cost expenditures in Canada (see Exercise 12), a researcher decides to add additional independent variables. (In the file **Health_Expenditures,** costs are in CDN$100,000.)
    a) Construct a regression equation for forecasting a year's drug expenditures, based on knowing the year's expenditures on hospital care and physicians.
    b) Is the regression significant? Justify your claim.
    c) What is the standard error of the regression?
    d) Use the regression to estimate expenditures for drugs in a year when expenditures for hospital care are $2,532,100,000 and for physicians' costs are $1,079,900,000.
    e) Suppose that in a year, costs for hospital care and physicians were as shown in (d) and the cost for drugs was $920,500,000. In relation to the above regression equation, what is the y-residual for the year?
    f) Do both of the independent variables in the above model appear to contribute significantly to the regression result? Explain how you determine your answer.
29. a) Repeat (a) through (c) of Exercise 28 but add capital costs as an *additional* independent variable.
    b) Based on adjusted R^2, is there any advantage to adding the extra variable to the model obtained in Exercise 28?
    c) In the model using the new variable, does the new variable appear to have a significant coefficient? Explain.
30. In her analysis of mineral waters around the world, the chemist in Exercise 14 hopes to improve estimates for the potassium content of the different waters by switching to multiple regression. (In the data file **Water_Mineral_Content,** all units are in mg/L.)
    a) Attempt three different regressions for estimating the level of potassium (K) in a water product, based on, respectively, its levels of:
        i) Sodium (Na) and calcium (Ca).
        ii) Sodium (Na) and chlorine (Cl).
        iii) Calcium (Ca) and chlorine (Cl).
    b) Which of these models seems preferable with respect to adjusted R^2?
    c) For which if any of the models do the residual plots seem consistent with the requirements for the regression? Explain.
    d) For which if any of the models do *both* the independent variables appear to have significant coefficients?
    e) Drawing on your previous answers, which models in (a), (i) through (iii), would you recommend? Explain.
31. The chemist in Exercise 30 decides to estimate the value of potassium (K) in a water product, based on its contents for all three of sodium (abbreviated Na), calcium (Ca), and chlorine (Cl). Show the regression equation. Is the adjusted R^2 notably different from the best of the regressions attempted in the previous question? Is the residual plot improved? Drawing on your previous answers, which of the models (the models in Exercise 30, (a), (i) through (iii), or the model for this exercise) would you recommend? Explain.

Basic Concepts 16.3

32. List five rules of thumb for constructing a good multiple regression model.
33. If you plan to construct a multiple regression model, why is it important to examine and analyze the residuals?
34. List several reasons why including more independent variables in the model is not always desirable.
35. What does it mean to say that a variable in a regression model is a proxy variable? What might cause this to occur? Why is this not desirable in a model?
36. What does it mean to say that there is collinearity among the variables in a regression model? Why is this not desirable in a model?
37. What does it mean to say that for a variable in a regression model there is autocorrelation?
38. What is an influential observation, and what problems can this cause for a regression model?
39. Can a model be highly significant and still not very useful in practice? If so, explain how and when this might occur.
40. What is the name of a formal approach to selecting independent variables, one by one, to include in the regression model?

Applications 16.3

For these problems, more than one right answer might be possible. Also, results may vary if your instructor recommends specific procedures (e.g., the use of stepwise regression with a particular software application). The common goal is to construct and apply the best linear regression model that you can—in keeping with the five rules of thumb and your instructor's guidelines—and to ensure, if possible, by residuals analysis, a reasonable conformity of the model with regression assumptions.

41. The regression model built in Exercise 26 to forecast the mean ages of dental assistants used only two independent variables. Using the data in the file **Age_Health_Occupation,** try to create a better model by using the ages for some combination of the variables ambulance attendant ("Amb_Att"), audiologist ("Audio"), dental hygienist ("Dent_Hyg"), medical laboratory technician ("ML_Tech1"), medical radiation technologist ("MR_Tech"), and other professional occupation ("Other"). All variables retained in the final model should have significant coefficients.
    a) What is the regression equation, based on your model?
    b) How much of the variance of the dependent variable is explained by the regression?

c) Use the regression equation to estimate the mean age of dental assistants in a year when the mean age of ambulance attendants, medical radiation technologists, and those classified as "Other" is 38 years, and the mean age of audiologists, dental hygienists, and medical laboratory technicians is 36.2 years.

d) Judging from the standard error of the model, as well as from the residuals analysis, how exact do you think your estimate was in (c)? Explain.

42. For the six variables originally considered when answering Exercise 41, test for collinearity. Then test for collinearity among just the variables that you retained in your solution for Exercise 41. Were any of the variables involved in collinearity removed from the final model?

43. The regression model built in Exercise 28 to estimate drug cost expenditures in Canada used only two independent variables. (In the file **Health_Expenditures,** costs are in CDN$100,000.) Try to create a better model by using some combination of the expenditures for hospitals, other institutions, physicians, other professionals, capital, public health & admin, and other health spending. All variables retained in the final model should have significant coefficients.

a) What is the regression equation, based on your model?

b) How much of the variance of the dependent variable is explained by the regression?

c) Use the regression to estimate expenditures for drugs in a year when expenditures (in CDN$100,000) are as shown in the following table.

Hospitals	Other Institutions	Physicians	Other Professionals
21,790.2	8,769.0	9,651.6	9,835.9

Capital	Public Health & Admin	Other Health Spending
2,339.7	4,082.7	8,389.1

d) Suppose that in year *T*, the costs for hospital care and physicians were as shown in (c) and the actual cost for drugs happened to be $1,050,000,000. Compare the actual drug costs in year *T* with the drug costs calculated in (c). What is the *y*-residual for year *T*?

44. In Exercise 43, the variable time was not included. Do all of the values of variables tend to be increasing with time? If so, do you expect that there will be a high collinearity among all of the *x*-variables? Explain.

45. Each of the regression models built in Exercise 30 to estimate potassium (K) content in bottled waters used only two independent variables. (In the data file **Water_Mineral_Content,** all of the data is in mg/L.) Try to create a better model by using some combination of the variables for Ca, Mg, Na, HCO_3, Cl, SO_4, F, B, P, Ti, and Cr. All variables retained in the final model should have significant coefficients.

a) What is the regression equation, based on your model?

b) How much of the variance of the dependent variable is explained by the regression?

c) Use the regression to estimate the potassium content in a mineral water when the other contents are as shown in the following table (in parts per billion: ppb):

Ca	Mg	Na	HCO_3	Cl	SO_4	F	B	P	Ti	Cr
45	66.1	16.1	171	9.9	0.01	0.07	119	444	2.5	0.01

d) Suppose that in a mineral water whose mineral contents are as shown in the table in (c), the potassium content is 2.77 ppb. In relation to your constructed regression equation, what is the *y*-residual for the year?

46. Try adding additional *x*-variables to the model created in Exercise 45 based on the other mineral contents in the data file that were not considered. In your judgment, are you able to improve very much on the significance or strength of the model by adding any additional variables? Explain.

Basic Concepts 16.4

47. Explain what is meant by a dummy variable. When can dummy variables be used?

48. Suppose a qualitative variable can take one of five values. The information in that variable could be coded using up to how many dummy variables?

49. Suppose a regression model for the price of an order of salad at a salad bar is:

$$\hat{y} = \$2.50 + [\$0.04 \times (\text{grams of salad})] + (\$1.50 \times \text{bottle})$$

where "bottle" takes the value 1 if you order a bottle of any available beverage or, otherwise, the value 0. How much does buying a bottle of beverage add to your bill?

50. Refer to Exercise 49. What is the expected cost of an order of 200 grams of salad plus a bottle of cream soda?

51. To include a dummy variable in the regression model, the values for the dummy variable should be (choose one):

a) Related to the values taken by the quantitative variables.

b) Independent from the values taken by the quantitative variables.

Explain your answer.

Applications 16.4

For these problems, more than one right answer might be possible. Construct and apply the best linear regression model that you can—in keeping with the five rules of thumb and your instructor's guidelines—and ensure, if possible, by residuals analysis, a reasonable conformity of the model with regression assumptions.

52. An economist is seeking variables to predict average rents of two-bedroom apartments in various Canadian cities. Using the data in the file **Housing_Facts,** create a multiple linear regression model by using some combination of the following information: ownership rates ("Owner," in percentages); rental vacancy rates ("Rental," in percentages); Ontario; far west (i.e., Alberta or British Columbia); and the Atlantic Provinces. The last three variables are dummy variables, to be coded based on the "Area" information in the data set. All variables retained in the final model should have significant coefficients.
    a) What is the regression equation, based on your model?
    b) How much of the variance of the dependent variable is explained by the regression?
    c) Use the regression equation to estimate the average rent for a city in Ontario that has an ownership rate of 65% and a rental vacancy rate of 2%.
    d) Repeat (c) but for a city in the far west.
    e) Judging from the standard error of the model, as well as from the residuals analysis, how exact do you think your estimate was in (c)? Explain.

53. A human resources manager in Welland, Ontario, is examining predictors of work sectors that offer high levels of full-time employment in the city. Using the data in the file **Welland_Employment,** create a multiple linear regression model by using some combination of the following x-variables: number of firms in the sector ("Firms"); number of part-time employees in the sector ("Part"); number of seasonal workers in the sector ("Season"); and sectors that focus on "Services," which is a dummy variable that is to be coded based on the name of the sector. All variables retained in the final model should have significant coefficients.
    a) What is the regression equation, based on your model?
    b) Did the dummy variable for "Services" remain in your model? Why or why not?
    c) How precise a prediction does your model offer? (Hint: Refer to the model R^2 and standard error.)

54. A power engineer noticed that large hydroelectric plants have not all reached their full design capacities. To confirm this, she collected the data in the file **Largest_Hydroelectric_Plants.** Create a regression model to estimate the current actual capacity of a plant based on its theoretical ultimate capacity. Is the model improved by adding a dummy variable that codes whether the plant is located in Canada? Try it, and explain your answer.

55. A nutritionist wants to estimate the calorie content of a food item, based on the food's contents of carbohydrates, fat, protein, saturated fat, and cholesterol. Using the data in the file **Nutrition,** construct a model based on as many of these variables as appropriate. Write the equation. Is the model improved by adding a dummy variable that codes whether the food item contains beef? Try it, and explain your answer.

56. The file **Foreign_Currency_Units** shows the relative value of the Canadian dollar against various foreign currencies at different points in time. An importer/exporter wants to use these data to estimate the Canadian/U.S. exchange rate based on the other currency exchange rates shown. Create the best model you can for making this estimate.
    a) What is the regression equation, based on your model?
    b) How much of the variance of the dependent variable is explained by the regression?
    c) Add a dummy variable that indicates whether the data are taken at a date after 1995. Is the coefficient for this new variable significant? If so, rewrite the model to include this variable.

Review Applications

For multiple regression problems, more than one right answer may be possible. Construct and apply the best linear regression model that you can—in keeping with the five rules of thumb and your instructor's guidelines—and ensure, if possible, by residuals analysis, a reasonable conformity of the model with regression assumptions.

57. According to data collected by the International Bicycle Fund, these are the comparative production numbers, worldwide, for bicycles versus automobiles.[3]

Year	1990	1991	1992	1993
Bikes (millions)	92	99	102	102
Autos (millions)	36	35	36	34
Year	**1994**	**1995**	**1996**	**1997**
Bikes (millions)	105	106	98	92
Autos (millions)	35	35	37	39
Year	**1998**	**1999**	**2000**	
Bikes (millions)	76	96	104	
Autos (millions)	38	40	41	

    a) Construct a linear regression equation to predict the numbers of autos produced in a year, given the number of bikes produced in that year.
    b) Is the apparent regression significant? How do you know?
    c) What percentage of the variation in annual numbers of autos can be explained in terms of the regression with the numbers of bikes?
    d) Based on your answers to (b) and (c), as well as on residuals analysis, would it be prudent to base an estimate for the numbers of autos produced on the regression equation constructed in (a)? Explain why or why not.

For Exercises 58–59, use data from the file **Canadian_Equity_Funds**. For all calculations, omit records that lack data for one of the variables being analyzed.

58. An investor is comparing the performance of several Canadian equity funds and in the file **Canadian_Equity_Funds,** has compiled related data on several variables for analysis.

a) Construct a linear regression equation to predict the one-month financial return for a fund ("Ret_1Mth") from its three-month financial return ("Ret_3Mth").

b) Is the regression in (a) statistically significant at the 0.05 level?

c) Use the regression in (a) to predict the one-month return for a fund whose three-month return was 2.5.

d) Fill in the blank: In general, predictions made from the regression equation in (a) will be accurate within ± ________, 95% of the time.

e) The regression equation from (a) explains what percentage of the variance in one-month returns compared to the mean for one-month returns?

59. The investor in Exercise 58 decides to expand her model for estimating the three-month return of a Canadian equity fund. Construct a regression model based on as many variables as appropriate from among (a) the one-month return, (b) the management expense ratio (MER), (c) the fund's assets (in millions of Canadian dollars), (d) the net asset value per share(NAVPS) in Canadian dollars as of the end of the previous month, and (e) whether the fund charges a sales fee for purchase, which is a dummy variable based on the "Load" column and equals 0 for the case of no load or fee. Write the equation. Is the dummy variable retained in the final model? Why or why not?

For Exercises 60–62, use data from the file **Weather_Data_Canada.**

60. After compiling the data in the file **Weather_Data_Canada,** a meteorology student began to explore the relationship between different weather variables. The file displays a number of various cities' weather norms for each of two months of the year.

a) Construct a linear regression equation for making estimates of this nature: If you know a city's typical maximum daily temperature for a certain month of the year ("Max_Temp"), estimate the typical *minimum* daily temperature for the same month/city combination ("Min_Temp").

b) Is the regression in (a) statistically significant at the 0.05 level?

c) Use the regression in (a) to estimate the typical minimum daily temperature for a city in July if the same city's maximum daily temperature in July is typically 24.5 degrees Celsius.

d) Your answer in (c) is a point estimate. Give an approximate 95% confidence interval for your answer to (c).

e) The regression equation from (a) explains what percentage of the variance in the dependent variable compared to the mean for the dependent variable?

61. In the **Weather_Data_Canada** file, the variable Ex_MaxT displays the maximum-ever temperature recorded (in degrees Celsius) for the indicated city and month of the year.

a) Construct a linear regression equation to predict the value of Ex_MaxT for a city/month combination if you know the usual average temperature ("Av_Temp") for that same city in the same month of the year.

b) Is the regression in (a) statistically significant at the 0.05 level?

c) Use the regression in (a) to predict the likely maximum-ever temperature for a city in a given month of the year if you know that the city's average temperature for that month of the year is 17.5 degrees Celsius.

d) Suppose that you know the average temperature for Edmonton for the January that has just passed. Can you now use the equation in (a) to estimate the maximum temperature that was reached in Edmonton in the January that just passed? Explain your answer.

62. Three of the variables for wind speed in the **Weather_Data_Canada** file are the normal wind speed in km/hr for a particular city in a particular month ("W_Speed") and the typical number of days in a given month that the wind in the city reaches at least 52 km/hr ("Wind_g52") or reaches at least 63 km/hr ("Wind_g63").

a) Construct a linear regression equation to predict the typical number of days in a month that the wind in a city will reach at least 63 km/hr, given that the average wind speed in that city for the month is known ("W_Speed"), as well as the typical number of days in a month that the wind in the city will reach at least 52 km/hr.

b) Is the regression in (a) statistically significant at the 0.05 level?

c) Use the regression in (a) to predict the typical number of days in a month that the wind in the city will reach at least 63 km/hr if the average wind speed in that city for that month is 21 km/hr, and the typical number of days in a month that the wind in the city will reach at least 52 km/hour is 5.

d) Note that the data set contains only the aggregate data for the months January and July. How confident would you feel about your answer to (c) if the month in question was March? Explain.

e) Create a dummy variable to represent whether the wind generally blows south ("S" or "SW" or "SE") versus in some other direction. Can the model in (a) be improved by including this dummy variable as an independent variable?

For Exercises 63–68, use data from the file **Worker_ Class.**

63. Concerned about employment equity, a researcher is studying the number of self-employed women across Canada. Using data in the file **Worker_Class,** construct a regression model to estimate the number of self-employed women in a province or territory if you are provided with the numbers of paid and unpaid women workers, respectively, in that region.

64. Using the model in Exercise 63, estimate the number of self-employed women in a province that has 700,000 paid women workers and 8,000 unpaid women workers.

65. Based on examining residuals and the standard error from your model in Exercise 63, how precise would you expect your estimate to be in Exercise 64? Explain.

66. Redo Exercise 63 but also create and consider, as a potential dummy variable, a column indicating whether the province for the observation was, or was not, in eastern Canada (i.e., one of the four Atlantic Provinces). Is the coefficient of this dummy variable significant in the model?

67. Redo Exercise 63 but also consider as potential x-variables the numbers in each province, respectively, of paid, unpaid, and self-employed *male* workers. Does the resulting model include all of the variables you have considered? Is the adjusted R^2 much improved from your model in Exercise 63?

68. The data set **Worker_Class** included several columns of totals (e.g., for all male workers of all classes and for all workers of either gender of a particular class). If you included these variables along with those considered in Exercise 67, do you think that collinearity could become a problem? Explain your answer.

Association with Time: Time Series Analysis

Chapter 17

• Chapter Outline

17.1 Linear and Nonlinear Trends

17.2 Seasonal Decomposition

17.3 Applying Dummy Variables for Time Series Analysis

• Learning Objectives

17.1 Describe what is meant by a time series, identify the four types of components of a time series, and construct equations to model linear or nonlinear trends in the data.

17.2 Perform seasonal decomposition on a time series that involves seasonality and combine these steps with regression to make a forecast for a time period *t*.

17.3 Construct and interpret a multiple regression equation that includes dummy variables to model the impact of qualitative variables on a time series.

17.1 Linear and Nonlinear Trends

Time series analysis is a general term that is applicable to a variety of techniques. A **time series** is an ordered set of observations for a variable made at regular intervals over time. Examples of time series include the monthly costs of maintaining a fleet of Boeing 767s, the annual shrinkage of the Helm Glacier in British Columbia, and annual expenditures on forest management in Canada. Time series data are often measured on a yearly basis. Depending on the purpose of the analysis, other time units, such as quarterly, monthly, weekly, or daily, can be used as well.

We analyze time series for two main reasons: First, to explain the changes of a variable over time. We may find that the values are decreasing, increasing, or varying in a regular pattern and at a quantifiable rate of change. Time series analysis per se does not give the causes for the identified patterns; for causes, we might look to factors such as increasing fuel costs or changes in global temperature. But whatever the underlying conditions, the analysis shows how these conditions impact over time on the value of the variable.

Second, time series analysis is an important tool in making forecasts. These forecasts can be used for planning for the future and also for process monitoring by allowing for ongoing comparisons of actual versus forecasted values.

Many techniques are used in time series analysis, and entire courses are offered on the topic. In this chapter, we focus on some basic methods in which regression plays a significant role. We will include some options for nonlinear, as well as linear, regression techniques.

A time series can be analyzed (or **decomposed**) into four types of components. These are: (1) the trend component, *T*; (2) the seasonal component, *S*; (3) the cyclical component, *C*; and (4) the irregular component, (*I*). If these components can be identified and quantified, then the time series can be modelled with an equation that combines those elements. For example, one approach is to interpret each value of the original variable as the product of four corresponding component values; for example, $Y_i = T_i \times S_i \times C_i \times I_i$.

Trend Component

The **trend component** of the time series represents those changes in the value of the variable that reflect a pattern of long-term growth or decline. As is the case for the annual shrinkage of British Columbia's Helm Glacier since 1930 (relative to the position of its leading edge in 1900), the growth or decline pattern may be linear (see Figure 17.1). On the other hand, the increasing spot prices for crude oil since 1998 show a curved, nonlinear trend (see Figure 17.2). In either case, the basic calculations follow the same principles as for simple linear regression, as described in Section 16.1. The regression line is the line of best fit, applying a least squares approach to the residuals to measure the degree of fit. The independent variables are based on the ordered sequence of time.

Seasonal Component

The **seasonal component** of the time series reflects short-term variations in the value of the variable that are associated with particular and recurring divisions of time. For example, unemployment rates tend to be higher in the winter than in the summer, due to work patterns in tourism, agriculture, construction, and other industries. Similarly, toy sales are highest in the months leading up to Christmas, due to the cultural buying (and advertising) patterns in North America. Seasonal fluctuations may occur for almost

Figure 17.1
Example of a Linear Trend

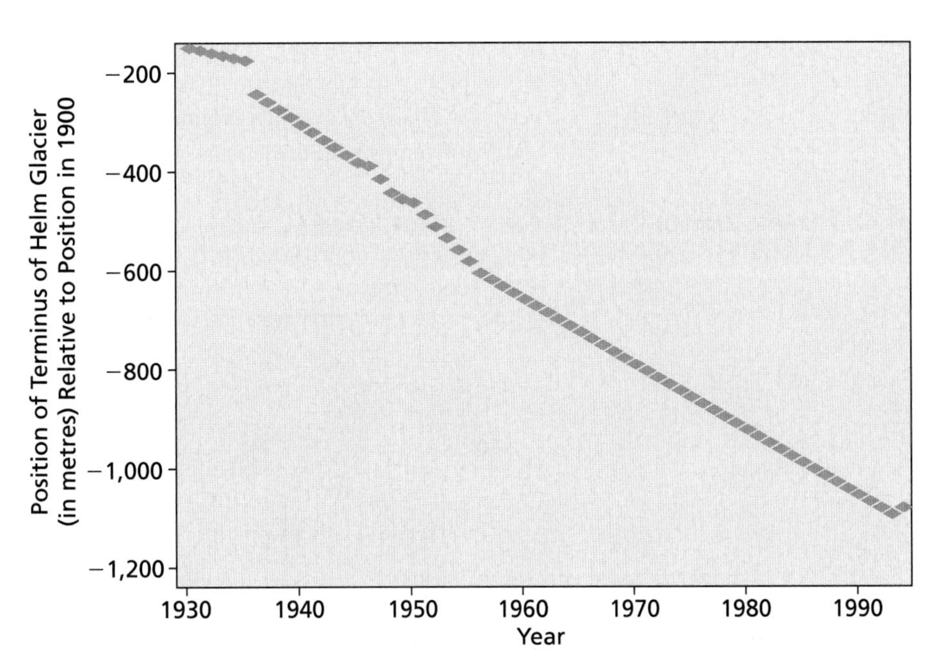

Source: Copyright © Province of British Columbia. All rights reserved. Reprinted with permission of the Province of British Columbia. www.ipp.gov.bc.ca. Found at: http://www.env.gov.bc.ca/air/climate/indicat/pdf/glacier_tdoc.pdf

Figure 17.2
Example of a Nonlinear Trend

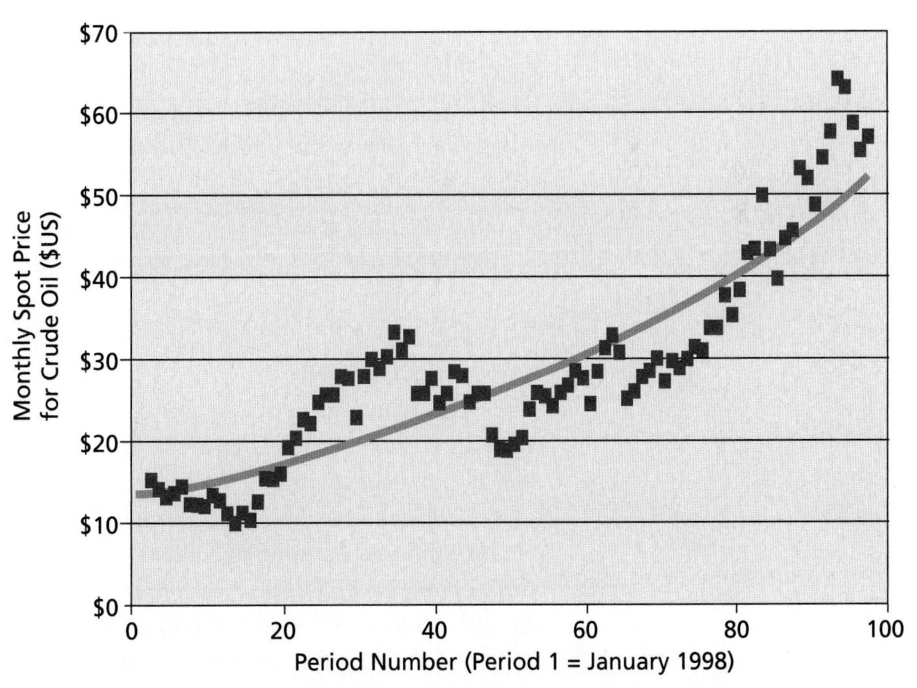

Source: Government of Newfoundland and Labrador, Newfoundland and Labrador Statistic Agency. Based on data in *Industry Statistics.* Original source: U.S. Energy Information Administration website. Found at: http://www.stats.gov.nl.ca/statistics/Industry/Oil_Crude_Prices.asp

any time division (day or part of day, week, month, and so on) and may be caused by regular changes in weather conditions, social and religious customs, institutional schedules (e.g., universities' dates for March break), or government regulations (e.g., deadlines for tax deductions on savings plan purchases). If data are collected and graphed for each season over several cycles, patterns of seasonal variation can often be visible by eye. This occurs in Figure 17.3, which originally accompanied a paper to show that hemoglobin levels in veterans with diabetes are variable with the time of year.

Figure 17.3
Example of Seasonal Fluctuations

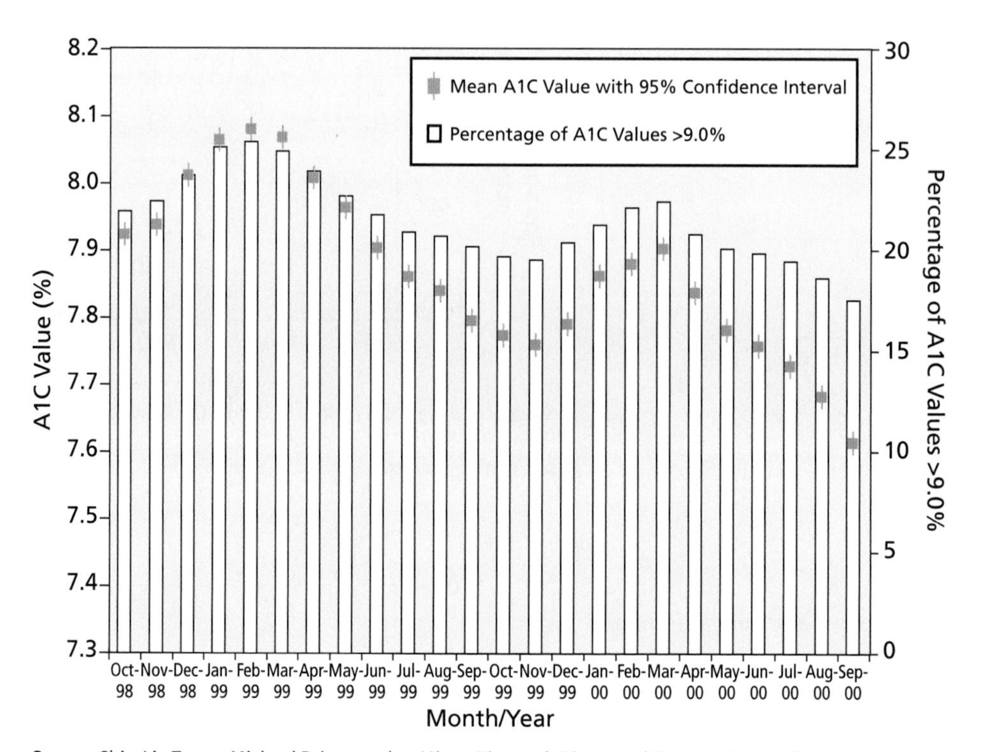

Source: Chin-Lin Tseng, Michael Brimacombe, Minge Xie, et al. "Seasonal Patterns in Monthly Hemoglobin A_{1c} Values." *American Journal of Epidemiology,* 161(6), (2005). Adapted from Figure 1, page 568. Used by permission of Oxford University. Found at: http://aje.oxfordjournals.org/cgi/reprint/161/6/565

Cyclical Component

The **cyclical component** of the time series reflects longer-term recurring changes in the value of the variable, compared to seasonal changes. In practice, the lengths of the cycles (i.e., the durations of time before the pattern repeats) are often less constant and harder to predict than those of a seasonal component. A well-known example is the cyclical pattern of abnormal warming (the El Niño effect) and cooling (the La Niña effect) of the surface of the Pacific Ocean. This cycle is associated with dramatic and extensive weather changes—from droughts in Australia to heavy rainfall in South America. The cycle of El Niño/La Niña occurrence is shown in Figure 17.4; the El Niño cycle is shown in red and the La Niña cycle in blue.

While there is clearly a rough alternation between El Niño and La Niña conditions in the time series, there is obviously no uniform distance between the occurrences. The case is similar for other important cycles, such as business cycles. Although it is meaningful to distinguish periods of recession or growth and to note that they alternate, the durations of these stages can vary considerably.

Figure 17.4
Example of Cyclical Fluctuations

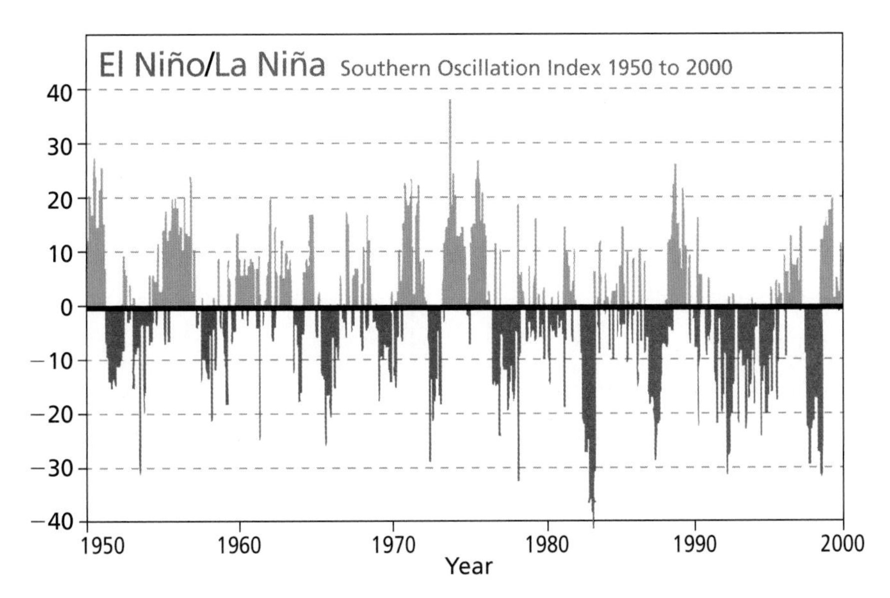

Source: John L. Daly, *The El-Niño Southern Oscillation (ENSO)*. Adapted figure. Web page as updated Sept. 30, 2006. Found at: http://www.john-daly.com/elnino.htm

Irregular Component

The **irregular component** of the time series reflects unanticipated changes in the value of the variable due to unusual or irregular events, such as a hurricane or tsunami, a crop failure, a major strike, or war. These events can also occur on a more local scale. For a retailer, the advent of a bold new advertising campaign by a competitor may play this role. The irregular component may be compared to the epsilon component in a regression model, which represents unaccounted-for residuals in the value of the response variable. However, we can sometimes predict what will happen to the variable of interest when a certain irregular event occurs, but we may not know in advance when it will occur or with what intensity, although there may be ways to include such components in a model. However, unexplained random residuals will always remain unexplained.

Consider, for example, the residuals shown in Figure 17.5. The figure is based on a study of fatal car crashes in Arizona and shows variations in the variable not explained by other variables in the time series analyses. Much of this variation may be just random noise but if, for example, we learned that there were widespread destructive storms during January 1990 and January 1993, this might explain the positive peaks during those two periods. Arizona might take preventive steps if other similar storms are forecast in the future.

Finding the Linear Trend

The basic procedures for this task are identical to those for linear regression, but some preliminary steps may be required. The independent variable is time—usually depicted as t rather than x. If a data set represents annual data from 1989 to 2005, this series of numbers *could* be used for the t-values. But such large values for the year numbers are somewhat artificial, and using them will distort the values of the coefficients. Instead, it is common to represent the time variable as the series beginning with "1" for the first year or time period recorded, "2" for the second time period, and so on.

Figure 17.5
Example of Irregular Fluctuations

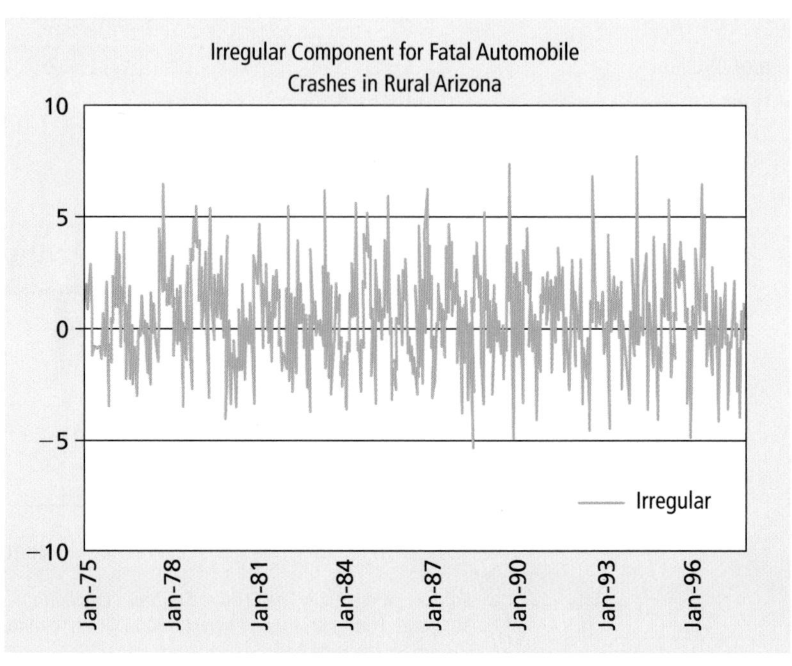

Source: Sandy Balkin, Ernst & Young LLP, and J. Keith Ord, Georgetown University. "Assessing the Impact of Speed Limit Increases on Fatal Interstate Crashes." *Consumers Union.* Adapted from Figure 3, "Irregular Component for Rural Arizona." Found at: http://www.consumersunion.org/other/speedlimits/speed031500b03.htm

Another modification from straight regression is the need to first adjust the data set if seasonality and possibly cyclicality are present. (This process will be discussed in more detail in the next section.) We have seen that these components in the data can be relatively orderly and predictable, yet, from the regression viewpoint, they represent extra variation of values around the trend line. If the seasonal or cyclical components are not removed before the analysis, the trend regression may appear less strong or significant than it actually is.

Once the regression is complete, future values for a variable can be estimated by inputting a future time value (as the x-value) into the trend (i.e., regression) equation. Estimates become less reliable as they extend far beyond the time range of the original data. The strength and significance of the trend can be assessed as was done in Chapters 15 and 16. If the regression was performed on adjusted data, then the regression estimate will need a follow-up adjustment to account for seasonality or cyclicality in the data.

Example 17.1

*(Based on the **Glacier_Retreats** file)*

An environmentalist is concerned with the phenomenon of retreating glaciers around the world. He compiled the data found in the file **Glacier_Retreats** and partially illustrated in Figure 17.1.

1. Using just the data for the Helm Glacier from 1930 to 1994 inclusive, find the regression equation that models the trend.
2. Assess the strength and significance of any apparent trend.

3. Assuming the trend equation remains valid until 2010, predict how far the Helm Glacier will have receded by that date from its level in 1900 (i.e., the baseline used for the dataset).
4. Do you think that your assumption in (3) is reasonable? Explain.

Solution

1. Because there is no evidence of seasonality or cyclicality in this data set (see Figure 17.1), we can perform the regression directly on the unadjusted data. (If we had monthly data, there could be seasonality within each year, but our data provide only one yearly value.) We therefore can apply the regression procedure directly to the Helm Glacier data for the specified years. Before inputting the t-values, convert the year numbers "1930, 1931, . . ., 1994" to the series "1, 2, . . ., 65," and use the latter for the independent variable. The computer output looks something like Figure 17.6.

Figure 17.6

Trend Line for the Helm Glacier Retreat

Regression Statistics	
Multiple R	0.996316323
R Square	0.992646215
Adjusted R Square	0.992529488
Standard Error	25.04641394
Observations	65

ANOVA

	df	*SS*	*MS*	*F*	Significance *F*
Regression	1	5334763.851	5334764	8504.016	6.26895E-69
Residual	63	39521.33963	627.3229		
Total	64	5374285.19			

	Coefficients	Standard Error	*t*-Stat	*p*-Value	Lower 95%	Upper 95%
Intercept	−151.5132692	6.285640439	−24.1047	1.66E-33	−164.0741192	−138.9524192
YearNum	−15.26966783	0.165583685	−92.2172	6.27E-69	−15.60056042	−14.93877524

Source: www.ipp.gov.bc.ca. Found at: http://www.env.gov.bc.ca/air/climate/indicat/ pdf/glacier_tdoc.pdf

Using the coefficients shown at the lower left of the figure, we construct the regression equation:

$$\hat{y} = -151.51 - 15.27t$$

2. From the output, we conclude that $r = -0.996$ (based on the value displayed as "Multiple R"—for the magnitude—and using the sign from the coefficient b_1); this indicates an exceptionally strong correlation between time and the retreat of the glacier. The regression explains 99.3% of the variation in the magnitudes of the variable based on r^2 for the simple linear regression. The p-value is virtually equal to zero (in scientific notation: 6.269×10^{-69}), so the regression is highly significant.

3. If 1930 is year 1, then 2010 is year 1 + 80 = 81. If the trend equation remains applicable by then, we can estimate:

$$\hat{y}_{81} = -151.51 - 15.27(81) = -1{,}388.38$$

In that year, the Helm Glacier will have retreated about 1.39 kilometres from its level in 1900.

4. The conclusion in (3) presumes that the regression equation, which was based on data ending in 1994, still applies 16 years later in 2010, but some scientists believe that the climate changes that impact glacier boundaries are accelerating. If so, conclusions based on old data may be not be reliable that far into the future.

Finding a Nonlinear Trend

There are two basic approaches to modelling a nonlinear trend. Statistical software might provide functions to create one or the other of these models but, if there is evidence of a seasonal or cyclical pattern, it is important to first adjust the data before proceeding, similar to the approach for a linear trend. Find the trend using data adjusted for seasons or cycles, if necessary, and then adjust your estimates, in turn, to account for any pre-adjustments.

A nonlinear trend for the spot prices of oil was illustrated in Figure 17.2. We can model the same relationships in a way that makes the trend line appear straight. This approach is illustrated in Figure 17.7, which reveals the underlying method for nonlinear regression: The raw data can be mathematically transformed in such a way that a particular type of nonlinear trend (such as exponential) would appear linear. In this example, each raw value of y has been replaced with the base-10 logarithm of y (that is, by a number p such that 10 raised to the power p (i.e., 10^p) equals the original y-value). Since the trend of the transformed values now appears straight, the least squares algorithm for linear regression can be applied. If the model is then used to make an estimate, a reverse transformation is applied to the output of the equation.

On the other hand, the version of a nonlinear equation produced by your software may look something like this: $\hat{y} = 13.198e^{0.0141t}$. (This corresponds to the curve in Figure 17.2.) Notice that the y-variable in the equation is calculated directly; the output is not a transformed value. The formula has dealt with the transformations. In this text, we allow for either approach since, for a given type of nonlinear model, either version of the formula can lead to the same estimates.

Also, we will not focus on p-values for these models, since regrettably not all software packages include them in their outputs for nonlinear trends. Fortunately, as a rule of thumb, if a regression explains much of the variation (i.e., r^2 is large) and there is a reasonably large sample size, then the regression is likely to be significant. Time series are generally analyzed for a practical goal—to make estimates or predictions. A low r^2 regression *can* sometimes be statistically significant and informative about the interactions of variables, but a trend equation for such a case is probably of limited use for making a practical estimate.

There are a variety of ways in which a trend can be nonlinear. Example trend types include exponential, power, and logarithmic. One software's selection of options is illustrated in Figure 17.8. Deciding which model best fits the data is called **curve fitting.**

Figure 17.7
Alternative View of a Nonlinear Trend

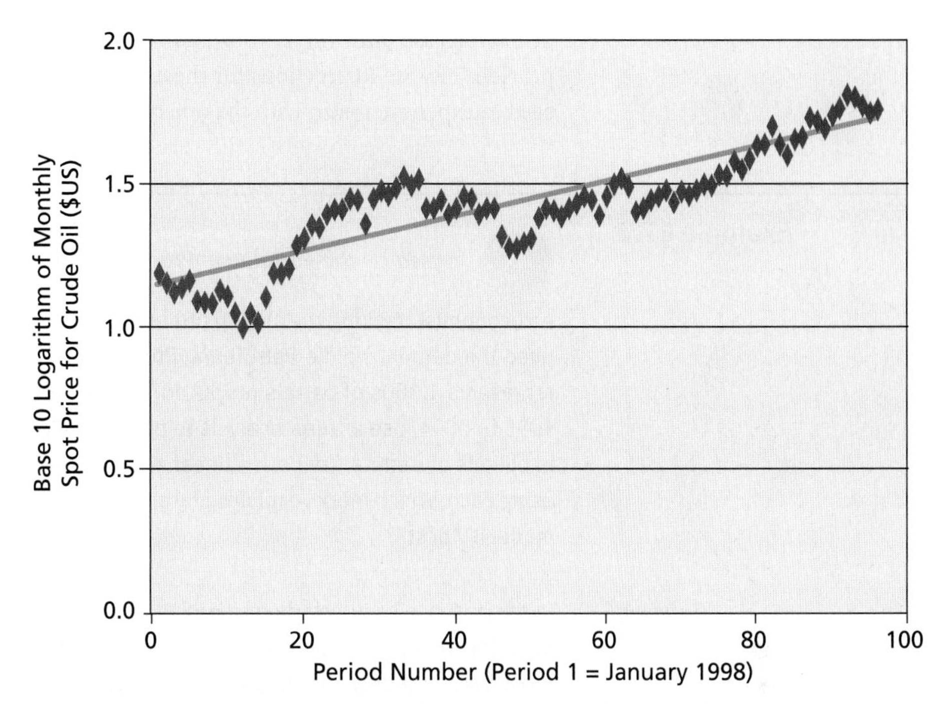

Source: Government of Newfoundland and Labrador, Newfoundland and Labrador Statistic Agency. Based on data in *Industry Statistics*. Original source: U.S. Energy Information Administration website. Found at: http://www.stats.gov.nl.ca/statistics/Industry/Oil_Crude_Prices.asp

Observe that in some models (e.g., polynomial) it is not even required that the data consistently increase or consistently decrease over time.

When selecting the trend model to use, keep in mind your practical goals. A linear model is relatively intuitive to understand and apply, and if you are communicating your results, more likely to be familiar to a broader audience. Ask yourself: How great is the increase in variation explained if you select a more complex model? What is the purpose

Figure 17.8
Some Trend Type Options

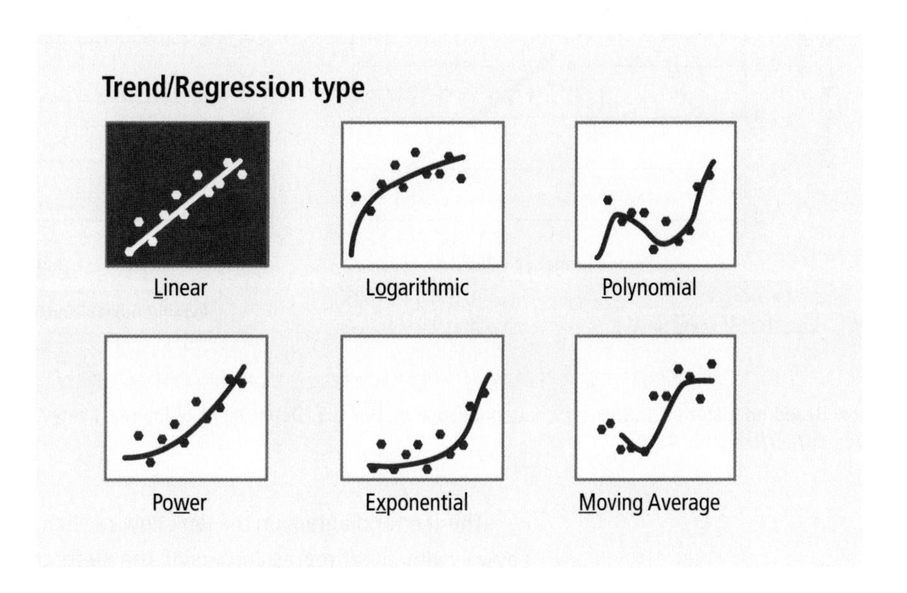

of the forecast, and who is your audience? In "In Brief 17.1" starting on page 716, we provide software instructions for the exponential model, in particular. You are encouraged to experiment also with the other options on your software.

Example 17.2

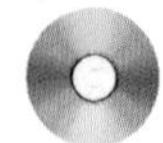

*(Based on the **Petroleum_Production_US** file)*

An economist tracking the annual production of petroleum in the United States has compiled the data in the file **Petroleum_Production_US.** (Note that each number in the file represents 1,000s of barrels produced.) Using only the data for crude oil production for 1985 to 2004, use a trend analysis to predict the quantity of crude oil produced in 2006 in billions of barrels. Try both a linear and an exponential model, and make a prediction using each. Which model explains more of the variation in the value of the variable? Which model is "best"?

Solution

In the absence of quarterly or monthly data to reveal possible seasonality, we can perform the regression directly on the unadjusted data for the specified years. Before inputting the time values (t), we convert the year numbers "1985, 1986, . . ., 2004" to the series "1, 2, . . ., 20." We can also convert the production data to units of billions of barrels, consistent with the desired units for the prediction. Depending on your software, the computer output will look something like parts (a) and/or (b) of Figure 17.9.

Figure 17.9
Trend Models for Crude Oil Production

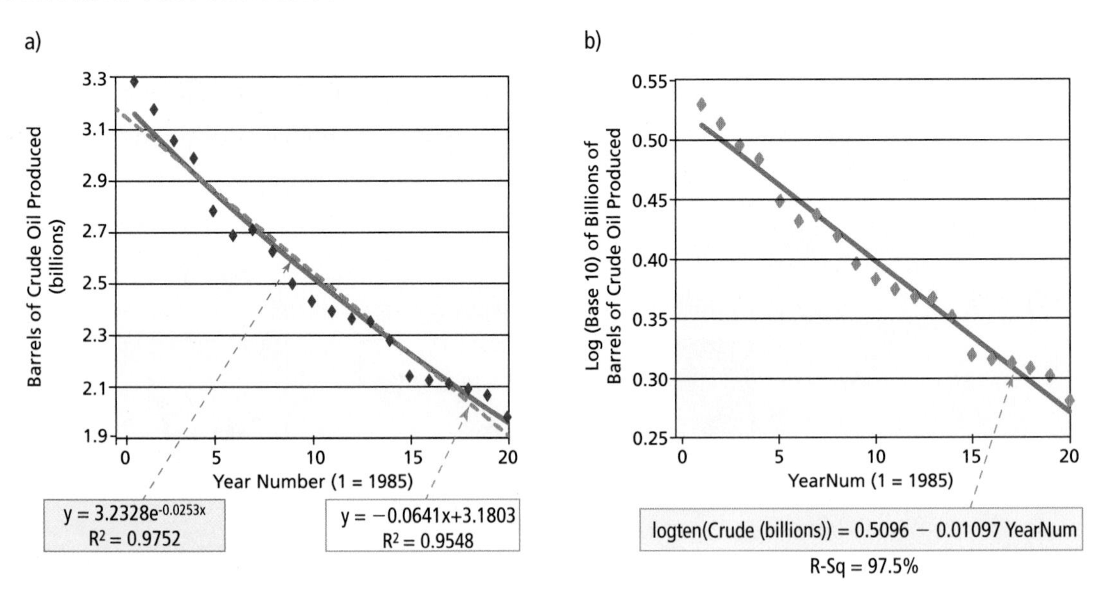

Source: Based on historical U.S. energy data published by the U.S. Department of Energy, Energy Administration. Found at: http://www.eia.doe.gov/emeu/aer/txt/stb0501.xls

The scatter diagram on the left shows a slight curve pattern in the trend. If we proceed anyway with linear regression (using the methods described earlier), we derive the trend equation shown at the bottom right of part (a) in the figure. The linear trend line is shown as green dashes. The r^2-value (typically displayed as "R-sq" or "R-square" by the software) tells us that 95.5% of the variation in production is explained by the linear model.

On the other hand, we can attempt an **exponential model for the regression.** (For the procedures, see "In Brief 17.1.") The trend equation can be shown in various equivalent ways. One version directly predicts the y-variable for each value of the time variable; this version is shown in the bottom left of part (a) in the figure. The symbol e represents a constant whose approximate value is 2.71828. The formula always takes this format: $\hat{y} = b_0 \cdot e^{(b_1 \cdot t)}$; where b_1 is the coefficient of t, e is raised to the $(b_1 \cdot t)$ power (using the "·" symbol for multiplication), and b_0 is an unchanging coefficient for $e^{(b_1 \cdot t)}$. (Your calculator probably has a key e^x for raising e to a specified power.)

An alternative way to represent the trend equation is shown at the bottom of Figure 17.9(b). This version estimates y indirectly: Directly, it estimates the base- 10 logarithm of y. Once this value is determined for a given value of t, raise 10 to the power (log 10 y) to predict the y itself. Note that, in structure, the right side of the equation is a standard linear model.

Regardless of the nonlinear equation approach taken, the r^2-value is the same; in this case, 0.975. Therefore, the exponential model explains 97.5% of the variation in production—a slight improvement over the linear model.

For making the prediction, convert year 2006 to year number 22 ($t = 22$) in the series.

- **By the linear model:**

$$\hat{y} = -0.0641(22) + 3.1803 = 1.77 \text{ billions of barrels of crude oil}$$

- **By the exponential model [version (a)]:**

$$\hat{y} = 3.2328 \times e^{(-0.0253 \cdot 22)} = 1.85 \text{ billions of barrels of crude oil}$$

- **By the exponential model [version (b)]:** Two steps are required.

$$\log \text{(base-10) } \hat{y} = 0.5096 - 0.01097(22) = 0.26826$$

$$\hat{y} = 10^{[\log \text{(base-10)} \hat{y}]} = 10^{0.26826} = 1.85 \text{ billions of barrels of crude oil}$$

(Note that the version (a) and version (b) solutions are the same.)

In general, there is no mechanical way to ascertain which model is best. Comparing r^2 can be informative, but in the oil production example, both approaches share very high values. Sometimes, the right side of a scatter diagram (which is "approaching" the future) seems to follow one model's curve more closely than the other's. This is not the case for the present example. Another consideration is your audience. For a technical report, the higher r^2 model may be best, but for a broader readership, the slight r^2 advantage of the exponential model in the example may not justify the extra complexity of presenting it.

Other comparisons of models can be based on the measurements of error that some software tools provide. Examples are **mean absolute deviation** (MAD) and **mean squared error** (MSE). For MAD, compare all of the actual values with what the model would have predicted for the corresponding time periods, and take the mean of the absolute values for all of these residuals. For MSE, take the mean of the squared values for all of the residuals. All things being equal, we prefer those models that have comparatively low values for MAD and MSE.

• *In Brief 17.1* *Nonlinear Regression*

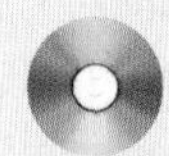

(*Based on the* ***Petroleum_Production_US*** *file*)

Among the software packages, there is more than the usual variation in their procedures. Only the exponential model is covered formally in this text. In the following, you will find references to other options that are available.

Excel

Begin by preparing two columns of data, one for the time-based independent variable and one for the response variable—either raw or adjusted (e.g., for seasonality). Using procedures in "In Brief 3.5(a)," construct a scatter diagram for the time series. Move the mouse pointer over one of the graphed points and right-click; in the resulting drop-down menu shown in the illustration, click on *Add Trendline*. In the dialogue box shown at the bottom left of the illustration, click on the desired type of trend model (e.g., *Exponential*) and then select the *Options* tab. On the *Options* page, select *Display equation on chart* and *Display R-squared value on chart*. Click *OK*. As shown in the illustration, the trend line, nonlinear equation, and r^2-value will all be added to the original chart.

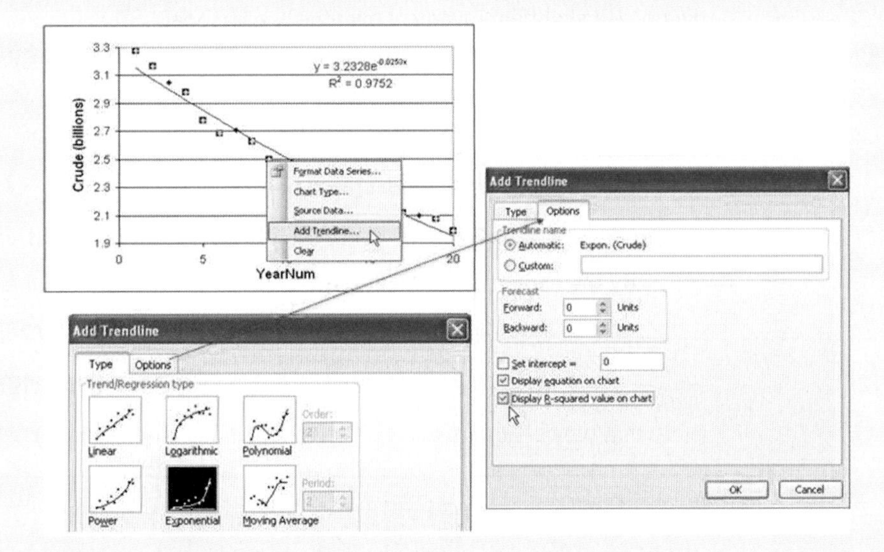

Minitab

Begin by preparing two columns of data, one for the time-based independent variable and one for the response variable—either raw or adjusted (e.g., for seasonality). Then use the menu sequence *Stat / Regression / Fitted Line Plot*. In the resulting dialogue box (the first one in the illustration), highlight and select, in turn, the response variable and the predictor (i.e., time) variable. Do not change the default option for a linear regression model. Next, click on *Options* and, in the next dialogue box, select the transformation *Logten of Y*. Then click *OK* twice. The output includes a graphic (lower left of the illustration) and some text output. The graphic includes the trend equation and the value of r^2. The text includes an ANOVA table, which gives the *p*-value for the regression.

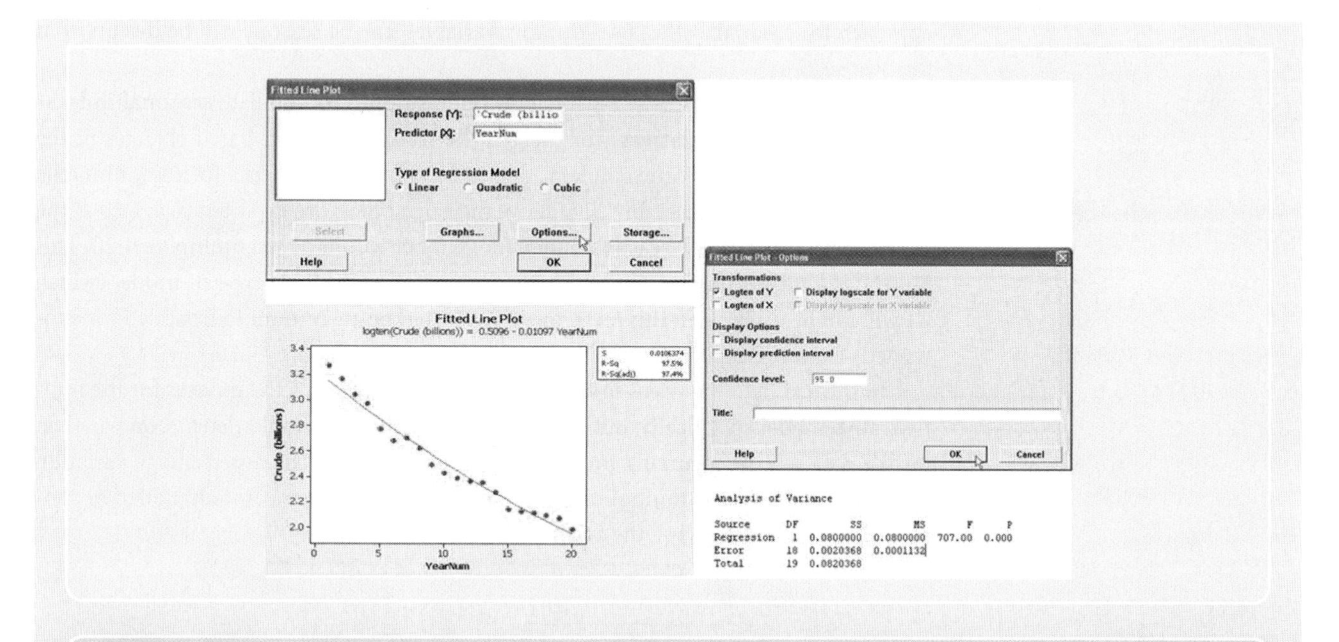

SPSS

Prepare a column with the response variable—either raw or adjusted (e.g., for seasonality). It is optional to also include a column for the time-based independent variable. Use the menu sequence *Analyze / Regression / Curve Estimation.* In the dialogue box (see the screen capture), highlight and select the response variable. If you then select *Time* as the Independent variable, SPSS will automatically use the series 1, 2, 3, etc., for the successive time periods. Alternatively, you can manually select another column for the independent variable. In the *Models* part of the dialogue box, select the names for one or more trend models to be created (such as *Exponential*). Then Click *OK.* The output includes a table, as shown in the illustration. This provides the b_0 and b_1 terms for the model, the r^2-value, and the p-value ("Sig.") of the regression.

Model Summary and Parameter Estimates

Dependent Variable: Crude_billions

Equation	Model Summary					Parameter Estimates	
	R Square	F	df1	df2	Sig.	Constant	b1
Exponential	.975	706.995	1	18	.000	3.233	-.025

17.2 Seasonal Decomposition

As noted in the previous section, data collected on a monthly, weekly, daily, or other basis, other than for a full year, may often exhibit seasonality. Waiting times between subway trains may differ for times during rush hours versus other times throughout the day. Numbers of umbrellas sold at a mall may vary based on month or season. If fewer umbrellas were sold last February than in the previous October, this is not good evidence that umbrella sales in general (i.e., the trend) are decreasing. With **seasonal decomposition,** we can isolate our analysis of seasonal patterns from our analysis of trend-based patterns. When making a forecast, we consider both of these elements separately, and later put them back together. The last step is called **reseasonalizing** the data. (An alternative approach to seasonality that uses dummy variables for the seasons will be discussed in the next section.)

The key to the process of seasonal decomposition is to calculate **seasonal indexes** (also called **seasonal relatives**) for each of the recurring season-based changes in the values of the variable. For example, if due to local weather patterns for rain, umbrella sales in October are predictably 20% above the value otherwise expected (i.e., 1.20 × the expected value), then the seasonal index for October would be the multiplier 1.20. This illustrates a **multiplicative model for the index;** for an additive type of model (which we will not be utilizing in this text), the index would be an amount to be added (or subtracted) from the expected value for a given season. (See Figure 17.10.)

In the text below, we will show how to calculate and apply the indexes for the multiplicative model specifically, but a user's choice of models should depend on whether, for the data set, the recurring impact of a season is more likely a percentage variation from the trend (use the multiplicative model) or simply an extra amount added or subtracted from the trend (use the additive model).

Figure 17.10
Additive and Multiplicative Models for Seasonality

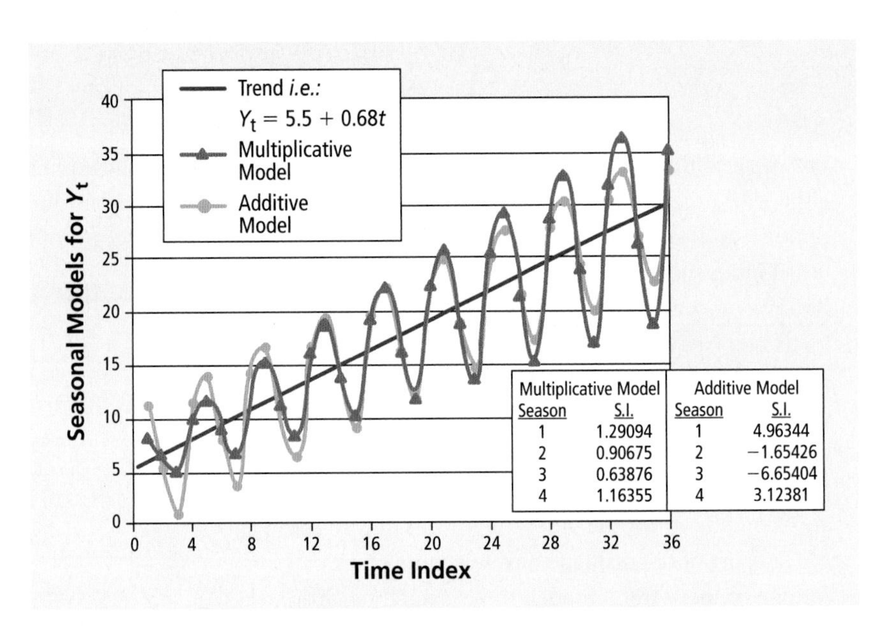

To create the multiplicative model, we calculate one seasonal index for each season in the full cycle of seasons (e.g., one for each month of the year or one for each hour in an operational day for the subway). The model can be applied whether the long-term average value is stable or whether there is a trend. The set of all of the seasonal indexes serves as a model for a predictable pattern of variation either around a stable mean or around a trend line of expected values over time.

The Analytical Sequence

For data that exhibit both seasonality and trend, the order of analysis is to:

1. Calculate all seasonal indexes (SIs) for the data set. (Details will follow.)
2. Create a column of **deseasonalized** values, also called **seasonally adjusted values.** For a multiplicative model, the procedure is:

$$Y_{\text{t (deseasonalized)}} = Y_{\text{t (actual)}} / SI_{\text{t}}$$

That is, divide each raw data value by the seasonal index that corresponds to the time period when the value was observed.

3. Use the deseasonalized data—not the original data—to determine the general trend, if any. Use the same procedures as in Section 17.1 but for the analysis, use the deseasonalized data as the y-values.
4. To make a forecast for any time period t:
   a) Determine the number of the time period (also called the *time index*) for which the estimate is required. For example, in Figure 17.10, if the next value of the variable is to be predicted, this would the time period $36 + 1 = 37$.
   b) Use the regression equation in step (3) to estimate the deseasonalized value for the time period t that is of interest. That is, find $Y_{t\,(\text{deseasonalized})}$.
   c) Determine for which season the estimate is required. For example, in Figure 17.10, there are four seasons, and data were provided for nine complete cycles of seasons. These are displayed (i.e., 9 years × 4 seasons/year = 36 time periods). So the season for the next value of the variable to be predicted (i.e., period 37) brings us back to season 1.
   d) Reseasonalize your answer from step (b) to account for the season [as in step (c)]. For a multiplicative model:

   $$Y_t\,(\text{reseasonalized estimate}) = Y_t\,(\text{deseasonalized estimate}) \times SI_t$$

   That is, multiply the raw estimate from the regression times the seasonal index corresponding to the time period for the period being estimated.

Example 17.3 Use the information provided in Figure 17.10 to forecast the next four values for the variable Y_t. Apply the multiplicative model for seasonality.

Solution Steps (1) through (3) have already been completed: The multiplicative seasonal indexes are displayed in Figure 17.10; the trend equation is also shown. The next four time periods to be estimated are t_{37}, t_{38}, t_{39}, and t_{40}. Given that there are four seasons in the seasonal cycle, it can be shown that $\text{Season}_{36} = 4$, and so the next four seasons in turn are 1, 2, 3, and 4. We can now perform the calculations:

For period 37: $Y_{37(\text{deseasonalized})} = 5.5 + 0.68(37) = 30.66$
$Y_{37(\text{reseasonalized estimate})} = 30.66 \times SI_{(\text{season} = 1)} = 30.66 \times 1.29094 \approx 39.58$

For period 38: $Y_{38(\text{deseasonalized})} = 5.5 + 0.68(38) = 31.34$
$Y_{38(\text{reseasonalized estimate})} = 31.34 \times SI_{(\text{season} = 2)} = 31.34 \times 0.90675 \approx 28.42$

For period 39: $Y_{39(\text{deseasonalized})} = 5.5 + 0.68(39) = 32.02$
$Y_{39(\text{reseasonalized estimate})} = 32.02 \times SI_{(\text{season} = 3)} = 32.02 \times 0.63876 \approx 20.45$

For period 40: $Y_{40(\text{deseasonalized})} = 5.5 + 0.68(40) = 32.70$
$Y_{40(\text{reseasonalized estimate})} = 32.70 \times SI_{(\text{season} = 4)} = 32.70 \times 1.16355 \approx 38.05$

Determining the Seasonal Indexes

Now that we have applied seasonal indexes to improve forecast accuracy, we will look at a standard textbook procedure for calculating the values of those indexes. The basic procedures are given below and will be illustrated later. You may find that your software uses a somewhat different model, yielding slightly different answers. These variations will be addressed, where applicable, in the "In Brief 17.2" section for your software.

1. **Collect and inspect the data for the variable of interest over several complete cycles of seasons.** For example, collect 12 monthly observations per year for 5 complete years or 8 hourly observations per 8-hour workday over 10 workdays. In Figure 17.11, five weeks of daily data are illustrated. Create a scatter diagram to confirm the appearance of seasonality and of how many seasons comprise the seasonal cycle. (If there is little seasonality, then calculating seasonal indexes may not be very useful.) In the figure, the pattern repeats after each seven days—which are the seasons.

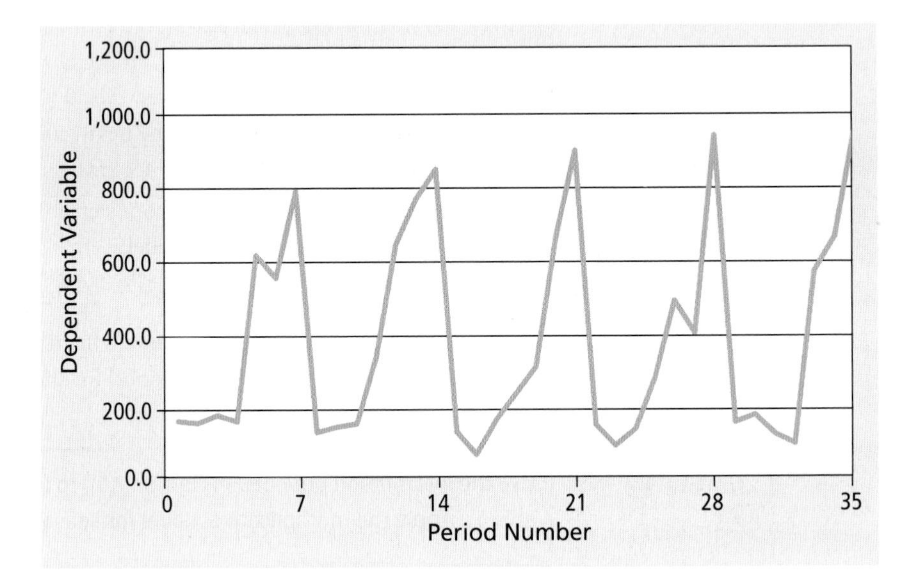

Figure 17.11
Scatter Diagram (with Connected Points) for the Time Series

2. **Next to each observation (where possible) find the centred moving average for one complete cycle of seasons.** We are looking for an average value for a complete cycle of seasons—to see if the value for an individual observation is higher or lower than that average. We take the centred moving average so that, even if there is a trend, we are finding the expected value in the middle of the line for that cycle—assuming no seasonality and ignoring possibilities of there being longer cycles or irregular variations. In Figure 17.12, the first entry for centred moving average is 380.06, lined up to the right of the observed value 169.4 (season 4). The mean of one complete week's data is 380.06, centred around the value for the first recorded season 4. Note that for the first three data entries, as well as for the bottom three entries, it is not possible to find a centred moving average based on seven seasons.

   When the number of seasons in a complete cycle is even, the above procedure must be slightly modified, as shown in Figure 17.13. Suppose the seasonality is four quarters per year. The centre for the centred moving average of the first four quarters is 172.93—but this does not correspond to any actual observation; the centre of the average falls between the second and third observations. Similarly, the centre for the next moving average—for observations two to five—falls

Figure 17.12
Textbook Model for Calculating Seasonal Indexes (Multiplicative)

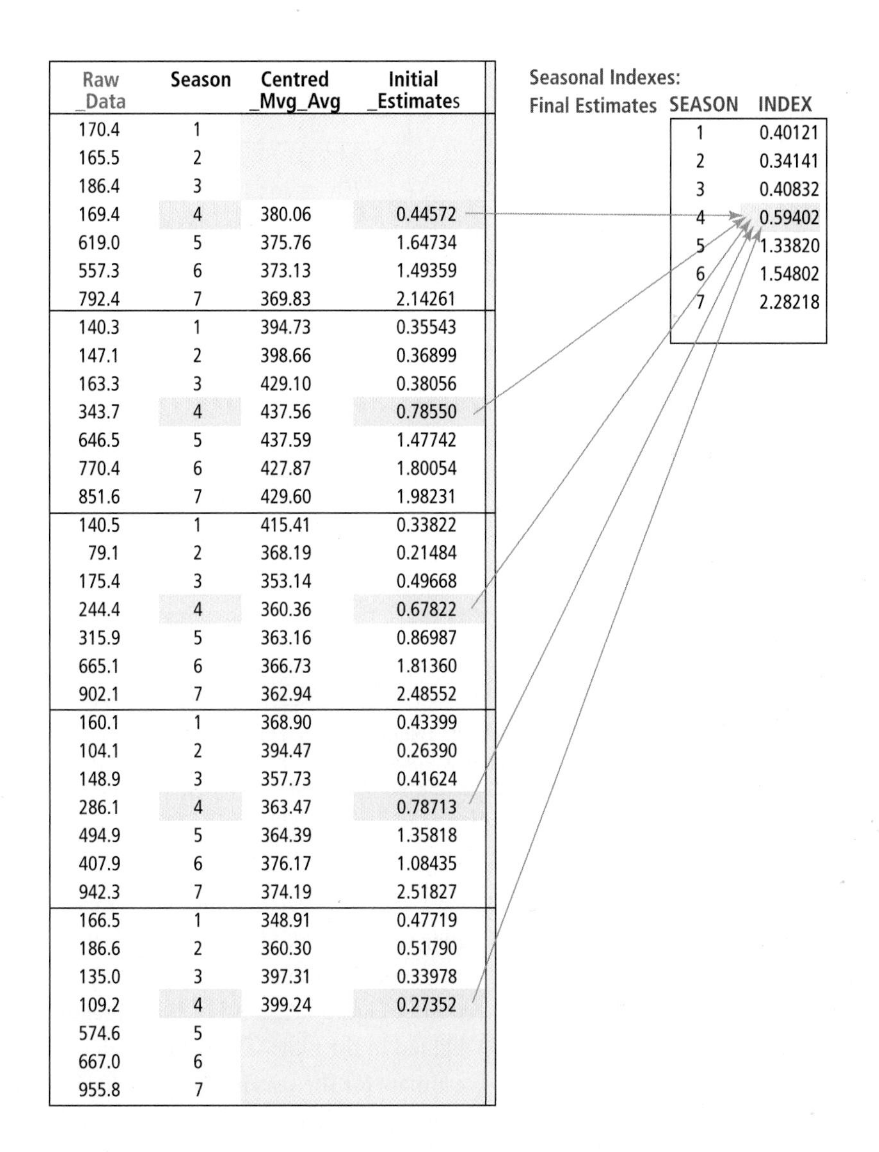

Raw _Data	Season	Centred _Mvg_Avg	Initial _Estimates
170.4	1		
165.5	2		
186.4	3		
169.4	4	380.06	0.44572
619.0	5	375.76	1.64734
557.3	6	373.13	1.49359
792.4	7	369.83	2.14261
140.3	1	394.73	0.35543
147.1	2	398.66	0.36899
163.3	3	429.10	0.38056
343.7	4	437.56	0.78550
646.5	5	437.59	1.47742
770.4	6	427.87	1.80054
851.6	7	429.60	1.98231
140.5	1	415.41	0.33822
79.1	2	368.19	0.21484
175.4	3	353.14	0.49668
244.4	4	360.36	0.67822
315.9	5	363.16	0.86987
665.1	6	366.73	1.81360
902.1	7	362.94	2.48552
160.1	1	368.90	0.43399
104.1	2	394.47	0.26390
148.9	3	357.73	0.41624
286.1	4	363.47	0.78713
494.9	5	364.39	1.35818
407.9	6	376.17	1.08435
942.3	7	374.19	2.51827
166.5	1	348.91	0.47719
186.6	2	360.30	0.51790
135.0	3	397.31	0.33978
109.2	4	399.24	0.27352
574.6	5		
667.0	6		
955.8	7		

Seasonal Indexes:
Final Estimates

SEASON	INDEX
1	0.40121
2	0.34141
3	0.40832
4	0.59402
5	1.33820
6	1.54802
7	2.28218

between the third and fourth observations. If we take the mean of those two preliminary centred moving averages, we get a value (169.16) that is lined up directly corresponding to the third observation. In the same way, if we take preliminary moving averages for observations 2 through 5, and then for observations 3 through 6, we can calculate a centred moving average (163.10) that lines up directly with the fourth observation. And so on.

3. **For each aligned pair of observed value *O* and corresponding centred moving average *MA*, calculate the ratio: *O/MA*.** These quotients are identified in Figure 17.12 as the *initial estimates* for the seasonal indexes, which are sometimes called the *raw estimates* for the indexes. Each estimate shows the extent to which an observed value for a particular season varies from the expected value for that season—*if* there was no seasonality. For example, in the fourth row, the ratio of observed to expected value is 169.4/380.06 = 0.44572. A ratio differing from 1.0 suggests that seasonality may be present for that season. (Note: For an additive model, the calculation for this step would be: $O - MA$.)

Figure 17.13
Centred Moving Averages—Given an Even Number of Seasons

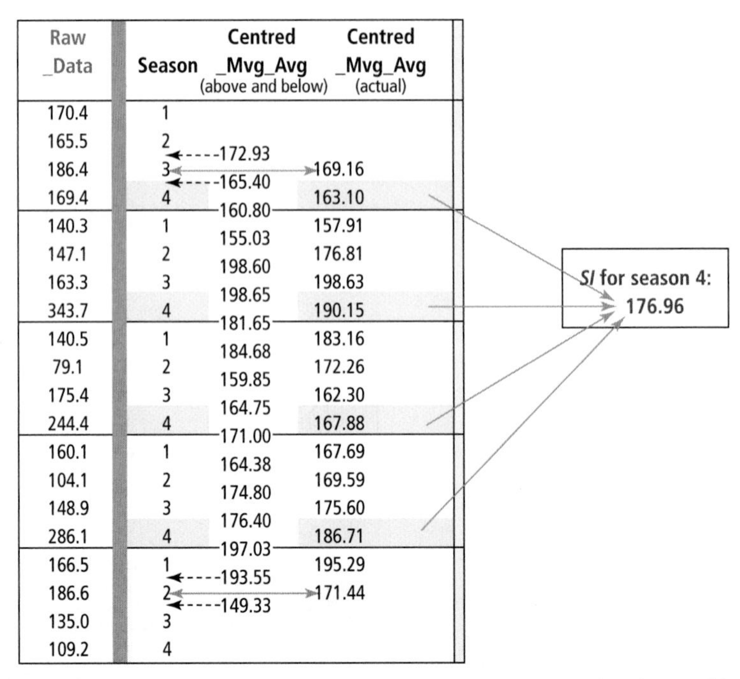

Raw _Data	Season	Centred _Mvg_Avg (above and below)	Centred _Mvg_Avg (actual)
170.4	1		
165.5	2	172.93	
186.4	3	165.40	169.16
169.4	4	160.80	163.10
140.3	1	155.03	157.91
147.1	2	198.60	176.81
163.3	3	198.65	198.63
343.7	4	181.65	190.15
140.5	1	184.68	183.16
79.1	2	159.85	172.26
175.4	3	164.75	162.30
244.4	4	171.00	167.88
160.1	1	164.38	167.69
104.1	2	174.80	169.59
148.9	3	176.40	175.60
286.1	4	197.03	186.71
166.5	1	193.55	195.29
186.6	2	149.33	171.44
135.0	3		
109.2	4		

Sources: Bank of Canada & Canadian Housing and Mortgage Corporation. Based on data in: (1) Mortgage interest column: Bank of Canada. Found at: http://www.bankofcanada.ca/cgi-bin/famecgi_fdps; (2) Canada Mortgage and Housing Corporation. Found at: http://www.cmhc-schl.gc.ca/en

4. **Calculate the seasonal indexes for each season in turn. To do this for a particular season (e.g., Thursday), take the mean for all of the initial estimates that correspond to that season.** This step is partially illustrated in the right-hand portion of Figure 17.12. All of the initial estimates for the season 4 index are highlighted in the table. The mean of all of the highlighted estimates equals the final estimate for the season, which is displayed in the upper right of the figure.

Following these procedures, we are a step closer to the process of finding a linear trend. If there is seasonality, finding the trend must be deferred until the seasonal indexes are determined. Then, **deseasonalize** each observation by the calculation:

$$\textbf{Seasonally adjusted (or deseasonalized) value}_i = \textbf{(Observed value)}_i / SI$$

where SI is the seasonal index for the season that corresponds to the observation. Perform the trend regression on the deseasonalized values—not the original values—and proceed as described earlier.

Example 17.4

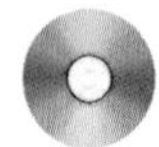

(*Based on the **Mortgages_Approved** file*)

A mortgage analyst wants to estimate the number of mortgages likely to be approved next month for new single-unit dwellings. Based on data from the Canada Mortgage and Housing Corporation, she compiled the data in the file **Mortgages_Approved.** Accounting for seasonality of the data, predict the number of approvals for April 2006.

Solution Data are found in the "NewSingle" column of the data set. Observations are recorded monthly, so if there is seasonality, each calendar month is one season and there are 12 seasons in a complete cycle (i.e., in a year). The nonadjusted time series data are displayed in Figure 17.14 (upper right), together with the basic calculations for seasonal adjustment and for solving the problem. Now many steps can be automated and the exact values for the seasonal indexes may vary with your type of software.

Figure 17.14

Model for Forecasting the Next Seasonalized Value

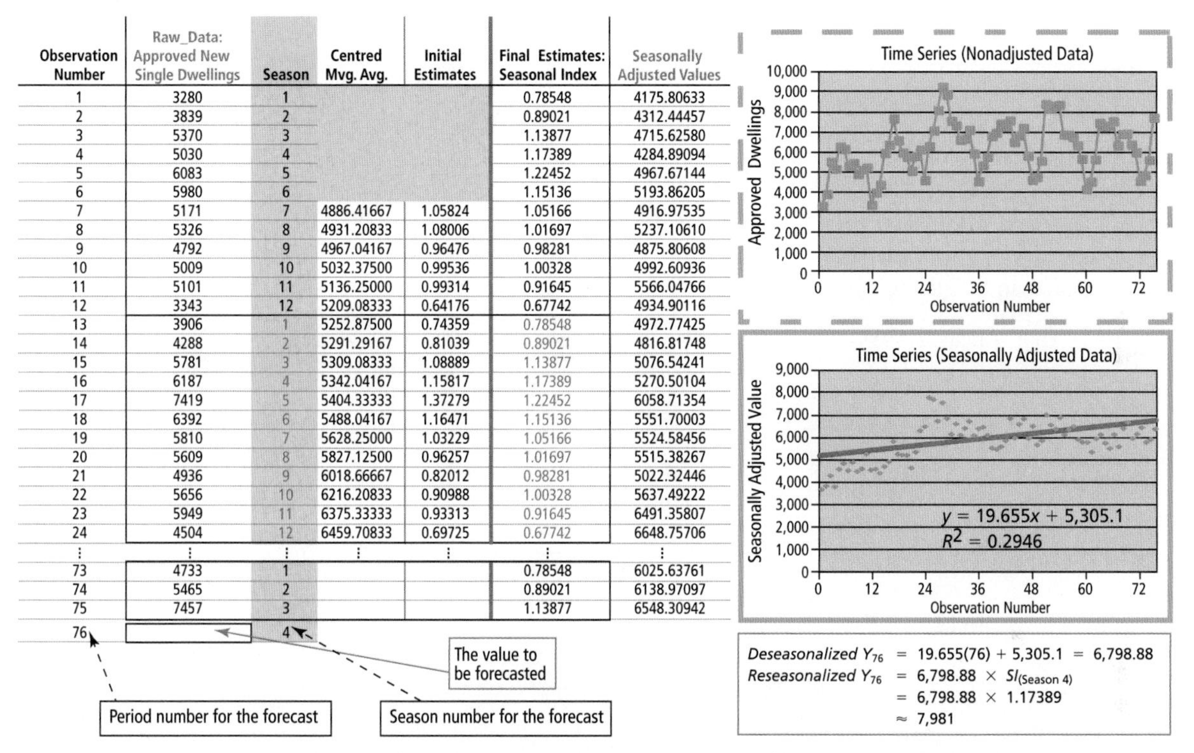

Observation Number	Raw_Data: Approved New Single Dwellings	Season	Centred Mvg. Avg.	Initial Estimates	Final Estimates: Seasonal Index	Seasonally Adjusted Values
1	3280	1			0.78548	4175.80633
2	3839	2			0.89021	4312.44457
3	5370	3			1.13877	4715.62580
4	5030	4			1.17389	4284.89094
5	6083	5			1.22452	4967.67144
6	5980	6			1.15136	5193.86205
7	5171	7	4886.41667	1.05824	1.05166	4916.97535
8	5326	8	4931.20833	1.08006	1.01697	5237.10610
9	4792	9	4967.04167	0.96476	0.98281	4875.80608
10	5009	10	5032.37500	0.99536	1.00328	4992.60936
11	5101	11	5136.25000	0.99314	0.91645	5566.04766
12	3343	12	5209.08333	0.64176	0.67742	4934.90116
13	3906	1	5252.87500	0.74359	0.78548	4972.77425
14	4288	2	5291.29167	0.81039	0.89021	4816.81748
15	5781	3	5309.08333	1.08889	1.13877	5076.54241
16	6187	4	5342.04167	1.15817	1.17389	5270.50104
17	7419	5	5404.33333	1.37279	1.22452	6058.71354
18	6392	6	5488.04167	1.16471	1.15136	5551.70003
19	5810	7	5628.25000	1.03229	1.05166	5524.58456
20	5609	8	5827.12500	0.96257	1.01697	5515.38267
21	4936	9	6018.66667	0.82012	0.98281	5022.32446
22	5656	10	6216.20833	0.90988	1.00328	5637.49222
23	5949	11	6375.33333	0.93313	0.91645	6491.35807
24	4504	12	6459.70833	0.69725	0.67742	6648.75706
⋮	⋮	⋮	⋮	⋮	⋮	⋮
73	4733	1			0.78548	6025.63761
74	5465	2			0.89021	6138.97097
75	7457	3			1.13877	6548.30942
76		4				

Sources: Bank of Canada & Canadian Housing and Mortgage Corporation. Based on data in: (1) Mortgage interest column: Bank of Canada. Found at: http://www.bankofcanada.ca/cgi-bin/famecgi_fdps; (2) Canada Mortgage and Housing Corporation. Found at: http://www.cmhc-schl.gc.ca/en

In the figure, centred moving averages have been used to estimate the seasonal indexes, as described previously. The full set of indexes is shown in the first 12 rows of the table under "Final Estimates: Seasonal Index." For convenience, the set of seasonal indexes has been copied and pasted into the column, so that the *SI* for every season depicted in the third column is shown explicitly in the "Final Estimates" column for that row. Then, using (Observed value)/*SI* for each row in the table, the row's seasonally adjusted (i.e., deseasonalized) value has been calculated.

The regression equation included in the second graph in Figure 17.14 is based on the column of deseasonalized values. We are reminded at the bottom left of the figure that the next period to be estimated equals $75 + 1 = 76$; the season for the next period will be $3 + 1 = 4$. On the lower right of the figure, the deseasonalized value for the 76th time period is estimated and then reseasonalized, based on the index for the 4th season. Our forecast is the resulting value: 7,981. However, given the low r^2 for the regression, do not expect the forecast to be exact.

• In Brief 17.2 Seasonal Decomposition

*(Based on the **Mortgages_Approved** file)*

Excel

Excel does not provide a function for seasonal decomposition. However, the steps illustrated in Figures 17.12 and 17.13 are easily implemented in Excel columns, using available formulas to model the concepts described in the text. For example, given a column "A" with raw data and a corresponding column "D" with the centred moving averages, a column "E" for initial estimates can be created with a copied and pasted formula such as =A3/D3.

As a convenience, a template file in Excel has also been created and stored on your data CD as **Template Centred Moving Avg.xls.** As shown in the illustration, only the column of raw data needs to be input. Through built-in formulas, the columns leading to the final estimates are produced. These are the seasonal indexes. The template includes separate tabs for cases of 4 seasons per cycle, 7 seasons per cycle, or 12 seasons per cycle.

	A	B	C	D	E	FG	H	I	J	K	L
1	Raw_Data		Season	Centred_Mvg_Avg	Initial_Estimates		Seasonal Indexes:	SEASON	INDEX		Enter data
2	3280		1				Final Estimates	1	0.78547704		in Column A only.
3	3839		2					2	0.89021434		
4	5370		3					3	1.1387672		
5	5030		4					4	1.17389219		
6	6083		5					5	1.22451738		
7	5980		6					6	1.15135904		
8	5171		7	4886.416667	1.058239678			7	1.05166279		
9	5326		8	4931.208333	1.080059823			8	1.01697386		
10	4792		9	4967.041667	0.964759372			9	0.98281185		
11	5009		10	5032.375	0.995355076			10	1.00328298		
12	5101		11	5136.25	0.993137016			11	0.91644921		
13	3343		12	5209.083333	0.641763586			12	0.67741985		
14	3906		1	5252.875	0.743592794						
15	4288		2	5291.291667	0.810388138						
16	5781		3	5309.083333	1.08888854						
17	6187		4	5342.041667	1.158171423						
18	7419		5	5404.333333	1.372787269						
19	6392		6	[illegible]	[illegible]						

Minitab

Minitab provides a function for seasonal decomposition. However, its underlying calculations differ in two ways from the standard textbook approach, leading to slightly different results. As described in the text, it calculates initial estimates of the seasonal indexes, based on each specific observation divided by a moving average centred on that observation. But for averaging the initial estimates to create final estimates, it uses the median rather than the mean. For example, based on Figure 17.12, Minitab's steps for the seasonal index for season 4 would produce 0.67822 (the median of the five initial estimates for the season) instead of 0.59402 (the mean of the initial estimates). The second difference for Minitab is that it proportionally adjusts all of the *SI* estimates, so that the sum of *SI* estimates always equals one.

To open the dialogue box shown in the illustration, use the menu sequence *Stat / Time Series / Decomposition.* Then highlight and select the time-based variable and specify the seasonal length (how many seasons in a complete cycle). Click *OK.* As shown in the illustration, Minitab produces not only the seasonal indexes but also the trend equation for the deseasonalized values. (See previous paragraph about its calculation assumptions.) A useful graphical version is also output, as well as measures of forecast accuracy.

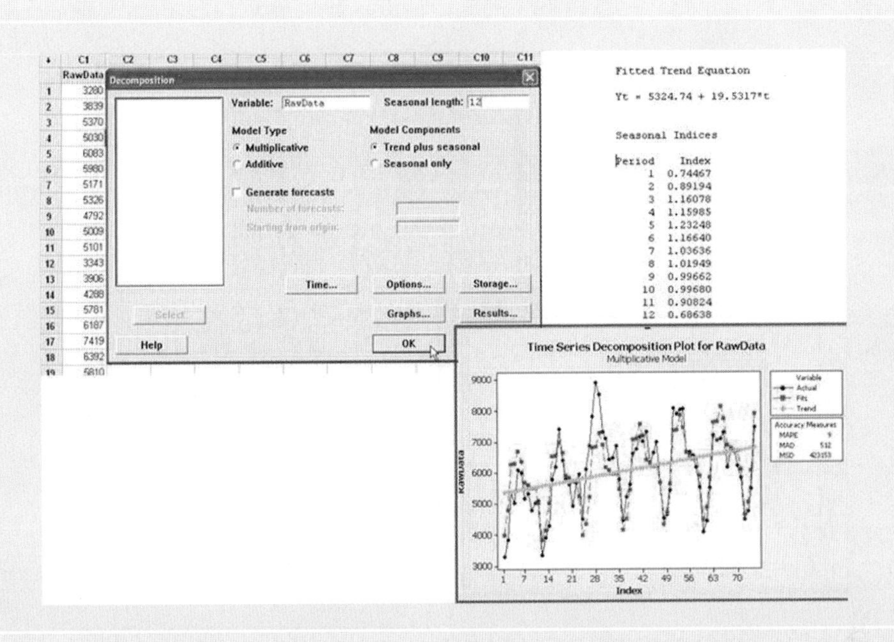

SPSS

SPSS provides a function for seasonal decomposition but its underlying calculations differ in three ways from the standard textbook approach, leading to slightly different results. The textbook method calculates initial estimates of the seasonal indexes, based on each specific observation divided by a moving average centred on that observation. But for averaging the initial estimates to create final estimates, SPSS uses a medial average rather than the conventional mean; that is, it omits the largest and smallest initial estimates from its calculations and takes the mean of only the remaining estimates. For example, based on Figure 17.12, SPSS's approach for the seasonal index for season 4 would produce 0.63648 (the mean of the three remaining estimates for that season, after trimming the two "extremes") instead of 0.59402 (the mean of all five initial estimates). The second difference for SPSS is that it proportionally adjusts all of the *SI* estimates, so that the sum of *SI* estimates always equals one.

The third difference applies when the number of distinct seasons is even. SPSS does not use the two-stage approach shown in Figure 17.13. Instead, to find the centred moving average for any rows t_1 to t_s (where *s* is the number of distinct seasons), SPSS finds the conventional mean for the values in those *s* rows and places this value directly in the row corresponding to $t_{(s/2+1)}$. For example, if there are four seasons, the first centred moving average is based on the mean of the first four entries and this value is placed in the same row as the third entry of the data.

To begin the analysis, input a column with the raw time series data into SPSS and then use the menu sequence *Data / Define Dates*. In the resulting dialogue box, specify the presumed nature of the seasonality. In the example (see the illustration), a complete cycle of seasons is a year, divided seasonally into months. Other possibilities are weeks divided into days, or days divided into hours, and so on. Where shown, input the identifier (in this case, year number) for the first data item and specify in which season you will start. (For example, if the first available data were for March, not January, you would input "3.") Click *OK*. Observe in part (b) of the illustration that SPSS will add three new columns to the working spreadsheet, which "remind" the program of the seasonal structure to

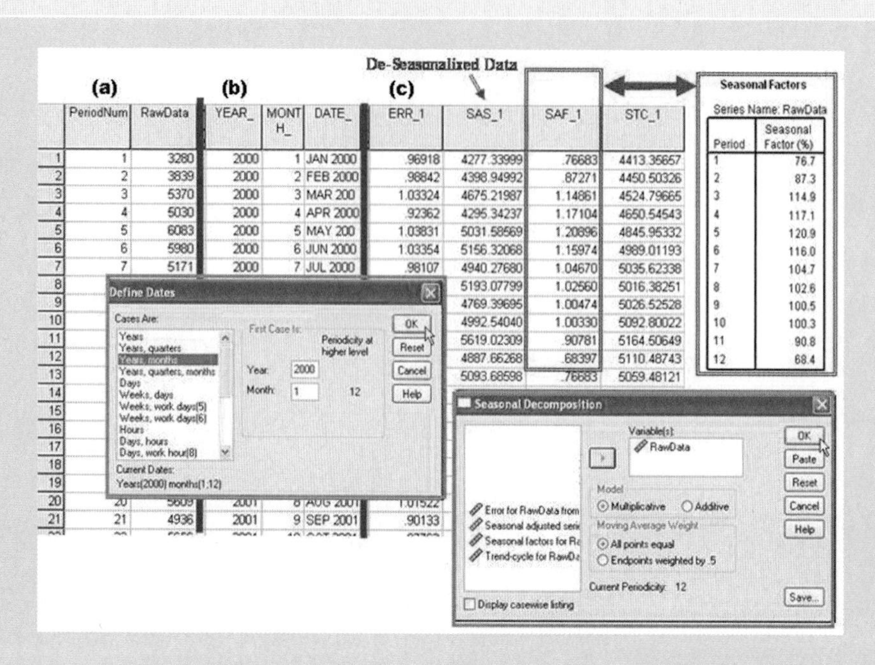

be used in the next steps. (Note that to simplify this complex illustration, most columns in the original data file are not displayed, and the data column of interest, "New Single," is here labelled "Raw Data.")

Next, use the menu sequence *Analyze / Time Series / Seasonal Decomposition.* Highlight, and then click the arrow to select, the name of the time series variable for analysis. SPSS will know, from the previous step, what seasonal structure to use. Click *OK.*

As shown in area (c) of the illustration, SPSS not only outputs the seasonal indexes but also adds four new columns to the working spreadsheet. Of interest here: The column "SAF_1" lists the seasonal factors corresponding to each row for the entire set. And in column "SAS_1," the deseasonalized value (based on the raw value divided by "SAF_1") for each row is provided. If trend analysis is to follow, you would use the "SAS_1" data as the inputs.

• Hands On 17.1 *Forecasting Practice*

Data for practising forecasting are widely available on the Internet. A good exercise is to input all but the last few records for a time series and try to predict those last items. See how close you can come to the actual records for those extra periods and, where you differ, try to understand some of the contributors to the variance.

To try this out, access the *Time Series Data Library* that Dr. Ron Hyndman of Monash University has made available at www-personal.buseco.monash .edu.au/~hyndman/TSDL. The site provides a list of topic areas such as agriculture, chemistry, and crime, for each of which a long list of site links is provided, along with a short data description. The following are three examples.

Topic: Meteorology

MELBMAX.DAT

Daily maximum temperatures in Melbourne, Australia, 1981–1990. Source: Australian Bureau of Meteorology.

Topic: Agriculture

COWS.DAT

Daily temperature of a cow at 6:30 a.m. measured by counting chirps from a telemetric thermometer implanted in the cow. Data are chirps per 5-minute interval -800. Source: Velleman (1981).

Topic: Miscellaneous

MCCLEARY8.DAT

Monthly average daily calls to directory assistance Jan. '62–Dec. '76. Source: McCleary & Hay (1980).

The downloaded data can be saved on your computer with the tag ".dat" and then opened into your statistics program. Check your software procedures for converting text data into numerical data and for aligning the data into a single column.

For at least 10 data sets, try making forecasts using holdout data (i.e., not using some of the more recent records) and then compare your predictions to the actual data for the holdout periods. Take into account some data sets that include potential seasonality (e.g., collected monthly or hourly) and other data sets that do not (e.g., collected annually).

17.3 Applying Dummy Variables for Time Series Analysis

We introduced dummy variables for multiple regression in Section 16.4. Such variables take only one of two numeric values (usually 0 or 1) and are used to represent ordinal or qualitative variables within a regression-based model. The **base case** (i.e., the default assumption) is that all dummy variables in the complete model have the value 0, so the y-estimate is based solely on the regression that uses the quantitative variables. If a dummy variable has the value 1, meaning that a particular qualitative characteristic applies, then the y-estimate is increased by exactly 1 times the coefficient of that variable. That is, the dummy variable coefficient models the effect when a particular quality is present.

Dummy variables can be very useful for a type of time series modelling called **event modelling.** For example, in tracking the weekly sales of a video game, there may be spikes that follow predictably from large TV advertising campaigns. In a time series model, a dummy variable could be added for "ad_campaign_effects." Suppose these effects usually happen in the second, third, and fourth weeks following the advertising. This would be called a **lagged variable** since it takes its value ("1") *following* the time period when the event actually occurs—because the impact on the dependent variable does not occur until that later time. For most weeks, the dummy variable's value would be set to zero. In a regression that included this variable, the variable coefficient would show the magnitude of the effects of the advertising campaign. For predictions, if you know an ad campaign is scheduled, the model will automatically make the adjustment.

Some of the irregular components of regression outputs can also be addressed with event modelling. For example, the occasional flooding of a local creek may impact on the population estimates for mosquito larvae. A time series model of the population could include a dummy variable for flooding. If, independently, the weather service now predicts a flood for next week, the model could help us to estimate the effect on the larvae population.

Moreover, as noted in the previous section, dummy variables can provide an alternative approach to modelling seasonality, compared to the decomposition methods described in that section. A regression with dummy variables is really a special case of multiple regression, so with dummy variables we can include variables based on time and variables not based on time in a single model to predict the response variable. Once the variables have been coded, perform the regression as described in Chapter 16.

Example 17.5

*(Based on the **Mortgages_Approved** file)*

The mortgage analyst in Example 17.4 is not satisfied with the low r^2 for the model that was obtained in that earlier solution. She believes that a much better model can be constructed if mortgage interest rate data are included as an independent variable. (Data for three-year mortgage interest rates, as of each recorded month, are included in the file **Mortgages_Approved.**) Construct a multiple regression-based model to project the number of new single-unit dwellings constructed in April 2006. As independent variables, use time-related variables, including dummy variables, as well as the variable for mortgage interest rates. The mortgage interest rate for April 2006 was 6.45%. Compare your model with the solution for Example 17.4.

Solution

A first step is to prepare the time variables for the regression analysis. In case there is a linear trend based on time, a time index variable (having consecutive values 1, 2, 3, . . ., n (where n is the number of recorded time periods)) should be added. This appears as the variable "TimeNum" to the right of the dependent variable in Figure 17.15(a). Then, dummy variables are added for each of the months (seasons) of February to December. The base case is January; for other months, when "1" is the value for the dummy variable for that month, that row of data corresponds to that particular month. The final column is the unaltered monthly data for the three-year mortgage interest rates.

To construct a model involving so many dummy variables, we need to balance some of the rules of thumb for multiple regression (see Chapter 16) (e.g., reduce the number of variables, if reasonable), bearing in mind the ease of interpreting the final model. We might test whether, collectively, the seasonal-based dummy variables have a significant impact on the response variable. If they do not, we could reject the whole set of them; if they do, these variables provide some useful information. According to the results shown in Figure 17.15(b), the seasonality variables are collectively significant, although two of them indi-

Figure 17.15
Preliminary Alternative Model for Forecasting the Next Seasonalized Value

a)

NewSingle	TimeNum	Feb	Mar	Apr	May	Jun	Jul	Aug	Sep	Oct	Nov	Dec	3-Yr Mort Int Rate
3280	1	0	0	0	0	0	0	0	0	0	0	0	8.3
3839	2	1	0	0	0	0	0	0	0	0	0	0	8.3
5370	3	0	1	0	0	0	0	0	0	0	0	0	8.15
5030	4	0	0	1	0	0	0	0	0	0	0	0	8.15
6083	5	0	0	0	1	0	0	0	0	0	0	0	8.55

b) ANOVA

	df	*SS*	*MS*	*F*	Significance *F*
Regression	11	66471378	6042853	8.375332	7.62025E-09
Residual	63	45454880	721506		
Total	74	1.12E+08			

	Coefficients	Standard Error	*t*-Stat	*p*-Value	Lower 95%	Upper 95%
Intercept	4616.57143	321.0487	14.37966	1.44E-21	3975.00673	5258.136127
Feb	664.714286	454.0315	1.464027	0.148159	−242.5952116	1572.023783
Mar	2285	454.0315	5.032691	4.3E-06	1377.690503	3192.309497
Apr	2345.59524	472.5709	4.963477	5.56E-06	1401.237572	3289.952904
May	2765.7619	472.5709	5.852586	1.89E-07	1821.404239	3710.119571
Jun	2392.92857	472.5709	5.063639	3.83E-06	1448.570906	3337.286237
Jul	1753.09524	472.5709	3.709697	0.000441	808.7375723	2697.452904
Aug	1550.09524	472.5709	3.280132	0.001693	605.7375723	2494.452904
Sep	1394.2619	472.5709	2.950376	0.004451	449.904239	2338.619571
Oct	1528.59524	472.5709	3.234637	0.001941	584.2375723	2472.952904
Nov	1021.7619	472.5709	2.162134	0.034413	77.40423898	1966.119571
Dec	−389.57143	472.5709	−0.82437	0.412842	−1333.929094	554.7862372

Sources: Bank of Canada & Canadian Housing and Mortgage Corporation. Based on data in: (1) Mortgage interest column: Bank of Canada. Found at: http://www.bankofcanada.ca/cgi-bin/famecgi_fdps; (2) Canada Mortgage and Housing Corporation. Found at: http://www.cmhc-schl.gc.ca/en

vidually (December and February) do not appear to have significant coefficients. Unless we have reason to focus on only some particular months, the variable set will be easier to interpret if all of the months (except the base case) are included in the model. A separate check, however, finds that the "TimeNum" variable makes no useful contribution.

Working with the remaining variables, apply multiple regression procedures and obtain the "final" results shown in Figure 17.16. Note the remarkably small *p*-values for most of the remaining dummy variables—suggesting highly significant seasonal effects for much of the year. The variable for mortgage rate is also highly significant and is the one predictor in the model that explains negative movement in the response variable.

For comparing the regression with the previous example, we use the adjusted R^2, since extra variables have been added. Even with this adjustment, R^2 has more than doubled compared to the simple regression model.

Figure 17.16
Final Alternative Model for Forecasting the Next Seasonalized Value

Regression Statistics	
Multiple R	0.894638749
R Square	0.800378491
Adjusted R Square	0.76174207
Standard Error	600.307562
Observations	75

ANOVA

	df	*SS*	*MS*	*F*	Significance *F*
Regression	12	89583369.44	7E+06	20.7156	2.60821E-17
Residual	62	22342888.48	360369		
Total	74	111926257.9			

	Coefficients	Standard Error	*t*-Stat	*p*-Value	Lower 95%	Upper 95%
Intercept	8382.287301	522.1012991	16.055	9.2E-24	7338.621686	9425.953
Feb	631.1667056	320.905232	1.9668	0.05368	−10.31370672	1272.647
Mar	2301.77379	320.8847251	7.1732	1.1E-09	1660.334371	2943.213
Apr	2435.055452	334.1670619	7.2869	6.8E-10	1767.06502	3103.046
May	2874.79154	334.2576715	8.6005	3.6E-12	2206.619982	3542.963
Jun	2428.572875	334.0099524	7.271	7.3E-10	1760.8965	3096.249
Jul	1759.385409	333.9812195	5.2679	1.8E-06	1091.766471	2427.004
Aug	1522.138921	333.9985394	4.5573	2.5E-05	854.4853608	2189.792
Sep	1332.0591	334.0706027	3.9874	0.00018	664.2614868	1999.857
Oct	1451.715367	334.1182374	4.3449	5.3E-05	783.8225334	2119.608
Nov	895.9584792	334.3495321	2.6797	0.00942	227.6032941	1564.314
Dec	−530.0519205	334.4406502	−1.585	0.11806	−1198.589248	138.4854
3-Yr Mort Int Rate	−587.0826528	73.30850714	−8.008	3.8E-11	−733.6242691	−440.541

Sources: Bank of Canada & Canadian Housing and Mortgage Corporation. Based on data in: (1) Mortgage interest column: Bank of Canada. Found at: http://www.bankofcanada.ca/cgi-bin/famecgi_fdps; (2) Canada Mortgage and Housing Corporation. Found at: http://www.cmhc-schl.gc.ca/en

Finally, to predict the numbers of approved mortgages for April 2006, we can apply this regression equation:

$$\hat{y} = 8382.3 + (631.17 \times \text{February dummy value})$$
$$+ (2301.8 \times \text{March dummy value})$$
$$+ (2435.1 \times \text{April dummy value})$$
$$+ \ldots$$
$$+ (-587.1 \times \text{mortgage rate for April 2006})$$

$$\hat{y} = 8382.3 + (631.17 \times 0)$$
$$+ (2301.8 \times 0)$$
$$+ (2435.1 \times 1)$$
$$+ \ldots (0 \times \text{all of the other dummy variable coefficients})$$
$$+ (-587.1 \times 6.45)$$

$$\hat{y} \approx 7031$$

This value is considerably lower than the forecast given for Example 17.4. If we scan the data, we see that interest rates are rising; since the earlier model failed to account for this, the estimate in this solution is likely more accurate.

Chapter Summary

17.1 Describe what is meant by a time series, identify the four types of components of a time series, and construct equations to model linear or nonlinear trends in the data.

Time series analysis is a general term, covering a variety of techniques. A **time series** is an ordered set of observations for a variable made at regular intervals over time, such as yearly, quarterly, monthly, weekly, or daily. A time series can be analyzed (or **decomposed**) into four types of components: (1) the **trend component,** for changes in the value of the variable that reflect a pattern of long-term growth or decline; (2) the **seasonal component,** which reflects short-term variations in the value of the variable that are associated with particular, and recurring, divisions of time; (3) the **cyclical component,** which reflects longer-term, recurring changes in the value of the variable that are often hard to predict; and (4) the **irregular component,** which reflects unanticipated changes in the value of the variable due to unusual or irregular events.

Deciding which model (such as a linear model or an **exponential model for the regression**) best fits the data is called **curve fitting.** For an exponential model, the general formula is $\hat{y} = b_0 \cdot e^{(b_1 \cdot t)}$, where e is a constant $\cong 2.71828$. The exponential model can also be shown as $\log_{\text{(base-10)}} \hat{y} = b_0 + b_1 \cdot t$. This version estimates y indirectly by estimating the base-10 logarithm of y; the base 10 must be raised to the power of this value (i.e., to log 10 y) to predict the y itself. **Mean absolute deviation** (MAD) and **mean squared error** (MSE) are two possible measures for assessing the model. All things being equal, we prefer those models that have comparatively low values for MAD and MSE.

17.2 Perform seasonal decomposition on a time series that involves seasonality and combine these steps with regression to make a forecast for a time period *t*.

With **seasonal decomposition,** we can isolate our analysis of seasonal patterns from our analysis of trend-based patterns. When making a forecast, we consider both of these elements separately and later put them back together. This last step is called **reseasonalizing** the data. The key to seasonal decomposition is to calculate **seasonal indexes** (also called **seasonal relatives**) for each of the recurring season-based changes in the values of the variable. In this text, we focus on a **multiplicative model for the indexes.**

For data that exhibit both seasonality and trend, the order of analysis is to (1) Calculate all seasonal indexes for the data set; (2) create a column of **deseasonalized** (or **seasonally adjusted**) **values** (for a multiplicative model, the procedure is $Y_{\text{t (deseasonalized)}} = Y_{\text{t (actual)}}/SI_{\text{t}}$); (3) use the deseasonalized data—not the original data—to determine the general trend, if any; (4) make a forecast for any time period t by using the regression from step (3), but reseasonalizing your preliminary answer to account for the season (for a multiplicative model: $Y_{\text{t (reseasonalized estimate)}} = Y_{\text{t (deseasonalized estimate)}} \times SI_{\text{t}}$).

Procedures for determining the seasonal indexes are also described in the text.

17.3 Construct and interpret a multiple regression equation that includes dummy variables to model the impact of qualitative variables on a time series.

Dummy variables are applied for a time series analysis similarly to the way they are used in multiple linear regression for modelling qualitative data (see Section 16.4). The **base case** (i.e., the default assumption) is that all dummy variables in the model have the value 0, so the y-estimate is based solely on the regression that uses time or other quantitative variables. The dummy variable coefficient models the effect when a particular quality is present. In a time series, the dummy variables can represent seasons, as an alternative to seasonal decomposition. Dummy variables can be useful for **event modelling,** such as modelling the effects on sales of an ad campaign. If the effects occur in periods following the actual event, then the dummy variable can be used as a **lagged variable;** that is, it takes its value (1) following the time period when the event actually occurs.

Key Terms

base case, p. 727
curve fitting, p. 712
cyclical component, p. 708
decomposed, p. 706
deseasonalized (*or* seasonally adjusted) values, p. 718
event modelling, p. 727
exponential model for the regression, p. 715
irregular component, p. 709
lagged variable, p. 727
mean absolute deviation (MAD), p. 715
mean squared error (MSE), p. 715
multiplicative model for the index, p. 718
reseasonalizing, p. 717
seasonal component, p. 706
seasonal decomposition, p. 717
seasonal indexes (*or* seasonal relatives), p. 718
time series, p. 706
time series analysis, p. 706
trend component, p. 706

Exercises

Basic Concepts 17.1

1. Define what is meant by a time series.
2. Describe two main reasons for analyzing time series.
3. Name and describe the four components into which time series data can be decomposed.
4. Which component of the time series appears to apply for each of the following cases?
   a) Sales of sunscreen predictably increase during the summer months compared to the winter months.
   b) A fruit orchard suffers periodical cicada damage about every 17 years, when the insects emerge from underground and begin to swarm.
   c) Over the past months, the price of gasoline has increased almost every week.
   d) Following a fire in an oil refinery, the price of gasoline has increased.
5. To perform a regression on time series data in which there is a seasonal component, why do the raw data have to be adjusted first? If the data have been adjusted, what should be done before interpreting an estimate that is based on the regression equation?
6. Suppose the equation below has been developed from a monthly time series for the price of an electronic device. Period 1 for the data set was January 2005.
   $$\hat{y} = \$165 - 2.5t$$
   a) Estimate the price of the electronic device in June 2006.
   b) Estimate the price of the electronic device in June 2012. Do you think this estimate is really plausible? Why or why not?
7. Suppose that the equation below has been developed from a monthly time series for the price of a litre of gasoline in a certain area. Period 1 for the data set was January 2006.
   $$\hat{y} = \$0.96 \cdot e^{(0.005t)}$$

a) What type of regression model has been used?
b) Estimate the price per litre of gasoline: for June 2007 and for June 2008.

8. Suppose that the equation below has been developed from a monthly time series for the price per share (in cents) of a popular penny stock. Period 1 for the data set was January 2006.

$$\log_{\text{base-10}} \hat{y} = 0.1 \cdot 0.055t$$

a) What type of regression model has been used?
b) Estimate the price per share (in cents): for June 2007 and for June 2008.

9. If an exponential regression model for a time series has a higher value for r^2 compared to a linear regression model for that time series, is it always best to use the exponential model? Explain your answer.

Applications 17.1

10. According to Canadian census data published by Statistics Canada, it appears that the median age of Canadians is increasing.[1]

Period Number	1	2	3	4
Year of census	1956	1961	1966	1971
Median age	27.2	26.3	25.4	26.2

Period Number	5	6	7	8
Year of census	1976	1981	1986	1991
Median age	27.8	29.6	31.6	33.5

Period Number	9	10	11
Year of census	1996	2001	2006
Median age	35.3	37.6	39.5

a) Construct a linear regression equation that relates the median ages of Canadians to the period number (where 1956 census = Period 1).
b) Assess the strength and significance of the relationship.
c) Use the model to forecast the median age of Canadians in 2011, which will be period 12.

11. Using the data in Exercise 10:
a) Construct an *exponential* regression equation that relates the median ages of Canadians to the period number (where 1956 census = Period 1) and assess the strength and significance of the relationship.
b) Use the model to forecast the median age of Canadians in 2011, which will be period 12.
c) Compare the results for (a) and (b) with the answers for Exercise 10. If you were presenting your forecast for 2011 in a newspaper article, which model (linear or exponential) would you recommend using? Why?

For Exercises 12–15, use data from the file **Age_Health_Occupation.**

12. A health insurance official wants to forecast the future mean ages of workers in various dental- and medical-related fields. Where possible, she hopes to use linear regression to model the age trends for the different groups. Using the data in the file **Age_Health_Occupation:**
a) Use linear regression for a trend analysis that relates the mean ages of dental assistants to time periods, starting from Period 1 = 1994. What is the equation?
b) Is the regression significant? Justify your claim.
c) What is the standard error of the regression? What is the coefficient of determination?
d) Use the regression equation to estimate the mean age for dental assistants in 2004. Would you feel confident using this prediction? Explain.
e) Use the regression to estimate the mean age for dental assistants in 2009. Would you feel confident using this prediction? Explain.

13. Repeat (a) through (d) of Exercise 12 for dental technicians.

14. Repeat (a) through (d) of Exercise 12 for medical laboratory technicians.

15. Repeat (a) through (d) of Exercise 12 for optometrists. For this group, is there any advantage to using the regression for estimating the 2004 value compared to simply using the mean of all prior observed values as an estimate? Explain.

16. A public health researcher needs to forecast, as accurately as possible, the next year's costs for drug expenditures. (See data in the file **Health_Expenditures;** costs are in CDN$100,000.)
a) Use linear regression for a trend analysis of drug costs, starting from Period 1 = 1990. What is the equation?
b) Is the regression significant? Justify your claim.
c) What is the standard error of the regression? What is the coefficient of determination?
d) Use the linear regression to estimate the drug costs, in Canadian dollars, for 2003.
e) What is the equation for an exponential model of the trend? Is the regression significant? What is the coefficient of determination?
f) Use the regression in (e) to estimate the drug costs, in Canadian dollars, for 2003.
g) Based on your answers above, combined with creating a scatter diagram for the time series data, would you recommend using the linear or the exponential model for the needed forecast? Explain.

Basic Concepts 17.2

17. Explain what is meant by seasonal decomposition and describe when this might be required as part of a forecasting process.

18. What does it mean to reseasonalize data and under what circumstances is this step required?
19. Explain the role of seasonal indexes in the forecast process.
20. What is another name for a seasonal index?
21. What is the difference between a multiplicative model for seasonality and an additive model?
22. What is another name for a deseasonalized value?
23. Suppose there are four seasons with seasonal indexes equal to 0.8, 1.1, 0.9, and 1.2, respectively. A raw data value that occurs in season 3 happens to equal 5.673. What is the seasonally adjusted value that corresponds to that raw data value?
24. If there are four seasons in a time series, during which of those four seasons would the 42nd time period occur?
25. Suppose there are four seasons with seasonal indexes equal to 0.8, 1.1, 0.9, and 1.2, respectively. An unreseasonalized forecast for period 42 happens to equal 7.991. What is the reseasonalized value that corresponds to that forecast?

Applications 17.2

For Exercises 26–30, use data from the file **Philippines Employment.**

26. The file saved as **PhilippinesEmployment** contains quarterly indexes for the levels of employment in different sectors of the economy over a number of years. (For each column, the index in a particular cell can be compared to the value "100"—which represents the level of employment in that sector in 1978.) Use scatter diagrams to see in which sectors of the economy the levels of employment showed seasonality in the time series from 2002 to 2006. Name the three sectors for which seasonality was most pronounced and consistent over that period.
27. Calculate and display the four seasonal indexes for quarterly employment levels for each of the following sectors: (a) mining, (b) manufacturing, (c) electricity and water, and (d) trade.
28. Based on the seasonal indexes calculated in Exercise 27, calculate the first four seasonally adjusted values for the employment levels in each of the following sectors: (a) mining, (b) manufacturing, (c) electricity and water, and (d) trade.
29. In such a way as to account for both trend and seasonality, forecast the employment index for the manufacturing sector for the first and second quarters of 2007.
30. In such a way as to account for both trend and seasonality, forecast the employment index for the electricity and water sector for the first and second quarters of 2007.

For Exercises 31 and 32, use data from the file **Yields.**

31. Having compiled the data in the file **Yields** during a job placement, you would like to forecast the Canada benchmark bond yields for the first six months of 2007. In the scatter diagram, you note that the time series for bond yields appears to have an approximately linear trend—starting from about 1982. Assuming that this trend will continue for at least the six months to be forecasted, use the data from January 1982 to December 2006. Calculate and display the seasonal indexes for each of the months from January to June.
32. Continuing Exercise 31, use the data for trend and seasonality to forecast the bond yields for the first six months of 2007, based on the data starting from January 1982.

Basic Concepts 17.3

33. If you use dummy variables in a regression model for a time series, what is meant by the "base case"?
34. Explain how you can use dummy variables for event modelling in a regression model for a time series.
35. What is a lagged variable? Describe how and why you might use a lagged dummy variable in a regression model for a time series.
36. Explain how you can use dummy variables to model seasonality in a regression model for a time series.
37. Suppose you have recorded sales data quarterly for umbrellas and it is clear from a scatter diagram that the time series exhibits seasonality. If you model the seasons with dummy variables, what is the largest number of dummy variables that you could require?
38. If you model the seasons with dummy variables, do you *necessarily* need to use the largest number of dummy variables that is possible? Give an example of when a smaller number of dummy variables might be sufficient.

Applications 17.3

For Exercises 39–41, use data from the file **Yields.**

39. In Exercise 32 you used the data from the file **Yields** to forecast the Canada benchmark bond yields for the first six months of 2007. Repeat that forecast but this time use dummy variables for the seasons, rather than seasonal

decomposition, and make the forecast using a regression that includes the period number and the dummy variables as the independent variables. What is the regression equation? How does your forecast compare with Exercise 32?

40. In your model from the previous question, are the coefficients of all of the dummy variables significant? If not, what does this imply?

41. Again referring to **Yields** data, suppose your goal is not to make a forecast but to quantify the reduction in yields that generally happens in November and December, as compared to the other months. Create a model that includes dummy variables for addressing this issue.

42. Data for the file **IceBakerLake** were ideally to have been collected weekly—as long as ice was present—but various weeks were missed. Even averaging the weekly data into monthly records left gaps. Despite this issue, create scatter diagrams for various blocks of time within the data set, using the period number as the *x*-variable and the average ice measurement (in cm) as the dependent variable. Do you detect seasonality?

43. As noted in Exercise 42, the data for **IceBakerLake** are not complete, so it would be very difficult to use seasonal decomposition unless this was done manually. Instead, create dummy variables based on the months for which data are generally available and omit records for which no data are available. Then run a multiple regression, using the principles discussed in Chapter 16. There may be more than one "right" answer.

a) What is the regression equation, based on your model?
b) How much of the variance of the dependent variable is explained by the regression?
c) What is the coefficient of the dummy variable for month 10 (October)?
d) What is the coefficient of the dummy variable for month 5 (May)?
e) What would you estimate as the ice thickness in February 2000? In October 2000?

44. Redo Exercise 43 but use the data in the file **IceCambridgeBay** and answer the questions in relation to ice thickness in that data set.

45. Redo Exercise 44 but answer the questions in relation to the variable for snow thickness in the data set for Cambridge Bay.

Review Applications

For questions involving multiple linear regression, construct and apply the best linear regression model that you can—in keeping with the five rules of thumb and your instructor's guidelines—ensuring, if possible, by residuals analysis, a reasonable conformity of the model with regression assumptions.

For Exercises 46–49, use data from the file **Landed_ Fish.**

46. Concerned about the decreasing fish stocks in the Maritimes region, an officer of the Department of Fisheries and Oceans collected the data in the file **Landed_Fish.**

a) Based on a scatter diagram, does the time series for the quantities of fish landed each year follow a linear pattern?
b) Construct a linear regression equation to forecast the quantities of fish landed in future years. Write the equation.
c) Is the apparent regression in (b) significant?
d) Use the model to estimate the quantity of fish landed in 2006.

47. a) Reuse the **Landed_Fish** data to construct an exponential regression equation to forecast the quantities of fish landed in future years. Write the equation and estimate the quantity of fish that were landed in 2006.
b) Is the apparent regression in (a) significant?
c) Of the models in Exercises 46(b) and 47(a), which explains a larger percentage of the variance in the quantities of landed fish?
d) What are some arguments for and against excluding the three earliest data points (1990–1992) from the model and revising the regression equation accordingly?

48. Redo all parts of Exercise 46, but use the value of the catches as the response variable, rather than the quantity of the catches.

49. Redo Exercise 47(a) and (b) but use the value of the catches as the response variable, rather than the quantity of the catches. Then compare the models for the values as instructed in Exercise 47(c).

For Exercises 50 and 51, use data from the file **PhilippinesEmployment.**

50. The file saved as **PhilippinesEmployment** contains quarterly indexes for the levels of employment in different sectors of the economy over a number of years (compared to a Base year 1978 = 100). Although the data are quarterly, a scatter diagram shows little evidence of seasonality for employment in the finance sector. Use a linear regression method to estimate the employment index for the finance sector in the first two quarters of 2007.

51. Although the data for the real estate sector are also quarterly, a scatter diagram shows little evidence of seasonality for employment in this sector. Use a linear regression method to estimate the employment index for the real estate sector in the first two quarters of 2007.

For Exercises 52–59, use data from the file **Philippines Revenue.**

52. The file saved as **PhilippinesRevenue** contains quarterly indexes for the levels of gross revenue received by different sectors of the Philippine economy over a number of years. (For each column, the index in a particular cell can be compared to the value "100"—which represents the level of revenue for that sector in 1978.) Using a scatter diagram, confirm that there is seasonality in the time series for revenue in the trade sector from 2002 to 2006. Calculate and display the four seasonal indexes for that variable.

53. Based on the seasonal indexes calculated in Exercise 52, calculate the seasonally adjusted values corresponding to all the values in the trade column.

54. In such a way as to account for trend and seasonality, forecast the revenue values for the trade sector for the second and third quarters of 2007.

55. Again forecast the revenue values for the trade sector for the second and third quarters of 2007 but, this time, use dummy variables for the seasons rather than using the seasonal decomposition method.

56. Next, using a scatter diagram, confirm that there is little or no seasonality in the time series for revenue in the real estate sector from 2002 to 2006. Does the trend appear to be linear or exponential?

57. For the real estate time series, find both the linear regression equation and the exponential regression equation; then compare the strength and significance of each model. Which model would you recommend to most accurately forecast the revenue for that sector for the first quarter of 2007? Justify your choice and use your selected model to make the forecast.

58. Next, using a scatter diagram, confirm that there is little or no seasonality in the time series for revenue in the private services sector from 2002 to 2006. Does the trend appear to be linear or exponential?

59. For the private services time series, find both the linear regression equation and the exponential regression equation; then compare the strength and significance of each model. Which model would you recommend to most accurately forecast the revenue for that sector for the first quarter of 2007? Justify your choice and use your selected model to make the forecast.

For Exercises 60 and 61, use data from the file **BMO_Stock.**

60. Historical closing prices for many stocks can now be downloaded at no charge from the issuing corporation's website. For example, more than five years of monthly closing prices for Bank of Montreal stocks are stored in the file **BMO_Stock.**
    a) Use the method of seasonal decomposition to find the seasonal indexes for each month.
    b) Based on (a), find the deseasonalized values corresponding to each of the data values for 2002.
    c) Construct the regression equation using the deseasonalized values for the closing prices.
    d) Accounting for both trend and seasonality, estimate the monthly closing prices for the remainder of 2007.

61. Accounting for both trend and seasonality, estimate the monthly closing prices for the remainder of 2007 for BMO stock but this time use the method of including dummy variables instead of using seasonal decomposition. Compare your forecast with your answer to 60(d).

Glossary

Term (Chapter) Interpretation

A

ABSOLUTE PROBABILITY (6) The **absolute** (or marginal) **probability** of event *A* is the probability of *A* occurring [in symbols: $P(A)$]—calculated without reference to other possible events.

ACCEPTANCE SAMPLING (7) **Acceptance sampling** is a statistical procedure used to decide whether to accept or reject a batch of merchandise or other output, based on the number of defects that are found in a sample.

ADJUSTED R^2 (16) **Adjusted R^2** is a recommended alternative measure to R^2 for comparing the relative strengths of multiple regression models that have different numbers of independent variables, taking into account the reduced degrees of freedom as variables are added.

α (11) **α** is the effective level of maximum risk for committing a **Type I error** when making the **statistical decision** of a hypothesis test.

ALTERNATIVE HYPOTHESIS (11) In a hypothesis test, the **alternative hypothesis** (in symbols: H_1 or H_a) is logically opposed to the **null hypothesis** and always includes the concept of inequality (i.e., the logical operator: $\neq$ or $<$ or $>$).

ANOVA TABLE (14) When conducting an analysis of variance (ANOVA) hypothesis test, the **ANOVA table** shows the basic elements that lead to the final calculation for the test statistic (the *F*-value).

A PRIORI PROBABILITY (6) [*See* **classical probability**]

ARITHMETIC MEAN (4) The **arithmetic mean** is a measure of centre that takes all values in the data set into account: Add all the values in the data set and divide by the number of values in the set.

ASSOCIATION (15) There is an **association** between two variables *A* and *B* if for some value *a* of *A*, the absolute probability $P(A = a)$ is *not* identical to the conditional probability $P(A = a \mid B = b)$ for all values *b* of *B*, and/or for some value *b* of *B*, the absolute probability $P(B = b)$ is *not* identical to the conditional probability $P(B = b \mid A = a)$ for all values *a* of *A*.

AUTOCORRELATION (16) In a regression where time is an independent variable, a **response variable** exhibits **autocorrelation** if the values of observations at times t_i are correlated with the values of observations at previous times t_{i-k}.

B

BAR CHART (3) In a **bar chart,** all bars have equal widths and their lengths are proportional to the frequencies (or relative frequencies) of occurrence of the different data values (i.e., categories) being graphed; the bars may be oriented vertically or horizontally.

BASE CASE (17) If dummy variables for qualitative data are included in a multiple regression, the **base case** (i.e., the default assumption) is that all dummy variables in the complete model have the value 0, so the *y*-estimate is based solely on the regression that uses the quantitative variables.

BAYES' RULE (OR BAYES' THEOREM) (6) **Bayes' Rule** (*or* **Bayes' Theorem**) was developed to facilitate *revision* of initial probability assignments (called the **prior probabilities**) based on new information, to develop revised **posterior probability** estimates.

BEFORE/AFTER METHOD (12) In the **before/after method** for obtaining a paired sample (also called the **pre-test/post-test method** or **repeated-measures method**), the same individuals are sampled both before and after an intervention or experiment, and results are compared.

BELL CURVE (8) ***Bell curve*** is an informal term that usually refers to the *normal curve.* See the **normal curve** definition.

BERNOULLI TRIAL (7) A concept that underlies the binomial distribution is the **Bernoulli trial** (*or* **trial**), which is a random experiment for which only two mutually exclusive outcomes are possible; one of the two outcomes is identified as a "success."

β RISK (11) In a hypothesis test, a **β risk** is the risk of committing a **Type II error.**

BIAS (2) **Bias** is a type of nonsampling error in which the data collected misleadingly tend more toward one possible conclusion than another.

BIMODAL (4) A data set that has two **modes** is called **bimodal.**

BINOMIAL EXPERIMENT (7) A **binomial experiment** is a random experiment, such as two coin flips in a row, consisting of a sequence of a fixed number of identical and independent **Bernoulli trials.**

BINOMIAL PROBABILITY DISTRIBUTION (*OR* BINOMIAL DISTRIBUTION) (7) The **binomial probability distribution** (or **binomial distribution**) is a discrete probability distribution for all possible values of a **binomial random variable.**

BINOMIAL RANDOM VARIABLE (7) A variable that counts the number of successes (regardless of exact sequence) among the trial results in a **binomial experiment** is called a **binomial random variable.**

BIVARIATE DESCRIPTIVE STATISTICS (3) In **bivariate descriptive statistics,** we compare the values for *two* variables that are observed for each individual in the population.

BIVARIATE NORMAL DISTRIBUTION (15) In a **bivariate normal distribution,** the deviations of actual y-values from the predicted y-values (as estimated by a regression line for the two variables) are normally distributed for all parts of the line.

BLOCKS (14) In a **randomized block design,** the **blocks** are the intentionally connected observations.

BOOTSTRAP METHODS (9) **Bootstrap methods** are **nonparametric** techniques that use repeated random sampling from the sample data themselves.

BOXPLOT (5) A **boxplot** (or box-and-whisker diagram) is a graphic method that can illustrate directly the **five-number summary.**

C

CASE 1 (12) **Case 1** in a t-test for the difference in population means presumes that the two unknown variances can reasonably be assumed to be equal.

CASE 2 (12) **Case 2** in a t-test for the difference in population means presumes that the two variances are unequal.

CATEGORICAL RANDOM VARIABLE (1) [*See* **qualitative random variable**]

CATEGORIES (1) **Categories** are the distinct traits or characteristics that are possible values for a **qualitative** (or **categorical) random variable.**

CAUSAL RESEARCH (2) [*See* **experimental research**]

CENSUS (1) The collection of data for *all* members of a population is called a **census.**

CENTRAL LIMIT THEOREM (9) According to the **central limit theorem,** when the sample size n is sufficiently large and sampling is done with replacement, then (1) the sampling distribution of the sample means is approximately normal, regardless of the distribution of the underlying population and (2) this approach toward the normal distribution becomes closer as the sample size n increases.

CHEBYSHEV'S INEQUALITY (5) Regardless of the **shape** of a distribution, **Chebyshev's Inequality** can provide estimates of the *minimum* proportions of values that can be expected to fall within specified distances of the mean.

CHI-SQUARE (χ^2) DISTRIBUTION (15) The **chi-square (χ^2) distribution** models the extent to which the observed and expected frequencies could be expected to differ (under the null hypothesis) for given distributions of values of categorical variables.

CLAIM (11) A **claim** is an assertion about facts, which can be formalized as a hypothesis for purposes of **hypothesis testing.**

CLASSES (3) In a frequency distribution table for continuous quantitative data, specified ranges of possible values are called **classes.**

CLASSICAL PROBABILITY (6) For **classical** (or **a priori**) **probability,** the probability of an event is determined from previously known definitions and properties of things; no actual observations (experiments) are required.

CLEAN (3) Data are **clean** and not dirty if the recorded numbers or codes are all correct and free of mistyped or poorly measured values.

CLINICAL TRIAL (12) [*See* **controlled experiment**]

COEFFICIENT OF CORRELATION (r) (15) [*See* **sample coefficient of correlation**]

COEFFICIENT OF DETERMINATION (r^2 or R^2) (16) The **coefficient of determination** (r^2 for simple regression and R^2 for multiple regression) indicates what proportion of the total variance of the dependent variable (which is the variance compared to simply the y-mean) is explained by, or attributable to, the variance due to the regression.

COEFFICIENT OF VARIATION (5) The **coefficient of variation** is a measure that expresses the relative size of the standard deviation as a percentage of the mean for a given population or sample.

COLLINEARITY (16) If there is **collinearity** among the x-variables in a regression model, then they are not totally independent from one another.

COMBINATIONS (6) **Combinations** are distinct sets of k objects that could be selected from a usually larger set of n objects; the order of selecting the objects does not matter.

COMPLEMENT (6) The compound event that occurs when a named simple event *does not* occur is called the **complement** of the named event; in notation, the complement of event A is the event $\overline{A}$.

COMPOUND EVENT (6) An event defined in terms of other simple events is a **compound event.**

CONCLUSIVE RESEARCH (2) **Conclusive research** can be conducted when the research objectives are clear and the problems are unambiguous.

CONDITIONAL PROBABILITY (6) The **conditional probability** of "event A, given event B" [in symbols: $P(A \mid B)$]

indicates that the occurrence of B may have an impact on the probability of A.

CONDITIONS OF PROBABILITY (6) These **conditions of probability** always apply: (1) The assigned value of probability is a number between 0 and +1, with 0 representing an impossible event and +1 signifying a certain event; (2) the sum of probabilities for all of the possible outcomes (if outcomes are discrete) equals 1 (or if possible outcomes are values on a continuous scale—divisible into arbitrarily small classes—the sum of probabilities for all of the classes of possible outcomes equals 1).

CONFIDENCE INTERVAL (10) A **confidence interval** is an estimate for a parameter that specifies both a range of possible values for the parameter, and also a degree of confidence that the computed confidence interval really does contain the true value of the parameter.

CONFIDENCE LEVEL (10) [*See* **degree of confidence**]

CONFIDENCE LIMITS (10) The lower and upper boundaries of the confidence interval estimate are called the **confidence limits.**

CONJUNCTIVE EVENTS (6) [*See* **joint events**]

CONSISTENT (10) A **consistent** estimator for a parameter has the property that, as sample size increases, the values of the statistic are more likely to be closer to the parameter being estimated.

CONSTANTS (1) **Constants** are a basic form of data whose values do not change over the course of an experiment.

CONTINGENCY TABLES (3, 6) **Contingency tables** (or **cross-tabulation tables**) display potential associations between two qualitative variables by displaying the frequencies or relative frequencies for both variables individually *and* for combinations of the two variables' values.

CONTINUOUS QUANTITATIVE FREQUENCY TABLE (3) A **continuous quantitative frequency table** is a frequency distribution table for continuous quantitative data that displays how many times the variable takes values that fall within specified **classes.**

CONTINUOUS RANDOM VARIABLE (1) A **continuous random variable** can potentially assume *any* value over a particular range of values.

CONTROL GROUP (12) The **control group** is a sample to which a particular treatment in a **clinical trial** is *not* applied.

CONTROLLED EXPERIMENT (12) For a **controlled experiment** (such as a **clinical trial**) a sample is taken from a single population and then a distinction is introduced by the experimenter—between an experimental group (or **treatment group**), to which an experimental procedure (or treatment) is applied, versus a **control group,** to which the treatment is not applied.

CONVENIENCE SAMPLING (2) **Convenience sampling** is a nonstatistical sampling technique that draws samples based on their convenience or availability.

CRITICAL REGION (11) The **critical region** is the portion of the hypothesized sampling distribution for which the probability that the **test statistic** will occur there (if H_0 is true) is so low that we will reject the **null hypothesis** if in fact the **test statistic** does fall in that region.

CRITICAL VALUE OR VALUES (10, 11) The **critical value** or **critical values** are values (given in standardized units, depending on the model used for the sampling distribution) that mark the boundaries of the **critical region.**

CROSS-TABULATION TABLES (3) [*See* **contingency tables**]

CUMULATIVE DISTRIBUTION FUNCTION (8) The **cumulative distribution function** (CDF) for a probability distribution (such as the normal curve) allows us to find the area under the curve (i.e., the cumulative probability) that is bounded on the right by a specified value of the variable.

CUMULATIVE PERCENT FREQUENCY DISTRIBUTION (3) Often displayed as a column in a frequency distribution table, the **cumulative percent frequency distribution** shows for each value of the variable (or **class**) the cumulative percentage of all cases having values up to or including the value (or upper bound of the **class**) displayed for that row.

CURVE FITTING (17) In time series regression modelling, **curve fitting** is a process for deciding which regression model (e.g., linear or exponential) best fits the data.

CYCLICAL COMPONENT (17) The **cyclical component** of a time series reflects longer-term recurring changes in the value of the variable compared to seasonal changes; cyclical patterns have less constant and harder-to-predict cycle lengths.

D

DATA (1) **Data** are sets of facts (numerical or nonnumerical) that record the results of observations and provide the raw materials for statistics.

DATA SET (1) The set of all observations for a given purpose is called a **data set.**

DATUM (1) A single observation is called a **datum.**

DECILES (4) **Deciles** are numeric boundaries that divide an ordered data set into 10 equal parts.

DECOMPOSED (17) To **decompose** a time series is to analyze it into its various components; the components are: (1) the **trend component, *T*;** (2) the **seasonal component, *S*;** (3) the **cyclical component, *C*;** and (4) the **irregular component, *I*.**

DEGREE OF CONFIDENCE (10) The **degree of confidence** (or **confidence level**) that is associated with a **confidence interval** conveys the confidence that the specified interval really does contain the true value of the parameter.

DEGREES OF FREEDOM (*DF*) (5, 10) **Degrees of freedom** indicate the number of observations that could freely change in value without changing the sample statistic (such as the mean) that has been calculated for the sample.

DENSITY SCALE HISTOGRAM (3) A **density scale histogram** is a type of histogram in which the area encompassed by each bar represents the percentage of the total number of observations that fall within the class boundaries for the bar.

DEPENDENT EVENTS (6) Two events are **dependent events** if the probability that one of the events will occur is affected by the occurrence or nonoccurrence of the other event.

DEPENDENT SAMPLES (12) In **dependent samples** (also called **related samples** or **paired samples**), the selection of elements in one sample is influenced by the selection of elements in the other sample.

DEPENDENT VARIABLE (16) In a regression model, the **dependent** (or **response**) **variable** is the variable whose values are to be estimated or predicted by the model.

DESCRIPTIVE RESEARCH (2) **Descriptive research** is used to describe the characteristics of a population.

DESCRIPTIVE STATISTICS (1) In **descriptive statistics,** the focus is on summarizing and presenting information.

DESEASONALIZE (17) To **deseasonalize** the observations in a time series is to rescale each raw data value based on the **seasonal index** that corresponds to the time period when the value was observed; for a multiplicative model, this is accomplished by dividing each raw data value by its corresponding seasonal index.

DESEASONALIZED VALUES (17) **Deseasonalized values** are data values that have been transformed by being **deseasonalized.**

DEVIATION (5) In general, the distance of a value from the mean is called its **deviation** from the mean.

DIRTY DATA (3) Data are described as **dirty** if the recorded numbers or codes are not all correct, possibly due to mistyped or poorly measured or recorded values.

DISCRETE QUANTITATIVE FREQUENCY TABLE (3) A **discrete quantitative frequency table** is a frequency distribution table for quantitative data that shows how many times the variable takes specific values from a finite list of alternatives.

DISCRETE RANDOM VARIABLE (1) A **discrete random variable** is a random variable for which the numerical values that could be observed are constrained to a limited set of values; e.g., to only whole numbers or to a limited set of real numbers.

DISJUNCTIVE EVENTS (6) **Disjunctive events** are compound events that occur if at least one of the named **simple events** occurs; in notation, "*A* or *B*" is a disjunctive event, based on the simple events *A*, *B*.

DISTRIBUTION-FREE (13) [*See* **nonparametric**]

DUE DILIGENCE (2) **Due diligence** is ensuring that you plan and execute each step of a statistical analysis very carefully, assess the limitations of the methods used, and report on any possible limitations and biases.

DUMMY VARIABLES (16) **Dummy variables** are additional independent variables added to a regression model, which by convention usually take a default value of 0 if some qualitative condition does not apply for an observation or a value of 1 if that condition *does* apply.

E

EFFICIENT (13) An (1) estimator *or* (2) hypothesis test is more **efficient** when, for the given sample size, it (1) gives a more precise estimate or (2) has more statistical **power.**

EMPIRICAL PROBABILITY (6) For **empirical probability** (also called **relative frequency** or **experimental probability**), we learn from experience how often a particular event has occurred in the past relative to all the other alternatives and take this as an estimate of the future proportion.

EMPIRICAL RULE (5) For an approximately bell-shaped distribution, the **empirical rule** (or **"68–95–99% rule"**) can be used to estimate the proportions of data that lie within certain distances of the mean.

EQUIPROBABLE DISTRIBUTION (7) An **equiprobable distribution** is a probability distribution such that each value in the sample space has the same probability of occurring as any other.

ESTIMATION (10) **Estimation** is a process of obtaining a single numerical value, or a set of values, intended to approximate the actual value of a parameter in the population.

ESTIMATOR (10) An **estimator** is a measured or calculated value based on a sample used to estimate a population parameter.

ETHICAL (2) **Ethical** research strives to follow legal and moral guidelines for respecting persons and **due diligence** when collecting and presenting data.

EVENT (6) For calculating probabilities, an **event** is an outcome or a combination of possible outcomes that the researcher judges can be classified as a single occurrence of interest.

EVENT MODELLING (17) In a time series model, **event modelling** is the use of a special variable to indicate the occurrence of a particular state of affairs during specified periods of time.

EXPECTED FREQUENCIES (15) In a chi-square test for two variables, **expected frequencies** are the frequencies that would be expected in each cell of the contingency table if the variables were independent; for goodness of fit of a single variable, they are frequencies expected in each class of the frequency distribution if the null hypothesis was true.

EXPECTED VALUE (7) If an **experiment** produces outcomes with probabilities consistent with a discrete probability distribution and the experiment is repeated many times, then the **expected value** is the mean for all of the outcomes that would be generated by the repeated experiments.

EXPERIMENT (2) In probability, an **experiment** is any procedure whose objective is to collect observations of outcomes. More formally in causal research, experiments are procedures to collect primary data in which the researcher attempts to control (keep constant) all factors except those factors being investigated.

EXPERIMENTAL GROUP (12) [*See* **treatment group**]

EXPERIMENTAL PROBABILITY (6) [*See* **empirical probability**]

EXPERIMENTAL RESEARCH (2) **Experimental** (or **causal) research** explores possible cause-and-effect relationships among the observed factors.

EXPLORATORY RESEARCH (2) **Exploratory research** is preliminary research, conducted when very little is known as yet about the problem being studied.

EXPONENTIAL PROBABILITY DISTRIBUTION (8) When the random variable relates to *time*, the distribution of possible values—such as times required for a randomly selected customer to be served in a restaurant, or the lengths of time between patients arriving at a hospital emergency ward—often follow the **exponential probability distribution.**

EXTERNAL SECONDARY DATA (2) **External secondary data** are data collected originally outside of one's own organization.

EXTRANEOUS VARIABLES (12) **Extraneous variables** are variables not included in the model that may be accounting for unexplained variation in the **response variable.**

EXTRAPOLATE (16) In regression, to **extrapolate** is to attempt a forecast or estimate based on an x-value that falls beyond the original range of data.

F

FACTOR VARIABLE (14) [*See* **grouping variable**]

FENCES (5) In a variation of the boxplot, **fences** (or **steps**) are calculated distances from the median used to identify and display possible outliers in the data.

FINITE POPULATION (1) In a **finite population,** it is possible to list or count every member of the population.

FINITE POPULATION CORRECTION FACTOR (9) The **finite population correction factor** is a multiplier used to adjust the estimate for the standard error in cases where sampling is done without replacement and the sample size is relatively large compared to the population size.

FIVE-NUMBER SUMMARY (5) The **five-number summary** is a set of five numbers intended to summarize a population distribution; it includes the minimum and maximum data values and the values for quartiles 1, 2, and 3.

FOCUS GROUPS (2) **Focus groups** are interactive sessions with a limited number of research participants; used in **exploratory research.**

FREQUENCY DISTRIBUTION TABLE (3) In a **frequency distribution table,** we count how many times the variable either (1) takes specific values from a finite list of possible values (in a qualitative or discrete quantitative frequency table) or (2) takes values that fall within specified **classes** of values (in a continuous quantitative frequency table).

FREQUENCY POLYGONS (3) **Frequency polygons** are alternatives to histograms that create a smoother image of the distribution's **shape.**

G

GEOMETRIC MEAN (4) The **geometric mean** is a measure of centre that is appropriate for data series that represent rates of change (e.g., interest rates) that accumulate their effects over the course of the data series.

GROUPING VARIABLE (14) In a one-way ANOVA model, the **grouping** (or **factor**) **variable** provides the basis on which the data for the **response variable** are divided into separate samples or groups (called **levels**).

H

HISTOGRAM (3) A **histogram** is a graph that uses the lengths of vertically oriented bars to represent the frequency (or relative frequency) of values in each **class** of a frequency distribution.

HOMOGENEOUS (15) Two or more populations are **homogeneous** in relation to a list of categories if each of the populations has the same distribution of proportions of individuals falling into each category.

HYPERGEOMETRIC EXPERIMENT (7) A **hypergeometric experiment** is an experiment that consists of randomly drawing a sample of size n from a population of size N, without replacement, and counting the number of items that have a certain property (i.e., are "successes").

HYPERGEOMETRIC RANDOM VARIABLE (7) The variable that counts the number of successes (regardless of exact sequence) among the sampled individuals in a hypergeometric experiment is called a **hypergeometric random variable.**

HYPOTHESIS (11) A **hypothesis** is a statement or claim made specifically in order to test that statement or claim, using a formal procedure for **hypothesis testing.**

HYPOTHESIS TESTING (11) **Hypothesis testing** is a set of procedures taken to formally evaluate a **hypothesis.**

I

INDEPENDENT EVENTS (6) Two events are **independent events** if the probability that one of the events will occur is not affected by the other event's occurrence or nonoccurrence, and vice versa.

INDEPENDENT SAMPLES (12) Samples are **independent samples** if the selection of elements for one sample is not influenced by the selection of elements for the other.

INDEPENDENT VARIABLE (16) In a regression model, the variable used as the basis for making an estimate or prediction is the **independent** (or **predictor**) **variable.**

INFERENCE (1) Making an **inference** is drawing a conclusion about a parameter based on the value of a statistic.

INFERENTIAL STATISTICS (1) In **inferential statistics,** we make **inferences** about the characteristics of a population—either *estimating* a characteristic of a population or *testing a claim or assumption* about a characteristic of a population—based on information in one or more samples.

INFINITE POPULATION (1) In an **infinite population,** it is either not possible or not realistic to count every individual member.

INFLUENTIAL OBSERVATIONS (16) **Influential observations** are outliers that if included in a regression model can have a major (and often misleading) effect on the slope or slopes of the model.

INTERACTION (14) There is **interaction** between the factors in an ANOVA model if some of the total variance can be explained by *combinations* of the values of the two factors.

INTERNAL SECONDARY DATA (2) **Internal secondary data** are data originally collected in one's own organization, but not for purposes of one's current research.

INTERQUARTILE RANGE (5) The **interquartile range** is a **measure of spread,** calculated as the distance between the first and third **quartiles** for the data set.

INTERVAL ESTIMATE (10) An **interval estimate** gives a range of values within which the true value of the population parameter is expected to lie.

INVERSE CORRELATION (15) If there is an **inverse correlation** between the values of two associated variables, then as the value of one variable increases, the corresponding value of the other variable decreases.

IRREGULAR COMPONENT (17) If a time series is **decomposed** into its components, the **irregular component** represents unanticipated changes in the value of the variable due to unusual or irregular events.

J

JOINT EVENTS (6) **Joint** (or **conjunctive**) **events** are compound events that occur only if all of the named simple events occur; in notation, "*A* and *B*" is a joint event, based on the simple events *A*, *B*.

JUDGMENT SAMPLING (2) **Judgment sampling** is a nonstatistical sampling technique that focuses on cases or individuals who the researcher believes can provide some key information.

K

KURTOSIS (5) **Kurtosis** is the relative degree to which a distribution is tall or flat, compared to a smooth bell-shaped curve drawn on the same scale.

L

LAGGED VARIABLE (17) For time series data, a **lagged variable** takes its value *following* the time period when an event actually occurs, because the impact on the dependent variable does not occur until that later time.

LAW OF LARGE NUMBERS (6) Suppose an experiment has a classical probability p that result R will occur. According to the **Law of Large Numbers,** if the experiment is repeated n times, the empirical probability of result R (which is the relative frequency of R in the n experiments) tends to approach the classical probability of R as the number of experiments n increases.

LEVEL OF MEASUREMENT (1) The **level of measurement** of data—on a scale from nominal to ratio data—is an indicator of how much potential information the data values contain.

LEVEL OF SIGNIFICANCE (11) **Level of significance (α)** of a hypothesis test expresses how small the probability is that a test statistic will fall into the critical region if the null hypothesis is really true.

LEVELS (14) In a one-way ANOVA model, the **levels** are the different groups into which the response variable data are divided, based on the values of the **grouping** (or **factor**) **variable.**

LINE OF BEST FIT (16) [*See* **regression line**]

LINEAR CORRELATION (15) A **linear correlation** is an **association** between two quantitative variables that follows a linear pattern in a **scatter diagram.**

LOCATION (4) Measures of **location** for a data set give a sense of where the centre or other portions of the data set lie in relation to a number line.

M

MAIN EFFECTS (14) In two-factor ANOVA, the **main effects** are the effects of each factor individually on the **response variable.**

MARGIN OF ERROR (10) In constructing a confidence interval for a parameter, the **margin of error** specifies the ± distance from the point estimate for the parameter to the respective boundaries of the confidence interval.

MARGINAL PROBABILITIES (6) **Absolute probabilities** of events are also called **marginal probabilities,** based on their relationship to numbers on the margins of contingency tables.

MATCHED-PAIRS APPROACH (12) The **matched-pairs approach** is one way to collect a pair of **dependent samples;** items are selected in pairs, one each from a predetermined set of categories.

MEAN ABSOLUTE DEVIATION (MAD) (5, 17) The **mean absolute deviation** is a measure of the effectiveness of a predictive model based on taking the mean of the absolute values of all of the **residuals** for the model.

MEAN SQUARED ERROR (MSE) (17) The **mean squared error** is a measure of the effectiveness of a predictive model based on taking the mean of the squared values of all of the **residuals** for the model.

MEASURE OF CENTRE (4) A **measure of centre** for a data set gives a sense of where the centre of the data set lies in relation to a number line.

MEASURE OF DISPERSION (5) [*See* **measure of spread**]

MEASURE OF SPREAD (5) A **measure of spread** (or **dispersion** or **variation**) is a single number that describes the degree of variety among the values of a distribution.

MEASURE OF VARIATION (5) [*See* **measure of spread**]

MEASUREMENT ERROR (2) **Measurement error** is the type of nonsampling error that occurs if one incorrectly reads or writes a measurement value, uses a poorly calibrated instrument, etc.

MEDIAN (4) The **median** is a measure of centre that identifies the middle value if the data are arranged into an ordered array.

MODAL CLASS (4) If continuous quantitative data are grouped into a frequency distribution, the class having the largest frequency is the **modal class.**

MODE (4) The **mode** is the particular value (or values—in the case of ties) that occurs most frequently in the data set.

MONOTONIC (15) The relationship between two variables is **monotonic** if, as you steadily increase the value (or rank) for one variable, the value (or rank) of the other variable increases (or decreases) consistently.

MULTIPLE-COMPARISON TESTS (14) **Multiple-comparison** (or **multiple-range) tests** are procedures to determine which mean or means are different from the others if in an ANOVA test, the null hypothesis of equal population means has been rejected.

MULTIPLE-RANGE TESTS (14) [*See* **multiple-comparison tests**]

MULTIPLICATIVE MODEL (17) When a time series is **decomposed,** a **multiplicative model** presumes that each value in the time series can be explained as the *product* of the four identified components.

MUTUALLY EXCLUSIVE (3, 6) Two events are **mutually exclusive** if there is no possible outcome that could count as both of the events having occurred.

N

NEGATIVE CORRELATION (15) [*See* **inverse correlation**]

NON-BIASED INDICATOR (9) An indicator or **estimator** is **non-biased** if the sample statistic has an equal probability of being larger *or* smaller than the parameter to be estimated.

NONPARAMETRIC (13) Hypothesis tests known as **nonparametric** (or **distribution-free**) require fewer assumptions than do parametric tests about the distribution of the population's data.

NONRESPONSE ERROR (2) **Nonresponse error** is the type of nonsampling error that occurs if cases that are not included in the sample tend to have different properties from cases that are included.

NONSAMPLING ERROR (2) **Nonsampling error** has occurred if the difference between information in the sample and in the population is due to missing data or incorrect measurement.

NONSTATISTICAL SAMPLING TECHNIQUES (2) **Nonstatistical sampling techniques** are ways to collect data that are not based on the randomizing strategies that are fundamental for **statistical sampling.**

NORMAL CURVE (8) The **normal curve** (often referred to informally as a ***bell curve***) is the familiar type of symmetrical probability density curve in which most values are clustered near the mean and the frequencies drop off after two or more standard deviations from the mean, in either direction.

NORMAL PROBABILITY PLOT (8) A **normal probability plot** compares the actual distribution of all data values to the distribution you would *expect* the values to follow *if* the data were normally distributed.

NULL DISTRIBUTION (11) The **null distribution** is the expected sampling distribution of the test statistic if the **null hypothesis** is true.

NULL HYPOTHESIS (H_0) (11) In a hypothesis test, the **null hypothesis** (in symbols: H_0) is logically opposed to the **alternative hypothesis** and always includes the concept of equality (i.e., the logical operator: $=$ or $\leq$ or $\geq$).

NUMERICAL RANDOM VARIABLE (1) [*See* **quantitative random variable**]

O

OBSERVATION (2) Personal or mechanical **observation** of what is actually happening is one of the ways to collect primary data.

OBSERVED FREQUENCIES (15) **Observed frequencies** are the actual number of cases that fall into each cell of a contingency table (or of a frequency distribution), regardless of what numbers were expected.

OGIVES (3) **Ogives** are line-drawing equivalents to cumulative frequency histograms that create a smoother image for the data pattern.

ONE-FACTOR REPEATED MEASURES (14) **One-factor repeated measures** is a variation of the randomized block design for ANOVA, except that the successive treatments are applied to the same individuals.

ONE-WAY ANOVA (ALSO CALLED ONE-FACTOR ANOVA) (14) **One-way** (or **one-factor) ANOVA** presumes that in the data set there are two variables: (1) the dependent variable and (2) a grouping (or factor) variable, on the basis of which the data in (1) are divided into separate samples or groups and the group means are tested for difference by analysis of variance.

OPERATING CHARACTERISTIC CURVE (11) An **operating characteristic curve** displays—over a range of possible distances between the actual and hypothesized population parameters—the values of β corresponding to each potential distance.

ORDINARY LEAST SQUARES (16) **Ordinary least squares** is the basic procedure to find a regression equation based on minimizing the sum of squared **residuals** between actual y-values and the corresponding predicted $\hat{y}$-values on the regression line.

OUTCOME (6) Each distinct observation that results from an **experiment** is called an **outcome.**

OUTLIERS (3) **Outliers** in the data are values that appear to be remote from all or most of the other values for the variable.

P

P-VALUE (11) The ***p*-value** is the conditional probability of obtaining a **test statistic** at least as far from the expected parameter as has occurred, if in fact the sampling distribution based on H_0 really applies.

PAIRED SAMPLES (12) [*See* **dependent samples**]

PARAMETER (1) A **parameter** is a value for summarizing the measurements of a quantifiable aspect of the population (such as a proportion).

PARETO DIAGRAMS (3) **Pareto diagrams** are similar to **histograms,** but designed to display all of the relative frequencies in decreasing order.

PERCENT FREQUENCY DISTRIBUTION (3) A **percent frequency distribution** shows for each distinct value (or class) of the variable its percentage of occurrence relative to the total number of observations.

PERCENTILES (4) **Percentiles** are values that divide an ordered data set into 100 equal parts.

PERMUTATIONS (6) **Permutations** are distinct sequences of k objects that could be selected from a usually larger set of n objects, wherein the order of selection does matter.

PIE CHART (3) A **pie chart** displays the relative frequencies of classes in a frequency distribution by means of the relative sizes of "slices" of a circle.

PILOT STUDIES (2) **Pilot studies** are preliminary forms of research, usually conducted with small samples as part of **exploratory research.**

POINT ESTIMATE (10) A **point estimate** is a single value intended to approximate the true corresponding value of a population parameter.

POISSON PROBABILITY DISTRIBUTION (7) The **Poisson probability distribution** is a type of model for the expected distribution of the number of *occurrences* of an object or circumstance over a stretch of time or space; for example, the number of arrivals at an Emergency Department in Ontario over the course of a minute or an hour.

POPULATION (1) A **population** is the total set of objects that are of interest to a decision maker.

POSITIVE CORRELATION (15) If there is a **positive correlation** between the values of two associated variables, then if the value of one variable increases, the corresponding value of the other variable increases as well.

POSTERIOR PROBABILITY (6) When applying **Bayes' Rule,** the **posterior probability** of an event A is its conditional probability $P(A \mid B)$—once it *becomes known* that the condition (B) has in fact occurred.

POWER (11) The **power** of a hypothesis test measures the test's ability to detect a difference of the actual population parameter from its presumed value in the null hypothesis; mathematically its value is $1 - \beta$, where β is the probability of a **type II error.**

POWER CURVE (11) The **power curve** is the complement of the **operating characteristic curve** and shows, for any distance between the true and hypothesized parameters, the corresponding probability $(1 - \beta)$ of correctly rejecting the **null hypothesis.**

PRECISION (1, 4) The **precision** of recorded measurements refers to the arbitrary limit that has been set for the number of decimal places or significant digits to be displayed for the numerical values.

PREDICTION INTERVAL (16) The **prediction interval** is an interval estimate for the value of the **dependent variable** in a regression, given a specific value for *x*.

PREDICTOR VARIABLE (16) [*See* **independent variable**]

PRE-TEST/POST-TEST METHOD (12) [*See* **before/after method**]

PRIMARY DATA (2) **Primary data** are source data for a research effort that are collected specifically for that research.

PRIOR PROBABILITIES (6) When applying **Bayes' Rule,** the **prior probability** of an event *A* is its **absolute** probability, assessed or estimated without reference to possible conditions.

PROBABILITY (6) The **probability** of an **event** is the proportion of times that the event is expected to occur—presuming that the random experiment that produces the event was repeated a large number of times.

PROBABILITY DENSITY CURVE (8) A **probability density curve** is a graphical model for the **probability distribution** of a continuous quantitative variable that looks similar to a smoothed frequency polygon, wherein the *relative area* under any part of the curve corresponds to the relative percentage of the total observations that are expected to fall within the boundaries for that area on the *x*-axis.

PROBABILITY DENSITY FUNCTION (8) The formula that determines the ordinate (*y*) value of the **probability density curve** for each *x*-value in the sample space is a **probability density function.**

PROBABILITY DISTRIBUTION (7) The distribution of all possible values of a discrete variable, together with the probabilities that each of these values will occur is a **probability distribution;** for a continuous variable, not all possible values can be listed, so probabilities may be given in relation to **classes** of *x*-values.

PROBABILITY SAMPLING (2) [*See* **statistical sampling**]

PROBABILITY TREES (6) **Probability trees** are tree-like diagrams that can help you visualize the conditional relationships among probabilities and guide calculations.

PROXY VARIABLE (16) If an important variable has been omitted from a regression analysis, a **proxy variable** is an included variable that may stand in for the missing variable (due to its connection with it) and possibly have an unexpected value for its coefficient.

Q

QUALITATIVE FREQUENCY DISTRIBUTION TABLE (3) A **qualitative frequency distribution table** displays how often each particular value or category of a qualitative variable has occurred.

QUALITATIVE RANDOM VARIABLE (1) For a **qualitative (or categorical) random variable,** the observed values are not numerical; they are traits or characteristics that can be classified into one of a number of categories.

QUALITATIVE STATISTICS (2) **Qualitative statistics** are methods based on **nonstatistical sampling techniques** that aim to provide a deeper view of how people are thinking in certain contexts and how they relate their ideas.

QUANTILES (4) **Quantiles** are values that divide an ordered distribution of data values into a specified number of equal parts.

QUANTITATIVE RANDOM VARIABLE (1) For a **quantitative (or numerical) random variable,** the observed values are numbers that can vary in magnitude.

QUARTILES (4) **Quartiles** are values that divide an ordered distribution of data values into four equal parts.

QUINTILES (4) **Quintiles** are values that divide an ordered distribution of data values into five equal parts.

QUOTA SAMPLING (2) **Quota sampling** is a **nonstatistical sampling** technique that focuses on cases or individuals who the researcher believes can provide some key information.

R

RANDOM EXPERIMENT (6) A **random experiment** is a planned process of observation to obtain a value for a random variable.

RANDOM SELECTION (2) **Random selection** refers to techniques for selecting samples such that you could not predict exactly which objects will be collected in the sample, as an attempt to ensure that objects of varying characteristics will be selected, by chance, in roughly the same proportions as in the full population.

RANDOM VARIABLE (1) A variable is a **random variable** if its next observed value is not known until it is observed.

RANDOMIZED BLOCK DESIGN (14) In ANOVA, **randomized block design** links related elements in the two samples in order to control for some of the variation among individuals.

RANGE (3, 5) The **range** is the simplest measure of variation, calculated as the difference between the largest and smallest data values.

RANK-ORDERING (OR RANKS) (13) **Rank-ordering** data that are originally measured at the interval or ratio level means to assign to each original value a number, called its **rank,** that indicates the relative position of the data value—presuming all values were sorted from lowest to highest (or vice versa).

RATE OF CHANGE (4) A **rate of change,** like an interest rate, tells by what proportion of an initial value the *next* value increases (or decreases).

RAW DATA (3) Data in the original form in which they are collected and recorded are called the **raw data.**

REGRESSION ANALYSIS (16) **Regression analysis** is the application of **ordinary least squares** procedures to produce a model, in the form of an equation, with which you can estimate or predict the value of one variable based on its (linear or nonlinear) relationship with one or more other variables.

REGRESSION COEFFICIENTS (16) In a linear **regression equation,** the **regression coefficients** ($b_1, b_2, \ldots, b_k$) are the values by which each of the x-variables is to be multiplied: $\hat{y} = b_0 + b_1x_1 + b_2x_2 + \ldots + b_kx_k$. (In *nonlinear* regression, the coefficients may play different roles in the model equation, such as $\hat{y} = b_0 \cdot e^{b1x1}$ in exponential regression.)

REGRESSION EQUATION (16) The output of a **regression analysis** is a **regression equation** that models algebraically the relationship between the dependent and independent variables in the **line of best fit** (i.e., the **regression line**).

REGRESSION LINE (16) The linear **regression line** (or **line of best fit**) is the one line for a given data set that minimizes the sum of squared **residuals** between actual y-values and the corresponding, predicted $\hat{y}$-values that fall on that regression line.

RELATED SAMPLES (12) [*See* **dependent samples**]

RELATIVE FREQUENCY PROBABILITY (6) [*See* **empirical probability**]

RELATIVE VARIATION (5) A measure of **relative variation** compares the sizes of the variation in each of two or more data sets in relative terms; e.g., as percentages of the respective data sets' means.

REPEATED MEASURES METHOD (12) [*See* **before/after method**]

REPRESENTATIVE (1) A sample is **representative** of the population it is drawn from if the statistics taken from the sample reasonably approximate the corresponding population parameters.

RESAMPLING (13) In **resampling,** the sample data are taken to provide the estimate for the population distribution, and the sampling distribution for the test statistic is approximated by repeated sampling from the original sample—each time recording the relevant statistic.

RESEARCH OBJECTIVES (2) The **research objectives** state the purpose for the study in clear, unambiguous terms.

RESEASONALIZE (17) If **deseasonalized** data have been used to estimate a future value, the final estimate (for a multiplicative model) requires that the raw output be **reseasonalized** by multiplying it times the seasonal index for the time period being estimated.

RESIDUAL (16) In regression, the distance between an actual value of y and the corresponding expected value ($\hat{y}$) for a given x-value is called a **residual.**

RESIDUAL PLOTS (16) **Residual plots** are a variety of graphic tools for assessing the conformity of a data set to the assumptions for regression; all are based on comparing **residuals** to the corresponding actual or expected values of other variables.

RESPONSE VARIABLE (16) [*See* **dependent variable**]

ROBUST (8) A statistical method is **robust** if it reaches reasonably accurate conclusions about the data, even if its assumptions about the data set do not fully apply.

S

SAMPLE (1) A **sample** is a subset of the population that is observed and studied to provide information about the overall population.

SAMPLE COEFFICIENT OF CORRELATION (r) (15) The **sample coefficient of correlation** (r) provides (1) a measure of the strength of the linear correlation (the strongest being -1 or $+1$) and (2) a sign ($+$ or $-$) to indicate a **positive correlation** or an **inverse correlation,** respectively.

SAMPLE SPACE (6) In probability for a given **experiment,** the set of all possible **outcomes** is called the **sample space.**

SAMPLE STATISTIC (1, 11) A **sample statistic** is a summary measure for the sample that corresponds to (and can help to estimate) a parameter of the population; for hypothesis tests, the sample statistic is converted into the appropriate standardized **test statistic.**

SAMPLE WITH REPLACEMENT (7) To **sample with replacement** means that after selecting one item from the population, you return it to the sample space so that it could potentially be selected again.

SAMPLING DISTRIBUTION (9) The **sampling distribution** is a probability distribution for a sample statistic, such that if every possible sample of a given size was taken from the population and the statistic was calculated for each sample, the sampling distribution would be the distribution of that statistic as taken from all the samples.

SAMPLING ERROR (2) **Sampling error** is the difference in value between a sample statistic and the corresponding population parameter that occurs merely because the sample is only a subset of the population.

SAMPLING FRAME (2) The **sampling frame** is the list of the individuals or objects from which the sample will be drawn.

SCATTER DIAGRAMS (3) **Scatter diagrams** are figures for comparing two quantitative variables, wherein the position of each dot displays simultaneously the values of *both* variables for *one* individual in the sample.

SCIENTIFIC APPROACH (2) A **scientific approach** to data analysis begins with the recognition of a problem to investigate and development of a research plan for obtaining the relevant data.

SEASONAL COMPONENT (17) The **seasonal component** of a **decomposed** time series reflects short-term variations in the value of the variable that are associated with particular, and recurring, divisions of time.

SEASONAL DECOMPOSITION (17) **Seasonal decomposition** refers specifically to **decomposing** a time series into its **trend** and **seasonal components** for purposes of making a forecast.

SEASONAL INDEX (ALSO CALLED SEASONAL RELATIVE) (17) A **seasonal index** (or **seasonal relative**) represents—for **deseasonalizing** a time series—the **seasonal component** corresponding to one particular "season"; for each of the seasons in the full cycle (e.g., for each month in a year) there are distinct **seasonal indexes.**

SEASONALLY ADJUSTED VALUES (17) [*See* **deseasonalized values**]

SECONDARY DATA (2) **Secondary data** are source data for a research effort that have been collected originally for some other purpose.

SHAPE (3, 5) The **shape** of a distribution refers to what characteristic pattern the data exhibit (such as **symmetrical** or **skewed**) if graphed in the form of a **histogram.**

SIGNIFICANT CORRELATION (15) A correlation is **significant** if there is sufficient evidence that the apparent correlation in the sample represents a true population parameter and not just a random sampling effect.

SIMPLE CLUSTER SAMPLING (2) **Simple cluster sampling** is a sampling technique for populations that are widely dispersed, wherein the population is divided into clusters, and samples can then be taken from randomly selected clusters.

SIMPLE EVENT (6) In probability, a **simple event** can be named or specified without reference to other defined events.

SIMPLE RANDOM SAMPLING (2) **Simple random sampling** is a way of sampling such that each object in the sampling frame has an equal chance of being selected, as does each possible *combination* of objects of a given size.

68–95–99% RULE (5) [*See* **empirical rule**]

SKEWED (3) The **shape** of a **skewed** distribution is such that one of the tails appears to retain thickness (i.e., frequencies greater than zero) for a greater distance from the mean compared to the other tail.

SKEWNESS (5) **Skewness** is a measure of the extent to which a distribution is **skewed.**

SLOPE (16) For a **regression line** in simple linear regression, the **slope** is the constant ratio between the rise (change in vertical distance on the y-axis) over the run (change in horizontal distance on the x-axis) between any pair of points on the line.

SPEARMAN (OR SPEARMAN'S) RANK CORRELATION COEFFICIENT, r_s (15) For bivariate data that are ordinal *or* have been transformed into ranks, the **Spearman** (or **Spearman's**) **rank correlation coefficient,** r_s, provides a measure of the strength of the **association** between the two variables' ranks (the strongest being -1 or $+1$), and provides a sign ($+$ or $-$) to indicate a **positive correlation** or an **inverse correlation** between their ranks.

SSB (14) [*See* **sum of squares between**]

SSW (14) [*See* **sum of squares within**]

STANDARD DEVIATION (5) The **standard deviation** is a common **measure of spread** based on the squares of the deviations of each data value from the mean.

STANDARD ERROR OF THE ESTIMATE (16) In regression, the **standard error of the estimate** is a measure of the variability of the **response variable** around the **regression line;** it is calculated as the root of [(the sum of squared residuals) $\div$ (n $-$ 2)].

STANDARD NORMAL DISTRIBUTION (8) A normal distribution for which the mean $= 0$ and the standard deviation $= 1$ is the **standard normal distribution.**

STATISTICAL DECISION FOR THE HYPOTHESIS TEST (11) The **statistical decision** for a hypothesis test is a formal step in which (based on either **critical values** or a ***p*-value**), you either *reject* or *fail to reject* the **null hypothesis.**

STATISTICAL SAMPLING (2) Methods of collecting data that are based on **random selection** in order to ensure that the collected samples are **representative** of the population are called **statistical sampling** (or **probability sampling**) methods.

STATISTICS (1) **Statistics** is the art and process of collecting, classifying, interpreting, and reporting numerical information in relation to a particular subject.

STEMPLOTS (OR STEM-AND-LEAF DISPLAYS) (3) **Stemplots** (or **stem-and-leaf displays**) group quantitative data into classes in a manner that retains information about the data within the classes; part of each number is identified as the stem and the next digit to the stem's right is the leaf.

STEPS (5) [*See* **fences**]

STEPWISE REGRESSION (16) **Stepwise regression** is a systematic approach for adding (or removing) variables in order to construct the optimum multiple regression model.

STRATIFIED RANDOM SAMPLING (2) **Stratified random sampling** is a method for collecting samples in which the population is divided into subpopulations (called *strata*) of interest (e.g., men versus women); simple random samples are then taken from each stratum.

STUDENT'S *T*-DISTRIBUTION (10) [*See* ***t*-distribution**]

SUBJECTIVE PROBABILITY (6) In **subjective probability,** a probability value is assigned to an event based on individual judgment, after taking into account experience and available evidence.

SUM OF SQUARES BETWEEN GROUPS (*SSB*) (14) In one-way ANOVA, the **sum of squares between groups** (***SSB***) (also called *sum of squares among* (*SSA*) or *sum of squares explained*) is a measure for the amount of variation explained by distinguishing groups of data based on the **grouping** variable.

SUM OF SQUARES BETWEEN TREATMENTS (*SSTr*) (14) In a **randomized block design** for ANOVA, the **sum of squares between treatments** (***SSTr***) is a measure for the amount of variation explained by distinguishing groups of data based on the experimental **treatment** that was applied.

SUM OF SQUARES BLOCKS (*SSBlocks*) (14) In a **randomized block design** for ANOVA, the **sum of squares blocks** (***SSBlocks***) is a measure for the amount of variation explained by distinguishing groups of data based on the **blocks** to which treatments were applied.

SUM OF SQUARES WITHIN (*SSW*) *OR* SUM OF SQUARES ERROR (*SSE*) (14) In ANOVA, the **sum of squares within groups** (SSW) (also called ***sum of squares error*** (***SSE***) or *sum of squares unexplained*) is a measure for the amount of variation left unexplained after the effects of all factors, blocks, and interactions are accounted for.

SURVEYS (2) **Surveys** are techniques such as telephone interviews or mail-in questionnaires for obtaining primary data.

SYMMETRICAL (5) The **shape** of a distribution is **symmetrical** when, if we could fold up a picture of the distribution (e.g., its histogram) along a centre line, the figure would line up with itself on each side.

SYSTEMATIC RANDOM SAMPLING (2) **Systematic random sampling** is a method of data collection based on taking every *k*th object (from a random starting point) from a list of members of the sampling frame.

T

TABLE OF OUTCOMES (6) For calculating probabilities, a **table of outcomes** is a display of the possible **outcomes** of the **experiment,** in which outcomes viewed as the **event** can also be identified and counted.

T-DISTRIBUTION (10) The ***t*-distribution** (or **Student's *t*-distribution**) is the appropriate model for the sampling distribution of the sample means when the population σ is not known—provided the sample is large or the population is normally distributed; there is family of *t*-distributions and each one is specified by its **degrees of freedom.**

TEST FOR GOODNESS OF FIT (15) A **test for goodness of fit** assesses the correspondence between an observed frequency distribution and our expectation of what that distribution might be.

TEST FOR HOMOGENEITY (15) A **test for homogeneity** determines whether two or more populations are **homogeneous** in relation to the proportions of their members that fall into various categories.

TEST FOR INDEPENDENCE (15) A **test for independence** determines whether samples presumably taken from two populations are **independent samples.**

TEST STATISTIC (11) The **test statistic** represents the value of the **sample statistic**—after its conversion to a standardized score, according to the model of the sampling distribution (e.g., a *t*- or *z*-value) assuming H_0 is true.

TIME SERIES (17) A **time series** is an ordered set of observations for a variable made at regular intervals over time.

TIME SERIES ANALYSIS (17) **Time series analysis** is a general term for techniques designed to make estimates or predictions based on the data in a **time series.**

TIME SERIES CHART (3) A **time series chart** is a type of **scatter diagram** wherein the variable measured horizontally is time (for example, in periods of months or years), and the second variable is either a discrete count of some quantity or a continuous value.

TREATMENT (12) A **treatment** is an experimental procedure that is applied to a single population in a **clinical trial.**

TREATMENT GROUP (12) A **treatment group** (or **experimental group**) is a sample within a **clinical trial** to which an experimental procedure or **treatment** is applied.

TREND COMPONENT (17) The **trend component** of the time series represents those changes in the value of the variable that reflect a pattern of long-term growth or decline.

TRIAL (7) [*See* **Bernoulli trial**]

TWO-FACTOR ANOVA (14) With **two-factor ANOVA,** we examine the effects on a **response variable** of two factors, each having several possible values or **levels.**

TYPE I ERROR (11) A **Type I error** occurs if H_0 happens to be true and yet the null hypothesis is rejected.

TYPE II ERROR (11) A **Type II Error** has occurred if the null hypothesis (H_0) is in fact false but we mistakenly failed to reject H_0.

TYPES OF PROBABILITY (6) The three basic types of probabilities are classical probability, empirical probability, and subjective probability.

U

UNBIASED (10) An estimator for a population parameter is **unbiased** if the mean of its sampling distribution is equal to the true parameter.

UNIFORM PROBABILITY DISTRIBUTION (8) A **uniform probability distribution** is a continuous probability distribution whose **probability density curve** appears rectangular; the **probability density function** returns the same value for any value of x.

UNIVARIATE DESCRIPTIVE STATISTICS (3) Univariate descriptive statistics is descriptive statistics for just one variable.

V

VARIABLES (1) **Variables** are that form of data whose values may change from observation to observation.

VARIANCE (5) The standard deviation squared is the variance.

W

WEIGHTED ARITHMETIC MEAN (4) The **weighted arithmetic mean** is the mean calculated for a set of data in which individual values are assigned unequal importance.

X

X-COEFFICIENT (16) For **simple linear regression,** the ***x*-coefficient** is the **regression coefficient** for the single independent variable.

Y

Y-INTERCEPT (16) For **simple linear regression,** the ***y*-intercept** (b_0) defines the expected value for the **response variable** if the value of $x = 0$.

Z

Z-SCORE (OR *Z*-VALUE) (5, 8) A ***z*-score** (or ***z*-value**) restates the value of a variable in terms of how many standard deviations the value is above or below the mean of the distribution for that variable.

Key Formulas

Chapter 4 Measures of Location

FORMULA 4.1 **Arithmetic Mean for a Population**

$$\mu = \frac{\sum_{i=1}^{N} x_i}{N}$$

where
μ = Arithmetic mean for a population (pronounced "mu")
x_i = Individual values of the population
N = Number of values in the population (population size)
$\sum_{i=1}^{N}$ is a mathematical operator for summation (adding).
$\sum_{i=1}^{N} x_i$ means: Add all of the x_i-values in a data set or column, from the first value ($i = 1$) to the last ($i = N$).

FORMULA 4.2 **Arithmetic Mean for a Sample**

$$\bar{x} = \frac{\sum_{i=1}^{n} x_i}{n}$$

where
$\bar{x}$ = Arithmetic mean for a sample (The symbol is pronounced "x bar." The x can be replaced with the name of any variable to be averaged.)
x_i = Individual values of the *sample*
n = Number of values in the sample (sample size)
$\sum_{i=1}^{n} x_i$ means: Add all of the values (in this case, the *sample* values) in a data set or column from the first value ($i = 1$) to the last ($i = n$).

FORMULA 4.3 **Weighted Arithmetic Mean**

3a) For a Population

$$\mu = \frac{\sum_{i=1}^{N} x_i w_i}{\sum_{i=1}^{N} w_i}$$

3b) For a Sample

$$\bar{x} = \frac{\sum_{i=1}^{n} x_i w_i}{\sum_{i=1}^{n} w_i}$$

where
μ = Arithmetic mean for a population ***or*** $\bar{x}$ = Arithmetic mean for a sample
x_i = Individual values of the population or sample
w_i = Corresponding weights assigned to the data values
N = Number of values in the population (population size) ***or*** n = Number of values in the sample (sample size)
$\sum_{i=1}^{N}$ *or* $\sum_{i=1}^{n}$ is the mathematical operator for summation (adding).

FORMULA 4.4 **The Geometric Mean**

$$GM = \sqrt[n]{x_1 \cdot x_2 \cdot x_3 \cdot \ldots \cdot x_n}$$

where GM = Geometric mean
n = Number of observations in the series, such as the number of years over which the growth rate is being assessed
x_i = Growth factor for a particular period in the series (e.g., x_2 = Growth factor for the second period)

Chapter 5 Measures of Spread and Shape

FORMULA 5.1 **Variance for a Population**

$$\sigma^2 = \frac{\sum_{i=1}^{N}(x_i - \mu)^2}{N}$$

where σ^2 is called the variance of the population.
x_i = Individual observation in the data set
μ = Mean of the population
N = Number of values in the population

FORMULA 5.2 **Standard Deviation for a Population**

$$\sigma = \sqrt{\sigma^2}$$

where σ (pronounced "sigma") is the population standard deviation, to be calculated.
σ^2 is called the variance of the population.

FORMULA 5.3 **Standard Deviation for a Population**

$$\sigma = \sqrt{\frac{\sum_{i=1}^{N}(x_i - \mu)^2}{N}}$$

FORMULA 5.4 **Variance for a Sample**

$$s^2 = \frac{\sum_{i=1}^{n}(x_i - \bar{x})^2}{(n - 1)}$$

FORMULA 5.5 **Standard Deviation for a Sample**

$$s = \sqrt{s^2}$$

FORMULA 5.6

Standard Deviation for a Sample

$$s = \sqrt{\frac{\sum_{i=1}^{n} (x_i - \overline{x})^2}{(n - 1)}}$$

where s is the *sample* standard deviation, to be calculated.
s^2 is called the *variance of the sample.*
x_i = Individual observation in the data set
$\overline{x}$ = Mean of the sample
n = Number of values in the sample

FORMULA 5.7

Converting a Data Value to a *z*-Score (for a population)

$$z = \frac{x - \mu}{\sigma}$$

FORMULA 5.8

Converting a Data Value to a *z*-Score (for a sample)

$$z = \frac{x - \overline{x}}{s}$$

FORMULA 5.9

Coefficient of Variation

$$CV = \frac{\sigma}{\mu} \times 100\%$$

where CV is the coefficient of variation.
σ is the population standard deviation (or use s if working from a sample).
μ is the population mean (or use $\overline{x}$ if working from a sample).

Chapter 6 Concepts of Probability

FORMULA 6.1

Calculation of a Classical Probability

$$P(A) = \frac{\text{number of ways A can occur}}{\text{number of different outcomes possible}}$$

where A = An event of interest
$P(A)$ = "The probability of the event A occurring," ***and***
$0 < P(A) < 1$, ***and***
"The number of ways A can occur" is the number of possible outcomes that would be considered cases of A occurring, ***and***
Both the number of ways A could occur (the numerator) and the number of different outcomes possible (the denominator) can be determined by the definitions of objects involved in the calculation.

FORMULA 6.2

Calculation of Empirical Probability

$$P(A) = \frac{(\text{number of ways } A \text{ did occur})}{(\text{number of different outcomes possible})}$$

where A = An event of interest
$P(A)$ = "The probability of event A occurring," ***and***
One "outcome" is the result of one observation or experiment, among the set of n observations or experiments that were conducted, ***and***
The number of ways A did occur (the numerator) is the number of outcomes that historically met the conditions for being considered cases of "A occurred," ***and***
The size of the sample space (i.e., the number of different outcomes possible) (the denominator) equals the sample size n (i.e., the number of observations or experiments conducted).

FORMULA 6.3

Complementary Event Rule 1

$$P(A) + P(\overline{A}) = 1$$

where A represents any event, and $\overline{A}$ is the complement of that event. Combined, the mutually exclusive events A and $\overline{A}$ represent every possible outcome in the sample space, so the sum of their two probabilities adds to 1.

FORMULA 6.4

Complementary Event Rule 2

$$P(A) = 1 - P(\overline{A})$$

FORMULA 6.5

Complementary Event Rule 3

$$P(\overline{A}) = 1 - P(A)$$

FORMULA 6.6

Multiplication Rule for *Independent* Events *A* and *B*

$$P(A \text{ and } B) = P(A) \cdot P(B)$$

FORMULA 6.7

Multiplication Rule for *Dependent* Events *A* and *B*

$$P(A \text{ and } B) = P(A) \cdot P(B|A)$$

FORMULA 6.8

Addition Rule for *Mutually Exclusive* Events *A* and *B*

$$P(A \text{ or } B) = P(A) + P(B)$$

FORMULA 6.9

Addition Rule for *Not Mutually Exclusive* Events *A* and *B*

$$P(A \text{ or } B) = P(A) + P(B) - P(A \text{ and } B)$$

FORMULA 6.10 **The Core Idea of Bayes' Rule**

$$P(A|B) = \frac{P(B|A) \cdot P(A)}{P(B)}$$

FORMULA 6.11 **Basic Bayes' Rule**

$$P(A|B) = \frac{P(B|A) \cdot P(A)}{P(B|A) \cdot P(A) + P(B|\overline{A}) \cdot P(\overline{A})}$$

FORMULA 6.12 **Extended Formal Version**

$$P(A_i|B) = \frac{P(B|A_i) \cdot P(A_i)}{P(B|A_1) \cdot P(A_1) + P(B|A_2) \cdot P(A_2) + \ldots + P(B|A_k) \cdot P(A_k)}$$

Chapter 7 Discrete Probability Distributions

FORMULA 7.1 **Expected Value of a Discrete Probability Distribution**

$$E(X) = \sum [xP(x)]$$

where $E(X)$ = Expected value of the random variable
x = A value of the random variable x
$P(x)$ = Probability of occurrence of the value x

FORMULA 7.2 **Variance of a Discrete Probability Distribution**

$$\sigma^2 = \sum [(x - \mu)^2 \times P(x)]$$

where μ is the expected value of the distribution

FORMULA 7.3 **Standard Deviation of a Discrete Probability Distribution**

$$\sigma = \sqrt{\sigma^2}$$

where σ^2 = Variance of a random variable
σ = Standard deviation of a random variable
x = A value of the random variable
μ = Mean, or expected value, of the random variable
$P(x)$ = Probability of occurrence of the value x

FORMULA 7.4 **Binomial Probability of an Outcome**

$$P(x) = {}_nC_x p^x (1 - p)^{(n-x)}$$

or

$$P(x) = \frac{n!}{x!\,(n - x)!} p^x (1 - p)^{(n-x)}$$

where $0 \leq X \leq n$
$0 < p < 1$

FORMULA 7.5

Expected Value of a Binomial Probability Distribution

$$E(X) = \mu = np$$

FORMULA 7.6

Variance of a Binomial Probability Distribution

$$\sigma^2 = np(1 - p)$$

FORMULA 7.7

Hypergeometric Probabilities

$$P(x) = \frac{\dfrac{r!}{x!\,(r-x)!} \times \dfrac{(N-r)!}{(n-x)! \times [(N-r)-(n-x)]!}}{\dfrac{N!}{n!\,(N-n)!}}$$

where
x = Specific number of successes (Example: "draw *2* diamonds")
n = Number of elements in the sample (Example: "draw *2* cards in total")
N = Number of elements in the population (Example: "a special deck with *8* cards")
r = Number of elements in the population with the characteristic of interest. (Example: "*2* diamonds are originally available in the deck")

FORMULA 7.8

Expected Value for a Hypergeometric Distribution

$$E(X) = \frac{nr}{N}$$

where
n = Number of elements in the sample
N = Number of elements in the population
r = Number of elements in the population with the characteristic of interest

FORMULA 7.9

Variance for a Hypergeometric Distribution

$$\sigma^2 = \frac{r(N-r) \cdot n(N-n)}{N^2(N-1)}$$

where
n = Number of elements in the sample
N = Number of elements in the population
r = Number of elements in the population with the characteristic of interest

FORMULA 7.10

Standard Deviation for a Hypergeometric Distribution

$$\sigma = \sqrt{\sigma^2} = \sqrt{\frac{r(N-r) \cdot n(N-n)}{N^2(N-1)}}$$

where
n = Number of elements in the sample
N = Number of elements in the population
r = Number of elements in the population with the characteristic of interest

FORMULA 7.11 **Poisson Probabilities**

$$P(x) = \frac{\mu^x e^{-\mu}}{x!}$$

where μ = Mean or expected number of occurrences per interval (alternatively, this parameter can be represented with the Greek letter lambda: λ)
x = Number of occurrences of a particular outcome
e = A constant, known as *Euler's constant*, which = 2.71828

Chapter 8 Continuous Probability Distributions

FORMULA 8.1 **Probability Density Function for the Normal Curve**

$$f(x) = \frac{1}{\sigma\sqrt{2\pi}} e^{\frac{-(x-\mu)^2}{2\sigma^2}}$$

where μ = Population mean
σ = Population standard deviation
π = The constant 3.14159
e = The constant 2.71828

FORMULA 8.2 **Calculating a *z*-Score**

$$z = \frac{x - \mu}{\sigma}$$

where μ = Population mean
σ = Population standard deviation
x = Actual value of a member of the population, to be converted to a *z*-score

FORMULA 8.3 **Probability Distribution Function for the Continuous Uniform Probability Distribution**

$$f(x) = \frac{1}{b - a}$$

where x = Any value within the sample space for the distribution
b = Upper bound of the distribution
a = Lower bound of the distribution

FORMULA 8.4 **Expected Value for a Continuous Uniform Probability Distribution**

$$E(X) = \mu = \frac{b + a}{2}$$

(Note that the formula simply finds the mean of the upper and lower values of the range—since the curve has a constant height over the range. Graphically, this value is always in the exact centre of the distribution.)

FORMULA 8.5 **Variance of a Continuous Uniform Probability Distribution**

$$\sigma^2 = \frac{(b - a)^2}{12}$$

FORMULA 8.6 **Standard Deviation of a Continuous Uniform Probability Distribution**

$$\sigma = \sqrt{\sigma^2} = \sqrt{\frac{(b - a)^2}{12}}$$

FORMULA 8.7 **Less Than Cumulative Probabilities for a Uniform Distribution**

$$P(X < x) = \frac{x - a}{b - a}$$

FORMULA 8.8 **Upper Bound of an Area Corresponding to a Cumulative Probability in a Uniform Distribution**

$$x = a + \{[P(X < x)] \times (b - a)\}$$

FORMULA 8.9 **Exponential Probability Distribution**

$$f(x) = \lambda e^{-\lambda x}$$

where x = Any value within the sample space for the distribution
λ = The mean or expected number of occurrences of an event within a given time interval (this is the *same parameter* as λ (or μ) encountered for the Poisson distribution)
e = The constant 2.71828

FORMULA 8.10 **Expected Value for an Exponential Probability Distribution**

$$E(X) = \mu = \frac{1}{\lambda}$$

FORMULA 8.11 **Variance for an Exponential Probability Distribution**

$$\sigma^2 = \frac{1}{\lambda^2}$$

FORMULA 8.12 **Standard Deviation for an Exponential Probability Distribution**

$$\sigma = \sqrt{\sigma^2} = \sqrt{\frac{1}{\lambda^2}} = \frac{1}{\lambda}$$

FORMULA 8.13

Less Than Cumulative Probabilities for an Exponential Probability Distribution

$$P(X < x) = 1 - e^{-\lambda x}$$

where
x = Any positive value of time within the sample space
λ = The mean or expected number of occurrences of an event within a given time interval (this is the same parameter as λ (or μ) encountered for the Poisson distribution)
e = The constant 2.71828

Chapter 9 The Distributions of Sample Statistics

FORMULA 9.1

Mean for the Sampling Distribution of the Sample Mean

$$\mu_{\bar{x}} = \mu$$

where
$\mu_{\bar{x}}$ = Mean of the sampling distribution of the sample mean, for a given sample size
μ = Mean of the full population from which the samples are drawn

FORMULA 9.2

Variance for the Sampling Distribution of the Sample Mean

$$\sigma_{\bar{x}}^2 = \frac{\sum_{i=1}^{N} (\bar{x}_i - \mu)^2}{N}$$

FORMULA 9.3

Standard Deviation for the Sampling Distribution of the Sample Mean

$$\sigma_{\bar{x}} = \sqrt{\sigma_{\bar{x}}^2} = \sqrt{\frac{\sum_{i=1}^{N} (\bar{x}_i - \mu)^2}{N}}$$

where
$\sigma_{\bar{x}}^2$ = Variance of the sampling distribution of the sample means
$\sigma_{\bar{x}}$ = Standard deviation of the sampling distribution of the sample means (also called the *standard error* of the sampling distribution)
$\bar{x}_i$ = For each value of i, the mean of the ith possible sample of size n from the population
μ = Mean of the sampling distribution, which is also the mean of the population
N = Number of sample means

FORMULA 9.4

Standard Deviation for the Sampling Distribution of the Sample Mean (Simplified Formula)

$$\sigma_{\bar{x}} = \frac{\sigma}{\sqrt{n}}$$

where
$\sigma_{\bar{x}}$ = Standard deviation of the sampling distribution of the sample means
σ = Standard deviation of the population from which the samples are drawn
n = Sample size for each individual sample in the sampling distribution

FORMULA 9.5 **Revised Version of Formula 9.4 Showing the Finite Population Correction Factor**

$$\sigma_{\bar{x}} = \frac{\sigma}{\sqrt{n}} \underbrace{\sqrt{\frac{N - n}{N - 1}}}_{\text{Finite population correction factor}}$$

where $\sigma_{\bar{x}}$ = Standard deviation of the sampling distribution of sample means
σ = Standard deviation of the population from which the samples are drawn
n = Sample size for each individual sample in the sampling distribution
N = Number of items in the population from which the samples are drawn

Chapter 10 Estimates and Confidence Intervals

$$\bar{X} - Z_{\alpha/2} \frac{\sigma}{\sqrt{n}} < \mu < \bar{X} + Z_{\alpha/2} \frac{\sigma}{\sqrt{n}}$$

FORMULA 10.1 **Formula for a *t*-Value**

$$t = \frac{\bar{x} - \mu}{\frac{s}{\sqrt{n}}}$$

FORMULA 10.2 **Sample-Based Estimate for Standard Deviation of the Sampling Distribution of the Sample Mean**

$$s_{\bar{x}} = \frac{s}{\sqrt{n}}$$

$$\bar{X} - t_{\alpha/2,n-1} \frac{S}{\sqrt{n}} < \mu < \bar{X} + t_{\alpha/2,n-1} \frac{S}{\sqrt{n}}$$

Chapter 11 One-Sample Tests of Significance

FORMULA 11.1 **Sample Size Determination to Construct a Confidence Interval for the Mean**

$$n = \frac{z_{\alpha/2}^2 \sigma^2}{E^2}$$

FORMULA 11.2 **Sample Size Determination to Construct a Confidence Interval for the Proportion**

$$n = \frac{z_{\alpha/2}^2 p(1 - p)}{E^2}$$

Chapter 12 Two-Sample Tests of Significance

FORMULA 12.1 **z-Value Calculation for the Difference in Sample Means**

$$z = \frac{(\bar{X}_1 - \bar{X}_2) - (\mu_1 - \mu_2)}{\sqrt{\frac{\sigma_1^2}{n_1} + \frac{\sigma_2^2}{n_2}}}$$

FORMULA 12.2 **t-Value Calculation for the Difference in Sample Means (Case 1)**

$$\text{a)}\ t = \frac{(\bar{X}_1 - \bar{X}_2) - (\mu_1 - \mu_2)}{\sqrt{\frac{s_p^2}{n_1} + \frac{s_p^2}{n_2}}}$$

where

$$\text{b)}\ s_p^2 = \frac{(\mathrm{n}_1 - 1)s_1^2 + (\mathrm{n}_2 - 1)s_2^2}{(\mathrm{n}_1 - 1) + (\mathrm{n}_2 - 1)}$$

FORMULA 12.3 **Welch's t-Test Calculation for the Test Statistic (Case 2)**

$$t = \frac{(\bar{X}_1 - \bar{X}_2) - (\mu_1 - \mu_2)}{\sqrt{\frac{s_1^2}{n_1} + \frac{s_2^2}{n_2}}}$$

where s_1 and s_2 are the individual variances of the two samples.

FORMULA 12.4 **Mean Difference in Pairs of Values**

$$\bar{d} = \frac{\sum_{i=1}^{n} d_i}{n}$$

FORMULA 12.5 **Variance in Pairs of Values**

$$s_d^2 = \frac{\sum_{i=1}^{n} (d_i - \bar{d})^2}{(n - 1)}$$

FORMULA 12.6 **Standard Deviation in Pairs of Values**

$$s_d = \sqrt{s_d^2} = \sqrt{\frac{\sum_{i=1}^{n} (d_i - \bar{d})^2}{(n - 1)}}$$

FORMULA 12.7 **Standard Error of the Differences**

$$s_{\bar{d}} = \frac{s_d}{\sqrt{n}}$$

FORMULA 12.8 ***t*-Value for a Specific Value Difference**

$$t = \frac{\bar{d} - \mu_D}{\frac{s_d}{\sqrt{n}}}$$

FORMULA 12.9 ***z*-Value for the Difference in Sample Proportions**

$$\text{a) } z = \frac{(p_{s_1} - p_{s_2}) - (p_1 - p_2)}{\sqrt{\frac{\bar{p}(1-\bar{p})}{n_1} + \frac{\bar{p}(1-\bar{p})}{n_2}}}$$

where p_{s_1} and p_{s_2} are the two *sample* proportions
p_1 and p_2 are the two *population* proportions
$\bar{p}$ is the pooled estimate of the proportion based on:

$$\text{b) } \bar{p} = \frac{x_1 + x_2}{n_1 + n_2}$$

x_1 and x_2 are the numbers of successes in each of the samples

Chapter 14 Analysis of Variance (ANOVA)

FORMULA 14.1 **Sum of Squares between Groups (*SSB*)**

$$SSB = \sum_{i=1}^{k} n_i(\bar{x}_i - \bar{\bar{x}})^2$$

where k = Number of groups

FORMULA 14.2 **Sum of Squares within Groups (*SSW*)**

$$SSW = \sum_{j=1}^{k}\sum_{i=1}^{n_j}(x_{ij} - \bar{x}_j)^2$$

where $\bar{x}_j$ = Sample mean of group j
x_{ij} = The ith observation in group j
k = Number of groups
n_j = Number of observations in group j

FORMULA 14.3 **Sum of Squares between Treatments (*SSTr*)**

$$SSTr = \sum_{i=1}^{k} b(\bar{x}_i - \bar{\bar{x}})^2 = b\sum_{i=1}^{k}(\bar{x}_i - \bar{\bar{x}})^2$$

where k = Number of treatment groups
b = Constant number of blocks

Chapter 15 Measuring and Testing Associations for Bivariate Data

FORMULA 15.1 **Value of the Correlation Coefficient *r***

$$r = \frac{\sum_{i=1}^{n} (x_i - \bar{x})(y_i - \bar{y})}{\sqrt{\sum_{i=1}^{n} (x_i - \bar{x})^2} \times \sqrt{\sum_{i=1}^{n} (y_i - \bar{y})^2}}$$

Chapter 16 Simple and Multiple Linear Regression

FORMULA 16.1 **Slope of the Regression Line**

$$b_1 = \frac{SSXY}{SSX} = \frac{\sum_{i=1}^{n} (x_i - \bar{x})(y_i - \bar{y})}{\sum_{i=1}^{n} (x_i - \bar{x})^2}$$

FORMULA 16.2 ***y*-Intercept of the Regression Line**

$$b_0 = \bar{y} - b_1\bar{x}$$

FORMULA 16.3 **Coefficient of Determination**

$$r^2 = \frac{SSregression}{SSTotal}$$

where *SSregression* is the sum of squared differences between the model-predicted *y*-values (i.e., $\hat{y}$) and $\bar{y}$.
SSTotal is the sum of squared differences between all the actual *y*-values and $\bar{y}$.

FORMULA 16.4 **Standard Error of the Estimate**

$$s_y = \sqrt{\frac{\sum_{i=1}^{n} (y_i - \hat{y}_i)^2}{n - 2}}$$

Solutions for Odd-Numbered Exercises

Note: Occasionally your exact solutions to some of the textbook problems may vary slightly from those displayed here, depending on (a) the software that you use, and/or (b) the round-off decisions that you make for intermediate calculations.

Chapter 1 Introduction to Statistical Data

Basic Concepts 1.1

1. In the first sense, the term *statistics* refers to numerical facts of interest in some context. As a course of study, *statistics* encompasses collecting, processing, and interpreting data by using a variety of mathematical approaches.
3. Descriptive statistics covers the collection and summation of data and its presentation in tables and graphs. Inferential statistics relies on measurements derived from samples to make conclusions about population parameters.
5. A table of NHL scoring leaders; a pie chart displaying the countries of origin of Canadian immigrants; a histogram with salary categories for faculty at Canadian universities.
7. (a) Descriptive, (b) inferential, (c) descriptive, (d) descriptive, (e) inferential.

Basic Concepts 1.2

9. No, it is not possible to tell whether a set of data—given without explanation—represents a sample or a population. The data set could be either, depending on whether values were taken from all, or only some, of the objects or measurements that are of interest to the decision maker.
11. No, because taking a full census for a large population could be affected by errors in data collection or coding, especially if many people (some possibly inexperienced) are working on the project. Also, for an infinite population, a census is not possible, so a sample provides the only realistic source of information.
13. a) In most of the cases that are reported, the data represent a sample, due to the cost and time constraints of conducting a census.
    b) In a typical article about obese children in Canada, for example, the data would be presented as descriptive statistics (with tables, graphs, averages), but inferences may be implied—for example, about the proportion of children *across Canada* who are obese.
    Data reported from the Canadian census would be descriptive statistics.
15. The most direct inference is that, based on a sample statistic, a conclusion can be made that 60% of *all* eligible voters (not just those sampled) support Franklin. This inference might be mistaken if the survey wasn't conducted properly (e.g., sample too small, sample not representative, errors in data entry and processing, etc.). If the poll was taken just before the election, it might also be inferred that Franklin will win (i.e., obtain over half the votes). But this inference has additional risks of error if, for example, many voters change their minds or fail to go to the polls.
17. a) Descriptive statistics.
    b) Approximately 35.9% of the British Columbians have no religious affiliation. This value is a parameter because the data were gathered from the 2001 Canadian census, not from a sample.
19. a) The population is the number of homes that are inspected.
    b) We cannot be confident that the sample is representative because no information is provided on how the inspector selected these houses, where these houses are located, and other characteristics (such as the ages of the houses).
    c) In addition to the factors already noted in part (b), his sample might be not representative of the population because the problems encountered with houses in a rundown part of the city may differ from those in newer or better-maintained areas.
21. a) The population of interest for the study is the set of all Canadians with declarable incomes.
    b) About 19% of Canadians with a declarable income would like to hide their income from the Canada Revenue Agency if they thought they could get away with it.
23. a) The population of interest consists of all beer drinkers residing in Canada.
    b) Infinite population. (Note: The number is not infinite in a pure mathematical sense; for example, the number is less than the known population of Canada. But statistically, the population is infinite in the sense that there is no realistic way to take a complete count.)
    c) It cannot be inferred that 61% of all Canadian adults would choose Canadian beer because the survey was

conducted only among Canadian beer drinkers; beer drinkers are not a representative subgroup of all Canadian adults.

Basic Concepts 1.3

25. A quantitative random variable takes a numerical value, whereas the values of a qualitative random variable are categories. The seniority of employees in a company expressed in years is a quantitative random variable, whereas people's favourite NHL hockey team is a qualitative random variable.

27. (a) Discrete; (b) continuous; (c) height is a continuous variable but the height of the CN Tower is really a constant—unless your measurements are precise enough to observe small changes due to corrosion); (d) discrete.

29. (a) Quantitative, (b) qualitative, (c) qualitative, (d) quantitative.

31. Data measured on the ordinal level are usually categorical and can be ranked. The categories of the ranking could be replaced by ordered numerical codes. A typical survey question could be: "Do you think the new cinema complex should be built?" with a set of answers like "fully agree," "slightly agree," "indifferent," "slightly disagree," and "fully disagree."

33. Data measured at the nominal level are qualitative because all items are classified into categories that can be identified by alphabetical (i.e., words) or numerical codes.

35. Both variables are measured at the ordinal level.

Applications 1.3

37. a) (i) Discrete, because the number can be counted; (ii) discrete, because the value for the difference between two counted numbers must itself be a whole number; (iii) continuous, because there is no restriction on the values that this ratio might take.
b) (i) Ratio, (ii) ratio, (iii) ratio.

39. (b) and (c).

41. The ratio of a tire's circumference to its diameter.

43. (a) Nominal, (b) ordinal, (c) ordinal, (d) ratio, (e) ratio.

Review Applications

45. Not a reasonable approach. Since a large country like Canada has diverse landscapes and environmental conditions, Ontario and its parks are not representative of all national parks in Canada.

47. "Year" (since only whole numbers are used for the values).

49. "Name" and "Province."

51. "Area" (Note: "Year" is just a counter on a scale representing time. If, instead, the variable was labelled "Years *elapsed* since the starting point of the calendar," then ratios between the values would become meaningful.)

53. A population, since the file lists *all* of the prize winners within the range of dates being studied.

55. "Winner," "Country," "Contin," (i.e., continent), "Rubric," and "Other."

57. None (since only whole numbers are recorded for the years).

59. None.

61. Not reasonable. Reasons include: Small sample size (only four listings for Canada) and sample not representative (all of the sampled, Canadian bridges were built in the sixties and technology has changed in the meantime; also the length of a bridge is dependent on the location, which is unique for each project.)

63. "Year" (since only whole numbers are used for the values). (Note: Although measurements for the length variables—Span_ft and Span_mtr—were also rounded to whole numbers, they can be viewed as continuous since their round-off "gaps" are trivially small compared to the large ranges of values taken by these variables.)

65. "Type," "TypeCode," "Name," and "Country."

67. "Span_ft" and "Span_mtr."

Chapter 2 Obtaining the Data

Basic Concepts 2.1

1. Customer lists, sales figures, accounting records, inventory data, payroll data.

3. Newspapers, like *The Globe and Mail;* journals, like *Nature;* the U.S. Department of Labor, Bureau of Labor Statistics (www.bls.gov/cps).

5. Define the objectives of the research; identify potential sources of the data (primary vs. secondary data); plan a method to obtain the data (e.g., face-to-face interview using a paper questionnaire, telephone interview with a computer-assisted questionnaire, use of secondary data published by Statistics Canada); develop a sampling plan; collect and analyze the data; then report the findings and draw conclusions with respect to the initial research objectives.

7. Survey method.

9. Exploratory research occurs when the research objectives have not yet been defined (often using qualitative approaches). Conclusive research encompasses descriptive and casual research. Research objectives are fixed, and data are collected to draw conclusions about the population of interest.

Applications 2.1

11. (a) Internal, (b) external, (c) external, (d) internal, (e) external.

13. Conclusive research.

Basic Concepts 2.2

15. A sample is a *random* sample when the objects or individuals of a population of interest are selected in a random way, such that any member of the population has an equal chance of being included in the sample. Using random samples helps to ensure that the sample gives a representative picture of the population when drawing conclusions about population parameters.

17. In a cluster sample, the members of the sampling frame are split up into subgroups, each of which should be roughly representative of the entire population. A subset of all of the clusters is selected by random sampling; then, within selected clusters, either all individuals or a random sample are selected. Example: For making a prognosis about Canada-wide election results, divide the country into reasonably similar election districts and then sample randomly from each of a random selection of the districts.

In stratified random sampling, the population of interest is split up into subgroups (strata) based on particular characteristics, and a simple random sample is drawn from each stratum. To make conclusions about the full population, the sample data may have to be weighted, depending on the sample sizes used. Example: For a study on different ethnic traditions practised in Canada, a stratified random sample plan could be used to ensure sufficient data about traditions involving only a small proportion of the population.

19. A random sample selects individuals or items by chance, so bias is less likely than in judgment samples, which are drawn based on the researcher's opinion of whom it would be useful to include.

21. The registrar could use software to randomly select from the list of ID numbers the students to be approached.

23. By definition, a census surveys all population members, so sampling error cannot occur. (In practice, some members may be missed or may not respond, in which case the "census" becomes a very large sample.) Even in a census, nonsampling errors or nonresponse errors could occur; for example, if some respondents refuse to answer all questions or print illegibly, or if data are transcribed incorrectly.

25. If possible, avoid self-completion questionnaires and hire well-trained, supervised interviewers instead. Be sure the questionnaire is clear and easily understood. When using a computer, check for consistency of data inputs. Follow-up letters or incentives could be used to motivate selected individuals to participate in a survey.

Applications 2.2

27. Conduct a face-to-face survey, using a stratified sampling procedure to include in the sample respondents from different districts of the city; or send out a questionnaire to each household by mail, asking the residents to complete it and send it back (be sure to include a stamped, self-addressed envelope); or carry out a qualitative or nonstatistical survey by setting up focus groups throughout the city.

29. Open-ended questions. Advantages: Provide opportunities to obtain information the researcher wasn't aware of, and avoid the bias of guiding respondents in a particular direction. Disadvantages: Often lead to nonresponse errors because respondents fail to answer, and measurement errors can arise because vague or complex responses need to be categorized.

Pre-categorized answers: Advantage: Easier and more accurate coding of the responses. Disadvantage: Nonresponse errors if respondents refuse to answer certain questions, perhaps because they're not comfortable with any of the choices; and measurement errors if data are recorded incorrectly.

31. Stratified sample.

33. In some university programs, such as engineering and computer sciences, women are underrepresented. Using judgment sampling, focus group discussions with female high schools students could be made to examine why most female university entrants choose nontechnical programs and to determine what steps have to be taken to encourage young women to choose technical programs.

Review Applications

35. a) Through a questionnaire asking various companies' security officials about the steps taken to reduce their companies' exposure to crime.
b) Through direct observation (which includes monitoring of appropriate instruments) of weather and snow-cover conditions at the site.
c) Through a customer survey (asking about customers' anticipated vehicle preferences, and when they expect to need their next vehicle) supported by observations during conversations with customers at the dealership.
d) Through surveys of workers about their perceptions of stress, supplemented by qualitative group discussions with employees.
e) Through an experiment (maybe as a simulation on the computer or in a laboratory).

37. Exploratory research.

39. Stratified random sample.

41. Judgment sampling.

43. One approach is to take a systematic random sample of the hits during a day. Given a time-ordered list of all hits throughout the day, start at a random point early in the list and sample every *k*th item in the list of the day's hits.

45. Stratified random sampling.

Chapter 3 Displaying Data Distributions

Basic Concepts 3.1

1. Yes. An outlier is a data value that is apparently inconsistent with the remaining observations. Sometimes the cause for such a value is a data entry or measurement error, which would be an example of dirty data.

3. Sturge's Rule is useful because it can help determine a reasonable number of classes to use for constructing a continuous frequency distribution.

5. a) Data value "02" could be dirty data because this value seems to be inconsistent with the remaining data.
b) Data value "02" could be considered an outlier because it is an extremely small value compared to the others.

7. Data set 1 (one possible solution):
a) Number of classes = 6.
b) Tentative class width = 5.
c) Note: Alternatively, the classes could be labelled "10–14"; "15–19", etc.

Classes	Frequency
10 up to 15	3
15 up to 20	4
20 up to 25	3
25 up to 30	2
30 up to 35	1
35 up to 40	2
Total	**15**

d)

Classes	Percent
10 up to 15	20.0
15 up to 20	26.7
20 up to 25	20.0
25 up to 30	13.3
30 up to 35	6.7
35 up to 40	13.3
Total	**100.0**

e)

Classes	Cumulative Percent
10 up to 15	20.0
15 up to 20	46.7
20 up to 25	66.7
25 up to 30	80.0
30 up to 35	86.7
35 up to 40	100.0

f) Note: This solution assumes discrete data so, for example, the upper bound of the class "10 up to 15" is exactly 14.

Classes	Midpoints
10 up to 15	12
15 up to 20	17
20 up to 25	22
25 up to 30	27
30 up to 35	32
35 up to 40	37

Data set 2 (one possible solution):
a) Number of classes = 6.
b) Tentative class width = 7.
c) Note: Alternatively, the classes could be labelled "0–6"; "7–13," etc.

Classes	Frequency
0 up to 7	6
7 up to 14	6
14 up to 21	8
21 up to 28	6
28 up to 35	5
35 up to 42	1
	$N = 32$

d)

Classes	Percent
0 up to 7	18.75
7 up to 14	18.75
14 up to 21	25.00
21 up to 28	18.75
28 up to 35	15.63
35 up to 42	3.13

e)

Classes	Cumulative Percent
0 up to 7	18.75
7 up to 14	37.50
14 up to 21	62.50
21 up to 28	81.25
28 up to 35	96.88
35 up to 42	100.00

f) Note: This solution assumes discrete data so, for example, the upper bound of the class "0 up to 7" is exactly 6.

Classes	Midpoints
0 up to 7	3
7 up to 14	10
14 up to 21	17
21 up to 28	24
28 up to 35	31
35 up to 42	38

9. Classes are not mutually exclusive; classes are not collectively exhaustive; class width is not identical for all classes; not all classes are showing convenient class limits; the total (= 245) does not correspond to the sum of the frequencies (= 246); the style of class labels is not identical for all classes; classes are not in the appropriate order.

11. a)

Eye Colour	Frequency
Brown	14
Blue	6
Grey	6
Green	4
Total	30

b)

Eye Colour	Frequency	Percent
Brown	14	46.67
Blue	6	20.00
Grey	6	20.00
Green	4	13.33
Total	30	100.00

Applications 3.1

13. a)

Number of Eruptions	Frequency
1	13
2	25
3	31
4	14
Total	**83**

b)

Number of Eruptions	Frequency	Percent
1	13	15.7
2	25	30.1
3	31	37.3
4	14	16.9
Total	**83**	**100.0**

c) Two or three eruptions of the Grotto Geyser per day seem to be "typical," since together these numbers occur two-thirds of the time (i.e., 30.1% + 37.3%).

15. a)

Number of Game-Tying Goals	Frequency
0	14
1	8
2	2
3	1
Total	25

b)

Number of Game-Tying Goals	Frequency	Percent
0	14	56.00
1	8	32.00
2	2	8.00
3	1	4.00
Total	25	100.00

c) No, the majority of the top goal scorers did not shoot a game-tying goal at all (56%).

17. a) One possible solution:

Densities	Count
0 up to 80	10
80 up to 160	3
160 up to 240	4
240 up to 320	0
320 up to 400	2

b)

Densities	Count	Percent	CumPct
0 up to 80	10	52.63	52.63
80 up to 160	3	15.79	68.42
160 up to 240	4	21.05	89.47
240 up to 320	0	0.00	89.47
320 up to 400	2	10.53	100.00

c) The table appears to back up the demographer's claim. Although about half (52.63%) of the observations fall with the range "0 up to 80," the remaining population densities are spread over ranges of considerably larger values.

19. a) One possible solution is shown below.

Note for parts (a) and (b): Depending on your software, the empty class "20 up to 22.50" may not be shown in the output.

Prices	Frequency
10.00 up to 12.50	1
12.50 up to 15.00	16
15.00 up to 17.50	8
17.50 up to 20.00	2
20.00 up to 22.50	0
22.50 up to 25.00	3
Total	**30**

b)

Prices	Frequency	Percent	Cumulative Percent
10.00 up to 12.50	1	3.3	3.3
12.50 up to 15.00	16	53.3	56.7
15.00 up to 17.50	8	26.7	83.3
17.50 up to 20.00	2	6.7	90.0
20.00 up to 22.50	0	0.0	90.0
22.50 up to 25.00	3	10.0	100.0
Total	**30**	**100.0**	

c) The "typical" prices are in the ranges between $12.50 and $17.49 (encompassing 80% of the cases). Prices above $17.50 are less common (only 17% of the wines), and wines priced between $22.50 and $24.99 are particularly "above the norm," with only 10% of the cases.

21. a)

State	Frequency	State	Frequency
Alabama	1	Nova Scotia	1
Alaska	1	Nunavut	1
Alberta	2	Ohio	1
Arizona	1	Oklahoma	1
Illinois	2	Ontario	5
Indiana	1	Quebec	9
Iowa	1	Saskatchewan	6
Kansas	1	Tennessee	2
Kentucky	1	Texas	3
Manitoba	2	Utah	1
Michigan	1	Virginia	1
Missouri	2	Wisconsin	2
Montana	1	Wyoming	1
Newfoundland/Labrador	1	Yucatan	1
		Total	**57**
North Dakota	2		
Northwest Territories	2		

b)

State	Frequency	Percent
Alabama	1	1.8
Alaska	1	1.8
Alberta	2	3.5
Arizona	1	1.8
Illinois	2	3.5
Indiana	1	1.8
Iowa	1	1.8
Kansas	1	1.8
Kentucky	1	1.8
Manitoba	2	3.5
Michigan	1	1.8
Missouri	2	3.5
Montana	1	1.8
Newfoundland/Labrador	1	1.8
North Dakota	2	3.5
Northwest Territories	2	3.5
Nova Scotia	1	1.8
Nunavut	1	1.8
Ohio	1	1.8
Oklahoma	1	1.8
Ontario	5	8.8
Quebec	9	15.8
Saskatchewan	6	10.5
Tennessee	2	3.5
Texas	3	5.3
Utah	1	1.8
Virginia	1	1.8
Wisconsin	2	3.5
Wyoming	1	1.8
Yucatan	1	1.8
Total	**57**	**100.0**

c) Quebec has been particularly prone to large meteor falls, followed by Saskatchewan and Ontario. Provinces and territories that do not have at least one known meteor crater are British Columbia, New Brunswick, Prince Edward Island, and the Yukon.

23.

Prize	Frequency	Percent
GMC Canyon	30	0.001
Panasonic plasma television	84	0.003
Cash prize of $1,000	460	0.015
Coleman camping package	6,644	0.221
Food prize	3,000,000	99.760
Total	3,007,218	100.000

The percentage of GMC Canyons that was awarded was 0.001%, cash prizes were awarded in 0.015% of the cases, and Coleman camping packages in 0.221%. If your "Roll Up the Rim" cup showed you won a prize in 2005, you most likely won a food prize and least likely won a GMC Canyon.

Basic Concepts 3.2

25. Data set 1 (one possible solution):

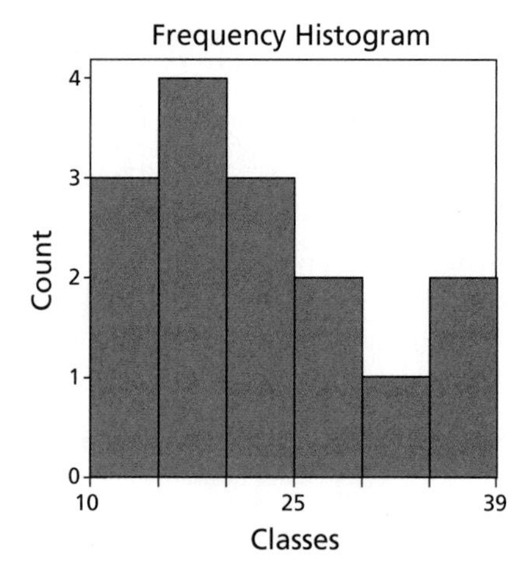

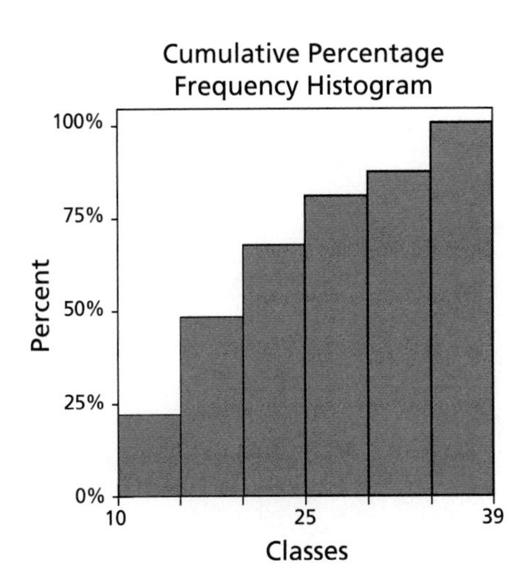

Data set 2 (one possible solution):

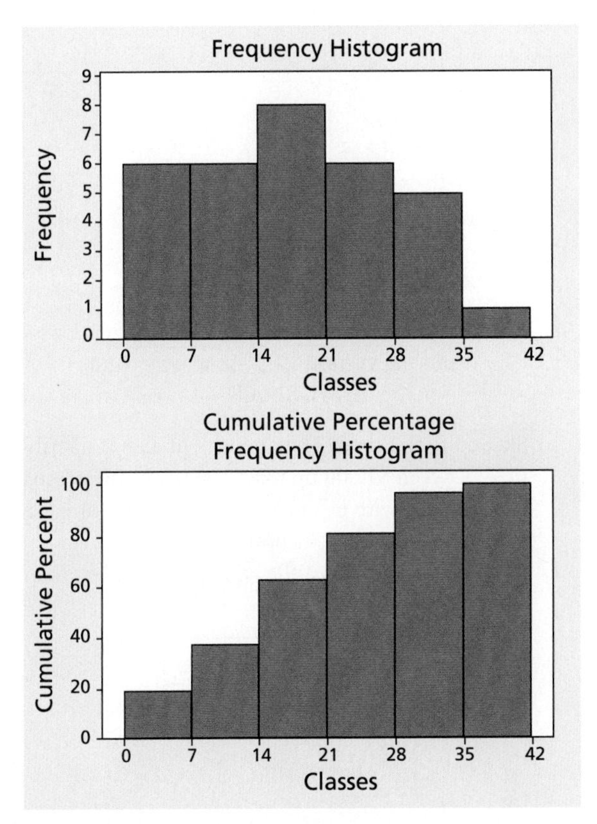

27.

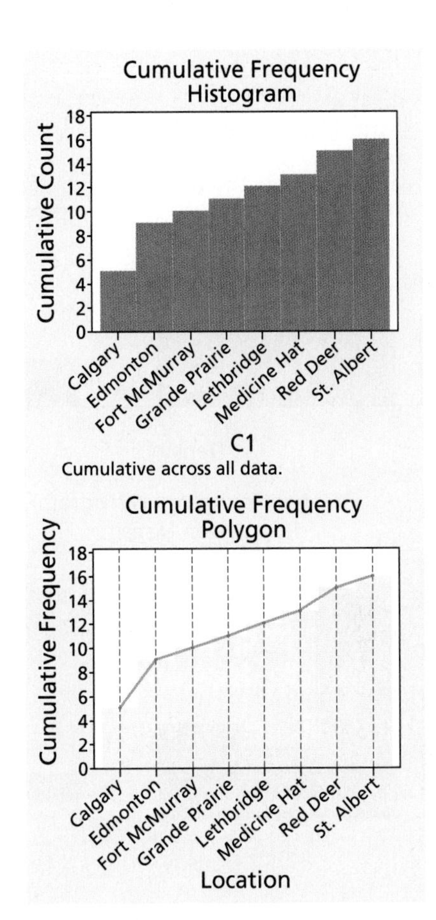

Applications 3.2

29. a) One possible solution:

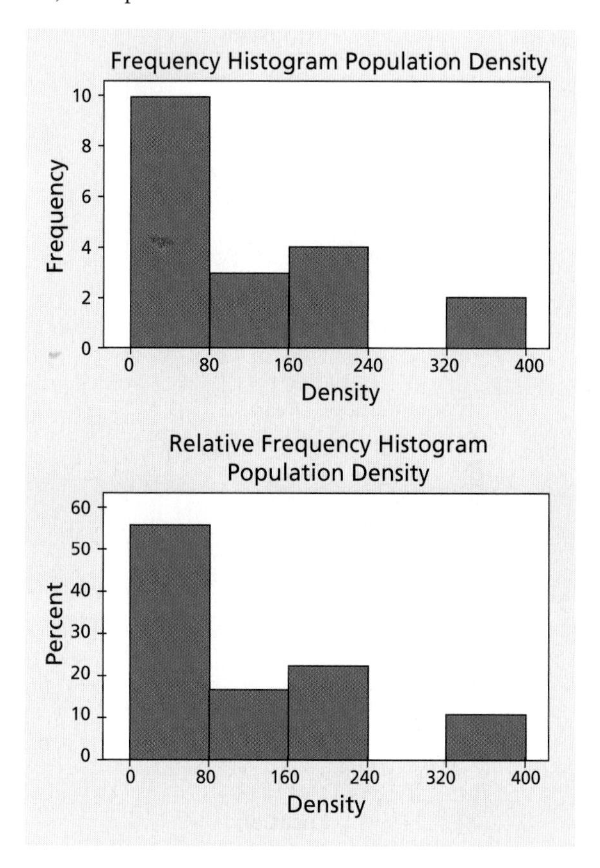

b) The distribution is right-skewed, not symmetrical.

c) Based on the figure, approximately 85% of countries have a population density less than 300 people per square kilometre.

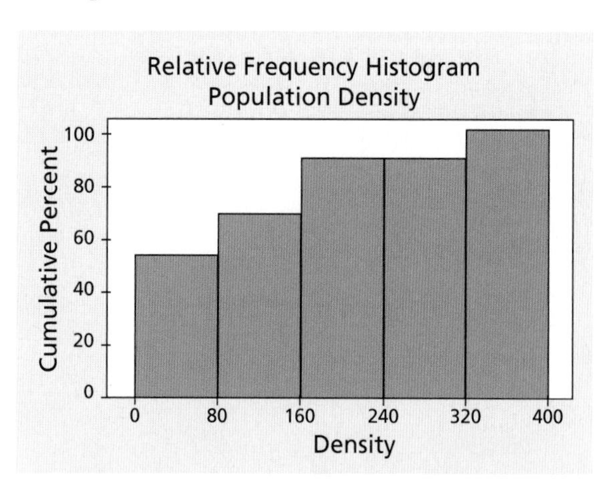

d) About 15% of countries have a population density of *at least* 300 people per square kilometre.

31. a) One possible solution:

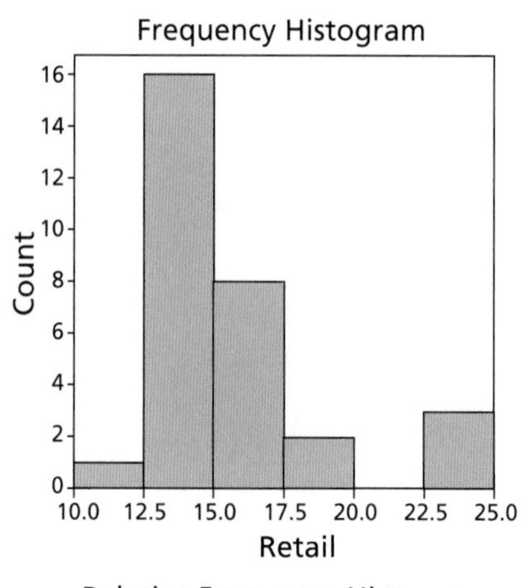

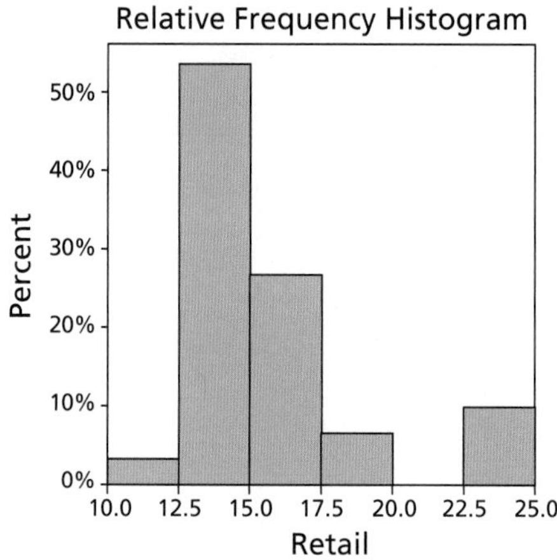

b) Skewed to the right. The majority of data is distributed between \$10.00 up to \$20.00 per bottle of Australian wine, with a concentration within a range of \$12.50 up to \$15.00 per bottle.

c) Approximately 90% of the wine prices are less than \$20.

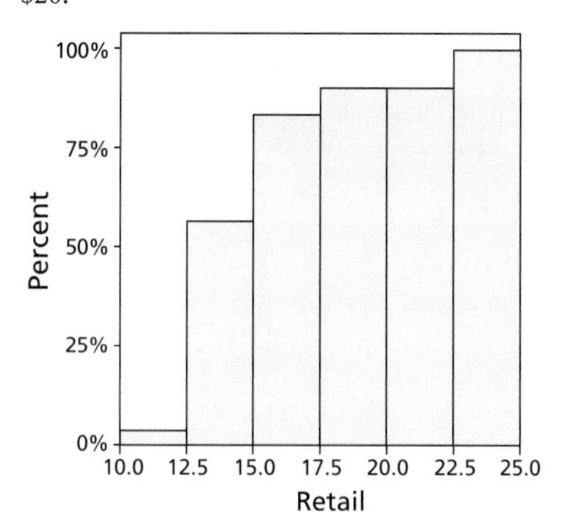

d) About 10% of the wine prices are *at least* \$20.

33. a) One possible solution:

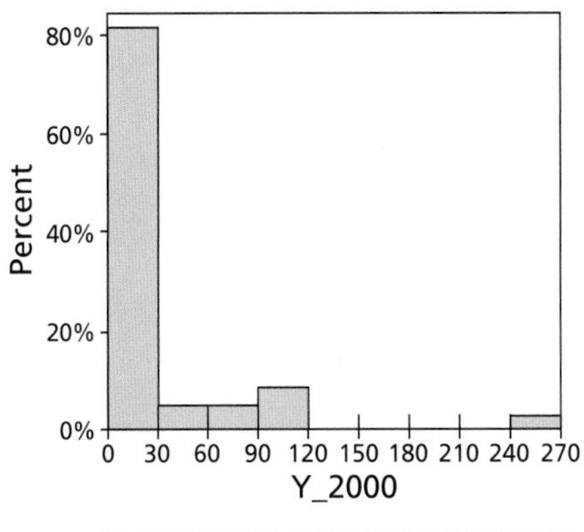

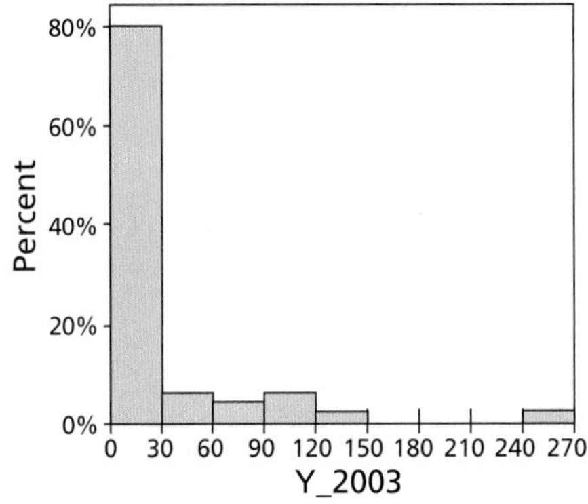

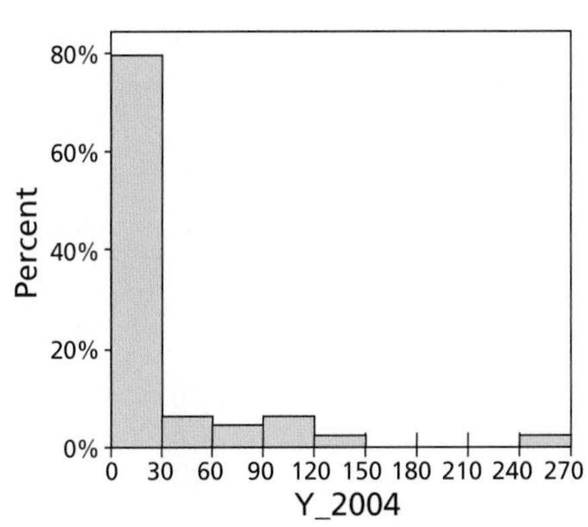

All three time periods are exhibiting a similar distribution shape that is highly right-skewed, with the largest concentration in the first class, below 30 billion barrels.

b) No, the distribution of the countries' oil production did not change significantly from 2000 to 2004.

c) Assuming that oil is produced in roughly the proportion of the size of a country's oil reserves, then Saudi Arabia, followed by Iran, are the countries producing most; a randomly selected barrel will more likely have been produced in one of those countries rather than elsewhere. A randomly selected barrel is least likely to have come from low-producing countries such as Uzbekistan, Turkmenistan, Thailand, and Romania.

d)

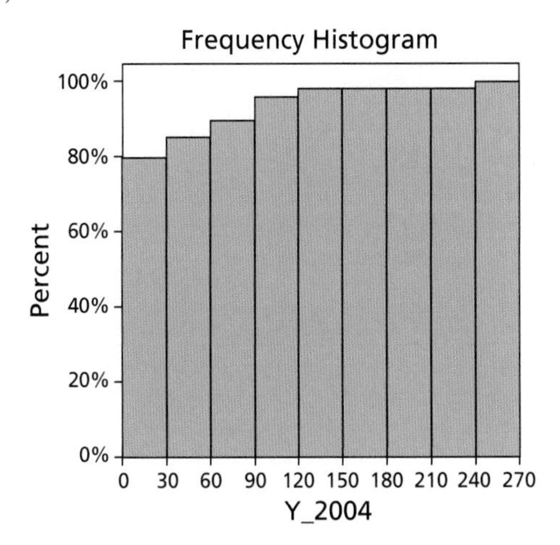

Approximately 80% of the countries produce less than 30 billion barrels.

e) About 20% of the countries produce *at least* 30 billion barrels.

Basic Concepts 3.3

35. a) **City**

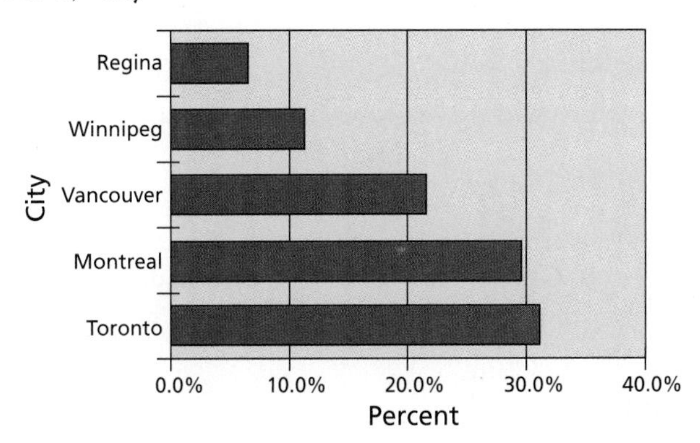

b) **Eye Colour**

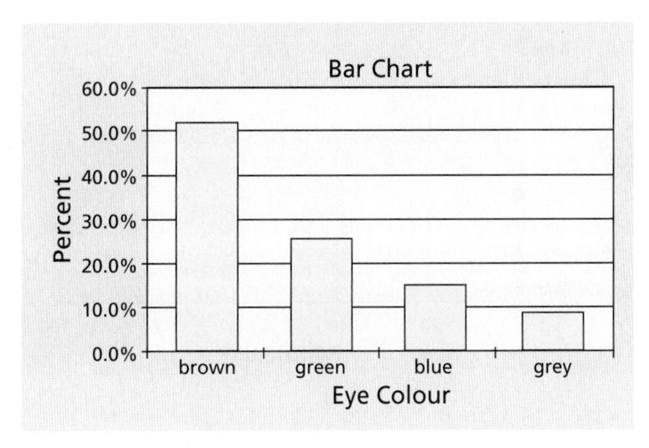

c) **Mode of Transportation**

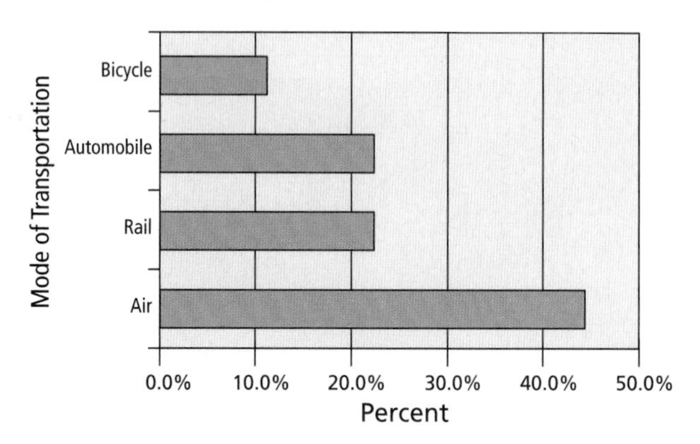

37.

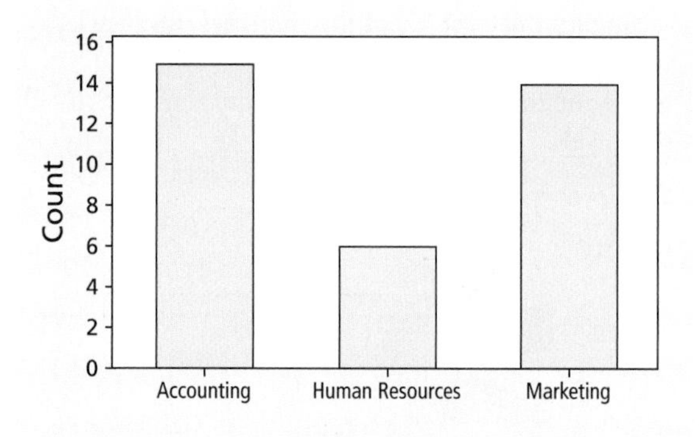

39. A Pareto diagram would represent the data in the best way, because the most common category is displayed first, followed by the second most common category, and so on.

Applications 3.3

41. a)

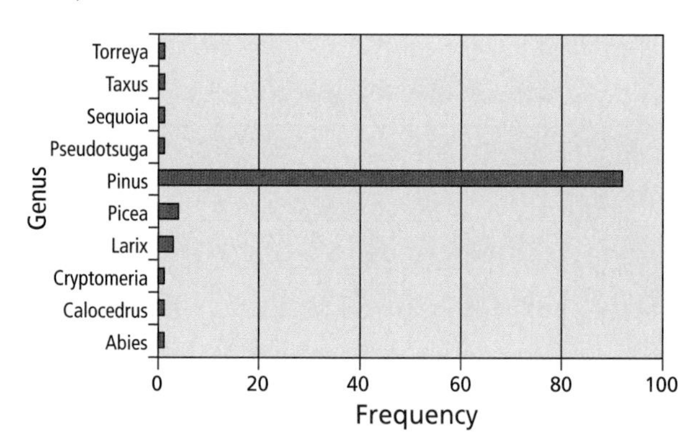

b)

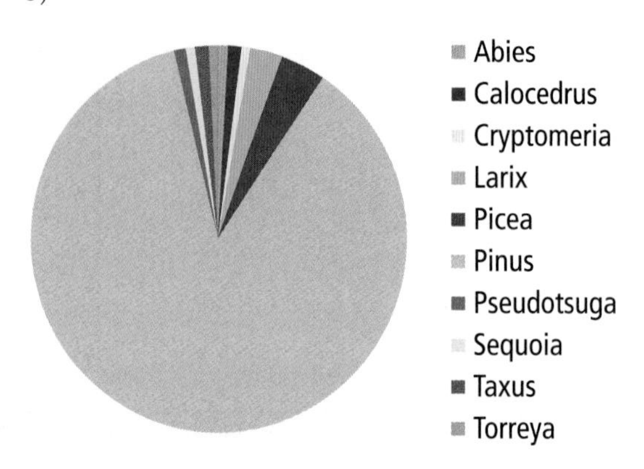

c) Both representations communicate clearly that the genus *Pinus* occurs much more frequently than the others. With the bar chart, it is much easier to read the frequencies for the non-*Pinus* cases.

43. a)

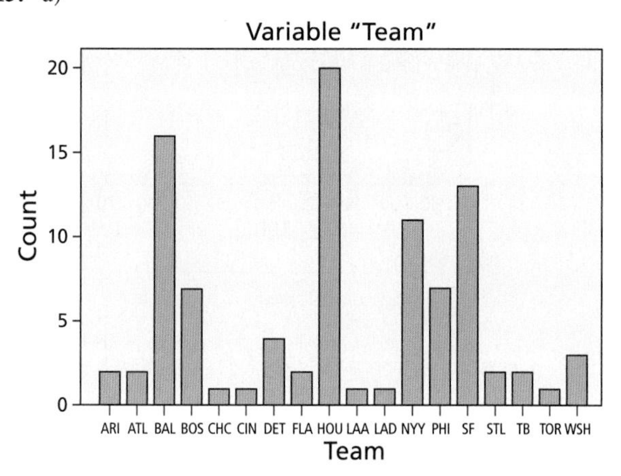

b) Members of the Houston and Baltimore baseball teams account for a comparatively large percentage of the milestones (20% and 16%, respectively). The event "Home Run" (HR) happens to occur most often. It is not very likely that these patterns will repeat the same way in future years, as individual baseball players change clubs or finish their professional careers.

Basic Concepts 3.4

45. a)

Stem	.	Leaf
2	.	133445
3	.	234
4	.	34567
5	.	56

b)

Stem	.	Leaf
5	.	46
6	.	7
7	.	88
8	.	6
9	.	5889
10	.	049
11	.	0

47. a)

Stem	.	Leaf
11	.	24679
12	.	2355
13	.	1248

b)

Stem	.	Leaf
3	.	2379
4	.	1245679
5	.	
6	.	56778
7	.	158

Applications 3.4

49. a)

Stem	.	Leaf
0	.	00011112579
1	.	02678
2	.	3
3	.	39

b) Both graphs are similar and show that population densities are concentrated at values smaller than about 100. For the histogram, there is more flexibility in setting class boundaries, so the histogram highlights that the concentration of values is actually below 80.

51. a)

Stem	.	Leaf
11	.	9
12	.	9
13	.	000467
14	.	019999999
15	.	1344899
16	.	5
17	.	
18	.	27

(Also: Three outliers, with values ≥ 22.7.)

b) The histogram shows different class boundaries from this stemplot, which in this case conveys a different impression of the shape. The choice of class boundaries is less flexible for the stemplot.

Basic Concepts 3.5

53.

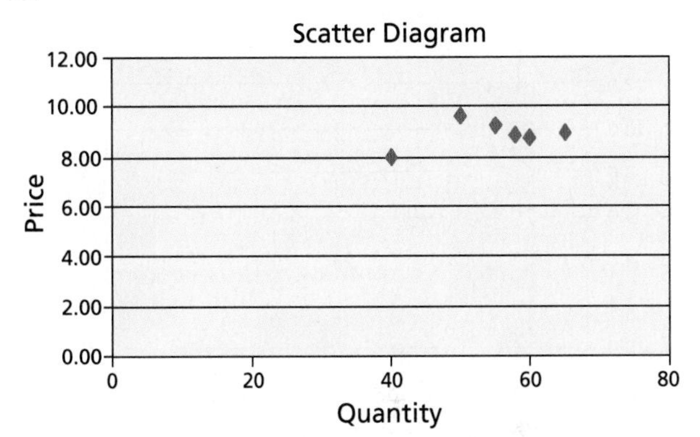

55.

Frequency of Churchgoing	Men	Women	Total
Regularly	40	60	100
Not regularly	35	15	50
Total	75	75	150

57.

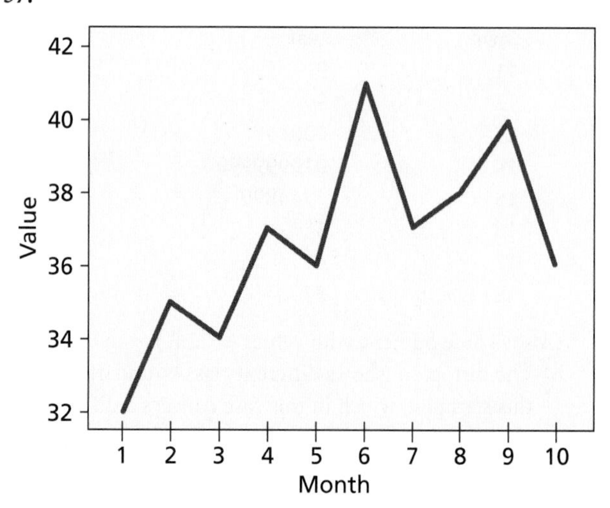

59.

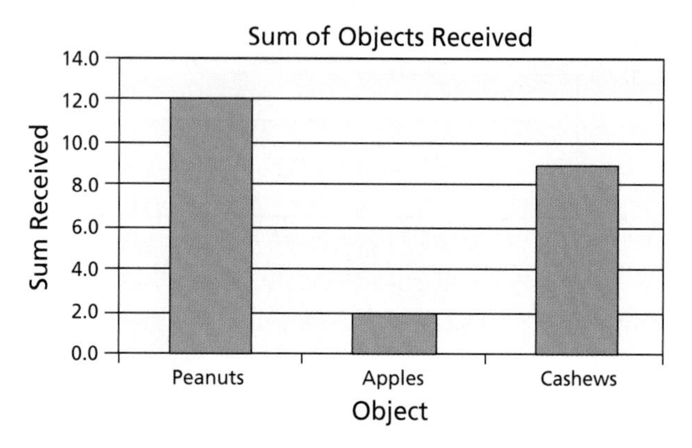

Applications 3.5

61. Yes. In general, for countries in which the infant mortality rate is low, the life expectancy is very high, whereas in countries that have a higher rate of infant mortality, the life expectancy tends to be lower.

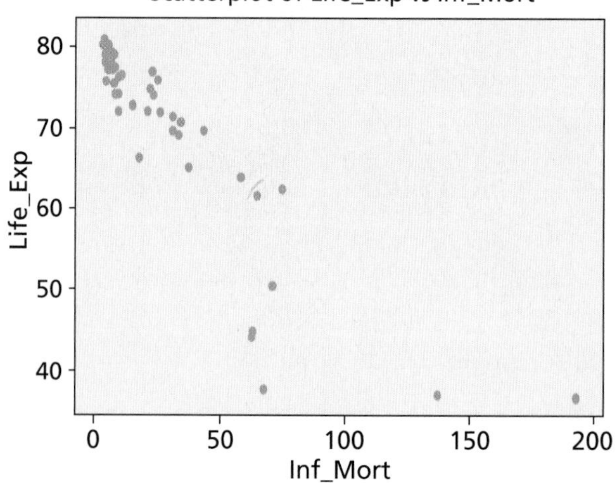

63. Yes, there does seem to be a positive relationship between the number of complaints about tickets/boarding passes and the perceived difficulties in receiving refunds. Both problems might have common causes, such as communication problems within the airline organization, or low employee–customer ratios, or an airline's focus on lower ticket prices as opposed to flexibility in handling ticket-change requests.

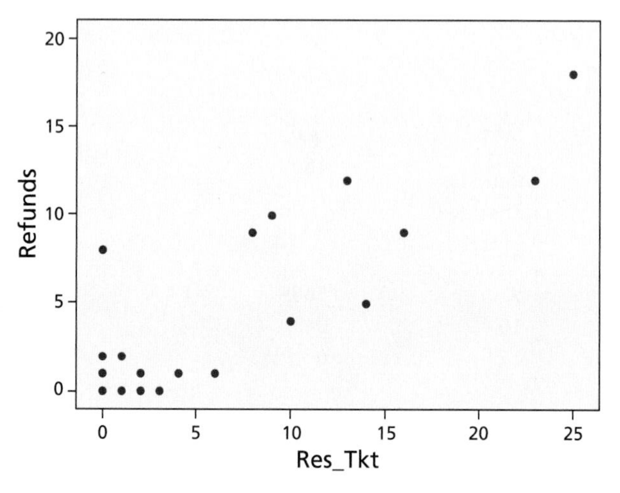

65. There is evidence of inflation for both sets of prices. Both food categories have become more expensive over time at roughly the same long-term rates, with variations for shorter-term periods.

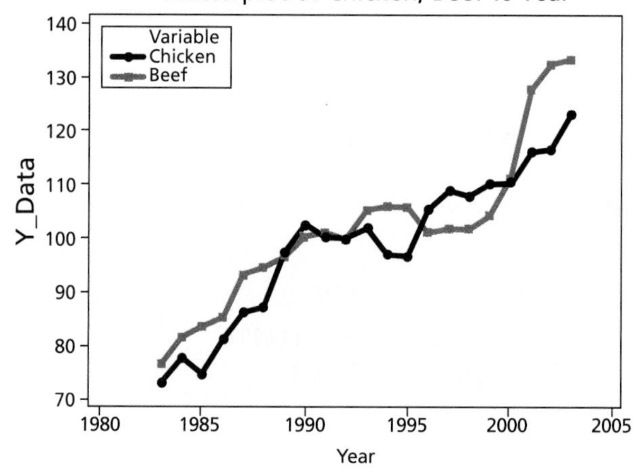

67. a) The minimum age most commonly suggested, overall, is 18 years and older.

Suggested Minimum Age	Ages of Respondents 18–29	30–39	40–49	50–59	60+	Total
12+ years of age	0	0	1	1	4	**6**
16+ years of age	21	13	7	6	5	**52**
18+ years of age	159	192	172	124	129	**776**
No age given	21	22	25	27	47	**142**
Unsure	8	3	1	4	8	**24**
Total	**209**	**230**	**206**	**162**	**193**	**1,000**

b) Although a majority of respondents in *all* age groups think that the minimum age for marijuana possession should be 18 years and older, the percentage of respondents who think the minimum age should be 16 is greater for those aged 18 to 29 (10%) than for those in other age groups.

69. a)

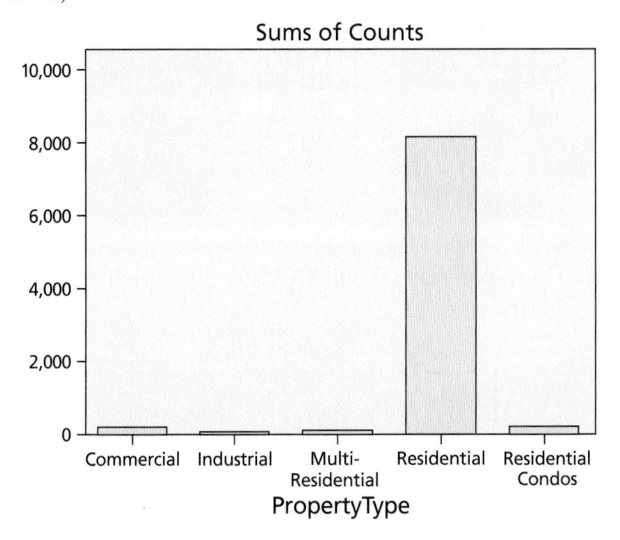

b)

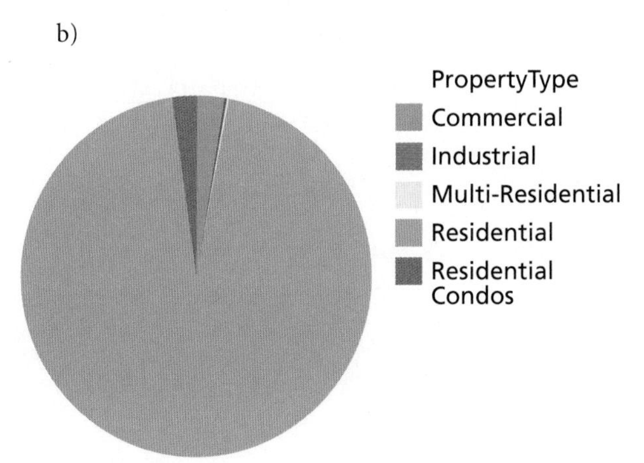

c) The total number of *residential*-code properties is about 8,000. The bivariate bar chart is the more helpful graph to answer this question, since the sum of counts for properties is clearly displayed on the y-axis.

Review Applications

71. One possible solution:

a)

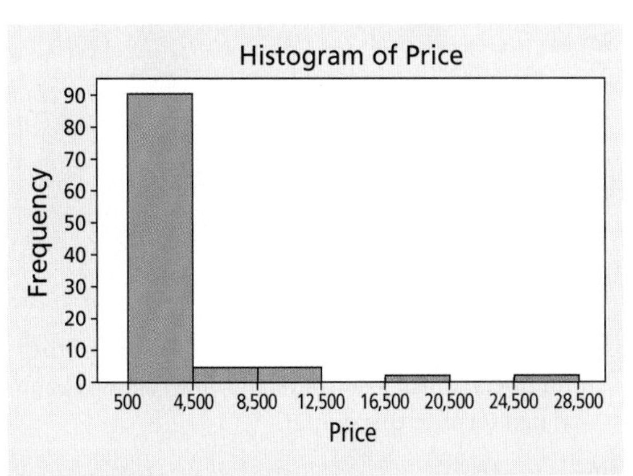

b)

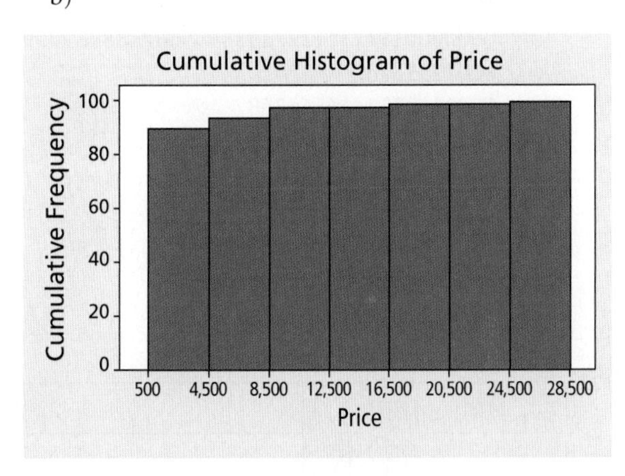

c)

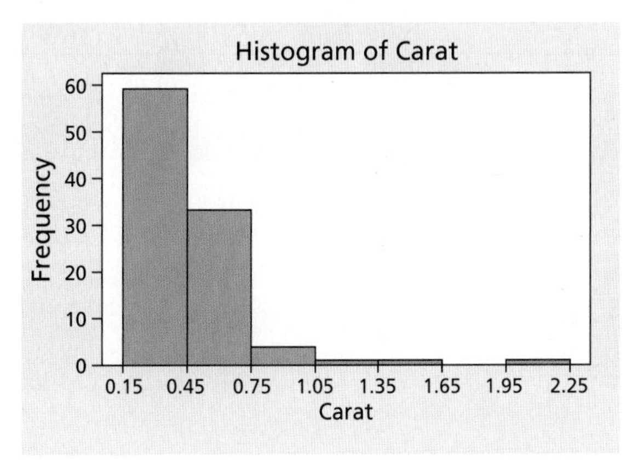

d)

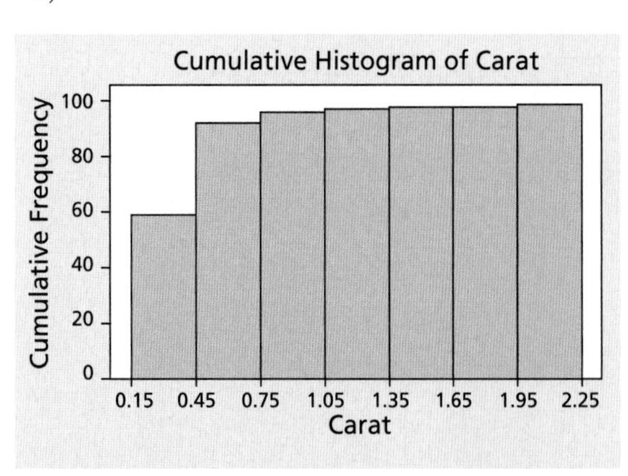

e) Note: The stems represent the tenths place of the numbers. If using software, they may show an abbreviated form of this output.

Stem	.	Leaf
1	.	889
2	.	1233445566789
3	.	00000001111111222233445566788888899999
4	.	0001235578
5	.	0001112233444556788
6	.	1
7	.	000011124
8	.	0

[Plus six outliers, with values ≥ 1.0]

73. a)

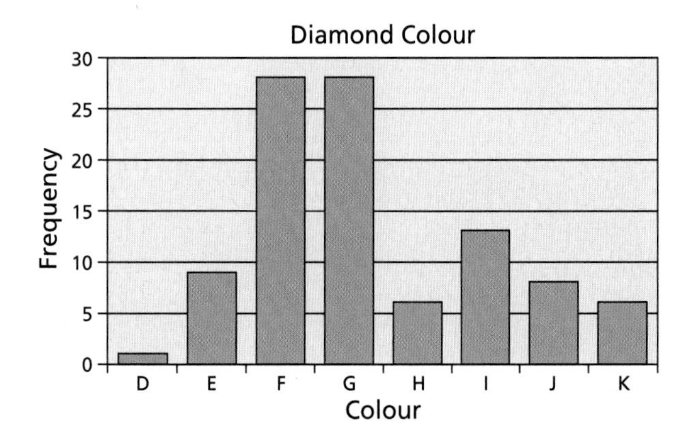

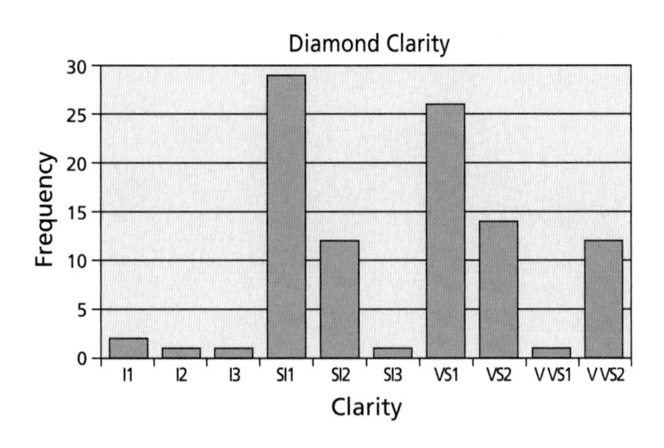

b) Note: Records without data for symmetry or polish are excluded.

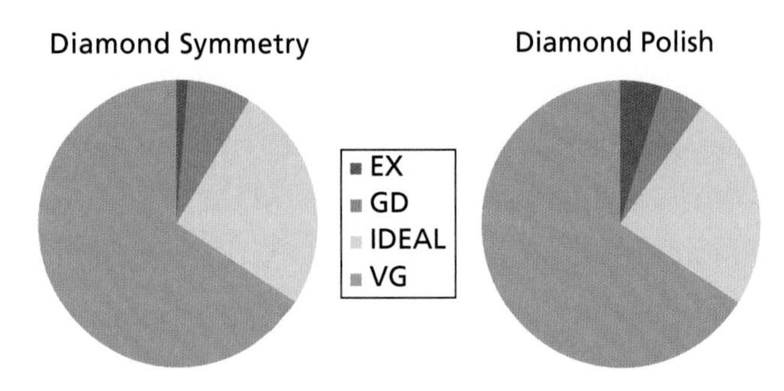

75. The colour values G and F account for most of the sales dollars.

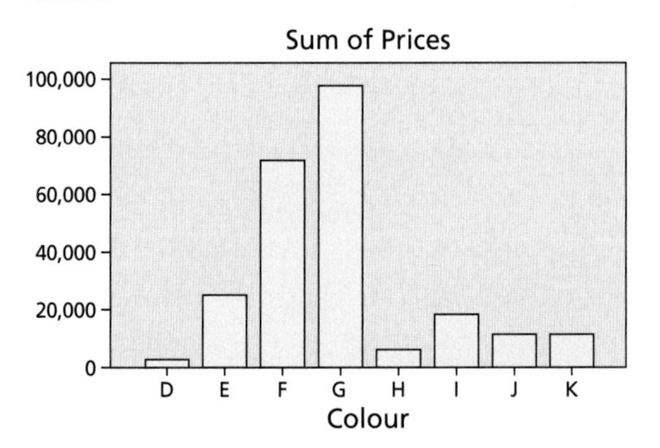

77. One possible solution:

a)

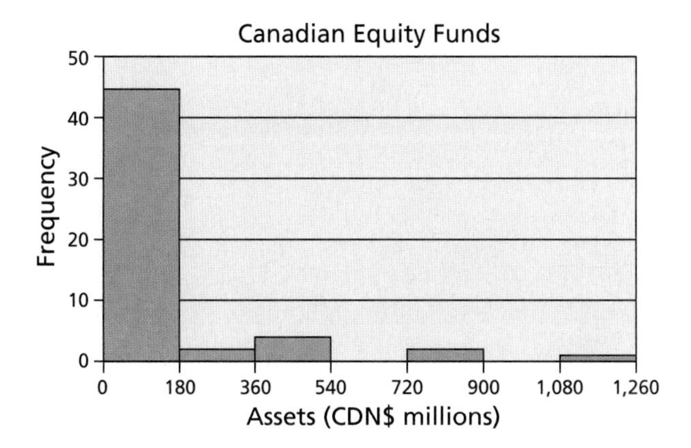

b)

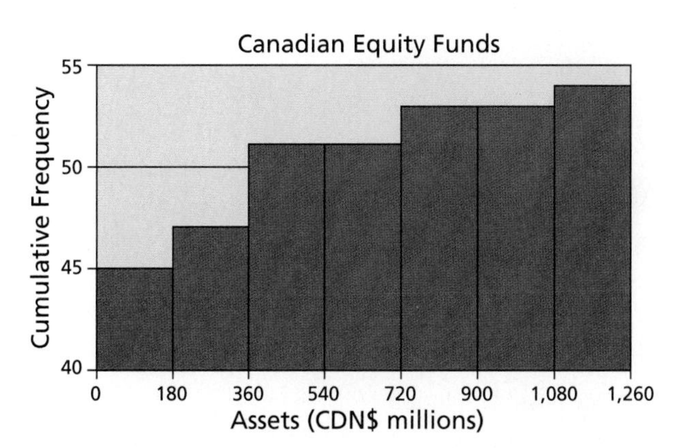

c)

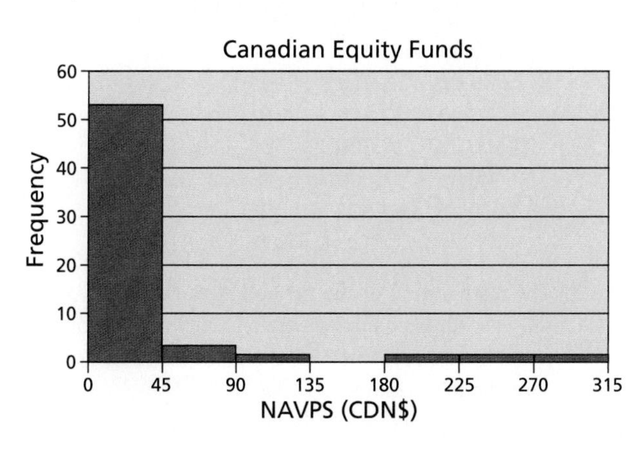

d)

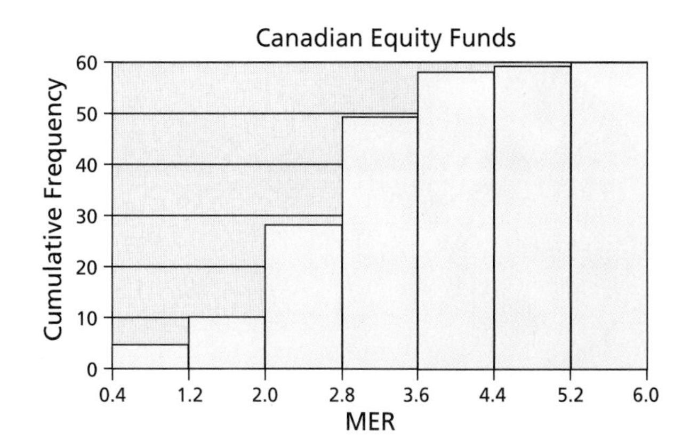

e)

Stem	.	Leaf
(And five extreme values ≤ 1.2.)		
1	.	2
1	.	5567
2	.	014444
2	.	5555566677778899999
3	.	001111122334
3	.	5666667
4	.	122
(And two extreme values ≥ 5.0.)		

79. a)

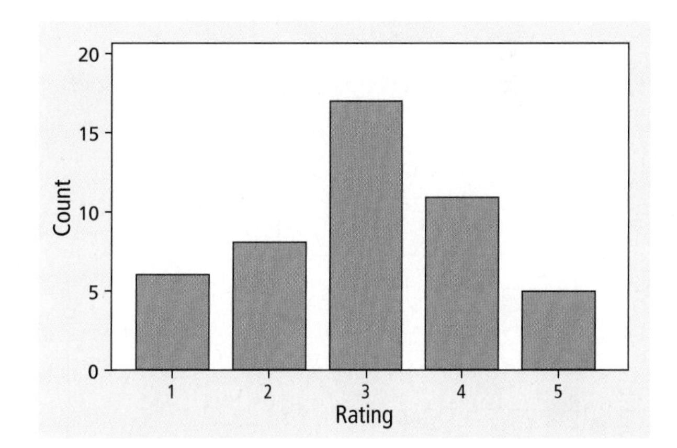

b)

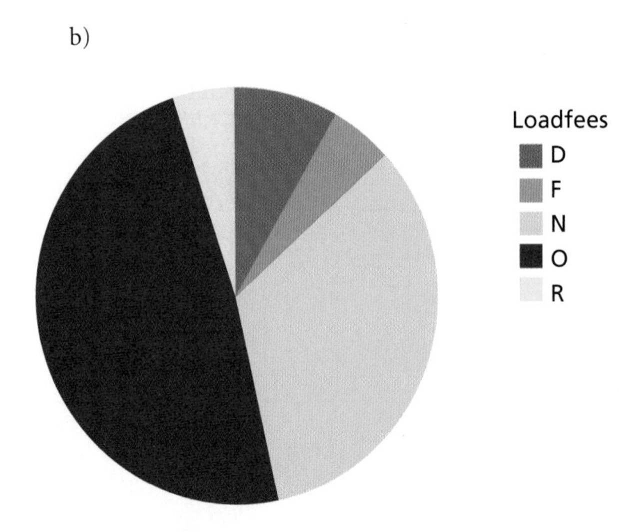

81. None of the ratings represent more than a third of the total assets.

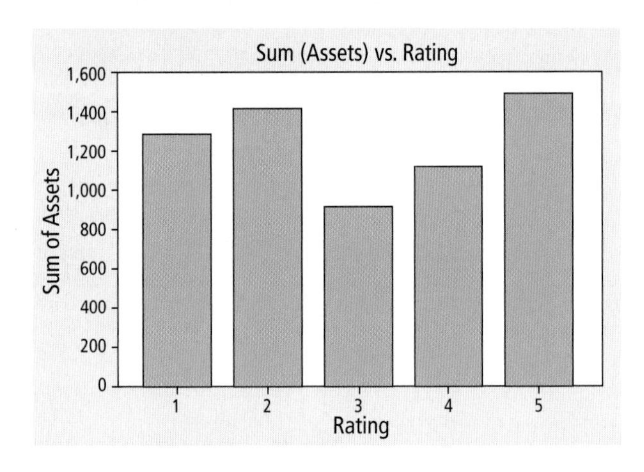

83. One possible solution:

a)

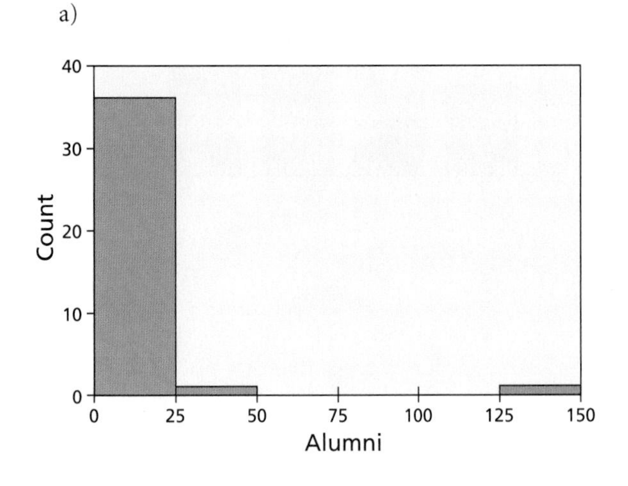

b)

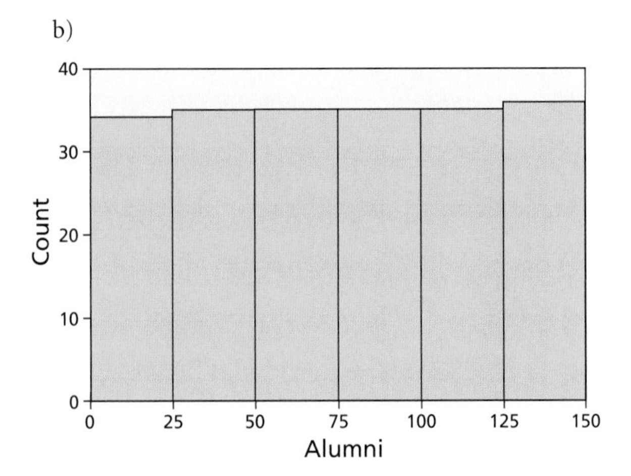

85.

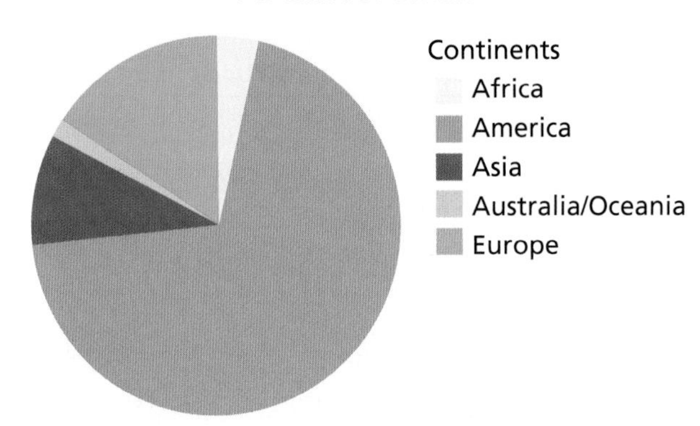

87. One possible solution:

a)

(i) AreaClass	(ii) Count	Percent	CumPct
0 up to 7,500	33	76.74	76.74
7,500 up to 15,000	4	9.30	86.05
15,000 up to 22,500	4	9.30	95.35
22,500 up to 30,000	0	0.00	95.35
30,000 up to 37,500	0	0.00	95.35
37,500 up to 45,000	2	4.65	100.00
	$N = 43$		

b) The distribution of the variable Area is not bell-shaped, but is positively skewed. Only two national parks have an area of 37,500 km^2 or more and can be considered as outliers.

89.

Stem	.	Leaf
0	.	000000000000000000001111111233446
1	.	00126
2	.	122
3	.	9
4	.	4

91. Based on this scatter diagram, there is no discernable relationship between the year a park was created and its area.

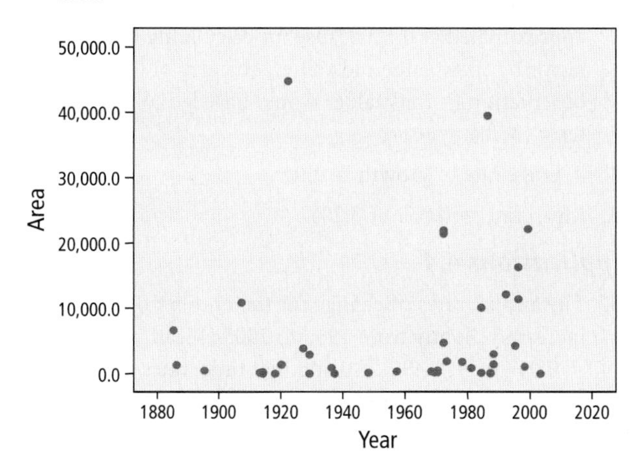

Chapter 4 Measures of Location

Basic Concepts 4.1

1. The arithmetic mean is located approximately in the middle (i.e., "centre") of the distribution.

3. The mean of Data set 2 is twice a high as the mean for Data set 1, due to including one much larger value in the set (i.e., replacing the value 2 with 25). Changing one number in a smaller data set has a greater impact on the mean than changing one number in a larger data set, which can be explained by the formula for the mean: With a small data set, the formula's denominator n is small, so the relative impact of adding one large number to the numerator is greater than if the denominator n is large.

5. Mean = 749.83.

Applications 4.1

7. a) Mean = 2.6.
   b) Yes. In the frequency distribution, we saw that "typically" there are about 2 to 3 eruptions per day. Similarly, the calculated mean of 2.6 lies roughly between 2 and 3 eruptions a day.

9. a) (i) Mean = −12.1; (ii) mean = 140.6.
   b) For melting points, a concentration can be observed in the class 0 up to 45°C. The corresponding mean of −12.1°C is misleading as a predictor for the majority of the melting points of the solvents, due to the 20% of solvents that have a melting point below −90°C. Boiling points are distributed more evenly and about 28% of the solvents have a boiling point in the range from 105°C to almost 175°C. The calculated mean of 140.6°C is located in the middle of this range. So if you randomly select one solvent, the mean of the boiling points appears to be a better predictor for the data than the mean of the melting points.

11. a) Mean (density) = 6.778; Mean (hardness) = 4.120.
    b) For the variable density, the calculated mean of 6.778 is a reasonable measure of location: 84% of the minerals have a density within a range of 3 up to 9. Also, the mean for hardness (4.120) is a reasonable measure for that variable, since over 80% of the minerals have a hardness factor between 1.5 up to 6. For both variables, only two or three cases have values at or more than twice the mean.

Basic Concepts 4.2

13. Although changing one number did impact the median, the impact on the calculation was not as strong as for the arithmetic mean, because the median is not affected by the presence of outliers.

15. a) Median = 0.262.
    b) Modes = 0.262, 0.269.
    c) For continuous data, the mode is not a reasonable measure of central location because identical continuous values will be observed only rarely—and possibly just by coincidence due to round-off.
    d) After aggregating continuous data into mutually exclusive groups, the modal class is the group with the highest frequency. Using ranges starting with "0.05 up to 0.10," "0.10 up to 0.15," etc., the modal class for the batting averages data would be "0.25 up to 0.30," with a frequency of 10.

17. Number of homes = 250.

Applications 4.2

19. a) Median = 12,430.
    b)

Diameters	Frequency
0 up to 25,000	6
25,000 up to 50,000	1
50,000 up to 75,000	1
75,000 up to 100,000	0
100,000 up to 125,000	1
125,000 up to 150,000	1

The mean for these diameters is 40,656.4 kilometres—which has been pulled up from more common values (in the range 0 up to 25,000) due to the high diameters for Saturn and Jupiter. The median gives a better picture of the real "centre" of this data set.

21. a) Median = 79.1.
    b) The median. We know (from Chapter 3, Exercise 17) that in 53% of countries (10 out of 19) the population is below 80 persons per km^2. The median (79.1) reflects this distribution. The mean with a value of

107.95 persons per km^2 was pulled higher by two outliers.

23. a) Mode = "With friends/family."
b) The single most common answer was "With friends and family."
c) (i) The category "With friends and family" would still be the mode.
(ii) The category "With friends and family" would still be the mode since it would still have the highest frequency.

25. a) Mode = 14.98.
b) The modal class is \$12.50 up to \$15.00, which in this case includes the actual mode. Both measures convey that there is a concentration of bottles of wine with a price close to \$15.00. The mode is more precise about an exact single value that recurs but it ignores information about the remaining bottles in the sample. The frequency histogram provides better information about the whole distribution of the data.

Basic Concepts 4.3

27. Weighted mean = \$953.50.

29. Weighted mean = 7,916.7 pbV.

31. Mean = 29.5. (Note: It has been assumed that the data are continuous.)

Applications 4.3

33. a) Average age of patients ≈ 61.
b) (i) For patients averaging 61 years old, it would be advisable not to choose music styles like rock, pop, hip hop etc., but rather choose classical music, jazz, or country music.
(ii) Select TV shows and magazines that are mostly seen or bought by older consumers.

35. a) Weighted mean ≈ 47°N (when calculating class midpoints for latitudes, note that the raw data for this variable are continuous.)
b) Without the raw data, you could not be sure. The mean falls near the middle of the class "45–49." If we assume that the data *within* that class are distributed evenly, then about 24 + 5 = 29 meteors fall below (to the south of) the mean, and 21 + 5 = 26 meteors fall to the north, so more would fall to the south.

37. a) Weighted mean = \$46.93.
b) A net increase. You could recalculate the weighted mean, which becomes \$47.53. Alternatively, note that red tickets are worth *more than* twice as much as white ones, so adding a number (200) of red tickets more than compensates in revenue for a loss of twice that number (400) of white tickets.

Basic Concepts 4.4

39. The mean rate of growth in speed (per two-second interval) is 3.98. The first entry in the data table is approximated at 0.1, because for calculating the growth factor by (new value/old value), an entry of 0 for the first observation (= old value) would have led to an error term (35/0).

41. Average rate of growth = 1.04.

43. Mean rate = 0.032 or 3.2%.

Applications 4.4

45. The annual rates of change for the number of wildfires increased slightly from 1950 to 2003 (Mean annual rate = 0.014 or + 1.4%). During that time, rates of change for the number of hectares destroyed decreased slightly rather than increasing (Mean annual rate = −0.024 or −2.4%).

47. a) Average growth rate = 1.0567.
b) Average annual return on investment = 0.0567 or 5.67%.
c) The average annual return on investment if the value for 1991 is omitted is 5.11%.
d) Not very sensitive. If the 1991 value in the data is replaced with 17, the annual return from 1991 to 2005 was 5.87%.

Basic Concepts 4.5

Section 4.5 Note: Answers for calculations of quantiles may vary, depending on your software.

49. Data set 1: 1st quartile = 3.25; 3rd quartile = 7.00.
Data set 2: 1st quartile = 4.0; 3rd quartile = 11.5.
Data set 3: 1st quartile = 13; 3rd quartile = 20.

51. a) 1st quartile = 0.2325; 2nd quartile = 0.262; 3rd quartile = 0.2725.
b) 5th percentile = 0.083; 50th percentile = 0.262; 95th percentile = 0.308

53. Data set 1: (a) 10, (b) 12, (c) 18, (d) 19.
Data set 2: (a) 5, (b) 5.75, (c) 8.25, (d) 9.
Data set 3: (a) 10, (b) 12, (c) 18.5, (d) 26.
Data set 4: (a) 1.1, (b) 3, (c) 6.5, (d) 9.

Applications 4.5

55. a) 1st quartile = 4,528; 3rd quartile = 68,472.5.
b) 10th percentile = 2,498.5; 90th percentile = 140,739.2.

57. a) (i) Melting points: 1st quartile = −64; 3rd quartile = 26.
(ii) Boiling points: 1st quartile = 81; 3rd quartile = 202.

b) (i) Melting points: 15th percentile = −95; 85th percentile = 34.2.
(ii) Boiling points: 15th percentile = 77; 85th percentile = 216.6.

c) The value for the randomly selected solvent is more likely to be bigger than the third quartile. By definition, about 25% of the observed values will fall above the third quartile, whereas only 15% of the observed values will fall above the 85th percentile.

59. a) (i) 1st quartile = 4.785; 3rd quartile = 7.785.
(ii) 1st quartile = 2.75; 3rd quartile = 5.5.
b) (i) 10th percentile = 4.008; 90th percentile = 10.311.
(ii) 10th percentile = 2.25; 90th percentile = 6.275.

Review Applications

Note: Answers for some calculations for quantiles may vary, depending on your software.

61. a) Median = 3,204.95
b) Mode = 3,202.50.

63. a) Mean = −1.968.
b) Median = −1.7; mode = −1.7.
c) 1st quartile = −2.975; 2nd quartile = −1.7; 3rd quartile = −1.050; 5th percentile = −5.060; 10th percentile = −3.4; 90th percentile = −0.230; 95th percentile = 0.190.

65. a) Mean = 30.694.
b) Median = 13.565; modes = 10.09, 12.16, 13.17.
c) 1st quartile = 11.393; 2nd quartile = 13.565; 3rd quartile = 19.793; 5th percentile = 6.453; 10th percentile = 7.532; 90th percentile = 54.088; 95th percentile = 209.752.

67. The mode for Loadfees is category "O" (i.e., Optional). This category occurs with the highest frequency, so if you randomly pick one of the funds, it is more likely to be in the O category than in any single other category.

69. (a) Median = 3; (b) mode = 1.

71. Smaller. If numbers of alumni in *all* countries outside the United States are included, there would be many values (equal to 0) added to the data set.

73. Median = 1,313.1 square kilometres.

75. Average rate of growth = 1.017.

77. Average rate of growth = 1.006.

79. Weighted mean = 35.3.

Chapter 5 Measures of Spread and Shape

Basic Concepts 5.1

1. Because without squaring, the sum of the positive and negative differences would always equal zero.

3. Data set 1: $s^2 = 16.554$, $s = 4.0686$.
Data set 2: $s^2 = 4.778$, $s = 2.1858$.
Data set 3: $s^2 = 9.361$, $s = 3.0596$.

5. $s = 0.98$.

Applications 5.1

7. a) $s = 0.953$.
b) With a mean of 2.55 eruptions per day and a standard deviation of 0.953 eruptions, the percentage of actual values between (mean − s = 1.60 eruptions per day) and (mean + s = 3.50 eruptions per day) = 56/83 ≈ 67.5%. *All* of the actual values lie between (mean − 2s = 0.64) and (mean + 2s = 4.46). Yes, it appears that the boundaries of the distribution can be estimated, based on adding or subtracting about two standard deviations from the mean.

9. a) (i) $s_{(\text{melting points})} = 62.787$; (ii) $s_{(\text{boiling points})} = 67.597$.
b) (i) The mean *melting point* is −12.1°C. For this variable, (mean − 2s) = [−12.1 − (2 × 62.787)] ≈ ×137. 7°C. No actual data points fall below this value. Also, (mean + 2s) = −12.1 + (2 × 62.787) ≈ 113.5°C. Only one actual data value (1/47 = 2.1% of the cases) is greater than this value.
(ii) The mean *boiling point* is 140.6°C. For this variable, (mean − 2s) = [140.6 − (2 × 67.597)] ≈ 5.4°C. No actual data points fall below this value. Also, (mean + 2s) = [140.6 + (2 × 67.597)] ≈ 275.8°C. Only two actual data values (2/47 = 4.3% of the cases) are greater than this value.
For both variables, it appears that we can estimate the "typical" spread of most values as being the within the range bounded by (mean ± 2s).

11. Note that in the question, the data are interpreted as a *population*.
a) Density: Range = 20.19, $\sigma^2 = 9.430$, $\sigma = 3.071$.
Hardness: Range = 8.25, $\sigma^2 = 2.768$, $\sigma = 1.664$.
b) Density: Two standard deviations above the mean = [6.778 + (2 × 3.071)] ≈ 12.92. Five percent of the actual data values falls above this value. Two standard deviations below the mean = 6.778 × (2 × 3.071) ≈ 0.64. No actual data values fall below this value. Hence, a randomly selected data value is more likely to be more than two standard deviations above the mean than two standard deviations below the mean. Given that the higher numbers of cases fall to the right of ($\mu + 2\sigma$) and none fall to the left of ($\mu - 2\sigma$) may suggest that the variable's distribution is skewed to the right.

Hardness: Two standard deviations above the mean = [4.120 + (2 × 1.664)] ≈ 7.45. Three of the actual data values fall above this value. Two standard deviations below the mean = $[4.12 - (2 \times 1.664)] \approx 0.792$. None of the actual data values fall below this value. Hence, a randomly selected data value is more likely to be more than two standard deviations above the mean than two standard deviations below the mean. Relatively few cases fall to the right of $(\mu + 2\sigma)$, and none fall to the left of $(\mu - 2\sigma)$, suggesting that the variable's distribution is nearly symmetrical but slightly skewed to the right.

Basic Concepts 5.2

13. (a) $z_{\$600} = 0.94$; (b) $z_{\$400} = -1.05$; (c) $z_{\$590.76} = 0.85$; (d) $z_{\$495.21} = -0.10$.

15. If the salaries are normally distributed, 95% are between approximately \$11.00 and \$21.00 per hour. If the distribution is very non-normal, at least 96% are between \$3.50 and \$28.50 per hour.

17. a) $s_{(\text{stock } 1)} = \$3.25$, $s_{(\text{stock } 2)} = \$2.04$.

b) Both stocks have about the same mean price but the variability of stock 1, as measured by *s*, is about 1.6 times larger than the variability of stock 2. So stock 2 represents a more stable investment since its price is less likely to vary from its mean.

Applications 5.2

19. (a) 75%; (b) between 45.9 and 130.8; (c) $z_{92} = 0.26$.

21. (a) 68%; (b) between 36.4 and 113.2 kg; (c) $z_{71.2} = -0.28$. (d) It is likely that the distribution of people's weights in the room will not be normally distributed, because it will become multimodal—with a cluster of low weights for the children and different clusters of weights for the football players and for the older males.

Basic Concepts 5.3

Section 5.3 Note: If you are using a computer program, different exact solutions may be displayed, depending on your software.

23. IQR = 4.

25. IQR = 6.

27. IQR = 1.3 millilitres.

Applications 5.3

29. a) IQR = 1.

b) The standard deviation (0.953) is very close to the IQR. Both measures provide a comparable measure for the degree of variation in the data.

31. a) IQR = 51.5.

b) One possible solution:

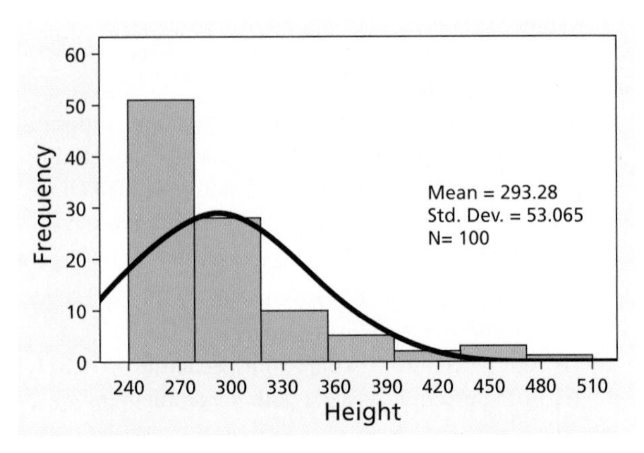

Just knowing the IQR does not tell us that the distribution of heights is very skewed, and from the IQR, it would be difficult to picture exactly how the data are spread. The storeys data are closer to being symmetrical, so you can better sense from the IQR how widely the (visually) middle set of classes in the distribution are spread.

33. a) Density: IQR = 3.00; hardness: IQR = 2.75.

b) The interquartile range is defined to show the spread in the middle 50% of the data. Hence, a randomly selected data value has a 50% chance of being within the interquartile range, regardless of the shape of the distribution.

Basic Concepts 5.4

Section 5.4 Note: If you are using a computer program, different exact solutions may be displayed, depending on your software.

35. Five-number summary: Min = 3, Q1 = 4, Median = 6, Q3 = 8, Max = 9.

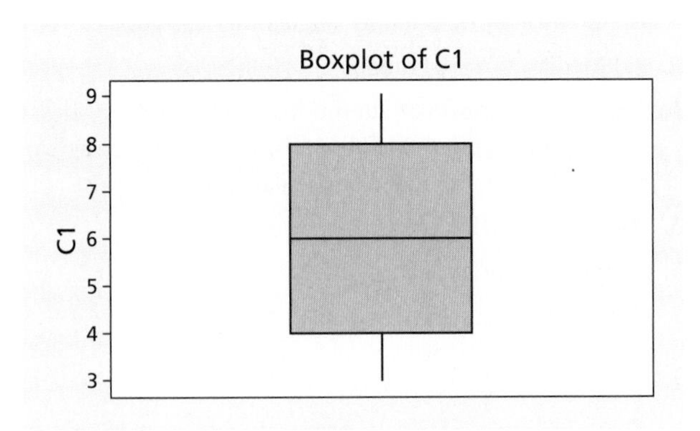

37. Five-number summary: Min = 101, $Q1$ = 123.75, Median = 139.5, $Q3$ = 153.25, Max = 194.

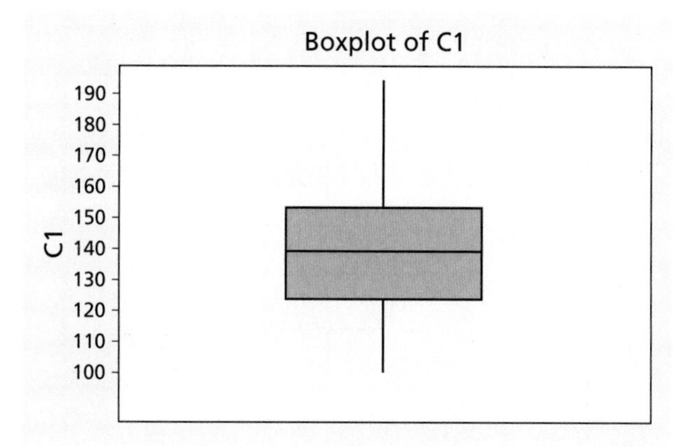

39. Five-number summary: Min = 196, $Q1$ = 218, Median = 262.5, $Q3$ = 293, Max = 314.

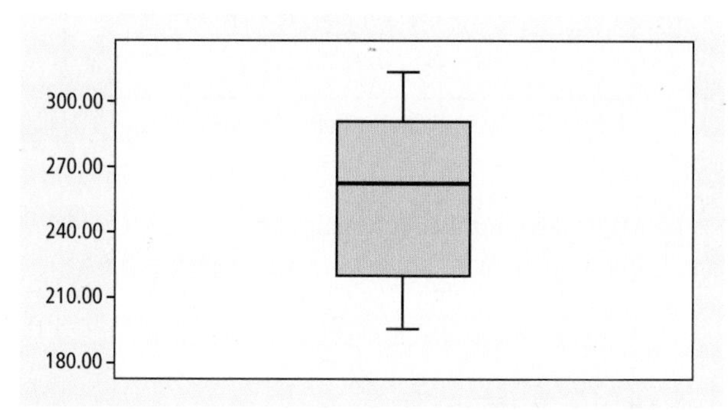

Applications 5.4

41. Five-number summary: Min = 30, $Q1$ = 56, Median = 64, $Q3$ = 72.75, Max = 110.

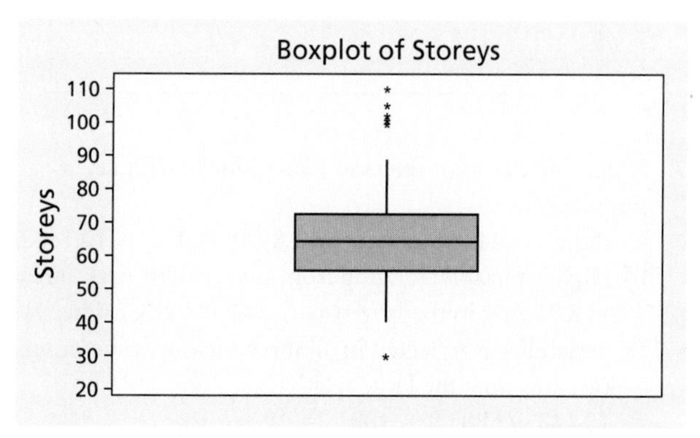

43. For *any* variable, the IQR lacks information about the *location* of the data. As far as spread, the distribution for variable *storeys* seems fairly symmetrical—if outliers are excluded—so the IQR conveys some information about the spread. The distribution for variable *height* has a very nonsymmetrical shape, so the five-number summary is much more informative about the distribution.

Basic Concepts 5.5

45. If two data sets with (a) different units or (b) very different measures of centre have to be compared, it would be misleading to compare their standard deviations directly; it would be better to rely on their coefficients of variation for comparison.

47. a) $CV = \frac{s}{\bar{x}} \times 100\% = 48.1\%$.

b) $CV_{\text{male}} = 47.6\%$.
$CV_{\text{female}} = 40.2\%$.
The variation around the average years of service is larger for male managers than for female managers.

49. Assuming that the data represent a *population:* $\sigma = 39.1$ and $\mu = 257.4$.

$$CV = \frac{\sigma}{\mu} \times 100\% = 15.2\%.$$

Applications 5.5

51. a) $CV_{\text{Xalostoc}} = \frac{s}{\bar{x}} \times 100\% = 56.4\%$.

$CV_{\text{Cero Estrella}} = 53.8\%$.
$CV_{\text{Netzahualcoyotl}} = 50.6\%$.

b) Xalostoc.

c) Xalostoc.

d) The standard deviation. Knowing the standard deviation and the mean, and also knowing that the PM_{10} level is normally distributed, he can use the empirical rule to estimate the likelihoods of upper levels of PM_{10} in Xalostoc.

53. a) $CV = 17.1\%$.

b) CV will become larger. By adding both heavier people (the football players) and lighter people (the children), the average weight may stay about the same but the standard deviation for weight will become larger due to more variation in weights. The ratio of the standard deviation to the mean will likely increase.

55. (The questions refer to *populations* of data. Note that data for Ontario include the ridings in the Greater Toronto Area.)

(a) $CV = 62.3\%$; (b) $CV = 77.9\%$; (c) the relative variability in the winning margins was greater in Ontario than in Quebec.

Basic Concepts 5.6

57. Kurtosis measures the extent to which a distribution is flat or tall. A positive kurtosis indicates a tall distribution with high peaks and "fat" tails. A flat-shaped distribution with "thin" tails shows a negative kurtosis.

59. The normal distribution.

61. Approximately symmetrical with a slightly negative kurtosis (i.e., flatter relative to a normal distribution).

63. Left-skewed distribution, with one extreme small value.

Applications 5.6

65. Density: Right-skewed with six outliers; if outliers are excluded, the distribution appears to be approximately normal. Hardness: Right-skewed distribution.

Review Applications

Note: If you are using a computer program, different exact solutions may be displayed, depending on your software.

67. Note: As worded in the question, the data can be interpreted as a *population.*

(a) Between 1,844,707 and 3,630,362; (b) between 2,291,121 and 3,183,948; (c) $z = 0.715$.

69. a) Crude oil: Five-number summary: Min = 1,841,940; Q1 = 2,359,316; Median = 2,695,863; Q3 = 3,154,128; Max = 3,517,450.

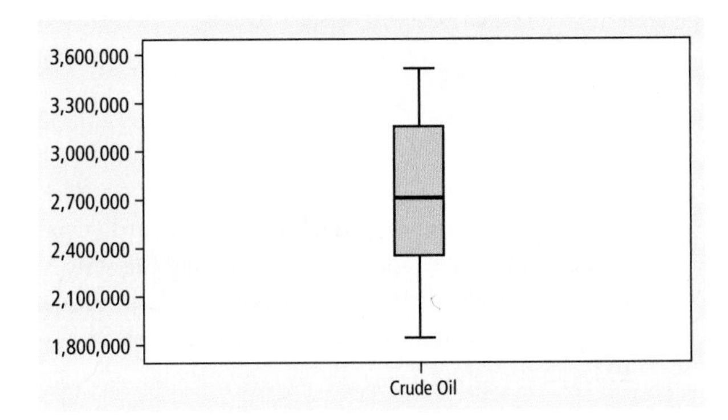

b) Natural gas: Five-number summary: Min = 157,086; Q1 = 379,750; Median = 579,132; Q3 = 625,927; Max = 699,415.

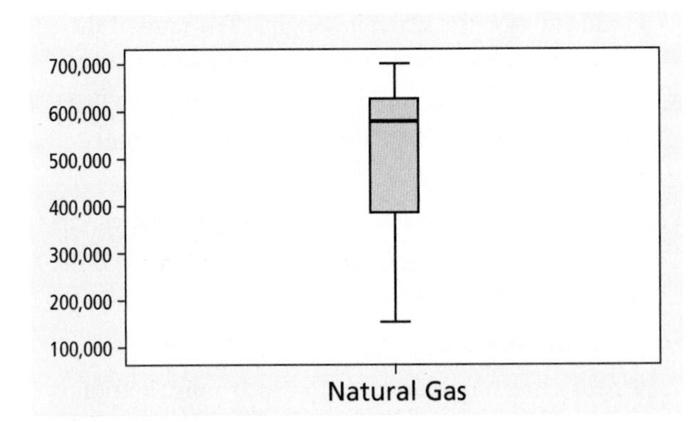

71. Range = 6.1, $s^2 = 1.73$, $s = 1.32$, $CV = 66.9\%$. (Note: Although the mean is negative, the ratio, as a percent, between the *magnitudes* of s and $\bar{x}$ is not negative.)

73. (a) Lower side: −4.60, higher side: 0.66; (b) 75%, 95%.

75. a) Ret_3Mth: Five-number summary: Min = −5.3; Q1 = 0.175; Median = 1.5; Q3 = 2.6; Max = 4.3.

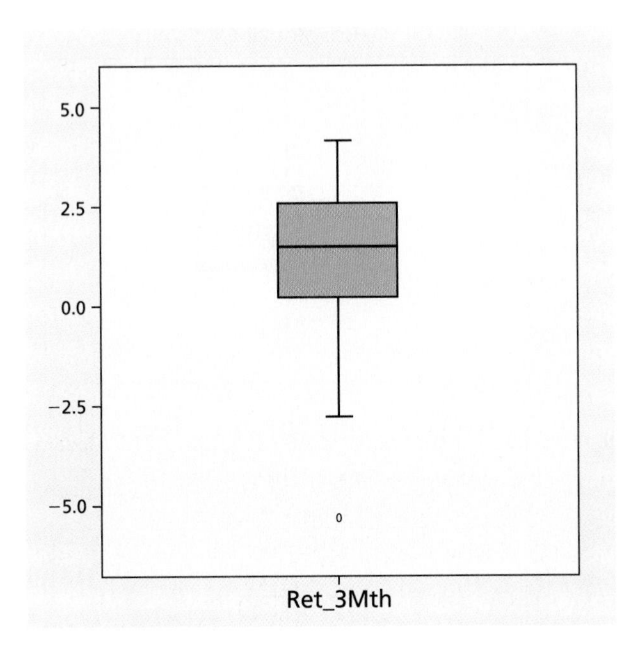

b) MER: Five-number summary: Min = 0.43; Q1 = 2.455; Median = 2.89; Q3 = 3.293; Max = 5.63.

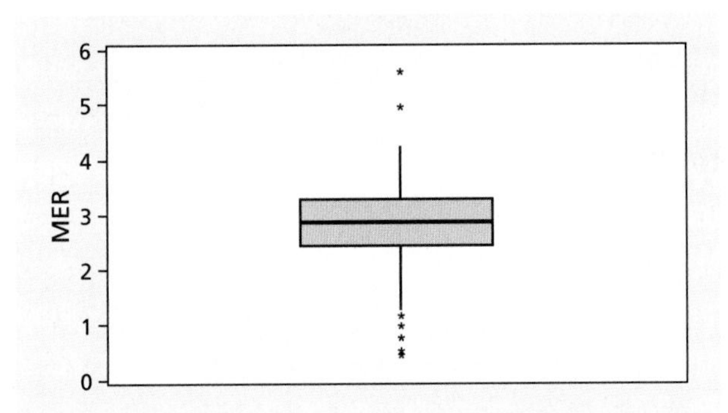

77. Note: The question refers to *all* national parks; i.e., a population.

a) Range = 44,798.3; $\sigma^2 = 102,892,095.9$; $\sigma = 10,143.6$.

b) Highly variable, ranging from the smallest park (area = 8.7 km^2) to the largest park (44,807 km^2). This variability is reflected in all three measures shown in (a); also note the large relative variation = 10,143.6/5,885.9 × 100% = 172%.

79. IQR ≈ 6,402.

81. Very right-skewed distribution, with large outliers.

83. a) Lower side: 0; higher side: 1,158; i.e., the actual values do extend to the theoretical boundaries for $(\bar{x} \pm 3s)$; (b) 89%, 99.7%; (c) $z = 2.25$.

85. a) Repair costs: Five-number summary: Min = 194.0; Q1 = 328.0; Median = 475.0; Q3 = 715.0; Max = 1,158.0.

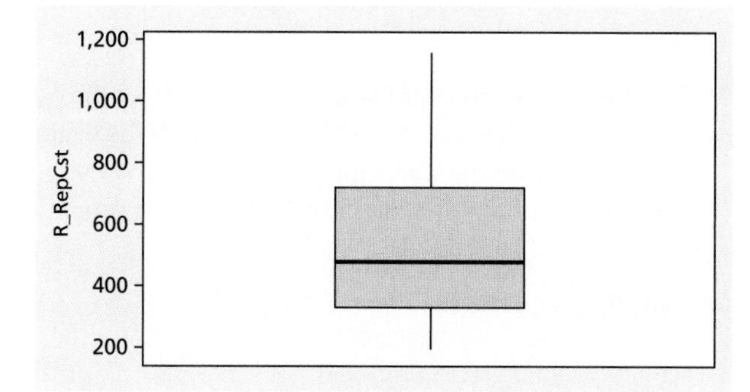

b) Fuel costs: Five-number summary: Min = 19.89; Q1 = 25.70; Median = 30.25; Q3 = 39.33; Max = 57.16.

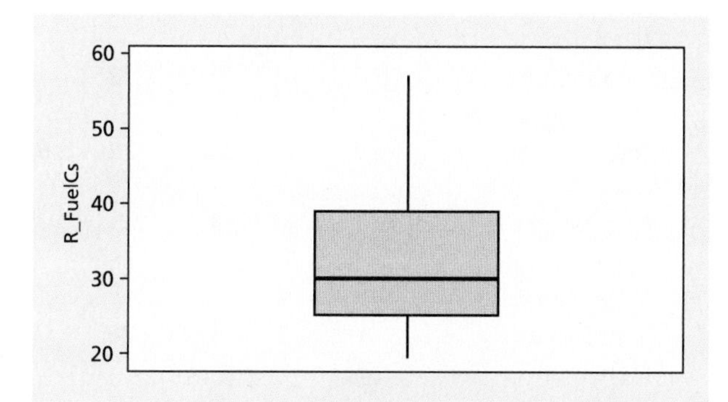

Chapter 6 Concepts of Probability

Basic Concepts 6.1

1. a) An experiment is a special observation under controlled conditions in order to achieve a value for a random variable.
b) The sample space is the set of all possible outcomes of an experiment.
c) Outcomes are the results from an experiment.
d) Events are outcomes or combinations of possible outcomes that the researcher judges can be classified as single occurrences of interest.

3. a) Collection of all 36 possible outcomes when the two dice were rolled.
b) Possible answers are if the first die and second die display, respectively, (i) 1 and 2; (ii) 3 and 6; (iii) 6 and 4; (iv) 5 and 5.
c) All possible outcomes where the two dice together add to 7 (e.g., a 1 and a 6).
d) 0.1667.

5. a) The (stop or don't-stop) actions taken by all 500 drivers.
b) Possible answers are:
- 200 drivers stop fully; 300 drivers do not stop fully.
- 400 drivers stop fully; 100 drivers do not stop fully.
- 125 drivers stop fully; 375 drivers do not stop fully.
- 430 drivers stop fully; 70 drivers do not stop fully.

7. Subjective probability.

9. a) *P*(red marble) = 0.5.
b) *P*(one blue and one white marble—in either order) = 0.125.

Applications 6.1

11. (a) *P*(sum = 58) = 0.0667; (b) *P*(sum < 60) = 0.4.

13. a) *P* = 0.61.
b) No. For nondrinkers, the historical proportions among beer drinkers are not representative. Their interests and preferences could be quite different.

15. (a) *P*(living in Manitoba) = 0.0372; (b) *P*(living in a province located west of Ontario) = 0.3017; (c) *P*(being Catholic) = 0.4365; (d) *P*(Buddhist living in the Yukon) = 0.000004.

Basic Concepts 6.2

17. a)

Book Return Status	Engi-neering	Business and Humanities	Health Sciences	Total
On Time	0.07	0.49	0.31	0.87
Late	0.03	0.06	0.04	0.13
Total	0.10	0.55	0.35	1.00

b) *P*(late) = 0.13; (c) *P*(late | Health Sciences student) = 0.1143; (d) *P*(Health Sciences student | on time) = 0.3563.

19. *P*(enjoyed concert | voiced an opinion) = 0.7074.

21. (a) *P*(low grade) = 0.2118; (b) *P*(low grade | frequently absent) = 0.4706; (c) *P*(frequently absent) = 0.20; (d) *P*(frequently absent | high grade) = 0.1053.

Applications 6.2

23. (a) *P*(atomic number < 21 | alkaline earth element) = 0.5; (b) *P*(atomic number of 12 | alkaline earth element whose atomic number is < 21) = 0.3333.

25. (a) *P*(individual chose maple syrup | individual did *not* choose Canadian beer) = 0.2974; (b) *P*(individual chose screech | individual chose a liquid beverage) = 0.0467.

27. a) *P*(spam) = 0.6036.
b) *P*(spam | 2005) = 0.7251; *P*(spam | 2004) = 0.4560. It was more likely that an e-mail received in 2005 was spam.

c) $P(2005 \mid \text{virus}) = 0.1768$; $P(2004 \mid \text{virus}) = 0.8232$.

Given that the e-mail was a virus, it was more likely to have come in 2004.

Basic Concepts 6.3

29. Independent, because $P(A \text{ and } B) = 0.14 = P(A) \times P(B) = 0.14$.

Dependent, because $P(A \text{ and } B) = 0.36 \neq P(A) \times P(B) = 0.40$.

31. $P(\text{not win the scholarship}) = 0.95$.

33. (a) $P(\text{both cards are aces}) = 0.0045$; (b) $P(\text{both cards are face cards}) = 0.0498$; (c) $P(\text{first card} = \text{ace } or \text{ second card} = \text{ace})$ (or both) $= 0.1493$; (d) $P(\text{neither card is an ace}) = 0.8507$.

35. (a) $P(\text{no} \mid \text{male})$ 0.4; (b) $P(\text{male and no}) = 0.198$; (c) $P(\text{female and yes}) = 0.4257$; (d) $P(\text{male and no } or \text{ female and yes}) = 0.6238$.

37. 0.0278.

Applications 6.3

39. a) $P(\text{home run and Boston}) = 0.0729$.

b) $P(\text{both selections are home run and Boston}) = 0.0046$.

c) Neither event is more likely than the other. Both events have the same chance to occur: $P(\text{RBI and Houston}) = P(\text{BB and Philadelphia}) = 0.0417$.

d) $P(\text{Houston or Baltimore}) = 0.375$.

e) $P(\text{RBI or Detroit}) = 0.1771$.

41. (a) $P(\text{at least one is struck by lightning}) = 0.00000064$; (b) $P(\text{both are struck by lightning}) = 0.0000000000001$; (c) $P(\text{neither are struck by lightning}) = 0.99999936$.

43. (a) $P(\text{2005 and spam}) = 0.3976$; (b) $P(\text{2005 or spam}) = 0.7543$; (c) $P(\text{virus received in 2005}) = 0.0015$; (d) $P(\text{virus received in 2004}) = 0.0068$.

Basic Concepts 6.4

45. $P(A \mid B) = 0.2893$.

47. $P(\text{night} \mid \text{defective}) = 0.6533$.

49. (a) $P(\text{the Siamese cat is the father of the litter} \mid \text{a kitten with a curly tail is born}) = 0.1927$; (b) $P(\text{the Ocicat is the father of the litter} \mid \text{a kitten with a curly tail is born}) = 0.7706$; (c) $P(\text{the Devon Rex cat is the father of the litter} \mid \text{a kitten with a curly tail is born}) = 0.0367$.

51. (a) $P(\text{telephone} \mid \text{payment}) = 0.5455$; (b) $P(\text{letter} \mid \text{payment}) = 0.1636$; (c) $P(\text{personal visit} \mid \text{payment}) = 0.2909$.

53. (a) $P(A \mid X) = 0.4444$; (b) $P(B \mid X) = 0.2963$; (c) $P(C \mid X) = 0.0741$; (d) $P(D \mid X) = 0.1852$.

Applications 6.4

55. (a) $P(\text{underground}) = 0.4318$; (b) $P(\text{Kentucky}) = 0.3088$; (c) $P(\text{underground} \mid \text{Kentucky}) = 0.5322$; (d) $P(\text{Kentucky} \mid \text{underground}) = 0.3806$.

57. $P(\text{banished} \mid \text{surviving}) = 0.3102$.

59. (a) $P(\text{the wine costs over \$14.00}) = 0.7333$; (b) $P(\text{the wine's sweetness is 1 on the scale}) = 0.6$; (c) $P(\text{the wine's sweetness is 1 on the scale} \mid \text{the wine costs over \$14.00}) = 0.6364$; (d) $P(\text{the wine costs over \$14.00} \mid \text{the wine's sweetness is 1 on the scale}) = 0.7778$.

Review Applications

61.

	Colour Code								
Carat Weight	D	E	F	G	H	I	J	K	Total
0 to 0.5 carats	1	4	20	19	5	9	5	3	66
More than 0.5 to 1 carat		5	6	6	1	4	3	3	28
More than 1 to 2 carats			2	2					4
More than 2 carats				1					1
Totals	1	9	28	28	6	13	8	6	99

63. (a) $P(\text{more than two carats and colour of category "G"}) = 0.0101$; (b) $P(\text{more than two carats or colour of category "G"}) = 0.2828$; (c) $P(\text{more than 1 carat}) = 0.0505$; d) $P(\text{faint yellow}) = 0.0606$.

65. (a) $P(\text{first diamond} = \text{"G" and second diamond} = \text{"G"}) = 0.0779$; (b) $P(\text{first diamond} \neq \text{"G" and second diamond} \neq \text{"G"}) = 0.5123$; (c) $P(\text{at least one} = \text{"G"}) = 0.4877$.

For the solutions to Exercises 67–70, it is assumed that the graduate herself is from the United States but that her travels will all be in non-U.S. countries.

67. (a) $P(\text{Canada}) = 0.4466$; (b) $P(\text{Cote d'Ivoire}) = 0.0097$; (c) $P(\text{Canada or Cote d'Ivoire}) = 0.4563$; (d) $P(\text{African continent}) = 0.0356$.

69. (a) $P(\text{both live in Africa}) = 0.0012$; (b) $P(\text{that one (in either order) lives in Africa and the other lives in Europe}) = 0.0113$.

71. (a) P(area is over 1,000 km^2) = 0.5349; (b) P(located in British Columbia) = 0.1860; (c) P(area is over 1,000 km^2 and located in British Columbia) = 0.1163; (d) P(area is less than 30 km^2 and located in Ontario) = 0.0698; (e) P(area is less than 30 km^2 or located in Ontario) = 0.1628.

73. P(located in British Columbia | area is over 1,000 km^2) = 0.2173.

75. (a) P(visits an Ontario park each year) = 0.0027; (b) P(never visits an Ontario park) = 0.6371.

77. (a) P(both are adults) = 0.6165; (b) P(neither are adults) = 0.0462; (c) P(exactly one is an adult) = 0.3374; (d) P(both live in Alberta) = 0.0102; (e) P(both live in Alberta *or* both live in Manitoba) = 0.0115.

79. (a) P(lives in Ontario) = 0.3855; (b) P(suffers from stress) = 0.2320; (c) P(suffers from stress | lives in Ontario) = 0.2310; (d) P(adult lives in Ontario | suffers from stress) = 0.3838.

81. a) P(no sales fee) = 0.3333.
b) P(no sales fee *or* is front end loaded) = 0.3833.
Note: For solutions (c)–(e), unrated funds have been included in the denominator.
c) P(rating = 4) = 0.1833.
d) P(rating between 2 to 4, inclusive) = 0.6000.
e) P(rating = 4 *or* charges no sales fee) = 0.4333.

83. (a) P(no sales fee | rating = 5) = 0.6000; (b) P(rating = 5 | no sales fee) = 0.15; (c) P(front end loaded | rating less than 4) = 0.0323.

85. (a) P(bridge is in Canada | steel arch bridge) = 0.3333; (b) P(steel arch bridge | located in Canada) = 0.5; (c) P(cable-stayed bridge | bridge is over 600 metres long) = 0.1429; (d) P(bridge is in Canada | it is under 200 metres long) = 0.0263.

87. (a) P(bridge was built before 1930) = 0.0513; (b) P(simple truss bridge) = 0.0769; (c) P(bridge was built before 1930 | simple truss bridge) = 0.3333; (d) P(bridge was a simple truss type | bridge was built before 1930) = 0.4996.

89. (a) P(well depth is greater than 50 feet) = 0.4286; (b) P(well depth is less than 20 feet) = 0.2857; (c) P(well depth is between 22 to 60 feet) = 0.2857.

91. (a) P(both wells used for domestic water supply) = 0.1429; (b) P(both wells used for public water supply) = 0.1429; (c) P(both wells have depths that, if added, add to less than 25 feet) = 0.0476; (d) P(both wells have depths that, if added, add to more than 100 feet) = 0.3333.

Chapter 7 Discrete Probability Distributions

Basic Concepts 7.1

1. (a) −3, 0, 3, 6; (b) $P(x > 0) = 0.5$; (c) Yes.

3.

Values	Probability
0	25/36 = 0.6944
1	10/36 = 0.2778
2	1/36 = 0.0278

5. (a) P(purchase Gator Gum) = 0.19; (b) $E(x)$ = \$0.315; (c) σ = \$0.656; (d) expected revenue from sales = \$141.75.

7. a)

Response	Probability
Yes, enjoyed the concert.	0.648
No, did not enjoy the concert.	0.268
Don't know or no response.	0.084

b) No, because the response variable is a categorical variable; i.e., no numerical values are assigned to the response categories.

Applications 7.1

9. a)

Atomic number	Probability
4	1/6 = 0.1667
12	1/6 = 0.1667
20	1/6 = 0.1667
38	1/6 = 0.1667
56	1/6 = 0.1667
88	1/6 = 0.1667

b) $E(x)$ = 36.3.

11. a)

Identified E-Mail Type	Probability
Spam	0.7251
Viruses	0.0027
Acceptable for delivery	0.2722

b) Estimated extra cost due to viruses or spam = \$789.90.

13. (a) P(next call cost \$0.084) = 0.127; (b) P(cost > \$0.084) = 0.586; (c) $E(x)$ = \$0.137; (d) Expected cost to make 200 calls = \$27.40.

Basic Concepts 7.2

15.
- Experiment consists of a fixed number of identical trials, n.
- Each trial has only two possible mutually exclusive results (success and failure).

- Outcome of any trial is independent of outcome of other trials.
- Probability, p, of a successful trial is constant for all trials.

17. (a) $P(X = 15) = 0.0072$; (b) $P(X = 5) = 0.0017$; (c) $P(X > 7) = 0.9662$; (d) $P(X \geq 7) = 0.9906$; (e) $P(3 \leq X \leq 7) = 0.0338$.

19. (a) $P(X = 10) = 0.0135$; (b) $P(X = 0) = 0.0000275$; (c) $P(2 \leq X \leq 4) = 0.0944$.

21. $\mu = 1.8$ (thus, between 1 to 2 days); $\sigma = 1.301$ days.

23. The mean and standard deviations for the binomial distributions:

(a) $\mu = 6, \sigma = 2.049$; (b) $\mu = 72, \sigma = 5.367$; (c) $\mu = 50, \sigma = 5.0$; (d) $\mu = 87.5, \sigma = 7.542$; (e) $\mu = 42, \sigma = 3.240$.

Applications 7.2

25. a) Between 0 and 1 pair of twins ($\mu = 0.6$).
b) Around 6 pairs.
c) $P(X > 1) = 0.1217$.
d) One of the conditions for using the binomial model would not apply if the chance of older mothers having twins is higher than for younger mothers; there would not be a constant probability for all trials.

27. a) $\mu = 39$.
b) A new-business survival rate of 9 out of 20 (i.e., 0.45) can be considered as unusually high because, according to the binomial distribution, the expected probability of at least nine Canadian businesses surviving is very small [$P(X \geq 9) = 0.00046$].
c) The chances of business survival may increase for new companies that accept good advice and have a budget for consultants. In that case, the overall estimated success rate of 13% may not be a constant success rate for the entire population of new businesses, which would violate the condition for a constant success rate for each trial.

29. a) The assumptions must be made that the risk of dog bites, as observed in Pittsburgh, will be the same in the city in question and will remain constant. These assumptions may not be valid, since the city of interest might have a different (or changing) mixture of dog breeds, dog-restraint rules, and different social attitudes and behaviours among dog owners. Also, the outcomes of trials (the biting or non-biting of individuals) are not dichotomous, since a single person can be bitten more than once.
b) Between 883 and 884 ($\mu = 883.5$); c) $P(10 \leq X \leq 15) = 0.0000$.

31. a) Between 20 and 21 students ($\mu = 20.7$), between 4 and 5 students ($\sigma = 4.340$).
b) $P(X = 0) = 0.0000$, $P(X \geq 20) = 0.5975$, $P(X \geq 40) = 0.00004$.
c) No. Observations regarding the depression rate could be made independently for several subgroups (e.g. for different ages, genders, ethnic backgrounds, social statuses). Binomial probabilities could then be calculated based on the group that is of interest.

Basic Concepts 7.3

33. To apply the binomial distribution, there has to be constant success rate from trial to trial. This condition does not apply for a hypergeometric distribution, which is based on drawing a random sample without replacement from a population—so the conditional probability of success changes with each trial.

35. (a) $P(X = 1) = 0.2679$; (b) $P(X \geq 2) = 0.7143$; (c) $E(x) = 1.875$; (d) $\sigma^2 = 0.502, \sigma = 0.709$.

37. $E(x) = 2$ defective bikes.

39. (a) $P(X = 10) = 0.0044$; (b) $P(X \geq 5) = 0.0080$.

41. $E(x) =$ between 4 and 5 Canada geese ($\mu = 4.541$).

Applications 7.3

43. $E(x) =$ about 2 or 3 left-handed pitchers ($\mu = 2.5$).

45. $E(x) =$ between 4 and 5 ($\mu = 4.67$), $\sigma = 1.31$.

47. (a) $P(X = 6) = 0.0000$; (b) $P(X = 0) = 0.0420$, (c) $P(X = 3) = 0.2398$, (d) $P(X \geq 3) = 0.2867$.

Basic Concepts 7.4

49. (a) 0.1680; (b) 0.1804; (c) 0.0595.

51. (a) 0.0404; (b) 0.2378; (c) 0.6375.

53. (a) $P(X < 5) = 0.4405$; (b) $P(X = 7) = 0.1044$.

55. (a) $P(X = 0) = 0.0025$; (b) $P(X = 6) = 0.1606$; (c) $P(X > 8) = 0.1528$; (d) $P(X \geq 5) = 0.7149$.

57. (a) $P(X = 0) = 0.0067$; (b) $P(X = 5) = 0.1755$; (c) $P(4 \leq X \leq 8) = 0.6669$.

Applications 7.4

59. (a) $P(X = 0) = 0.1353$; (b) $P(X = 5) = 0.0361$; (c) $P(2 \leq X \leq 4) = 0.5413$; (d) $P(X \geq 2) = 0.5940$.

61. (a) $P(X < 150) = 0.0346$; (b) $P(X > 180) = 0.2814$; (c) $P(150 \leq X \leq 180) = 0.6840$.

63. a) $P(X = 0) = 0.0016$.
b) $P(X < 4) = 0.1146$.
c) $P(X > 8) = 0.2037$.
d) Not very likely. The rate is probably higher when the services of family physicians and other clinics are not available (e.g., late afternoon, evenings, nights, and weekends). To adjust the calculations, determine the

average arrival rate for different time slots during the day and for the weekend, and base your calculations only on comparable times.

Review Applications

65. a)

Number of Candidates	Probability
4	2/14 = 0.1429
5	3/14 = 0.2143
6	5/14 = 0.3571
7	4/14 = 0.2857

b) $E(x)$ = Between 5 and 6 candidates (= 5.79); $\sigma = 1.013$.

c) 0.2857.

d) If there is an alternative location in Provencher where at least 7 candidates can have their debate, I might prefer not take the over 1 in 4 (28.6%) chance that the town hall stage will too small. But if there is no alternative, the risk might be acceptable: The expected number of candidates is 5 or 6, and possibly fewer. Also note that in 2004, Provencher had among the fewest numbers of candidates (4) for ridings in Manitoba, so its local probability of over 6 candidates may be less than the federal or Manitoba averages.

67. a) Between 192 and 193 times ($\mu = 192.6$).

b) $\sigma = 10.496$.

c) $P(X \geq 4) = 0.2201$; $P(X \leq 2) = 0.4864$.

d) The probability that Cory will reach base on his sixth time at bat is 0.428, because if the conditions of the binomial model apply, then Cory's on-base percentage is constant from trial to trial.

e) His success rate could vary, depending on his general fitness, events in his home life, and whether he suffers from an injury. Other factors could include whether a game is a home match, and the strength of the opposing pitchers or fielders.

69. Between 1 and 2 individuals ($\mu = 1.35$); $\sigma = 1.14$.

71. (a) $P(X > 6 \mid \mu = 1.258) = 0.0003$; (b) $P(X = 0 \mid \mu = 1.258) = 0.2842$; (c) $P(X > 6 \mid \mu = 0.364) = 0.0000$; (d) $P(X = 0 \mid \mu = 0.364) = 0.6949$.

73. Between 0 and 1 complaints ($\mu = 0.80$); $\sigma^2 = 2.509$; $\sigma = 1.584$.

75. Between 4 and 5 days ($\mu = 4.6$); $\sigma = 1.576$.

77. (a) $P(X > 50) = 0.8979$; (b) $P(X > 66) = 0.0240$.

79. (a) $P(X \geq 20) = 0.5396$; (b) $P(X = 0) = 0.0000$; (c) $P(14 \leq X \leq 17) = 0.2255$.

81. Between 5 and 6 victims ($\mu = 5.376$); $\sigma = 1.723$.

Chapter 8 Continuous Probability Distributions

Basic Concepts 8.1

1. (a) *P*(between 220 and 260 days) ≈ 0.68; (b) *P*(between 200 and 280 days) ≈ 0.95; (c) *P*(fewer than 180 days or more than 300 days) ≈ 0.003.

3. (a) 0.2778; (b) 0.1667; (c) −0.2222; (d) −0.0833.

5. (a) −1.3165; (b) −0.7089; (c) −0.4810; (d) 1.5190; (e) 0.0000.

7. (a) $P(X < 9) = 0.2980$; (b) *P*(two DVDs *both* cost < \$9.00) = 0.0888.

9. (a) $P(X > 23) = 0.2839$; (b) $P(22 < X < 23) = 0.2729$; (c) 20.41; (d) between 20.59 and 23.81.

Applications 8.1

11. (a) 1.09; (b) 0.79.

13. a) $P(5 < X < 8) = 0.1186$.

b) 16.60.

c) Mean is larger than median, indicating a right-skewed distribution.

d) An overestimate. Since the values are concentrated in the lower part of the distribution, 90% of the area under the curve will be represented by a smaller *x*-value than for a normal distribution.

15. (a) 1st quartile = −3.25, 3rd quartile = 2.95; (b) $P(X < 6.2) = 0.9167$; (c) $P(X \geq -6.2) = 0.9063$; (d) $P(-6.2 < X < 6.2) = 0.8230$.

Basic Concepts 8.2

17. • Compare mean and median (for normality they need to be almost identical).

• Check whether approximately 95% of the data fall within ± 2 standard deviations from the mean.

• Check whether nearly all values fall within ± 3 standard deviations from the mean.

Sometimes there are outliers that cannot be detected by these numerical measures, but they can be visualized using different plots.

19. A normal probability plot is a graph that displays for each actual value of *X* on the *x*-axis the expected value of *X* (based on the assumption that *X* is normally distributed) on the *y*-axis. The ideal fit of these two variables is displayed is a positively sloped line. If the plotted values are arranged close to the straight line, the distribution could be considered as normal. For skewed distributions, some of the plotted values diverge significantly from the straight line. Outliers appear relatively far from the straight line.

21. Only some checks are listed:

- $\bar{x} = 749.83 \approx \text{median} = 749.9$.
- $\bar{x} \pm 3s = 746.88$ to 752.78; no data value is outside this range ($s = 0.9832$).
- See below

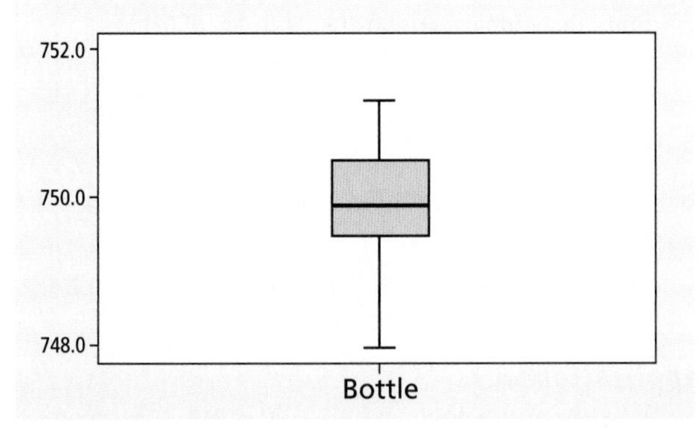

- The distribution of volumes is approximately normal. However, there are two lower values that skew the distribution slightly to the left.

23. For each data set, only some checks are listed.

Data set 1:

- $\bar{x} = 1{,}992.0 > \text{median} = 1{,}959.0$
- See below

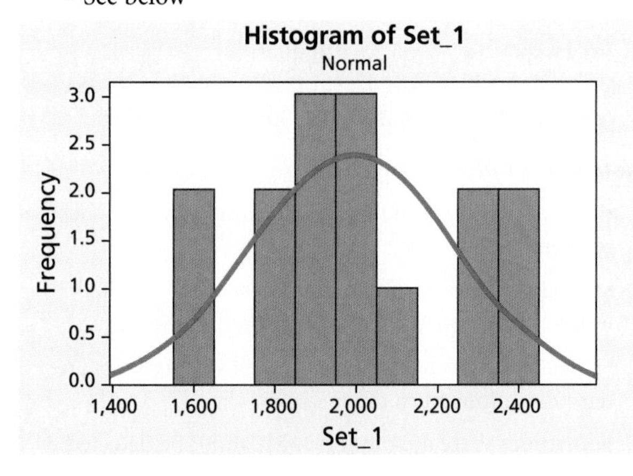

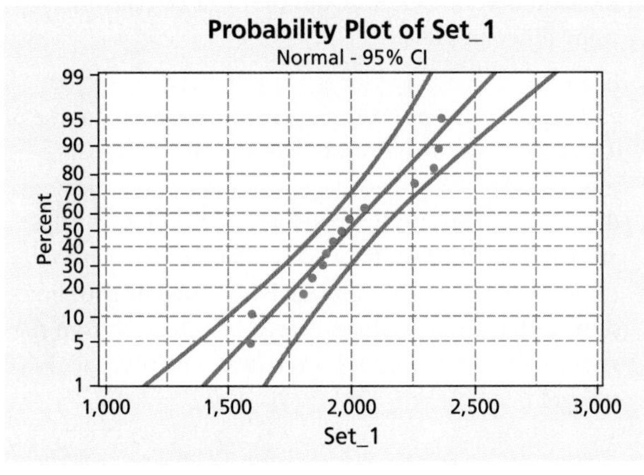

- Based on the comparison of mean and median, the distribution seems to be right-skewed. According to the histogram and normal probability plot, the distribution seems to be approximately normal; yet, in the normal probability plot, both tails of the distribution deviate from the normal line. Overall, you can conclude that the underlying distribution loosely approximates a normal distribution.

Data set 2:

- $\bar{x} = 50.61 > \text{median} = 47.00$.
- $\bar{x} \pm 2s = 27.67$ to 73.55; one data value (6.67% of cases) outside ($s = 11.47$).
- See below

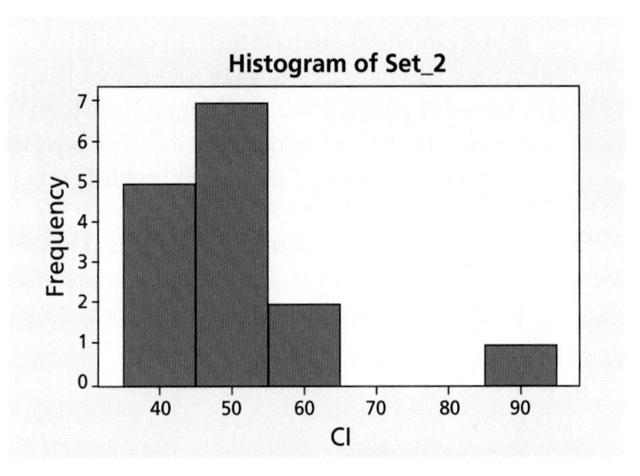

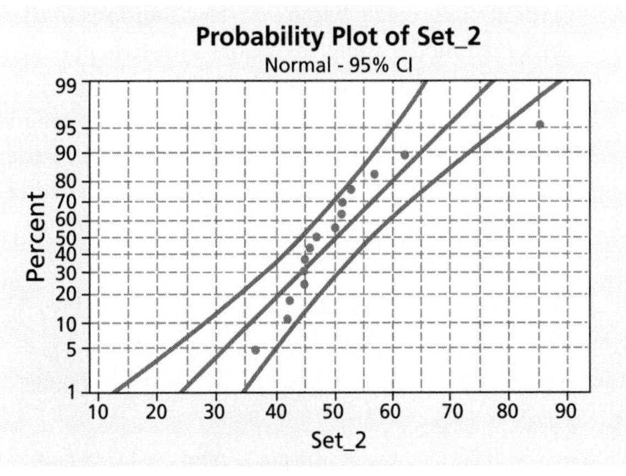

- The mean of the distribution is larger than its median, indicating a right-skewed distribution, which the histogram also suggests. There is one outlier at the upper tail (visible on the upper right of the normal probability plot), which falls above the boundary of +3 standard deviations from the mean. Combined with the fluctuation of the values on the probability plot, it appears that the underlying population is not very normally distributed.

Data set 3:

- $\bar{x} = 18.33 \approx$ median $= 18.00$.
- See below

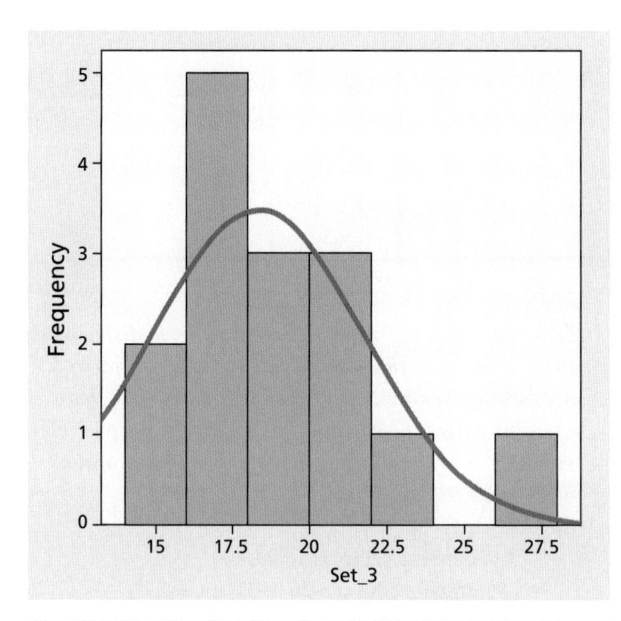

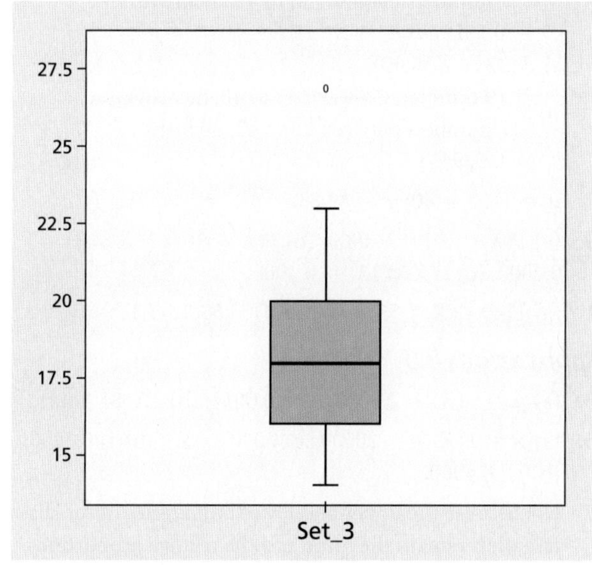

- The mean and median of the distribution are almost identical. An outlier at the upper tail affects the distribution and skews it to the right. Thus, it depends on the interpretation of the outlier whether you could conclude that the underlying population is normally distributed.

25. This example shows why it is useful to test normality with a variety of techniques: The mean is very close to the median, suggesting normality. The boxplot shows a perfect symmetrical shape, but with short whiskers, but the histogram reveals a bimodal, not normal, distribution. The normal probability plot by itself is difficult to interpret. Overall, this sample follows a roughly symmetrical, but not normal, distribution.

Applications 8.2

27. a) The distribution of solubilities of gases in water is strongly right-skewed (the median is much smaller than the mean). Three outliers on the upper tail of the distribution stretch the curve to the right.

b) No, because the shape of this data set is far from normal.

29. a) Not exactly; it appears to be right-skewed, with outliers.

b) If we use the convention described in Chapter 5 (in the text referring to Figure 5.16), there are two outliers (114, 802) in the data set. (The numbers of identified outliers may vary, depending on your procedures and software.) With outliers excluded, the distribution is still only approximately normal.

c) $P(X > 250 \mid \bar{x} = 372.067, s = 106.511) = 0.8741$ (The exact value will vary if different outliers are identified in (b).)

d) Yes. Even with outliers excluded, more than the required 80% of months meet the criterion of at least 250 messages; some months had considerably higher numbers of messages.

31. a) The data appear not to be normally distributed, but right-skewed (the mean is somewhat greater than the median), possibly with an outlier on the upper tail. The skewing is visible in both a histogram (shown below) and a boxplot, and non-normality is also suggested by the normal probability plot (shown below).

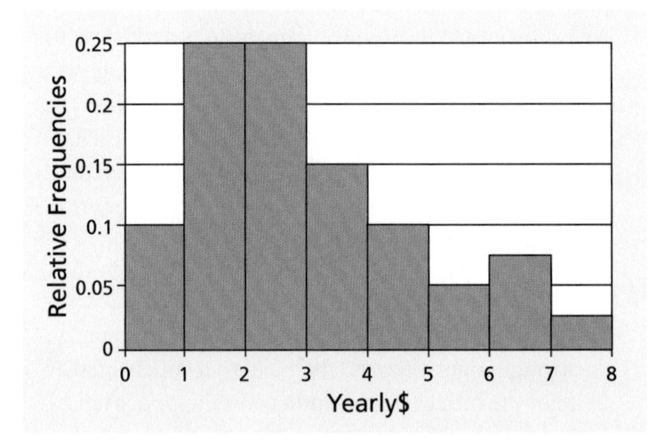

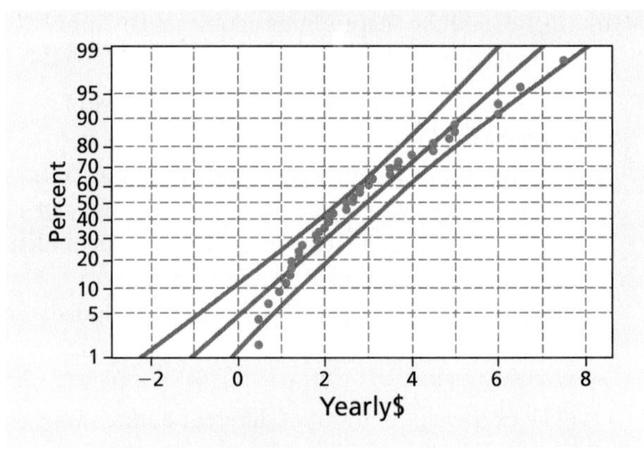

b) As there are different possible patterns for right-skewed distributions, it is difficult to answer this question with a general rule. For a normal distribution with this sample's mean (2.895) and standard deviation (1.749), 1 (million) is only about one standard deviation below the mean, so we'd expect more of the cases to be below that value than the 10% shown in the distribution; our estimate would be an overestimate.

Basic Concepts 8.3

33. (a) $\mu = E(X) = 22.5, \sigma = 4.33$; (b) $P(X > 25) = 0.3333$; (c) $P(18 < X < 25) = 0.4667$.

35. a) Discrete uniform distribution.
b) 26.
c)

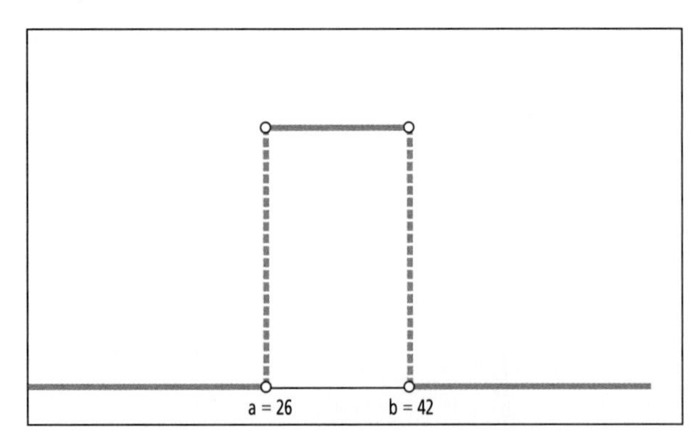

Note for (d) and (e): For a discrete uniform distribution, the probabilities that can be calculated by Formula 8.7 are only approximate. Especially for a case like this exercise, where the variable has relatively few possible values, it is preferable to calculate using basic probability rules.

d) $P(X < 32)$
= $P(X$ between 26 to 31 inclusive)
= (6 numbers between 26 to 31, inclusive)/ (17 numbers between 26 to 42, inclusive)
= 0.3529

e) $P(32 \leq X \leq 40)$
= (9 numbers between 32 to 40, inclusive)/ (17 numbers between 26 to 42, inclusive)
= 0.5294

37. $\mu = E(X) = 40; \sigma = 8.083$.

39. (a) $P(X < 10) = 0.4000$; (b) $P(X \leq 8) \approx P(X < 8) = 0.2000$; (c) $P(X \geq 12) = 0.4000$.

41. (a) $P(40 < X < 55) = 0.2500$; (b) 55; (c) 83.

Applications 8.3

43. (a) $\mu = E(X) = 2.1238, \sigma = 0.0054$; (b) 2.1181 hours.

45. (a) $\mu = E(X) = -2000$ (i.e., 2000 B.C.E.); (b)(i) 0.7000; (b)(ii) 0.4000.

47. No, because the upper and lower boundaries of the distribution remain the same and the relative proportions of water falling in the different ranges do not change.

Basic Concepts 8.4

49. (a) 0.0025; (b) 0.0498; (c) 0.9889.

51. (a) 0.2231; (b) 0.7769; (c) 0.4226; (d) 0.6667.

53. (a) 1 car / minute; (b)(i) 0.3935; (b)(ii) 0.3679; (b)(iii) 0.4237.

55. (a) 0.5 customers / minute; (b)(i) 0.3935; (b)(ii) 0.2212; (b)(iii) 0.1723.

57. (a) $\mu = 0.2$ weeks between articles; (b) $\sigma = 1.4$ days; (c) 0.4895; (d) 0.3722.

Applications 8.4

59. (a) $\mu = 0.5$ minutes between calls; (b) $\sigma = 0.5$ minutes; (c)(i) 0.3935; (c)(ii) 0.2231; (c)(iii) 0.3834.

61. a) $\mu = 75{,}187.97$ miles between accidents.
b) $\sigma = 75{,}187.97$ miles.
(Hint for (c)–(e): Convert numbers of train miles into units of *millions* of train miles.)
c) 0.5498.
d) 0.3451.
e) 0.0491.

63. (a) $\lambda = 0.4$ battery failures per year per laptop; (b)(i) between 6 and 7 battery failures (i.e., 6.594); (b)(ii) approximately 11 battery failures (i.e., 11.014); (b)(iii) approximately 14 battery failures (i.e., 13.976).

Review Applications

65. (a) $\mu = 45{,}693.71$; $\sigma = 5{,}340.80$ (Note: Data are given for the full *population*.); (b)(i) $P(X < 43{,}000) = 0.3070$; (b)(ii) $P(X > 49{,}000) = 0.2679$; (b)(iii) $P(43{,}000 < X < 49{,}000) = 0.4251$.

67. a) The numbers of votes cast in Quebec ridings were somewhat normally distributed. The mean (45,693.71) and median (46,140) are relatively close. The histogram appears skewed to the left, although the boxplot reveals that if four outliers on the left tail were excluded, the distribution is reasonably symmetrical. The normal probability plot shows some non-normality.
b) The answer to Exercise 65(b)(i) would be an overestimate because the estimate does not account for the fact that for a left-skewed distribution, fewer than half of the cases are spread to the left of the mean. This would also cause the X-value chosen for Exercise 66 to be smaller than necessary. As there are different possible patterns for left-skewed distributions, it is difficult to anticipate the effects on the estimates for the right tail, such as in Exercise 65(b)(ii). (As it happens the estimate is close to the actual proportion.) The answer to Exercise 65(b)(iii) may be an overestimate because the estimate does not account for the fact that for a left-skewed distribution, more than half of the cases are spread to the right of the mean—and most of the data range for the intended estimate is on that side.

69. 451.12 minutes (or 7.52 hours).

71. 57.34°N. (Note: Data are considered as a population.)

73. The distribution of latitudes is close to normal but slightly skewed to the right. The mean (46.68) is very close in value to the median (46.6) and the boxplot is quite symmetrical with no outliers. However, skewing appears in the histogram, and in the normal probability plot, values at latitudes above 70°N appear as possible outliers.

75. 3.30 decimal minutes. (Note: Data are given for the full *population* of songs on the CD.)

77. The set of fees in the range \$2.00 to \$3.50 seems to be relative evenly spaced. But since this is a fairly small sample with 14 different bank accounts, the match with the uniform distribution is not perfect in detail.

79. Note: Because the data are discrete, with very few specific values, the estimates that follow (based on a continuous distribution) may be approximate:
(a) 0.67; (b) 0.70; (c) 0.37.

81. (a) $\lambda = 8.667$ heat alert arrivals per month; (b) $\mu = 0.115$ months between arrivals, $\sigma = 0.115$ months; (c)(i) $P(X < 3) = 0.5760$; (c)(ii) $P(X < 1.5) = 0.3489$.

Chapter 9 The Distributions of Sample Statistics

Note: Depending on round-off decisions you make for intermediate values, your exact solutions to some of this chapter's problems may vary slightly from those displayed.

Basic Concepts 9.1

1. No. The standard error of the mean is the standard deviation of all possible sample means for a given sample size in relation to the true population mean.

3. (a) Not necessary; (b) necessary; (c) necessary; (d) not necessary.

5. Population mean.

7. (a) $\mu_{\bar{x}} = 300$; (b) $\mu_{\bar{x}} = 300$; (c) $\sigma_{\bar{x}} = 4.564$; (d) $\sigma_{\bar{x}} = 2.635$.

9. (a) $\mu_{\bar{x}} = 11.4$.; (b) $\sigma_{\bar{x}} = 0.693$.

Applications 9.1

11. $\sigma_{\bar{x}} = 0.0121$.

13. a) $\mu_{\bar{x}} = 4.10$.
b) $\sigma_{\bar{x}}^2 = 3.17$, $\sigma_{\bar{x}} = 1.78$.
c) No; No. The central limit theorem does not inform about the shape or variance of a sampling distribution unless the population itself has a normal distribution or else the size of the samples is at least equal to about $n = 30$.

15. (a) 0.9506; (b) $\sigma_{\bar{x}} = 23.17$.

Basic Concepts 9.2

17. Applied to large samples, the central limit theorem ensures that the sampling distribution will be normal, regardless of the distribution of the population. If the sampling distribution is normal, then the formulas for calculating probabilities for a normal distribution (see Section 8.1), can be applied.

19. (a) $P(\bar{x} < 294) = 0.0537$; (b) $P(\bar{x} \geq 303) = 0.2104$; (c) $P(294 < \bar{x} < 303) = 0.7359$.

21. (a) $P(\bar{x} < 12.3) = 0.9030$; (b) $P(\bar{x} \leq 13.1) = 0.9929$; (c) $P(10.9 < \bar{x} < 11.1) = 0.0972$.

23. $P(\bar{x} < 3.4) = 0.1030$.

25. Because, without knowing the population distribution, the sample size of $n = 7$ is too small to conclude that the sampling distribution is normally distributed.

Applications 9.2

27. a) $P(196 < \bar{x} < 248) = 0.7075$
b) Because the exercise stated that the distribution of trial delays is normal, we know that even for a small *n*, the sampling distribution is normal.

29. a) $P(1141 < \bar{x} < 1212) = 0.7973$.
b) 0.8810.
c) No. In both cases, the sample sizes were larger than 30 so, according to the central limit theorem, the sampling distributions would be approximately normal.

31. a) $P(1.2 < \bar{x} < 1.3) = 0.5117$.
b) The mean of the original sample can be viewed as one possible result from the set of all possible means in the sampling distribution for samples of that size. If *n* is very large, then we know that (1) the standard error is small—so there is not likely to be a large difference between the sample mean and the population mean, which is at the centre of the distribution, and (2) the estimate is unbiased, so it is no more likely to be larger or smaller than the true population mean.

Basic Concepts 9.3

33. (a) $\mu_{(p_s)} = 0.55$; (b) $\sigma_{(p_s)} = 0.0908$; (c)(i) $P(X > 0.75) = 0.0138$; (c)(ii) $P(X < 0.40) = 0.0493$.

35. (a) $\mu_{(p_s)} = 0.09$; (b) $\sigma_{(p_s)} = 0.0342$; (c) $P(X > 0.09) = 0.5000$.

37. $P(X < 4) = 0.2578$.

39. (a) $P(X > 6) = 0.8152$; (b) $P(X < 12) = 0.8194$; (c) $P(6 \leq X \leq 12) = 0.7970$.

41. a) (i) $P(X \geq 5) = 0.7320$; (ii) $P(X \leq 7.5) = P(X \leq 7) = 0.7533$; (iii) $P(5 \leq X \leq 12.5) = P(5 \leq X \leq 12) = 0.7269$.
b) The calculations in (a) assumed that every new enrollee has an equal chance of reaching and completing Level 10. We are not told about the qualifications of most new enrollees in the school, but *if* the Level 8 transfers are more qualified than the usual new enrollees, then this batch of new students would have a higher than usual probability of completing Level 10; so the estimate in (a)(i) is an underestimate.

Applications 9.3

43. a) $P(X \geq 75) = 0.0000$.
b) $P(X \geq 9) = 0.0005$.
c) No. 9 out of 20 clients would be a success rate of 45%, which on the assumption that the consultant's clients succeed no better than the usual 13% would be highly unlikely to occur by chance (probability = 0.0005). It is more likely that this consultant's clients have a success rate higher than 13%.

45. (a) $P(X \geq 300) = 0.0025$; (b) $P(200 \leq X \leq 300) = 0.9980$.

47. a) $P(X \leq 393) \approx 0.0000$.
b) $P(X \geq 459) \approx 0.0000$.
c) Yes. Parts (a) and (b) looked at how likely it would have been for a sample from each of those two years to have produced the *other* year's proportion. For both cases, the probabilities are virtually 0.0000—i.e. very unlikely to have occurred due to sampling error. More likely, the reported years do have different proportions.

Basic Concepts 9.4

49. False.

51. One example of sample statistics for which no formal mathematical model is available is illustrated by the ratios between the counts of two possible values of a discrete variable.

53. A sampling distribution that was created by a bootstrap model can be displayed as a relative frequency distribution histogram. Based on the histogram, probabilities of drawing samples of a certain size with sample statistics of specific values can be estimated.

Review Applications

55. (a) $P(\bar{x} > 55) = 0.0127$; (b) $P(\bar{x} < 35) = 0.0526$; (c) $P(35 < \bar{x} < 55) = 0.9347$.

57. a) Lefke and Yeni İskele. (Note: The population correction factor should be used when $n \geq 5\%$ of the *adult* population. Determine the adult population per location by multiplying the whole population times 0.67.)
b) For Lefke = 0.9709; for Yeni İskele = 0.9316.

59. (a) $\mu_{\bar{x}} = 19.0$, (b) $\sigma^2_{\bar{x}} = 0.7938$, $\sigma_{\bar{x}} = 0.8910$.

61. (a) $\mu_{\bar{x}} = 21.0$; $\sigma^2_{\bar{x}} = 1.0658$, $\sigma_{\bar{x}} = 1.0324$.
(b) $\mu_{\bar{x}} = 24.7$; $\sigma^2_{\bar{x}} = 0.8192$, $\sigma_{\bar{x}} = 0.9051$.
(c) $\mu_{\bar{x}} = 29.9$; $\sigma^2_{\bar{x}} = 0.5618$, $\sigma_{\bar{x}} = 0.7495$.
(d) $\mu_{\bar{x}} = 35.6$; $\sigma^2_{\bar{x}} = 0.5000$, $\sigma_{\bar{x}} = 0.7071$.

63. a) $\mu_{\bar{x}} = 15.5$.
b) $\sigma_{\bar{x}} = 0.3227$.
c) No. According to the central limit theorem, the sample size ($n = 60$) is large enough that the sam-

pling distribution will be normally distributed, regardless of the overall population distribution.

65. (a) $P(\bar{x} > 16) = 0.0607$; (b) $P(\bar{x} > 16.5) = 0.0010$; (c) $P(16 < \bar{x} < 16.5) = 0.0597$.

67. (a) $P(X \geq 12.5) = P(X \geq 13) = 0.5539$; (b) $P(X \leq 15) = 0.7925$; (c) $P(10 \leq X \leq 17.5) = P(10 \leq X \leq 17) = 0.7958$.

69. (a) $P(X \leq 25.2) = P(X \leq 25) = 0.6879$; (b) $P(X \geq 22.5) = P(X \geq 23) = 0.6573$; (c) $P(23.4 \leq X \leq 24.3) = P(24 \leq X \leq 24)$ **or** $P(X = 24) = 0.1184$.

71. (a) $P(X \geq 41.25) = P(X \geq 42) = 0.7959$; (b) $P(X \leq 50.25) = P(X \leq 50) = 0.9037$; (c) $P(41.25 \leq X \leq 50.25) = P(42 \leq X \leq 50) = 0.6997$.

73. $P(60 \leq X \leq 63.75) = P(60 \leq X \leq 63) = 0.4123$.

Chapter 10 Estimates and Confidence Intervals

Note: Depending on round-off decisions you make for intermediate values, your exact solutions to some of this chapter's problems may vary slightly from those displayed.

Basic Concepts 10.1

1. A *point estimate* is a single value based on a sample that is used to estimate or predict the value of a population parameter. **Examples:** A bottling plant estimates that the mean amount of cola in each bottle is 2.0 litres. A student predicts that his final mark will be 91%.

An *interval estimate* is a range of values within which the true population parameter is expected to fall. **Examples:** A bottling plant estimates that the mean amount of cola in each bottle falls within the range from 1.95 to 2.05 litres. A student predicts that his final mark will be A+—i.e., somewhere in the range from 90% to 100%.

3. (a) Point estimate; (b) interval estimate; (c) interval estimate; (d) 67% by itself is a point estimate but if the information given in parentheses is also included, it is a confidence interval; (e) confidence interval.

5. An estimator is called *unbiased* if the mean of its sampling distribution is equal to the true parameter.

7. No. An estimator is called *consistent* when—with increasing sample size—it becomes more precise. But for nonsymmetrical populations, the value of

$$\frac{\text{Sample maximum} + \text{Sample minimum}}{2}$$ will tend *not* to be a precise estimate for the mean—even if we increase the sample size.

9. Although the (range/6) = $\dfrac{\text{Sample maximum} - \text{Sample minimum}}{6}$ is sometimes used as an estimator for the standard deviation of a *normally distributed* population, it is not consistent as a general estimator for *s* because it is too dependent on the precise values of the largest and smallest numbers.

Applications 10.1

11. Confidence interval.

13. Point estimate.

15. Interval estimate.

Basic Concepts 10.2

17. $10.04 < \mu < 12.76$.

19. (a) 1.645; (b) = 2.326; (c) = 1.960.

21. (a) $3.37 < \mu < 3.63$; (b) $3.32 < \mu < 3.68$; (c) $3.38 < \mu < 3.62$.

23. No, because given $n = 45$, we can assume that the sampling distribution will be normal or close to normal.

25. With a small sample size ($n = 7$, which is less than 30), the conditions for assuming a normally distributed sampling distribution, based on the central limit theorem, are not satisfied.

Applications 10.2

27. a) $77{,}429 < \mu < 220{,}669$.
b) Changes in the climate, with higher temperatures and fewer rainfalls, could possibly result in drier forests, posing a greater fire risk from lightning and human activity than previously. On the other hand, technologies to extinguish forest fires may become more efficient than in the past.
c) No. The *z*-based model presumes that σ is known. If instead σ is estimated from *s*, then additional error is introduced into the estimate, which is not modelled by the normal distribution.

29. (a) 2.0537; (b) ±13.7961; (c) $500.5 < \mu < 528.0$.

31. $1.223 < \mu < 1.401$.

Basic Concepts 10.3

33. a) $9.99 < \mu < 12.81$.
b) Slightly narrower. A larger sample reduces standard error and so reduces the margin of error. Thus, the interval boundaries will lie closer to the mean.

35. (a) $7.50 < \mu < 11.10$; (b) The variable has to follow a normal distribution; (c) $t_{\alpha/2,n-1} = 2.0462$, $E = \pm 1.7963$.

37. $0.339 < \mu < 0.368$.

39. a) $292.0 < \mu < 352.3$.
b) The confidence interval could be too large *or* too small—depending on whether the sample includes one of the relatively few values from the extended tail of the population distribution (which would inflate the variance) or whether it does not include values

from the extended tail (which would make the variance look artificially small). Put another way: The estimate of a *symmetrical* confidence interval does not accurately reflect that, for such a small sample size, the sampling distribution is skewed.

41. 10,628.2 $< \mu <$ 11,399.2.

43. (a) 2.5168; (b) ±16.7022; (c) 500.0 $< \mu <$ 533.5.

Applications 10.3

45. 1.203% $< \mu <$ 1.395%.

47. 2.1519 $< \mu <$ 2.1564. (Note: The data file contains the "original list" with N = 200 values; a sample of $n = 100$ is taken from that list. Therefore use the Finite Population Correction Factor.)

Basic Concepts 10.4

49. (a)(i) 0.2250; (a)(ii) ±0.0154; (b)(i) = 0.2250; (b)(ii) ±0.0183; (c)(i) 0.7954; (c)(ii) ±0.0300; (d)(i) 0.2674; (d)(ii) ±0.0210; (e)(i) 0.8348; (e)(ii) ±0.0088.

51. (Note: The "7%" in the problem must be an approximation—since 7% of 75 cats would be 5.25 cats; not meaningful. Therefore, calculate based on *five* cats, which is, more accurately, 6.667% of *n*.)
(a) 0.012 $< p <$ 0.1231; (b) ±0.0565.

53. (a) 0.1100; (b) 0.0530 $< p <$ 0.1670; (c) ±0.0570.

55. 0.0597 $< p <$ 0.2403.

57. (a) 0.0395 $< p <$ 0.2005; (b) ±0.0805.

Applications 10.4

59. a) 0.8000.
b) 0.6854 $< p <$ 0.9146.
c) No, because a proportion of 0.9 is well within the confidence interval estimate for the population proportion.

61. a) 0.7232 $< p <$ 0.7768.
b) 0.7025 $< p <$ 0.7575.
c) Yes. The two confidence intervals overlap. If the true proportion of those very concerned about air pollution falls at the lower end of the interval estimate for that proportion, and the true proportion of those very concerned about water is at the upper end of its interval estimate, then it could in fact happen that the population proportion of those very concerned about water is larger than the population proportion of those very concerned about air pollution.

63. (a) 0.7091; (b) 0.6901 $< p <$ 0.7281.

Review Applications

65. (a) 1.96; (b) 4.6386; (c) ±9.0915; (d) 36.01 $< \mu <$ 54.19.

67. 17.91 $< \mu <$ 19.10.

69. a) 14.37 $< \mu <$ 17.23.
b) It is possible but not very likely. According to part (a), we are 95% confident that the true average attenuation falls within a range *that does not include* the government's standard of 17.9 dB.

71. (a) 1.9600; (b) 0.3953; (c) ±0.7747; (d) 14.93 $< \mu <$ 16.47.

73. (a) 2.5706; (b) 10.0789; (c) ±25.9086; (d) 2.09 $< \mu <$ 53.91.

75. (a) 2.0452; (b) 51.0718; (c) ±104.4535; (d) 206.8 $< \mu <$ 415.7.

77. 324.3 $< \mu <$ 404.8.

79. 618.0 $< \mu <$ 663.0.

81. 0.0404 $<$ p $<$ 0.1915. Not likely. The two confidence intervals do not overlap at all.

83. 0.3433 $<$ p $<$ 0.4505. Not likely. The two confidence intervals do not overlap at all.

85. (Note: Presume, as in Exercise 84, that $n = 75$.)
(a) 63; (b) 0.7659 $< p <$ 0.9141.

87. 27.27%. The 95% confidence interval is 0.0866 $< p <$ 0.4588.

89. a) 0 $< p <$ 0.2608. (Note: The calculated value for the lower boundary was actually −0.0790.)
b) 0.3521 $< p <$ 0.9206.
c) Not very likely. The confidence intervals do not overlap and *all* values within the interval estimate for the higher-ranked companies are *lower* than all values within the interval estimate for the lower-ranked companies.

Chapter 11 One-Sample Tests of Significance

Basic Concepts 11.1

1. In a hypothesis test, the null hypothesis (in symbols: H_0) is logically opposed to the alternative hypothesis and always includes the concept of equality (i.e., the logical operator = or ≤ or ≥).

3. No, claims about ideas or assumptions can be made in various ways. The alternative hypothesis is a special formal type of claim—specifically used for testing a hypothesis—that is not based on equality and that is mathematically the opposite of the null hypothesis.

5. (a) and (d).

7. A Type I error occurs if the null hypothesis is true but is rejected. A Type II error occurs if the null hypothesis is false but is not rejected. Both error types can lead to undesirable wrong decisions. For example, a Type I error could be the rejection of a true statement that a person is innocent of a crime—resulting in the punishment of an innocent person. A Type II error could be to accept (i.e., not reject) the innocence of someone who is in fact guilty of a serious crime.

9. The sampling distribution can be divided into two regions: a non-critical and a critical region (for a two-tail test, there are two critical regions). The critical region is an area under the curve where the sample statistic could occur, but only with a low probability if the null hypothesis is true. If the test statistic falls into that region, H_0 is rejected.

11. A sample statistic is a number calculated on the basis of sample information (e.g., the sample mean or variance). For a hypothesis test, we convert the sample statistic to a standardized score for the relevant sampling distribution (e.g., *z*-score, *t*-score). The value of the test statistic (relative to the critical value) determines whether to reject or not reject H_0.

13. Reject H_0.

15. The *p*-value displays directly the probability of obtaining the test statistic (or a value at least that extreme) under the assumptions of the null hypothesis. This probability can be compared to α—or even to several levels of α (such as significant "at the 0.01 level" or "at the 0.001 level"). When both raw data and statistical software are available, the *p*-value can generally be obtained easily. The critical value approach is sometimes more practical if you are working from summary statistics instead of raw data. For a problem requiring a *z*-test for the mean, your statistical software may not generate a *p*-value solution.

17. (a) Do not reject H_0 (because *p*-value $= 0.062 > \alpha = 0.05$); (b) the data do not support the claim that Farmer A has the larger proportion of cows that yield top-grade milk.

19. Type I error: Conclude that the evidence supports J. J. Jackson's claim, even though he did *not* in fact obtain over 50% of the votes. Type II error: Conclude that the evidence does not support J. J. Jackson's claim, although in fact he *did* win more than half of today's vote.

21. The null hypothesis would be rejected because the *p*-value is less than α.

23. Claim ($= H_1$): $p_{Head} > p_{Tail}$; H_0: $p_{Head} \le p_{Tail}$. Alternatively: H_1: $p_{Head} > 0.5$; H_0: $p_{Head} \le 0.5$.

25. In general, a researcher states a conclusion in the context of a specific claim. If the claim becomes H_0 and it is not rejected, the null hypothesis is not proved true or supported. We can state only that there is not sufficient evidence to reject the claim. If the claim becomes H_1 and the null hypothesis is not rejected, then we still cannot say that the null hypothesis is proved true or supported; instead, we would say that the evidence does not support the claim H_1.

Basic Concepts 11.2

27. Use the normal distribution for hypothesis testing (of the mean) if the population standard deviation σ is known and either the sample size is large (30 or greater) or the population is known to have normal distribution.

29. a)

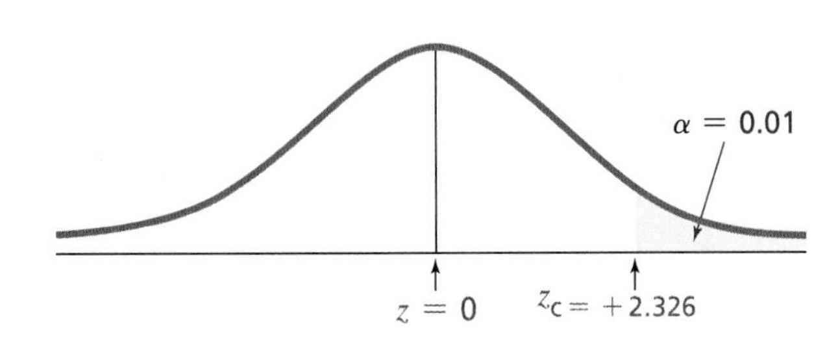

b)

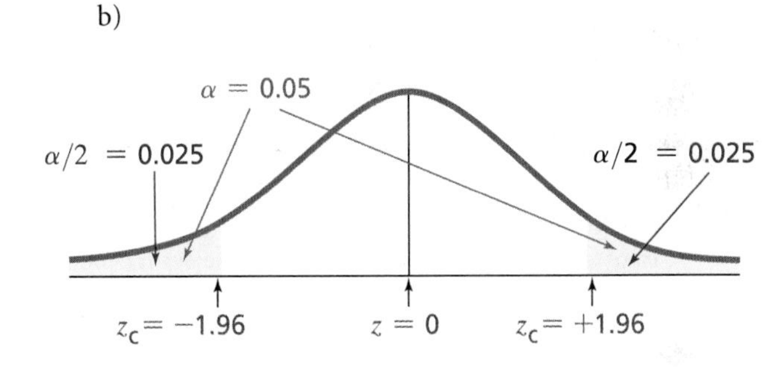

c)

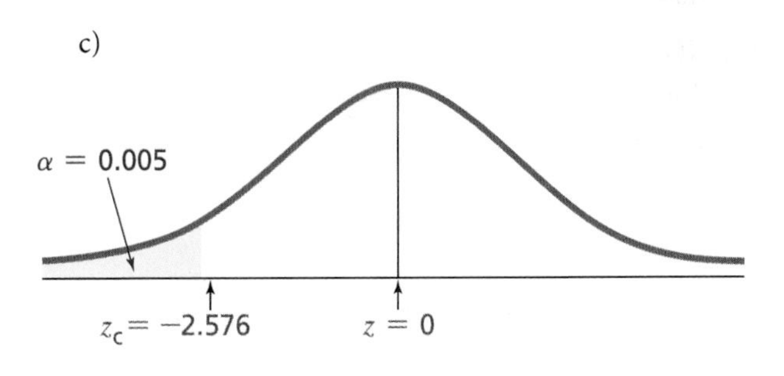

d)

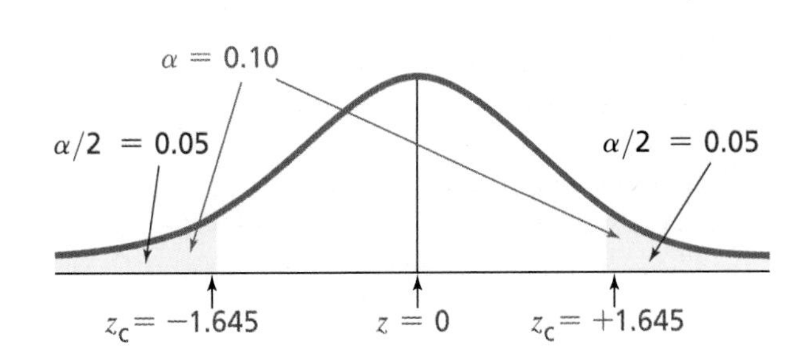

e)

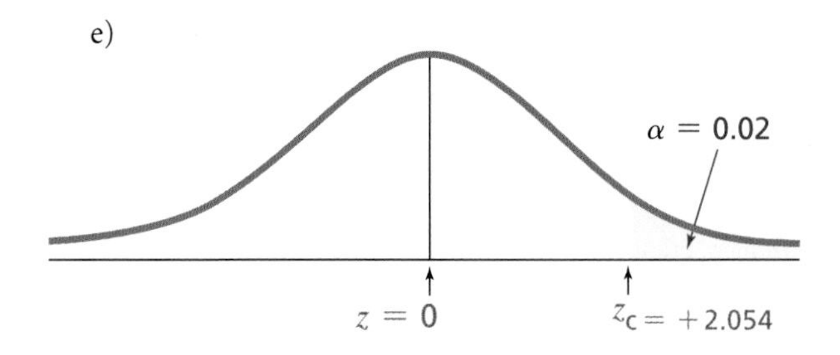

f)

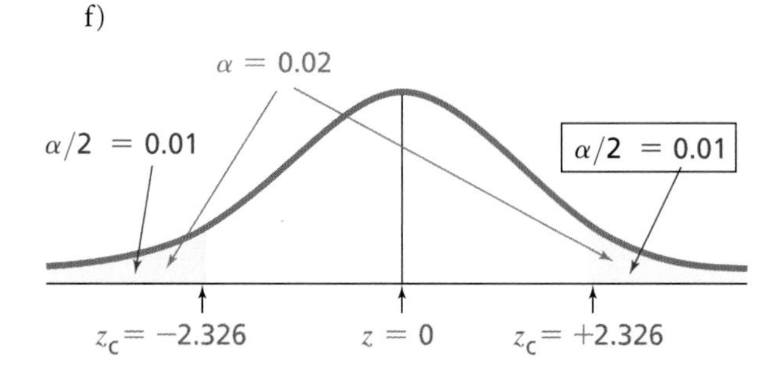

31. (a) 2.0706; (b) −2.9693; (c) 3.6683; (d) −2.1224; (e) −2.0000.

33. (a) 2.7616; (b) −3.6204; (c) 2.0971; (d) −2.5696; (e) 0.2974.

35. a) p-value = 0.0092; for $\alpha = 0.05$: Reject H_0 (no difference from Exercise 34); for $\alpha = 0.01$: Reject H_0 (no difference from Exercise 34).
  b) p-value = 0.0028; for $\alpha = 0.05$: Reject H_0 (no difference from Exercise 34); for $\alpha = 0.01$: Reject H_0 (no difference from Exercise 34).
  c) p-value = 0.0200; for $\alpha = 0.05$: Reject H_0 (no difference from Exercise 34); for $\alpha = 0.01$: Do not reject H_0 (no difference from Exercise 34).
  d) p-value = 0.0078; for $\alpha = 0.05$: Reject H_0 (no difference from Exercise 34); for $\alpha = 0.01$: Reject H_0 (no difference from Exercise 34).
  e) p-value = 0.3839; for $\alpha = 0.05$: Do not reject H_0 (no difference from Exercise 34); for $\alpha = 0.01$: Do not reject H_0 (no difference from #34).

Applications 11.2

37. **Step 1.** Claim: $\mu < \$1,175$.

Step 2. H_0: $\mu \geq \$1,175$; H_1: $\mu < \$1,175$.

Step 3. $\alpha = 0.02$.

Step 4. z-test for the mean, $n = 10$.

Step 5.

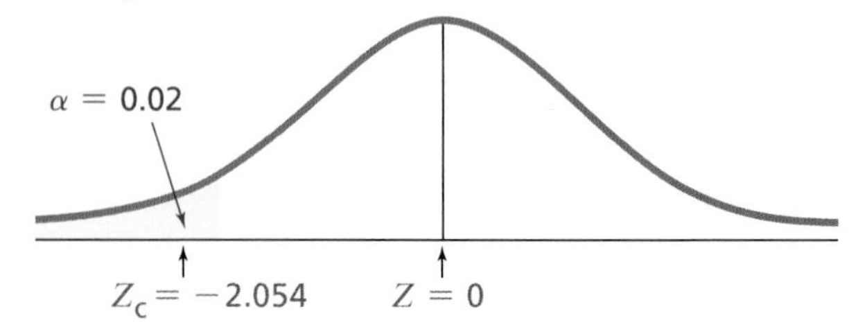

Steps 6 and 7. Test statistic $z = -0.4133$, p-value = 0.3397; do not reject H_0.

Step 8. The evidence does not support the claim that the prices for subwoofers are lower than the online average.

39. **Step 1.** Claim: $\mu \neq 2.667$.

Step 2. H_0: $\mu = 2.667$; H_1: $\mu \neq 2.667$.

Step 3. $\alpha = 0.01$.

Step 4. t-test for the mean, $n = 45$, $df = 44$.

Step 5.

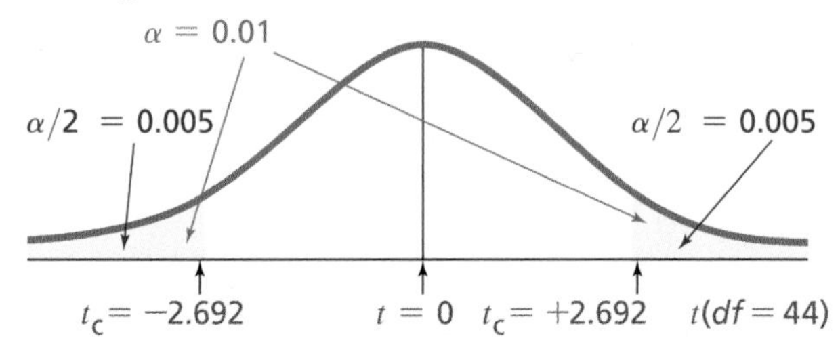

Steps 6 and 7. Test statistic $t = -2.1184$, p-value = 0.0398, do not reject H_0.

Step 8. The evidence does not support the claim that the mean bulk density is different from its earlier value of 2.667 g/cm³.

The above conclusion (based on a sample standard deviation) differs from the conclusion in Exercise 38 (based on a population standard deviation). The answer that uses the sample standard deviation is more reliable because the sample is based on more recent observations.

Note: In subsequent solutions for hypothesis tests, the diagrams corresponding to step 5 in each case will not be provided, except when they are different in format from the examples illustrated in this section.

41. **Step 1.** Claim: $\mu = 130$.

Step 2. H_0: $\mu = 130$; H_1: $\mu \neq 130$.

Step 3. $\alpha = 0.05$.

Step 4. t-test for the mean, $n = 54$, $df = 53$.

Step 5. Draw the null distribution for a two-tail t-test.

Steps 6 and 7. Test statistic $t = -0.2843$, p-value = 0.7777; do not reject H_0.

Step 8. There is not sufficient evidence to reject a claim that the mean asset value for the full population of GlobeFund listings was \$130 million.

43. a) **Step 1.** Claim: $\mu > 20$.
  Step 2. H_0: $\mu \leq 20$; H_1: $\mu > 20$.
  Step 3. $\alpha = 0.05$.
  Step 4. t-test for the mean, $n = 60$, $df = 59$.
  Step 5. Draw the null distribution for right-tail t-test.

Steps 6 and 7. Test statistic $t = 1.4517$, p-value = 0.0762; do not reject H_0.

Step 8. The evidence does not support the claim that the mean asset value per share for the full population of GlobeFund listings was greater than \$20.

b) **Step 1.** Claim: $\mu > 16$.

Step 2. H_0: $\mu \leq 16$; H_1: $\mu > 16$.

Step 3. $\alpha = 0.05$.

Step 4. t-test for the mean, $n = 60$, $df = 59$.

Step 5. Draw the null distribution for a right-tail t-test.

Steps 6 and 7. Test statistic $t = -1.9947$, p-value = 0.0256; reject H_0.

Step 8. The data support the claim that the mean asset value per share for the full population of GlobeFund listings was greater than \$16.

Basic Concepts 11.3

45. (a) Larger; (b) smaller; (c) larger; (d) larger.

47. The β risk is the probability of committing a Type II error; i.e., the risk that a false null hypothesis will mistakenly fail to be rejected. This risk is decreased if the range of values in the non-rejection area is decreased. It can be shown that this occurs if the sample size is increased—thus decreasing the standard error and the range of values in the non-rejection area.

49. An operating characteristic curve displays—over a range of possible distances between the actual and hypothesized population parameters—the values of β corresponding to each potential distance. The power curve is the complement of the operating characteristic curve and shows, for any distance between the true and hypothesized parameters, the corresponding probability $(1 - \beta)$ of correctly rejecting the null hypothesis.

51. $n = 136.1 \rightarrow$ use 137.

Applications 11.3

53. $n = 51.7 \rightarrow$ use 52.

55. $n = 135.4 \rightarrow$ use 136.

57. $n = 173.8 \rightarrow$ use 174.

59. $n = 209.3 \rightarrow$ use 210.

Basic Concepts 11.4

61. (a) Do not reject H_0 (p-value $= 0.4032 > \alpha = 0.01$); (b) do not reject H_0 (p-value $= 0.0219 > \alpha = 0.01$); (c) do not reject H_0 (p-value $= 0.2375 > \alpha = 0.01$).

The statistical decision for test (b) changes from "reject H_0 at the 0.05 level of significance" (Exercise 60) to "do not reject H_0 at the 0.01 level of significance" (Exercise 61). A smaller level of significance reduces the risk of committing a Type I error because it is harder to reject the null hypothesis.

63. (a) Do not reject H_0 (p-value $= 0.3980 > \alpha = 0.01$); (b) do not reject H_0 (p-value $= 0.0328 > \alpha = 0.01$); (c) do not reject H_0 (p-value $= 0.0692 > \alpha = 0.01$).

The statistical decision for test (b) has changed from "reject H_0 at the 0.05 level of significance" (Exercise 62) to "do not reject H_0 at the 0.01 level of significance" (Exercise 63).

65. Given a probability distribution of proportions, each possible proportion is equal to a number of successes; i.e., (frequency of cases that meet a certain condition) divided by (sample size). The denominator (sample size) is a constant for all proportions in the distribution, so the distribution pattern is unchanged if we multiply all proportions times the sample size. The result is a distribution of possible numbers of successes (from 1 to n) that meet the condition, with a probability assigned to each. The distribution described is a binomial distribution—provided the other conditions for binomial distributions (see Exercise 66) are also satisfied.

67. **Step 1.** Claim: $p < 0.10$.

Step 2. H_0: $p \geq 0.10$; H_1: $p < 0.10$.

Step 3. $\alpha = 0.02$.

Step 4. Binomial test for a single proportion, $n = 64$.

Step 5.

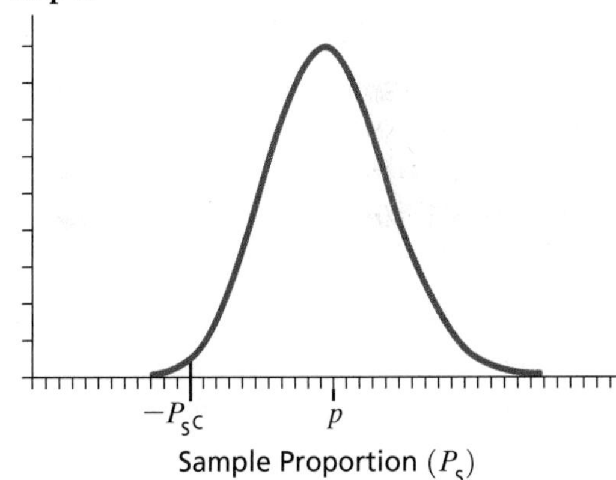

Steps 6 and 7. p-value $= 0.3727$; do not reject H_0.

Step 8. The evidence does not support the claim that the quality of the jackets has improved.

Applications 11.4

69. **Step 1.** Claim: $p = 0.085$.

Step 2. H_0: $p = 0.085$; H_1: $p \neq 0.085$.

Step 3. $\alpha = 0.05$.

Step 4. Binomial test for a single proportion, $n = 86$.

Step 5. Draw the null distribution for two-tailed binomial test.

Steps 6 and 7. p-value $= 0.7041$; do not reject H_0.

Step 8. There is not sufficient evidence to reject the claim that about 8.5% of adult male frogs migrate from their breeding ponds to a separate summer habitat.

71. **Step 1.** Claim: $p < 0.25$.

Step 2. H_0: $p \geq 0.25$; H_1: $p < 0.25$.

Step 3. $\alpha = 0.05$.

Step 4. Binomial test for a single proportion, $n = 142$.

Step 5. Sketch the null distribution for left-tailed binomial test.

Steps 6 and 7. p-value $= 0.3544$; do not reject H_0.

Step 8. The evidence does not support the claim that less than a quarter of all avalanche victims in Canada were riding snowmobiles at the time of the tragedy.

73. For $\alpha = 0.10$:

Step 1. Claim: $p < 0.25$.

Step 2. H_0: $p \geq 0.25$; H_1: $p < 0.25$.

Step 3. $\alpha = 0.10$.

Step 4. Binomial test for a single proportion, $n = 500$.

Step 5. Draw the null distribution for left-tailed binomial test.

Steps 6 and 7. p-value $= 0.053$; reject H_0.

Step 8. The data support the claim that less than a quarter of the population were unsure about how they would spend the holiday.

For $\alpha = 0.05$:

Step 1. Claim: $p < 0.25$.

Step 2. H_0: $p \geq 0.25$; H_1: $p < 0.25$.

Step 3. $\alpha = 0.05$.

Step 4. Binomial test for a single proportion, $n = 500$.

Step 5. Draw null distribution for left-tailed binomial test.

Steps 6 and 7. p-value $= 0.053$; do not reject H_0.

Step 8. The evidence does not support the claim that less than a quarter of the population were unsure about how they would spend the holiday.

The two tests lead to different statistical decisions. When α is smaller (in the second scenario), the risk of committing a Type I error is smaller, but it is also more difficult to reject H_0.

75. a) **Step 1.** Claim: $p \leq 0.50$.
Step 2. H_0: $p \leq 0.50$; H_1: $p > 0.50$.
Step 3. $\alpha = 0.01$.
Step 4. Binomial test for a single proportion, $n = 500$.
Step 5. Draw the null distribution for right-tailed binomial test.
Step 6 and 7. p-value $= 0.0000$; reject H_0.
Step 8. There is sufficient evidence to reject the claim that, at most, half of the population did not intend to be with friends and/or family during the holiday.

b) No, because many who answered that they intended to spend the holiday fishing, or at the cottage, for example, may also have intended to share these activities with friends or family members. However, we have no information about how often this occurred.

Review Applications

77. **Step 1.** Claim: $\mu < 43.4$.

Step 2. H_0: $\mu \geq 43.4$; H_1: $\mu < 43.4$.

Step 3. $\alpha = 0.05$.

Step 4. t-test for the mean, $n = 50$, $df = 49$.

Step 5. Draw the null distribution for left-tail t-test.

Steps 6 and 7. Test statistic $t = -0.4562$, p-value $= 0.3251$; do not reject H_0.

Step 8. The evidence does not support the claim that the mean formaldehyde content in direct cigarette smoke from the "light" brands is less than 43.4 μg.

No, the same conclusion is reached as in Exercise 76.

79. **Step 1.** Claim: $\mu < 17.9$.

Step 2. H_0: $\mu \geq 17.9$; H_1: $\mu < 17.9$.

Step 3. $\alpha = 0.01$.

Step 4. t-test for the mean, $n = 50$, $df = 49$.

Step 5. Draw the null distribution for left-tail t-test.

Steps 6 and 7. Test statistic $t = -1.4731$, p-value $= 0.0736$; do not reject H_0.

Step 8. The evidence does not support the claim that the EasyEar product provides less mean attenuation at the frequency 125 Hz than is required.

81. **Step 1.** Claim: $\mu < 24.7$.

Step 2. H_0: $\mu \geq 24.7$; H_1: $\mu < 24.7$.

Step 3. $\alpha = 0.01$.

Step 4. t-test for the mean, $n = 50$, $df = 49$.

Step 5. Draw the null distribution for left-tail t-test.

Steps 6 and 7. Test statistic $t = -3.8777$, p-value $= 0.0002$; reject H_0.

Step 8. The data do support the claim that the EasyEar product provides less mean attenuation at the frequency 1,000 Hz than is required.

83. $n = 11.5 \rightarrow$ use 12.

85. **Step 1.** Claim: $\mu > 15.5$.

Step 2. H_0: $\mu \leq 15.5$; H_1: $\mu > 15.5$.

Step 3. $\alpha = 0.05$.

Step 4. z-test for the mean, $n = 60$.

Step 5. Draw the null distribution for right-tail z-test.

Steps 6 and 7. Test statistic $z = 2.1689$, p-value $= 0.0150$; reject H_0.

Step 8. The data support the claim that this population has a higher than normal mean eye pressure.

87. (Note: The sampling method has been applied directly to the 1285 records in the file, resulting in 43 selected records—of which only $n = 40$ contain data for the Domestic Water use variable. A different sample size would be obtained if the method were applied only to the records for which Domestic Water Use has a value.)

Step 1. Claim: $\mu > 6{,}000$.

Step 2. H_0: $\mu \leq 6{,}000$; H_1: $\mu > 6{,}000$.

Step 3. $\alpha = 0.04$.

Step 4. t-test for the mean, $n = 40$, $df = 39$.

Step 5. Draw the null distribution for right-tail t-test.

Steps 6 and 7. Test statistic $t = 0.785$, p-value $= 0.219$; do not reject H_0.

Step 8. The evidence does not support the claim that the mean domestic water use in all municipalities of this size is greater than 6,000 cubic metres per day.

89. **Step 1.** Claim: $\mu \neq 47$.

Step 2. H_0: $\mu = 47$; H_1: $\mu \neq 47$.

Step 3. $\alpha = 0.04$.

Step 4. t-test for the mean, $n = 43$, $df = 42$.

Step 5. Draw the null distribution for two-tail t-test.

Steps 6 and 7. Test statistic $t = 1.653$, p-value $= 0.106$; do not reject H_0.

Step 8. The evidence does not support the claim that the mean latitude for Canadian municipalities with populations over 1,000 is not equal to 47°N.

91. $n = 15.4 \rightarrow$ use 16.

Chapter 12 Two-Sample Tests of Significance

Basic Concepts 12.1

1. Samples are independent if the selection of objects for any one sample is not affected by the selection of objects for any of the other samples. Samples are dependent if the selection of objects for one sample *is* related to the selection of objects in another sample.

3. In repeated-measures sampling, the same individuals are examined before and after receiving a treatment (e.g., a weight-loss program). In matched-pairs sampling, dependent samples are created by selecting comparable individuals from different populations and examining each individual for a particular characteristic.

5. Yes. For both samples, the initial skills of participants are not known. If by chance the "before" group and "after" group had different levels of initial skills, then any apparent difference in skills between the "before" and "after" groups may reflect this extraneous variable, rather than the impact of the treatment.

7. Extraneous variables that might influence the house prices in each sample may include residential location, ages of the houses, the number of bedrooms, sizes of the backyards, and numbers of appliances.

9. Test for two dependent samples (matched pairs); controlled experiment.

Basic Concepts 12.2

11. (a) H_0: $\mu_1 = \mu_2$; (b) reject H_0; (c) there is sufficient evidence to reject a claim that the two means are equal.

13. (a) H_0: $\mu_1 \leq \mu_2$; (b) $t = 1.6633$; (c) 0.0539; (d) do not reject H_0; (e) the evidence does not support a claim that the first population has a larger mean than the second population.

15. (a) H_0: $\mu_1 = \mu_2$; (b) do not reject H_0 (p-value $= 0.011 > \alpha = 0.01$); (c) there is *not* sufficient evidence to reject the claim that the two population means are equal.

17. (a) H_0: $\mu_1 = \mu_2$; (b) 0.159; (c) 0.160; (d) (using Case 1): do not reject H_0; (e) there is not sufficient evidence to reject the claim that the two population means are equal.

19. (a) H_0: $\mu_1 - 3 = \mu_2$; (b) 0.996; (c) 0.996; (d) (using Case 1): do not reject H_0; (e) there is not sufficient evidence to reject the claim that the two population means differ by 3.0.

21. $-31.97 < \mu_1 - \mu_2 < -14.03$.

23. $-33.70 < \mu_1 - \mu_2 < -12.30$

25. $-2.87 < \mu_1 - \mu_2 < 8.85$

Applications 12.2

27. $-10{,}886.2 < \mu_1 - \mu_2 < 20{,}286.2$

29. $-32{,}366.2 < \mu_1 - \mu_2 < 13{,}366.2$

31. Group 1: Use subscript "1"; group 3: use subscript "3."

Step 1. Claim: $\mu_1 < \mu_3$.

Step 2. H_0: $\mu_1 \geq \mu_3$; H_1: $\mu_1 < \mu_3$.

Step 3. $\alpha = 0.01$.

Step 4. z-test for difference in population means, two independent samples, $n_1 = 100$, $n_3 = 90$.

Step 5.

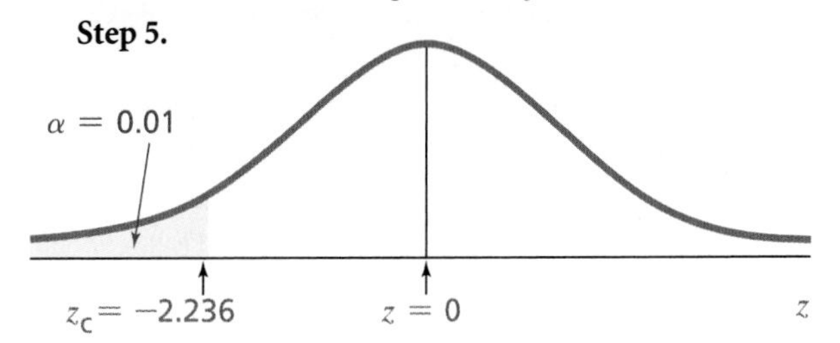

Steps 6 and 7. Test statistic $z = \times 0.7919$ (p-value = 0.2142); do not reject H_0.

Step 8. The evidence does not support the claim that the durations of depressive incidents are longer for women aged 65 or older than for women aged 12–24.

33. -1.0 months $< \mu_1 - \mu_3 <$ 0.4 months.

35. a) Subscript 1 = previous (1950 to 1980); subscript 2 = current (1981 to 2004).
Step 1. Claim: $\mu_1 < \mu_2$.
Step 2. H_0: $\mu_1 \geq \mu_2$; H_1: $\mu_1 < \mu_2$.
Step 3. $\alpha = 0.05$.
Step 4. t-test for difference in population means, two independent samples (Case 2: unequal variances), $n_1 = 31$, $n_2 = 24$, $df \approx 31$.
Step 5.

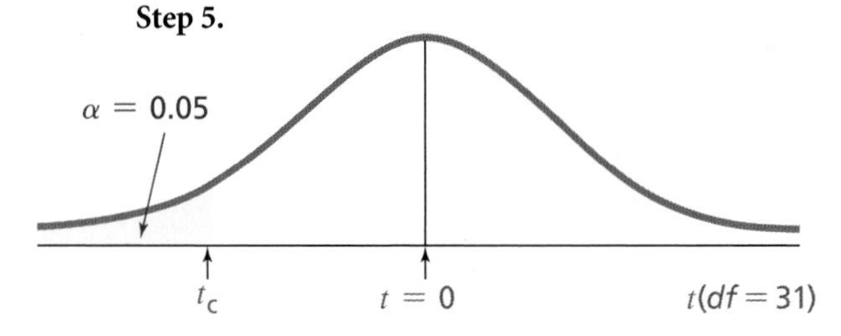

Steps 6 and 7. Test statistic $t = -4.332$, p-value (unequal variances) ≈ 0.000; reject H_0.
Step 8. The data support the claim that since 1981, lightning has caused more forest fires per year than was previously the case.

b) The final conclusion does not differ if Case 1 is used: p-value = 0.000; $t = -4.714$, $df = 53$.

c) Check of variable "lightning" for previous fire losses (1950 to 1980):
- Mean (33.81) > median (21).
- Boxplot shows a right-skewed distribution.
- Normal probability plots show that the distribution is not normal.

Check of variable "lightning" for more recent fire losses (1981 to 2004):
- Mean (89.75) > median (76.50).
- Boxplot shows a right-skewed distribution.
- Normal probability plots show that the distribution is not normal.

It appears that the data for lightning-caused fires do not meet the condition of normality for either time period. While the sample size for the earlier data is 31—large enough to apply a t-test, the sample size for the current data is only 24. Because the conditions to apply a t-test are not formally met, a conclusion based on the t-test may not be reliable. Nonetheless, the apparent difference in groups is so great in part (a) that the tentative conclusion might still be worth confirming; e.g., by applying a nonparametric method (see Chapter 13).

37. Assuming unequal variances: $-91.41 < \mu_1 - \mu_2 < -20.47$.

Basic Concepts 12.3

39. Mechanically, yes, it could be done, but by treating the two samples' means as independent when they are not, you lose the opportunity to control for extraneous variables by pairing the individuals from the dependent samples.

41. (a) H_0: $\mu_D = 0$, where D = (Group A values) − (Group B values); (b) $t = -1.8469$; (c) 0.1072; (d) do not reject H_0; (e) there is not sufficient evidence to reject the claim that there is no mean difference between the paired values.

43. (a) Reject H_0; (b) the data support the claim that prices are generally higher for the new-formula version.

45. $-27.93 < \mu_D < 3.43$.

47. $-0.164 < \mu_D < 1.664$.

Applications 12.3

49. $-0.1566 < \mu_D < -0.0074$, where D = (March 2005 values) − (April 2005 values).

51. a) D = (College diploma/degree values) − (Trades education values).
Step 1. Claim: $\mu_D = 0$.
Step 2. H_0: $\mu_D = 0$; H_1: $\mu_D \neq 0$.
Step 3. $\alpha = 0.05$.
Step 4. t-test for the mean difference, two dependent samples, $n = 13$, $df = 12$.
Step 5.

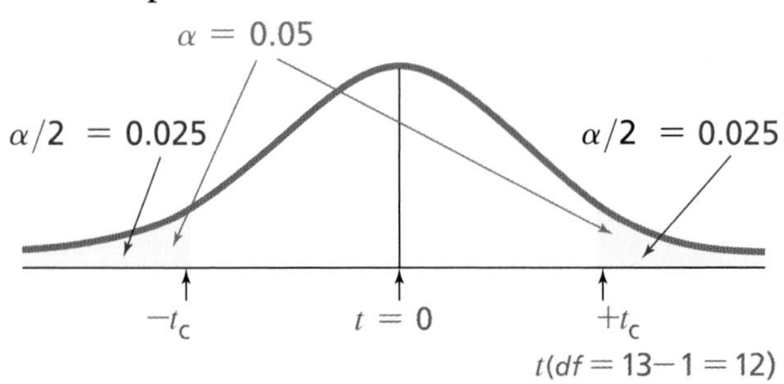

Steps 6 and 7. Test statistic $t = -0.440$, p-value = 0.668; do not reject H_0.
Step 8. There is not sufficient evidence to reject the claim that there is not any difference in salaries based on a college diploma/degree versus a trades education.

b) $-1{,}163.62 < \mu_D < 1{,}751.78$.

Note: In subsequent solutions for hypothesis tests in this chapter, the diagrams corresponding to step 5 will not be provided except where different in format from the examples that precede it.

53. a) $D = (2004 \text{ values}) - (2005 \text{ values})$.

Step 1. Claim: $\mu_D < 0$.

Step 2. H_0: $\mu_D \geq 0$; H_1: $\mu_D < 0$.

Step 3. $\alpha = 0.05$.

Step 4. t-test for the mean difference, two dependent samples, $n = 10$, $df = 9$.

Step 5. Draw the null distribution for left-tail t-test.

Steps 6 and 7. Test statistic $t = -0.686$, p-value $=$ 0.255; do not reject H_0.

Step 8. The evidence does not support the claim that the average number of dairy cows being kept in each province has increased between 2004 and 2005.

b) For the goal of forecasting into the future, the data set used in (a) represents a small sample of just two years of observation. Before making a conclusion about a long-term trend, it would be advisable to include more time periods.

Basic Concepts 12.4

55. For both samples—condition 1: $n \times \overline{p} \geq 5$ and condition 2: $n \times (1 \times \overline{p}) \geq 5$ (where $\overline{p}$ = pooled estimate of the proportion based on the two samples combined).

57. Subscript 1 = fourth-year students; subscript 2 = first-year students.

Step 1. Claim: $p_1 > p_2$.

Step 2. H_0: $p_1 \leq p_2$; H_1: $p_1 > p_2$.

Step 3. $\alpha = 0.01$.

Step 4. Conditions to apply normal distribution are valid; z-test for difference in population proportions, $n_1 = 130$, $n_2 = 300$.

Step 5.

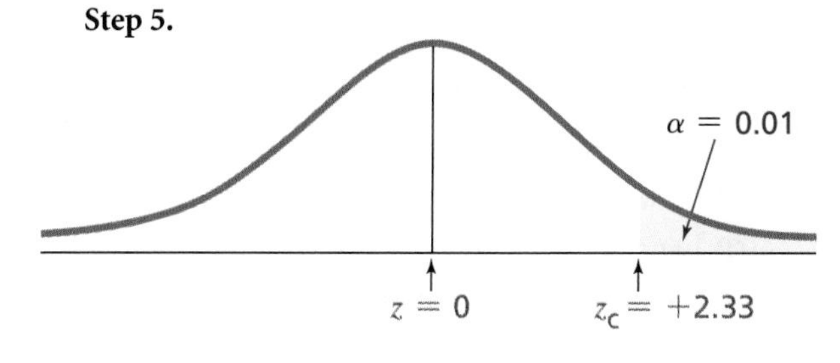

Steps 6 and 7. Test statistic $z = 7.7070$, p-value $= 0.000$; reject H_0.

Step 8. The data support the claim that the proportion of fourth-year students who buy at least one secondhand textbook is larger than the proportion of first-year students.

59. $-0.196 < p_1 - p_2 < 0.070$.

61. Subscript 1 = day shift; subscript 2 = evening shift.

Step 1. Claim: $p_1 < p_2$.

Step 2. H_0: $p_1 \geq p_2$; H_1: $p_1 < p_2$.

Step 3. $\alpha = 0.02$.

Step 4. Conditions to apply normal distribution are valid; z-test for difference in population proportions, $n_1 = 150$, $n_2 = 150$.

Step 5. Draw the null distribution for left-tail z-test.

Steps 6 and 7. Note: Given the discrete nature of the data "5% of the 150 trials" must equal exactly seven cases (4.67%) or exactly eight cases (5.33%). The following assumes seven cases: Test statistic $z = -1.77$, p-value $=$ 0.038; do not reject H_0. (The statistical decision is the same if *eight* cases are assumed.)

Step 8. The evidence does not support the claim that proportionally more flawed jackets are produced during the evening shift than during the day shift.

63. Subscript 1 = university A; subscript 2 = university B.

Step 1. Claim: $p_1 > p_2$.

Step 2. H_0: $p_1 \leq p_2$; H_1: $p_1 > p_2$.

Step 3. $\alpha = 0.05$.

Step 4. Conditions to apply normal distribution are valid; z-test for difference in population proportions, $n_1 = 73$, $n_2 = 65$.

Step 5. Draw the null distribution for right-tail z-test.

Steps 6 and 7. Test statistic $z = 0.7379$, p-value $= 0.2303$; do not reject H_0.

Step 8. The evidence does not support the claim that university A has recruited a higher proportion of female students for its engineering program than university B.

Applications 12.4

65. Subscript 1 = people who have *never* had over four drinks; subscript 2 = people who *at least once* have had over four drinks.

Step 1. Claim: $p_1 = p_2$.

Step 2. H_0: $p_1 = p_2$; H_1: $p_1 \neq p_2$.

Step 3. $\alpha = 0.05$.

Step 4. Caution: The conditions to apply the normal distribution model are not fully valid: $n_2 \times \overline{p} = 30 \times 0.086 = 2.57 < 5$; z-test for difference in population proportions, $n_1 = 926$, $n_2 = 30$.

Step 5. Draw the null distribution for two-tail z-test.

Steps 6 and 7. Test statistic $z = -4.2543$, p-value $=$ 0.000; reject H_0.

Step 8. According to the z-test, there is sufficient evidence to reject the claim that for the two groups, the same proportions of respondents think that the reduced limits would be too strict.

67. $-0.153 < p_1 - p_2 < 0.201$.

69. $0.014 < p_1 - p_2 < 0.026$.

Review Applications

71. **Step 1.** Claim: $\mu_A < \mu_B$.

    Step 2. H_0: $\mu_A \geq \mu_B$; H_1: $\mu_A < \mu_B$.

    Step 3. $\alpha = 0.05$.

    Step 4. z-test for difference in population means, two independent samples, $n_A = 50$, $n_B = 45$.

    Step 5. Draw the null distribution for left-tail z-test.

    Steps 6 and 7. Test statistic $z = -0.3858$, p-value = 0.3498; do not reject H_0.

    Step 8. The evidence does not support the claim that the smoke from brand B cigarettes has a higher mean formaldehyde content than the smoke from brand A cigarettes.

73. **Step 1.** Claim: $\mu_A < \mu_B$.

    Step 2. H_0: $\mu_A \geq \mu_B$; H_1: $\mu_A < \mu_B$.

    Step 3. $\alpha = 0.05$.

    Step 4. t-test for difference in population means, two independent samples (equal variances assumed), $n_A = 50$, $n_B = 45$, $df = 93$.

    Step 5. Draw the null distribution for left-tail t-test.

    Steps 6 and 7. Test statistic $t = -0.3819$, p-value = 0.3528; do not reject H_0.

    Step 8. The evidence does not support the claim that the smoke from brand B cigarettes has a higher mean formaldehyde content than the smoke from brand A cigarettes. Not surprisingly, given the large sample sizes, the t-test comes to the same conclusion as the z-test.

75. Subscript 1 = communities that use chlorine; subscript 2 = communities that do not use chlorine.

    Step 1. Claim: $\mu_1 = \mu_2$.

    Step 2. H_0: $\mu_1 = \mu_2$; H_1: $\mu_1 \neq \mu_2$.

    Step 3. $\alpha = 0.05$.

    Step 4. z-test for difference in population means, two independent samples, $n_1 = 1038$, $n_2 = 179$.

    Step 5. Draw the null distribution for two-tail z-test.

    Steps 6 and 7. Test statistic $z = 0.5097$, p-value = 0.6103; do not reject H_0.

    Step 8. There is not sufficient evidence to reject the claim that both groups of communities consume the same mean quantity of water per day in the domestic sector.

77. $-3{,}066.7 < \mu_1 - \mu_2 < 5{,}222.0$.

79. $-3{,}072.0 < \mu_1 - \mu_2 < 5{,}227.3$

81. Subscript 1 = fee; subscript 2 = no fee.

    Step 1. Claim: $p_1 < p_2$.

    Step 2. H_0: $p_1 \geq p_2$; H_1: $p_1 < p_2$.

    Step 3. $\alpha = 0.03$.

    Step 4. Conditions to apply normal distribution are valid; z-test for difference in population proportions, $n_1 = 30$, $n_2 = 17$.

    Step 5. Draw the null distribution for left-tail z-test.

    Steps 6 and 7. Test statistic $z = 0.6195$. Observe that for the samples, $p_1 > p_2$, which is opposed to H_0; p-value = 0.733; do not reject H_0.

    Step 8. The evidence does not support the claim that the proportion of no-fee funds rated at least 3 by GlobeFund was actually better than the proportion of other funds that achieved that rating.

83. Subscript 1 = British Columbians; subscript 2 = Albertans.

    Step 1. Claim: $p_1 = p_2$.

    Step 2. H_0: $p_1 = p_2$; H_1: $p_1 \neq p_2$.

    Step 3. $\alpha = 0.05$.

    Step 4. Conditions to apply normal distribution are valid; z-test for difference in population proportions, $n_1 = 1{,}000$, $n_2 = 1{,}000$.

    Step 5. Draw the null distribution for two-tail z-test.

    Steps 6 and 7. Test statistic $z = -1.6512$, p-value = 0.0987; do not reject H_0.

    Step 8. There is not sufficient evidence to reject the claim that the same proportions of British Columbians and Albertans suffer from obesity.

85. Subscript 1 = Quebecker; subscript 2 = New Brunswicker: $0.023 < p_1 - p_2 < 0.097$.

87. D = (Scores for obesity) − (Scores for stress).

    Step 1. Claim: $\mu_D = 0$.

    Step 2. H_0: $\mu_D = 0$, H_1: $\mu_D \neq 0$.

    Step 3. $\alpha = 0.05$.

    Step 4. t-test for the mean difference, two dependent samples, $n = 13$, $df = 12$.

    Step 5. Draw the null distribution for two-tail t-test.

    Steps 6 and 7. Test statistic $t = -0.6316$, p-value = 0.5395; do not reject H_0.

    Step 8. There is not sufficient evidence to reject the claim that in general, for all provinces and territories, the two scores for obesity and stress have about the same mean.

Chapter 13 Nonparametric Tests of Significance

Basic Concepts 13.1

1.

(a) Original Number	Rank	(b) Original Number	Rank
1946	1	22	1
1999	2	32	2
2058	3	34	3
2206	4	41	4
2408	5	44	5
		47	6
		58	7
		59	8
		65	9
		89	10
		96	11

(c) Original Number	Rank	(d) Original Number	Rank
41	2.5	1974	1
41	2.5	1999	2
41	2.5	2058	4
41	2.5	2058	4
46	5	2058	4
59	7	2111	6
59	7	2206	7
59	7		
63	9		
65	10		
71	11		

3. The test statistic is the number of plus signs (+) in the sample. (A plus sign is assigned to every value in the sample that falls above the hypothesized median.)

5. Raw data must be at least at the ordinal level; observed values are from a random sample; data are continuous or drawn from a very large population; the underlying distribution is symmetrical.

7. Raw data must be at least at the ordinal level. Higher measurement levels would also work because it is possible to rank the data in order to validly perform the Wilcoxon signed-ranks test.

9. (a) H_0: median = 100, H_1: median ≠ 100; (b) $T = 56.5$; (c) p-value = 0.858 (Note: The p-value solution appears slightly different if your software uses the z-approximation.); (d) do not reject H_0 because there is not sufficient evidence to reject the claim that the median weight of luggage is 100 kg.

Applications 13.1

11. **Step 1.** Claim: median = 40.

    Step 2. H_0: median = 40; H_1: median ≠ 40.

    Step 3. $\alpha = 0.05$.

    Step 4. One-sample Wilcoxon signed-ranks test for median, $n = 54$.

    Step 5. Test is nonparametric; no parametric null distribution needs to be displayed.

    Steps 6 and 7. Test statistic $T = 705$, p-value = 0.750; do not reject H_0.

    Step 8. There is not sufficient evidence to reject the claim that the median asset value for the full population of GlobeFund listings in 2005 was $40 million.

13. **Step 1.** Claim: median < 16.

    Step 2. H_0: median ≥ 16; H_1: median < 16.

    Step 3. $\alpha = 0.05$.

    Step 4. One-sample Wilcoxon signed-ranks test for median, $n = 60$.

    Step 5. Test is nonparametric; no parametric null distribution needs to be displayed.

    Steps 6 and 7. Test statistic $T = 801.5$, p-value = 0.203 (Note: The p-value solution may appear slightly different if your software uses the z-approximation.); do not reject H_0.

    Step 8. The evidence does not support the claim that the median net asset value per share for the full population of GlobeFund listings in 2005 was less than $16.

15. **Step 1.** Claim: median = 42.

    Step 2. H_0: median = 42; H_1: median ≠ 42.

    Step 3. $\alpha = 0.02$.

    Step 4. One-sample Wilcoxon signed-ranks test for median, $n = 11$.

    Step 5. Test is nonparametric; no parametric null distribution needs to be displayed.

    Steps 6 and 7. Test statistic $T = 30$, p-value = 0.719 (Note: The p-value solution appears slightly different if your software uses the z-approximation.); do not reject H_0.

    Step 8. There is not sufficient evidence to reject the claim that the median threshold is 42 hours per week.

Basic Concepts 13.2

17. If the data are not normally distributed and the sample size is not large enough; if the data are not at least at the interval level.

19. H_0: median$_1$ = median$_2$ (or ≤ or ≥); H_1: median$_1$ ≠ median$_2$ (or < or >).

21. a) **Step 1.** Claim: median$_1$ ≠ median$_2$.

Step 2. H_0: median$_1$ = median$_2$; H_1: median$_1$ ≠ median$_2$.
Step 3. $\alpha = 0.01$.
Step 4. Wilcoxon rank-sum test for two independent samples, $n_1 = 10$, $n_2 = 9$.
Step 5. Test is nonparametric; no parametric null distribution needs to be displayed.
Steps 6 and 7. $W_{\text{sample 1}} = 99$ (or you could use $W_{\text{sample 2}} = 91$), p-value = 0.953 (Note: The p-value solution appears slightly different if your software uses the z-approximation.); do not reject H_0.
Step 8. The evidence does not support the claim that the medians of the two distributions are different.

b) **Step 1.** Claim: $\mu_1 \neq \mu_2$.
Step 2. H_0: $\mu_1 = \mu_2$, H_1: $\mu_1 \neq \mu_2$.
Step 3. $\alpha = 0.01$.
Step 4. t-test for difference in population means, two independent samples (equal variances assumed), $n_1 = 10$, $n_2 = 9$, $df = 17$.
Step 5.

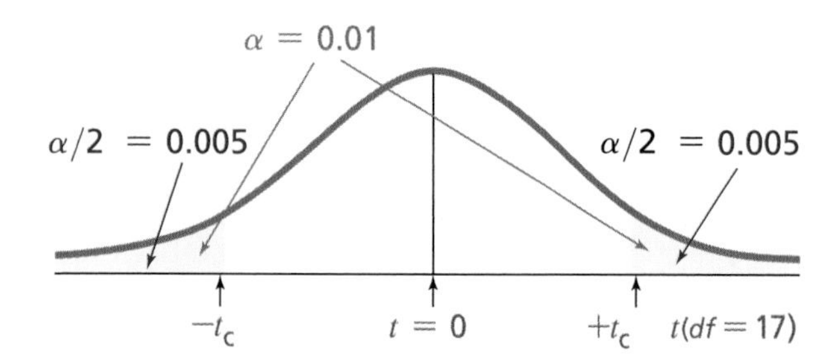

Steps 6 and 7. Test statistic $t = 0.066$, p-value = 0.948; do not reject H_0.
Step 8. The evidence does not support the claim that the means of the two distributions are different.

23. H_0: median$_{\text{sample 1}}$ = median$_{\text{sample 2}}$ (or ≤ or ≥); H_1: median$_{\text{sample1}}$ ≠ median$_{\text{sample2}}$ (or < or >).

25. a) **Step 1.** Claim: median$_1$ ≠ median$_2$.
Step 2. H_0: median$_1$ = median$_2$; H_1: median$_1$ ≠ median$_2$.
Step 3. $\alpha = 0.01$.
Step 4. Wilcoxon signed-ranks test for two dependent samples, $n = 10$.
Step 5. Test is nonparametric; no parametric null distribution needs to be displayed.
Steps 6 and 7. Test statistic $T = 21$, p-value = 0.539 (Note: The p-value solution appears slightly different if your software uses the z-approximation.); do not reject H_0.
Step 8. The evidence does not support the claim that the medians of the two distributions are different.

b) D = (set1 values) − (set2 values).
Step 1. Claim: $\mu_D \neq 0$.
Step 2. H_0: $\mu_D = 0$; H_1: $\mu_D \neq 0$.
Step 3. $\alpha = 0.01$.
Step 4. t-test for the mean difference, two dependent samples, $n = 10$, $df = 9$.
Step 5.

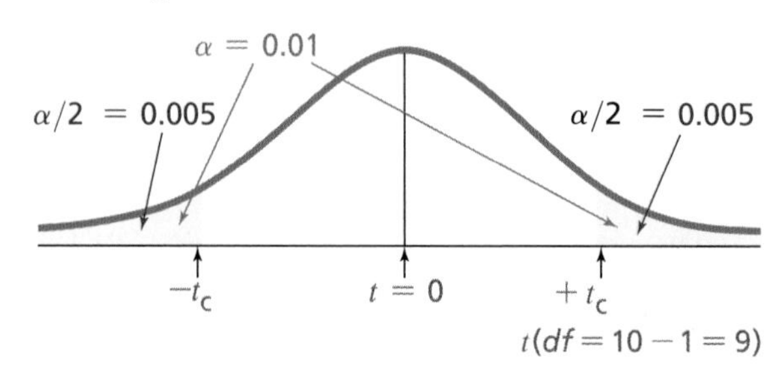

Steps 6 and 7. Test statistic $t = 0.287$, p-value = 0.781; do not reject H_0.
Step 8. The evidence does not support the claim that the mean difference of the two distributions is unequal to zero.

Applications 13.2

27. Subscript 1: previous patterns (1950 to 1980); subscript 2: current patterns (1981 to 2004).
Step 1. Claim: median$_1$ < median$_2$.
Step 2. H_0: median$_1$ ≥ median$_2$, H_1: median$_1$ < median$_2$.
Step 3. $\alpha = 0.05$.
Step 4. Wilcoxon rank-sum test for two independent samples, $n_1 = 31$, $n_2 = 24$.
Step 5. Test is nonparametric; no parametric null distribution needs to be displayed.
Steps 6 and 7. $W_{\text{sample\_1}} = 617.5$ (or $W_{\text{sample\_2}} = 922.5$), p-value = 0.000; reject H_0.
Step 8. The data support the claim that, based on the median, more of the territories' forests have been destroyed yearly by lightning since 1981 than was previously the case.

29. a) Subscript 1 = books on education theory; subscript 2 = books on home schooling.
Step 1. Claim: median$_1$ < median$_2$.
Step 2. H_0: median$_1$ ≥ median$_2$; H_1: median$_1$ < median$_2$.
Step 3. $\alpha = 0.05$.
Step 4. Wilcoxon signed-ranks test for two independent samples, $n_1 = 9$, $n_2 = 9$.
Step 5. Test is nonparametric; no parametric null distribution needs to be displayed.

Steps 6 and 7. $W_{sample\_1} = 63.5$ (or $W_{sample\_2} = 107.5$), p-value $= 0.026$; reject H_0.
Step 8. The evidence supports the claim that the median price for books on education theory is less than the median price for books on home schooling.

b) Subscript 1 = books on education theory; subscript 2 = books on home schooling.
Step 1. Claim: $\mu_1 < \mu_2$.
Step 2. H_0: $\mu_1 \geq \mu_2$; H_1: $\mu_1 < \mu_2$.
Step 3. $\alpha = 0.05$.
Step 4. t-test for the mean, two independent samples (equal-variances assumption not rejected), $n_1 = 9$, $n_2 = 9$, $df = 16$.
Step 5. Draw null distribution for left-tail t-test.
Steps 6 and 7. Test statistic $t = -1.807$, p-value $= 0.045$; reject H_0.
Step 8. The evidence supports the claim that the mean price for books on education theory is less than the mean price for books on home schooling.

c) Yes, although the p-value was just barely less than α for the t-test approach. This might be because, with small samples from populations that appear to be right-skewed, the parametric t-test is a less reliable test for comparing the measures of centre.

31. a) **Step 1.** Claim: $\text{median}_{March} < \text{median}_{April}$.
Step 2. H_0: $\text{median}_{March} \geq \text{median}_{April}$; H_1: $\text{median}_{March} < \text{median}_{April}$.
Step 3. $\alpha = 0.03$.
Step 4. Wilcoxon signed-ranks test for two dependent samples, $n = 10$.
Step 5. Test is nonparametric; no parametric null distribution needs to be displayed.
Steps 6 and 7. Test statistic $T = 11$, p-value $= 0.052$ (Note: The p-value solution appears slightly different if your software uses the z-approximation.); do not reject H_0.
Step 8. The evidence does not support the claim that the median price for commodities purchased by the inspector increased between March and April 2005.

b) **Step 1.** Claim: $\text{median}_{Feb} < \text{median}_{March}$.
Step 2. H_0: $\text{median}_{Feb} \geq \text{median}_{March}$, H_1: $\text{median}_{Feb} < \text{median}_{March}$.
Step 3. $\alpha = 0.03$.
Step 4. Wilcoxon signed-ranks test for two dependent samples, $n = 10$.
Step 5. Test is nonparametric; no parametric null distribution needs to be displayed.
Steps 6 and 7. Test statistic $T = 19$, p-value $= 0.216$ (Note: The p-value solution appears slightly different if your software uses the z-approximation.); do not reject H_0.
Step 8. The evidence does not support the claim that the median price for commodities purchased by the inspector increased between February and March 2005.

33. **Step 1.** Claim: $\text{median}_{1993} = \text{median}_{2003}$.

Step 2. H_0: $\text{median}_{1993} = \text{median}_{2003}$; H_1: $\text{median}_{1993} \neq \text{median}_{2003}$.

Step 3. $\alpha = 0.05$.

Step 4. Wilcoxon signed-ranks test for two dependent samples, $n = 6$.

Step 5. Test is nonparametric; no parametric null distribution needs to be displayed.

Steps 6 and 7. Test statistic $T = 10$, p-value $= 1.000$; do not reject H_0.

Step 8. There is not sufficient evidence to reject the claim that the median consumption of non-chicken meat did not change between 1993 and 2003.

Basic Concepts 13.3

35. When the conditions for applying a parametric test *are* met and data information would be lost if a nonparametric test was used instead; when another test version would be more efficient—i.e., had more power and therefore would have a higher probability of correctly rejecting a false null hypothesis.

Review Applications

37. **Step 1.** Claim: median = 500.

Step 2. H_0: median = 500; H_1: median $\neq$ 500.

Step 3. $\alpha = 0.05$.

Step 4. Sample Wilcoxon signed-ranks test for median, $n = 12$.

Step 5. Test is nonparametric; no parametric null distribution needs to be displayed.

Steps 6 and 7. Test statistic $T = 26$, p-value $= 0.562$ (Note: The p-value solution appears slightly different if your software uses the z-approximation.*); do not reject* H_0.

Step 8. There is not sufficient evidence to reject the claim that the median weight for the domestic animals in the collection is 500 grams.

39. Subscript 1 = livestock animals; subscript 2 = domestic animals.

Step 1. Claim: $\text{median}_1 > \text{median}_2$.

Step 2. H_0: $\text{median}_1 \leq \text{median}_2$, H_1: $\text{median}_1 > \text{median}_2$.

Step 3. $\alpha = 0.01$.

Step 4. Wilcoxon rank-sum test for two independent samples, $n_1 = 16$, $n_2 = 12$.

Step 5. Test is nonparametric; no parametric null distribution needs to be displayed.

Steps 6 and 7. $W_{\text{sample 1}} = 237$ (or $W_{\text{sample 2}} = 169$), p-value $= 0.414$ (Note: The p-value solution appears slightly different if your software uses the z-approximation.); do not reject H_0.

Step 8. The evidence does not support the claim that livestock animals consume a greater median amount of water per kilogram of body weight than do domestic animals.

41. Since the distribution of the variable domestic water use is not symmetrical, the sign test for the median is the appropriate test model.

(Note: The sampling method has been applied directly to the 1285 records in the file, resulting in 43 selected records—of which only $n = 40$ contain data for the Domestic Water use variable.)

Step 1. Claim: median > 1500.

Step 2. H_0: median ≤ 1500, H_1: median > 1500.

Step 3. $\alpha = 0.04$.

Step 4. Sign test for the median, $n = 40$.

Step 5. Test is nonparametric; no parametric null distribution needs to be displayed.

Steps 6 and 7. Test statistic (number of plus signs) $= 21$, p-value $= 0.438$; do not reject H_0.

Step 8. The evidence does not support the claim that the median domestic water use in all municipalities with a population over 1,000 is greater than 1,500 cubic metres per day.

43. Since the distribution of the variable longitude is not symmetrical, the sign test for the median is the appropriate test model.

Step 1. Claim: median $\neq 72$.

Step 2. H_0: median $= 72$; H_1: median $\neq 72$.

Step 3. $\alpha = 0.05$.

Step 4. Sign test for the median, $n = 43$.

Step 5. Test is nonparametric; no parametric null distribution needs to be displayed.

Steps 6 and 7. Test statistic (number of plus signs) $= 20$, p-value $= 0.761$; do not reject H_0.

Step 8. The evidence does not support the claim that the median longitude in municipalities with a population over 1,000 is not equal to 72°W.

45. Subscript 1= population in 1999; subscript 2 = population in 1996.

Step 1. Claim: $\text{median}_1 > \text{median}_2$.

Step 2. H_0: $\text{median}_1 \leq \text{median}_2$; H_1: $\text{median}_1 > \text{median}_2$.

Step 3. $\alpha = 0.05$.

Step 4. Wilcoxon signed-ranks test for two dependent samples, $n = 43$. (Note: Although the individual populations are highly skewed, the population of *differences* for paired values is roughly symmetrical, as this method requires.)

Step 5. Test is nonparametric; no parametric null distribution needs to be displayed.

Steps 6 and 7. Test statistic $T = 52$, p-value $= 0.001$ (Note: The p-value solution may appear slightly different if your software uses the z-approximation.); reject H_0.

Step 8. The data support the claim that in general the median municipalities' populations have increased from 1997 to 1999.

47. Subscript 1 = obesity scores; subscript 2 = stress scores.

Step 1. Claim: $\text{median}_1 = \text{median}_2$.

Step 2. H_0: $\text{median}_1 = \text{median}_2$; H_1: $\text{median}_1 \neq \text{median}_2$.

Step 3. $\alpha = 0.05$.

Step 4. Wilcoxon signed-ranks test for two dependent samples, $n = 13$.

Step 5. Test is nonparametric; no parametric null distribution needs to be displayed.

Steps 6 and 7. Test statistic $T = 35$, p-value $= 0.497$ (Note: The p-value solution appears to be slightly different if your software uses the z-approximation.); do not reject H_0.

Step 8. There is not sufficient evidence to reject the claim that the median obesity scores are the same as the median stress scores for the various regions.

49. For actual revenue tons *performed,* use the variable All_RevP.

Subscript 1 = regional; subscript 2 = national.

Step 1. Claim: $\text{median}_1 < \text{median}_2$.

Step 2. H_0: $\text{median}_1 \geq \text{median}_2$; H_1: $\text{median}_1 < \text{median}_2$.

Step 3. $\alpha = 0.05$.

Step 4. Wilcoxon rank-sum test for two independent samples, $n_1 = 10$, $n_2 = 17$.

Step 5. Test is nonparametric; no parametric null distribution needs to be displayed.

Steps 6 and 7. $W_{\text{sample 1}} = 95$ (or $W_{\text{sample 2}} = 283$), p-value $= 0.012$; reject H_0.

Step 8. The data support the claim that regional airlines tend to service fewer median revenue tons than national airlines do.

Chapter 14 Analysis of Variance (ANOVA)

Basic Concepts 14.1

1. To test for a difference among more than two means.
3. There are two main reasons why using a series of t-tests is not always recommended:
   - For each pair of means, a t-test has to be set up. As the number of means becomes larger, the number of possible paired combinations can become very large.
   - The risk of committing a Type I error increases as more tests are performed.
5. Answers are based on $df_{\text{between}} = k - 1$ and $df_{\text{within}} = n - k$, where k = number of groups and n = number of distinct observations.

   (a) $df_{\text{between}} = 4$, $df_{\text{within}} = 45$; (b) $df_{\text{between}} = 2$, $df_{\text{within}} = 15$; (c) $df_{\text{between}} = 3$, $df_{\text{within}} = 44$; (d) $df_{\text{between}} = 5$, $df_{\text{within}} = 42$; (e) $df_{\text{between}} = 3$; $df_{\text{within}} = 32$.
7. Answers are based on $MS_{\text{within}} = SSW/df_{\text{within}}$ and $df_{\text{within}} = n - k$.

   (a) 2.1; (b) 1.12; (c) 1.733; (d) 0.368.
9. The null hypothesis would be rejected at the 0.01 level of significance.

Source of Variation	SS	df	MS	F	P
Between groups	51.67	2	25.835	21.176	0.000
Within groups	30.5	25	1.22		
Total	82.17	27			

11. In the case of rejecting the null hypothesis in an ANOVA test, we conclude that the tested means are not all the same, but this conveys no information about *which* of the means was different. A follow-up test can help to answer this question.

Applications 14.1

Notes: (1) For the hypothesis tests problems in this chapter, complete ANOVA tables are not reproduced for all solutions; however, your own tables can be confirmed by checking against the displayed degrees of freedom, test statistic, p-value, etc. (2) Also for hypothesis tests in this chapter, the diagrams corresponding to step 5 will not be provided unless different in format from the examples that precede it.

13. Subscript 1 = Asia; subscript 2 = Africa; subscript 3 = Europe.

    Step 1. Claim: $\mu_1 = \mu_2 = \mu_3$.

    Step 2. H_0: $\mu_1 = \mu_2 = \mu_3$; H_1: not all the means are equal.

    Step 3. $\alpha = 0.05$.

    Step 4. One-way ANOVA test, $n_1 = 4$, $n_2 = 4$, $n_3 = 13$ ($df_{\text{between}} = 2$, $df_{\text{within}} = 18$, $df_{\text{total}} = 20$).

    Step 5.

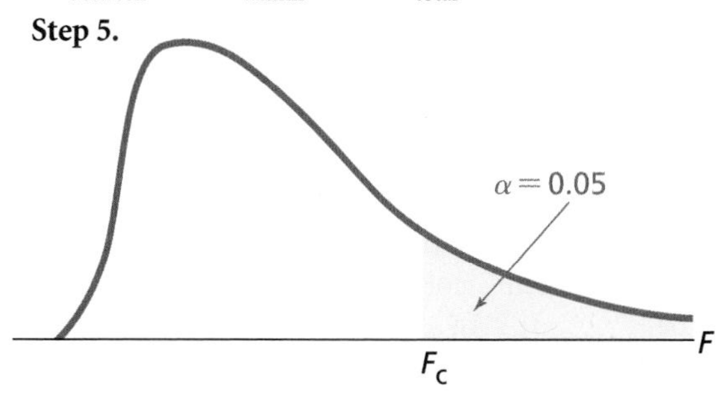

    Steps 6 and 7. Test statistic $F = 1.332$, p-value $= 0.289$; do not reject H_0.

    Step 8. There is not sufficient evidence to reject the claim that the mean numbers of Grace alumni per country listed in each of Asia, Africa, and Europe are the same.

15. Subscript 1 = clarity "V"; subscript 2 = clarity "S"; subscript 3 = clarity "I."

    Step 1. Claim: $\mu_1 = \mu_2 = \mu_3$.

    Step 2. H_0: $\mu_1 = \mu_2 = \mu$; H_1: not all the means are equal.

    Step 3. $\alpha = 0.10$.

    Step 4. One-way ANOVA test, $n_1 = 53$, $n_2 = 42$, $n_3 = 4$ ($df_{\text{between}} = 2$, $df_{\text{within}} = 96$, $df_{\text{total}} = 98$).

    Step 5. Draw the null distribution for the F-test.

    Steps 6 and 7. Test statistic $F = 0.553$, p-value $= 0.577$; do not reject H_0.

    Step 8. There is not sufficient evidence to reject the claim that the mean prices for diamonds in all three categories are the same.

17. Subscript 1 = Alberta; subscript 2 = British Columbia; subscript 3 = Manitoba; subscript 4 = New Brunswick; subscript 5 = Newfoundland & Labrador; subscript 6 = Nova Scotia; subscript 7 = Ontario; subscript 8 = Prince Edward Island; subscript 9 = Quebec; subscript 10 = Saskatchewan.

    Step 1. Claim: All of the means are equal.

    Step 2. H_0: all the means are equal, H_1: not all the means are equal.

    Step 3. $\alpha = 0.05$.

    Step 4. One-way ANOVA test, $n_1 = 2$, $n_2 = 7$, $n_3 = 1$, $n_4 = 2$, $n_5 = 1$, $n_6 = 2$, $n_7 = 17$, $n_8 = 1$, $n_9 = 5$, $n_{10} = 2$ ($df_{\text{between}} = 9$; $df_{\text{within}} = 30$; $df_{\text{total}} = 39$).

    Step 5. Draw the null distribution for the F-test.

    Steps 6 and 7. $F = 1.105$, p-value $= 0.389$; do not reject H_0.

Step 8. There is not sufficient evidence to reject the claim that the mean living costs are the same for the cities in all of the provinces listed.

Basic Concepts 14.2

19. Prefer the randomized block design when the samples are *dependent;* by linking related elements in the two samples, you control for some of the variation among individuals.

21. Blocks in the randomized block design are the intentionally introduced factors that connect the observations.

23.
- The treatment observations are independent and ideally taken from a random experiment.
- The observations of block/treatment combinations are approximately normally distributed.
- The variances of the distributions are approximately equal.
- The data have to be at least at the interval level of measurement.

25. p-value (for treatment) $= 0.013$. Reject H_0. The data support the claim that the mean wear is different for different loads.

27. p-value (for treatment) $= 0.000$. Reject H_0. The data support the claim that the mean pollen yield is different based on treatment.

Applications 14.2

29. **Step 1.** Claim: All of the treatment means are equal.

Step 2. H_0: all of the treatment means are equal; H_1: not all of the treatment means are equal.

Step 3. $\alpha = 0.05$.

Step 4. ANOVA test, with randomized block design, five treatments, three blocks.

Step 5. Draw the null distribution for the F-test.

Steps 6 and 7. Test statistic F (for treatment) $= 0.884$, $df_{\text{numerator}} = 4$, $df_{\text{denominator}} = 8$, p-value (for treatment) $= 0.515$; do not reject H_0.

Step 8. There is not sufficient evidence to reject the claim that there is no difference in the mean yields for different types of rootstocks.

31. **Step 1.** Not all of the treatment means are equal.

Step 2. H_0: all of the treatment means are equal; H_1: not all of the treatment means are equal.

Step 3. $\alpha = 0.05$.

Step 4. ANOVA test, with randomized block design, six treatments, three blocks.

Step 5. Draw the null distribution for the F-test.

Steps 6 and 7. Test statistic F (for treatment) $= 1.465$, $df_{\text{numerator}} = 5$, $df_{\text{denominator}} = 10$; p-value (for treatment) $= 0.284$; do not reject H_0.

Step 8. The evidence does not support the claim that there is a difference in the mean potassium level in the harvested leaves for the different types of rootstocks.

33. **Step 1.** Claim: Not all of the treatment means are equal.

Step 2. H_0: all of the treatment means are equal; H_1: not all of the treatment means are equal.

Step 3. $\alpha = 0.05$.

Step 4. ANOVA test, with randomized block design, three treatments, three blocks.

Step 5. Draw the null distribution for the F-test.

Steps 6 and 7. Test statistic F (for treatment) $= 5.967$, $df_{\text{numerator}} = 2$, $df_{\text{denominator}} = 4$, p-value (for treatment) $= 0.063$; do not reject H_0.

Step 8. The evidence does not support the claim that the mean resistances resulting from the treatments are not the same in relation to genotypes that originated in Syria.

Basic Concepts 14.3

35. Use a two-factor ANOVA to test a hypothesis when there are two factors used as grouping variables. The two-factor ANOVA tests whether there is an interaction between the two grouping variables, and if there is no interaction it can test whether each of the factor variables has an effect on the response variable.

37.
- The treatment observations are independent and ideally taken from a random experiment.
- The observations of each combination of factors are approximately normally distributed.
- The variances of the distributions are approximately equal.
- The data are at least at the interval level of measurement.

39. Yes, because the p-value for the test of interaction between the two factors ≈ 0.000, so reject the implied null hypothesis that the impact of the interaction on the total variance is not significant.

41. The p-value for the test of interaction between the two factors ≈ 0.000, so reject H_0. Since the claim corresponds to H_1, the conclusion is that the data support the claim that brand and load have an interaction effect on the mean durability of tires.

43. The p-value (for brand) ≈ 0.000, so reject H_0. Since the claim corresponds to H_1, the conclusion is that the data support the claim that tires of different brands do not all have the same mean durability.

Applications 14.3

45. a) **Step 1.** Claim: All of the group means based on sector levels are equal.
Step 2. H_0: all of the group means based on sector levels are equal; H_1: not all of the group means based on sector levels are equal.
Step 3. $\alpha = 0.01$.
Step 4. Use the two-factor ANOVA test.
Step 5. Draw the null distribution for the *F*-test.
Steps 6 and 7. Test statistic *F* (for sector factor) = 0.022, $df_{\text{numerator}} = 2$, $df_{\text{denominator}} = 60$, *p*-value (for sector factor) = 0.978; do not reject H_0.
Step 8. There is not sufficient evidence to reject the claim that means for water use are the same for all of the listed sectors; none of the sectors appear to have a greater water use.

b) **Step 1.** Claim: Not all of the group means based on community size levels are equal.
Step 2. H_0: all of the group means based on community size levels are equal; H_1: not all of the group means based on community size levels are equal.
Step 3. $\alpha = 0.01$.
Step 4. Use the two-factor ANOVA test.
Step 5. Draw the null distribution for the *F*-test.
Steps 6 and 7. Test statistic *F* (for the size factor) = 14.061, $df_{\text{numerator}} = 4$, $df_{\text{denominator}} = 60$, *p*-value (for the size factor) = 0.000; reject H_0.
Step 8. The data support the claim that means of water use vary for the different sizes of communities. When community size increases, the water use also increases.

47. **Step 1.** Claim: The interaction between the factors has a significant impact on the mean span of bridges.

Step 2. H_0: the impact of the interaction on the mean span is not significant; H_1: the impact of the interaction on the mean span is significant.

Step 3. $\alpha = 0.05$.

Step 4. Use the two-factor ANOVA test.

Step 5. Draw the null distribution for the *F*-test.

Steps 6 and 7. Test statistic *F* (for interaction factor) = 5.862, $df_{\text{numerator}} = 2$, $df_{\text{denominator}} = 12$; *p*-value (for interaction factor) = 0.017; reject H_0.

Step 8. The data support the claim that the mean span of a major bridge is affected by the interaction of its era and type.

49. a) **Step 1.** Claim: The interaction between the factors has a significant impact on the mean gas prices.
Step 2. H_0: the impact of the interaction on mean gas prices is not significant; H_1: the impact of the interaction on mean gas prices is significant.
Step 3. $\alpha = 0.05$.
Step 4. Use the two-factor ANOVA test.
Step 5. Draw the null distribution for the *F*-test.
Steps 6 and 7. Test statistic *F* (for interaction factor) = 2.060, $df_{\text{numerator}} = 4$, $df_{\text{denominator}} = 18$; *p*-value (for interaction factor) = 0.129; do not reject H_0.
Step 8. There is not enough evidence to reject the claim that mean gas prices are not affected by the interaction of community and brand of gas.

b) **Step 1.** Claim: All of the group means based on brand levels are equal.
Step 2. H_0: all of the group means based on brand levels are equal; H_1: not all of the group means based on brand levels are equal.
Step 3. $\alpha = 0.05$.
Step 4. Use the two-factor ANOVA test.
Step 5. Draw the null distribution for one-tail *F*-test.
Steps 6 and 7. Test statistic *F* (for brand factor) = 0.761, $df_{\text{numerator}} = 2$, $df_{\text{denominator}} = 18$; *p*-value (for interaction factor) = 0.481; do not reject H_0.
Step 8. There is not sufficient evidence to reject the claim that mean gas prices are no different for groups based on the brand of gas.

Review Applications

51. a) Subscript 1 = Alberta; subscript 2 = British Columbia; subscript 3 = Manitoba; subscript 4 = Newfoundland & Labrador; subscript 5 = Northwest Territories; subscript 6 = Nunavut; subscript 7 = Ontario; subscript 8 = Quebec; subscript 9 = Saskatchewan; subscript 10 = United States.
Step 1. Claim: Not all of the mean vertical drops are equal.
Step 2. H_0: all of the means are equal; H_1: not all of the means are equal.
Step 3. $\alpha = 0.05$.
Step 4. One-way ANOVA test, $n_1 = 1$, $n_2 = 7$, $n_3 = 2$, $n_4 = 4$, $n_5 = 7$, $n_6 = 3$, $n_7 = 15$, $n_8 = 6$, $n_9 = 1$, $n_{10} = 1$ ($df_{\text{between}} = 9$; $df_{\text{within}} = 37$; $df_{\text{total}} = 46$).
Step 5. Draw the null distribution for the *F*-test.
Steps 6 and 7. Test statistic $F = 5.859$, *p*-value = 0.000; reject H_0.
Step 8. The data support the claim that the mean vertical drops of the waterfalls are not the same in all of the listed provinces and territories.

b) Alberta and British Columbia.

53. Subscript 1 = Brazil; subscript 2 = China; subscript 3 = Czech Republic; subscript 4 = Egypt; subscript 5 = France; subscript 6 = Germany; subscript 7 = Italy; subscript 8 = Poland; subscript 9 = South Africa; subscript 10 = Sweden; subscript 11 = Thailand; subscript 12 = Turkey.

Step 1. Claim: All of the means for calcium content are equal.

Step 2. H_0: all of the means for calcium content are equal; H_1: not all of the means for calcium content are equal.

Step 3. $\alpha = 0.05$.

Step 4. One-way ANOVA test, $n_1 = 3$, $n_2 = 1$, $n_3 = 3$, $n_4 = 1$, $n_5 = 2$, $n_6 = 5$, $n_7 = 3$, $n_8 = 2$, $n_9 = 3$, $n_{10} = 1$, $n_{11} = 1$, $n_{12} = 5$ ($df_{\text{between}} = 11$, $df_{\text{within}} = 18$; $df_{\text{total}} = 29$).

Step 5. Draw null distribution for the *F*-test.

Steps 6 and 7. Test statistic $F = 0.849$, *p*-value $= 0.599$; do not reject H_0.

Step 8. There is not sufficient evidence to reject the claim that the mean contents of calcium are the same in the mineral water products for all the countries listed.

55. Subscript 1 = weights less than 90 kg; subscript 2 = weights from 90 to 100 kg; subscript 3 = weights greater than 100 kg.

Step 1. Claim: The mean penalties in minutes are not equal for all weight groups.

Step 2. H_0: the mean penalties in minutes are equal for all weight groups; H_1: the mean penalties in minutes are not equal for all weight groups.

Step 3. $\alpha = 0.05$.

Step 4. One-way ANOVA test, $n_1 = 10$, $n_2 = 12$, $n_3 = 5$ ($df_{\text{between}} = 2$, $df_{\text{within}} = 24$; $df_{\text{total}} = 26$).

Step 5. Draw the null distribution for the *F*-test.

Steps 6 and 7. Test statistic $F = 7.585$, *p*-value $= 0.003$; reject H_0.

Step 8. The data support the claim that the mean time spent in the penalty box is different for the different weight categories.

57. Subscript 1 = weights less than 90 kg; subscript 2 = weights from 90 to 100 kg; subscript 3 = weights larger than 100 kg.

Step 1. Claim: The mean points earned are not equal for all weight groups.

Step 2. H_0: The mean points earned are equal for all weight groups; H_1: The mean points earned are not equal for all weight groups.

Step 3. $\alpha = 0.05$.

Step 4. One-way ANOVA test, $n_1 = 10$, $n_2 = 12$, $n_3 = 5$ ($df_{\text{between}} = 2$, $df_{\text{within}} = 24$; $df_{\text{total}} = 26$).

Step 5. Draw the null distribution for the *F*-test.

Steps 6. and 7. Test statistic $F = 1.658$, *p*-value $= 0.212$; do not reject H_0.

Step 8. There is not sufficient evidence to reject the claim that the mean number of points earned by players in the season was the same for all coded weight categories.

59. Step 1. Claim: All of the means for revenue tons are equal (regardless of performance category).

Step 2. H_0: all of the means for revenue tons are equal (regardless of performance category); H_1: not all of the means for revenue tons are equal (regardless of performance category).

Step 3. $\alpha = 0.05$.

Step 4. ANOVA test, with randomized block design, three treatments, three blocks.

Step 5. Draw the null distribution for one-tail *F*-test.

Steps 6 and 7. Test statistic *F* (for performance) $= 1.180$, $df_{\text{numerator}} = 2$, $df_{\text{denominator}} = 4$, *p*-value (for performance) $= 0.396$; do not reject H_0.

Step 8. There is not sufficient evidence to reject the claim that the mean values for revenue tons carried are about the same, regardless of the performance category of the airline.

61. Subscript 1 = dairy products; subscript 2 = fruit; subscript 3 = grain products, legumes, nuts; subscript 4 = meat, poultry, fish; subscript 5 = vegetables.

Step 1. Claim: Not all of the means are equal.

Step 2. H_0: all of the means are equal; H_1: not all of the means are equal.

Step 3. $\alpha = 0.05$.

Step 4. One-way ANOVA test, $n_1 = 4$, $n_2 = 10$, $n_3 = 5$, $n_4 = 9$, $n_5 = 13$ ($df_{\text{between}} = 4$, $df_{\text{within}} = 36$; $df_{\text{total}} = 40$).

Step 5. Draw the null distribution for the *F*-test.

Steps 6 and 7. Test statistic $F = 11.778$, *p*-value $= 0.000$; reject H_0.

Step 8. The data support the claim that the mean calories of the foods within different groups are not the same.

63. Step 1. Claim: The interaction between the factors has no significant impact on the mean density.

Step 2. H_0: the impact of the interaction of factors on mean density is not significant; H_1: the impact of the interaction of factors on mean density is significant.

Step 3. $\alpha = 0.05$.

Step 4. Two-factor ANOVA test.

Step 5. Draw the null distribution for the *F*-test.

Steps 6 and 7. Test statistic *F* (for interaction factors) $= 0.162$, $df_{\text{numerator}} = 2$, $df_{\text{denominator}} = 18$, *p*-value (for interaction of factors) $= 0.852$; do not reject H_0.

Step 8. There is not sufficient evidence to reject the claim that there is no interaction between the variables' main colour and hardness code in relation to density.

65. **Step 1.** Claim: All of the group means for density based on colour levels are equal.

Step 2. H_0: all of the group means for density based on colour levels are equal; H_1: not all of the group means for density based on colour levels are equal.

Step 3. $\alpha = 0.05$.

Step 4. Two-factor ANOVA test.

Step 5. Draw the null distribution for the *F*-test.

Steps 6 and 7. Test statistic *F* for colour factor = 0.678, $df_{\text{numerator}} = 1$, $df_{\text{denominator}} = 18$, *p*-value for colour factor = 0.421; do not reject H_0.

Step 8. There is not sufficient evidence to reject the claim that mean densities of minerals grouped by main colour are equal.

67. a) **Step 1.** Claim: The interaction between the polling organization and party has a significant impact on the preference score.
Step 2. H_0: the interaction between polling organization and party has no significant impact on the preference score; H_1: the interaction between the polling organization and party has a significant impact on the preference score.
Step 3. $\alpha = 0.05$.
Step 4. Two-factor ANOVA test.
Step 5. Draw the null distribution for the *F*-test.
Steps 6 and 7. Test statistic *F* for interaction of factors = 2.487, $df_{\text{numerator}} = 24$, $df_{\text{denominator}} = 108$, *p*-value for interaction of factors = 0.001; reject H_0.
Step 8. The data support the claim that the mean preference scores do show an interaction effect between polling organization and party preference.

b) Yes. *If* in fact polling organizations were biased in how they collected or reported preference scores—e.g., preferring certain parties or certain causes associated with parties—it would be consistent to discover that reported preference scores were affected by the combinations of polling organization and party preference.

c) No, the results of (a) do not *prove* that some organizations are exhibiting bias in their collection or reporting. Other influencing factors, for which we have no information, may be affecting the results—e.g., time frame in which interviews were conducted, sample sizes, sampling frames, method used to obtain preference scores, and return rate.

69. **Step 1.** Claim: Not all of the group means based on age levels are equal.

Step 2. H_0: all of the group means based on age levels are equal; H_1: not all of the group means based on age levels are equal.

Step 3. $\alpha = 0.05$.

Step 4. Two-factor ANOVA test.

Step 5. Draw the null distribution for the *F*-test.

Steps 6 and 7. Test statistic *F* for age factor = 4.978, $df_{\text{numerator}} = 2$, $df_{\text{denominator}} = 18$, *p*-value for age factor = 0.019; reject H_0.

Step 8. The data support the claim that the mean prices of children's books, when grouped by the target age, are not the same.

71. a) **Step 1.** Claim: All of the group means for numbers of hits based on the numbers of search words are equal.
Step 2. H_0: all of the group means based on numbers of search words are equal; H_1: not all of the group means based on numbers of search words are equal.
Step 3. $\alpha = 0.05$.
Step 4. Two-factor ANOVA test.
Step 5. Draw the null distribution for the *F*-test.
Steps 6 and 7. Test statistic *F* for words factor = 8.178, $df_{\text{numerator}} = 2$, $df_{\text{denominator}} = 18$, *p*-value for words factor = 0.003; reject H_0.
Step 8. There is sufficient evidence to reject the claim that the mean number of hits is the same regardless of the numbers of words in the search phrase.

b) **Step 1.** Claim: Not all of the group means based on different search engines are equal.
Step 2. H_0: all of the group means based on different search engines are equal; H_1: not all of the group means based on different search engines are equal.
Step 3. $\alpha = 0.05$.
Step 4. Two-factor ANOVA test.
Step 5. Draw the null distribution for the *F*-test.
Steps 6 and 7. Test statistic *F* for search engine factor = 0.926, $df_{\text{numerator}} = 2$, $df_{\text{denominator}} = 18$, *p*-value for interaction of factors = 0.414; do not reject H_0.
Step 8. The evidence does not support the claim that the mean number of hits is different for the different search engines.

Chapter 15 Measuring and Testing Associations for Bivariate Data

Basic Concepts 15.1

1. An association between two variables *A* and *B* exists when the conditional probability $P(A \mid B)$ (for particular values *a* of *A*) has varying results for different values of *B*, and/or the conditional probability $P(B \mid A)$ (for particular values *b* of *B*) has varying results for different values of *A*.

3. Range: −1 to +1; negative r suggests an inverse relationship between the correlated variables (e.g., if values of A increase, the values of B decrease).

5. • Data must be numerical and at the interval or ratio level of measurement.
   • The set of data pairs needs to be drawn as a random sample.
   • The sample of data pairs needs to be drawn from a bivariate normal population.

7. The strength of a correlation between two variables in a sample is measured as r; the corresponding population parameter is ρ. By itself, r gives no information as to whether the correlation is significant; to test significance, the implied null hypothesis H_0 is $\rho = 0$ (i.e., no correlation). We interpret r as a sample statistic, and assess whether to reject the null hypothesis.

9. (a) 0.954; (b) yes, because r is close to 1; (c) yes (p-value = 0.012); (d) r = value 0.943 (strong), p-value = 0.005 (significant).

Applications 15.1

11. a) The scatter diagram does not suggest a linear correlation between the two variables. The scatter of actual points around an imagined line clearly does not exhibit a bivariate normal distribution.

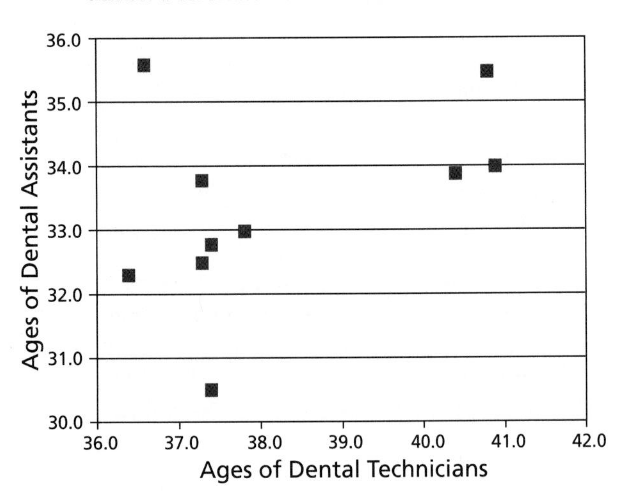

b) Based on $r = 0.417$, any apparent correlation is weak.
c) Based on p-value = 0.231, the weak apparent correlation is not significant.

13. a) Overall, the conditions are not ideal for calculating a linear correlation coefficient because of the clearly nonlinear pattern when drug expenses are less than about $11,000. However, the conditions for calculating a linear correlation look very good for the portion of data where drug expenses are more than about $11,000.

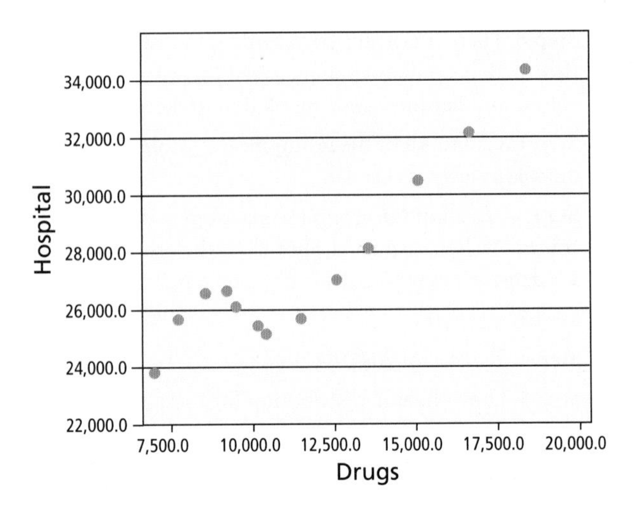

b) Based on $r = 0.929$, the correlation appears to be strong.
c) Based on p-value = 0.000, the strong apparent correlation is significant.

15. Yes. If variables for which the costs have both increased proportionally over time are paired, there will appear to be a positive linear correlation between them—even if the two variables are otherwise unrelated and have no causal connection between them.

Basic Concepts 15.2

17. The appropriate correlation measure is r if the raw data are at the interval or ratio level of measurement; r_s is the appropriate measure for either ordinal data or data that have been ranked (i.e., transformed into ordinal rank values).

19. False. Even if the raw data are measured at the ratio level, it is always possible (e.g., if the paired data do not follow a linear pattern) to transform all of the values for each variable into ranks; then it would be valid to calculate r_s based on the ranked values.

21. If the value or rank of one variable is steadily increased, then the value or rank of the other variable generally must increase or decrease consistently.

23. r_s is significant if an implied null hypothesis H_0: ($\rho_s = 0$) can be rejected, based on the value of the sample statistic r_s.

Applications 15.2

25. (a) $r_s = 0.359$; (b) weak; (c) significant (p-value = 0.011).

27. a) Weak ($r_s = 0.470$), significant (p-value = 0.001).
b) Weak ($r = 0.395$), significant (p-value = 0.005).
c) Yes, the correlation is somewhat stronger for the ranked values. Differences are not unusual, because

the two measures do not reflect exactly the same information. r_s shows only how the relative ranks of the two variables are correlated. r tells whether there is a correlation between the unranked original values of the two variables.

29. (a) $r_s = 0.924$; (b) significant (p-value $= 0.000$), very strong.

31. a) Very weak, close to no correlation ($r_s = -0.005$), not significant (p-value $= 0.963$). (Note: $r_s = -0.005$ is actually a weak *positive* correlation—since '1' is the *highest* rank for change, and '100' is the smallest rank.)

b) r_s is measured for variables that are ordinal or ranked. The data for the original variable "Change" were not ranked, whereas total sales data were already ranked in the variable "Rank2004." To assess the correlation between sales and change, the variable "Change" also had to be transformed into a ranked variable.

Basic Concepts 15.3

33. A test for independence determines whether samples presumably taken from two populations are independent samples. A test for homogeneity determines whether two or more populations are homogeneous in relation to the proportions of their members that fall into various categories.

35. The test statistic chi-square (χ^2) measures the squared distances between actual and expected values.

37. Test for goodness of fit.

39. a)

	Plain	Cherry	Total
Male	22.5	22.5	45
Female	27.5	27.5	55
Total	50	50	100

b) 9.091.

c) 1.

d) p-value $= 0.0026$; (assuming $\alpha = 0.05$ or $\alpha = 0.01$) reject H_0. The data support the claim that the gender of a customer is related to the type of cheesecake that is purchased.

Applications 15.3

41. **Step 1.** Claim: All of the corresponding proportions are equal.

Step 2. H_0: all of the corresponding proportions are equal; H_1: not all of the corresponding proportions are equal.

Step 3. $\alpha = 0.05$.

Step 4. Chi-square test for homogeneity, $df = 2$.

Step 5.

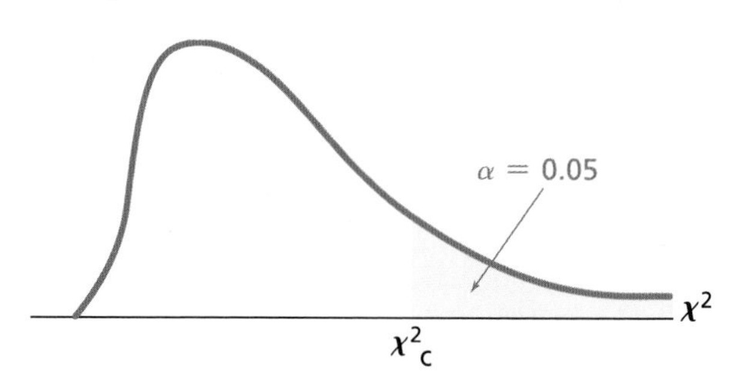

Steps 6 and 7. Test statistic $\chi^2 = 188{,}941.5$, p-value $= 0.000$; reject H_0.

Step 8. There is sufficient evidence to reject the claim that the same proportions of different e-mail types were received in July 2005 as in March 2004.

Note: In subsequent solutions for hypothesis tests in this chapter, the diagrams for Step 5 will not be provided.

43. **Step 1.** Claim: The variables "Province" and "Subdivision of Christianity" are independent.

Step 2. H_0: the variables "Province" and "Subdivision of Christianity" are independent; H_1: the variables "Province" and "Subdivision of Christianity" are related (not independent).

Step 3. $\alpha = 0.05$.

Step 4. Chi-square test for independence, $df = 3$.

Step 5. Draw the null distribution for χ^2-test.

Steps 6 and 7. Test statistic $\chi^2 = 3{,}490{,}821$, p-value $= 0.000$; reject H_0.

Step 8. There is sufficient evidence to reject the claim that for residents of Ontario and Quebec, their stated branch of Christianity is independent of which province they live in.

45. **Step 1.** Claim: Not all of the corresponding proportions are equal for the two groups.

Step 2. H_0: All of the corresponding proportions are equal; H_1: Not all of the corresponding proportions are equal.

Step 3. $\alpha = 0.05$.

Step 4. Chi-square test for homogeneity, $df = 2$.

Step 5. Draw the null distribution for χ^2 test.

Steps 6 and 7. Test statistic $\chi^2 = 7.837$, p-value $= 0.020$; reject H_0.

Step 8. The data support the claim that in the populations of people aged 18–29 and 40–49, respectively, the proportions who prefer the various minimum-age options are not the same.

47. **Step 1.** Claim: The observed and expected proportions are equal for all five countries.

Step 2. H_0: the observed and expected proportions are equal for all five countries; H_1: the observed and expected proportions are not all equal for all five countries.

Step 3. $\alpha = 0.05$.

Step 4. Chi-square test for goodness of fit, $df = 4$.

Step 5. Draw null distribution for χ^2-test.

Steps 6 and 7. Test statistic $\chi^2 = 6.223$, p-value $= 0.183$; do not reject H_0.

Step 8. There is not sufficient evidence to reject the claim that the recorded production numbers of bicycles in 1998 follow the predicted distribution of bicycle numbers as shown in the bottom row of the table.

Review Applications

49. (a) $r = -0.296$ (weak correlation); (b) not significant (p-value $= 0.377$).

51. (a) For Assets ↔ MER: $r = -0.011$, for MER ↔ NAVPS: $r = -0.098$; (b) neither; (c) neither (p-value for Assets ↔ MER $= 0.938$, p-value for MER ↔ NAVPS $= 0.455$).

53. **Step 1.** Claim: The variables "Fees" and "Rategroup" are not related (i.e. they are independent).

Step 2. H_0: The variables "Fees" and "Rategroup" are not related (i.e., they are independent); H_1: The variables "Fees" and "Rategroup" are related (i.e., not independent).

Step 3. $\alpha = 0.05$.

Step 4. Chi-square test for independence, $df = 2$.

Step 5. Draw the null distribution for χ^2-test.

Steps 6 and 7. Test statistic $\chi^2 = 1.370$, p-value $= 0.504$; do not reject H_0.

Step 8. There is not sufficient evidence to reject the claim that there is no relationship between the variables for fees and rate group for these funds.

55. (a) $r = -0.216$; (b) very weak and not significant (p-value $= 0.288$).

57. (a) $r = 0.947$; (b) very strong, significant (p-value $= 0.000$); (c) there is a significant strong positive correlation between the maximum-ever recorded temperatures in selected Canadian cities during the months of January and July and the corresponding minimum-ever recorded temperatures in those cities; that is, for cities with higher maximum-ever temperatures, the minimum-ever temperatures also tend to be higher values.

59. (a) $r_s = 0.886$; (b) strong, significant (p-value $= 0.000$); (c) there is a significant strong positive correlation between the *ranks* of the maximum-ever recorded temperatures in selected Canadian cities during the months of January and July and the *ranks* of the corresponding minimum-ever recorded temperatures in those cities; that is, for cities with higher-ranked maximum-ever temperatures, the ranks for the minimum-ever temperatures also tend to be higher values.

Both correlations were concluded to be strong and significant. The strength of the correlation is somewhat weaker for the ranked pairs of values for these two variables.

61. This exercise can be interpreted as: We expect (in the null hypothesis) that the airlines' numbers of reports would follow the same proportions as the numbers of passengers sampled from each airline. It is claimed that this is not the case.

Step 1. Claim: The corresponding observed and expected proportions of reports filed are not all equal.

Step 2. H_0: all of the corresponding observed and expected proportions are equal; H_1: the corresponding observed and expected proportions are not all equal.

Step 3. $\alpha = 0.05$.

Step 4. Chi-square test for goodness of fit, $df = 4$.

Step 5. Draw the null distribution for χ^2-test.

Steps 6 and 7. Test statistic $\chi^2 = 0.144$, p-value $= 0.998$; do not reject H_0.

Step 8. The evidence does not support the claim that the proportion of reports filed in relation to passengers sampled is not the same for all of these airlines.

63. **Step 1.** Claim: The observed and expected proportions are equal for all airlines.

Step 2. H_0: the observed and expected proportions are equal for all airlines; H_1: the observed and expected proportions are not all equal for all airlines.

Step 3. $\alpha = 0.05$.

Step 4. Chi-square test for goodness of fit, $df = 4$.

Step 5. Draw the null distribution for χ^2-test.

Steps 6 and 7. Test statistic $\chi^2 = 2.667$, p-value $= 0.615$; do not reject H_0.

Step 8. There is not sufficient evidence to reject the claim that the number of reports filed by the airlines followed the expected distribution.

Chapter 16 Simple and Multiple Linear Regression

Basic Concepts 16.1

1. The independent variable is used to make predictions or estimates for the dependent variable.
3. The regression line (or line of best fit) is a graphical display of the relationship between the dependent and independent variables; it is the one line for a given data set that minimizes the sum of squared residuals between actual y-values and the corresponding, predicted $\hat{y}$-values that fall on that regression line. The regression equation is an algebraic expression of the same relationship as is modelled in the regression line.
5. (a) Size of this year's tree-ring growth; (b) average winter temperature (in 1893); (c) number of housing starts.
7. a) Graphical display of the regression relationship between a dependent (y) and independent (x) variable.
   b) The procedure by which the regression equation/line are developed—by comparing the actual and expected values of a variable and attempting to minimize the sum of squared distances between those values.
   c) For each value for x, the distance between the actual corresponding value of y and the expected value of y (i.e., $\hat{y}$), based on the regression.
   d) The process of using the regression equation to predict values for the dependent variable that correspond to values of the independent variable (x-values)—that lie beyond the range of x-values in the data set that was used for constructing the equation.
   e) A proportion that shows how much of the variance of the dependent variable could be explained by the variance due to the regression.
   f) r^2 is the abbreviation for the coefficient of determination; see answer (e).
   g) A standard deviation measure for the distances between all actual values of y in comparison to all the corresponding expected values $\hat{y}$.
   h) A confidence interval estimate for the value of the dependent variable, given a *specific* value of the x-variable.
   i) A regression is significant when the amount of variance explained by the regression is significantly different from zero.
9. a) $\hat{y} = 171 + 3.12x$.
   b) 91%; based on $r^2 = 0.910$.
   c) \$311.40.
   d) "... within ± \$35.92" [based on $\pm(2 \times$ standard error)].
   e) (a. $\hat{y} = 171 + 3.18x$; (b. 89% ($r^2 = 0.889$); (c. \$314.10; (d. "...within ± \$35.79." [based on $\pm(2 \times$ standard error)].

Applications 16.1

11. Dependent variable (y) = mean ages of dental technicians, independent variable (x) = mean ages of dental assistants.
    a) $\hat{y} = 22.222 + 0.479x$; the regression is not significant (p-value of 0.231).
    b) Because the linear regression is not significant; also, only 17.4% of the variation in the mean age of dental technicians is explained by the regression. There might be other factors that have more impact on the dependent variable that are not considered in this regression model.
13. Dependent variable (y) = physician expenditures; independent variable (x) = expenditures on hospital care.
    a) $\hat{y} = -2{,}905.364 + 0.523x$ (for xs in units of \$100,000); (b) significant (p-value = 0.0000); (c) 417.94; (d) \$1,033,752,000.
    b) Residual = (actual physician costs) − (regression-estimated costs for physicians) = 14,321 − 14,485.96 = −164.96 (in CDN\$100,000s)
15. Dependent variable (y) = level of potassium (K); independent variable (x) = level of calcium (Ca).
    a) $\hat{y} = 6.427 + 0.00043x$.
    b) *Not* significant (p-value = 0.990).
    c) Regression is very weak ($r = 0.002$).
    d) 6.47 mg/L. Because the regression is not significant (and in any case, extremely weak), simply use the mean of the dependent variable.

Basic Concepts 16.2

17. In simple linear regression there is only one independent variable x to explain the dependent variable y, whereas in multiple linear regression, there are at least two independent variables x_i.
19. R^2 (similar to r^2 for simple linear regression) is a proportion that shows the relative variation in y that is explainable by the regression equation. But for comparing alternative multiple regressions, R^2 can be misleading as a measure, since adding new independent variables into the model reduces degrees of freedom, which reduces the usable information in the calculation. The calculations for adjusted R^2 take these considerations into account.
21. The slope of a single variable x_i is interpreted as the effect on the dependent variable of increasing that x_i-value by one unit, providing that all other independent variables are kept constant.
23. t-test.

25. Yes.

Applications 16.2

27. Dependent variable (y) = mean ages of dental assistants; independent variables: x_1 = mean ages of dental hygienists, x_2 = mean ages of dental technicians, x_3 = mean ages of denturists.
    a) $\hat{y} = 8.796 + 0.568x_1 + 0.127x_2 + 0.002x_3$.
    b) Not significant (p-value $= 0.134 > \alpha = 0.05$).
    c) R^2 (= 58%) stayed the same.
    d) Adjusted R^2 (= 37.1%) decreased. When the lost degrees of freedom factor is taken into account, the model got *less* reliable by adding the extra variable.

29. Dependent variable (y) = drug expenditures; independent variables: x_1 = expenditures on hospital care, x_2 = expenditures on physicians, x_3 = capital costs.
    a) $\hat{y} = -6854.169 - 0.639x_1 + 3.028x_2 + 0.406x_3$; regression is significant (p-value = 0.000); 419.652 (in CDN$100,000s).
    b) No, because the adjusted R^2 value has not increased; it has actually dropped slightly.
    c) No, the p-value for an implied t-test of significance of the variable's coefficient equals 0.348.

31. Dependent variable (y) = level of potassium; independent variables: x_1 = level of sodium, x_2 = level of calcium, x_3 = level of chlorine.

    Regression equation:
    $\hat{y} = 1.861 + 0.016x_1 - 0.026x_2 + 0.067x_3$. Adjusted R^2 (= 69.1%) is not notably different from the best of the regressions attempted in the previous question. The residual plot did not improve. Since the regression with all three independent variables (which is actually the addition of the variable chlorine into model (a)(ii) did not show a significant impact on the regression, it is recommended to use model (a)(ii).

Basic Concepts 16.3

33. Because with multiple variables, it is not feasible to use a scatter diagram to decide graphically whether the assumptions for regression (e.g., the normal distribution of errors) are met.

35. A proxy variable in a regression model is a variable that replaces a missing, but more causally relevant variable, perhaps because data for the more suitable variable are not available or (worse) because the researcher is unaware of the other variable. If included in the regression model, the proxy variable may take on unexpected values and possibly lead to a wrong conclusion or cause the entire model to appear to be not significant.

37. In a regression where time is an independent variable, a response variable exhibits autocorrelation if the values of observations at times t_i are correlated with the values of observations at previous times t_{i-k}.

39. Yes. One or more coefficients of the independent variables or the y-intercept may be not significant—not just increasing error, but also adding to the burden of data collection. The validity of the multiple regression model as a whole could be negatively affected if there is collinearity among independent variables, or if the distribution of the residuals is not reasonably close to normal, or if there are influential outliers. If the model contains proxy variables, these may distort the results of the regression model. A model that includes too many independent variables may be difficult to interpret.

Applications 16.3

41. Dependent variable (y) = mean age of dental assistants; independent variables (mean ages of . . .): x_1 = ambulance attendant, x_2 = audiologist, x_3 = dental hygienist, x_4 = medical laboratory technician, x_5 = medical radiation technologist, x_6 = other professional occupation. (Note: Records missing at least one value were excluded.)
    a) **One possible model:**
       $\hat{y} = -29.813 + 0.561x_1 + 1.101x_4$.
       Considerations: *Rejected variables:* Other (not significant and collinear with three other variables), Audio, MR_Tech, and DentHyg (not significant). *Strength of the model:* Overall p-value = 0.001, all x-coefficients are significant.
    b) $R^2 = 87.5\%$.
    c) Note: Only the inputs that have a role in the model are utilized.
       $\hat{y} = -29.813 + (0.561 \times 38) + (1.101 \times 36.2)$
       $= 31.36$ years
    d) The standard error of the model is 0.6107 and thus quite small in comparison to the average value for the dependent variable (= 33.4 years). The residual plot shows that the residuals are fairly normally distributed. Overall, based on these observations, the estimate in (c) seems to be quite reliable.

43. Dependent variable (y) = drug expenditures; independent variables (expenditures for . . .): x_1 = hospitals, x_2 = other institutions, x_3 = physicians, x_4 = other professionals, x_5 = capital, x_6 = public health & admin, x_7 = other health spending.
    a) **One possible model:**
       $\hat{y} = -7223.536 + 0.212x_1 + 1.374x_4$.
       Considerations: Rejected variables: Oth_Inst (x_2), Physic (x_3), Capital (x_5), Pub_Heal (x_6), and Oth_Heal (x_7) (not significant); all variables are collinear with the other variables. *Strength of the model:* Overall p-value = 0.000, all x coefficients are significant; standard error is fairly small (= 182.81);

the distribution of residuals is reasonably approximate to normal. *Weakness of the model:* Variables in the final model are correlated, *so use the model with caution.*

b) R^2 almost 100% (99.8%).

c) Note: Only the inputs that have a role in the model are utilized.

$$\hat{y} = -7223.536 + (0.212 \times 21790.2) + (1.374 \times 9835.9) = 10910.5 \text{ (in CDN\$100,000s)}$$

d) −$410.51 (in CDN$100,000s) (based on actual drug costs − estimated drug costs).

45. Dependent variable (y) = K; independent variables: x_1 = Ca, x_2 = Mg, x_3 = Na, x_4 = HCO_3, x_5 = Cl, x_6 = SO_4, x_7 = F, x_8 = B, x_9 = P, x_{10} = Ti, x_{11} = Cr.

a) **One possible model:**

$$\hat{y} = -3.799 + 0.021x_2 + 0.017x_4 + 7.807x_7 - 0.002x_8$$

Considerations: Rejected variables: Na (x_3), P (x_9), Ti (x_{10}), Cr (x_{11}) (not significant); omission of SO_4 (due to collinearity with Mg) and Ca (due to collinearity with HCO_3 and B). *Strength of the model:* Overall p-value = 0.000, all x-coefficients are significant; residuals are almost perfectly normal distributed. *Weakness of the model:* HCO_3 and B are moderately correlated (r = 0.671) and the standard error is fairly high (= 5.214).

b) R^2 = 92%.

c) Note: Only the inputs that have a role in the model are utilized.

$$\hat{y} = -3.799 + (0.021 \times 66.1) + (0.017 \times 171) + (7.807 \times 0.07)(0.002 \times 119) = 0.805$$

d) 1.965 (= 2.77 − 0.805 = actual potassium content − estimated potassium content).

Basic Concepts 16.4

47. A dummy variable is an additional independent variable added to a regression model, which by convention usually takes a default value of 0 if some qualitative condition does not apply for an observation or a value of 1 if that condition *does* apply. A dummy variable can be used to enable a categorical variable to be included in a linear regression model.

49. $1.50.

51. Answer (b) (independent) because the slopes for the quantitative variables should not be affected by the values of the dummy variable; otherwise, the impact of a dummy variable could not be modelled by simply adding or subtracting its coefficient from the regression output based on the quantitative variables.

Applications 16.4

53. Dependent variable (y) = level of full-time employment (Full); independent variables: x_1 = number of firms (Firms), x_2 = number of part-time employees (Part), x_3 = number of seasonal workers (Season), x_4 = dummy variable: sector that focuses on service (Service, where 0 = not service sector and 1 = service sector).

a) **One possible model:** $\hat{y} = 217.382 + 5.229x_2$.
Considerations: Rejected variables: Firms (x_1), Season (x_3), Service (x_4) (not significant). *Strength of the model:* Overall p-value = 0.000. *Weakness of the model:* Although the (one) x-variable coefficient is significant (p-value = 0.000), the constant (b_0) is not significant (p-value = 0.076); the regression explains only about one-third of the variance in y (R^2 = 33.1%); the residuals do not follow the expected distribution well; the standard error is very high (644.7) in comparison with the y-mean.

b) No, it was not significant.

c) The model does not offer a very precise prediction. See "*Weaknesses of the model*" in the solution to (a).

55. Dependent variable (y) = calorie content (Calories); independent variables (content of . . .): x_1 = carbohydrates (Carbohyd), x_2 = fat (Fat), x_3 = protein (Protein), x_4 = saturated fat (Sat_Fat), x_5 = cholesterol (Cholest), in the second step, x_6 = dummy variable (Beef, where 0 = no beef content and 1 = some beef content).

Model without dummy variable:

One possible model:
$\hat{y} = 156.975 - 10.174x_1 + 14.124x_4$.

Considerations: *Rejected variables:* Cholest (x_5) (not significant); Fat (x_2) and Protein (x_3) due to collinearity with remaining variables. *Strength of the model:* Overall p-value = 0.002, significant x-coefficients, adjusted R^2 is high with 93.5%, standard error is fairly small, residuals are fairly reasonably distributed. *Weakness of the model:* The two retained variables Carbohyd and Sat_Fat are moderately correlated (r = −0.709).

Model with dummy variable: No, the added dummy variable was not significant and adding the variable had little positive impact on adjusted R^2.

Review Applications

57. Dependent variable (y) = numbers of autos; independent variable (x) = number of bikes.

(a) $\hat{y} = 44.633 - 0.079x$; (b) no ($p$-value = 0.377); (c) 8.8% ; (d) no, because the model is not significant and the regression , in any case, explains just a small percentage of the variation in the dependent variable. Also, the distribution of the residuals is not close to being bivariate normal.

59. Dependent variable (y) = three-month return (Ret_3Mth); independent variables: x_1 = 1-month return (Ret_1Mth), x_2 = management expense ratio (MER), x_3 = fund's assets (Assets), x_4 = net asset value per share (NAVPS), x_5 = dummy variable (Fee; code "no fee" as 0 = no fee and fee as 1). (Note: Records missing at least one value were excluded.)

One possible model: $\hat{y} = 1.457 - 0.003x_3$.

Considerations: *Rejected variables:* Ret_1Mth (x_1), MER (x_2), NAVPS (x_4), and Fee (x_5) (not significant). *Strength of the model:* Overall p-value = 0.010, significant x-coefficient, residuals are fairly reasonably distributed. *Weakness of the model:* Adjusted R^2 is low with 10.5%, standard error is very high; the resulting model is only a simple linear regression.

The dummy variable is not retained in the final model, because it did not contribute a significant coefficient to the regression model.

61. Dependent variable (y) = Ex_MaxT; independent variable (x) = Av_Temp.

a) $\hat{y} = 21.249 + 0.755x$.
b) Yes (p-value = 0.000).
c) $\hat{y} = 21.249 + (0.755 \times 17.5) = 34.5°C$.
d) No, because of the aggregate nature of the data that was used to construct the regression (i.e., each data value is based on summarizing many months of observations for a given city); therefore, the regression cannot be used to input or output a temperature statistic for one single month of a particular year for a city.

63. Dependent variable (y) = number of self-employed women (F_Self); independent variables: x_1 = number of paid women workers (F_Paid), x_2 = number of unpaid women workers (F_Unpaid).

$$\hat{y} = -432.984 + 0.047x_1 + 3.534x_2$$

65. The standard error of the regression (7,378.4) is comparatively small in comparison to the average number of self-employed women (35,809.2), indicating that the predictions may not be too far from the actual values. However, analysis of the residuals shows that the residuals are not very normally distributed. Also, both independent variables are highly correlated ($r = 0.955$). Therefore, although the model is significant, it should be applied with some caution.

67. Dependent variable (y) = number of self-employed women (F_Self); independent variables: x_1 = number of paid women workers (F_Paid), x_2 = number of unpaid women workers (F_Unpaid), x_3 = number of paid male workers (M_Paid), x_4 = number of unpaid male workers (M_Unpaid), x_5 = number of self-employed male workers (M_Self).

One possible model:

$$\hat{y} = -219.386 - 0.180x_1 - 12.126x_2 + 0.149x_3 + 31.501x_4 + 0.714x_5$$

Considerations: *Rejected variables:* None. *Strength of the model:* Overall p-value = 0.000, significant x-coefficient, residuals are fairly reasonably distributed, high adjusted R^2 and R^2; low standard error. *Weakness of the model:* There are strong significant correlations between all independent variables; the y-intercept is not significant; numerous independent variables—yet omission of any independent variables results in nonsignificance of other independent variables and weaker results for R^2, the standard error and the residual plot.

In this solution, all variables are kept in the model and the adjusted R^2 has improved.

Chapter 17 Association with Time: Time Series Analysis

Basic Concepts 17.1

1. A time series is an ordered set of observations for a variable made at regular intervals over time.
3. The trend component (changes in the value of the variable that reflect a pattern of long-term growth or decline); the seasonal component (changes in the value of a variable that are associated with particular, recurring divisions of time); the cyclical component (longer-term recurring changes in the value of the variable compared to seasonal changes; these longer-term recurring changes generally have less constant and harder-to-predict cycle lengths than seasonal changes); and the irregular component (changes in the value of the variable due to unexpected or unanticipated events).
5. Because the seasonality in the data could reduce the apparent strength or significance of the regression model. Readjust the estimate that is based on the regression equation to account for the seasonal component.
7. (a) Exponential model; (b) for June, 2007 ($t = 18$): $\hat{y}$ = \$1.050. For June, 2008 ($t = 30$): $\hat{y}$ = \$1.115.
9. No. If the difference of r^2 between the models is only slightly different, sometimes it is preferable to rely on the linear model because it is less complex and so its results are easier to communicate (and to calculate) than those of the exponential model.

Applications 17.1

11. (a) $\hat{y} = 23.393 \cdot e^{(0.045t)}$; the relationship is very strong ($r = 0.948$); 89.8% of the variation in the median ages is explained by the regression; regression is significant (p-value $= 0.000$); (b) $\hat{y} = 40.14$; (c) both models fit the data well—both are significant and both explain a nearly identical high proportion of the variation in the median age. For the newspaper article, it would be advisable to use the less-complex linear model, which is likely to be familiar to more readers.

13. (a) $\hat{y} = 36.547 + 0.306t$; (b) no ($p$-value $= 0.117$); (c) standard error $= 1.582$, $r^2 = 0.279$; (d) for year 2004 ($t = 11$): $\hat{y} = 39.91$. No, because the linear model has these weaknesses: The regression is not significant, r^2 is quite low, and the data are not bivariate normally distributed.

15. (a) $\hat{y} = 39.993 + 0.170t$; (b) no ($p$-value $= 0.618$); (c) standard error $= 2.981$, $r^2 = 0.033$; (d) for year 2004 ($t = 11$): $\hat{y} = 41.86$. No, because the linear model has these weaknesses: The regression is not significant, r^2 is quite low, and the data are not bivariate normally distributed.

No, the variation in the y data from the mean y is mostly *not* explained through the regression.

Basic Concepts 17.2

17. For data that show a seasonal pattern, seasonal decomposition is the process of analyzing the data to distinguish its seasonal components from its trend components. To prepare the data for a trend analysis, the seasonal component is removed prior to the analysis.

19. In seasonal decomposition, the expected change of value for an observation due to seasonality, compared to its simply lying on the trend line, is modelled by a seasonal index (SI) that corresponds to the season when the observation is made. For forecasting (with a multiplicative model for seasonality), divide every observed value by its SI to obtain the deseasonalized values; then apply trend analysis to make a (deseasonalized) prediction or forecast, and for the final prediction, multiply this estimate times the SI for the corresponding season.

21. In a multiplicative model, the trend-based (deseasonalized) predictions are multiplied times their respective seasonal indexes to make the final estimates, whereas in an additive model, the trend-based predictions are *added* to their respective seasonal indexes to make the final estimates.

23. 6.303 (= 5.673/0.9; assume multiplicative model).

25. 8.790 (= 7.991 × 1.1; period 42 would occur during season 2; assume multiplicative model).

Applications 17.2

Note: Due to variations in software procedures, your answers for seasonal indexes in this section may vary slightly, depending on your software. Where this occurs, subsequent calculations that use the seasonal indexes may vary slightly as well.

27. (a) Q1: 0.961, Q2: 0.777, Q3: 1.278, Q4: 0.984; (b) Q1: 0.947, Q2: 1.301, Q3: 0.949, Q4: 0.803; (c) Q1: 1.098, Q2: 0.920, Q3: 1.007, Q4: 0.975; (d) Q1: 0.980, Q2: 0.995, Q3: 0.999, Q4: 1.025.

29. Linear and exponential model have about the same r^2; the less-complex model is chosen: $\hat{y} = 232.787 + 0.127t$.

Forecast I ($t = 21$): $\hat{y}_{(\text{deseasonalized})} = 232.787 + (0.127 \times 21) = 235.454$; $\hat{y}_{(\text{reseasonalized estimate})} = 235.454 \times 0.947 = 222.97$.

Forecast II ($t = 22$): $\hat{y}_{(\text{deseasonalized})} = 232.787 + (0.127 \times 22) = 235.581$; $\hat{y}_{(\text{reseasonalized estimate})} = 235.581 \times 1.301 = 306.49$.

31. January: 0.987, February: 0.987, March: 1.003, April: 1.011, May: 1.008, June: 1.008.

Basic Concepts 17.3

33. If dummy variables for qualitative data are included in a multiple regression, the base case (i.e., the default assumption) is that all dummy variables in the complete model have the value 0, so the y-estimate is based solely on the regression that uses the quantitative variables.

35. A lagged variable in a time series regression model is a variable that takes its value at some time period after the event that is represented by the value of the variable. For example, a dummy variable that takes a value "1" when an advertising event occurs may not be assigned "1" until March, although the advertising was placed in February—so as to account for the time-delayed effects that the advertising has on the dependent variable (e.g., the quantities of product sold).

37. 3.

Applications 17.3

39. (Recall that only data from 1982 onward are utilized.) $\hat{y} = 12.752 - 0.030t + 0.010(\text{Feb.}) + 0.134(\text{Mar.}) + 0.176(\text{Apr.}) + 0.137(\text{May}) + 0.187(\text{June}) + 0.216(\text{July}) + 0.090(\text{Aug.}) + 0.122(\text{Sept.}) - 0.012(\text{Oct.}) - 0.109(\text{Nov.}) - 0.163(\text{Dec.})$.

Forecast Jan. 2007 ($t = 301$): $\hat{y} = 12.752 - (0.030 \times 301) + (0.010 \times 0) + (0.134 \times 0) + (0.176 \times 0) + (0.137 \times 0) + (0.187 \times 0) + (0.216 \times 0) + (0.090 \times 0) + (0.122 \times 0) - (0.012 \times 0) - (0.109 \times 0) - (0.163 \times 0) = 3.72$. **Forecast Feb. 2007 ($t = 302$):** $\hat{y} = 3.70$. **Forecast March 2007 ($t = 303$):** $\hat{y} = 3.80$. **Forecast April**

2007 (t = 304): $\hat{y}$ = 3.81. **Forecast May 2007 (t = 305):** $\hat{y}$ = 3.74. **Forecast June 2007 (t = 306):** $\hat{y}$ = 3.76.

For January and February, the forecasts are slightly smaller than in Exercise 32; for March to June, the forecasts are higher than observed in Exercise 32.

41. A regression model is constructed, using the variable "Period" and dummy variables for November and December. The months January to October are considered as the base case: $\hat{y}$ = 12.857 − 0.030(Period) − 0.215(Nov.) − 0.269(Dec.). (Caution, however: Neither of the coefficients of November or December test as significant.

Conclude that neither November nor December has a predictable effect on the yield.)

43. Months for which data were (generally) available: January to July and October to December.

One possible model. Considerations: Attempt the forecast for only the generally available months and let January be the base case. Then drop the period number variable, which is found to be nonsignificant.

a) $\hat{y}$ = 142.038 + 28.559(Feb.) + 57.097(Mar.) + 75.670(April) + 76.054(May) + 46.496(June) − 34.204(July) − 120.671(Oct.) × 83.590(Nov.) − 38.991(Dec.).
b) R^2 = 93.6%.
c) −120.671.
d) 76.054.
e) **Forecast February 2000 (period = 494):** $\hat{y}$ = 142.038 + (28.559 × 1) + (57.097 × 0) + (75.67 × 0) + (76.054 × 0) + (46.496 × 0) − (34.204 × 0) − (120.671 × 0) − (83.590 × 0) − (38.991 × 0) = 170.60. **Forecast October 2000 (period = 502):** $\hat{y}$ = 21.37.

45. Months for which data were generally available: January to July and October to December.

One possible model. Considerations: Attempt the forecast for only the generally available months and let January be the base case. A collinearity check shows no strong significant correlations between the independent variables. Dropping the nonsignificant dummy variable for December results in the significant model below (p-value = 0.000), which has significant x-coefficients and y-intercept. The residuals are fairly reasonably distributed.

a) $\hat{y}$ = 10.315 − 0.010t + 2.478(Feb.) + 2.893(Mar.) + 4.237(April) + 3.025(May) − 6.081(June) − 9.048(July) − 5.598(Oct.) − 3.713(Nov.).
b) R^2 = 40.1%.
c) −5.598.
d) 3.025.
e) **Forecast February 2000 (period = 494):** $\hat{y}$ = 10.315 − (0.010 × 494) + (2.478 × 1) + (2.893 × 0) + (4.237 × 0) + (3.025 × 0) − (6.091 × 0) − (9.048 − 0) × (5.598 × 0) − (3.713 × 0) = 7.853. **Forecast October 2000 (period = 502):** $\hat{y}$ = −0.30 ≈ 0.00. (Negative thicknesses are not defined.)

Review Applications

47. a) $\hat{y}$ = 204.141 • $e^{(-0.088t)}$; forecast: $\hat{y}_{(t=17)}$ = 204.141 • $e^{(-0.088 \times 17)}$ = 45.73.
b) Regression is significant (p-value = 0.000).
c) Exponential model explains a larger percentage (r^2 = 70.9%) vs. the linear model (r^2 = 63.2%).
d) The three earliest data points (1990–1992) have much higher values than the data points for the following years. A data set that includes those three early data points may not be representative for predictions beyond 2005. However, if the three points are excluded from the two models (linear and exponential), the proportion of variance in the dependent variable that is explained by the regression becomes *smaller* than before (54.1% for the exponential model and 45.8% for the linear model). Thus, the exclusion of the data points does not improve the models.

49. (a) $\hat{y}$ = 157.594 • $e^{(-0.052t)}$; forecast: $\hat{y}_{(t=17)}$ = 157.594 • $e^{(-0.052 \times 17)}$ = 65.11; (b) regression is significant (p-value = 0.000); (c) the exponential model explains a larger percentage of the variance (r^2 = 61.9%) compared to the linear model (r^2 = 57.8%).

51. Equation: $\hat{y}$ = 45.509 + 0.367t. **Forecast 1st quarter 2007 (t = 21):** $\hat{y}$ = 45.509 + (0.367 × 21) = 53.22. **Forecast 2nd quarter 2007 (t = 22):** $\hat{y}$ = 53.58.

Note: Due to variations in software procedures, your answers for seasonal indexes in this section may vary slightly, depending on your software. Where this occurs, subsequent calculations that use the seasonal indexes may vary slightly as well.

53. Period: Seasonally adjusted values for the "Trade" column:

2002 Q1: 4,566.4; **2002 Q2:** 4,637.5; **2002 Q3:** 5,188.6; **2002 Q4:** 5,050.4; **2003 Q1:** 4,950.1; **2003 Q2:** 5,013.3; **2003 Q3:** 5,359.1; **2003 Q4:** 5,702.8; **2004 Q1:** 5,532.4; **2004 Q2:** 5,968.7; **2004 Q3:** 6,206.6; **2004 Q4:** 6,473.2; **2005 Q1:** 6,962.6; **2005 Q2:** 6,940.2; **2005 Q3:** 7,205.4; **2005 Q4:** 7,515.3; **2006 Q1:** 8,280.7; **2006 Q2:** 8,353.2; **2006 Q3:** 8,224.3; **2006 Q4:** 8,440.6

55. Equation: $\hat{y}$ = 3719.799 + 220.356(Period) + 64.544(Q2) + 96.849(Q3) + 1,073.033(Q4); overall the model is significant, R^2 = 96.7%, but the coefficients for the 2nd and 3rd quarters are not significant. **Forecast 2007 Q2 (t = 22):** $\hat{y}$ = 3,719.799 + (220.356 × 22) +

$(64.544 \times 1) + (96.849 \times 0) + (1{,}073.033 \times 0) =$ 8,632.18. **Forecast 2007 Q3 ($t = 23$):** $\hat{y} = 8{,}884.84$. (An alternative, not shown, would be to rerun the regression, excluding all quarters but Q4; the results would be similar.)

57. Linear equation: $\hat{y} = 293.343 + 35.884t$; exponential equation: $\hat{y} = 367.230 \cdot e^{(0.052t)}$

Both regression models are strong ($r_{linear} = 0.941$, $r_{exponential} = 0.968$) and both models are significant (p-value $= 0.000$). However, because 93.7% of the variation in the revenues in real estate is explained by the exponential regression (compared to $r^2 = 0.886$ for the linear model), the most likely accurate forecast for the 1st quarter 2007 index is $\hat{y} = 367.230 \cdot e^{(0.052 \times 21)} =$ 1,094.43.

59. Linear equation: $\hat{y} = 2{,}028.394 + 61.800t$; exponential equation: $\hat{y} = 2{,}081.141 \cdot e^{(0.023t)}$. Both regression models are equally strong ($r \approx 0.97$) and equally significant (p-value $= 0.000$). Thus, both models would appear to give an equally accurate forecast. Applying the simpler linear model to forecast the 1st quarter 2007 value of the index: $\hat{y} = 2{,}028.394 + (61.8 \times 21) = 3{,}326.19$.

61. Equation: $\hat{y} = 37.010 + 0.573t - 0.885(\text{Feb.}) - 1.723(\text{March}) - 3.090(\text{April}) - 3.676(\text{May}) - 4.803(\text{June}) - 2.708(\text{July}) - 2.759(\text{Aug.}) - 3.070(\text{Sept.}) - 2.052(\text{Oct.}) - 1.147(\text{Nov.}) - 0.256(\text{Dec.})$; overall the model is significant and $R^2 = 0.096$, but the coefficients for February, March, July, August, September, October, November, and December are not significant. **Forecast July 2007 ($t = 67$):** $\hat{y} = 37.010 + (0.573 \times 67) - (0.885 \times 0) - (1.723 \times 0) - (3.09 \times 0) - (3.676 \times 0) - (4.803 \times 0) - (2.708 \times 1) - (2.759 \times 0) - (3.070 \times 0) - (2.052 \times 0) - (1.147 \times 0) - (0.256 \times 0) = 72.69$. **Forecast Aug. 2007 ($t = 68$):** $\hat{y} = 73.22$. **Forecast Sept. 2007 ($t = 69$):** $\hat{y} = 73.48$. **Forecast Oct. 2007 ($t = 70$):** $\hat{y} = 75.07$. **Forecast Nov. 2007 ($t = 71$):** $\hat{y} = 76.55$. **Forecast Dec. 2007 ($t = 72$):** $\hat{y} = 78.01$. Overall, the forecasts determined by the different approaches are more or less equivalent.

Notes and Data Sources

Chapter 1

1. Definition for the word *information* in this context obtained from the NATO-sponsored site *MERREA* for the project *Managing Effective Risk Response: An Ecological Approach.* Retrieved from http://www.merrea.org/glossary%20i.htm
2. *Dan's World,* "Famous Heights." Retrieved from http://www.famousheights.com
3. Canadian Football League. *2007 CFL Statistics.* Retrieved July 27, 2006, from http://www.cfl.ca/CFLStatistics05/ssk_reg1.pdf
4. Statistics Canada, press release, November 3, 2006. *Latest Release from the Labour Force Survey.* Retrieved from http://www.statcan.ca/english/Subjects/Labour/LFS/lfs-en.htm
5. Statistics Canada. Based on 2001 census data. Retrieved from http://www40.statcan.ca/l01/cst01/demo30a.htm
6. Based on data published in a Letter by the Secretary General of the United Nations, January 2006, "Top 10 Financial Contributors of Assessed and Voluntary Contributions to the United Nations Budget, Funds, Programs, and Agencies in U.S.$." At URL: UN_Financial-ContributorList_16Jan06.pdf or at: http://www.reformtheun.org/index.php?module=uploads&func=download&fileId=1241
7. Professional Home Inspector Wayne C. Wright, A.H.I. Found at: http://www.realestatelink.ca/homeinspection/wwbulltn4.htm
8. Natural Resources Canada. Based on *Canadians' Attitudes Towards Natural Resources Issues, 2002.* Retrieved from http://www.nrcan.gc.ca/inter/survey/publicConcerns_e.htm
9. Peter Landry, *A Blupete Essay: Taxes.* Retrieved from http://www.blupete.com/Literature/Essays/BluePete/Taxes.htm
10. Based on data reported in "National Hospital Discharge Survey: 2001 Annual Summary with Detailed Diagnosis and Procedure Data," published by U.S. Department of Health and Human Services, in Series 13 #156 of *Vital and Health Statistics.* June 2004.
11. Problem cited the name and business of Engineering Dynamics Corporation: http://www.edccorp.com/index.html
12. Rodrigo de la Jara, *IQ Comparison Site,* "IQ Basics." Used for general reference information in constructing the question. Retrieved from www.iqcomparisonsite.com/IQBasics.aspx
13. Veterans Affairs Canada. Document: PEN 57E (2006-02). Found at: http://www.vac-acc.gc.ca/content/dispen/pdfs/pen57e.pdf

Chapter 2

1. Canadian Avalanche Centre. *Annual Report 2004/2005.* Retrieved from http://www.avalancheinfo.net/CAC/Canadian%20Avalanche%20Centre%20Annual%20Report%202004-2005.pdf
2. Canada Health Portal. Retrieved from http://chp-pcs.gc.ca/CHP/index_e.jsp
3. Statistics Canada. Retrieved from http://www.statcan.ca/start.html
4. Statistics Canada, *The Daily,* August 8, 2005. Retrieved from http://www.statcan.ca/Daily/English/050629/d050629d.htm
5. Rosanna L. Hamilton, *Views of the Solar System.* Retrieved from http://www.solarviews.com/eng/moon.htm
6. Natural Resources Canada, *Towards an Idle-Free Zone in the City of Mississauga,* "Baseline Resident Telephone Survey: Summary Report," October, 2001. Retrieved from http://oee.nrcan.gc.ca/transportation/idling/material/reports-research/Mississauga-survey-report.cfm?attr=16#2
7. Ontario Universities Application Centre, *Highlights from the 2004-2005 Ontario University Graduate Survey,* July 2005. Retrieved from http://www.ouac.on.ca/news/2002_Survey.pdf
8. Thompson Rivers University. Quote is from a questionnaire for its agents who promote the institution internationally. Retrieved from http://www.truworld.ca/questionnaire.asp?PageID=217

Chapter 3

1. Government of Alberta, Alberta Economic Development, *Study of Visitors to Alberta during Summer/Fall of 2000: Telephone Survey Report.* Submitted by Bruce A. Campbell.
2. Canadian Firearms Centre, *Quick Facts* as of January 31, 2005. Retrieved from http://www.canadianfirearms.com
3. Canada.com, Canadian retirees spend average of $822 on presents, survey finds, Tuesday, Dec. 28, 2004.
4. Re/Max Realtron Realty Inc., Lu and Stefan Hyross, *MLS Listings* for Richmond Hill, March 14, 2005. Retrieved from http://www.torontohomescanada.com/mypages7.html
5. Red Label Vacations Inc., red tag.ca Vacations, *Flight Deals.* Retrieved March 15, 2005, from http://www.redtag.ca/flights.php
6. University of New Brunswick, *Earth Impact Database, 2003.* Retrieved April 8, 2005, from http://www.unb.ca/passc/ImpactDatabase/NAmerica.html
7. *Canadian Business Online,* "Best Cities for Business in Canada, 2003/2004." Retrieved from http://www.canadianbusiness.com/bestcities/ranking.jsp
8. Tim Hortons, *Frequently Asked Questions.* Question asked: "How many people win prizes during the Roll Up The Rim To Win promotion?" Answer: "In 2005, we gave away 30 2005 GMC Canyons, 84 Panasonic Plasma Televisions, 460 cash prizes of $1,000 and 6,644 Coleman Camping Packages, as well as millions of Tim Hortons food prizes including coffee, donuts, muffins and cookies." Retrieved from http://www.timhortons.com/en/about/faq.html#fifteen

Chapter 4

1. Statistics Canada. 2001 Census. Published as part of a CBC report on the 2001 Aboriginal People's Survey. Found at: http://www.cbc.ca/news/background/aboriginals/facts_figures.html. Accessed Aug. 10, 2007.
2. Statistics Canada. 2001 Census. Published as part of a CBC report on the 2001 Aboriginal People's Survey. Found at: http://www.cbc.ca/news/background/aboriginals/facts_figures.html. Accessed Aug. 10, 2007. http://www.phac-aspc.gc.ca/publicat/dic-dac99/d06_e.html
3. *The Globe and Mail.* Based on data in "Federal Election Results 2004," June 30, 2004, pp. A12 and A13
4. University of New Brunswick, *Earth Impact Database, 2003.* Retrieved April 8, 2005, from http://www.unb.ca/passc/ImpactDatabase
5. Transport Canada. Based on data in *Weight and Balance of Aircraft.* Retrieved from http://www.tc.gc.ca/mediaroom/backgrounders/b04-a003e.htm
6. Calgary Stampeder Football Club. Found at: http://www.stampeders.com/index.php?module=ContentExpress&func=display&ceid=541
7. Syncrude Canada. *2004 Sustainability Report,* Section: "Air: Sulphur Emissions." Found at: http://sustainability.syncrude.ca/sustainability2004/environment_health_safety/air.shtml
8. Statistics Canada. Table 358-00013. "Gross domestic expenditures on research and development, by science type and by funder and performer sector, annual (Dollars × 1,000,000)" [Country: Canada; Funder: All sectors; Performer: All sectors]. Found at: http://estat.statcan.ca.uproxy.library.dc-uoit.ca/cgi-win/CNSMCGI.EXE. Accessed Aug. 10, 2007.
9. "15 Year Mutual Fund Review—November 11, 2005" published by Investors Group powered by GlobeFund.com. Found at: http://www.igf.globefund.com/globefund/investorsgroup/review/20070216/TABRET_1.html February 16, 2007.
10. American Public Power Association and the Sacramento Municipal Utility District. Found at: http://www.appanet.org/treeben/data/growthdata.asp
11. Based on data published by the Minnesota Crop Improvement Association. URL: http://www.mncia.org/pub_newvarieties2004.pdf

Chapter 5

1. Katharine B. Perry, Department Extension Leader, Department of Horticultural Science, North Carolina State University. *Average First Fall Frost Dates for Selected North Carolina Locations.* Retrieved from http://www.ces.ncsu.edu/depts/hort/hil/hil-708.html
2. Katharine B. Perry, Department Extension Leader, Department of Horticultural Science, North Carolina State University. *Average First Fall Frost Dates for Selected North Carolina Locations.* Retrieved from http://www.ces.ncsu.edu/depts/hort/hil/hil-708.html
3. P. Cicero-Fernandez, V. Torres, A. Rosales, et al. *Evaluation of Human Exposure to Ambient PM10 in the Metropolitan Area of Mexico City Using a GIS-Based Methodology.* Found at: http://www.idrc.ca/IMAGES/ecohealth/100205-1.pdf
4. J. J. Ray. Based on data in "Attitude to Authority" scale. Originally published in *Australian Psychologist,* 6(1), March 1971, pp. 31–50.
5. Based on data published on the Learning Outcomes Assessment website for the two private online institu-

tions that comprise the American Public University System. Retrieved from http://www.apus.edu/Learning-Outcomes-Assessment/Reports/Major-Field-Test-Results/MFT-Criminal-Justice-htm

6. Based on data from U.S. Environmental Protection Agency, National Center for Health Statistics, 1987. Retrieved from http://www.epa.gov/ncea/pdfs/efh/sect-7.pdf
7. Dr. Alice H. Suter. Based on data in *Noise and Its Effects* (November 1991). Prepared for the consideration of the Administrative Conference of the United States. Retrieved from http://www.nonoise.org/library/suter/suter.htm#aircraft
8. P. Cicero-Fernandez, V. Torres, A. Rosales, et al. *Evaluation of Human Exposure to Ambient PM10 in the Metropolitan Area of Mexico City Using a GIS-Based Methodology.* Found at: http://www.idrc.ca/IMAGES/ecohealth/100205-1.pdf
9. Based on data published on the Learning Outcomes Assessment website for the two private online institutions that comprise the American Public University System. Retrieved from http://www.apus.edu/Learning-Outcomes-Assessment/Reports/Major-Field-Test-Results/MFT-Criminal-Justice-htm
10. Based on data from U.S. Environmental Protection Agency, National Center for Health Statistics, 1987. Retrieved from http://www.epa.gov/ncea/pdfs/efh/sect-7.pdf

Chapter 6

1. Public Service Human Resources Management of Canada. Based on data in *Minority Populations by First Official Language Spoken (2001 Census Data).* Retrieved from http://www.hrma-agrh.gc.ca/ollo/reimplementation-r%E9application/MP-PM200101_e.asp#pres
2. Public Service Human Resources Management of Canada. Based on data in *Minority Populations by First Official Language Spoken (2001 Census Data).* Retrieved from http://www.hrma-agrh.gc.ca/ollo/reimplementation-r%E9application/MP-PM200101_e.asp#pres
3. Yinon Bentor, *Chemical Elements* website. Based on data in "Periodic Table: Atomic Numbers." Retrieved from http://www.chemicalelements.com/show/atomicnumber.html
4. Integrated Publishing, *Electrical Engineering Training Series.* Adapted information in "Magnetic Poles." Retrieved from http://www.tpub.com/neets/book1/chapter1/1g.htm
5. Yinon Bentor, *Chemical Elements* website. Based on data in "Periodic Table: Atomic Numbers." Retrieved from http://www.chemicalelements.com/show/atomicnumber.html
6. Drexel School of Education, *Ask Dr. Math.* Based on an estimate in "The Math Forum@Drexel." Retrieved from http://mathforum.org/dr.math/faq/faq.prob.world.html
7. Pathway Communications. Based on data published in a reference essay "The Gray Jay, Algonquin Park (and 'The Raven')" by an amateur naturalist. Retrieved from http://www.pathcom.com/~wgbz/grayjay.htm

Chapter 7

1. Unspam Technologies, Inc. *Quick Information.* Retrieved June 2004 from http://www.unspam.com/fight_spam/information/spamstats.html?start=0
2. Natural Sciences and Engineering Research Council Canada. Based on data in the newsletter *NSERC Contact,* January 2004, p. 7.
3. Adoption Council of Canada. *Canadians Look Favourably on Adoption: Ipsos-Reid Survey* (July 12, 2005). Copyright 2005 Adoption Council of Canada. Retrieved from http://www.adoption.ca/news/050712ipsos.htm
4. Canadian Association of Emergency Physicians (CAEP). Based on data in *CAEP Backgrounder: Emergency Department Overcrowding in Canada, Myth and Reality.* Retrieved from http://www.caep.ca/002.policies/002-05.communications/2004/040614.bgnd-overcrowding.pdf
5. Published on the website for *The Science of Algonquin's Animals.* © Copyright 2004–2005. Presented by The Friends of Algonquin Park in cooperation with Ontario Parks and the Ontario Ministry of Economic Development and Trade. (Website developed and maintained by Hyperweb.ca). Retrieved from http://www.sbaa.ca/assets/attachments/cms/wildlife_and_technology.pdf
6. Keith G. Calkins, Andrews University, Berrien Springs, MI, *Queuing Theory and the Poisson Distribution.* Retrieved from http://www.andrews.edu/~calkins/math/webtexts/prod10.htm
7. Yinon Bentor, *Chemical Elements* website. Based on data in "Periodic Table: Atomic Numbers." Retrieved from http://www.chemicalelements.com/show/atomicnumber.html
8. Concordia University. Found at: http://iits.concordia.ca/statistics/email
9. Connecticut Department of Environmental Protection. Based on data in the *Wildlife in Connecticut* series. Retrieved from http://dep.state.ct.us/burnatr/wildlife/factshts/wdduck.htm

10. Based on data provided in April Holladay's *Wonderquest* website column. Retrieved from http://www.wonderquest.com/TwinsTrigger.htm

11. M. L. Green, Ocean Mammal Institute. Based on data in *The Impact of Parasail Boats on the Hawaiian Humpback Whale.* Retrieved from http://www.oceanmammalinst.org/w91.html

12. Powerhouse International Inc. Based on data in a promotional document. Retrieved from http://www.powerhouseinc.ca/html/boards/empowered.html

13. Health Care Corporation of St. John's. Based on data in *MRSA Infection Rates.* Retrieved from http://www.hccsj.nl.ca/about/MRSA_Rates.html

14. Public Health Agency of Canada, *Chronic Diseases in Canada Abstract Reprints,* Yue-Fang Chang, Joan E. McMahon, Deidre L. Hennon, Ronald E. LaPorte, and Jeffrey H. Coben, "Dog bite incidence in the city of Pittsburgh: A capture–recapture approach." Published in the *American Journal of Public Health,* 1997; 87(10), pp. 1703–1705. Retrieved from http://www.phac-aspc.gc.ca/publicat/cdic-mcc/18-4/i_e.html

15. Asia Pacific Foundation of Canada. Found at: http://www.asiapacific.ca/analysis/pubs/pdfs/invest_survey/invest_intentions2006.pdf.

16. Government of Nova Scotia. Based on data in *Community Health Survey Topics,* "A First Look at Depression in Nova Scotia," 2004. Retrieved from http://www.gov.ns.ca/heal/downloads/CCHS_Depression.pdf

17. *Space Daily.* Based on data in "Heavy Metals Rich Stars Tend To Harbor Planets." Retrieved from http://www.spacedaily.com/reports/Heavy_Metals_Rich_Stars_Tend_To_Harbor_Planets.html

18. *Sports Illustrated.* Based on data in the Blue Jays' team roster as of June 23, 2005. Retrieved from http://sportsillustrated.cnn.com/baseball/mlb/rosters/teams/blue_jays.html

19. Toronto Zoo. Based on data in *Vancouver Island Marmots—Conservation Program.* Retrieved from http://www.torontozoo.com/Conservation/mammals.asp

20. Statistics Canada, *The Daily.* Based on data in "Health Reports: Deaths involving firearms," June 28, 2005. Retrieved from http://www.statcan.ca/Daily/English/050628/d050628a.htm

21. Boeing, *Boeing Frontiers* (employee newsletter), "We never sleep: Newly consolidated Enterprise Help Desk gives all employees 24-hour support," August 2004. Retrieved from http://www.boeing.com/news/frontiers/archive/2004/august/i_ssg.html

22. FastWebServer.com. Based on promotional material for Global Crossing. Retrieved June 2006 from http://www.fastwebserver.com/data_center_and_connection.htm

23. F. A. Baker and K. R. Knowles. Based on data in *Case Study: 36 Years of Dwarf Mistletoe in a Regenerating Black Spruce Strand in Northern Minnesota.* Retrieved from http://www.cnr.usu.edu/departments/forest/fab/bdale_2_03.pdf

24. D. Lam, D. Cox, and J. Widom. Based on data in *Teletraffic Modelling for Personal Communications Services.* Retrieved from http://www.cs.ucl.ac.uk/staff/S.Bhatti/teaching/3g-4g/lcw1997.pdf

25. A. Forster, I. Stiell, G. Wells, A. Lee, and C. van Walraven. Based on data in "The effect of hospital occupancy on emergency department length of stay and patient disposition," *Annals of Emergency Medicine* 2003, 10(2), pp. 127–133. Retrieved from http://www.ices.on.ca/webpage.cfm?site_id=1&org_id=32&morg_id=0&gsec_id=1217&item_id=1217&category_id=35

26. United States Department of Agriculture (Animal and Plant Health Inspection Service, APHIS 11-55-009). Based on data in *Bird Predation and Its Control at Aquaculture Facilities in the Northeastern United States,* issued June 1997. Retrieved from http://www.aphis.usda.gov/ws/birdpred.html

27. Toronto Blue Jays. Based on data retrieved from http://toronto.bluejays.mlb.com.

28. National Eye Institute. Based on data retrieved from http://www.nei.nih.gov

29. Public Health Agency of Canada. CHIRPP injury reports. Found at: http://www.phac-aspc.gc.ca/injury-les/chirpp/injrep-rapbles/inline_e.html

Chapter 8

1. Itron of Spokane, Washington. Promotional material. Retrieved from http://www.itron.com/asset.asp?path=support/whitepaper/pdf/itr_000494.pdf

2. Niagara Falls Thunder Alley. Based on data in *Origins of Niagara: A Geological History.* Retrieved from http://www.iaw.com/~falls/origins.html

3. Niagara Falls Thunder Alley. Based on data in *Origins of Niagara: A Geological History.* Retrieved from http://www.iaw.com/~falls/origins.html

4. Environment Canada. Data based on a specific query related to *Water Level and Streamflow Statistics,* specifically for station (01AF011) Little River Near Grand Falls. Additional information retrieved from http://www.wsc.ec.gc.ca/staflo/index_e.cfm?cname=flow_daily.cfm

5. University of British Columbia. Based on information in *Process Flow Management,* p. 46, related to the Vancouver Airport. Retrieved from http://coe.ubc.ca/users/marty/batl510_05/mbpf3.pdf

6. Ledtronics, Inc. Based on definitions found in *Common Light Measurement Terms* on Light Resource.com, Light Research Center, Light Board, *IES Lighting Handbook,* 5th ed. Retrieved from http://www.ledtronics.com/pages/tech4.htm

7. Canadian Institute for Health Information. Based on data *Ontario Trauma Registry Analytic Bulletin,* July 2003. Retrieved from http://www.icis.ca/cihiweb/en/downloads/bl_otrJul2003_e.pdf

8. Roger Boe, Larry Motiuk, and Mark Nafekh, Correctional Service Canada. Based on data in *An Examination of the Average Length of Prison Sentence for Adult Men in Canada: 1994 to 2002,* March 2004. Retrieved from http://www.csc-scc.gc.ca/text/rsrch/reports/r136/r136_e.shtml#t4

9. Roger Boe, Larry Motiuk, and Mark Nafekh, Correctional Service of Canada. Based on data in *An Examination of the Average Length of Prison Sentence for Adult Men in Canada: 1994 to 2002,* March 2004. Retrieved from http://www.csc-scc.gc.ca/text/rsrch/reports/r136/r136_e.shtml#t4

10. Yoichi Watanabe, Tomohide Akimitsu, Yutaka Hirokawa, Rob B. Mooij, and G. Mark Perera. Based on data in "Evaluation of dose delivery accuracy of gamma knife by polymer gel dosimetry," *Journal of Applied Clinical Medical Physics,* 6(3), Summer 2005.

11. Publications International, Ltd., *Consumer Guide.* Based on U.S. dollar data on *HowStuffWorks* website. Retrieved July 2006 from http://shopper.howstuffworks.com/products/Velodyne+SPL_1000+Subwoofer/SF-1/PID-20182150

12. United States Faceters Guild (USFG). Based on message history data from December 2000 to June 2006. Retrieved from http://groups.yahoo.com/group/usfgfaceterslist

13. TBX, the Building Exchange. Based on data in *Canadian Housing Observer, 2003.* The average rental data are from "Table 12: Average Rent for Two-Bedroom Apartments, Canada: Provinces and Metropolitan Areas, 1992–2002." Retrieved from http://www.tbxhome.com/public/newsresearch/Canadian_Housing_Observer.pdf

14. *The Globe and Mail.* Based on data on Canadian equities at GlobeFund.com. Retrieved from http://www.globefund.com/static/romf/generic/tabceq.pdf

15. Based on data in source file prepared by Dr. John Andraos, Department of Chemistry, York University. Retrieved from http://www.careerchem.com/NAMED/Elements-Discoverers.pdf1

16. The Weather Doctor. Based on data published in *Weather Almanac for April 2003,* "The Energy of a Rainshower." Retrieved from http://www.islandnet.com/~see/weather/almanac/arc2003/alm03apr.htm

17. Government of Nova Scotia. Based on data in *Fall Protection and Scaffolding Regulations Made under Section 82 of the Occupational Health and Safety Act,* S.N.S. 1996, c. 7 O.I.C. 96-14 (January 3, 1996), Nova Scotia. Reg. 2/96. Retrieved from http://www.gov.ns.ca/just/regulations/regs/ohs296f.htm

18. H. Paul Barringer and Michael Kotlyar. Based on data in *Reliability Of Critical Turbo/Compressor Equipment.* Originally presented at the Fifth International Conference on Process Plant Reliability, Houston, Texas, October 2–4, 1996. Retrieved from http://www.maintenanceworld.com/Articles/barringerkotlyar/Rel_Crit_Turbo_Comp.pdf

19. Boeing, *Boeing Frontiers* (employee newsletter), "We never sleep: Newly consolidated Enterprise Help Desk gives all employees 24-hour support," August 2004. Retrieved from http://www.boeing.com/news/frontiers/archive/2004/august/i_ssg.html

20. FastWebServer.com. Based on promotional material for Global Crossing. Retrieved June 2006 from http://www.fastwebserver.com/data_center_and_connection.htm

21. Treasury Board of Canada Secretariat. Based on data in *Transport Canada Departmental Performance Report for the Period Ending March 31, 2001.* Retrieved from http://www.tbs-sct.gc.ca/rma/dpr/00-01/TC00dpr/TC00dpr-PR_e.asp?printable=True

22. D. Lam, D. Cox, and J. Widom. Based on data in *Teletraffic Modelling for Personal Communications Services.* Retrieved from http://www.cs.ucl.ac.uk/staff/S.Bhatti/teaching/3g-4g/lcw1997.pdf

23. Vidtek Battery Canada.com. Based on data in *Battery Facts.* Retrieved from: http://www.batterycanada.com/Battery_Facts.htm

24. A. Forster, I. Stiell, G. Wells, A. Lee, and C. van Walraven. Based on data in "The effect of hospital occupancy on emergency department length of stay and patient disposition," *Annals of Emergency Medicine* 2003, 10(2), pp. 127–133. Retrieved from http://www.ices.on.ca/webpage.cfm?site_id=1&org_id=32&morg_id=0&gsec_id=1217&item_id=1217&category_id=35

25. Goran Lazac. Based on data in *School Speed Zones: Before and After Study. City of Saskatoon.* Presented at the 2003 Annual Conference of the Transportation Association of Canada. Retrieved from http://www.tac-atc.ca/English/pdf/conf2003/lazic2.pdf

26. Environment Canada. Based on data in *Top Ten Weather Stories for 2005.* Retrieved from http://www.msc-smc.ec.gc.ca/media/top10/2005_e.html

27. *South Florida Sun-Sentinel.* Based on data in "Average Hurricane Dates." Retrieved from http://www.sun-sentinel.com/news/weather/hurricane/sns-hc-canedates,0,5453803.htmlstory

Chapter 9

1. Natural Sciences and Engineering Research Council Canada. Based on data in the newsletter *NSERC Contact,* January 2004, p. 7.

2. Natural Sciences and Engineering Research Council Canada. Based on data in the newsletter *NSERC Contact,* January 2004, p. 7.

3. Travel Alberta. Based on a page of facts about Alberta. Retrieved from http://www1.travelalberta.com/content/albertafacts

4. Wikipedia, *Village of Alix* [Alberta]. Retrieved from http://en.wikipedia.org/wiki/Alix,_Alberta

5. *About.com: Geography.* Based on data in "The number of countries in the world." Retrieved from http://geography.about.com/cs/countries/a/numbercountries.htm

6. Canada-City.ca. Based on "List of ALL cities in Canada." Retrieved from http://www.canada-city.ca/all-cities-in-canada.php

7. Ledtronics, Inc. Based on definitions found in *Common Light Measurement Terms,* based on Light Resource.com, Light Research Center, Light Board, *IES Lighting Handbook,* 5th ed. Retrieved from http://www.ledtronics.com/pages/tech4.htm

8. Travel Alberta. Based on a page of facts about Alberta. Retrieved from http://www1.travelalberta.com/content/albertafacts

9. Wikipedia, *Village of Alix* [Alberta]. Retrieved from http://en.wikipedia.org/wiki/Alix,_Alberta

10. *About.com: Geography.* Based on data in "The number of countries in the world." Retrieved from http://geography.about.com/cs/countries/a/numbercountries.htm

11. *Canada-City.ca.* Based on a "List of ALL cities in Canada." Retrieved from http://www.canada-city.ca/all-cities-in-canada.php

12. Roger Boe, Larry Motiuk, and Mark Nafekh, Correctional Service Canada. Based on data in *An Examination of the Average Length of Prison Sentence for Adult Men in Canada: 1994 to 2002,* March 2004. Retrieved from http://www.csc-scc.gc.ca/text/rsrch/reports/r136/r136_e.shtml#t4

13. Roger Boe, Larry Motiuk, and Mark Nafekh, Correctional Service Canada. Based on data in *An Examination of the Average Length of Prison Sentence for Adult Men in Canada: 1994 to 2002,* March 2004. Retrieved from http://www.csc-scc.gc.ca/text/rsrch/reports/r136/r136_e.shtml#t4

14. Publications International, Ltd., *Consumer Guide.* Based on U.S. dollar data on *HowStuffWorks* website. Retrieved July 2006 from http://shopper.howstuffworks.com/products/Velodyne+SPL_1000+Subwoofer/SF-1/PID-20182150

15. Roger Boe, Larry Motiuk, and Mark Nafekh, Correctional Service Canada. Based on data in *An Examination of the Average Length of Prison Sentence for Adult Men in Canada: 1994 to 2002,* March 2004. Retrieved from http://www.csc-scc.gc.ca/text/rsrch/reports/r136/r136_e.shtml#t4

16. Roger Boe, Larry Motiuk, and Mark Nafekh, Correctional Service Canada. Based on data in *An Examination of the Average Length of Prison Sentence for Adult Men in Canada: 1994 to 2002,* March 2004. Retrieved from http://www.csc-scc.gc.ca/text/rsrch/reports/r136/r136_e.shtml#t4

17. Publications International, Ltd., *Consumer Guide.* Based on U.S. dollar data on *HowStuffWorks* website. Retrieved July 2006 from http://shopper.howstuffworks.com/products/Velodyne+SPL_1000+Subwoofer/SF-1/PID-20182150

18. Saskatchewan Highways and Transportation. Based on data in *Canadian Asphalt Mix Exchange Historical Data.* Retrieved from http://www.highways.gov.sk.ca/docs/reports_manuals/reports/History2000-2004.pdf

19. Saskatchewan Highways and Transportation. Based on data in *Canadian Asphalt Mix Exchange Historical Data.* Retrieved from http://www.highways.gov.sk.ca/docs/reports_manuals/reports/History2000-2004.pdf

20. The Tolly Group. Based on data in Newsletter 202117 (April 2002) evaluating the Octasic OCT6100. Retrieved from http://www.octasic.com/priv/oct6100sl2000.php

21. M. L. Green, Ocean Mammal Institute. Based on data in *The Impact of Parasail Boats on the Hawaiian Humpback Whale.* Retrieved from http://www.oceanmammalinst.org/w91.html

22. Powerhouse International. Based on data in a promotional document. Retrieved from http://www.powerhouseinc.ca/html/boards/empowered.html

23. City of Kingston. Based on data in *Solid Waste Services Situation Analysis Review,* 2002. Retrieved from http://www.cityofkingston.ca/pdf/waste/situation-analysis_may.pdf

24. Asia Pacific Foundation of Canada. Based on data in *2006 Asian Investment Intentions Survey.* Retrieved from http://www.asiapacific.ca/analysis/pubs/pdfs/invest_survey/invest_intentions2006.pdf
25. Government of Nova Scotia. Based on data in *Community Health Survey Topics,* "A First Look at Depression in Nova Scotia," 2004. Retrieved from http://www.gov.ns.ca/heal/downloads/CCHS_Depression.pdf
26. Statistics Canada. Based on data in *Proportion of labour force and paid workers covered by a registered pension plan (RPP).* Retrieved from http://www40.statcan.ca/l01/cst01/labor26a.htm
27. *Space Daily.* Based on data in "Heavy Metals Rich Stars Tend To Harbor Planets." Retrieved from http://www.spacedaily.com/reports/Heavy_Metals_Rich_Stars_Tend_To_Harbor_Planets.html
28. Health Canada. Based on data in *Environmental & Workplace Health,* "Proposed residential indoor air quality guidelines for formaldehyde." Retrieved from http://www.hc-sc.gc.ca/ewh-semt/pubs/air/formaldehyde/sources_e.html
29. Health Canada. Based on data in *Environmental & Workplace Health,* "Proposed residential indoor air quality guidelines for formaldehyde." Retrieved from http://www.hc-sc.gc.ca/ewh-semt/pubs/air/formaldehyde/sources_e.html
30. Highline Air Tours. Based on data in *About North Cyprus.* Retrieved from http://www.highlineparagliding.com/HL-northcyp.html
31. University of Illinois Eye & Ear Infirmary. Based on data in "Corneal Thickness and Glaucoma," *The Eye Digest.* Retrieved from http://www.agingeye.net/mainnews/glaucomapachymetry.php

Chapter 10

1. Rosanna L. Hamilton, *Views of the Solar System.* Retrieved from http://www.solarviews.com/eng/moon.htm
2. Phyllis M. Pineda and Boris C. Kondratief. Based on data in *What Do Tiger Beetles Do When It's Hot?* 2003. Retrieved from http://www.nps.gov/grsa/resources/curriculum/mid/insects/images/activity_graph.htm
3. *Straight Goods.* Based on data in "NDP Losing Ground," December 14, 2005. Retrieved from http://www.straightgoods.ca/Election2006/ViewNews.cfm?Ref=3
4. John Daly. Based on statistics and opinion in *Global Warming.* Retrieved from http://www.john-daly.com
5. Colorado Department of Natural Resources, Colorado Avalanche Information Centre. Retrieved from http://geosurvey.state.co.us/avalanche/Default.aspx?tabid=176
6. Colorado Department of Natural Resources, Colorado Avalanche Information Centre. Retrieved from http://geosurvey.state.co.us/avalanche/Default.aspx?tabid=176
7. University of Waterloo Weather Station Archives. Retrieved from http://weather.uwaterloo.ca/data.htm
8. Pacific Salmon Commission, Vancouver. Based on News Release No. 6, Aug. 5, 2005. Retrieved from http://www.psc.org/NewsRel/2005/NewsRelease06.pdf
9. Fisheries and Oceans Canada. Based on data in *Information to Media,* "Harp Seal Population Assessment Results," April 2000. Retrieved from http://www.dfo-mpo.gc.ca/media/imedia/2000/im1_e.htm
10. Fisheries and Oceans Canada. Based on data in *Information to Media,* "Harp Seal Population Assessment Results," April 2000. Retrieved from http://www.dfo-mpo.gc.ca/media/imedia/2000/im1_e.htm
11. Fisheries and Oceans Canada. Based on data in *Information to Media,* "Harp Seal Population Assessment Results," April 2000. Retrieved from http://www.dfo-mpo.gc.ca/media/imedia/2000/im1_e.htm
12. Canada Border Services Agency. Based on data in *Border Wait Times* as of August 14, 2006, 2:15 p.m. EDT. Retrieved from http://www.cbsa-asfc.gc.ca/general/times/menu-e.html
13. Canada Border Services Agency. Based on data in *Border Wait Times* as of August 14, 2006, 2:15 p.m. EDT. Retrieved from http://www.cbsa-asfc.gc.ca/general/times/menu-e.html
14. BC Ministry of Education. Based on data in *Estimated Instructional Time for Main Components of Grade 12 Calculus Courses.* Retrieved from http://www.bced.gov.bc.ca/irp/math1012/calc12est.htm
15. Petroleumshow.com. Based on promotional materials for the Oils Sands Trade Show and Conference, 2006, Fort McMurray, Alberta. Retrieved from http://www.petroleumshow.com/Files/ContentVersionFile/44055/Section_6.pdf
16. Roger Boe, Larry Motiuk, and Mark Nafekh, Correctional Service of Canada. Based on data in *An Examination of the Average Length of Prison Sentence for Adult Men in Canada: 1994 to 2002,* March 2004. Retrieved from http://www.csc-scc.gc.ca/text/rsrch/reports/r136/r136_e.shtml#t4
17. Publications International, Ltd., *Consumer Guide.* Based on U.S. dollar data on *HowStuffWorks* website. Retrieved July 2006 from http://shopper.howstuffworks.com/products/Velodyne+SPL_1000+Subwoofer/SF-1/PID-20182150

18. Saskatchewan Highways and Transportation. Based on data in *Canadian Asphalt Mix Exchange Historical Data.* Retrieved from http://www.highways.gov.sk.ca/docs/reports_manuals/reports/History2000-2004.pdf

19. Saskatchewan Highways and Transportation. Based on data in *Canadian Asphalt Mix Exchange Historical Data.* Retrieved from http://www.highways.gov.sk.ca/docs/reports_manuals/reports/History2000-2004.pdf

20. The Tolly Group. Based on data in Newsletter 202117 (April 2002) evaluating the Octasic OCT6100. Retrieved from http://www.octasic.com/priv/oct6100sl2000.php

21. Saskatchewan Highways and Transportation. Based on data in *Canadian Asphalt Mix Exchange Historical Data.* Retrieved from http://www.highways.gov.sk.ca/docs/reports_manuals/reports/History2000-2004.pdf

22. Saskatchewan Highways and Transportation. Based on data in *Canadian Asphalt Mix Exchange Historical Data.* Retrieved from http://www.highways.gov.sk.ca/docs/reports_manuals/reports/History2000-2004.pdf

23. PromoCycle Foundation, Quebec. Based on data in *Task analysis for intensive braking of a motorcycle in a straight line,* January 2004. Retrieved from http://www.fmq.qc.ca/pdf/amorce-freinage_eng.pdf

24. *The Globe and Mail.* Based on data in "On-line banking is a habit that's really starting to click," August 16, 2006, p. B1. The article, in turn, cites Statistics Canada.

25. M. L. Green, Ocean Mammal Institute. Based on data in *The Impact of Parasail Boats on the Hawaiian Humpback Whale.* Retrieved from http://www.oceanmammalinst.org/w91.html

26. Northstar Research Partners. Based on data in the report *Stormwater Pollution Poll: Public Attitude Survey,* presented in March 2000 to the City of Toronto (Works and Emergency Services). Retrieved from http://www.toronto.ca/water/protecting_quality/stormwater_pollution/pdf/respres.pdf

27. Northstar Research Partners. Based on data in the report *Stormwater Pollution Poll: Public Attitude Survey,* presented in March 2000 to the City of Toronto (Works and Emergency Services). Retrieved from http://www.toronto.ca/water/protecting_quality/stormwater_pollution/pdf/respres.pdf

28. Statistics Canada. Based on data in *Proportion of labour force and paid workers covered by a registered pension plan (RPP).* Retrieved from http://www40.statcan.ca/l01/cst01/labor26a.htm

29. CBC. Based on data in *Children and men lead the way as Americans get fatter: Survey,* April 4, 2006. The article, in turn, references a report by the U.S. Centers for Disease Control in the then-current issue of the *Journal of the American Medical Association.*

30. CBC. Based on data in *Children and men lead the way as Americans get fatter: Survey,* April 4, 2006. The article, in turn, references a report by the U.S. Centers for Disease Control in the then-current issue of the *Journal of the American Medical Association.*

31. Health Canada. Based on data in *Environmental & Workplace Health,* "Proposed residential indoor air quality guidelines for formaldehyde." Retrieved from http://www.hc-sc.gc.ca/ewh-semt/pubs/air/formaldehyde/sources_e.html

32. Canadian Museum of Nature. Based on data in *Rideau River Biodiversity Project—1999 Water Quality Data.* Retrieved from http://www.nature.ca/rideau/pdf/1999_wqdata_e.pdf

33. Canadian Museum of Nature. Based on data in *Rideau River Biodiversity Project—1999 Water Quality Data.* Retrieved from http://www.nature.ca/rideau/pdf/1999_wqdata_e.pdf

34. University of Illinois Eye & Ear Infirmary. Based on data in "Corneal Thickness and Glaucoma," *The Eye Digest.* Retrieved from http://www.agingeye.net/mainnews/glaucomapachymetry.php

35. Environment Canada. Based on weather data for the Iqualuit A weather station in Nunavut. http://www.climate.weatheroffice.ec.gc.ca/climateData/monthlydata_e.html?timeframe=1&Prov=XX&StationID=1758&Year=2006&Month=8&Day=16

36. Environment Canada. Based on weather data for the Iqualuit A weather station in Nunavut. http://www.climate.weatheroffice.ec.gc.ca/climateData/monthlydata_e.html?timeframe=1&Prov=XX&StationID=1758&Year=2006&Month=8&Day=16

37. United Nations. Economic Commission for Latin America and the Caribbean. *Statistical Yearbook for Latin America and the Caribbean.* 2004. Found at: http://www.eclac.cl/publicaciones/xml/1/21231/p2_1.pdf

38. United Nations Economic Commission for Europe. Based on searchable data c. 2002. Retrieved from http://w3.unece.org/stat/

39. Lottery Canada. Based on lottery results for various Canadian lotteries. Retrieved from http://www.lotterycanada.com/lottery/?job=frequency_chart&lottery_name=qc_banco&order=0&years=0

Chapter 11

1. Katharine B. Perry, Department Extension Leader, Department of Horticultural Science, North Carolina State University. *Average First Fall Frost Dates for Selected North Carolina Locations.* Retrieved from http://www.ces.ncsu.edu/depts/hort/hil/hil-708.html

2. Environment Canada. Based on data in spreadsheet file "Municipal Water Use & Pricing Data, All Municipalities With Population > 1000, 1998." *Data for All Canada, 1999 Survey.* Retrieved from http://www.ec.gc.ca/water/en/manage/data/Use_DB_83-99_DB.xls
3. Source: U.S. Environmental Protection Agency, National Center for Environmental Assessment. Based on data in "Technical Support Document for Exposure Assessment and Stochastic Analysis," Table 10, *Exposure Factors Handbook Volume 1: General Factors,* August 1997, EPA/600/P-95/002Fa. Retrieved from http://www.oehha.ca.gov/air/hot_spots/pdf/chap10.pdf
4. G. Cayrel de Strobel, Observatoire de Paris. Based on data in *The Study of Stars with Planets,* GEPI/CNRS UMR 8111, 92195 Meudon cedex, France. Retrieved from http://www.rssd.esa.int/SA/GAIA/docs/Gaia_2004_Proceedings/Gaia_2004_Proceedings_267.pdf
5. Roger Boe, Larry Motiuk, and Mark Nafekh, Correctional Service Canada. Based on data in *An Examination of the Average Length of Prison Sentence for Adult Men in Canada: 1994 to 2002,* March 2004. Retrieved from http://www.csc-scc.gc.ca/text/rsrch/reports/r136/r136_e.shtml#t4
6. Publications International, Ltd., *Consumer Guide.* Based on U.S. dollar data on *HowStuffWorks* website. Retrieved July 2006 from http://shopper.howstuffworks.com/products/Velodyne+SPL_1000+Subwoofer/SF-1/PID-20182150
7. Saskatchewan Highways and Transportation. Based on data in *Canadian Asphalt Mix Exchange Historical Data.* Retrieved from http://www.highways.gov.sk.ca/docs/reports_manuals/reports/History2000-2004.pdf
8. The Tolly Group. Based on data in Newsletter 202117 (April 2002) evaluating the Octasic OCT6100. Retrieved from http://www.octasic.com/priv/oct6100sl2000.php
9. A. N. Hamilton, D. C. Heard, and M. A. Austin, prepared for the British Columbia Ministry of Water, Land and Air Protection. Approximate values are drawn from statistics in *British Columbia Grizzly Bear (Ursus arctos) Population Estimate 2004,* June 14, 2004. Retrieved from http://wlapwww.gov.bc.ca/wld/documents/gb_bc_pop_est.pdf
10. Association of American Publishers. Reference is to data in *Textbook Facts vs. Rhetoric,* April 2006. Retrieved from http://www.publishers.org/highered/pdfs/Textbook%20Facts%20Versus%20Rhetoric%20April%202006.pdf
11. Powerhouse International Inc. Based on data in a promotional document. Retrieved from http://www.powerhouseinc.ca/html/boards/empowered.html
12. Steven D. Weisbord, et al. Based on data in "Prevalence, Severity, and Importance of Physical and Emotional Symptoms in Chronic Hemodialysis Patients," *Journal of the American Society of Nephrology,* June 23, 2005, 16, pp. 2487–2494. Retrieved from http://jasn.asnjournals.org/cgi/content/full/16/8/2487
13. David S. Pilliod, Charles R. Peterson, and Peter I. Ritson. Based on data in *Seasonal migration of Columbia spotted frogs (Rana luteiventris) among complementary resources in a high mountain basin.* Retrieved from http://article.pubs.nrc-cnrc.gc.ca/ppv/RPViewDoc?_handler_=HandleInitialGet&journal=cjz&volume=80&calyLang=eng&articleFile=z02-175.pdf
14. Alberta Gaming and Liquor Commission. Based on the report *Casino Gaming (Table Games).* Retrieved from http://www.aglc.gov.ab.ca/pdf/lpr/LPR_Report_09-Casino.pdf
15. Canadian Avalanche Centre. *Annual Report 2004/2005.* Retrieved from http://www.avalancheinfo.net/CAC/Canadian%20Avalanche%20Centre%20Annual%20Report%202004-2005.pdf
16. Health Canada. Based on data in "Proposed residential indoor air quality guidelines for formaldehyde," *Environmental & Workplace Health.* Retrieved from http://www.hc-sc.gc.ca/ewh-semt/pubs/air/formaldehyde/sources_e.html
17. University of Illinois Eye & Ear Infirmary. Based on data in "Corneal Thickness and Glaucoma," *The Eye Digest.* Retrieved from http://www.agingeye.net/mainnews/glaucomapachymetry.php

Chapter 12

1. Fraser Institute. Based on data in *Prescription Drug Prices in Canada and the United States—Part 3, Retail Price Distribution.* Retrieved May 2006 from http://oldfraser.lexi.net/publications/pps/50/section_04.html
2. Environics Research Group. Based on data in a press release dated April 6, 2006. Retrieved from http://erg.environics.net/news/default.asp?aID=603
3. American Industrial Hygiene Association. Based on data in *The 1999 Survey of Salary Levels among BC Occupational Hygiene Professionals.* Retrieved from http://www3.bc.sympatico.ca/aiha/salary.doc
4. Scott B. Patten. Based on data in "The Duration of Major Depressive Episodes in the Canadian General Population," Public Health Agency of Canada journal *Chronic Diseases in Canada,* 22(1), 2001. Retrieved from http://www.phac-aspc.gc.ca/publicat/cdic-mcc/22-1/b_e.html
5. Scott B. Patten. Based on data in "The Duration of Major Depressive Episodes in the Canadian General Population," Public Health Agency of Canada journal *Chronic Diseases in Canada,* 22(1), 2001. Retrieved from http://www.phac-aspc.gc.ca/publicat/cdic-mcc/22-1/b_e.html

6. David S. Pilliod, Charles R. Peterson, and Peter I. Ritson. Based on data in *Seasonal migration of Columbia spotted frogs (Rana luteiventris) among complementary resources in a high mountain basin.* Retrieved from http://article.pubs.nrc-cnrc.gc.ca/ppv/RPViewDoc?_handler_=HandleInitialGet&journal=cjz&volume=80&calyLang=eng&articleFile=z02-175.pdf

7. SES Research. Based on data in one of the three monthly *National Omnibus Surveys.* Retrieved from http://www.sesresearch.com/news/press_releases/MADD%20September%2029%202003.pdf

8. Canadian Avalanche Centre. *Annual Report 2004/2005.* Retrieved from http://www.avalancheinfo.net/CAC/Canadian%20Avalanche%20Centre%20Annual%20Report%202004-2005.pdf

9. Statistics Canada. Based on data in *Proportion of labour force and paid workers covered by a registered pension plan (RPP).* Retrieved from http://www40.statcan.ca/l01/cst01/labor26a.htm

10. Health Canada. Based on data in "Proposed residential indoor air quality guidelines for formaldehyde," *Environmental & Workplace Health.* Retrieved from http://www.hc-sc.gc.ca/ewh-semt/pubs/air/formaldehyde/sources_e.html

Chapter 13

1. Government of Nova Scotia, Labour Standards Division. Based on data in *Report to the Minister: Averaging Review,* February 2005. Retrieved from http://www.gov.ns.ca/enla/employmentrights/docs/AveragingReview.pdf

Chapter 14

1. Jaladhi Chaudhury, Uttam Kumar Mandal, K. L. Sharma, H. Ghosh, and Biswapati Mandal. "Assessing Soil Quality Under Long-Term Rice-Based Cropping System," *Communications in Soil Science and Plant Analysis,* issue 36: pg. 1141–1161. www.informaworld.com. © Copyright Taylor & Francis, Inc.

2. Toronto Blue Jays website. Found at: http://toronto.bluejays.mlb.com/NASApp/mlb/stats/sortable_player_stats.jsp?c_id=tor. Accessed August 30, 2006.

3. Adapted from data in Table 21—"Number of all-listed diagnoses for discharges from short-stay hospitals, by geographic region and diagnosis: United States, 2001," Table 21, *National Hospital Discharge Survey: 2001 Annual Survey,* June 2004. U.S. Dept. of Health and Human Services, Centers for Disease Control and Prevention.

4. László Kocsis and Éva Lehoczky. "The Significance of Yield Production and Sugar Content of the Grape Juice with Macronutrients in Grape Rootstock–Scion Combinations on Dry Climatic Condition." *Communications in Soil Science and Plant Analysis,* Vol. 33 Nos. 15–18, pp. 3159–3166, 2002. www.informaworld.com. © Copyright Taylor & Francis.

5. László Kocsis and Éva Lehoczky. "The Significance of Yield Production and Sugar Content of the Grape Juice with Macronutrients in Grape Rootstock–Scion Combinations on Dry Climatic Condition," *Communications in Soil Science and Plant Analysis,* Vol. 33 Nos. 15–18, pp. 3159–3166, 2002. www.informaworld.com. © Copyright Taylor & Francis.

6. László Kocsis and Éva Lehoczky. "The Significance of Yield Production and Sugar Content of the Grape Juice with Macronutrients in Grape Rootstock–Scion Combinations on Dry Climatic Condition," *Communications in Soil Science and Plant Analysis,* Vol. 33 Nos. 15–18, pp. 3159–3166, 2002. www.informaworld.com. © Copyright Taylor & Francis.

7. M. Arabi. "Inheritance of Partial Resistance to Spot Blotch in Barley." *Plant Breeding,* Vol. 124, 2005, pages 605–607. Blackwell Publishing.

8. *GasTips.com.* Based on data posted as of February 25, 2007. Specific data were found in "Greater Vancouver, Burnaby, Richmond, Surrey Gas Tips" and selected links. Retrieved from http://van.bc.gastips.com

9. Point 2 Homes. Based on random selection from the data at http://homes.point2.com/CA/Ontario-Real-Estate.aspx

10. Based on data collected using randomized inputs, as of February 25, 2007, from the three search engine websites: Google, Ask.com, and AltaVista.

11. Based on data posted as of February 25, 2007, on Amazon.ca. Specific data were found by an interactive search of "Children's Books" and then by age of child.

Chapter 15

1. Matthew Horne. *Incorporating Preferences for Personal Urban Transportation Technologies into a Hybrid Energy-Economy Model,* p. 36. Diss., Simon Fraser University, School of Resource and Environmental Management, 2003.

2. International Bicycle Fund. Based on data in "World Bicycle and Automobile Production, 1950–2000," *Bicycle Statistics: Usage, Production, Sales, Import, Export.* Historical data series compiled by Worldwatch Institute, Vital Signs 1996, 2002, 2005 (New York: W. W. Norton & Company, 1996, 2002). Retrieved from http://www.ibike.org/library/statistics-data.htm

Chapter 16

1. Government of British Columbia, Ministry of Environment, Lands and Parks, Water Management Branch, Environment and Resource Division. Based on data in *Animal Weights and Their Food and Water Requirements,* 1996 (minor updates 2001). Retrieved from http://wlapwww.gov.bc.ca/wat/wq/reference/foodandwater.html#table1
2. Virtual Agent Network. Based on data on Toronto real estate listings as of October 13, 2006. Retrieved from http://on.virtual-agent.com/toronto-real-estate.html
3. "Figure: World Automobile Production" and "Figure: World Bicycle Production," from Vital Signs 2002: The Trends that Are Shaping Our Future by Lester R. Brown, et al. Copyright © 2002 by Worldwatch Institute. Used by permission of W.W. Norton & Company, Inc.

Chapter 17

1. *The Globe and Mail.* Based on data in "Aging Population Set to Alter Landscape," July 18, 2007, p. A10. Research for the figure: Rick Cash and Mike Bird of *The Globe and Mail.*

Index

T

U

V